The Aquarist's Encyclopaedia

THE AQUARIST'S ENCYCLOPAEDIA

Günther Sterba

English Editor Dick Mills
Translated by Susan Simpson

BLANDFORD PRESS
POOLE DORSET

List of Contributors and their Subject Areas

Dr. rer. nat. GERT BRÜCKNER

Genera of the Anabantoidei, Gobiidae, Goodeidae, Poeciliidae; anatomy, embryology and genetics of fish; science of feeding freshwater fish; contributions to general biology.

Dr. rer. nat. HANNS-JOACHIM FRANKE

Genera of the Characoidei, Siluriformes; ethology; fish breeding; catching fish; ornamental fish trade.

Dr. rer. nat. UDO JACOB

Genera of the Centrarchidae, Centropomidae, Cichlidae, Cyprinodontidae Nandidae; invertebrates in the freshwater aquarium; ecology; water chemistry; zoogeography; general geography; taxonomy.

Dr. rer. nat. JOACHIM KORMANN

Genera of the Tetraodontidae and marine fish; science of feeding marine fish; invertebrates in the marine aquarium.

Dr. rer. nat. HELMUT MÜHLBERG

Systematics, anatomy and physiology; advice on the cultivation of aquatic plants; plant diseases; aquatic plant cultures.

Dr. rer. nat WILFRIED NAUMANN

Genera of the Atherinidae, Cyprinoidei, Gymnotoidei, Mastacembelidae, Melanotaeniidae, Mormyridae, Percidae, Polypteridae; fish diseases; fish transportation; techniques of freshwater and marine aquarium-keeping; history of aquarium-keeping and ichthyology; aquarium photography; foreign words.

Dr. sc. nat. Dr. h.c. GÜNTHER STERBA

Families and higher taxa of the Agnatha, Chondrichthyes and Osteichthyes.

First published in the German Democratic Republic in 1978 by Edition Leipzig as *Lexikon der Aquaristik und Ichthyologie*
 First published in the U.K. in 1983 by Blandford Press, Link House, West Street, Poole, Dorset BH 15 1LL

British Library Cataloguing in Publication Data

The Aquarist's encyclopaedia.
1. Aquarium fishes—Dictionaries
I. Sterba, Gunther II. Mills, Dick
III. Lexicon der Aquaristik und Ichthyologie. *English*
639.3'4'03 SF457

ISBN 0 7137 1146 9

Printed in the German Democratic Republic

Foreword by Günther Sterba

The main aim when conceiving and producing this handbook was to put together a reference work for freshwater and sea water aquarium-keeping that provided basic information about general and specialist ichthyology, hydrology, fish economy, the biology of freshwater animals and marine biology. With such an aim in mind, the hope was to ensure a multipurpose use of the book. A secondary intention was to make clear to anyone interested in aquarium-keeping just how his hobby relates to the various biological disciplines, and perhaps to deepen his potential interest in all things biological by explaining their inter-relationships. This daunting task could only be realised by a careful choice of entry words, by treating them in a generally understood and scientifically exact manner, and by achieving a good degree of continuity within the information given under each entry.

When discussing the individual groups of organisms, such as the fishes, invertebrate water animals and water plants, the genera were given greatest attention, followed by very brief descriptions of the individual species. This was a deliberate attempt to separate this book from specialist aquarium-keeping literature. From the higher taxonomic groups of the cartilaginous and bony fishes, as well as the jawless fishes, this book provides a comprehensive survey of the modern extant orders and families that are important in aquarium-keeping or in fish economy. Among the aquatic plants, mainly those groups were chosen that can be cultivated under normal aquarium conditions. Among the invertebrate animals, not only have we included those species that can occasionally be kept in the freshwater or marine aquarium, but also those that are come across quite often in ponds or streams, or even when diving offshore. Other broad areas of study included in the book and that should be mentioned here include: water chemistry; technical aquarium equipment; biology and behaviour of fishes; anatomy and embryology of fishes; how to care for, breed and rear aquarium-fishes; fish diseases; taxonomic, ecological and geographical terms; important aspects of general biology and the history of ichthyology.

The printing and general appearance of the book make clear just how much, in addition to the authors, the artists, photographers and publishing colleagues have contributed to the realization of the project. I would like to thank all concerned for their individual efforts and for their understanding and forbearance with regard to the wishes of the editor, not all of which were easy to fulfil.

I hope that the people who use this work will find that it really does extend their knowledge of the subjects and that it makes clear how aquarium-keeping fits in with biological fields of study as a whole. No book is without errors, and this is especially true of a lexicon that attempts to pursue such a broad-based aim. Along with my co-authors, I already extend my thanks to you, the readers, for any corrections or additional remarks you may wish to make.

Leipzig, July 1977 — Günther Sterba

English Editor's Preface

The name GÜNTHER STERBA is enough to send ripples of anticipation throughout the fish-keeping hobby, no matter where it is practised; news of any publication from Dr Sterba's pen is eagerly awaited by aquarists the world over.

To be associated with the production of a new 'Sterba' can only be regarded, by the hobbyist especially, as a privilege and an honour. It may be asked what more is there to be added to such authoritative text: in truth, the answer should be 'nothing', but such are the diversities within the hobby owing to language and fish popularity differences that it was felt that an anglicised edition should be produced, so that a more complete work could be available to English-speaking aquarists. Any such additional information has been limited to the provision of English common names of species where known, together with brief notes on the availability and culture of some species where differences arise with the original text.

For my own part, it has been a labour of love, and the task was made even less burdensome by the quality of the translation provided by Susan Simpson. I hope that our efforts have done nothing to impair the reputation of Dr Sterba and his associates, and that readers will be tolerant of our labour.

Dick Mills 1983

How to Use this Book

Cross references are indicated either by the word *'see'* or by an asterisk above the relevant word. Main entries will be found under the scientific name rather than the common name. Descriptions of the higher systematic taxa are accompanied by generic descriptions, and descriptions of species that are of interest in aquarium-keeping.

Abbreviations

Fish		*Systematics*			
D	= dorsal fin	Ph	= phylum	Sub-O	= sub-order
D1	= 1st dorsal fin	Sub-Ph	= sub-phylum	Sup-F	= super-family
D2	= 2nd dorsal fin or adipose fin	D	= division	F	= family
		Sub-D	= sub-division	Sub-F	= sub-family
C	= caudal or tail fin	Sub-Cl	= super-class	G	= genus
An	= anal fin	Cl	= class	Ga	= genera
V	= ventral or pelvic fin	Sub-Cl	= sub-class	Sub-G	= sub-genus
P	= pectoral fin	Coh	= cohort	Sp	= species
LL	= lateral line	Sup-O	= super-order	Sub-Sp	= sub-species
		O	= order		

Abalistes JORDAN and SEALE, 1904. G of the Balistidae. Sub-F Balistinae. Well-known species in Indo-Pacific, including the Red Sea.

– *A. stellaris* (BLOCH and SCHNEIDER, 1801). Speckled or Starry Triggerfish, Flat-tailed Triggerfish or Sea-vark whose habits and shape are typical of the family. To 75 cm. The caudal peduncle is flattened dorso-ventrally which makes it look very thin side-on. Grey-white with white spots on its darker dorsal surface.

Abdomen *see* Anatomy

Abdominal Cavity *see* Body Structure

Abraminae, Sub-F, *see* Cyprinidae

Abramis CUVIER, 1817. G of the Cyprinidae*, Sub-F Abraminae. Bream are distributed throughout Europe and N Asia where they live in lakes and slow-flowing waters with vegetation. Also found in brackish waters. Large, sociable fish that like to keep close to the bottom. Fairly deep, elongated body which is much compressed. Keel-shaped belly. The tall, narrow dorsal fin is easily distinguishable from the long anal fin, and the lateral line is complete. The mouth has no barbels. Ground colour whitish-grey to greenish-grey with a darker back and shining scales. Young, healthy specimens will survive well when kept in a spacious tank with a sandy substrate. Temperature should not exceed 20 °C. Food is generally taken from the bottom and consists of molluscs, small crustaceans, insect larvae and worms.

Fig. 1 *Abramis brama*

– *A. brama* (LINNAEUS, 1758); Common Bream. Found in central, N and E Europe, but not south of the Alps. Up to 75 cm. Males acquire tubercles at spawning time (*see* Spawning Coloration). Spawning takes place at night among dense vegetation in shallow water. An important sporting and edible fish. Close relatives: Zobel or Danube Bream *(A. sapa)* PALLAS, 1811; and *A. ballerus*, LINNAEUS, 1758, both of which are more difficult to keep in aquaria than the Common Bream (fig. 1).

Abramites FOWLER, 1906. G of the Anostomidae*. Found in the Amazon region and Guyana, these fish live in shoals among the vegetation close to the riverbank. Between 10 and 14 cm long. Deep body, small head, pointed snout. Long An (24–34 fin rays), D2 well developed with irregular transverse bars. Differences between the sexes and breeding habits unknown. Temperature 25–27 °C. Ideal aquarium fish, now imported more frequently than ever before, since it is only recently that the main distribution area was firmly established. Breeding experiments (*see* F Chilodus). Happy to eat plant food from time to time (eg lettuce leaves). Intolerant of other individuals of their own species. Only 2 species, similar in shape and requirements to members of the Chilodus family.

Fig. 2 *Abramites hypselonotus hypselonotus*

– *A. eques* STEINDACHNER, 1875. Rio Magdalena. Up to 12 cm. 4 vertical bar.

– *A. hypselonotus hypselonotus* GÜNTHER, 1868. Headstander. Amazon and Guyana. Up to 14 cm. Irregular, slightly oblique bars. Yellow fins, the An with a broad, blackish-brown base (fig. 2).

– *A. microcephalus* NORMAN, 1926. Marbled Headstander. Another name for *A. hypselonotus hypselonotus.*

Abscesses are characterised by a localised loss of external skin or of the mucous membrane surrounding internal organs. What causes them is uncertain in many cases. They are often a sign of well-known diseases, like *Ichthyosporidium**, Fish Tuberculosis*, Stomach Dropsy*, *Pleistophora** and Nematode Disease*, as well as other unspecified bacterial infections. With diseases that can be treated successfully, the abscesses will clear up too.

Absorption. The absorption of light, gases, fluids and nutrients plays an important role in the metabolism of living organisms. Absorption is observed when light rays pass through a substance. A certain amount of the radiation is converted into heat and other forms of energy. For example, when light is passed through glass or water, certain wavelengths are absorbed. Depending upon the thickness and type of glass or on the depth of the water and the materials it contains, certain characteristic absorption spectra will be produced. At sea, as the depth of the water increases, first the red and lastly the blue part of the light spectrum will be absorbed. Gases are dissolved by being absorbed by liquids and the amount of gas dissolved decreases as the temperature rises. At 0 °C and 760 Torr, 1 litre of water can, for example, dissolve 20 ml oxygen. The ability of oxygen and carbon dioxide to dissolve in water enables plant and animal life to exist in this medium.

Abudefduf FORSKÅL, 1775; G of the Pomacentridae*. Members of the family are found in all tropical seas particularly on coral reefs. In most species, a single individual or a pair occupies a territory which is aggressively defended against other members of the same species or similar fishes. Even species that live in shoals when young become territorial in later life (eg *A. saxatilis*). These are small forms whose shape is typical of the

Fig. 3 *Abudefduf cyaneus*

family. As in *Chromis** none of the opercular bones is serrated and the scales are large. However *Chromis* has sharp pointed teeth, whereas *Abudefduf* has flatter rounder ones. Most species of *Abudefduf* do not have the deeply forked C typical of *Chromis*. The G *Abudefduf* has been revised by ALLEN (1975) and split up into several Ga that are not discussed here for the moment. Almost all imported species of *Abudefduf* have proved good aquarium fish which can be fed a wide variety of foods. All that needs to be remembered is that they are intolerant of their own species. Some have even spawned in the aquarium, but the young have not been reared successfully. The following species are often imported.

– *A. annulatus* (PETERS); Yellowtail Sergeant Major. Found in the Red Sea, East Africa, in algal zone, not on coral reefs. Up to 8 cm. Resembles *Dascyllus* aruanus*.

– *A. biocellatus* (QUOY and GAIMARD, 1824); Blue-banded Damselfish; Ocellated Sergeant Major. Red Sea, Indo-Pacific, lives among reefs. Up to 10 cm. Steel-blue back, yellowish body. D with 1–2 eye spots. Becomes very aggressive.

– *A. oxyodon* (BLEEKER, 1850); Blue-velvet Damselfish, Blue-banded Sergeant Major, Emerald Perch, Neon Reef Perch. Indo-Australian Archipelago, lives among reefs. Up to 11 cm. In young fish the head has bright blue-green, longitudinal stripes and the belly has a white band on a velvety-black or dark brown background. The bright colours disappear with age.

– *A. hemicyaneus* (WEBER 1913). Indo-Australian Archipelago, Philippines, lives among reefs. About 4 cm. Brilliant blue, caudal peduncle and C base bright yellow. Often confused with *Chromis* xanthurus.*

– *A. saxatilis* (LINNAEUS, 1758); Sergeant Major, Coralfish, Five-banded Damselfish. In all tropical seas. Up to 20 cm. Sociable when young. Tall, whitish fish with 5 black transverse bars. Deeply forked C.

– *A. sexfasciatus* LACÉPÈDE, 1802. Six Banded Sergeant Major. Indo-Pacific. Like *A. saxatilis*, but slightly smaller, with 2 dark longitudinal stripes on the outer edge of the C.

Abudefdufidae, F, name not in current usage, *see* Pomacentridae

Abyss. That part of the sea with depths greater than 1000 m (*see* Benthal), where no light penetrates and no plants can grow, but where great numbers of sponges (Porifera), coelenterates (Coelenterata), crustaceans (Crustacea), molluscs (Mollusca) and echinoderms (Echinodermata) live. Abyssal animals feed either on organic matter falling down from above, eg Sea Lilies (Crinoidea); or they live as predators, eg Cuttlefish (Cephalopoda). The term abyss is also used for inland waters with great depths, eg Lake Baikal* and Lake Tanganyika*. It is also quite usual to describe the pitch-dark depths of all waters (including the deep pelagic* zone) as the abyss.

Acanthaceae. G of the Magnoliatae*, 250 Ga, 2,600 species. Tropical to subtropical, with only a few species in temperate regions. Mostly herbs, rarely trees or shrubs, with decussate leaves. Flowers hermaphroditic, zygomorphic with lips. Acanthaceae prefer localities with constant water supply. Only a few genera contain marsh-dwelling species which will tolerate submersion, eg *Hygrophila** and *Nomaphila**.

Acanthaster GERVAIS, 1841 (fig. 4) F Asteroidea* (Starfishes). Large Starfish from tropical seas, with

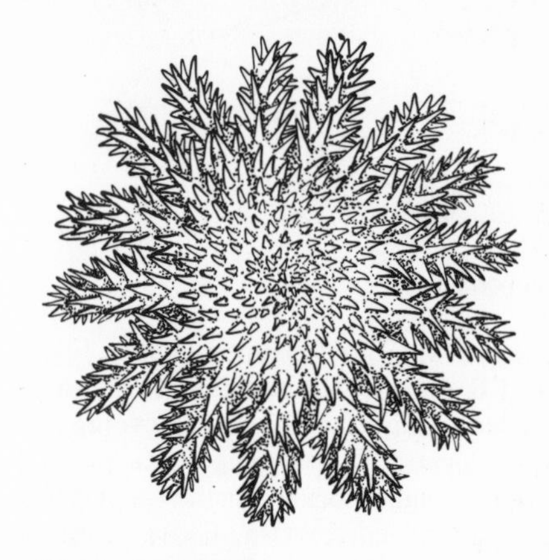

Fig. 4 *Acanthaster*

many arms beset with needle-sharp spines. For some years *A. planci* (LINNAEUS, 1758), the Crown of Thorns, had been a pest in some parts of the Indo-Pacific coral reefs. The animals evert their stomach over the coral polyps and digest them. The remaining dead, calcareous skeleton soon becomes overgrown by other organisms and is destroyed. A broad band of *Acanthaster* move over the reefs at a speed of about 1 m a day totally destroying the living communities on the reefs. It has been suggested that the huge increase in starfish populations is associated with environmental changes caused by man. In the mean time, a further spread has been prevented.

Acanthocephala (fig. 5) *see* Nemathelminthes, of which there are several genera and species. Worm-like, intestinal parasites, of fish and other vertebrates, which range in size from a few mm to several cm. They possess well-developed gonads, but no mouth, gut or anus, and absorb food through the whole body surface (*see* Osmosis). They attach themselves firmly to the host's gut wall by means of hooks surrounding a proboscis. Their eggs pass down the fish's gut into the water, where they hatch into larvae (Acanthella) which live in *Gammarus** and other small crustaceans, insect larvae or fishes, and which therefore serve as intermediate hosts (*see* Paras-

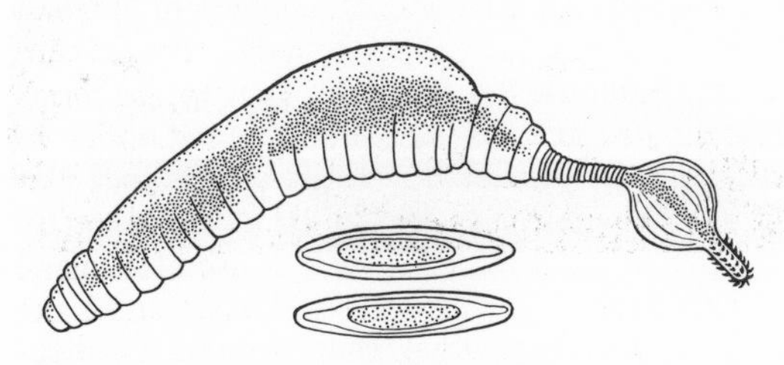

Fig. 5 *Acanthocephalus*. Below, eggs

itism). When such an infected animal is eaten by a fish the larva develops inside its intestine into a sexually mature acanthocephalan. These parasites damage the host by depriving it of food and by wounding the gut wall, which encourages other infections. The fish becomes emaciated and in severe cases dies. Under a microscope a smear taken from the gut will reveal the spindle-shaped inner shell of the acanthocephalan eggs, but the parasite itself can only be seen when the gut is opened up. In fish, the larvae are usually found in the body cavity. In the aquarium, these parasites usually only occur in imported specimens. There is no known cure.

Acanthocephalans, Cl, *see* Acanthocephala

Acanthodi, Sub-Cl, *see* Aphetohyoidea

Acanthodians, Sub-Cl, *see* Aphetohyoidea

Acanthodoras BLEEKER, 1863. G of the Doradidae* (Catfishes). Widely distributed in northern S America, particularly in central Amazon region. Squat, flattened, thorny Catfishes, 12–15 cm long. Body and head covered with bony plates, which, like the spiny first fin rays, are armed with rows of numerous spiny outgrowths. Head much flattened, with broad mouth and 3 pairs of barbels. Like the pectorals, D1 starts far forwards and has a strengthened, serrated 1st fin ray. Rounded C, D2 present. Hardy, long-living, crepuscular fish that like to bury themselves or hide away by day. They feed on all kinds of live food, eg *Daphnia*, various worms and insect larvae, and occasionally prepared food. Breeding habits of these peaceable fish unknown.

– *A. spinosissimus* EIGENMANN and EIGENMANN, 1888. Talking or Spiny Catfish. Central Amazon. Up to 15 cm. Flanks coffee-brown with dark brown markings and a white longitudinal band formed by the white spines of the bony plates. Female uniformly brown, male has brown and white spots on the belly. C rounded, with pale transverse bars, An has a white edge. When re-

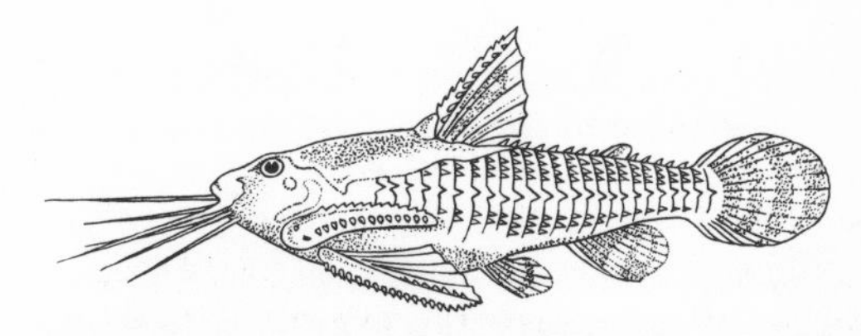

Fig. 6 *Acanthodoras spinosissimus*

moved from the tank, the breast fins become hooked to the net. You may have to cut the net to free the fish. The fish will grip your finger between its body and the breast fin and can inflict painful wounds which may bleed. May emit croaking sounds when removed from the water. The species like a tank with subdued lighting (fig. 6).

Acanthophthalmus VAN HASSELT, 1823. G of the Cobitidae*. Found in slow-flowing streams with soft bottoms in tropical SE Asia, Sumatra and Borneo. Small, bottom-dwelling fish up to 12 cm long. Worm-like, elongated body with circular cross-section, slight lateral compression in tail region. Fins small with soft rays. D starts far behind the Ps. Body covered with very small scales, head has no scales. No LL. Mouth ventral with 2 pairs of barbels on upper mandible, 1 pair on lower. Eyes small, protected by transparent skin. Beneath each eye is a bifid spine which can be erected and locked in position. Coloration very characteristic: pale with blackish-brown transverse bands, the arrangement of which varies even within a single species. *Acanthophthalmus* are crepuscular. They are suitable for a small tank but not for a community tank. Soft sand with a layer of fibrous peat is recommended for the substrate, as the fish like to burrow. They dislike light so provide hiding-places (stones, branches etc). The water should be clear, not too hard, and fresh water should be added every so often. Temperature 24–30 °C. Food is taken from the bottom and consists of worms and other live food as well as dried food. There are conflicting reports about breeding habits. They are said to coil themselves tightly round one another and spawn just beneath the surface.

– *A. cuneovirgatus* RAUT, 1957. Coolie Loach. Up to 5.5 cm. Yellow to whitish-yellow body with 13–17 short, blackish-brown transverse bands, ending above the LL.

– *A. kuhli kuhli* (CUVIER and VALENCIENNES, 1846). Coolie Loach, Giant Coolie, Leopard Eel. Malaya, Borneo, Sumatra, Java. Up to 10 cm. Yellow-orange to red with 15–20 broad, black transverse bands which do not extend to the belly. A few have been bred successfully.

– *A. kuhli sumatranus* FRASER-BRUNNER, 1940. Sumatra. Up to 10 cm. Golden-yellow to salmon-pink with 12–15 broad black transverse bands which extend over the LL (fig. 7).

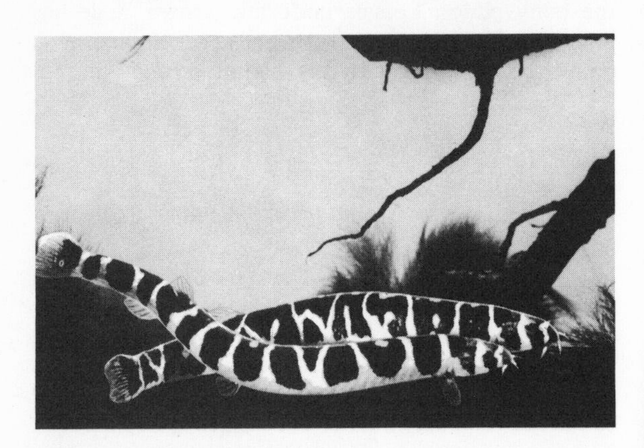

Fig. 7 *Acanthophthalmus kuhli sumatranus*

– *A. myersi* HARRY, 1949. Myer's Loach. Slimy Myersi. Thailand. Up to 8 cm. Yellow to salmon-pink with 10–14 very broad black transverse bands, which extend to belly, sometimes forming complete rings (fig. 8).

– *A. robiginosus* RAUT, 1957. W Java. Up to 5 cm. Yellow-brown to rust-red with 16–22 irregular, branched, dark brown transverse bands which end slightly below the middle of the body (fig. 8).

– *A. semicinctus* FRASER-BRUNNER, 1940. Half-banded Coolie Loach. Malaya. Up to 8 cm. Yellowish-white to salmon-pink with 12–17 dark brown, irregular transverse bands.

Fig. 8 *Acanthophthalmus myersi, A. robiginosus, A. shelfordi*

– *A. shelfordi* POPDA, 1901. Borneo. Shelford's Coolie Loach. Up to 8 cm. Reddish-yellow to salmon-pink with a row of black transverse bars, which meet up with a second row extending upwards fron the belly (fig. 8).

Acanthopsis VAN HASSELT, 1823, *see* Cobitidae, F. SE Asia, Greater Sunda Islands, Thailand, Malaysia, in fast-flowing streams with gravel bottoms. Medium-sized, bottom-living fish with elongated bodies, flattened beyond the D. Small scales. Large pointed head with a ventral mouth, surrounded by 3 pairs of barbels. Lower lip fringed. Double-pointed spines below the eyes can be erected. Nostrils elongated and form tubes. Very hardy fish, several of which should be kept together. They require medium-hard, clear, oxygen-rich water, parts of which should be renewed periodically. As these fish like to burrow, it is necessary to provide shaded hiding places and above all, a soft substrate. Temperature 24–28 °C. Food taken from bottom and consists of worms, insect larvae, small crustaceans and occasionally plant matter. Not yet bred in captivity.

– *A. choirorhynchus* BLEEKER, 1854. Horsefaced Loach, Long-nosed Loach. Thailand, Malaysia, Indonesia. Up to 18 cm. Grey-brown to yellowish-ochre with variable markings of dark-brown spots and bands. Sex differences not known (fig. 9).

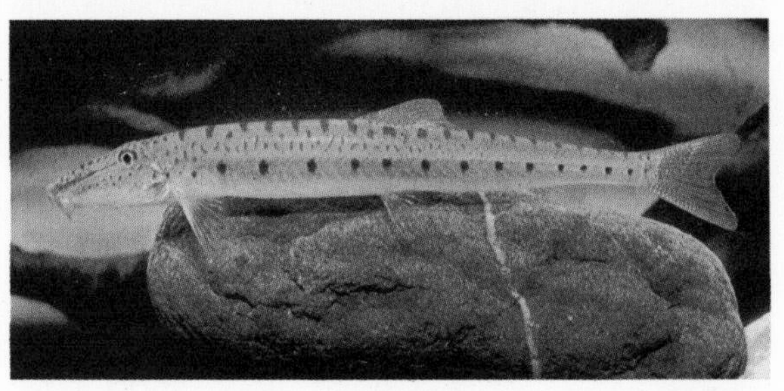

Fig. 9 *Acanthopsis choirorhynchus*

Acanthopterygii *see* Osteichthyes

Acanthostracion BLEEKER, 1865, *see* Ostraciontidae, F. A few species live in tropical parts of the Atlantic, where they are found near the sea bed on coral reefs and rocky coasts. They feed on small animals and marine growths. Back of the armour is arched, belly more or less straight. Characterised by 2 forward-pointing spines above the eyes and 2 backward-pointing spines below the caudal peduncle. Like all Boxfish,

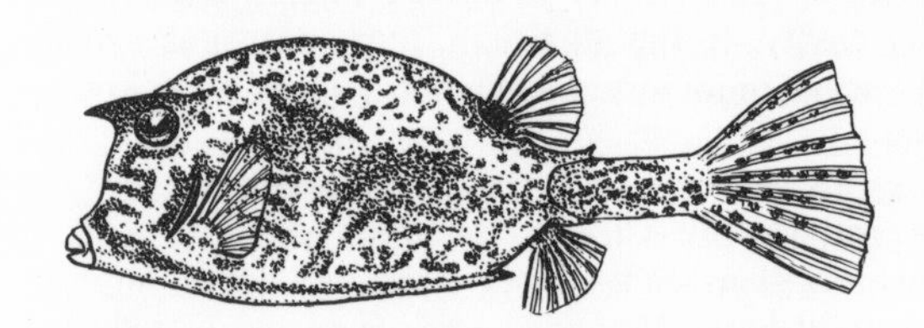

Fig. 10 *Acanthostracion quadricornis*

these specimens are delicate aquarium fishes, difficult to keep. The following species is occasionally imported.

– *A. quadricornis* (LINNAEUS, 1758). Scrawled Cowfish. Tropical W Atlantic. Up to 50 cm. Pale brownish to greenish with a pattern of dark spots and stripes (fig. 10).

Acanthuridae, F, *see* Acanthuroidei

Acanthuroidei. Related to Surgeonfishes. Sub-O of Perciformes*, Sup-O Acanthopterygii. Marine fish almost exclusively tropical, living near the coasts, particularly on reefs. The oldest fossils date from the Early Tertiary. Of the 2 families, the first is most important to the aquarist.

– *Acanthuridae.* Surgeonfishes, Tangs (fig. 11). Body tall, laterally compressed, the long and broad D and An giving it an almost disc-like appearance. Mouth small, snout often protruding, a few species with frontal outgrowths in the form of a swelling or knob. Scales small and rough. Most characteristic is the scalpel-like spine on each side of the caudal peduncle. In the genera *Acanthurus* and *Zebrasoma* the spine lies in a groove and is only raised occasionally when it is held diagonally forwards. In other genera the spine is fixed and in others it is reduced to a bony knob. The spine is used chiefly in defence and when used in conjunction with powerful movements of the tail it can inflict deep wounds on its adversary. The popular names in German and in other languages are derived from this particular behaviour.

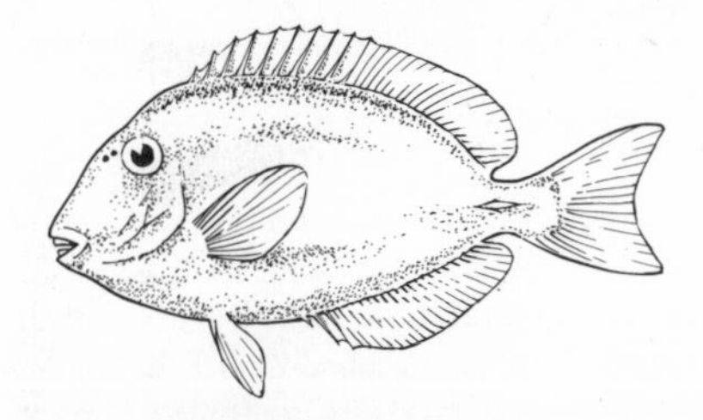

Fig. 11 Acanthuridae-type

Many species are exceptionally brightly coloured. Their distribution range is restricted to rocky coasts or coral reefs, for only there can they find sufficient algae, their main diet. They swim mainly with their Ps, so their movements are interrupted, almost jerky. They are on the move almost continuously. Only a few species are longer than 35 cm.

Surgeonfishes breed throughout the year and in some species the spawning periods appear to be linked to the phases of the moon. The eggs contain a large oil globule, so they float at the surface. The eggs hatch in 1–2 days and the fry develop into pelagic larvae up to 20 mm long. At this stage they are disc-shaped with a silvery belly, no scales, but densely packed, vertical, rib-like ridges. They feed on small crustaceans. They drift along with the current, then sink to the bottom, provided they have reached a suitable habitat. Here they turn into small Surgeonfishes. This larval stage that drifts along with the current ensures a wide distribution for the different species. The young Surgeonfishes begin to feed on algae and develop the long alimentary tract needed for the digestion of plant food. At the same time the wonderful, velvety, shimmering colours start to appear, which only start to fade in old specimens to a grey or greenish colour. Acanthuridae require a large aquarium and lots of plant food. If several specimens are to be kept together, keep a look out for signs of intolerance. Some species make an excellent food for man, although from time to time fish poisoning has been encountered. Recently, the Tobies (Moorish Idols) (Zanclidae F) have been grouped with the Surgeonfishes. For further details *see Acanthurus, Naso, Paracanthurus, Zanclus, Zebrasoma.*

– *Siganidae.* Rabbitfishes or Spinefeet (fig. 12). Distributed in coastal regions of the Indian and Pacific

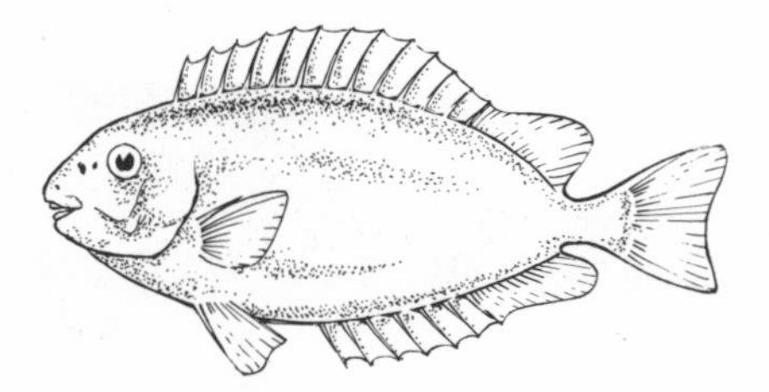

Fig. 12 Siganidae-type

Oceans. Body relatively tall, in side view has a longish-oval appearance. The mouth area looks a little like the snout of a rabbit, and when it feeds the similarity becomes more apparent. These fish browse among marine growth, eating algae mainly. Rabbitfishes are recognised by the 1st spine of the long D which is relatively short and directed obliquely forwards. C slightly indented or straight, An long, Vs each have a spine on the outer and inner edge with 3 soft rays in between. All the spines are very hard and pointed, those on the D and An containing poison glands. Many Rabbitfishes have a labyrinthine pattern of worm-like lines on a coloured background. The colour itself changes remarkably. Approximately 30 species, of which only a few are longer than 35 cm. Some species may live for a while in brackish water. Large aquaria may occasionally house the Reticulated Rabbitfish or Scribbled Spinefoot *(Siganus vermiculatus)* (fig. 13), from the coasts of

Fig. 13 *Siganus vermiculatus*

Indonesia. It is a brownish species with blue lines on its head and body. Another large aquarium species from the same area is *Lo vulpinus* (fig. 14), Foxface or Badgerfish, which has a long snout and is yellow-black in colour.

Fig. 14 *Lo vulpinus*

Acanthurus FORSKÅL, 1775. Surgeonfishes, Tangs. F Acanthuridae, Sub-O Acanthuroidei*. Found in all tropical seas, where they live in shoals on reefs and rocky shores. They browse among marine growth, particularly algae. Tall body, almost oval in profile. Characterised by the steep forehead and the high position of the eyes, as well as the spines on the caudal peduncle typical of the family. Nearly all are brightly coloured. Spawning behaviour has been observed in *A. triostegus.* Within a feeding shoal, small groups come together for a brief courtship session, then shoot upwards several m away from the shoal to release the pelagic eggs which are then fertilised. Unfortunately, Surgeonfishes are often intolerant of other fish in the aquarium, especially their own species. They can inflict grievous wounds with their caudal spines. They require some plant food and the aquarium should not be too small. Commonly imported species are:

– *A. coeruleus* BLOCH and SCHNEIDER. Blue Tang. Tropical W Atlantic. Up to 30 cm. Bright blue with dark, longitudinal stripes, grey when unwell. Juveniles yellow.

– *A. glaucopareius* CUVIER and VALENCIENNES, 1856. Philippine Surgeonfish, White-cheeked Surgeonfish, Gold-rimmed Surgeonfish, Lipstick Surgeon. Eastern half of Indian Ocean, Tropical Pacific. Up to 17 cm. Keel-shaped black marking runs from eye to mouth. Ground colour yellow, becoming black or violet when excited.

– *A. leucosternon* BENNETT, 1856. Powder-blue or White-breasted Surgeonfish. Indo-Pacific. Up to 30 cm. Head black, breast white, body sky-blue.

– *A. lineatus* LACÉPÈDE, 1853. Clown Surgeonfish, Blue-lined Surgeonfish, Striped Surgeon. Indo-Pacific. Up to 18 cm. Ground colour brown, with blue longitudinal stripes edged with black.

Fig. 15 *Acanthurus achilles*

Acari. Mites. An order of the arachnids (*see* Arachnida) with more than 10,000 species. Originally exclusively terrestrial, a considerable number have, in the course of evolution, become adapted to life in the sea or in inland waters. The marine-dwelling family Halacaridae, together with over 30 freshwater families, are collectively called the Hydracarina. The Acari are characterised by their unsegmented, usually flattened and strikingly coloured body which is never more than a few mm long and has 4 pairs of legs. By far the majority are predators and they either have a parasitic larval stage or as adults they feed on the tissues of suitable hosts (animals and plants), thereby transmitting disease.

The following account relates primarily to aquatic forms (fig. 16). Although these have retained the system of breathing through the open tracheae of their land-dwelling ancestors, they do not, like other secondarily aquatic animals, have to come to the surface to take in air. All aquatic mites have separate sexes and frequently show sexual differences (*see* Sexual Dimorphism). Eggs are yellow, red or brown and are attached to a substrate

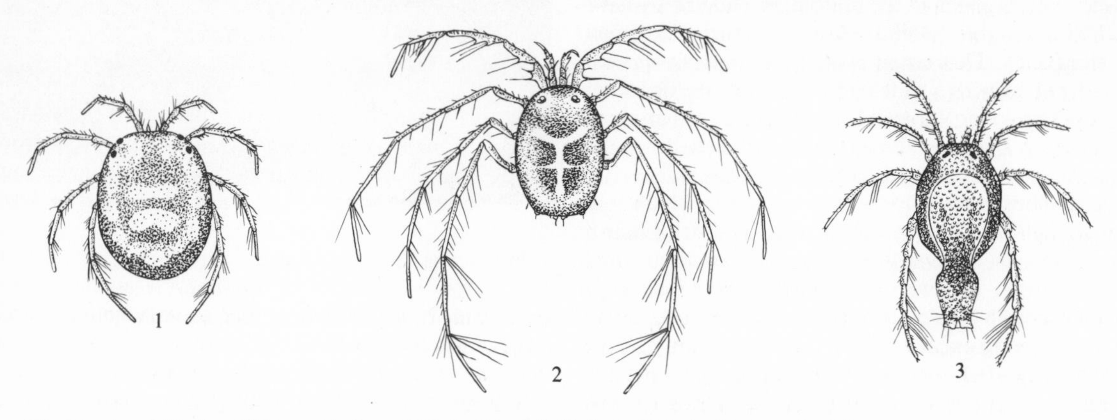

Fig. 16 *Acari.* Various Water-mite Ga. 1 *Hydrodroma* 2 *Unionicola* 3 *Arrhenurus* (male)

(water plants, rocks). Usually water mites have 6 different stages of development, including a parasitic larval stage with 3 pairs of legs and an immobile, sac-like pupal stage. Hosts are usually soft-bodied animals (eg Mollusca*) or various kinds of aquatic insect.

In spite of their small size, aquatic mites make interesting observation studies in specially designed aquaria. Their mating behaviour and life-cycles have largely been studied in specimens kept in aquaria.

Acaronia MYERS, 1940 (alternative name for *Acaropsis* STEINDACHNER, 1875). F Cichlidae* with only one species distributed from Orinoco to Northern Amazon region, in quiet streams with plenty of vegetation. Up to 25 cm. Large, robust fish with typical cichlid shape. Tall body, laterally much compressed, mouth wide and projecting forwards. Eyes very large. Soft rayed parts of D and An pointed, C rounded and fan-shaped. Under aquarium conditions these fish become quarrelsome and territorial, so should be kept in very large tanks with a lot of dense vegetation, rocks and roots. Well-suited to public aquaria, less so the amateur aquarist. These specimens need warmth, and temperature should not fall below 23–24 °C. To match their natural habitat, use very soft water. Should be fed only on larger kinds of live food, mostly small fish. Not yet bred in captivity.
– *A. nassa* HECKEL, 1840. Big-eyed Cichlid. In males, body greenish with distinct, dark longitudinal markings. In the centre of the flank is a large, prominent, darker spot. Scales have a pale, green or silvery-blue iridescence. Fins greenish to brownish. Females paler.

Acaropsis, name not in current usage, *see Acaronia*

Acclimatisation. Adaptation to changes in climate, environment or way of life. Acclimatisation is an important factor when organisms are taken into captivity and it should also be given much consideration by the aquarist. As a general rule, it is a slow process, brought about gradually. The ability to adapt is developed to a different degree in different species and is to a greater or lesser extent determined genetically. Readily adaptable fishes are easy to keep and care for. Species that live in a highly specialised way in their natural habitats find it more difficult to acclimatise; caring for them is only possible for a certain length of time, and they rarely breed in captivity. *See also* Temperature Damage, Temperature Control in the Aquarium.

Acentronichthys EIGENMANN and EIGENMANN 1889. F Pimelodidae*. Small naked Catfishes from eastern and south-eastern Brazil, scarcely more than 12 cm in length. Body elongated, cylindrical. 3 pairs of barbels. D1 is short, beginning above the Vs, D2 very long, low and fused with the deeply forked C, An long. Crepuscular fish, so aquaria should have subdued lighting and soft substrate, and roots or hollow bits of wood. Omnivores that particularly like live food as well as raw, scraped meat. Sex differences and breeding unknown.
– *A. leptos* EIGENMANN and EIGENMANN, 1889. Eastern and south-eastern Brazil. Up to 11 cm. Back dark brown, flanks pale, belly yellowish-white to brownish. A narrow black longitudinal band runs from the tip of the snout to the base of the C, a second one running diagonally downwards from the operculum to the end of the An, and a third down from the operculum and along the ventral line to the beginning of the An. In between, the flanks are dotted with irregular, dark marks. Fins colourless, snout blunted (fig. 17).

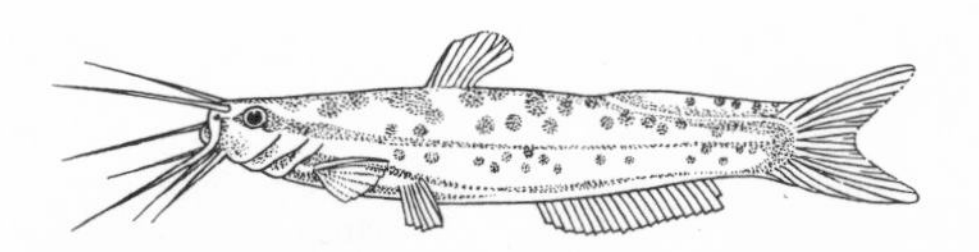

Fig. 17 *Acentronichthys leptos*

Acerina GÜLDENSTADT, 1774. F Percidae*. N and central Europe, Danube area, Siberia. Medium-sized, sociable fish that often inhabit the bottom of rivers and lakes, and also brackish waters. Body elongated with dense ctenoid scales. Large, conical head with a terminal or ventral mouth and high eyes. Operculum has spiny processes. The D is supported at the front by hard rays, at the rear mainly by soft rays. LL complete. Keep these specimens in large tanks standing in a cool place, with plenty of oxygen. Gravelly substrate and good aeration. Temperature not higher than 16–17 °C. Eats live food of all kinds, especially insect larvae, worms, snails and small fish, mostly taken from the bottom. Very productive fish which spawn from April to May, laying large strings and clumps of eggs on rocks.
– *A. cernua* (LINNAEUS, 1758). Ruffe or Pope. Central and N Europe, eastern Baltic Sea. Up to 25 cm. Back olive to grey-green with dark spots which extend onto the paler flanks. Belly yellowish, fins yellowish-grey. If kept cool over winter, this species may breed in the aquarium. In some places this species has some economic importance.
– *A. schraetzer* (LINNAEUS, 1758). Danube and tributaries. Up to 25 cm. Back and flanks dirty yellow with 3–4 black longitudinal stripes that are often broken. Rare, somewhat delicate fish (fig. 18).

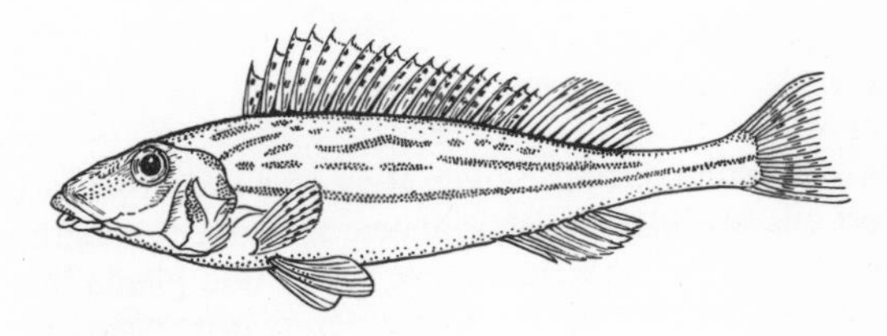

Fig. 18 *Acerina schraetzer*

Acestrorhamphus EIGENMANN, 1903. G of the Characidae*. Voracious predatory Characins that come from SE Brazil. They live among the reeds, and are about 20 cm long. Elongated body, short head, wide mouth. Habits similar to species *Roeboides**. Feed with live fish, insect larvae, raw fish etc in the aquarium. Breeding habits unknown. *Acestrorhamphus* specimens are interesting to keep in large public aquaria but unsuitable for normal aquaria.
– *A. hepsetus* (CUVIER, 1842). Sharp-toothed Tetra. SE Brazil. Up to 20 cm. Flanks silvery, back brownish-olive. A longitudinal stripe along the body begins

with a black mark at the shoulder and ends with a diamond-shaped mark edged with pale yellow on the C. Fins colourless (fig. 19).

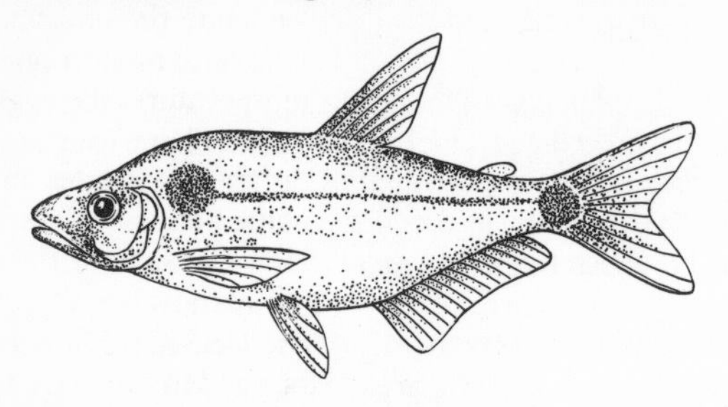

Fig. 19 *Acestrorhamphus hepsetus*

Acestrorhynchus EIGENMANN, 1903. G of the Characidae*. Predatory Characins from Amazon region and Guyana, about 30 cm long. Elongated body, pike-shaped. D1 stars in rear third of body, C deeply indented. Interesting specimen in public aquaria, that preys on small fish and is difficult to acclimatise. Breeding habits unknown.

– *A. microlepis* (SCHOMBURGK, 1841). Pike Characin. Amazon, Guyana. Up to 30 cm. Adults very colourful. According to Rachow the back is greenish, the rear of the head red, breast and belly region reddish, flanks blue-green with a longitudinal row of dark spots, ending in a dark splodge on the caudal peduncle. D1, D2 and upper lobe of C brick red, lower lobe of C orange, remaining fins bluish, An marked with dark spots. Only rarely imported. The tank should be well supplied with plants (fig. 20).

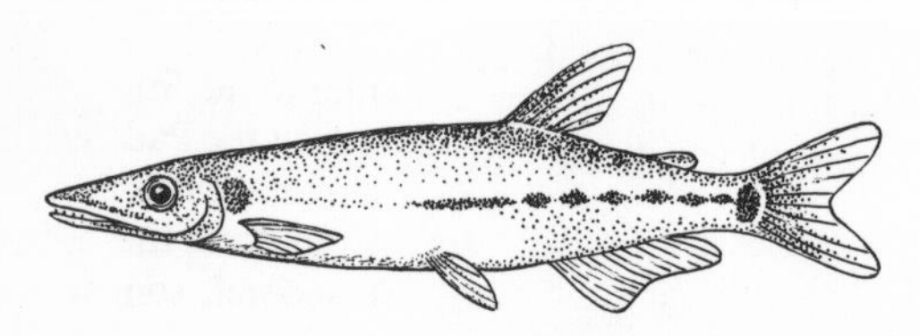

Fig. 20 *Acestrorhynchus microlepis*

Acetabularia. G of the green algae (*see* Chlorophyta). Umbrella-shaped, whitish in appearance due to calcium deposits. In the aquarium these rather odd plants live quite well, in contrast to most green marine algae, although they become overgrown by *Vaucheria**. The following species is common in the Mediterranean.

– *A. mediterranea* LAMOUROUX. In shallow depths, sometimes many together. About 5 cm high, umbrella diameter 1 cm.

Achiridae, F, name not in current usage, *see* Pleuronectiformes

Achirus LACÉPÈDE, 1802. G of the Soleidae, O Pleuronectiformes*. In warm seas, sometimes entering fresh water. Shape typical of the family. Food consists of bottom-living animals. So far only one species has been imported for aquaria:

– *A. fasciatus* LACÉPÈDE, 1803. Tropical and sub-tropical Atlantic coast of N America, penetrating far into fresh waters. Up to 15 cm. Colour changes according to background. With age, narrow dark transverse bars appear. These specimens mostly spend their days buried. They eat worms, mosquito larvae and sometimes organic waste. Add a small amount of sea water to the aquarium water.

Achlya, G, *see* Saprolegniacea

Acid Combining Capacity. An identical term for alkalinity*.

Acid Reaction *see* pH Value

Acid Sickness. Aquatic animals have different requirements of the pH of the water. Species that are adapted to slightly alkali to neutral values can be damaged in water that is only slightly acidic. But even organisms from slightly acidic water should not be kept at values below pH 5.5. A lowering of the pH value is to be feared under certain circumstances, especially in very soft fresh water. Even sea water, that usually has a pH value of 8.2–9.5, is inclined to suffer a lowering of pH value under aquarium conditions. In sea water, although it is almost impossible to achieve values that lie within the acid range, a pH below 8.0 can be harmful for sea-water animals. Fish react to a sudden reduction in pH by dashing around jerkily, jumping out of the water, and by quivering in the body. They usually die very quickly in a natural body position and with an intense body colouring. A slow decline in pH value causes a heavy mucus to form all over the epidermis; dot-like skin inflammations appear, the gills appear brighter because of brownish flecks and they experience a shortage of breath. To prevent this happening, the pH value must be checked regularly; if the situation is acute, suitable measures must be taken to effect an immediate remedy (pH Value*).

Acid Value *see* pH Value

Acidity is the presence of carbonic acid and humic acids* in certain natural waters that give it the ability to bind bases. The only aspect of interest to the aquarist is the determination of the humic acid content in the use of peat filters or extracts. To do this: boil away the carbon dioxide in a 200-ml water sample, leave to cool, then titrate the humic acids with 0.05 N caustic soda solution, using phenolphthalein as an indicator. The acidity determined by humic acids can be calculated mathematically: the used up ml of caustic soda solution minus $0.3 \times 0.25 =$ mval humic acids/l.

Acids. Substances that yield hydrogen ions in aqueous solution. Thus, for example, hydrochloric acid dissociates in water into hydrogen ions and chloride ions, and back again into hydrochloric acid:

$HCl \rightarrow H^+ + Cl^-$

Depending on the capacity to yield only a few or many hydrogen ions, a distinction is made between weak acids (eg most organic acids, carbonic acid, hydrogen sulphide) and strong acids (many inorganic acids).

Of the inorganic acids, the carbonic acid equilibrium is of direct importance in aquarium-keeping (Temporary Hardness*), phosphoric acid is occasionally used to lower the pH value*, and strong mineral acids (hydrochloric acid, sulphuric acid) are used, for example, in the regeneration of ion exchangers*. Among the organic acids, tannins (Tannic Substances*) and humic acids* are particularly important because of their properties as chelatores* and their antiseptic and astringent qualities. Chelaplex* prevents the formation of salts that are difficult to dissolve and so helps plant growth in the aquarium.

Acipenseridae, F, *see* Acipenseriformes

Acipenseriformes. O Osteichthyes*, Coh Chondrostei. Primitive, shark-like fish found exclusively in the north temperate zone. About 25 species, $^{2}/_{3}$ of which occur in Eurasia, $^{1}/_{3}$ in N America. Some live only in freshwater, others in the sea, but the latter come into freshwater to spawn (so-called anadromous fishes). The order contains the following families.

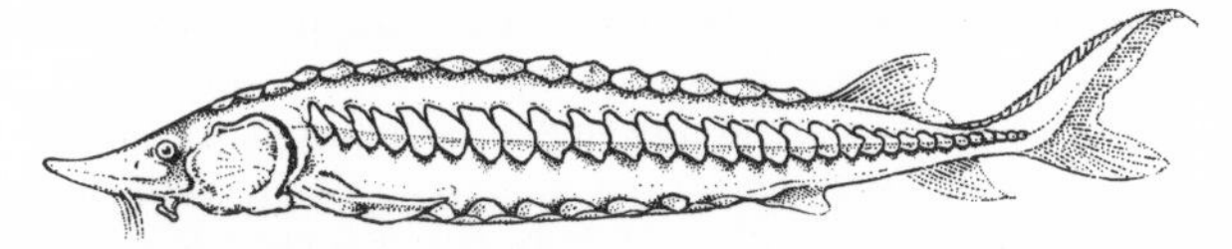

Fig. 21 Acipenseridae-type

– *Acipenseridae.* Sturgeons (fig. 21). Body elongated, head at the front drawn into a long snout, mouth ventral and protrusible, with no teeth. A single transverse row of 4 barbels lies in front of the mouth. Some forms have spiracles. D set back well to the rear, C has a very elongated upper lobe as in sharks – the vertebral column extends into this (heterocercal tail). No scales, but there are 5 longitudinal rows of strong, rhomboid bony plates. One runs along the dorsal ridge on each side of the body, and one on both the right and left edges of the belly. Bone around scull incomplete. Sturgeons move slowly

Fig. 22 *Acipenser ruthenus*

but continuously over the bottom using their barbels to search for food. This consists of soft-bodied animals, crustaceans, insect larvae and small fish and these are taken into the mouth along with sand and mud. The largest member of the family is the Beluga *(Huso huso)*, which reaches lengths of up to 9 m and can weigh over 1,000 kg. Its territory ranges from the Adriatic through the Black and Caspian Seas to the Volga. The Atlantic Sturgeon *(Acipenser sturio)* lives off the coasts of Europe and N America and the Russian Sturgeon *(Acipenser güldenstaedti)* in SE Europe. Together, they are the main sources of caviare. The white Sturgeon *(A. transmontanus)* from the N American Pacific coasts can reach similar lengths. In public aquaria, one sees the Sterlet *(A. ruthenus)* (fig. 22) more often. It measures up to 1 m in length, and is distributed from the Danube to Siberia. Another Danube species, *A. glaber,* has been introduced into various parts of Europe in recent years. Sturgeons spawn mainly in spring in relatively shallow rivers with gravelly bottoms (at the same time marine species travel up into fresh water). Several million eggs are laid. Depending on the water temperature, the eggs hatch in 14–50 days. The fry are about 12 mm long and grow very fast to begin with, but rather slowly later on. Sturgeons are of great economic value because of their tasty, boneless flesh and particularly because of their roe, which is known as caviare. They are now being bred very successfully, particularly in the USSR. Caviare is a product of unripe, green eggs, which are either stripped from the fish or taken from a dead female. The eggs are mechanically freed from the egg-cases and mixed with salt, which turns them a blackish colour.

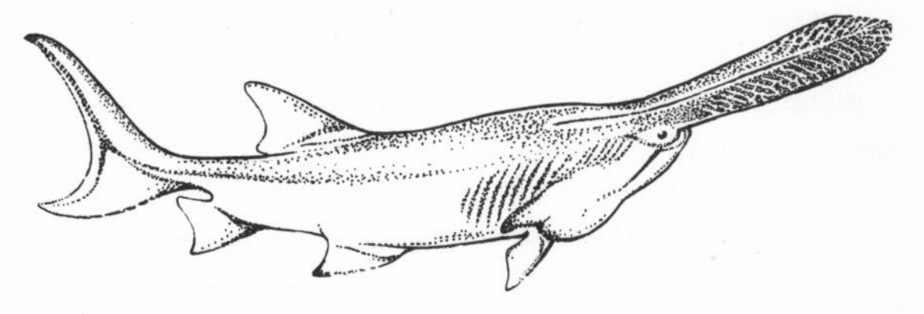

Fig. 23 Polyodontidae-type

After pouring off the brine, the caviare is ready for despatch. Malossal caviare comes from *A. sturio* (= *oxyrhynchus*) and *A. güldenstaedti,* Beluga caviare from *Huso huso* (up to 100 kg from a single female). Sturgeons caught in British rivers are traditionally offered to the reigning monarch. Young sturgeons are easy to keep in aquaria but give them clear water and a substrate of sand, not mud.

– *Polyodontidae* (fig. 23). Very similar to the Sturgeons but with a quite unusually long snout. In the Paddlefish *(Polyodon spathula)* (fig. 24), which can measure up to 1.5 m and comes from the Mississippi area, the snout is also flattened at the front. The second, somewhat

Fig. 24 *Polyodon spathula*

larger species of this family is *Psephurus gladius,* or Chinese Paddlefish, which is a very rare fish from the Yangtze-Kiang region of China. Neither species has bony plates in the skin, nor do they possess spiracles. The small crustaceans and plankton which form their food are not taken from the substrate but are sifted from the water by long gill rakers.

Acnodon EIGENMANN, 1903. G of the Serrasalmidae*. Shoaling fishes that live in open water in the Amazon region and Guyana. Piranha-like shape. Length 13–15 cm, with 6–9 serrations on the belly between the Vs and the An; fin rays absent, except for a single ray in front of the D1. To date rarely imported. Breeding habits unknown. Can be kept in the aquarium like the species of *Serrasalmus**. Only two species:

– *A. normani* GOSLINE, 1951. South America, Rio Santa Teresa, a western tributary of the upper Rio Tocantins. Up to 15 cm. Flanks silvery, with bluish stripes, which extend from the dorsal ridge to the middle line. Fin membrane between the rays is a black colour. As in Catoprion the first rays of the D1 and An are elongated. C with a dark border.

– *A. oligacanthus* MÜLLER & TROSCHEL, 1844. Guyana. Up to 13 cm. 8–9 serrations between the Vs and the An. D2 small. This species has not yet been imported alive.

Acontia. Stinging tentacles found in many sea-anemones (*see* Actiniaria). When stimulated, the tentacles are projected through the mouth opening and through pores in the side walls (eg *Adamsia**, *Aiptasia**, *Metridium**).

Acorn Barnacles, G, *see Balanus*

Acorus LINNAEUS, 1753. G of the Araceae*. Two known species, but there are probably more. Eastern Asia, Europe, N America, sometimes introduced. Herbs with a relatively large rhizome that come in a variety of sizes. Leaf-like spathe does not enclose the club-shaped inflorescence. Flowers inconspicuous, hermaphroditic. Perianth consists of 6 short, greenish scales. 6 stamens, 3 carpels, closely packed in the flower spike. Fruit a reddish berry. All the species of *Acorus* are marsh plants, growing along the banks of lakes and ponds rich in nutrient, or on the edges of ditches. Sometimes found growing in great quantities. *A. gramineus* also thrives in drier places, eg meadows.

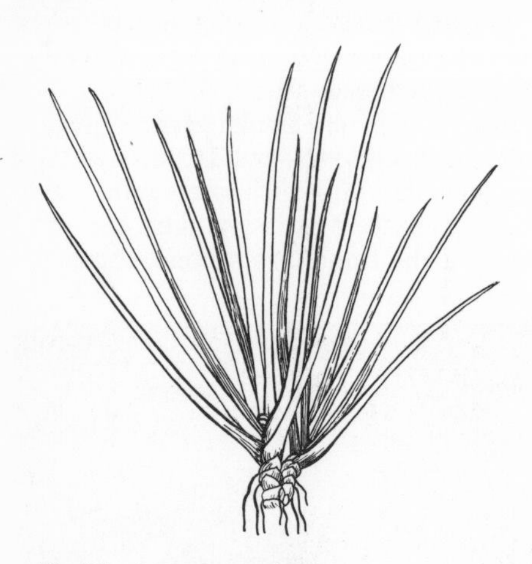

Fig. 25 *Acorus gramineus*

– *A. calamus* LINNAEUS. Sweet Flag. Eastern Asia, Europe, N America. Up to 1 m high, leaves up to 15 mm broad, each with a distinct mid-rib. There is a strong aromatic smell about the plant. Introduced into Europe and N America. In Europe it has been grown as a medicinal plant, and the rhizome has various uses. If not cultivated, the species tends to run to seed. The European plants are sterile, propagated by division of the rhizome. These plants need little attention and are suitable for the shallower parts of open-air ponds.

– *A. gramineus* SOLANDER. Japanese Rush (fig. 25). Eastern Asia. Up to 40 cm high, leaves up to 10 mm broad, without a mid-rib, growing in the form of a fan. One variety has green leaves, but there is a cultivated green and white striped variety which has been developed as a garden plant in Asia. Although available since aquarium-keeping first started, this plant is not really suitable as an aquarium plant because it does not like being submersed all the time. The dwarf variety, *A. pusillus* (ENGLER), which only grows up to 10 cm, is better suited for the aquarium.

Acrania. Sub-Ph of Chordata* (fig. 26). Small marine chordates with an elongated, laterally compressed body, pointed at each end. The body can be divided into two parts: the long main torso and the short tail (there is no area corresponding to the head in higher chordates). The elastic notochord* *(Chorda dorsalis)* forms the only skeletal element and it extends right along the body. Above the notochord lies the tubular nerve cord which widens out at the front into a bubble-shaped vesicle. The pharynx forms the main part of the body. It consists of numerous oblique gill arches separated by gill slits, the whole forming a kind of net trap. Like the Tunicata* and Agnatha*, the Acrania are mucus filter-feeders. The water containing the food particles passes over the front filter which strains the coarser particles before the water enters the pharynx. Here a continuously renewed mucus filter lying on the inside of the gill arches retains the food particles, but lets the water through, so that it can pass through the gill slits. However, the water does not flow directly to the outside, but goes first into a chamber surrounding the pharynx (the atrium). This chamber opens to the outside via the atriopore which lies near the middle of the body. The intestine has a liver sac and the vascular system is similar to that of a fish, except that there is no heart. A narrow fin runs from the front end of the body backwards along the mid-dorsal line, forming a small lanceolate C at the rear end. It then continues forwards along the belly as far as the atriopore. The primitive, but in essence typical, arrangement of some of these organs has led some people to the conclusion that the Acrania living today are the ancestors of the Agnatha and Gnathostomata*, but this is certainly not correct. It is

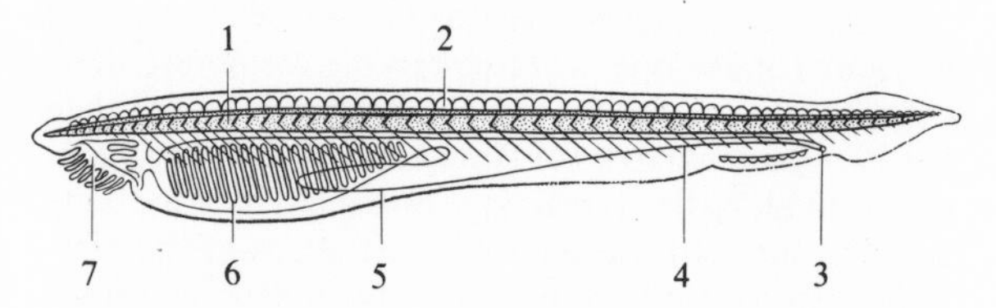

Fig. 26 Anatomy of *Branchiostoma*. 1 Chorda dorsalis 2 Neural tube 3 Anus 4 Intestine 5 Blind sac of liver 6 Pharynx with gill slits 7 Mouth

more true to say that Acrania are secondarily modified and specialised descendants of the ancestral form.

The Acrania live in soft sandy substrates into which they burrow. There, they lie on their backs with only the mouth in contact with the open water, so that they can filter-feed continuously. When disturbed they spin rapidly out of the substrate, but only swim for a short distance before quickly disappearing into the substrate once more. The sexual products are released into the water. The eggs hatch into free-swimming larvae which rotate on their own axis. The 13 species are all small and very similar. The best known is the pale flesh-coloured Lancelet or Amphioxus *(Branchiostoma lanceolatum).* It is about 60 mm long and is found in relatively shallow depths in the N Atlantic and adjacent seas such as the Mediterranean and North Seas. Around Taiwan some 30–60 tonnes of *B. belcheri* are caught every year for human consumption.

Acriflavin *see* Trypaflavin

Acroloxus BECK, 1837. F of aquatic pulmonate snails (*see* Basommatophora). Similar in structure to the lake

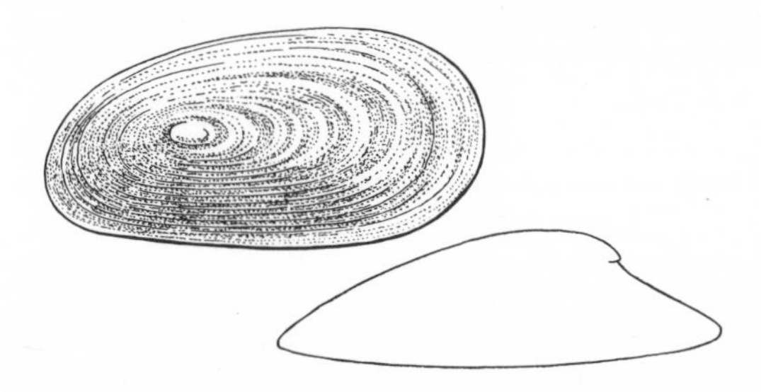

Fig. 27 *Acroloxus lacustris.* Top, view from above shell

limpets (*see* Ancylus), but much flatter, narrower, and with the tip of the shell displaced to the left. Up to 6 mm long. These snails are mostly sessile*, living in still waters on aquatic plant stems or on the undersides of floating leaves (fig. 27).

Acropora OKEN, 1815. F with many species of stony corals (*see* Madreporaria), to which *c.* 25% of all tropical reef-building corals belong. Colonies of Acropora are usually much branched and they grow relatively fast. A single coral stock may cover an area of several square metres. *A. prolifera* is a typical species. It is shrub-like, with cylindrical pointed branches, and comes from the Caribbean. The enormous Elk's Antler Coral, *A. palmata,* comes from the same area. It forms stocks a metre high and has thick stems and broad branches.

Actinia BROWNE, 1756. G of the Anthozoa*. Cl Actiniidae. *Actinia* are sea-anemones with a world-wide distribution, often occurring in very large numbers in the intertidal zone on rocky sea coasts. Medium-sized and squat, usually with strikingly coloured knob-like structures below the crown of tentacles. In some species the young develop inside the mother until they are fully formed small sea-anemones; in others eggs are laid. *Actinia* are hardy in the aquarium and live for many years. *A. equina* is known to have lived for 66 years! They like plenty of water movement and are best fed on small crustaceans.

– *A. equina* LINNAEUS, 1766. Beadlet Anemone. Cosmopolitan. Diameter of disc *c.* 5 cm, height 3–4 cm, length

of tentacle 1–2 cm. Bright red, brown, blackish or green. There are various races some of which lay eggs, others are livebearing. Easy to keep in the aquarium.

–*A. cari* DELLE CHIAJE, 1825. Mediterranean. Size and form as for *A. equina.* Olive-green with dark rings on the column. Egg-laying.

Actiniae, O, *see* Actiniaria

Actiniaria. Sea-anemones. O of Anthozoa*. About, 1,000 species found in all seas. The polyps do not form colonies. They have a diameter ranging from a few mm to 1.5 m, and they have no skeleton. Most live attached by the base to a firm substrate. The side walls form what is usually known as the column. Below the ring of tentacles some have acrorhagi, or knob-like structures, which are a particularly bright blue in *Actinia* equina.* The mouth is surrounded by 6 to over 1,000 tentacles, all more or less identical in form, which carry sting cells. In those species which catch relatively large prey the tentacles are strong with powerful stings (eg *Actinia*, Anemonia*, Tealia**). They are used to hold the prey fast and transfer it to the mouth. Species that feed on tiny prey or particles have numerous delicate tentacles (*Metridium**), and the food is carried into the mouth by ciliary currents. Many Actiniaria have defensive organs known as Acontia* which can be pushed out from inside the body through the mouth or through pores in the column wall. In spite of this they are still eaten by some starfish (*Marthasterias*, Asterina**) and fish. All Actiniaria can creep slowly along by expanding and contracting the foot, some even detach themselves for a time from the substrate and hop *(Stomphia),* and others are pelagic. Forms without a true base live buried in the substrate. Most species have separate sexes which release eggs or sperm into the open water. In some the young develop within the female's body (*Actinia*, Bunodactis**). There may also be asexual reproduction by longitudinal fission (*Anemonia**) or by laceration (*Aiptasia*, Metridium**). Various sea-anemones form an association with gastropods or hermit crabs (*Adamsia*, Calliactis**), with prawns of the G *Periclimenes**, with the crustacean *Neopetrolisthes* oshimae* or with Anemone-fishes*. Many sea-anemones have symbiotic algae (Zooxanthellae*) in the endoderm.

Most sea-anemones live well in the aquarium, some for decades. They can be put in with other marine animals provided the tank is not too small. They are best fed on small crustaceans which can be dropped onto the tentacles from a feeding tube. Those that eat large prey can be given pieces of mussel or prawn from time to time, but they must not be overfed. Both the species that produce live young and those that reproduce asexually will breed regularly. Sea-anemones firmly fixed to a rock or other substrate are best removed with a fingernail. Many species can be transported long distances if packed in damp material without water. The systematics of the Actiniaria are based primarily on the number and form of the internal septa.

Actinistia, O, name not in current usage, *see* Coelacanthiformes

Actinopterygia, Sub-Cl, *see* Osteichthyes

Actinothoe Fischer, 1890. Sea-anemone family (Actiniaria*). Cl Sagartiidae. Small sea-anemones with up to 200 marginal tentacles and no tentacles on the column wall. Hardy in the aquarium.

Fig. 28 *Actinothoe clavata*

– *A. clavata* Ilmoni, 1830. North Sea to Black Sea, below the intertidal zone. Column 5–10 cm, with brownish longitudinal stripes and translucent tentacles which are as long as the body (fig. 28).

Activated Charcoal is produced by the carbonisation of plant matter. By this means the cell wall structure is retained as a carbon skeleton and the numerous cavities form a large internal surface area. This structure gives the charcoal its ability to adsorb* numerous substances. For aquarium purposes its use depends on its quality. In the carbonisation process certain residues (eg tars) remain behind in the charcoal and alter its properties, decreasing its adsorptive and reducing properties and releasing harmful substances. The complete removal of these residues is technically expensive. The amount of residue in each type of charcoal can vary from one lot to the next, so test each one before using it in the aquarium. The adsorptive capacity can be tested with water-soluble pigments, most easily with dyes. Water with a known concentration of dye is passed through a known quantity of the charcoal. The colour starts to disappear and this goes on until the adsorptive capacity of the charcoal is exhausted, at which point it can be calculated approximately. However, even if the adsorptive properties are good, this does not mean that the charcoal is suitable for aquarium use. It is, in fact, necessary to load the water artificially with organic matter (*Tubifex, Daphnia*, mussel flesh etc) and then filter it through the charcoal. The water should immediately become clear and should not develop any cloudiness later. After 24 hours the amount of ammonia*, nitrite* and nitrate* in the water should be tested – it should not have risen. Insufficient adsorption and a diffusion of ions out of the charcoal render it unfit for aquarium purposes and usually indicate insufficient cleansing properties. Tar residues can be tested by soaking a sample of charcoal in distilled water for 1–2 days, and then determining the phenol* content of the water. If the reaction is positive the charcoal must not be used in the aquarium. A good charcoal can be used as a filter medium (*see* Filtration) and it will adsorb colloids, eg protein, dyes, organic acids, water-soluble drugs, and also gases such as chlorine* and ozone*. Amonia, nitrite and nitrate are not absorbed. Before being used as a filter medium the charcoal should be boiled for a short time in order to drive out air from the pores. Activated charcoal can also be used for the removal of impurities in the air (see Air Filtration). Exhausted charcoal should not be regenerated for use in the aquarium.

Active Substances. Various organic substances which are necessary in very small amounts as metabolic products of plant cells for the life processes to take place. They belong to the groups of substances known as the hormones* and vitamins*.

Adamsia Forbes, 1840. Sea-anemone G (*see* Actiniaria), F Hormathidae. *Adamsia* has close symbiotic relationship with hermit-crabs. The crab's shell is enveloped by the very broad base of *Adamsia* and enlarged by the secretion of a horny membrane.

– *A. palliata* Bohadsch, 1781. North Sea to Mediterranean. Base 5–7 cm across, disc up to 2 cm broad, with 500 short tentacles. Body whitish with red spots. Close relationship with hermit-crabs of the G *Eupagurus**. Unfortunately very delicate in the aquarium and will not survive for any length of time without the hermit-crab (fig. 29).

Fig. 29 *Adamsia palliata*

Adaptation. The process whereby organisms become suited to specific environmental factors, both quantitative and qualitative. This may involve short-term or long-term reactions which are capable of being regulated by the organism (for example, the way in which flatfish change colour rapidly to suit their background).

The reaction of a sense organ* to a given stimulus is also known as adaptation. Depending on the strength of the stimulus, the threshold* of stimulation of the sense cells (ie their ability to respond) may become adjusted to an average. This prevents the organ from being overworked. In certain vertebrates, the eye adapts in very different ways (*see* Sight, Organs of). In darkness the threshold of stimulation is very low and the sense cells react to very tiny amounts of light; they become dark-adapted. In bright light, on the other hand, the threshold rises sharply and the eyes only react to considerably greater amounts of light; they become light-adapted.

Adaption. Gradual alteration of the anatomy and way of life of organisms in relation to conditions in their environment. In the course of evolution (*see* Phylogenesis) brought about by changes in the ancestral form (*see* Mutation) and natural selection (*see* Selection), the ability of organisms to develop and adapt has led to today's rich variety of species and forms in the different environments (*see* Evolution). Constant adaption is necessary if a group of organisms is to survive for any lengthy period.

Alongside adaptions to ensure the continued existence of an organism in a certain environment (eg specialisation in feeding, reproduction etc), there are whole series of adaptions designed to protect an organism from its natural enemies. These protective adaptions are important among fish. Many bottom-dwellers, eg Flatfishes, Pleuronectiformes*, can change colour* to match their surroundings – the outline of their bodies seems to disappear completely. In *Mimese*, a certain feature in the environment is mimicked. The well-known Leaf-fish *Monocirrhus* polyacanthus* looks like a dead leaf, the stalk consisting of a barbel-like growth on the fish's lower jaw. Such imitation mimicry) is also found in animals. If the organism being mimicked is poisonous or very aggressive, the imitator will seldom be attacked by enemies. One fish, *Aspidontus* taeniatus*, benefits in another way from mimicry. It is so like the cleaner fish *Labroides* dimidiatus* that it is invited to set about cleaning by other fish adopting a particular mode of behaviour. It is never eaten. But instead of ridding the fish of external parasites or cleaning the teeth and gills, the impostor bites into the victim with sharp sword-like teeth and tears off bits of fin or skin. Such deception can only work for a long time if the imitators are less numerous than the organism being mimicked. This shows that adaption is always associated with the maintenance or reintroduction of a biological equilibrium*.

Adenome *see* Tumour Diseases

Adipose Fin (fig. 30). An additional fin that many fish possess, usually found on the back between the D and C. In rare cases, it may be located on the central surface in front of the An. Those fish with an adipose fin include the Salmoniformes*, most Characins (Characoidei*), many Siluriformes* and the Trout-perches (Percopsidae*). Usually, the adipose fin is kept rigid only by fatty tissue or connective tissue although there may also be supporting filaments (actinotrichia) and in a very few cases fin rays. In many armoured Catfishes (Loricariidae*), there is a mobile spine at the front edge of the adipose fin. In this book the adipose fin is often simply referred to as the D2.

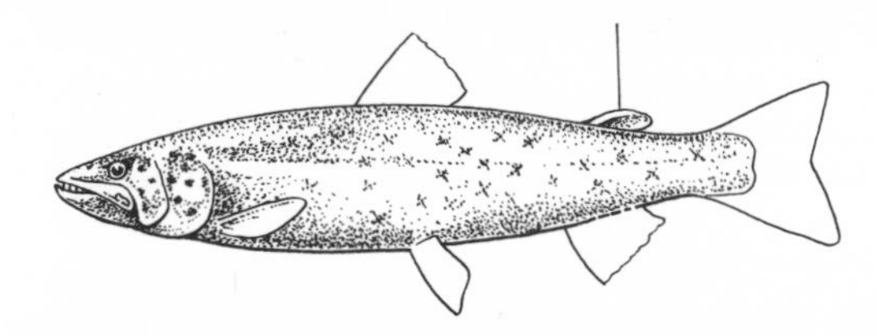

Fig. 30 Adipose fin (arrowed) in the salmon

Adrianichthyidae, F of the Atheriniformes*, Sub-O Cyprinodontoidei (fig. 31). A small family with only 3 species, all from the Celebes. Characterised by the relatively long D, the even longer An and by the enlarged mouth which is similar to a duck's bill. Vs very small. The largest species, which is up to 20 cm long, is of local economic importance. There is no information about keeping these fish in the aquarium.

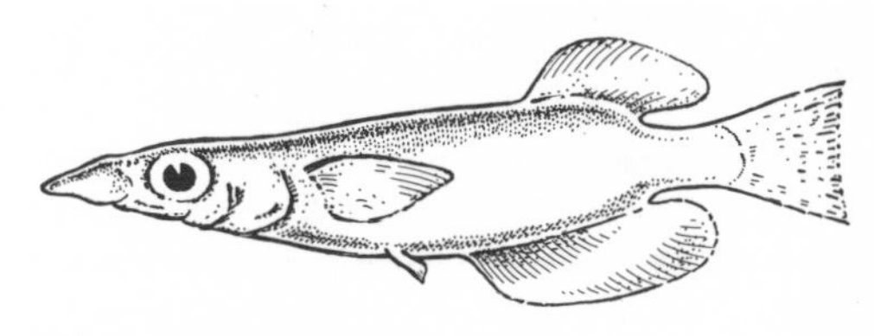

Fig. 31 Adrianichthyidae-type

Adriatic. An area of the Mediterranean* measuring 132,000 km^2 and lying between the Apennines in Italy and the Balkan peninsula. It runs roughly N-S for about 800 km and at its narrowest point W-E it is 76 km. The northern part is shallow and considerably diluted by the streams flowing down from the Italian Alps. Some parts in the southern area of the sea reach 1,251 m in depth, and there the salinity is 39 %. The varied topography of the Yugoslav coastline provides numerous habitats where marine animals and plants can be observed and collected. So far more than 6,000 species are known from this region, which makes it one of the most rigorously studied areas of the world's seas. As a rule, the composition of the flora and fauna in the Adriatic is similar to that of the Mediterranean.

Adsorbent Resins. Substances with large internal and external surface areas that enable them to bind or adsorb (*see* Adsorption) colloidal or dissolved substances from the water. If it is to be used in the aquarium, an adsorbent resin must have a broad adsorption spectrum – it must be able to adsorb all the organic impurities that occur, in particular proteins and their breakdown products. At present there are no absolutely ideal resins suitable for the aquarium. The chief disadvantages of the available adsorbent resins are: leaching out of poisonous impurities (phenols), insufficient ability to adsorb all the metabolic products present, too rapid exhaustion of the adsorptive capacity, and the impossibility of regenerating these substances for subsequent use. In many cases adsorbed proteins are broken down by bacteria and the poisonous substances thus produced are released into the water, because the resin is unable to adsorb them. Many substances that are suggested as adsorbent resins are, in fact, large-pore ion exchange resins, and their adsorptive capacities become exhausted very quickly. A filter will do much the same job for much less money. The use of ion exchange resins (Ion Exchangers*) as a form of 'filter' is not really possible for any length of time, as ions are only exchanged for one another. If metabolic by-products are removed in this way the natural ionic relationships in the water are altered, and eventually conditions become hostile to life.

Adsorption is the combining and adhesion of gases or dissolved substances (adsorbate) to the surface of a solid body (adsorbent), without a chemical reaction. The greater the external and or internal surface of the adsorbent the greater its adsorptive capacity; this capacity decreases with rising temperatures, increases at lower temperatures. Porous substances with a large internal surface used as adsorbents include silica gel, clay and activated charcoal which has a surface area of over 100 m^2 per gram. The adsorbed substances will be released (desorbed) when suitable solvents are added or when the temperature gets very high. Adsorption plays an important part in the uptake of nutrients by plants (*see* Plant Nutrition). In the aquarium certain adsorbents, such as activated charcoal* or adsorbent resins*, are used to remove unwanted substances from the water.

Adult. The adult phase of a living organism is associated with the onset of sexual maturity and the full development of the characteristic anatomy. Opposite to juvenile*. The distinction between adult and juvenile is particularly marked in animals with a larval phase and metamorphosis*, eg insects, crustaceans, echinoderms, cyclostomes, amphibians. In many species of fish the adult also differs considerably from the larva or juvenile, eg. in form (eels, herrings, flatfish) or in coloration *(Astronotus ocellatus).*

Adventitious Plants. 1) In the morphological sense these are young plants which arise asexually on leaves (eg in *Ceratopteris*), on inflorescences (eg in *Echinodorus*) or on other parts of the plant. These play an important part in the propagation of aquarium plants*. 2) In the geobotanical sense plants which are not originally native to an area, eg *Elodea canadensis in Europe.*

Aedes MEIGEN, 1818. G of mosquitoes (Culicidae, Nematocera*). As in *Culex** and *Anopheles**, the females of *Aedes* bite mainly by day. In many cases the sucked blood is a prerequisite for the ripening of the eggs. The larvae and pupae resemble those of *Culex* and provide an excellent food for fish. In contrast to *Culex* the females usually lay their eggs out of the water, and they can only develop when the water level rises. This explains why *Aedes* likes to live in marshland or flooded areas, and in particular tundra, where they can sorely irritate man and animals.

Aegagropila, G. *see Cladophora*

Aegidae, F, *see* Nerocila

Aeolidia, G, *see* Nudibranchia

Aequidens EIGENMANN & BRAY, 1894. G of the Cichlidae*, with numerous species. Mainly distributed in tropical S America, with a few species, eg *A. portalegrensis*, occurring in the southern part of this subcontinent. Moderately elongated, laterally compressed fishes, 8–25 cm long, depending on the species. Forehead profile arched, older individuals may even be hump-backed. Species of *Aequidens* can be distinguished from fishes with a similar shape, such as *Cichlasoma**, by the very elongate D and An and by the small number of spiny rays (3) in the An. Unlike most species of *Cichlasoma* or even the African cichlids, the species of *Aequidens* are relatively peaceful, especially in large tanks. *A. curviceps, A. itanyi, A. maronii* and other species can therefore be kept in a community tank, without danger to the other larger fish or to plants. During breeding periods, of course, even many of the *Aequidens* species have the unfortunate habit of unsettling the substrate when digging spawning pits. This makes the water cloudy, uproots plants and causes rocks to fall. The G consists mainly of species that spawn out in the open (*see* Cichlidae), which means that they are relatively primitive forms and are much less specialised in comparison with the species that spawn in hidden locations. Basic information on care and breeding can be found under the family description (Cichlidae). *Aequidens* require the same temperature as in their place of origin, ie 23–25 °C, rising to *c.* 28 °C when breeding. *A. portalegrensis* can be kept at a temperature 3–4 °C lower. Some species are sensitive to old water and are then prone to certain diseases, such as protruding eyes and cloudiness of the cornea. Frequent changes of a proportion of the water are therefore advisable.

– *A. curviceps* (AHL, 1924). Flag Cichlid, Sheepshead Acara, Amazon region. Up to 8 cm. Body silvery-grey to silvery-green. Scales with darker edges. An has indistinct black marking above the insertion of the P, another at the upper insertion of the C, and a third in the centre of the body. Bright, pale blue dots on the cheeks, opercula, An and C. Upper part of iris blood-red (fig. 32).

Fig. 32 *Aequidens curviceps*

– *A. itanyi* PUYO, 1942. Dolphin Cichlid. River Itany in eastern Guyana. Up to 15 cm. Body red-brown to coffee-brown above, sea-green to olive below, with a dark brown longitudinal band (frequently broken up into spots) running from the mouth, over the eye, down to below the soft-rayed part of the D. A broad dark stripe between the eyes.

– *A. maronii* STEINDACHNER, 1882. Keyhole Cichlid. Guyana, Surinam. Up to 10 cm. Beige to pale brown, the flanks having 12–13 longitudinal rows of darker dots. Two vertical dark brown to blackish bands, one from the nape through the eye to the throat, the other in the rear third of the body (fig. 33).

Fig. 33 *Aequidens maronii*

– *A. portalegrensis* Hensel, 1870. Port, or Black, Brown or Green Acara. S Brazil, Bolivia, Paraguay. Up to 25 cm, but considerably smaller in the aquarium. Body and fins brownish or greenish, the scales with darker edges. A dark longitudinal band runs from behind the eye to the upper insertion of the C, where it ends in a black marking with a pale border.
– *A. pulcher* Gill, 1858. Blue Acara. Central and northern S. America. Up to 17 cm. Body yellow-brown or grey-brown, with 5–8 indistinct, broad, dark transverse bands. Each scale has an iridescent blue or green spot. In fine, sexually mature individuals the C is wine-red. Eyes golden, with a red ring, lips pale blue. Also known as *A. latifrons* (Steindachner, 1878).

Aerenchyma *see* Tissue

Aerobes. Organisms which depend upon oxygen in the air and satisfy their energy requirements by the oxidation of organic matter, using free oxygen. Most plants and animals live as aerobes. In the process of respiration* hydrogen is split off from various organic nutrients and gradually, through reactions with various

enzymes, the hydrogen becomes bound with oxygen to form water. The energy thus released is bound chemically and stored in this form. Opposite to anaerobes*.

Aeromonas, G, *See* Fin-rot

Aeschna Fabricius, 1775. G of Dragonflies (*see* Odonata), with numerous species. The larvae are very

Fig. 34 *Aeschna*

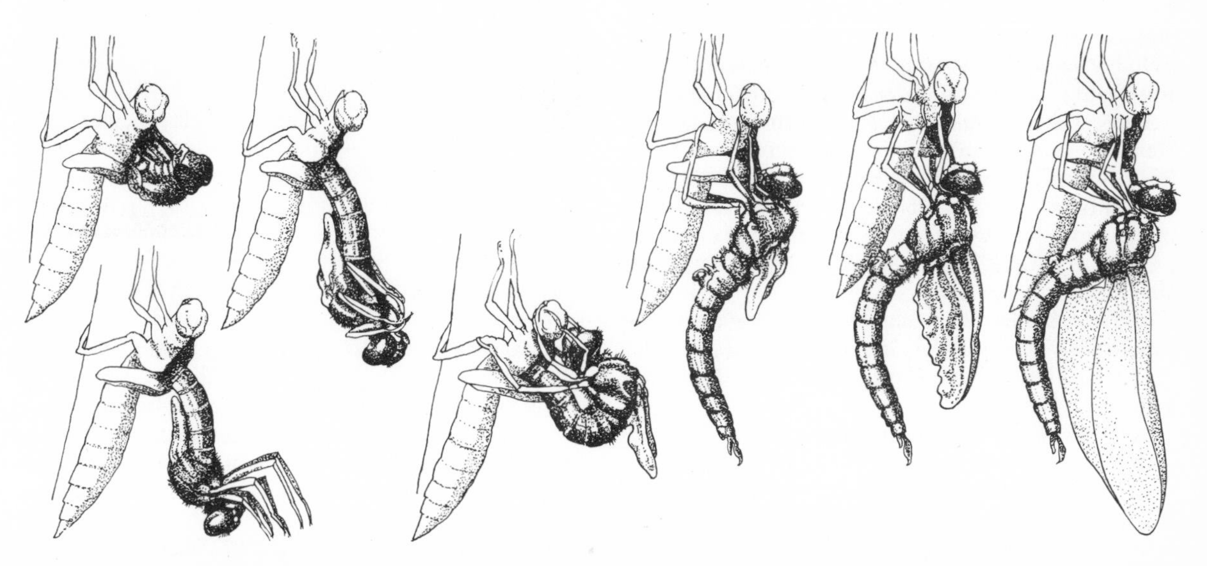

Fig. 35 *Aeschna.* Changing into imago

big (up to 5 cm). If you dip a net into a pond, you will frequently scoop up some of these larvae. If kept in a small tank, you will be able to observe how they catch their prey, how they moult, and the emergence of the imago (figs. 34–36). The larvae must be kept separately, as they are cannibals. If larvae of their own size are moulting, they will even eat them.

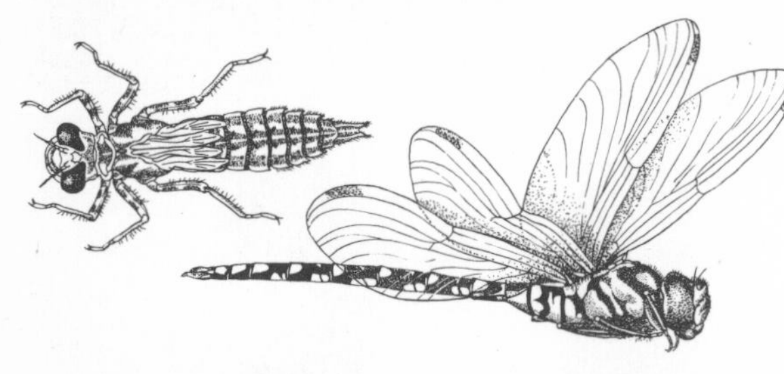

Fig. 36 *Aeschna.* Top left, larva; below right, imago

Aeshnid Dragonflies, G, *see Aeschna*

Africa. Second largest continent, covering 30 million km^2. Connected with Asia* only by the Sinai Peninsula, which makes the continent very isolated biogeographically, and as a result it is rich in endemic* species. For example, of the 44 Fs of freshwater fish represented in Africa, 16 occur only there. Those of interest to the aquarist include Protopteridae, Polypteridae, Pantodontidae, Mormyridae, Cichlidae, Gymnarchidae, Citharinidae, Malapteruridae and Mochokidae. Of the 280 freshwater fish Ga occurring in Africa, 263 are endemic. Together with India and Australia*, the central and southern parts of Africa belong to the earth's oldest land masses. This can be proven by, among other things, the wealth of primitive forms, or relict species*, found in these places. They include fish such as the Protopteridae, Polypteridae, Clupeidae, Denticipitidae, Cromeriidae, Kneriidae and Phractolaemidae. (These Fs are described under the relevant entries.) Africa has the largest continuous tropical land mass in the world, only the north and south being subtropical. Africa also has extensive, and very varied, inland waters, ranging from ponds in savanna country and tropical rain forest to very large rivers (eg Nile*, Congo*, Niger*, Zambezi*) and enormous lakes (eg Malawi*, Tanganyika*, Victoria*). There is a total of 2,510 species of freshwater fish, including 675 Cichlids, 453 Catfishes and 122 Characins (figures valid as at 1973). From the zoogeographical viewpoint, with the exception of the northern part, Africa forms the Ethiopian region*.

African Ammannia *(Ammannia senegalensis) see* Ammannia

African Banded Barb *(Barbus fasciolatus) see Barbus*

African Blockhead *(Steatocranus casuarius) see Steatocranus*

African Butterfly Cichlid *(Haplochromis thomasi) see Haplochromis*

African Catfishes, F, *see* Amphiliidae

African Climbing Perches, G, *see Ctenopoma*

African False Featherback *(Xenomystus nigri) see* Mormyriformes

African Freshwater Pipefish *(Syngnathus pulchellus) see Syngnathus*

African Glass Catfish *(Physailia pellucida) see Physailia*

African Knifefish *(Notopterus afer) see* Mormyriformes

African Leaf-fish *(Polycentropsis abbreviata) see Polycentropsis*

African Long-finned Characin *(Alestes longipinnis) see Alestes*

African Lungfishes (Protopteridae, F) *see* Ceratodiformes

African Mudfishes (Phractolaemidae) *see* Gonorynchiformes

African One-striped Characin *(Nannaethiops unitaeniatus) see Nannaethiops*

African Pike Characin *(Phago maculatus) see Phago*

African Red-eyed Characin *(Arnoldichthys spilopterus) see Arnoldichthys*

African Red-finned Barb *(Barbus camptacanthus) see Barbus*

African Spot-scale Barb *(Barbus holotaenia) see Barbus*

Agamyxis COPE, 1878. G of the Doradidae*. E Ecuador. Up to 16 cm. Similar in form to members of the G *Acanthodoras**, but only the bony plates on the rear half of the body carry spines. 3 pairs of barbels. D1 short, inserted in front of the Vs, D2 small, C laterally rounded, the rear edge being truncated. Crepuscular fish, which require large live food of all kinds, also dried food and lean meat (heart). Breeding habits unknown.

Fig. 37 *Agamyxis pectinifrons*

– *A. pectinifrons* COPE, 1878. E Ecuador. Up to 16 cm. Dark brown to blue-black, the head and body with pale spots, the spotted belly coffee-brown. Fins dark with pale streaks and dots, which often merge to form transverse bands. Old individuals are uniformly dark, with white spots on the belly. Very hardy in the aquarium, but unfortunately rarely imported (fig. 37).

Agassiz, Louis (1807–1873). Zoologist and distinguished ichthyologist, born in Motier, Switzerland. Studied philosophy and medicine in Zurich, Heidelberg and Munich. In 1829, after working on a collection of Brazilian fish made by MARTIUS and SPIX, he published in Munich his work *Selecta genera et species piscium, quae in itinere per Brasiliam collegit JB de Spix,* in which he gave a detailed description of the fish and their habitats. At the age of 23 he went to Paris and, inspired by BRONN and CUVIER, continued his studies on fossil fish. The results appeared in an important 5-volume work, *Recherches sur les poissons fossiles,* published in Neuchatel, Switzerland, in 1833–42. During his stay in Switzerland from 1832 to 1846 he lectured on zoology and geology as Professor of Natural History. His geological studies also led to publications on fossil echinoderms (*Monographie d'echinodermes vivants et fossiles,* in 1838, and *Description des echinodermes de la Suisse,* in 1839–40). In 1846 he went to the USA, and in 1847 became Professor of Geology and Zoology at Harvard University, Cambridge, Mass. Here he carried out numerous studies on the fish fauna of America, went on collecting expeditions for the Museum of Comparative Zoology in Cambridge, Mass., and investigated the coral reefs of Florida. In 1865 he travelled to Brazil, the results appearing in *A Journey in Brazil* (1868). In 1871 he made a voyage round Cape Horn to California, carrying out deep-sea research on the way. Agassiz recognised the value of fish scales in systematics and described the ganoids, a group no longer regarded as uniform (ganoid scales, *see* Scales). His work was carried on by his son Alexander Agassiz who, amongst other things, was interested in deep-sea fauna.

Agassiz's Dwarf Cichlid *(Apistogramma agassizi) see Apistogramma*

Age *see* Life Span

Ageneiosus LACÉPÈDE, 1805. G of the Siluridae*, Sub-F Ageneiosinae. Northern S America. About 12 species of true catfishes, 20–100 cm long. Swimbladder small, snout much longer than the distance between the eyes and broadly compressed. Mouth has a large gape. One pair of short barbels on the upper lip. Head flattened. Dorsal profile straight. D1 set well forwards, C forked, An long. These are voracious predators which will eat fish and lean meat, so they are best kept alone. Breeding habits unknown. Some species have very attractive colouring when young which makes them popular with the amateur aquarist and public aquarium alike.

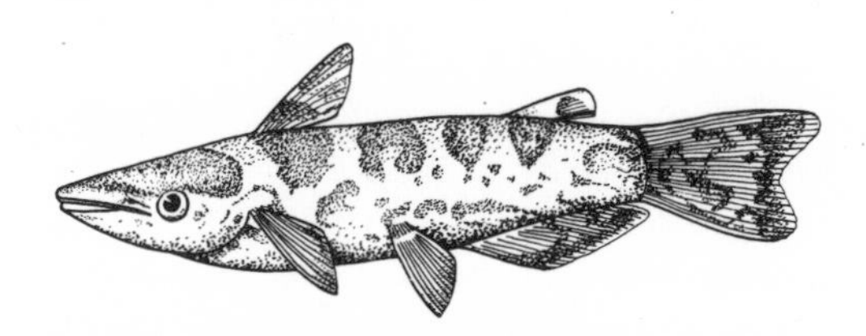

Fig. 38 *Ageneiosus marmoratus*

– *A. marmoratus* EIGENMANN, 1912. Northern S America. Up to 20 cm. Flanks pale, with slate-grey spots and stripes. Along the back there are 5 large, dark markings, not extending to the LL. D1 black with 2 pale bands, D2 pale with one black marking. An and C pale with a broad black band at the base. One of the most striking and colourful catfishes, but unfortunately only rarely imported (fig. 38).

Aggressive Fighting Behaviour. An ethological term. In contrast to symbolic fighting behaviour* this type may lead to true aggressive fighting and damage to the skin and fins. Sometimes it becomes a fight to the death. During courtship one of the partners, usually the male, may suddenly start fighting aggressively. This is because if the courtship drive* of one partner is too great and the other is not ready to mate, then the courtship and fighting behaviour may easily overlap. If a fish is kept in a tank that is too small and has no chance of escape, it may also be killed.

Aglaonema, G, *see* Araceae

Agnatha (meaning 'jawless'). Sub-Ph of the Chordata*. Fish-like aquatic animals which reached their peak from the Ordovician to the Upper Devonian. Only a few, highly specialised genera today. Distinguished from the Acrania* mainly by the primitive skull which surrounds the brain itself and to a certain extent also the larger sense organs, such as sight, smell and balance. Distinguished from the Gnathostomata* by the absence of jaws working one against the other. Many Agnatha do have skeletal traces in the region of the pharynx which give some support. However, they cannot make snapping movements. These skeletal structures serve to expand the lumen of the pharynx – a function necessary to enable the Agnatha to feed. All Agnatha are mucus filterers, that is, they draw water and its content

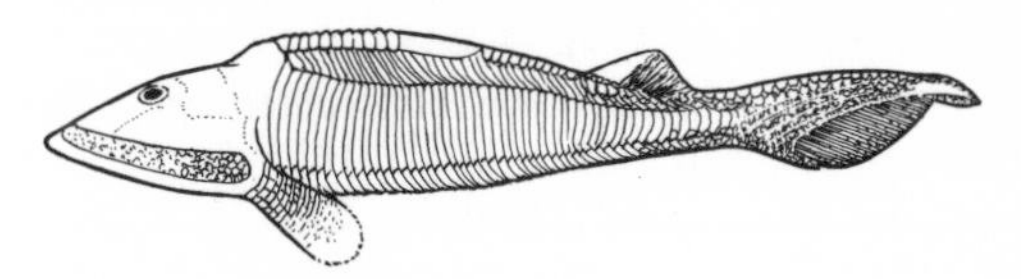

Fig. 39 *Hemicyclaspis murchisoni* after Stensiö (Osteostraci)

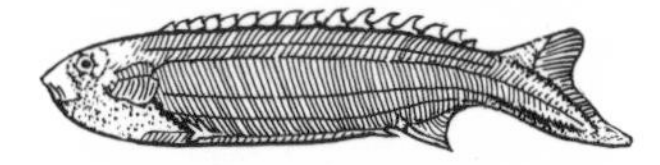

Fig. 40 *Pterolepis nitidus* after Stensiö (Anaspida)

of microscopic organisms into the pharynx where it is pressed against its inner wall by a continuously renewed mucus layer. The particles of food remain attached to the mucus, but the water flows out through the gill slits. The mucus and food particles then pass to the digestive tract.

The Agnatha were mainly small bottom-living animals found in fresh waters, with a body not unlike that of an armoured catfish. On the upper side of the head lay the paired eyes and the single entrance to the organ of smell. There were numerous round gill slits on the underside of the head. In contrast to the Gnathostomata the organ of balance had only 2 semicircular canals. Paired appendages resembling pectoral fins often present, the lower side of the C usually enlarged. Most Agnatha had a strong armour of dermal bony plates. Fossil groups are the Osteostraci (fig. 39), Anaspida (fig. 40), Heterostraci, Thelodontidae and Phlebolepida. Only one group has survived to the present day – the Cyclostomata*. Early on in the course of evolution the bony skeleton of the cyclostomes was completely lost, and so their ancestors have left few, if any, fossil remains which might have provided some clues to their evolution.

Agoniatinae, Sub-F, *see* Characoidei

Agonidae, F, *see* Cottoidei

Agonus BLOCH & SCHNEIDER, 1801. G of the Agonidae, Sub-O Cottoidei*. Slow-moving, bottom-dwelling fish of northern seas, heavily armoured. Numerous short barbels on underside of the pointed head. They feed on various bottom-living animals. Like many cold-water fish they do not tolerate an aquarium temperature above 18–20 °C and are very delicate.

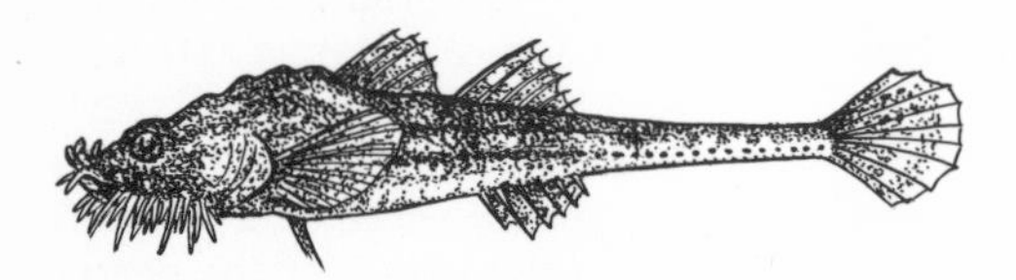

Fig. 41 *Agonus cataphractus*

– *A. cataphractus* LINNAEUS, 1758. Hooknose or Pogge. N Atlantic to Baltic Sea, on soft bottoms. Up to 20 cm. Brown, with 4 transverse bands across the back (fig. 41).

Ahl, Ernst (1898–1943). Ichthyologist. Director of the Ichthyological and Herpetological Department of the Zoological Museum, Berlin. From 1927–34 he was Editor of the journal *Das Aquarium.* In 1936, together with J. P. Arnold*, he published published *Die fremdländischen Süsswasserfishe* ('Exotic Freshwater Fishes').

Aiptasia GOSSE, 1858. G of sea-anemones (*see* Actiniaria) F Aiptasiidae. Medium-sized sea-anemones with acontia* (stinging tentacles) which are not fully retractable. Usually associated with the algae Zooxanthellae*. Interesting animals for the aquarium.

– *A. diaphana* RAPP, 1829. Mediterranean. Height 5 cm. Tentacles 2–4 cm long. Translucent and usually brownish owing to the presence of Zooxanthellae. Likes warmth and light, and reproduces regularly by laceration*. Has a strong sting, so care should be taken when keeping it in a tank with other sea-anemones or with other delicate animals (fig. 42).

Air Breathing *see* Respiration

Air Filtration. If atmospheric conditions are particularly bad, it may be necessary to remove impurities from the air that is being pumped (Pumps*) in to the aquarium (*see* Ventilation) and filter systems. In rooms with a high amount of tobacco smoke, exhaust or chemical fumes, or in places with high levels of industrially impure air, it may not always be possible to place the inlet tubing of the pump somewhere else, so that mainly pure air is taken in. In such a case, a wash-bottle must be fitted into the air stream (fig. 43). The simplest thing to do is to fill the bottle with water so that poisonous, soluble gases will be filtered off. The water must be renewed at regular intervals. Filtering air through lye to remove carbon dioxide is of no value. Disturbances in the biological equilibrium of an aquarium will scarcely be

Fig. 42 *Aiptasia diaphana*

eliminated in this way. If a piece of equipment is being driven with large compressors, it is essential that the air has any traces of oil removed. In such cases, it is advisable to fit in an oil separator (as used industrially) in the form of a wash-bottle of appropriate size, filled with activated charcoal*. This contraption will bind even the smallest traces of oil through adsorption*.

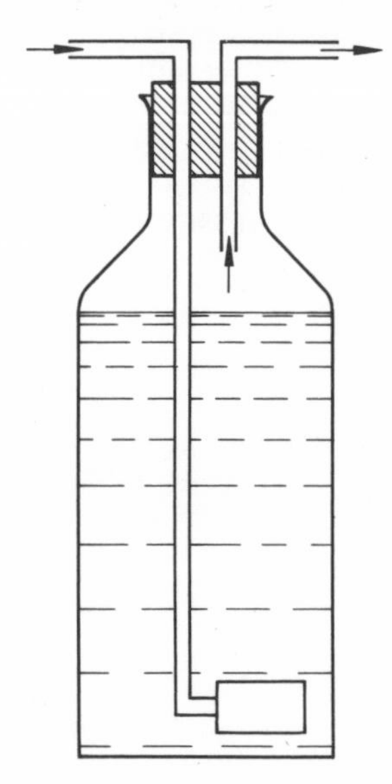

Fig. 43 Wash-bottle

Alarm Substances. Secretions produced by special gland cells in a fish's skin, particularly common among Carp-types (Cypriniformes). They are released when the fish is injured and they cause a flight reaction among like-species.

Albinism. Loss or lack of pigment. In animals and man albinism is due to malfunction of the process of pigmentation (*see* Colour), which is caused by a change in hereditary material (*see* Mutation). In contrast to melanism*, albinism is usually a recessive characteristic (*see* Heredity). It will only appear in the offspring of two animals that both show the appropriate recessive changes in hereditary material. This can happen frequently with aquarium fish. However, albino races of eg Guppy or Swordtail can only result after repeated cross-breeding of genetically pure albino animals, or by back-crossing. In the wild, albino animals do not usually survive long, because their camouflage is poor. Cave-dwelling animals sometimes prove the exception; in fish, for example, some of the unpigmented species are cavefishes (Amblyopsidae, see Percopsiformes, the Mexican Blind Cave Characin *(see Astyanax)* or the Blind Cave Barb *(see Caecobarbus geertsi)* from the Congo.

Albino Helleri *(Xiphophorus helleri helleri) see Xiphophorus*

Albula vulpes (Bonefish, Ladyfish) *see* Elopiformes

Albulidae, F, *see* Elopiformes

Alburnoides, G, *see Alburnus*

Alburnus RAFINESQUE, 1820. G of the Cyprinidae*, Sub-F Abraminae. Europe. Medium-sized shoaling fishes that live near the surface in slow-flowing and still waters. Elongate body, slender and laterally much compressed. Base of An longer than that of D. LL complete. Mouth facing upwards. Ground colour grey to greenish-grey with beautiful silvery iridescence. These fish should be kept as a small shoal in a long tank with a large surface area and good aeration. The vegetation should not be too dense and should leave enough room for swimming. Temperature not above 18–20 °C, in winter down to 4 °C. The water must be clear and the occasional renewal of a proportion is advisable. Omnivorous fish that prefer to feed at the surface. They spawn in shoals in shallow, marginal areas with a gravelly bottom. Males develop spawning tubercles on the head. Numerous subspecies.

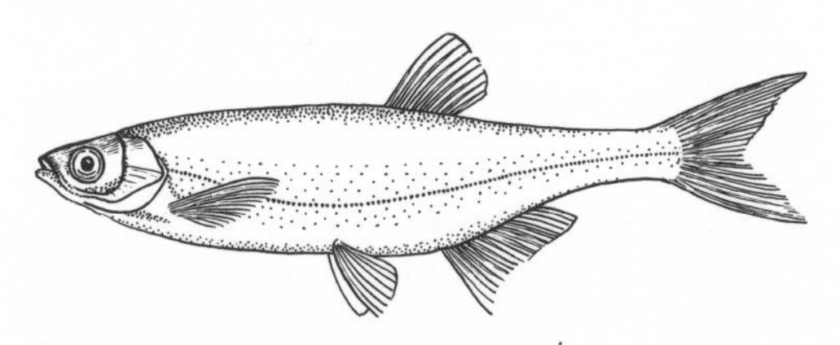

Fig. 44 *Alburnus alburnus*

– *A. alburnus* LINNAEUS, 1758. Bleak (fig. 44). Europe north of the Alps, eastern Baltic Sea. Up to 25 cm. Flanks have a particularly beautiful iridescence. The scales yield a pearl essence used in the manufacture of artificial pearls. Closely related Ga are *Chalcalburnus* BERG, 1933, and *Alburnoides* JEITTELES, 1861, with *A. bipunctatus* (BLOCH, 1782); Schneider.

Alcyonaria. Leathery or soft corals. O of the Anthozoa*, Sub-Cl Octocorallia, containing roughly 800 species. Main distribution area is the tropical Indian Ocean, but a few live in colder seas. Alcyonaria are colourful, usually soft forms that come in a variety of shapes. The skeleton is composed of small calcareous structures (sclerites) as in *Alcyonium*. In the Red Coral (Corallium*) and the Organ-pipe Coral (*Tubipora**), the sclerites are cemented together to form a firm, coloured skeleton.

Alder Flies, O, *see* Megaloptera

Aldrovanda LINNAEUS, 1753. G of the Droseraceae*, containing only one species. Europe, N Africa, Asia, Australia. Rootless aquatic plants with a mechanism for catching prey. Shoots with elongated internodes, leaves in whorls. Petioles wedge-shaped, with a few bristles at the insertion of the blade. Leaf blades roundish to kidney-shaped, up to 7 mm long and 10 mm broad, and thickened in the middle. The two halves of the leaf blades can spring shut, and have trigger hairs at the edges and glands on the upper surface. Small, hermaphroditic flowers, growing singly in the leaf axils, and reaching the water surface. Perianth consists of 5 sepals and petals. 5 stamens, 5 carpels, fused and pointing upwards. *Aldrovanda* is found very rarely in soft, acidic, still waters in low-lying country, particularly in shallow places among the rushes, where summer temperatures can rise to 30 °C. The plant is short-lived in temperate regions, although winter buds may be formed. Migrating waterfowl from the subtropics are probably re-introducing the plant to these regions continuously. The mechanism for trapping prey is released by touching the hairs at the edge of the leaf blade; the 2 halves shut together in 0.2 seconds. Their prey consists of various small insects and their larvae, and above all, small crustaceans. Proteolytic enzymes help with digestion. Each leaf blade can function several times. There is no reliable information about cultivating *Aldrovanda.*

– *A. vesiculosa* LINNAEUS.

Alectis RAFINESQUE, 1815. G of the Carangidae*. An extremely tall carangid living in the littoral* of tropical

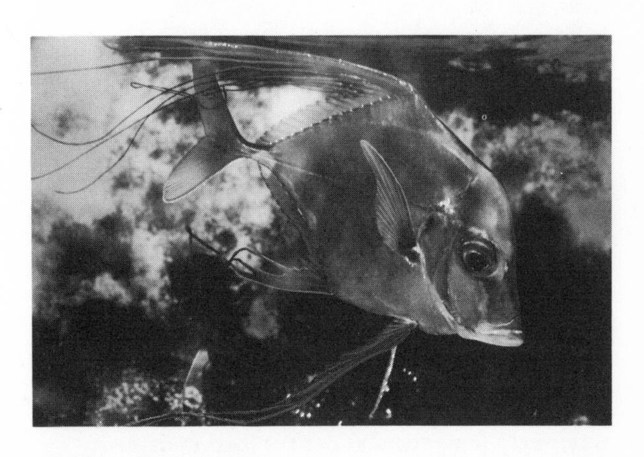

Fig. 45 *Alectis ciliaris*

and subtropical seas and feeding on plankton. The D and An have trailing streamers which in young fish are longer than the body, and so give the fish an unusual appearance. A delicate fish in the aquarium. It requires plenty of room for swimming about in and a diet of small live crustaceans. The following species is often imported:

– *A. ciliaris* BLOCH, 1788; Pennant Trevally. Tropical Indo-Pacific. Up to 30 cm. Body is a silvery colour (fig. 45).

Alepisauridae, F, *see* Myctophoidei

Alestes MÜLLER & TROSCHEL, 1844. G of the Characidae*. According to TRAVASSOS, *Alestes* is an old name for *Myletes* CUVIER, 1814. However, this view cannot be supported. GÉRY placed the G *Alestes* in its own family, the Alestidae (a group of characins with numerous species, widely distributed in Africa and mostly brightly coloured). They are closely related to *Micralestes**. Length *c.* 7–15 cm for the smaller and *c.* 25–45 cm for the larger species. Elongate body, laterally compressed. D1 usually inserted in the middle of the body, D2 well developed. These are elegant shoaling fish that jump well (the tank must be well covered). They have good vision, and will remain timid if the tank is kept in the light with insufficient vegetation. Soft, slightly acid water is essential for breeding. They spawn communally (*see* Selection of Breeding Fishes). The actual spawning takes place after vigorous driving among feathery plants. The 300 or more eggs are only slightly sticky and they swell considerably. They hatch after *c.* 6 days and the young can be fed immediately on rotifers and very small *Cyclops* nauplii. At the present time *A. humilis* and *A. longipinnis* have been bred successfully. Breeding individuals will spawn more readily if given plenty of live food, such as flies, fruit-flies, small cockroaches, etc. The species of *Alestes* are ideal aquarium fish.

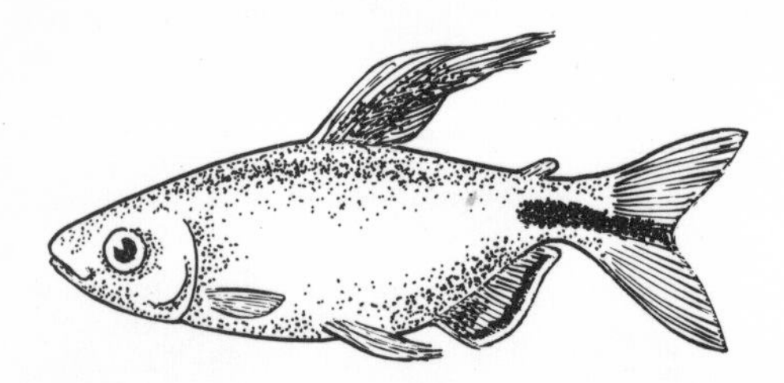

Fig. 46 *Alestes longipinnis*

GÉRY maintains that the species *A. chaperi, A. humilis*, and *A. longipinnis* belong to a little known G *Brycinus* VALENCIENNES in CUVIER and V., 1849.

– *A. chaperi* SAUVAGE, 1882; Chaper's or Green Characin. W Africa, Ghana to Nigeria. Up to 7 cm. A brightly coloured fish similar in general to the well-known *A. longipinnis.* Flanks grass-green, with deep black longitudinal stripes on the caudal peduncle. D1 and C brick-red with a broad yellow border. Eye red. So far only rarely imported and not yet bred in captivity.

– *A. humilis* BOULENGER, 1905. Angola, area of River Kwanza. Up to 9 cm. Males translucent, the upper half of the body covered with pale blue iridescent scales, the lower half with golden-yellow scales. Caudal peduncle and the base of the C have a black rhomboid spot. D1 colourless, with several long filaments. An has two lobes, is pointed, and, like the tips of the Vs white in colour. The other fins are yellowish. Females translucent with silvery iridescence, the fins not elongated. There is a dark, rhomboid marking on the C. A very pretty species, unfortunately only imported once into Europe; a few were bred by FRANCKE in 1968.

– *A. longipinnis* GÜNTHER, 1864; Long-finned Characin. Tropical W Africa, Sierra Leone to Congo (Zaire). Up to 13 cm. Body olive-green to yellow-green with strong silvery iridescence. A broad black band extends from the caudal peduncle into the C and it has a bright golden-yellow zone above it. Iris brilliant reddish-gold. D1 much elongated in the male, the An has a white border. The base of the C is a purply-red. The fins of the female do not have the elongated filaments as in the male. A very attractive fish for the larger community tank; it has been bred frequently in recent years (fig. 46).

Alestidae, F, name not in current usage, *see* Characidae

Alestopetersius HOEDEMAN, 1956. G of the Characidae*. According to GÉRY, however, this G is a Sub-G of *Hemigrammopetersius* PELLEGRIN, 1925. Africa, River Congo (Zaire) and its tributaries. Up to 6–8 cm. These fish live in the open water of tropical forest rivers. Elongate body, laterally much compressed. D1

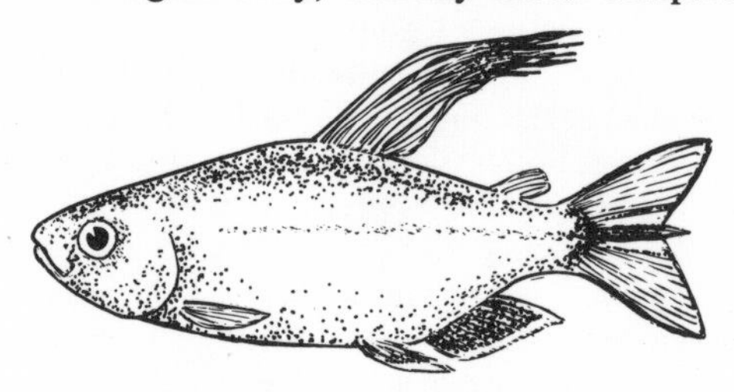

Fig. 47 *Alestopetersius caudalis*

located around the middle of the back, with long filaments in the male. D2 is present. An is longer than D1 and in the male there is a broad margin. C has two lobes, the central rays elongated to form a short point.

– *A. caudalis* BOULENGER, 1899, Yellow Congo Characin. Lower Congo and tributaries. Up to 7 cm. Males translucent yellowish-grey, with delicate pale blue iridescence. White belly. Fins yellow, the rays of D1 and An elongated, the latter with a whitish border. Central rays of C slightly elongated and a rich black colour. Sometimes you can pick out an indistinct dark longitudinal band. Females uniformly translucent without elongated fin rays. Frequently bred in captivity, using the method given for *Alestes** (fig. 47).

Alfaro MEEK, 1912. G of the Poeciliidae*, Sub-F Poeciliinae. Caribbean side of Guatemala, Honduras, Nicaragua, Costa Rica, Panama and the northern part of the Brasilian Amazon. These fish live in the lower zones of small streams. Males up to 9 cm, females up to 10 cm. Body greatly compressed. Lower part of the caudal peduncle has a sharp-edged keel make up of scales. An, forming the gonopodium in males, is situated just in front of the keel of the caudal peduncle. Vertebral column transparent. Can be kept in a well-planted tank at 24–28 °C, with some sea salt added. This G of livebearing toothcarps has been of little importance in the aquarium up to now. The following species has been bred occasionally.

– *A. cultratus* REGAN, 1908. Knife Livebearer. In addition to the main central American form there is also a smaller Brasilian form. Males up to 9 (5) cm, females up to 10 (6) cm. Male pale brown to greenish-brown, with blue iridescence on the flanks, and a pattern of fine black dots, particularly on the back. Fins colourless to delicate yellowish-green. C with a dark border. Female inconspicuously coloured. Live food preferred. Gestation period 8–10 weeks.

Alfonsino *(Beryx splendens) see* Beryciformes

Algae. Plant protobionts (*see* Eukaryota). 30,000 species. The typical characteristics of algae are: the possession of cell nuclei; the presence of chlorophyll, and with it the ability to photosynthesise (*see* Photosynthesis); the occurrence of cysts* during reproduction; and a vegetative body which has not evolved beyond the stage of a thallus*. Nevertheless, algae are not systematically totally alike – they consist of a number of independent groups which are isolated from one another to a greater or lesser extent. The vegetative body may be a mobile or immobile single cell, or a colony of cells; have branched or unbranched cell filaments; or sheets of cells with one or more layers. The most differentiated forms have tissues and they sometimes resemble higher plants superficially by having shoots and leaves. The thallus is usually made up of cells, but it may be a non-cellular structure with numerous nuclei. The photosynthetic chloroplasts, like those of higher plants, may be lens-shaped and there may be large numbers of them in each cell; sometimes there are only 1–2 per cell, however. The form of the chloroplasts varies, eg they may be cup-shaped, ribbon-like, lobed, cylindrical or mesh-like. Likewise, there are more different types of chlorophyll than in higher plants. In many algae the chlorophyll is masked by other yellow, brown, red or blue pigments. In addition to intensive vegetative reproduction, algae also reproduce asexually by means of spores* and sexually by gametes*. Most species show an alternation of generations. The vast majority of algae are entirely aquatic, with only a few occurring on land. The different groups of algae are distinguished by their content of pigment, by the type of cilia or flagella of the mobile cells and by the chemistry of their reserve substances. The most important groups are the Euglenophyta*, the Dinophyta*, the Crysophyta*, the Bacillariophyta*, the Xanthophyta*, the Green Algae or Chlorophyta*, the Phaeophyta*, the Rhodophyta* and the Charophyta*. The Blue-green Algae, or Cyanophyta*, do not belong to the algae, as they lack nuclei. (*See also* Filamentous Algae.)

Algae, Control of. Thick growths of algae are very unusual in a well-balanced, properly planted aquarium with a moderate stock of fish. There are no universally applicable or reliable methods of controlling algae. This is because, on the one hand, the algae in an aquarium can adapt to a wide range of living conditions, and on the other they sometimes react to the slightest change in the environment and suddenly die off. If freely mobile one-celled algae, or those with a few cells, start to accumulate, the water may become cloudy (*see* Water Bloom). After 2–3 days such water usually becomes clear again. If the cloudiness persists, it may help to darken the aquarium tank or even better introduce *Daphnia*. Other one-celled and few-celled algae are sessile, living on the plants, on the bottom and on the glass, or on any material for that matter. These are actually beneficial, as they produce oxygen in light and are eaten by many fish. Unfortunately, an algal growth on the front glass must be removed from time to time. This can be done with a glass cleaner*, available from aquarium dealers. Blue-green Algae, on the other hand, can be dangerous when it settles on the substrate or on plants. A slimy blue-green or black deposit forms which may seriously damage the higher plants. Here the only remedy is to remove the biological conditions which have allowed them to form in the first place; alter the strength and duration of the lighting, the pH, the hardness of the water, or remove any excess waste matter that may be present. New water should not be added. Chemical preparations should only be used with great care as these usually damage the higher plants. Finally, thick growths of filamentous algae* may occur. In such cases it is best to attempt their removal mechanically; the coarser types can easily be drawn out by a piece of rough stick. This type of algae can also be controlled by certain fish. *Gyrinocheilus aymonieri* and various species of *Otocinclus* and *Hypostomus* (formerly *Plecostomus*) use their broad lips to rasp at these algal growths. Even a badly infested tank will be clean again in no time. Dense growths of Blue-green Algae will not be eaten, however. To a limited extent, some types of algal growths are eaten by Ramshorn and other snails. Otherwise the best remedy is to use the method suggested for Blue-green Algae.

Algae-eaters *see* Gyrinocheilidae, Loricariidae

Algae-scrapers. Pieces of equipment for removing algal growth from aquarium sides. They usually consist of a long rod, at the end of which is a clamp for fixing in razor blades, plastic scrapers, felt, hard sponge tissue or rubber. Recently, equipment has come onto the market where a small scraper can be moved on the inside of the side of the aquarium by means of a magnet. It is important that the material is inert, particularly if it is to be used in a sea-water tank, and also that it is constructed in such a way that it will not scratch the pane of glass. Tanks made of Plexiglass cannot be cleaned with such equipment; instead, use a sponge at regular intervals to get rid of any algae growing on the sides (this will prevent the surface from becoming scratched).

Algal Growth in the Marine Aquarium. The only plants suitable for a marine aquarium are algae*. A tank with a good growth of algae means that the water conditions are healthy and stable. Unfortunately, only a few of the many extremely pretty algae can be kept. Brown algae (*see* Phaeophyta), found on the shore, will quickly die and pollute the water. Attempts can be made to grow certain of the algae that live in quiet, shady places in depths of a few metres, eg the Red Algae (Rhodophyta*). However, only those algae that grow well will improve the quality of the water. The most reliable in this respect are some of the usually self-sown green-coloured *Vaucheria** and *Caulerpa**. Algal growth is an important indication of water quality: healthy water conditions with a high redox potential are always associated with a good growth of green algae, whereas growths of slimy, blue-green algae (*see* Cyanophyta) or of dark-brown diatoms (*see* Bacillariophyta) show that the water is in poor condition. Sometimes green algae that have been growing well over a period of time start to turn pale green. This means that there is a deficiency of trace elements, and the remedy is to add such substances in solution or to renew a proportion of the water. A newly established tank will often produce a growth of diatoms, but these will be replaced by green algae after a few weeks. For good algal growth the following is recommended. Inoculate the tank with material (water, substrate) from an established healthy tank, wait for some green algae to appear and then introduce the fish. Avoid overstocking, feed sparingly, keep the water moving, and possibly add Chelatores*. Fertiliser should not need to be added when fish are present. The tank should be lit by daylight or by white artificial light. Dense growths of algae must be given a regular supply of light.

Alisma LINNAEUS, 1753. G of the Alismataceae*, containing 4 species. N America, Europe, N Asia, and introduced elsewhere. Perennial herbs with tubers and radical leaf rosettes. Submerged leaves pointed when young, or also later. Aerial leaves have stalks with lanceolate to ovate blades. Flower-stem has a number of branches. Hermaphroditic flowers, polysymmetric. Perianth with 3 sepals and 3 whitish or pale violet petals. Numerous stamens and carpels. These are common plants in marshland and in areas of shallow water. They are suitable for planting in an outdoor pond, but they produce large amounts of seeds and may easily become weeds. The flowers should always be cut off as they appear.

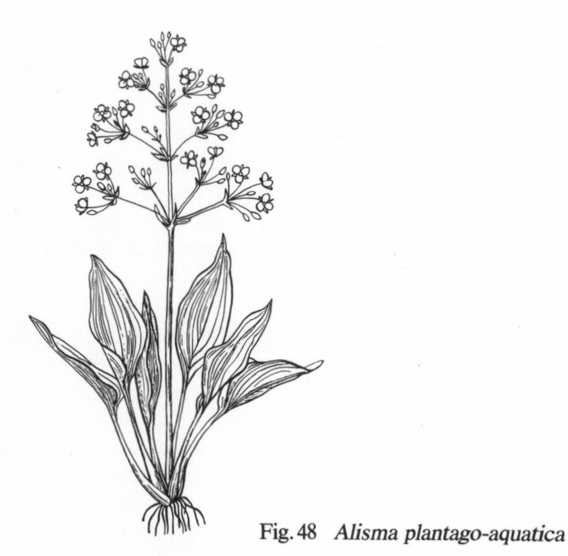

Fig. 48 *Alisma plantago-aquatica*

– *A plantago-aquatica* LINNAEUS; Common Water Plantain. N America, Europe, N Asia, introduced elsewhere. A marshland plant, but also found growing submerged when young. Leaves broadly ovate, sometimes with a cordate base (fig. 48).

– *A. gramineum* C. C. GMELIN. N America, Europe, N Asia, rare. Leaves almost always submerged, narrow, linear.

– *A. lanceolatum* WITHERING. Europe, N Asia. A marshland plant, but also submerged when young. Leaves lanceolate.

Alismataceae. F Liliatae*. 10 Ga with 100 species. Tropical to temperate, mainly in the N Hemisphere and in S America. Perennials, rarely annuals. Leaves in radical rosettes. Flowers hermaphroditic or unisexual, and then usually monoecious, polysymmetric. Stamens and carpels usually numerous. The A are exclusively marsh and water plants. Important Ga are *Alisma**, *Echinodorus** and *Sagittaria**.

Alkali reaction. Means of identification for a base reaction, rarely used today (*see* pH).

Alkali sickness. If the pH value* of fresh water is over 8–9, it will harm the fish. In particular you will notice milky cloudiness of the skin and gills and frayed fins. How prone individual species of fish are to alkali sickness is variable. In well-planted tanks a rise in pH value until it reaches basic conditions may be caused by biogenic decalcification*. In the presence of ammonium compounds, a basic pH value may enrich the amounts of highly poisonous ammonia*. To prevent alkali sickness it is necessary to keep a regular check on the pH value and take immediative corrective action should it be required.

Alkalinity. The presence of hydro-carbons, carbons and hydroxides in natural water, which make it able to bind with acids. As alkalinity is in direct proportion to the temporary hardness* of the water, it is an important point to consider when testing the quality of aquarium water. Titrate 100 ml of the water to be tested with 0.1 N hydrochloric acid. Suitable indicators are methyl orange and a mixture of 3 parts bromocresol green and 1 part methyl red. At the equivalence point the methyl orange will turn from yellow to orange and the mixed indicator from green to red. The acid binding capacity (in ml) thus determined gives the alkalinity. Multiply this figure by 2.8 and you can calculate the temporary hardness.

All-glass Aquaria *see* Aquaria

Alligator Gar *(Lepisosteus tristoechus) see* Lepisosteiformes

Allis Shad *(Alosa alosa) see* Clupeiformes

Allochthonic. Originally not native to a country; introduced into a specific region, or spread by other means. For example, *Salvelinus** *fontinalis* or Brook Trout is native or autochthonic* to N America, but allochthonic to Europe.

Allochthonic Movements. Term used in ethology to describe movements which lead to no conclusion, and which therefore bear no relation to the function being performed at the time. For example, darting movements often start up in conflict situations, but there appear to be two opposed spheres of activity going on at the same time – attack and flight. An example might be two male Cichlids which are neither fighting nor fleeing from one another; instead, they do something quite meaningless: they start pecking on the bottom for food although there is no food there. Making the fins or body quiver are other examples of allochthonic movements. In the course of evolution, these darting movements can become ritualised through a change in

significance; they may come to signify courting or threatening behaviour. So, whenever the function is performed, it conveys a specific and appropriate meaning, subsequently becoming autochthonic* movements.

Allogamy *see* Pollination

Allonatry. Word used to denote a certain type of range in biogeography*, in this case the habitats (*see* Areal) of forms that are distinguished from one another by comparison with one another, ie the areals do not overlap. If the areals of closely related forms come into contact they can be regarded as geographical races (compare concept of species). On the other hand if a biographical comparison shows an areal disjunction, it is not possible to draw a safe conclusion about the relationship between the forms being compared. (Are they different species? One species? Sub-species?) *See* Sympatry.

Allotriognathi, O, name not in current usage, *see* Lampridiformes

Alopiidae. Thresher Sharks. F Chondrichthyes*, O Lamniformes. 4–5 species known from all tropical and temperate seas. The torpedo-shaped Alopiidae are instantly recognisable by the immensely long, scythe-like upper lobe of the tail which may be as long again as the body. The Alopiidae lash their tail when chasing prey, herding shoals of fish into a tighter and tighter mass to give themselves easy targets. Spiracle is present. Alopiidae are excellent, fast swimmers that live in

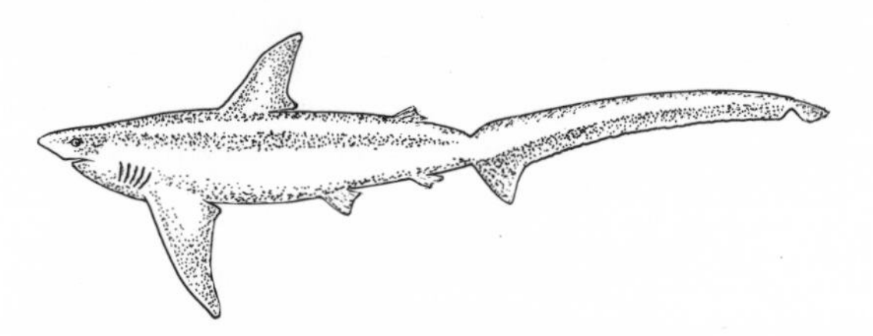

Fig. 49 *Alopias vulpinus*

the upper part of the seas. They are not a danger to people. All Thresher Sharks give birth to live young. Usually only 2–4 pups form a litter, but even at birth they may be 1 m in length. The best-known species, the slate-grey, greenish or blue-black *Alopias vulpinus* (fig. 49), can reach more than 5 m. It is commonly found in the temperate and subtropical regions of the Atlantic and eastern Pacific Oceans.

Alosa alosa *see* Clupeiformes

Alosa fallax *see* Clupeiformes

Alpheus WEBER, 1795. F Decapoda*. Smallish, shrimp-like, burrowing crustaceans that live on the sea bottom. A trigger mechanism on the enlarged right or left claw can produce a noise sounding like a shot and a strong explosion. *A. californensis* (HOLMES) can stun

Fig. 50 *Alpheus djiboutensis*

prey in this way. *A. djiboutensis* (fig. 50) lives in the Red Sea in symbiotic partnership with members of the *Cryptocentrus* goby family, laying claim to the burrows they make in the sandy bottom.

Alternanthera FORSKÅL, 1775. F Amaranthaceae*. 170 species. Tropical-subtropical America. Some species also found in other parts of the world. Small perennial herbs or annual shrubs. Stems upright or creeping, with decussate leaf arrangement. Leaves partially coloured or spotted. Very small, hermaphroditic flowers, dense inflorescences. Perianth consists of 5 free scales. 5–2 stamens, 3–2 fused carpels. Nut for a fruit. On account of its pretty foliage, several species were used as ornamental flowers around the turn of the century. In their natural habitat they are found in a wide variety of places. A few can grow in marshy areas where they endure submersion to a greater or lower degree. It is these species that can be used in the aquarium, but they take a long time to adjust and grow slowly. The most common species in aquaria is:

– *A. reineckii* BRIQUET. Tropical S America.

Alternate *see* Leaf Arrangement

Alternation of Generations in Plants (fig. 51). It is a typical feature of the majority of plants and fungi to have an alternation of generations; ie a regular alternation between two generations whose methods of reproduction differ from one another. It is possible to

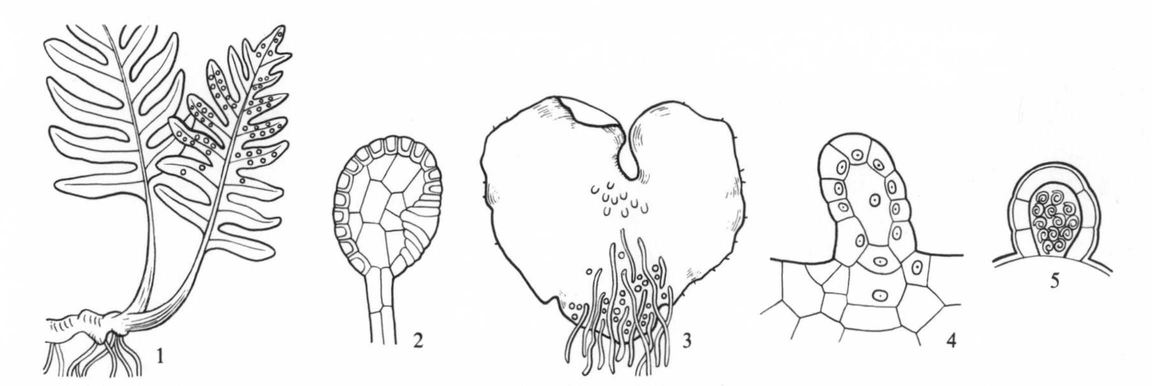

Fig. 51 Alternation of generations in Ferns. 1 Fern plant 2 Sporangium 3 Prothallium 4 Archegonium 5 Antheridium

distinguish sexual and asexual generations. The asexual generation, the *sporophyte*, is diploid and forms haploid spores (*see* Spore) by reduction division as a conclusion to its development. The sexual generation, the *gametophyte*, develops from the spores, and as a conclusion to its development it forms sex cells (the gametes*). After successful fertilisation, the asexual generation develops once more from the zygote. In complicated cases, the generations occur in several phases, as, for example, the two sporophyte phases in the Red Algae. In most algae and ferns, both generations occur as separate entities. The 2 generations in such cases are clearly distinguishable; in some algae, however, they have the same shape. Other algae and mosses have generations that are closely bound together. Fungi have multi-faceted and complicated alternations of generations. Some, as in the seed-plants, have both generations united in the one plant-body, making it impossible to recognise the alternation of generations without further examination. For information on alternation of generations in animals, *see* Reproductive Cycles.

Amanses, G, *see Monacanthus*

Amaranth, related species, F, *see* Amaranthaceae

Amaranthaceae F Magnoliatae*. 60 Ga, 900 species. Tropical to temperate, especially in America and Africa. Rarely found as small trees and bushes, more common as perennial herbs and annual shrubs. Leaves alternate or decussate. Flowers mostly small, hermaphroditic, polysymmetrical. At all costs Amaranthaceae avoid dry places, on the other hand some Ga contain marsh plants which tolerate submersion. Some species prefer nitrogen-rich areas. Cultivate in the same way as *Alternanthera**.

Amaryllidaceae. F Liliatae*. 65 Ga, 850 species. Tropics and subtropics of southern hemisphere, especially S Africa. A few species are found in northern temperate regions. Amaryllidaceae rhizomes and bulbs and their leaves come in rosettes. Flowers mostly conspicuous, hermaphroditic and polysymmetrical. Amaryllidaceae mostly colonise areas that are dry periodically. They include several types of ornamental plants. For care, *see Crinum*.

Amaryllis Plants, F, *see* Amaryllidaceae

Fig. 52 Tropical rain forest on the Amazon

Amazon (fig. 52). Largest river in S America*, situated near the equator, over 6,000 km long, up to 100 m deep and over 300 km wide where it meets the Atlantic. The drainage area of the Amazon is almost as big as Europe*, and for the greater part it has a tropical climate. Its source rivers the Marañón and Ucayali begin in the Peruvian Cordillera Mountains and altogether it takes in over 500 tributaries, including the Rio Xingú, the Rio Tapajóz, the Rio Madeira and the Rio Negro. The area around the River Amazon is the most important in S America for the richest variety of species and life-forms. Nearly all aquarium Characins and many Ga of Catfishes and Cichlids are typical inhabitants of the Amazon and its tributaries. *See Hemigrammus, Hyphessobrycon, Corydoras, Apistogramma, Cichlasoma, Geophagus, Pterophyllum* and *Symphysodon.*

The tributaries of the Amazon can be put into 3 groups according to the condition of the water.

1) White water rivers, so called on account of their high content of loam and clay particles. The water is yellow-grey in colour and sluggish, eg Rio Madeira and large stretches of the Amazon itself.
2) Clear water rivers. Water that is very transparent and yellow-green to olive-green in colour, eg Rio Tapajóz and Rio Xingú.
3) Black water rivers. Water is similarly very transparent, but is coloured dark brown because of high concentrations of tannic* and humic* acids (compare dystrophic waters), eg Rio Negro.

In general the Amazon has very clean water and an extremely low electrolyte* content. The clear water rivers are even comparable with distilled water in purity. Its total hardness is less than 0.5°, its pH value is extremely low (4.5–5), on account of a high content of free carbonic acid and a deficiency of buffer-capacity. Of all the other matter common in water, such as ions of calcium, magnesium, sodium, hydrogen carbonate, phosphate and nitrate, only the nitrates occur in such great concentrations as the free carbonic acid (this is caused by the chemical reactions of the living organisms in the water). With certain Amazon fish, such as many species of *Hyphessobrycon**, *Paracheirodon**, *Symphysodon** and *Uaru**, it is essential that the conditions of their natural habitat be reproduced, if they are to be cared for and bred succesfully (compare Water Preparation).

Amazon Dwarf Cichlid *(Apistogramma pertense) see Apistogramma*

Amazon Molly *(Poecilia formosa) see Poecilia*

Amazon Plants (various species) *see Echinodorus*

Amazon Sword Plant *(Echinodorus amazonicus) see Echinodorus*

Amazon Sword Plant *(Echinodorus bleheri) see Echinodorus*

Ambassidae, F, name not in current usage, *see* Centropomidae

Ambassis lala, name not in current usage, *see Chanda*

Amblopites RAFINESQUE, 1820. G of the Centrarchidae*. Spread throughout eastern N America. *Amblopites* is outwardly very similar to the *Lepomis** species, but is distinguished from it and other Sunfishes by the large number (6) of hard rays in the An. Up to 30 cm long. *Ambloplites* are tasty fish and are popular with anglers in N America, as well as being commercially important. Only young specimens are suitable for the aquarium. For advice on care *see Lepomis.*

– *A. rupestris* RAFINESQUE, 1817; Rock Bass, Goggle-eye. Mouth and eyes particularly large. Body and fins yellowish-brown to olive, scales partly covered in dark spots which merge into large clusters near the dorsal ridge. Iris red.

Amblycirrhites GILL, 1862. Family Cirrhitidae*. Hawkfishes. Small, solitary fish that live among the coral reefs and rocky coasts of tropical seas where they lie in wait for their prey. The tufts on the dorsal fin spines are typical of the family. From time to time the one west-Atlantic species of the family is imported:

Fig. 53 *Amblycirrhites pinos*

– *A. pinos* MOWBRAY; Red-spotted Hawkfish. Up to 8 cm. Brown-white bars with orange spots on the head and front portion of the body (fig. 53).

Amblydoras BLEEKER, 1863. G of the Doradidae*. Lowlands of the Amazon basin, Peru, Bolivia and Guyana. Thorny Catfish roughly 15 cm long. Sides with heavy, spiny scutes. 3 pairs of barbels. Body compressed, flattened anteriorly. The large Ps have a strong, toothed first fin ray. In contrast to the G *Acanthodoras**, the first ray of the D1 is not toothed. The As are prettily patterned, very peaceful fish that mostly stay hidden in the daytime. They prefer small live food, such as *Tubifex*, White worm *(Enchytraeus)* and insect larvae. They are hardy in the aquarium, not yet bred successfully.

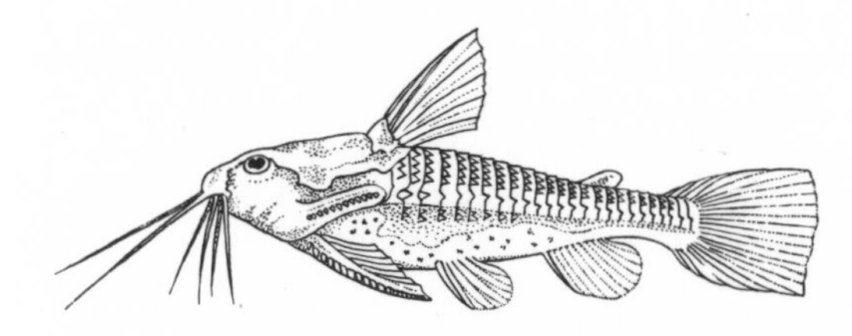

Fig. 54 *Amblydoras hancocki*

– *A. hancocki* (CUVIER and VALENCIENNES, 1840); Hancock's Amblydoras. Lowlands of the Amazon basin, Peru, Bolivia, Guyana. Up to 15 cm. Body and fins yellow-brown with irregular black-brown markings. A bright streak made up of numerous silvery dots is visible through the white thorns of the scutes. Belly of the female is dull white, that of the male is covered with brown spats. Barbels with brown and white rings. Eyes gleaming blue. These fish are said to build nests of leaves in which they lay their eggs. If removed from the water, the fish make purring noises (fig. 54).

Amblyopsidae, F, *see* Percopsiformes

Amblyopsis spelaeus *see* Percopsiformes

Ambulacrum *see* Echinodermata

Ambulia, G, name not in current usage, *see Limnophila*

Ameiuridae, F, name not in current usage, *see* Ictaluridae

American Black-nosed Dace *(Rhinichthys atratulus atratulus) see Rhinichthys*

American 'Brook Trout' *(Salvelinus fontinalis) see* Salmonoidei

American Flag Fish *(Jordanella floridae) see Jordanella*

American Frogbit, G, *see Limnobium*

American Knifefishes (Apteronotidae and Rhamphichthyidae, Fs) *see* Gymnotoidei

American Lizard's Tail *(Saururus cernuus) see* Saururaceae

American Lotus Flower *(Nelumbo lutea) see* Nymphaeaceae

American Marsh Wild Thyme *(Peplis diandra) see Peplis*

American Minnow *(Phoxinus neogaeus) see Phoxinus*

American Perch *(Perca flavescens) see Perca*

American Red Shiner *(Notropis lutrensis) see Notropis*

American Synbranchoid Eel *(Synbranchus marmoratus) see* Synbranchiformes

Amiidae, F, *see* Amiiformes

Amiiformes. Bowfins. O of the Osteichthyes*, Coh Holostei. Only one F (Amiidae) containing a single species – the 60 cm long, sometimes larger, Bowfin or Mudfish *(Amia calva)* (fig. 55). Found in eastern N America. Ancestors of Amiiformes flourished during the Jurassic and Cretaceous, and representatives of the F Amiidae are known in Europe throughout the Tertiary

Fig. 55 *Amia calva*

(67–20 million years ago). *Amia* is of particular interest to the academic ichthyologists, since, in spite of the fact that it looks like the higher bony fish (Sub-Cl Teleostei), it retains various features possessed by its ancestors. For example, the heavy bony plates surrounding the skull and backbone, and the bony gular plate between the lower jaws. Upperside olive-green, belly yellowish to orange. Body elongate, slightly compressible, the main characteristic being the very long, soft-rayed D, which stretches from the front end of the dorsal ridge almost as far as the rounded C. Short An. Just in front of the C, on the upper part of the caudal peduncle, there is a dark blotch. In the males, this blotch is much more pronounced and usually with a bright border. Thick rows of hard scales (cycloid). The inner walls of the net-like swim-bladder can be used as a lung, enabling the fish to live in oxygen-poor waters. Eats anything it can manage to overpower.

Amia spawns in May to June in areas of the bank with much vegetation. The male builds a nest out of parts of plants and guards the eggs and the newly hatched fish. In southern N America *Amia calva* is valued as a food. In Europe, the species attracts a great deal of interest in large aquaria. In general it should be treated like our native Carps. *Amia* will grow quickly if supplied with lean meat, fish or animal fish-feed.

Amitose *see* Cell Division

Ammannia LINNAEUS, 1753. G of the Lythraceae*. 20 species. Tropics and Subtropics throughout the world, in isolated cases found in temperate regions. Small perennial or annual plants with horizontal or upright stems. Decussate leaf arrangement. Leaves sessile. Flowers hermaphroditic, small, polysymmetrical. Double perianth, 4 sepals and 4 petals or corolla absent. Stamens 8–2. Carpels 4–2, fused, pointing upwards. Fruit is a capsule. All the species like flooded areas, so they are quite happy living submerged. Only recently, some species were tried out in the aquarium with varying degrees of success. Propagation is possible with cuttings.

– *A. latifolia* LINNAEUS, Broad-leaved Ammannia. America. Leaves lanceolate, pointed, bright green. Flowers have no petals.

– *A. senegalensis* LAMARCK. African Ammannia. Africa. Leaves lanceolate, rounded off, bright green. Very small petals on the flowers.

– *A.* sp. Olive-green Ammannia. Origin unknown. Leaves lanceolate to inverted oval shape, rounded off, olive-green to reddish in colour. Flowers have relatively large petals.

Ammocoete Larva *see* Cyclostomata

Ammodytes LINNAEUS, 1758, Sand Eel. F Ammodytidae, Sub-O Ammodytoidei*. Eel-like, shoaling fish widespread along the sandy coastlines of northern seas. Can burrow into the sand with the aid of a horny process on the lower jaw. Lives on plankton. Used as bait. Not very suitable for the aquarium, eg:

– *A. tobianus* LINNAEUS, 1758. N Atlantic to Baltic Sea, on sandy substrates. Up to 30 cm. Silver.

Ammodytidae, F, *see* Ammodytoidei

Ammodytoidei. Relatives of the Sand Eels (fig. 56). Sub-O Perciformes*, Sup-O Acanthopterygii. Marine fish spread throughout the world. Eel-like form with fossils dating back to the early Tertiary. Of the 2 Fs, we are only dealing with the Ammodytidae (Sand Eels). Long pointed head, prominent lower jaw, fairly wide gape to the mouth but no teeth. Fins unpaired without

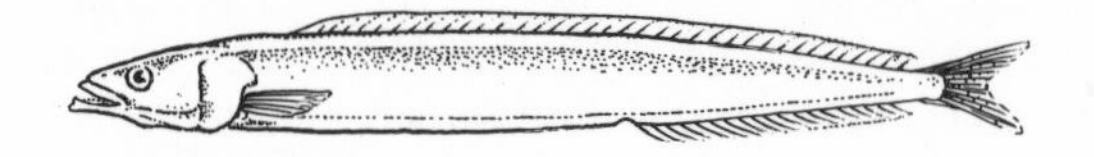

Fig. 56 Ammodytoidei-type

spines. D very long, C deeply lobed, An about half as long as the D, no Ps. The skin makes delicate folds and a longitudinal furrow near the belly. No cycloid scales, LL straight, near the dorsal ridge. No swimbladder. The small Ammodytoidei live mostly in schools or huge shoals and form an important part of the diet of larger predatory fish. They can disappear into the sand with astonishing speed. For further details see *Ammodytes**.

Ammonia. Chemical formula NH_3. A combination of nitrogen and hydrogen. Under normal conditions, ammonia is a colourless, pungent gas that is highly poisonous. Easy to dissolve in water where it may in part react to give ammonium ions and hydroxyl ions: $NH_3 + H_2O \rightarrow NH_4^+ + OH^-$ (ammonia + water = → ammonium + hydroxyl ion). In contrast to ammonia, ammonium ions are only poisonous to fish in high concentrations. Too much ammonium in the aquarium can be a problem, however, as the chemical equilibrium between ammonia and ammonium is very dependent on the pH value*. In acidic conditions, only ammonium ions are pressent, in neutral conditions (pH 7) 1% ammonia and 99% ammonium, at pH 8, 4% ammonia and 96% ammonium, at pH 9, 25% ammonia and 75% ammonium. Ammonia and ammonium ions build up in the aquarium as a result of the bacterial decay of urea and protein, ie, there are too many fish and too much food. In running water nitrogen-fixing bacteria oxidise the ammonia, changing it from nitrite to nitrate, and so rendering it largely harmless. On the other hand, ammonia poisons accumulate relatively easily in the aquarium, if ammonia is allowed to build up and the nitrogen-fixation process is not very efficient (newly-established marine aquaria) or if the redox potential* is so low that nitrogen-fixation cannot take place. In freshwater aquaria, provided the pH does not go above the neutral point, there is no real danger. When the water is changed, however, there may be the very slightest increase in pH, and suddenly large amounts of ammonia will be released from the ammonium ions. A mere 0.2–0.5 mg ammonia in 1 l water is enough to kill most fish. To prevent this happening take an ammonia reading at regular intervals. Working with a strong base medium, all the ammonium is converted into ammonia. The results obtained will show how much ammonium can be converted into ammonia, after an increase in pH. Use an alkaline solution of potassium mercuric iodide $K_2(HgJ_4)$ (Nessler's reagent). The mercuric acid amides that results colours the water being tested yellow to brown in relation to the ammonia content.

Instructions: 10 ml of the water being tested should be placed in a white cup along with 5 drops of 50% refined salt solution, 6 drops of the Nessler's reagent and 6 drops of 25% caustic soda. If the water remains clear, the ammonia concentration is lower than 0.15 mg/l; if, after 2–3 mins, a pale yellow colour appears, this indicates a level of 0.15–0.25 mg/l; if a yellow colour appears immediately, the ammonia concentration is 0.5–1.0 mg/l. A brown colour indicates a level of 3 mg/l.

Ammonia Poisoning *see* Poisoning

Ammonium *see* Ammonia

Amoebae (collective name) *see* Rhizopoda

Amphibia. The amphibians are a class of backboned animals (Vertebrata) comprising approximately 2,800 species. There are 3 orders, the Caecilians (Gymnophions), Newts and Salamanders (Caudata) and Frogs and Toads (Salientia). With a worldwide distribution, the Amphibia inhabit very different biotopes, above all water, but also dry land, shrubs and tall trees. The smallest amphibian is the Cuban Dwarf Frog *(Sminthillus limbatus)* at 1 cm long, and the largest are the Giant Salamanders of E Asia *(see Andrias)*, which can measure up to 1.5 m. Most amphibian larvae have to live in water, but the adults are nearly all suited to life on

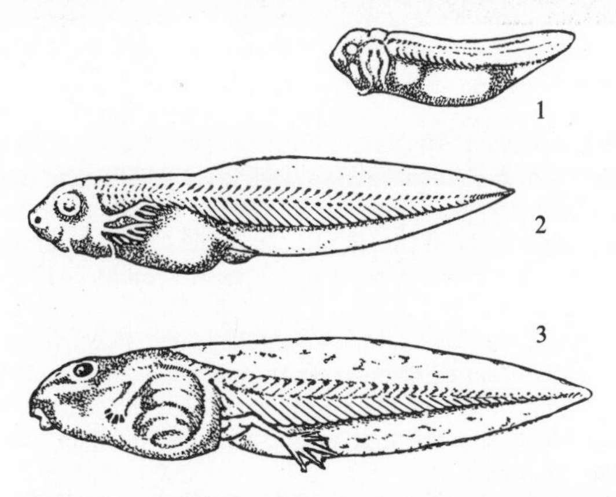

Fig. 57 Amphibia. Development of Common Frog: larval stages. 1 with gill slits 2 with external gills 3 front legs beginning to show and hind legs already formed.

land. The life-cycle involves radical changes in anatomy which take place during a metamorphosis*. The larvae possess external gills, a caudal tail with fins, and a lateral line to help orientate themselves in the water (fig. 57). After the metamorphosis, limbs appear, as well as lungs, salivary glands, multi-celled skin glands, a connecting tube between the nose and mouth and other features required for life on land. In a few species, the metamorphosis is incomplete, ie they retain

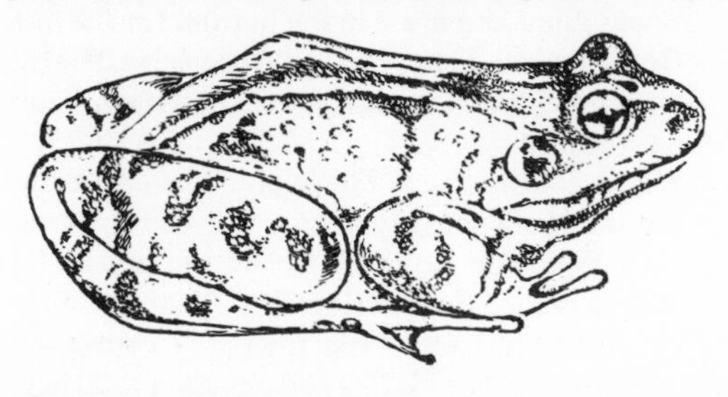

Fig. 58 Common Frog, adult

the outward appearance of the larva even when mature, eg the well-known Mexican Salamander, the Axolotl *(Ambystoma mexicana)*. All Caudata (Newts and Salamanders) retain the tail during metamorphosis, whereas Frogs and Toads lose theirs. Other changes, such as the formation of a larynx that enables the croaking noise to be produced and an 'ear' with ear-drum are peculiar to the Frogs and Toads. Their larvae, or tadpoles, feed on plants, while the adults, like nearly all amphibians, are predators. The same is true of *Xenopus laevis* (a tropical African Frog often kept in aquaria) and the dwarf variety, *Hymenochirus, Pseudohymenochirus.* They belong to the tongueless Frogs (Pipidae) which eat underwater and have a kind of lateral line for detecting the slightest water movements. As a rule, amphibians reproduce by laying eggs in water which are then fertilised. The Caecilians and various other amphibians, eg the Fire Salamander *(Salamandra salamandra)*, give birth to live young. Some species take care of their young, eg the European Midwife Toad *(Alytes obstetricans)* or the colourful tree-climbing frogs (Dendrobatinae) of the American Tropics. In these species the eggs, and sometimes the larvae, are carried by the male.

Amphibians, Cl, *see* Amphibia

Amphibious. From the Greek *amphi* meaning twofold. Used to describe animals that live in two different habitats, usually attributed to the water-land combination. For example, the word can be applied to most amphibians, to the Mudskippers (Pteriophthalmidae*) and the Lungfishes (Dipnoi*) among the fish. Plants are made up of both aquatic and terrestrial plants, which may look very different from one another but nevertheless belong to one and the same species, eg Amphibious Bistort *(Polygonum amphibium)*.

Fig. 59 *Phractura ansorgei*

Amphibious Bistort *(Polygonum amphibium) see* Polygonaceae

Amphiliidae, F, Siluriformes*, Sup-O Ostariophysi. Small Catfishes that live in flowing waters of tropical Africa with a body similar to that of the armoured Catfishes, Loricariidae*. Very elongate body. The round trunk of the body ends abruptly in a very long thin tail region. Head has a fairly pointed snout, mouth ventral with 2–3 pairs of short barbels on either side. 1–2 rows of bony scutes can form along the sides. Adipose fin present. In contrast to the Loricariidae, the pupils are round, ie without lids. One member of the family, *Phractura ansorgei* (fig. 59), was imported from the lower Niger and observed in the aquarium by W. FOERSCH. It is up to 6 cm long, fawn-coloured, with dark markings. The fish likes to hide among the vegetation and eats live food as well as soaked spinach. During the breeding season the male becomes more red-brown in colour and makes chirping noises. Pairing takes place in open water, the male bending his body round the female's head in a U-shape. In this way, the female's mouth is placed near the male sexual orifice. The sperm may be transported to the eggs (which have already been ejected by the female along with the water being breathed in via the mouth and gill slits. The eggs are a green colour, about 1 mm across, and they hatch after 3 days. These fish are difficult to breed.

Amphioxus *(Branchiostoma lanceolatum) see Acrania*

Amphipnoidae, F, *see* Synbranchiformes

Amphipnous cuchia *see* Synbranchiformes

Amphipoda. Amphipods Large order of higher crustaceans (Malacostraca*), with many species. Most species live in the sea, but a few make their way into ponds, streams and underground water as well as fresh water. Other species live as parasites on jellyfish and whales, and in the tropics some even live on land. Amphipods range in size from a few mm to several cm; most are compressed to some extent and have long plates on the sides of the body. This produces a sunken-in, groove-like belly along which flows the water being breathed in and out. The eggs and newly hatched young are kept here too. The head and first segment of the thorax are fused into a short cephalothorax, then come 7 free thorax segments and 6 abdomen segments, all of them bearing limbs. Light-sensitive organs are sessile, complex eyes lying next to the antennae. Most Amphipods are good swimmers, whilst the very compressed species move around by contraction of the abdomen on one side of the body. Economically their value as a foodstuff for many fishes and birds is not to be underestimated. In the aquarium, too, Amphipoda are a popular food, eg members of the well known G *Gammarus**, Freshwater Shrimps.

Amphiprion BLOCH and SCHNEIDER, 1801. Anemonefishes, Clownfishes. F Pomacentridae*. These fish live in the tropical Indo-Pacific and the Red Sea. All the species live on coral reefs in close association with large sea-anemones (Actiniaria*, Stoichactis*), which is where the name 'Anemonefishes' comes from. The fish are ungainly swimmers and never stray far from the protection of the sea-anemone. They form binding relationships and defend a territory. The spawn is laid very close to sea-anemones on a rock that has been cleaned. There, it is guarded. The newly-hatched fish feed on plankton and are not looked after. Like many territorial coral fish, the *Amphiprion* are brilliantly coloured. The colouring of many species is not always accurate. In the aquarium most *Amphiprion* live well with or without sea anemones; *A. percula* has survived for up to 9 years! They will also nestle inside Mediterranean sea-anemones. Many have spawned in the aquarium and with some the minute plankton-eating larvae have survived. The following species are frequently imported.

– *A. bicinctus* RÜPPEL, 1828, Two-banded Anemonefish, Red Sea Clownfish. Indo-Pacific, Red Sea. Up to 12 cm. Orange-brown with a broad, white band on the head and a narrower one in the middle of the body. In juveniles there is a third band on the caudal peduncle.

– *A. clarki* (BENNETT, 1830); Chocolate Clownfish. Indo-Pacific. Up to 13 cm. 3 white bands on a black-brown background, the second band not running across the D. Fins yellow to blackish (fig. 60).

Fig. 60 *Amphiprion clarki*

– *A. ephippium* BLOCH, 1790, Tomato Clown; Fire Clown; Saddle Anemonefish, Black-backed Anemonefish. Indo-Pacific. Up to 12 cm. Dark red with an indistinct dark spot on the flanks. Juveniles have a white band on the head. Very similar to *A. frenatus* (BREVOORT, 1856).

– *A. ocellaris* CUVIER and VALENCIENNES, 1830; Common Clownfish or Orange Anemonefish. Indo-Pacific. Up to 9 cm. Male smaller. Orange, with 3 broad, white bands. The species that is imported most often, but rather delicate.

– *A. perideraion* BLEEKER, 1855; Pink Skunk Clown, False Skunk-striped Anemonefish. Indo-Pacific. Up to 7.5 cm. Bright flesh colour with white stripes on the back and a white band on the head.

– *A. sebae* BLEEKER, 1853; Black Anemonefish; Black Clown. Indo-Pacific. Up to 12 cm. Dark brown with 2 broad white bands, the second one continuing to the D. In juveniles there is a third band on the caudal peduncle. C yellow.

Amphiprionidae, F, name not in current usage, *see* Pomacentridae

Amphiura FORBES, 1843. G of the Ophiuroidea* (Brittlestars), possessing very long arms equipped with short spines. They lie buried in the bottom for the most part. A few are luminous, eg *A. filiformis* (O. F. MÜLLER), a species which is found from the North Sea to the Mediterranean.

Ampullaria LAMARCK, 1799; Apple Snail, Infusoria Snail. G of the Mollusca (Molluscs), Sub-Cl Prosobranchia*. Very large (up to 7 cm) freshwater snails that come from tropical S America and Asia. In shape, colour and way of life, not unlike the native Freshwater Winkle (*Viviparus**) with darker bands on a horn-coloured shell. Ampullaria possess both gills and lungs, so they lead an amphibious life. In addition, these animals possess a long breathing tube which can be stuck up into the air and used like a snorkel. Ampullaria make interesting specimens and should be kept either as a small group in a single-species aquarium or as a single specimen in a correspondingly large community tank. The water is liable to cloud quickly as a result of the high metabolic rate of Ampullaria or because it has a voracious appetite for aquatic plants. The latter problem occurs if there is not enough algae or remains of food, but it can be avoided by giving the snails lettuce leaves which they love to eat. Ampullaria reproduce happily in small tanks, the pink eggs being laid out of water. These snails are often kept in fry tanks and fed on lettuce leaves. Their droppings produce natural infusoria which the young fish eat.

Amur Pike *(Esox reicherti) see* Esocoidei

Anabaena azollae *see Azolla*

Anabantidae, F, *see* Anabantoidei

Anabantoidei. Relatives of the Climbing Perches. Sub-O Perciformes*, Sup-O Acanthopterygii. Small to medium-sized freshwater fishes from Africa, S and E Asia as far as Korea, the Philippines and Indonesia. Sometimes found in estuaries, and fossils date from the Lower Tertiary. All members of the Sub-O possess a comprehensive accessory respiratory organ, the so-called labyrinth, which lies in the gill chamber. This is why these fish are also known as Labyrinth Fishes, a term that is sometimes used in a wider sense to include the Channiformes*, which possess a similar organ. The labyrinth is situated in a protruding part of the gill chamber that points in the direction of the nape. It is formed after the embryonic stage from the top-most bone (epibranch) of the first gill arch and consists of lamellae which are folded like the leaves of a loose cabbage. Blood vessels criss-cross the lamellae. Most Anabantoidei can no longer obtain enough oxygen with their gills alone, and must regularly take in air from the surface. They suffocate even when conditions in the water are good if access to air is blocked off (eg as an experiment, netting placed in the water). Only the African species are more independent from the need to breathe air. On the other hand the possession of such an organ allows the Anabantoidei to live in waters which are oxygen-poor, either because the water is too warm or because of impurities (eg flooded rice fields). There are even reports that in the early hours of the morning, when the humidity level is high, or after downpours, *Anabas testudineus* leaves the water and slithers across the ground to reach other waters. The Anabantoidei are elongate, moderately tall (or less commonly, very tall) and compressed. In part, like perches to look at. D very variable, long with numerous spines and soft rays, or short and like a pennant with only one spine. C mostly rounded, in a few cases indented, occasionally with elongated middle rays in the males. An long, most reaching the breast region, with several spines in the front part. The D, C and An can border so closely on one another that a complete fin corona is formed (eg in many *Ctenopoma* species). In other Ga, the C and An are elongated into a point. Vs long, in some Ga drawn out into separate elongated fin rays, in others the first soft ray is drawn into a whip-like appendage, which can be moved in all directions. It serves as an organ of touch, and possible, of taste. Ps strong. Thick covering of ctenoid scales, swimbladder elongated at the back and often divided into two in the tail region. The Anabantoidei prefer waters that are often muddy, with vegetation. As a rule, these fish do not form shoals but do establish territories. The larger species at least of the F Anabantidae (Climbing Perches) live entirely as predators. The other individuals either live on small animals or are omnivores.

The way these fish breed is very interesting. In the typical case the males build floating nests of foam from bubbles of air containing spittle and parts of plants. Pairing takes place beneath the nest when the male twists himself round the female from the side and turns her onto her back. As a rule, the eggs are lighter than water, so they float upwards and into the foam nest. Until the young fish start to eat, the male continues to repair the foam nest. Independently of one another, several Ga (eg the *Betta* species and *Sphaerichthys osphromenoides*) have developed into fish that brood the eggs in the mouth. Only a very few Anabantoidei do not care for their young at all.

The tall, 60-cm-long Gourami *(Osphronemus goramy)* is most important economically. In SE Asia this excellent food-fish often takes the place of Carps. It grows very quickly and brings profitable returns even in unfavourable living conditions. Other species are used as a source of protein, but today only occasionally, eg during times of famine. In some places, fights between males of the *Betta splendens* species are staged in public, with bets being wagered.

A separate discussion of each family is unnecessary as all of them are very similar. For similar reasons, all the species of this group of fish are sometimes included under the one F, the Anabantidae. GREENWOOD and his colleagues (1966) distinguished 4 families. A great many Anabantoidei make ideal aquarium fish. Their brilliant colouring, undemanding nature and, at least in some cases, the ease with which they can be bred, make them popular throughout the world. For further details turn to the particular G from the following Fs:

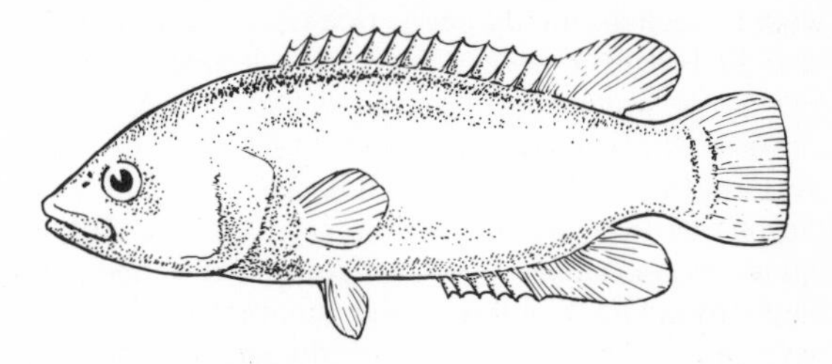

Fig. 61 Anabantidae-type

Anabantidae (Climbing Perches (fig. 61)): *Anabas, Ctenopoma.* Belontiidae (Bettas or Fighting Fishes (fig. 62)): *Belontia, Betta, Colisa, Ctenops, Macropodus, Parosphromenus, Sphaerichthys, Trichogaster, Trichopsis* (Gouramies). Helostomatidae (Kissing Gouramies (fig. 63)): *Helostoma.* Osphronemidae (fig. 64): *Osphronemus.*

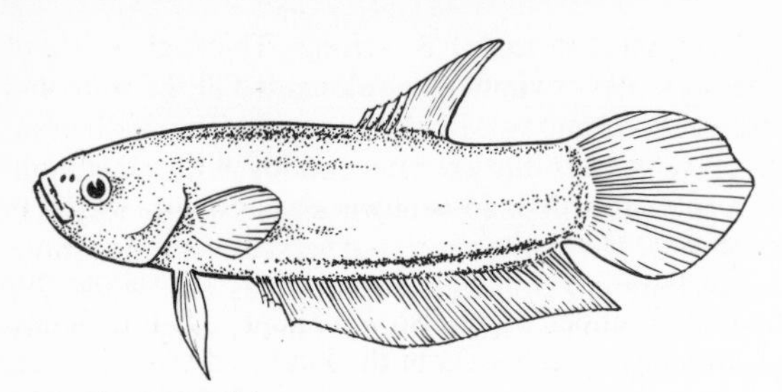

Fig. 62 Belontiidae-type

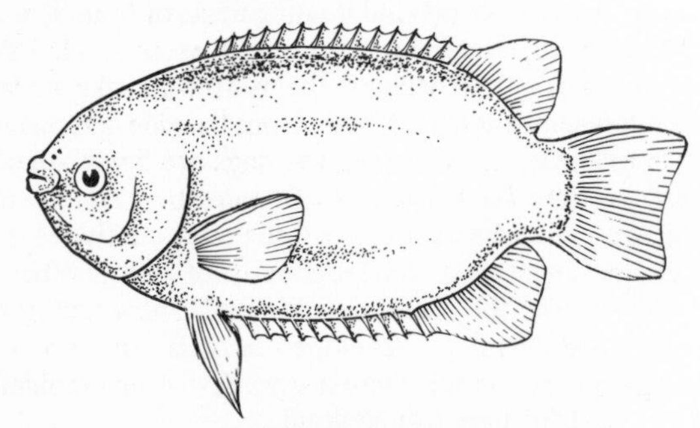

Fig. 63 Helostomatidae-type

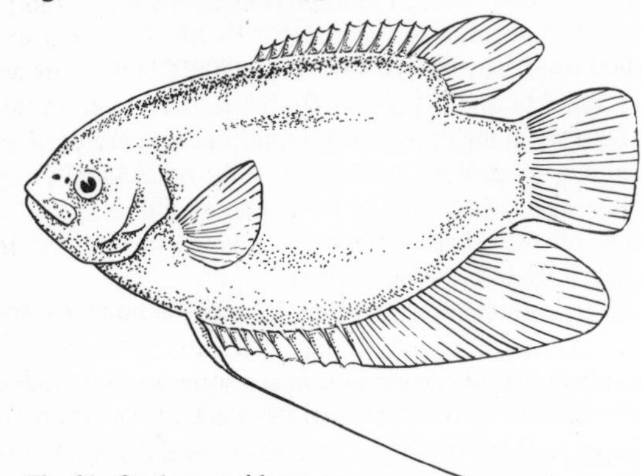

Fig. 64 Osphronemidae-type

Anabas CUVIER, 1817; Climbing Perch. G of the Anabantidae, Sub-O Anabantoidei*. Tropical Asia, from India and Sri Lanka (Ceylon) to S China, Malaysia, Indonesia, the Philippines. In slow-flowing and stagnant waters, such as muddy pools, rice fields, weedy rivers, also in brackish water. Especially at night, *Anabas* leaves the water and, if the air is humid, it can move considerable distances over land with the help of its tail, pectoral fins and operculum. It is capable of surviving periods of drought in the mud. Up to 25 cm. Robust, elongated, perch-like body, with large scales and especially strong Ps. Relatively large mouth, the jaws filled with teeth. The outer edge of the operculum bears numerous spines. The ground colour of older specimens is grey-brown to grey-green, the edges of the scales and fins being brighter. A dark spot appears on the operculum and at the root of the tail. These fish are somewhat intolerant, and should be kept in a roomy, well-planted tank at a temperature of 20–25 °C. Feed with live food and plant matter. For breeding, temperature should be 25–30 °C. Not a builder of foam-nests. Just one species:

– *A. testudineus* BLOCH, 1795; Climbing Perch. An is not rounded in the male. In its native home, a food fish.

Anablepidae. Four Eyes. F of the Atherinomorpha, Sub-O Cyprinodontoidei*. A very small, very interesting, livebearing F (2 species) spread throughout S Mexico down to northern S America. Typical surface fish with very elongate bodies and a flattened head. Very large, protruding eyes like a frog. The cornea, pupil and retina, but not the lens, of each eye is divided across the centre by a band of tissue. Usually, the fish swim with the upper part of the eye above the surface and the lower part below. The asymmetrical nature of the lens allows an image of distant objects above the surface to be formed in the lower part of the upper eye, and near objects under the water to be seen in the upper part of the retina of the lower eye. Through observation,

Fig. 65 *Anableps anableps*

it appears that the upper eye is used in the recognition of enemies. If trying to escape, these fish leap right out of the water. The sexual organs are arranged in a similar way to those of the Jenynsiidae*. In Anablepidae populations, the ratios of sinistral and dextral males and females are roughly 25 % each. The females give birth twice a year to a small number of babies, which are already 3–5 cm long at birth. The well-known species Four Eyes *(Anableps anableps)* (fig. 65) can reach up to 20 cm in length and is often kept in aquaria. These fish need a large area of surface water and mainly insects for food. Keep well covered.

Anacanthini, O, name not in current usage, *see* Gadiformes

Anadromous Fish. Species that breed in fresh water, therefore classified as part of freshwater fauna, but as juveniles they return to the sea to reap the benefits of the richer supply of food that that environment has to offer. Anadromous fish return to fresh water to breed, sometimes returning to the same place (river or stream) that their egg development took place. It has been proved that Salmon (*see* Salmonoidei) use their sense of smell to trace their old habitats. Salmon and some Lampreys (Petromyzoniformes*) only spawn once. Other anadromous fish, like certain Sturgeons (Acipenseriformes*), return to breed in fresh water every so often. Opposite to catadromous fish.*

Anaerobes. Organisms that live permanently or partially without free oxygen from the air. Many fungi, bacteria and numerous parasites are anaerobes. They get their energy, for example, through fermentation. This is a process whereby the organic foodstuffs are not entirely used up as energy, and as well as carbon dioxide other characteristic metabolic end products, such as alcohol, lactic acid and butyric acid, are produced. Insufficient air supply to the substrate in an aquarium encourages the existence of anaerobic micro-organisms, and their metabolic products are poisonous to fish and plants. Opposite to aerobes*.

Anaesthetics. These substances bring about a reversible, temporary 'switching off' of the central functions of the nervous system, in particular the centres of pain, movement and consciousness. The metabolic processes in the organism are severely limited at the same time. There is a large number of substances, with very different chemical structures, that are known to produce an anaesthetising effect. Anaesthetics are being used on fish increasingly. There are lots of circumstances under which it is a good idea, if not essential, to put a fish to 'sleep' for a short while. These include marking and determining the exact species a fish represents, finding out about disease processes or during operations, experimental work in the study of spawning specimens in relation to the artificial breeding of fish. MS 222-Sandoz® (methylsulphonate of the m-aminobenzoic-acidic-ethylester) has proved of particular value in fish. This substance dissolves very well in water (fresh and sea), takes effect quickly and does not produce any disadvantageous side-effects. MS 222 is applied in a concentration of 1 g to 2–4 litres of water. It takes effect after 15–60 seconds; obviously, its speed will depend on the water temperature, the concentration applied, and the size of the fish (as well as its species). With very large fish, a solution of 1 g MS 222 to 1 litre of water can be sprayed in through the gill cavity. Weak concentrations (1 g to about 25 litres of water) reduce metabolism, especially oxygen requirements, and so they are used when transporting fish, to cut down on losses (Transport of Aquatic Animals*). Other anaesthetics that are used include ethyl urethane (3–10 g to 1 litre of water) and diethylether (use with care!) (10–15 g to 1 litre of water). Whatever is used, the fish must be observed afterwards to note their reactions, and if necessary, individual doses of the anaesthetic should be calculated. Invertebrate animals must also be anaesthetised before being preserved (Conservation of Aquatic Animals*) to avoid too violent contractions in the preserving fluid. Most Coelenterates (Coelenterata*) can also be anaesthetised by carefully adding, drop by drop, a 30% solution of magnesium chloride ($MgCl_2$) to the water in which they are being kept; add sufficient until the water contains about 1% of the salt. Several hours may pass before the effect is fully achieved. Mollusca*, Echinodermata* and other invertebrate animals from fresh and sea water can be anaesthetised in the same way, or by carefully adding alcohol or nicotine. Some invertebrates can only be anaesthetised with very special substances, that are usually extremely toxic for the human organism. Giving an animal 3 to 5 times the dose of an anaesthetic will kill it safely and painlessly (Fish, Killing of*).

Anal Fin *see* Fins

Analogous Organs. Parts of the body with the same function and often similar structure that have arisen during phylogenesis in unrelated plant or animal groups and that develop in dissimilar ways during ontogenesis* from different embryonal hereditary factors. For example, the wings of insects and birds, and the dorsal fins of fish and whales, are analogous organs. *See also* Homologous Organs.

Anamnesis. Pre-illness history. Details about the conditions under which fish were kept prior to an illness give important clues about the kind of disease it might be. In many instances it is only possible to give an exact diagnosis* only after a carefully conducted anamnesis, taking into account all the possible factors. Consideration of the following is essential: behaviour and appearance of the fish before the illness or death; progress of disease over a certain period and the period of greatest damage; conditions of water, temperature and lighting, as well as type of feed and contact with newly-introduced fish.

Anarrhichadidae, F, *see* Blennioidei

Anarrhichas, G, *see* Blennioidei

Anatomy is the science of the external and internal structure of animals and man. Macroscopic anatomy is the study of the outer form of organs or parts of the body, microscopic anatomy is the study of their internal structure (with the aid of a microscope). Included in microscopic anatomy are histology* and cell structure. Comparative anatomy examines corresponding organs and organ systems in different systematic categories, eg the structure of the skeleton or brain in vertebrates from the Lampreys to the mammals. In this way, the underlying principles of organisation of the individual body organs as well as the whole body is made clear. Similarities or disparities between certain structural features can tell you how closely related living, as well as fossil, organisms are. Comparative anatomy is therefore very important in taxonomy* and palaeontology*. In botany, anatomy is generally taken to mean the inner structure of plants.

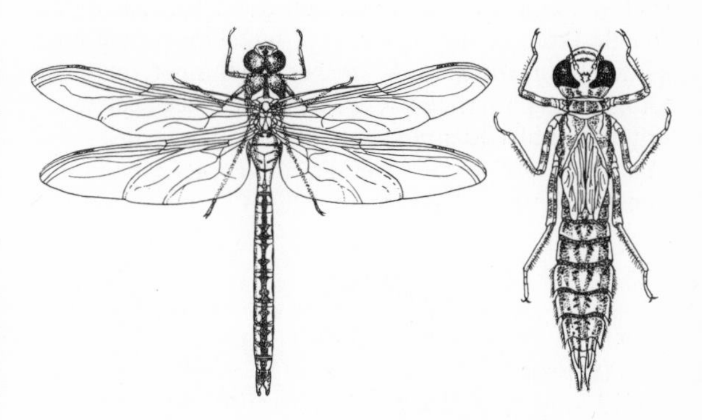

Fig. 66 *Anax imperator.* Left, imago; right, larva

Anax LEACH, 1830; Dragonflies (fig. 66). Widespread G of the Dragonfly O (Odonata*). In Europe, 2 species are common in still waters. Their 6 cm long larvae are not unlike those of *Aeschna**, and they make equally interesting specimens in pond aquaria.

Anchovies (Engraulidae, F) *see* Clupeiformes

Anchovy (*Trissocles*, sometimes spelt *Thrissocles, setifer*) *see* Clupeiformes

Ancistrus KNER, 1853. G of the Loricariidae*. Armoured Catfishes of the Amazon region and Guyana. Bottom-dwelling fish that like to cling to rocks and plants with their sucking mouths. Can be distinguished from the closely related G *Hypostomus** by the numerous antennae-like processes on the edge of the snout and by the structure of the operculum, especially the movable interoperculum with its hooked, pointed spines. The body is covered with bony scutes which overlap like roof-tiles. Head and front half of the body are much flattened. D1 is large, fan-shaped, but mostly laid against the body, D2 small, C broad, cut obliquely from top to bottom, An very small. The Vs and Ps are broad and bent outwards at the side. Recently, *A. dolichopterus* has been bred with more success, and FRANCKE has had a number of successes with *A. cirrhosus* or *A. temmincki*. After much forceful swimming by the male, the fishes spawn in narrow plastic tubes. The eggs are very large, about 3 mm in diameter, and a brilliant orange colour. They are laid in a clump and intensively cared for by the male. After 5 days, up to 200 babies hatch, each with a large, orange-coloured yolk sac. In the next few days, the fish darken, but they mostly hang in a cluster on the roof of the inside of the tube under the watchful eye of the male. About 7–9 days after hatching they leave the tube. By this time they are about 14 mm long. They are easy to rear on finely ground powdered food and tender lettuce leaves. Adults and juveniles require a lot of vegetable food as well as animal food; wood, in particular as ballast for the digestion; and pieces of leached wood for rasping. Otherwise, deficiency diseases may result, causing the deaths of most of the fish. All the species are relatively unaffected by low temperatures. Hiding-places will allow the fishes to feel more at home. Differences between the sexes: in the male, tentacles on the forehead and on the edge of the snout are very long and often forked; in the female there is only one row of tentacles on the edge of the snout, and they are shorter and thinner, or absent altogether. The *Ancistrus* are by far the best destroyers of algae of all aquarium fish, even *Gyrinocheilus**. One *Ancistrus* will keep a well-lit, 100-cm-long aquarium completely free of algae.

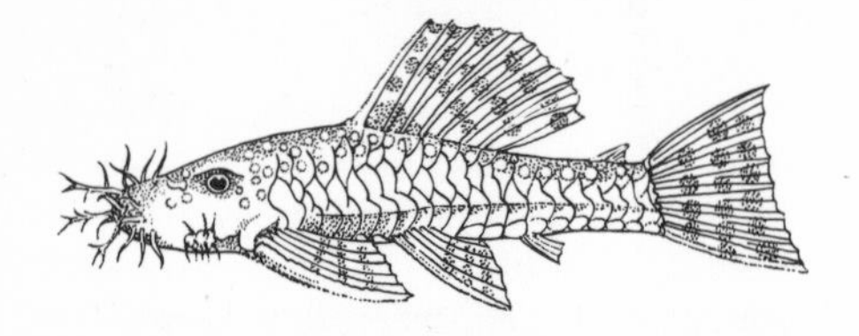

Fig. 67 *Ancistrus cirrhosus*

– *A. cirrhosus* CUVIER and VALENCIENNES, 1840; Bristlemouth. Up to 14 cm, most smaller. Body dark olive-brown, belly bright grey-green to fawn-coloured, with a large number of irregular bright spots. D1 large, with dark speckles, and a large black spot at the base. There are also 3–4 rows of dark speckles on the C, Vs and Ps, and the small An is colourless (fig. 67).

– *A. dolichopterus* KNER, 1854; Bushmouth or Blue Antenna Catfish. Amazon region and Guyana. Up to 13 cm, most remaining smaller. Body dark brown to blackish with a blue-black shimmer and dark speckles. Belly brighter. Fins black-blue with white speckles. D1 and C have bright spots and a white border. The juveniles are more vividly coloured. Several successes with breeding (fig. 68).

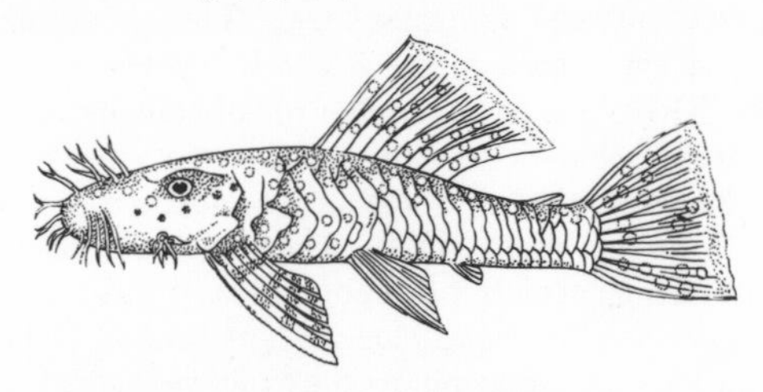

Fig. 68 *Ancistrus dolichopterus*

Ancylus O. F. MÜLLER, 1774 (fig. 69). G of the Basommatophora*. Limpets with uncoiled, basin-like carapace. Up to 0.5 cm, rarely up to 1 cm long. *Ancylus* lives in clear, sub-Alpine streams, but also occurs in flowing waters on plains and in the surf of lakes. Lives on stones (in contrast to *Acrolexus**), almost sessile and feeds on blankets of algae.

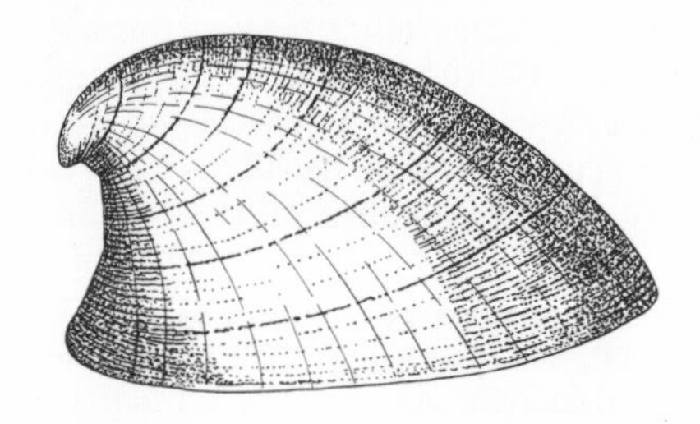

Fig. 69 *Ancylus fluviatilis*

Anemogamy *see* Pollination

Anemone Crab *(Petrolisthes oshimae) see* Porcellanidae

Anemone Hermit Crab *(Eupagurus prideauxi) see Eupagurus*

Anemonefishes. Tropical fish of the F Pomacentridae* that live in close association with sea-anemones (Actiniaria*). In particular, the fish of the Ga *Amphiprion** and *Premnas**, as well as young *Dascyllus* trimaculatus*, live in giant sea-anemones of the G *Stoichactis** in the tropical Indo-Pacific. While all other organisms are stung by the slightest brush against the anemones, the Anemonefishes can move about unharmed between the tentacles and so enjoy an effective protection. It is not clear what benefit the sea-anemone gets out of this association, but perhaps the Anemonefishes keep the anemone-eating fishes (Chaetodontidae*) at bay. The fish are not immune to the poison, however, but they do not trigger off the explosion that releases the stinging cells*. The release of these cells is suppressed by mucus on the fish. It may not be just the fish's own mucus, but that of the sea-anemone as well that combine to prevent the explosion of the cells whenever the tentacles are touched. The fish renew this protective layer continually by bathing in the tentacles; they lose the protection if they are separated from the anemone for any length of time. The Anemonefishes either live alone or in pairs in a particular sea-anemone. They do not stray far and lay their eggs nearby. In the aquarium anemones that are too small will mostly be exhausted by several fish. The fish will accept larger Mediterranean sea-anemones as substitutes.

Anemones, O, *see* Zoantharia

Anemonia RISSO, 1826. G of the Actiniaria* (Sea-anemones), F Actiniidae. Large sea-anemones with long tentacles that cannot be fully retracted, powerful sting, very sticky. Often found with Zooxanthellae*. Only one species is found in Europe:

Fig. 70 *Anemonia sulcata*

– *A. sulcata* (PENNANT, 1766); Snakelocks or Opelet Anemone. Mediterranean, E Atlantic. Often found in large numbers in top part of the littoral, the base being attached not all that firmly below stones or in crevices. Up to 20 cm. Brown to greenish, often with tentacles tipped with violet. Loves warmth and light, and hardy in the aquarium (fig. 70).

Angelfish *(Pterophyllum scalare) see Pterophyllum*

Angelfishes, G, *see Angelichthys*

Angelichthys JORDAN and EVERMANN, 1896; Angelfishes. G of the Chaetodonitidae*. Sub-F Pomacanthinae. This generic name is sometimes applied to some of the Angelfishes of tropical W Africa (eg *A. ciliaris,* LINNAEUS, 1758). However, they should all probably be placed under the G *Holacanthus**.

Angelsharks, F, *see* Squatinidae

Angiospermae, Sub-D, *see* Spermatophyta

Angiosperms (Angiospermae, Sub-D, *see* Spermatophyta

Angium. A structure found in mosses (Bryophyta*) and ferns (Pteridophyta*) in which the reproductive cells are made. It is made up of the large number of cells, so, in contrast to the algal cyst*, the outer wall consists of a single layer of sterile cells. If the structure produces asexual reproductive cells (spores*), it is called a *sporangium*; if it produces sexual reproductive cells (gametes*), it is known as a *gametangium.* The female gametangium is referred to as the *archegonium*, the male as the *antheridium.*

Anglerfish *(Lophius piscatorius) see* Lophiiformes

Anglerfishes, G, *see Antennarius*

Anglerfishes (Antennariidae, F) *see* Lophiiformes

Anglerfishes (Lophiidae, F) *see* Lophiiformes

Anglerfish-types, O, *see* Lophiiformes

Anguillidae, F, *see* Anguilliformes

Anguilliformes; Eel types. O Osteichthyes*. Coh Teleostei. Elongated, snake-like fish that live only in the sea, with the exception of the Anguillidae F. In most species, the head is pointed and the mouth stretches to below the small eyes, or even farther. Nasal openings often tubular, elongated on the outside. Operculum often reduced. All modern species lack ventral fins. D is very long. It often unites with the C and long An to form a fin corona of uniformly low height. Ps present. None of the fins has hard rays. The shoulder region is joined to the long vertebral column, not the skull. Scales either absent or tiny and deeply embedded. LL complete. Many Anguilliformes have a willow-leaf-shaped, pelagic larva or leptocephalus that turns into a young eel (metamorphosis). There are approximately 350 modern species of eel belonging to 22 Fs, of which nearly all belong to the Sub-O Anguilloidei. Only 3 of the Fs confined to deep waters are members of the Sub-O Saccopharyngoidei. No-one is entirely sure how to classify them. At one time they were placed in their own O (Lyomeri). The following descriptions refer only to the most important Fs.

– *Anguillidae.* True Eels. Characteristics of the well-known F are well-formed Ps, long ventral fin reaching

Fig. 71 *Anguilla anguilla*

to the anus and tiny, deeply embedded scales. The members of this F are the only ones to live for a time in fresh water. The life-cycle of the European or Common Eel *(Anguilla anguilla)* (fig. 71) is typical of all the species. It was described around 1920 after much observation, most importantly by the Dane, J. SCHMIDT. In European waters there are 2 types of eel, the predominantly fish-eating broad-nosed eel and the pointed-nosed eel, which eats water insects, crustaceans and molluscs. The sex of young eels is barely detectable. Only when they are about 30 cm long do they develop into males and females. Males are fully grown at about 50 cm long, females can grow up to 1 m long and have strong pectoral fins. They stay in fresh water for between 9 and 12 years, until they reach sexual maturity. This sexually mature phase is induced by hormones and is characterised by the eyes getting bigger, by the belly becoming a bright silver, and by the intestine regressing to a useless cord. On moonlit evenings in late summer these so-called Silver Eels swim down-river to the sea, then journey westwards from the European coasts to the Sargasso Sea in the area of Bermuda. They find their way with the help of a sun compass mechanism. This 5,000 km trip takes them about 18 months, during which time they live on their own body substance. About 400 m down in the seaweed-strewn Sargasso Sea lies a layer of water which has a higher salt content and a temperature of around 17 °C. This zone is the real spawning

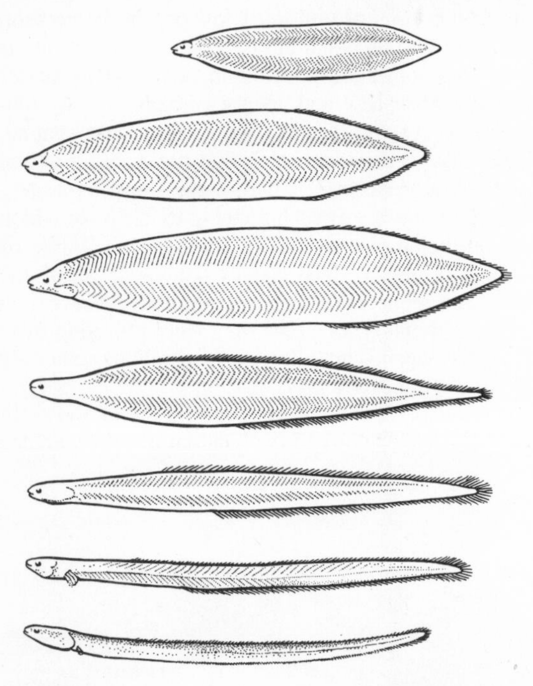

Fig. 72 Leptocephalus and how it changes into a young eel

ground as well as the grave of our eels, for they die directly after spawning. The eggs float to the surface where they hatch into small, willow-leaf-shaped, transparent larvae. In about 3 years they are once more carried to the European coasts by the Gulf Stream, during which time they feed on plankton. In Europe the leptocephali (fig. 72) turn into Glass-eels until as Elvers they ascend the rivers and reach further inland. From time to time the Elvers may even travel short distances across marshy ground or use underground waters. Eels are typical bottom-dwelling fish that hunt for food at night. Eel blood contains a nerve poison (neurotoxin) that is destroyed when cooked or smoked. Other species occur in N America *(Anguilla rostrata)*, around Japan *(Anguilla japonica)* and around Australia. All are very important economically. Only in N America is the demand relatively small. The fishing industries of many countries have a proportionately high eel catch. In the aquarium, eels live well for many years, although they are rarely seen on account of their tendency to lie hidden and become active at night. Feeding is no problem and the tank must be well covered.

– *Congridae;* Marine Eels. Anguilliformes that are very similar to the True Eels (Anguillidae) but which can be differentiated by various details of the skull, the teeth, and the gill cover slits, which can reach fairly close to the throat. Besides this, the Marine Eels have no scales

Fig. 73 *Conger conger*

at all, unlike the Anguillidae. On the other hand, their method of reproduction is largely the same as in the True Eels. Upon reaching sexual maturity, food intake ceases, the intestine regresses and the eggs are reputedly laid at great depths where they develop into leptocephali, ie the willow-leaf-shaped larvae of many Anguilliformes. The best-known species, the Conger Eel *(Conger conger)* (fig. 73), is found in the northern E Atlantic, especially around the British Isles, often in the English Channel, but is not known in the Mediterranean and the western Black Sea. Females can be up to 2 m long, sometimes even bigger, but the males remain smaller. Notable features are the darker dorsal surface and bright ventral surface, as well as the black edge of the tips of the fins. The Conger Eel prefers rocky places at depths of 40–100 m. It feeds on fish, squids and crustaceans. In SW Europe it is very important economically, but is nowhere near as good a quality fish as the European Eel. According to NORMAN, extracts from the Conger Eel flesh are used in turtle soup. In nets, congers can cause considerable harm; they are prized by anglers. *Conger oceanicus* is found off the coasts of N America and *Conger esculentus* in the Caribbean Sea.

– *Moringuidae;* Worm Eels (fig. 74). Worm-like eels that live in sand and mud of tropical regions. Distinguished from the closely related Moray Eels by the reduced set of teeth, the small Ps, the straight edge to

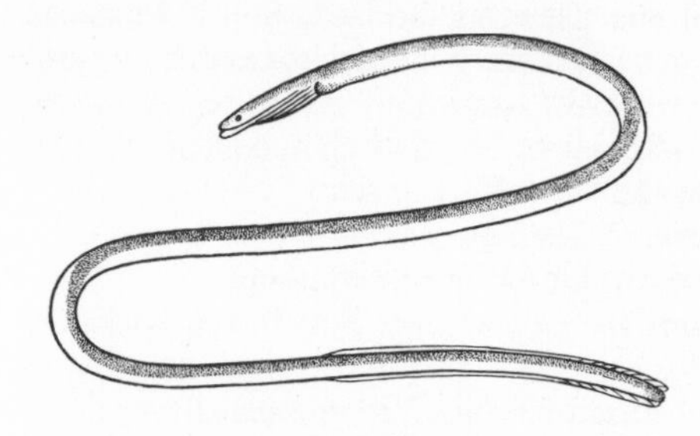

Fig. 74 *Moringus javanicus*

the tail, and the far-back position of the anus. One lip, usually the upper one, is enlarged and extended in front of the other like a flap. The way the lips thicken can vary between male and female. Worm Eels bury themselves head-first in the bottom. According to GOSLINE, the females, but not the males, undergo a marked transformation as they become sexually mature. Their eyes enlarge, the height of their fins increases and their colour changes from yellow to a silvery iridescence. It is thought that the sexually mature fish leave the bottom and live pelagically from then on.

– *Muraenidae;* Moray Eels (fig. 75). Cylindrical or laterally flattened fishes without paired fins. Head pointed, mouth opening small and very deep, reaching far behind the eyes. Jaws and roof of the mouth are usually beset with rows of pointed teeth that are easily bent backwards. 2 nasal openings on each side, which are tubular in many species, elongated on the outside. External gill opening forms a hole. The D, C and An often form a single low fin which often lies completely hidden in the skin. The skin is leathery and without scales. The 80 or so species almost all live in coral reefs and rocky coastal areas of warm and temperate seas. Moray Eels are typical crepuscular and nocturnal fishes, lying hidden in cracks and crevices during the day. Divers mostly spot the heads of the eels as they are poked out of the hiding places. They may look a little threatening as the mouth is always slightly open for breathing, thus revealing the teeth to some extent. However, Morays only bite when disturbed in their hiding places. The bites are painful and often poisonous, though for a long time it was unclear where this poison came from. Then poison glands were discovered in the roof of the mouth in some species; the poison was released from the glands when the fish bit. The normal diet of the Moray Eels consists of fish and squids, which they detect in the darkness with their excellent sense of smell. Many species are prettily patterned, some, on account of their meagre size and lovely colouring, make

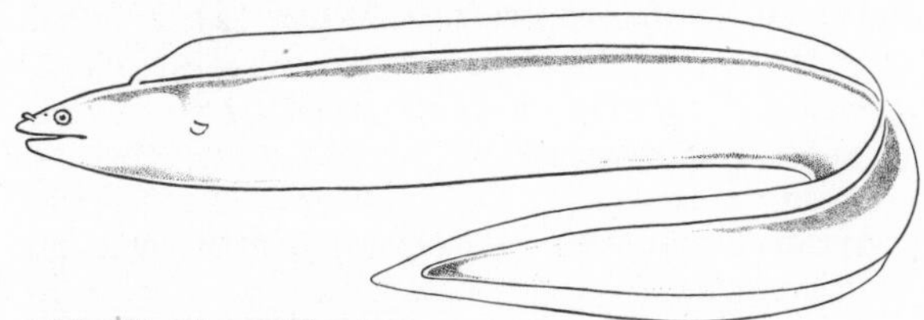

Fig. 75 Muraenidae-type

interesting subjects in the marine aquarium, *see* Ga *Echidna, Gymnothorax, Lycodontis, Muraena, Myrichthys, Rhinomuraena.* In fishing economy, Moray Eels are of some regional importance. One Mediterranean species, the poisonous 1 m long, several kg heavy *Muraena helena,* was considered a delicacy in Ancient Rome and was kept in specially built stone tanks.

– *Nemichthyidae;* Snipe Eels (fig. 76). These Anguilliformes are very bizarre members of the O, which live in very deep water and which are only rarely brought to the surface intact. The head is especially characterised by the very elongated, often needle-like jaws which can be bent in such a way that they can no longer be closed together. Large gill covering slit, small Ps present. Dorsal and ventral fins begin directly behind the head and continue all along the very long, flattened body, along the rear whip-like section to the thread-like tail fin. Little is known about its biology.

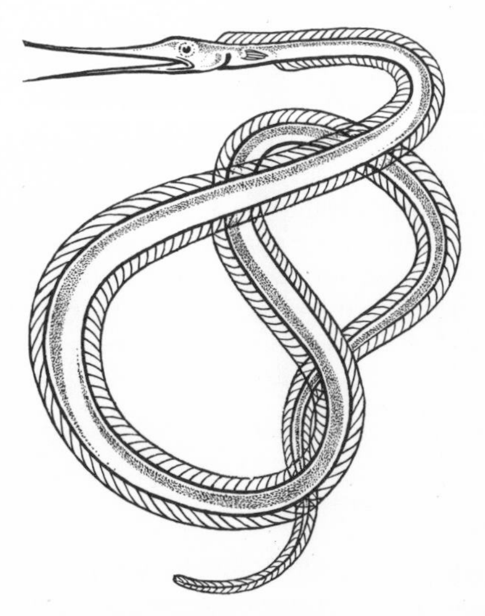

Fig. 76 *Nemichthys scolopacea*

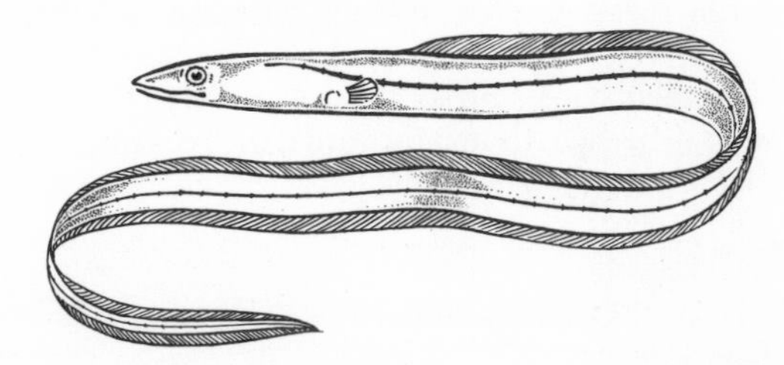

Fig. 77 *Ophichthys unicolor*

– *Ophichthyidae*; Snake Eels (fig. 77). Smallish Anguilliformes, often beautifully patterned and coloured. Very elongate and slender. Usually found in shallow areas of warm seas. Typical of the Snake Eels are the relatively far back, slit-like openings of the gill covers, the absence of scales, and the pointed tail. The latter is naked, ie it has no fin, and sharp; it is probably an adaption to life in the sandy bottom which is where these fish bury themselves, tail-first, at lightning speed. However, you also find naked pointed tails in species that prefer rocky places. Small Ps may or may not be present. Captive Snake Eels are extremely lively, slipping through the smallest hole and cracks. Occasionally, mummified Ophichthyidae have been found inside large predatory fish. They had been swallowed live and apparently still had the strength to bore through the intestine wall. Although they have only small teeth, they are supposed to deliver a painful bite.

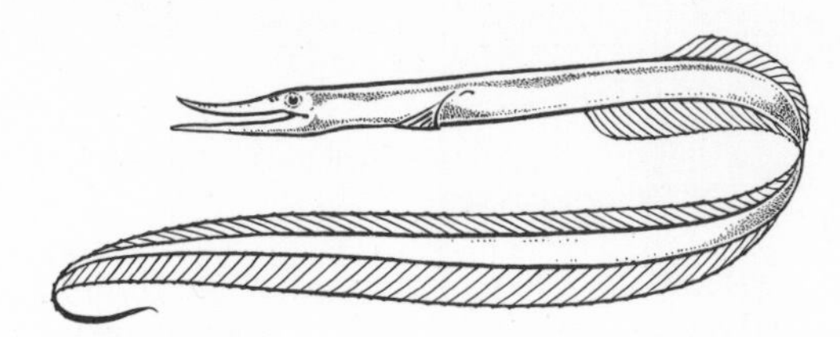

Fig. 78 *Serrivomer beani*

– *Serrivomeridae;* Deep-sea Eels (fig. 78). Typical Anguilliformes with a conical head and a mouth opening that reaches to behind they eyes. Slit-like gill openings positioned on the throat between the Ps. Small, rudimentary scales embedded deep in the skin. Known in great depths of all oceans. *Synaphobranchus pinnatus* (fig. 79) is widespread in the E Atlantic, *Synaphobranchus infernalis* in the W Atlantic. Both species spawn in the Sargasso Sea, the larvae drifting along with

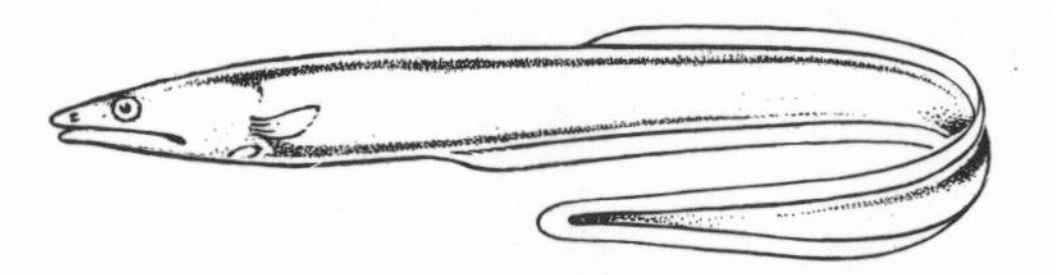

Fig. 79 *Synaphobranchus pinnatus*

the currents. In this respect there are remarkable parallels with the European and American Eel.

There is a whole host of other families, some with interesting forms besides those discussed so far, eg the Tube Eels (Heterocongridae) which live together in colonies, their bodies projected well out into the mud and moving backwards and forwards; the Parasitic Eels (Simenchelyidae) (fig. 80), which can eat into other fish

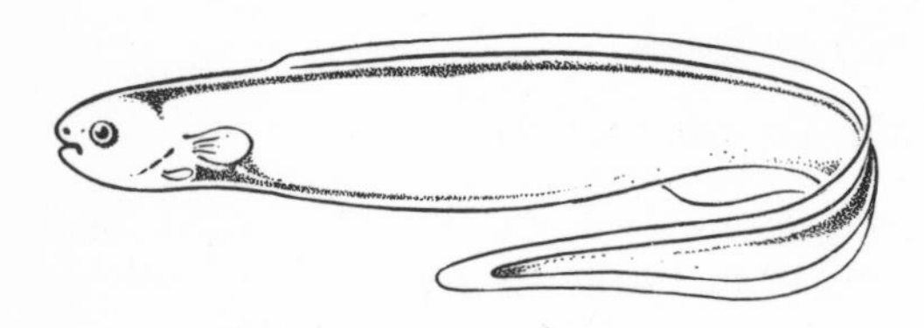

Fig. 80 *Simenchelys parasitica*

with their sharp set of teeth; or the Muraenesocidae with a set of teeth like that of a Pike. The families grouped under the Sub-O Saccopharyngoidei (Gulper Eels) (fig. 81) have a very strange appearance scarcely recognisable as belonging to the Anguilliformes. They were formerly placed in their own O; TSCHERNAVIN even put them in their own Class. The jaws of all the species are packed with teeth and so greatly enlarged

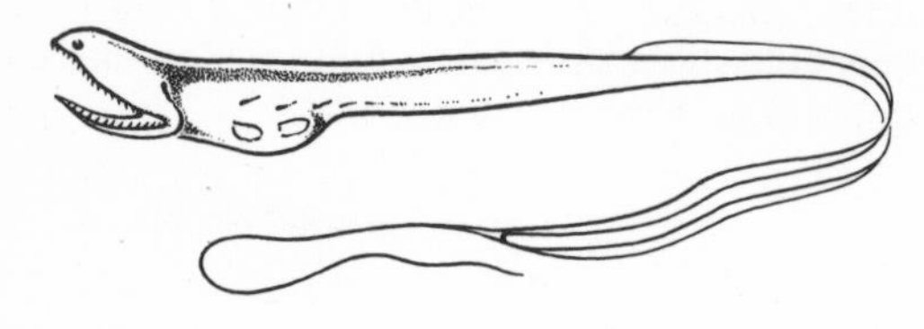

Fig. 81 *Saccopharyngus* sp.

that they look like a large bag, open at the front and hanging beneath the small skull. They can devour enormous prey which are apparently caught not by active hunting but by swimming along with the mouth open as described. The skeletal elements of the skull can vary enormously, eg the entire upper jaw is absent in Monognathidae (fig. 82). The eyes are very small, the 5 or 6 gill slits open usually well to the rear and the fringe of fin begins on the dorsal surface either behind or at the head and reaches almost as far forwards on the

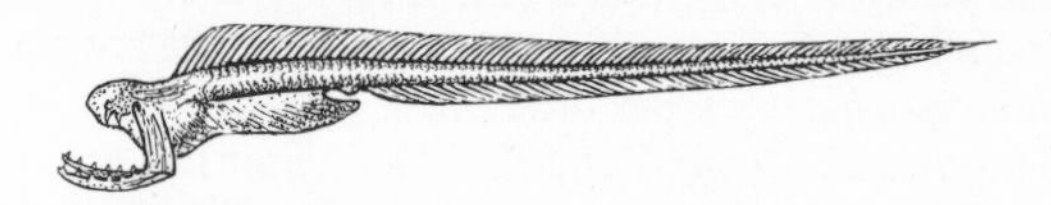

Fig. 82 *Monognathus taningi*

ventral surface. Like other deep-sea fish, many species have light organs. *Eurypharynx palecanoides* (fig. 83) is up to 60 cm long and lives in depths of about 4,000 m. This species may occur in all oceans.

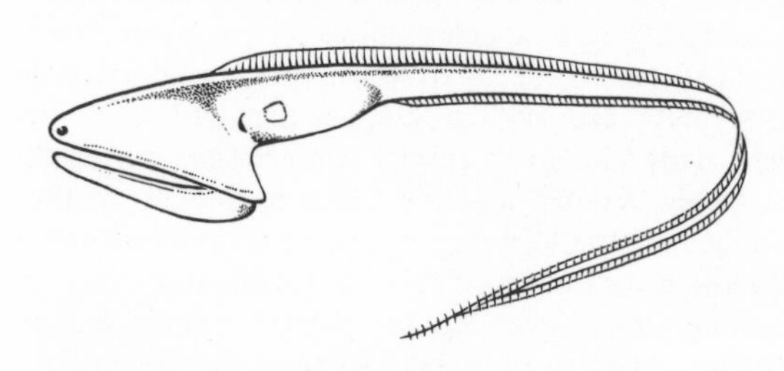

Fig. 83 *Eurypharynx palecanoides*

Anguilloidei, Sub-O, *see* Anguilliformes

Anilocra physodes LINNAEUS, 1758. Parasitic woodlouse (*see* Isopoda) found in the sea and brackish waters. Very common in the Mediterranean where it settles in the gill slits or mouths of fish. On account of its size (females up to 24 mm) it can do no damage. Numerous other Ga and species of parasitic isopods are known, eg *Livoneca symmetrica* VAN NAME, 1925, found on South American freshwater fish or *Nerocila**.

Animal Geography *see* Zoogeography

Animal Pollination *see* Pollination

Anion *see* Ions

Anisitia EIGENMANN, 1903. Name not in current usage, *see Hemiodus*

Anisophyllous. Having differently formed leaves on one shoot, eg *Salvinia**.

Anisotremus GILL, 1861; Grunts. G of the Pomadasyida*. From tropical W Atlantic. Relatively tall, medium-sized, sociable fish, perch-like, with striking colouring. Soft D and An, in contrast to *Haemulon** without scales. By grinding the teeth in their pharynx

Fig. 84 *Anisotremus virginicus*

they can produce purring noises. No problem about keeping them in aquaria. An occasionally imported species is:
– *A. virginicus* (LINNAEUS, 1758). Porkfish. Tropical W Atlantic. Up to 30 cm. Body with blue stripes on a yellow background, Head with 2 dark bars. Nocturnal hunter that eats small animals; juveniles act as cleaner fish (fig. 84).

Annelida. Segmented Worms. Ph of articulated animals (*see* Articulata). The 3 main classes of annelids are the Polychaeta*, the Oligochaeta* and the Hirudinea*. The Annelida have many primitive features, eg their uniform segmentation; so they can be regarded as the original type of articulated animal. All Annelida have a secondary body cavity, ie the primary body cavity is almost completely displaced by an extensive chamber lying within the mesoderm called the coelom. The closed circulatory system consists in essence of two vessels running length-wise; a dorsal vessel which carries blood anteriorly and a ventral vessel through which blood flows posteriorly. There are also circular vessels at the point where the segments meet. Another feature all Annelida have in common is a rope-ladder-effect nervous system situated on the ventral side. Only the 'brain' – a paired knot of nerves in the head (cerebral ganglion) – is positioned above the digestive tract.

Annelids range in size from microscopically small to several m long. How worm-like they are can vary, but, almost exclusively, they are typical inhabitants of the bottom of seas and fresh waters, or the ground on dry land. Some annelids, particularly among the marine variety, are very colourful, and so are kept in aquaria.

Annelids, Ph, *see* Annelida

Annual. Annual fish complete their life-cycle from egg to sexual maturity and death within a year (*see* Cyprinodontidae). Many plants are also annuals. They germinate, flower and die within one period of growth. Aquatic plants are only rarely annuals.

Annual Plant. In contrast to a perennial plant*, an annual takes only a few weeks (up to a year at most) to complete the cycle from germination through to seed maturity and death. Some aquarium plants, eg *Limnophila aquatica*, were described as annuals in scientific literature but proved to be perennials when cultivated.

Anodinae, Sub-F, *see* Characoidei

Anodonta LAMARCK, 1799. Pond Mussels (fig. 85). G

Fig. 85 *Anodonta cygnea*

of the Bivalvia*. Two species are found in middle Europe. The largest, *A. cygnea* or Swan Mussel, is up to 25 cm long and is very common in still waters with a muddy bottom. In contrast to the River Mussels *(see Unio)*, the shells of *Anodonta* are higher, thinner walled, and have no shutting teeth. The reproductive period of *Anodonta* occurs in winter; the eggs remain for many months in the gills where they grow into larvae (glochidia). The larvae hook themselves to a fish's skin and in this way are spread. *Anodonta* live well in the aquarium, but require a thick substrate which should be turned over regularly. This will upset any plant growth; fish may also suffer if subjected to too many glochidia. Bitterlings *(see Rhodeus)* are particularly suited to living alongside *Anodonta*, as they need the mussels to complete their reproductive cycle.

Anomalopidae, F, *see* Beryciformes

Anomalops, G, *see* Beryciformes

Anomura, Sub-O, *see* Reptantia

Anopheles MEIGEN, 1818; Malarial Mosquito (fig. 86). G of the Mosquitoes (Culicidae, Nematocera*). Widespread in the tropics and the temperate zone, feared as a carrier of malaria in warm regions. Only the females bite, for without the sucked blood the eggs will

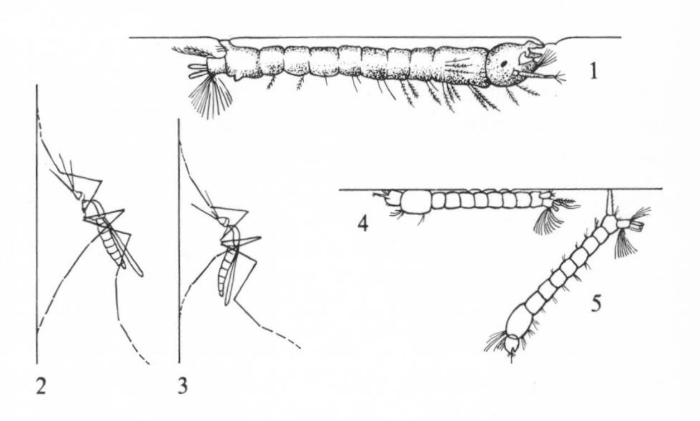

Fig. 86 *Anopheles* and *Culex*. 1, 4 Larva of *Anopheles* 2 Imago of *Anopheles* 3 Imago of *Culex* 5 Larva of *Culex*

not ripen. Larvae and pupae of *Anopheles* are an important food of many fishes, so one way of combating mosquito plagues is to introduce certain Tooth Carps (*Gambusia**). In contrast to related Ga, *Anopheles* larvae swim horizontally beneath the surface of the water; the resting position of the adult insect imago is different from that of other mosquitoes (*Culex**, *Aedes**). Apart from still waters of all kinds, particularly marshes and areas liable to flood, some species are also found in flowing waters*.

Anoptichthys jordani HUBBS and INNES, 1936. Name not in current usage, *see Astyanax*

Anostomidae. F of the Cypriniformes*, Sub-O Characoidei* (fig. 87). Relatively uniform group of Characoidei, which are found mostly in S America but also on some W Indian islands. Elongate body, like a spindle, only slightly compressed laterally. Head conical, mouth small, often with strong lips. In the Ga *Anostomus, Synaptolaemus* and *Gnathodolus* the mouth is directed upwards, rarely ventral. Very strong teeth, either truncated or with many small points, and usually firmly lodged in the jaws. Arrangement of teeth varies. Intermaxillary bone and lower jaw are toothed, whereas the upper jaw has no teeth. D roughly in the middle of the body, virtually opposite the Vs, adipose fin present. C forked, An short. Cycloid scales. In the G *Leporellus* scales even cover the part of the C nearest to the body.

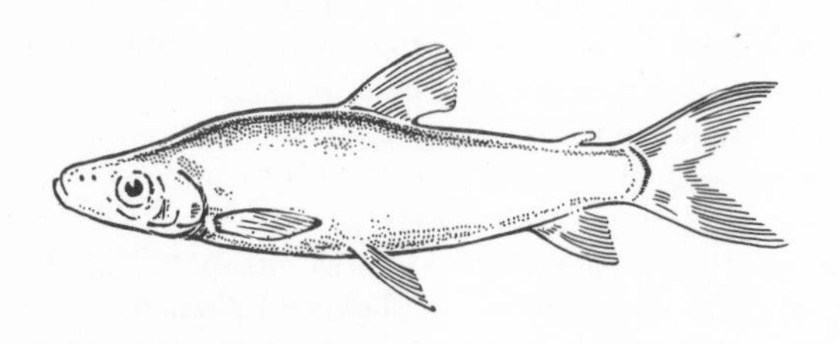

Fig. 87 Anostomidae-type

Many Anostomidae are Headstanders (eg Ga *Anostomus, Abramites, Laemolyta*), ie they stand head-down amongst the water plants or roots. Even as they undulate gently towards the plants they are positioned head-down, at a slight angle to the bottom. Only when escaping do they shoot horizontally through the water. Standing on the head is first and foremost a protective function. They melt imperceptibly into their surroundings when they match their bodies to the oblique or vertical lines of their natural habitat. This advantage was acquired because of difficulties when feeding. To be able to win food from the bottom Headstanders must in effect dislocate their upward-pointing mouths or turn on their backs. However, in the wild, they seem to feed on the undersides of leaves or roots chiefly – a method of finding food that can be easily observed in *Anostomus* in the aquarium. Those Anostomidae that do not stand on their heads are to a large extent bottom-feeders, living on both animal and algal food. Many species are resident fish, others live in small shoals. It has been said of *Leporinus* that during the breeding season they travel up small streams to the spawning grounds where they congregate in such numbers that they can hardly move. All species appear to spawn in the open. Most members of the family are nicely patterned with pretty colours at least in the juvenile phase, but with the onset of sexual maturity many species lose their youthful colouring. Most Anostomidae reach lengths of between 20 and 30 cm and are prized delicacies in their native lands. In the aquarium usually only young specimens are kept. However, many species are intolerant of like species – behaviour which is understandable when you consider that they defend territories in the wild. For further details see *Abramites, Anostomus, Leporinus.*

Anostominae, Sub-F, *see* Characoidei.

Anostomus SCOPOLI, 1777. G of the Anostomidae*. Amazon and Orinoco basins, Guyana. Torpedo-shaped fishes, 14–20 cm long. All species have an upward-pointing mouth, so they turn onto the back when feeding. Plants and stones are grazed for animal or plant food. D1 and An roughly the same length, D2 present. Nasal openings elongated like stalks. They swim obliquely, propelling themselves along head-down. Very colourful, and in comparison with other species they are peaceable, although territorial, aquarium fish. Water hardness should be 4–6°, slightly acidic pH. This, together with a varied diet, should ensure that they survive for many years. Lighting should be obscured by surface vegetation and the substrate should be dark. The *Anostomus* are often quarrelsome among each other (especially *Anostomus anostomus*), so keep at least 4, preferably 6–8, fish in a large, well-planted tank with hiding-places. Like this the fish will swim mostly as a shoal and the aggression of high-ranking fish will be contained. Reproduction unknown. For the aquarist these are interesting and sought-after specimens that will hopefully be bred soon. 3 species are often imported:

– *A. anostomus* (LINNAEUS, 1758); Striped Anostomus. Amazon, Orinoco, Guyana. Up to 18 cm. Ground colour ochre to yellow with 3 dark length-wise stripes. The scales at the edges of these stripes jut into the yellow zones, so they make a jagged edge. Red base to all fins; the base of the An and C is blood-red especially in the males, becoming brown-red with age. Adult males are more slender and somewhat smaller than the more compact females. Young specimens are among the most beautiful of S American aquarium fish (fig. 88).

Fig. 88 *Anostomus anostomus*

– *A. ternetzi* FERNANDEZ-YEPEZ, 1949 (fig. 89). Guyana. Up to 12 cm. Body similar to *A. anostomus.* 3 black, sharply delineated length-wise stripes against a yellow background. Adults are black on the ventral surface, mouth area reddish, fins colourless. Differences in shape and size between the sexes as for *A. anostomus.* More peaceable than *A. anostomus.*

– *A. trimaculatus* KNER, 1858); Three-spot Anostomus

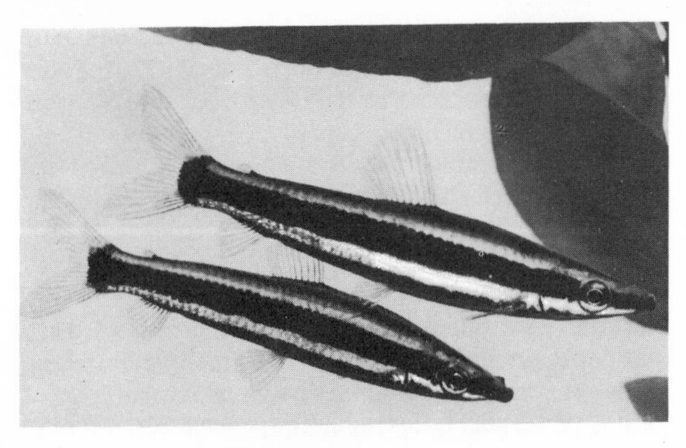

Fig. 89 *Anostomus ternetzi*

(fig. 90). Lower Amazon, Guyana. Up to 20 cm. 3 black spots or an olive-grey to walnut-brown background with fine dark transverse bars. The spots are located on the operculum, near the middle of the body and on the

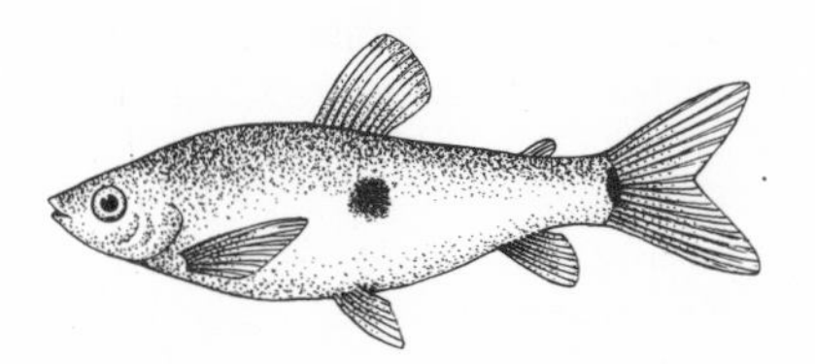

Fig. 90 *Anostomus trimaculatus*

caudal peduncle. Unpaired fins bright red on the free edges, yellow on the base. Taller and more robust than either *A. anostomus* or *A. ternetzi*.

Anostraca. O of the Crustacea* restricted to inland waters and having many primitive features, eg elongated body, large number of body segments, numerous pairs of legs each bearing gills, and no shell-like extension of the last segment of the head (carapace). The Anostraca are up to 4 cm long and swim slowly on their backs in open water. They are almost always only found in waters without fish, especially in pools and puddles which only receive water from time to time. The life-cycle of the Anostraca is astonishingly well adapted to the extreme conditions under which it lives. The eggs can desiccate for years without losing their ability to develop, and from hatching the nauplius larvae become

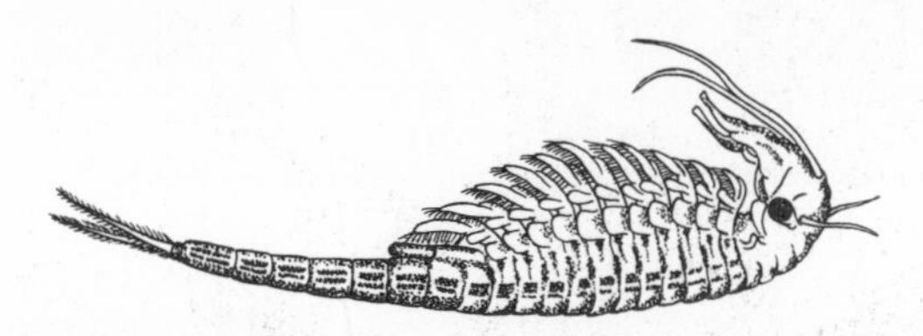

Fig. 91 *Chirocephalus* sp., member of the Anostraca

mature adults in only a few days. Such decidedly seasonal forms can be found in the Ga *Branchipus* SCHAEFFER, 1766, and *Chirocephalus* PRÉVOST, 1803 (fig. 91). If inhabitable water is free of fish for other reasons (eg too high a salinity level) Anostraca may well colonise it. A well known example of the latter is *Artemia**.

Ansorge's Characin *(Neolebias ansorgei) see Neolebias*

Antagonism. In higher organisms numerous organs and their systems are made to work by opposite effects. Such antagonistic effects can be released by mechanical, nervous or substantial factors. For example, a muscle functions as a rule in relation to the effect of an antagonist which counteracts the muscle contraction in the form of other muscles, connective tissue or skeletal elements. The intestines and glands of vertebrates are acted upon by two antagonistic parts of the autonomic nervous system* (sympathetic and para-sympathetic nervous system) which either increase or inhibit the organ's effect. Hormones* are also known to produce antagonistic effects upon certain functions of the organism.

Antarctic Cods, relatives of, *see* Notothenioidei

Antedon DE FRÉMINVILLE, 1811; European Feather-star. G of the Sea-lilies (*see* Crinoidea). Echinoderms (*see* Echinodermata) that live below the shore line. The 5 arms are broken up into 10 slender, feathery branches. Sessile when young, and in later life it uses its arms to climb about or swim among the vegetation on the bottom. Particle-eater, the pieces of food being driven by cilia into a food funnel running along the arms, and from there to the mouth. Very delicate, sensitive creatures, inclined towards autonomy; will survive in quiet aquaria if looked after carefully, eg:

– *A. mediterranea* LAMARCK. Rosy Feather-star. Found at less than 10 m depth. Length of arms 6–8 cm. Colour yellow, red or brown (fig. 92).

Fig. 92 *Antedon mediterranea*

Antenna Catfishes, F, *see* Pimelodidae

Antennariidae, F, *see* Lophiiformes

Antennarius LACÉPÈDE, 1798; Anglerfishes, Frogfishes (fig. 93). G of the Antennariidae, O Lophiiformes*. These fish live along tropical coasts where they lie in wait for prey. They entice their victims with a movable lure on the head which looks like a piece of

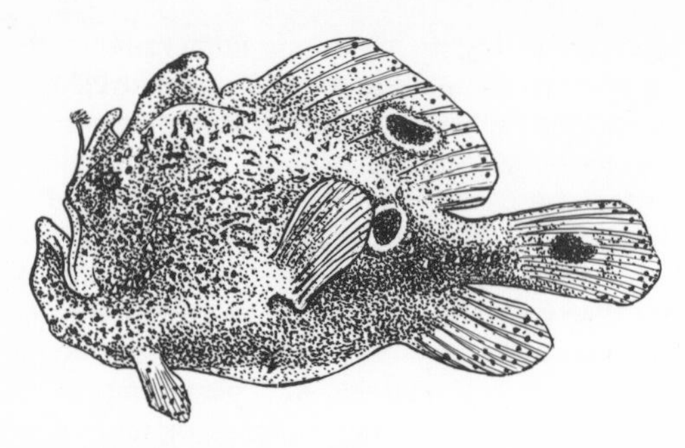

Fig. 93 *Antennarius ocellatus*

food. These graceless fish are perfectly camouflaged by their cryptic colouring and ugly outgrowths of the skin. The Ps have become adapted for crawling. Although the Anglers eat well in the aquarium, they cannot be kept for long. Several species are often imported on account of their decidedly odd appearance.

Anther (part of stamen) *see* Flower

Antheridium *see* Angium

Anthias BLOCH, 1792. G of the Serranidae*, to which small, sociable, colourful fish of tropical and subtropical seas belong. Many are typical of fish that live on coral reefs. Some occur in large shoals *(A. squamipinnis).* If danger threatens they hide among the corals. Their food consists largely of small crustaceans. Elongate body, perch-like mouth can be projected far forwards. Especially in the males, the fins are very much elongated, the C is to some degree forked. The species so far imported have proved to do well in the aquarium, if several fishes of the same species are kept together.

– *A. anthias* (LINNAEUS, 1758). Mediterranean, E Atlantic, found in shadowy crevices and caves. Up to 24 cm. V drawn far out. Brilliant pink-red colouring.

Fig. 94 *Anthias squamipinnis*

– *A. squamipinnis* (PETERS, 1855); Lyretail Coralfish, Orange Sea Perch, Wreckfish. Indo-Pacific, Red Sea, very common on reefs. Bright red with pale band between the eye and P (fig. 94).

Anthozoa; Sea-anemones, Corals. Ph Coelenterata*, Sub-Ph Cnidaria*. The Anthozoa are exclusively sea creatures, living alone in colonies. Frequently, very large polyps are formed with a gut cavity traversed by at least 6 septa (partitions) deep into which a gullet penetrates. In contrast to the Hydrozoa* and Scyphozoa* the Anthozoa do not reproduce by budding. There are 2 distinct Sub-Cl, the Octocorallia, in which the polyps possess always 8 feathery tentacles, and the Hexacorallia with up to 1,000 unfeathered tentacles. The Alcyonaria*, Gorgonaria* and Pennatularia* all belong to the Octocorallia. The Sea-anemones (Actiniaria*), Madreporaria*, Ceriantharia*, Zoantharia*, Antipatharia*, as well as the Corallimorpheria belong to the Hexacorallia. Many Anthozoa particularly the Sea-anemones and Ceriantharia, are well suited for the aquarium as they are very pretty and durable.

Anthozoans, Cl, *see* Anthozoa

Antibiotics are metabolic products produced by micro-organisms, which have the capacity to inhibit the growth of another micro-organism. Already applied to human medicine with great success. In the treatment of bacterial diseases of fish, especially stomach dropsy*, antibiotics are beginning to play an increasingly effective role. Antibiotics that might be used include Aureomycin*, Chloromycetin*, Penicillin*, Streptomycin*.

Antibody *see* Immunity

Antigene *see* Immunity

Antigonie's Boarfish (Antigoniidae, F, name not in current usage) *see* Zeiformes

Antigoniidae, F, name not in current usage, *see* Zeiformes

Antiparallel Converging Behaviour *see* Display Behaviour

Antipatharia. Black Corals. O of Anthozoa* with approximately 100 species. Often found growing in large polyps in deep parts of the sea. Much branched with a horny black skeleton that is characterised by a thorny edging. No information from aquarium observations.

Fig. 95 *Anubias congensis*

Anubias SCHOTT, 1857 (fig. 95). G of the Araceae*. 15 species. W Africa. Small to medium-sized perennial plants with rhizomes. Stalked leaves in radical rosettes, with rough, mostly dark green blades. Flowers small and

inconspicuous, unisexual and monoecious, surrounded by a greenish spathe. Perianth absent, 5 stamens. 1 carpel, pointing upwards. Fruit is a green berry. The species of the G grow in shallow streams in woods or along the sides, preferring lightly shaded areas. They either stand emersed from the water or their rhizomes grow in the water with the leaves jutting out. When the rivers become swollen, the plants are totally submerged for a time. In the aquarium, they require a nutrient-rich substrate and they grow very slowly. They are seldom kept in aquaria. Well-known species are:
– *A. barteri* ENGLER. W Africa. Leaf blades spear-shaped.
– *A. lanceolata* N. E. BROWN; Water Aspidistra. W Africa. Leaf blades broadly lanceolate, rough.
– *A. nana* ENGLER. W Africa. Leaf blades narrow and spear-shaped, small.

Aorta *see* Blood Vascular System

Aphaniinae, Sub-F, *see* Cyprinodontidae

Aphanius NARDO, 1827. G of the Cyprinodontidae*. About 10 species which prefer to live in brackish water and salt-retaining inland waters near the coast in S Europe and N Africa, across Asia Minor to India. Build is typical of the family. Small to medium-sized fish, laterally compressed. Fins not brought to a point, D practically placed at the tail, C rounded or marginally indented. Males are always smaller than the females, transverse stripes typical, females mostly spotted. If the

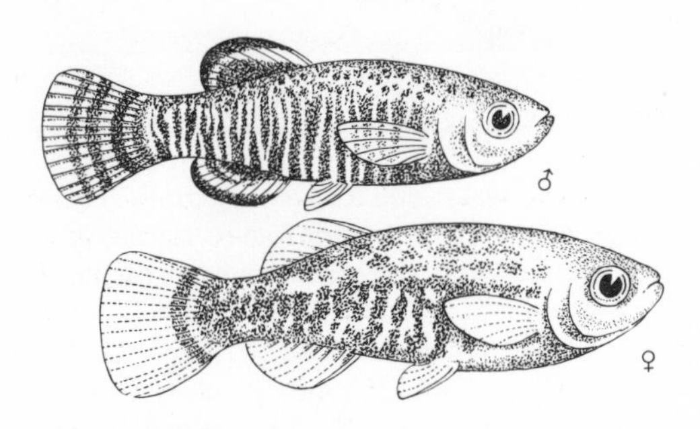

Fig. 96 *Aphanius iberus*

salt content climbs too high, the individual populations can be confined to the smallest, less highly concentrated areas of distribution. Such extreme isolation of the communities has led to marked genetic differences. *A. fasciatus* has been employed successfully in the fight against *Anopheles** in tropical and subtropical regions. *A. dispar* has been able to overcome the limitations of its distribution by travelling through the Suez Canal and so increasing its range through India to the Red Sea and Palestine. Another characteristic of the *Aphanius* is that they only reach their full potential in a species tank. Best of all are large, shallow tanks with very hard water or addition of sea salt (3–6 g/l). Give sufficient opportunity for hiding places with plenty of rocks, aquatic and floating plants. Temperature 15–25 °C, 10 °C cooler than this is still tolerable, at least for the subtropical species. Live food of all kinds as well as supplementary plant food (algae, cooked spinach or softened lettuce leaves). In the species aquarium most of the fish will reproduce without any special preparation, but you should separate the sexes first or change part of the water. Court-

ship is hectic and the eggs are laid on finely feathered plants. On a few occasions spawn is eaten. The eggs hatch after 6–14 days, depending on the species and the temperature of the water. Their care is easy. Species that are often kept in aquaria include:
– *A. fasciatus* (VALENCIENNES, 1811); Banded Minnow. Coasts of middle and eastern Mediterranean, in fresh and brackish waters. Up to 6 cm. Male olive to blue-green ground colour with 10–15 dark transverse bands. Fins yellow, D with a dark border. Female grey-green and transverse bars indistinct.
– *A. iberus* (CUVIER and VALENCIENNES, 1846); Spanish Minnow, Spanish Killi. Coasts of western Mediterranean in fresh and brackish waters. Up to 5 cm. Male blue-green to aquamarine, with about 15 narrow, pale blue transverse bars. C also barred. Female blue-green or brown-green with dark streaks and blotches (fig. 96).

Aphetohyoidea. Cl Gnathostomata*. These ancient groups of fish are very varied in form. They look very strange to us because the shape of their bodies and the arrangement of their fins are very different from those of the typical fish of today. Their features are reminiscent in part of the Agnatha*, Chondrichthyes* and also the Sarcopterygia*. Common to all is the structure of the jawbones. In contrast to the Agnatha, the Aphetohyoidea have an upper jaw and a lower jaw. Both jaws have developed from the 1st gill arch and carry teeth. However, the 2nd gill arch – and this is where the Aphetohyoidea differ from the cartilaginous fishes (Chondrichthyes) and the bony fishes (Osteichthyes*) – does not form a maxillary stalk (Hyomandibular) and a bone at the base of the tongue (Hyoid); instead, it is a normal gill arch. Therefore, a perfectly normal gill slit is positioned between the 1st and 2nd gill arches. The scales are primitive ganoid scales (*see* Scales) or similar. Paired fins are usually present, some have spines and additional fins. The Aphetohyoidea were thought to be primarily freshwater fish that only ventured into the sea in the Devonian. There are Sub-Cl of Aphetohyoidea.

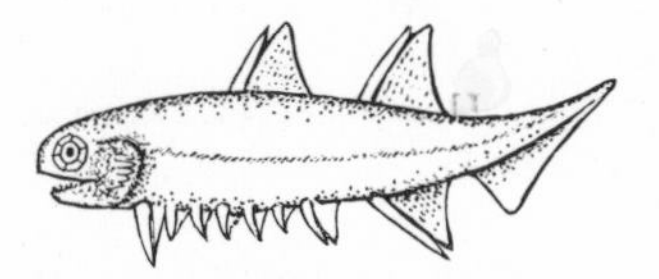

Fig. 97 *Climatus* sp. after Traquair and Watson

– *Acanthodi* (fig. 97). The most primitive of fish from the Upper Silurian and Lower Devonian. Body like a shark's but there is no long snout. The skull surrounding the brain and jaws partly covered with thin bone. Gills protected by a primitive operculum. Fins have strong spines, the Ps and Vs are joined both sides of the ventral suface by a row of additional spiny fins. Scales are formed out of small ganoid plates arranged like a mosaic. Both the cartilaginous and bony fishes probably evolved from this Sub-Cl.
– *Placodermi.* Armoured Fishes (fig. 98). Primitive fish from the Devonian and Lower Carboniferous, the head

and front part of the body being covered with large bony plates like a suit of armour. An articulated joint almost always links the armour of the head and torso. The Ps are reduced to spines or have become specialised as arm-like, armoured extension. Rear of the body naked or covered with scales.

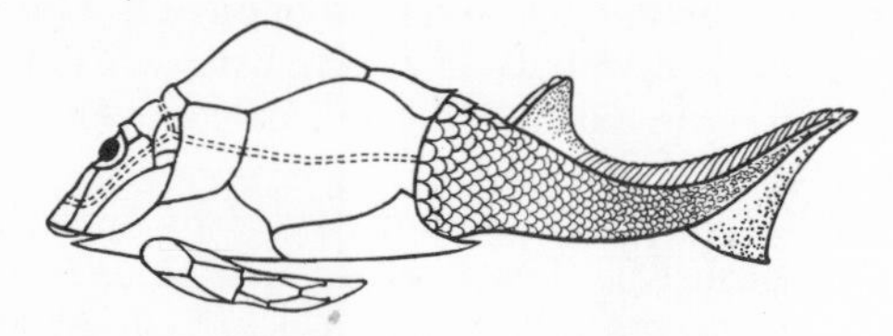

Fig. 98 *Pterichthys* sp. after Traquair

– *Rhenanida* (fig. 99). Primitive fish from the Lower-Upper Devonian which are reminiscent of the Angelsharks (*see* Squatinidae). Bony armour considerably reduced. Body much flattened, Ps large. Small ganoid scales.

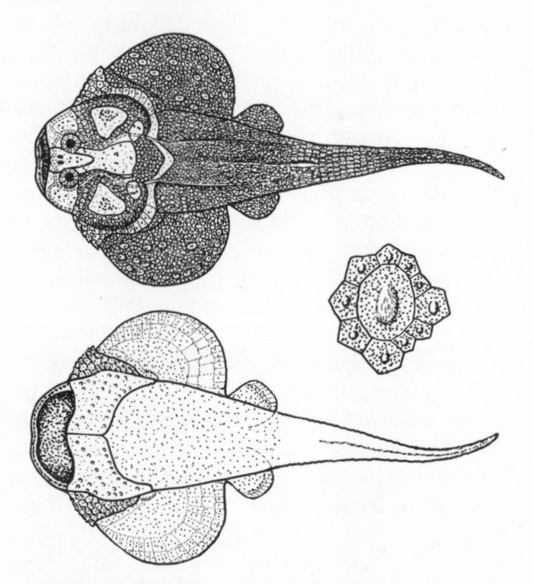

Fig. 99 *Gemundina stuertzi.* Upperside, underside and cyclomorial scale. After Gross

Aphredoderidae, F, *see* Percopsiformes

Aphredoderus sayanus *see* Percopsiformes

Aphrodita LINNAEUS, 1758. G of the Polychaeta*. These worms live buried in the mud on the bottom of the sea where they eat small animals and detritus. The fluffy chaetae on the back give a beautiful multi-coloured iridescence. The Sea Mouse *(A. aculeata)* LINNAEUS, 1758, lives along the coasts of Europe. Up to 20 cm long and 6 cm wide. This species lives well in the aquarium.

Aphyocharacinae, Sub-F, *see* Characidae

Aphyocharax GÜNTHER, 1868. G of Characidae*. S America. About 20 species. The majority are elongated Characins 3–8 cm long. D extremely small, An long, C deeply forked, adipose fin present. Peaceable, lively shoaling fish that live in the upper and middle levels. Best suited to a temperature of 22–26 °C, and soft water is needed. A lot of room for swimming is required in the tank. Females can be recognised by their stouter ventral surface. The males tend to get caught by the hooks on the anal fin when fished out in a net. Be careful when you free them. Otherwise the hooks are ripped out and mating can no longer take place; without the hooks they cannot cling to the female during mating. The *Aphyocharax* eat live food of all kinds, as well as dried food. The Bloodfin, *A. anisitsi,* is one of the most popular Characins; other species are also imported occasionally.

– *A. alburnus* (GÜNTHER, 1869); Golden Crown Aphyocharax. Rio Marañó, Rio Paraná. Up to 7 cm. Elongate body, delicate blue-green with silvery iridescence and a small dark spot at the shoulder. Fins yellowish to colourless. C has a broad red stripe at the base; the free edges are colourless. The males are thinner and more colourful. Not yet bred successfully.

– *A. anisitsi* EIGENMANN and KENNEDY, 1903; the Bloodfin (fig. 100). Rio Paraná, Argentina. Up to 5.5 cm. Body elongate, silvery, with a suggestion of blue. The lower parts of D1, An, C and Vs are blood-red, Ps colourless. Breeding takes place in pairs at a temperature of 24–26 °C. The water should be soft, mildly acidic

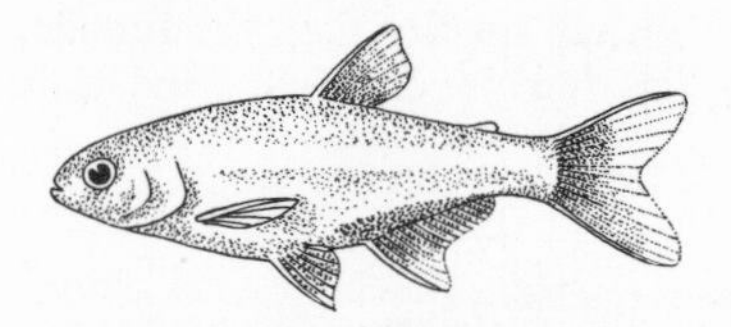

Fig. 100 *Aphyocharax anisitsi*

with only 5–6 cm water and a spawning grate. This reduces the tendency of the parent fish to eat the spawn as it floats downwards. The male drives at the female with a dorsal swimming action (*see* Leading Posture). They often break water when pairing. Up to 800 eggs are released. After about 20 hours the glass-clear eggs hatch. The young grow quickly; they are easy to rear on rotifers or *Cyclops* nauplii. According to GÉRY, *A. rubropinnis* PAPPENHEIM, 1921, is an invalid synonym for *A. anisitsi.*

– *A. axelrodi* TRAVASSOS, 1960; Calypso Tetra, Red Pristella. Trinidad. Up to 3 cm. Body untypically small for the genus, upper half pink, lower half silver. D1 and An delicate pinky-red, the former with a dark stripe. C blood-red. Females paler. Breeding is simple; the spawn is laid on thick clumps of vegetation. For care of the young *see A. anisitsi.* According to GÉRY, this species belongs to the G *Megalamphodus.*

Aphyocypris pooni, name not in current usage, *see Tanichthys*

Aphyoplatys CLAUSEN, 1967. G of the Cyprinodontidae*. Only 1 species, occurring in Central Africa. Small, very elongate fishes with round fins, except for the C; D, C and An noticeably elongated. Best kept in a small species aquarium with soft, mildly acidic water and dense vegetation. Temperature 22–26 °C. These fish like to stay near the surface but do not swim much. They feed on small live food as well as small insects floating on the surface (eg Leaflice). Breeding is simple. The eggs are laid on finely feathered foliage and they de-

velop without delay (*see* Diapause). Finely ground nauplii and Radiolaria make the best food.
– *A. duboisi* (POLL, 1953). Central Congo near Stanley Pool. Up to 3 cm. Male has a metallic blue iridescence, body and fins dappled with red. Female grey or brownish.

Aphyosemion MYERS, 1924 (fig. 101). G of the Cyprinodontidae*. At present comprising about 60 species, most of which have already been bred and cared for in the aquarium. Closely related to *Roloffia**. All members

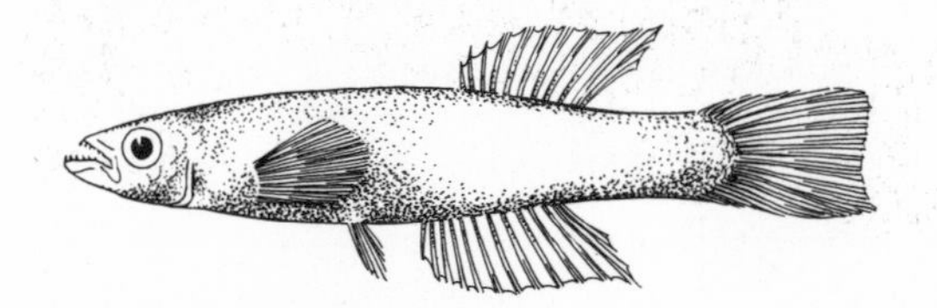

Fig. 101 *Aphyosemion* sp.

of *Aphyosemion* live exlusively in tropical W and Central Africa where they inhabit still-water ponds or slowly flowing waters in the tropical forests and the savannah. Much rarer in large rivers. Without exception, the G consists of very slim, slightly compressed fish with large eyes and a broad, upward-pointing mouth. About 5 cm long, a very few species reaching up to 10 cm. The fins and coloration in the males are always more gorgeous than in the females; the unpaired fins are lengthened into pennant-like tips with bold, sometimes fantastic, colour combinations – blood-red bands or lyre-shaped patterns; bright yellow, skin white or sky-blue edges; red blotches on an emerald green background etc. In contrast, the females are all a brown colour with the occasional red-brown spots. Data regarding the state of the water in the places the *Aphyosemion* live is now well recorded and should be borne in mind in the aquarium. Temperature relatively low (22–24 °C), pH slightly acidic (6.3–6.7) and the total hardness remarkably minimal (0.2–3 °). Some of the areas inhabited by the *Aphyosemion* only contain water during the rainy season, and they dry out quickly afterwards. Some *Aphyosemion*, the so-called annual species (*see* Cyprinodontidae), have adapted to these extreme conditions. They live through the dry period as embryos inside hard eggs, hatching only at the beginning of the next rainy season. Then they grow quickly and lay more eggs after only a few weeks or months. Even in the aquarium, some species are extremely short-lived and will only develop completely from their eggs if conditions in the dry period are imitated (*see below* and *Cynolebias*). The general biology (*see* Systematics) of the G has only recently been fully described. Taxonomically important features are eg the chromosome count and the scales on the head, neither of which need much concern the aquarist. Experiments in cross-breeding also play an important role. They have proven, for example, that *A. bivittatum* is a superspecies (*see* Species), ie all the different populations* of *A. bivittatum* are not necessarily able to be crossbred with one another.

For the purposes of aquarium-keeping, the G can roughly be divided into 2 sub-genera, *Fundulopanchax* MYERS (species type *A. sjoestedti*) and *Aphyosemion* (in the narrow sense). The former mostly have comblike spines on the scales and fin rays, while the latter have none. The sub-genus *Aphyosemion* can be further divided into several related groups.
– *Bivittatum* Group. C extended into long points, in a few cases much shorter or rounded. No lyre-like pattern, but there is usually a red line in the lower part of the fins in both sexes with two distinct, dark, lengthwise stripes. Member of the group is the superspecies *A. bivittatum.*
– *Calliurum* Group. C often extended into long points, pronounced lyre pattern well to the edge, with dark red spots in between. An has a broad red band across its lower part, beyond which it is often very pale, white or yellow. Flanks of the males very dark, as is the throat. Includes *A. ahli, A. australe, A. calliurum, A. cinnamomeum, A. gardneri.*
– *Cameronense* Group. C usually rounded, rarely extended into short points, lyre pattern absent or indistinct, central part of the fins mostly fine red speckles. Yellow throat with no spots. D and An short, rounded, with no distinct red or black spots. Includes *A. cameronense, A. obscurum, A. lujae, A. louessense* etc.
– *Elegans* Group. C usually extended into long points, lyre pattern present, very regular in shape, with rows of red spots in between. D and An similarly extended. Spots on the flanks do not usually form length-wise or cross-wise rows. Spots on the throat vary. Includes *A. christyi, A. cognatum, A. elegans* etc.
– *Exiguum* Group. C with or without pennant-like extensions, lyre pattern present, very dense on the upper and lower edges; red bands very narrow, the middle part of the fins usually with red streaks in the male, brown in the female. D and An have pennant-like extensions, mostly without a definite pattern of colour. Includes *A. exiguum, A. bualanum.*

Keep in a species tank, as only there will their full beauty be revealed. The tank can be small, should not stand in too bright a light, and should contain water plants. As the males are very aggressive to one another, either put only 1 or 2 males in with several females or have a sufficiently dense foliage to allow escape routes. A layer of well-boiled peat is a good substrate as it helps prevent bacterial growth. (Most species are very susceptible to bacteria.) In the wild, members of the G *Aphyosemion* live chiefly on small insects and their larvae; in the aquarium they will also eat small crustaceans. The G includes both adhesive or plant-spawners and substrate-spawners. The former are the more primitive form, the eggs being laid in batches on fine-leaved plants during the breeding season that normally lasts for several weeks. The embryos hatch in 10–20 days. If the water dries up, the eggs laid by plant-spawners will not survive, so the species that lay their eggs in this way are only found where water levels are constant. Those species of the G that lay their eggs in the substrate are completely adapted to the temporary presence of water. Before the dry period, the eggs are laid in the substrate. After a while, the embryos stop developing, ie there is a diapause* of up to several months. When the next rain comes, the egg finishes its development and the brood hatches. This type of annual life-cycle is also found in *Cynolebias** and *Nothobranchius**. Of course the division is not quite as neat as that: certain species can be either plant-spawners or substrate-spawners depending on the ecological factors, and are often referred to as switch-spawners by aquarists. To be successful with breeding those species that have annual life-cycles, it is necessary to replicate the natural conditions. Plant-spawners may utilise the aquarium plants, or artificial nylon mops can be substituted for the plants. A good spawning substrate is flaky, well-boiled peat. Either make a layer of it 3–6 cm thick or simply use a peat-filled spawning bowl. The eggs are either fanned by the tail fins into the peat or mating takes place in the substrate. If you separate the sexes for a while before breeding, mating often takes place as soon as male and female are brought together again. The spawning period lasts for weeks; it works well if the partners are separated at times and the fish are kept well fed. If enough eggs are laid after a few days or weeks, the peat containing the spawn can be removed from the tank and dried. To do this let the water run out through a sieve and finish off by drying it on layers of newspaper until it is just damp. Then put the peat into jars or plastic bottles and ventilate regularly. After about 4 weeks you can try to make the embryo hatch out by soaking the peat in rain or aquarium water. Some embryos will start hatching immediately, others after a few days. Blowing through some carbon dioxide or adding some sugar or dried food appears to speed up the process. Nevertheless, only some of the embryos will hatch; others will need the whole procedure repeated, ie another dry period. The young fish can eat right away. They require large amounts of live food such as *Artemia* nauplii and *Cyclops* nauplii as well as rotifers. They appear to suffer if the water quality deteriorates, so change part of the water regularly for a short while after hatching occurs.

Fig. 102 *Aphyosemion australe*

– *A. australe* (RACHOW, 1921); Lyretailed Panchax, Lyretail, Cape Lopez Lyretail. Congo to Gabon. Up to 5.5 cm. Plant-spawner (fig. 102).

– *A. bivittatum* (LÖNNBERG, 1895); Two-striped Aphyosemion, Red Killi. Togo to Guinea. Up to 5 cm. Superspecies with several, in part biologically separate, populations. Lays spawn on plants (fig. 103).

– *A. bualanum* (AHL, 1924). E Cameroun to Central African Republic. Up to 5 cm. Plant-spawner, easy to keep and breed in hard water.

Fig. 103 *Aphyosemion bivittatum*

– *A. calliurum* (BOULENGER, 1911); Red-chinned Aphyosemion, Banner Lyretail. Nigeria. Up to 5 cm. Lays spawn on plants. Resistant species.

– *A. cameronense* (BOULENGER, 1903); Cameroun Round-tailed Aphyosemion. Cameroun. Up to 5.5 cm. Not introduced as an aquarium fish so far, it is mentioned here because of the group description above.

– *A. christyi* (BOULENGER, 1915); Christy's Lyretail. Central Congo. Up to 5 cm. Superspecies. Care and breeding simple (fig. 104).

Fig. 104 *Aphyosemion christyi*

– *A. cinnamomeum* (CLAUSEN, 1963); Cinnamon Lyretail. W Cameroun. Up to 5 cm. By choice substrate-spawner. Shy, sensitive species.

– *A. cognatum* (MEINKEN, 1951); Red-spotted Aphyosemion. Middle and lower Congo. Up to 5.5 cm. Plant-spawner, lively, difficult to breed (eggs very sensitive to bacteria and moulds) (fig. 105).

– *A. elegans* (BOULENGER, 1899). Central Congo. Up to 4 cm. As yet not introduced to the aquarium, included because of group description above.

– *A. exiguum* (BOULENGER, 1911). Cameroun, Gabon. Up to 4 cm. Plant-spawner. Care and breeding simple.
– *A. filamentosum* (MEINKEN, 1933); Lyretail from Togo, Plumed Lyretail. Nigeria. Up to 5.5 cm. Along with another species – *A. arnoldi* (BOULENGER, 1908) – very isolated within the G. Body and fins sky-blue to green, red spots. C with short pennant-like

Fig. 105 *Aphyosemion cognatum*

points. Along with the An, C occasionally has a red band. Predominantly a substrate-spawner.
– *A. gardneri* (BOULENGER, 1911); Steel-blue Aphyosemion. Nigeria, Cameroun. Up to 6 cm. Can choose to be either plant or substrate-spawner. Easy to keep and breed (fig. 106).
– *A. louessense* (PELLEGRIN, 1931). Congo. Up to 5.5 cm. Plant-spawner. Sensitive species.
– *A. lujae* (BOULENGER, 1911); Red-striped Lyretail. Central Congo. Up to 5.5 cm. Plant-spawner.
– *A. obscurum* (AHL, 1924). Cameroun. Up to 4.5 cm. Plant-spawner, sensitive to bacteria and moulds.
– *A. sjoestedti* (LÖNNBERG, 1895); Blue Gularis. Nigeria, Cameroun. Up to 12 cm. Belongs to sub-genus *Fundulopanchax*. A frequently used synonym* for this

Fig. 106 *Aphyosemion gardneri*

species is *A. coeruleum* (BOULENGER, 1915). Body of the males metallic blue-green, front part consisting of lengthwise stripes made up of red dots, the back part of similar cross-wise stripes. C has three flaps, with a red band at the bottom, like the An. Substrate-spawner (fig. 107).
– *A. walkeri* (BOULENGER, 1911); Walker's Killifish. Ghana, Ivory Coast. Up to 6.5 cm. Most closely related to *A. gardneri*, especially in old age fairly compact. D, An and C drawn out into short points, middle region of C rounded towards the outside. Body vibrant blue-green and iridescent, red spots. Unpaired fins have broad, dark red border. Substrate-spawner. Care and breeding simple.

Aphyosemion sjoestedti (ARNOLD, 1911). Name not in current usage, *see Roloffia*

Apiaceae. (Umbelliferae, name not in current usage.) F of the Magnoliatae*. 300 Ga, 3,000 species. Northern subtropical areas, as well as northern and southern temperate areas. Mostly perennials with rhizomes or taproots. Alternate leaves, usually feathered, rarely with simple form or scaly leaves. Flowers small, hermaphroditic, predominantly polysymmetrical, arranged in simple or compound umbels. Essential oils in almost all parts. Besides carrots, several other root vegetables

Fig. 107 *Aphyosemion sjoestedti*

belong to the Apiaceae. Found in places ranging from the sea-shore to high mountains, as well as marshland and water. Species of the Ga *Sium* with their distinct heterophylly* and *Oenanthe* (Water Dropwort) are commonly encountered. Growing alongside European waters you will also find the highly poisonous *Cicuta virosa* or Cowbane. The G *Hydrocotyle** is cultivated, and probably species of the G *Lilaeopsis* are also suitable for the aquarium.

Apistogramma REGAN, 1913. G of the Cichlidae*. S America, with over 25 species. The majority of the species live in tropical regions of this sub-continent, especially Guyana, Surinam, Peru, Bolivia and Brazil; a few species have also penetrated into the subtropical south. The G comprises small fish with a shape typical of the family, and together with the related Ga *Nannacara** and *Apistogrammoides** are appropriately described as Dwarf Cichlids. All *Apistogramma* are very elongate and laterally compressed. In addition they always have 3 spiny rays in the An and the LL is broken up and displaced. The shape of the head and the arrangement of the fins can vary enormously. Some species may have particularly well arched forehead profiles, eg *A. sweglesi* and *A. trifasciatum*. Especially in the males the D may have elongated spiny rays, as in *A. ornatipinnis* and *A. ramirezi*, for example. The simplest form of C is in the shape of a fan (*A. commbrae*,

A. reitzigi, A. pertense), has elongated middle rays *(A. agassizi)* or the rays at the edge are extended like pennants *(A. borelli, A. kleei, A. ortmanni, A. wickleri).* Often the Ps are much elongated too (eg *A. kleei, A. ornatipinnis, A. trifasciatum*). Three striking dark bands running from the eye to the mouth, round underneath the edge of the operculum to the C are typical of the G.

All the *Apistogramma* hide their eggs in caves and incubate them (*see* Cichlidae). The differences between the sexes are typical of this method of reproduction. The males are almost always larger, more boldly coloured and have more strongly developed fins. The social structure of the *Apistogramma* is very interesting and well researched. Each individual adopts a territory

Fig. 108 *Apistogramma agassizi*

and defends it against any other fish of its own sex. In a male's territory there may be several small female territories, which means that not only does he defend the territories of 'his' females but he also runs a veritable 'harem'. Under the unnatural conditions of a hobbyist's aquarium, with its lack of space, the size of the tank represents one male territory, ie the strongest male takes possession. For this reason *Apistogramma* males of the same species should only be kept together in very large aquaria. This G needs a greater degree of care than *Cichlasoma** or *Tilapia**, for example. As far as possible the water should correspond to natural conditions, ie very soft, sightly acidic and reasonably fresh. To this end filter over peat for a time. Temperature according to the origin of the species; most *Apistogramma* like warmth and should be kept at a temperature not below 25 °C, for breeding not below 28 °C. Places of refuge can be created with plants, rocks and roots, even fine leaved plants are not harmed by *Apistogramma.* Holes in the rocks make good spawning places as well as flower heads that have fallen off, or broken-up coconut shells. The male mates with a female that is ready to spawn, after which they both clean the spawning area. 50–200 eggs are laid at a time, usually on the roof of the cave. They are an intense pink or red colour. While the female devotes herself entirely to the care of her brood, the male defends the territory where the spawn has been laid. The eggs hatch about 2–4 days later, or they are chewed out of their protective egg-cases by the female. Finally they are ferried to a safe place. Even 5–6 days after hatching, when the young are swimming around freely, the female still acts as a nurse-maid. Contact is maintained by certain signals of movement. It should be mentioned that females with the instinct to care for a brood, but which do not have a brood, adopt water fleas and carefully keep them together.

On account of their small size, beautiful colouring and the care taken of the young, the *Apistogramma* make ideal fish for the seroius aquarist. Already they are a well-known and popular G. The many facets to their way of life, coupled with their colouring, make them interesting fish to observe, something which is far more practical in the aquarium than in the wild. To experience it at its best, always leave the brood to be brought up by the female, even if the majority of the young are eaten. If fed well, a female can be ready to lay again in 3–4 weeks, so the patience of the aquarist should not be too greatly taxed.

Nearly all species of *Apistogramma* are imported, but only the best-known ones can be described here.

– *A. agassizi* (STEINDACHNER, 1875). Agassiz's Dwarf Cichlid. Brazil, Bolivia. Up to 8 cm. Male's body blue-green, front orange, with a dark band running form the mouth to the C. Fins smoke grey on the outside, partly edged in bright red. Female yellow, with a similar dark length-wise band (fig. 108).

– *A. borelli* (REGAN, 1906); Borell's Dwarf Cichlid, Umbrella Cichlid. Tropical and subtropical S America. Up to 8 cm. Males iridescent blue or yellowish, with dark length-wise bands, particularly on the underside. The bands running along the middle of the body are particularly broad and intense. The back is covered with spots stretching into the D. Tips of the elongated D rays red, 1–3 red spots at the top of the C. Female stronger yellow with similar markings.

– *A. commbrae* (REGAN, 1906); Corumba Dwarf Cichlid. Correct species name *corumbae.* Rio Paraná. Up to 5 cm. Yellowish to yellow-brown, with a dark length-wise band ending in a spot.

– *A. kleei* MEINKEN, 1964. Thought to come from upper Amazon. Up to 7.5 cm. Slim, very thick-lipped, ventral profile almost straight. Dark-grey, lower half of the body with two broad, dark, length-wise bands. Fins ochre with wine-red rays and tips.

– *A. ornatipinnis* AHL, 1936; Ornate Dwarf Cichlid. Guyana. Up to 7 cm. Yellow-brown, very dark oblique band across the operculum. Dark spot in the middle of the body and at the root of the tail fin. D, C and An orange, rear portion of which are crossed by dim bands.

– *A. ortmanni* (EIGENMANN, 1912); Ortmann's Dwarf Cichlid. Guyana, Brazil. Up to 7 cm. Very similar to *A. commbrae.*

– *A. pertense* (HASEMAN, 1911); Yellow Dwarf Cichlid; Amazon Dwarf Cichlid. Amazon River. Up to 5 cm. Variable colouring, grey-brown to yellow-brown ground colour with one dark length-wise band and 7–8 cross-wise bands. Very intense, oblique band across the operculum. Front rays of P and D black, the latter running into orange.

– *A. ramirezi* MYERS and HARRY, 1948, Ramirez's

Dwarf Cichlid; Ram; Butterfly Dwarf Cichlid. Venezuela, Bolivia. Up to 7 cm. Beautiful species. Delicate purply-red with brilliant blue-green spots. Large dark spot under the D. First rays of D black. Sensitive, demanding Cichlids (fig. 109). A 'Gold' variety is also available (not found in nature).

Fig. 109 *Apistogramma ramirezi*

– *A. reitzigi* AHL, 1939; Yellow Dwarf Cichlid, Reitzig's Dwarf Cichlid. Rio Paraguay. Up to 5 cm. Yellowish-grey, throat and belly yellow. When excited a dark longitudinal bar and several transverse bars appear. Female becomes lemon-yellow during the spawning period (fig. 110).

Fig. 110 *Apistogramma reitzigi*

– *A. sweglesi* MEINKEN, 1961. Peru. Up to 7 cm. Brownish-yellow, scales with a dark border. A dark lengthwise band typical of the genus and another band on the operculum bent forward in the characteristic manner of the species.
– *A. taeniatum* (GÜNTHER, 1862). Peru. Up to 7.5 cm. Olive-yellow, with a dark longitudinal band and oblique bar on the operculum. A suggestion of some broad dark transverse bars near the dorsal surface. Tips of fins orange-red.
– *A. trifasciatum* (EIGENMANN and KENNEDY, 1903); Blue Apistogramma. Brazil. Up to 5 cm. The subspecies *A. trifasciatum heraldschultzi* MEINKEN, 1960, is most often introduced as an aquarium fish. Gleaming blue-green, with characteristic bar patterns, but often only the length-wise bar is clearly visible.
– *A. wickleri* MEINKEN, 1960; Wickler's Dwarf Cichlid. Northern S America. Up to 8 cm. Grey-blue markings depend very much on the mood, usually with a black spot at the centre of the body and at the root of the rail fin.

Apistogrammoides MEINKEN, 1965. G of the Cichlidae*. At present only includes one species, which lives in small Peruvian streams and was imported into the USA as an aquarium fish in the 1960s. As far as Cichlids go, these are very small fish, barely 4 cm long. Similar shape to the G *Apistogramma**, but unlike them, the *Apistogrammoides* have 8 spiny rays in the An and the front part of the LL is positioned lower down.
– *A. pucallpaensis* MEINKEN, 1965. Male olive-green, with a black longitudinal band running from the operculum to the C. C red, with gleaming blue tips. Female brownish-green, with a similar dark longitudinal band that often breaks up into individual spots.

Aplocheilichthys BLEEKER, 1863. Thought to be a polyphyletic* G of the Cyprinodontidae*, with many species. With a few exceptions (eg *A. pumilus*), all *Aplocheilichthys* are typical fish living in small areas of flowing water in Africa. They swarm at the surface looking for insect food (insects, spiders etc). All are small, very elongated surface fish, laterally only slightly compressed and with noticeably large eyes. Unpaired fins well developed, D lying over the back part of the An, both often tapering. C long rather than tall, droplet-shaped or fan-shaped. Females mostly smaller than the males. As a group very agile swimmers, difficult to catch. Like most fish that inhabit flowing waters, the *Aplocheilichthys* are more difficult to care for and breed than still-water fish. Above all, consideration must be given to the social behaviour of the *Aplocheilichthys* (shoaling fishes!). For this reason, the G is not popular in the aquarium, except for *A. macrophthalmus* which can be delightful to watch if given the correct care (ie large enough shoal, subdued overhead lighting). The first requirements for the care of these fish are a well-ventilated tank, not too high, with enough room for free swimming. Soft, slightly acidic water is equivalent to the natural conditions and is essential for successful breeding, but all the same it increases considerably the tendency that *Aplocheilichthys* have to contract fish tuberculosis*. Therefore, harder, neutral, or slightly alkaline water is often preferred. Care is needed when transferring the fish or when changing the water, as damage can easily be done to the gills if there are large differences in ion concentration. Moreover, it is important to remember that most *Aplocheilichthys* lose their sense of direction in complete darkness and will rush about helplessly, easily damaging themselves. The same thing can happen if the lighting is suddenly changed (turning it on and off). The remedy for this is to keep an 'emergency light' burning all the time. All *Aplocheilichthys* are non-annual species and plant-spawners (compare Cyprinodontidae). In a breeding tank there are several pairs, as the members of *Aplocheilichthys* like to spawn in a shoal. The eggs are relatively large and are laid on fine-leaved plants or on artificial spawning substrates (Perlon yarn). Every 3–4 days all the eggs can be collected from the plants and substrate medium and transferred to another tank for rearing. Newly-hatched young already lead a pelagic* life-style like their parents. They swim with eel-like movements, need a great deal of food and grow slowly. The following species are often kept in aquaria.

Fig. 111 *Aplocheilichthys katangae*

– *A. katangae* (BOULENGER, 1912); Katanga Lampeye. Congo region. Up to 4.5 cm. Male silvery with a blue-green lustre and one dark longitudinal band. Head and back yellow, eyes turquoise blue. Female less brightly coloured (fig. 111).
– *A. macrophthalmus* (MEINKEN, 1932); Lamp-eyed Panchax. Tropical W Africa. Up to 4 cm. Male brass-coloured with two beautiful blue iridescent, longitudinal bands. C dotted blue and orange. Female duller (fig. 112).

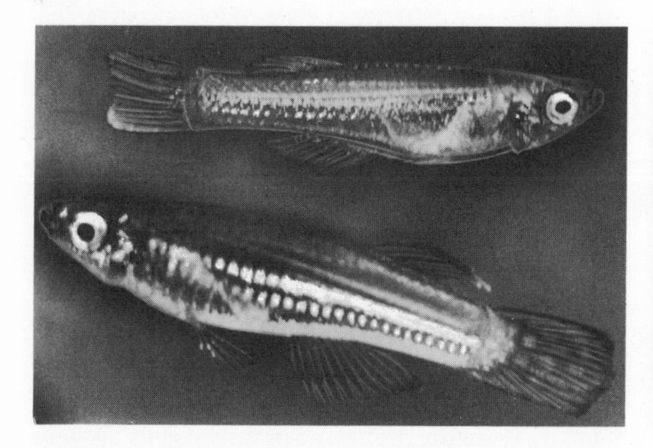

Fig. 112 *Aplocheilichthys macrophthalmus*

– *A. myersi* POLL, 1952; Myer's Lampeye. Central Africa. Up to 2.2 cm. One of the smallest vertebrate animals. D is pennant-shaped in the male, body iridescent blue-green. Female does not have D elongated to a point.
– *A. pumilus* (BOULENGER, 1906); Tanganyika Lampeye. Large lakes of Central Africa (Tanganyika, L. Kiwu, L. Edward etc). Up to 5.5 cm. Male an intense, iridescent green-blue with one indistinct dark band. Fins orange-brown. Female less intensive colouring, colourless fins.

Aplocheilus MCCLELLAND, 1839. G of the Cyprinodontidae*. About 7 species, found in S Asia, most confined to S India and Sri Lanka. All members of the G are small to medium-sized, pike-like, elongated fish with a broad, upward-pointing mouth and remarkably large eyes. Forehead almost horizontal. Arrangement of fins much the same throughout the G, D well to the rear, An attached by long base, often elongated like a pennant, C large and either rounded or ovate. Especially the larger species of *Aplocheilus* are predatory, so large, flat, well-planted tanks are recommended. A thick layer of floating plants suits the fish's life-style, as it offers plenty of cover. The *Aplocheilus* do not swim about

Fig. 113 *Aplocheilus dayi*

much, preferring to lie in wait for their prey, be it fish fry or flying insects (eg Leaflice). Midge larvae, large *Daphnia* and several types of worm are not refused either. The water temperature should be about 24–28 °C. There are no particular recommendations for

Fig. 114 *Aplocheilus lineatus*

preparation of the water although soft water, filtered through peat, is best, especially for breeding purposes. After an interesting courtship display, the spawn is laid on fine-leaved plants. The spawning season may last several weeks. The serious breeder should transfer the spawning substrate containing the eggs to a separate rearing tank. The young hatch after about 2 weeks and soon lie in wait for prey at the surface, just like their parents. So the young are best reared by floating the food (*Artemia* and *Cyclops* nauplii, rotifers) along with the current so that it passes over their mouths.
– *A. blocki* (ARNOLD, 1911); Dwarf Panchax, Green Panchax. S India and Sri Lanka. Only 5 cm (smallest species). Male metallic yellow-green, body and fins dotted red. Female paler.

– *A. dayi* (STEINDACHNER, 1892); Day's Killifish, Ceylon Killifish. Sri Lanka. Up to 7 cm. Male shining metallic green, body scattered with red spots. An has fine red stripes, and C has similar ones at the lower part. Female, less brilliantly coloured, caudal peduncle with 6 or 7 dark transverse bars (fig. 113).
– *A. lineatus* (CUVIER and VALENCIENNES, 1846); Striped Panchax, Panchax Lineatus. India and Sri Lanka. Up to 10 cm. Male olive-brown, scales have a shiny gold colour, breast and belly dotted red. Edges of fins red. Female darker, with noticeable transverse bands. Base of the D has a dark spot (fig. 114).
– *A. panchax* (HAMILTON-BUCHANAN, 1822); Blue Panchax. S Asia, from India to the Malaysian Archipelago.

Fig. 115 *Aplocheilus panchax*

Fig. 116 *Aplocheilus panchax siamensis*

Up to 8 cm. Male grey-yellow, scales have an iridescent blue zone. D has black spot at the base. Female has similar colouring with fins a stronger orange (figs. 115 and 116).

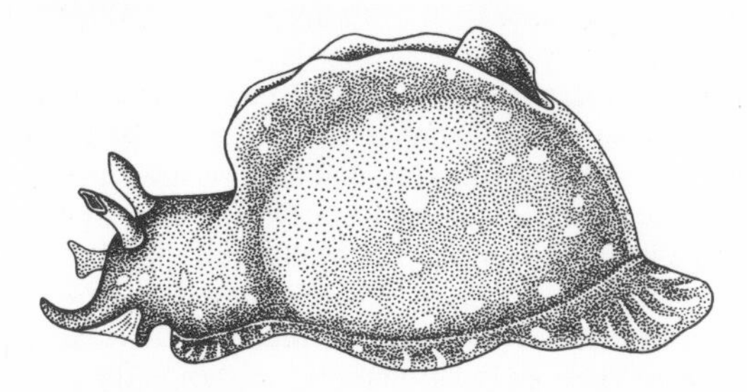

Fig. 117 *Aplysia*

Aplodinotus grunnieus *see* Sciaenidae

Aplysia LINNAEUS, 1758; Sea Hare (fig. 117). G of the Opisthobranchia (*see* Euthyneura), with a small shell covered by the mantle (O Tectibranchia). These Sea Hares live in the sea, weigh up to 1 kg and can be 40 cm long. Can swim by means of wing-like fleshy extensions (parapodia); when threatened secrete a purple fluid. They are herbivores, and will eat lettuce as well as algae in the aquarium. They live quite well in the aquarium. In certain places these molluscs are very common. They are used in experiments, particularly the neurophysiological kind, for which their large nerve fibres are particularly suited. Several species live in the Mediterranean.

Apodes, O, name not in current usage, *see* Anguilliformes

Apogon LACÉPÈDE, 1802. G of the Apogonidae* with numerous species. In tropical seas, isolated sightings in temperate regions. Sociable fish that eat small animals. Very prettily coloured, small perch-like fish with a compressed body, large mouth and big eyes. As far as we know, the males carry the brood in the mouth. After a long courtship, fertilisation probably takes place internally. In *A. imberbis*, for example, the female releases about 20,000 tiny eggs in a clump, which the male takes into his mouth immediately. The young fish are not looked after. The *Apogon* are perfect for the aquarium.

Fig. 118 *Apogon nematopterus*

They like to hide and are primarily active at night. Feed with small crustaceans. Do not put them in with lively large fish. So far only a few of the many species have been kept in the aquarium, eg:
– *A. imberbis* (LINNAEUS, 1758). Mediterranean in shaded areas. Sociable. Up to 13 cm. Gold-red.
– *A. nematopterus*. BLEEKER, 1856; Cardinalfish, Pyjama Cardinalfish, Pennant Cardinalfish. Tropical Indo-Pacific. Up to 9 cm. Head, D and P yellow, eye red, a dark band around the middle of the body. Posterior pale with brown spots (fig. 118).

Apogonidae. Cardinalfishes (fig. 119). F Perciformes*, Sub-O Percoidei*. About 10 to a maximum 20 cm long. These fish are mostly red and brown and from tropical seas, where they live primarily in shallow water, especially mangroves, as well as reefs. A few species live in deeper waters, and a very small number

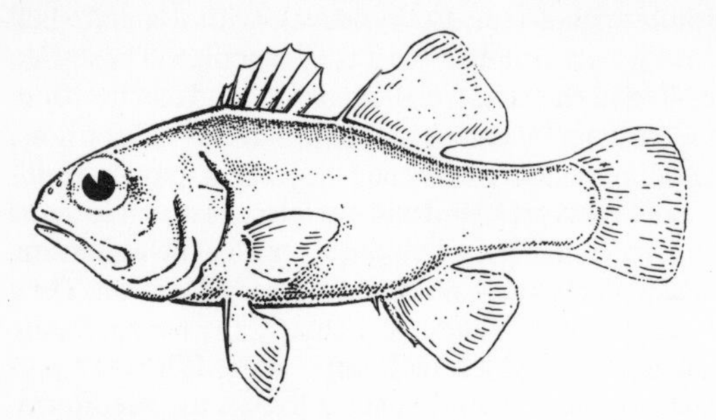

Fig. 119 Apogonidae-type

enter fresh water in some South Sea Islands. They often occur in large numbers and are an important part of the diet of larger predatory fish. Body mostly elongated, large eyes. 2 dorsal fins, D1 has 5–9 spines, An has only 2, like the Percidae*. Relatively large comb-edged (ctenoid) scales. The eggs and young fish are taken into the mouth by the male, and occasionally by the female, if danger threatens. Several species play dead when threatened, ie they lie on their side. *Apogon ellioti* from the Gulf of Tonking has 3 light organs in the intestinal canal. Their function is unknown. Several species are quite often kept in marine aquaria; *see* Ga *Apogon* and *Siphamia*.

Apolemichthys FRASER-BRUNNER, 1933. G of the Chaetodontidae*. Sub-F Pomacanthinae. 3 species live in the tropical Indo-Pacific. Territorial fish that live on reefs and eat animal and plant food. FRASER-BRUNNER originally described the G as a Sub-G of *Holacanthus**. Very hardy, long-lasting species if kept in large aquaria with regular changes of water and plant food. The following species is often imported.

– *A. trimaculatus* (CUVIER and VALENCIENNES, 1831); Three-spot Angelfish. Bright shiny yellow, mostly with a double black spot high above the eyes and a brown spot behind each of the upper edges of the gill opening. Lower half of the An black, mouth mostly bright blue.

Apollo Shark *(Luciosoma spilopleura) see Luciosoma*

Apomorphic. Word used to describe a feature that is a deviation from the original plesiomorphic* feature. In fish, for example, this word could be applied to the fin rays in sticklebacks that have evolved into spines, to the spatula-shaped teeth of many L. Malawi Cichlids, or to the different ways of caring for the young. The more apomorphic features groups of organisms have in common, the surer you can be of their natural, objective relationship. In systematics*, common apomorphic features form the basis of a reconstruction of the phylogenetic relationship.

Aponogeton LINNÉ, fils, 1781. The only G of the Aponogetonaceae*. 40 species, found in Africa, Madagascar, S Asia, Australia. Perennials with compressed, slightly branched rhizomes. In some species there may also be true tubers. Leaves in radical rosettes, of many shapes. Inflorescences have 1, 2 or sometimes more spikelets (fig. 120), partially encased by a spathe. Flowers usually hermaphroditic, but can also be unisexual and split between two plants. Monosymmetrical. Perianth usually formed of 2 scales, which, like the axil of the inflorescence, is white, yellow, pink or violet, becoming green as the fruit ripens. 6 stamens, 3 carpels, free, superior. Carpels unite to form fruit. Occasionally adventitious plants form near the inflorescences. The species of the G are aquatic plants, living in a wide variety of waters. There are also some marsh species. They grow in a very uniform manner, but because of the variety of leaves they have become important as aquarium plants. Those species that only form floating leaves, with the exception of the rarely cultivated *A. distachyus*, are not suitable for the aquarium. Better suited are species that produce both floating and sub-

Fig. 120 *Aponogeton*-inflorescence

merged leaves. The submerged leaves are formed after a resting period, only to be cut off later by the floating leaves. For the aquarist, those plants that produce submerged leaves only are the most important. The leaves of these species are distinguished by colour, the length and breadth of the blades and their consistency, as well as their shape. In *A. undulatus* and some other species, dark, translucent patches appear on the pale surfaces of the leaves in varying quantities. These patches have a different anatomical structure to the rest of the plant tissue. In *A. madagascariensis* the leaf blades become latticed when leaf tissue is lacking between the nerves. Some marsh species have round leaves like rushes.

We know little about the growth cycle. Some African species are said to live through a resting period lasting several months when the waters dry up. The rhizomes lie in the bottom, leafless. Other species probably have resting periods which are related to temperature. They do not grow at these times, but keep some of their old leaves.

Apart from *A. madagascariensis*, the species cultivated so far give no problems in the aquarium. A strong substrate and fairly good to good lighting have proved worthwhile. Occasional transplanting seems to help plant growth. Propagation is usually only possible by cultivating seedlings. Many plants produce seeds without assistance; others must be pollinated artificially. In many cases cross-pollination is necessary. As the flowers of one spikelet open gradually, they must be dusted with pollen several times. When the fruits are nearly ripe, they should be cut off and transferred to another container, so that when the seeds fall out you can have them under control. These have no resting period, and germination starts after only a few days. Only plant the seedlings in the substrate after they are sufficiently developed.

– *A. boivinianus* BAILLONER JUMELLE. N Madagascar. Leaves dark green, as if beaten by a hammer. Inflorescence containing 2–3 spikes, white or pink. V decorative, large species, but rarely cultivated (fig. 121).

– *A. crispus* THUNBERG; Ruffled Sword Plant. Sri Lanka (Ceylon). Leaves variable, pale to dark green, leaf edges wavy or curled. Inflorescence with 1 spike, white. Seldom cultivated (fig. 122).

Fig. 121 *Aponogeton boivinianus*

Fig. 122 *Aponogeton crispus*

– *A. distachyus* LINNÉ, fils. S Africa, introduced into S France, California, Australia. Strong plants that only form floating leaves. Inflorescence with 2 spikes. Flowers have only one large scale. Not suitable for aquaria. Can be kept in the open over winter if frost-free; better results achieved if cultivated in a greenhouse. Winter flowers form with a 6-month summer resting period. Cultivated in Europe since 1788.

– *A. echinatus* ROXBURGH. S India. Submerged leaves pale green, delicate, wavy, reminiscent of *A. ulvaceus*. Floating leaves coarser, rare in aquaria. Inflorescence with 1 spike, white or pink. Grows well. (Several species collected in Asia and kept in aquaria exhibit features like those of *A. crispus* as well as *A. echinatus*.) However, it is not satisfactory to regard these plants as hybrids, which arose out of cultivation or already existed in their place of origin. In fact a study of the general biology of these related plants needs to be made. *A. crispus* and *A. undulatus* are the most important species in this respect (*see* earlier literature on aquaria).

– *A. madagascariensis* (MIRBEL) VAN BRUGEN; Madagascar Lace Plant. This name is given to all Madagascan plants with leaves either partially or completely in the form of a lattice-work, eg *A. fenestralis* (POIRET) HOOKER fils, *A. fenestralis* var. *guillottii* (HOCHREUTINER) JUMELLE; *A. henkelianus* FALKENBERG and BAUM and *A. bernerianus* (all forms of aquarium plants not entirely separate from one another). *A. madagascariensis* is found in central Madagascar. Leaves of different shape and lattice-work. Inflorescences with 2–6 spikes, white or pink. Difficult to cultivate; so far, this species has only survived for a short time in aquaria. Successes

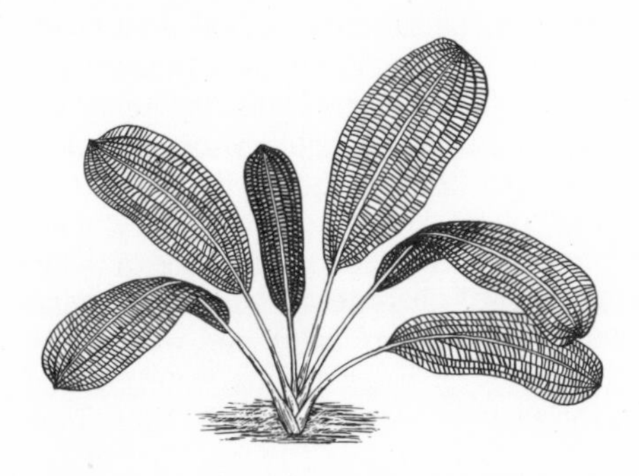

Fig. 123 *Aponogeton madagascariensis*

were obtained in botanical gardens in the first half of the century, eg in Kiev, Leningrad, Rostock. The plants were mostly kept in oak containers with a moderate amount of light, temp. 18–25 °C, pH values below 7 and a water hardness of 10 °, with a regular change of water. Water movement also helps. The substrate appears to be of less importance (fig. 123).

– *A. rigidifolius* VAN BRUGGEN. Sri Lanka (Ceylon). Long rhizome. Leaves dark olive green, coarse. Inflorescence with 1 spike, white. Undemanding species.

– *A. ulvaceus* BAKER. N and central Madagascar. Leaves transparent, pale green to reddish, flat to very wavy. Inflorescence with 2 spikes, yellow. Large undemanding species. Cross-pollination necessary.

– *A. undulatus* ROXBURGH. India. Occasionally with floating leaves. Submerged leaves have dark, translucent patches. Adventitious plants form near the inflorescence. 3 types are cultivated and can be distinguished by the width of the leaf blades. *A. stachyosporus* DE WIT belongs here too. Undemanding species.

Aponogetonaceae, F of the Liliatae*. The only G is *Aponogeton**.

Appendicularia, Cl, *see* Tunicata

Appetence Behaviour. Term used in ethology. An external stimulus will induce a fish to release energy stored within it for a specific reaction. This results in 'search' behaviour. It is important for the successful unfolding of very different modes of behaviour, however it is not essential in every case. In fish, appetence behaviour can mean going off in search of many things, for example: 1) a male searching for a female in breeding condition; 2) a female seeking a suitable spawning ground during courtship; 3) searching for a suitably aroused combatant when battle is the action needed to complete the behavioural cycle; 4) searching for prey etc. Every part of behaviour can be regarded as appetence behaviour for the type of behaviour to follow or as the completion of the previous type of behaviour. Appetence behaviour requires no external stimulus to activate a sphere of action. Once the specific energy is released and the final act of behaviour has run its course, appetence behaviour is complete.

Apteronotidae, F, *see* Gymnotoidei

Apteronotus albifrons *see* Gymnotoidei

Aquaria. The term 'aquarium' was first used by the Englishman GOSSE (1853) to describe transparent containers used for keeping aquatic animals (*see* Aquarium-keeping). Today's aquarium technology means that the technical problems of keeping aquatic life forms have in principle been solved. We now know all about water preparation* and the constant monitoring of water quality (its pH value*, hardness*, redox potential*, oxygen content*, temperature* etc), and we can also make use of automatic heating elements, water circulating pumps (*see* Pumps), filtration*, UV light, foam filters* etc.

Modern technology has had to face certain demands in quality, stability and reliability in the construction of modern aquarium equipment, eg complete neutrality in respect of both fresh and sea water, great stability and security against breakage, ease of maintenance and a colour-correct, undistorted reflection of the contents.

There are various types of aquaria that meet these requirements to a greater or lesser degree. 'Type' is defined by the manufacturing process:

1) All-glass aquaria consist of moulded glass. They are cheap, corrosion-free and need no maintenance. Disadvantages are limited size, little stability and flaws in the glass, produced during manufacture. But because they are easy to sterilise these aquaria are well suited to breeding, keeping fish broods, treating disease and keeping infected animals in quarantine. Another use is as filter tanks.

2) Framed aquaria consist of a metallic or plastic frame with an angular profile, into which sheets of glass (*see* Aquarium 'Glass'), or plastic and asbestos cement matting can be sealed (Aquarium Sealants*) to give greater stability. Because of this higher degree of stability, metal-frames are to be preferred, but they MUST be well sealed in every case (by coating with plastic, epoxy resin or by enamelling). For marine aquaria the top edge of the frame is turned to the outside to get rid of possible overhanging corrosive surfaces. These tanks are extremely stable and can be made quite large. By inserting plate glass, you will be able to achieve a colour-correct, undistorted view of the contents. Poisonous substances can get into the aquarium water if unsuitable sealant is used. Leaks may occur when the tank is drained or moved. Framed aquaria have proved their worth in practice and are found everywhere.

3) Glued glass aquaria combine the advantages of all-glass and framed aquaria. With the aid of silicon india-rubber, sheets of glass can be stuck together to make any kind of aquarium you like. Even one of considerable size will have a high level of stability and the glue will not deteriorate with age, either in fresh or sea water.

4) Asbestos cement aquaria (eternit aquaria) have as their foundation material asbestos powder mixed with cement, bonded under high pressure. These tanks are completely watertight but very heavy. Tanks made out of asbestos cement as well as pure cement can be made into large aquaria, but they must be rinsed thoroughly or isolated with epoxy resin.

5) Plexiglass aquaria consist of the raw material Polymethylmetacrylic and can be bought either as a portable, ready-made tank, or as separate components for construction yourself. Plexiglass is easy to shape by machine; it is less brittle but more flexible than silica glass. When completely neutral, a 3 mm thick sheet of Plexiglass will still let through roughly 70% UV light. Piacrylic glass can be used for fresh and sea water. The only disadvantage is that the material is not very hard and so scratches easily, making it difficult to clean. With suitable raw materials, it is possible to make an aquarium any size and shape you like and still meet all the requirements. Any stable material (wood, cardboard, etc) can be sealed with some kind of resin and used to contain your aquarium frame.

Table for angle-iron and bottom sheet

Size of tank cm	Angle-iron mm	Thickness of bottom sheet mm
50 × 25 × 25	15 × 15 × 3	1.5–2
60 × 25 × 30	15 × 15 × 3	1.6–2
60 × 30 × 40	15 × 15 × 3	1.7–2
80 × 30 × 30	20 × 20 × 3	2.0–2.6
80 × 30 × 40	20 × 20 × 3	2.2–2.6
80 × 40 × 45	25 × 25 × 4	2.6
100 × 40 × 40	25 × 25 × 5	2.7–2.9
100 × 35 × 40	25 × 25 × 4	2.9
100 × 40 × 60	30 × 30 × 5	2.9
120 × 50 × 50	30 × 30 × 5	3.4
150 × 60 × 80	45 × 45 × 9	5
200 × 80 × 100	80 × 80 × 8	8

The choice of fresh or sea water aquaria depends on the requirements of the animals and plants you are going to keep, and whether your tank contains warm or cold water depends on their temperature requirements. A community tank tries to unite plants and animals of different origins but with virtually the same needs for life; a biotope or species aquarium tries to replicate geographical, ecological and ethological factors. Quite different in purpose is the hygiene-aquarium. It usually dispenses with the whole idea of an aquarium shape (at least in as far it is not necessary to keep the aquarium animals alive) and devotes its attention to water quality. Tanks that are only used for breeding will be set up entirely from the point of view of how efficiently they will fulfil that function.

Aquaria in zoological gardens or animal parks are provided for the public to observe water animals. Besides these, there are also large aquaria in numerous marine biology stations where special research is carried out. Independent enterprises such as well equipped oceanaria or dolphinaria are increasingly being devoted to research into marine biology, in addition to their obvious attraction as public show-places.

Aquarium *see* Aquaria

Aquarium, Care of. There are certain basic rules that need to be adhered to when caring for your aquarium. The work needs to be carried out regularly but it should involve the absolute minimum of disturbance for the specimens in the tank. Great changes of any kind can alter the fragile equilibrium in an aquarium, and possible consequences can only be guessed at. The tools used to care for your tank should be made out of non-toxic material, and in some circumstances they should be seawaterproof. If the same tools are used in several tanks, you must take relevant hygienic precautions (disinfect them, for example). Before you tackle any big jobs, you should interrupt the flow of water to the aquarium and *switch off* all equipment. This will prevent accidents. Your daily routine will include feeding* the aquarium animals; the removal of remains of food, mould on the substrate (*see* Mud-extracting Tube) and any dead animals; observation of the animals (behaviour*, method of feeding*, colour and state of health, *see* Fish Diseases); monitoring of the water (colour, smell, cloudiness); and checking that all the technical pieces of apparatus are working (temperature-control*, aeration*, filtration* units, lighting*). Periodically, the following should be carried out: clean the front and lid of the aquarium; clean the filter unit; observe plant growth (in seawater look for algal growth, which is a sensitive indicator of a change in water quality) and notice any addition of plant manure or trace elements; check the pH* and water hardness* as well as the salinity level in seawater (*see* Areometer); replace any evaporated water with distilled or completely purified water at the same temperature (*see* Desalination). Depending what you have in the tank, $^1/_5$–$^1/_4$ of the water should be siphoned off every 10–14 days and replaced by fresh water of the same quality. If you suspect irregularities it may be wise to check the amounts of ammonia*, nitrates, nitrites, phenols, oxygen and carbonic acid as well as the redox potential*, in addition to the routine testing of the pH, water hardness and salinity. A falling redox potential*, lack of oxygen, high concentration of carbon dioxide (*see* Carbon Dioxide Poisoning), or an increase in toxic metabolic products all point to a change in the equilibrium of the tank. The causes may often be traced to technical apparatus that is not working properly, or it may mean that toxins are present that have interfered with the metabolic processes in the aquarium. Other causes might be a build-up of organic waste through dead animals, remains of food and rotting vegetation or over-feeding, and the introduction of too many animals may, among other things, overload the system to such an extent that no amount of technical apparatus can compensate. In such cases the cause of the problem must be found and dealt with as quickly as possible if the equilibrium of the tank is to be re-established to allow the aquarium specimens to go on living.

Aquarium 'Glass'. In framed aquaria (see Aquaria), sheets of silica glass, or more rarely Plexiglass, are puttied together with plastic or asbestos-cement matting (*see* Aquarium Sealants). The type of silica glass that is most used comes in the form of standard glass and is suitable for the floor of the aquarium, as well as its back and side walls. Neutral-coloured plate-glass is more suitable for the front wall. The glass should be as free of all impurities and can be slotted into the frame if there is just 4 mm tolerance in both planes of expansion. To reduce the chances of breakage in a large tank it is advisable to file down the edges of the sheets of glass and place pieces of cork in the frame to form the boundary. Glaze the frame in the following order: the sheet forming the floor followed by the longer then the shorter sides. Look at the table to see how strong the glass in your aquarium must be. In tanks made out of pieces of silica glass stuck together, the sheets should be 2–3 mm thicker than the figures in the table. The vertical sheets of glass must stand on the bottom sheet and the side walls should be inserted between the walls at the front and back. If using other raw materials, choose how strong it will have to be by comparing it with the pressure of water you will expect to have.

Aquarium, Installation of. How an aquarium is set up is always determined by the particular use it will have. That is why aquaria used for breeding purposes*, or for quarantine* or experimental purposes, will be very different from community tanks or biotope and species tanks (*see* Aquaria). The first thing to be considered is the natural demands of life of your specimens. As well as numerous other factors, relevant decorative mat-

Thickness of glass in mm

Length cm	30	40	50	60	70	80	90	100/110	110/130	130/150
Height										
30	2.8	3.3	3.8	4.1	4.2	4.4	4.6	6.3	6.9	9.1
40	3.4	4.3	5.1	5.6	6.0	6.3	6.5	6.9	7.1	9.2
50	4.4	5.1	5.8	6.5	7.2	7.7	8.2	8.7	9.1	11.1
60	–	6.0	6.5	7.5	8.5	9.3	9.7	10.7	11.4	11.7
70	–	6.6	7.3	8.2	9.0	10.0	10.9	12.2	13.1	13.6
80	–	7.4	8.2	8.8	9.3	11.0	12.2	13.7	14.9	16.1

erial* and aquatic plants should contribute to the best possible living conditions. When setting up an aquarium you should follow some basic rules and in particular keep to a certain order of tasks. Firstly, clean the tank thoroughly, if necessary disinfect it (*see* Disinfection) and test for leaks. Place large rocks and corals directly on the bottom before introducing any substrate. In this way you will prevent layers of sand from becoming stuck under the rocks where it will decay. It also means that the decorations will not be overturned by fish rooting around on the bottom. If you stick rocks together with cement or some other suitable glue to make large areas of decoration, or even stick them to the back walls, the danger of upsetting them becomes very slight, but you may end up having difficulties with cleaning the tank (*see* Aquarium, Care of). Now you can lay down a suitable substrate. It should be damp, and make sure it is pressed down firmly, slanting the sand a little towards the front. Soak away excess water carefully with a sponge. Complete your layout by adding smaller rocks, roots etc. Now the necessary technical pieces of equipment (for aeration*, filtration*, temperature control*) can be put in. Before pouring in the water it is best to lay a piece of paper over the whole set-up and carefully run the water through a hose into the tank. Of course, your water should already be prepared and at the right temperature (*see* Water Preparation). This way of filling your tank prevents the substrate from being churned up which clouds the water. When filled, you can add plants to the tank. What you plant will depend on the fish you are going to keep. In community tanks and ornamental aquaria you can use several species of plants that blend well in growth pattern, size, leaf formation and colour, either as groups of plants or as solitary specimens (fig. 124). When selecting plants, bear in mind how much light they need and later growth. Any roots present should be removed as far as possible, because otherwise they die after planting and start to decay. They are quickly replaced by new, shoot-bearing roots. Plants with rhizomes or tubers can be pressed into the substrate with the fingers, when the depth of the plant is not a significant factor. Use pincers to insert small shoots. Secure with plant pins*. Lastly, connect the electrical apparatus to the mains. Switch on the filter circuit, put on the lid (*see* Aquarium Lids) and install lighting*. Under no circumstances should fish or any other animals be placed into the tank immediately. Leave standing for 10–14 days until a biological equilibrium is arrived at, then put your first animals in. Newly installed aquaria should be provided with animals gradually and carefully, even after they have been left standing for 2 weeks. This is to avoid upsetting the delicate biological balance. Animals that are particularly sensitive, and these include some fish and many marine invertebrates, should only be placed in tanks with 'established' stable conditions, if you want to avoid losing many of your specimens.

Fig. 124
A well-planted aquarium

Aquarium Lids. Adding a lid to an aquarium is necessary for different reasons. Above all, a lid should reduce evaporation and stop the fish from jumping out. Sheets of silica glass or plexiglass are most often used. Plexiglass lets through UV light and does not alter the spectrum (*see* Lighting). The lids are fixed inside the upper edge of the aquarium frame by a corrosive-free support. Condensation can easily be removed by tipping the lid up slightly. Casings for lights, made out of stable material, can also be useful in aquaria which are completely covered. A lid keeps dust off the surface of the water and creates a zone with the right temperature and humidity for floating plants or plants with aerial leaves. Such a zone is also essential for fish that supplement their oxygen supply from the air (Labyrinth fish). Small flaps in the lid make it easier to care for your specimens, eg feeding, without having to take the whole lid off. It is important that the lid does not prevent the right amount of light from reaching the surface of the water. There should also be enough openings for air circulation, especially for the exchange of the carbon dioxide that is released (*see* Ventilation).

Aquarium Plant. Marsh or aquatic plant, suitable for keeping in an aquarium.

Aquarium, Rear Wall of, *see* Decorative Material

Aquarium Sealants. To seal the walls of an aquarium with a plastic or metal frame (*see* Aquaria) you need a pliable sealant that will remain elastic, waterproof and non-toxic. According to old recipes only certain putties made out of linseed oil and varnish with a dash of red lead and 'filler' substances were used. 1 kg varnish putty, 1 tablespoon red lead, dry chalk (calcium carbonate). 1 kg putty, 1 tablespoon linseed-oil, 1 tablespoon red lead. 2 parts linseed-oil varnish, 4 parts finely ground glass, 4 parts lead oxide.

More suitable are the very long-lasting, usually completely non-toxic synthetic putties. Thoroughly recommended are those products based on neoprene, silicon india rubber and thiocol. The latter is often made hard with the addition of lead, so must be isolated against reaction with the water. Silicon india rubber is better suited than traditional fluid materials for the subsequent sealing of conventionally puttied tanks. It is also good as a glue in unframed glass aquaria.

Aquarium-keeping. The observation, care and breeding of aquatic animals and plants in aquaria*. Essentially it serves to popularise scientific understanding particularly in the field of biology. It has developed from a simple hobby into a serious one. Increasingly, aquarium-keeping has had an influence on related scientific fields of study, and a great deal of our knowledge of ichthyology has been instigated by aquarists. To a very large extent, aquarium-keeping has contributed information about native and exotic species of fish and aquatic plants. The collected experiences of people who have kept and bred aquatic life forms down through many centuries have enabled all the technical problems to be solved. Such advances in technique have increased the amount of biological research carried out successfully by the aquarist's methods. In addition to information regarding care, breeding and reproductive behaviour in aquatic life forms, research regarding the fishing industry, fish disease, chemical and microbiological problems of water, ecological problems and environmental pollution is also important. Recently, basic information in the field of ethology has been arrived at by watching fish in aquaria.

Keeping aquatic animals, particularly fish, is an ancient activity. There are accounts dating back to ancient Egypt. In Greece and Rome too, containers for fish (piscinae) have been found in the gardens of large palaces. According to reports from the time of the conquest of Mexico, there were man-made containers for fresh and sea-water fish in Montezuma's gardens. In China, fish were already kept in the tenth century. In the twelfth century they bred goldfish, which were later kept as abnormal breeds in gigantic porcelain tubs (called dragon tubs) which stood in the palace grounds of the Chinese emperor. Goldfish-breeding reached Japan in the fifteenth century and by the beginning of the seventeenth century it was also known in Europe. Only with the increase of knowledge regarding the mutual exchange between aquatic plants and animals, and the importing of exotic fish, did aquarium-keeping receive an upsurge of popularity. Large sections of the population were won over, and aquarium-keeping became a hobby, practised worldwide. Large industries sprang up to breed ornamental fish, import and export companies were established and whole branches of industries became involved in making equipment needed by the aquarist. Since aquarium-keeping is now a worldwide industry, it represents a not insignificant factor in the economy. The literature in this field has grown into an extensive specialist study with numerous journals appearing periodically. Today, there are many clubs and societies concerned with aquarium-keeping, and it has become a valuable source of education, developing the minds of all those interested in the subject.

Aquarium-keeping, Science of. Started as a hobby, when aquatic life forms were kept and cared for in aquaria*. In Germany aquarium-keeping was founded chiefly by a natural historian called Roßmäßler (1806–67) who came from Leipzig. At first people concentrated on keeping native freshwater fish and invertebrate aquatic animals. As the import trade improved, more and more tropical and subtropical species were brought in. This began in the last quarter of the nineteenth century, and in the twentieth century aquarists began to keep marine fish and other animals. Very large public aquaria were built in geological gardens, which also enabled large aquatic animals to be kept and shown. (The supreme examples today are the dolphinaria.) Alongside the care of animals, people began to cultivate aquatic plants as it became more important to create a natural-looking biotope in the aquarium. As aquarium-keeping developed however scientists also began to take a keener interest. It began to develop from a simple hobby into a scientific method of keeping aquatic animals and, as such, numerous scientific experiments came to be carried out. There have been contributions to taxonomy*, distribution ethology*, anatomy*, genetics*, and disease, in fish, in the higher aquatic vertebrates, in invertebrates and in aquatic plants. On the other hand, scientific institutions increasingly make use of observations made in the aquarium during experiments with aquatic animals and plants. In most industrialised nations aquarium-keeping is the most widespread type of hobby to impart basic principles about biology. In schools, an aquarium will provide ample opportunity for observing biological specimens. The technique of aquarium-keeping has grown up along with the science, and has much in common with the science of terrestrial aquarium-keeping. (*See* also Aquarium, Care of; Aquarium-keeping; Terrarium-keeping.)

Aquatic. Life-forms that live in a watery environment are described as aquatic. In nature this includes all fish, all truly aquatic plants and by far the majority of crustaceans, to name but a few. The term 'aquatic' is commonly used when there are both aquatic and terrestrial forms in a single group, eg of the web-spinning spiders, only one G, *Argyroneta**, is aquatic. Opposite to terrestrial.

Aquatic Plant Diseases *see* Chlorosis, Cryptocoryne Disease

Araceae. F of the Liliatae*. 100 Ga, 1,800 species. Tropical-subtropical, few representatives in temperate regions. Perennials with rhizomes and tubers, climbing plants, epiphytes. Leaves alternate or with double rows (distichous), lattice-work of nerves. Flowers small, mostly without a perianth, unisexual and monoecious, standing on a spadix and surrounded by a sheathing bract (the spathe) that is usually strikingly coloured. There are often calcium oxide crystals in the cells. The Araceae are plants of the forest, marsh and water. *Calla palustris,* or Bog Arum, has a white spathe and red berries and is found in N America, Europe and N Asia. It is a poisonous plant that grows in places with a low pH. Difficult to cultivate. In shallow waters and flooded areas of the eastern part of N America you will find *Orontium aquaticum* which has gold-yellow flowers and no spathe. Various species of the G *Peltandra* grow in the same region on the edge of the water. They have a whitish or greenish spathe. Both *Peltandra* and *Orontium* are suitable for open-air ponds, but in Europe they should be protected with a lid in winter. The well-known *Zantedeschia aethiopica* is originally a S African marsh plant. In aquarium literature another G that is often mentioned is *Lagenandra* from S Asia. Its species are small to medium-sized marsh plants which look like *Cryptocoryne ciliata* but which are not suitable for keeping in aquaria. Species of the Ga *Aglaonema* from S Asia and *Spathiphyllum* from the tropics of America and Asia have also been kept in aquaria. However, often they did not survive for long and so proved unsuitable. As they are naturally marsh plants they are only used to being submerged for a short period. Important Ga in the aquarium are *Acorus**, *Anubias**, *Cryptocoryne** and *Pistia**.

Arachnida. Spiders. Very abundant Cl of arthropods (*see* Arthropoda) with over 30,000 species. No antennae. The best known groups are the Scorpions, the web-spinning Spiders (Araneae) and the Mites (*see* Acari). The scissor-like or blade-like second pair of legs (chelicera) carry food to the mouth (often the third pair, or pedipalpa, are also used in this way). External digestion is very common in Arachnida, ie the prey is sprayed with digestive juices which breaks it down so that it can be absorbed. The Arachnida are among the oldest land animals. Their body was originally long and split into many segments (Scorpions), but in the course of evolution it has tended to become much smaller (web-spinning Spiders). In the forms that have changed most (*see* Apomorphic), the body has been reduced to a short, sack-like structure (eg Mites). Relatively speaking, only a few Arachnida have changed to a watery existence, including about 3,000 species of Mites and just one G of web-spinning Spiders (the Argyroneta*).

Arapaima gigas *see* Osteoglossiformes

Arbacia Gray, 1835. G of small to medium-sized Sea-urchins (Echinoidea*) found in warm seas. They possess uniformly long, pointed spines. *A. lixula* (Linnaeus, 1758) and *Paracentrotus** live together on rocks along the shores of the Mediterranean.

Archegonium *see* Angium

Archerfishes, F, *see* Toxotidae

Architheutis, G, *see* Cephalopoda

Arctoscopus japonicus *see* Trachinoidei

Areal (fig. 125). Area of distribution, in the narrow sense, ie the area where a particular systematical group lives and breeds. Usually, the word refers to a species. The migratory area is not part of the areal, even although this might be considerably larger than the areal itself (eg anadromous fish*). The shape and size of the areal can vary enormously but they are characteristic for the systematical group concerned. In general sea creatures have much larger areals than freshwater life forms, for example. A freshwater organism might only occur in one river system or just one lake (*see* Endemic). The most important factors determining the size of the areal are the ecological potence* and geographical/ecological barriers preventing more widespread distribution. Compare also Distribution and Zoogeographical Regions.

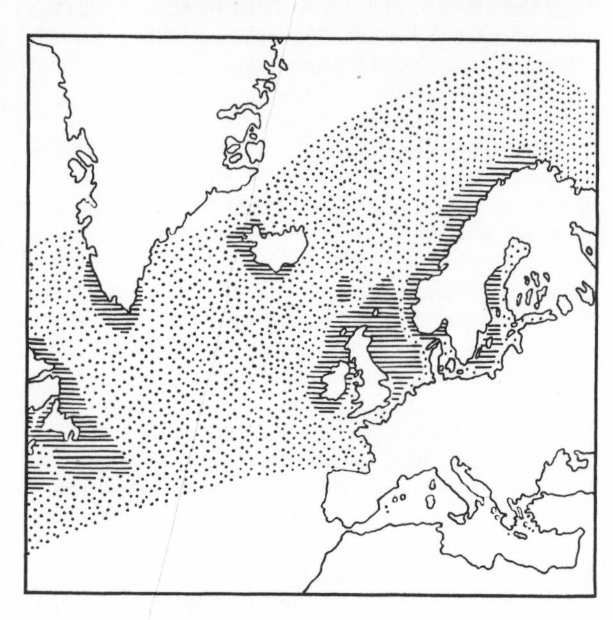

Fig. 125 Areal (hatched) and migratory area (dotted) of the Cod *(Gadus morrhua)*

Arenicola Lamarck, 1801; Lugworms. G of the Polychaeta*. A rather inactive G found in the sandy bottom of the sea. Casts betray their presence. The Arenicola build U-shaped burrows. A collapsible funnel develops on the side of the mouth where sand is swallowed. At the rear end of the burrow piles of sausage-shaped excrement build up. Not possible to be kept in aquaria.

– *A. marina* (Linnaeus, 1758). Very common along coasts of Europe. Up to 20 cm. Olive-green to blackish with red gill bristles.

Areometer (fig. 126). An instrument used to measure the density of a liquid. As a rule, an increase in the concentration of dissolved substances will increase the density of the liquid. Liquid areometers are used in aquarium-keeping to determine the salt concentration of sea water*. The instrument consists of a calibrated glass tube, the lower end of which widens into a spindle-like shape and is weighted with lead or mercury. When this end is floated into a liquid, the instrument stands up vertically and the depth which it has reached can be read off on the scale. The deeper the tube sinks into the liquid, the less the liquid density, and the volume displaced equals the weight of the areometer. Some areometers are calibrated in such a way as to give the specific gravity, so the corresponding concentration has to be calculated mathematically. The aquarist can make use of special ones which are specifically calibrated

Fig. 126 Areometer (hydrometer)

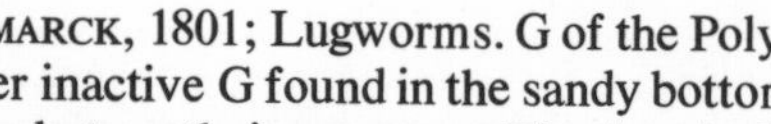

for a particular liquid. With these he can read off the concentration in % or ‰ directly from the instrument. As density is dependent on temperature, comparative readings should always be taken at the same liquid temperature or mathematical adjustments made to the reading to compensate for temperature differences.

Argentine Pearlfish *(Cynolebias bellotti) see Cynolebias*

Argentinian Amazon Sword Plant *(Echinodorus argentinensis) see Echinodorus*

Argonauta argo *see* Cephalopoda

Argulidae. Fishlice. F Brachyura*. Shield-shaped, flattened crustaceans which are parasitic on fish. About 50 known species. About 5–8 mm long. Part of the mouth is shaped like a hook or it is modified into two suckers. By these means they attach themselves to fishes from time to time; but they can also move freely in the water with the help of 8 legs. Usually, Argulidae possess a poison spine which pierces the skin of the fish. Surrounding the spine is a tube mouth through which cell tissue and blood is sucked from the host. Stricken fish suffer a great deal of discomfort; the fins are clamped and where the skin is pierced it is red and inflamed, providing easy access for disease-carriers.

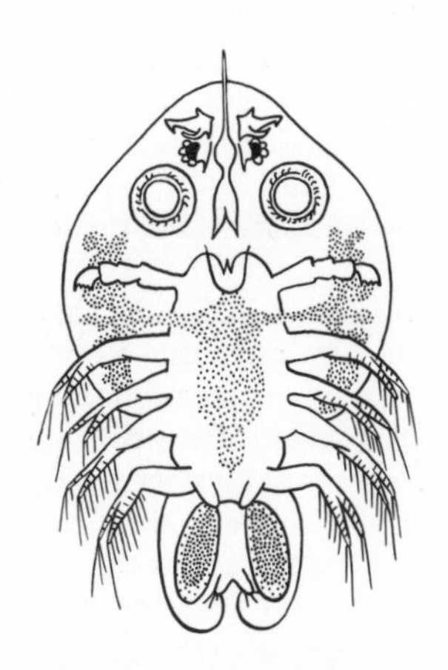

Fig. 127 *Argulus foliaceus*

Small fish may die as a result of the poison from the spine. Besides this, the Argulidae transmit stomach dropsy* and *Cryptobia** LEIDY, 1846. *Argulus foliaceus* LINNAEUS, 1758 (fig. 127), most frequently turns up in the aquarium, but tropical and subtropical species have also made an appearance, eg *Argulus chromidis* KROYER and *Argulus indicus* WEBER. Native species can only withstand higher temperatures for a short period. A good therapy is to dip or bathe infected fish in lysol* and potassium permanganate*, but it is also possible to pick the parasites off carefully with tweezers.

Argulus, G, *see* Argulidae

Argus Fish *(Scatophapus argus) see* Scatophagidae

Argus Fishes, 'Scats', F, *see* Scatophagidae

Argyroneta LATREILLE, 1804; Water Spiders (fig. 128). G of the Arachnida* with very few species. Widespread in Europe and Asia, also known in New Zealand. It is thought that Argyroneta is the only G of the very abundant web-spinning Spiders (Araneae) to spend its whole life underwater. Found chiefly in still waters with dense vegetation, in particular, ponds and bogs. The Argyroneta not only carry around a bubble of air with them, like many other arthropods (Arthropoda*) that once lived an land, they also fill their underwater webs with air. This creates a kind of diving-bell into which the spider puts its abdomen and breathes. Natural hollows of a suitable size, eg empty snail shells, are also filled with air for the same purpose. Male *Argyroneta* are bigger than the females, in particular, they have longer legs. They like to wander around without a permanent home, merely renewing their air supply at the surface now and again and searching out females in their web domes with which to mate. Eggs are laid in mid-summer, and at this time the web dome is made into two storeys. The eggs sit in the upper storey and the female sits below. *Argyroneta* make good specimens in pond aquaria, especially as they are not aggressive towards other members of their G, unlike so many of their terrestrial relatives. For this reason, several can be kept together. Feed with woodlice *(Asellus*)* which also form the main food in the wild.

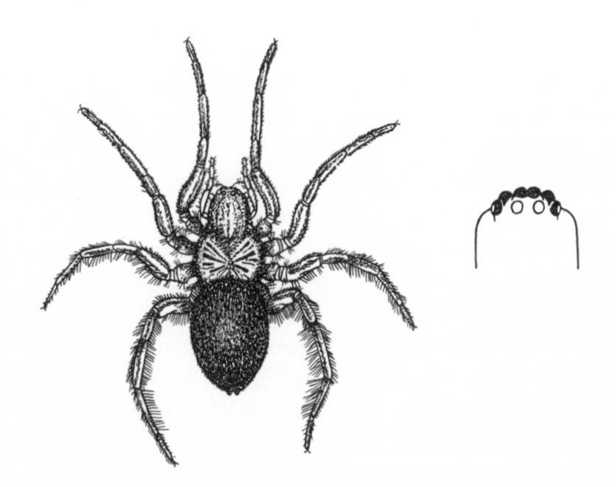

Fig. 128 *Argyroneta aquatica.* Right, head and eyes (enlarged)

Argyropelecus, G, *see* Stomiatoidei

Ariidae (fig. 129). F Siluriformes*, Sup-O Ostariophysi. In body shape these Catfishes are reminiscent of the Bagridae*, ie elongated, laterally compressed, but they have only 1–2 pairs of barbels that are often very short. Broad head, flattened dorso-ventrally, large mouth, positioned obliquely. The front and back nasal opening are placed very close together. D often shaped like a pointed sail and with a strong spine, like the Ps. Adipose fin present, C deeply indented. Many Ariidae are marine fish living in the coastal waters of tropical

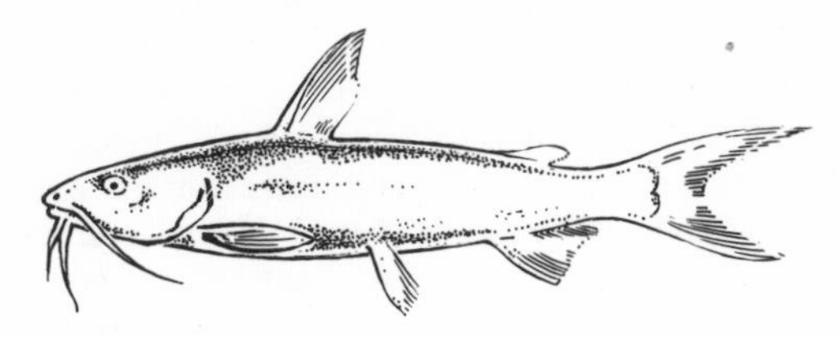

Fig. 129 Ariidae-type

and subtropical areas. Others are found in fresh water, eg S America and E Asia. All the marine forms and some freshwater forms carry their eggs in the mouth. The males incubate the cherry-sized eggs in their mouths, even the young fish may find shelter there for a while. During the breeding season, the females develop a fold of skin on the Vs which is certainly used to catch the large eggs. How they then get into the male's mouth remains a mystery. (The German term *Kreuzwelse* [Crucifix Catfishes] is a reference to the bony skull of the species *Arius proops* found in S America and in the West Indies. Seen from below, the skull resembles a crucifix.) The dissected skulls are made into souvenirs of various kinds. Many Ariidae are valuable food fish.

Aristotle's Lantern *see* Echinoidea

Armoracia aquatica, name not in current usage, *see Rorippa*

Armoured Catfish *(Callichthys callichthys) see Callichthys*

Armoured Catfishes (Corydoras, G) *see* Callichthyidae

Armoured Catfishes, F, *see* Callichthyidae

Armoured Fishes (Placodermi, Sub-Cl) *see* Aphetohyoidea

Armoured Fishes (Scleroparei, O, name not in current usage) *see* Scorpaeniformes

Armoured Phago *(Phago loricatus) see Phago*

Armoured Sea Robins (Peristediidae, F, name not in current usage) *see* Scorpaenoidei

Arnold, Johann Paul (1869–1952). Journalist. One of the best-known and most distinguished vivarium-keepers of his time. He gave the first descriptions of a great many aquarium fish. In 1904 he co-founded the *Wochenschrift für Aquarien- und Terrarienkunde* ('Weekly Magazine for Vivarium-keepers'). In 1936, together with ERNST AHL*, he published a book *Fremdländische Süßwasserfische* ('Exotic Freshwater Fishes'). Each of the entries was illustrated and supplemented with a German name.

Arnoldichthys MYERS, 1926. G of the Characidae*. Tropical W Africa. Very elongated Characins, 10 cm in length. The D1 emerges above the Vs and the D2 is well-developed. The complete LL runs through the lower-half of the body, and its scales, like the rows of scales beneath it, are essentially smaller than the scales of the upper half of the body. Only one species:

– *A. spilopterus* (BOULENGER, 1909); Arnold's Red-eyed Characin, African Red-eyed Characin, Big-scaled African Characin. Tropical W Africa, Lagos to the mouth of the Niger. Up to 10 cm. Flanks iridescent green to blue-green. A dark band runs from the operculum to the middle rays of the C with a rainbow-coloured, shimmering longitudinal band above. Ventral surface yellow. D1 has a large black spot, An has dark-green and yellow longitudinal stripes. The males of one local species have a blood-red blotch on the An. Eye bright red. The females have a black blotch at the base of the An and are a less intensive green colour. Shy shoaling fish that need a lot of room for swimming. They can be kept with other peaceful fish (eg *Micralestes interruptus*). They should be kept at a temperature of 27 °C, 2–6° water hardness with a slightly acidic pH. Food should be live, particularly fly larvae and surface insects, but dried food is also acceptable. First bred succesfully in 1967 by BECK. After hectic driving, up to 1,000 eggs, 1.2 mm long, are produced. The embryos hatch after 30–35 hours and swim free on the 7th day. So far, it has proved difficult to rear the young fish

Fig. 130 *Arnoldichthys spilopterus*

which grow quickly in ideal conditions. A beautifully coloured species, unfortunately very delicate. Particularly prone to fish tuberculosis* if kept in water that is too hard (fig. 130).

Arnold's Red-eyed Characin *(Arnoldichthys spilopterus) see Arnoldichthys*

Arothron MÜLLER, 1852. G of the Tetraodontidae* with over 20 species in tropical Indo-Pacific. Very intolerant, territorial fish that can give a painful bite with their ridged teeth which are arranged like a beak. No notice is taken of large fish belonging to other species. The set of teeth are used to overpower their hard-shelled prey, such as molluscs and crustaceans. As individuals, the *Arothron* make good aquarium specimens but they should not be put in with invertebrates or smaller fishes. One species often imported is:

– *A. hispidus* (LINNAEUS, 1758); Broad-barred Toadfish, Stars and Stripes Toad. Indo-Pacific, Red Sea. Up to 50 cm. Brown with white and brownish spots (fig. 131).

Fig. 131 *Arothron hispidus* with Cleanerfish *(Labroides dimidiatus)*

Arowana *(Osteoglossum bicirrhosum) see* Osteoglossiformes

Arrow-arum (*Peltandra,* G) *see* Araceae

Arrowhead, G, *see Sagittaria*

Arrowhead *(Sagittaria sagittifolia) see Sagittaria*

Artedi, Peter (1705–35). Ichthyologist. Born in Anundsjö in Sweden. In 1728 he was studying in Uppsala with LINNÉ, with whom he formed a deep friendship. In London he made ichthyological studies in the collections of SLOANE, WILLOUGHBY and RAY, which he continued later in Amsterdam. Died after an accident at the age of 30. His manuscript was published by CARL VON LINNÉ in Leiden in 1738. The 5-volume *Bibliotheca Ichthyologica* contains a critical evaluation of the work of other ichthyologists, a detailed morphological description of fish, a key for the recognition of 45 Ga and a discription of 72 species. The division of fishes into 4 orders was later adopted by LINNÉ in his work.

Artefact. Structures or substances that can only be produced in an organism artificially, after a special method of examination, eg the colouring of a fish after preservation in alcohol or formulin.

Artemia salina LINNAEUS, Brine Shrimp. Free-swimming crustaceans, with no shell, found in salt-retaining inland waters in all parts of the earth, not in the sea. Size up to 15 mm, colour mostly red, like raw meat, both of which are influenced by the salt content of the water. Along the body are 11 pairs of leaf-shaped swimming legs with gill processes. Reproduction by eggs, mostly without the males, through a process called parthenogenesis*. The long-lasting eggs are produced at intervals, have thick protective cases, and can survive periods of drought lasting 2–3 years. The larvae (nauplii*) that hatch out are about 0.5 mm long. The larvae and the adults make valuable foodstuffs. Care and breeding: to obtain the nauplii, the brown eggs obtained in the commercial world (approx. 0.3 mm long) are used. Place $1 1^1/_2$ level teaspoons of eggs into 1.5–3 % table-salt solution or the equivalent concentration of sea water. Provide with good ventilation to keep eggs moving. Suitable containers are colourless bottles. At 24 °C the larvae hatch after 1 day. Recommended as food for the rearing of young fish (*see* Fry, Feeding of). If you want to rear adult specimens keep them in large, ventilated bowls standing in a bright spot. The nauplii can be fed with naturally-occurring algae in suspension or small amounts of yeast or artificial food. Shining a light on one side of the container makes it easier to fish the larvae out, as they gather around light. Use nets of gauze to sieve out and separate unusable egg cases. Both nauplii and adults will only live for a short time in fresh water.

Artery *see* Blood Vascular System

Arthropoda. About three-quarters of all known animal forms belong to the Arthropoda, among them the Crustacea*, Arachnida* and Insecta*. The Arthropoda stemmed from the ancient segmented worms (Annelida*) during the course of evolution, and together they also form the phylum Articulata* (articulated animals). While the Annelida have at most stump-like limbs, all Arthropoda possess jointed legs. The number of legs varies from 6 in insects to a staggering 680 in some tropical millipedes. Another characteristic of the Arthropoda is a hard, armour-like exoskeleton made of chitin. This skeleton does not grow and so must be shed every so often. Judged by the number of species and variety of forms, the Arthropoda are the most successful animal group ever. There is no habitat that has not been colonised by them. You will find arthropods from the Arctic to the tropics, from the highest mountain to the deepest ocean, and even in other organisms.

Arthropods, Sub-Ph, *see* Arthropoda

Articulata. The most extensive phylum in the animal kingdom with an extraordinary variety of forms and an enormous number of species, comprising most importantly the segmented worms (Annelida*) and the arthropods (Arthropoda*) as well as some small, largely unknown groups. The most fundamental feature all Articulata have in common is segmentation (metamery). In the primitive forms of the Articulata the basic structure is repeated in each segment, eg the secondary body cavity formed from the third germ layer (mesoderm), paired excretory organs (nephridia), a pair of nerve cords (ganglia) on the ventral side, and limbs. In many arthropods, for example, the exactly repeated structure of each segment may be abandoned completely or greatly modified.

Artificial Food *see* Feeding

Artificial Materials *see* Materials

Artificial Pollination *see* Propagation of Aquarium Plants

Arulius Barb *(Barbus arulius) see Barbus*

Arum Lily *(Zantedeschia aethiopica) see* Araceae

Ascidia LINNAEUS, 1758. G of solitary Sea-squirts (Ascidiacea, *see* Tunicata) found in cool and warm seas, with several species along the European coasts. The Sea-squirts of this G are imposing creatures, that may reach up to 15 cm in length. They are characterised by a straight pharynx and papillae on the pharynx and lengthwise ribs. They survive very well in the aquarium. Another species, *Phallusia mammilata* (CUVIER), which is common in the Mediterranean, has similar requirements.

Ascites. The build-up of watery, bloody fluid in the body cavity. Occurs in fish after infection with stomach dropsy*, and occasionally with fish tuberculosis*.

Asellus GEOFFROY St. HILAIRE, 1764. A freshwater G of the Isopoda*. Water lice belonging to an O of the higher crustaceans (Malacostraca*). They are distributed widely across the palaearctic realm*, and there are many forms within the general species. The best known and most common is *A. aquaticus* (LINNAEUS, 1758) (fig. 132). This water louse lives in still or slowly flowing waters where there is rich vegetation. It feeds on decaying matter, so may become abundant in polluted water. Its reproductive biology is very interesting. Before mating, the males, which may be 20 mm long, ride about on the smaller females. The eggs, as well as the newly hatched offspring, are kept in a brood chamber under the female's belly, which is covered at the bootom by broad, overlapping plates projecting from the middle of the legs. This undemanding and tenacious species is well suited for observation in pond aquaria.

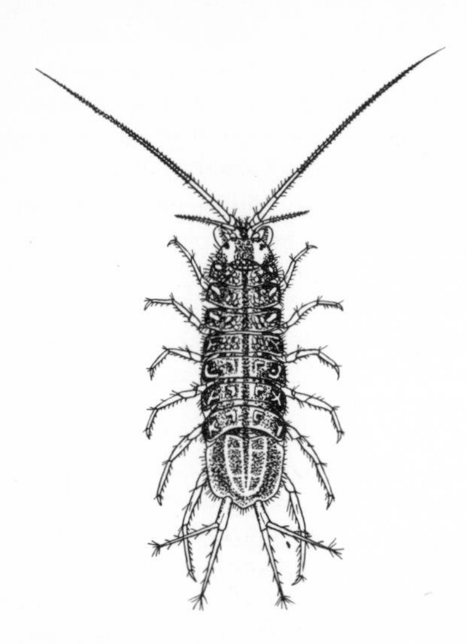

Fig. 132 *Asellus aquaticus*

Asexual Reproduction *see* Reproduction

Asia measures some 44 million km^2, and together with the Asian islands it makes up one-third of the total land mass of the earth. The south of Asia has a tropical climate and zoogeographically it belongs to the Oriental realm*. As this region has close associations with the Ethiopian zoogeographical region, the two regions are grouped together in biogeography in the palaeotropic. Palaeotropically distributed taxa among the fish are the G *Barbus* and the F Mastacembelidae*.

The south of Asia, ie the Near East and Indo-China as well as the Greater Sunda Islands (Java, Sumatra), is the home of many well-known aquarium fish. Their families and the most important Ga are named below:
F Cyprinidae* – *Brachydanio**, *Danio**, *Barbus** and *Rasbora**.
F Cyprinodontidae* – *Aplocheilus** and *Oryzias**.
F Cobitidae* – *Acanthophthalmus** and *Botia**.
Sub-O Anabantoidei* – *Belontia**, *Betta**, *Colisa**, *Macropodus** and *Trichogaster**.

The central and northern regions of Asia belong to the palaearctic* in biogeography, which is a much less interesting region from the aquarist's point of view. On the other hand, there are also some important waters there as far as the ichthyologist is concerned, eg the Amur* and Lake Baikal*.

Asian Lizard's Tail *(Houttuynia cordata) see* Saururaceae

Aspidontus Quoy and Gaimard, 1835. G of the Blenniidae, Sub-O Blennioidei*. Tropical Indo-Pacific. These fish have massive, sword-like teeth with which they tear off bits of skin and fins from other fish. *A. taeniatus* has evolved into a perfect copy of the Cleaner Wrasse, *Labroides** *dimidiatus,* particularly in colour and behaviour. It uses this resemblance to approach unsuspecting larger fish. *A. taeniatus* even imitates local colour variations in *Labroides*. It is often sold erroneously as a Cleanerfish, as its large teeth are concealed in the mouth, which is ventral rather than terminal as in the true Cleanerfish *L. dimidiatus.*

– *A. taeniatus* Quoy and Gaimard, 1835; False Cleanerfish. Found on Indo-Pacific coral reefs, particularly in tube-worm cases. Up to 10 cm. In contrast to *Labroides* the mouth points downwards (fig. 133).

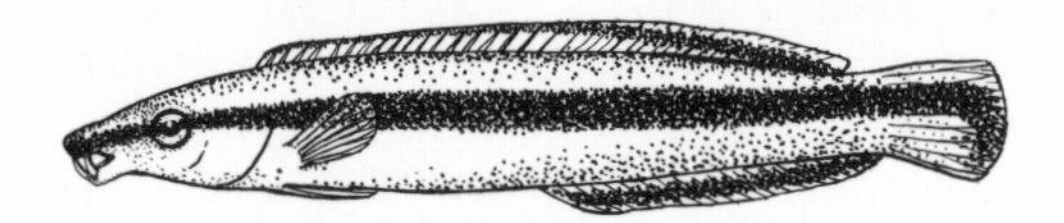

Fig. 133 *Aspidontus taeniatus*

Aspredinidae. F Siluriformes*, Sup-O Ostariophysi. Widespread S American Catfishes with a much flattened, foreshortened, anterior part of the body and a slender, often very long posterior region. The representatives of the Sub-F Bunocephalinae (Banjo Catfishes) (fig. 134) have a rhomboid head/body crosssection which continues into the long, stalk-like tail, ending with a straight-edged C. Mouth relatively small, surrounded by 3 pairs of barbels, of which the top-lip barbels are fairly long. D and An short, the Vs small, The Ps on the other hand are not only very large, they also have a strong serrated spine. Adipose fin absent. The skin has no scales; it is wrinkled and punctuated with gritty warts. All the species of this Sub-F are freshwater, bottom-dwelling fish from the Rio Magdalena region to Argentina. Active at night. Species of the Sub-F Aspredininae (Banjo Catfishes) are found in the brackish waters of estuaries in NE South America or even at sea. They can be distinguished from the Bunocephalinae by their rounder body, shorter caudal peduncle, and, above all, by the very long An. A very unusual method of caring for the young has been observed in some species of this Sub-F. Stalk-like processes develop at the rear of the female's belly during the breeding season. The eggs become attached there. The Bunocephalinae on the other hand spawn in flat hollows.

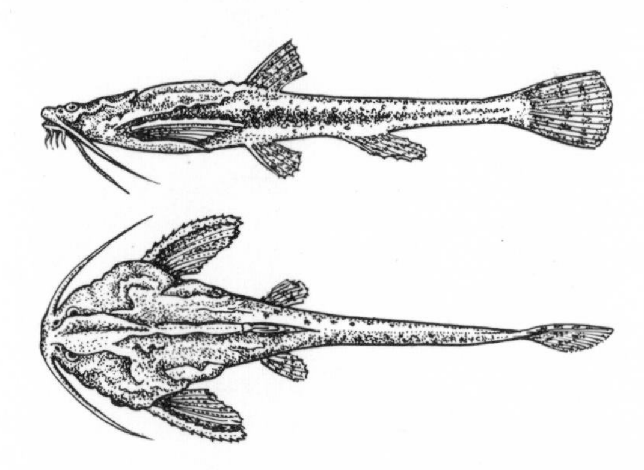

Fig. 134 Aspredinidae-type of the Sub-F Bunocephalinae

Unfortunately, little is known about this F. Observations in the aquarium could provide valuable information in this respect. However, these Catfishes are only imported very rarely. *See also* the G *Bunocephalus.*

Assimilation *see* Carbon Dioxide Assimilation

Association. In ethology and animal psychology, association is taken to mean the linking of various experiences. Associations can arise because of a similarity or contrast in experiences, or because experiences are linked in time or space. It is part of learning. Taste and smell, pain, hearing and facial stimuli are associated, for example. In the Pavlovian sense, associations can be retained as an association chain even after the stimulus for a particular reaction has ceased. Herein lies the basis of achievements in animal training, and it is most successful in the case of dolphins.

In chemistry, association is the word used to describe the way in which several, identical, solitary molecules gather into complex molecules. Such molecule associations can possess different characteristics from the solitary molecules. Water is an example of such an association.

In plant geography, the term is used to describe plant communities that have the same species structure as well as the same habitat conditions and appearance. Eg in lakes, Myriophyllo-Nupharetem is the belt of floating plants in which the Ga *Nuphar**, *Nymphaea**, *Potamogeton**, *Myriophyllum** regularly occur.

Astacura, Sup-F, *see* Reptantia

Astacus FABRICIUS, 1793; Crayfishes. G of the ten-legged crustaceans (Decapoda*) confined to fresh and brackish waters. Some species, measuring 10–25 cm long, are spread throughout Europe and western Asia; in N America they are replaced by related Ga (*Cambarus**, *Orconectes*).

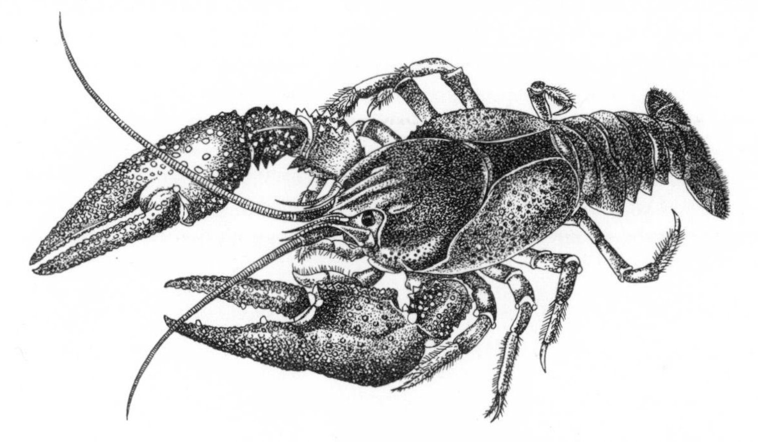
Fig. 135 *Astacus astacus*

The best-known species, *A. astacus* (LINNAEUS, 1758) (fig. 135) (identical to *A. fluviatilis* FABRICIUS, 1793) used to be bred in great quantities as a delicacy. Today, the populations of *Astacus* have been sorely decimated by crab pest and water pollution. With the correct care (cool, oxygen-rich water, hiding-places), crayfishes make good aquarium specimens and will live for years. They can be fed with animal food such as worms and freshly killed fish, as well as additional plant food.

Asterias LINNAEUS, 1758. G of the Starfishes (Asteroidea*). Several species found in the N Atlantic and the N Pacific. The *Asterias* have five arms and a small mouth. The skin is leathery and the arms have no large, calcareous plates. They eat mussels mainly, in particular *Mytilus**, which they prize open in a characteristic way. The starfish settles on top of its victim, trying for hours or even days, with its tube-feet, to get a pull on the two halves of the mussel's shell. Eventually, the adductor muscles of its prey will become weaker and the starfish can force its stomach between the shells. With special mussel traps, the *Asterias* have reached weights of 2–3 kg. The soft-bodied prey is digested and absorbed outside the body. In the aquarium *Asterias* make lively and voracious specimens. They are best kept at a temperature under 20 °C and the water quality must be well-maintained. They will accept different kinds of animal food. The following species has been kept in the aquarium for 7 years.

– *A. rubens* LINNAEUS, 1758; Common Starfish. Norway to the African Atlantic coast, also in brackish water of the Baltic Sea. Often found in large numbers. Up to 20 cm. Yellow, red or violet.

Asterina NARDO, 1834. Gibbons Starlet. G of the Starfishes (Asteroidea*). In warm seas. These Starfish have short arms; often they are almost like a five-cornered shape. No conspicuous scales on the sides of the arms. The following details refer to *A. gibbosa*. *Asterina* live at shallow depths, under stones and among grass weed. They are hermaphrodites, able to fertilise their own eggs. Unlike most echinoderms, the larvae remain on the bottom, developing into 1 mm long young starfish without absorbing any food. Having grown this big, they start to eat hydroid polyps. *A. gibbosa* can also be kept in the aquarium. They will eat all kinds of animal food and are not over-delicate.

– *A. gibbosa* (PENNANT); Small Gibbons Starlet, Cushion Star. Mediterranean, E Atlantic. Up to 5 cm. Yellowish, olive-green or brown (fig. 136).

Fig. 136 *Asterina gibbosa* with clutch of eggs

Asteroidea. Starfishes. Cl of echinoderms (*see* Echinodermata) with about 1,500 species. They have colonised all seas and some have penetrated brackish water. Most are predators, attacking almost all sessile or slow-moving animals, and some Sea-anemones and Sea-urchins. Some species, like *Astropecten**, transfer their prey to the stomach; others digest it outside the body *(Asterias*)*. Some species also eat tiny particles which they waft into the mouth by ciliary currents, and some *only* eat in this way *(Henricia)*. The Asteroidea possess 5 or more arms which are not separated sharply from the mouth (as in the Brittlestars (Ophiuroidea*)), but join up with the mouth at the broad base of the arms. The mouth faces the bottom and the anus lies on the upper side of the body, near the middle. To the side of this you will be able to see the madreporic plate (*see* Echinodermata). The upper side of the body is covered with small bumps called papulae. The sexual openings are usually situated in the corners of the arms. The sucker-feet lie in grooves on the underside of the arms. One arm will lead the way when moving along. A well-developed chemical sense enables the Asteroidea to seek out their prey. Light is perceived along the entire surface of the body as well as with special sense pads at the tips of the arms. The majority of the Asteroidea are separate sexually, and the eggs and sperm are usually released into the water. Plankton larvae (Bipinnaria, Brachiolaria) equipped with cilia are the outcome, but these rarely survive in the aquarium. In some species the eggs and larvae remain on the bottom where they live off the rich yolk, without the need for other food. These grow directly into small Starfish, even in the aquarium *(Asterina*, Echinaster*)*. Many Asteroidea reproduce asexually by binary fission *(Coscinasterias*)* or by an arm splitting off *(Linckia*)*. In the aquarium, arms have sometimes been discarded *(Asterias)*, but usually when conditions are poor, and this has nothing to do with reproduction. The Asteroidea are very interesting aquarium specimens. Feed them on mussel flesh in particular, and there should be no problems. Only bear in mind that they may attack many other animals. It is not possible to keep those species that are particle-eaters only. Like all echinoderms, the Asteroidea need excellent water conditions.

Astrodoras BLEEKER, 1863. G of the Doradidae*. Central Amazon region. Thorny Catfishes 10–12 cm long. Body compressed, head flattened. The flanks are covered with bony plates that have strong spines. 3 medium length barbels. 1st ray of the D1 straight, with a row of simple teeth on the front edge and 2 prominent lateral ridges. This makes the G easy to distinguish from the closely related Ga *Acanthodoras** and *Amblydoras*. For care and feeding *see Acanthodoras*. Not yet bred in captivity.

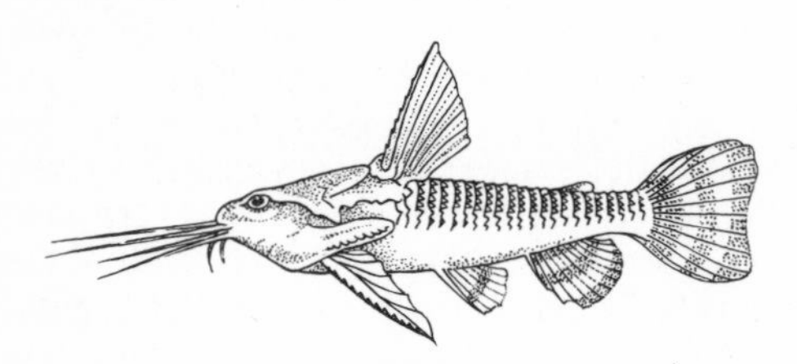

Fig. 137 *Astrodoras asterifrons*

– *A. asterifrons* (HECKEL, 1855). Central Amazon region. Up to 11 cm. Body yellow-brown to blackish. Ventral surface delicate pale brown to white. Several dark, indistinct transverse bars in the rear part of the body, but they do not reach the belly. The D1 and D2 are colourless. C, An and Ps have a transparent ground colour with dark transverse bars (fig. 137).

Astronotus SWAINSON, 1839, G of the Cichlidae* with few species. Widespread in tropical S America. Robust, moderately flattened fish, up to 35 cm long. Upper and lower profiles about equally convex, body fairly deep. C fan-like. Unlike most other cichlid genera, the soft rays of the D and An are rounded. Thick-lipped mouth, eyes relatively small. Coloration extremely variable, dependent on age and locality of origin, but there is also great variation between individuals. Very generally speaking, they have a dark ground colour with irregular pale bands and spots. Hardly any differences between the sexes, which is typical of fish (such as this G) that spawn in open sea (*see* Cichlidae). There should be no problems keeping them in the aquarium, but the size of the tank should be suitable for the size of the fish. Although most *Astronotus* are relatively peaceable, it is recommended that they be kept with tolerant fish of similar size, such as *Cichlasoma* severum* or *Cichlasoma** festivum and *Pterophyllum**. The tank should contain a thick layer of clean sand with roots for decoration. Either do without vegetation altogether or have large, strong potted varieties. Water temp. about 25 °C,

Fig. 138 *Astronotus ocellatus*

higher for breeding, in winter about 20 °C. *Astronotus* needs a lot to eat, but these fish are not choosy and devour all types of suitable food such as large insect larvae, small fish, earthworms. After a period of adjustment they will also eat minced meat and small pieces of meat, cooked rolled oats and dried food. Reproductive behaviour is absolutely typical of all spawners in the open sea. The spawn is attached to a piece of ground that has been cleaned, and the parents care for it assiduously. After 3–4 days, the brood hatches and they are placed in sandy hollows for the next week. Occasionally you will observe that the young fish cling

to their parents even after they can swim freely (compare *Symphysodon**). It is not yet fully understood whether or not they also eat a secretion from the skin when they cling in this way.
– *A. ocellatus* (CUVIER, 1829); Oscar, Marbled Cichlid. S America from the Orinoco to the Rio Paraguay. Up to 35 cm. Body usually very dark, olive-brown to deep blue-black with pale yellow markings. Particular scales have orange-coloured spots. Base of the C has a deep black peacock-tail spot (ocellus), bordered with red. There may be similar ocelli on the D as well. Fins very dark like the body (fig. 138).

Astropecten BLAINVILLE, 1830. G of the Asteroidea*. The *Astropecten* are widespread in sandy substrates. Their tube-feet are not suckered, but they can move surprisingly quickly with their cone-shaped ambulacral feet. They often burrow into the sand. The *Astropecten* devour their prey whole, which mainly consists of molluscs. Typical of the G are the comb-like spines on the sides of the arms. There are no anus or pedicellaria (*see* Echinodermata). These lively starfish are good specimens for marine aquaria, eg:
– *A. aurantiacus* LINNAEUS, 1758, Red Comb Starfish. Mediterranean, E Atlantic. Radius up to 28 cm. Coloration red-orange.

Astyanax BAIRD and GIRARD, 1854. G of the Characidae* with about 75 species. Distributed from New Mexico and Arizona across Central America to Patagonia. 5–20 cm long *(A. maximus).* These fish live mostly in the open waters of lakes, rivers and streams right up to the mountainous areas. Hardy, undemanding aquarium fish. Elongated, laterally compressed body, with relatively short D1, long An and deeply forked C. D2 present. The complete LL allows you to distinguish between this G and the otherwise similar Ga *Hemigrammus** and *Hyphessobrycon**. The G *Astyanax* is divided into 3 Sub-Ga, of which only the Sub-G *Astyanax* has any importance in aquarium-keeping. The remaining Sub-Ga *Poecilurichthys* GILL, 1858, and *Zygogaster* EIGENMANN, 1913 comprise relatively uncolourful species. Almost all the species are relatively peaceable, to a large extent not dependent on the water quality and insensitive to sharp changes in temperature. These fish are omnivorous, and eat live and dried food as well as plant food. Almost all the species have a silver background with a slight tinge of colour and a dark spot on the shoulder. There is often a dark silver or dark green longitudinal band and a dark spot on the caudal peduncle. Fins yellowish or reddish, often with white tips. Females in breeding condition are easy to tell apart from the slimmer, smaller males by the more robust ventral surface. Breeding is usually easy. After hectic driving by the male the spawn is laid at random in and on top of fine-leaved plants. The eggs do not stick. Fish hatch after about 24 hours, swim freely after 5 days. They grow quickly with good amounts of powdered food. For the aquarist, only the smaller species are important.
– *A. (Poecilurichthys) bimaculatus* (LINNAEUS, 1758); Two-spot Astyanax. N eastern and eastern S America, south to the Plate basin. Up to 15 cm. Flanks silver to yellowish, dorsal surface olive. Behind the operculum and above the LL lies an oblong black spot, with another one at the root of the trail fin. Fins yellowish. *A. bimaculatus* is very productive (fig. 139).
– *A. (Astyanax) fasciatus fasciatus* (CUVIER, 1819); Banded Astyanax. Widespread from Central America to Argentina. Up to 17 cm. Body silver, dorsal surface olive-brown. A black oblong spot at the shoulder is only just visible. The greenish-silvery longitudinal band in the middle of the body becomes black at the caudal peduncle and stretches as far as the middle rays of the C. D1 and C are red, the remaining fins colourless.

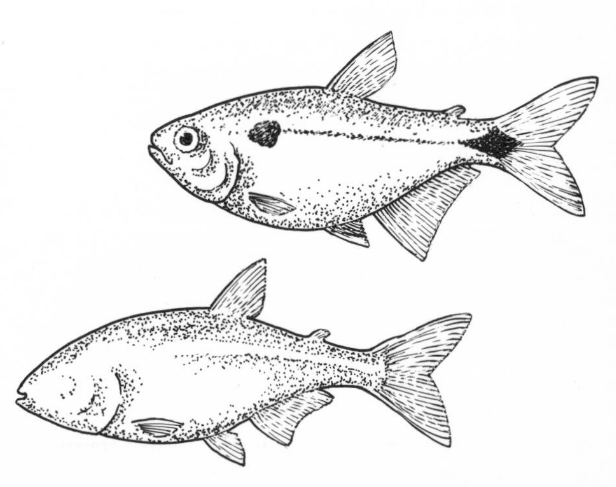

Fig. 139 *Astyanax bimaculatus, A. mexicanus,* blind cave form

– *A. (Astyanax) kennedyi* (GÉRY, 1964. Upper Amazon, near Iquitos. Up to 5 cm. Body silver with a strong pale green iridescence. A black, rhomboid spot on the caudal peduncle which spreads out into the middle rays of the An. The dark, vertical spot at the shoulder is often unclear. A silver longitudinal band becomes tinged with gold at the caudal peduncle. All fins yellowish. A very pretty species, easy to keep and breed.
– *A. mexicanus* (FILIPPI, 1853). Texas to Panama. Up to 9 cm. The body of this species is silvery to brassy in colour, dorsal surface olive. An indistinct greenish-silvery longitudinal band becomes dark at the height where the An begins. The dark spot at the base of the C is bordered by pale yellow. The middle rays of the C are dark. All fins are yellow to reddish, the An with a white tip. *Anoptichthys jordani* HUBBS and INNES, 1936, is the scientific name given to the well-known Blind Cave Characin, but as SADOGLU was to prove, it is just a cave-dwelling form of the species from the province San Luis Potosi. The blind fish populations breed without difficulty with the main form, producing intermediary offspring. Young fish of the cave forms are often still able to see. Sight deteriorates only with age, as the eyes sink deeper into the eye sockets and become covered with numerous layers of skin. By this stage the fish orientate themselves only by means of the sense of smell and sense of distance in the LL. It is also the means by which they capture prey. The body of the blind form is flesh-coloured with a silvery iridescence, the fins reddish or colourless. Easy to breed, including the blind form (fig. 139).

Atebrin is a derivative of acridine. Easily soluble in water and used by aquarists to combat skin parasites in fish. Atebrin works as a mitotic poison. 0.1 g Atebrin is used per 10 l water, and is left in the aquarium for several days. The effect is all the greater if the temperature is raised and there is good aeration. If applied for too long, the plants may suffer. Atebrin can be removed from water by filtering through activated charcoal* (and *see* Adsorption).

Atherina. LINNAEUS, 1758. G of the Atherinidae, Sub-O Atherinoidei*. Small, much elongated fish of tropical and temperate seas, sometimes found in brackish and fresh water. Plankton-eaters. Live near the coasts in large shoals and rarely kept in aquaria. Some are fairly important as food fish, eg:

– *A. hepsetus* LINNAEUS, 1758. E Atlantic to the Black Sea. Up to 14 cm. Green-grey with a silvery to dark longitudinal band.

Atherinidae, F, *see* Atherinoidei

Atheriniformes. O of the Osteichthyes*, Coh Teleostei. The system described here is based on that defined by GREENWOOD and colleagues (1966). Numerous groups of fish are placed together in the O Atheriniformes. In other systematic classifications, these fish groups are arranged very differently and are grouped under O, Sub-O and F. Eg the O Synentognathi in the REGAN system, which largely corresponds to the O Beloniformes in the BERG system, the O Microcyprini (= Cyprinodontiformes) without the F Amblyopsidae, the F Atherinidae from the Sub-O Percesoces (= Mugiliformes) and the O Phallostethiformes. There are very specific facts that brought about the new scientific grouping. There is not room here to go into them in detail, but they prove that all the groups now placed together in the Atheriniformes have a common ancestral form (*see* D. E. ROSEN: Bull. Amer. Mus. Nat. Hist. 127,217–268, 1964). Almost all the fish described here are relatively small to very small surface fish, or they live in shallow waters. Many feed almost exclusively on food that floats by. The Atheriniformes occur in both fresh and brackish water, as well as in the sea. Nearly all the freshwater forms show a clear distinction between the sexes, in size, marking, coloration as well as fin arrangement. In the livebearing species, spermatophores (cased or uncased) are transported by an adapted An, V or P. GREENWOOD and colleagues (1966) distinguish the 3 Sub-Os Exocoetoidei*, Cyprinodontoidei* and Atherinoidei*. For details on the Fs belonging to the Sub-Os, *see* the systematics under Osteichthyes* or the particular entry for each of the Sub-Os.

Atherinoidei. Sub-O of the Atheriniformes*, Sup-O Atherinomorpha. D. E. ROSEN (1964) places 5 Fs in the Sub-O, of which the Atherinidae, Melanotaeniidae and Phallostethidae are the most important. Moreover, the Sub-O, along with the Exocoetoidei* and the Cyprinodontoidei*, are grouped together in the O Atheriniformes*. This approach was completely new, especially as far as the Fs grouped in the Sub-O Atherinoidei were concerned, as they had been grouped very differently for a long time. For example, the Atherinidae (including the Melanotaeniidae) had always been regarded as close relatives of the Grey Mullets (Mugilidae), and were grouped along with the Barracudas (Sphyrnidae) in a Sub-O (Mugiloidea) and sometimes in an O (Peresoces or Mugiliformes). The Phallostethidae were mostly placed alongside the Cyprinodontidae. It is an important point to note that D. E. ROSEN has been able to show, on the basis of an exhaustive analysis of general and specific features, that the named Sub-O and Fs are fairly closely related and probably stem from common ancestors in the early Tertiary. Details are given in the family descriptions below.

– *Atherinidae;* Silversides, Sand Smelts, Whitebait (fig. 140). Mostly small, spindle-shaped, laterally compressed fish. Head has scales, mouth usually pointing slightly upwards. Large eyes, jaws toothed (as well as the palatinum and the vomer in many cases). 3rd and 4th upper pharynx bones on each side merge into one another. 2 clearly separated Ds. D1 supported by elastic,

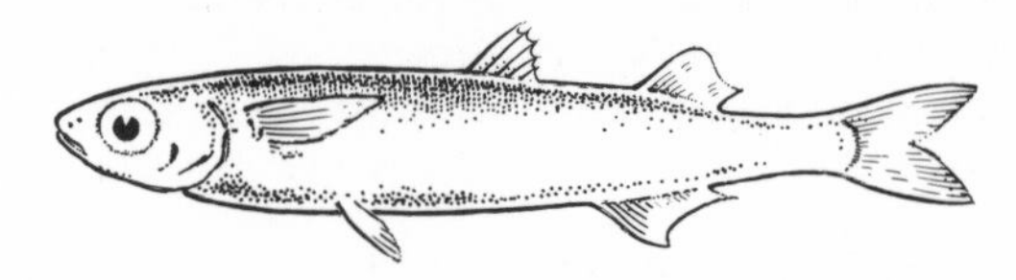

Fig. 140 Atherinidae-type

spine-like rays, D2 by soft rays; often with a spine at the front end. C indented, An with 1–3 spines, about as long as D2 and positioned almost directly opposite it. Vs have 1 spine and 5 soft rays and are positioned fairly far forwards, but clearly behind the Ps. The latter are relatively high up. There are bands between the pelvic girdle and the shoulder area. LL much reduced or lacking. Swimbladder not joined to the intestine. Cycloid scales. Eggs mostly have long filamentous outgrowths. The approximately 160 species live near the coasts, above all in tropical and subtropical seas; a few species live in fresh water. Most species have a silvery metallic ground colour and a strongly iridescent stripe may be visible running lengthwise. Many small types of Atherinidae are translucent. The shoal-forming Sand-smelt *(Atherina presbyter)* is found on the W and NW coasts of Europe, as well as in the Mediterranean. It grows up to 15 cm long. The California Grunion *(Leuresthes tenuis)* grows up to 15 cm long and has a very interesting spawning behaviour. The fish spawn 1–2 days after a new or full moon at spring flood tide. In their thousands the males and females wriggle out of the water, burrow into the sand and lay their sexual products at a depth of 3–5 cm. 14 days later, at the next spring flood, the young hatch out and are swept into the sea with the water. However, most species spawn in open water among grass weed. The spawning period can last several weeks. There are some prettily coloured species, suitable as aquarium specimens. *See* the Ga *Alepidomus, Atherina, Austrominidia, Bedotia, Pseudomugil, Telmatherina.*

– *Melanotaeniidae*; Rainbowfishes (fig. 141). These fish are usually considered as a Sub-F of the Atherinidae. The majority are restricted to Australia. They are distinguishable from the Atherinidae by, among other things, the deeper body and above all by the very long

D2 and the even longer An, which practically reaches past the middle of the body towards the front. Certain species are wel-known aquarium fishes, G *Nematocentrus**.

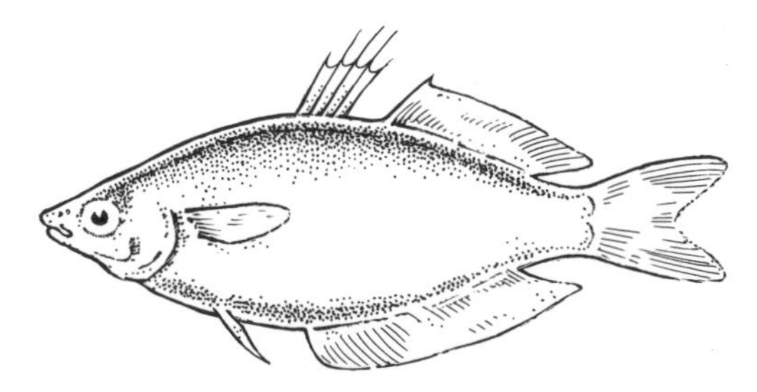

Fig. 141 Melanotaeniidae-type

– *Phallostethidae* (fig. 142). This F contains few species and consists of small fishes reminiscent of the Toothed Carps or Barbels. They occur regionally in SE Asia and in the Philippines both in fresh and brackish water. They have become known particularly for the copulatory organ (priapium) in the throat region of the male. It consists of a main section with 2 forward-springing processes that look like sabres. This curious organ is formed from the 1st pair of ribs and specialised elements of the pelvic and shoulder girdles. The anus, as well as the openings to the ureter and the sex organs, are also found in the priapium. In the females these openings are located between the Ps. The eggs are either fertilised inside the female or after they have been

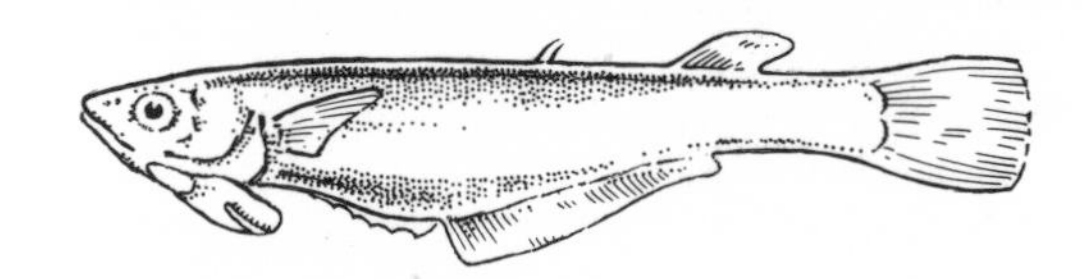

Fig. 142 Phallostethidae-type

released. The two Ds are like those found in the other Fs of the Sub-O – D1 is very small and represented only by 1–2 spines. C indented, An relatively long and inserted far forwards, Vs absent. One of the smallest fish belongs to this family. A male *Phenacostethus smithi* is only 17.8 mm long, and a female 18.7 mm. These glass-like fish live in cloudy waters of the Malayan peninsula. The Neostethidae are a closely related F, also with a priapium.

Atherinomorpha, Sup-O, *see* Osteichthyes.

Atipa *(Hoplosternum thoracatum) see Hoplosternum*

Atlantic Salmon *(Salmo salar) see Salmonoidei*

Atlantic Saury *(Scomberesox saurus) see* Exocoetoidei

Atoll *see* Coral Reefs

Atom. The smallest part of a chemical element. There are 88 naturally ocurring atoms, each with different characteristics. In solids atoms lie close to one another and occur in a regular order; in liquids they can move about in relation to one another, whereas in gases the atoms move about freely and there is a large gap between each one. The nucleus of an atom consists of positive charge carriers (the protons) and neutrons which have no charge. It contains the total mass, but its diameter is up to 100,000 times smaller than the diameter of the whole atom. According to the number of protons, which is the same as the charge on the nucleus, atoms have been arranged in a periodic table Negatively charged particles, the electrons, circle the nucleus at a relatively large distance in circular or wave-like orbits. The number of electrons is exactly the same as the number of protons, a feature which maintains the neutrality of the atom. A certain number of electrons is linked with a characteristic chemical property. Altering the number of electrons in relation to the stable nucleus will create a positively or negatively charged atom which is then described as an ion*. Such changes in the number of electrons is the basis of chemical reactions.

Atrophy. The wasting away of organs or parts of organs until they disappear completely. Causes, such as lack of use, eg muscular atrophy, and under-nourishment, can be regarded as the natural ageing process or as the consequences of illness. For example, parasites may obstruct or seal off the gill vessels of a fish, preventing it from getting sufficient food and producing atrophy of the gill epithelium.

Attached Spawners *see* Substrate Spawners

Auchenipteridae. Scaleless Catfishes (fig. 143). F of the Siluriformes*, Sub-O Ostariophysi. Small, compressed to elongate Catfishes with an area of distribution that stretches from the east coast of mid South America to the R. Magdalena in the west. Closely related to the Thorny Catfishes (Doradidae*), but with noticeably smaller fins, no bony plates on the flanks of the body and a swimbladder that lies unattached in the body cavity. Also unlike the Doradidae, the C is indented. At least some species are lively by day. They live in shoals in their natural habitats, often joining up

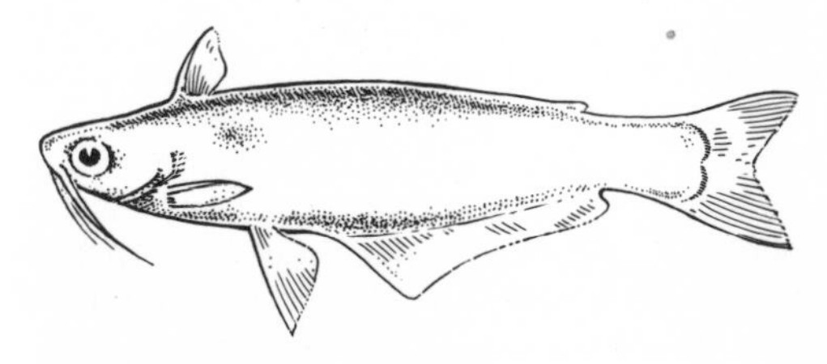

Fig. 143 Auchenipteridae-type

with Armoured Catfishes. There are no known details about reproduction. Look under the G *Centromochlus* for information about observations in the aquarium.

Auchenoglanis occidentalis *see* Bagridae

Auditory Organ *see* Balancing Organ

Aulonocara REGAN, 1921, G of the Cichlidae*, Endemic* to Lake Malawi, with 3 species represented. Similar in form to the G *Haplochromis**, ie fairly tall. Laterally moderately flattened Cichlids. Particularly characteristic of *Auloconara* and the related G *Trematocranus** are the 2 rows of clearly visible depressions in the head (sense organs?), which stretch from the corner of the mouth, around the eye to the upper edge of the operculum; eyes remarkably large. Forehead sunken in older male fish. D has 15–16 hard rays and 10–11 soft ones, An has 3 hard and 9 soft rays; the rear portions of the D and An are rounded. C slightly indented. An has 3–6 gold-coloured oval markings. Like related Ga *(Haplochromis, Pseudotropheus*)*, *Aulonocara* are mouth brooders.

Care is similar to that required by other Cichlids from L. Malawi. The tank should contain rocky structures and the water should be medium hard with a temperature up to 27 °C. Feeding should cause no problems especially as the *Aulonocara* are much less specialised feeders than, for example, *Labeotropheus** and *Pseudotropheus* particularly with regard to the eating of algal growth. The *Aulonocara* are also much more peaceable fish than the above-mentioned Ga. One thing they do have in common is that only the highest ranking male exhibits intensified coloration, whilst lesser ranking males have insignificant colouring similar to the females. Behaviour during breeding is similar to that of *Pseudotropheus.* After a lively courtship the female lays her eggs in batches and takes them immediately into the mouth. The oval markings on the An (mentioned above) help to ensure fertilisation because the female bites at them, believing them to be eggs, which brings her mouth close to the male's genital region. The spawn takes 3 weeks to mature and like other highly specialised mouth brooders the egg count is low (about 35 in all).

– *A. nyassae* REGAN, 1921. L. Malawi. Up to 13 cm. Males have a beautiful steel-blue ground colouring with black oblique bands. D has a blue-white edge, V rust colours. Females and lesser ranking males are an insignificant brown colour with darker oblique bands.

Aulopodidae, F *see* Myctophoidei

Aulorhynchidae, F, *see* Gasterosteoidei

Aulorhynchus flavidus *see* Gasterosteoidei

Aulostoma LACÉPÈDE, 1803; Trumpetfishes. G of the Aulostomidae, Sub-O Aulostomoidei*. Found in the tropical Indo-Pacific and the W Atlantic. Similar to the Pipefishes (*see* Syngnathoidei) with their large tube mouth. The *Aulostoma* feed on small fish and shrimps. They lie in wait either among the plants and corals or hide behind peaceful fishes, which enables them to get close to their prey. Not easy to keep in the aquarium. The same can be said of the related Cornetfishes or Flute-mouths (Fistulariidae, F *see* Aulostomioidei). The following species are often imported.

– *A. maculata* VALENCIENNES. Tropical W Atlantic. Up to 50 cm. Yellow to brown stripes and dots (fig. 144).

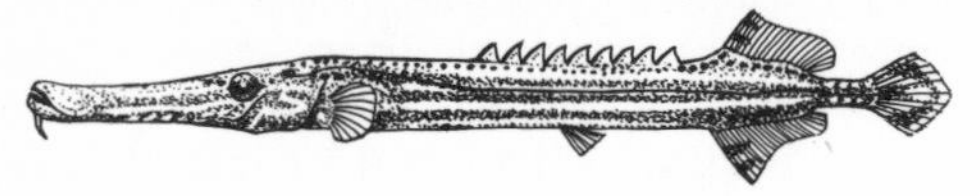

Fig. 144 *Aulostoma maculata*

– *A. valentini* (BLEEKER, 1853). Indo-Pacific. Up to 50 cm. Yellow or brown with a dark marking and bright spots on the rear part of the body below the D.

Aulostomidae, F, *see* Aulostomioidei

Aulostomioidei Trumpetfishes, related species; Sub-O of the Gasterosteiformes*, Sup-O Acanthopterygii. All the fish grouped under this Sub-O have an elongated, tube-like snout with a small, terminal mouth opening. Food is sucked in jerkily and mostly consists of small plankton animals. Various skull bones are elongated because they are adapted to the tube snout. Teeth either weak or absent. Most characteristic is the way in which the first 4–6 vertebrae are elongated and have grown into one another. Ribs and connecting muscles are missing. Comb-edged gills. The Aulostomioidei are primarily marine fish, often found near coral reefs. The oldest fossils date back to the Lower Eocene.

– *Aulostomidae;* Trumpetfishes (fig. 145). These fish can be distinguished from other members of the Sub-O by the free-standing spines in front of the D, by a barbel

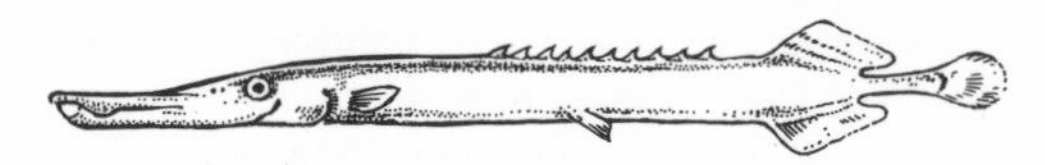

Fig. 145 Aulostomidae-type

on the chin and by the small ctenoid scales. The dermal bony plates characteristic of almost all the other members of the O Gasterosteiformes* are absent in the Aulostomidae. Individual members of the one G *Aulostoma** are occasionally kept in aquaria.

– *Centriscidae;* Razorfishes, Shrimpfishes (fig. 146). Laterally, the body is flattened in the extreme, keel of the belly sharp-edged, head has a long tube snout, ie the fish's shape is reminiscent of a knife-blade. Body covered with thin dermal bony plates. The position of

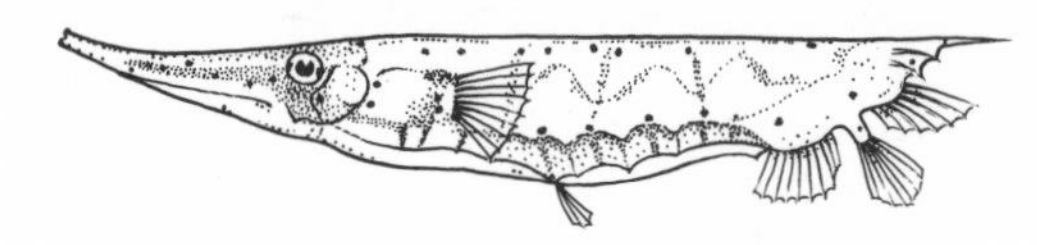

Fig. 146 Centriscidae-type

the fins is very strange. The D1 is placed right back at the rear end of the body, its elongated 1st spine practically forming an extra length of the body. The D2 and C lie on the ventral side of the rear end of the body, directly behind the small An. The Centriscidae stand vertically, ie they swim or stand head-down mostly in small schools, and so provide excellent camouflage for themselves among the branches of coral, among the spines of sea-urchins or algae. They can rush backwards and forwards very well, using the spine of the D1 or the sharp edge of the belly as a defence weapon. Confined to the tropical Indo-Pacific. 2 Ga, 4 species. Aquarium observations chiefly concern *Aeoliscus strigatus,* which can reach a length of 15 cm and has a movable spine on the D1.

– *Fistulariidae;* Cornetfishes, Flutemouths (fig. 147). Very elongated fish with a flute-shaped, elongated

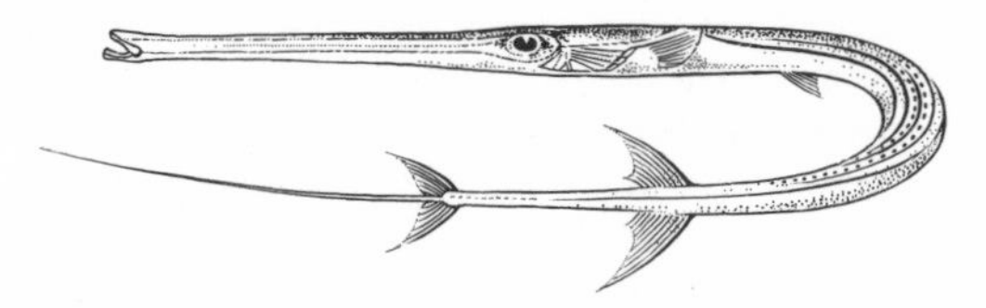

Fig. 147 Fistulariidae-type

mouth which works like a pipette. Small fins. Middle section of the C drawn out into a long thread. Bones traversed by small narrow canals, a peculiarity almost unique in fish. No scales, but in some places there are deeply embedded dermal bony plates. Only 1 G with 4 species. The Fistulariidae are found chiefly among coral reefs. They swim in small schools, and also have excellent camouflage. The best-known species, the red-brown Blue-spotted Cornetfish *(Fistularia tabaccaria)* has length-wise rows of bluish spots and can grow up to 2 m in length.

– *Macrorhamphosidae;* Snipefishes (fig. 148). Viewed from the side, the body has a long oval shape, much compressed, with a very long tube snout, large eyes and an unmistakable fin arrangement. D1 inserted only behind the middle of the body, fin rays like spines, the second one of which is particularly strong and long. D2 short, C small. Two rows of bony plates on each side of the dorsal ridge. The Macrorhamphosidae are chiefly inhabitants of deep waters of warm seas. They remain relatively small, stand mostly vertically in the water and can swim equally well backwards or forwards. The

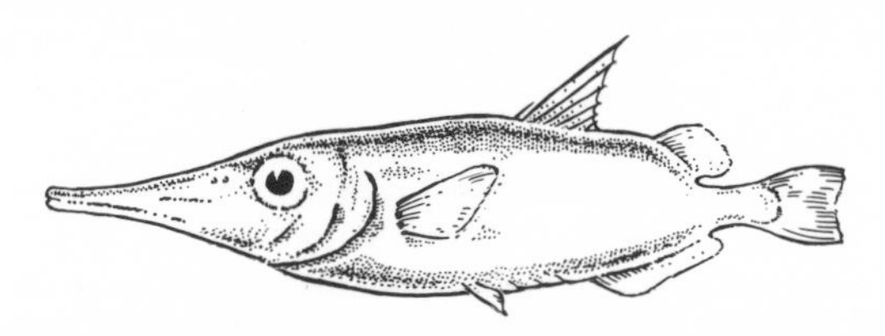

Fig. 148 Macrorhamphosidae-type

most beautiful species, the Banded Snipefish *(Centriscops obliquus),* is found around New Zealand – it has 5 dark oblique bands running across an orange ground colouring. Various species have been kept separately in aquaria.

Aurelia LINNAEUS, 1758. G of the Jellyfish (Scyphozoa*), distributed worldwide. Characterised by a flat umbrella-shape with short tentacles around the edge. The 4 arms of the mouth are short. The violet circles or arcs visible through the glassy umbrella are in fact the gonads. The eggs are fertilised inside the females and the larvae develop on the mouth-arms of the female parent. (They look like a brown deposit.)

Small *Aurelia* can sometimes survive for some time in the aquarium if transported carefully. Small polyps are often introduced and it is possible to observe the formation of medusae. It is very difficult to rear these animals to the sexually mature phase. To start with they eat small crustaceans, later on Enchytraea and the flesh of Crabs and Mussels. The commonest medusa of the North Sea and Baltic is:

– *A. aurita* LINNAEUS, 1758. Atlantic, often found in great clusters. Up to 40 cm. In addition, often found on the bottom where they find food.

Aureomycin. Highly effective antibiotic* often applied to combat infections by bacteria (*see* Bacteriophyta), viruses (*see* Viral Infections) and protozoans. In aquarium-keeping the use of Aureomycin is particularly effective against *Ichthyosporidium**, Oodinium* and all bacterial fish diseases. In marine fish its use is recommended for skin tumours (*see* Skin Diseases) and 'spot' diseases*. Aureomycin is applied as a soaking solution (concentration 13 mg/l 1 freshwater or 15–20 mg/l 1 sea water). Special tanks should be used. Aureomycin can be removed from the water by filtering through activated charcoal*.

Australia. The smallest continent, measuring about 7 million km^2. Large parts of the west and south have no rivers, likewise the interior. So the Australian freshwater animal kingdom is confined to the north and east, as the distribution of the Fs Melanotaeniidae* and Ceratodidae* clearly shows. These 2 families are also the only important freshwater genera of Australia as far as the aquarist is concerned, The exact opposite can be said of the animals that live in the coastal region. Coral reefs stretch along the entire northern coast of Australia and along large sections of the eastern coast. The diversity of forms and species to be found there is exceptional. Zoogeographically, Australia forms the largest part of Australis*.

Australian Gudgeon *(Mogurnda mogurnda) see Mogurnda*

Australian Lungfish *(Neoceratodus forsteri) see* Ceratodiformes

Australian Lungfishes (Ceratodidae, F) *see* Ceratodiformes

Australian Sleeper Goby *(Carassiops compressus) see* Carassiops

Australis. Zoogeographical* region of Australia including the islands of Tasmania, New Guinea, New Zealand and the Polynesian Islands. This region is rich in reptiles, birds and mammals, but has extremely few freshwater fish. Most of the fish are descendants from marine groups and have only secondarily migrated into fresh water, eg the important aquarium fish family Melanotaeniidae*. Primarily freshwater fish are a rare thing in Australia, but one should mention *Neoceratodus* forsteri* and one *Scleropages* species (F Osteoglossidae*). The groups of freshwater fish we know today established themselves in the Tertiary, ie at a time when Australis was already isolated from the other continental land masses, and the sea presented an insuperable barrier to migration for the majority of freshwater fish.

Austrofundulus MYERS, 1932. G of the Cyprinodontidae*. About 5 species. Confined to north-east S America, where they are mostly found in small, shaded waters with soft, slightly acidic water of 23–27 °C. Such waters may even dry up from time to time. Features found in them relate to *Pterolebias* and *Rachovia** in particular, the latter of which is barely distinguishable to the layman. Some species have the strongly extended D, C and An fins found in *Pterolebias.* Medium-size body (up to more than 10 cm), tall and compressed, laterally flattened, reminiscent of *Nothobranchius**, but not so brilliantly coloured. D and An almost opposite each other. In accordance with their natural habitat, only annual species belong to the G (compare Cyprinodontidae). There are no special requirements as far as size of tank is concerned, although a dark, soft substrate (perhaps boiled peat mould) is essential. The aquarium should be well planted and should not be in direct light. If breeding, place 1 strong male in with 1–2 females. The spawning period lasts a few weeks. As in *Cynolebias** and *Pterolebias* the eggs are fanned into the substrate and from then on should be treated as described in the G *Cynolebias*. The eggs lie dormant for 4–6 months, and only then are the embryos ready to hatch. The following species is often kept in aquaria.

– *A. dolichopterus* (WEITZMAN and WOURMS, 1967); Sabrefin. Venezuela. Up to 5 cm. D and An of the male extended like a long pennant, likewise the upper and lower tail fin rays. Body and fins a delicate metallic blue, with red or dark brown spots. Female uniformly grey-olive, with shorter fins. Recently, this species was transferred to the new G *Terranatus* (fig. 149).

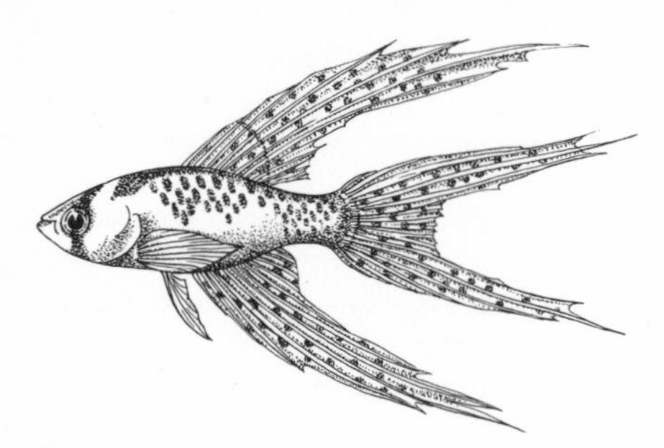

Fig. 149 *Austrofundulus dolichopterus*

Autecology. A branch of ecology* which examines the inter-relationships between the individual organism (representative of a species) and its environment, eg the preferred temperature of a fish species, the dependence of the growth of aquatic plants on the redox potential and the influence of the salt content on the size of the shell of a brackish water mussel.

Autochthonic. Indigenous; originally resident in a particular area. For example, Rainbow Trouts (*see* Salmonoidei), Sunfishes (*Lepomis**, *Micropterus**) and Bullheads (*Ameiurus**) are indigenous to N America, ie autochthonic, and introduced into Europe (allochthonic*) (fig. 150).

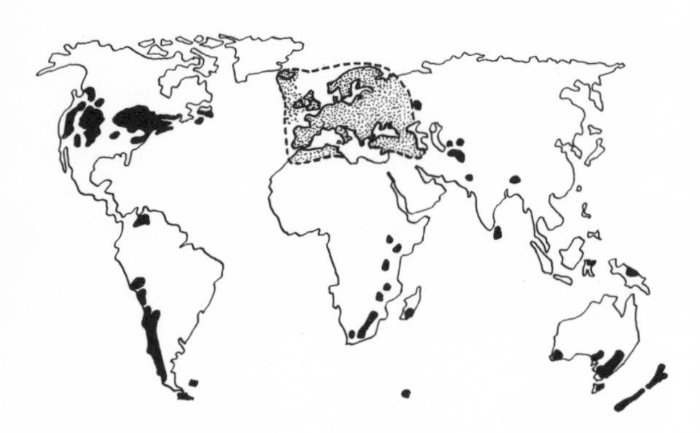

Fig. 150 Areal of the Brown Trout. Where it occurs autochthonically, the area is dotted, allochthonic areas are in black.

Autochthonous Movements. Term used in ethology. They are modes of behaviour that are part of the sphere of function of the movement, in contrast to allochthonic movements*. Eg if two battle-ready *Nannostomus-beckfordi-aripirangensis* males strike up a horizontal battle position (symbolic fighting behaviour*) then switch to fighting vertically (symbolic fighting behaviour) and finally start circling each other with the aim of ramming one another (*see* Ramming), the sequence of movements is autochthonous, as they are part of the battle sphere of action.

Autogamy *see* Pollination

Autumnal Starwort *(Callitriche autumnalis) see Callitriche*

Aves. Birds. A class of vertebrate animals encompassing about 8,600 species whose systematic classification is in part still disputed. The most extensive O, with 5,100 species, is that of the passerines (Passeriformes). The largest living bird is the African Ostrich *(Struthio camelus)* which is 2.5 m tall and weighs 100–150 kg. The smallest is the Cuban Hummingbird *Calypte helenae,* at 6 cm and weighing 3–8 g. Birds inhabit the most varied habitats on earth: dry land and water, the tropics and temperate regions, the islands of the northern seas and the Arctic.

Apart from the mammals, the birds are the only class of animals with warm blood. Their body temperature is 38–44 °C, and their feathers protect them from loss of heat. Feathers are moulted and replaced once a year. Their body is adapted in various ways for flying—the bones are filled with air sacs which are linked with the lungs and eyesight and hearing are well developed. In birds that can fly well, the skeleton possesses a well-defined breast-bone, to which large flight muscles are attached. Some birds, eg the Swifts (Apodiformes) remain in the air almost continuously, reaching speeds of up to 100 km/h. Others journey for up to 6,000 km without a break during migration. Some birds have become unable to fly once again, eg the O of Penguins (Sphenisciformes) which are distributed in the Southern Hemisphere. Penguins are completely adapted to an aquatic way of life, as can be seen by their body shape, fin-like wings, scale-like feathers and colouration (fig. 151). Birds reproduce by eggs which have a calcareous shell. The eggs are fertilised in the females oviduct and are incubated by the parents. The chicks are mostly looked after with great care for a long period.

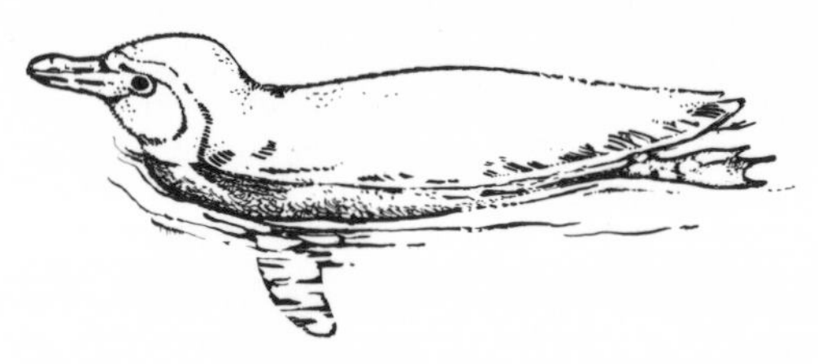

Fig. 151 Adaptation of the penguin to water, source of its food

Avitominosis *see* Deficiency Diseases

Awl-wort, G, *see Subularia*

Axelrodia GÉRY, 1965. G of the Characidae*. Very small (scarcely 2 cm long) South American Characins of the Rio Meta basin. In habits very similar to the *Hyphessobrycon** species. Only recently imported to any great extent. They are easily the smallest Characins known so far, so should only be put together with small, peaceful species or kept separately. Shoaling fish, which have not yet been reared successfully. Feed with the smallest of live food and keep in very soft, slightly acidic water, otherwise they are very delicate fish. Two known species.

– *A. riesei* GÉRY, 1966; Ruby Tetra. Upper Rio Meta basin near Villavicencio Columbia. Up to 2 cm. Body coloration inky red in natural habitat, which has given it its American name 'Red Ruby Tetra'. A dark brilliant pink marking at the root of the C. Fins delicate reddish colour with white tips. Upper half of the iris blood red, the lower half gold with a blue iridescence. Unfortunately, these lovely shades of colour take on an uninteresting pink colour after a short while in captivity. The species is similar in colour and form to the well-known Georgette's Tetra *(Hyphessobrycon* georgettae)*, but it is even smaller (fig. 152).

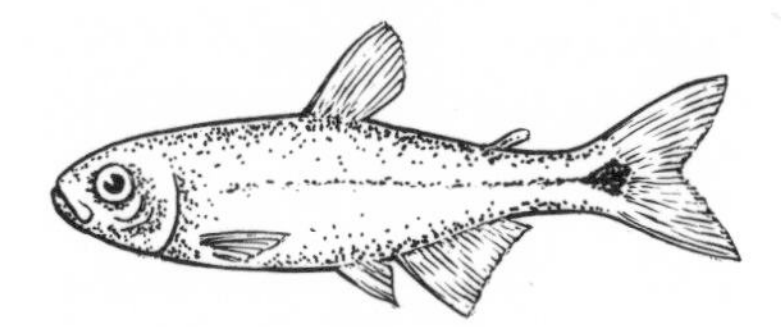

Fig. 152 *Axelrodia riesei*

– *A. stigmatias* (FOWLER, 1914). Lower Amazon basin, Igarapé, Preto, Purus. Up to 3 cm. Body coloration silver, dorsal surface olive. A deep black streak on the caudal peduncle has a gold iridescent spot above it, similar to that found in *Hemigrammus ocellifer*. The male is slimmer than the ripe female. *Hyphessobrycon stigmatias* FOWLER, 1914, is another name for this species.

Azolla LAMARCK, 1783. G of the Pteridophyta*. 5 species. Tropical-subtropical, occasionally occurring in temperate regions. These plants are ferns with many branches and they reproduce rapidly by an asexual method. Distichous, scaly leaves divided into an upper and lower lobe. Complicated reproduction through spores. The species of the G float on the water surface or take root in damp substrate. Occasionally transported to temperate regions by migrating waterfowl. They survive well there throughout the summer. The blue alga *Anabaena azollae*, which binds atmospheric nitrogen, lives in depressions in the upper leaf lobe, so *Azolla* is used in rice fields as a fertiliser. These plants are not very suitable for the aquaria as they shut out light from above; on the other hand they can be put to good use in greenhouses where they will provide natural shade. The plant may be endangered if illumination lasts under 12 hours. Most common species is:

– *A. caroliniana* WILLDENOW. America (fig. 153).

Fig. 153 *Azolla caroliniana*

Bacillariophyta. Diatoms. Sub-D of the algae*. 10,000 species. Yellow-green to brown algae with chlorophyll a and c, each cell containing numerous lentil-shaped or 2 plate-like chromatophores. No flagella. Found in the sea, freshwater and on the bottom. The Bacillariophyta are single-celled or colony-forming algae. They possess two silicic acid capsules which enable them to grip onto one another – like the lid of a small box. The capsules have extremely fine and complicated structures. The immobile species in the marine plankton mostly have a circular or rounded triangular outline. The freshwater forms are often shaped like a rod or carina. They creep along on substrates, on the bottom, on rocks, aquatic plants, aquarium glass etc.

Backboned Animals (collective term) *see* Vertebrata

Backswimmers, G, *see Notonecta*

Bacopa AUBLET, 1775. G of the Scrophuleriaceae*. 20 species. Southern Hemisphere and N America. Small perennials with elongated stem axes. A creeping or stiffly vertical growth. Decussate leaf arrangement or whorled. Sedentary leaves. Flowers hermaphroditic, polysymmetrical. Double perianth, sepals and petals in fives, the latter growing together into a bell shape. 4 stamens. 2 carpels, fused, superior. Fruit is a capsule. The species of the G are small marsh plants, a few of which can also withstand being flooded for a long time,

Fig. 154 *Bacopa amplexicaulis*

and so are suitable for the aquarium. The following is an undemanding species, satisfied with a temperature of 15–20 °C.

– *B. amplexicaulis* (PURSH) WETTSTEIN (*B. caroliniana*, name not in current usage). South-east N America. Ovate leaf blades. Blue flowers (fig. 154).

Bacteria, D, *see* Bacteriophyta

Bacteria Filter *see* Filtration

Bacterial Changes to Spawn can be recognised during egg-laying in fish by a cloudiness and grey-white spots (White Spot Disease) or a softening of the eggs (Soft Egg Disease). Later, egg-cells can become coated with cotton-wool-like layers of fungus (Saprolegniacea*). Bacterial infections can often be observed in a clutch of eggs whose natural habitat would consist of soft water enriched with tannins* or very deficient in bacteria and which are kept in aquaria under different conditions. Add malachite green*, formalin*, antibiotics* or Sulphonamides* to the water to prevent the disease.

Bacterial Fin-rot *see* Fin-rot

Bacterial Infections are the cause of many illnesses in fish. The characteristics of bacteria (Bacteriophyta*) (their tiny size, their ability to spread rapidly etc) make a diagnosis very difficult, especially as specific signs of illness may not be apparent. To identify the pathogen correctly it is necessary to grow a sub-culture on special nutrient mediums and examine it through a microscope using different staining methods. Often fish fall ill from infection by several types of bacteria. The best-known illnesses are probably stomach dropsy*, fish tuberculosis*, fin-rot*, spot diseases*, Columnaris-disease. In addition, diseases of the organs may often be attributable to bacteria. These might include damage to the kidneys and the liver, the intestine, the swimbladder, as well as changes to the skin, in particular disease of the scales. Bacterial infection is usually combated with antibiotics* and sulphonamide*. The alternative of immunisation (*see* Immunity) is very rarely applied to fish.

Bacterial Poisons *see* Toxins

Bacteriophyta. Bacteria. Sub-D of 1,600 species. Primitive, extremely small, single-celled organisms with no cell nucleus. Spherical, rod-like or spiral in shape. During certain stages of development many of them possess delicate plasma flagella which enable them to move about. In certain Bacteriophyta the cells stay loosely connected with one another after they have divided, so forming piles of cells, packets of cells or cell networks. If conditions are poor many species develop winter-spores*. There are enormous numbers of bacteria in water, in the ground, in dust particles as well as everywhere in the atmosphere and on every object. This can be explained by the great resistance of their cells and winter-spores and the very high rate of reproduction (brought about by the high level of metabolic activity). Bacteriophyta have a great number of different modes of nutrition. Some can form organic substances from inorganic ones. The dark red purple bacteria and the red and green sulphur bacteria possess a special chlorophyll, and obtain the necessary energy for building up organic matter like plants through photosynthesis*. However, no oxygen is produced. Other bacteria obtain the energy for building up organic matter by the oxidation of inorganic compounds (chemosynthesis). Those that can do this include the colourless sulphur bacteria. They are able to oxidise substances which are harmful to most other organisms – eg Hydrogen sulphide, sulphides, sulphites and other sulphurous compounds. They are an important factor in nature for the self-purification of water as well as the purification of waste water. The nitrifying bacteria also live chemosynthetically, oxidising partly ammonia, partly nitrite. Iron (II) carbonate is oxidised by iron bacteria. You can quite often see their brown excretory products in ditches and puddles, but they also live in water pipes. Methane bacteria turn methane (marsh gas) into carbon dioxide and water while the oxyhydrogen gas bacteria form water out of molecular hydrogen and atmospheric oxygen. In the oxidation process, the de-nitrifying and de-sulphurising bacteria use the oxygen obtained by the reduction of nitrates and sulphates. Sulphate reduction, next to protein decomposition, is one of the chief causes of the build-up of hydrogen sulphide, especially in stagnant water or aquaria. The black colour of the bottom which is often observed with hydrogen sulphate build-up is caused by iron being precipitated. There are also many bacteria which cannot build organic compounds from inorganic ones. They produce their own body substances by the restructuring of any dead or living organic matter that may be to hand. In fact almost every natural substance can be broken down by bacteria, even oil, paraffin, asphalt. Occasionally some artificial resins and plasters can resist bacteria decay to a large extent. Decomposition occurs as a result of the release of enzymes. Fermentation may also be caused by bacteria. Some bacteria are poisonous because of the toxins they release and are pathogens in plants, animals and people. The products released by certain bacterial species are used in medicines as antibiotics. Many bacteria, which live symbiotically with plants, are able to bind atmospheric nitrogen. Luminescent bacteria produce a substance which emits light through oxidation. Many marine animals possess luminescent bacteria in special organs.

Bade, Ernst. Ichthyologist. Author of the book *Die mitteleuropäischen Süßwasserfische* ('Middle European Freshwater Fishes') which was published in two volumes in Berlin in 1901/02 and is still of importance today. His second, very popular book *Das Süßwasser-aquarium* ('The Freshwater Aquarium') was the first important work to cover the entire subject matter, including plants and invertebrates that can be kept in the aquarium. Although the ichthyological section became quickly out of date, the book lasted several editions.

Badis BLEEKER, 1853. G of the Nandidae*. A single species, widespread in India. *Badis* differs in many respects from the other Ga of the F. Body very much like that of the Cichlids, elongated, fairly short and only slightly compressed. Mouth remarkably small, unpaired fins have no glassy, translucent, soft-rayed parts. Behaviour essentially more peaceable than other Nandidae and cave-brooders. The male cares for the brood. Unlike the other Nandidae, *Badis* can be cared

for in community tanks, but there should be hiding-places (root stumps, thick groups of plants, holes in the rock, flowerpots). If there are not enough places of refuge *Badis* remains very shy. Temperature 25–28 °C, varied diet of live food.
– *B. badis* (HAMILTON-BUCHANAN, 1822); Badis, Dwarf Chameleon Fish. India. Up to 8 cm, female remains smaller, male liable to have a hollow stomach. Coloration extremely changeable. Ground colour clay-yellow, greenish or bluish. Especially fine male specimens have a mosaic-like arrangement of differently coloured scales. Juveniles often have dark oblique bands. Females more plainly coloured (fig. 155).

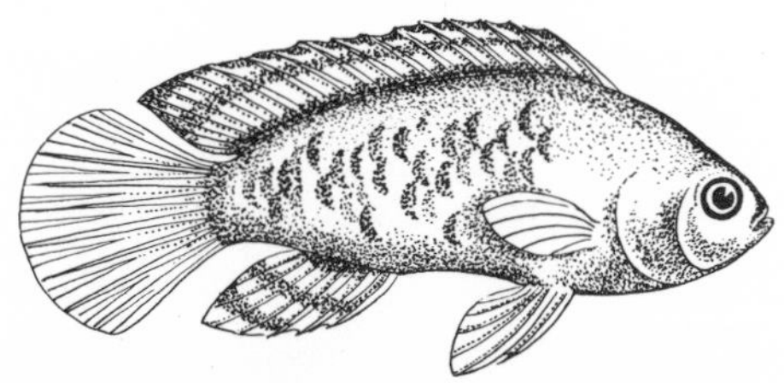

Fig. 155 *Badis badis*

Baetis LEACH, 1815 (fig. 156). Widespread G of the Mayfly insect O (Ephemeroptera*). Their larvae can be found in streams and rivers in every season of the year, sometimes in great numbers on stones and aquatic plants. On each side of the larva are seven simple, ovoid, tracheal gills along the segments of the abdomen. The

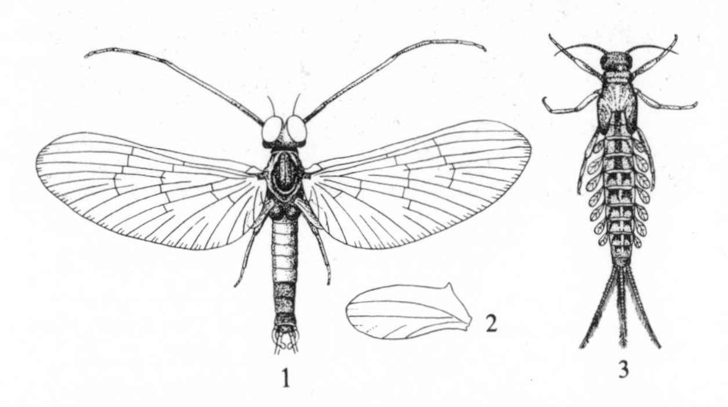

Fig. 156 *Baetis*. 1 Imago (no tail threads) 2 Rear wing (much enlarged) 3 Larva

larvae also have three tail threads, of which the middle one is often shortened. *Baetis* larvae are very suitable as food for large aquarium fish eg Cichlids; transporting them is the only difficulty (water warms up, lack of oxygen). The male adult insects (Imago*) swarm in the morning and evening sunlight at a low height on the banks where they live.

Bagridae; Naked Catfishes (fig. 157). F of the Siluriformes*. Sup-O Ostariophysi. These fish first developed in the early Tertiary and today they are spread across Africa, the Near East, South and East Asia as well as the islands of the Indo-Malaysian Archipelago. Compressed to elongated body shape. Mouth usually at an angle, often pointing upwards slightly. The jaws always have teeth, and usually the palatal bones too. As a rule there is 1 pair of barbels by the rear nasal opening, 1 pair of very long barbels on the upper jaw and 2 pairs of short barbels on the chin. The relatively short D

inserted in the front of the Vs, has a strong spine. This can reach 30 cm long in *Bagrichthys hypselopterus* ($^3/_4$ of the length of the body!), which is a species from Kalimantan and Sumatra. Large, much elongated adipose fin, C indented, An almost as long as the D. The eyes are often covered by a transparent skin. Many species are very important economically in their homelands. Different kinds of small Bagridae or juveniles of the larger species are regularly imported for the

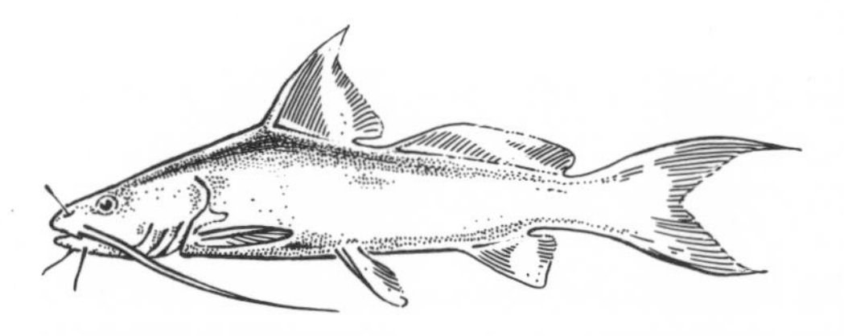

Fig. 157 Bagridae-type

aquarium, eg in particular members of the Ga *Leiocassis** and *Mystus**. The 18-cm-long Bumble-bee Catfish *(Leiocassis siamensis)* has 4–5 irregular, grey-yellow oblique bars on a dark brown background (at least in young fishes). *Mystus vittatus* from flowing and still waters of SE Asia has several very dark longitudinal bands, which contrast sharply with the silvery matt to golden ground colour. In its homeland this species grows up to more than 20 cm long, but remains smaller in the aquarium. The up to 50 cm long *Auchenoglanis occidentalis* is widespread in Africa and makes a striking aquarium specimen with its honeycombed or latticed markings. In captivity almost all the species make no special demands. However, they should be provided with places of refuge and a varied diet of live food. Some members of the G are lively free-swimming fish throughout the day. *See also* the Ga *Leiocassis, Mystus.*

Bait Killi *(Fundulus heteroclitus) see Fundulus*

Bala Shark *(Balantiocheilus melanopterus) see Balantiocheilus*

Balancing Organ (fig. 158). A sense organ that together with the auditory organ forms a functional unit in vertebrates. This acoustic organ (or inner ear) is paired and is located at the back of the cranium. It is a membranous system of interconnected canals and sacs filled with a fluid known as endolymph which is surrounded by a fluid layer known as the perilymph. In fish too, the balancing organ consists of 3 semicircular canals at right angles to one another which open out from above, via an ampulla-like extension of the canals, into a larger cavity known as the otosaccus. The otosaccus is divided into an upper and a lower part (the utriculus and the sacculus). A mass of sense cells receptive to stimuli lie at the end of each canal, embedded in the inner wall. The cells are covered by a jelly-like mass. With every active or passive body movement this jelly-like mass, together with the endolymph, alters its position. This in turn causes the sense cells in the variously orientated semicircular canals to be stimulated in different ways. The corresponding signals are transmitted to the brain via the organs of balance and of hearing where they are interpreted. Further information is transmitted from the sensitive areas of the utriculus and sacculus which contain small ear stones or otoliths made of calcium carbonate (*see* Otoliths).

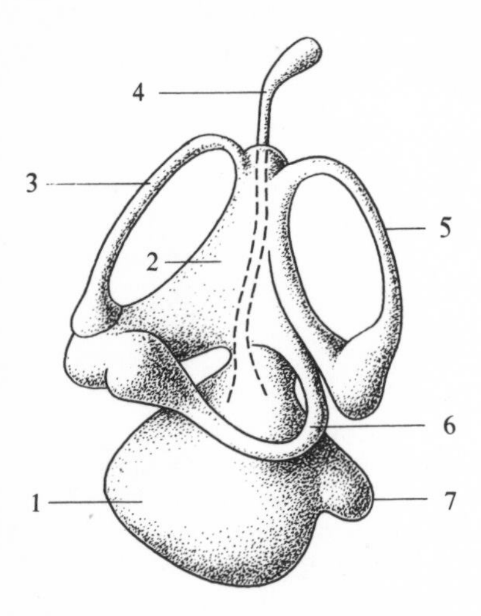

Fig. 158 Inner ear of a bony fish. 1 Sacculus 2 Utriculus 3, 5, 6 Semicircular canals 4 Endolymphatic duct 7 Lagena

The auditory organ of a fish is made up of groups of sense cells in the sacculus. One of these groups, the lagena, undergoes great development in higher vertebrates, at the same time protruding out of the rear wall of the sacculus like a piece of hose. In mammals, this hose is shaped like a spiral and has a very large number of sense cells.

Many groups of fish are also able to detect mechanical waves by means of the sense cells in their lateral lines*.

Balantiocheilus BLEEKER, 1859. G of the Cyprinidae*, Sub-F Cyprininae. SE Asia, Thailand, Vietnam, Malaya, Kalimantan (Borneo), Sumatra. Large fish that live in flowing waters. Slim, elongated body, laterally much compressed. The back part of the D and An are concave. The large C is deeply indented and drawn out into two points. LL complete. Mouth on the ventral side with a thickened, granular upper lip and a lower lip containing a groove which forms a pouch that opens to the back. No barbels. Lively, sociable fish which are only suitable for the aquarium when young. They require large tanks with a lot of horizontal room, soft sand and a medium amount of foliage. Good jumpers, which also take their food from the bottom. Preferably feed with large live food, and occasionally plant food as well. Temperature 24 °C. Not yet bred successfully. Only one species:

– *B. melanopterus* (BLEEKER, 1850); Silver Shark, Bala Shark. Up to 35 cm. Dorsal surface blue-grey, flanks a beautiful silvery iridescence. Yellow fins with a black edge at the rear end (fig. 159).

Fig. 159 *Balantiocheilus melanopterus*

Balanus DA COSTA, 1778. Best-known G of the Barnacles (Cirripedia*) (fig. 160). The G is represented by several species along the coasts of Europe. The most common species in the North Sea is *B. balanoides* BRUGIERE. Its place is taken by *Chthamalus stellatus* (POLL) along southern-European coasts.

Fig. 160 *Balanus improvisus*

Baldellia PARLATORE, 1854. G of the Alismataceae*. 2 species. Western N Africa, W Europe. Small perennials with compressed stem axes; leaves form rosettes. Aerial leaves dark green, pedunculate, up to 25 cm long. Submerged leaves light green, linear, up to 30 cm long and 5 mm wide. Flowers polysymmetrical, perianth in threes. In contrast to *Echinodorus* the calyx drops off when the fruit is ripe; the corolla is white or pale violet, 6 stamens. Many carpels, free, superior. Collective fruit. The species of the G grow at various points along riverbanks, in ditches or in flooded fields. Ideal for the aquarium if kept at a temperature above 20 °C. At one time they were counted among the G *Echinodorus*.

– *B. ranunculoides* (LINNÉ) PARLATORE. Upright inflorescence, with no adventitious plants*. Submerged with no runners.

– *B. repens* LAMARCK. Inflorescence horizontal, with adventitious plants*. Submerged with runners.

Baldner, Leonhardt (1612–94). An angler, from Strasbourg, who was the first person to keep fish and plants in glass containers as specimens. In 1666 he published a book about his observations entitled *Vogel-, Fisch- und Tierbuch* (‘A Book about Birds, Fishes and Animals’).

Balistapus Tilesius, 1820. G of the Balistidae*. A single unmistakable species from the tropical Indo-Pacific. Its bony shape is typical of the family, as are all its other characteristics. *Balistapus* is extremely intolerant, territorial and predatory. They survive well in the aquarium.
– *B. undulatus* (Mungo-Park, 1797). Undulated or Red-lined Triggerfish. Up to 30 cm. Green with orange pattern of stripes (fig. 161).

Fig. 161 *Balistapus undulatus*

Balistes Linnaeus, 1758. G of the Balistidae*, Sub-F Balistinae. Representatives in all tropical seas and some in the temperate zone. These fish live on coral reefs or rocky coasts. They hide away in crevices at night or when danger threatens by clinging firmly with the spine on their backs which they can erect and the knob-like ventral fins. Most species form territories and are very intolerant of one another. They eat echinoderms, molluscs, pieces of coral and calcareous algae which they bite through with their strong teeth. They also attack other fish. The *Balistes* species are shaped in the typical way of the family and like other Ga they have a deep furrow before the eyes. The scales are moderately large and smooth even on the laterally compressed stalk of the tail fin. The soft D and An are elongated at the front. The species of *Balistes* are suited to the aquarium; just remember that they are intolerant. Many are too big for normal aquaria. Known species include:
– *B. capriscus* (Melin); Grey Triggerfish. Warm parts of the Atlantic, Mediterranean. Up to 40 cm. Grey with blue flecks. Relatively peaceable.
– *B. vetula* Linnaeus, 1758; Cochino, Conchino, Old Wench, Old Wife, Peje Puerco, Queen Triggerfish. Tropical Atlantic. Up to 50 cm. Olive green with blue stripes on the head and edges of the fins. C much elongated. Relatively peaceable, grows very quickly (fig. 162).

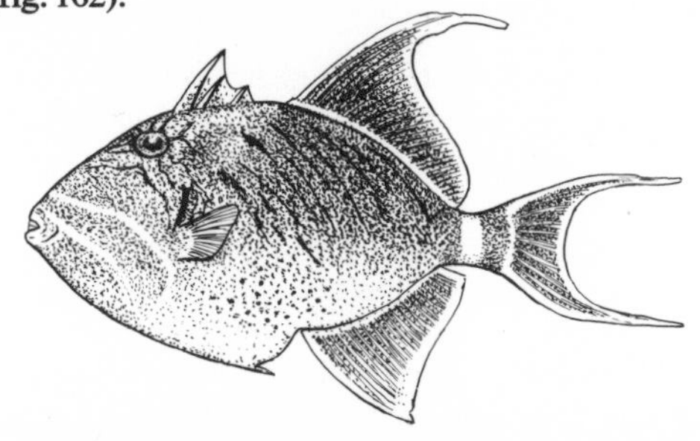

Fig. 162 *Balistes vetula*

Balistidae; Triggerfishes. F of the Tetraodontiformes*, Sub-O Balistoidei. Tall, laterally compressed fish with very large heads that are found in the tropical Atlantic and Indo-Pacific. Members of the Sub-F Balistinae have a long oval shape when viewed in profile. Their small mouth often looks very large because of characteristic facial markings. Teeth not fused, very strong. The D1, which is inserted far forwards, is supported by 3 spines, the first and strongest of which can be locked into place by the second spine. This short second spine rests in a groove along the rear edge of the first spine, and 'triggers' the front spine into its upright position (trigger mechanism). The dorsal fin can also be collapsed so that it rests in a groove (fig. 163). The long, broad D2 and An are used for propulsion, during which they are moved in an undulating fashion. Only when swimming fast is the C ever employed. This fin is rounded or elongated by the fin rays on the outside edge. Vs reduced. Strong scales, which may be dotted with tubercles in the rear part of the body. Purring noises, increased in volume by the swimbladder, and clicking and grinding sounds are made with the teeth or pharyngeal teeth. Many species have beautiful colouring and many

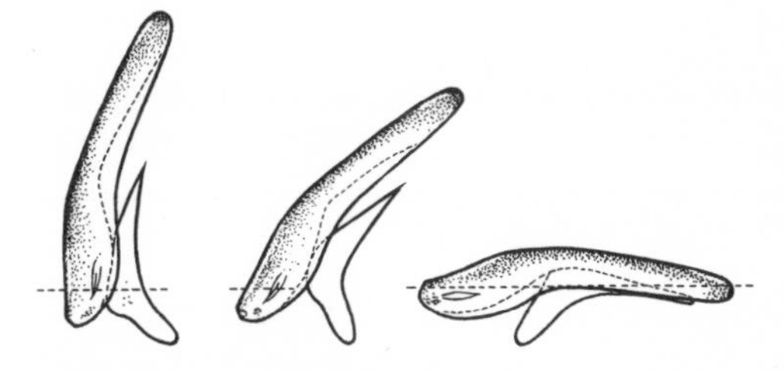

Fig. 163 Trigger mechanism

lay on their sides at night. They eat crustaceans and mussels mainly, which they pull out of crevices or blow out of the sand. Chiefly found among coral reefs

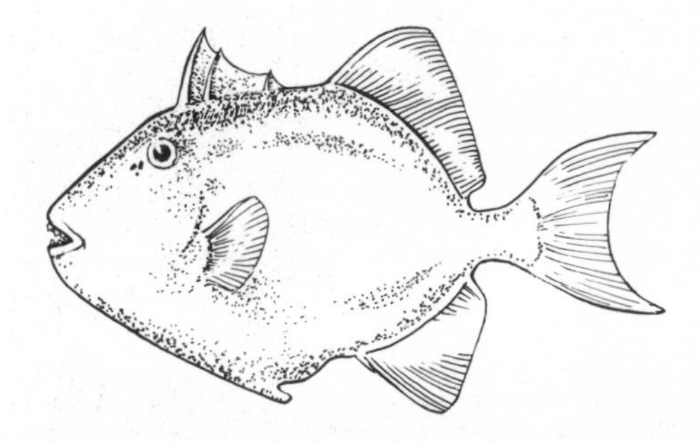

Fig. 164 Balistidae-type of the Sub-F Balistinae

(fig. 164). In profile, members of the Sub-F Monacanthinae (Filefishes) look similar to a kite (fig. 165). The D1, which is positioned at the back of the head, consists entirely of a large spine held erect by a second, much reduced spine which can lock it in place. Tubercles in the skin make it as rough as a file. True scales are absent. Some Filefishes also produce sounds. The flesh of different members of both Sub-Fs is very poisonous;

other species are eaten by local people. When poisoned there is often a reversal of the feelings of warmth and cold. Many Balistidae are well-known aquarium fish, big attractions in show aquaria. *See also* the Ga *Abalistes,*

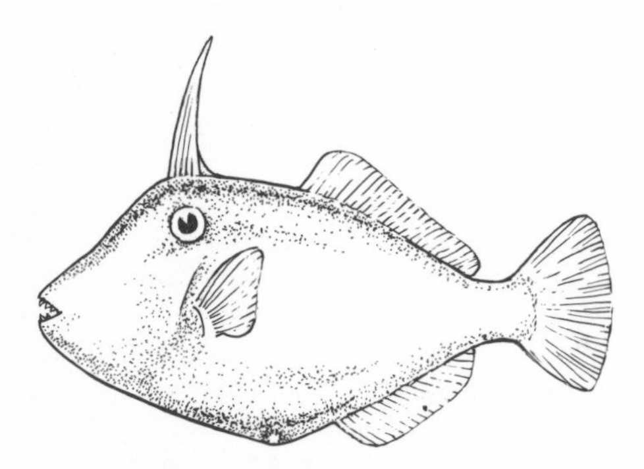

Fig. 165 Balistidae-type of the Sub-F Monacanthinae

Balistapus, Balistes, Balistoides, Melichthys, Monacanthus, Odonus, Oxymonacanthus, Paraluteres, Pseudobalistes.

Balistoides FRASER-BRUNNER, 1901. G of the Balistidae*, Sub-F Balistinae. Contains some very large species found in the tropical Indo Pacific. Way of life, shape and behaviour are typical of the family. Head profile convex, soft D and An that are low down, C rounded. Scales at the back of the body rough, smaller on the cheeks than on the body. These fish are 50–75 cm long, and should be cared for as *Balistes**. The best-known species is the brilliantly coloured:

– *B. conspicillum* (BLOCH and SCHNEIDER, 1801); Clown Triggerfish, Leopard Triggerfish. Tropical Indo-Pacific, rarely a loner. Up to 50 cm. Black with white spots on the belly. A yellow patterned marking on the back (a bit like a saddle). Mouth yellow-orange, D and An have a brilliant blue base (fig. 166).

Fig. 166 *Balistoides conspicillum*

Baltic Herring *(Clupea harengus membras) see* Clupeiformes

Baltic Prawn *(Palaemon squilla) see Palaemon*

Bambus, G, *see* Poaceae

Banded Barb *(Barbus vittatus) see Barbus*

Banded Characin *(Characidium fasciatum) see Characidium*

Banded Cichlid *(Cichlasoma severum) see Cichlasoma*

Banded Climbing Perch *(Ctenopoma fasciolatum) see Ctenopoma*

Banded Clown *(Amphiprion bicinctus) see Amphiprion*

Banded Coral Shrimp *(Stenopus hispidus) see Stenopus*

Banded Fighter *(Betta taeniata) see Betta*

Banded Humbug *(Dascyllus aruanus) see Dascyllus*

Banded Jewel Fish *(Hemichromis fasciatus) see Hemichromis*

Banded Knifefish *(Gymnotus carapo) see* Gymnotoidei

Banded Leporinus *(Leporinus fasciatus fasciatus) see Leporinus*

Banded Loach *(Botia hymenophysa) see Botia*

Banded Long-snout Butterfly-fish *(Chelmon rostratus) see Chelmon*

Banded Minnow *(Aphanius fasciatus) see Aphanius*

Banded Panchax *(Epiplatys macrostigma) see Epiplatys*

Banded Sea-perch *(Serranellus scriba) see Serranellus*

Banded Sunfish *(Enneacanthus obesus) see Enneacanthus*

Bandtail Barb *(Barbus stigma) see Barbus*

Banjo Catfish *(Bunocephalus kneri) see Bunocephalus*

Banjo Catfishes, F, *see* Aspredinidae

Bank Zone *see* Littoral

Barbel *(Barbus barbus) see Barbus*

Barbel Region *see* Potamal

Barbels (fig. 167). Thread-like or bristle-like processes in the region of the head of many different fish.

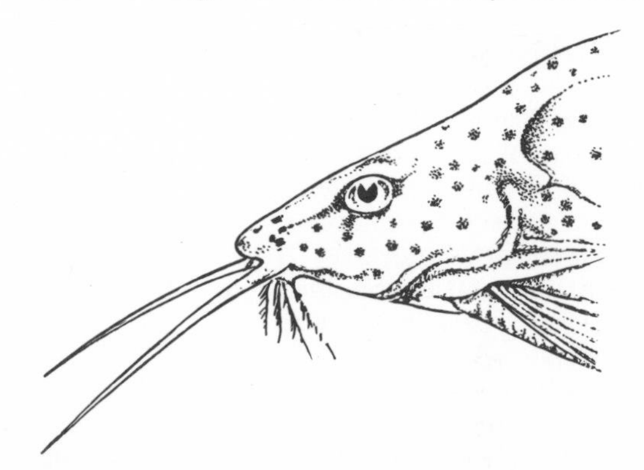

Fig. 167 Barbels. Head of the Catfish *Synodontis nigrita*

In particular, those fish that live near the bottom have barbels, eg Catfish types (Siluriformes), Loaches (Cobitidae), Carps (Cyprinidae), Cods (Gadidae), Goatfishes (Mullidae) and Sturgeons (Acipenseridae). They are used as organs of taste and touch for finding food. The barbels can be stiffened by cartilage or bones and in many species they can be moved by muscles inserted at the base of the barbels. The scaleless Catfishes (Mochokidae) possess branched barbels on the lower jaw and chin, the barbels on the upper jaw are smooth.

Barber's Tetra *(Mimagoniates barberi) see Mimagoniates*

Barbus CUVIER, 1817. G of the Cyprinidae*, Sub-F Cyprininae. The G Barbus is a collective G, under which a large number of species have been grouped, although they are not directly related if you look at their evolutionary history. Attempts at compiling a system according to natural relationships has so far proved unsuccessful. Breaking the G down according to the number of barbels in the Ga *Puntius* HAMILTON-BUCHANAN, 1822, with no barbels, *Capoeta* VALENCIENNES, 1842, with 2 barbels and *Barbodes* BLEEKER, 1860, with 4 barbels is not an adequate solution. So, after the recommendation of MYERS (1956), the term G *Barbus* has continued to be used. Species are spread throughout Europe, SE Asia, India, Sri Lanka, the Indo-Australian archipelago and Africa. They range in size from very small to large, are peaceable and sociable fish, some of which may form large shoals. Their body is moderately to much elongated and laterally compressed. Elongations of the fins are rare, the scales are large. LL complete or reduced. The mouth is terminal to inferior, without barbels or with 1–2 pairs of barbels. Only young specimens of the European species can be kept in well aerated cold-water aquaria with the occasional addition of fresh water. The adults are only suitable for public aquaria. The SE Asian and African species are peaceable shoaling fish, which are almost without exception ideal for the community aquarium. They need large, very long tanks with a soft substrate. The vegetation must leave enough free swimming area, and a shady covering of floating plants will ensure that the fish settle down. Most of the species require no particular water quality, but the occasional addition of fresh water is advised. Temperature 20–25 °C although a temporary drop is tolerated well. These fish are easily satisfied omnivores, which will accept all types of live and dried food. Rearing these often very productive species is not difficult and sometimes it is very easy. Use a medium-sized tank (20–30 l) with soft cold water (total hardness 4–8°) and a little fresh water. A few fine-leaved plants should be added along with a layer of gravel on the bottom as the majority of the species will take spawn as prey. Put the females in first (recognisable by their stouter belly), and 1–2 days later, and in the evening, add the more slender, and more strikingly coloured males.

After a very turbulent driving the fish mate several times, particularly at sunrise. In a lot of cases, the male pushes his C or caudal peduncle behind the D over the back of the female, during which his body is at an oblique angle head-down. The great majority are open-water spawners*; only a few are substrate spawners*. The brood hatches after 24–26 hours and still live on their yolk sac for 1–2 days more. The fast-growing fry are easy to rear, and they will have already achieved sexual maturity after 9–12 months.

– *B. arulius* (JERDON, 1849); Longfin Barb, Arulius Barb. S and SE India, Travancore, Cauvery. Up to 12 cm. Flanks of the body greenish to reddish with a silvery iridescence and an indistinct grey longitudinal band. Some black-blue oblique bands run over the back to the lateral line. An and C yellowish with a red tip.

Males have a red D, the fin rays of which are elongated beyond the edge of the fin.

– *B. barbus* (LINNAEUS, 1758); European Barbel. Europe. Up to 90 cm. Typical of the European Barbels. Sociable fish that live near the bottom of fast-flowing, oxygen-rich rivers with sand and gravel. Crepuscular fish which take their food from the bottom. Dorsal surface brown-green to black-green, flanks paler. Some have irregular spotting. At spawning time in May–June they migrate up river to a suitable headwater. Males acquire spawning tubercles on the head and dorsal surface. Numerous other species and their sub-species are spread across Europe.

– *B. bariloides* BOULENGER, 1914; Orange-fin Minnow. S Africa, Angola, Southern Zimbabwe, Katanga. Up to 5 cm. Flanks of the body orange to reddish with 12–16 fine, dark, oblique stripes, the third and last of which may broaden out into a spot. Fins yellowish, D has a crimson red spot (fig. 168).

Fig. 168 *Barbus bariloides*

– *B. bimaculatus* (BLEEKER, 1864). Only found in Sri Lanka (Ceylon). Up to 7 cm. Flanks pale coppery colours with a silvery iridescence. A dark net-like patterning in the upper half of the body. The base of the D and the root of the tail each have a black spot. The males have a dark red longitudinal band.

– *B. binotatus* CUVIER and VALENCIENNES, 1842; Spotted Barb. Common in Indo-China, Malaya, Indonesia, the Philippines. Up to 18 cm. Flanks bluish with a silvery iridescence, with 5 dark spots in young fish. D and C reddish with dark edges.

– *B. callipterus* BOULENGER, 1907; Clipper Barb. W Africa, Niger to the Cameroun. Up to 9 cm. Flanks have a strong silvery iridescence, scales in part have dark tips and the base of the lateral line scales have a dark spot. D reddish with a black triangular spot. So far not reared in captivity.

– *B. camptacanthus* (BLEEKER, 1863); African Red-finned Barb. Common in tropical W Africa, Niger to the Congo. Up to 16 cm. Dorsal surface olive-green, flanks pale yellowish to reddish with a silvery iridescence. Along the flanks there is a brown longitudinal band which is indistinct in places. In the male, the fins are a vivid red, in the female orange-yellow. The male

acquires spawning tubercles. Not yet bred in captivity.
– *B. candens* (NICHOLS and GRISCOM, 1917). Congo Basin. Up to 4 cm. Flanks pink with 3 black spots, the last of which lies below the middle of the caudal peduncle. D has a red base and a black band. V and An have red and black tips. For breeding, soft, slightly acidic water (total hardness 1–2°, pH 5–6) is needed. The eggs are laid in the substrate which should consist of a layer of peat. The baby fishes hatch after 1 week.
– *B. chola* HAMILTON-BUCHANAN, 1822; Swamp Barb. E India, Burma. Up to 15 cm. Dorsal surface olive, flanks yellowish with a silvery iridescence. An ill defined golden-yellow spot on the operculum. Base of the D and tail root each have a black spot. Fins yellowish, reddish in the male.
– *B. conchonius* (HAMILTON-BUCHANAN, 1822); Rosy Barb. NE Indo-China. Up to 14 cm. Dorsal surface dark olive, flanks yellowish or reddish with a silvery iridescence. A black spot, edged in golden-yellow, on the caudal peduncle. At spawning time the males are coloured deep red with a black tip on the D. Very easy to breed.
– *B. cumingi* GÜNTHER, 1868; Cuming's Barb. Only found in Sri Lanka (Ceylon). Up to 5 cm. Dorsal surface grey-brown, flanks grey-white with a silvery-golden iridescence. Each scale has a dark edge. A black, obliquely oval spot on the nape and at the beginning of the caudal peduncle. Fins yellow to reddish.
– *B. dunckeri* E. AHL, 1929; Duncker's Barb. Singapore, Malacca. Up to 30 cm, but sexually ripe at 8–10 cm. Dorsal surface brown to greenish, flanks with a golden iridescence and large black-green spots. Orange belly. Fins red.
– *B. everetti* BOULENGER, 1984; Clown Barb. Singapore, Kalimantan (Borneo). Up to 10 cm. Dorsal surface brown-red, flanks silvery to reddish with a golden iridescence and irregular blue-black spots. Fins red. Keep the sexes separate before breeding and feed adequately.
– *B. fasciatus* (JERDON, 1849); Striped Barb, Zebra Barb. Malaysia, Indonesia. Up to 6 cm. Dorsal surface green-golden, flanks wine-red with net-like markings and irregular black spots near the dorsal region.
– *B. fasciolatus* GÜNTHER; African Banded Barb. Angola. Up to 7 cm. Dorsal surface blue-grey, flanks silvery with a blue-green iridescence and 12 narrow, short, oblique bands. Fins colourless to pale brownish. To some extent intolerant. Not yet bred in captivity.
– *B. filamentosus* (CUVIER and VALENCIENNES, 1844); Filament Barb, Featherfin Barb. S and SW India, Sri Lanka (Ceylon). Up to 15 cm. Very variable colouring, dorsal surface brown to greenish, flanks with a silvery iridescence. A large lack spot on the caudal peduncle above the An. Fins mostly red. Young fish have broad black oblique bands on the flanks. Males have elongated dorsal fin rays.
– *B. gambiensis* SVENSSON, 1933; Brassy Barb. W Africa (R. Gambia). Up to 10 cm. Flanks with a yellowish-silvery iridescence and a black-brown longitudinal band which is edged with a golden iridescence in the dorsal region. D has a black tip.
– *B. gelius* (HAMILTON-BUCHANAN, 1822); Golden Dwarf Barb. Indo-China, Bengal, Assam. Up to 4 cm. Dorsal surface brown-olive, flanks a translucent gold colour with irregularly distributed black spots and a red-gold longitudinal band. Fins yellowish, C pale reddish.
– *B. holotaenia* BOULENGER, 1904; African Spot-scale Barb. Congo Basin, Cameroun to Angola. Up to 12 cm. Flanks yellowish with silvery iridescence and a blue-black longitudinal band. A dark half-moon spot on the base of the scales. Unpaired fins red, D with a black tip. Not yet bred in captivity (fig. 169).

Fig. 169 *Barbus holotaenia*

– *B. hulstaerti* POOL, 1945; Butterfly Barb. Lower Congo. Up to 3.5 cm. Flanks a coppery to brown colour with 3 black patent, to some extent elongated, spots. Fins yellow, D, An and V with black tips. In the male the front body spot is shaped like a half moon. For breeding, place in a darkened tank, temp. not more than 22 °C.
– *B. lateristriga* CUVIER and VALENCIENNES, 1842; Spanner Spanner Barb. Thailand, Malaysia, Indonesia. Up to 18 cm. Dorsal surface greenish orange, flanks yellowish-brown with a golden iridescence, orange belly. In the front part of the body, 2 blue-black, pointed, oblique bands run across the dorsal surface; in the rear part a longitudinal band of the same colour but narrowing towards the C. A blue-black spot above the An. Fins red, especially in the male. Elderly fish are paler.

Fig. 170 *Barbus lineatus*

– *B. lineatus* DUNCKER, 1904; Striped Barb, Zebra Barb. Southern part of the Malaysian Peninsula. Up to 12 cm. Flanks yellowish-brown to silvery, 4–6 blue-black longitudinal stripes (fig. 170).
– *B. nigrofasciatus* GÜNTHER, 1868; Black-ruby Barb, Purple-headed Barb. Only found in Sri Lanka (Ceylon). Up to 7 cm. Flanks yellowish-grey with 3–4 dark oblique bands, head reddish. Scales with silvery iridescent edges. At spawning time the males become a deep purply-red in the front part of their bodies, caudal peduncle and C black with a green iridescent dorsal surface. D black, An black-red. V red.
– *B. oligolepsis* (BLEEKER, 1853); Checker Barb. Sumatra. Up to 5 cm. Flanks brown-yellow with a bluish-green iridescence and numerous dark spots in the youthful phase. Fins reddish-yellow. The unpaired fins of the male have black tips (fig. 171).

Fig. 171 *Barbus oligolepis*

– *B. orphoides* CUVIER and VALENCIENNES, 1842; Orange-cheek Barb. Indonesia, Thailand. Up to 25 cm. Dorsal surface olive, flanks bluish with a silvery iridescence. An ill-defined red spot, on the operculum. Fins red, the C with a broad black edge at the top and at the bottom. Only young specimens are suitable for a room aquarium.
– *B. pentazona pentazona* BOULENGER, 1894; Five-banded Barb. Indonesia, Malaysia. Up to 5 cm. Flanks yellow to orange with 5 blue-black oblique bands. Fins have a vivid red base.
– *B. pentazona hexazona* WEBER and BEAUFORT, 1912; Six-banded Barb. Sumatra. Has one more band than the aforementioned. The first band is broad and completely encircles the eye. Several sub-species, which are essentially distinguished by the patterns of the bands.
– *B. phutunio* (HAMILTON-BUCHANAN, 1822); Dwarf Barb, Pygmy Barb. E India, Sri Lanka (Ceylon). Up to 8 cm. Dorsal surface brown-green, flanks a silvery iridescence with some dark spots, which fade away with age. Fins yellowish to red.
– *B. roloffi* (KLAUSEWITZ, 1957); Roloff's Barb, Streak-scale Barb. S Thailand. Up to 4 cm. Dorsal surface olive, flanks have a silvery iridescence and each scale has a black streak. Root of the tail has a black spot with a silvery-white one on top of it.
– *B. schwanenfeldi* BLEEKER, 1853; Tinfoil Barb, Gold Foil Barb. Indonesia, Thailand, Malaysia. Up to 35 cm. Flanks have a silvery iridescence, with a yellow-bluish shimmer in part. Fins yellow. D red with a black spot, C red and black at the base. Differences between the sexes not known.
– *B. semifasciolatus* GÜNTHER, 1868; Half-striped Barb, Green Barb. SE China. Up to 10 cm. The brown dorsal surface gives way to green and finally brassy-yellow flanks. 5–7 narrow, black, oblique bands. Fins yellow to reddish. Scales dark-edged. A xanthic (yellow) form (*see* Colour) of this species has been called *B. schuberti* (name not now in current usage). It is more orange in colour with dark spots in certain parts.
– *B. stigma* CUVIER and VALENCIENNES, 1842; Band-tail Barb. Dotted Barb. Indo-China, Burma. Up to 15 cm. Dorsal surface brown-green, flanks with a silvery iridescence. The base of the D and the root of the tail each have a black spot. At spawning time the male acquires a reddish longitudinal band.
– *B. terio* (HAMILTON-BUCHANAN, 1822); One-spot Barb. Bengal. Up to 9 cm. Flanks have a silvery iridescence. An indistinct black longitudinal band runs from the middle of the body to the cuadal peduncle. At each end of this band is a black spot, edged in gold.
– *B. tetrazona tetrazona* (BLEEKER, 1855); Sumatra Barb, Tiger Barb. Sumatra, Kalimantan (Borneo). Up to 7 cm. Flanks yellowish with a golden iridescence and 4 black oblique bands. The third oblique band runs up into the red coloured D and An (fig. 172).

Fig. 172 *Barbus tetrazona tetrazona*

– *B. tetrazona partipentazona* FOWLER, 1934. Thailand, Malaysia. Distinguished from the aforementioned by a fifth oblique band which reaches from the base of the D to the middle of the body.
– *B. ticto ticto* (HAMILTON-BUCHANAN, 1822); Ticto Barb, Two Spot Barb. India, Sri Lanka (Ceylon). Up to 10 cm. Dorsal surface dark, flanks with a silvery iridescence. Above both the P and An in the middle of the body is a black spot, edged in gold. In the male, the V is reddish, the D with black spots. Can be overwintered at a temp. of 14–16 °C.
– *B. ticto stoliczkae* (DAY, 1869). S Burma, in the river basin of the Irawaddy. Introduced as a sub-species, and although it is uncertain it is distinguished from the aforementioned by a smaller number of scales in each oblique row.

– *B. titteya* (DERANIYAGALA, 1929); Cherry Barb. Sri Lanka (Ceylon). Up to 5 cm. Flanks yellow to reddish with a silvery iridescence. A broad black band runs from the snout into the C. Above it lies a golden, turquoise iridescent band and below it a double row of dark spots.
– *B. vittatus* (DAY, 1865); Banded Barb. India, Sri Lanka (Ceylon). Up to 6 cm. Body greenish with a silver iridescence, scales have a dark base. A small black spot in front of the C on the caudal peduncle and above the An. Fins yellow, D with a black band, edged in orange at the top.
– *B. viviparus* WEBER, 1897; Livebearing Barb, Silver Barb. S Africa (Natal). Up to 6 cm. Flanks with a silver iridescence and a dark longitudinal band at the middle of the body. A second stripe, bowing, down towards the ventral surface, corresponds to the LL. Recent observations suggest the species is not viviparous*.

Barclaya WALLICH, 1827. G of the Nymphaeaceae*. 3 species. S Asia. Rhizomed perennials with short runners. Leaves in basilar rosettes, stalked, with circular to lanceolate blades and heart-shaped bases. The upper surface is olive-green, the lower surface reddish. Flowers hermaphroditic, polysymmetrical. Perianth consists of a five-part calyx and a multi-part corolla. The petals are green on the outside, dark red on the inside. Many stamens. Many carpels, with a medial, fused, floral axis. The fruit is a green berry.

The species grow scattered on flat areas of tropical forest streams. In the aquarium dimmed lighting and a temperature over 20 °C have proved best. The substrate should not be cold. The flowers either open on the surface of the water or remain submerged. Not every flower produces seeds. Propagation either by separation of plants growing runners or by sowing of seeds. Most often kept in aquaria is:
– *B. longifolia* WALLICH. SE Asian Mainland. Leaves up to 60 cm long and 5 cm wide, but most are smaller (fig. 173).

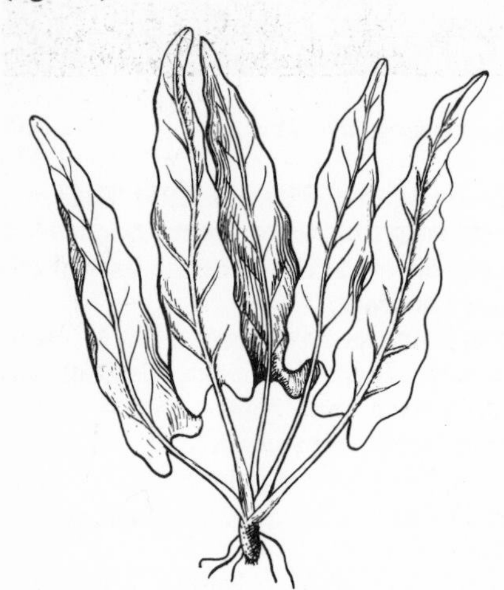

Fig. 173 *Barclaya longifolia*

Barclayaceae, F, name not in current usage, *see* Nymphaeaceae.

Barilius HAMILTON-BUCHANAN, 1822. G of the Cyprinidae*, Sub-F Cyprininae. Japan, SE China, S Asia, Indo-Australian archipelago up to Africa. These small to medium-sized shoaling fish live in mountain rivers and those of flat lands. Elongated shape, like a torpedo and flattened laterally. LL complete. The large mouth has a wide gape with 4 barbels or none at all (very rarely there are 2). The D is inserted behind the start of the V. Flanks usually silver with characteristic pattern of spots and bands. These peaceable species are kept as a shoal in large, long tanks with a lot of swimming area and medium-hard to soft water. They prefer the upper levels of water. Temperature 23 °C. Live food of all sorts is accepted, but expecially insects. Of the numerous species only a few have been introduced.
– *B. christyi* BOULENGER, 1920; Gold-lipped Barb. Congo Basin. Up to 13 cm. Dorsal surface dark olive, sides green-yellow with a silver iridescence and 8–18 narrow black-brown oblique bands. Root of the tail has a black blotch. On the upper jaw is a gold-red blotch, the lower jaw is blue. Fins yellow-orange with black edges. Not yet reared in captivity.
– *B. neglectus* STIELER, 1907; Japanese Barb. Japan. Up to 7 cm. Compressed body and dull brownish to coppery colours with a silver iridescence. A bluish, silvery iridescent longitudinal band on the flanks with indistinct dark spots. Temperature about 20 °C. Breeding is simple. These fish will spawn in the morning in small tanks with some plant growth and fresh water, the pH and hardness of which is immaterial. The fry hatch after 20–24 hours and should be reared on the finest powdered food.

Barnacles, Sub-Cl, *see* Cirripedia

Barnea, G, *see* Boring Molluscs

Barracudas (Sphyraenidae, F) *see* Sphyraenoidei

Barracudas, related species, Sub-O, *see* Sphyraenoidei

Barramundi Cod *(Cromileptis altivelis) see Cromileptis*

Barred Loach *(Noemacheilus fasciatus) see Noemacheilus*

Barred Millions Fish *(Phalloptychus januarius) see Phalloptychus*

Barred Pencilfish *(Nannostomus espei) see Nannostomus*

Barrier Reef *see* Coral Reefs

Bases. Matter, very varied in structure, but with the property in common of splitting off hydroxyl ions (OH^-) or binding hydrogen ions (H^+) in an aqueous solution. The size of the imbalance between the known ion concentrations is expressed in terms of pH value*, which is always greater than 7 in a basic reaction. According to the strength of the possible fluctuation in pH value, weak and strong bases can be distinguished. Strong bases, eg the caustic potash or caustic soda used in the regeneration of negative ion exchangers, must be handled with extreme care, particularly in concentrated form, because they may burn skin or clothing (hence the description 'caustic'). The cornea of the eye is extremely sensitive to bases, so when handling caustic solutions wear some protective glasses. If any of the solution gets onto the skin or clothes, wash off immediately with a lot of water.

Basic Reaction *see* pH value

Basilar *see* Leaf Arrangement

Basking Shark *(Cetorhinus maximus) see* Cetorhinidae

Basking Sharks, F, *see* Cetorhinidae

Basommatophora. O of the Euthyneura* (Sub-Cl of the Mollusca). Freshwater and brackish snails found worldwide, in which the eyes do not sit at the end of stalks (tentacles) but at the base of them in the skin on the head. A shell is always formed; it may form a spiral as in *Planorbis** or it may not be coiled as in *Acroloxus**. All the snails of this O can retreat fully into their shells. The German name for these snails *Wasserlungenschnecken* (water-lung snails) refers to the fact that the Basommatophora can breathe in air, although they almost exclusively live in water. They come to the surface of the water from time to time to exchange the air supply in the lung cavity. Well-known Ga are *Helisoma**, *Lymnaea**, *Physa** and *Planorbarius**.

Bassalm Tetra *(Hemigrammus marginatus) see Hemigrammus*

Batanga lebretonis, name not in current usage, *see Eleotris*

Batfishes (Ogcocephalidae, F) *see* Lophiiformes

Batfishes (Platacinae, Sub-F) *see* Ephippidae

Bathing Treatments. For effective treatment of sick fish, it is important to diagnose correctly whatever is wrong. Bathing treatments have proved to be simple but effective methods of treating the skin and gill parasites (Parasitism*) which so often attack fish. In a lot of cases raising the water temperature to 32–34 °C proves successful. This method increases the metabolism which may activate the natural defence mechanisms of the fish. Many parasites require a relatively narrow range of temperature for their development and will die under these altered conditions. Even more effective are bathing treatments with the addition of medicaments. These must be completely dissolved before being placed in the treatment tank. Instructions about concentration, length of treatment, water temperature, pH values and water hardness must be rigidly adhered to. A combination of different medicaments may severely damage the fish. Whenever fish are treated by the bathing method, it is best to do this in a separate glass container with good aeration. Filters must be operated using fresh filter wadding only, as activated charcoal* in particular will adsorb the medicaments quickly, with the result that the specimens will not be bathed at the correct concentrations for a sufficient length of time. Adding medicaments directly to the tank where the fish are kept is not desirable, although often unavoidable in practice. When dealing with parasites that can also survive for a while without their host, it is best to keep fish out of the aquarium for a certain period, or to disinfect it in an appropriate manner (Disinfection*). Fish react to medicaments or to particular concentrations in very different ways, so it is important to observe them carefully during the treatment period. If the fish is observed to lie at an unusual angle, or to start reeling or to be short of breath, it should be transferred immediately to a tank that has already been prepared with suitable water. Small concentrations of medicaments can be tolerated by fish for several days or weeks without harm. Parasites that have very hardy stages of development can only be overcome effectively by such long-term treatment. Fish can also be treated in short-term immersion baths, whereby they are subjected to much higher concentrations of medicaments for only a few minutes. Such bathing treatments must only be carried out in separate glass tanks and under the strictest observation. The infected specimens are suspended in a net in the tank so that they can be transferred immediately to another container with normal water if need be. Highly concentrated baths are more tiring for the fish and must usually be repeated several times at intervals. Internal diseases (*see* Fish Diseases) will only rarely be affected by bathing treatments. Medicaments have to be applied in such cases along with the food or as an injection*.

Bathygobius BLEEKER, 1878. G of the Gobiidae, Sub-O Gobioidei*. America, from Florida to SE Brazil, including the Bahamas and Antilles, W and E Africa, SE Asia. Found in shallow coastal waters in the tidal zone in mangrove swamps or even on sandy or rocky bottoms. Can live in sea water as well as brackish water. Small to medium-sized fish (6.5–15 cm). Elongated body, cylindrical at the front, flattened laterally at the back. Large, thick lipped mouth. Several rows of teeth, the outermost row in the upper jaw being larger. V fused with one another. The hallmark of the G is the thread-like, free-standing, upper rays of the P. Colouring various shades of brown with small or large dark spots, some of which may form vague outlines of oblique bands. Male often darker, D2 and An longer. Keep the brackish water forms in 50–100 g of sea salt to 10 l water. Large, live food. Breeding conditions not known. The following used to be imported frequently.

– *B. fuscus* (RÜPPEL, 1828). E coast of Africa, the Red Sea, SE Asia. Sea-water and brackish water. Up to 12 cm. Colour very variable as is typical of the fish's wide area of distribution. Upper surface of body olive-green, flanks brownish, paler below. Head and body have irregular dark spots, which form rows in places. Fins grey with some brown spotting of many of the rays. D1 often tipped in blue.

Bathypteroidae, F, *see* Myctophoidei

Bathysaurus ferox *see* Myctophoidei

Batoidea, O, name not in current usage, *see* Chondrichthyes

Batrachoididae, F, *see* Batrachoidiformes

Batrachoidiformes (fig. 174). Toadfishes. (Haplodoci is a name not currently used.) O of the Osteichthyes*, Coh Teleostei. In appearance tadpole-like, flattened, bottom-dwelling fish of rocky coastal regions of tropical and subtropical seas. About 30 modern species and

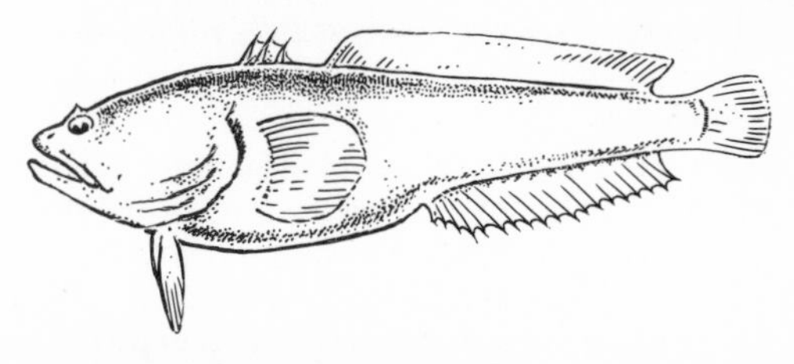

Fig. 174 Batrachoidiformes-type

1 fossil species from the Neogene Period in N Africa. The powerful, broad head with its large, well-toothed mouth forms the biggest part of the body. Adjoining it is the torso-tail region which gets increasingly narrower away from the head. The Vs are situated at the throat, ie in front of the muscular Ps. D1 small, situated at the nape of the neck and supported by 2–4, partially hollow, spines. D2 and An very long, C small. Only 3 pairs of gills. Many species possess poison glands with outlets either through the hollow spines of the D1 or through hollow spines of the operculum. Wounds caused by these spines are very painful but not dangerous. Many Batrachoidiformes have the ability to exist in extremely oxygen-poor waters or even to survive for hours out of water. Many Toadfishes are known to produce grunting or whistling noises with the aid of the swimbladder. The largest member of the single F (Batrachoididae) is the Plainfin Midshipman *(Porichthys notatus).* It is found in the Pacific and reaches up to 40 cm in length. Like many other members of this F, it possesses numerous luminescent organs on the ventral surface. Most Batrachoidiformes are much smaller, however. In the marine aquarium these odd, mostly dark or cryptically coloured fish usually refuse food.

Beacon Fish *(Hemigrammus ocellifer) see Hemigrammus*

Beadlet Anemone *(Actinia equina) see Actinia*

Beaked Butterfly-fish *(Chelmon rostratus) see Chelmon*

Beaked Coralfish *(Chelmon rostratus) see Chelmon*

Beaked Leatherjacket *(Oxymonacanthus longirostris) see Oxymonacanthus*

Beau Gregory *(Eupomacentrus leucostictus) see Eupomacentrus*

Beaufort's Loach *(Botia beauforti) see Botia*

Bedotia REGAN, 1903. G of the Atherinidae, Sub-O Atherinoidei*. Area of distribution confined to the fresh waters of Madagascar. Small to medium-sized shoaling fish that live in the upper surface region. The much elongated body is flattened laterally and is covered with cycloid scales (*see* Scales). The head is pointed, ending in a superior mouth. D split, D2 extending far to the back, C straight. These fish are very peaceful and undemanding, and should be kept in large tanks with a medium amount of plants and room for swimming. Temperature 23–25 °C. Every sort of live and dried food is accepted, provided it floats on the surface. There are no particular points to bear in mind when getting the water ready, although calcium-rich water appears to be most favourable. Eggs are released continuously over long periods of time (*see* Breeding Season). They have thread-like processes and hatch at 25 °C after 5–8 days. Easy to rear. Only 1 species of this G has been introduced so far:

– *B. geayi* PELLEGRIN, 1907; Madagascar Rainbow. Madagascar. Up to 15 cm. Yellow to yellow-brown ground colouring. A dark, grey-black longitudinal band stretches from the mouth to the C. A less distinct band runs parallel to it to the belly region. The unpaired fins vary enormously in colour but as a rule they are yellow to yellow-brown with a black band and red edges (fig. 175).

Beebe, William (1877–1962). Ichthyologist and deep-sea explorer. Born in New York, USA. He was Director

Fig. 175 *Bedotia geayi*

of the Department of Tropical Research in New York. In 1923 he led an expedition to the Galapagos Islands. The records of this trip were published under the title *Galapagos, the End of the World.* In 1925, on behalf of the Zoological Society of New York, he explored the upper levels of the sea with the aid of a diving helmet. Even at that time Beebe triggered off greater interest in diving with simple aids in his book *Arcturus Adventure* (1928). His most successful undertaking was in 1934, when he dived to a depth of 923 m off the islands of Bermuda in a bathysphere built by himself. In 1935 he described his impressions and experiences of this in his book *923 Metres beneath the Surface.* Beebe discovered numerous new deep-sea fish and was the first to publish his personal observations about their way of life.

Beetles, O, *see* Coleoptera

Behavioural Inventory. Term used in ethology. The behavioural inventory of a species of animal takes into account all its modes of behaviour and is divided by ethologists into various spheres of function*. During ethological studies, a behavioural inventory is presented as an ethogram*, which is afterwards compared with the behavioural inventory of species that appear to be related. Important clues may be drawn from these comparisons as to the relationships of the species concerned within the natural system of species.

Behavioural Sciences *see* Ethology

Behavioural Syndrome (fig. 176). Term used in ethology. A behavioural syndrome is the combining in time or on a hierarchical basis of various fixed action patterns* of a sphere of function*. Although the basic coordination of these fixed action patterns is different, they usually occur again and again in the same sequence and, just like other behavioural elements in the ethogram*, they can be taken into account in the characterisation of a species. A typical male behavioural syndrome is the courtship behaviour of Nannostomids. Here, the following combinations take place: head, dorsal or ventral display swimming (undulating display swimming) pausing, body vibrating, on plants (*see* drawing), whereby the females are led to the spawning ground and there induced to mate. This male behavioural behavioural syndrome is a chain of actions following on one from the other (Initial Action*), and the subsequent mating is regarded as the consummatory act*.

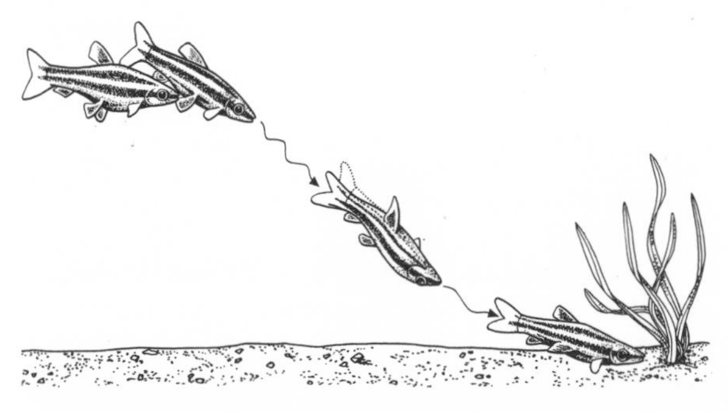

Fig. 176 Behavioural syndrome of the courtship of *Nannostomus marginatus*; from the left: release of cranial courtship, undulating enticement swimming (horizontal oscillation), pausing routine with body quivering

Beirabarbus palustris, name not in current usage *(Barbus palustris), see Barbus*

Belgian Flag Tetra *(Hyphessobrycon heterorhabdus) see Hyphessobrycon*

Belone belone *see* Exocoetoidei

Belonesox KNER, 1860; Pike Top Minnows. G of the Poeciliidae*, Sub-F Poeciliinae. Eastern Central America from S Mexico to Honduras (Nicaragua). Lives in the upper water levels of small dirty waters and lakes, as well as brackish water. Male up to 10 cm, female up to 20 cm. Elongated body, head pointed with a deeply-cleft mouth and a long snout. The jaws are slightly bent together and are beset with pointed teeth. Other typical features: the D is inserted far back and the scales are fine; male has a gonopodium; body is brownish, breast and belly pale during the day, dark at night; when light falls on it, a bronze shimmer shows through; flanks have up to 5 complete longitudinal rows of dark spots. The C has a black, pale-bordered blotch at the base. Otherwise, the fins are colourless or pale yellow. These predatory fish should be kept in roomy, well-planted tanks, as they also attack their own species. A tiny amount of sea salt should be added. 20–30 °C. Mostly, only the larger live food is accepted, particularly fish, but also dragonfly larvae and worms. Breed at 25–30 °C. It is necessary to isolate the female to give birth to the youngsters. Gestation period 30–50 days. Up to 100 young in a brood. The only species is:

– *B. belizanus* KNER, 1860; Livebearing Pike Top Minnow (fig. 177).

Fig. 177 *Belonesox belizanus*

Belonidae, F, *see* Exocoetidae

Beloniformes, O, name not in current usage, *see* Atheriniformes

Belontia MYERS, 1923. G of the Belontiidae, Sub-O Anabantoidei*. Sri Lanka (Ceylon), Sumatra, Java. Lives in rivers and still waters. Medium-sized fish, with a body similar to that of the Cichlids. Both front rays of the V somewhat elongated. D and An, particularly in the male, drawn out like threads at the back. Should be kept in spacious tanks with hiding places. 22–30 °C, higher for breeding. Omnivorous. A foam nest is built. Often kept in the aquarium is the following.

– *B. signata* (GÜNTHER, 1861); Combtail Paradise Fish. Only found in Sri Lanka (Ceylon). Up to 13 cm. Body moderately elongated. Ground colouring greenish, and occasionally dark-brown oblique bands. In older fish the colour becomes brick-red. The scales have a red-brown edge. Fins reddish, darker on the edges. C has blue rays and tip. There is a clear dark blotch at the base of the soft-rayed D. Female less conspicuous, fins less elongated. Eggs 1.2 mm. About 200 young, that all grow quickly (fig. 178).

Fig. 178 *Belontia signata*

Belontidae, F, *see* Anabantoidei

Belted Sandfish *(Serranellus scriba) see Serranellus*

Benthal, Benthic. The bottom zone of waters (in contrast to the free-swimming zone or pelagial*). It is split up vertically into a border zone or littoral* which is exposed to light and a deeper zone or profundal* which receives little light. Depending on the substrate (mud, sand, loam, gravel, stones, rock etc), characteristic living communities live there, which are described as benthos because of their common link with the bottom zone.

Benthos. Aquatic organisms which are bound to the bottom zone (*see* Benthal). They are either sessile, like most aquatic plants, sponges, bryozoans and many coelenterates, or they are free-moving (vagile) like the higher crustaceans, many insect larvae etc. Fish which have a close relationship with the bottom zone on account of their behavioural and reproductive biology may also be benthos (eg Flounders, and many Cichlids). The opposite to benthos are those forms of life that always live in the free-swimming zone (Pelagial*), such as plankton* and necton*.

Berg, Lejew Semjonowitsch (1876–1950). Important geographer, natural historian and ichthyologist. Born in Benderach in Russia. He studied zoology at Moscow University and managed matters to do with fishing on Lake Aral till 1903. As an extremely versatile natural historian he worked in numerous institutes in his homeland. From 1904 to 1914 he was a zoologist at the Zoological Museum of the Academy of Sciences in Petersburg and in 1914 he became Professor of Ichthyology at the Moscow Agricultural University. In the years that followed he became head of the Department of Applied Ichthyology at the State Institute for Experimental Agronomy and head of the Marine Department at the Hydrological Institute. From 1934 he worked as ichthyologist on fossil fish at the Zoological Museum of the Academy of Sciences of the USSR in Leningrad. In 1946 he became a member of the Academy of Sciences of the USSR. From his limnological studies on the lakes of Siberia, Lake Aral, Balkasch, Issyk-Kul, Ladoga etc he put together works about the fish of Russia: *The Fishes of Turkestan* (1905), *The Fishes of the Amur Basin* (1909), and the major work *The Freshwater Fishes of the USSR and Neighbouring Countries* (1948–49). In conjunction with his geological-geographical studies, Berg carried out palaeontological research on fish, which he published in his major work *The Systematics of Modern and Fossil Fish Types and Fishes* (1955). In addition to his work on phylogenesis and systematics, Berg tackled the problems of evolution, particularly anti-Darwinism.

Bering Wolf-fish *(Anarrhichas orientalis) see* Blennioidei

Bernhard's Hermit Crab *(Eupagurus bernhardus) see Eupagurus*

Berycidae, F, *see* Beryciformes

Beryciformes. O of the Osteichthyes*, Coh Teleostei. Called the Berycomorphi in the Regan system, these fish form a characteristic and very important group, which combines primitive and special features. They belong to the Sup-O Acanthopterygii which means that, like all higher bony fish, their fins possess spine-like hard rays along with soft rays, although the number of hard rays is relatively small. In non-Acanthopterygii, only soft rays are present. They may become like spines at the beginning of the fins, although true hard rays are rare. On the other hand, as in nearly all non-Acanthopterygii, part of the neurocranium is formed of an orbitosphenoid set of bones (which is usually absent in higher bony fish). The Beryciformes date back to quite an early time in the Earth's history, the oldest fossils being from the Upper Cretaceous, ie 100–120 million years ago. In the Tertiary, the O underwent vigorous development. The German term *Schleimkopfartige* (Mud head types) refers to depressions beneath the thin skin on the head, which are usually filled with mud. They are found in many fish belonging to the O. The body is mostly relatively tall to very tall and flattened laterally. Deeply cleft mouth, the upper edge of which is formed only of premaxillary bones. The presence of supramaxillae must be regarded as primitive. Most species live near the coasts of tropical and subtropical seas, where they are particularly common among coral reefs. Several species are found in deep sea (bathypelagic). 3 Sub-Os with 12 Fs in total. Only 4 Fs of the Sub-O Berycoidei are considered here.

– *Anomalopteridae*, Lantern Eyes (fig. 179). This is a small F with 3 species. They look a little like Barbs with a large bean-shaped organ of light beneath each eye. The light organs can be hidden by an eyelid-like fold (G

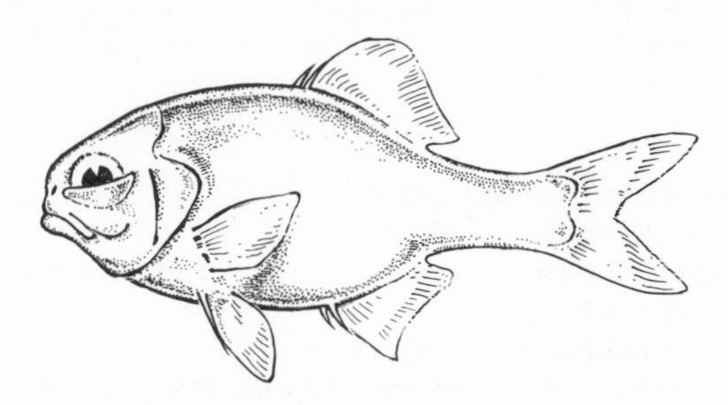

Fig. 179 Anomalopteridae-type

Photoblepharon) or they can be turned inwards by a special muscle (G *Anomalops*). In both cases the light is, as it were, switched off. The source of light is luminous bacteria which live in the gland canals of the light organs. As Lantern Eyes do not live in deep sea, the reason for these organs remained unclear for a long time. Today we believe they act as light lures to attract prey. But the way in which the lights can be turned on and off also suggests that they act as signals between like species. The light is very bright.

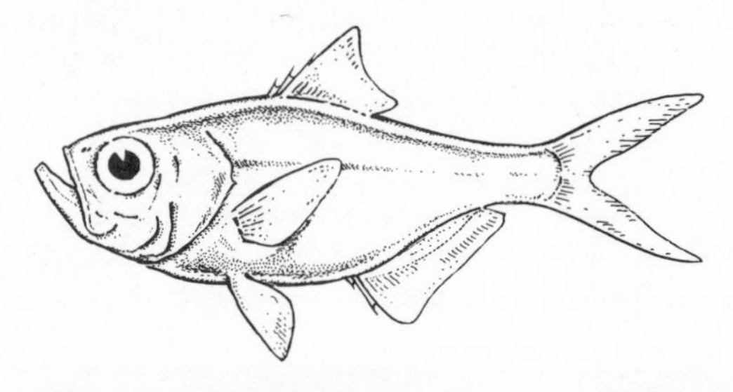

Fig. 180 Berycidae-type

– *Berycidae;* Berycoids (fig. 180). Very primitive F of the O Beryciformes. In body shape the Berycidae look like relatively tall Soldierfishes (*see* below) but they are distinguished from them by the slimmer caudal peduncle and only 4 spiny rays in the D. There are numerous mud depressions on the head and some species have spines above the snout. The F is distributed worldwide, but many species are quite rare. Most are reddish in colour. Red Alfonsino *(Beryx splendens)* is a relatively common species, growing up to 60 cm long and living at depths of 500–600 m. The Bercdidae have a very tasty flesh, but because most are rare they cannot be exploited commercially.

– *Holocentridae;* Soldierfishes or Squirrelfishes (fig. 181). A relatively uniform group of fish, characterised by a moderately tall, elongated, laterally compressed body. The height of the body remains roughly the same until the middle of the An, and then it suddenly narrows into a very slim, short caudal peduncle. Very

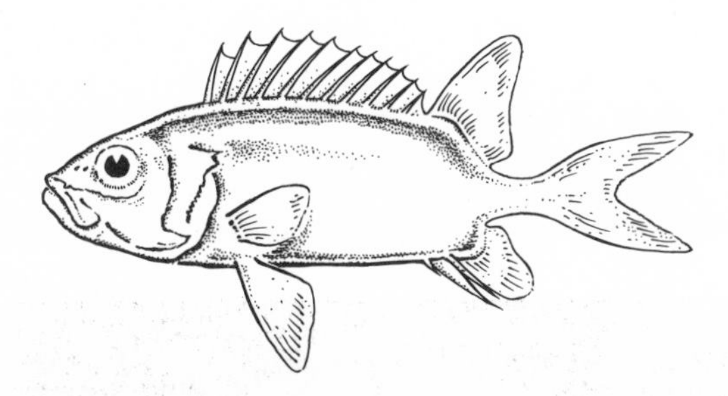

Fig. 181 Holocentridae-type

large eyes. The D consists of a long section supported by 11–12 spines and a short, soft-rayed section. The An contains 4 spines, the third of which is usually very strong. Various Ga have a long spine at the corner of the operculum. Heavy ctenoid scales on the head and body. A great number of species are coloured a lovely red. Most Holocentridae typically live on coral reefs or rocky coasts of tropical and subtropical seas. During the day they lie hidden in crevices and holes and only become active at night. They eat bottom-dwelling animals mainly. Some members of the G *Myripristis* have some economic importance. This is because they are the only ones to form shoals in open water, so can be fished efficiently. Soldierfishes are often kept in marine aquaria. For their needs and peculiarities turn to the Ga *Holocentrus, Myripristis.*

– *Monocentridae;* Pine-cone Fishes (fig. 182). Short, compressed fish whose large head is a third the size of

Fig. 182 Monocentridae-type

the oval-shaped body. Very big scales which together form a rough body armour. The dorsal fin spines stand separately in front of the soft-rayed part of the fin and are inclined alternately to the right and to the left. Beneath the chin are 2 luminous organs. The best-known species, the up to 16-cm-long *Monocentrus japonicus* (fig. 183) is found in the W Pacific and Indian Ocean far as E Africa. Mostly caught at depths of 100–200 m. Some Monocentridae are occasionally kept individually in aquaria.

Berycoidei, Sub-O, *see* Beryciformes

Berycomorphi, O, name not in current usage, *see* Beryciformes

Fig. 183 *Monocentrus japonicus*

Beryx splendens *see* Beryciformes

Betta BLEEKER, 1850; Fighting Fishes. G of the Belontiidae, Sub-O Anabantoidei*, Sub-F Macropodinae. SE Asia, Indo-China, Malaysian Peninsula, Sumatra, Java, Kalimantan (Borneo). Biotope varies with the species. Found in clear, weedy streams, small rivers and ditches, as well as marshy waters. From flat land up to 1500 m high in the mountains. These are small fish (5–11 cm) with an elongated body and slim build. LL reduced. Pointed, conical teeth in the jaws. D short, An long, and both may be elongated. P pointed but not elongated like a thread. In the wild, most forms have a brown ground colouring with often red and blue-green tones on the fins. The sides of the body often have metallic greenish speckles. In the aquarium these fish should be kept as pairs in well-planted, not too small tanks. Temperature 22–28 °C, depending on species, and there should be enough hiding-places for the female. The males are particularly aggressive amongst themselves. For centuries in Thailand *B. splendens* has been used in fish fights, during which 2 males are put together until one of them submits to the other. The species has been bred for this purpose for over 100 years. All fighting fish eat larger varieties of live food, in particular mosquito larvae, although they will also accept dried food. Before breeding also add *Enchytraeus* to the diet. The mouth-brooding species *B. brederi, B. picta* and *B. pugnax* lay their eggs in a trough in the sand. The bubble-nesting varieties can be placed in the smallest of containers for breeding. The ripe female need only be brought in for laying of the eggs. The male takes care of the brood, and should be removed from the breeding tank after 2–3 days. The tiny fry hatch after 24–40 hours. Rear them in a shallow depth of water, feeding at first with powdered food. The following make interesting and often colourful aquarium specimens.

– *B. bellica* SAUVAGE, 1884; Slim Fighting Fish. Malaysian Peninsula. Up to 11 cm. Very slim. Dark blue-grey to brown. Upper side of head marbled. There are some transverse bars on the body, only some being distinct. Scales on the flanks brilliant green at the front. P red, C and An red in part. Female less conspicuously coloured (fig. 184).

Fig. 184 *Betta bellica*

– *B. brederi* MYERS, 1935; Javan Mouth-breeding Fighting Fish. Java, Sumatra. Up to 8 cm. Body very robust, relatively tall. Middle rays of the C elongated. Ground colouring brown with some distinct irregular, pale transverse bands. Upper side of head and nape marbled. A dark longitudinal band runs from the mouth, across the eye and gill cover. Above it is a paler stripe. Mouth-brooders. The fry hatch from the large eggs after aout 40 hours.

– *B. fasciata* REGAN, 1909; Striped Fighting Fish. Sumatra. Up to 10 cm. Elongated, hardly compressed at all laterally. Middle rays of the C elongated. Body blue-black to dark blue-green. Some distinct transverse bands on the sides with rows of iridescent green spots. Fins dark. D and in part the upper section of the C have black speckles. Female has smaller, less elongated fins. Needs warmth, 26–28 °C.

– *B. picta* (CUVIER and VALENCIENNES, 1846); Javanese Fighting Fish. Sumatra, Java, Singapore, Bangka, Belitung. Found in Java at heights up to 1600 m. Up to 5 cm. Body elongated, only slightly compressed laterally. Brownish. 3 dark longitudinal bands, the longest of which (the middle one) stretches from the root of the tail to the snout. Fins colourless to yellowish with dark spots and rows of speckles. Male has a blue-black edge to his An. Female often has no markings on the fins. Does not require warmth. 22 °C, for breeding higher (26–29 °C). Mouth-brooder, with 60–130 per brood.

– *B. pugnax* (CANTOR, 1849); Mouth-brooding Fighting Fish, Penang Mouth-breeding Fighter. Penang Island, west of the Malaysian Peninsula. Up to 9 cm. Robust build. D inserted relatively far forwards – about midway between the head and the root of the tail. Colour variable. Grey-blue to red-brown with some distinct transverse bands and a dark longitudinal band, edged above and below in a paler colour. The scales all have metallic blue-green dots. In the male the throat, V and the edge of the An may be blue-green. Female always paler. Likes warmth to some degree, 22–25 °C. Mouth-brooders.

– *B. splendens* REGAN, 1909; Siamese Fighting Fish. Indo-China, above all in Thailand and the Malaysian Peninsula. In the wild grows up to 5 cm, otherwise up to 6.5 cm. Body elongated, D inserted far back and very broad. C rounded. Even the wild forms vary in colour. Mostly brown with a blue-green iridescence, but may appear red or blue when excited. Body has metallic, usually green dots, often in rows. Fins red-brown with a green pattern. The C has a fan-like pattern of green stripes. V red with white tips. Female a dirty brown with pale transverse bands and some faintly indicated longitudinal ones. Fins, small, greenish, tipped in pale red. A warmth-loving species, 25–28 °C. 400–500 eggs (fig. 185).

Long-finned varieties of *B. splendens* have been bred, and are available in a number of colours – red, blue, green, black, and multi-coloured. They are very short-lived fish. The males should be reared individually in small vessels lined up next to one another so the fish can see each other.

Fig. 185 *Betta splendens*, veil form

– *B. taeniata* REGAN, 1909; Banded Fighter. Sumatra, Kalimantan (Borneo). Up to 8.5 cm. D positioned nearer the head than the root of the tail. Middle rays of C elongated. Body brown with 2 dark longitudinal bands which unite into a large blotch at the root of the tail. The band running above the P begins at the snout, the one running below the P begins on the head. An and C have some black spots.

Bicarbonate Hardness, *see* Temporary Hardness

Bichir types, O, *see* Polypteriformes

Bichirs (Polypteridae, F) *see* Polypteriformes

Bicolor Blenny *(Escenius bicolor) see Escenius*

Bicolour Chromis *(Chromis dimidiatus) see Chromis*

Big-eye Squirrelfish *(Myripristis murdjan) see Myripristis*

Big-eyed Cichlid *(Acaronia nassa) see Acaronia*

Big-eyes, F, *see* Priacanthidae

Bigmouth Sucker *(Ictiobus cyprinella) see* Catostomidae

Big-spot Rasbora *(Rasbora kalochroma) see Rasbora*

Billfishes (Istiphoridae, F) *see* Scombroidei

Biocenesis. The biotic community of plants and animals in a biotope*. The way in which it is made up depends on the given ecological factors*. If the structure of the biotope and its biocenesis correspond with one another, there is a biological equilibrium*. The relationships between the organisms of a biocenesis rest essentially on the food system, ie the sum of all the food-chains that are bound up with one another: green plants – plant-eaters – small predators – large predators – scavengers (necrophage) – decomposers.

Whilst bioceneses rich in species scarcely show any fluctuation in population, those that have few species are very prone to sharp fluctuations. Monocultures (eg tanks containing fish for breeding or for fish foodstuffs) are particularly vulnerable to disturbances, such as diseases. The more variable the living conditions, and the more these conditions are suited to the majority of organisms, the greater the number of species in a biocenesis. The more extreme these environmental factors, the more characteristic, but also lacking in species, the biocenesis becomes. All the same, the species that then remain occur in large numbers.

Biochemistry. Biological discipline which studies living matter with chemical methods. Alongside the study of the nature of the material of which micro-organisms, plants, animals and people are made up, biochemists also carry out research into the chemical reactions of the life processes and the principles that regulate them. This can include a very broad spectrum of interest, from heredity, growth and the biological methods of acquiring energy to the specialised functions of particular cells or organs and the way they change during illness. Besides this, biochemistry may also provide information about the closeness of the relationship between organisms by comparing their chemical structure, eg their protein substances. Such information may be useful in taxonomy* and for research into evolution.

Biocybernetics is a modern branch of biology dealing mathematically with the control of functions and methods of information transfer within living beings (eg questions relating to heredity, the regulation of metabolism and the processes of learning and memory). Terms and methods used (eg the technique of calculation and imparting of information) have found a place in biology.

Biodemography, also Population Ecology. A branch of ecology*, which studies the correlation between environment* and population* as well as the norms of reaction of a population as a result of internal and external influences. The central questions with which biodemography is concerned include the periodical fluctuations in population density on account of variable rates of multiplication, changes in the relationship between predator and prey, different immigration and migration patterns of members of a population, varying food supplies etc.

Biogenetic Basic Laws. These laws state that during development (Ontogenesis*) every organism passes in quick succession through stages of its phylogenetic development. This does not mean that every part of the anatomy of an organism's ancestors is formed—it is restricted to a few characteristic features. However, important information about the development of different systematic groups during the Earth's history can be concluded from such features. For example, for a while, the embryos of amphibians, reptiles, birds, mammals and humans possess gill slits and a simple blood circulatory system like fish. This proves that all land vertebrates, including man, are descended from gill-bearing, and therefore aquatic, ancestors. In a lot of cases the larval form suggests a close relationship originally. This is true of annelids and molluscs, many of which form a ciliate larva or trachophore. In many organisms such links between ontogenesis and phylogenesis are no longer apparent, because the development has been modified owing to adaptation to the environment. Nevertheless, knowledge of the biogenetic basic laws is essential to explain natural relationships. Biogenetic basic laws offer one way of proving the accuracy of descent theory*.

Biogenic Decalcification. Term used to describe the process during which aquatic plants (under certain circumstances) extract chalk from the water which becomes deposited as a crust on the plants and can impair the plant's growth. It occurs both in nature and in the aquarium and is caused by a shift in balance from calcium-carbonic acid in the water to hydrogen carbonate.

$CaCO_3$	+ H_2O	+ CO_2	$Ca(HO_3)_2$
chalk (practically insoluble)	water	carbon dioxide	calcium bicarbonate solution (more easily soluble than chalk)

Obviously, there is a magnesium carbonate-carbonic acid balance too. This undesired shift in balance towards the virtually insoluble calcium or magnesium carbonate can be brought on by various factors.

1) An increase in concentration of those substances that cause carbonate hardness as a result of water evaporation, and making up the deficit with water also containing carbonate hardness substances.

2) Insufficient aeration, therefore not enough CO_2 got rid of.

3) High temperatures, which have an adverse effect on the ability of the water to dissolve CO_2.

4) Too high a concentration of chalk in the substrate, which primarily increases the hydrogen carbonate concentration and brings about a shift in balance to the left.

5) Too much vegetation (coupled with deficiency, insufficient water movement and high carbonate hardness), which as a result of assimilation can reduce the CO_2 concentration so much that the carbonates are precipitated.

The problem can be helped by using soft water (make up the deficit of any evaporated water with distilled or completely de-salted water), by installing an efficient aerator, by lowering the temperature and by using a more suitable substrate ie one containing less calcium).

Biogeography. The study of the distribution of different groups of organisms, divided into plant geography (geobotany*) and animal geography (zoogeography*). Chorology is largely a descriptive science within biogeography and concerns the distribution ranges (*see* Areal) of groups of organisms. In addition, ecological and historical biogeography tries to discover causal connections between areal* and ecology* or between areal and historical-geographical events (eg ice ages, continental drift, mountain folding).

Bio-indicators. Organisms, whose presence or absence enables conclusions to be drawn about chemical and physical (climatic) situations. Bio-indicators are very important in the characterisation of biotopes* and environmental pollution.

Biological Equilibrium. Balanced relationship between conditions of the living space (*see* Biotope) and its biotic community (*see* Biocenosis). Numerous environmental factors, for example the pH value, salt content, oxygen content and sunshine factor in waters, create the living conditions for organisms which are bound together in complex food chains. As a result, plants (including plant plankton*) are directly or indirectly the basis of fish nutrition. If there becomes a shortage of food in the water it will cause a decline in the fish population.

Usually, a biotope is continually subject to minor changes, eg climatic influences, but the biological equilibrium always reasserts itself. This is also often true of artificially created changes, such as pollution of the water through industrial waste. However, if the waste builds up to such a degree that one of the interdependent organism communities can no longer survive, the biological equilibrium is lost. In the aquarium there is no biological equilibrium, it can only be approximated. Even in the so-called biotope aquarium, the simplest action such as feeding makes it impossible for a self-regulating equilibrium to exist.

Biological Filter *see* Filtration

Biology. The science of life, whose aim is the study of all the characteristics of living matter. The studies of micro-organisms, plants, animals and people are called micro-biology, botany, zoology and anthropology respectively. These 4 fields of research have similarly been broken down into a whole range of individual sciences. Morphology*, biochemistry*, physiology*, genetics*, taxonomy*, ecology*, ethology* and palaeontology* are just some of these specialist areas. They tackle biological research with different methods and from very different viewpoints, but they do have much common ground. Biology was a total science in the nineteenth century, when MATTHIAS SCHLEIDEN and THEODOR SCHWANN discovered the cell, GREGOR MENDEL* created the basis of the heredity laws, and CHARLES DARWIN* helped bring about the recognition of descent theory*. The methods of research in the natural sciences were particularly improved at that time. Biological experiment came to count for as much as simple description of the objects under study. Modern biology to a large extent makes use of mathematics, physics and chemistry, as in biochemistry*, biocybernetics* and biometry*. Research is not just aimed at the discovery of basic principles, it is very much concerned with the solving of practical problems, eg regarding medicine, the production of food and environmental pollution.

Biometry. Biostatistics. Range of application of mathematics in biology* and related fields. Biometry provides methods with which observations and experiments can be carried out on a mathematical basis. The results can also be interpreted mathematically. So, for example, the minimum number of necessary measurements in an experiment can be determined. By taking measurements, recording weights, or taking counts, exact information can be given regarding the mean value, the significance of any differences that have been observed or the variability of a feature. Decisions can be facilitated by probability calculation. The risk of error should be calculated and therefore avoided. Biometry has become an indispensable scientific aid in all the disciplines of basic biological research and application, eg medicine and food production.

-biont. Last syllable of a word which denotes that a species is bound to a particular biotic or abiotic environmental factor. Therefore, the Brown Trout *(Salmo trutta fario)* for example, or many coral fish, are described as polyoxybiont because they need very oxygen-rich water (poly = much; oxy = short for oxygen). The Brine Shrimp *Artemia* can only develop in salt-rich water, and so is halobiont (halos = salt).

Biostatistics *see* Biometry

Biotope. Well-definable living spaces of a biotic community (*see* Biocenesis), eg a mountain stream, a savanna pool, or a coral reef. The structure of the biotope has vital influence upon the composition of the biocenesis, as well as the other way round (*see* Biological Equilibrium). If a biotope has many separate units and offers the most favourable living conditions (for the majority of the organisms), it tends to be rich in species (eg coral reefs). In contrast, a biotope with extreme living conditions is usually home for only a few species that are adapted in characteristic ways (eg a mountain stream), or it has a small amount of flora and fauna (eg a brackish water biotope or a high moor pool). Biotope and biocenesis together form the ecosystem.

Bipinnaria larva *see* Asteroidea

Bird Wrasses, G, *see Gomphosus*

Birds, Cl, *see* Aves

Birgus latro *see* Paguridea

Bishop Fish *(Brachyrhaphis episcopi) see Brachyrhaphis*

Bispira KROYER, 1856. G of the sessile Bristle Worms (Polychaeta*). Found in warm seas. Paired crown of tentacles and a short, soft tube-body. Bispiras have simple eyes on the tentacles and react to light stimuli

Fig. 186 *Bispira volutacornis*

by retreating quickly into their tube. For specimens in the aquarium *see Spirographis.* The following is a species commonly kept.

– *B. volutacornis* MONTAGU. Mediterranean, from 2 m deep on a firm substrate. Diameter of crown of tentacles 3–4 cm. White or greenish with dark eyes (fig. 186).

Bitter Cress *(Cardamine lyrate) see Cardamine*

Bitterling *(Rhodeus sericeus amarus) see Rhodeus*

Bitterlings, Sub-F, *see* Cyprinidae

Bivalves, Cl, *see* Bivalvia

Bivalvia; Bivalves (fig. 187). A class of molluscs (Mollusca*) with a great many species and forms (25,000). They are without exception water-dwelling animals, most of them being marine. The body is bilaterally symmetrical, with no head, and covered by two valves. The edges of the valves are almost always near the point of concrescence (the ligament) and their hinge consists of teeth which stops the dislocation of the valves against one another. The structure of this hinge is an important systematical feature. The animal's foot extends from the two valves. Bivalves burrow into the ground with this foot, and move slowly along, although many species are sessile (*Mytilus**, *Dreissena**). Types

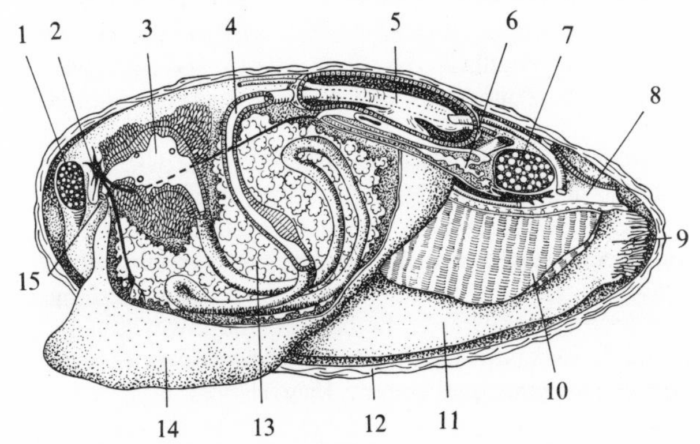

Fig. 187 Bivalvia. Anatomy of a Bivalve. 1 Anterior adductor muscle 2 Head ganglion 3 Stomach 4 Intestine 5 Intestine continuation in the heart 6 Kidney 7 Posterior adductor muscle 8 Cloaca 9 Respiratory opening 10 Gill 11 Mantle 12 Shell 13 Gonad 14 Foot 15 Mouth

that burrow deeper possess much elongated incurrent and excurrent siphons (for the transport of respiratory water, food and excrement). The gills inside bivalves have several functions. They produce a strong current of water with their cilia which filter out any microorganisms and detritus* for food. Besides this, in bivalves that live in cold seas as well as in freshwater bivalves, the egg development takes place in the gills. In *Anodonta** and *Unio** the eggs develop as far as the parasitical stage known as the *glochidium,* in other species they develop into complete small bivalves. In contrast to this, bivalves from warm seas and the G *Dreissena* pass through a planktonic veliger larval stage. Many bivalves are hardy in the aquarium. However, the large freshwater species *(Anodonta, Unio)* should only be kept as an object of study in specially arranged tanks (exception, the rearing of Bitterlings, *see Rhodeus*).

Bivibrachiinae, Sub-F, *see* Hemiodidae

Black Acara *(Aequidens portalengrensis) see Aequidens*

Black Anemonefish *(Amphiprion sebae) see Amphiprion*

Black Angelfish *(Pomacanthus arcuatus) see Pomacanthus*

Black Clown *(Amphiprion sebae) see Amphiprion*

Black Corals, O, *see* Antipatharia

Black Echinodorus *(Echinodorus parviflorus) see Echinodorus*

Black Flies, G, *see Simulium*

Black Goby *(Gobius niger) see Gobius*

Black Line Tetra *(Hyphessobrycon scholzei) see Hyphessobrycon*

Black Marlin *(Istiompax marlina) see* Scombroidei

Black Molly *(Poecilia sphenops) see Poecilia*

Black Neon Tetra *(Hyphessobrycon herbertaxelrodi) see Hyphessobrycon*

Black Paradisefish *(Macropodus opercularis concolor) see Macropodus*

Black Phantom Tetra *(Megalamphodus megalopterus) see Megalamphodus*

Black Sea. An inland sea between Europe* and Asia*, joined to the Mediterranean* via the Bosphorus and the Dardanelle Straits, which means that is is connected in this way to the world's oceans. For lengthy periods of time, however, the Black Sea has had no maritime connection, and at those times it was a largely edulcorated inland sea. This has given the sea a brackish water character that is still much in evidence today, eg through the lack of species that live there, the absence of stenohaline (Steno-*) marine creatures such as Corals, Cephalopods (Cephalopoda*), Echinoderms (Echinodermata*) on the one hand, and on the other the presence of Black Sea brackish and freshwater species such as *Dreissena**, various *Acipenser** species and *Lucioperca* marina*. The present connection to the Mediterranean only came about around 10,000 years ago. Nevertheless, the dominant elements in the Black Sea fauna today are migrants from the Mediterranean – eg among the fish, species of *Blennius**, *Crenilabrus**, *Mugil** and *Hippocampus**; plus, from the invertebrates, certain Jellyfishes, Sea-anemones, Hermit crabs and many Molluscs (Mollusca*). The formerly typical brackish water and freshwater species are now concentrated along the NW and N coasts (the Danube delta, the Sea of Azow). Another important feature of the Black Sea is that hydrogen sulphide builds up at depths of around 150 m downwards, which means that no animal life can survive.

Black Sea-urchin *(Arbacia lixula) see Arbacia*

Black Shark *(Morulius chrysophekadion) see Morulius*

Black Tetra *(Gymnocorymbus ternetzi) see Gymnocorymbus*

Black Triggerfish *(Odonus niger) see Odonus*

Black Triple-tail *(Lobotes surinamensis) see* Lobotidae

Black Water *see* Dystrophic Waters

Black Wedge Tetra *(Hemigrammus pulcher) see Hemigrammus*

Black Widow *(Gymnocorymbus ternetzi) see Gymnocorymbus*

Black and Gold Angelfish *(Centropyge bicolor) see Centropyge*

Black and White Damsel *(Dascyllus aruanus) see Dascyllus*

Blackamoor *(Gymnocorymbus ternetzi) see Gymnocorymbus*

Black-backed Anemonefish *(Amphiprion ephippium) see Amphiprion*

Black-banded 'Pyrrhulina' *(Copella metae) see Copella*

Black-banded Sunfish *(Enneacanthus [Mesogonistius] chaetodon) see Enneacanthus*

Black-barred Limia *(Poecilia nigrofasciata) see Poecilia*

Black-barred Livebearer *(Quintana atrizona) see Quintana*

Blackbellied Limia *(Poecilia melanogaster) see Poecilia*

Black-blotched Turretfish *(Tetrosomus gibbosus) see Tetrosomus*

Black-blotched Wrasse *(Halichoeres gymnocephalus) see Halichoeres*

Black-blotched Wrasse *(Halichoeres marginatus) see Halichoeres*

Black-dotted Trunkfish *(Ostracion tuberculatus) see Ostracion*

Black-eyed Thicklip *(Hemigymnus melapterus) see Hemigymnus*

Black-fin Goodea *(Goodea atripinnis) see Goodea*

Black-finned Pearlfish *(Cynolebias nigripinnis) see Cynolebias*

Black-finned Triggerfish *(Paraluteres prionurus) see Paraluteres*

Black-ruby Barb *(Barbus nigrofasciatus) see Barbus*

Black-saddled Leatherjacket *(Oxymonacanthus longirostris) see Oxymonacanthus*

Black-spotted Boxfish *(Ostracion tuberculatus) see Ostracion*

Black-spotted Corydoras *(Corydoras melanistius) see Corydoras*

Black-striped Rasbora *(Rasbora taeniata) see Rasbora*

Black-tailed Footballer *(Dascyllus melanurus) see Dascyllus*

Black-tailed Humbug *(Dascyllus melanurus) see Dascyllus*

Black-winged Hatchet Fish *(Carnegiella marthae) see Carnegiella*

Bladder Snails, Ga, *see Physa* and *Ampullaria*

Bladderwort, G, *see Utricularia*

Bladderworts, F, *see* Lentibulariaceae

Bladderwrack *(Fucus vesiculosus) see Fucus*

Bleak *(Alburnus alburnus) see Alburnus*

Bleeding Heart Tetra *(Hyphessobrycon erythrostigma) see Hyphessobrycon*

Bleeker, Pieter (1819–78). Dutch natural historian and doctor. For many years he lived in India and Java where he built up extensive collections of fish. A large number of first descriptions were published by him. His 9-volume *Atlas icthyologique des Indes Orientales Néerlandaises, publié sous les aspices du Gouvernement colonial néerlandais* ('Ichthyological Atlas of Dutch Oriental India, published under the auspices of the Dutch Colonial Government') which appeared in Amsterdam in 1862–77, was the first comprehensive survey of the fish of SE Asia.

Blennies, related species, Sub-O, *see* Blennioidei

Blennies, Scaled (Clinidae, F) *see* Blennioidei

Blenniidae, F, *see* Blennioidei

Blennioidei. Sub-O of the Perciformes*, Sup-O Acanthopterygii. These fish live on the bottom of the sea from the tropics to the Arctic. Elongated to eel-like form, with Vs positioned at the throat to the chin. They are much reduced, and occasionally totally absent. D usually uniform in height and very long. As a rule, it begins at the nape and reaches almost to the C. An similarly long. In some Fs the D, C and An form a uniform fin corona. The Ps are of normal shape, in part much reduced in the eel-like members of the O. The Blennioidei live primarily in shallow waters near the coast where there are lots of places to hide. A few types are found in deeper waters, and there are one or two freshwater forms. Their way of life, feeding and reproduction are very variable. A few fossil forms from the Neogene belong to this Sub-O. As in the system used here which was devised by GREENWOOD and colleagues (1966), the Zoarcidae* are discussed under another entry.

– *Anarrhichadidae;* Sea Wolf-fishes (fig. 188). Large Blennioidei from northern cold seas, these types have a huge, compressed head and an elongated body. Mouth very large, always slightly open, and surrounded by thick lips. In the front of the jaws there are sharp fang-like teeth, and towards the back of the lower jaw and along various bones at the base of the cranium there are strong crushing teeth. Very long D beginning at the nape. Hard rays weak and flexible. An long, Ps large and Vs missing. Small scales embedded deep in the thick skin. Of the 9 species, almost all live on the sea bottom at depths of 20–50 m, where they feed on crabs and

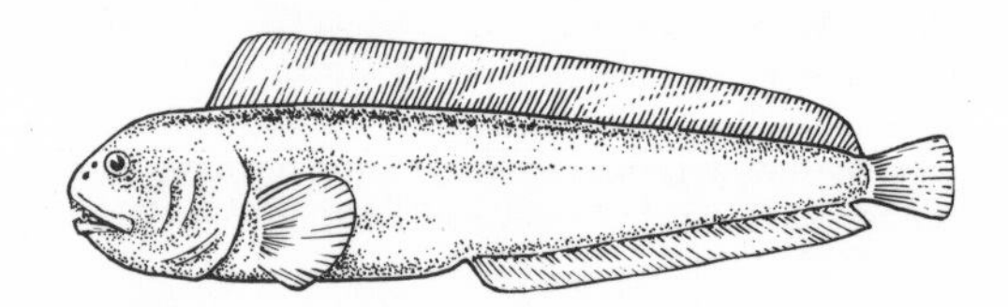

Fig. 188 Anarrhichadidae-type

Fig. 189 *Anarrhichas lupus*

shellfish. The large eggs are laid in deeper water and guarded for a while by the male. Several species are commercially exploited, in particular the up to 125 cm long Atlantic Sea Wolf *(Anarrhichas lupus)* (fig. 189) which is often sold (minus the head) as fish cutlets. Also known as Catfishes, the blue-grey Atlantic Sea Wolf sometimes wanders into the Baltic Sea. The more elongated Thin-tailed Sea Wolf or Bering Wolf-fish *(Anyrrhichas orientalis)* occurs in the Pacific. Because of their unusually strong bite, Sea Wolf-fishes are handled with extreme care by fishermen.

– *Blenniidae;* Scaleless Blennies (fig. 190). Small, elongated fish, laterally only moderately compressed. Eyes positioned high up on a bulldog-like head. Small mouth usually with small, densely packed maxillary teeth. A few parasitical species possess a tusk-like canine tooth

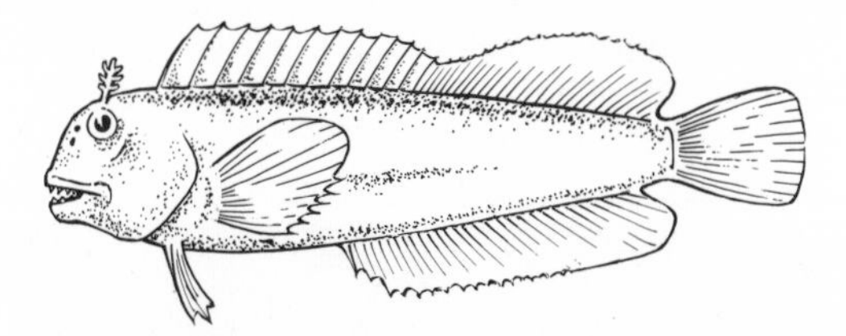

Fig. 190 Blenniidae-type

in both the right and left side of the lower jaw (fig. 191), but they are only visible when the lower jaw falls down fully. Several species have a frontal comb or outgrowths between the eyes which have the shape of a feather or paintbrush tufts. D is long and even the spinous rays are soft and flexible. The fin is uniform in appearance or curving inwards near the middle, or it may have a short anterior section separated from the main part of the fin. An long, Vs positioned at the throat. A thick skin with a great many mucous glands, but lacking scales. Most Blenniidae live among reefs and in stony, shallow waters. There are a few freshwater species, eg *Blennius fluviatilis* which is found in Europe south of the Alps.

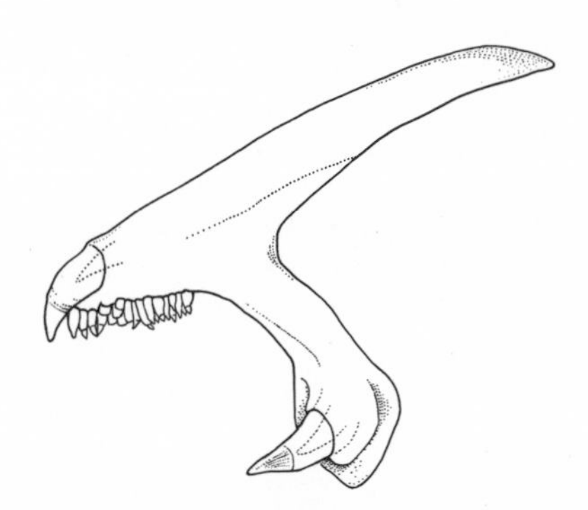

Fig. 191 Teeth of *Aspidonotus*

All the species hide away in places of various sorts. The males are often bigger and have a stronger colouring. Many Blenniidae can survive long periods out of the water, and some search for food in the intertidal areas. *Dialommus fuscus* lives in the Galapagos Islands in perpendicular holes. It has eyes like those of the Anablepidae*, ie divided by a septum. However, in *Dialommus fuscus* this partition runs vertically which enables the fish to see above the water with the front half of the eye and below it with the rear half, as it pushes itself out of its hole. Many Blenniidae make interesting aquarium specimens. Of the 13 Ga, 4 are treated under separate entries where you will also find information about their biology: *Aspidonotus, Blennius, Petroscirtes, Runula.*

– *Clinidae;* Scaled Blennies (fig. 192). Primarily found in warm seas in the southern hemisphere, these fish are small to medium-sized, similar to the Blenniidae. Body elongated, head compressed or pointed. Mouth usually small, surrounded by thick lips, teeth may vary. Many Ga have short tentacles between the often highly mobile eyes. *Cirrhibarbis* has barbs on the snout and chin. D very long, chiefly supported by hard spines. It is uniform in appearance or curving inwards well to the rear, and in some Ga a short anterior section is separated from the rest of the fin. C rounded, An long, Vs situated at the throat and much reduced. Mucilaginous skin with small scales or without scales. The family comprises many species and several of them are livebearing. During the breeding season, the males possess a conical, forwards-facing copulatory organ located in front of the An. The Clinidae live primarily in the intertidal zone and

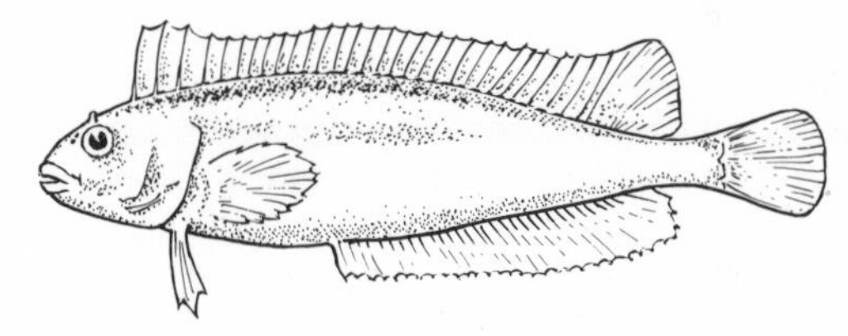

Fig. 192 Clinidae-type

hide under stones, among corals or seaweeds. Some live in estuaries and a few in deep water. The majority have a pretty colouring, often with mother-of-pearl blotches. In a lot of cases males and females have very different colouring. The largest species is *Heterostichus rostratus* which lives in the tropical eastern Pacific and can grow up to 60 cm long. Closely related to the Clinidae are the Tripterygiidae, which are the only F of the Sub-O Blennioidei to have 3 separate dorsal fins *(see Tripterygion).* Many Clinidae are well-known aquarium fish. See under *Cristiceps* and *Ecsenius,* where you will also find biological details.

– *Pholididae;* Gunnels (fig. 193). These live in northern and temperate seas and are very similar to the next group of fish to be described, the Stichaeidae. The differences between them lie in the anatomical features of the skeleton. D, C, and An are usually united as one, the Vs are absent or consist of 1 spine and 1 soft ray. LL short or absent. The Pholididae also live hidden beneath stones or among carpets of seaweed. They can

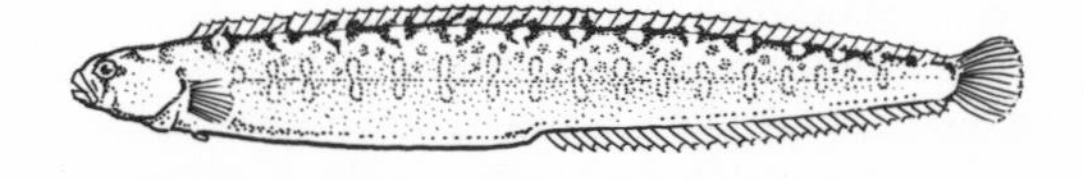

Fig. 193 Pholididae-type

survive for 2–3 days in the intertidal zone living under washed-up sea grass, for example. The eggs are laid in bundles on the bottom and are guarded by the female. They hatch after about 2 months and the fry are pelagic, drifting over a wide area. One species, the Butterfish or Gunnel *(Pholis gunellus),* lives in the shallow water of the eastern and western North Atlantic. It grows up to 30 cm long. Economically the Pholididae are of little importance.

– *Stichaeidae* (fig. 194). Very elongated, ribbon-like Blennioidei from northern temperate and cold seas. In some Ga the head has a frontal comb, there are

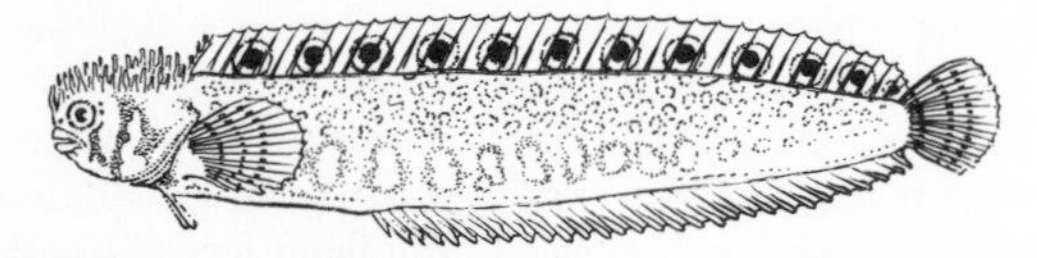

Fig. 194 Stichaeidae-type

branched tentacles between the eyes or moss-like carpets of tentacles along the entire top of the head to the dorsal fin. The D itself is very long, only supported by spines. Occasionally, like the long An, it is fused with the rounded C. Vs positioned at the throat or absent. LL indistinct or incomplete or divided into 3 main branches which may have side branches extending vertically from them. The LL is rarely absent. The Stichaeidae live near the bottom in shallow, coastal waters, although a few have been caught at greater depths. They eat crustaceans and other invertebrates mainly. *Chirolophis ascanni,* which can grow up to 30 cm long, is found from the Arctic waters to the Channel area at depths of 30–100 m. The eggs are guarded by the female and the fry are pelagic, drifting across

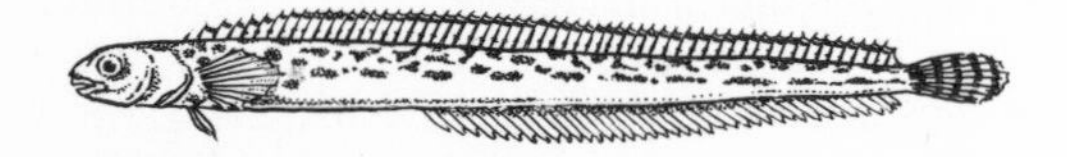

Fig. 195 *Lumpenus* sp.

great distances. The largest species is the 75-cm-long *Cebidichthys violaceus* which lives along the coast of California. The Snake-eels of the G *Lumpenus* (fig. 195) also belong to this family. One species, *Lumpenus lampretiformes,* from the Barents Sea, has also been caught in the Baltic from time to time.

Blennius Linnaeus, 1758. G of the Blenniidae, Sub-O Blennioidei*. There are numerous species found along the coasts of warm seas, and a few in fresh water. They are small, lively, bottom-dwelling fish which are among the most common forms found in the upper rocky littoral along the southern European coasts. Elongated fish with no swimbladder, they have to make constant winding movements when swimming in open waters. They support themselves on the bottom with their Vs which are located at the throat. D and An are both long. There are various kinds of outgrowths on the head, usually taking the form of tentacles above the eyes. They are usually more noticeable in the male. Many species have beautiful colouring. They eat small animals, although some prefer plant food, while others are fairly predatory. *Blennius* is territorial and the males guard the eggs in rock crevices. These fish are highly recommended for the aquarium and are easy to care for. Courtship and spawning may well be observed. The following are well-known species of the southern European coasts.

– *B. galerita* Linnaeus, 1758. Atlantic to the Black Sea. Up to 8 cm. There are unpaired rows of bristles on the head and folds of skin seem to make the mouth slit look bigger. Patterned in grey-white.

– *B. pavo* Risso; Peacock Blenny. E Atlantic to the Black Sea. Up to 12 cm. Mature males have a helmet-like head-dress. An olive colour with bluish-white flecks.

– *B. sphinx* Valenciennes, Sphinx Blenny or Mudfish. Mediterranean, Black Sea. Up to 8 cm. Yellowish to olive-brown oblique bands and blue iridescent streaks. Small eye tentacles (fig. 196).

Fig. 196 *Blennius sphinx*

– *B. tentacularis* Brünnich. E Atlantic to Black Sea. Up to 15 cm. Males have high eye tentacles. Brown marble colouring. Robust, predatory species.

Blepharoceridae (fig. 197). F of midges (Nematocera*) comprising about 170 species in total, about 28 of which are found in Europe. With the exception of the imago phase, the Blepharoceridae are totally aquatic. They are typical creatures of the stream region (*see* Rhithral). The adult insects resemble mosquitoes *(Culex*, Aedes*)* but with longer legs and a system of small folds (omentum) alongside the wing venation. The

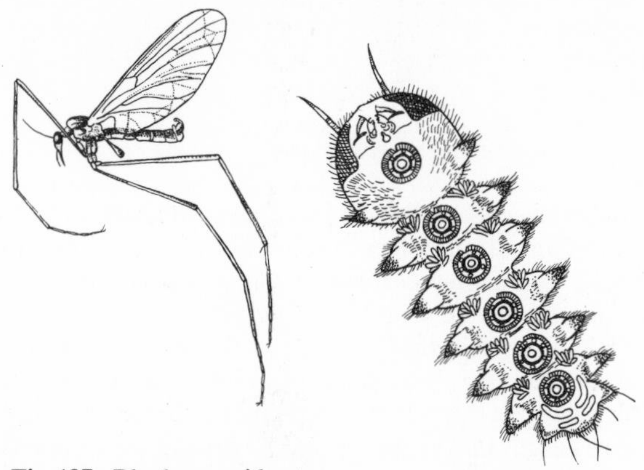

Fig. 197 Blepharoceridae-type

vae could never be confused with that of any other on account of the 6 characteristic suckers arranged one after the other. These stop the larvae drifting away even in a strong current. Although the Blepharoceridae are of little interest to the aquarist, they command our attention in any survey of a mountain stream, which is why they are included here.

Blicca HECKEL, 1843. G of the Cyprinidae*, Sub-F Abraminae. Europe, N Asia, Trans-Caucasia. Large fish that live in lakes and large rivers, near the bottom. Sociable during the breeding season. The body is tall, laterally much compressed, similar to *Abramis* brama*. Ground colouring grey-green, the paired fins being pale reddish. Their food consists of bottom-dwelling insect larvae, crustaceans and molluscs, as well as plants. Keep in a large, cool tank with a sandy substrate. Temperature not above 18–20 °C. Only one species:

– *B. bjoerkna* (LINNAEUS, 1758); Silver Bream. Europe, north of the Alps and Pyrenees, SE Scandinavia. Up to 30 cm. In places it is a prized food fish, but contains a great many bones (fig. 198).

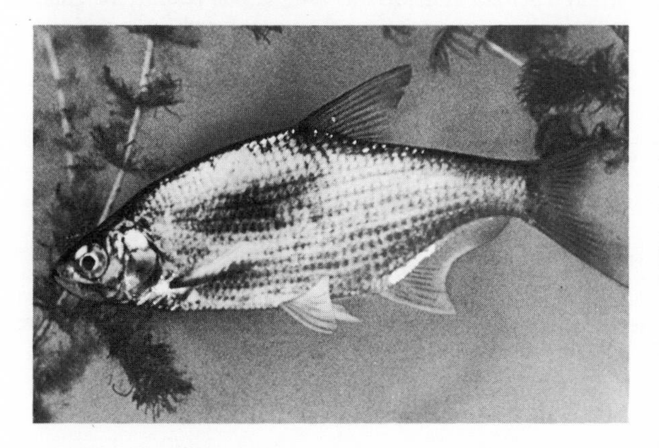

Fig. 198 *Blicca bjoerkna*

Blind Cave Barb *(Caecobarbus geertsi) see Caecobarbus*

Blind Cave Fish *(Astyanax mexicanus) see Astyanax*

Blindfishes (Amblyopsidae, F) *see* Percopsiformes

Blindness *see* Eyes, Diseases of the

Bloch, Marcus Elieser (1723–99). Ichthyologist, born in Ansbach, Bavaria. Studied medicine in Berlin and worked there as a surgeon for many years. At the same time he pursued natural history. He only began to study fish fauna at a relatively late date. Between 1782 and 1784 in Berlin he published his observations on native fish under the title *Oeconomische Naturgeschichte der Fische Deutschlands* ('Economic Natural History of Germany's Fishes'). It represents Bloch's most important work, combining scientific thoroughness with a high aesthetic value on account of its unique copperplate engravings. The work achieved worldwide recognition. In 1785 the first volume of *Naturgeschichte der Ausländischen Fische* ('Natural History of Exotic Fishes') appeared, to be followed by 8 more volumes up to the year 1795. This work contains numerous first descriptions of many of the aquarium species popular today, such as *Corydoras punctatus* (BLOCH, 1794), *Hoplias malabaricus* (BLOCH, 1794), *Leporinus fasciatus* (BLOCH, 1794), *Etroplus maculatus* (BLOCH, 1795). However, the work remains a fragment. After his death, the systematical work *M. E. Blochii Systema ichthyologiae iconibus CX illustratum* was published by J. G. SCHNEIDER in 1801. Bloch's fish collection forms a part of the collections of the Zoological Museum in Berlin.

Blood. Circulating body fluid which serves to transport substances. It is a participant in several of the body's processes: eg respiration*, nutrition*, the regulation of bodily functions with hormones* and defence against diseases and toxins. It also plays a part in sealing a wound, through coagulation. The basic component of a vertebrate animal's blood is the watery *blood plasma*, which contains primarily salts, proteins and sugar. The *blood cells* are transported in it round the body. They consist of red and white corpuscles (erythrocytes and leucocytes) and platelets (thrombocytes) which produce the coagulation of the blood. Blood gets its colour from *blood pigments*. In vertebrates a substance called haemoglobin can increase the capacity of the red blood corpuscles to transport oxygen and carbon dioxide through chemical combining. Crustaceans and many molluscs possess a blue pigment. The components of blood are constantly renewed. Erythrocytes form primarily in the bone marrow; they are broken down in the liver (*see* Digestive organs) and the spleen. Any change in the composition of the blood, in particular an increase in the amount of leucocytes, can give important indications about diseases.

The blood of vertebrates is in principle the same, although the fully developed erythrocytes of mammals do not contain a nucleus. In fish the amount of haemoglobin and the number of erythrocytes (and therefore the oxygen capacity) is less than in the mammals. Selachians and marine bony fish have a higher content of salt and protein in the blood plasma than freshwater fish, on account of the relative concentration of the surrounding water. Migrating fish which leave the fresh water or return to it do not alter the concentration of their blood fluid.

Blood, Examination of. Examinations of the blood can give an early warning of disease in an organism. For example, the early diagnosis of stomach dropsy* in the rearing of carps is possible through the examination of blood serum. However, this can only be carried out in special laboratories. A blood deficiency can be diagnosed by the layman too – the fish will look paler in colour, and it could be a concomitant symptom of stomach dropsy, gill parasites (*see* Gills, Diseases of the), or blood parasites*. The well-equipped aquarist may be able to prove the existence of blood parasites with a microscopic examination. The fish are anaesthetised (*see* Anaesthetics), dried thoroughly, and then an incision is made with a knife at the caudal peduncle. A few drops of blood are caught on a microscope slide. In uncoloured preparations, with great magnification, it may be possible to recognise the highly mobile, single-celled members of the Ga *Trypanosoma** and *Hexamita**. But, in general, microscopic blood examinations are not possible for the layman to carry out, nor are they worthwhile.

Blood Flagellates (fig. 199). There are numerous Flagellata* which live as parasites in a fish's blood. The single flagellum species of the G *Trypanosoma* GRUBY, 1843, occur quite often in marine and freshwater fish but are not pathogenic. For example, a high percentage of eels are stricken with *Trypanosoma granulosum* LAVERAN and MESNIL. The species of the G *Cryptobia* LEIDY, 1846 (*Trypanoplasma*, name not in current usage), which all have 2 flagella, cause much harm to fish,

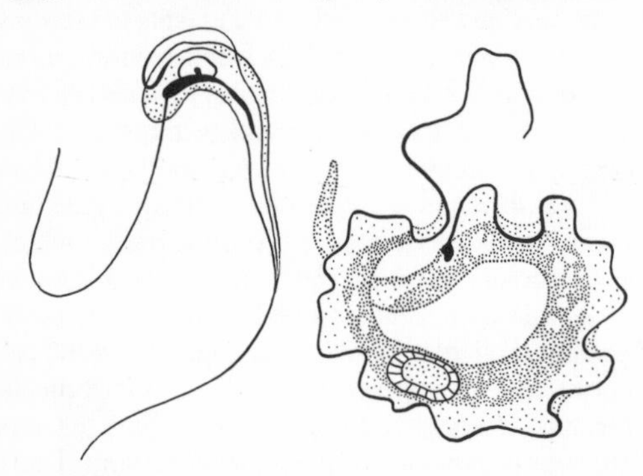

Fig. 199 Blood flagellates. Left, *Cryptobia;* right, *Trypanosoma*

eg extreme emaciation, sluggishness, sunken eyes, pale gills. In severe cases, the fish may die. The propagation of both Ga is only possible in leeches (*see* Hirudinea), which also transfer the parasites to the fish. Aquarium fish are rarely stricken with blood flagellates. Diagnosis can be made from microscopic examination of blood smears (*see* Blood, Examination of), but there is no known treatment.

Blood Fluke *(Sanguinicola) see* Trematodes

Blood Parasites. Characteristic of the blood parasites which attack fish are members of the Ga *Trypanosoma* and *Cryptobia* (*see* Blood Flagellates); *Sanguinicola** and *Hexamita**. Many Bacteriophyta* and Sporozoa* may also live in the blood, some permanently, others for certain periods. Numerous parasites are only found in the blood during particular stages of development and their presence is difficult to prove. It is only in later stages that they live as parasites in a particular organ system, such as various Nematoda (*see* Nematode Diseases), Trematoda*, Cestoda*, Sporozoa* etc.

Blood Vascular System. In many invertebrates (insects, molluscs) blood circulates through gaps in the tissue and through the body cavity. In others (annelids) and in all vertebrates, blood is contained inside a closed vascular system. This system determines the route of the blood in the body and is the basis of the direction and regulation of its circulation (fig. 200). Pumped by the heart, it flows through muscular *arteries* to a network of fine *capillaries* where oxygen exchange with the tissues* takes place. The capillaries open out into a system of *veins* which take the blood back to the heart again. The respiratory organs are linked in with this system in different ways. In fish the gills are a stage along the route of the simple circulation (heart–gills–body–heart). In vertebrates with lungs a separate lung circulation develops in which the blood flows once more to the heart before it is pumped into the body (heart–

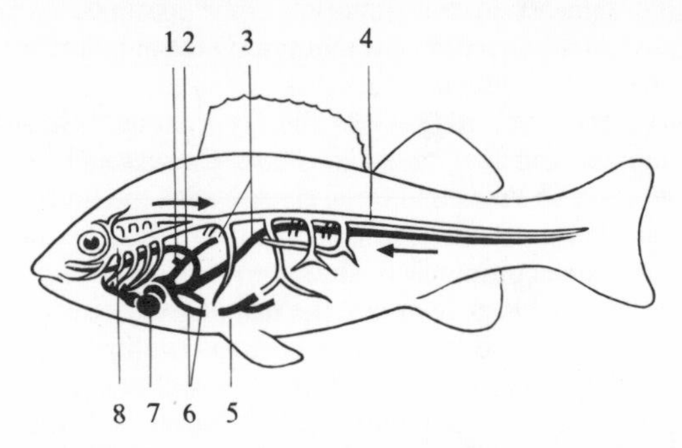

Fig. 200 Blood circulation. 1, 2 Cranial veins 3 Cardinal veins 4 Aorta 5 Portal vein 6 Hepatic veins 7 Heart 8 Gill blood vessels

lungs–heart–body–heart). The heart (fig. 201) of a fish lies in a pericardium on the ventral side behind the gills. It can be compared with an S-bend tube which is divided into different parts. The first chamber, the *Sinus venosus*, collects the blood flowing from the body. From there it passes into the atrium. Here, rhythmical muscle contractions from the neighbouring heart chamber cause the blood to be sucked through and forced into the system of arteries. Valves lying between the different heart sections prevent the blood flowing the wrong way. The blood flows first through the arteries and the gill arch arteries into the capillary network of the gills. Here it is enriched with oxygen. One lot of

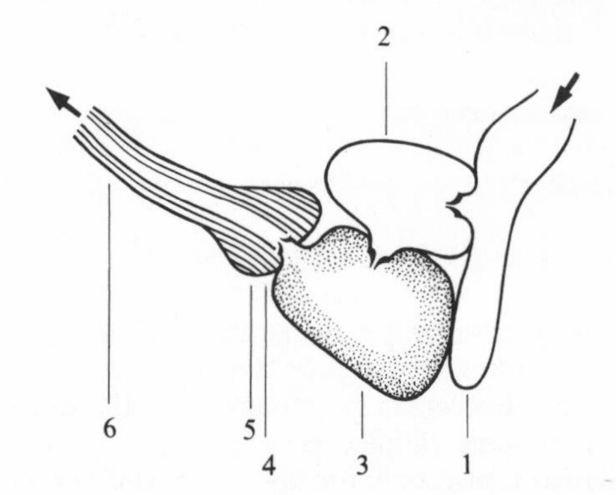

Fig. 201 Heart of a bony fish. 1 Sinus venosus 2 Atrium 3 Ventricle 4 Valves 5 Conus arteriosus 6 Truncus arteriosus

blood then flows into the carotid arteries, the largest part going to the aorta. The latter blood vessel is protected by the vertebral column and it reaches to the tail region. Numerous arteries branch off from the aorta taking oxygen-rich blood to all parts of the body apart from the head. After passing through the tissue capillaries and smaller veins, the now oxygen-poor blood goes back to the heart via networks of large veins. Blood flowing from the stomach, intestine, spleen and pancreas is sent through the liver beforehand so that the nutrients taken from the digestive organs* may be released here.

Blotched Hawkfish *(Cirrhitichthys aprinus) see Cirrhitichthys*

Blotcheye *(Myripristis murdjan) see Myripristis*

Blue Acara *(Aequidens pulcher) see Aequidens*

Blue Algae, D, *see* Cyanophyta

Blue Antenna Catfish *(Ancistrus dolichopterus) see Ancistrus*

Blue Crab *(Callinectes sapidus) see Callinectes*

Blue Damsel *(Chromis chromis) see Chromis*

Blue Damselflies (collective term for several Ga) *see Coenagrion*

Blue Danio *(Brachydanio kerri) see Brachydanio*

Blue Demoiselle *(Chromis xanthurus) see Chromis*

Blue Discus *(Symphysodon discus) see Symphysodon*

Blue Eyes *(Oryzias javanicus) see Oryzias*

Blue Gambusia *(Gambusia puncticulata) see Gambusia*

Blue Gourami *(Trichogaster trichopterus sumatranus) see Trichogaster*

Blue Gourami *(Trichogaster trichopterus trichopterus) see Trichogaster*

Blue Gularis *(Aphyosemion sjoestedti) see Aphyosemion*

Blue Keel-belly *(Oxygaster atpar) see Oxygaster*

Blue Limia *(Poecilia melanogaster) see Poecilia*

Blue Marlin *(Makaira audax) see* Scombroidei

Blue Mosquitofish *(Gambusia puncticulata) see Gambusia*

Blue Panchax *(Aplocheilus panchax) see Aplocheilus*

Blue Poecilia *(Poecilia caudofasciata) see Poecilia*

Blue Puller *(Chromis coeruleus) see Chromis*

Blue Sac Disease *see* Yolk Sac Dropsy

Blue Shark *(Prionace glauca) see* Carcharhinidae

Blue Sharks, F, *see* Carcharhinidae

Blue Starfish *(Coscinasterias tenuispina) see Coscinasterias*

Blue Streak *(Labroides dimidiatus) see* Labroides

Blue Surgeon *(Paracanthurus hepatus) see Paracanthurus*

Blue Tang *(Acanthurus coeruleus) see Acanthurus*

Blue Tang *(Paracanthurus hepatus) see Paracanthurus*

Blue Tetra *(Glandulocauda inaequalis) see Glandulocauda*

Blue Tetra *(Mimagoniates [Coelurichthys] microlepis) see Mimagoniates*

Blue-banded Angelfish *(Pygoplites diacanthus) see Pygoplites*

Blue-banded Damselfish *(Abudefduf biocellatus) see Abudefduf*

Blue-banded Sea Perch *(Lutjanus sebae) see Lutjanus*

Blue-banded Triggerfish *(Pseudobalistes fuscus) see Pseudobalistes*

Blue-fin Top Minnow *(Lucania goodei) see Lucania*

Blue-fin Tuna *(Thunnus thynnus) see* Scombroidei

Bluefish *(Pomatomus saltatrix) see Pomatomidae*

Bluefishes, F, *see* Pomatomidae

Blue-green Chromis *(Chromis coeruleus) see Chromis*

Blue-green Puller *(Chromis coeruleus) see Chromis*

Bluehead *(Thalassoma bifasciatum) see Thalassoma*

Blue-line Rasbora *(Rasbora einthoveni) see Rasbora*

Blue-lined Surgeonfish *(Acanthurus lineatus) see Acanthurus*

Blue-spot Copella *(Copella nattereri) see Copella*

Blue-spot Keel-belly *(Chela caeruleostigmata) see Chela*

Blue-spot Rock Cod *(Cephalopholis miniatus) see Cephalopholis*

Blue-striped Angelfish *(Pygoplites diacanthus) see Pygoplites*

Blue-velvet Damselfish *(Abudefduf oxyodon) see Abudefduf*

Blyxa NORONHA ex THOUARS, 1908. G of the Hydrocharitaceae*, with 10 species. Found in tropical regions of Africa, Asia, Australia. Small perennial or annual plants. Stem axes elongated with alternate leaf arrangement or compressed with basilar leaf rosettes. Pointed, linear leaves. 1 or several inflorescences surrounded by a spathe. Hermaphroditic or unisexual, and therefore monoecious or dioecious, polysymmetrical. Double perianth, in threes. 9–3 stamens, 3 carpels, fused, inferior. Fruit is a berry. The species of the G are found in shallow parts of flowing and still waters. There is little information regarding their care in the aquarium, but they are thought to require soft, acidic water and a lot of light. Temperature 20–25 °C. The following species has been cultivated many times, but has not survived for long.

– *B. echinosperma* (CLARKE) HOOKER fil. Found in S Asia and Australia.

Boarfish *(Capros aper) see* Zeiformes

Boarfishes (Caproidae, F) *see* Zeiformes

Bodianus BLOCH, 1790. G of the Labridae, Sub-O Labroidei*. Found in the tropical and subtropical Indo-Pacific and Atlantic. *Bodianus* lives on reefs and rocks and eats small animals. The G comprises moderately tall Wrasses with a pointed head and a drawn-out C in old age. So far only one species has been imported now and again:

– *B. rufus* LINNAEUS 1758; Spanish Hogfish. Found on both sides of tropical parts of the Atlantic. Up to 60 cm maximum, most much smaller. Head and dorsal surface violet, belly and rear part of the body brilliant yellow. The juveniles, like many Wrasses, are Cleaner Fish (fig. 202).

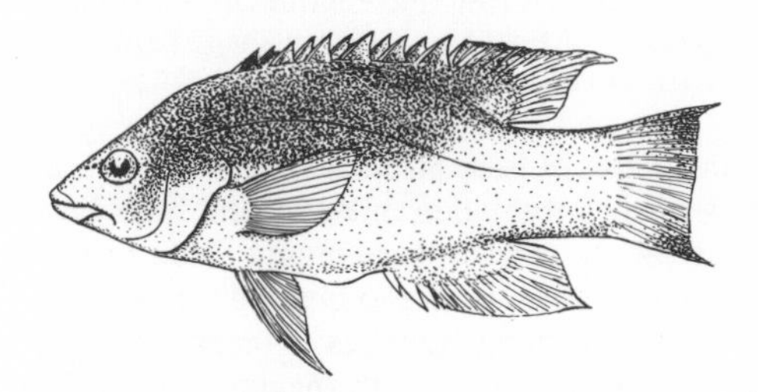

Fig. 202 *Bodianus rufus*

Body Care Behaviour *see* Grooming

Body Rubbing *see* Grooming

Body Shapes *see* Body Structure

Body Structure. The basic structure of organisms and how closely related it is to generally valid principles of structure and function within the various systematic groups. The body structure of fish contains many of the basic features found in vertebrates: internal skeleton with vertebral column and cranium (*see* Skeleton); a nervous system* with brain, spinal cord, nerves and sense organs*; enclosed blood-vascular system*; typical digestive organs*, urinary organs*, sex organs*, musculature*, multi-layered skin* and a tail region.

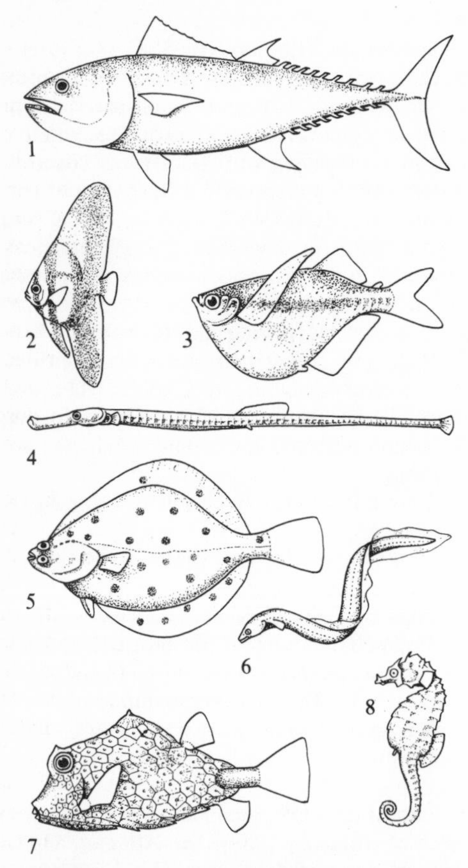

Fig. 203 Variety of body shapes in bony fish. 1 Tuna Fish *(Thunnus)* = torpedo-shaped 2 Batfish *(Platax)* and 3 Flying Hatchetfish *(Thoracocharax)* = bilaterally symmetrically compressed 4 Pipefish *(Syngnathus)* = needle-shaped 5 Flat-fish *(Pleuronectes)* = asymmetrically compressed 6 Eel *(Anguilla)* = eel-shaped 7 Boxfish *(Tetrosomus)* 8 Sea-horse *(Hippocampus)*

The peculiarities of body structure found in fish is the expression of their evolution in water and their lower position in the system of vertebrates. This is most clearly seen in the use of gills to breathe (*see* Respiration), the lack of connecting tubes (choanae) between the nose and mouth cavity (with the exception of the Sarcopterygia), the formation of unpaired fins*, the presence of ganoid or bony scales*, and, in bony fishes, the presence of a swimbladder*. The shape of the body often looks like a spindle that has been flattened laterally. Depending on the habitat, however, there are a great number of deviations from the norm (fig. 203). Examples are the torpedo-shaped bodies of sharks, mackerel and tuna; the dorso-ventrally flattened body of the rays; the bilaterally compressed, symmetrical bodies of *Chaetodon* and *Pterophyllum*; the asymmetrically compressed body of *Pleuronectes*; the eel-shaped bodies of *Anguilla* and *Nemichthys*; the ribbon-shaped bodies of *Regalecus* and *Trichiurus*; the arrow-shaped bodies of *Belone* and *Lepidosteus*; the spherical bodies of *Diodon* and *Tetraodon*; or the needle-shaped body of *Syngnathus*. There are also many fish with greatly altered body shapes, such as *Hippocampus* and *Ostracion*.

Fig. 204 shows how a fish's body is divided up externally.

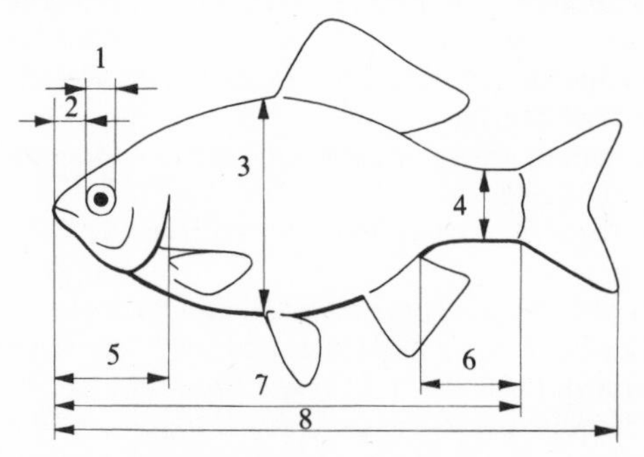

Fig. 204 The most important body measurements in bony fish. 1 Diameter of the eye 2 Length of snout 3 Height of body 4 Height of caudal peduncle 5 Length of head 6 Length of caudal peduncle 7 Length of body 8 Total length

Body Temperature. In contrast to the mammals and birds, fish have poikilothermal* blood. Their body temperature depends very largely on the temperature of their surroundings. With species from tropical, temperate or polar regions, the metabolism is geared to the corresponding ranges of temperature, which of course must be adhered to in the aquarium. Because muscle movement produces heat, the body temperature of very muscular species is often greater than that of the water temperature (eg *Thunnus*).

Boehlkea GÉRY, 1966. G of the Characidae*, Sub-F Tetragonopterinae. Small South American Characins from the Rio Marañón from Iquitos to Leticia. 4–5 cm. Very similar in body to the closely related species of the G *Glandulocauda**, although they do not possess any gland-like formations on the root of the C. Body elongated and laterally much compressed. Colourful, lively Characins which are peaceable and undemanding. These fish live in the upper water levels and fertilisation is internal as in *Gephyrocharax, Glandulocauda* and *Mimagoniates*. The fertilised females lay the spawn on leaves without the attendance of the male. Rearing is not difficult—feed with rotifers and later on *Cyclops* nauplii.

– *B. fredcochui* GÉRY, 1966. Rio Marañón from Iquitos to Leticia. Up to 5 cm. In the male, the flanks are a brilliant beautiful violet, but sea-green near the dorsal surface. The unpaired fins have pastel-white tips: the An is relatively long. An indistinct, dark, longitudinal band ends in a dark blotch at the root of the C. The body coloration of the somewhat larger females is pale blue to blue. The species was originally wrongly referred to as *Microbrycon cochui* (fig. 205).

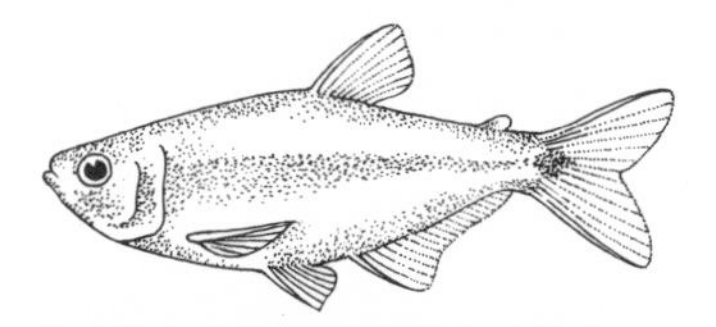

Fig. 205 *Boehlkea fredcochui*

Bog Arum *(Calla palustris) see* Araceae

Bogbean *(Menyanthes trifoliata) see* Menyanthaceae

Bolbitis SCHOTT; Water Fern (fig. 206). G of the Pteridophyta*. 85 species. Tropical, primarily S Asia. Small to medium-sized ferns with elongated rhizomes and simple or feathery leaves. The species of the G are ferns of tropical forests and grow in places with high humidity. Found on riverbeds, on rocks in mountain streams or on tree trunks along the banks. The following species can be kept continually submerged in the aquarium if cultivated attached to wood or stones. Grows very slowly. A gentle water flow is said to suit the plant.

– *B. heudelottii* (BORY ex FÉE) ALSTON. Tropical W Africa. Dark green.

Fig. 206 *Bolbitis*

Boleosoma, G, name not in current usage, *see Etheostoma*

Bombay Ducks (Harpodontidae, F) *see* Myctophoidei

Bonellia ROLANDO, 1821 (fig. 207). G of the Echiurida*, a small marine animal Ph closely associated with the true worms (Annelida*). The G is biologically interesting because of its special form of sex determination, which only takes place in the larval stage. On hard substrates in the Mediterranean, it is possible to spot the very long dark, T-shaped proboscis of the females. Detritus is conveyed to the mouth via the proboscis. *Bonellia* is dark green and spherical. It sits in a hole which it rarely leaves, or beneath stones. Occasionally kept in aquaria.

Bones *see* Skeleton

Bonito *(Euthynnus pelamis) see* Scombroidei

Bony Fish, Cl, *see* Osteichthyes

Bony Fish, Higher (Teleostei, Coh) *see* Osteichthyes

Bony Tongues (Osteoglossidae, F) *see* Osteoglossiformes

Boops CUVIER, 1817. G of the Sparidae*. Several species found in warm seas where they live in shoals near the coast. Mainly feed on animal and plant growths. Unfortunately these attractive fish are not really suitable for the aquarium. A well-known species is:

– *B. salpa* (LINNAEUS, 1758). Mediterranean, E Atlantic. Up to 45 cm. Gold longitudinal stripes on a bluish-silver background.

Borell's Dwarf Cichlid *(Apistogramma borellii) see Apistogramma*

Boring Molluscs. Marine molluscs of different systematic groups which bore into wood, firm sediment and stone-work. *Lithophaga lithophaga*, LINNAEUS, 1758, 'bores' holes in calcareous rock chemically, with acid. The majority of boring molluscs, however, bore mechanically with the scraper-like front end of their shell (*Zirfaea*, GRAY, *Barnea* LEACH, 1826, *Pholas* LINNAEUS, 1758). To do this, the molluscs turn on their lengthwise axis, opening and shutting the shell valve rhythmically. In the wood-boring *Teredo navalis* LINNAEUS, 1758, the shells have entirely lost their protective function and are only used for boring. The well-known giant clams (*Tridacna**) of tropical seas bore through soft coral with movements of the shell.

Boring Sponges, G, *see Cliona*

Bosmina BAIRD, 1845. G of the Water Fleas (Cladocera*). Distributed worldwide in still freshwaters. Even the individual species are mostly typical cosmopolites*. *Bosmina* can be up to 1.5 mm long, but is often no more than 0.4–0.6 mm. A typical feature is the flat head drawn out into a long proboscis. Other than that the species are extremely varied, depending on the conditions of the environment. The G is of some importance as a food for many freshwater fish and regularly occurs among plankton*.

Botanical Nomenclature. Although botanical and zoological nomenclatures* have much in common, they are independent of one another. The basis of botanical nomenclature forms the *International Code of Botani-*

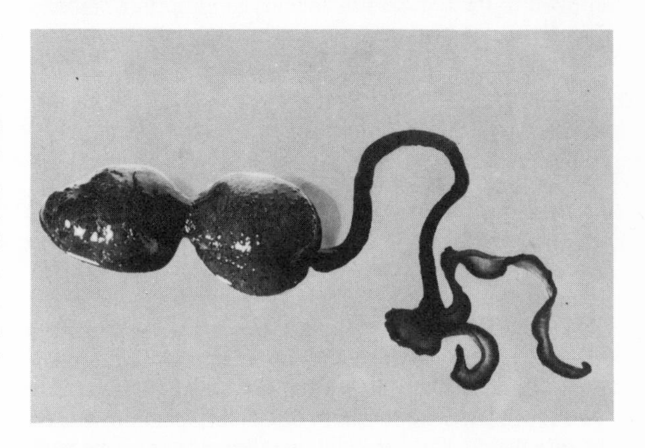

Fig. 207 *Bonellia viridis*

cal Nomenclature which is available in English, French and German. The most recent issue of the code as discussed and agreed at an international botanical congress supersedes all earlier issues. The code contains principles, rules and recommendations.

The systematical categories* of the plants and fungi are laid down in the code. For the aquarist, the species and the specific name are the most important items. This consists of the genus name followed by a specific name (epitheton) and lastly the author's name is added (*Echinodorus osiris* RATAJ). Each species bears only one correct name – ie the oldest one which corresponds to the rules of the code. All names adopted later are synonyms. A valid name must have appeared in a publication which is available across the counter, through exchange or donation, and likely to be accessible to botanists. At the very least it must be available in libraries. Publication in popular scientific journals is very ill-advised. As well as naming the name, a description or diagnosis must follow. Since 1. 1. 1935 this has been written in Latin. By diagnosis we mean those features which distinguish one species from other species. It is also important that each name is linked to a type.

The author's name is the name of the person who first officially published the specific name. Many authors' names are abbreviated in particular ways (CHAMISSO = CHAM., LINNÉ = L.). If a name has been published jointly by 2 or more authors, their names are linked by the word *et* or the symbol & (*Hydrocotyle leucocephala* CHAM. et SCHLECHT). When an author publishes a name on the recommendation of another person or would like to ascribe the name to another person, the name of that other person can be inserted with the linking word *ex* in front of the name of the actual author (*Mayacca vandellii* SCHOTT ex ENGLER). If father and son have carried out species descriptions jointly, the word *filius* appears after the author's name, in the case of the son (HOOKER filius). If a species is transferred to another G, the author of the first description is placed in brackets, whilst the name of the author who has proposed the change is placed after the brackets: *Alisma tenellum* MART. = *Echinodorus tenellus* (MART.) BUCH. In ranks below species, no author's name is applied to the typical sub-species or variety (*Cabomba caroliniana* A. GRAY var. *caroliniana*), only to the deviant (*Cabomba caroliniana* A. GRAY var. *paucipartita* RAMSH. et FLORSCH).

Botany is a branch of biology*. It concerns the science of plants – their structure, life processes and distribution. There are many disciplines within botany, morphology*, physiology*, taxonomy*, geobotany* etc, each of which sets out to answer particular questions regarding plants.

Botia GRAY, 1831; Loaches. G of the Cobitidae*. Tropical SE Asia. Small to medium-sized fish which live in small groups in streams. Body compressed, flattened laterally and some have a high caudal peduncle. The dorsal profile is arched more noticeably than the ventral profile which, like the belly, is often straight. The long, often pointed head ends in an inferior mouth with 3–4 pairs of barbels. In contrast to many other Ga the eyes are not covered by skin. Beneath the eyes are bifid spines which can be erected and locked in position. Scales much reduced. The unpaired fins are often more strongly defined than in other Cobitidae. The D begins directly above or slightly in front of the V. Several specimens of the same species can be kept together in large tanks, where they may well exhibit territorial behaviour*. Very peaceable behaviour as well as intolerance to like species will be observed. The tank should stand in a shady spot with not too coarse a sand substrate, and there should be plenty of hiding-places, eg wood, rocks or sturdy plants. The water, which can be replaced with fresh water, should not be too hard, clear and oxygen-rich. Temperature 24–26 °C. Many species are crepuscular and can utter characteristic cracking noises. All species require live food, especially worms, as well as dried and plant food. Sexual maturity achieved very late.

– *B. beauforti* SMITH, 1931; Beaufort's Loach. Thailand, Laos. Up to 25 cm. Grey-green to yellowish in colour, with rows of dark spots on the flanks. Towards the dorsal surface, behind the eye, there are small spots which unite at the front into fine lines. D and An orange-yellow.

– *B. berdmorei* BLYTH, 1860; Berdmore's Loach. Thailand, Burma. Up to 25 cm. Yellowish-white to ochre with 10–11 broad transverse bars. The black colouring of the barbels continues in a line to the eye. Behind it there are two black longitudinal bands which continue towards the rear of the fish as rows of speckles. D and C have transverse bars.

– *B. horae* SMITH, 1931; Hora's Loach, Skunk Loach. Thailand. Up to 10 cm. Yellowish-green, occasionally with indistinct transverse bars. A black longitudinal band extends from the head and across the dorsal surface, ending in a transverse bar each side of the C (fig. 208).

Fig. 208 *Botia horae*

– *B. hymenophysa* (BLEEKER, 1852); Banded Loach, Tiger Loach, Tiger Botia. Thailand, Malaysia, Singapore, Greater Sunda Islands. Up to 20 cm. Brownish-yellow with 12–15 grey-blue bands, which run from the dorsal surface slightly at an angle to the rear of the fish. Fins pale yellow.

– *B. lecontei* FOWLER, 1937, Leconte's Loach. Thailand. Up to 8 cm. Greenish to blue-grey with a large blotch at the root of the tail. Fins orange to reddish (fig. 209).

– *B. lohachata* CHAUDHURI, 1912; Pakistani Loach, Reticulated Loach. N India, Bangladesh. Up to 10 cm. Body yellowish-grey with a very variable pattern of

Fig. 209 *Botia lecontei*

black-grey bands. This is a sociable species which is peaceable.

– *B. lucas bahi* FOWLER, 1937; Barred Loach. Thailand. Up to 8 cm. Brown to grey-brown ground colouring with some indistinct transverse bands and numerous dark spots. D has a black edge.

Fig. 210 *Botia macracantha*

– *B. macracantha* (BLEEKER, 1852); Clown Loach, Tiger Botia. Sumatra, Kalimantan (Borneo). Up to 30 cm. Orange ground colouring with three black, wedge-shaped transverse bands. P, V and C red, An and D yellowish with a black band (fig. 210).

Fig. 211 *Botia sidthimunki*

– *B. modesta* BLEEKER, 1864; Orange-finned Loach. Thailand, Vietnam, Malaysia. Up to 10 cm. Grey-green to grey-blue. Indistinct blotch at the root of the tail. C yellow.

– *B. sidthimunki* KLAUSEWITZ, 1959; Dwarf Loach, Chained Loach. Thailand. Up to 3.5 cm. Dorsal surface brownish, flanks yellowish-silver with black-brown longitudinal bars, which run along the dorsal ridge on the back and along the vertebral column on the sides. A mesh-like patterning results from the transverse bands. Usually a free-swimming species (fig. 211).

Botrulidae, F, name not in current usage, *see* Gadiformes

Botryllus PALLAS, 1774. G of the colony-forming Sea-squirts (Ascidiacea*). Found in cool and warm seas. *B. schlosseri* (PALLAS) is widespread along the European coasts. The jelly-like colonies take over firm substrates. Each individual animal is only a few mm long. They arrange themselves in groups of 3–12 individuals in a star formation around a common cloaca. The colonies came in a variety of colours, including yellow, blue, violet, red or white. *B. schlosseri* is often placed in large aquaria. Take care the colonies are not squashed!

Bottom Sword (*Poecilia reticulata*, standard species) *see Poecilia*

Bottom Zone *see* Benthal

Boulenger, George Albert (1858–1937). Important Belgian ichthyologist and botanist. He was successor to A. GÜNTHER* as Director of the Ichthyological Department of the British Museum. Boulenger was concerned with the examination of tropical African fish. He made a large number of first descriptions and his chief work *Catalogue of the Freshwater Fishes of Africa in the British Museum* is still of great interest today. It appeared in London in 1909–1916 and comprised 4 volumes.

Boulengerella EIGENMANN, 1903. G of the Ctenoluciidae*. Pike-like predatory fish about 35–60 cm long, which live in the Amazon Basin. They have an elongated snout with an outgrowth at the tip which points upwards. D1 inserted very far back and small, D2 present, C deeply forked, An short. LL complete and numerous cycloid-ctenoid scales are present. *Boulengerella* lives in the reedy area and feeds on fish and large water insects. A dark longitudinal band stretches across the greenish-white colouring of the body. *B. lucia* has a deep black blotch on the C. The fins are colourless with dark bar markings. The various species need warmth and are only suitable for the aquarium when young. Reproduction unknown. The following are imported:

– *B. (Spixostoma) lucia* (CUVIER, 1817). Amazon. Up to 60 cm.

– *B. (Boulengerella) lateristriga* (BOULENGER, 1895). Amazon. Up to 50 cm (fig. 212).

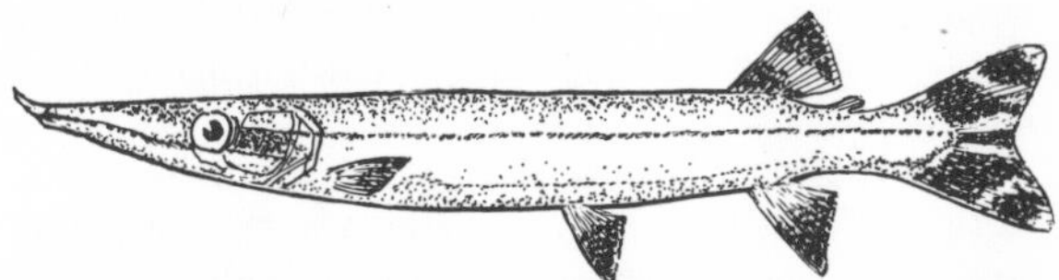

Fig. 212 *Boulengerella lateristriga*

– *B. (Boulengerella) maculata* (VALENCIENNES in CUV. and VAL., 1849). Amazon. Up to 35 cm.

Bowfin types, O, *see* Amiiformes

Bowfins (Amiidae, F) *see* Amiiformes

Bowfins (Amiidae, F) *see* Osteichthyes

Box Crab *(Calappa granulata) see Calappa*

Box Jellyfish, O, *see* Cubomedusae

Brachiolaria-larva *see* Asteroidea

Brachiopoda. Lamp-shells. Cl of the Tentaculata*, to which the better known bryozoans (Bryozoa*) belong. Evolutionarily speaking, the Brachiopoda are very ancient marine animals that reached their peak in the Palaeozoic, about 500 million years ago. They possess a two-valved calcareous shell, which externally reminds one (confusingly) of molluscs, although they are not related to one another. Brachiopods are filter-feeders, living attached to firm substrates. There are no reports of aquarium observations.

Brachydanio WEBER and DE BEAUFORT, 1916. G of the Cyprinidae*, Sub-F Rasborinae. The whole of SE Asia, Burma, Malaysian Archipelago, Sumatra, but *not* in N India. Small fish that are often found in large shoals in still and flowing waters, occasionally also in flooded rice fields. Body elongated, slim and much flattened laterally. The terminal or slightly superior mouth has 2 pairs of barbels round it. *Brachydanio* is distinguishable from the G *Danio* by 7–9 separate soft rays in the D. LL much reduced or completely absent. These are beautifully coloured, peaceful fish which are excellent swimmers and good jumpers, making them ideal for the community aquarium. They should always be kept as a shoal in large, lengthy tanks. Any plants must leave enough room for swimming. The water should not be too old, but there are no particular requirements as far as the water is concerned. Temperature 22–24 °C, somewhat cooler in winter. Live and dried food of all kinds is eaten. For rearing shoaling spawners, smaller glass tanks are used (10–30 l) filled with fresh water. The hardness and pH value of the water does not matter. The bottom is covered with gravel and some groups of fine-leaved plants are placed in the corners away from the light. The more robust female is put in first, followed by the much slimmer males 1–2 days later, in the evening. Some breeders have 2 females to every 1 male or they let the fish spawn in the shoal. Spawning takes place in the early morning hours and is helped along by the sun streaming in. The partners entwine or press against one another, and the large eggs are released into the open water in the course of several matings. These fish are very prone to eating the spawn. The young hatch after 20–36 hours, living for another 48–72 hours on their yolk provision. Afterwards, the fast-growing youngsters should be provided with fine live food or dried food as well. These fish are highly productive and will spawn again after 3–4 weeks.

– *B. albolineatus* (BLYTH, 1860); Pearl or Golden Danio. Burma, Thailand, Malaysia, Sumatra. Up to 6 cm. Ground-colouring grey-green, flanks with greenish, blue or violet reflective colours. A reddish, blue-edged

Fig. 213 *Brachydanio albolineatus*

longitudinal band begins above the V, and reaches to the root of the tail (fig. 213).

– *B. kerri* (SMITH, 1931); Kerr's or Blue Danio. The islands of Koh Yao Yai and Koh Yao Noi (N Thailand). Up to 5 cm. The species is more compressed and possesses 2 pairs of barbels. Dorsal surface grey-blue, flanks blue to an iridescent blue-green. A gold iridescent longitudinal line on the flanks which has another line below it running parallel.

– *B. nigrofasciatus* (DAY, 1869); Spotted or Dwarf Danio. Burma. Up to 4 cm. Dorsal surface olive-brown, belly yellowish-white to faint orange. A brown-gold longitudinal stripe on the flanks edged in blue-black bands and reaching to the C.

– *B. rerio* (HAMILTON-BUCHANAN, 1822); Zebra Danio. E India, Bangladesh. Up to 5 cm. Dorsal surface olive-brown, flanks an iridescent silvery-gold with 4 vivid blue longitudinal bands which continue into the C (fig. 214).

Fig. 214 *Brachydanio rerio*

– *'B. frankei'* MEINKEN, 1963; Leopard Danio. Up to 5 cm. Flanks bronze to gold in colour with numerous irregularly distributed dark markings. As these fish can be crossed with *B. rerio*, it is doubtful whether this is a proper species.

Brachygobius BLEEKER, 1874. G of the Gobiidae, Sub-O Gobioidei*. Tropical Asia, India to Thailand, Malaysia, Indonesia, Philippines, They live as bottom-dwellers in rivers, canals and ditches of the brackish water zone and in different types of small waters (perhaps even polluted). Small fish up to 4.5 cm. Body cylindrical at the front, flattened laterally at the rear. Compressed shape. D1 consists of hard rays. Ctenoid scales in part on the operculum and head, naked otherwise. The nape has a furrow in the middle. Ground colouring of the body yellow with different kinds of brown to black blotches which also cover the D and C. The female is paler in places. Best kept in a species tank with hiding-places, at a temperature of 24–30 °C. An addition of sea salt (1–2 tsp to 10 l water) is recommended. Live food, depending on size, is necessary. Spawning takes place under rocks or other hiding-places. Up to 150 eggs are laid. At the beginning the young fishes are pelagic. All the species are very similar and the following are imported under the name Golden-banded Gobies or Bumble-bee Fish:

– *B. aggregatus* (HERRE, 1940). N. Kalimantan (N Borneo), Philippines. Up to 4.4 cm. 22–26 scales per longitudinal row. Sides of the head and throat have dark-brown spots. D2 and C yellow.

– *B. nunus* (HAMILTON-BUCHANAN, 1822). Indo-China, Thailand, Malaysian Peninsula, Greater Sunda Islands. The commonest and most widespread species. Up to 4.2 cm. 25–27 scales per longitudinal row. D1, D2 and An have a broad yellow tip. Base of the P and V black.

– *B. xanthozona* (BLEEKER, 1849); Bumble-bee Fish. Sumatra, Java, Kalimantan (Borneo). Up to 4.5 cm. More than 50 scales per longitudinal row. D1, D2 and An black. Base of the P and V black (fig. 215).

Fig. 215 *Brachygobius xanthozona*

Brachyrhaphis REGAN, 1913. G of the Poeciliidae*, Sub-F Poeciliinae. Central America, Panama, Costa Rica, 1 species known from S Mexico. Found in rivers, still zones near the banks of large waters, in ditches, pools and ponds as well as in brackish water. Small fish, males 2–4.5 cm, females 4–6.5 cm. Slim-bodied. Ground colouring of the body often brownish, often with darker, transverse bands and spots on the sides. D and C likewise partially covered in dark stripes and dots. Males have a relatively short gonopodium. Keep at 25–30 °C with adequate vegetation in the tank. Live food preferred. Rearing conditions largely unknown. So far, the following is the only species of interest to the aquarist.

– *B. episcopi* (STEINDACHNER, 1878); Bishop Fish. Panama. Male 3–3.5 cm, female 4–5 cm. Male pale olive-green to yellowish-brown, dorsal surface darker. A row of dark flecks or transverse bars along the sides. D and C yellow to yellow-orange, with dark brown tips. The An in the female is red and elongated into a point, otherwise the female is less conspicuous. Young fish have only been reared occasionally (fig. 216).

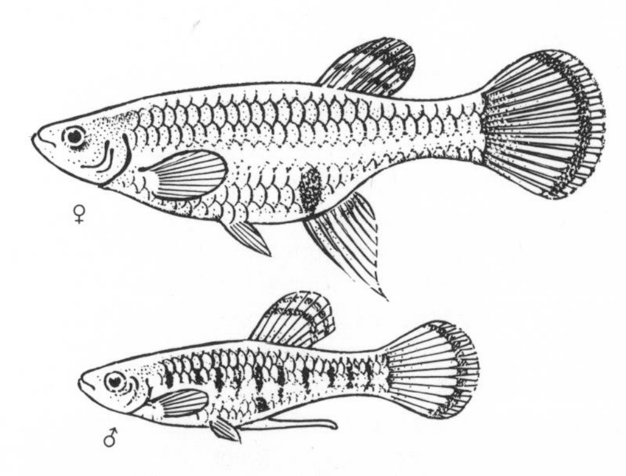

Fig. 216 *Brachyrhaphis episcopi*

Brachyura; Crabs. With about 4,500 species this group of Decapoda* has the largest number of species. Sub-O Reptantia. The majority live in seas in all climatic zones, some also in fresh water; many leave the water at times *(Grapsus, Pachygrapsus*)* and some have changed to living totally on land (F Gecarcinidae). There are several predatory species of crab, many eat carrion, some eat plants and detritus. The compressed Cephalothorax, usually triangular, rectangular or round in shape, is typical of the crabs. Beneath it, the tongue-like, symmetrical abdomen is hinged towards the front. In contrast to the similar Porcellanidae* there are no appendages (uropoda) on the last segment of the abdomen. The abdomen is narrow in the males, much broader in the females, since they carry the eggs beneath it until the larvae hatch. The crab's first pair of walking legs possesses strong pincers which are used like hands. There are often depressions in the carapace into which the stalk-eyes can be folded for protection. The front edge of the carapace contains teeth, spines or notches of different kinds, depending on species. It is therefore an important feature of identification. The average width of the carapace is 2–10 cm, up to 30 cm in the largest European species *(Cancer*)*; the largest species of all is the Japanese Giant Crab *Kaempfferia kaempfferi* DE HAAN *(= Macrocheira)* which has pincers up to 1.50 m long. Many crabs run sideways. The swimming crabs (eg *Portunus**) can swim by beating with their last paddle-like pair of walking legs. *Maja* uses camouflage, allowing plant or animal growths to settle on its upper surface; *Dromia* covers itself with living sponges, *Ethusa* keeps a mollusc shell over the body, *Pinnotheres* lives inside large molluscs and ascidians, and some tropical crabs *(Lybia)* carry a sea anemone with a sharp sting in each of the pincers. Whilst the larval development in the marine and land crabs takes place in the marine plankton, the pelagic larval stage (zoea) does not exist in the freshwater crabs (Potamonidae). Instead, their young develop in a similar way to those of the crayfishes–they grow into small crabs beneath the female's abdomen. Crabs are lively and often very colourful animals which are usually easy to care for in the aquarium. They make quite charming specimens. However, care must be taken with the various predatory forms. A good lid for the aquarium is very important when keeping amphibious species. Most crabs eat animal food of all sorts; the tropical Fiddler Crabs *(Uca*)* eat fine organic particles which they obtain by scavenging on the sea shore. Amphibious crabs can also be kept in aqua-terraria without ventilation. Large crabs can nip painfully *(Eriphia*, Cancer)*. Some crabs are edible *(Callinectes*, Cancer)*. Apart from the ones already mentioned, the following Ga are also considered suitable for the aquarium: *Calappa*, Carcinides*, Hyas*, Macropodia*, Pilumnus*, Stenorhynchus*, Xantho**.

Brackish Water. Sea-water that contains a high proportion of fresh water. Salinity level between 30 and 0.5‰. Large parts of the Baltic and Black Seas, Lakes Caspi and Aral consist entirely of brackish water. There is a poor number of species in a brackish water habitat. Compare Brackish Water Species, Biotope, Biocenesis.

Brackish Water Species. The living communities of brackish water* are largely composed of those species that are marine, but can tolerate a lower salinity level, or are freshwater and can withstand a higher salt content. There are also 'true' brackish water species which are tied to the habitat or prefer it. However, in comparison with the former, 'true' brackish water species are very much in the minority. As a result there is a noticeable lack of species in a brackish water biotope in contrast to marine and freshwater biotopes (fig. 217).

Bract. This is a leaf in the axil of which a side-shoot, inflorescence, part inflorescence or simple flower has developed.

Brain *see* Nervous System

Brain Corals, G, *see* Madreporaria

Branching in higher plants is the development of one or several buds* on a stem, which are usually found in the leaf axils* in seed-plants (Spermatophyta*). As a result of branching a stem system* develops. With branching Thread Algae, either the topmost cell divides or lateral growth takes place from a cell towards the rear.

Branchiomyces-fungi *see* Gill Rot

Branchiostoma belcheri *see* Acrania

Branchiostoma lanceolatum *see* Acrania

Branchipus, G, *see* Anostraca

Branner's Livebearer *(Poecilia branneri) see Poecilia*

Brasenia SCHREBER, 1789. G of the Cabombaceae*, with only one species. Found in N and Central America, Africa, S and E Asia, Australia. Small perennials with underground creeping stems and upright stems and alternately arranged peltate leaves 2–10 cm long. Flowers hermaphroditic, polysymmetrical. Double perianth, parts in multiples of 4 to 3. Stamens 18–12. Carpels 20–2, free-standing, superior. A nut forms a small fruit. The species is a common aquatic plant in many areas. Seems to prefer slightly acidic water, but so far has been cultivated only rarely.

– *B. schreberi* GMELIN; Water Shield.

Brassicaceae (Cruciferae, name not in current usage). F of the Magnoliatae* with 350 Ga, 3,000 species. Found in temperate and boreal regions of the northern hemisphere, but only up in the mountains in the tropics and sub-tropics. Perennial and annual herbs with alternate leaf arrangement. Flowers hermaphroditic, polysymmetrical.

Numerous cultivated plants belong to the Brassicaceae, eg European cabbage plants. Only a few species grow in wet places and only a few can withstand submersion. In aquarium-keeping the Ga *Rorippa** and *Cardamine** are of importance.

Brassy Barb *(Barbus gambiensis) see Barbus*

Brazilian Milfoil *(Myriophyllum brasiliense) see Myriophyllum*

Bread-crumb Sponge *(Halichondria panicea) see Halichondria*

Bream *(Abramis brama) see Abramis*

Bream Region *see* Potamal

Breathing Root. A vertical root growing upwards (negatively geotropic), which becomes aerial as it emerges from the soil or water. Contains a great deal of ventilating tissue and provides air for the root system in the ground. Breathing roots are found in mangrove plants, for example, as well as in many species of the G *Ludwigia*.

Breeding Aquarium; spawning aquarium. All-glass or frame aquaria of varying extent, in each case corresponding to the size and swimming habits of the fish

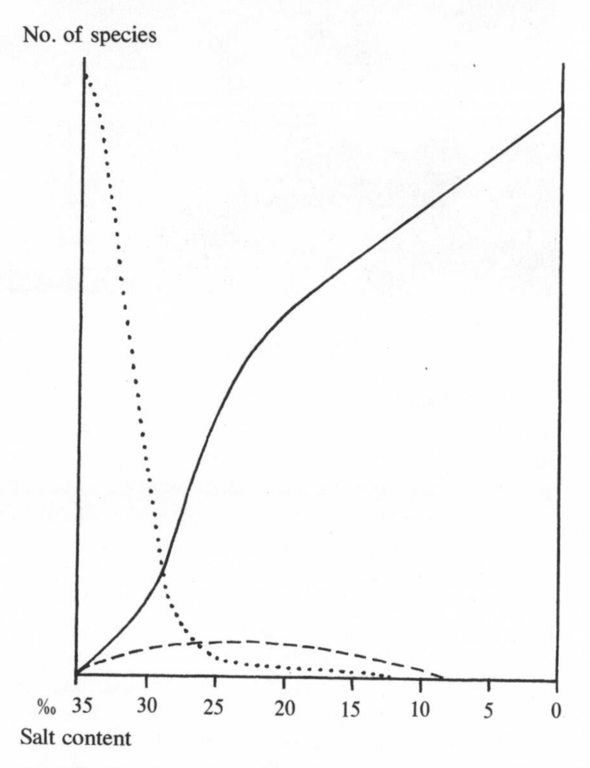

Fig. 217 Number of species related to salt content. Note the lack of species in the brackish water range. (Dotted curve: marine species; broken curve: brackish water species; unbroken curve: freshwater species)

being bred, and filled with suitable breeding water (Water, for Breeding Purposes*). For many smaller fish species of the *Hyphessobrycon*-, *Hemigrammus*-, *Nannostomus*- or *Barbus*-type, normally all-glass aquaria measuring 20 × 25 × 15 cm or 25 × 18 × 15 cm are adequate. For larger species needing greater swimming areas during courtship and mating, such as *Micralestes*, *Metynnis*, larger kinds of *Barbus* and *Danio* species, Atherinids of various Ga, larger Labyrinth Fishes and Catfishes, the tank should measure between 60 × 30 × 25 cm and 100 × 35 × 30 cm (especially if the fish concerned are shoaling spawners [Selection of Breeding Fish*]). Species that practise brood-care, eg Sunfishes, Leaf-fishes, Cichlids, do not need a separate breeding tank; they are usually bred in the biotope aquarium. The same goes for many Egg-laying Tooth-carps, where the parent fish are removed after a certain time, and the fry then hatch out and are reared in the biotope aquarium. With the smaller Characin and Barb species mentioned at the beginning, the fry are transferred a few days after feeding begins, or even after hatching, from the smaller breeding aquarium into a larger *rearing aquarium**. To give a better over-all view and to keep the tank clinically clean, a breeding aquarium has no substrate, the only exceptions being species that lay their spawn in the substrate. To protect the spawn from being eaten by the parent fishes (Spawn, Eating of*), a spawning grill made of synthetic gauze should be placed about 1–2 cm above the bottom of the container. The modern use of silica-rubber to stick aquaria together means that the aquarist can now make his tank any size he likes, and he can also give free rein to his imagination for constructing special forms of breeding tank to see if that helps in any way with the breeding of species that have so far never been bred or only rarely. Eg, for species that have to remain for quite some time in the breeding tank, a narrow feeding area for Enchytreids or *Tubifex* can be erected, simply by sticking a strip of glass 2–3 cm tall onto the bottom pane, such that it is not covered by the spawning grill. The food animals are thus not able to crawl under the spawning grill, dead food animals can be sucked away effortlessly, and the egg-cells can still be seen quite clearly underneath the spawning grill through this ‘double window’. A breeding aquarium can also be fitted with a bottom pane that slopes forwards, forming a collecting bay at the front pane where non-sticky eggs can collect (eg those of *Micralestes*-, *Alestes*- and *Danio*- species). This means that the eggs are safe from being eaten by their parents, and that they can also be sucked out easily through a thin piece of hose fixed into a suitable glass tube. If plants are brought into the breeding tank as a spawning substrate, make sure they are absolutely clean. The addition of a good lid will spare the breeder from any unpleasant surprises with jumping species such as *Thayeria*, *Carnegiella* and *Copella*, as well as Egg-laying Tooth-carps etc. At the same time it also prevents the build-up of any undesirable deposits on the surface.

Breeding Condition. This expression is taken to mean breeding fish in the best possible physical condition, achieved by the species concerned having been kept under its *optimal living conditions*. By this means the maturing of the sex cells is promoted, and the fish reach the necessary mood for reproduction. This presumes knowledge of or research into the abiotic and biotic factors which favourably influence these processes. In the first instance, the following abiotic factors are all important: composition of the water, its movement, temperature, hardness and pH value, light intensity, the angle of light rays, the length of the daily illumination during the spawning period, the extent of the breeding aquarium*, level of the water etc. Included in the biotic factors are the right food basis, suitable spawning plants (for Substrate Spawners*), the right choice of substrate (eg coarse peat for *Nothobranchius* species), with species that practise brood-care, possibly the hostility factor* too. The study of any literature on the subject of our knowledge of the reproductive biology of the species concerned assumes in the first instance a good breeding condition. If this knowledge is only slight or lacking altogether, the breeder simply has to carry out his own experiments in an attempt to achieve the optimal conditions. This demands a great deal of ‘feel’ for the subject if, instead of the hoped-for breeding condition, constitutional damage is not to be caused.

Breeding Season. All fish have a breeding season. Some, such as the Atherinidae (*see* Atherinoidei) spawn almost daily, although only a small number of eggs ripen at any one time. In other kinds of fish the intervals between spawning may vary greatly. In captivity, many Characins, Barbs and Cichlids (eg *Pterophyllum scalare*) spawn in roughly weekly cycles, but after some time there will be longer gaps between each spawning. Especially in imported fish, the time when the eggs mature and the breeding season begins coincides with the onset of the rainy season in their tropical homelands (where it is brought on by rainfall coupled with a rise in oxygen levels). The females of livebearing fish undergo a particularly rigorous reproductive period. Various types of fish (eg the Eel, Salmon and some brackish-water fish) travel great distances during their breeding season to reach their spawning grounds. They may journey from fresh or brackish water to the sea, or upstream to the source of a freshwater river, at which time the sexual cells become mature.

Brehm, Alfred Edmund (1829–84). Zoologist and field researcher, born in Rentendorf in Thuringia. From 1853–56 he studied natural sciences and zoology at Jena and Vienna. At 18, he went on his first, highly successful field trip to N Africa and in 1862 he visited Eritrea (Ethiopia). His experiences on these journeys were described in *Reiseskizzen aus Nordafrica* (‘Sketches of a Journey to North Africa’) (1855) and *Ergebnisse einer Reise nach Habesch* (‘Results of a Journey to Habesh’) (1863). He recorded his observations of the bird world made during trips through northern Europe, in his book *Das Leben der Vögel* (‘The Life of Birds’) (1861). The various experiences accumulated on his field observation trips were to form the basis of his most popular work *Illustriertes Tierleben* (‘Illustrated Animal Life’) which he published in 6 volumes in 1863–69 along with E. Taschenberg and O. Schmidt. From 1863 Brehm was Director of the Zoological Gardens of Hamburg and in 1869 he founded the Berlin Aquarium. There followed a very active publishing period: 1875 – new edition of ROSSMÄSSLER'S *Süßwasser-Aquarium* ('Freshwater Aquarium'), 1876 – the handbook *Gefangene Vögel* ('Captive Birds') and the second edition of *Animal Life*. In 1877 he went on more trips taking him from W Siberia to W Turkestan, as well as to Hungary, Spain and Portugal. He travelled across N America in 1883 giving lectures. Brehm's most valuable contribution lies in the way he could convey the life of animals at a popular level, stimulating the interest of a very broad spectrum of the population. For several decades his books were the only comprehensive reference works on the animal kingdom. In recent times there have been many imitations of the approach offering modern scientific concepts.

Brevoortia tyrannus *see* Clupeiformes

Brichard's African Dwarf Cichlid *(Teleogramma brichardi) see Teleogramma*

Bridled Beauty *(Labroides dimidiatus) see* Labroides

Brill *(Scophthalmus rhombeus) see* Pleuronectiformes

Brilliant Rasbora *(Rasbora einthoveni) see* Rasbora

Brine Shrimp *(Artemia salina) see Artemia*

Brissopsis *see* Echinoidea

Bristle Teeth, F, *see* Chaetodontidae

Bristle Worms, Cl, *see* Polychaeta

Bristlemouths (Gonostomatidae, F) *see* Stomiatoidei

Brittanichthys GÉRY, 1965. G of the Characidae*, Sub-F Tetragonopterinae. Small S American Characins from the Rio Itu, a tributary of the Rio Negro. 3.5–4 cm long. Elongated body, LL incomplete. The males are dinstinguishable from the females by having small hooks on the first 4 anal rays and an S-shaped middle ray on the caudal fin. The first third of the caudal fin lobes have scales on them. The care and rearing of the 2 species is not known. Not yet imported alive.

– *B. axelrodi* GÉRY, 1965. Rio Itu, 80 km before the river meets the Rio Negro. Up to 3.5 cm. Body colouring pale yellowish, belly blood-red in the area of the Vs, with a blue zone above it. A longish, blood-red blotch on the root of the C. The front rays of the D1 are blood-red, the other fins being yellowish (fig. 218).

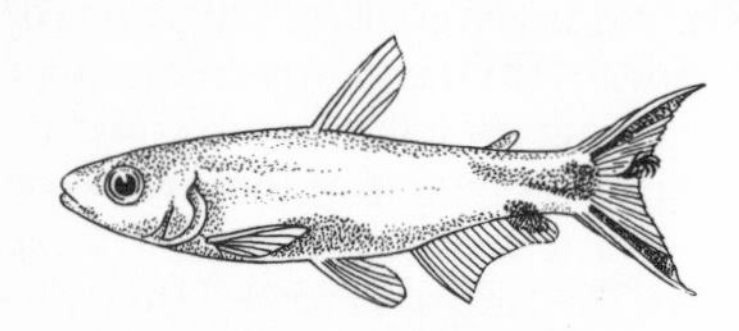

Fig. 218 *Brittanichthys axelrodi*

– *B. myersi* GÉRY, 1965. Inlet on the Rio Negro, 13 km west of its confluence with the Amazon at Manaus. Up to 3.5 cm Body a little less elongated than that of *B. axelrodi*, pale grey-olive with a silvery longitudinal band. 2 black dots are placed behind the operculum. Upper half of the head and body, as well as a zone above the An, are citrus-yellow. There are several blood-red, elongated speckles between the Ps and the An on the caudal peduncle and on the middle rays of the D1. A brilliant blue zone runs obliquely downwards above the red ventral speckles. There is a protrusion on the edge of the lower caudal lobe which has a complicated structure.

Brittle-stars, Cl, *see* Ophiuroidea

Broad-barred Toadfish *(Arothron hispidus) see Arothron*

Broad-bill Swordfish *(Xiphias gladius) see* Scombroidei

Broadside Display *see* Display Behaviour

Brochis COPE, 1872. G of the Callichthyidae*. Armoured Catfishes, up to 8 cm long, found in the Amazon Basin. The shape of the body is very similar to that of the well-known *Corydoras** species, but they can be distinguished because the body is more flattened laterally and the D1 is longer. *Brochis* also has bony plates round the mouth unlike *Corydoras*. These fish live in shoals near the bottom. They are undemanding, omnivorous fish that can survive for a long time in the aquarium. Of the few species of this G, *B. splendens* has been imported.

– *B. splendens* (DE CASTELNAU, 1855); Short-bodied Catfish. Upper Amazon near Iquitos. Up to 7 cm. The flanks are a lovely shining emerald green: the belly ochre-yellow, fins colourless. Successfully bred several times, according to KNAACK the species is not very productive. For information on reproductive behaviour see *Corydoras*. *B. coeruleus* is a synonym for this species (fig. 219).

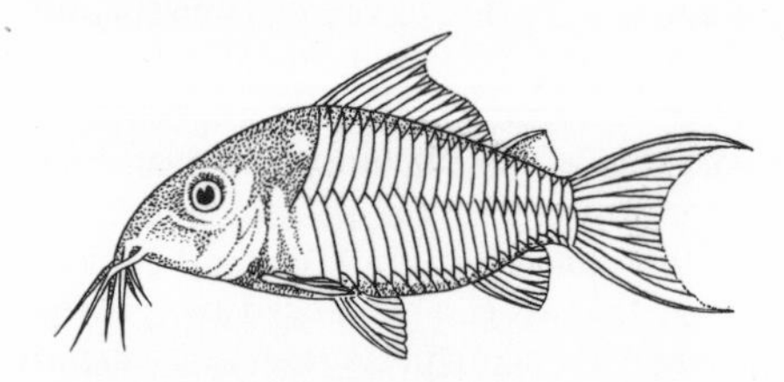

Fig. 219 *Brochis coeruleus*

Brook Lamprey *(Lampetra planeri) see* Cyclostomata

Brooklime *(Veronica beccabunga) see* Scrophulariaceae

Bronze Corydoras *(Corydoras aeneus) see Corydoras*

Brotulids (Ophidiidae, F) *see* Gadiformes

Brown Acara *(Aequidens portalegrensis) see Aequidens*

Brown Algae, D, *see* Phaeophyta

Brown Bullhead *(Ictalurus nebulosus) see* Ictaluridae

Brown Discus *(Symphysodon discus) see Symphysodon*

Brown Rivulus *(Rivulus cylindraceus) see Rivulus*

Brown Triggerfish *(Pseudobalistes fuscus) see Pseudobalistes*

Brown Trout *(Salmo trutta fario) see* Salmonoidei

Brown Water *see* Dystrophic Waters

Brycinus, G, *see Alestes*

Brycon MÜLLER and TROSCHEL, 1844. G of the Characidae*. Sub-F Bryconinae. Distributed from Guatemala to La Plata. Characin-type fish of typical *Hemigrammus** shape, growing to between 10–60 cm long. The *Brycon* species are robust, omnivorous fish that grow quickly when kept well-fed. Only suitable for the aquarium when young. Not yet bred successfully. Valuable edible fish in their home waters.

– *B. whitei* MYERS and WEITZMANN, 1958. Around Los Micos on the Rio Guaviare in Colombia. Up to 35 cm. Silver body with broad black longitudinal band, reaching from the operculum to the middle ray of the C and broadening out into a rhomboid shape on the root of the C. The dark edges of the scales form fine, dark, lengthwise lines. The scales themselves produce warm, metallic reflections, unpaired fins red; Ps, Vs and D2 dark (fig. 220).

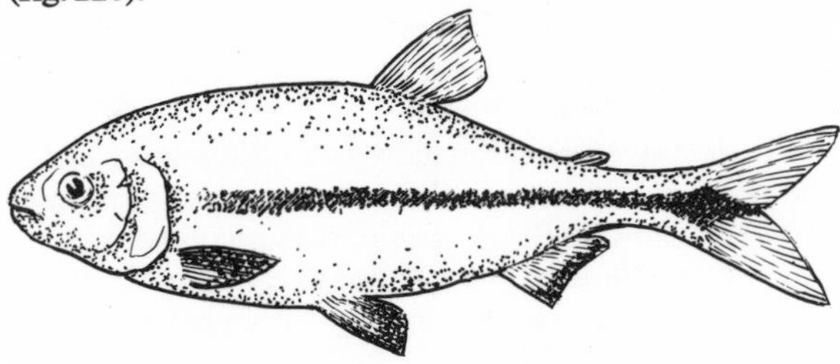

Fig. 220 *Brycon whitei*

Bryconinae, Sub-F, *see* Characoidei

Bryophyta; Bryophytes. Division of the Phytobionta with 26,000 species. Small-sized plants with a thallus* for a vegetative body. This can be simple in shape, eg flat with a lobed edge or band-like with forked branches, but it may also be differentiated into small stems and small leaves. Ingestion and the intake of water takes place over the entire upper surface of the thallus and by means of specially shaped hairs (rhizoids*), which also anchor the plant to its substrate. There is no vascular system for water or nutrients. Distribution via spores.

Bryophyta have an alternation of generations* in which, unlike the pteridophytes (Pteridopyta*), the sexual generation is brought about. First, a proembryo, looking like filamentous algae, grows out of a germinating spore. Later, the actual bryophyte, the sexual generation, develops from this proembryo. Egg-cells or spermatozoids are formed in special containers (gametangium, *see* Angium) on the bryophyte. The stalked capsule, which is brown when ripe, grows out of the fertilised egg-cell. A large number of spores develop inside the capsules which represents the asexual generation. The sexual and asexual generations remain inextricably bound to each other – in fact the capsule is nourished by the gametophyte.

Apart from the sea and places with extreme desert conditions, the Bryophyta can colonise almost anywhere. They are most suited to areas with a high humidity, however, as found in woods, moors, on rocks in shady gorges. Only a few species are adapted to life in fresh water. Bryophyta achieve their greatest variety of forms in the tropics. They are divided into distinct classes:

– *Hepaticae;* Liverworts. 10,000 species. Thallus* flat or band-like, or consisting of small stems and leaves. Unistratose leaves with no mid-rib and parts arranged in twos. Only the Ricciaceae F is of interest to the aquarist, as it contains the Ga *Riccia* and *Ricciocarpus* which include water-dwelling species.

– *Musci;* Mosses. 16,000 species. Thallus always divided into stems and leaves. Unistratose leaves with multi-layered mid-rib, arranged in spirals. Of interest to the aquarist are the Fs Fontinalaceae, with the G *Fontinalis**, and the Hypnaceae, with the G *Vesicularia**. Members of the G *Sphagnum** may crop up in marshy and acidic places, particularly moors. *Sphagnum* is the only G of the Sphagnaceae F.

Bryozoa; Moss Animals. Cl of the Tentaculata with about 4,000 species, the overwhelming majority of which live in the sea, with a dozen or so species having come to have lived in fresh water in the course of their evolution. Moss animals range in size from between

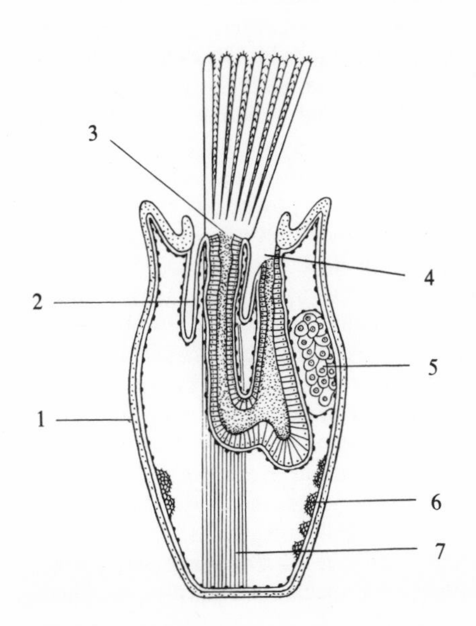

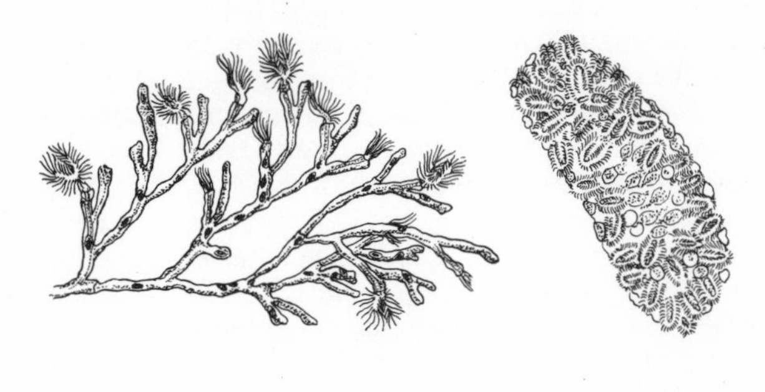

Fig. 221 Bryozoa. Left, the anatomy of a 'moss animal'; Middle, *Plumatella fruticosa*; Right, *Cristatella mucedo*. 1 Body wall 2 Anterior section with crown of tentacles 3 Mouth 4 Anus 5 Ovary 6 Testis 7 Retractor muscle

fractions of 1 mm to several mm, depending on species. However, with the exception of a marine G, the Bryozoa form colonies of considerable size at times (fig. 221). Such a colony may form from sexually produced individuals that propagate themselves by budding, ie asexually. The individual organism, often numbering thousands, can either all be the same shape, or they are built differently to correspond to their special functions within the colony. There is considerable variation in form between the different species of bryozoan colonies, but most sit like a crust on rocks, aquatic plants, fences and other objects. However, there are sponge-like or antler-like formations, too. Like most sessile animals (at least the marine varieties), the Bryozoa have a free-swimming larval stage which helps the animals to spread. In freshwater forms, small spores form that can survive the winter. They contain barbs which become attached to water birds and such like and so are carried from one water to another. Bryozoans eat the finest plankton* which they swish towards them with their crown of tentacles. The beat of the tentacles in the larger freshwater forms can be observed on the aquarium even with the naked eye.

Bubble-nests. Many Labyrinth-fishes (Anabantoidei*) and a few other species (eg *Hoplosternum**) build bubble-nests to protect their eggs and young fish (a form of Brood Care Behaviour*). The nest is usually made by the male from air-bubbles which he fills with saliva. It floats on the water or under plant leaves, for example, at a small distance from the surface. The size and shape of the nest varies with species. The eggs are laid under the nest and often climb upwards because of their oil content; or they are gathered together by the male and spat into the nest. Many Labyrinth-fishes build bubble-nests out of air bubbles and plants.

Buckbean, F, *see* Menyanthaceae

Buckbean *(Menyanthes trifoliata) see* Menyanthaceae

Bucktoothed Tetra *(Exodon paradoxus) see Exodon*

Bud. The rudimentary shoot or rudimentary flower. Buds usually develop in a leaf axil*, more rarely on other parts of the plant. They may start forming right away or they may remain in a dormant state for some time. We know of this latter fact from the reserve buds found on rearward rhizome sections in species of *Cryptocoryne* and *Echinodorus*.

Budding. An asexual means of reproduction found in some Sea-anemones (Actiniaria*), eg *Aiptasia**, *Metridium**. Small pieces of the anemone's foot break loose and develop into new specimens (Laceration*).

Buenos Aires Tetra *(Hemigrammus caudovittatus) see Hemigrammus*

Buffer. A mixture made of a weak acid* and its salt or one made of a weak base* and its salt. In the aquarium, the carbonate-carbonic acid equilibrium in particular works as a buffer. In a somewhat simplified form, this is how the buffer system works.

Carbonic acid dissociates noticeably in water into ions*; it can be seen in the following equation that this dissociation leads to hydrogen ions and hydrogen carbonate ions:

H_2CO_3	$\rightarrow H^+$	$+ HCO_3^-$
carbonic acid	hydrogen ions	hydrogen carbonate ions

Through the salt that belongs to it (in this case, likewise a hydrogen carbonate), the amount of hydrogen carbonate ions is dramatically increased. According to the law of mass effect, all known concentrations of substances stand in a certain relation:

$$\frac{H^+ \cdot HCO_3}{H_2CO_3} = K$$

The quotient of the known concentration of substance is therefore constant. If the hydrogen ion concentration increases through the addition of an acid, carbonic acid will arise immediately from H^+ and HCO_3^- ions—ie a lowering of the pH value* is prevented. By analogy, when a base is added, H^+ ions form immediately from carbonic acid, the same system having also prevented an increase in pH. From this can be explained why buffered water is unaffected by the addition of bases and acids. In practice in the aquarium, therefore, one must not allow the hydrogen carbonate content to go below a certain minimum (Water Preparation*, Standard Water*). The accompanying carbonic acid is introduced by the ventilation.

Bugs, O, *see* Heteroptera

Bulb. A bulb consists of a very compressed stem axis, usually underground, containing tightly packed cup-shaped or scale-like cataphyllary leaves in which reserve substances are stored. Only found in the G *Crinum** among marsh and water plants.

Bullhead *(Cottus gobio) see Cottus*

Bullhead Sharks, F, *see* Heterodontidae

Bullheads (Cottidae, F) *see* Cottoidei

Bullheads, related species, Sub-O, *see* Cottoidei

Bumble-bee Catfish *(Leiocassis siamensis) see Leiocassis*

Bumble-bee Fish *(Brachygobius xanthozona) see Brachygobius*

Bunocephalinae, Sub-F, *see* Aspredinidae

Bunocephalus KNER, 1855. G of the Aspredinidae*. Found in tropical S America, east of the Andes from Venezuela to Argentina. 12–15 cm long, naked, flattened Catfishes, like a paper kite in shape, with a very long caudal peduncle. The head has bony outgrowths and ridges, and the caudal peduncle is three or four cornered. The praedorsal ray of the short D1 has saw-teeth which are bent backwards and which are longest at the tip. The Ps are very powerfully developed with a strengthened and toothed first ray. An short, C cut off almost straight, Vs small, D2 absent. Of the 3 pairs of barbels, the pair on the upper jaw are the longest. Coloration very variable, upper side grey-brown with marbled effect or rows of dark blotches. Dark longitudinal bands also sometimes present. Ventral side whitish with dark spotting, often close together. *Bunocephalus* is an inactive bottom-dwelling fish which lies motionless for hours buried in the ground. According to RÖSSEL they swim jerkily by pressing water out through their gill slits, following the jet principle. They are long-lasting omnivores which have not yet been bred successfully. All species are hard to distinguish.

– *B. kneri* STEINDACHNER, 1882; Banjo Catfish. Western Amazon to Ecuador. Up to 12 cm (fig. 222).

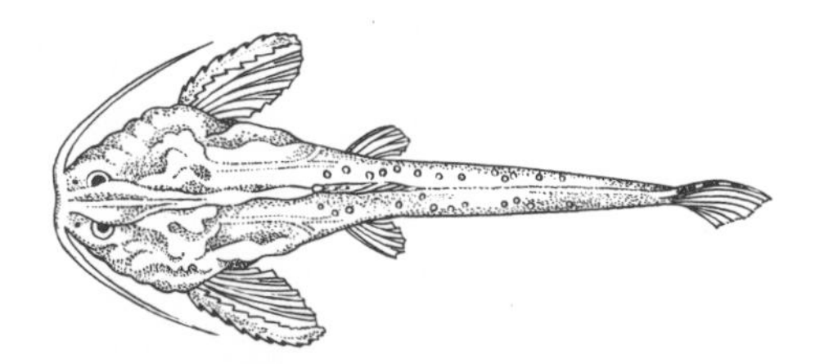

Fig. 222 *Bunocephalus kneri*

Bunodactis VERRILL, 1899; Gem Anemone. G of the Sea-anemones (Actiniaria*). Lives in the littoral* of warm and temperate seas, attached inside crevices and holes. *Bunodactis* is a small, strikingly coloured anemone without little sacs round the edge but with rows of suckers on the column and strong tentacles. The tentacles are fully retractable. One species in particular is kept in the aquarium which is livebearing and regularly reared:
– *B. verrucosa* (PENNANT, 1766); Gem Anemone. Mediterranean, European Atlantic coast. Diameter of the crown of tentacles up to 4 cm. Tentacles blackish with shining white flecks, mouth red-violet, colouring variable.

Burbot *(Lota lota) see* Gadiformes

Bur-reed (*Sparganium,* G) *see* Sparganiaceae

Bur-reeds, F, *see* Sparganiaceae

Bushmouth Catfish *(Ancistrus dolichopterus) see Ancistrus*

Butomaceae. F of the Liliatae*, with 4 Ga, 15 species. Primarily tropical. Perennials which can grow in very different forms, with a rosette leaf arrangement. Flowers hermaphroditic, polysymmetrical. Perianth comes in various forms, but the majority are strikingly coloured. All members of the Butomaceae are marsh or aquatic plants. In temperate parts of Europe and Asia *Butomus umbellatus* LINNÉ is found. It grows up to 1.5 m tall and has red umbels (fig. 223). In places, it may grow among the reeds in still water or slow-flowing waters. More rarely, it grows completely submerged with leaves streaming. Butomaceae are suitable for open-air ponds. In aquatic plant greenhouses in botanical gardens you will often see the S American *Hydrocleis nymphoides* (WILLDENOW) BUCHENAU with its brilliant yellow flowers floating on the surface.

Fig. 223 *Butomus umbellatus*

Butomus umbellatus *see* Butomaceae

Butter Hamlet, G, *see Hypoplectrus*

Buttercup Family *see* Ranunculaceae

Butterfish *(Pholis gunellus) see* Blennioidei

Butterfishes (Nomeidae, F) *see* Stromateoidei

Butterfishes (Pholididae, F) *see* Blennioidei

Butterfishes, related species, Sub-O, *see* Stromateoidei

Butterfly Barb *(Barbus hulstaerti) see Barbus*

Butterfly Cod *(Pterois volitans) see Pterois*

Butterfly Dwarf Cichlid *(Apistogramma ramirezi) see Apistogramma*

Butterfly Tetra *(Gymnocorymbus ternetzi) see Gymnocorymbus*

Butterfly-fish *(Pantodon buchholzi) see* Osteoglossiformes

Butterfly-fishes (collective term) *see Chaetodon*

Butterfly-fishes (Pantodontidae, F) *see* Osteoglossiformes

Butterfly-fishes or Coralfishes, G, *see Chelmon, Forcipiger*

Cabomba AUBLET, 1775; Fanwort. G of Cabombaceae*; 10 species. Tropical and temperate America. Small perennial plants with spreading stem-axes. Underwater leaves decussate or arranged in whorls of 3 with the pinnules often spreading out like a hand. The floating leaves near the flower are small, round or sagittate in alternate arrangement. The flowers are hermaphroditic, polysymmetric. Double perianth without a separate calyx or corolla, three-part, corolla leaves partly lobed. Stamens 6–3. Carpels 4–2 free, superior. Carpels unite to form a nut-fruit. The species of this G are exclusively water plants, many of which have already been used as aquarium plants for a long time. They need plenty of light and are sensitive to disturbed detritus. They can be propagated by cuttings.

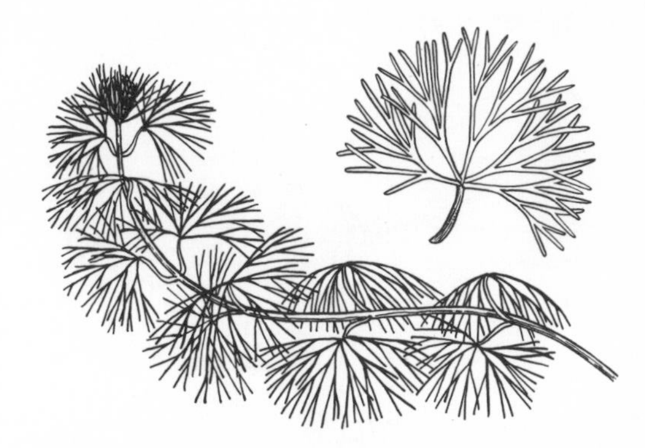

Fig. 224 *Cabomba caroliniana*

– *C. aquatica* AUBLET. Central America, northern S America. Underwater leaves green. Leaf-blade edge is roundish. Floating leaves round, flowers yellow.
– *C. caroliniana* A. GRAY; Carolina Water Shield (fig. 224). America. Underwater leaves green. Leaf-outline kidney-shaped. Floating leaves sagittate. Flowers white. To this species belong the var. *paucipartita* RAMSHORST et FLORSCHUTZ with broader pinnules and var. *tortifolia* MÜHLBERG (fig. 225) with pinnules which turn in on themselves.
– *C. pulcherrima* (HARPER) FASSETT. Southern N America. Leaves reddish, floating leaves sagittate. Flowers red-violet.

Fig. 225 *Cabomba caroliniana* var. *tortifolia*

Cabombaceae. Fanwort family. F of Magnoliatae*. 2 Ga, about 10 species. Southern hemisphere and N America. Small perennial plants with spreading stem-axes and varied leaf arrangement. Pinnules of underwater and floating leaves form a hand-shape. Flowers hermaphroditic polysymmetric, 3 part. Exclusively water-plants, the Ga are *Brasenia** and *Cabomba**.

Cactus Algae, G, *see* Halimeda

Caddis Flies, O, *see* Trichoptera

Caecobarbus BOULENGER, 1922. G of Cyprinidae*, Sub-F Cyprininae. Small fish found in the waters of the cave regions of Ngungu (Thysville) in the lower reaches of the Congo and other caves as far as the northern border of Angola. The body is elongate and hardly flattened at the sides. 2 long barbels in upper jaw. The eyes are completely non-existent. The skin which is completely lacking in pigment is given a pinkish hue by the vessels which shine through. The fish live in the waters of caves. They shun light and find their way about exclusively by sense of touch and use of the lateral line organ. Tanks should stand in the dark and should be spacious enough for these restless fish. Temperature 18–20 °C. This fish will eat any kind of live or dried food. Reproductive habits unknown. There is only one species in the G.
– *C. geertsi* BOULENGER, 1921; Blind Cave Barb. Up to 9 cm. Sexual differences unknown.

Caenis STEPHENS, 1835 (fig. 226). Widespread and numerous G of the insect order of Mayflies (Ephemeroptera*). The larvae live in lakes but also in rivers and streams where they creep around sluggishly on the plant growth on stones and wood. Their body, legs and tail-threads are thickly covered with bristles, and mud which get caught in these provides them with an excellent camouflage. *Caenis* larvae are particularly good as food for the larger bottom-feeding fish. The imagos fly at dawn and dusk in thick swarms and only live for a few hours.

Caesio LACÉPÈDE, 1802; Fusilíer (fig. 227). G of Lutjanidae*. Different species found in tropical and subtropical Indo-Pacific and in the Red Sea. *Caesio* species

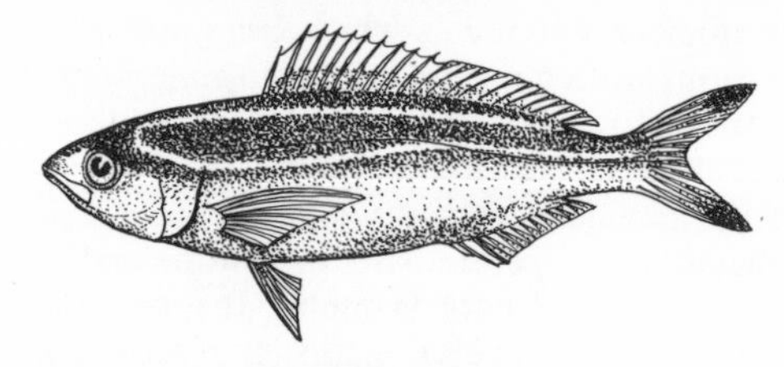

Fig. 227 *Caesio diagramma*

are spindle-shaped free-swimming fish with a deeply indented C. They live in shoals near reefs but do not conceal themselves in the coral when danger threatens.

Fig. 226 *Caenis*. Left, imago (without tail-threads); right larva

They grow to 20–35 cm long. They are sometimes treated as a separate F (Caesiodidae) from the snappers who live as bottom-dwelling predators and hunters. There are no recorded aquarium observations.

Calabar Lyretail *(Roloffia liberiensis) see Roloffia*

Calamoichthys Smith, 1865. G of Polypteridae, O Polypteriformes*. Found in W Africa in the Niger delta near Calabar and in Cameroun. They live in muddy or brackish waters. These large eel-like fish possess a strikingly broad head and a mouth with a large gape. Both nasal orifices are extended by short, tubular projections. LL present. The upper surface of the body is covered with ganoid scales (*see* Scales). The D consists of 10–13 separate little fins which are each formed from one jointed fin-ray. The Vs are absent. The fan-shaped Ps are positioned at the base of a muscular stump and serve as an additional locomotive organ and as a support when resting. The swimbladder is directly connected to the rear oral cavity and functions as a supplementary respiratory organ. These tranquil fish are very robust if kept in tanks where the water is not too hard and there is plenty of cover. Temperature 23–28 °C. The fish are active at night and need a substantial diet of live food, particularly worms, insect larvae and crabs. Their growth is relatively slow. Only one species.

– *C. calabaricus* Smith 1865; Reedfish. Up to 37 cm long. Olive-green back, sides yellowish-green, belly yellow. They have a black spot on the Ps, male 12–14 and female 9–12 fin-rays. Breeding habits unknown.

Calanus Leach, 1816. G of the Copepods (Copepoda*). Sea plankton with long waving antennae. The *C. finmarchicus* Gunn which is found from the Arctic to the Indian Ocean grows up to 5 mm long. It forms large swarms which sometimes make the water look red. It is an important food for fish, such as the herring.

Calappa Weber, 1795. G of Crabs (Brachyura*), F Calappidae. The *Calappa* species live in the littoral and sub-littoral waters of all the warmer seas. They prey on every kind of slow-moving and immobile creature. Characteristic of them are the strikingly broad claws. When these are together they can fit exactly, like a shield, over the front part of the body, including the oral organs and the respiratory openings. *Calappa* lives in sand and buries itself up to its eyes. The claws which are held above the body then allow an unimpeded circulation of water for breathing. The single European species is sometimes also kept in aquaria.

– *C. granulata* (Linnaeus, 1758); Box Crab. Found in the Mediterranean in depths of 30–100 m. Width of the body up to 9 cm. Sandy yellow with red spots. Upper side of claws like a coxcomb.

Calcareous Algae *see* Coral Reefs

Calcareous Tube-worm, G, *see Spirorbis*

Calcium-Carbolic Acid Equilibrium *see* Biogenic Decalcification

Calico Goldfish *(Carassius auratus auratus) see Carassius*

California Halibut *(Paralichthys californicus) see* Pleuronectiformes

Caligus, G, *see* Copepod Infestation

Calla palustris *see* Araceae

Calliactis Verrill, 1869. G of Sea Anemones (Actiniaria*). Found in warm and temperate seas, as a rule on the shells of hermit crabs and snails. The *Calliactis* species are, however, not totally dependent on these and exhibit no particular adaptation of body-form as do *Adamsia**. These are smaller Actinia with numerous relatively short, thin tentacles. They have stinging threads (Acontia*), which are emitted through the mouth and slits in the body wall when stimulated. The following species is often kept in aquaria.

– *C. parasitica* (Couch, 1838). Found in the Mediterranean and the European Atlantic coast. Height up to 10 cm. Dirty grey in colour with dark longitudinal stripes on the body wall.

Callichthyidae; Mailed Catfishes (fig. 228). F of Siluriformes*, Sub-O Ostariophysi. The Callichthyidae which are widespread in South America are also found in Trinidad. They are easily recognised by the smooth bony plates arranged in double rows on the flanks. The individual plates of each row are arranged from front to back like tiles. In addition those of the upper and

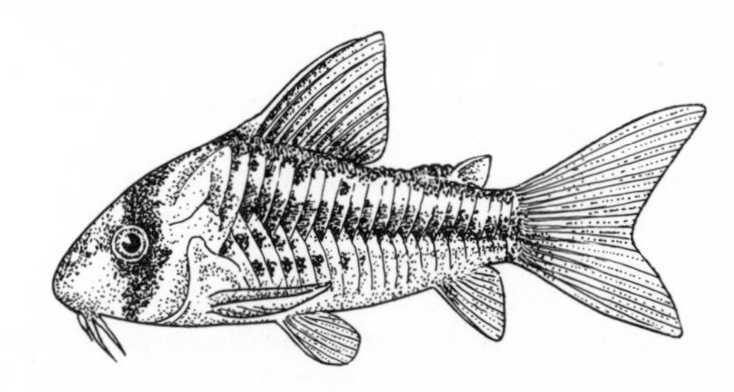

Fig. 228 Callichthyidae-type

lower rows overlap so that the total impression is one of a herringbone pattern. Furthermore, part of the head and dorsal ridge may also be armoured. The shape of the body is more or less elongated with only slight lateral compression and the Ga *Corydoras* or *Brochis* are characterised by a relatively flat underside and a curving high-arched dorsal profile. The slightly pointed, terminal and somewhat protractile small mouth is surrounded by two pairs of barbels each originating from the side of the upper and lower lip. In addition many barbel-like extensions have developed between the two barbels of the lower lip. The length of the barbels varies greatly between individual Ga, for example, those of the Ga *Callichthys* and *Dianema* are long and those of the G *Corydoras* are short. The jaws may be toothed or toothless. Eyes large and moveable. The large D, which is often sail-shaped, as well as the Ps, begin with a spiny ray, and the small adipose fin is also supported by a spine. C rounded or indented. An short, swimbladder is in two parts and encased in bone. The representatives of the well-known G *Corydoras*, in particular, have supplementary hind-gut respiration. Air taken from the surface in the mouth is swallowed and carried by intestinal peristalsis to the hind-gut, as with the Cobitidae*. Here the exchange of gas takes place, the used air leaving the intestine via the anus. This supplementary respiration is essential for many species even in good aquatic conditions. The representatives of most Ga live in shoals in weeded waters. They feed mainly on worms and insect larvae which they seek out, constantly rummaging in the soft sandy bottom. Using the spines of their pectoral fins the Armoured Catfishes at least can cross short flat stretches of land such as small sandbanks at low water. Some species are eaten by man in areas where they are commonly found. They are roasted and broken open like sweet chestnuts; in the case of the heavily armoured *Callichthys callichthys*, however, a stick or stone must be used for this purpose. Many Ga, but above all the numerous varieties of *Corydoras*, are very popular and interesting aquarium fish. For information about the individual Ga, their breeding and their needs when in captivity see under *Brochis, Corydoras, Dianema*, and *Hoplosternum*.

Callichthys LINNAEUS 1754. G of the Callichthyidae*. Found in East Brazil to La Plata. Elongated torpedo-shaped Mailed Catfish of 8–18 cm with only slight lateral compression. The armour of the dorsal hump between D1 and D2 is characteristic of this fish. This is comprised of bony plates which are not connected to the scutes of the upper lateral row. C is rounded. There are two pairs of long barbels in the upper jaw extending both forwards and downwards. They are a dull brown colour with a delicate blue sheen. The transparent fins have fine dark spots. The first ray of the Ps is thickened and orange-coloured in the male. The females are somewhat larger and more compact. The fish are indiscriminate omnivores, which build bubble nests among the surface leaves of plants. The eggs are attached to the undersides of these and are guarded by the female. The young are reared on rotifers, then later with *Enchytraeus* and freshly killed *Cyclops* as well as minced *Tubifex*.

– *C. callichythys* (LINNAEUS, 1758); Armoured Catfish, Hassar. Found in Eastern Brazil up to La Plata. Grows up to 10 cm (fig. 229).

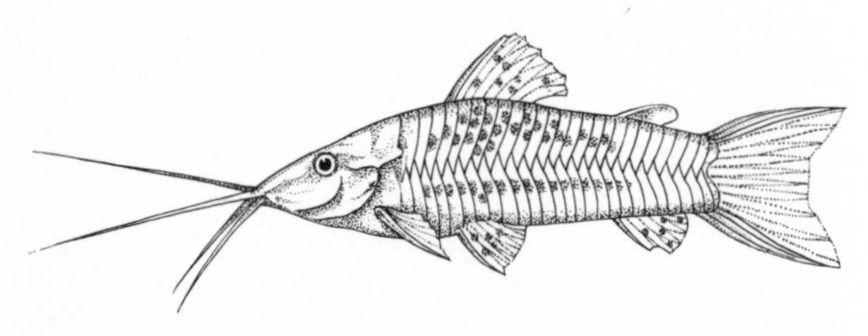

Fig. 229 *Callichthys callichthys*

Callinectes STIMPSON, 1860. G of Crabs (Brachyura*). Several species are found on the Atlantic and Pacific coasts of America. The carapace is broader than it is long and has two long lateral spines, the front edge is covered with a sharp point. The claws are large and sharp. In the USA *C. sapidus*, the Blue Crab, is of greater economic importance and is caught and eaten in huge quantities. It has also been imported by ship to Europe.

Callionymidae, F, *see* Callionymoidei

Callionymoidei; Dragonets, related species. Sub-O of Perciformes*, Sup-O Acanthopterygii. Small bottom-dwellers which are found in warm and temperate seas, fossils of which can be shown to date back to the late Tertiary (Miocene). It is comprised of one single F in which the Dragonets are included. These were formerly sometimes treated as an independent F. Elongated body, head pointed and flat. Upward-facing frog-like eyes. Operculum with backward pointing spines, in the case of the Sub-F Callionyminae (fig. 230) there is only on spine on the praeoperculum. Wide gill-slits (Sub-F Draconettinae) or like a hole and set on the upper side (Sub-F Callionyminae). D1 short, and set far forward, in the case of the male often enlarged and with ex-

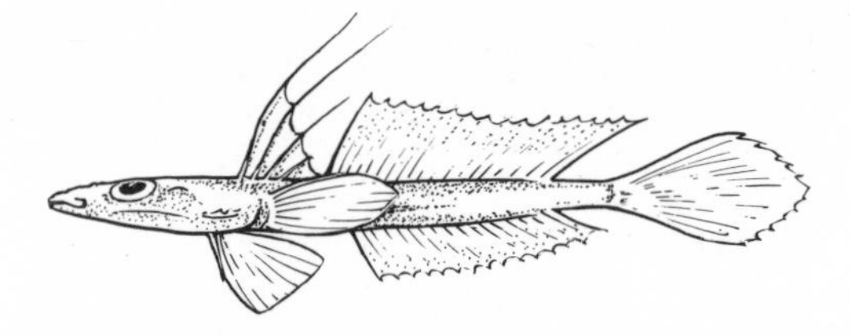

Fig. 230 Callionyminae-type

tremely elongated spines. D2 longer and in the case of the male also appreciably broader. C rounded. An approximately as long as D2. Vs set in front of the Ps, large and fan-shaped. The Cs are brightly coloured as a rule. Further details are to be found under the interesting Ga *Callionymus**, *Synchiropus**.

Callionymus LINNAEUS, 1758; Dragonets. G of the Callionymidae. Sub-O Callionymoidei*. Bottom-dwellers of warm and temperate seas, which can bury themselves in sand up to their protruding eyes. They have a broad flattened head with spines at the side and a small pointed mouth. Their food consists of various small animals. As sand-dwellers their colouring matches that of the sea-bed. They are, however, equipped with a visual means of communicating with members of their own species which is in the form of erectile, coloured Ds. The males are usually significantly larger than the females. They have greatly elongated Ds and are brillantly coloured in the breeding season. In the case of *C. lyra* spawning has been observed: after vigorous mating the pairs swim together to the water surface while the male links himself to his mate with the large P. The pelagic eggs are laid near the surface. Dragonets are somewhat delicate when kept in aquaria. They need water of high quality and require a great variety of live food. The G, however, includes fish which are interesting and delightful to keep, eg:

– *C. lyra* LINNAEUS, 1758; Dragonet. Found on European coasts to the Baltic, in shallow waters and up to 300 m deep. The male grows up to 25 cm long and displays shining blue markings during the breeding period.

– *C. festivus* PALLAS. The most common species found in shallow waters of southern European coasts. The male grows up to 14 cm long. Sandy in colour, with

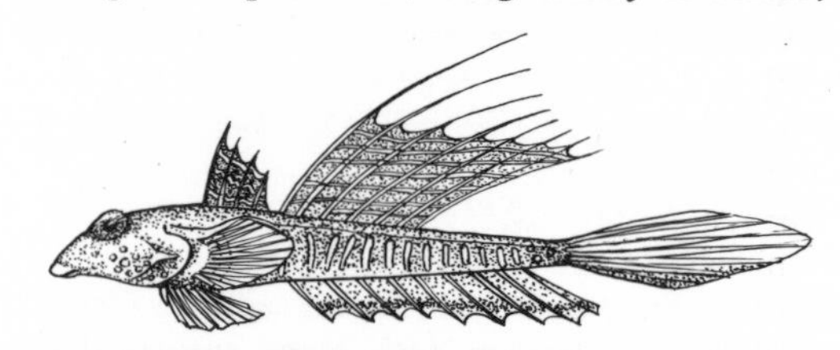

Fig. 231 *Callionymus festivus*

shining blue horizontal stripes during the breeding period (fig. 231).

Callistus Tetra *(Hyphessobrycon callistus) see Hyphessobrycon*

Callitrichaceae; Water Starworts. F of Magnoliatae*. The only G is *Callitriche**.

Callitriche LINNÉ, 1753; Water Starworts. Only G of the Callitrichaceae*. 25 species. Found almost throughout the world apart from S Africa, but may be rare in some areas. Small perennial or annual plants with many-branched spreading stem-axes. The emerse plants form a thick mat, the submerged growth being in branching tufts with upright shoots. The leaves are loosely opposite, often compressed together in rosettes on the surface. These are linear, spatula-shaped, rounded, truncated, or double-pointed. Flowers inconspicuous, polysymmetric, unisexual and dioecious. Perianth absent. One stamen. Two carpels, fused and superior. Winged fruit. All the species are marsh and water plants. Emerse plants rooting in mud usually die off in winter, whereas those growing submerged in still or flowing water exhibit weak growth. Some endure a very low temperature. The species are to some extent very variable and hard to define. There is hardly any practical knowledge concerning their cultivation. Indigenous species could be considered for cold water aquaria. Since the plants are very attractive, experiments with higher temperatures should also be undertaken. Widespread species are as follows:

– *C. autumnalis* LINNÉ; Autumnal Starwort. Found in N America, N Europe, N Asia. They exist only as a submersed plant with no rosettes.

– *C. vernalis* LINNÉ; Vernal Starwort. N America, Europe, Africa, Asia. Emerse and submerse plants with rosettes.

Callyodon (BLOCH and SCHNEIDER, 1802). G of the Scaridae, Sub-O Labroidei*. There are many species found in the tropical Indo-Pacific and Atlantic where they live in shoals, mostly on coral reefs. These fish, which are about 30–60 cm long, bite off little pieces of coral and plant matter with their beak-like dental plates. Their powerful longitudinally elongated oval bodies are covered with large scales and there are three rows of scales on the cheek beneath the eye. They are brightly coloured. The sexes often look quite different from each other. Many cover themselves in mucus at night. These fish which require special diets are seldom imported, but it is quite possible to keep them in an aquarium.

Calopteryx LEACH, 1815 (fig. 232). G of the insect family of Dragonflies (Odonta*). They are to be found throughout the northern temperate hemisphere (Holarctic*), mainly near flowing water. Their bodies are a deep metallic blue colour. So too are certain wing areas of the male. The larvae (fig. 233) have markedly long legs and are immediately recognised by the three-cornered external appendages of their hind-quarters. They can be used as food and varieties from slow-flowing water are also suitable for observation in the aquarium.

Calyx *see* Flower

Camallanus, G, *see* Nematode Diseases

Cambarus ERICHSON, 1846; American River Crabs. G of ten-legged crustaceans with many species (Decapoda*). Indigenous to various freshwater biotopes of N America. Members of the *Cambarus* G dig channels near to water but their actual hole is always found beneath the surface of the water (ground water, water seepage). The following species is very well-known and imported into Europe (allochthonic*):

Fig. 232 *Calopteryx* mating

– *C. affinis* SAY, 1817. Now also known as *Orconectes limosus* RAFINESQUE, 1817. This river-crab is well suited to the cold-water aquarium by virtue of its modest demands in respect of water temperature, water purity and oxygen content.

'Camerunensis' (*Pelvicachromis* sp.) *see Pelvicachromis*

Canadian Waterweed *(Elodea canadensis) see Elodea*

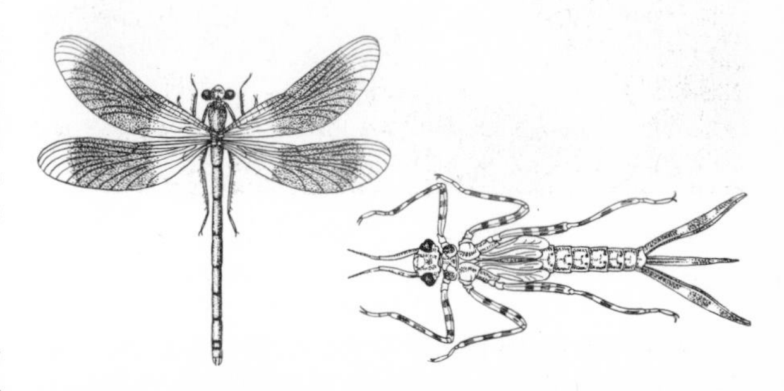

Fig. 233 *Calopteryx splendens.* Left, imago; right, larva

Cancer LINNAEUS, 1758; Edible Crab. G of Crabs (Brachyura*). Found in northern seas where they live on hard ground. They are clumsy, large, predatory crabs with broad carapaces, indented at the front. In common with other large crabs they are not really suited for domestic aquaria. In the North Sea the following species is often found.
– *C. pagurus* LINNAEUS, 1758; Edible Crab. Found from Lofoten to the Black Sea but not in the Baltic Sea. Up to 30 cm wide. Yellowish-red with black tips to the claws.

Candiru (*Vandellia*, G) *see* Trichomycteridae

Candlefish *(Thaleichthys pacificus) see* Salmonoidei

Candy-striped Basslet *(Chorististium carmabi) see Chorististium*

Canthigaster SWAINSON, 1839. G of Tetraodontidae*. Found in the tropical Indo-Pacific and Atlantic oceans. These are territory-forming fish which live on reefs and rocks. Their food consists of small creatures including coral polyps. At night they cling to the bottom with the skin of their undersides. They are small, long-nosed and brilliantly coloured Pufferfishes whose Vs have completely degenerated. When they display and inflate their bodies they raise a ventral comb. All species of the G have shown themselves to be very easy to keep in aquarium and are very peaceable except towards members of their own species. The following species are imported in great numbers.
– *C. margaritatus* RÜPPEL; Ocellated Pufferfish, Peacock-eyed Puffer, Diamond-flecked Pufferfish. Indo-Pacific. Up to 15 cm long with shining blue pearl spots on a brown-red background and a dark eye-spot on the D.
– *C. valentini* BLEEKER; Minstrel Pufferfish. Indo-Pacific. Up to 20 cm long. Yellow with two large dark-brown saddle markings on the back and sides of the body (fig. 234).

Fig. 234 *Canthigaster valentini*

Canthigasteridae, F, name not in current usage, *see* Tetraodontidae

Capillaria, G, *see* Nematode Diseases

Caprella LAMARCK, 1801. G of Skeleton Shrimps (Caprellidae), which belong to the Amphipods (Amphipoda*). These are extremely elongated marine Shrimps with reduced abdomens. They live as predators on algae and colonies of Hydrozoa and the like. They cling with their claw-like rear legs and wait without moving for their prey which they seize with their two pairs of claws. Because of their behaviour and their good camouflage they are often overlooked.
– *C. linearis* (LINNAEUS, 1758). North Sea, Baltic Sea. Up to 32 mm.

Caproidae, F, *see* Zeiformes

Capros aper *see* Zeiformes

Caracanthidae, F, *see* Scorpaenoidei

Carangidae; Jacks, Scads, Pompanos (fig. 235). F of Perciformes*, Sub-O Percoidei*. F with many species, fossils of which have been found in the upper layers of chalk. The Carangidae have relatively high backs or are elongated and laterally much compressed. Caudal peduncle is very slim, D1 small and short, D2 long with

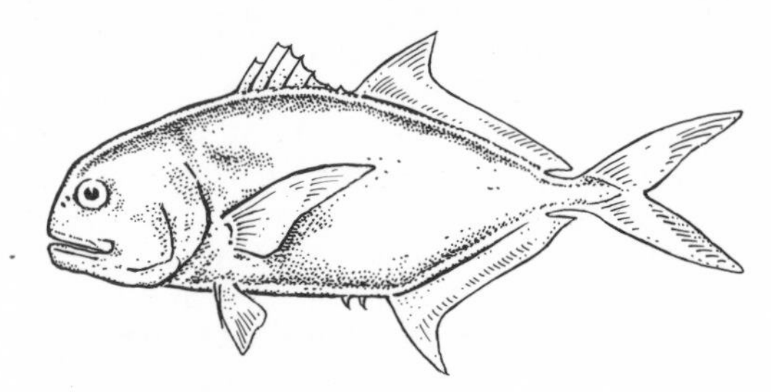

Fig. 235 *Carangidae*-type

the front section often lengthened like a pennant. The C is deeply forked, the An shaped similarly to the D2. In front of the An there are two free-standing spines. Some species have small fins behind the D2 and An. Tiny round scales. There may be enlarged spiny scales along the whole lateral line or along the rear part of the fish. The Carangidae are predominantly deep sea fish of warm and temperate seas. A few species penetrate into fresh water from time to time. Almost all species are free spawners, the eggs floating in the water. These fish usually grow up to 60–100 cm long and in rare cases reach 2 m. They are excellent to eat and are much loved by sea anglers. One of the best known species, the dark-banded bluish Pilotfish *(Naucrates ductor)* (fig. 236), accompanies large fish, above all sharks, and

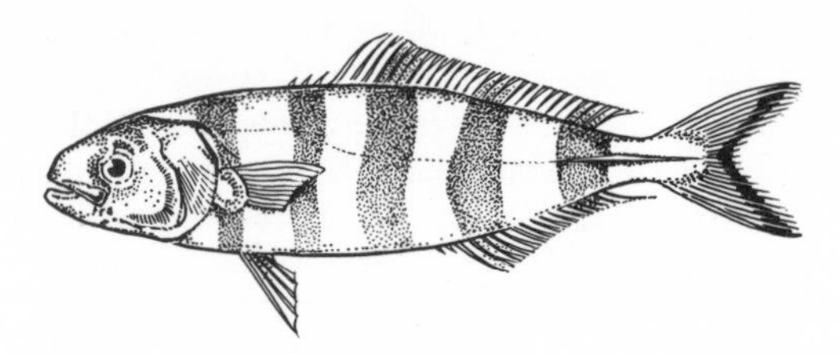

Fig. 236 *Naucrates ductor*

also ships – a habit that was already known about in ancient times but has still not been satisfactorily explained. Many species can be kept quite easily in marine aquaria when they are young. For information about this see under the Ga *Alectis, Caranx, Trachurus.*

Caranx LACÉPÈDE, 1802. G of Carangidae*. There are many species of these found in tropical and moderately warm seas; a few also penetrate into river estuaries. Found near coasts where they feed on small fish and crabs. Longish to oval in shape. LL is accompanied by scutes in the rear section of the body, C deeply indented. The young fish sometimes live under the protection of large jellyfish. One rarely sees *Caranx* in aquaria because of their great mobility and rapid growth. An exception to this rule is the Kingfish or Golden Trevally, *C. speciosus* (FORSKÅL, 1755) (fig. 237), from the Indo-Pacific to the Red Sea. This fish, which grows up to 90 cm long, is silvery white and has broad black and golden lateral bands. Head and fins reddish-yellow. Very rapid growth.

Fig. 237 *Caranx speciosus*

Carapidae, F, *see* Gadiformes

Carassiops OGILBY, 1897. G of Gobiidae, Sub-O Gobioidei*. Australia. Bottom-dwellers of fresh water and brackish coastal waters. They can survive dry periods in mud. Medium-sized fish with laterally compressed bodies. Head and back relatively high. P not fused at the base. Colouring of those species which have been kept so far greenish to blue with red spotted markings. Can be kept at 20–25 °C in tanks with hiding places and soft, fine, sandy bottoms. The addition of salt (one to two teaspoons to 10 litres of water), is recommended. They need live food, including fish. There are no observations on breeding habits.

– *C. compressus* (KREFFT, 1864); Australian Sleeper Goby. Up to 15 cm. Body relatively strongly compressed and elongated. Greenish to blue, red spots and stripes on the body which become little dots at the base of the unpaired fins. Well defined blue spot in front of the P. Female paler, D2 and An smaller (fig. 238).

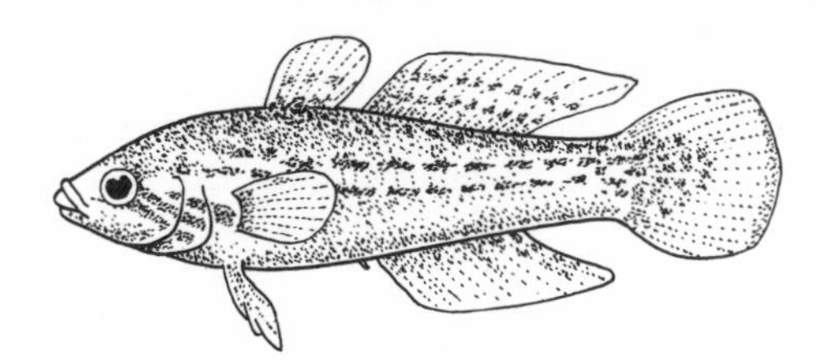

Fig. 238 *Carassiops compressus*

Carassius JAROCKI, 1822; Goldfish. G of Cyprinidae*. Sub-F Cyprininae. Europe, N Asia to China. Bottom-dwelling, medium-sized fish, which live in both smaller and larger groups in all waters though predominantly still ones. Their body is elongate, laterally compressed, in general similar to the Carp. Body often becomes very deep when older. The terminal mouth has no barbels. High, long D, shorter, slightly indented C. LL complete. The colour is very variable. These undemanding fish thrive in large cold water aquaria with sandy base material and a moderate number of plants. They need clear, clean water which is partially renewed weekly. They eat all kinds of live and dry food and plant food.

– *C. auratus gibelio* (BLOCH, 1822); Prussian or Gibel Carp (fig. 239). Up to 20 cm. Colour yellowish to silver-grey. Spread widely from E Asia to Central Europe. In the more western regions there are often only females which spawn with males of other species of carp. The egg-cells are stimulated to develop by the sperm from a different species and without exception produce only female offspring.

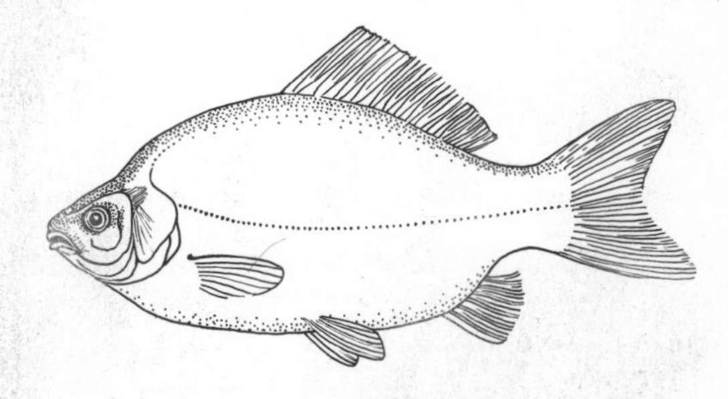

Fig. 239 *Carassius auratus gibelio*

– *C. auratus auratus* (LINNAEUS, 1758). Goldfish; Veiltail, Shubunkin, Calico Goldfish (fig. 240). Sub-species of *C. auratus* which was developed by special breeding in China and Japan about 1,000 years ago. It was introduced into Europe as an ornamental fish in the seventeenth century and became established in the wilds in some parts of south-east Europe. Very variable colouring. Back red-gold, sides golden coloured, underside yellowish. Fins red-gold to yellow. In addition some are pale pink, red and white, also speckled black to blue-black and yellow to brown. The young are uniformly brown. Further breeds have evolved from the principal form with magnificently developed fins, such as Veiltail, Comet, Shubunkin, Telescope-eyes, Oranda and Lionhead amongst others. Their main impact is visual and it was to this end that they were bred. The principal form is well able to survive the winter outside in suitable conditions. In temperatures under 10 °C they no longer feed. The specially bred species are more delicate and should be kept at higher temperatures (not less than 15 °C). The males are slimmer with an indented anal region; in the female this can be everted. Breeding of these prolific fish is possible at 22 °C in large tanks. It is not difficult when the fish are well-fed and in suitable pairs. Because of the high cost involved in the sorting of young fish, however, breeding is carried out almost exclusively in large installations in climatically suitable regions.

– *C. carassius* (LINNAEUS, 1758); Crucian Carp. Up to 45 cm. Europe, Siberia to the Lena. Back olive-green. Sides yellowish to grey-green with a bronzed sheen. Young have a dark spot on the base of the tail. Very undemanding, hardy fish, the young of which are well-suited for beginners. Endures both very low and relatively high temperatures. Eaten in certain parts of East Europe.

Fig. 240 *Carassius auratus auratus* (Lionhead)

Carbasone. A common name for 4-Carbamylaminophenilarson acid, also known as Ameban, Fenarsone, and Leucarsone. Carbasone is a white powder which dissolves in water and alcohol with difficulty and is used for combatting amoebic illnesses (Rhizopoda). Aquarists can use it to treat Discus Disease* and in the case of infection with *Hexamita** DUJARDIN, 1838. It can be used as an additive to food, mixed in the proportions of 1 g Carbasone to 500 g dried food or agar, and given to the diseased fish on 3–4 consecutive days.

Carbohydrates. A group of natural substances which consist of carbon, hydrogen and oxygen and which usually contain no nitrogen. To a very large extent, carbohydrates are produced by green plants by their utilising the energy of the sun's rays. The build-up of higher carbohydrates begins with glucose which is obtained during the assimilation of carbon dioxide (*see* Carbon Dioxide Assimilation) from the carbon dioxide in the air or in the water. Glucose is extremely important to organisms as the form in which carbohydrates are transported. In plants, carbohydrates are stored in the form of starch and sucrose, in animals and in humans in the form of glycogen. Among vertebrates, glycogen is concentrated in the liver and in the muscles mainly. During carbohydrate metabolism, glycogen is broken down as it is needed and energy is released; provided there is sufficient glucose or other primary material (such as fats), glycogen will then be built up again. There are many regulating factors involved in these metabolic processes, eg the hormones in the pancreas (*see* Digestive Organs). Carbohydrates make up a very important part of energy-giving foods (*see* Nutrition).

Chemically, carbohydrates are divided up according to the number of their constituent parts into monosaccharides, oligosaccharides and polysaccharides. The first two groups consist of sugars in their broadest sense, eg glucose and fructose, sucrose and lactose. Starch, cellulose and glycogen all belong to the third group as they are classed as high-molecule carbohydrates. Chitin is a carbohydrate containing nitrogen.

Carbon Dioxide Assimilation. Certain groups of organisms can form organic substances from carbon dioxide (CO_2) by metabolic processes (*see* Metabolism). By this process the carbon dioxide is reduced through energy being used up. Organisms containing chlorophyll use light as the energy source. This is known as *photosynthesis**. Many bacteria (Bacteriophyta*) attain the energy for assimilation through the oxidation of anorganic compounds – ie they carry out a *chemosynthesis*.

Carbon Dioxide Poisoning. Sometimes, too much uncombined carbon dioxide (CO_2) can build up in aquarium water. This can be caused by the water becoming too stagnant, by the pH values being around the neutral point and below, and by a lack of buffer capacity. Fish will react to such conditions by speeding up their respiration (this can be observed when the fish holds open its gill covering every so often). The same reactions can be observed when there is a lack of oxygen (Oxygen, lack of*) or too high a concentration of ammonia* (*see* Poisoning), so the cause is not always due to the toxic effect of CO_2. Rather, it can be traced to a disturbance of the balance between the amount of oxygen and carbon dioxide dissolved in the water. A one-sided increase in uncombined CO_2 can completely block the capacity of the fish to take in oxygen. The taking in of oxygen and the release of CO_2 via the gills is a complicated process, controlled by enzymes, and it is very much subject to the properties of the water – pH value, temperature, buffer capacity etc. If CO_2 does build up in the aquarium, do not try to increase the oxygen content by a corresponding amount; instead, create turbulence (*see* Ventilation) which will help to release the excess CO_2. In very serious situations, when a test for ammonia* proves negative, the pH value can be raised by a unit and so, via the combining of carbonic acid, a lowering of the uncombined carbon dioxide can be achieved. Fish damaged in this way usually recover quickly once conditions are normalised again.

Carbonate Hardness *see* Temporary Hardness

Carcharhinidae; Blue Sharks. F of Chondrichthyes*, O Carcharhiniformes. Found primarily in tropical seas. A few species are also found in temperate latitudes, with one freshwater example. The elongated spindle-shaped body of the Carcharhinidae tapers off towards the rear to a slender caudal peduncle which bends upwards into the greatly enlarged upper lobe of the tail-fin. The head is characterised by a long conical rostrum which is usually somewhat flattened and a large crescent-shaped mouth equipped with rows of powerful teeth. The nictitating membrane or, at the very least, a lower eye-fold is always present. All Carcharhinidae have 5 gill-slits on each side, of which the 4th and 5th or only the 5th is situated over the P. Blow-holes may be absent. Of the two dorsal fins the D1 is substantially larger, the D2 is usually directly above the small anal fin. With a total of some 65 species the Carcharhinidae are the largest F of the Chondrichthyes. These very greedy and rapacious animals primarily hunt fish, and often smaller members of their own species, but also snap at any other quarry. When seizing their prey the head is raised and moved from side to side after biting. In this way pieces are torn from the catch and swallowed in lumps. In the stomachs of larger blue sharks, not only have large fish been found but also small seals, dolphins, seabirds, and turtles, amongst other animals. They often hunt in groups and gather together around ships and boats. Some Carcharhinidae, but by no means all, also attack humans. The very common, relatively slender Great Blue Shark *(Prionace glauca)* (fig. 241) belongs to the man-eating sharks. It has a deep blue to grey-blue upper side and a light underside. Smaller examples of this pelagic cosmopolitan of the tropical and temperate seas

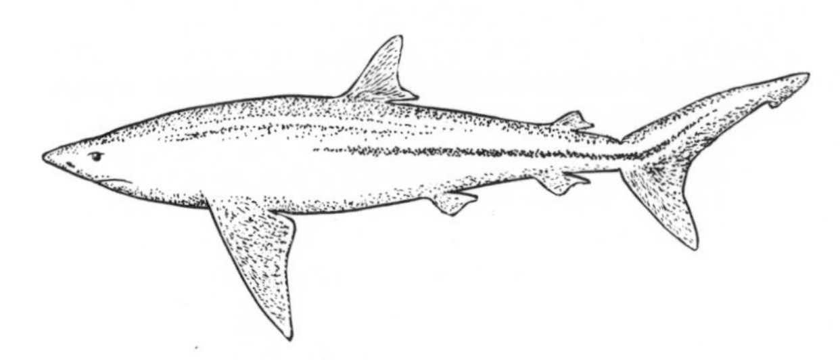

Fig. 241 *Prionace glauca*

sometimes appear in the North Sea and even very rarely in the middle Baltic. It follows shoals of fish mainly and as a rule remains in the open sea. A female can give birth to 45 live young, each of which is connected by an umbilical cord to a simple placenta in the uterus and is nourished in this manner. The Tiger Shark *(Galeocerdo cuvieri)* (fig. 242), which grows up to 5 m long, and very occasionally even larger, is also feared as a man-eating shark. The younger ones have dark spots on a light

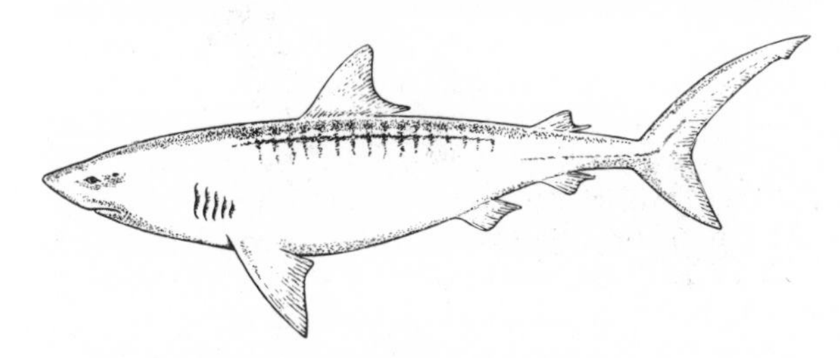

Fig. 242 *Galeocerdo cuvieri*

brown background and also have narrow crossbands. Older ones, on the other hand, are fairly uniformly grey-brown. They are found in all tropical seas and only occasionally do a few individual ones penetrate colder waters. The Tiger Shark eats anything in the main, with a predilection for other sharks and carrion. Further dangerous man-eating sharks of this family are, amongst others, the oceanic Whitetip Shark *(Carcharhinus longimanus)* of the deep seas of the Pacific and Atlantic oceans. A few species of this F enter river estuaries from the sea. Thus the Zambesi Shark *(Carcharhinus zambezenis)* has already been observed 200 km inland; the Ganges Shark *(Carcharhinus gangeticus)* is found not only in the lower reaches of the Ganges but also in the Tigris and other rivers. In the large Lake Nicaragua lives a shark which has accustomed itself completely to fresh water. This Nicaragua Shark *(Carcharhinus nicaraguensis)* which grows up to a length of 2.5 m is one of the members of this family which is most dangerous to man. The Carcharhinidae are of relatively small economic significance. Sometimes the liver oil which is fairly rich in vitamins can be used. Other species provide good leather or are processed to make fishmeal. In certain parts Carcharhinidae attract the interest of game fishermen.

Carcharhinus (various species) *see* Carcharhinidae

Carchariidae, F, name not in current usage, *see* Odontaspididae

Carcharodon carcharias *see* Lamnidae

Carcinides RATHBUN, 1897 (not actual name *Carcinus*). G of Crabs *(Brachyura*)*, F of Shore Crabs (Portunidae). They leave the sea and penetrate far into brackish water. The front edge of the carapace has saw-like indentations. In contrast to the other Swimming Crabs, the last pair of legs is only slightly broadened so that the *Carcinides* species cannot swim. These active crabs carry off any creatures they can. They are easy to keep in an aquarium. One must, however, keep them on their own if possible. If they have the opportunity to leave the water one does not even need an aerator.

– *C. maenas* (LINNAEUS, 1758); Shore Crab. Warm and temperate seas of both hemispheres. It is the most common crab of the flats of the coast of the Northern Baltic seas, and it penetrates river estuaries. Carapace up to 8 cm side, olive-green to reddish brown.

Carcinoma *see* Tumour Diseases

Cardamine LINNÉ, 1753. G of Brassicaceae*. 130 species. Temperate regions of almost the whole world. Small perennial and annual plants with leaves in a rosette and spreading stem-axes with alternate leaves on these. Leaf blades vary from undivided to pinnate. Hermaphroditic flowers, double perianth. Calyx and corolla absent, four-part. 6 stamens, 2 carpels, fused, superior. Husked fruit. The species of the G establish themselves in damp to wet environments and can sometimes tolerate even prolonged flooding. The following species is a plant which is easy to keep in aquaria but the temperatures should not be too high.

– *C. lyrata* BUNGE; Chinese Ivy, Bitter Cress. Temperate E Asia. Submerse leaves with roundish blades.

Cardinal Tetra *(Cheirodon axelrodi) see Cheirodon*

Cardinalfishes, F, *see* Apogonidae

Cardium LINNAEUS, 1758; Cockles. G of Mussels (Bivalvia*). Widely found, sea-dwelling mussels with heart-shaped, high-domed and ribbed shells and with a moveable folded foot with which some can hop forwards. Mostly they lie buried in the bottom just below the surface. Like most other sand-dwellers and filterers, they are not very suitable for aquaria. The following species, amongst others, live on European coasts:
– *C. edule* LINNAEUS, 1758; Edible Cockle. Baltic to the Black Sea. To 5 cm. White shell, yellowish or brown patterning.
– *C. aculeatum* LINNAEUS, 1758; Spiny Cockle. Atlantic coast, Mediterranean. To 9 cm. Ribbed shell set with spines. Onion-coloured.

Carex, G, *see* Cyperaceae

Caribbean Sea. A neighbouring sea of the Atlantic, located in Central America. Bounded on one side by the Greater Antilles, and linking with the Gulf of Mexico at the north-west corner via a channel between the Yucatan Peninsula and Cuba. The Caribbean is more than 3,000 km long and has a maximum depth of 7,241 m. Its coastal waters are very clearly defined. Salt content 34–37‰. Biologically, the main feature of the Caribbean is the high concentration of tropical and subtropical species. This is the result of, among other things, the changeable history of this area in recent geological times which has caused the breaking up of areals* and the isolation of organisms. The Caribbean is a popular destination for many marine aquarists and many diving and fishing expeditions are carried out.

Caribbean Scorpion *(Pterois volitans) see Pterois*

Carinotetraodon BENL, 1957. G of Tetraodontidae*. One well-known species in fresh water of South East Asia. The males of these Pufferfish (which in other respects have the typical shape of their family) erect during display a comb-like crest along their back and a sharp ventral keel like the marine species of the G *Canthigaster**. They are territory-forming and therefore incompatible with members of their own species and also, to some extent, with other fish. The small fish are very suitable for keeping in aquaria. Breeding has been successful. Both parents care for the young.
– *C. loreti* (TIRANT, 1885) (*C. somphongsi*, name not in current usage); Comb Pufferfish. Thailand, in fresh water. To 6.5 cm. Males greyish yellow with blue C; females more varied in colour, mainly plain yellowish with irregular dark spots.

Carmabi Fish *(Chorististium carmabi) see Chorististium*

Carnegiella EIGENMANN, 1909. G of Gasteropelecidae*. Widely found in the Amazon area. Small surface shoaling fish with hatchet shape typical of its family. Only 2.5–4.5 cm in length. The missing D2 distinguishes the G from the *Gasteropelecus*-species. Lateral line short or absent. The Vs are formed from 5 unbranched soft rays. All species are splendid jumpers because of their wing-like Ps, and therefore the tank must be well-covered. They eat almost exclusively flying insects such as the Vinegar-fly, small *Daphnia* on the water surface, Mosquito larvae and *Cyclops*, as well as dried food. The *Carnegiella* species are fish which are mainly active at dusk. There have been one or two successful attempts at breeding *C. strigata* and *C. marthae.* The act of spawning occurs amongst thick roots at the water surface presumably in the evening or by moonlight. According to KLUGE, when mating, the males do a fluttering dance parallel to the females on the shoots of finely feathered plants under the water surface. The eggs cling to them. The rearing of the fish, which hatch after 24–30 hours, is not difficult with rotifers and later *Cyclops* nauplii. Up till now 4 species have been imported which are interesting to keep in an aquarium.
– *C. marthae* MYERS, 1927; Black-winged Hatchet Fish. Venezuela, Peru, Brazilian Amazon region. Rio Negro, Rio Orinoco. To 3.5 cm. The sides of the body are silver with a longitudinal stripe which is dark and delicately edged in gold at the top. The Ps have a dark stripe and the ventral keel is edged with black. They are very delicate creatures which need to be tended carefully and kept in soft, slightly acidic water. They only take food from the surface, and not when it is sinking (fig. 243).

Fig. 243 *Carnegiella marthae*

– *C. myersi* FERNADEZ-YEPEZ, 1950. Peru, Upper Amazon, tributaries of Rio Marañón. To 2.5 cm. Smallest species of the G. Body almost transparent and colourless; occasionally there is a hint of a dark longitudinal stripe. Fins colourless. This species is hardier than *C. marthae.* They pick up food as it is sinking, including even *Enchytraeus*. Very elegant swimmers and good jumpers. Not yet bred in the aquarium.
– *C. strigata* GÜNTHER, 1864; Marbled Hatchet Fish. Found in small forest streams of the Amazon and Guyana. To 4.5 cm. Body sides greenish-silver to purple silver. With 3 irregular curved black cross bands. Fins colourless. Contrary to the previously widely-held opinion that they were a delicate species, they are in fact not very susceptible to illness and eagerly eat any kind of small live food and also dried food. Breeding

Fig. 244 *Carnegiella vesca* (at the top) and *Carnegiella strigata*

has been successful on one occasion. Active at dusk, prettily coloured and hardy shoaling fish which love shady positions (fig. 244).

– *C. vesca* FRASER-BRUNNER, 1950. Lower Amazon region, West Guyana. To 4.5 cm. Very similar to *C. strigata* in colour and shape but the stripes are broader. GÉRY maintains that this is a sub-species of *C. strigata*, as hybrids are said to occur in areas near their principal habitats. This fish has not yet been imported. It is the most colourful of its G. Breeding has not yet been successful (fig. 244).

Carolina Yellow Pond-lily *(Nuphar sagittifolium) see Nuphar*

Carousel Swimming *see Hippocampus*

Carp *(Cyprinus carpio) see Cyprinus*

Carp, G, *see Carassius*

Carpel *see* Flower

Carpet Shark *(Orectolobus barbatus) see* Orectolobidae

Carpet Sharks, F, *see* Orectolobidae

Carpiodes cyprinus *see* Catostomidae

Carps (Cyprininae, Sub-F) *see* Cyprinidae

Carps, F, *see* Cyprinidae

Carps, related species, Sub-O, *see* Cyprinoidei

Carp-types, O, *see* Cypriniformes

Cartilaginous Fish, Cl, *see* Chondrichthyes

Caryophylla, G, *see* Madreporaria

Caryophyllaeus laticeps *see* Cestodes

Cassiopeia, G, *see* Scyphozoa

Cat Sharks, F, *see* Scyliorhynidae

Catadromous Fish. Fish that live for a while in fresh water, although their spawning grounds are in the sea which means that they are classed as part of the marine fauna. Probably the best known example is the eel (*see* Anguilliformes).

Cataphyll (cataphyllary leaf) *see* Leaf Succession

Catfish *(Anarrhichas lupus) see* Blennioidei

Catfishes, Electric, F, *see* Malapteruridae

Catfishes, True, F, *see* Siluridae

Catfish-types, O, *see* Siluriformes

Cation *see* Ions

Catoprion MÜLLER and TROSCHEL, 1844. G of Serrasalmidae*. 14 cm long. Living in the Amazon region and Guyana. Not to be confused with the Piranha species by the flag-like elongated first rays of the D, and An. Very powerful lower jaw. Mouth superior. The Catoprions are relatively peaceable and eat only small fish and water insects. In their true habitat they are said to live largely from the scales of larger fish. They are splendidly coloured, sturdy, shoaling fish which need plenty of room. Only one species. There are no observations on breeding habits.

– *C. mento* MÜLLER and TROSCHEL, 1844. Winged or Wimple Piranha. Amazon region, Guyana. 14 cm. The disc-shaped body is silver with a delicate green sheen. For the shape of D and A see the description of the G. The first ray of the D is white, the following 3 black; the D2 is well developed. The An is brick red with white first ray. The C is deeply forked and widely curved, its base and outer rays as well as its edge are black and the rest yellowish. There is an orange-red spot on the operculum and the lateral line is marked by a silver stripe. They are aquarium fish which are much in demand but seldom imported (fig. 245).

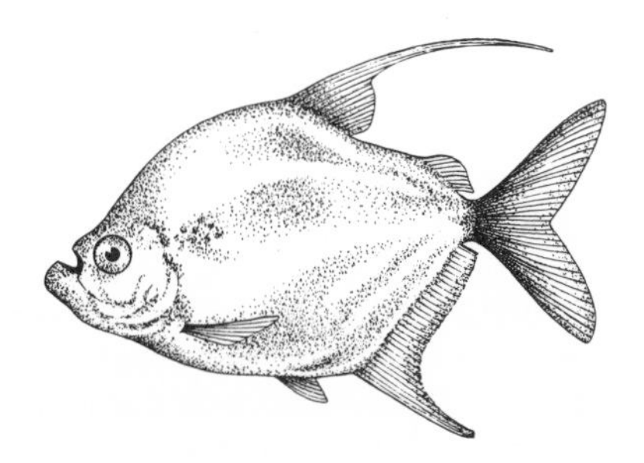

Fig. 245 *Catoprion mento*

Catoprioninae, Sub-F, *see* Serrasalmidae

Catostomidae; Suckers. F of Cypriniformes*, Sub-O Cyprinoidei. The members of this family are very similar to the Cyprinidae*. Many Ga can hardly be distinguished from the similarly formed Cyprinidae by external examination alone. The most important distinguishing characteristics between the two closely related families are the following: the Catostomidae have a short, broad head and a downward-pointing protrusile mouth, surrounded by large thick lips, generally with papillae. In addition the lower lip, which is usually larger, has a central cleft. It is almost split in half. The pharyngeal teeth are significantly more numerous, smaller and always arranged only in single rows. The first fin-ray of the D is never developed into a spine. Finally it should be mentioned that in the case of the Suckers the ratio of the distances 'leading edge of the An to mouth: leading edge of the An to C' is at least 2.5:1; in the Cyprinidae (with the exception of the Carp) this ratio is lower, usually about 2:1. The main regions inhabited by the Catostomidae extend from Mexico through the whole of N America to the Arctic. A limited number of species are found in NE Asia. In their native areas they are the main representatives of the Cyprinidae. NIKOLSKI points out that one can distinguish two groups of species. The one group has a relatively high back, small close pharyngeal teeth and a relatively long thin intestine. These are typically found in the large rivers of the lowlands and in lakes. Many grow fairly large, such as the Buffalo Fishes which are reminiscent of the Carps. The best-known species is either the Bigmouth Sucker *(Ictiobus cyprinella)* from central N America which grows up to 90 cm long (fig. 246) or the spiny-backed Quillback *(Carpiodes*

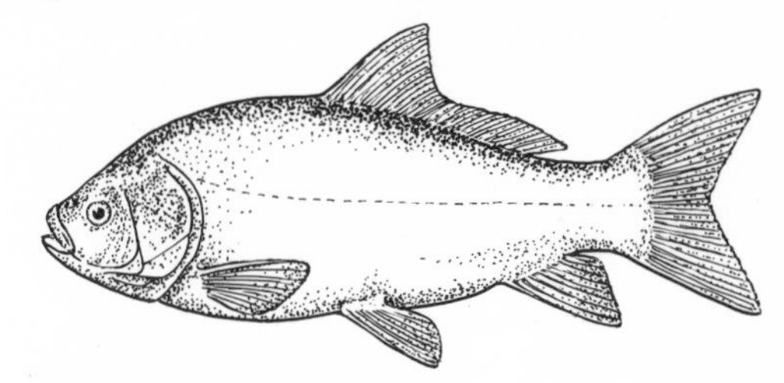

Fig. 246 *Ictiobus cyprinella*

cyprinus), which can grow up to 60 cm long and is found in the large lakes of N America, amongst other places. The Catostomidae of the second group, on the other hand, are typically found in faster flowing waters. They generally stay smaller and are an elongated spindle-shape. They also have stronger pharyngeal teeth and a shorter intestine. The Spotted Sucker *(Minytrema melanops)* from the eastern United States, which grows to 45 cm long, belongs to this group, as do members of the Ga *Catostomus* (fig. 247) and *Moxostama*. All Catostomidae live mainly on a diet of invertebrates. The high-backed species also eat plants. The males are

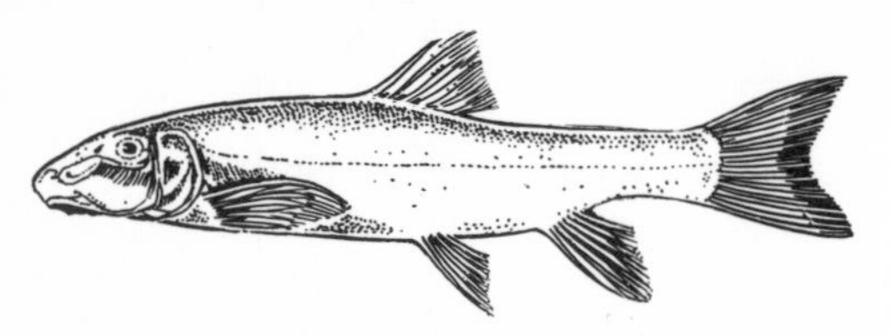

Fig. 247 *Catostomus catostomus*

smaller as a rule and during spawning have tubercles in the form of light-coloured nodes as well as a display of strong colours. The eggs are deposited on the bottom or on plant substrata. There are about 100 different varieties, some of which are of economic importance. Young fish of the Catostomidae from still waters are relatively easy to keep in the domestic aquarium.

Catostomus, G, *see* Catostomidae

Caudal. Situated near the tail. From the Latin *cauda* the 'tail'. Anatomical term used to describe the position of parts of the body. Opposite: cranial*.

Caudal Fins *see* Fins

Caudata, O, *see* Amphibia

Caudo *(Phalloceros caudimaculatus) see Phalloceros*

Caudovittatus *(Hemigrammus caudovittatus) see Hemigrammus*

Caulerpa (fig. 248). G of marine green algae (Chlorophyta*) with creeping stem, root-like holdfasts (Rhizoids) and leaf-life or bush-like 'leaves'. *C. prolifera*, found in the Mediterranean, grows well even under normal aquarium conditions (like some other representatives of the G). It flourishes at temperatures

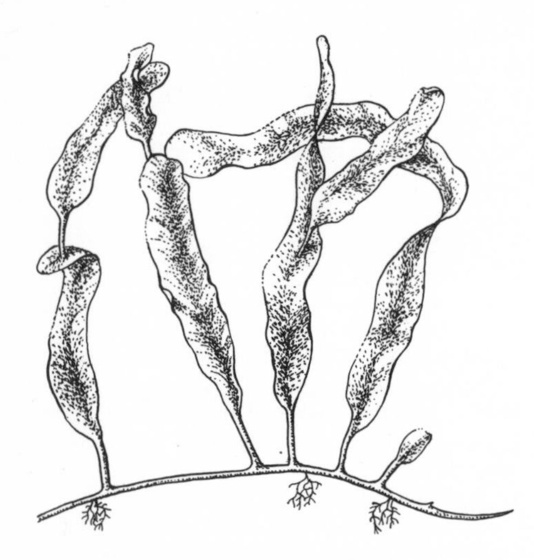

Fig. 248 *Caulerpa prolifera*

up to 28 °C, and good water circulation and lighting encourage its growth. Sudden fluctuation in salt content can make the multinucleate cytoplasm (which in these algae is not divided into cells) explode and gush out.

Cavalla *(Caranx speciosus) see Caranx*

Cavefishes, North American (Amblyopsidae, F) *see* Percopsiformes

Cebidichthys violaceus *see* Blennioidei

Celebes Goby *(Stigmatogobius hoeveni) see Stigmatogobius*

Celebes Rainbou Fish *(Telmatherina ladigesi) see Telmatherina*

Cell (fig. 249). The smallest structural unit in organisms that is capable of independent life. In single-celled organisms all the life processes of an individual are contained herein, such as metabolism, locomotion, irritability and reproduction. The cells of multi-celled plants and animals, however, are component parts of the tissue* and as such have specialised functions of various kinds. As a result, their shapes, sizes and internal structures can vary. A cell's size usually lies in the

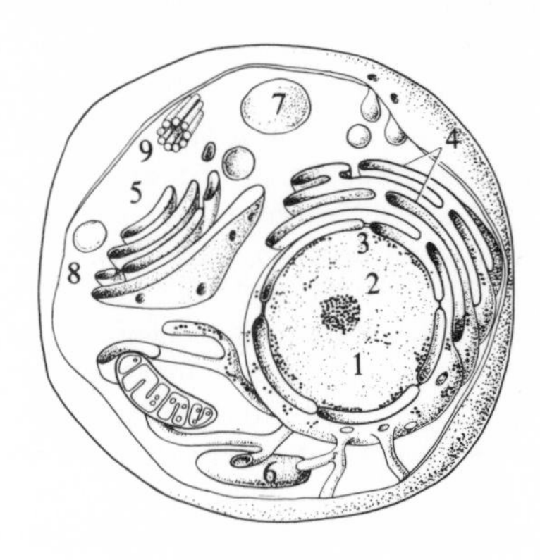

Fig. 249 Section of an animal cell under an electron microscope. 1 Cell nucleus 2 Nucleolus 3 Nuclear pores in the nuclear membrane 4 Endoplasmic reticulum 5 Golgi bodies 6 Mitochondria (the one on the left of the cell is shown in section) 7 Lipid drop 8 Lysosome 9 Centrosome

region of 0.01 to 0.2 mm, although there are a large number of far bigger cells. Some cells form very long processes or are very elongate, eg the processes of nerve cells can be more than a metre long and plant fibre cells more than 5 m long. Nevertheless, all cells are constructed according to a basic principle. They consist of a *cell body* and a *cell nucleus* (bacteria and blue algae have a nucleus equivalent). The cell body is surrounded by a thin cell membrane which apart from forming the outside boundary also performs a host of other functions (especially exchange of substances with the outside medium, eg through pinocytosis). In addition, plants have a firm cell wall that usually consists of cellulose. The cell body is made up of unstructured *cytoplasm* in which numerous proteins, carbohydrates and salts are dissolved, and various structural elements whose finer detail can only be detected under an electron microscope. Included in it is the *endoplasmic reticulum*, a network of very fine tubes and cisterns, which is particularly important for the metabolism and protein synthesis of the cell. The energy-producers of the cell are the *mitochondria*, small formations of variable shape with a complicated membranous internal structure. Other important organelles in the cytoplasm are the *golgi bodies* which carry out various synthetical procedures, and the *lysosomes* which are involved in the breakdown of cellular components. For animal cells, the centrosome is also typical (sometimes present as a double quantity known as a diplosome). During cell divisions, the division spindle is formed from the centrosome. Many plant cells contain plastids which are involved in the carrying out of photosynthesis*.

The *cell nucleus* is a fairly large area of the cell, often spherical in shape, which is divided off from the rest of the cell by a double membrane containing pores. Its essential constituents are nucleic acids*: DNA (deoxyribose nucleic acid) as the carrier of genetic information, and RNA (ribonucleic acid) as an important component-part in the realisation of this information; the nucleolus is particularly rich in RNA. In the Eukaryota the DNA lies in the *chromosomes*. These very fine thread-like structures are scattered irregularly in the cell nucleus and warped, when not in the cell division phase. In the microscopic picture at this stage they only form a chromatin substance. But during cell division* they become compact formations, often shaped like small rods. When this happens it also becomes apparent that the shape and size of the individual chromosomes, as well as their number, are species-specific (eg. Vinegar-fly 8, Human 46, Carp 104). Animal body cells (except the Flagellata and Sporozoa from the single-celled organisms) always have pairs of identical (homologous) chromosomes, ie a *diploid* set of chromosomes (symbol 2n). Apart from paired autosomes, this set contains usually 2 separate heterosomes (X and Y chromosomes) which are unlike in one sex; as sex chromosomes they are involved in sex determination*. The sex cells on the other hand are *haploid*, usually as a result of the special cell division* (meiosis) they were previously subjected to. During fertilisation*, 2 haploid sets of chromosomes are once again united into a diploid set. The cells of many organisms have a multifarious, *polyploid* set of chromo somes.

Cell Division. The division of cells, ie the development of new cells from old ones, is the basis of reproduction in all cellular organisms and it is also essential for the development and viability of multi-celled individuals. With each cell division the genetic information rooted in the hereditary material (DNA) is passed on to the daughter cells (Cell*, Nucleic Acids*). In multi-celled organisms it is the cells of the embryonal tissue that have the greatest amount of cell division activity. But cell divisions do take place throughout an organism's lifetime, in animals particularly in tissue that wears out a lot (eg in the skin, in the mucous membranes of the digestive canal or in the formation of new blood cells). In growing seed-plants cell multiplication takes place primarily in special meristematic tissue located in the stem and in the vegetative points of the tips of the stem and root.

In both plants and animals, the typical form of cell division is indirect cell division or *mitosis* which is characterised by a change in form of the chromosomes in the cell nucleus (fig. 249). The division process is

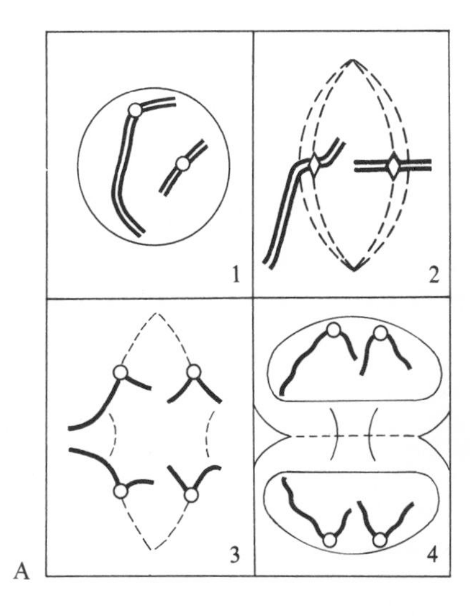

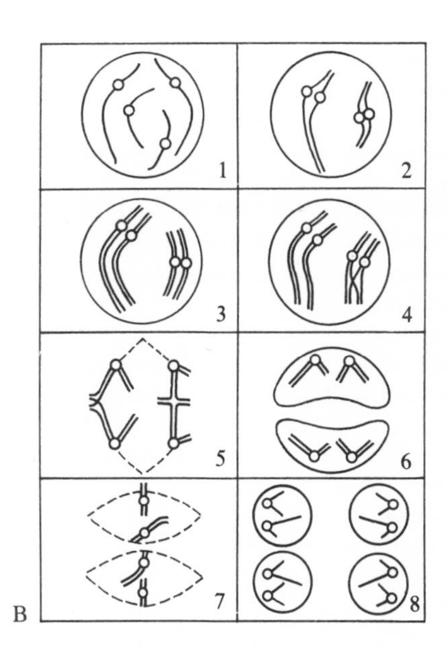

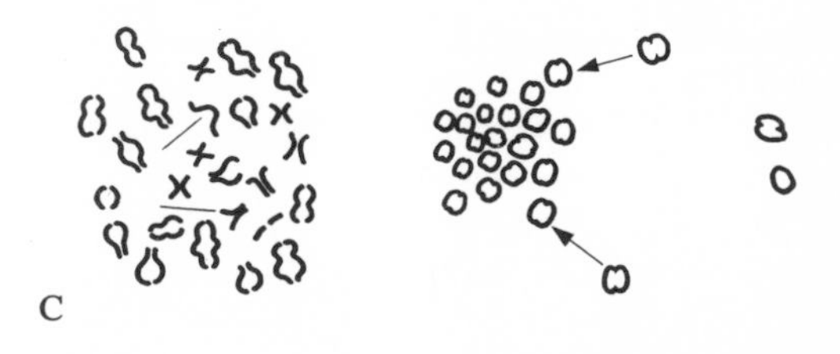

Fig. 250 Cell division. A Mitotic division with 2 chromosomes. 1 Prophase 2 Metaphase 3 Anaphase 4 Telophase B Most important stages of the 1st (1–6) and 2nd (7 and 8) meiotic division. 1–4 Prophase I with pairing of the homologous chromosomes. In 4, chromatids cross over and there is segment exchange. 5 Metaphase I 6 Telophase I 7 Metaphase II 8 Telophase II with 4 haploid cell nuclei. C Chromosomes of the metaphase-stage in man (left), and in the Stickleback (right)

divided into four stages. In the prophase the chromosomes are clearly visible as thread-like structures (*mitos* = Greek for 'thread'). They become increasingly more compact and can be seen to have a longitudinal dehiscent split into 2 identical halves (chromatids). After the nucleus membrane has broken up, the nuclear spindle (the apparatus of division for the chromosomes) forms from the finest of protein tubuli. In the metaphase the chromosomes line up in one plane in the middle of the cell (equatorial plane). In the anaphase the split halves of the chromosomes move apart and, with the aid of the nuclear spindle, they go to the cell poles. In the final telophase, the chromosomes return to the resting condition (fine threads) and so become invisible again. The nuclear membrane of the daughter nuclei forms and, finally, the division of the cell body takes place. Direct cell division or *amitosis* is more rare. Here, the nuclear division is a simple case of peeling off into two – there is no nuclear spindle and no change of phase with the chromosomes. The cells of ageing tissue can divide in this way.

Maturation division or *meiosis* occurs with diploid animals during the formation of the germ cells (Gametes*). It leads to the new combination of the genetic material of the cells and to the reduction of the diploid set of chromosomes to a simple (haploid) set of gametes. In haploid animals (eg Sporozoa*) meiosis only takes place after the gametes have joined (during zygote reduction) (fig. 250). There are 2 steps in meiotic division, the first and second meiotic division. During the long-lasting prophase of the first meiotic division important chromosomal changes take place. The corresponding (homologous) paternal and maternal chromosome threads of a nucleus lie next to one another and divide along their length-wise axis, so that 4 identical parts lie together in a tetrad. The pairing is so intense that it is not uncommon for pieces of chromosomes to be exchanged between the chromatids. After the metaphase (corresponding to the mitosis) the tetrads are divided during the anaphase into 2 pairs. The second-stage miotic division now follows on in which the haploid sets of chromosomes develop by separation of these two pairs. The new distribution of the originally paternal and maternal chromosomes is fortuitous and, alongside the exchange of pieces of chromosomes, it represents a second possibility of new combinations of the hereditary material. During the spermiogenesis of animals, four spermatozoa develop from the four division products of the original cell. During the meiotic division of the egg-cells, however, only one of the four daughter cells contains the yolk supply. The other three remain small and die (abortive egg-cells). The maturing into fertilizable gametes follows on from the meiosis.

Cellophane Plant *(Echinodorus berteroi) see Echinodorus*

Cement Poisoning *see* Poisoning

Central Nervous System *see* Nervous System

Centrarchidae; Sunfishes and Basses (fig. 251). F of Perciformes*, Sub-O Percoidei. Found predominantly in the warmer waters of central and eastern N America. These fish are similar to Perches, with elongated to relatively deep laterally compressed bodies. Mouth with a fairly wide gape, and on the jaws and other bones of the oral cavity, as well as on the tongue, there are fine teeth arranged in bands. D and An almost the same length in the case of the Ga *Pomoxis* and *Centrarchus*; in all other Ga the An is shorter. D uniform or almost split in two with 6 to 13 spines. An has 3–9 spines. C is rounded or slightly indented. The lateral line is almost complete but absent in the G *Elassoma*. Cycloid scales or ctenoid. The oldest fossils are from the Miocene age.

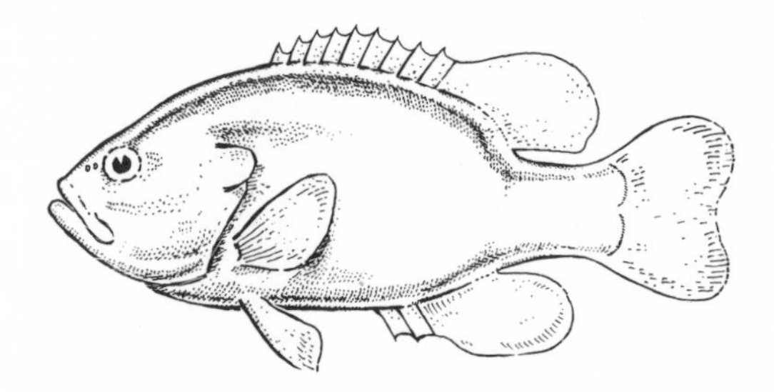

Fig. 251 Centrarchidae-type

Many of the 30 or so species are brightly coloured, particularly in the breeding season, the females often more so than the males. The Centrarchidae prefer warmer, weedy, still or slowly flowing waters. The eggs are mostly deposited in shallow hollows. As a rule, the males practise very intensive brood-care. Various species are now distributed throughout the world and have increased their numbers greatly, in southern Europe, for example. Some of the larger species can be eaten and become the quarry of sport fishermen. The smaller species are popular aquarium fish; *see* under Ga *Ambloplites, Centrarchus, Elassoma, Enneacanthus, Lepomis, Micropterus, Pomoxis.*

Centrarchus CUVIER, 1829. G of Centrarchidae*. Lives in the south-eastern states of the USA. Largish, laterally much flattened fish, oval in shape. D only marginally longer than An, with 11–12 hard rays (in the case of the similar G *Pomoxis**, however, only 5–8 hard rays). One of the most rewarding specimens for keeping in unheated indoor aquaria. Their behaviour is quiet, they recognise their keepers (thus giving the impression of true 'intelligence'. For the care of these fish one needs large tanks in a sunny position with abundant plant-life and water which has been allowed to stand. Take care when there is a sudden change of environment, as old fish in particular are prone to develop fungus. When in good health, these fish will reproduce in a species aquarium with no difficulty. Open-air tanks however are particularly good for breeding *Centrarchus.* The male digs a hollow in which, after mating, the eggs are laid and fertilised. The male alone then takes over the care of the spawn and the newly-hatched young.

– *C. macropterus* (LACÉPÈDE, 1802); Flier, Peacock-eyed Sunfish. To 16 cm. Body bright olive, shimmering bluish silver, coloured with fairly clear dark bands and spots. Head with dark vertical stripe over the eye, black eye-spot with light edge at the base of the soft-ray part of the dorsal fin (fig. 252).

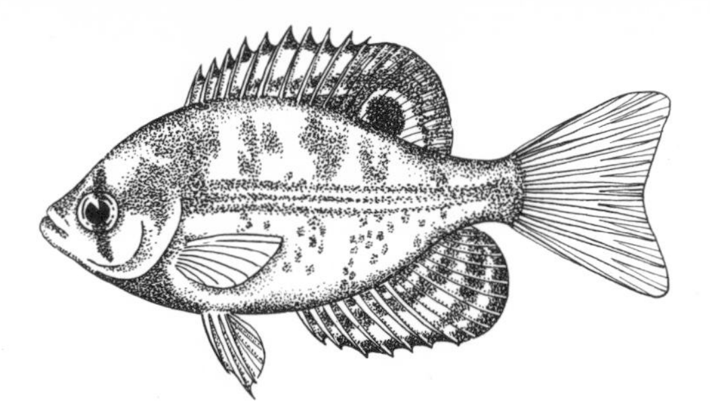

Fig. 252 *Centrarchus macropterus*

Centrifugal Pumps *see* Pumps

Centriscidae, F, *see* Aulostomioidei

Centromochlus KNER, 1857. G of Auchenipteridae. 8–15 cm long. Naked Catfishes living north of the Amazon in eastern S America. They have an elongated shape, equally tall at front and back. There is one pair of long barbels on the upper jaw and two pairs of fine, very short barbels on the lower jaw. The D1 and D2 are relatively small. The former has a powerful spine and is positioned far forward. The C is deeply forked. The An is small and short. The mouth is terminal. Very active species which should, therefore, be kept in larger aquaria, which are not too dark. Omnivores which will accept raw meat and dried food. There is no information about breeding habits. Many representatives of the G *Centromochlus* have recently been placed in the G *Tatia* RIBEIRO, 1912.

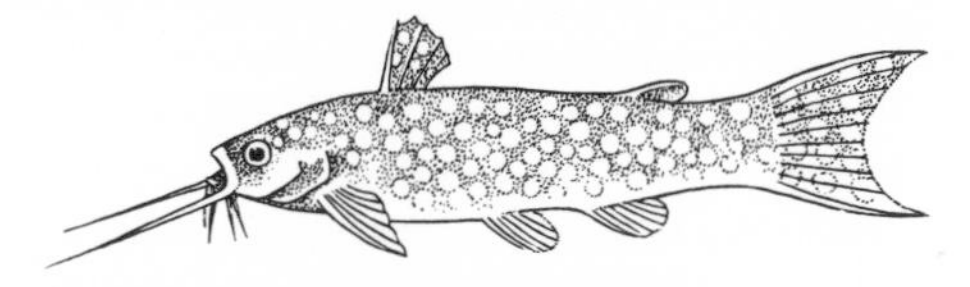

Fig. 253 *Centromochlus aulopygius*

– *C. aulopygius* KNER, 1857. Eastern S America, north of the Amazon. To 8 cm. The sides of the body as well as D1 and D2 and C are chocolate brown with bright spots. The other fins are colourless (figs. 253 and 254).

Fig. 254 *Centromochlus aulopygius*

Centropomidae; Glassfishes. F of Perciformes* Sub-O Percoidei. Many differently formed fish belong to this family, of which only the very small species merit the name Glassfishes, on account of their transparent bodies. The large members of this family on the other hand are sturdy fish of the Perch or Pike-Perch type. Body relatively short to elongated; more or less laterally compressed, lateral line as a rule continuing uninterrupted to the base of the C. D in one or two parts. C rounded or indented. Scales ctenoid or cycloid. Head either completely or only partly covered in scales. The Centropomidae are found in coastal regions of the warmer parts of the Atlantic and Indo-Pacific. Various species sometimes make their way into fresh water; others are pure freshwater fish. The large mouthed Snook *(Centropomus undecimalis)* of the east and west coast of tropical America, at over a metre long, and the still longer Nile Perch *(Lates niloticus)* (fig. 255) are some of the species of economic interest to fisheries. Ga of interest to the aquarist are discussed separately; *Chanda**, *Gymnochanda**.

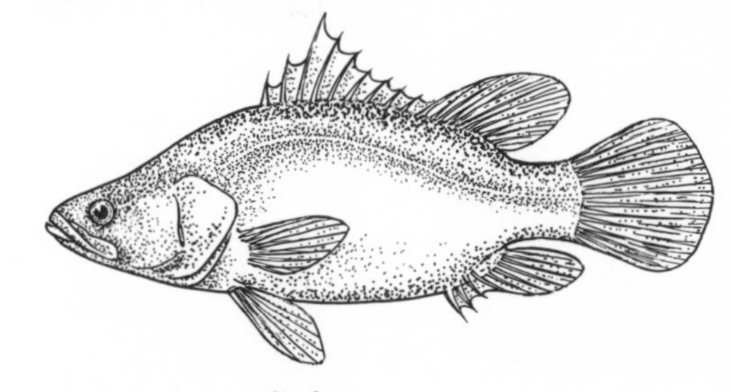

Fig. 255 *Lates niloticus*

Centropomus undecimalis *see* Centropomidae

Centropyge (KAUP, 1860) (figs. 256 and 257). G of the Chaetodontidae* Sub-F Pomacanthinae. Small Angelfishes of the tropical Indo-Pacific and Atlantic, where they live as browsing fish on reefs and rocks. They eat

Fig. 256 *Centropyge eibli*

algae and probably animal matter as well. The shape and behaviour of *Centropyge* are reminiscent of Damselfishes (Pomacentridae*). They are, however, easy to recognise by the spine on the operculum which is typical of the Angelfishes. Unlike the large Pomacanthinae, the colour of the young fish of this G is similar to that of the older ones. *Centropyge* species are easy to keep in the aquarium, eg:

– *C. argi* WOODS and KANAZAWA, 1951; Pygmy Angel, Caribbean Dwarf Angelfish, Cherubfish. Tropical W Atlantic. Up to 6.5 cm. Deep blue, lower half of head and breast goldish red (fig. 257).

Fig. 257 *Centropyge argi*

– *C. bicolor* BLOCH, 1787; Oriole Angelfish, Black and Gold Angelfish, Bicolor Cherub, Bicolor Angelfish. Tropical, Indo-Pacific. To 12 cm. Front section of body yellow, back blue black.

Cephalochordates, Sub-Ph, *see* Acrania

Cephalopholis BLOCH and SCHNEIDER, 1801; Jewel Grouper. G of Serranidae*. The *Cephalopholis* species are tropical sea fish, which live on reefs and rocks where they prey mainly on other fish. These are medium-sized, relatively elongated Serranidae which are covered with a spotty pattern which acts as a camouflage and is very useful when they are catching fish. These extremely attractive fish are easy to keep in the aquarium. Various species are frequently imported:

– *C. argus* BLOCH and SCHNEIDER, 1790; Jewel Grouper; Jewel-spotted Grouper. Red Sea, Indo-Pacific. About 40 cm. Gleaming blue spots on brown background.

– *C. miniatus* (FORSKÅL, 1775); Blue-spot Rock Cod, Jewel Bass, Coral Trout. Red Sea, Indo-Pacific. About 40 cm. Gleaming blue flecks on a shining red background colour (fig. 258).

Fig. 258 *Cephalopholis miniatus*

Cephalopoda; Cephalopods. Cl of Invertebrates (Mollusca*) with over 700 species. Cephalopods live only in the sea, in all types of habitat. They are predominantly predators including both the *Sepiola* species which are only a few centimetres long and the Giant Squids of the G *Architheutis* which grow to 18 m. The body consists of two main sections – the head with suckered tentacles and the body. A small neck section connects the two. In most forms living today the shell is reduced to parts which are not externally visible. The main sense organs of the Cephalopods are the large efficient eyes which give colour vision. As they also possess a well-developed central nervous system the speed of reaction to optical stimuli approaches that of the vertebrates. Tentacles surround the mouth which is equipped with sharp beak-like horny jaws which can bite very powerfully. Anal, renal and sexual orifices as well as the gills are to be found on the underside in the large mantle cavity which is surrounded by a muscular wall. The respiratory water flows in through the mantle opening, which can be closed, and out through a cylindrical orifice, the so-called funnel. When in danger, many Squids can expel a black secretion from a large ink-gland which opens into the hind-gut. This distracts enemies and covers the squid's flight. Slow movement of bottom-dwelling Cephalopods is achieved by using the arms. In the case of free-swimming varieties fins on the body moving in wave shapes are used. They can move extraordinarily quickly with the help of the funnel. Through this, water is forced out of the mantle cavity with great power, and the creatures are propelled by the backwards thrust. Cephalopoda possess a striking ability to change their colour, which is due to contractile chromatophores* in the skin. The ability to change colour is used to match the surroundings (which, for example, is mastered quite amazingly by *Octopus**) as well as for communication between members of a given species, such as during the courtship display of *Sepia*. Cephalopods are unisexual. A specialised tentacle (hectocotylus) is used by many to transfer the sperm capsules (spermatophores). The eggs are attached to the substratum in horny capsules *(Loligo, Sepia, Octopus)*, are hidden in the shell *(Argonauta)* or are contained in gelatinous rolls (Oegopsida). The class Cephalopoda embraces the Sub-Cl of Tetrabranchiata, with 4 gills, many short arms and a well-developed snail-like shell, as well as the Sub-Cl Dibranchiata, with 2 gills and ten arms (O Decabrachia) or 8 arms (Octobrachia). Of the Tetrabranchiata there are only 3 species of the G *Nautilus* alive today and these are found in the Indo-Pacific region. The Decabrachia contain the well-known free-swimming squids with a stream-lined shape, such as *Sepia* and *Loligo*. They have two arms which have developed into catching tentacles which can be shot far forwards. Predominantly bottom-living varieties with clumsy, sack-like hind-quarters such as *Octopus* or *Eledone** belong to the Octobrachia but so too does the floating Paper Nautilus *Argonauta argo* LINNAEUS, 1758, the female of which forms a thin secondary shell with two specialised arms. There are great difficulties involved in keeping Cephalopods in aquaria, since the quick-growing short-lived creatures need extremely large quantities of food and possess a correspondingly high metabolic rate. Since they require water of a high quality, even small animals need large aquaria, intensive filtration and continuously changing water. Only small bottom-living Cephalopods, which are best kept singly, come into consideration (*Eledone, Sepia,* possibly *Octopus*). These highly-developed Molluscs are some of the most fascinating sea creatures.

Cephalopods, Cl, *see* Cephalopoda

Ceramium. G of marine red algae (Rhodophyta*) with finely branched bushy growth. The cylindrical, coiled branches have a fork-shaped tip. The species *C. rubrum* (HUDSON), which is plentiful everywhere, dies very quickly in the aquarium.

Ceratioidei (Sub-O) *see* Lophiiformes

Ceratodidae, F, *see* Ceratodiformes

Ceratodiformes; Lungfishes. Order of Osteichthyes*. Sub-Cl Sarcoptherygia. As shown under the entry Osteichthyes both the Sup-Os Dipnoi (Lungfishes) and Crossopterygii (Stump-fins) belong to the Sub-Cl of Sarcopterygia. The recent Dipnoi are included in the O Ceratodiformes. The F Ceratodidae (Amphibious fishes, Australian Lungfish: 1 species), Protopteridae (African Lungfishes: 1 G, 4 species) and Lepidosirenidae (South American Lungfishes: 1 species) belong to this order. As the popular name implies, the fish concerned are similar to the Newts and Salamanders, and are closer to them in many anatomical characteristics than the bony fish (which is of course not to say that they are the direct ancestors of the Amphibia). The paired fins are formed as *Archipterygium* (Osteichthyes*), and *Neoceratodus* has retained them in the most basic form as paddle fins. In the case of the other species these fins are simplified and have the shape of conical appendages which are moved alternatively like limbs. The lungs are ventral sack-like outgrowths of the gut which in the case of *Neoceratodus* are sited over the gut. The gills are usually reduced. The air, which is taken in through the nose, passes via the nasal, palantal passage into the oral cavity and is forced into the lungs from here. Dermal as well as pulmonary respiration is highly developed. The circulation of the blood and the heart have adapted, to a certain extent, to pulmonary respiration which is perfected by the land-living vertebrates. The eyes are hardly developed. No lateral line but numerous skin sense organs. The skeleton is partly cartilaginous. No vertebrae.

The typical jaw-bones of the Osteichthyes such as the praemaxillary, maxillary and dental bones are missing from the skull. The jaw-bone is directly attached to the cranium. The jaw-bone (hyomandibular bone) is reduced (*see also* Osteichthyes). Jaw-edge teeth are absent, but in their place strong dental plates have developed, the radial branches of which constitute fused teeth. Dental plates of this type are found in the lower jaw and in the roof of the mouth above. They are for grinding animals with shells, crabs and parts of plants (fig. 259). The gut has a spiral valve. Newt-like larvae hatch from the richly yolked egg which are sometimes encapsulated in gelatine. In the case of *Protopterus* and *Lepidosiren* they have four pairs of feathery external gills which do not disappear until between the first and sixth month of their lives during their metamorphosis. The amphibious fish are a very old fish-group which had their heyday in the world's infancy (Devonian to Triassic), that is 310–280 million years ago. Their de-

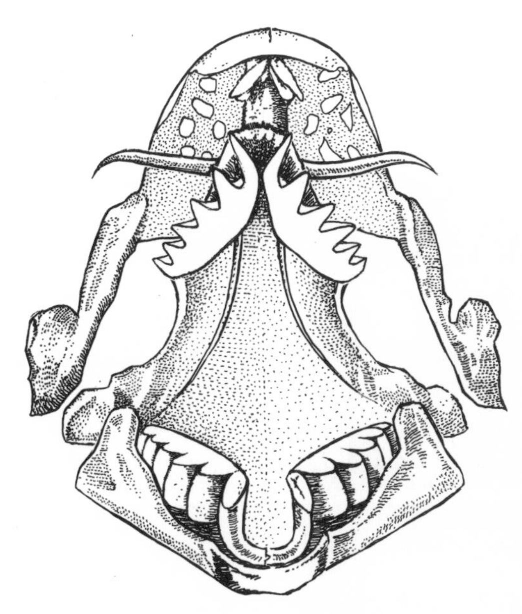

Fig. 259 Dental plates of *Neoceratodus*

velopment can be traced back almost without a break through fossils. The Australian Lungfish *(Neoceratodus forsteri)* (fig. 260) which grows up to 1.70 m long and weighs up to 50–60 kg has only one unpaired lung and

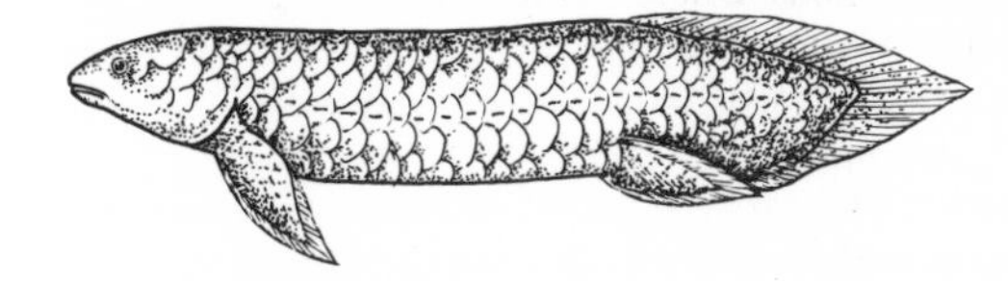

Fig. 260 *Neoceratodus forsteri*

cannot live out of water. The large scales cover not only the body but also the muscular paired fins with their lateral horny rays. The African Lungfishes, such as, for example, the West African brown spotted *Protopterus annectens* which grows up to 70 cm long, or the East African Grey-Flecked or Marbled Leopard Lungfish *(Protopterus aethiopicus)* (fig. 261) which is almost

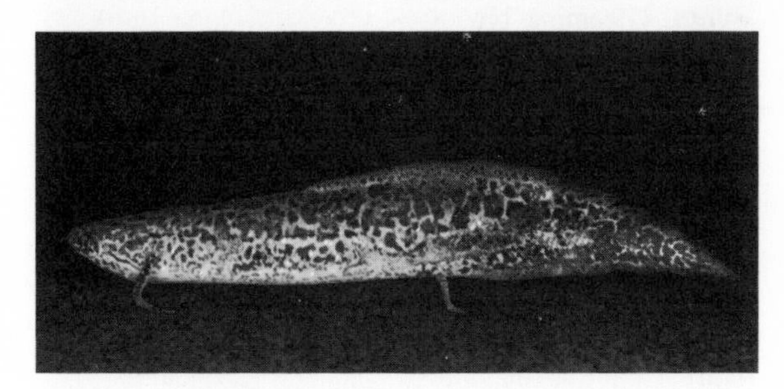

Fig. 261 *Protopterus aethiopicus*

twice as long, are able to survive dry periods without difficulty due to their reasonably efficient lung respiration. They withdraw further and further into the mud as the water level recedes, roll up in a hole and secrete a mucus which forms a cocoon in which they become dormant. An air shaft to the surface provides fresh air (fig. 262). Breeding takes place after the dry period. The eggs which are 3–4 mm in size are laid in a U-shaped channel and are watched over by the male which fans

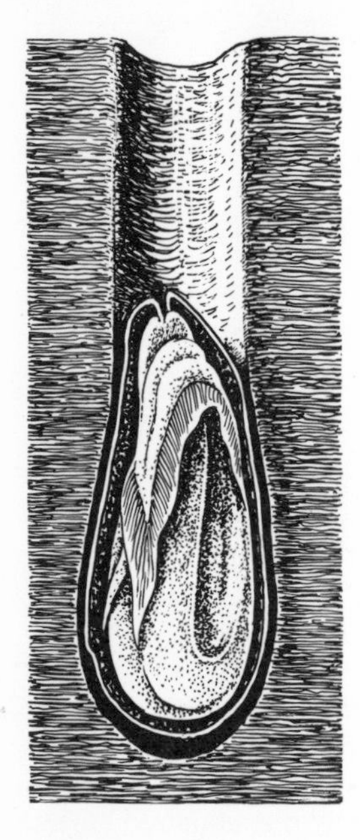

Fig. 262 *Protopterus* in mucus cocoon

them with fresh water. The South American Lungfish *(Lepidosiren paradoxa)* (fig. 263) lives in the swampy regions of the northern Gran Chaco and attains a length of 125 cm. It also withdraws into holes in the ground during dry periods but does not form a cocoon of mucus.

Fig. 263 *Lepidosiren paradoxa*

During brood-care the male develops many branched feathery filaments on the pectoral fins which act as supplementary dermal breathing organs. Equipped with these it can do without air from the surface and so does not need to leave the brood. The African Lungfishes are regarded as special delicacies to eat and are dug out by Africans during the dry season. All Ceratodiformes are great attractions in zoo aquaria. In recent years larvae and young fish have been kept by many enthusiasts. Indeed these sluggish creatures prove to be very robust, but it is not always easy to get them to feed. This can be attempted by holding out Earthworms, Guppies, small Mussels or Cow's heart. Only fish of the same size should be kept together.

Ceratophylliaceae; Hornworts. F of Magnoliatae*. The only G is *Ceratophyllum**.

Ceratophyllum LINNÉ, 1753; Hornwort. Only G of the Ceratophyllaceae*. 3 species. Found in all parts of the world. Submerse rootless perennials. Leaves arranged in whorls anchored in the ground or free-floating. Stem-axes spreading, much-branched with numerous leaf whorls. Leaves with many pinnules, spreading out like a hand. Flowers inconspicuous and submerse, single in the leaf axils, polysymmetric unisexual and monoecious. Perianth a greenish perigonium. Stamens 6–16, carpels 1, superior. Nut-fruit. The species of this G form a thick mat in still and gently flowing waters, some distance from the bank on muddy bottoms. Sometimes shoots which have been torn off may float freely in the water. Propagation is mostly achieved asexually by removing side-shoots. Sexual propagation depends on water pollination. The pollen has approximately the same specific gravity as water and thus can float to the stigmae of the female flowers. In autumn all shoots sink to the bottom and pass the winter there. The shoots which develop after autumn are more compact and their leaves are more robust. All species are suited to life in outdoor pools and cold water aquaria. However, they often also survive in high temperatures. They tend to form local varieties.

– *C. demersum* LINNÉ; Common Hornwort. Cosmopolitan. Leaves stiff. Fruit with 2 spines at the base.

– *C. echinatum* GRAY; Spiny Hornwort. Eastern N America. Leaves stiff. Fruit with several spines.

– *C. submersum* LINNÉ; Spineless Hornwort. Europe, South Asia. Leaves soft, fruit with no spines.

Ceratopteris BRONGNIART, 1822. G of Pteridophyta*. 4 species. Found in the tropics and subtropics of the whole world. Fern with leaf arrangement in rosettes, leaves light green with many variations in size, varying between lobed and compound pinnate. Spore-bearing leaves are emerse, very narrow pinnules. The *Ceratopteris* species are marsh and water ferns. Their rosettes either float on the water-surface with their bushy roots hanging down in the water, or they take root in the marsh and withstand flooding well. A good deal of light and a temperature of over 20° is recommended for their cultivation in the aquarium. Propagation is no problem because of the numerous plantlets which develop along the edge of the leaves. The following species and varieties are cultivated.

– *C. cornuta* (BEAUVOIS) LE PRIEUR; Floating Fern. Central Africa, SE Asia, and N Australia. Roots in the ground. Broad pinnules.

Fig. 264 *Ceratopteris pteridoides*

– *C. pteridoides* (HOOKER) HERON; Floating Water Sprite. S America to Southern N America, West Indian Islands. Floats on the water surface (fig. 264).
– *C. thalictroides* (LINNÉ) BRONGNIART. Indian or Sumatra Fern, Water Sprite. All tropics and subtropics. Species with many varieties. Roots in the ground. Narrow pinnules (Sumatra fern) or very narrow pinnules (Filigree fern).

Cercaria *see* Trematodes

Cereus OKEN, 1815. G of Sea Anemones (Actiniaria*). Found in the Mediterranean and on the European Atlantic coast, where they cling tightly to holes in the rocks or to stones under sand and gravel in shallow water. When danger threatens they withdraw far back into their holes. The following beautiful species is particularly well-suited for keeping in the marine aquarium.
– *C. pedunculatus* (PENNANT, 1776); Daisy Anemone. Mediterranean to North Sea. Diameter up to 10 cm. Very variable in colour, one colour or patterned in brown, yellow and red tones (fig. 265).

Fig. 265 *Cereus pedunculatus*

Ceriantharia; Cylinder Roses or Tube Anemones; O of Anthozoans (Anthozoa*). Approximately 50 species, found predominantly in the warmer seas. They do not form colonies. They are extremely elongated polyps without foot-discs and without a skeleton, which live on soft bottoms in leather-like tubes of mucus into which they can withdraw quickly when in danger. The presence of two different crowns of tentacles is characteristic of them: short brush-like mouth or labial tentacles and long thread-like edge or marginal tentacles with which they catch plankton. Reproduction occurs almost exclusively sexually via freely discharged egg and semen cells. The *Ceriantharia* species, like the Actiniaria*, are extremely well suited to being kept in the aquarium.

Fig. 266 *Cerianthus membranaceus*

They require a deeper soft bottom to burrow into. The best-known G is *Cerianthus* DELLE CHIAJE 1832, with the following species.
– *C. membranaceus* (SPALLANZANI, 1784); Mediterranean Cylinder Rose. Found in the Mediterranean on soft bottoms in shallower waters. Diameter of the crown of tentacles up to 20 cm. Body length to 35 cm. White or coloured brown or violet. Tentacles often curled. These splendid animals can survive for many decades in an aquarium (fig. 266).

Cerianthus, G, *see* Ceriantharia

Cerrithium BRUGUIÈRE, 1792. G of snails with mantle cavity opening at the front of the body (Prosobranchia*). Some found on South European coasts where they lead a concealed life on the rocky bottom on a diet of algae. They have long, pointed, conical shells which are covered with small tubercles. The spirals of the shells are not much in evidence. One finds shells occupied by Hermit-crabs *(Clibanarius*)* more often than live *Cerrithium* species. Best known species is the following.
– *C. vulgatum* LINNAEUS, 1758. Mediterranean, E Atlantic. To 7 cm.

Cestodes; Tapeworms. The Cestodes are tape-like, flattened, intestinal parasites of vertebrates. Intestine, mouth and anus are absent. The Tapeworm absorbs its food over the total surface of its skin (Osmosis*). Its small head is anchored into the intestinal wall with the help of sucker cups or crowns of hooks. In many species the elongated body is jointed in single sections (proglottidae) which are continually reformed behind the head and are cast off from the end of the worm, filled with ripe egg-cells. After leaving the host's intestine the larvae hatch and do not reach the intestine of the final host until they have travelled through one or more intermediate hosts (Parasitism*). Fish can be final and intermediate hosts. *Caryophyllaeus laticeps* PALLAS (fig. 267), Clove Worm, of up to 3 cm, inhabits the intestine of fish, and its larvae can be transferred via *Tubifex**. More often the fish is intermediate host, the larval stages (plerocercoid) are whitish knots or larger larvae which live in the body cavity, musculature or liver. *Ligula intestinalis* LINNAEUS, 1758 (fig. 267), *Diphyllobothrium latum* LINNAEUS, 1758, examples of the G *Schistocephalus* CREPLIN, 1829 (fig. 267), are found more frequently along with numerous other species and Ga. The infestation often occurs via *Cyclops** MÜLLER and other Copepoda, which act as intermediate hosts. Aquarium fish are seldom infested with Tapeworms. Only imported fish and of these above all the sea-fish, occasionally carry parasites. A minor infestation with larval stages generally stays unnoticed and is harmless; only mass infestations lead to damaged organs. A course of therapy in this case is impossible. Intestinal Cestoda cause severe emaciation in fish in spite of a good intake of food. If valuable fish are affected, a treatment with Thiabendazol may be attempted.

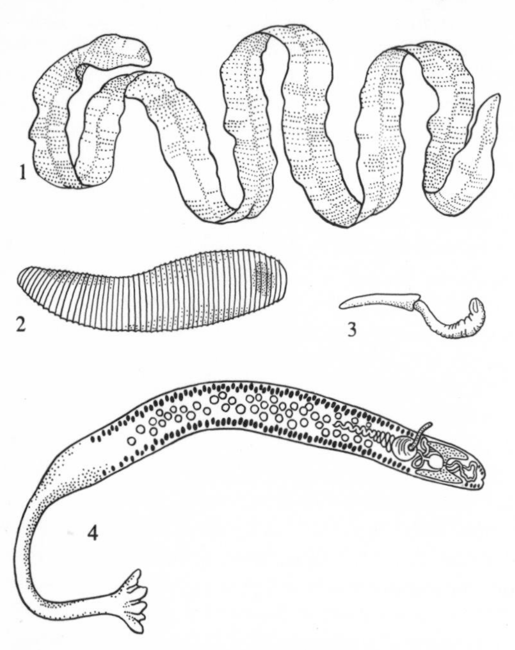

Fig. 267 Cestodes. 1 *Ligula* 2 *Schistocephalus* 3 *Diphyllobothrium* 4 *Caryophyllaeus*

Cestus, G *see* Ctenophora

Cetacea; Whales. O of mammals (Mammalia*) The Cetacea are classified in the Sub-O of the Odontoceti (Toothed Whales) and Mystacoceti (Whalebone Whales) with several Fs in each case. With the exception of some species which are found solely as freshwater forms in the great river-systems of the Ganges, Indus, Yangtse-kiang, Amazon and La Plata, the Cetacea inhabit the seas of the world. The whales belong to those mammals which are best adapted to life in water. They never leave it and are not able to move on land. The body is more or less elongate and spindle-shaped. The head is connected directly to the body region. The heavily compressed neck vertebrae are partly fused. The front limbs have been transformed into organs similar to pectoral fins, the characteristic skeletal elements of the limbs of the vertebrates are much shortened and mostly rigidly fused together. These 'fins' serve as body-stabilisers. The rear limbs are almost completely rudimentary. Due to their sponge-like loose structure and abundant fatty deposits the bony elements of the whole skeletal system have a relatively low weight. The end of the body has differentiated to become a horizontal fin-like organ which acts as a locomotive organ. Many Cetacea have in addition a 'dorsal fin' which, like the tail fin, contains no supporting skeletal elements, and the skin formation of which is constructed simply of connective tissue and fat. All Cetacea lack body-hair and the skin is smooth and without glands. The paired milk-glands are sunk in deep skin pockets and lie far back at the side of the sexual orifice. The olfactory sense has been almost completely lost and the nose is used by these lung-breathing animals solely to take in air. The nasal channel which begins with a paired or unpaired nasal opening (blow-hole) is directly connected to the trachea, that is there is effectively no connection with the throat area, so that the entry of water into the lungs via the oral cavity is made impossible. When diving the nasal openings are shut tight by a system of pocket-flaps. When exhaling the highly compressed air, which is expelled with great force, a cloud of steam of up to several metres high and made of condensed droplets of water develops—the blow. Inhalation is extraordinarily quick. Periods of submersion can extend up to nearly two hours in individual species of whales. This is achieved not only by the large capacity of the lungs, but also by an enlarged blood-volume with an increased count of red blood corpuscles, a heavily enriched muscle-haemoglobin and the ability to retain high concentrations of CO_2 in the blood. The eyes are usually small. Between these and the setting of the pectoral fins are situated the small aural orifices, which no longer have any form of ear-flaps. Although the outer hearing passage is closed, and the conduction of sound is taken over by bones of the skull, all Cetacea hear very well. Cetacea are warm-blooded animals which in a medium of such intensive contact as water must maintain a constant body temperature against temperatures which are generally substantially lower than those experienced by others. A fibrous layer, very rich in oil, the blubber, is formed under the skin as heat insulation. Besides helping to regulate the body temperature, this blubber is also important for this large animal's buoyancy and ability to swim. Some large species of whale, such as the Sperm-whale and *Hyperodon ampullatus* (Bottle Nosed Whale) among others, are decidedly deep divers, which seek depths of more than a thousand metres. The oil cavities in the head region of these animals are understood to be connected with the equalising of pressure which is necessary at these depths.

There are six Fs in the Sub-O of the Odontoceti (Toothed-whales), characteristic of which are the jaws, in which are found a large number of similarly formed teeth. The skull is asymmetric which is shown amongst other things by the differing number of teeth in each half of the jaw and in the single nasal opening which is positioned to one side. The number of fingers in the pectoral fins is not reduced. The members of the F Platanistidae (River Dolphins) are seen as very early forms of whale and are freshwater dwellers of large tropical river-systems. The largest Toothed-whale, the Sperm-whale *(Physeter catodon)* belongs to the F of Physeteridae which has only 2 Ga with one species in each. The male Sperm-whale grows to 20 m. The head represents up to a third of the total body length and has a snub end. The relatively narrow and short lower jaw has teeth, whereas the upper jaw has corresponding horny sheaths. These whales, which are found all over the world, feed mainly on squid (Cephalopoda*) even on the larger ones which measure over 10 m. One secretion of the gut of the Sperm-whale is ambergris, a grey-white substance which is used as a perfume in the cosmetic industry. Together with *Hyperodon ampulla-*

tus (F Ziphiidae), which grows up to 10 m long, the Sperm-whale is one of the most economically important Toothed-whales. Oil is obtained from the blubber and the head cavities. One particularly striking Toothed-whale is the Narwhal *(Monodon monoceros)* which belongs to the F Monodontidae. The males of this type which grow to 4–5 m long develop a pair of teeth in the upper jaw. The left one is quite rudimentary, but the right one grows into a corkscrew and stretches forward horizontally up to a length of 2.7 m. The function of this extremely elongated tooth has not yet been properly explained, and forms the basis of the fables of the unicorn. Those Cetacea which in a strict sense are regarded as dolphins belong to two Fs: the Phocoenidae (Porpoises), and the Delphinidae (Dolphins). They differ amongst other things in the shape of their teeth which is conical in the Delphinidae and spatula-shaped in the Phocoenidae. For the most part these whales reach a length of 1.5 m to little more than 3 m. Some species are found all over the world, others inhabit strictly defined areas of the sea. These whales which are all without exception agile and fast swimmers feed mainly on fish but also on squids and larger crabs. Two of the best known and most numerous of the Delphinidae are the Common Dolphin *(Delphinus delphis)* and the large 'Porpoise' *(Tursiops truncatus)*, which are found all over the world in many breeds. The G *Sotalia* includes smaller dolphins which live in the brackish waters around estuaries of large tropical rivers. One species, *Sotalia tenszii*, is worth mentioning. It lives in the estuary of the Cameroon and is the only whale to live exclusively on a vegetable diet. The Killer Whale *(Orcinus orca)*, which also belongs to the F Delphinidae, is a predator. The large pectoral fins and a raised sword-like dorsal fin show this Toothed-whale to be an extremely fast swimmer. It grows to 9 m. It can reach speeds of up to 37 kph. It does not have the beak-like elongated mouth. As a predatory whale it feeds on large fish, penguins, seals and other whales. In large groups they can even attack the giant Whalebone whales (Mysticeti) successfully.

The small porpoise *(Phocoena phocoena)*, which does not grow more than 2 m, is one of the most common species of the F Phocoenidae. It lives in the coastal waters of not only the North Sea and Baltic but also the Black Sea. It is also often found in large rivers.

In the Sub-O Mysticeti (Whalebone Whales) teeth are only developed in the early embryonic stage. In the adult they are completely rudimentary. In the upper jaw or palate respectively there are 2 rows of about 300–400 horny plates – the baleen or whalebone. The outer edge of the triangular whalebone is smooth whereas the inner side finishes with brush-like rough threads. The inner sides work together as a filter system. The whales feed predominantly on pelagic crabs of the G *Euphausia*, the so-called krill*. Large numbers of these can be taken into the roomy mouth of the whale. When the lower jaw is shut, the edges of which stretch far beyond the upper jaw, the water is pressed out to the side and the food remains in the filter system of the barbs. The mass of food is carried into the gullet with a push from the tongue. The gullet of whalebone whales is very narrow so that only in exceptional cases can creatures as large as a herring be swallowed. The Mysticeti are mostly extremely large creatures which grow to 12 m and longer. The Sub-O comprises 3 Fs, the Balaenidae (Right Whales), the Balaenopteridae (Rorquals and Humpbacks) and the Eschrictiidae (Grey Whales). With the exception of *Neobalaena marginata*, the Balaenidae have no dorsal fins. The baleen of the powerful mouth is very long and pliable and supplies the whalebone which was once much coveted. Besides other anatomical characteristics this family can be distinguished from others by the coalescence of the cervical vertebrae. The F of Balaenidae, which are relatively slow swimmers and tend to stay near the surface of the sea, also includes the Greenland Right Whale *(Balaena mysticetus)* which grows to about 18 m and the Atlantic Light Whale *(Eubalaena glacialis)* which grows to over 20 m long. Various breeds of these inhabit the polar seas predominantly. Both species were the main prey of the whalers up to the beginning of the twentieth century and were nearly completely exterminated. Only a few remain and are now protected by strict international regulations. The smaller *Neobalaena marginata* which only grows to 6 m keeps to the southern seas and is less important economically. The members of the F Balaenopteridae can be distinguished from the Balaenidae mainly by the presence of a row of neck and stomach furrows extending along the body axis and also by the triangular dorsal fin. In general they look more elongated and flat, the head is smaller and the pectoral fins are pointed. They are very fast swimmers. The baleen is short and the cervical vertebrae always separated. The whales of the greatest economic importance nowadays belong to the F of Balaenopteridae (Rorquals). The Blue Whale *(Balaenoptera musculus)* is found worldwide, and with its length of over 30 m and weight of over 150 tonnes is the largest creature on earth. It feeds exclusively on Krill, and the areas where it can be found are defined by their occurrence and movements. The slimly built Fin Whale *(Balaenoptera physalus)* only grows to 25 m but is an extraordinarily agile and fast swimmer considering its size. It can reach speeds of up to 22 kph. Besides its normal food of small crabs this whale sometimes eats fish up to the size of a cod. The Sei Whale *(Balaenoptera borealis)* which grows to 18 m is also found all over the world, but prefers the temperate zones and does not penetrate the ice regions of the pole. The smallest of the Balaenopteridae is the Little Piked Whale *(Balaenoptera acutorostrata)* This slimly built whale only grows to 9 m and is found in all cooler seas but is not so important economically as its larger relation. The Humpbacked Whale *(Megaptera novaeangliae)* is of economic use. It grows up to 15 m but looks plump and squat because of its large girth. The furrows which are wide apart stretch from the chin to navel. The head and pectoral fins are covered with numerous nodes. The third F of the Mysticeti, the Eschrictiidae, is only represented by one G with one species the Grey Whale *(Eschrictius glaucus)*. They only have 2–4 furrows in the neck area and never grow longer than 15 m. They are found in northern seas and prefer the coastal waters. This species which was once of economic importance but is now rare and completely protected.

Whaling is very old and was already carried on by coastal dwellers 4,000 years ago. At first they only hunted whales in the areas around the coast. It was only when the coastal waters became depleted at the beginning of the seventeenth century that the whales were hunted all over the northern seas. The first whalers chased almost exclusively the slower Balaenidae in order to acquire the whalebone and oil from the blubber. They used small rowing boats (scullers) and harpooned the whales by hand. The more intensive capturing methods and the use of large whaling fleets led finally at the end of the nineteenth century to almost complete extermination of the Balaenidae. With the improvement of the capturing techniques, especially faster whaling boats with harpoon cannons and ships able to negotiate the high seas, even the fast Balaenopteridae could be successfully hunted. The Antarctic waters also became a new whaling ground. In 1937–38 here alone 46,000 whales were captured, which represented 80% of the world whale catch. Of those whales which are still economically important today the Blue Whale, the Fin Whale, the Humpbacked Whale, the Sei Whale and the Sperm-whale play the largest part. There has been an international whaling agreement since 1937 which has been revised and improved in the following years. It banned hunting of Eschrictiidae and Balaenidae (Grey Whales and Right Whales) worldwide. It fixed minimum sizes at which each species could be caught and the total animal catch is limited as well as having a closed season. Whole branches of industry have been built up on whaling, mainly for the extraction of the valuable oils, but which also endeavour to exploit commercially every part of these animals. In spite of intensive worldwide whaling, the biological data about this group of animals, particularly concerning the larger species, is still incomplete. It was with the possibility of keeping smaller whales in large marine aquaria that some interesting facts were discovered. In particular, dolphins *(Delphinus delphis)* and porpoises *(Tursiops truncatus)*, together with other species of the Delphinidae and Phocoenidae (Porpoises), are often kept in large pools, Dolphinaria, and besides providing attractive exhibits are the objects of scientific research. They are soon tame, are very quick to learn and develop a marked love of play. In contrast to their sight their sense of touch is extremely well developed. Their hearing encompasses ultra-sonic levels and together with the ability to emit a wide range of sounds including ultra-sonic waves, provides the animals with a system of orientation, communications and navigation. Dolphins and porpoises have a relatively large brain which enables them to achieve extraordinarily high levels of intelligence which are in every way comparable with the abilities of primates. Their social behaviour is particularly noteworthy and reports of their helping wounded members of their species and saving drowning men have been confirmed again and again by recent investigations.

Cetomimidae, F, *see* Cetomimiformes

Cetomimiformes. O of Osteichthyes*, Coh Teleostei. GREENWOOD and colleagues (1966) have combined in this O different fish groups which appear in other fish systems partly as separate Os, Sub-Os or Fs. These include, for example, the Giganturiformes, the Chondrobrachii and the Mirapinniformes. The O is grouped in the Sub-Os Cetomimoidei, Ateleopodoidei, Mirapinnatoidei and Giganturoidei. Nevertheless many uncertainties still remain in this new classification which are simply caused by the fact that morphological knowledge of these groups is still too fragmentary and does not permit any true analysis. As we are dealing almost entirely with small and sometimes rare deep-sea fish this situation can only be changed slowly. Only 2 of the numerous families are dealt with here because of their biological peculiarities.

– *Cetomimidae*; Flabby Whalefishes (fig. 268). These are true fish whereas it is a well-known fact that the true Whales themselves are marine mammals. The Cetomimidae are deep-sea fish which are 5–15 cm long and whose body shape is reminiscent of that of whales. The

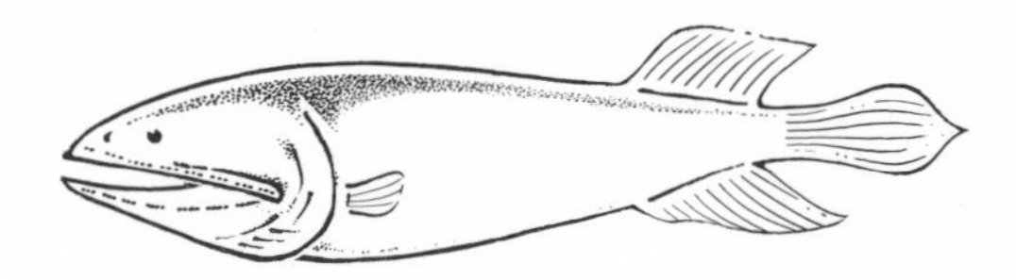

Fig. 268 Cetomimidae-type

Vs are absent, the D and An are positioned far back and opposite each other. They have a large head, a mouth with a wide gape and small rudimentary eyes. They have no swimbladder. The hydrostatic functions are achieved by the tube-like specialised lateral lines. The Cetomimidae are predators which can seize prey as big

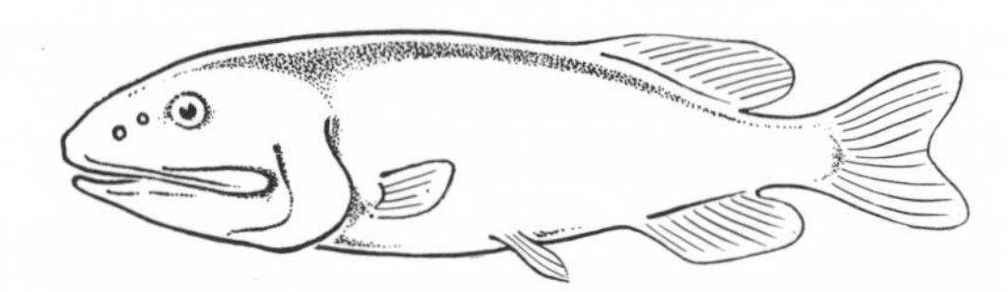

Fig. 269 Barbourisiidae-type

as themselves in their large mouth and carry them in their extensible stomachs. Of the ten or so well-known species most are brightly coloured although they live in deep and dark waters. The shining areas of the D and An are presumed to be caused by glandular secretions. The Cetomimidae together with the similarly shaped

Fig. 270 Rondeletiidae-type

Barbourisiidae (fig. 269) and Rondeletiidae (fig. 270), which however are equipped with sighted eyes, form the Cetomimoidei.

– *Giganturidae* (fig. 271); Giganturids. The few species which have been named up to now are 6–11 cm long and were caught in depths of 450–1,800 m. The relatively elongate Giganturidae are characterised above all by the various peculiarities of the skull bones, the telescopic, forward-pointing eyes, the missing Vs and greatly enlarged Ps, as well as a long lower lobe to the caudal fin.

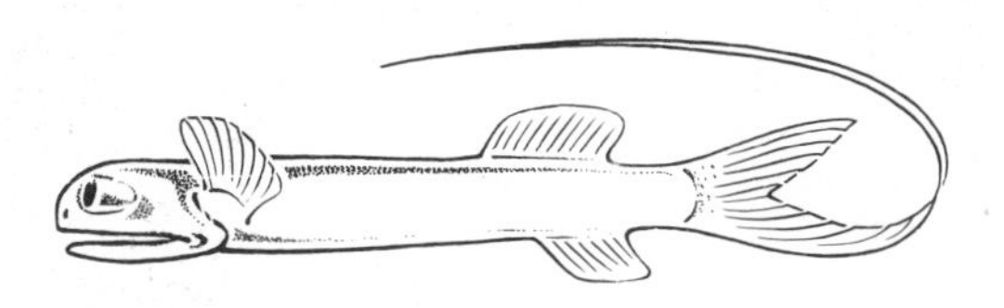

Fig. 271 Giganturidae-type

In addition there is no swim-bladder and the fin-rays are jointed but not branching. The telescope eyes can be pictured as a cylinder-shaped section from a highly enlarged eye which is bordered by the lens at the front and the retina at the back. This type of eye is very sensitive to light but has a very limited field of vision. Giganturidae can swallow prey which is larger than they are. Together with the Rosauridae, which are perhaps only larval forms of the Giganturidae, they form the Sub-O Giganturoidei.

Apart from the two Fs already dealt with the Mirapinnidae (Hairyfishes) and the Eutaeniophoridae (Tapetails or Ribbonbearers) which are related to them should be mentioned. A 4 cm-long fish found in the Azores formed the basis of the former of these two Fs and led to a whole series of surprises.

The whole body surface of this *Mirapinna esau* (fig. 272) is covered with hair-like projections which give the impression of velvety skin. The extremely large

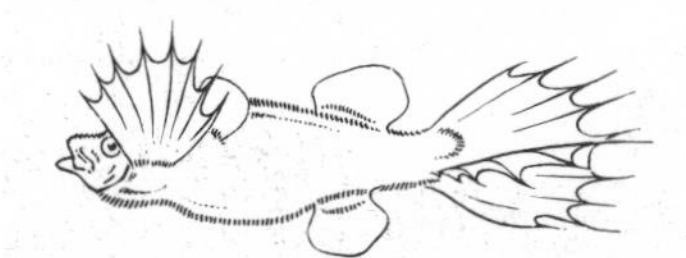

Fig. 272 *Mirapinna esau*

winglike Vs are positioned in front of the Ps and have extended fin-rays. The upper lobe of the C overlaps the lower laterally and the rays are also extended. It is presumed that this is a deep-sea fish type. The Eutaeniophoridae (fig. 273) have not been satisfactorily investigated either. Perhaps the known specimens are older larval stages of deep-sea fish which have remained undiscovered up to now. The special feature of the Eutaeniophoridae is the middle section of the C. This is elongated and becomes a tape-like appendage which

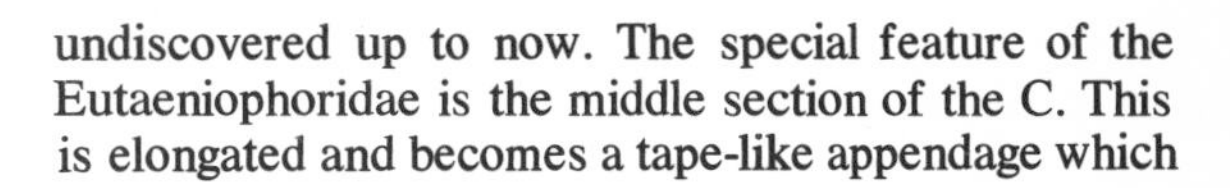

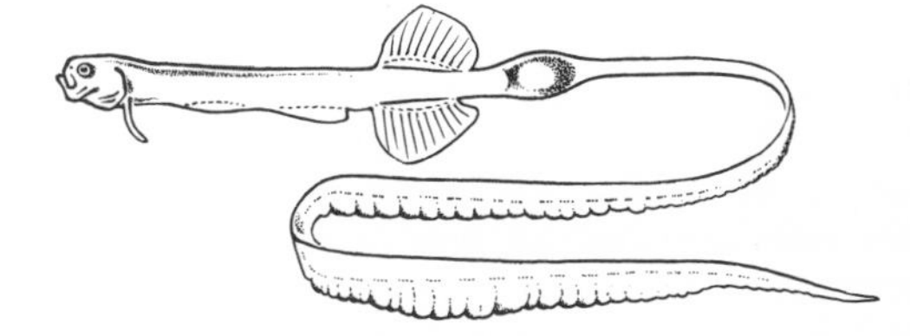

Fig. 273 Eutaeniophoridae-type

is many times longer than the body and which floats behind it. Perhaps it is intended to draw enemies away from the body itself. Both cases should serve as a reminder of how little we know about some fish and what surprises the deep sea still holds for us.

Cetorhinidae; Basking Sharks. F of Chondrichthyes. The only species of this family, the Basking Shark *(Cetorhinus maximus)* (fig. 274) is a powerful creature which usually grows to between 9–11 m and occasionally up to 14 m long. It is found mostly in moderate seas.

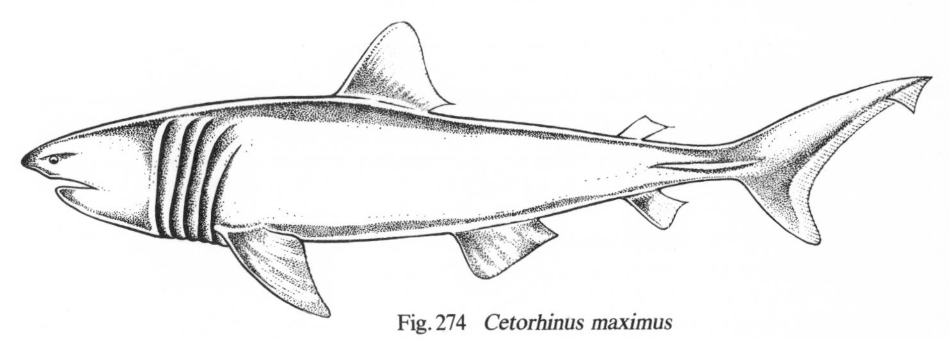

Fig. 274 *Cetorhinus maximus*

Apart from its torpedo-shaped body the 5 very large gill-slits which stretch like belts from the back to the under surface are characteristic of this shark. The first of the two Ds is very large and set high in the shape of a spinnaker. The C is in the shape of a half-moon. At each side of the caudal peduncle are keels which like the large Ps help stabilise the fish. The Basking Shark is livebearing—the young are between 1.50 m and 1.80 m at birth. Like the huge Whale Shark (Rhiniodontidae*), the Basking Shark often feeds on plankton which it filters out of the water with thread-like gill-arch appendages which together form a huge weir-basket. This shark usually floats directly under the surface and is completely harmless. Divers often dare to ride on its tail. The liver-oil of the Basking Shark contains no vitamins but is used in the leather industry and the meat is sometimes made into fish-meal. They are caught with harpoons from boats.

Cetorhinus maximus *see* Cetorhinidae

Ceylonese Fire Rasbora *(Rasbora vaterifloris) see Rasbora*

Chaca chaca *see* Chacidae

Chacidae; Frog-mouthed Catfishes. F of Siluriformes*. The only species of this family, *Chaca chaca* (fig. 275), which grows to 20 cm, is found in SE Asia (India, Burma, Sumatra, Borneo). This strange tadpole-like fish is broad and flattened in the head and body both dorsally and ventrally, almost right-angled in appearance only becoming increasingly laterally compressed

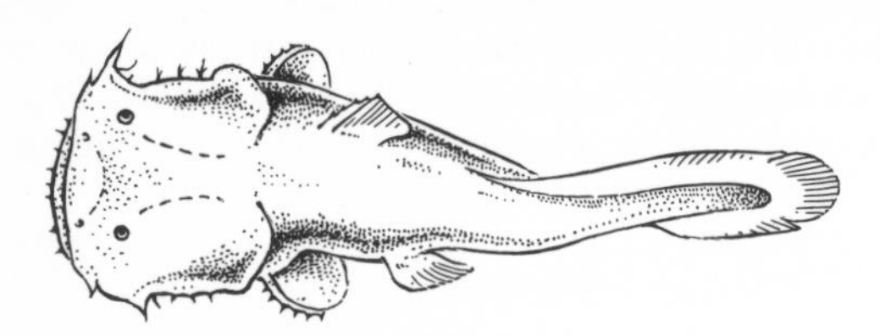

Fig. 275 *Chaca chaca*

in the caudal peduncle towards the C. Skin horny without bony scales. Mouth transverse across the whole body width, in each corner of the mouth are small barbels. D is short, the C surrounds the whole tip of the tail and extends very far forward on the upper side. No adipose fin. On adult specimens arborescent appendages are to be found on the head. *Chaca chaca*, whose top side is dark and spotted and the belly paler, is definitely one of the stranger freshwater fish and is somewhat reminiscent of the Ogcocephalidae*. When found in the aquarium it often looks like a piece of wood particularly as it does not give up its camouflaged position even when it is touched. Hardly anything is known of its biology.

Chaenocephalus aceratus *see* Notothenioidei

Chaetobranchiopsis STEINDACHNER, 1875. G of Cichlidae*. Has 3 species which live in the Amazon basin and in the upper River Paraguay. Medium-sized, tall, laterally much compressed fish with 4 hard rays in the An and sabre-shaped extended Ps. C rounded like a fan. They need warmth, are quiet, peaceable fish, sometimes splendidly coloured. They have not often been kept up to now and have not yet been bred. Probably open-water breeders.

– *Chaetobranchiopsis bitaeniatus* AHL, 1936; Two-striped Cichlid. Found in Amazon Basin. 12 cm. Splendidly coloured with golden glint. Two bold dark longitudinal lines, the upper of the two being widened by a dark blotch in the middle of the body. D, C an An yellowish, with wine-red patterns.

Chaetodipterus faber (Ephippinae, Sub-F) *see* Ephippidae

Chaetodon LINNAEUS, 1758; Butterfly-fishes (figs. 276, 277 and 278). G. of Chaetodontidae*. Sub-F Chaetodontinae. The main habitat of these enchanting fish are the coral reefs and rocky coasts of the tropical Indo-Pacific including the Red Sea. Some also live in the tropical Atlantic and the Caribbean Sea. Most of them have territories which they hold individually or in pairs and which they defend against members of their species. Some live in small groups. They are grazing fish which spend the whole day amongst the coral seeking out food in the form of small creatures. They often eat Coral Polyps or Sea-anemones. There are probably many which eat specialised food. The Chaetodons have high backs and are very strongly compressed laterally, sometimes being almost flat like a leaf. Their mouth is small and conical and not noticeably elongated like *Chelmon** and *Forcipiger**. The LL does not stretch to the tail fin. Most species are extremely brightly coloured and can

Fig. 276 *Chaetodon austriaca*

easily be distinguished for this reason alone. They grow hardly any larger than 20 cm. With the exception of a few species they are hard to keep in an aquarium. They need a very good quality of water and need a very varied diet with vegetable food and Sea-anemones if possible. The aquarium must have hiding-places and be brightly lit. They should not be kept with large robust fish. They attack Sea-anemones and Polychaetes. They are most unsociable towards members of their own species. There are no findings about their propagation. A few species which can be kept are imported a great deal, such as the following.

Fig. 277 *Chaetodon fasciatus*

– *C. auriga* FORSKÅL, 1775; Thread-fin Butterfly-fish. Indo-Pacific, Red Sea. 18 cm. White, D, C and An yellow, black bands on head and black eye-marking on the extended D of older specimens.

– *C. capistratus* LINNAEUS, 1758; Four-eye Butterfly-fish. Tropical Atlantic, Caribbean Sea. To 15 cm. Whit-

ish yellow with large white-edged eye spot in front of the tail.
– *C. collare* BLOCH, 1787; Collared Coralfish; Collared Butterfly-fish. Indo-Pacific. Up to 15 cm. Varying tones of brown, black eye band and white collar.

Fig. 278 *Chaetodon unimaculatus*

Chaetodontidae; Coralfishes, Bristle Teeth (fig. 279). F of Perciformes*. Sub-O Percoidei*. Small to medium-sized; seen from the side it is oval lengthwise or disc-shaped. Found in tropical and subtropical coral reefs and seaweed mats. The head is usually relatively small; some species have a long extended nose region. The mouth is narrow with fine brush-like teeth. D undivided, long, sometimes with very elongated spiny rays, rounded or pointed like a lobe at the back. C never indented. An shorter than D and mostly corresponding in shape with the back section of the latter. Fairly small comb-edged scales which extend onto the unpaired fins, particularly onto the soft-rayed sections. Lateral line arched upwards, not reaching as far as the C in some species. Post-temporal bone firmly attached to the shoulder girdle. The Chaetodontidae are known above all for their magnificent colours. The colours found together are full of contrast and have a velvet-like rather than a shining quality. They serve primarily to warn of

Fig. 279 Chaetodontidae-type

the territorial boundaries but are also used for camouflage and contact between members of the species. The colours of the young fish are in general more brilliant and can easily be distinguished from those of the mature fish. Some species live in pairs. The Coralfishes are substrate feeders, ie they feed on small stationary or crawling organisms like Polyps, Worms and Opisthobranchs. Their agile way of swimming is completely

suited to the crevices of the coral structures. Little is known about their reproductive habits. The young fish often have armour-like bony plates in the head region. Some larger species are highly prized delicacies. The Chaetodontidae are of particular significance for marine aquarists although keeping them and feeding them is difficult in many respects. In relation to this see under Ga *Centropyge, Chaetodon, Chaetodontoplus, Chelmon, Euxiphipops, Forcipiger, Heniochus, Holacanthus, Parachaetodon, Pomacanthus, Pygoplites.*

Chaetodontoplus BLEEKER, 1876. G of Chaetodontidae*. Sub-F Pomacanthinae. Indo-Pacific. G of smallish Angelfishes with small scales; D, An and C are not extended even in older fish in contrast to those of *Pomacanthus*. Eats Algae and small animals. Sometimes imported.
– *C. mesoleucus*, BLOCH, 1787. Indo-Pacific. To 16 cm. Black head bands, white front body section, slate-grey patterned rear body section. C yellow (fig. 280).

Fig. 280 *Chaetodontoplus mesoleucus*

Chaetognatha; Arrow Worms. The Ph comprises mostly small plankton-like, glassy transparent creatures with fish-like appearance. They are symmetrical on two sides and have fins at the sides and at the end of the body. They dash suddenly through the water, and seize their prey with their large curved head-spines. The best known genus is *Sagitta*. There are no aquarium observations.

Chaetognaths, Ph, *see* Chaetognatha

Chained Top Minnow *(Fundulus caternatus) see Fundulus*

Chalcalburnus, G, *see Alburnus*

Chalceidae (fig. 281). F of Cypriniformes*, Sub-O Characoidei*. This family has very few species and is

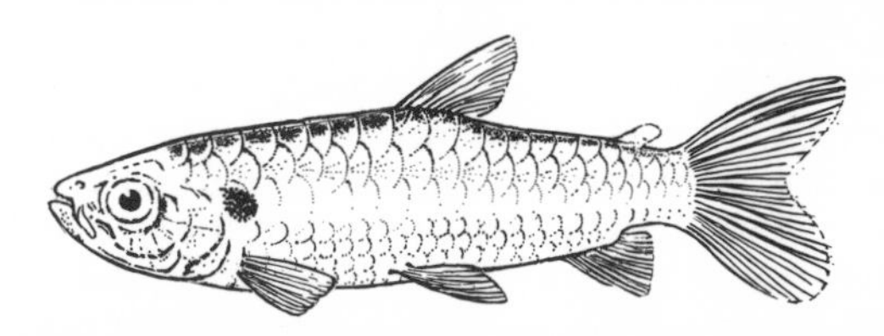

Fig. 281 Chalceidae-type

only found in tropical S America. The family embraces elongate, laterally compressed fish which are characterised above all by scales of varying sizes. There are three rows of very large scales above the lateral line which runs fairly low beneath the middle of the fish's flank. The scales of the LL and below it are considerably smaller. The mouth is terminal. Premaxilla (middle jaw) with three rows of many pointed teeth, Maxilla (upper jaw) relatively large with one row of teeth. Pterygoid has no teeth, the back row of teeth of the Dental (lower jaw) consists only of one pair of small conical teeth. D positioned slightly behind the middle of the back. Adipose fin present. Lower caudal lobe somewhat larger. An short. The Chalceidae are shoal-fish. See under G *Chalceus* for details.

Chalceus CUVIER, 1818. G of Chalceidae*. Elongate Charlacins growing up to 25 cm. They are found in the Amazon area and Guyana. D1 short and positioned behind the middle of the body, adipose fin present. An short. C deeply forked. The lateral line runs along the lower half of the body and consists of small scales, the row above is made up of surprisingly large scales. This shoaling fish which swims elegantly and jumps well, but is unfortunately predatory, needs to be able to move around a great deal. They are fed with small fish, insect larvae, small worms, lean beef, and also dried food. Breeding of these brightly coloured but unfortunately seldom imported Characins has only been successful on rare occasions.

– *C. macrolepidotus* CUVIER, 1818; Pink-tailed Characin. Guyana. To 25 cm. Silvery flanks with deep green or violet sheen according to the way the light falls. Back dark, belly silvery white. The colouring of the fins is variable depending on where they are caught. The D, C and V are mostly coloured an intense wine-red but also greyish yellow. An red or yellowish red, the shoulder spot is dark brown.

Champsocephalus esox *see* Notothenioidei

Chanchito *(Cichlasoma facetum) see Cichlasoma*

Chanda HAMILTON-BUCHANAN, 1822. G of Centropomidae*. Found from E Africa through S Asia and the Malysian Archipelago to Australia. The *Chanda* inhabit coastal and brackish waters there but sometimes penetrate into fresh water. Species which are only found in freshwater are rare; amongst them are *Chanda wolffi* and *Chanda buruensis*. The G includes small, to middle-sized, fairly high-backed fishes with laterally very compressed, silver white frequently glassy transparent bodies. Mouth upturned, neck-line in part clearly depressed. The D consists of a pointed front section with 7–8 hard rays and a back section with one hard ray and countless soft rays. Both sections are widely separated by a wedge-shaped cleft. C deeply forked.

Where they are found the larger species are a cheap food; the smaller ones have sometimes been kept by the aquarist, particularly *Chanda ranga* which was earlier also known as *Ambassis lala*. The freshwater species should be kept in long established, richly planted aquaria with hard water and temperatures of 20–25 °C. For species from brackish waters it is advisable to add about 5–20 g of salt per litre of water, in any case one would then as a rule have to do without the plants. The Chanda are not really suited for a communal aquarium, they are either shy or quarrelsome. If the aquaria are in sunny position, the species *Chanda ranga, Chanda buruensis* and *Chanda commersoni* breed well. When rearing the young fish one must make sure young Glassfishes do not have to look for food but can snap it up from right in front of their mouths. They must be fed plentifully and the tank must have slight ventilation to provide a current which will always bring the food to the fishes.

– *C. buruensis* (BLEEKER, 1856); East-Indian Glassfish. Freshwater species of SE Asia. Grows to 7 cm. Yellowish with very strong iridescent zone along the middle of the side. D and A partly blackish (fig. 282).

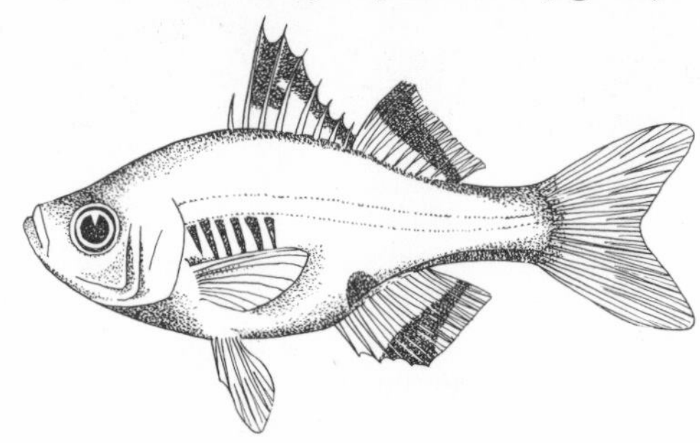

Fig. 282 *Chanda buruensis*

– *C. commersoni* (CUVIER and VALENCIENNES, 1828). Found from E Africa to Australia, predominantly in sea and brackish water. To 10 cm. Delicate yellowish colour, shining metallically. C, D and An orange, in males the tips of D1 and C are black.

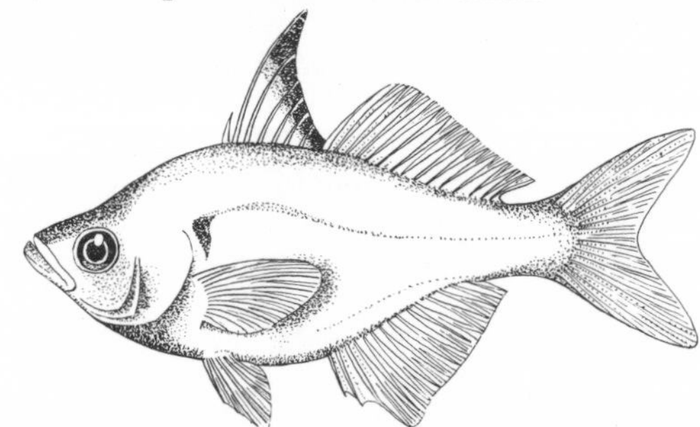

Fig. 283 *Chanda nana*

– *C. nana* HAMILTON-BUCHANAN, 1822. S Asia in brackish and freshwater. To 11 cm. Greenish yellow, fins of the males orange, sometimes with black tips (fig. 283).

Fig. 284 *Chanda ranga*

– *C. ranga* HAMILTON-BUCHANAN, 1822; Indian Glassfish. S and E Asia in fresh and brackish waters; to 7 cm. Greenish yellow, shining gold and iridescent bluish-green with dark horizontal stripe and transverse lines of pale dots. D and An edged with pale blue (figs. 284 and 285).

Fig. 285 *Chanda ranga*

– *C. wolffi* (BLEEKER, 1851). SE Asia. Malayan Archipelago. To 20 cm. Freshwater species. Pale yellow, with a distinct yellow sheen. 2nd fin-ray of the An very long.

Chanidae, F, see Gonorynchiformes

Channa (various species) *see* Channiformes

Channalabes GÜNTHER, 1873. G of Clariidae*. In the Congo Basin and Angola. Over 30 cm. Eel-like elongated catfishes with compressed head and the typical air-breathing organ of the family. The D1, C and An form a uniform fringing fin. The Ps are under-developed. Vs absent. Voracious omnivore which is very difficult to keep and which needs hiding-places and likes to dig hollows for itself. Only the young of these unsociable and predatory fish are suited for keeping in aquaria. Method of reproduction unknown.

– *C. apus* GÜNTHER, 1873. Congo Basin and Angola. Up to 30 cm long. Monotone dark-brown species.

Channel Catfish *(Ictalurus punctatus) see* Ictaluridae

Channichthyidae, F, *see* Notothenioidei

Channidae, F, *see* Channiformes

Channiformes; Snake-heads (fig. 286). O of Osteichthyes*, Coh Teleostei. Sturdy predatory fishes of Africa and SE Asia. Elongated body at the front, almost round in cross-section. The large head recalls the head of a snake, an impression which is strengthened further by the large plate-like scales on the upper side. Deeply

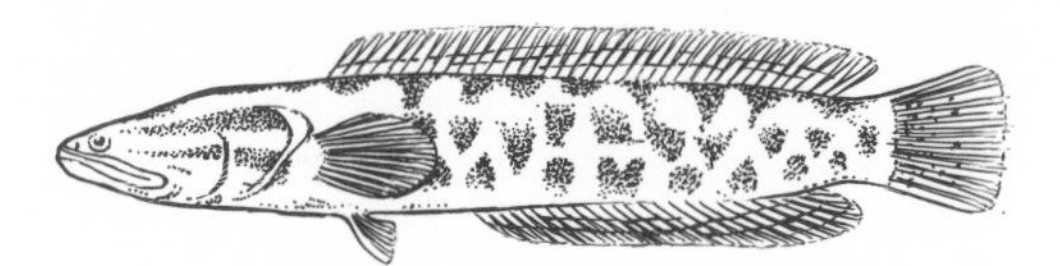

Fig. 286 Channiformes-type

gaping mouth, capable of extending and protruding a long way. Slightly projecting lower jaw, fully toothed. Front nasal opening extended externally and tube-like. Very long D and An, supported by soft rays. In *Channa orientalis* the Vs are absent. Cycloid or ctenoid scales. Swimbladder stretching into the tail and in two parts at the rear. Particularly characteristic is a supplementary air-breathing organ in the form of a sinus over the gill cavity into which two lamellae reach (fig. 287). Mainly because of this organ the Channiformes were often grouped together with the Anabantoidei*. The oldest fossils are from the Pliocene period. The snake-heads inhabit many different waters and even survive in water with very low oxygen content. They are greedy predators which eat everything which they can overcome and this includes fish which are hardly any smaller than they are. Many species are held to be excellent food varieties and because of this have been released into various regions. Little is known about their reproductive habits. Some lay their eggs in nests of plant matter, which are built by the male, others are free spawners whose eggs rise to the surface. The incubation period is short (2–3 days). All species jump well. Various specimens of the single F *Channidae* (Snake-heads) have already

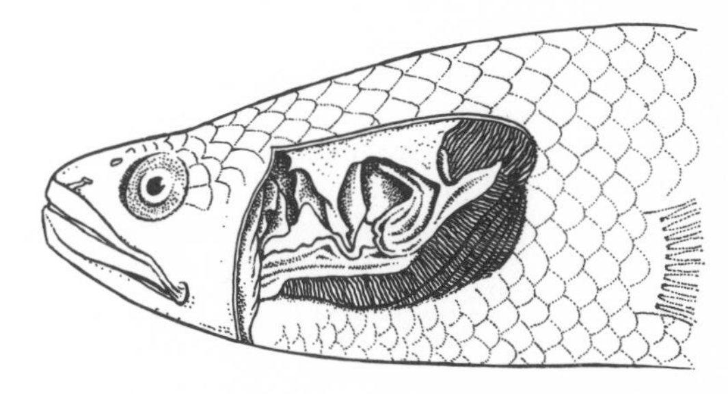

Fig. 287 Respiratory organ of the Channiformes

been kept in aquaria and sometimes have even been bred (fig. 288). These include *Channa orientalis* from S and SE Asia which grows up to 30 cm and *Channa africana* from tropical W Africa which grows to ap-

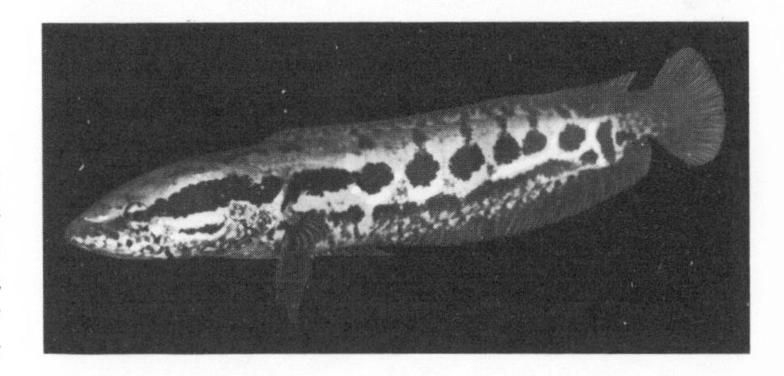

Fig. 288 *Channa* sp.

proximately the same length. The large species of up to 120 cm, such as *Channa marulia* from SE Asia, are sometimes exhibited in aquaria. It is easy to keep these hardy creatures and they also learn to recognise their keeper. Only fish of approximately the same size, however, should be kept together. They take lean meat (cow's heart) as well as larger forms of live food. Most have interesting markings and are beautifully coloured in ochre tones.

Chanos chanos *see* Gonorynchiformes

Chaper's Characin *(Alestes chaperi) see Alestes*

Char *(Salvelinus salvelinus) see* Salmonoidei

Chara, G, *see* Charophyta

Characidae; Characins (fig. 289). F of Cypriniformes*, Sub-O Characoidei*. F of Characoidei which has many species, whose habitat stretches from S and central America to Texas and also includes Africa. Many authorities put the African specimens in a separate F Alestidae. With regard to general features the Characidae conform to the details specified for the Sub-O Characoidei. The following further characteristics are typical of the Characin family. They are mainly small or tiny fish which have a typical fish-shape, less frequently their bodies are elongated or droplet spindle or disc-shaped. Mouth pointing forward or slightly upturned. Premaxilla (middle jaw) with 1–2 or more rarely 3 or 4 *(Brycon, Creagrutus)* rows of teeth, Maxilla (upper jaw) normally toothed or at least with a few teeth. On the Dentral (lower jaw) usually one, and less frequently two, rows of teeth. The teeth themselves are simply conical or chisel-shaped with several points (fig. 290). Full set of fins. Adipose fin generally present. D short, more or less positioned exactly in the middle of the body, when spread out normally spinnaker-shaped. C deeply indented, in some species with elongated middle rays or elongated lower lobe. An longer than D, in the males occasionally with specialised fin rays in the front section. Gill-skin is free, not fused with the isthmus. Most specimens are shoal-fish, many species are brightly patterned and coloured. The presence of strongly iridescent coloured body parts, lines, patches or spots is typical of Characins from darker tropical rain-forest waters of the black water type. These are not light organs such as those found in deep sea fish but are specialised skin areas with reflecting guanophores as a base layer and specialised chromatophores above them which have coloured pigments. These iridescent parts do however fulfil the same function as light organs. They provide a means of recognition between members of the same species in the half-light of the tropical rain-forest waters and thus for example keep the shoal together. Most species are passive when out of the sight of men; a few however must be numbered amongst predators of the S American waters, such as *Raphiodon vulpinus.* This elongated fish has a slanting mouth and a terrifying bite in which two needle-sharp fangs in the lower jaw are particularly noticeable. These are so long that when the mouth is shut they have to be kept in sheaths in the lower jaw. The eggs of the Characidae are as a rule expelled in open water between water plants and adhere to substrata. In the case of some Ga *(eg Glandulocauda* and *Corynopoma*) the males transfer capsules of milt whereby the An acts as a clasping organ. The females store these in the oviduct; fertilisation follows, however, shortly before expulsion of the egg (Ovoviparity). There are no livebearing species of this family. The Characidae do not look after the brood as a rule. It is often observed in captivity that the parents even neglect the eggs. This is definitely not the rule in the spacious spawning grounds of their habitat. The pairing behaviour of the free-spawning species has been described for vaious species under laboratory conditions particularly by STALLKNECHT. According to this research a relatively uniform basic pattern seems to hold for the mating and pairing. Partners swim together in parallel lines or in circles. Following on from this, actual mating begins with butting which can be done by either by the male or the female. The luring display of the male follows this. The female who is ready for spawning follows the fluttering male and eventually overtakes him and is then led by the male from below in a turning movement, around on imaginary vertical or horizontal axis. The ventral fins interlock during this. Egg and sperm cells are expelled at the same time. The duration of the individual phases and the nature of their execution can vary from G to G and also within a G. STALLKNECHT thought it probable that the direction of the turning depended on which ovary was more developed; eg if the right were strongest the fish would turn to the left, and vice versa. There are supposed to be males which are specialised in turning in one direction. The Characidae do not have any particular economic significance for fisheries. In some parts of S America abundant species are said to be used as food. Many Characidae are numbered among the best known aquarium fish. Millions of fish are bred annually for amateurs throughout the world; unfortunately, however, not always under good conditions. Neither is enough care always taken to use only selected breeding stock. As a result many originally splendid species are traded as colourless bulk merchandise. More and more newly imported species of Characins come onto the market, sometimes including specimens which have not yet been scientifically described. In this family in particular the number

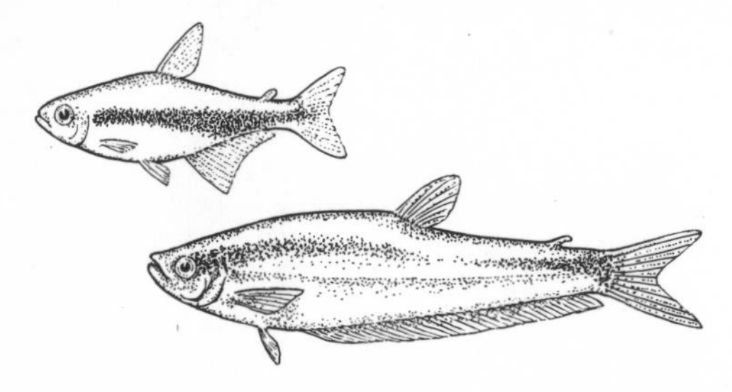

Fig. 289 Characidae-type. Top. G *Hyphessobrycon;* bottom, G *Leptagoniates*

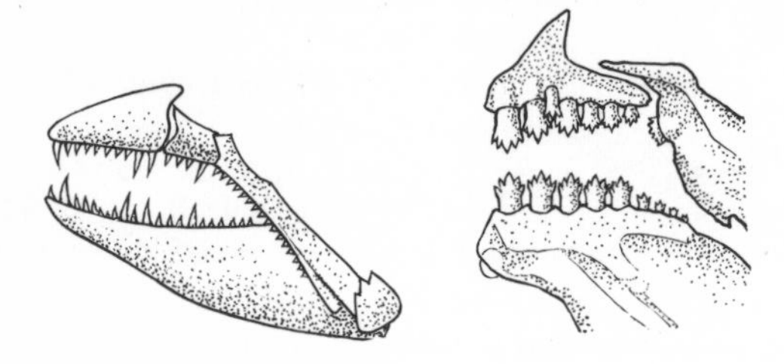

Fig. 290 Jaws and dentition of Characins. Left, jaws of Predatory Characin with conical teeth; right, jaws of those Characins which eat small creatures, showing many-pointed chisel teeth

of species which are of interest to the aquarist is very large. The following particularly important Ga are treated as separate entries. Under these, detailed information on care and breeding is to be found. *Acestrorhamphus, Acestrorhynchus, Alestes, Alestopetersius, Aphyocharax, Arnoldichthys, Astyanax, Axelrodia, Boehlkea, Brittanichthys, Brycon, Charax, Cheirodon, Ctenobrycon, Exodon, Gephyrocharax, Glandulocauda, Gymnocorymbus, Hemigrammus, Hyphessobrycon, Megalamphodus, Micralestes, Mimagoniates, Moenkhausia, Nematobrycon, Paracheirodon, Petitella, Poptella, Prionobrama, Pristella, Pseudocorynopoma, Roeboides, Stevardia, Thayeria.*

Characidiidae (fig. 291). F of Cypriniformes*. Sub-O Characoidei*. These S American Characoidei are small elongated, fairly round in cross-section, somewhat flattened on underside, and keep near to the bottom. They do not form shoals but live more in fixed positions in small territories. Small mouth forward-pointing or

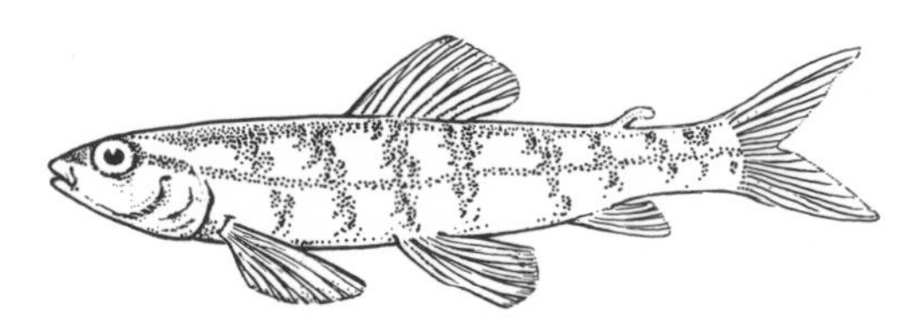

Fig. 291 Characidiidae-type

slightly inferior. Premaxilla (middle jaw) with one simple row of narrow, mostly three-pointed, close-set teeth, maxilla (upper jaw) small toothed at the front or toothless. Dentral (lower jaw) with 2 rows of teeth – in the front row close-set, three-pointed teeth and in the back very small, conical ones. D positioned in middle of the body, adipose fin present, C indented, An short, paired fins relatively large. The Characidiidae have a characteristic way of swimming. Like many fish living on the bottom they swim short stretches jerkily and keep stopping again and again. When they stop they support themselves on their large pectorals *(Characidium)* and look round with their head slightly raised. All species are small. Various specimens are kept as interesting aquarium fish. Compare *Characidium**.

Characidiinae, Sub-F, *see* Characidiidae

Characidium Reinhardt, 1866. G of Characidiidae*. Amazon Basin. G of bottom-dwelling Characins with many species. 6–10 cm long. Body elongate, laterally compressed. Lateral line complete. D1 short, beginning in middle of body. D2 absent. C forked. An short, Ps and Vs very developed. Ps are used to support the body on the bottom. They are mostly inconspicuously coloured with spots or band patterns. They are peaceful quiet fish which swim with jerky movements. According to previous reports breeding is not difficult. After much passionate activity amongst the finely leaved plants, the spawn is set down on the bottom. It does not stick well. The young fish hatch after 24–30 hours and must be fed with rotifers or *Cyclops* nauplii after 4 days. *C. fasciatum* has been imported alive in recent years several times. They have died, however, without having been propagated.

– *C. fasciatum* Reinhart, 1866; Darter Characin, Banded Characin. S America from Orinoco area to La Plata

Basin. Grows to 10 cm. The fish is yellowish-brown with a line along its body from its snout to the end of the tail stalk, crossed by several undefined bands, and ending in a black caudal blotch. The fins are colourless. The females which are ready to spawn can be recognised by their rounded stomach area (fig. 292).

Fig. 292 *Characidium fasciatum*

– *C. pellucidum* Eigenmann, 1910. Guyana. To 8 cm. They are almost white with several lines of small black spots along their body. A large dark patch covers the head area and behind it across the line of the back are countless irregular black transverse brands reaching to where the C begins. Fins colourless or with faintly dark spots (fig. 293).

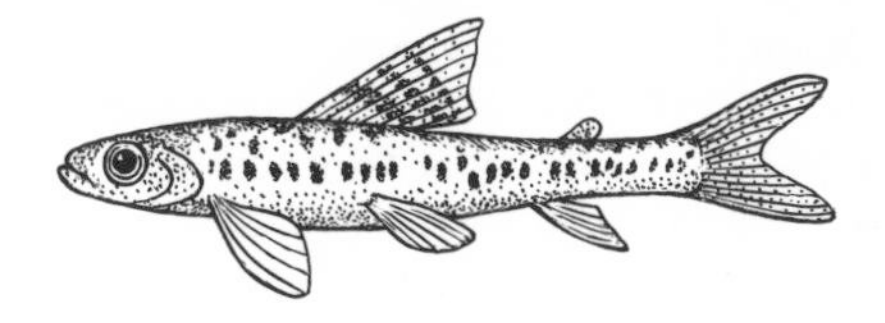

Fig. 293 *Characidium pellucidum*

Characidoidea, Sup-F, *see* Characoidei

Characinae, Sub-F, *see* Characoidei

Characins, F, *see* Characidiidae

Characins, related species, Sub-O, *see* Characoidei

Characins, True, F, *see* Characidae

Characoidei; Characin related species. Sub-O of Cypriniformes*. Sub-O Ostariophysi. About 1,200–1,300 present-day species of which there are about 200 in Africa, and 900 to 1,100 in S America, Central America and in southern N America up to the Rio Grande. The earliest fossils supposed to belong to this group are from the Late Cretaceous in Brazil and Africa. They have also been found from the Eocene in Europe. Most Characoidei have a form similar to *Barbus* but although there are great deviations from this there are no specimens which are dorso-ventrally compressed. The body has scales but the head does not. The mouth is not extensible (with the exception of the Hemiodidae), no barbels. Jaw generally toothed, in the upper jaw the teeth are generally limited to the Premaxilla. The Maxilla has hardly any or no teeth. Besides this the Palatinum and Pterygoid in the roof of the oral cavity can also have teeth. The form of the teeth is very varied, in some Fs they are reduced, in others they are completely missing. The bones of the skull, the vertebral column and the paired fins show some peculiarities. The Weberian Apparatus* which is typical of the whole O Cypriniformes, is of fairly simple construction. In rare instances, the 2nd and 3rd vertebrae are fused. The set of fins is usually completed by a small adipose fin behind the D. The fin-rays are of the soft ray type. The anal opening lies at the back in front of the An. The latter can show secondary sexual characteristics in the males in the form of specialised fin-rays which can usually only be recognised with a microscope. Swimbladder in two sections, openly connected to the gut, and functioning as an additional respiratory organ in some species eg *Erythrinus*. The Characin family are with few exceptions active throughout the day. Many species stay together in shoals and also spawn in shoals. Brood-care is only seen rarely, normally the parent fish do not bother about the abundant spawn which they set down on plants.

In captivity at least they often prey on their own spawn. The Characoidei include predatory fish, omnivores and also plant-eaters. Many of the larger species are a significant food for man in the areas where they are found despite being full of fish-bones. That is particularly true in S America. Because of their liveliness and their patterning and colouring which is frequently pretty, many of the small species are standard aquarium specimens and many of them are also relatively easy to breed. Opinions differ greatly about the number of Fs and Sub-Fs, ie the classification of the Characoidei. This situation is caused mainly by the fact that few fossils are available and that there are far too few exact results of research of present-day species, quite apart from the fact that knowledge of the areas of distribution is far too imperfect and new species are continuously being discovered. Here, the system of classification suggested by GÉRY is applied. GÉRY differentiates between the two super-Fs Characidoidea and Erythrinidoidea mainly on the basis of the length of the anal fin. The Fs and Sub-Fs belonging to each are listed below. They are treated under separate headings because of their significance to the aquarist.

Super-F: Characidoidea (Characoidei with long anal fin)
- F: Characidae
  - Sub-F: Agoniatinae
  - Sub-F: Raphiodontinae
  - Sub-F: Characinae
  - Sub-F: Bryconinae
  - Sub-F: Clupeacharacinae
  - Sub-F: Paragoniatinae
  - Sub-F: Iguanodectinae
  - Sub-F: Aphyocharacinae
  - Sub-F: Glandulocaudinae
  - Sub-F: Stethaprioninae
  - Sub-F: Tetragonopterinae
  - Sub-F: Rhodsiinae
  - Sub-F: Cheirodontinae
- F: Serrasalmidae
  - Sub-F: Myleinae
  - Sub-F: Catoprioninae
  - Sub-F: Serrasalminae
- F: Gasteropelecidae
  - Sub-F: Thoracocharacinae
  - Sub-F: Gasteropelecinae

Super-F: Erythrinidoidea (Characoidei with short anal fin and groups which can be included here because of other characteristics)
- F: Erythrinidae
- F: Ctenoluciidae
- F: Chalceidae
- F: Crenuchidae
- F: Characidiidae
  - Sub-F: Elachocharacinae
  - Sub-F: Characidiinae
- F: Lebiasinidae
  - Sub-F: Lebiasininae
  - Sub-F: Pyrrhulininae
  - Sub-F: Naunostomidae
- F: Anostomidae
  - Sub-F: Leporellinae
  - Sub-F: Anostominae
- F: Hemiodidae
  - Sub-F: Parodontinae
  - Sub-F: Hemiodinae
  - Sub-F: Bivibranchiinae
- F: Curimatidae
  - Sub-F: Chilodinae
  - Sub-F: Prochilodinae
  - Sub-F: Curinatinae
  - Sub-F: Anodinae
- F: Citharinidae

Charax GRONOVIUS, 1763. G of Characidae*. Sub-F Characinae. Found in Central and lower Amazon, River Paraguay and in Guyana. Elongate Characins which swim with their head pointing downwards. To 15 cm. They are not very active swimmers, and inhabit reedy waters. Hump-shaped back, long caudal peduncle. D1 short, D2 present, deeply forked C, An very long. Very similar to the better known, near related *Roeboides** species in habits and behaviour. Has been successfully bred, for breeding experiments and care see *Roeboides*.

– *C. gibbosus* (LINNAEUS, 1758); Humpbacked Headstander. Found in Guyana and Central and Lower Amazon, River Paraguay. To 15 cm. Body yellowish brown with a silver sheen. Upper part of body thickly sprinkled with brass coloured or iridescent green spots. Behind the gill cover there is a long, dark, shoulder blotch, the fins are colourless. This interesting fish is unfortunately seldom imported (fig. 294).

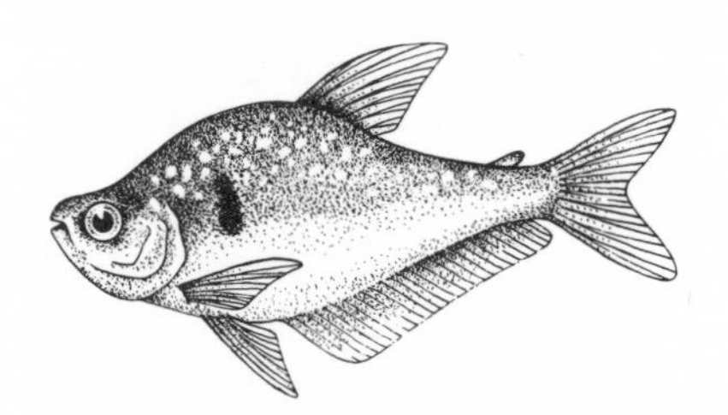

Fig. 294 *Charax gibbosus*

Charonia GISTL, 1848; Trumpet-snails (= *Tritonium* LINK 1807) (fig. 295). G of predatory Prosobranchs (Prosobranchia*) of warmer seas which feed on Echinoderms and molluscs. Their high-domed shells of up to 40 cm have a blow-hole and are used in Africa, Asia and also Mediterranean countries as a signalling device.

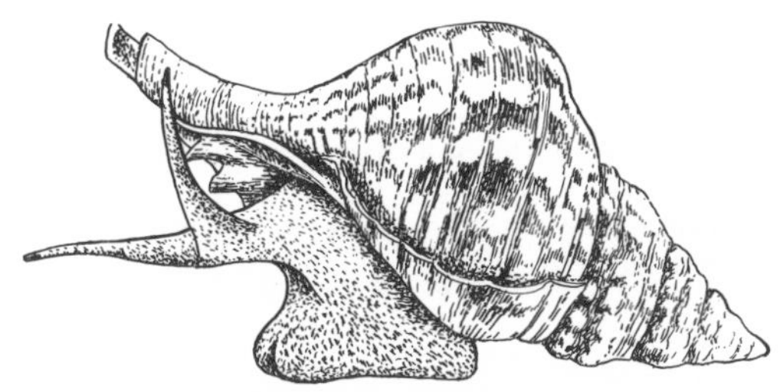

Fig. 295 *Charonia* sp.

Smaller specimens which would suit an aquarium are rarely obtainable. As natural enemies of the predatory *Acanthaster** (Crown-of-Thorns Starfish), they have an important function in the community life of the coral reef.

Charophyta; Stoneworts. Division of Algae* with 300 species. They are green or grey through calcareous incrustation, have chlorophyll a and b, lentil-shaped chromatophores. They have a highly differentiated thallus, which is deceptively similar to the shoots of higher plants. The branches are grouped in elongate sections and in nodes, on which long or short side branches are then arranged in whorls. The plants are anchored by rhizoids in soft substrate, such as mud or sand. The round yellow-red male reproductive organs and the egg-shaped female ones are formed on nodes of side-branches and can even be seen by the naked eye. The Charophyta are dispersed over the whole world and can appear in all types of fresh and brackish waters. In the early days aquarists occasionally used them as a spawning substrate, nowadays they have no significance. An exact classification is difficult even for the specialist and is only possible by microscopic examination. *Chara* and *Nitella* are widespread varieties.

Chars (various species) *see* Salmonoidei

Checked Snapper *(Lutjanus decussatus) see Lutjanus*

Checker Barb *(Barbus oligolepis) see Barbus*

Checkerboard Cichlid *(Crenicara maculata) see Crenicara*

Cheirodon GIRARD, 1854. G of Characidae*, Sub-F Cheirodontinae. Widespread throughout tropical S America. 4–8 cm long. Elongate, laterally compressed Characins which resemble species of the Ga *Hemigrammus* and *Hyphessobrycon.* Dorsal and ventral profiles are approximately equally convex; lateral line incomplete, D2 always present. The males have small plates on the underside of the caudal peduncle. *Cheirodon* species live in shoals at middle depths. Their base colour is silver with a blue or green sheen and a dark longitudinal band on the rear half of the body. Apart from *C. (Lamprocheirodon) axelrodi,* this G includes less attractive aquarium fish, the reproductive habits of which are not well-known. Contented omnivores which are active swimmers. Breeding should be attempted in the same way as for the Ga *Hemigrammus** and *Hyphessobrycon**.

– *C. (Lamprocheirodon) axelrodi* SCHULTZ, 1956; Cardinal Tetra. Found in the tributaries of the Rivers Negro and Orinoco. 4–4.5 cm. Shape typical of the genus. A broad, blue-green, iridescent, longitudinal band stretches from the tip of the snout through the eye to the root of the tail. Underneath it runs a broad shining cardinal red band of the same length and leaves only a narrow silvery edge to the belly. Red-brown back, fins colourless, D1 and An have a white tip, D2 present. These are shoal-fishes which prefer less-bright positions. Spawning takes place near finely pinnate plants in dark twilight, or at night by moonlight. For attempts at breeding it is best to leave a 25 watt lamp burning in a distant corner of the room at night. The pairs do not spawn as freely as the well-known *Paracheirodon** *innesi.* The rearing of the young fish is also more problematical. Soft, slightly acid water (2–4 °dH) is a requirement for successful breeding. The young fish which develop slowly must be fed with rotifers or the smallest nauplii. *C. axelrodi* is probably the most attractive and most sought after of all American Characins (fig. 296).

Fig. 296 *Cheirodon axelrodi*

– *C. meinkeni* AHL, 1928. Coastal waters of Brazil, from Bahia to Rio de Janeiro. To 5 cm. Flanks silvery with an olive tone; a longitudinal band with a metallic sheen which ends in a dark caudal blotch is visible from time to time. Fins a delicate yellow, the tips of the D and An are white. This species has not yet been bred successfully.

Cheirodontinae, Sub-F, *see* Characoidei

Chela HAMILTON-BUCHANAN, 1822 (fig. 297) G of

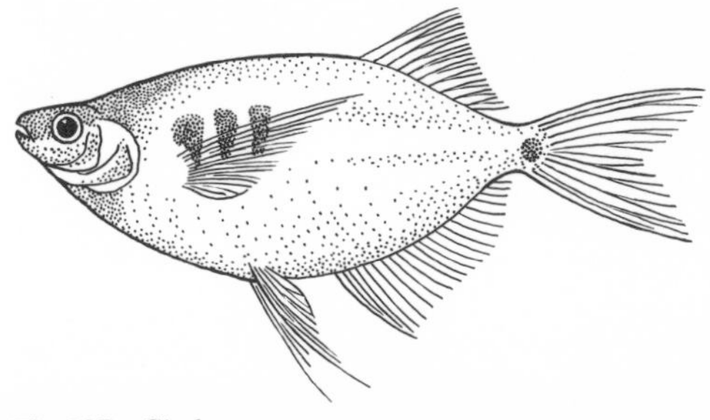

Fig. 297 *Chela*

Cyprinidae*, Sub-F Abraminae. Widespread in lakes and rivers of SE Asia. Characteristically small surface fish, the bodies of which are strongly compressed laterally. The strongly bowed belly profile is sharply keeled. Mouth upward-pointing. The many-rayed An is longer than the D, Ps elongate like wings, outer ray of the Vs also elongate. Colouring is delicately translucent, with shiny silver sides. Peaceable shoaling-fishes which are kept in well covered aquaria, having a large surface area and well-stocked with plants. The quality of the water is not important, but the addition of fresh water from time to time is advisable. Temperature not lower than 24 °C. Eats all kinds of live and dried foods, but especially insects from the surface. Elongated aquaria with fresh water and finely pinnate leaves are required for breeding, pH and hardness of the water not important. The fish begin to spawn at twilight when the male clasps the female briefly. The young hatch after 20 hours (at 35–26 °C) and should be reared with the finest rotifers.

– *C. caeruleostigmata* SMITH, 1931; Blue-spot Keel-belly. Thailand. Up to 6 cm. Back reddish-brown, sides silvery, with a blue to greenish sheen. Black blotch behind the gill-cover to which 4–12 short, lateral bands are attached. An and C a delicate yellowish colour.

– *C. laubuca* (HAMILTON-BUCHANAN, 1822); Indian Glass Barb. SE Asia. To 6 cm. Back grey-green, sides delicate greenish-blue with a silver or gold sheen. Dark, golden-edged bands in the region of the first half of the body and a blotch on the caudal peduncle (these may also be absent). A longitudinal band (at the level of the vertebral column) which changes in colour.

– *C. mouhoti* SMITH, 1945. Thailand. To 6 cm. Colouring the same as *C. caeruleostigmata*, but with only a black, golden-edged blotch behind the gill-cover.

Chelaplex (Komplexon, Titriplex). Trade-names for a group of Chelatores* of which only the easily soluble Chelaplex III (Sodium-salt of Ethylendiamintetraacetic acid) is of aquaristic significance. This substance has water-softening characteristics, hinders the formation of insoluble salts and helps plants and animals to absorb vital metal ions. Because of these characteristics water plants, for example, grow better after the addition of Chelaplex III than before. The quantity to be used depends on the hardness of the water*, because Chelaplex III binds metal ions quantatively in a perticular order. A mixture of 0.1 g of Chelaplex III per degree of total hardness per 100 litres of water has proved to be particularly effective. It should, however, be observed that Chelaplex III can also effect the pH value, which should, therefore, be checked before and after the addition of Chelaplex III.

Chelatores. Weak organic acids*, which because of certain chemical combining relationships keep polyvalent metal ions in solution even in the presence of precipitants. Such precipitants in the aquarium are mainly phosphate, carbonate and sulphate ions. Thus chelatores practically prevent, through their very presence, the build-up of insoluble salts and such important substances as magnesium, iron, calcium, manganese etc, remain available in the water for plants and animals, or are often only fully useable in conjunction with chelatores. Because of this valuable characteristic chelatores are used deliberately by aquarists and among these Chelaplex and humic-acids are particularly important.

Chelmon CLOUQUET, 1817; Butterfly-fish. G of Chaetodontidae* Sub-F Chaetodontinae. From the Indo-Pacific, where the fish live on coral reefs. With their remarkably elongated snout they are able to pick small animals out of holes and crevices. The territory-forming fish are unsociable with their fellows and are not easily kept in the aquarium. *Artemia, Mysis* and *Tubifex* can be used as food. They are particularly discriminating with regard to water-quality. Only one certain species is known, which varies within its wide area of dispersion.

– *C. rostratus* LINNAEUS, 1758; Copper-banded Butterfly-fish; Beaked Coralfish; Beaked Butterfly-fish; Banded Long-snout Butterfly-fish. E Africa to the Philippines. To 17 cm. Silver with broad orange lateral stripes (fig. 298).

Fig. 298 *Chelmon rostratus*

Chelon oligolepis *see* Mugiloidei

Chemosynthesis see Bacteriophyta

Cherry Barb *(Barbus titteya) see Barbus*

Cherubfish *(Centropyge argi) see Centropyge*

Chewing Plate. A horny pad found in the roof of the pharynx in Carp-type fish (Cyprinidae). It is fixed to the base of the skull in such a way that the pharyngeal teeth can bite against it (*see* Teeth). The food is either crushed or torn apart by this method.

Chiasmodentidae, F, *see* Trachinoidei

Chilodinae, Sub-F, *see* Curimatidae

Chilodonella. G of Ciliata*. Some species are ectoparasites (Parasitism*) in fish of which the best known is the large heart-shaped *C. cyprini* MOROFF, 1902. It is 40–60 μm and is given a heart-shaped appearance by a small indentation on the cell's surface, which has an uneven covering of hairs. Propagation occurs by simple cell division. The host fishes can be actively sought out so that a direct transfer is possible. *Chilodonella cyprini* (fig. 299) is a typical debilitating-parasite which can only reproduce on previously injured fish. Affected fish rub themselves, hold in their fins and have a whitish blue skin-discolouration particularly in the region of the

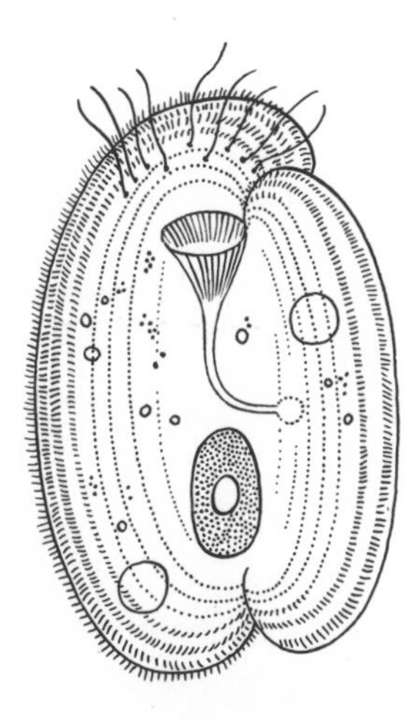

Fig. 299 *Chilodonella cyprini*

nape. In the advanced stages, the skin falls away in shreds. If the gills are badly affected, it results in shortness of breath. All species of fish suffer. A serious infection can be fatal. Skin and gill smears from living fish must be examined under a microscope for diagnosis since the parasites leave dead fish very quickly. Affected fish should be kept in a Trypaflavin-bath whilst maintaining a temperature of 28 °C. Empty aquaria are free of contamination after 3–5 days at 28 °C–30 °C since *Chilodonella* can only exist for a short time without a host and are sensitive to high temperatures.

Chilodus MÜLLER and TROSCHEL, 1844. G of Curimatidae*. Found in the Amazon area and Guyana. Elongate Characins of up to 10 cm. Shoaling fish which swim with their head downwards in reedy areas where they are well camouflaged by their posture and markings. The dorsal profile is more convex than the ventral. Mouth terminal with powerful lower lip. D1 spotted, D2 present, deeply forked C. An of moderate length. Nearly all the fins are colourless and transparent. Every scale has a dark blotch at the base on a silvery background giving rise to rows of spots. A small longitudinal band may be present. As well as types of live food, vegetable matter is readily taken. Breeding has been frequently successful in latter years. Two dark spots as big as peas behind the gill-cover and the paling of the longitudinal band show that breeding fish are ready to spawn. Mating follows after vigorous activity and after tapping with the mouth the spawning area which is in fine-leaved Java moss, Perlon Yarn and the like near the ground.

The breeding fish barely look after the swelling eggs. Artificial assistance is usually necessary for hatching the young fish. FRANKE (1963) helps the young fish by cutting the egg membranes with two scalpel needles under a magnifying glass after 48 hours. BECK burst these by letting a glass plate fall onto the eggs in the water. He had previously placed the eggs between two glass strips about 1.5 mm thick. The young fish which swim freely after about 5 days stand in the water even more steeply than their parents and even lie on their backs sometimes. Feeding with rotifers and *Cyclops* nauplii is not without problems. Soft, slightly acidic water is indispensable for keeping and breeding these fish.

– *C. punctatus* MÜLLER and TROSCHEL, 1844; Spotted Headstander (fig. 300). From Guyana and Loreto in Peru. To 10 cm. A narrow longitudinal band is sometimes visible on the body. D1 with a broad, black blotch

Fig. 300 *Chilodus punctatus*

on the end and an orange-yellow zone beneath it, as well as dark spots. For other colouring see above. The subspecies *C. punctatus zunevei* (PUYO, 1945) from the Rio Maroni, has, in contrast to the principal form, no black longitudinal band.

Chilomycterus DE BARNEVILLE, 1846. G of Diodontidae* with some species in tropical W Atlantic and Indo-Pacific. Lives in shallow waters and eats small creatures. Similar to *Diodon**, however spines stand

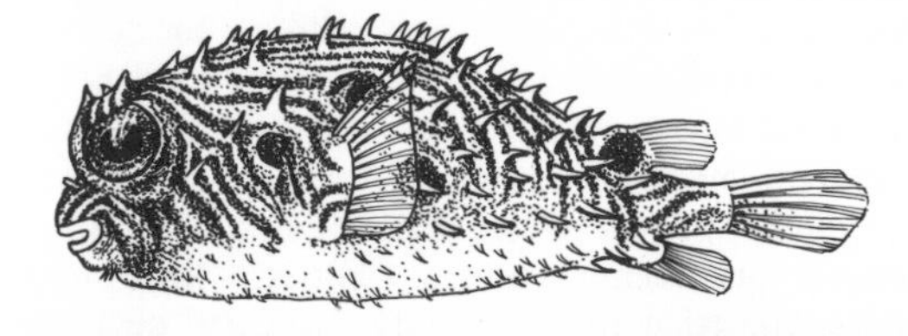

Fig. 301 *Chilomycterus schoepfi*

away from body constantly. These bizarre fishes need hard-shelled food—at least occasionally—such as mussels, shrimps etc. Often imported is:

– *C. schoepfi* WALBAUM; Spiny Burrfish or Boxfish. From tropical parts of the W Atlantic. To 25 cm. Young fish have elongated soft horn above the eyes. Body has olive-coloured, wavy patterns (fig. 301).

Chimaera monstrosa *see* Chondrichthyes

Chimaeras (Holocephali, Sub-Ph) *see* Chondrichthyes

Chimaeridae, F, *see* Chondrichthyes

Chimaeriformes, O, *see* Chondrichthyes

'Chinese Arrowhead' *(Sagittaria graminea) see Sagittaria*

Chinook Salmon *(Oncorhynchus tschawytscha) see* Salmonoidei

Chirocentridae, F, *see* Clupeiformes

Chirocentrus dorab *see* Clupeiformes

Chirocephalus, G, *see* Anostraca

Chirolophis ascanii *see* Blennioidei

Chironomidae; Non-biting Midges (fig. 302). F of the Flies (Nematocera*). This is the insect family of inland waters that contains most species. There are more than 1,500 species in Europe alone. Also in terms of quantity the Chironomidae are often the most common water fauna, so that they are of the greatest importance as fish food. As the individual species of Chironomidae are found in very specific biotopes (Biotope*) they are an important means for the ecological characterisation of waters. There is scarcely any fresh or brackish water

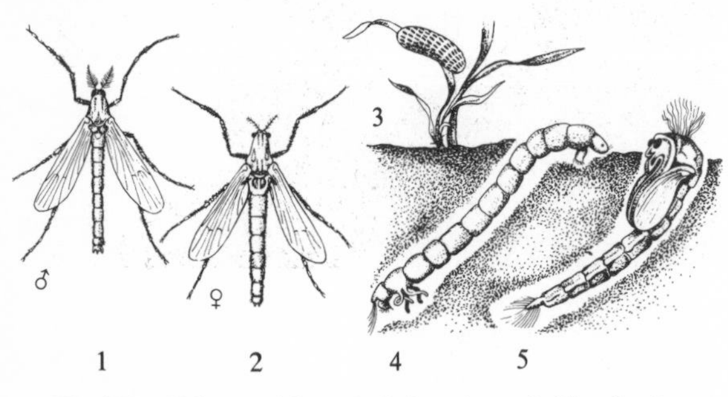

Fig. 302 Chironomidae. 1, 2 Imagines 3 Clutch of eggs 4 Larva 5 Pupa

which is not a home for Chironomidae. The adult insects are between 2 and 14 mm long, do not bite and noticeably twitch their raised front legs whilst they are sitting. The males have thickly feathered feelers, those of the females are thread-shaped. During their short adult life which only lasts a few days, the males form thick swarms into which the females then fly to mate. The larvae of the Chironomidae are relatively uniformly built and are generally known to aquarists in the form of the 'Red Mosquito Larvae'. As well as the species which are red because of the blood colour substance haemoglobin, there are also colourless, yellow, green and grey Chironomidae larvae. Almost all live on the bottom and dart around catching things or they build tubes and webs. The pupae are of the general fly pupa type; when disturbed the mobile forms flee 'tumbling' into deep waters. There are also others that live in shells. It should be mentioned that amongst the Chironomidae themselves there are parasitic larvae and pupae which live under the wing sheaths of other insect larvae. These larvae, which are used as fish food, mostly live in the mud of very dirty waters and should therefore be rinsed before use. Shallow dishes filled with a little water are recommended for keeping the larváe. These should be kept in a refrigerator. The larvae can be kept for several weeks at temperatures between 4–6°C without any appreciable change and thus hardly need any attention.

Chiton LINNAEUS, 1758 (fig. 303). G of Coat-of-nails shells (Placophora*). Found on the rocky coast of nearly all seas, and well-suited to life in the surf. It has

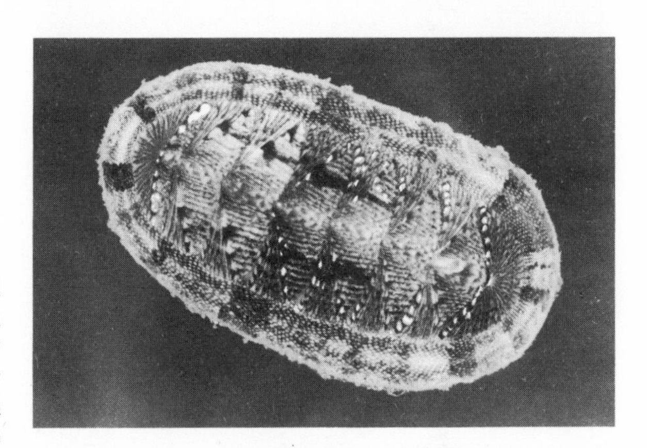

Fig. 303 *Chiton olivaceus*

a flat oval body and clings extremely tightly with its large foot. If chitons are dislodged they curl up. They live on algae and animal growth. They are occasionally kept in the aquarium.

Chitons, Cl, *see* Placophora

Chlamydoselachidae; Frill Sharks. F of Chondrichthyes*. O of Hexanchiformes, which is only represented by the species *Chlamydoselachus anguineus* (fig. 304). This elongate shark which is found all over the world has many primitive characteristics. The 6 large gill slits, the first of which is connected at the throat with its opposite, are partly overlapped by the curly gill arches which are extended externally. The mouth is terminal and equipped with powerful teeth.

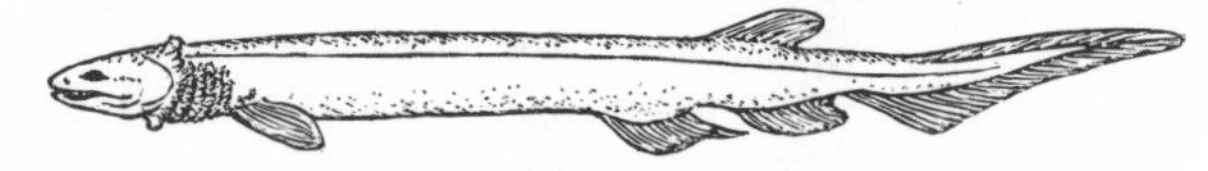

Fig. 304 *Chlamydoselachus anguineus*

The An consists simply of the very long upper lobe. The fish grow up to 1.5 m, and live in depths of 450 m and lower. Their diet consists primarily of squids. They are ovoviviparous, ie the young hatch while the eggs are stored in the mother's body. In a gestation period which lasts about 24 months, 10–15 eggs develop; the eggs themselves have a diameter of 12 cm.

Chloramine *(p-Toluolsulphon chloramide-Na)* is a salt, easily soluble in water, and contains about 25% active chlorine. It is used to disinfect wounds. It is used by the aquarist in a concentration of 0.1 g to about 5–10 litres of water, to combat Turbellaria* and skin parasites. It can be very poisonous to fish when used in soft water. Treatment should not exceed 2–4 hours.

Chloramphenicol *see* Chloromycetin

Chlorine. Chemical element. Symbol Cl. A gas which in its normal state is a pale yellowish-green colour and has a pungent smell. Because of its toxicity, it is used amongst other things to reduce the bacteria content of drinking water. Domestic water which has been considerably chlorinated can be detected by its smell and should be prepared first (Water Preparation*) before it is used in the aquarium.

Chlorine Poisoning *see* Poisoning

Chlorohydra, G, *see Hydra*

Chloromycetin, also called Chloramphenicol or Chloronitrin (D-Threo-1-p-nitrophenyl-2-dichloracetylamino-1.3-propandiol). Chloromycetin is a highly effective antibiotic* with low solubility in water which is used in medicine against Typhus and other bacterial diseases. Its effect is based on its ability to impede the growth of protein in the bacteria-cells. Stomach dropsy* and Fin-rot* amongst other bacterial infections of fish can be successfully treated with chloromycetin. Treatment can be effected by putting the fish in a separate aquarium containing the chloromycetin and should not exceed 6 hours. It can be repeated several times if necessary. 15 mg of chloromycetin are added to 1 litre of fresh water or alternatively 20 mg chloromycetin to 1 litre of sea water. For injections into the body cavity the dose should be 0.1 mg of chloromycetin per ten grams of the fish's body weight. As an additive to food 1 mg of chloromycetin is mixed with one gram of a suitable food.

Chloronitrin *see* Chloromycetin

Chlorophyll *see* Plant Colouring Matter

Chlorophyta; Green Algae. Class of Algae*. 10,000 species. Green algae with chlorophyll a and b, chloroplasts dish-shaped, ribbon-shaped, cylindrical and filamentous or lens-shaped and in large numbers. These actively mobile cells have at their front edge 2 or 4 flagella of the same length. 90 per cent of the species belong to the benthos* and plankton of fresh water. The number in marine plankton is small, a few species live in a fixed position in coastal waters of tropical and temperate seas. Some others have adapted to life on land. The higher plants such as mosses, ferns and seed plants have developed from Chlorophyta. The simplest forms are microscopically small and unicellular, and live a floating existence immobile in the water, or move by using their flagella. They are common plankton-organisms of freshwater, such as *Chlamydomonas,* which are found in rain puddles, ponds but also in aquaria. A mass concentration of them leads to a green colouration of the water—water bloom*. In some species consumption of organic nutrients furthers their development and they multiply fast in water enriched with organic material. Some immobile unicellular green algae such as *Chlorella* float in the water or are attached to stones and plants, on the sides of the aquarium and on technical equipment or they form green coverings out of the water on damp ground, on walls and on tree bark. Some live in infusorians and in Hydra. One special group, which is distinguished by its multiplicity of outer form and includes some very bizarre forms is that of the Desmids (Desmidiaceae). They are found predominantly in slightly acidic water. Cell-colonies and thread-like algae attach themselves to the unicellular Chlorophyta. The colonies are either actively mobile or passively floating in the shape of a ribbon or hollow sphere. They can be constructed of just four or of thousands of cells. In the case of the mobile hollow spheres of the G *Volvox,* which can comprise 20,000 cells and in which a certain division of labour amongst individual cell groups is recognisable, daughter spheres develop inside other spheres and are only freed when the mother sphere decays. In the case of *Hydrodictyon,* the Water-net, the cylindrical cells join end to end in the shape of a star and form a cell community in the shape of a fine-meshed, hollow net. The thread-like Chlorophyta may be branched. They grow free in the water or attached to objects such as stones or plants. When they grow in great concentration they can overgrow and damage higher forms of plants. They belong to very varied Ga such as *Ulothrix, Cladophora*, Oedogonium.* One of the best known Ga is *Spirogyra,* species of which are often plentiful in still water. More highly developed thalli occur in Chlorophyta which live in coastal regions. Here we encounter flat, ribbon-shaped and large leaf-like, double-layered bodies of vegetation, as with *Ulva**. Tuba algae are found predominantly in the coastal waters of warm seas. Their thallus is often highly differentiated and consists of rhizoids as well as stem-like and leaf-like parts. It has many nuclei but is not arranged in cells. *Caulerpa** is an example of this. Some Chlorophyta such as *Halimeda** secrete calcium and, like the corals, are involved in the formation of reefs. Other chlorophyta of importance for marine aquarists are: *Codium**, *Enteromorpha* Udotea**, *Vaucheria**.

Chlorosis. Anaemia of plants as a result of inadequate formation of chlorophyll. The cause is a lack of iron or insufficient absorption of iron which is part of the structure of green chloroplasts. Chlorosis occasionally occurs in newly established aquaria and disappears in time without treatment. In persistent cases the addition of Chelatores* helps.

Chocolate Gourami *(Sphaerichthys osphromenoides) see Sphaerichthys*

Chologaster cornutus *see* Percopsiformes

Chondrichthyes; Cartilaginous fish. Primitive Cl of the Sub-Ph of animals with jaw-bones (Gnathostomata*). Two Sub-Cl belong to the Chondrichthyes: the Elasmobranchii (plate-jaws) and the Holocephali (Chimaeras). Elasmobranchii such as sharks and rays are generally well-known fish, but it is mainly ichthyologists who know of the few species of Holocephali. The skeleton of Chondrichthyes living today is cartilaginous and in places is strengthened by layers of calcium. The vertebrae, for example, can have centres of calcium. The ancestors of the Chondrichthyes, the highly developed Placoderms of about 300 million years ago, had a bony skeleton that was reduced during the evolution of the cartilaginous fishes.

The most essential elements of the cartilaginous skeleton are formed around the Chorda dorsalis (Skeleton*). They are the vertebral column, which consists of vertebral elements (this is lacking in the Holocephali), the capsule-like cranium, the facial skull (fig. 305) which is made up of cartilaginous joints like the upper jaw (Palatoquadratum), the lower jaw (Mandible), the jaw process (Hyomandible), the tongue cartilage (Hyoid) and gill-arch cartilages; other elements are the shoulder and pelvic girdle and the skeleton of the paired fins. In

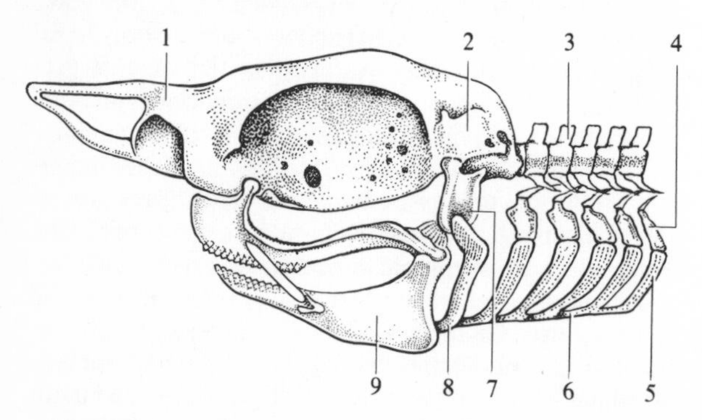

Fig. 305 Skeleton of head of shark. 1 Capsule for olfactory organ 2 Capsule for the balancing organ 3 Vertebrae 4 Epibranchials 5 Hypobranchials 6 Cardiobranchials 7 Hyomandibular (jaw process 8 Hyoid 9 Mandibular (lower jaw)

the case of the Holocephali there ist a firm connection between the upper jaw and the cranium. This is similar in the case of the primitive Elasmobranchii but the upper jaw is also connected to the jaw-process cartilage, in the higher Elasmobranchii however, the upper jaw has isolated itself from the cranium and is only connected to it indirectly by ligaments and jaw-process cartilage. In the Elasmobranchii the gill-slits open outwards individually and are uncovered; in the Holocephali, as in the Osteichthyes*, a gill-cover slides over the gillslits.

The fin-rays of the paired and unpaired fins are horny, the C is always clearly heterocercal (*see* Fins).

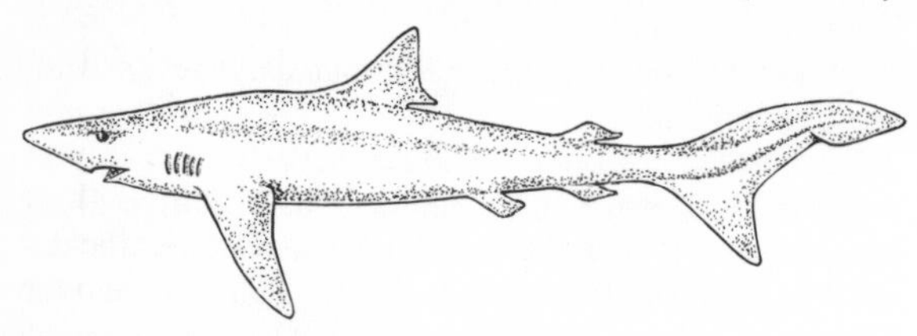

Fig. 306 Basic shape of sharks

Both Sub-Cls have inner fertilisation of the eggs. The transferrence of the spermatozoa is carried out by the Vs which in the males have become copulatory organs (Myxopterygia). The swimbladder is always absent in the Chondrichthyes. In the gut there is a characteristic spiral fold which acts as an enlargement of the surfaces. The following features are characteristic of both Sub-Cls.

– *Elasmobranchii.* The Sharks and Rays included in this Sub-Cl form a relatively uniform systematic group despite their very different body shape (figs. 306 and 307). The main distinguishing characteristic is the position of the gill-slits. In the Shark they are always on the body sides, slightly in front of or partly over the start of the Ps; in the Ray, on the other hand, they are on the underside of the flatterned body. This positioning of the gill-slits on the underside is causally linked with a second adaptation.

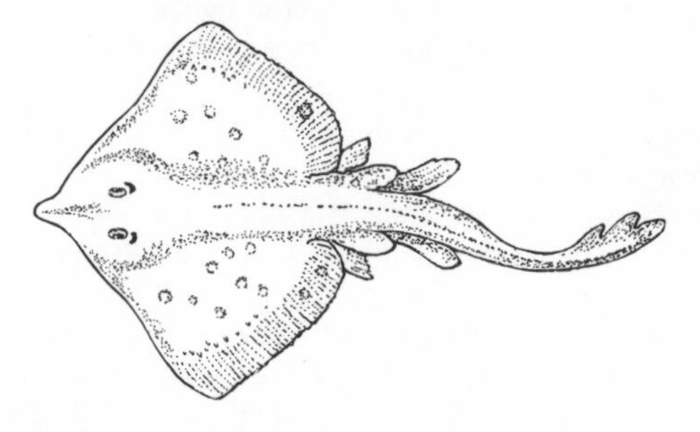

Fig. 307 Basic shape of rays

When the Rays are lying on the bottom or have dug themselves in, they suck in breathing water through their spray-holes and let it out again the same way. When swimming, however, they normally breathe through the mouth. A further difference is to be found in the Ps which are free-moving in the shark and function as steering and stabilising organs. In the Ray, the Ps are completely fused with the body and form the side parts of the body disc. In this form they are also the main propulsion element, which, particularly in the pelagic forms, beat up and down like wings and in bottom forms effect wave-like movements. Sharks often have mobile lid-like eyefolds (nictitating membrane) whereas such differentiations are absent in the Ray. The same is true of the An, which no Ray has. Their common characteristics are far more numerous. We need only mention the skin and dentition. With a few exceptions in the case of the Rays, the skin of both groups is very firm and above all rough. It feels like rough sandpaper and owes this characteristic to small dermal denticles (placoid scales) which are anchored in great numbers in the skin, often in slanting rows. The placoid scales are a completely different skin derivative from the familiar scales of the bony fish but are however phylogenetically precursors of the teeth of vertebrates. Every placoid scale is composed of a plate-like bony socket hidden in the skin and a cone-shaped tooth projecting quite a way out of the skin (fig. 308). The

Fig. 308 Placoid scales

tooth consists of dentine and a covering of enamel (Scales*). On the jaw edges the placoid scales are formed as teeth and flattened like a blade, sharf-edged with main and side points. They form rows of teeth. There are many rows one behind the other. The place where the teeth are formed is a sunken, groove-shaped strip of teeth from where row after row of teeth are formed and continually pushed forward depending on the rate at which the front rows are worn out. The rows of teeth move up just like the unrolling of a rush mat. In this way the teeth always remain new and sharp (fig. 309). In certain cases in species which eat creatures with hard shells the teeth have become bony-plates. A few plankton-

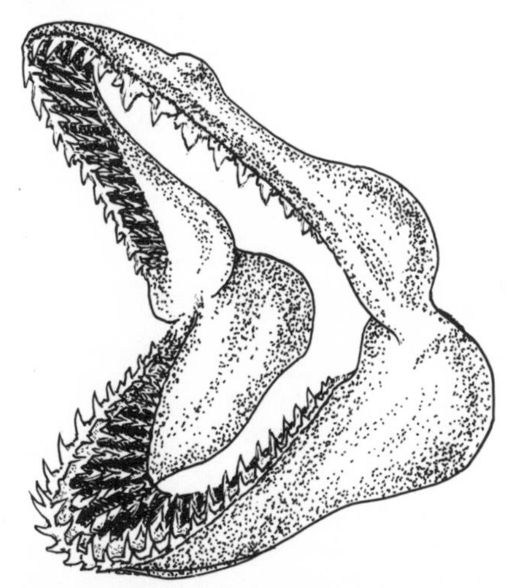

Fig. 309 Jaws of a shark

eaters amongst the Sharks and Rays dispense completely with their teeth and instead they filter small creatures from the water using their pharynx which acts like a weir-basket. Fertilisation is always internal. The males transfer the spermatozoa with the adapted Vs (Myxopterygia) (fig. 310). The eggs are either deposited in capsules provided with projections or develop in these capsules in the mother's body and only burst from these shortly before the birth (fig. 311). In some cases the embryos

Fig. 310 Vs of the ray. Female and male (on the right). The Vs of the male are also used in mating (myxopterygian).

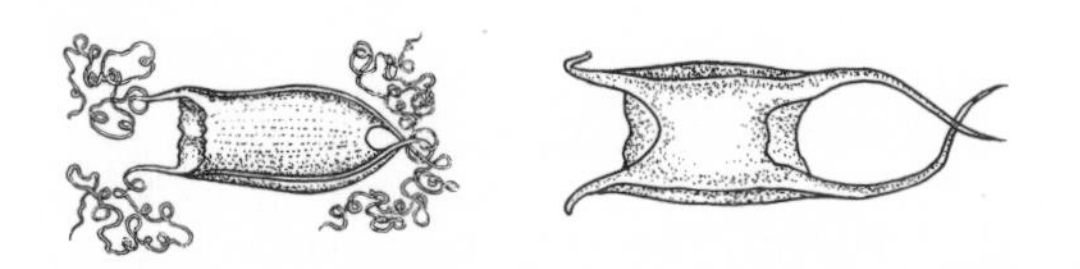

Fig. 311 Shape of the horny egg capsules of sharks (left) and rays (right)

are fed by the mother by means of a primitive placenta. The gestation period of these embryos is in all cases long or very long – eg 2 years in the case of the Basking Sharks. Sharks and Rays are creatures of the sea. They live in the deep sea as well as shallow water areas near the coast.

Some species penetrate far into the freshwater of large rivers, but only a very few species have adapted themselves completely to freshwater. Sharks are of greater importance economically than other Chondrichthyes. The meat of various shark species is tasty and, when prepared in a special way, is in many cases offered as a speciality. In China, sharks' fins are used to make soup. The shark liver which is rich in vitamins was of great value for a time before vitamins could be synthesised. The skin of some sharks is made into leather when the placoid scales have been removed. Some shark leathers are special patented commodities. The economic importance of Rays is, however, slight. In the last few decades, Rays have become increasingly popular as objects for sport fishing. They are reputed to be tenacious and tough fighters. Finally a few words about the danger of sharks. We must first remember that man does not form part of the normal diet of the shark but there are species which indiscriminately attack and devour everything they can. All those species which are called Man-eating Sharks belong to this category. Apart from these other less aggressive species can also be incited to attack under certain circumstances, eg when bathers or divers annoy them unintentionally or when there is blood or fish meat in the water. The individuals of one species can also behave very differently from each other. Sometimes there are a few from a species which normally flees from man, which might attack men. Basically they seem more ready to attack in higher temperatures. Thus there are considerably fewer accidents with sharks in the temperate seas than in the tropical areas of the oceans. Furthermore the time of day or degree of brightness seems to influence their readiness to attack. Finally, it should be mentioned that considerably more men are attacked than women. Despite the many methods described up to now, there is no sure way of defending oneself against a shark. After considering many possibilities S. Harald advised that you should act like a trainer, ie show no fear and swim bravely towards the shark (provided that you see it in time!). Hans Hass held sharks away from him by shouting loudly and using a shark-stick. Amongst the chemicals used for defence against sharks, copper acetate has been particularly useful.

– *Holocephali;* Chimaeras (Rabbitfishes). This Sub-Cl of the Chondrichthyes is mostly of scientific importance. The Holocephali which are known from the late Devonian have common ancestors with the Elasmobranchii but developed a few characteristics which are recognised as specialisation and are similar to features of the bony fish. Amongst these are the gill cover which covers the outlets of the 4 gill slits, the upper jaw (Palatoquadratum) which is firmly connected to the cranium and the separated outlet from the gut and ureter. It was previously thought that because these fish had characteristics of both cartilaginous and bony fish they must be a cross breed and thus they came to be called the Chimaeras. In all, there are 28 living species known. The body is deepest just behind the large head and tapers towards the back into a long (in some species very long), thin tail. The large head is characterised by the mouth which is set underneath and equipped with chisel-shaped dental plates, by the club-like brow projection which is only found in the males and can also be tucked into a groove in the skin, by the easily visible mucus channels and by the snout. 3 Fs are distinguished by the shape of their snout. The Ratfishes (Chimaeridae) (fig. 312) have a snub rounded snout, the Long-nosed Chimaeras (Rhinochimaeridae) have a

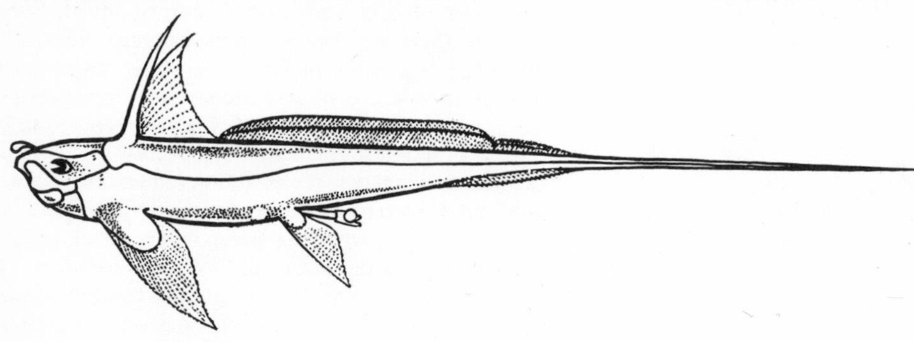

Fig. 312 Chimaeridae-type

long, pointed, stiletto-like snout and the Callorhynchidae (fig. 313) have a ragged, elongate snout turned down towards the mouth. The best known species, the Ratfish *(Chimaera monstrosa)*, grows to 1.5 cm long and is found in the deep water levels from Norway to the Mediterranean and around S Africa. It is not found in

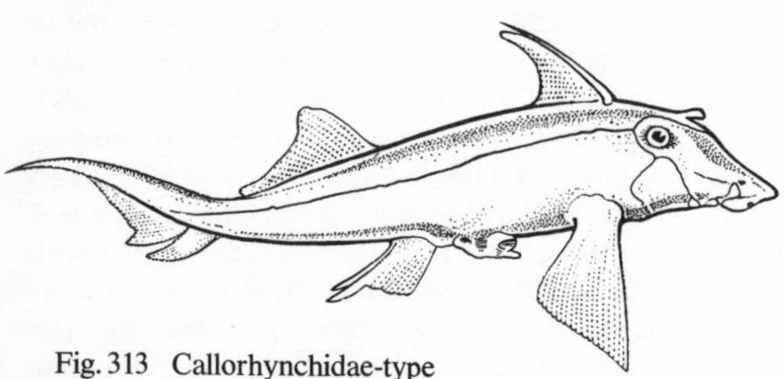

Fig. 313 Callorhynchidae-type

the equatorial areas of the E Atlantic. On the front edge of the D1, which lies directly behind the head, there is a poisonous ray. The economic value of the Chimaeras is very slight. The classification of the Chondrichthyes is still not absolutely clear, although great progress has been made in recent years regarding the living species. One of the best authorities on cartilaginous fishes, L. J. V. Compagno, suggested a new classification in the Zoological Journal of the Linnean Soc. Vol. 53 Suppl. 1 (1973) and this has also been taken as a basis in this work. It is briefly repeated here. The numbers by the Os give the number of Fs in the respective Os. The representatives of the Ga mentioned below are mentioned under the corresponding family entry.

Cl Chondrichthyes (Cartilaginous Fishes)
- Sub-Cl Elasmobranchii (plate-jaws)
  - Super-O Squalomorphi
    - O Hexanchiformes 3
      - F Chlamydoselachidae (Frill Sharks) Chlamydoselachus
      - F Hexanchidae (Cow Sharks) Hexanchus
    - O Squaliformes 2
      - F Squalidae (Dogfish Sharks) Squalus, Etmopterus, Somniosus, Isistius
    - O Pristiophoriformes 1
      - F Pristiophoridae (Saw Sharks) Pristiophorus
  - Super-O Batoidea
    - O Rajiformes (Ray-types) 8
      - F Rhinobatidae (Guitar-Fish) Rhinobatos.
      - F Rajidae (True Rays) Raja
    - O Pristiformes 1
      - F Pristidae (Sawfishes) Pristis
    - O Torpediniformes (Electric Ray types) 4
      - F Torpedinidae (Electric Rays) Torpedo
    - O Myliobatiformes (Eagle-Ray types) 6
      - F Dasyatidae (Sting Rays) Dasyatis
      - F Myliobatidae (Eagle Rays) Myliobatis, Aetobatis
      - F Mobulidae (Devil Fish) Manta, Mobula
  - Super-O Squatinomorphii
    - O Squatiniformes (Angel-Shark types) 1
      - F Squatinidae (Monkfish Angel Shark) Squatina
  - Super-O Galeomorphii
    - O Heterodontiformes 1
      - F Heterodontidae (Bullhead, Horn or Port Jackson Sharks) Heterodontus
    - O Orectolobiformes 7
      - F Orectolobidae (Carpet Sharks) Orectolobus
      - F Ginglymostomatidae (Nurse Sharks) Ginglymostoma
      - F Rhiniodontidae (whale sharks) Rhiniodon
    - O Lamniformes 7
      - F Odontaspididae (Sand Tigers) Odontaspis
      - F Scapanorhynchidae (Goblin Sharks) Scapanorhynchus
      - F Alopiidae (Thresher Shark) Alopias
      - F Cetorhinidae (Basking Sharks) Cetorhinus
      - F Lamnidae (Mackerel Sharks) Carcharodon, Lamna, Isurus
    - O Carcharhiniformes (Blue-Shark types) 8
      - F Scyliorhinidae (Dogfish) Scyliorhinus, Cephaloscyllium, Haploplepharus
      - F Triakidae (Smooth Dogfishes) Mustelus, Triakis, Galeorhinus, Triaenodon
      - F Carcharhinidae (Blue Sharks) Carcharhinus, Galeocerdo, Prionace

F Sphyrnidae (Hammerhead Sharks) Sphyrna
Sub-Cl Holocephali (Chimaeras)
O Chimaeriformes (Chimaera types) 3
F Chimaeridae (Chimaeras) Chimaera

Chondrostean Fishes (Chondrostei, Coh) *see* Osteichthyes

Chondrostei, Coh, *see* Osteichthyes

Chorda dorsalis *see* Skeleton

Chordata. Ph to which the highest developed animals belong and from which man developed. The Chordata are characterised by a flexible supportive rod, the Chorda dorsalis (notochord) which forms in the early embryonic stage from the mesoderm and remains unchanged for life or is superseded by a vertebral column. The flexible notochord passes through the body longitudinally as an axis and consists of special chord cells which are enclosed in a firm cover. The central nervous system which develops as a tube-like structure of nerves is situated over the notochord, below it is the body cavity. The fore-gut of the Chordata is extended into a pharynx; all Chordata have a tail region. In the Gnathostomata paired limbs only appear in the form of paired fins and the front and back limbs which evolve from them. The higher Chordata have conquered the land; a development which became possible by the acquisition ot respiration using lungs. Various air-breathers developed the ability to fly (birds and bats), others returned to the water (eg turtles and whales). The Chordata are generally grouped in 4 Sub-Pha: Tunicata* (animals with mantles), Acrania* (animals without skulls), Agnatha* (animals without jaws), Gnathostomata* (animals with jaws). The classes Aphetohyoidea*, Chondrichthyes*, Osteichthyes*, Amphibia*, Reptilia*, Aves* and Mammalia* belong to the Sub-Ph Gnathostomata. The characteristics of Chordata are greatly reduced or substantially altered in the Tunicata. The archetype of the higher Chordata is not represented by the Lancet fish (Acrania*); rather, among modern Chordata, the larvae of Cyclostomata* come substantially closer to this hypothetical archetype.

Chordate animals, Ph, *see* Chordata

Chorisochismus dentex *see* Gobiesociformes

Chorististium Gill, 1862. G of Serranidae*. Found in tropical W Atlantic. Small predatory fish which are active at night. They live in deep waters such as on the edge of reefs. These large-eyed fish with pointed snouts grow to 10 cm and have a D in two parts. D1 has eight spines. They are colourful fishes and are sometimes imported, eg:

– *C. carmabi* Randall, 1963; Candy-striped Basslet, Carmabi Fish. This species varies between salmon-pink and orange and has 4 lavender-blue, longitudinal stripes edged with orange. It has been kept in an aquarium for up to 2 years (fig. 314). This species is also described in the literature of marine fish as *Liopropoma carmabi.*

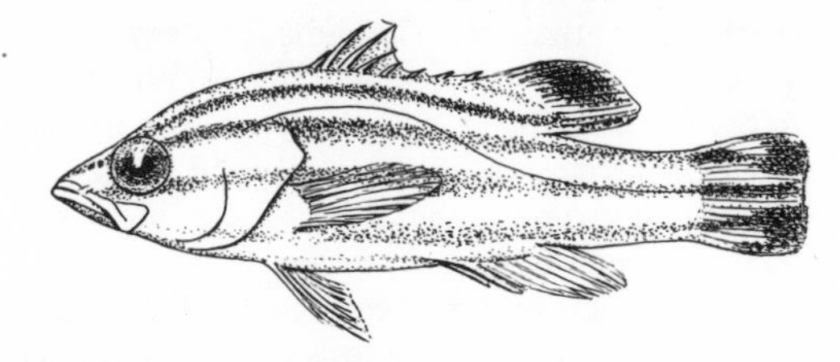

Fig. 314 *Chorististium carmabi*

Chromatophores. Pigment cells. In plants these are *plastides* in the cell plasma of various tissues; in animals, however, they are specialised pigment cells mainly in the skin. Individual colouring is dependent on the specific arrangement and combination of the chromatophores. The chromatophores of fish lie in the true skin (dermis). They are mostly flat expanded cells with branched projections. Several types are distinguished according to their chemical nature or the colour of their pigment. Melanophores contain brown, grey or black melanin (fig. 315). In the lipophores, predominantly yellow to red colours are found (flavine, carotin) when the cells are known as xanthophores or erythrophores. The guanophores or iridozytes have no pigments as a rule but have light crystals which consist of guanine or similar end-products of proteometabolism. The guanophores have varying affects according to the structure and distribution of their crystals. When they reflect almost all the direct light they produce a silver sheen. They can also split light into the colours of the rainbow and accentuate particular colour tones. The shining blue colours arise predominantly through the dispersion effect of finely distributed crystals through which only the blue part of the light is visible – the so-called Tyndall-blue. The matt white elements in the fins of some fish are due to the guanophores. Metallic colours are produced when the guanophores are positioned over yellow or red pigment cells, for example. If there is a layer of melanophores beneath them their effect is either changed or considerably increased. Blue pigments are relatively rare. In some species, blue-green derivatives of blood pigments (bilichromes) play a certain role.

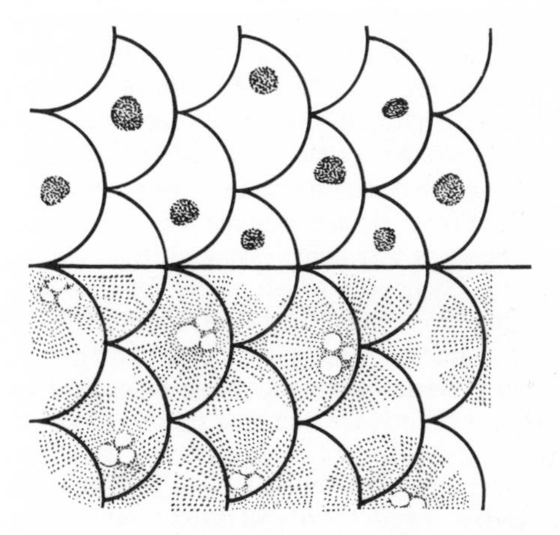

Fig. 315 Melanophores in fish-skin. The melanin pigment is concentrated in balls in the top section, and spread out in the bottom part.

The ability of many fish to change colour is linked to a displacement of pigment in the chromatophores which is governed by nerves or hormones. Thus the skin becomes increasingly dark when the melanin expands in the melanophores, when it is concentrated in the centre of the cell, however, the skin grows lighter. In contrast the colour-change of Cephalopods results from the distortion of the whole pigment cell by the finest muscle fibres.

Chromis CUVIER, 1815. G of Pomacentridae*. Has many species in tropical seas. Some species also penetrate temperate zones. The fish live mainly in large shoals, or in smaller groups in shallow waters of coral reefs and rocks, and feed on various small animals. When danger threatens they take refuge in crevices in between the coral branches. They grow to a maximum length of 10–17 cm. Characteristics of the G are the deeply forked, pointedly elongated C, the large scales and the protrusile mouth; the bones of the gill-covers are densely serrated. One must certainly look very closely, however, in order to notice the fine serration of the front gill-cover of the G *Dascyllus**. The colours of most *Chromis* species are very beautiful, those of many young fish being particularly bright. As far as is known the fish are substrate breeders. The males of *Chromis chromis* occupy small territories in dense colonies at breeding time. They clean the stone which has been chosen as the spawning place and with their display entice passing females which are ready for spawning. The fish spawn in the manner of the Cichlidae*. However, they do not form pairs, and the male's care of the eggs ends with the hatching of the young which then live like plankton. *Chromis* are amongst the most popular aquarium fish and are regularly imported. One should always keep several fish of one species together. Feeding with live food of all kinds and dried food presents no difficulties. Breeding has not yet been successful, particularly because of the sensitive plankton-larvae. Some important species are:

– *C. coeruleus* (CUVIER and VALENCIENNES, 1830); Blue or Blue-green Puller, Blue-green Chromis, Green Damsel. Indo-Pacific, Red Sea and on coral reefs. To 12 cm with a green or blue sheen according to the way the light falls. Very delicate but some people keep them well. They like to hide amongst bushy coral.

– *C. chromis* (LINNAEUS, 1758); Blue Damsel, Mediterranean Brown Forktail Damselfish. Mediterranean, E Atlantic on rocky coasts. To 12 cm. Brown or blue-black. Young have shining azure-blue stripes.

– *C. dimidiatus* (KLUNZINGER, 1871); Bicolour Chromis, Half and Half Puller. To 11 cm. Front part of body dark brown, rear part of body yellowish white.

– *C. xanthurus* (BLEEKER, 1854); Blue Demoiselle, Yellowtail. Indo-Pacific on coral reefs. To 16 cm. Blue tail-stalk and C sharply set apart with a yellow sheen. This species is often confused with *Abudefduf* parasema*, but can be distinguished by deeply forked tail fin.

Chromosomes *see* Cells

Chrosomus RAFINESQUE, 1820. G of Cyprinidae*, Sub-F Leuciscinae. S Canada, N and Central USA to the Rocky Mountains. The small shoaling fish live in clear running waters and lakes. The body is elongate, only slightly compressed laterally and covered with very small scales. The mouth is terminal and has no barbels. Lateral line incomplete. These tranquil fish are best kept in a shoal and need large tanks with plenty of swimming space. They eat live and dry food of all types. The temperature of the water should not be higher than 18–22 °C in summer; although the fish can tolerate high temperatures they need high oxygen content. Breeding is difficult and is only successful in large aquaria after a cool winter season at 6–8 °C.

– *C. erythrogaster* RAFINESQUE, 1820; Redbelly Dace. N USA, S Canada. To 8 cm. The back is dark olive brown. Two dark bands run from the mouth to the roof of the tail which border a yellow-gold stripe on the middle of the body. Unpaired fins are yellowish to brown with red at the base. In the spawning season the middle stripe becomes goldish red to red and the bands which border it become deep black and the throat and the belly carmine red. The males are more deeply coloured with tubercles on the gill-cover at spawning time (fig. 316).

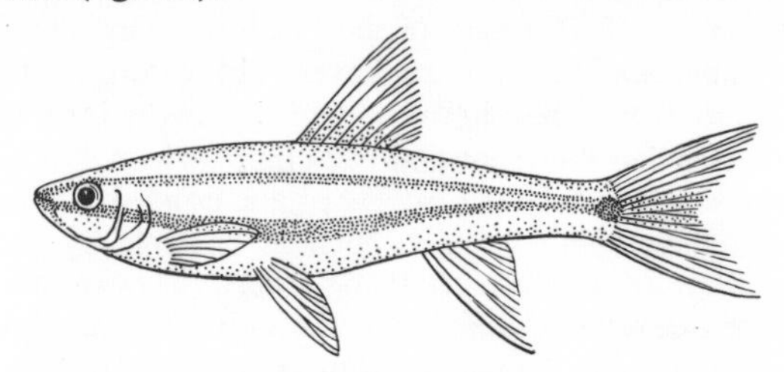

Fig. 316 *Chrosomus erythrogaster*

Chrysophyta; Gold Algae. D of Algae*. Algae of goldish brown to brown colouration with chlorophyll a and c., mostly with 2 chloroplasts per cell. The flagellate cells have 2 flagella of varying lengths at the front end. The Chrysophyta form a substantial proportion of marine plankton but are also found in fresh water. The simplest forms are either mobile or immobile and unicellular. There are, however, also mobile spherical and static bushy cell-colonies and more rarely simple threads. The single cells may have supplementary supportive elements such as external armour of silicic acid platelets, silicic acid shells or conical cellulose shells or internal calcium skeletons. Their corresponding fossil forms played a significant part in the formation of calcium sediments, eg drawing chalk.

Chthamalus stellatus *see Balanus*

Chub *(Leuciscus cephalus) see Leuciscus*

Chum Salmon *(Oncorhynchus keta) see* Salmonoidei

Cichla BLOCH and SCHNEIDER, 1801. G of Cichlidae*. Tropical and subtropical S America. Medium-sized to large. *C. ocellaris* grows up to 60 cm. Their slender form, the small scales and the deep indentation between the hard and soft rayed part of the D recall the true Perch (Percidae*) rather than the Cichlids. The soft rayed parts of the D and An are rounded, the An positioned closer to the tail than the D. Fan-shaped C. The fish are out-and-out predators, and at best only the young fish are suitable for domestic aquaria. Up to now the following species has been imported and kept:

– *C. ocellaris* BLOCH and SCHNEIDER, 1801; Eye-spot Cichlid. Tropical S America. Variable colouring, has indistinct transverse stripes when young, and is uniformly silver at maturity. Base of C has black spot with a lighter edge. Needs water with a high oxygen content. Has not yet been bred.

Cichlasoma SWAINSON, 1839. G of Cichlidae* with many species. Mainly found in tropical S America, various species are however also found in middle and southern N America as well as on the Greater Antilles. The members of *Cichlasoma* are medium-sized (10–30 cm) and have the typical shape of the Cichlids, ie longitudinally oval and laterally compressed. A few species, *C. festivum* and *C. severum* for example, develop a high back with age. In general the appearance of the *Cichlasoma* is particularly similar to the G *Aequidens**. However, in the case of the *Cichlasoma*, the D and An are less extended and pennant-like. Furthermore the An has 5 or more hard rays (*Aequidens* has 3 at the most). The markings are very variable; many species have dark transverse bands and particularly prominent spots in the middle of the sides and on the caudal peduncle. Being typical open-water brooders (Cichlidae*), the sexes are difficult to distinguish from external characteristics and sex can only be determined from the shape of the genital papilla which appears a few days before spawning and is pointed in the case of the male and has a blunt end in the case of the female. Most species are territory-forming and in maturity will only tolerate their selected partner in their vicinity. In the aquarium this means that the pairs should be kept alone as a rule and should have come together independently from a shoal of young fish. Alternatively they should be allowed to become acquainted and accustomed to one another through a glass partition. In spite of this, where space is restricted, serious quarrels often develop which tend to end in the death of the weaker partner if roots and stones are not available to offer opportunities for concealment. It is best to cover the bottom with a thick layer of clean sand in which the fish can form large hollows when they are ready to reproduce. Because the fish work the bottom in this way, only hardy potted plants can hold their own and even these cannot survive in the tanks of the particularly pugnacious species (eg *Cichlasoma biocellatum*) which will not tolerate any form of plants. The high-backed species (*C. festivum* and *C. severum*) do not however behave in the manner described above, except perhaps during the reproductive period. These Cichlids are much more passive than other members of their G and are particularly at home in tanks of a suitable size which are well-stocked with plants. They swim around there peacefully and cautiously. They are similar to the well-known Angelfishes (*Pterophyllum**) and will even tolerate much smaller fish near them.

Because of the points mentioned above only large roomy aquaria are suitable for the care and breeding of *Cichlasoma*. The condition of the water is not usually especially important but one should consider that most members of the G live in very soft water in their natural state. The temperature of the water is adjusted to suit the origin of the fishes. N and Central American varieties do well in 20 °C with still lower temperatures in winter. On the other hand, tropical species need temperatures of 23–25 °C. The temperature should be raised by 3 °C for the breeding of all species. Further information on care and breeding is to be found under the description of the family. It should be observed, furthermore, that many *Cichlasoma* learn to recognise their keeper personally and become very friendly. In many cases the fish give a decidedly intelligent impression. Frequently kept species are the following.

– *C. bimaculatum* (LINNAEUS, 1758); Two-spot Cichlid. Northern S America. To 20 cm. Body grey-brown with darker indistinct transverse bands, underside light silvery-white, with a large black spot on the middle of the sides and the caudal peduncle.

– *C. biocellatum* REGAN, 1909; Jack Dempsey. Central Amazon region. To 18 cm. Body dark blue, with 7 to 8 indistinct dark transverse bands especially in the young fish. Black spot with a light border on the middle of the body and another on the caudal peduncle. Scales with blue-green spots (fig. 317).

Fig. 317 *Cichlasoma biocellatum*

– *C. cutteri* FOWLER, 1932; Cutter's Cichlid. Honduras. To 12 cm. 7 to 8 black transverse bands on a blue to olive-green background, with a dark bridle-band (over the eye) from the mouth to the nape. D, C and An wine-red. Passive species.

– *C. cyanoguttatum* (BAIRD and GIRARD, 1854); Texas Cichlid. Up to 30 cm. Body blue-grey, with light blue or green spots. D, C and An similarly marked. Very beautiful but unsociable species. High oxygen requirement. Sensitive to old water (fig. 318).

Fig. 318 *Cichlasoma cyanoguttatum*

– *C. facetum* (Jenyns, 1842); Chanchito. Subtropical S America. To 30 cm. Base colouration very variable. Body sides with several deep black transverse bands which spread into the D and An.
– *C. festivum* (Heckel, 1840); Flag or Festive Cichlid. Tropical S America. To 15 cm. V, An and D extended further than in many other *Cichlasoma*. Yellowish in colour. Back very dark with a broad black band from the mouth to the soft-rayed part of the D. Black golden-edged spot to the caudal peduncle. Tranquil fish outside the breeding period, usually somewhat shy (fig. 319).

Fig. 319 *Cichlasoma festivum*

– *C. hellabrunni* Ladiges, 1942. Presumed to come from Upper Amazon basin. To 30 cm. Forehead strikingly arched and humped in older specimens. Body and fins wine-red, forehead with yellowish transverse bands. They have a blotch in the middle of the side and on the caudal peduncle. It is a peaceful species.
– *C. meeki* (Brind, 1918); Fire-mouth Cichlid. Found in SE, S and Central America and also in subterranean waters. To 15 cm. Bluish-grey with 5–7 indistinct dark transverse bands. The throat is brick red, belly yellow to orange. The scales have a red edge. The gill cover and caudal peduncle have a shiny black blotch which is edged with gold. Mostly peaceful (fig. 320).

Fig. 320 *Cichlasoma meeki*

– *C. nigrofasciatum* (Günther, 1869); Zebra or Convict Cichlid. Found in some lakes in Guatemala. To 10 cm. Grey with 8 or 9 dark transverse bands. Dark blotch on gill-cover and caudal peduncle, the fins are a metallic grey. They are very unsociable and need supplementary plant food (fig. 321).

Fig. 321 *Cichlasoma nigrofasciatum*

– *C. severum* (Heckel, 1840); Banded Cichlid. Found in tropical S America and have also moved into Nevada. To 20 cm. Variable colouring but mostly yellowish brown. The scales have red-brown spots. There is a large dark spot on the soft rayed sections of both the D and An and running between these is a dark band. In young specimens more dark transverse bands are to be found. The male has vivid red-brown patterning on the head and strongly produced D and An. Very peaceful except at spawning time (fig. 322).

Fig. 322 *Cichlasoma severum*

Cichlidae; Cichlids (fig. 323). F of Perciformes* with many species. Sub-O Percoidei*. The Cichlidae are found in S and Central America, Africa and SW and S Asia. The greatest number of species are found in Africa, particularly in the large E African lakes. There are about 650 known species. The body is elongate to arched, often sturdy and compressed laterally. The head is large, and the forehead section often bulges out forwards in older males. Unlike many other Percoidei the Cichlidae only have one nostril on each side which is both an exit and entrance at the same time. The mouth often has a wide gape, has swollen lips and is protrusile.

Maxilla teeth 1–3 pointed, formed as incisors or molars. D uniform with powerful spiny-rays in front section often finishing in a point at the back or sometimes extended like a sail. C rounded off or straight, outer rays sometimes extended. An shorter than D but similar in form to the soft-rayed section of D. Vs sometimes elongated moderately or a great deal. Ctenoid or cycloid scales, lateral line often divided and occasionally greatly reduced. Physoclists*. The Cichlidae have inhabited nearly all types of water in the tropics and subtropics and occupied countless ecological niches.

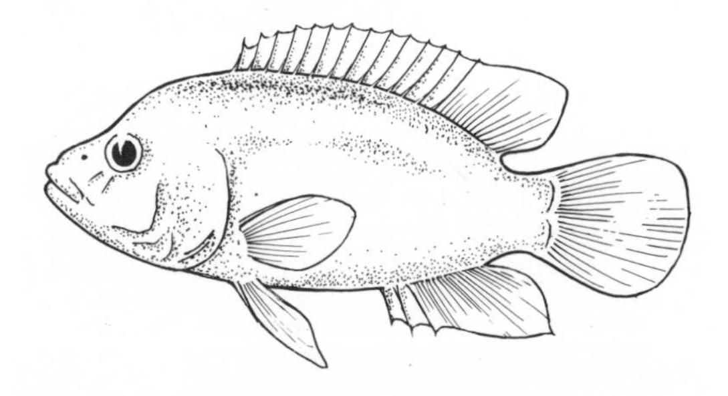

Fig. 323 Cichlidae-type

The Asiatic species also live in brackish coastal waters. There are omnivores and plant or algae eaters as well as the typical meat eaters. All Cichlidae have distinct behaviour patterns when reproducing and they care for the eggs and brood in some way or another. Because of this they have often been the object of physiological behaviour studies. There are open-water brooders and secret brooders. The former only show slight differences between the sexes and generally deposit their eggs on soft substrates; in the latter the males are larger and more brightly coloured and the eggs are laid in hollows. A special form of the concealed brooders is the mouth brooder – they hatch the eggs in their mouth. The females usually perform this task, rarely the males, although they occasionally share the responsibility. The eggs of the mouth-brooders are fewer, more brightly coloured and larger. They are not adhesive. Many species of Cichlidae are brightly coloured and some are very important regional food fish (eg *Tilapia* species). Most are ideal aquarium fish. For biological details about all breeding and aquarist information see under Ga *Acaronia, Aequidens, Apistogramma, Astronotus, Chaetobranchiopsis, Cichla, Cichlasoma, Crenicara, Crenicichla, Eretmodus, Etroplus, Geophagus, Haplochromis, Hemichromis, Hemihaplochromis, Julidochromis, Labeotropheus, Lamprologus, Leptotilapia, Lobochilotes, Nannacara, Nannochromis, Pelmatochromis, Pelvicachromis, Petrotilapia, Pseudotropheus, Pterophyllum, Sarotherodon, Simochromis, Spathodus, Steatocranus, Symphysodon, Teleogramma, Telmatochromis, Tilapia, Tropheus, Uaru.*

Cichlids, F, *see* Cichlidae

Cicuta virosa *see* Apiaceae

Cidaris; LESKE, 1778; Lance Sea Urchin. G of Sea Urchins (Echinoidea*).
Found in warmer seas where it lives underneath the surf zone and in greater depths. They feed on slow-moving and static creatures. The name Lance Sea Urchin refers to 60–90 long powerful spines which are situated far apart and on which they move as if on stilts. The large spines are surrounded at the base by small flat neighbouring spines. Often they are overgrown with various organisms. The *Cidaris* species can certainly be kept in an aquarium with good water quality. Feed them with different kinds of animal matter, as well as algae.

Ciguatera Poisoning *see* Poisonous Fish

Cilia. Actively mobile, thread-like processes of a cell. In the single-celled Ciliates (Ciliata*), the upper surface is completely or partially covered in cilia which make synchronised, beating movements to push the cell body along, or which produce eddy currents to pull food particles into the cell mouth. In multi-celled animals ciliated cells are often joined together into broad-surfaced ciliated epithelia which fulfil various functions of movement and transport (eg in the respiratory passages of air-breathing vertebrates). Flagella* are similarly built, but much longer cell organelles.

Ciliata; Ciliates; Infusoria (fig. 324). Most highly developed of the primeval animals (Protozoa*) rich in species and forms. A particular feature of the Ciliata is the border of cilia which stretches either over the whole cell surface or in isolated spots. The cilia beat in a co-ordinated manner and thus provide a means of locomotion and gathering food by making eddies. The Ciliates are distinguished further from other Protozoa by

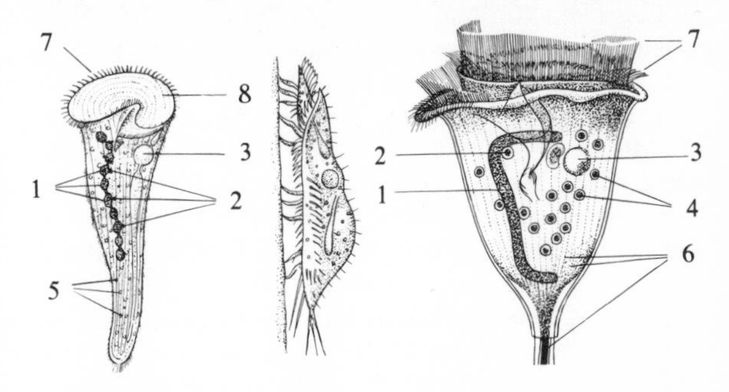

Fig. 324 Ciliata. Left, G *Stentor*; Centre, G *Stylonychia*; Right, G *Vorticella* (without stalk). 1 Large nucleus 2 Small nucleus (nuclei) 3 Pulsating vacuole 4 Feeding vacuoles 5 Zoochlorellae 6 Contractile fibrils 7 Spiral of Cilia 8 Oral region

a thin elastic outer layer of plasma (pellicula) and at least two nuclei per individual. The large nucleus (macro-nucleus) controls the whole metabolic process, while the small nucleus (micro-nucleus) plays a decisive role in the extremely complicated processes in the exchange of hereditary material. Such well-known forms as *Paramaecium** (Slipper Animalcule), *Vorticella* (Bell animals) and *Stentor* are Ciliates. Species of the G *Stylonychia* happily establish themselves in the aquarium. Their 'ventral' cilia have become stuck together to form some rod-like organelles with which the animals appear to walk on water, plants and other substrates. The Ciliates live in fresh water as well as in the sea. Some are important as foodstuffs to be used in the aquarium. Infusoria are thus used as food, although not of expecially high nutritional value, on which particularly small fish embryos are reared. A few Ciliates are fish parasites *(Ichthyophthirius**, *Chilodonella**, *Trichodina**, *Cryptocaryon**).

Ciliates, Ph, *see* Ciliata

Ciona FLEMING, 1822. G of solitary Sea-squirts (Ascidiacea*) with the cosmopolitan *Ciona intestinalis* (LINNAEUS, 1758), which is also frequently found in the western Baltic Sea at depths of 5 m and below. This glassy translucent yellowish-red slender Sea-squirt of up to 17 cm in length occasionally develops in the aquarium from imported larvae. It is difficult for large animals to accustom themselves to the aquarium.

Circular Display Swimming *see* Display Behaviour

Cirrhitichthys BLEEKER, 1856. Hawkfishes. G of Cirrhitidae*. Tropical Indo-Pacific. These fish are predatory fish which grow up to 10 cm in length, lurk on coral reefs and rocky coasts. They are heavier than water and lie still, resting on their Ps, on corals or in hiding, while they attentively inspect their surroundings. Their main food consists of small crabs and fish broods. The fish are territory-forming and live either alone or in fixed pairs *(C. aprinus).* All have lively colouration. They are easy to keep in the aquarium and take all forms of live food. In confined aquaria they fight amongst themselves. The following are frequently imported:

– *C. aprinus* (CUVIER and VALENCIENNES, 1829); Blotched or Spotted Hawkfish. Indo-Pacific, Red Sea. To 10 cm. Large dark-brown patches on a light background.

– *C. aureus* (SCHLEGEL); Spotted Hawkfish. Indo-Pacific. To 10 cm. Golden brown with darker patches. First ray of the soft D elongated.

Cirrhitidae; Hawkfishes (fig. 325). F of Perciformes*. Sub-O Percoidei*. Small, brightly coloured, territorial fish of the coral reefs from the tropical Indo-Pacific. The body is elongate and laterally compressed. The

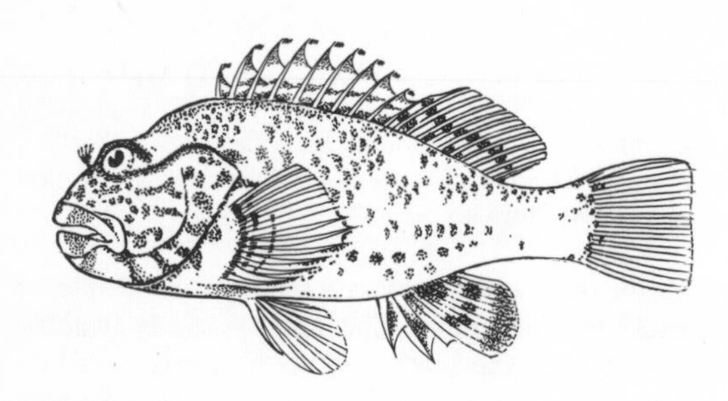

Fig. 325 Cirrhitidae-type

head is often bullish, eyes protruberant. The jaw and vomer are always filled with pointed conical teeth and the palatal bone sometimes has them. The D is uniform, the soft-rayed section is often broader, the C is rounded or straight. An short. The small pennant-like or tuft-like extensions on the points of the dorsal rays are characteristic of some species. The Cirrhitidae hardly move throughout the day. They remain for hours in one place in their territory, eg in a niche in the coral, and only start looking for food when it is dark. They are of no economic importance. There are some aquarists' observations about various species. See under Ga *Amblycirrhites, Cirrhitichthys, Paracirrhites* for information.

Cirrhosis *see* Liver, Diseases of

Cirripedia; Barnacles. Sub-Cl of Crustaceans (Crustacea*). About 800 species. They are with a few exceptions sea creatures. When mature, they live a sessile, partly parasitic, life and are barely recognisable as crustaceans. In fact, their true place in the classification was first recognised from the larva, which swims freely and is a typical nauplius*. The stalked Goose Barnacles (fig. 326) belong to the O Thoracica, as do the crater-shaped *Balanus,* which often build huge colonies. The body is completely enclosed in white calcium plates.

Fig. 326 Cirripedia (Barnacle)

They are filterers which strain the water with their protrusile hand-like feathered filtering legs. Most are hermaphrodite (bi-sexual). The G *Balanus** is equipped with a long penis-tube with which the adjacent animals can fertilise themselves. The larvae prefer to settle near other *Balanus.* The Cirripedia of the O Rhizocephala *(Sacculina)* are parasites and still more profoundly modified. The body has no limbs, is like a sack, and attacks the host animal with numerous root-like projections. The ten-footed higher crabs (Decapoda*) are the most susceptible. Thus one often finds the sack-like *Sacculina* on the back section of the Shore Crabs *(Carcinides**). South European Hermit Crabs *(Diogenes**) is parasitised by *Peltogaster.* The *Balanus* species and Goose Barnacles are interesting aquarium animals and can be kept for a long time. As these animals settle in large crowds on ships' hulls underneath the water line, they are a nuisance for shipping.

Ciscoes (Coregonidae, F, name not in current usage) *see* Salmonoidei

Citharichthys sordidus *see* Pleuronectiformes

Citharinidae; Moonfishes (fig. 327). F of Cypriniformes*. Sub-O Characoidei*. All specimens of the F are found only in Africa, and apart from this there are very few characteristics in this group which are common to all its members. These include the straight lateral line and the cyclo-ctenoid scales. The members of the individual Ga are very varyingly formed, some are reminiscent of Barbs and others of the true Characins and some have a slim elongate form with a long snout mouth *(Phago, Phagoborus).* There are, in the same family, both medium-sized and large species which grow to 60 cm and are relatively high-backed (some *Citharinus* and *Distichodus*) and dwarfs of 3–4 cm long. Many attempts have of course been made to divide the Citharinidae into several families and give some account of the natural relationship of the Ga, but up to now the various suggestions have not received any general recognition. In their natural habitat many of these fish

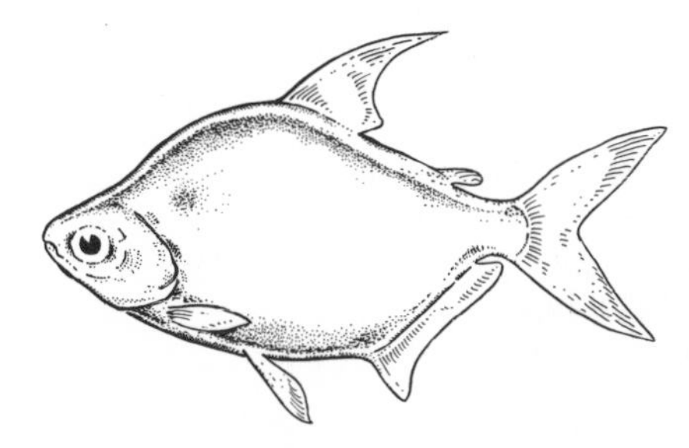

Fig. 327 Citharinidae-type

are very greatly used; some *Distichodus* are supposed to provide a delicious meat. Some small species of the Ga *Neolebias** and *Nannaethiops** have often been kept in aquaria because of their brilliant colouring, especially when spawning. The interesting members of the G *Phago** have unfortunately only been imported rarely and have proved to be very sensitive to injuries and difficult to feed. The young of the G *Citharinus** and some smaller brightly coloured members of the G *Distichodus** are only seen very occasionally in amateurs' aquaria, although they can be admired more often in aquaria for public display.

Citharinus. CUVIER, 1817. G of Citharinidae*. Found in Central Africa. Similar to the Leuciscinae – deep, laterally much compressed Characins of 35–50 cm. D triangular, well-developed adipose fin, C deeply forked, An long. They are usually uniformly silver with a darker back section. The fins are usually colourless. The lateral line is to be found in the middle of the body. They are active shoaling fish which will eat plant food (oats) as well as live food of all kinds. They are particularly suited for large display aquaria. They have hardly any significance for aquarists. Sometimes the following species are imported.

– *C. (Citharus) citharus* (GEOFFROY, 1809). Tropical Central Africa. To 50 cm (fig. 328).

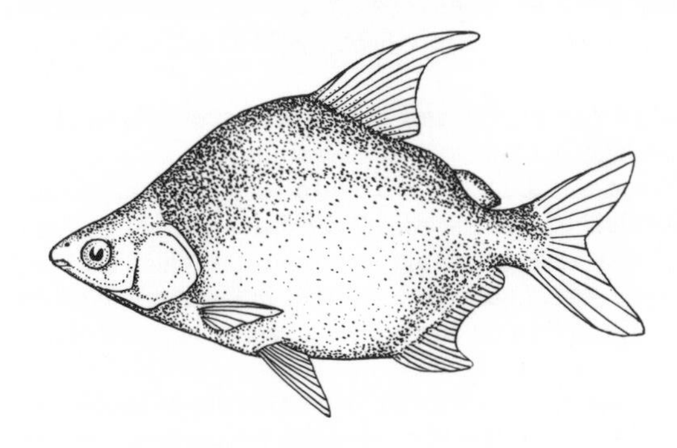

Fig. 328 *Citharinus citharus*

– *C. (Citharus) congicus* BOULENGER, 1901. Lake Malebo (Stanley Pool) Upper Congo. To 45 cm.

– *C. (Citharinoides) latus* MÜLLER and TROSCHEL 1845. Central and East Africa, lower Nile. To 42 cm.

Cladocera; Water-fleas (fig. 329). Highly specialised Fairy Shrimps (Phyllopoda*). With the exception of a few Ga they are found in fresh water and frequently near the banks of still waters (Littoral*). There are also other Cladocera which usually live in open water (Pelagial*) and even parasitic forms on Coelenterates (Coelenterata*) The body of the Cladocera is generally anything up to a few mm long, if segmented then only indistinctly. Apart from the head it is covered with a jointed shell (carapace). The second pair of extremities (antennae) which are particularly well developed provide a means of locomotion. The sole function of the remaining extremities (including up to six pairs of legs) is to gather food. The eyes which were originally in pairs in the Phyllopoda have been fused into a single moveable complex eye. A further characteristic of many

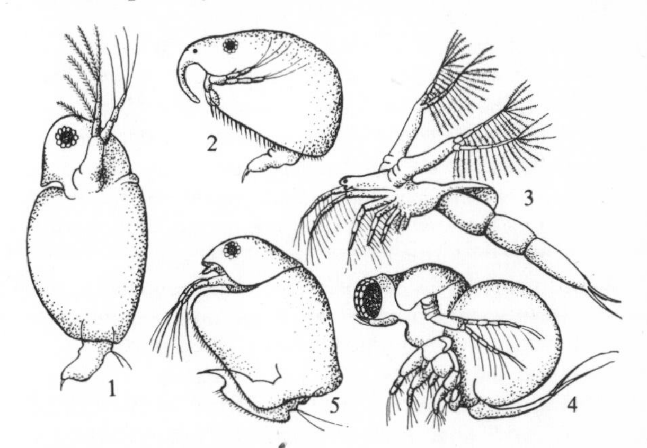

Fig. 329 Cladocera. Various Ga. 1 *Daphnia* 2 *Bosmina* 3 *Leptodora* 4 *Polyphemus* 5 *Eurycercus*

Cladocera is the changing of their shape according to the season (cyclomorphosis). The heads of *Daphnia* species for example are hood-like in summer but round in the colder seasons. Reproduction is very interesting. Basically both bisexual reproduction and parthenogenesis occur within one year. They are able to reproduce quickly, and rapidly increase the population* by parthenogenesis* when conditions are favourable (in summer). Males reappear in autumn and fertilise eggs which are stored in large numbers in a capsule (ephippium). These eggs are very resilient and are easily carried by water-fowl or strong wind.

Cladocera usually feed on the smallest particles (bacteria, algae, protozoa etc) but there are also some which are decidedly predatory and attack and overpower other small crustaceans. Predatory Water-fleas of this type are mostly oddly formed and grow up to 10 mm. In general, Cladocera form one of the most important basic foods of many freshwater fish. Members of the Ga *Daphnia** and *Bosmina**, amongst others, are used by aquarists for fish-food, both live and in dried form.

Cladocora EHRENBERG, 1834. G of Madreporaria*. In warm seas the colonies are either grassy, bushy, or mat-like, since the individual polyps (produced by

budding) demonstrate parallel upward growth but are not laterally fused together. The only European species which can be kept fairly well in the aquarium is the following.
– *C. caespitosa* (LINNAUS, 1767). Mediterranean. Often in shadowy shallower water. Polyps are brown because of the Zooxanthellae*. Diameter up to 5 mm. Colonies up to 30 cm.

Cladonema DUJARDIN, 1843. G of Hydrozoa*. Chiefly found in warm areas, with tiny, hardly branched polyp colonies and some creeping medusae. *C. radiatum* from the Mediterranean can immediately be designated as an aquarium medusa. The polyps are often inadvertently introduced and overlooked until the medusae (which are a few mm long) appear. They usually sit with their tentacles clinging fast and only swim away when disturbed. The whole process of alternation of generations takes place even in the smallest tanks without ventilation.

Cladophora. G of Green Algae (Chlorophyta*) with countless species both in fresh water and the sea. They form tufts of finely branched threads on firm substrates near the water surface. When examined under a microscope these show a cellular structure. Cladophora does not grow well in marine aquaria. The green thread-like algae which are so often seen are mostly *Vaucheria**. Cladophora is usually not wanted in the freshwater aquarium, with the exception of *C. (Aegagropila) sauteri*, from around Moscow, which has now become quite well known. It forms nice delicate green mats.

Clarias GRONOVIUS, 1781. G of Clariidae*. Many species, spread over almost all Africa, as well as Sri Lanka and East Indies. Torpedo-shaped Catfishes. From 15 to over 120 cm long. D1 and An long, C small and round, Ps and Vs well developed, D2 absent, 4 pairs of well developed barbels. Dark brown or brownish olive, sometimes bluish-green, and often with some degree of mottling. Thanks to their accessory respiratory organs the *Clarias* can live in dirty waters with a low oxygen content. As they are rather unsociable and voracious, tend to be active at night and are often of a considerable size, they are really only suited for large aquaria. Their reproductive behaviour is not known. The following species are often imported.
– *C. angolensis* (STEINDACHNER, 1866). Tropical W and Central Africa. To 35 cm. Coffee-brown or blackish with light spot. C has dark edge.
– *C. dumerili* STEINDACHNER, 1866; Barbel Catfish. Angola and Old Calabar. To 16 cm. Dark brown with greenish lustre. Fins brown.
– *C. mossambicus* PETER, 1852. E Africa. To 70 cm. Body and D olive brown to coffee brown to light brown, whitish marbled patterns. C and An dark brown and often with dark edges (fig. 330).

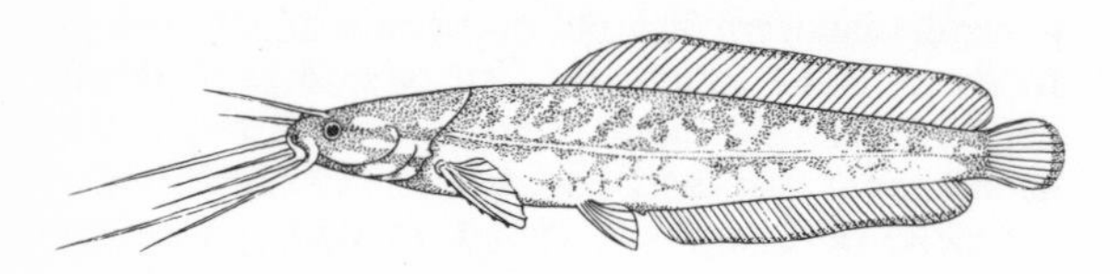

Fig. 330 *Clarias mossambicus*

Clariidae; Airbreathing Catfishes (fig. 331). F of Siluriformes*. Sup-O Ostariophysi. Elongate, plump, almost round when seen in cross-section, sometimes eel-shaped. The Clariidae, which have existed from the early Neocene are now spread from Africa over the whole of S and SE Asia including Indonesia. The characteristic feature of the F is a sac which extends backwards and upwards from the gill-chamber and into which much branched cauliflower-like processes project from the 2nd and 4th gill-arch. These processes are absent in the G *Heteropneustes*. It has instead a particularly large sac which extends back to the tail region in the form of two tubular blind sacs on either side of the vertebral column. These provide an additional respiratory organ which helps these fish to survive in warm waters with a low oxygen content. The Clariidae come to the surface to breathe. Apart from the P, the fins have no spiny soft ray. Head flattened, bullish with very wide gape. 4 pairs of barbels, usually long. The very long D and An form a border of fin which is fused with the C in the West African eel-shaped Ga *Channalabes* and *Gymnalabes*. The D is only short in *Heteropneustes*. The Ps and Vs are fairly small and the latter are absent on *Channalabes*. Some of the Clariidae grow to nearly 1 m long and they are unusually greedy predators. They hunt mainly at night but seek out hiding places during the day. In captivity they often eat so much that they lie around sluggishly on the bottom, balancing on their spherical stomach. Almost all species are very tasty food-fish. They have been released and bred in several subtropical and tropical areas. Some smaller species and young fish are sometimes kept in aquaria. Larger specimens are constantly exhibited in public aquaria. For advice on care see *Channalabes**, *Clarias**, *Heteropneustes**.

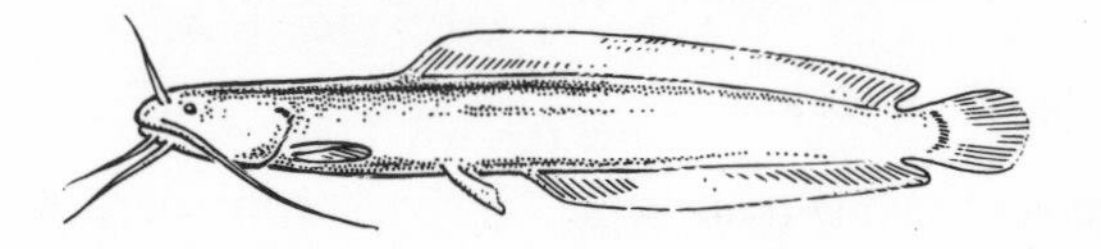

Fig. 331 Clariidae-type

Class *see* Systematic Categories

Clavelina, G, *see* Tunicata

Clay *see* Substrate

Cleaner Shrimp *(Hippolysmata grabhami) see Hippolysmata*

Cleanerfish *(Labroides dimidiatus) see Labroides*

Cleavage *see* Embryonic Development

Clibanarius DANA, 1852. G of small Hermit Crabs (Paguridea*) from the rocky coastal waters of warmer seas. They have brush-like hairy pincers which are both the same size. One species found in large numbers on South European coasts which is very suited to life in the marine aquarium is:
– *C. misanthropus* (RISSO). Mediterranean, Black Sea in shallow water. To 2 cm. Reddish brown with red feelers and red-blue striped feet.

Climate. The weather conditions, taken as a whole, over an average year or day. Climate has a very great effect on the vegetation and fauna* in a particular geographical region. The most important climatic factors are temperature and rainfall. Rainfall depends on air currents as well as the temperature of neighbouring seas or the location of high mountain ranges. It is possible to distinguish different climatic zones (*see* Vegetation) that are determined by the macroclimate. The distribution of an organism species, however, is often not merely dependent on the macroclimate but also on the particular microclimate in a certain habitat*.

Climbing Perch *(Anabas testudineus) see Anabas*

Climbing Perches (Anabantidae, F) *see* Anabantoidei

Climbing Perches, related species, Sub-O, *see* Anabantoidei

Cling-fishes (Gobiesocidae, F) *see* Gobiesociformes

Clingfish-types, O, *see* Gobiesociformes

Clinidae, F, *see* Blennioidei

Cliona GRANT, 1826; Boring Sponges. G of Porifera*. Found in warmer seas where they bore through into limestone, corals and mussel shells so that their only contact with the water is through small openings. The limestone probably becomes crumbly with the help of the carbonic acid from their breath and is pushed forward in microscopic fragments. The orange-coloured *C. celata* is found in great numbers on south European coasts. The material containing the sponge can often cause damage in small aquaria when the sponge dies.

Clipper Barb *(Barbus callipterus) see Barbus*

Cloaca. A common point of exit for faeces, urine and sexual products. Amphibians, reptiles, birds and monotremes (phylogenetically, older kinds of mammals) all have a cloaca. Sharks, rays and lungfishes also have one. In bony fish, the anus and the openings for the urinary and sex organs are separate. Normally, these openings lie next to one another in front of the anal fin, but in exceptional cases, the anal opening is very much further forwards (Gymnotidae, *Electrophorus*).

Cloeon LEACH, 1815 (fig. 332). Widespread G of Insect Order, Mayflies (Ephemeroptera*). The larvae of *Cloeon* often live in hordes in small thickly grown pools. They clamber around between the water plants. They can be distinguished from the *Baetis* larvae, which are similar, by the doubled tracheal gills on the abdomen. They are easy to catch and to keep and are excellent food for larger aquarium fish. The sub-imagines of *Cloeon* make their way into houses in autumn in order to shed their skin to become imagines.

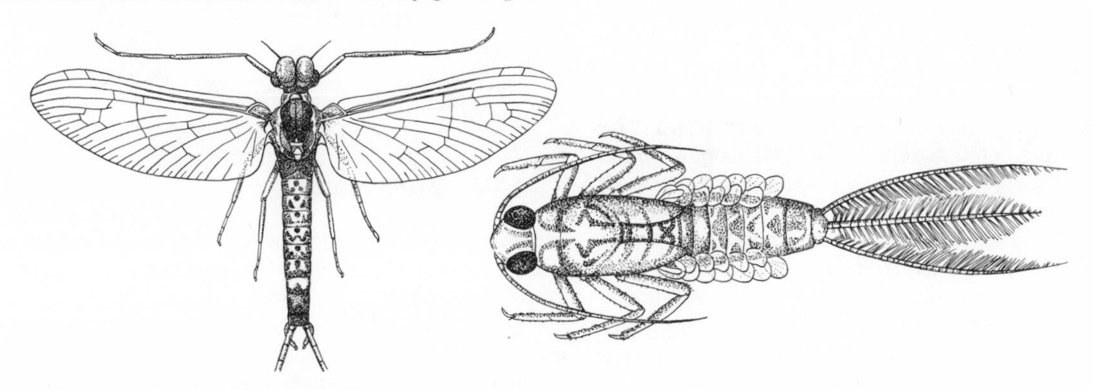

Fig. 332 *Cloeon.* Left, imago (without tail-threads); right, larva

Cloudiness of the Skin in fish can be caused by single-celled skin parasites (Parasitism*). They destroy the epidermis (*see* Skin), causing a bluish-white cloudiness. The principal parasites concerned are species of the Ga *Chilodonella**, *Costia**, *Trichodina** and *Oodinium**. All are typical weak parasites.

Cloudy Moray *(Echidna nebulosa) see Echidna*

Clown Barb *(Barbus everetti) see Barbus*

Clown Killie *(Epiplatys annulatus) see Epiplatys*

Clown Knifefish *(Notopterus chitala) see* Mormyriformes

Clown Labrid *(Coris angulata) see Coris*

Clown Labrids *(Coris formosa, Coris gaimardi) see Coris*

Clown Loach *(Botia macracantha) see Botia*

Clown Rasbora *(Rasbora kalochroma) see Rasbora*

Clown Surgeonfish *(Acanthurus lineatus) see Acanthurus*

Clown Triggerfish *(Balistoides conspicillum) see Balistoides*

Clown Wrasse *(Coris gaimardi) see Coris*

Clownfish *(Amphipirion ocellaris) see Amphiprion*

Club Mosses (Lypcopsida, Cl) *see* Pteridophyta

Clupea, G, *see* Clupeiformes

Clupeacharacinae, Sub-F, *see* Characoidei

Clupeidae, F, *see* Clupeiformes

Clupeiformes; Herring types. O of Osteichthyes*, Coh Teleostei. Predominantly marine fish with a uniform herring-like shape and a marked silver sheen. With the exception of the Denticipitidae, all Clupeiformes have the following common characteristics: The swimbladder forms a sac which extends forward on either side; it penetrates the cranium and usually forks here. Each end piece finishes with an extension into particular bones of the hearing or equilibrium organ capsules. The skeleton of the root of the caudal fin is further characterised by the lack of a flexible joint between the 2nd hypural and the urostyle bones. Finally, the special features of the lateral line system must be mentioned – such as the additional channels on the operculum and the lateral line on the body which is greatly or completely reduced. Apart from these characteristics, the Clupeiformes share various common features in the structure of the cranium, the vertebrae, the shoulder girdle and the ribs amongst other things. The following families are of special interest.

– *Chirocentridae*; Wolf Herrings. This family deserves to be mentioned as it differs from the other Clupeiformes in a few ways. In contrast to most other members of the O, the 2–3 m long fish, *Chirocentrus dorab* (fig. 333), does not filter plankton or eat small creatures, but is one of the greediest predatory fish. It has strong

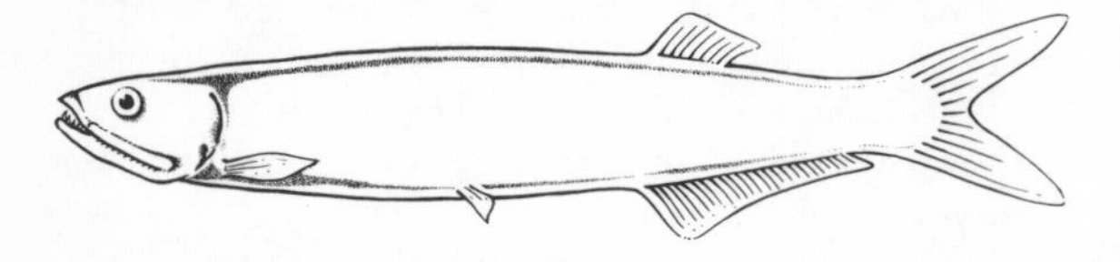

Fig. 333 *Chirocentrus dorab*

jaws with fangs. The short D lies very near the back and begins level with the long An. V small and lying approximately in the middle of the body. The back section of the body has a sharp underside edge. Swimbladder divided. Gut has no blind guts. This fish is to be found in almost all seas and feeds predominantly on fish.

– *Clupeidae*; Herrings. The 120 or so species of this interesting family differ only slightly from the well-known Herring *(Clupea harengus)* (fig. 334). The fossils belonging to the early Crustacean period are no exceptions to this rule. The characteristic appearance is primarily defined by the elongate and elegant body which is laterally compressed, and the relatively small

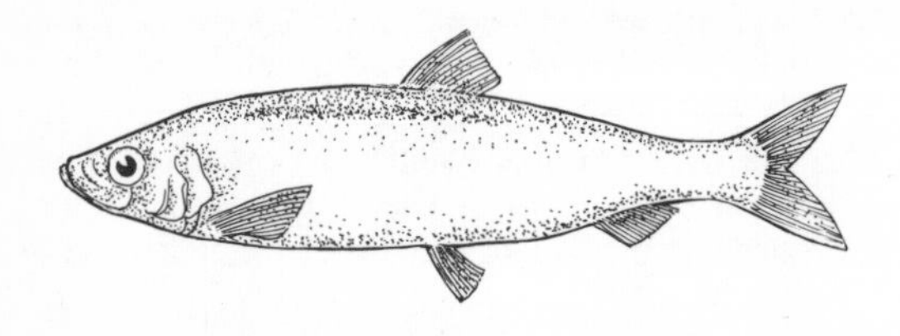

Fig. 334 *Clupea harengus*

fins. The belly (underside) is often sharp-edged. The short D lying in the middle of the back is opposite the Vs, C is deeply forked, An short. Amongst the most important characteristics are the lateral line bordered by 2–5 scales, the swimbladder which is connected to the hearing capsules and the fact that the swimbladder often has a channel-like extension to the rear which opens near the anus in many species. All species are physostomes, ie they have a swimbladder passage which joins the swimbladder to the gut. The fins are only supported by soft rays and the cycloid scales sit loosely in the skin. The number of vertebrae can vary greatly within one species; the specimens in cooler areas invariably have more vertebrae than members of their species in warmer regions. The pharynx is distinguished by numerous gill-rakers which protrude inwards from the gill-arch and are arranged like the prongs of a rake. Each gill-arch can have up to 200 such processes and all the gill-rakers together form a weir-basket with which they can filter the smallest food out of the water which they suck in with their extended mouth. This method of feeding has led to a great reduction in their jaws in the course of their phylogenesis. Most species are typical shoal fish. When watching shoals, the harmony of the movements of the individuals with one another is always surprising. Many species are amongst the most important of all the fish we use, and of these the Herring *(Clupea harengus)* is again of primary importance. When alive, it is blue green on its upper side and has a splendid silver sheen often with shimmers of violet, and can grow to 50 cm long and live to 20 years old. It is found in the whole N Atlantic and the seas near it. When searching for areas with plenty of plankton or combing them, the fish form giant swarms which are often several kilometres long. They go in shoals along the traditional routes to their spawning grounds which are usually, but not always, in shallow waters fairly near the coast. These spawning shoals provide ideal conditions for catching herrings. These routes and spawning grounds are however not absolutely fixed but change after a certain period so that the herring fisheries cannot always count on sure yields. Thus it happens that after a few years of overfishing the stock is apparently considerably reduced. The eggs which are expelled and fertilised in the open water sink to the bottom and stick fast. Every female produces about 20–30,000 eggs. After a shoal of herring has spawned, the bottom of the spawning ground is often covered with eggs up to several cm deep. The young hatch after 9–20 days depending on the temperature and sexual maturity is reached after 3–5 years. The Baltic Herring *(Clupea harengus membras)* only grows to 20 cm. It has developed various local forms which have spawning seasons in spring or autumn. In the N Pacific is found the Pacific Herring *(Clupea pallasi)*. A small version of the Herring is the Sprat of up to 15 cm *(Sprattus sprattus)* (fig. 335) which with various sub-species is found in the Baltic, the Atlantic and the Mediterranean as well as the Black Sea. This is of great economic importance

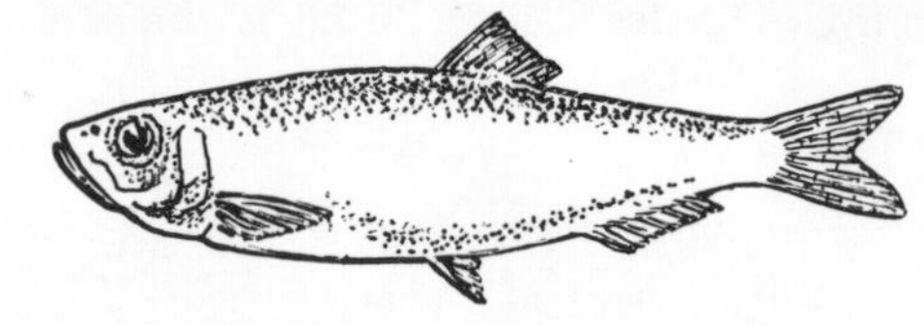

Fig. 335 *Sprattus sprattus*

to fisheries everywhere. The Baltic is one of the main areas in which it is caught. The Sardine and Pilchard *(Sardina pilchardus)* (fig. 336) are somewhat larger than the Sprat. They are shoal fish with a protruding lower jaw and fairly large scales. They are mainly found in the Mediterranean and the neighbouring coastal waters of

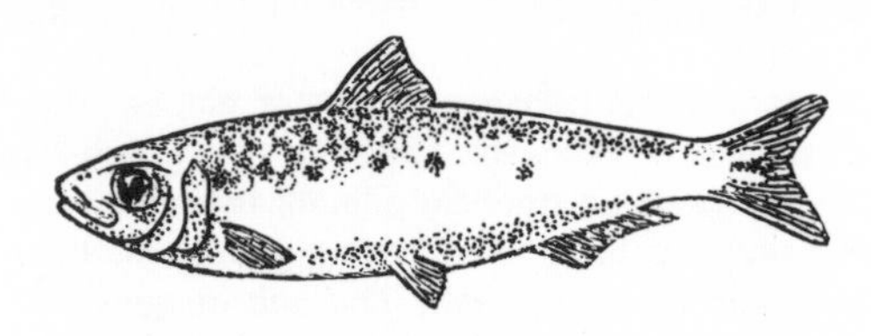

Fig. 336 *Sardina pilchardus*

the Atlantic. They are spread from here to the Canary Isles in the South and Norway in the North. The young fish, about as long as a finger, are those which are tinned. They are caught in trawls or purse-seine nets and the shoal is often enticed into the net by lights. The species *Sardinops sagax* WECCE which is found on the west and East coast of the Pacific is also economically important for fisheries. A less well-known member of the F is American Menhaden or Mossbunker *(Brevoortia tyrannus)* which forms the main raw material for American fish-meal factories. There are also various species of Herring which live in fresh water, some are anadromous and go up the rivers from the sea to spawn. On the European coast these are the Twaite Shad *(Alosa fallax)* (fig. 337) and the Allis Shad *(Alosa alosa)*. The numbers of these have however dwindled greatly in the last 30 years due to the river pollution. The Allis Shad *(Alosa alosa)* usually spawns in May in the lower courses of the river and then returns to the sea. The young fish follow in autumn. In N America the 50 cm long *Pomolobus chrysochloris*, which tastes very

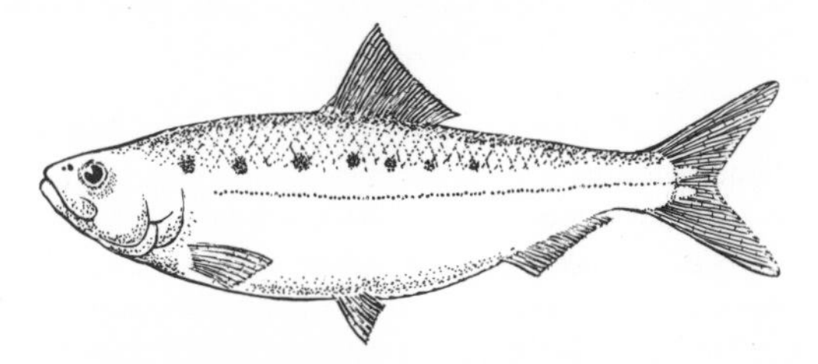

Fig. 337 *Alosa fallax*

good, comes up over the Mississippi Delta into the fresh water to spawn. Herrings have many enemies; some of them, such as the Tuna fish, follow the shoals on their wanderings. All members of the family can only be kept in large tanks under special conditions. For this reason they are only rarely seen, even in public marine aquaria.

– *Denticipitidae*; Freshwater Herrings. There is only one known present-day species of this very primitive family and that is *Denticeps clupeoides* (fig. 338) from the slow-flowing waters of the West African Rain Forest area (SW Nigeria). Fossil species have been found in E Africa. This differs from other Fs in various basic characteristics. These include peculiarities of the skeleton of the C-root, the complete lateral line on the head, the gill covers and in the front section of the scaly

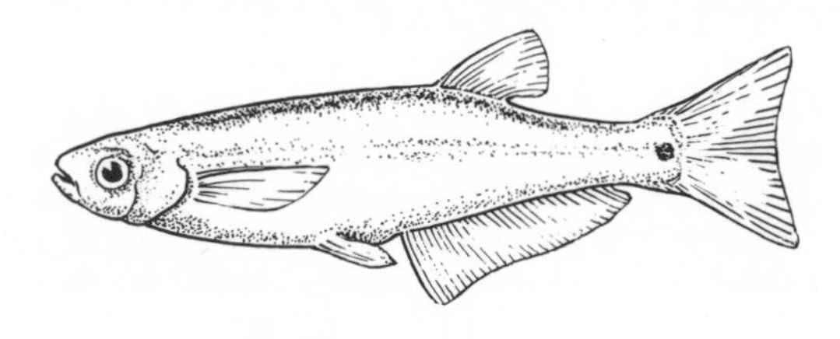

Fig. 338 *Denticeps clupeoides*

body. This shoal fish grows to 5 cm, has a greenish sheen on its upper side and a brownish yellow sheen underneath and is otherwise silver. It has been imported quite often in recent years. When keeping this fish in an aquarium it is important to know that it should have plenty of room to move, since it can only achieve a sufficient flow of water in the gills by constant swimming. It is always timid and susceptible to skin parasites. It always remains in the central area of water and avoids the bottom except when searching for food.

– *Engraulidae*; Anchovies. Silver shoaling fish from the size of a finger to that of a pencil. They differ from other Fs primarily in the underslung mouth which extends far behind the eyes. Fossils have been found from the Tertiary period; most of the 100 species alive now live in the Pacific Ocean. The Anchovies are plankton-filterers like the Clupeidae. When they are collecting food they chase through the water with their mouths wide open. The edges of the gills form the frame of the net and the pharynx the weir-basket. Every couple of seconds the mouth is shut and the filtrate is swallowed. The European Anchovy *(Engraulis encrasicholus)* (fig. 339) occurs in many races over the large area where it is found, which includes the European and the West African Atlantic coasts, the Mediterranean, the Black Sea and the Sea of Azov. Their oval eggs float on the surface of the water and mature relatively quickly. In the N Pacific is found the 20 cm long *Engraulis mordax*. Both species are important food fish which are usually caught with drag-nets on their extensive wanderings and made into pastes, sauces and also fillets. Some species live in brackish water and even occasionally in fresh water. Some quite monstrous species also belong to this family, such as the Whiskered Anchovy *(Trisocles setifer)*, the upper jawbone of which is so greatly ex-

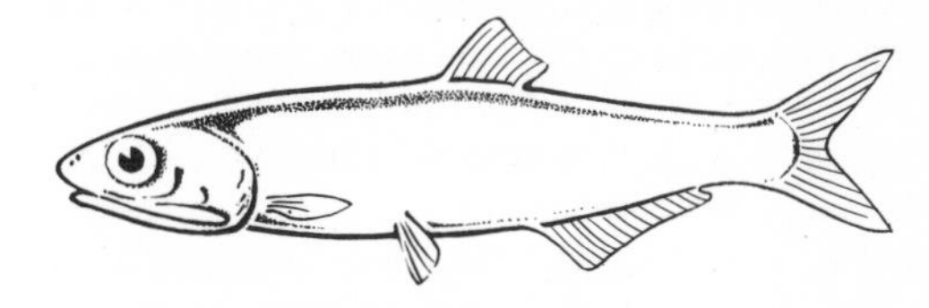

Fig. 339 *Engraulis encrasicholus*

tended that it almost reaches to the An. Specimens of the Indo-Pacific G *Coilia* have in addition to the greatly extended upper jawbone a long pointed tail with a small C (the underside of which is bordered by a very long An like a fin-fringe).

Clupeomorpha, Sup-O, *see* Osteichthyes

Clypeaster LAMARCK, 1801. Sand Dollars. G of Sea-urchins (Echinoidea*) Various species in tropical and subtropical seas. The disc-like flat *Clypeaster* live on soft bottoms. They mostly lie buried and sometimes move along the bottom. They eat little particles. There are hardly any laboratory observations of them.

Cnesterodon GARMAN, 1895. G of Poeciliidae. Sub-F Poeciliinae. S America, S Brazil, Paraguay, Uruguay, NE Argentina. Found in ditches, streams and other fairly still waters. Very small fish with slim shape. Males to 2.5 cm. Females 4 cm. Basic colouring yellowish with a row of dash-like dots along it. Male has fairly long gonopodium. Should be kept at 20–24 °C; lower temperatures are also possible. It is advisable to keep them with plenty of plants. They eat algae as well as live food and some dried food. So far only the following has been bred.

– *C. decemmaculatus* (JENYNS, 1843); Ten-spotted Livebearer. Argentina, Chaco region. Male to 2.5 cm. Female to 4 cm. Gonopodium has hooked appendage. The male has 6–12 grey spots along its central line and the female has 10. These are not limited to the rear section of the body in the female and can sometimes be black. Male has violet sheen. C with dark edge. Gestation period 4–6 weeks. Up to 40 young (fig. 340).

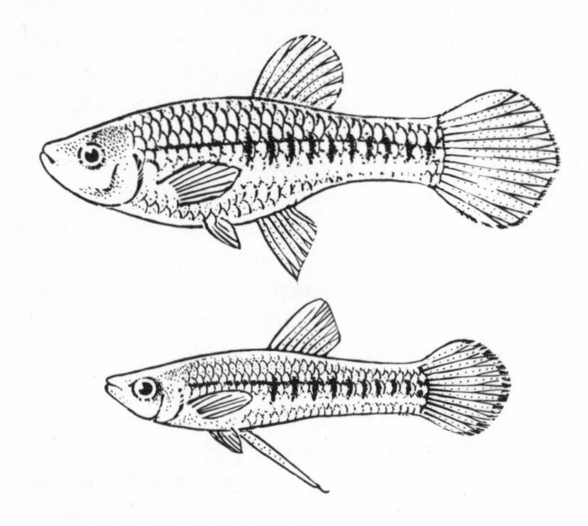

Fig. 340 *Cnesterodon decemmaculatus*

Cnida *see* Stinging Capsules

Cnidaria; Stinging Animals. Sub-Ph of Coelenterata*. They are distinguishable because they possess stinging cells*. Because of these they are different from the second Sub-Ph, the Acnidaria (Sea Gooseberries, Ctenophora*). Both groups are often regarded as separate Phyla. The classes Hydrozoa*, Scyphozoa* and Anthozoa* belong to the Cnidaria. Many Anthozoans (Corals) such as Sea Anemones (Actiniaria*) or *Cerianthus* (Ceriantharia*) are often kept in marine aquaria.

Cnidocil *see* Stinging Capsules

Cnidom *see* Stinging Cells

Coalfish *(Pollachius virens) see* Gadiformes

Coastal Reefs *see* Coral Reefs

Cobalt Blue Cichlid *(Pseudotropheus zebra) see Pseudotropheus*

Cobitidae; Loaches (fig. 341). F of Cypriniformes*, Sub-O Cyprinoidei*. Small to very small, elongate to eel-shaped; laterally compressed bottom-living freshwater fish. Found from N Africa (Morocco, Ethiopia) across Europe and Asia to the islands of the Malay Archipelago. Most of the species are native to SE Asia.

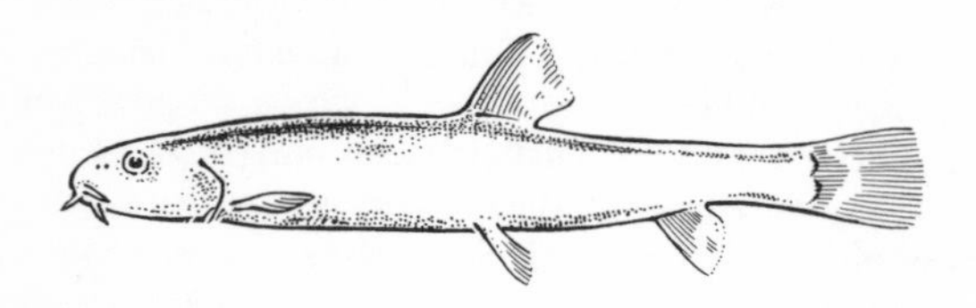

Fig. 341 Cobitidae-type

The Cobitidae can be distinguished from other Fs of the Sub-O Cyprinoidei* by their large number of barbels (3–6 pairs), their relatively weak pharyngeal bones, the large number of pharyngeal teeth arranged in one row and the missing grinding-stone on the process of the rear base of the skull (basioccipital bone). Other features are the tiny round scales which lie deep in the skin (and thus can hardly be seen) and the underslung, partly extensible mouth, the relatively small gill-slits and the specialised swimbladder. The latter, of course, consists of two parts, as is the case in most Cypriniformes, but the front section is surrounded by a bony capsule which belongs to the Weberian apparatus (*see* Cypriniformes for information about this). The rear section is normally developed in some species but small or involuted in bottom-dwelling species. A further peculiarity of some Loaches (Sub-F Cobitinae) is the inner movement of the neurocranium. By this is meant their ability to move the front section of the skull actively forwards and upwards. Finally, the simple or bifid spine near the eye must be mentioned, which has given rise to the name Spined Loach (fig. 342). In the case of the *Botia* species, for example, these spines are very long, forked, erectile and can be locked in place by a special hook. Through aquarists, we know that the Cobitidae often hang by their spine in the mesh of nets. Injuries through these spines, which grow away from the bones of the skull, are very painful. In some species, the eyes themselves are covered with a transparent film of skin. Loaches also have an unusual form of additional respiration. They rise to the surface to swallow air and this is passed by intestinal peristalsis to the hind gut which is rich in capillaries. Here the gas exchange takes place. The used air is passed out through the anus. Most species search for food on the bottom using their barbels as touch and taste organs. They suck their food in

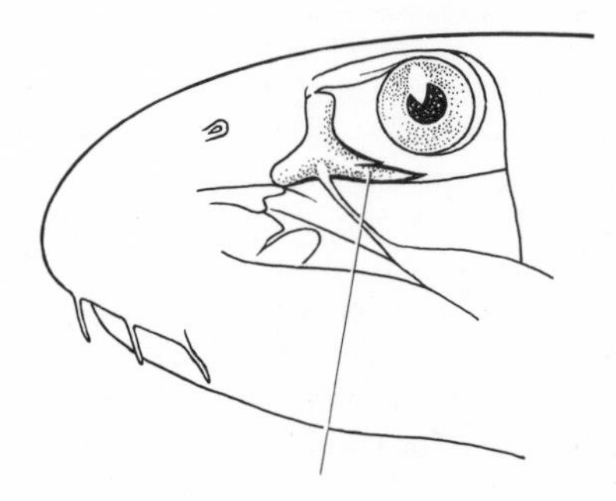

Fig. 342 Eye-spine (as indicated) of *Cobitis*

rather jerkily. Many species are inveterate muddy water-dwellers, eg the Weather-fish *(Misgurnus fossilis)*; others are found in streams and rivers with sandy and gravelly bottoms (G *Cobitis**). Some can dig themselves into the bottom, many species spend the whole day buried under stones or in holes in the bank. It is known that several Cobitidae make cracking noises which perhaps serve to define territories or form part of the mating behaviour. When pairing and mating they entwine themselves or press close to each other at the side, the pectoral fins of the males often have a special function as a directing organ and occasionally have special morphological equipment. The eggs cling firmly to various substrates. The young of the Weather-fish mentioned above are especially suited to living in the waters of the spawning ground with their low oxygen content owing to external gills which later regress. This species also reacts to lowering of the air pressure by frequently going to swallow air – a mode of behaviour that used to help in forecasting periods of bad weather. As prey of many predatory fish the Cobitidae are of indirect economic importance to fisheries. Many tropical and subtropical species are brightly patterned and coloured and thus popular with aquarists. The following species are dealt with under their own entry: *Acanthophthalmus, Acanthopsis, Botia, Cobitis, Lepidocephalus, Misgurnus, Noemachilus.*

Cobitis LINNAEUS, 1758. G of Cobitidae*. Europe, N Africa, Central Asia to the Pacific. These small territory-forming, bottom-dwelling fish live in clear streams and lakes with sandy and gravelly bottoms. The body is elongate and laterally compressed. Their tiny scales are buried in the skin. The very short head with a blunt snout finished in an inferior mouth which has 3 pairs of barbels. Under the eyes are 2 pointed erectile spines. During the day these nocturnal fish like to bury themselves in sand. They must be kept in tanks which are not too small, with sandy bottom-soil and good ventilation as the supplementary intestinal respiration is not very well developed. Stones and roots should be placed so as to provide dark hiding places. The water should be clear and partly renewed occasionally. The temperature should not be higher than 15–18 °C. All species will pick up live food from the bottom, but they prefer to dig up worms and insect larvae from the sand. After a cold overwintering these fish will also spawn in the aquarium from April to June. Rearing the young is not difficult.

– *C. taenia* LINNAEUS, 1758; Spined Loach. Europe and large parts of Asia excluding N Scandinavia and N Asia. To 12 cm. Back and body sides sand-coloured, brownish-yellow with dark roundish spots and kinds of stripes partly with cross-bands. Numerous other species endemic to Europe (fig. 343).

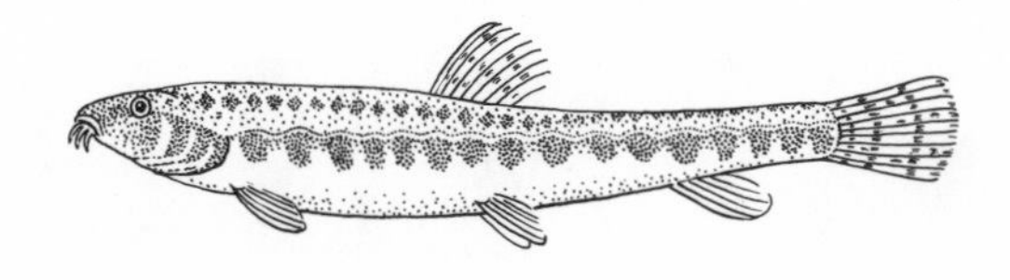

Fig. 343 *Cobitis taenia*

Cobra Fish *(Pterois volitans) see Pterois*
Cochino *(Balistes vetula) see Balistes*
Cockles, G, *see Cardium*
Cod *(Gadus morrhua) see* Gadiformes
Cod Types, O, *see* Gadiformes

Codium. Marine G of Green Algae (Chhorophyta*). Bulky Green Algae with thick-twigged or tuberous growth. Found at depths of 5–40 m. The potato-like, dark-green tuberous plants form a sea ball. *C. bursa* (L), from the Mediterranean, lives for years in a marine aquarium.

Cods (Gadidae, F) *see* Gadiformes
Coelacanthidae, F, *see* Coelacanthiformes

Coelacanthiformes; Coelacanth-types. O of Osteichthyes*; Sub-Cl Sarcopterygia, Sup-O Crossopterygii. For a systematic representation compare Osteichthyes* and Ceratodiformes*. When a living representative of the Crossopterygii (Stump-fins) was found in 1939, it caused a great sensation amongst Ichthyologists and Zoologists. This group of fish which was so important to the history of the phylum of vertebrates had been believed extinct for 50 million years. In 1939 L. B. Smith called this fish, which was discovered in 1938 by Mrs COURTENAY-LATIMER, *Latimeria chalumnae* (fig. 344). Despite great efforts it was 14 years before a second example was found in the Comoro Islands and its area of distribution established, 3,000 km from where the first was caught. *Latimeria* is a fish of up to 1.50 m in length and 80 kg in weight, grey-blue with white spots and an archaic perch-like appearance. The paired fins,

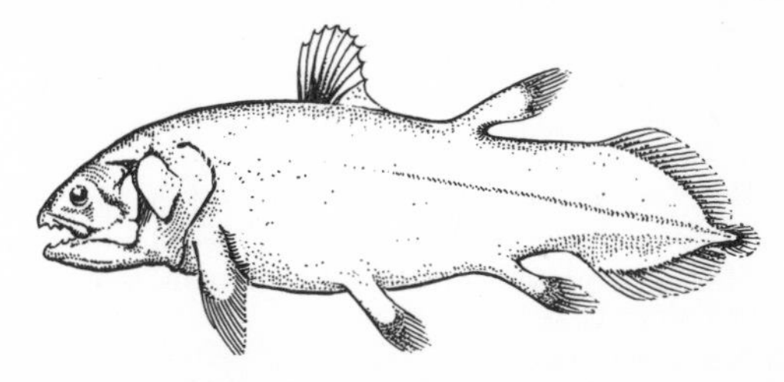

Fig. 344 *Latimeria chalumnae*

the D2 (which lies far back) and the An opposite it have muscular scaled stalks. D1 normal with 8 rays and situated more or less in the middle of the body. The spinal chord (chorda) extends to the far end of the C and draws the middle out into a point. The skin is very slimy. Scales large; brain very small. Heart primitive. Lungs reduced to a sack filled with fatty tissue. *Latimeria* lives in great depths over rocky bottoms and preys on fish. So far, it is the only species of the F Coelacanthidae; the second specimen caught was first of all described as *Malania anjouanae,* but very soon became known as Latimeria. The Coelacanthiformes recorded by fossils since the Carboniferous period are not the ancestors of the Amphibia but specialised Crossopterygii and this is of course obviously true of the living species. The Crossopterygii of the O Osteolopiformes (fig. 345), which lived 370–280 million years ago,

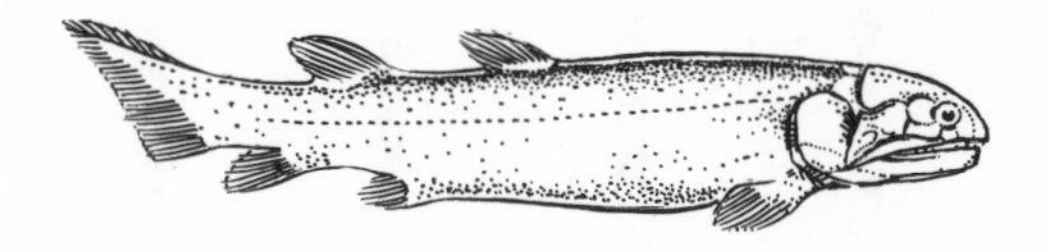

Fig. 345 Osteolepis-type from Devonian period

(15 Ga of which are known in fossilised forms), must rather be regarded as the phylum group of the Tetrapods.

Coelacanths (Coelacanthidae, F) *see* Coelacanthiformes

Coelacanth-types, O, *see* Coelacanthiformes

Coelenterata (fig. 346). Scientific name for the Ph of Coelenterates. Some specimens have rather confusingly been given plant names, eg Sea-Anemone, a reference to their plant-like appearance. The Coelenterates, to which such well-known forms as jelly-fish and corals belong, are very simply organised. The great complexity of the body forms is traced to a uniform

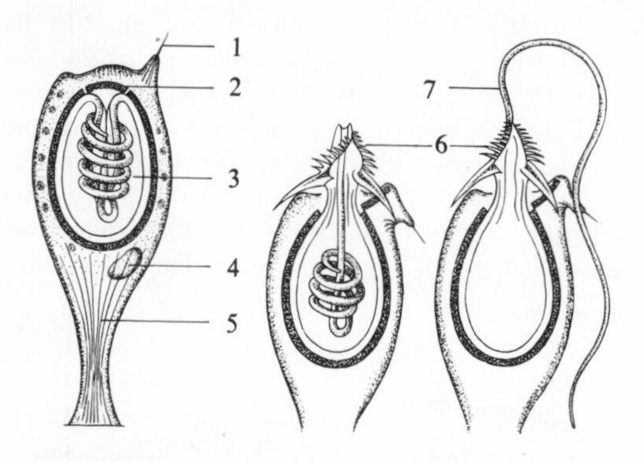

Fig. 346 Coelenterata. Left, anatomy of a stinging cell; centre and right, discharge of a stinging cell. 1 Cnidocil 2 Cover 3 Stinging capsule 4 Cell nucleus 5 Stinging cell 6 Barbed hook 7 Stinging thread

gastrula-like basic form. A large stomach or gastric cavity is surrounded by two layers of cells which developed from ectoderm and entoderm*. Here, in contrast to the Porifera*, true tissue with specialised cells appears for the first time in the animal kingdom, eg muscle cells and sense cells. The latter are connected to a simple nervous system, which for its part makes co-ordinated movements of the epithelium muscle cells possible. Between the two layers of cells is a gelatinous supporting lamella which can be either thin, as in the *Hydra**, or thick, as in many jelly-fish. In a typical case there are tentacles around the general oral anal orifice with which the food can be seized and overpowered. Even large prey are swallowed whole and only gradually broken up in the gastric chamber. Many species, eg Corals, only eat the smallest plankton which they filter from the water with cilia, some have Algae in their body cells which form nutrients through assimilation. Most of the Coelenterata live in the sea, partly remaining sessile and partly swimming around freely as typical dwellers of the pelagial*. Sexual and asexual reproduction can alternate (Reproductive Cycles*). If the case of asexual reproduction by budding or fission separation of individuals remains incomplete then clusters of animals are formed. The separate individuals can either be exactly the same as each other (eg as in many kinds of Anthozoa*) and then occur in very large numbers, or, as in the case of the Siphonophora, they can be variously formed corresponding to their special function. The Coelenterata are grouped into Cnidaria* and Ctenophora* according to whether or not they have stinging cells*.

Coelenterates, Ph, *see* Coelenterata

Coenagrion Kirby, 1890; Damselflies (figs. 347 and 348). G of insect order Libellae (Odonata*). Numerous species in Europe and Asia (palaearctic Realm*), some also in N America. The larvae of *Coenagrion* are characteristic inhabitants of pool biotopes with plenty of plants and are interesting aquarium objects. They are also good food for larger predatory fish. The male imagines of the *Coenagrion*, as well as the related G *Enallagma* which

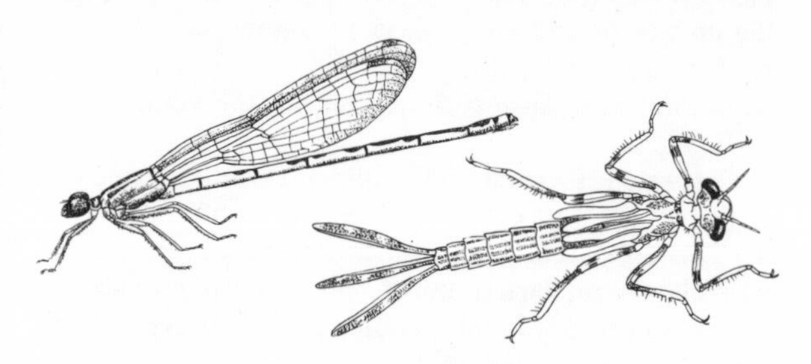

Fig. 347 *Coenagrion.* Left, imago; right, larva

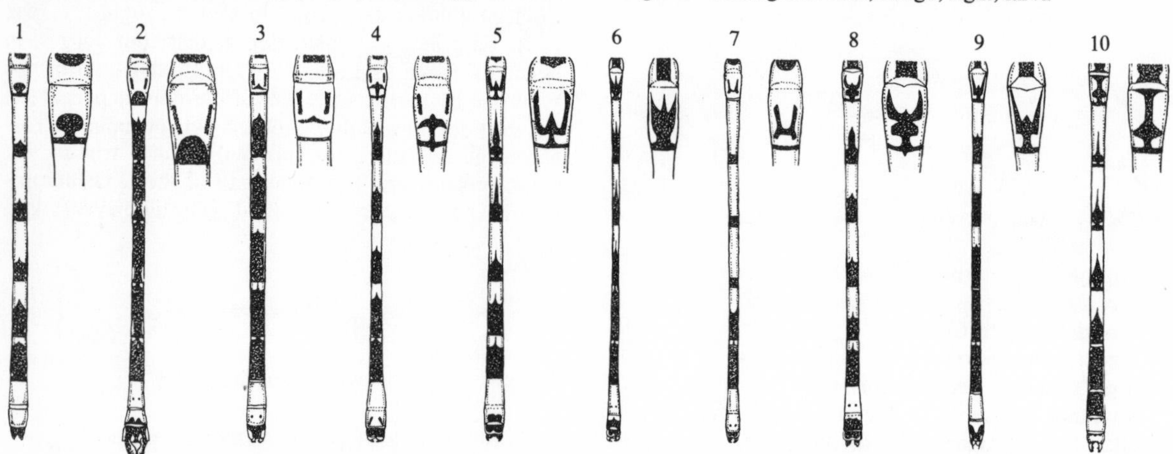

Fig. 348 Illustration of abdominal section of the Coenagriidae males (G *Coenagrion*, G *Enallagma*, G *Cercion*), 1st and 2nd abdominal ring enlarged more in each case. 1 *Enallagma cyathigerum* (Common Blue Damselfly) 2 *Coenagrion armatum* 3 *Coenagrion lunulatum* 4 *Coenagrion hastulatum* 5 *Coenagrion ornatum* 6 *Coenagrion pulchellum* 7 *Coenagrion puella* (Common Coenagrion) 8 *Coenagrion mercuriale* 9 *Coenagrion scitulum* 10 *Cercion lindeni*

is found in N America, are immediately recognisable by their azure bodies with shiny black markings.

Coenobita LATREILLE, 1829; Terrestrial Hermit Crabs. G of Paguridea* of tropical sea coasts which have gone over to living on the land and only pass their larval stage in the sea. They are omnivores that also eat plant food. The gills now only play a small role in their breathing, the wrinkled rear body (which is richly supplied with blood) functions as a respiratory organ. In adapting themselves to living on the land these creatures shut the former snail shell quite precisely with their pincers and front feet and thus make a shield against evaporation. This exact fit is achieved after moulting while the limbs are still soft. These lively, clambering *Coenobita* are to be recommended for amphibious displays.

Cold, Damage Caused by, *see* Temperature Damage

Coleoptera; Beetles. This animal order belonging to Insecta* has more species than anything on earth. So far more than half a million beetle species are known, of which by far the greatest number are purely land-dwelling. However several families can adapt themselves fairly well to life in fresh water. The most important of these are Dytiscidae* (Water Beetles), Hydrophilidae* (Great Silver Beetles) Gyrinidae* (Whirligig Beetles) and the Haliplidae* (Crawling Water Beetles. Apart from these, members of many other families prefer damp biotopes* such as river banks and wet carpets of moss. There is, however, no species of beetle which has settled in the sea. Furthermore, despite the fact that the freshwater beetles are thoroughly acclimatised to the water they cannot completely renounce their terrestrial* origins. Thus many species have to come up to the water surface for air in their larval and imaginal stages and the pupal stage is always spent on land.

Colisa CUVIER, 1831. G of Belontiidae, Sub-O Anabantoidei*. Asia, India, Bangladesh, Burma, Malaysian Peninsula, in the extensive river basins, eg of the Ganges and Brahmaputra. 15–12 cm. Females somewhat smaller. Strongly compressed laterally. More or less egg-shaped. D and An the same length, beginning before the centre of the body and extending to the tail stalk. Second ray of the V extended, thread-like. Body and fins brightly coloured mostly with cross bands running from the front upwards. Pectoral region and throat of a strong blue or green to black colour, particularly in the spawning season. Females plainer, mostly more brownish. These fishes need warmth. 24–28 °C up to 32 °C for breeding. Not demanding eaters, preferably live food of suitable size. All species build a fairly large bubble nest, mostly using parts of plants. The males care for the brood. The following are recommended for keeping in the aquarium.

– *C. chuna* (HAMILTON-BUCHANAN, 1832); Honey Gourami. NE India, Bangladesh. To about 7 cm. Territory-forming or mating males goldish-yellow to brick-red. Throat, breast and hard-rayed An blue-green to black. D has broad bright yellow edge. Threads of V white to red. Female plainer and brownish with fairly well marked longitudinal stripe which young males also have. This species has a small incomplete bubble nest (fig. 349).

Fig. 349 *Colisa chuna*

– *C. fasciata* (BLOCH and SCHNEIDER, 1801); Giant Gourami, Striped or Banded Gourami. India to Malaya. To 12 cm. Upper lip thick, especially in old males. Basic colour of the upper body sections is greenish-brown, becoming blue-green mixed with red towards the stomach. Blue-green bands on the body. D has white, An red edge. The inner section of An is metallic green. Females less intensely coloured. Vertical fins smaller. Large bubble nest, often over 10 cm wide. 600–1,000 eggs.

– *C. labiosa* (DAY, 1878); Thick-lipped Gourami. Burma. To 9 cm. The lips of the males are thick. The sides of the body are brown and the bands which cross at a fairly steep angle are blue-green. An blue in the male in the soft-rayed section and reddish in the female. D drawn out to a point in the male.

– *C. lalia* (HAMILTON-BUCHANAN, 1822); Dwarf Gourami. NE India, Bangladesh, Assam. To 5 cm. Ground colour of males red and metallic blue-green alternating as stripes on the body sides. On the D, An and C, blotches more or less form stripes. This species loves warmth. Prefers to breed in tanks with plenty of algae and sunshine (fig. 350).

Fig. 350 *Colisa lalia*

Colisa Disease *see Oodinium*

Collared Butterfly-fish *(Chaetodon collare) see Chaetodon*

Collargol. In finely dispersed amounts, silver has a strong inhibitive effect on fungi and bacteria. In fish aquarium-keeping, it is used against bacterial infections* of the skin and against saprolegniacea*. Silver preparations that can be dissolved as colloids (Colloid*) are used – eg collargol (colloidal silver), protargol (silver proteinate)–in the proportions of 1 mg to 10 litres of water. Sick fish should be treated with short-term baths (*see* Bathing Treatments) lasting a maximum 20 mins.

Collembola; Springtails (fig. 351). O of Insecta*. Comprising 3,500 species spread worldwide. The Collembola always live in a damp environment in earth, in rotton foliage, in old wood, on plants or on the water (eg *Podura aquatica*). Because they are so small (mostly 1–2 mm) the Collembola are sometimes used by aquarists for rearing fish, especially surface fish. Collembola are primitive wingless insects, which have 3 metamorphosed pairs of legs on their abdomen as well as the 3 pairs of walking legs on the thorax. In several species the last pair of legs has grown together to form a springfork which is tucked up under the abdomen. It can be jerked against the ground and helps the creatures to make jumps of several centimetres. Depending on habitat, the Collembola are mostly whitish, some species are however coloured or patterned. They eat dead pieces of plants, algae, fungus or micro-organisms and thus, as they appear in such vast numbers, they play a part in forming humus. Some of them are looked upon as plant pests. One can isolate the Collembola most easily in damp earth in woods. They enter a sieve with a mesh size of 1–2 mm, this is laid over a funnel under which there is a glass. Then the earth must be lit and dried out from above with an incandescent lamp. The Collembola withdraw underneath and fall through the funnel into the glass. They can then be put into a culture vessel – a covered shallow dish filled with damp leaf mould. They can be fed with a little sprinkling of flour, boiled potato or moistened white bread. They should be kept at not less than room temperature. The time taken for the development of the larvae which hatch out of the eggs after 1–3 weeks depends on the temperature. After a further 2–4 weeks they reach sexual maturity. The larvae and adult Collembola are hard to distinguish from external appearances. In order to take them out to feed them to your specimens, damp cloths or tree bark should be laid down on which the Collembola will collect.

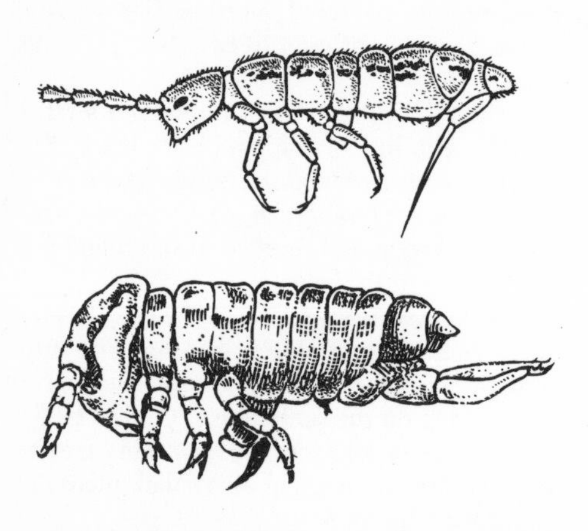

Fig. 351 Collembola. Top, *Isotoma*; bottom, *Podura*

Colloid. A substance whose particles measure between 10^{-5} and 10^{-7} cm, and therefore can no longer form true solution in water; on the other hand, it is not yet suspended either (precipitated). There are various colloidal substances in organisms (many proteins*, nucleic acids and polysaccharides), as well as in naturally occurring waters (eg humic substances and tannins).

Cololabis saira *see* Exocoetoidei

Colomesus FOWLER, 1911. G of Tetraodontidae*. The habitat of *Colomesus* is tropical S America, where they live in fresh water and brackish waters. They reach a size of up to 20 cm. Their shape is typical of their family. They are fairly timid fish in the aquarium and need plenty of oxygen. Nothing is known about their reproduction. The species imported most is:

– *C. psittacus* (SCHNEIDER, 1801); Parrot Pufferfish. West Indies, Amazon Basin. To 20 cm. Back olive-green with 6 black saddle markings. Stomach whitish.

Colour. In fish colour is mainly caused by colour cells (chromatophores*) lying in the skin. Within a certain range of variation, colour is typical for a species. It protects the body from being damaged by ultra-violet light particles. The importance of individual colouring is manifold. Certain markings, blotches of colour of reflective colours are signals for shoaling fish or they may help with recognition among those species that care for their young. Other colour details may indicate sex differences or the particular age of a fish. Some species have a 'nuptial dress' to indicate that they are ready for mating. Colour may also be a form of adaption*, since it may allow a fish to blend with its background. This is especially true of fish that can actually change colour. The markings on some fish sometimes work in the same way, breaking up the outline of the body. For example, the length-wise stripes of shoaling fish make it very difficult to pick out an individual within the shoal. Colour is also part of mimicry, when a creature takes on the appearance of parts of plants or some other animals. The fact that many fish larvae are colourless and transparent gives them good protection from their enemies. In nature, any freak individual that lacks pigments (*see* Albinism) is particularly vulnerable.

Colour Change. This is taken to mean both a complete and partial colour change. In fish, colour change can be a form of adaptation to the environment. For example, Flatfishes (Pleuronectiformes) and other bottom-dwelling fish can take on both the colour and structure of the substratum. A change in colour is often associated with particular behaviour, eg when a fish is frightened, when it is adopting a threat posture, or when it is ready to mate. Many species (eg *Nannostomus*, some Cichlids) change colour at night as soon as darkness falls. Colour change is regulated by the brain and is usually influenced by the organs of vision (*see* Sight, Organs of). Rapid colour change, which may take place in a matter of seconds, is due to a direct innervation of the chromatophores*. Slower changes of colour are controlled chiefly by hormones in the pituitary body of the brain. These hormones reach the chromatophores in the blood. As an example, the melanophore hormone causes the melanin pigment to be dispersed among the chromatophores, so turning the particular regions of the body dark.

Columnaris Sickness; 'Cotton-wool' Disease. Bacterial infection* of fish which is caused by *Chondrococcus columnaris* and *Cytophaga columnaris*. The visible symptoms are greyish-white film, especially in the head region but also on the fins, gills and upper body. The diagnosis can be proved only after microscopic examination. *Columnaris* is found widely and Poeciliidae* are particularly susceptible. It has been treated with varying degrees of success with Aureomycin*, Chloromycetin*, Sulphonamides* or medications containing Nitrofuran*.

Comb Pufferfish *(Carinotetraodon loreti) see Carinotetraodon*

Comb-edged or Ctenoid Scale *see* Scales

Combtail Paradisefish *(Belontia signata) see Belontia*

Comb-toothed Sharks, F, *see* Hexanchidae

Commensalism. A harmless taking advantage of the food of other animals. The Pilotfishes *(Naucrates ductor)*, for example, have a commensal association with Sharks (Carcharidae) and the Giant Devil Ray *(Manta birostris)*; they like to remain close by at all times and snatch at any bits of the Sharks' prey that falls in their direction.

Common Cuttlefish *(Sepia officinalis) see Sepia*

Common Dragonfish *(Pterois volitans) see Pterois*

Common Hatchetfish *(Gasteropelecus sternicla) see Gasteropelecus*

Common Name. The popular name given to an organism, although it is not necessarily binding, which is how it differs from the scientific name*. Common names can be very different in different languages, and even in the one language there are often several common names for the one object (eg Pearl Gourami; Lace Gourami; Mosaic Gourami; Leeri Gourami *[Trichogaster leeri]*). In aquarium-keeping, all the well-known fish and fish groups have common names, and these are usually enough to get by on for most people interested in the subject. But if scientific demands are to be met, perhaps when making observations or when anyone is concerned with the fish trade, the exact scientific names are indispensable.

Common Octopus *(Octopus vulgaris) see Octopus*

Common Oyster *(Ostrea edulis) see Ostrea*

Common Pike *(Esox lucius) see* Esocoidei

Common Pufferfish *(Tetraodon cutcutia) see Tetraodon*

Common Shore Crab *(Carcinides maenas) see Carcinides*

Common Shrimp *(Crangon crangon) see Crangon*

Common Squirrelfish *(Holocentrus rubra) see Holocentrus*

Common Starfish *(Asterias rubens) see Asterias*

Common Water Hyacinth *(Eichhornia crassipes) see Eichhornia*

Communication. Term used in ethology. In this context, the word is taken to mean contact (occasional or constant) between animals of one species (a pack or shoal, for example) and those of other species, or between one species and humans, or between sex partners for the purpose of reproduction etc. There are various *means of communication* that can come in the form of chemical, tactile, acoustic and visual stimuli. The individual sending the stimulus is known as the expedient, the one receiving the signal is known as the recipient. *Pheromones* are a particularly important *chemical* means of communication. Like the organism's hormones, pheromones act as chemical transmitters of information between the individuals. Female fish emit pheromones before and during the mating phase, thus stimulating the reproductive drive of the males and controlling their subsequent *appetitive behaviour** which promotes contact between the two individuals. Other kinds of chemical means of communication include the release of secretions from certain glands and the use of scent marks to indicate territories (by spraying trees and shrubs with dung). This is done by Rhinoceroses, for example, when they rotate their tails at the same time as they are defaecating; dogs are also doing it when they urinate against objects. Different types of *tactile* means of communication include the way in which the male and female in many species of fish touch sides during courtship swimming*; the grooming behaviour found among many species of birds and mammals; or the way in which social insects touch certain parts of each other's bodies. *Acoustic* forms of communication comprise the greatest variety of sounds and they are emitted by almost every class of animal. *Visual* means of communication are usually ways of moving that are coupled with particular patterns of shape or colour. They are a highly species-specific and extremely varied means of communication, and have the effect of reducing distance. In such a context, the ethologist will refer to affine systems of communication that serve to overcome certain ethological limitations. As imitative modes of behaviour, such communication systems are more differentiated and therefore usually effective only at even less distances. *Diffuse* means of communication on the other hand are often used as threatening gestures, and tend to increase distances. As a rule, they are more noticeable than the affine types since they are usually meant to be effective at greater distances. All types of communication are an important basis for the social behaviour* of animals.

Complete Desalination *see* Desalination

Compressors *see* Pumps

Compsopogon, G, *see* Rhodophyta

Concept of Species. A species is the unit of taxonomic classification for all forms of life. In contrast to all higher systematic categories, the concept of species reflects objective reality directly and undisputedly, since species today are defined as closed genetic systems on the basis of the fact that they can only interbreed with each other. Populations are said to belong to the same species when, because of their common ancestry, they exhibit extremely small variations in feature. Their genetics, physiology, biological behaviour and usually their morphology are so alike that under natural conditions all the members of a species can interbreed, geographical and ecological limitations apart. Within a species it can be guaranteed (at least potentially) that there will be an unlimited exchange of genetic material. Biologically, the species will become isolated from all other species – this may be manifested in cross-bred sterility*, different courtship behaviour, reproductive periods that take place at a different time or place, different demands on the biotope etc. Every species has a characteristic area of distribution (*see* Areal), depending on its historical development and the specific demands made by it on its environment. If all the populations of a species throughout the areal exhibit the same features, the species is called 'monotypic'. However, particularly in very large areals, the individual populations tend to deviate from one another, and this may lead to the formation of geographical races and subspecies within the species. Those species that are split up into races are known as 'polytypic'.

In practice, determining a species is fraught with problems. Working with museum material or allopatrically distributed populations (*see* Allopatry), it can be difficult to judge whether or not breeding communities with similar features are biologically isolated. In such cases the distinction between the species is mostly based on morphological differences; the term 'morphospecies' can here be taken to mean that doubt must remain about whether or not species is a justified description. In fact, examples are becoming more numerous where such morphospecies often encompass whole groups of biologically isolated species, eg within the Ga *Hyphessobrycon** or *Aphyosemion**. But the opposite is also true: subspecies are sometimes wrongly accorded the status of species. Even greater complications arise when races that are near neighbours interbreed freely, ie form mixed populations (a criterion that determines a species), while races that live further apart have developed a biological isolation mechanism (a criterion that determines different species). The terms 'racial group' and 'superspecies' were coined to describe this phenomenon; this state of affairs has arisen among the fish, eg *Aphyosemion* bivittatum*. A further restriction on the use of the biological concept of species can be found in the plant kingdom, where hybrids among species or even Ga are quite common in the wild. The biological concept of species can certainly not be applied to those groups of plants and animals that never reproduce bisexually. In such cases there can be no unequivocal statement about what makes up the species. The term 'agamospecies' is used when life forms are assumed to be equivalent to a biological species.

Conchostraca. Tribe of Crustacea* belonging to the Phyllopoda* (Fairy Shrimps), most nearly related to the Water-flea, but they are more primitive in many of their characteristics. Thus the body has many segments and has 10–32 pairs of legs. The Conchostraca grow up to 17 mm and are almost always completely covered by a hinged shell. They inhabit freshwater ponds which dry up at times and live there preferably in or on the bottom.

Concrete Poisoning *see* Poisoning

Concurat®; (2,3,5,6 Tetrahydro-6-phenyl-imidazo (1.1-b)thiazol-hydrochloride). A widely effective anti-helminthic medicament which can be used in fish to combat intestinal worms (Helminthiasis*). In order to treat the fish, food (mosquito larvae, shrimps and *Daphnia*) should be soaked in a solution of Concurat (1–4 g Concurat to 1 litre water) until they begin to die and then they must be fed to the fish immediately.

Conductivity. By measuring the conductivity of water, the total content of truly dissolved, electrically charged particles (ie ions*) can be determined. The ability of water to conduct current rests with this ion content, otherwise known as electrolytes (Electrolyte Content*). Conductivity is inversely proportional to the electrical resistance. The unit of measure is siemens (S) per cm. Since conductivity is altered in relation to temperature, any measurements must be taken under defined conditions. Two electrodes are suspended in the water. Alternating current is used to prevent any changes being caused to the water through electrolytical fission around the electrodes. Conductivity meters that are suitable for the aquarist's purposes can be bought from shops. Water that does not contain large amounts of electrolytes (which is the condition of the water where most freshwater species of aquarium fish live) can be measured for electrolyte content by determining its conductivity. The level of conductivity of the water is very close to the electrolyte concentration. Measuring conductivity levels is of particular value when it is important to keep a check on breeding water that has a low electrolyte content and when ion-exchangers are being used in desalination processes. The method is no good for higher electrolyte concentrations, such as those found in brackish water and sea water. As the concentration climbs higher, the conductivity is reduced again through ion hindrance. In such cases, the electrolyte content is determined with greater accuracy by means of density, using an areometer*.

Conductivity Meters *see* Conductivity

Condylactis DUCHASSAING and MICHELOTTI, 1866. G of Sea-anemones (Actiniaria*). The Condylactis are fairly large Actiniae of the littoral regions of warmer seas. They have a well-developed foot and suckers on their body column, without little sacs at the edge. They have powerful tentacles which are, however, much

Fig. 352 *Condylactis aurantiaca*

shorter than those of *Anemonia**. These brightly coloured creatures are somewhat more delicate in the aquarium than many other Sea-Anemones. The following species is often kept:

– *C. aurantiaca* (DELLE CHIAJE, 1825); Golden Sea-rose. Found only in the Mediterranean sitting firmly on stones underneath sand and gravel. The ring of tentacles is about 15 cm wide. The column is goldish-red, tentacles yellowish with violet tips. Livebearing (fig. 352).

Cone Shells, G, *see Conus*

Conflicting Drive Situation. Term used in ethology. This is a commonly observed phenomenon during which the animal suddenly changes its situation and so finds itself in the midst of two spheres of function*, although it is unable to adapt without any transition phase to the new factors in existence. It reacts to this situation with so-called displacement activity* which develops from a third sphere of function, but under the situation that exists this activity seems senseless. If, for example, a rival suddenly pops up in front of a courting male Cichlid, the Cichlid neither attacks immediately nor does it flee. Instead, it does something completely different, which at the time appears ridiculous – it starts pecking about on the substrate for food animals that are not even there. Only after this pause will the Cichlid launch into an attack on its opponent. The conflicting drive situation between the courtship and aggression spheres of function was in this case released by the detour of taking in food. During this, the action-specific energy for the courtship behaviour that was probably still in existence but which had become functionless would have been broken down.

Congo. At 4,374 km long, the Congo is the second longest river in Africa*, including the Ethiopian* region. It is also the largest in terms of volume of water. The most important source river, the Lualaba, gives the river its name up as far as Kisangani (Stanleyville). In the 350-km-long lower river section between Lake Malebo (Stanley Pool) and the rivermouth at the Atlantic, the Congo drops 265 m. This section is marked by tremendous rapids (Livingstone Falls) and it contains a very interesting fish fauna that is amazingly well adapted to very fast-flowing water. Examples include *Leptotilapia**, *Teleogramma**, *Steatocranus**. Since the Congo passes through areas of tropical rain forests, it has been fed with water continuously throughout long epochs of geological history. The rich tropical fauna contained in it could therefore develop undisturbed over long periods, a process that is almost unique in Africa in this form. It also explains why there are a great many endemic* species here that are only found in the Congo river system. Among aquarium fish this includes the above-mentioned Ga and *Lamprologus** *congolensis*.

Congo Backswimmer *(Synodontis nigriventris) see Synodontis*

Congo Climbing Perch *(Ctenopoma congicum) see Ctenopoma*

Congo Salmon *(Micralestes interruptus) see Micralestes*

Congo Tetra *(Micralestes interruptus) see Micralestes*

Congridae, F, *see* Anguilliformes

Conjugation. A method of reproduction* found in ciliates (Ciliata*). It does not result from the formation of sex cells, but from the exchange of haploid sex nuclei between two individuals lying next to one another.

Connective Tissue *see* Tissue

Conservation of Aquatic Animals. Many animals that have been kept in aquaria can be conserved for lengthy periods in a more or less life-like state, to be used as scientific exhibits, as collector's items or as part of a museum collection. It is vitally important that the animal is treated as soon after death as possible to avoid any bodily decay. Specimens that are killed specially for conserving should be treated with suitable anaesthetics* so that too strong a contraction in the conserving medium is avoided. Invertebrates such as Coelenterata and Echinodermata are particularly difficult to prepare in this way; to anaesthetise them, alcohol must be added to the aquarium water very carefully, drop by drop, or a 30%, watery magnesium chloride ($MgCl_2$) solution can be used (applied in the same way). Hours may pass before the reaction stops completely. Anaesthetised or dead animals should be placed in a good amount of conserving fluid, such as 4% formaldehyde solution or 70–80% ethyl alcohol. After some time has elapsed, the solution must be changed completely at least once. With larger types of fish, the conserving fluid will have to be injected into the body as well – into the body cavity, via the anus into the intestine, and into the oral cavity. Conserved specimens can be transported either in the conserving fluid itself or in cloths that have been soaked in the fluid. But whichever method of transport is used, it is essential that the container into which the specimen is put is absolutely airtight. Fish lose their colouring quickly when placed in conserving fluid. There are a few simple ways of slowing down this process: after the fish has been anaesthetised, keep it for 10 days in a mixture of 200 ml formaldehyde (35–40%), 1,000 ml distilled water, 15 g potassium nitrate (KNO_3) and 30 g potassium acetate (CH_3COOK). After that, to keep the colours at their best, transfer the fish for a maximum of 5 hours to a 70–80% ethyl alcohol solution. The final step in the preservation of the specimen uses 200 ml distilled water, 200 g potassium acetate, 400 ml glycerin and some crystals of thymol. For specimens that are to be exhibited, there are a great many different ways of conserving them, for example, freeze-drying, encasing in artificial resins or making casts. But all such techniques require highly specialised knowledge and equipment. Every conserved animal must be furnished, with exact details of species, origin, habitat and time caught, as well as notes about its colour, age, weight and sex. If such information is not supplied along with a conserved specimen, it is useless.

Consummatory Act. Term used in ethology. The consummatory act is usually introduced by *appetitive behaviour**. *Key stimuli** then bring about the release of stored specific-reaction energy which will satisfy the drive to complete the consummatory act. A period of relative quiet follows the act. Eg, courtship swimming* whereby the male fish leads the female to the spawning ground and the action of seeking a suitable place for spawning by the females are both examples of appetitive behaviour. The subsequent mating and the simultaneous release of the sexual products make up the consummatory act, brought on by the female adopting a mating position (the key stimulus).

Continual Spawner *see* Breeding Season

Controlled Breeding. The formation of new organisms under the influence of Man. Certain features of plants, animals or micro-organisms are developed further according to a set plan or are newly created in order to achieve a particular aim with breeding. Genetics* provides the scientific basis of selective breeding.

The oldest kind of artificial breeding is *selective breeding*, in which over many generations the most suitable individuals are selected for breeding. Gradually, groups of organisms develop that differ clearly from the original form and that correspond to the aim in mind. In order to obtain from them genetically and externally uniform (homozygous) lines, sorts of plants or animal races, a consecutive mating of very closely related individuals (In-breeding*) is necessary. When artificially breeding aquarium fish too (eg the hybrids of the Guppy, Swordtail, Black Molly or the breeding of Veil-tail forms), selective breeding (which may last years) in conjunction with consequent in-breeding is the most important method. *Combination breeding* is also important. By crossing unrelated breeding strains, new combinations of hereditary factors can be achieved that will form the basis of new features. Here too, the material must be made genetically uniform by subsequent selection and in-breeding to consolidate the course of inheritance and to prevent the new features dispersing once again (Heredity*). *Mutation breeding* is based on the alteration of the hereditary substance (Mutation*) through experimental intervention. In order to increase the frequency of hereditary changes, various treatments with mutagenes are carried out in the laboratory. (Examples of mutagenes are X-rays, radioactive rays or certain chemicals.) From a large number of offspring that deviate in some way, there will also be individuals with a positive selective value, ie those that have new, usable features that allow them to be used as the original form for selective and combination breeding.

Conus LINNAEUS, 1758; Cone Shell. G of Prosobranchia*. Found chiefly in warmer seas. The snails live as predators on coral and rocky coasts and are easily recognisable by the elongated conical shell. The mouth opening extends almost the whole length of this. The radula has become pointed darts with which the snails stab their prey. At the same time they discharge a poison which immediately paralyses small creatures. Tropical species are also of danger to man. *C. mediterraneus* BRUGUIÈRE from the Mediterranean is harmless.

Convergence (fig. 353). Among plants and animals it has been noticed that organisms with a different phylogenetic history will nevertheless adapt to a similar environment in similar ways. Whales, turtles and penguins, for example, all of which had land-dwelling ancestors, have formed very similar fins from their forefathers' legs or wings. Mammals that live in the water have a body that is once more becoming like that of a fish (in shape and in the kind of currents it creates). The formation of analogous organs that fulfil the same functions despite having different origins is also a form of convergence.

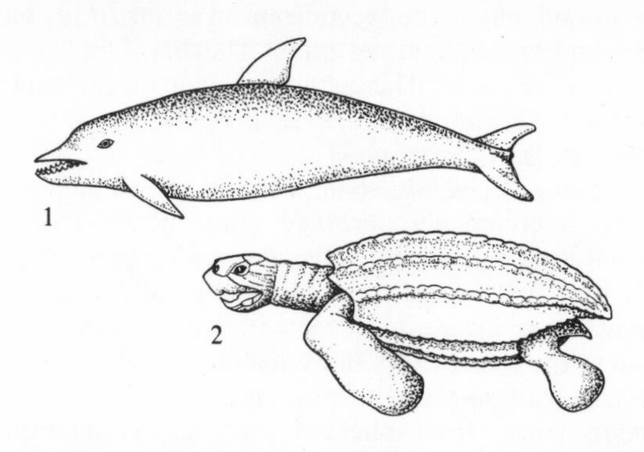

Fig. 353 Convergence. Formation of fin-like extremities in the habitat water 1 Mammal–Dolphin *(Delphinus delphis)* 2 Reptile–Leather-backed Turtle *(Dermatochelys coriacea)*

Convict Cichlid *(Cichlasoma nigrofasciatum) see Cichlasoma*

Coolie Loach *(Acanthophthalmus cuneovirgatus) see Acanthophthalmus*

Coolie Loach *(Acanthophthalmus kuhli kuhli) see Acanthophthalmus*

Coolie Loaches, G, *see Acanthophthalmus*

Cope, Edward Drinker (1840–97). North American palaeontologist and ichthyologist. Born in Philadelphia (USA). He studied medicine and was Professor of Natural History at Haverford College in Pennsylvania and later Professor of Comparative Anatomy and Zoology at the University of Philadelphia. He made countless geological and fossil studies in N and S America. His extensive collections of fossil vertebrates include more than 1,000 unknown species, families and orders. One of the most important books on the subject was published in 1882 in three volumes *The Vertebrata of the Cretaceous Formation of the Palaeozoic and Mesozoic Formations and of the Tertiary Formations.* His publications about the phylogenetic connections between amphibia, reptiles and mammals and primary factors of evolution are also worth mentioning. As an ichthyologist he particularised countless species of fish of southern N America and S America for the first time.

Copeina FOWLER, 1906. G of Lebiasinidae*. Sub-F Pyrrhulininae. Found in the tributaries of the central Amazon. Characin living in the middle and lower water levels. 8–15 cm long. Body slim to compact. All the fins are fairly small and short, D2 absent, C forked. According to GÉRY, the G *Copeina* only includes the species *C. guttata*, according to MYERS (from information in letters) it probably also includes *C. arnoldi* REGAN 1912. The latter is, however, probably not identical to the well-known Spraying Characin, as generally supposed up to now, but a species which has not yet been imported. The Spraying Characin itself perhaps belongs to the G *Copella**, and there are further particulars under that entry. *C. guttata* spawns in shallow pits in the sand which they make by circling round and

shoving and pushing at the bottom or occasionally by fanning movements. When they mate, the male first rides on the female then glides sidewards near the female and pushes the anal fin sac under the genital opening, catches the spawn with it, fertilises it and drops it into the pit. Finally the male fans the eggs and guards the territory until the young fish hatch and swim freely after 24–26 hours. They are easy to rear with *Cyclops* nauplii and then with small *Cyclops*. When attempting to breed them keep them in soft, slightly acidic water.
– *C. guttata* (STEINDACHNER, 1875); Red-Spotted Copeina. Middle Amazon and tributaries. To 8 cm. Back of the males brownish green, body sides with light blue to greenish-blue sheen, stomach white. The red to violet spots at the base of each scale give the appearance of several rows of spots. D yellowish with black spot. C, An and Vs yellow with orange edges. The females are generally paler. They are interesting and robust aquarium fish (fig. 354).

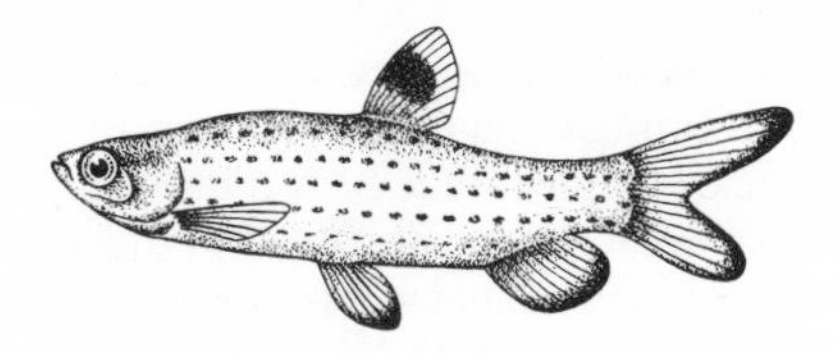

Fig. 354 *Copeina guttata*

Copeina arnoldi, name not in current usage, *see Copella*

Copelata, Cl, *see* Tunicata

Copella MYERS, 1956. G of Lebiasinidae*, Sub-F Pyrrhulininae. Amazon basin. S American Characin with many species. 5–8 cm long. Clear dimorphism between the sexes in both size and fin arrangement. The species here were taken by Myers from the Ga *Copeina** and *Pyrrhulina** (because of their divergent form of upper jaw) and united in the newly formed G *Copella*. The upper jaw bone of the males has two semi-circular curvatures sticking out towards the front, which are lacking in specimens of the Ga *Copeina* and *Pyrrhulina**. This characteristic is also absent in females of the G. The well-known Spraying Characin probably belongs to this G and, as Myers intimated in his letters, is presumed not to be one and the same as *Copeina arnoldi* REGAN 1912 as was previously believed. In his opinion the Spraying Characin is a species of *Copella* which is very like or even identical to *C. compta* MYERS, 1927, and *C. vilmae* GÉRY, 1963. The Pyrrhulininae are in urgent need of a thorough revision, as many species are wrongly designated. These elegant creatures have an elongated body shape which is hardly compressed laterally. The head is flattened at the top and the mouth is slightly superior. All fins are well developed. D, An, Vs and the upper lobe of the C are much elongated in the male and substantially shorter in the female. The *Copella* need slightly acidic water of 2–6° dH and a temperature of 25–27 °C. They need a spacious aquarium well supplied with large-leaved species of plants which reach up to the water surface. They are timid when they are kept somewhere which is too bright, they do not show their most brilliant colouring and they hide themselves away and do not spawn. They eat prey that falls onto the surface and are particularly fond of small insects. They should be kept only with other peaceful fish or best of all only with fish like them. Apart from the Spraying Characin all the species kept so far spawn on large horizontal leaves in the water, such as those of *Echinodorus, Aponogeton* and also leaves which have come off the *Aspidistra*. The male first cleans the spawning ground for hours by moving his body and fanning with his fins. Then the mate or mates appear on the leaf ready for mating. Using his anal fin sac the male catches the 6–10 eggs which she releases, fertilises them and finally drops them onto the leaf. There are about 300 eggs in total which cling well to the leaf. Until the young hatch the male fans them and he also defends the breeding territory. The young fish remain just under the water surface once they have hatched and should be fed rotifers in the first few days and then *Cyclops* nauplii. They are not very easy to rear and one must give them just the right amount of food and know exactly when to change the water. The brood also hatches in medium-hard water. The young fish grow relatively slowly. The reproduction of *C. metae*, *C. nattereri* and *C.* sp. follows a similar pattern. The Spraying Characin, on the other hand, spawns on the underside of leaves growing horizontally from 4–7 cm over the water surface. In order to do this, the male and female rise vertically together through the water to the surface, bend their tail stalks and jump together to the spawning place above the water surface. 5–10 eggs are laid and fertilised each time until several hundred eggs have been laid. In the aquarium, the eggs are usually stuck to the underside of the aquarium cover and the distance between the surface and the cover should thus not be too great. The male then sprays the eggs with a drop of water at regular intervals using his caudal fin. The young, when hatched, fall into the water when sprayed by several drops one after another. All species of the G *Copella* are ideal aquarium fish although they are unfortunately somewhat problematic at times.
– *'C. arnoldi'* Spraying Characin. Lower Amazon, Rio Para. Male to 8 cm. Female to 6 cm. This species was previously wrongly known as *Copeina* or *Copella ar-*

Fig. 355 *'Copella arnoldi'*; Spraying Characin

noldi (see above). The male is very brightly coloured and has a shark-like appearance. The body sides of the male are yellowish with dark edges to their scales, becoming very dark with silver spots during excitement when fighting or mating. The gill-cover has a green-gold spot. They have a black band across their snout, extending from eye to eye. The long extended triangular D is yellow with a black triangular spot which is white at the base, the first fin ray and the tip are red. All the other fins are greatly elongated and yellow red. The lobes of the C have blood red tips, the upper lobe is longer than the lower one. The females are more plainly coloured and have no extended fins. The Spraying Characin is one of the best loved and most elegant of aquarium fish (fig. 355).

Fig. 356 *Copella* sp.

– *C.* sp. Upper Amazon, for a wide distance round Iguitos and Requena. To 8 cm. A decorative species which is probably still not properly defined. The back is fawn-coloured, body sides silverish with a touch of brown or blue. A wide blue-black zig-zag longitudinal band extends to the tip of the mouth through the eye to the base of the C where it finishes in a black spot. The fins of the males are extended as is typical of this G. The females are altogether paler. They have already been successfully bred (fig. 356).

– *C. metae* (EIGENMANN, 1914); Black-banded 'Pyrrhulina', Red-dotted 'Pyrrhulina'. Peruvian Amazon. Rio Meta. To 6 cm. This species was previously known as *Pyrrhulina nigrofasciata* MEINKEN, 1952. Back of male fawn-coloured, body sides brownish, stomach section whitish. A coffee-brown broad longitudinal band extends from the tip of the mouth to the start of the C. This can at times be paler and give place to a zig-zag band, since the dark rows of scales in this area are then visible. However, according to FRANKE who bred both species, the *Copella metae* are in no way interchangeable with the *Copella* sp. as the zig-zag band is much broader and deeper in colour in the latter and is always visible. The dark longitudinal band is bordered by a cream band at the top. There is also a blood red spot on every scale. The reddish brown fins are not quite so extended in the males as they are in the other species. The females are altogether paler and have short fins. It is a splendid species but unfortunately rather delicate (fig. 357).

– *C. nattereri* (STEINDACHNER, 1875); Blue-spot Copella, Natterer's 'Pyrrhulina'. Central Amazon, Rio Negro? To 6 cm. Body sides olive-yellow, each scale has blue-brown spot at the bottom. A dark stripe runs from the tip of the snout through the eye to the edge of the gill-cover. The back is a pale olive-brown and underneath runs a cream-coloured longitudinal band which is particularly marked in the females. The fins are a pale yellow. The D has a dark spot in the lower third and in the male a bright red spot is visible under this. The fins of the male are longer than those of the female. This species has also already been bred. According to GÉRY and MYERS, *Copella callolepis* (REGAN, 1912) is the same species as this (fig. 358).

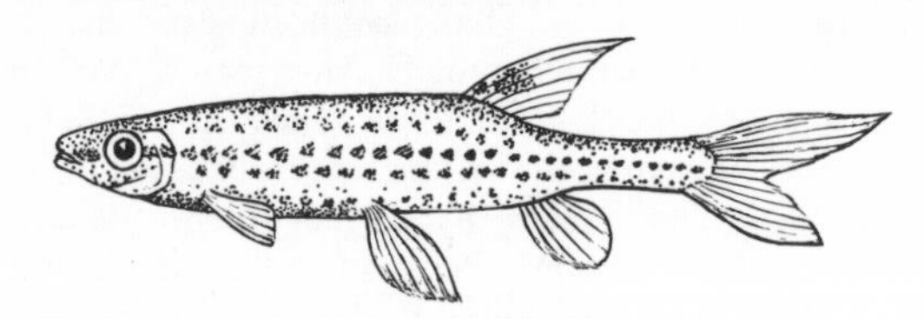

Fig. 358 *Copella nattereri*

– *C. vilmae* GÉRY, 1969; Rainbow Copella. Upper Amazon near Leticia. To 6 cm. Body sides pale olive-yellow each scale with dark edge and light brown spot on the base. Back gold-brown with a brown band occasionally visible along the body sides from the mouth to the caudal peduncle. The fins are pale yellow, D has bright red base and a large black spot. An and C bordered with red. All fins on the male are greatly extended as is typical of this G. According to the report of

Fig. 357 *Copella metae*

HARALD SCHULTZ when he first discovered the fish, it had a red gleam on the body and fins, but unfortunately this red colouration soon disappeared completely on the imported fish and did not return. This extremely elegant species should be very popular once breeding has been successful (fig. 359).

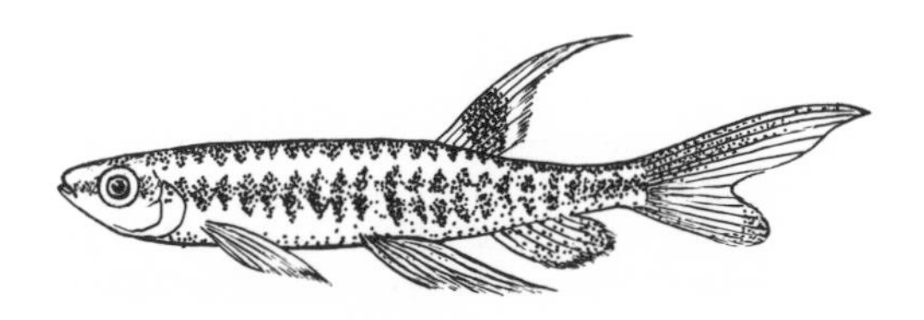

Fig. 359 *Copella vilmae*

Copepod Infestation (fig. 360). Parasitic Copepods only rarely affect aquarium fish. It is usually exclusively the females which are suited to the parasitic life in the skin or on the gills of fish. The development of clasping apparatus and organs in the mouth for taking in brood or tissue fluid is characteristic. Individual parasites usually remain unnoticed and only a strong infestation leads to the emaciation* of the affected fish. Examples of the Ergasilidae are gill parasites 0.5–2 mm in length to 1 mm long egg-sacks and these are visible to the naked eye. Herrings and Tenches are particularly prone to infestation by the females of *Ergasilus sieboldii* NORDMANN, 1832, whilst tropical ornamental fish are particularly susceptible to *Ergasilus boettgeri*

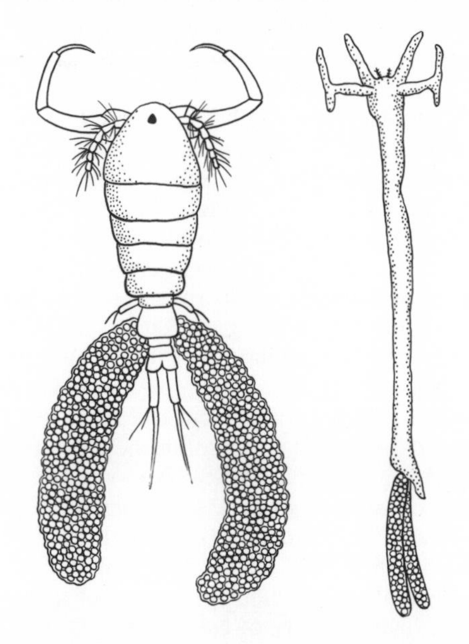

Fig. 360 Parasitic Copepods. Left, *Ergasilus*; right, *Lernaea*

REICHENBACH-KLINKE, 1958. An extreme modification of the body shape is noticeable in species of the G *Lernaea*. These elongated creatures grow to 22 mm long and are anchored by the head end in the musculature of the fish whilst the rear end with its egg sacks projects out of the fish. Development takes place without change of host via a nauplius larva (Copepoda*).

Parasitic Copepods are also found on sea fish especially on imported animals. The species of the G *Caligus* should be mentioned which infest the oral cavity and also those of the G *Sphyrion* which reach a length of up to 60 mm. Where fish are badly infested, short baths in potassium permanganate*, sodium chloride* Lindan* and Formalin* are recommended. The last two are also suitable for treating marine fish.

Copepoda; Copepods (fig. 361). O of Crustaceans (Crustacea*) with many species and forms found all over the world. The original and principal habitat of the Copepoda is the sea, although many species have become true freshwater dwellers in the course of development. The typical Copepod has a body of a few

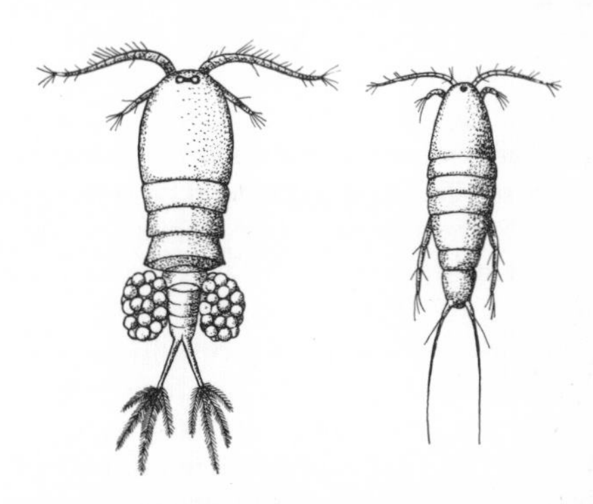

Fig. 361 Various Copepods. Left, *Cyclops* sp. (female); right, *Canthocamptus* sp. (male)

mm long, which consists of a pear or egg-shaped cephalothorax and a narrow abdomen which ends in a fork (furca). The most striking of their parts is the first pair of feelers (antennulae) with which the characteristic hopping swimming movement, is achieved. The simply formed nauplius eye functions as a photo-sensitive organ. The two egg-sacks on the abdomen are typical of the female, a characteristic which allows even greatly modified forms to be identified as belonging to the Copepoda. The larval stage of the Copepods is the nauplius*, a larval form with 3 pairs of limbs. The Copepoda are grouped as free-living or parasitic forms according to their way of life. The latter live on fish and molluscs (Mollusca*) and can do considerable damage. There are also predators amongst the independently living Copepoda. These even attack creatures which are bigger than themselves, such as a brood of fish. For this reason great care should be taken when using Copepods as food for rearing fish. In this case only the nauplii are not dangerous and thus well suited for this purpose. Apart form this limitation the Copepoda form a considerable part of natural fish food and thus are amongst the creatures most used by aquarists as food – namely the species of the Ga *Diaptomus** and *Cyclops**.

Copepods, O, *see* Copepoda

Copper Sulphate ($CuSO_4 \cdot 5H_2O$). Blue copper sulphate crystals dissolve easily in water. The copper ions that result are very poisonous to many organisms which is why copper sulphate is used in the treatment of fish parasites. Invertebrate animals and algae suffer most damage, but so too do the higher water plants. In sea water and in fresh water, copper ions are precipitated as insoluble compounds after a fairly short time, so that it is difficult to maintain the necessary concentration of ions that is effective against the parasites. A number of different factors contribute to the precipitation of the copper ion compounds: the salt content of the water, its pH value, the carbonic acid content, the amount of magnesium salts etc. In aquarium tanks, under certain circumstances, it is possible for the precipitated copper compounds to go back into solution again, in which state they may produce uncontrollable reactions and effects upon the physiological equilibrium in the tank. Because of these reasons, it is best to treat infected fish in separate tanks. The copper concentration can then be tested every so often to make sure that it is the correct strength and, if necessary, a second dose can be added. Fish that have been attacked by Saprolegniacea*, *Dactylogyrus* or *Gyrodactylus* (*see* Trematodes) should be treated in short-term baths (Bathing Treatments*) for 10–30 mins. The concentration needed is 1 g to 10 litres of water. To combat *Oodinium**, prolonged bathing treatments are needed that last for 3–10 days. Use a concentration of 1.5 mg to 1 litre of water. To make sure the dosage is correct, make up the following quantity: 1 g copper sulphate ($CuSO_4 \cdot 5H_2O$) to be dissolved in 1 litre water, and, if necessary, add 60 mg crystalline citric acid. For a prolonged bath, you will need 1.5 ml of this solution for every 1 litre of water in the treatment tank; for short-term baths, 100 ml of the solution for every 1 litre of water. For further doses in fresh sea water, the following rule of thumb can apply: After every 24, 32 and 48 hours, the treatment water should be topped up with 0.4 ml of the original solution for every 1 litre of water in the tank. After 3 days, the treatment tank must be reset-up. Copper sulphate is one of the most important disinfectants used in ponds (*see* Disinfection).

Copper-banded Butterfly-fish *(Chelmon rostratus) see Chelmon*

Copper-striped Rasbora *(Rasbora leptosoma) see Rasbora*

Copulatory Organs *see* Sex Organs

Coral Butterfly-fishes, G, *see Chaetodon*

Coral Catfishes, F, *see* Plotosidae

Coral Fish Disease *see Oodinium*

Coral Fishes (collective term) *see* Coral Reefs

Coral Formations *see* Decorative Material

Coral Gobies, C, *see Gobiodon*

Coral Reefs. Enormous animal structures found in tropical seas and made up mainly of the limy skeletons of polyps (Madreporaria*), but also including a great many other organisms such as Fire Corals (Milleporidae*), Organ Pipe Corals (Tubipora*), Tube Worms (*see* Polychaeta), calcareous algae etc. There are various types of reef: coastal reefs lie directly alongside rocky coastlines; barrier reefs rise like an embankment out of the sea just offshore; and finally ring-shaped atoll reefs. Barrier reefs and atolls enclose an area of still water known as a lagoon. The lagoon, the inner side of the reef, the reef plate near the surface and the often very steep seaward incline of the reef abound with a whole array of living communities. Coral reefs form best at a temperature of 25–29 °C, the absolute lowest limit being 20 °C. They also require strong sunlight. Because the growth of coral is to a large extent dependent on the zooxanthellae* found there, which are dependent on light penetrating through, a reef can only extend to a depth of about 60 m. Another factor which influences the growth of a coral reef is the purity of the sea water; a reef will not flourish near where a river enters the sea, for example. In the history of our planet, reefs have led to the formation of islands and calcareous mountain chains. One of the largest areas of reef formation in the world today is the Great Barrier Reef off Australia's east coast, which covers an area of 260,000 km^2. The chief threat to coral reefs is sea pollution, but in some places they are also under threat from collectors and other people who raid them for aquarium-keeping purposes. The plague of coral-eating Crown of Thorn starfishes *(Acanthaster*)* in the Indo-Pacific is probably caused by human activity in the first instance. Under natural, healthy conditions, the only fishes that eat coral are some Parrot-fishes (Scaridae*) and Pufferfishes (Tetraodontidae*), and the amount they consume will not harm the coral.

Tropical coral reefs abound with the greatest variety of living communities. In aquarium-keeping, the term 'Coral Fishes' includes every type of colourful tropical marine fish. But in the ecological sense the term should really only be used for those fishes that actually live in close association with coral reefs. Coral fishes belong to a very wide range of Fs. There is an especially large number of reef-dwellers among the Damselfishes (Pomacentridae*), the Butterfly-fishes (Chaetodontidae*), the Wrasses (Labridae, Sub-O Labroidei*), the Trigger-fishes (Balistidae*), the Trunkfishes (Ostraciontidae*), the Surgeonfishes (Acanthuridae, Sub-O Acanthuroidei*), the Cardinal-fishes (Apogonidae*), the Soldier-fishes (Holocentridae, O Beryciformes*), the Porcupine-fishes (Diodontidae*), the Sea Perches (Serranidae*), the Scorpion-fishes (Scorpaenidae, Sub-O Scorpaenoidei*), the Hawkfish (Cirrhitidae*) and the Morays (Muraenidae, O Anguilliformes*). The number of different invertebrates among the coral reefs is almost indescribably large. Apart from the main reef-builders that we have already mentioned, there are innumerable Sponges (Porifera*), Sea-anemones (Actiniaria*), Horny Corals (Gorgonaria*), Leathery or Soft Corals (Alcyonaria*), Anemones (Zoantharia*), brilliantly coloured Flatworms (*see* Polycladia), Bristle Worms (Polychaeta*) with colourful circlets of tentacles, an extraordinary variety of crustaceans (Crustacea*), Molluscs (Mollusca*), Starfishes (Asteroidea*), Brittle-stars (Ophiuroidea*), Sea Urchins (Echinoidea*), Sea Cucumbers (Holothuroidea*), graceful Sea-lilies (Crinoidea*), Bryozoans (Bryozoa*) and Tunicates (Tunicata*).

The relationships between the animals that live on the coral are also astonishingly varied. Some can be observed in the aquarium. A few examples that spring immediately to mind are the actions of Cleaner Fishes *(Labroides*, Gobiosoma*)* and Cleaner Shrimps *(Hippolysmata*, Stenopus*)* as well as Anemonefishes*. With today's technology, keeping coral fishes in the aquarium presents no particular problems. The food specialists, such as many Butterfly-fishes, can be a little tricky. Another particular problem related to coral fishes is that many of the very colourful species are intolerant of like species. This is because they form territories, their poster-colour markings to a certain extent acting as a demarcation. So, in a tank of normal proportions it is not possible to keep 2 specimens of the same species, instead examples of various species can be kept together. The reef-dwelling invertebrates can provide great pleasure for the aquarist. However, since it would be impossible to give information regarding their upkeep that would be generally valid, it is best to look for details under the separate entries for the different animal groups. Live reef-building coral can also be kept in the aquarium (*see* Madreporaria). If a reef is going to be formed, however, you will first of all have to provide coral skeletons. Before using them, bathe them for several days in diluted potash or soda lye (about 300 ml 15% lye to 10 litres water – handle with care as it is corrosive); then rinse them thoroughly, brush them down, and finally boil them for half an hour in tap water. In this way, the skeleton will end up completely free of organic remains. The gleaming white of the cleaned coral is completely unnatural. Before long, however, algae will begin to settle on it, and then the clumps of coral will better resemble the brown, green or red colouring of live coral which is caused by the polyp layer. Although every effort is made to achieve a natural effect, the first priority is to have a clearly arranged decoration which can be dismantled if necessary. Compact corals are more practical than the finely branched forms which get dirty all too easily. Coral sand is highly recommended as a substrate and filter medium. For the aquarist, keeping organisms that live on coral reefs is one of his most delightful pursuits.

Coral Sand *see* Substrate

Coral Skeleton *see* Coral Reefs

Coral Trout *(Cephalopholis miniatus) see Cephalopholis*

Coralfish *(Abudefduf saxatilis) see Abudefduf*

Corallium LAMARCK, 1801. G of Alcyonaria*. Horny Corals. Various species found in warmer seas. The elements of the skeleton are cemented to a firm central axis by calcite which is covered by a soft crust out of which the polyps project. The shrub-like colonies and the skeleton are red, yellow or white in colour. Only one species lives in European seas which can be kept in the aquarium if fed with small crustaceans, providing it is protected from being covered with Algae. This species is:

– *C. rubrum* (LINNAEUS, 1758); Precious Coral. Found in Mediterranean in depths of 30–200 m. Colonies of up to 1 m. Blood red with white polyps. The skeleton is used to make jewellery.

Coregonidae, F, name not in current usage, *see* Salmonoidei

Coregonus (various species) *see* Salmonoidei

Corethra MEIGEN, 1803; Phantom Midges (fig. 362). G of Mosquitoes (Nematocera*). Either grouped in their own family or added to the Culicidae. The larvae of the *Corethra* are an important fish food. Characteristic qualities of these larvae are their glassy transparency and their horizontal hovering in the water whereas the pupae stand vertically. The stages mentioned in contrast to other species of mosquitos *(Aedes*, Anopheles*, Culex*)* can breathe through the skin and thus are not dependent on the water surface. The preferred habitat

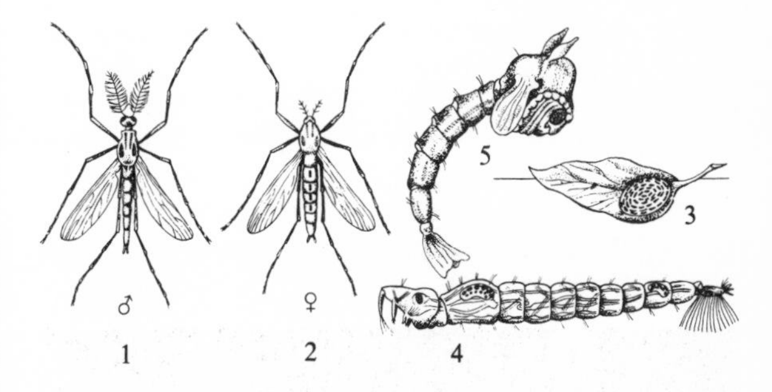

Fig. 362 *Corethra.* 1, 2 Imagines 3 Clutch of eggs 4 Larva 5 Pupa

of *Corethra* is clear still waters where the larvae can stop in open water and lay in wait for their prey. It must also be mentioned that the imagines do not bite and that the *Corethra* larvae are most often to be found in pools surrounded by birch trees.

Coris LACÉPÈDE, 1802; Wrasses. G of Labridae. Sub-O Labroidei*. Found in tropical and subtropical Indo-Pacific in the Mediterranean and Atlantic. The *Coris* species live alone or in loose groups in the littoral*. They feed on Crabs, Molluscs and Echinoderms which they crunch with their powerful teeth. These lively fish search for food all day long. At night and when in danger they often bury themselves in sand. Most grow to 20–30 cm, *C. angulata* to 120 cm. The *Coris* species have a slim, almost cylindrical, pliable body and a pointed head. They are all magnificently coloured. Young and old fish are usually so different from each other in colouring that many were at first described as separate species. This is how the species *C. julis* and *C. giofredi* from the Mediterranean were shown to be one and the same. They first pass through a female phase in the plain 'giofredi'-garb and then change to the brilliant coloured 'julis' males. The picture is even more complicated by the fact that some of the 'giofredi' are mature males and there are a few females amongst the 'julis'. The *Coris* species do not practise brood-care; their eggs are pelagic. These attractive fish are good to keep in the aquarium, although often quarrelsome with members of their own species. They need aquaria which are not too small with a layer of sand for them to dig into. *Coris julis* generally disappears into the bottom for quite a while after being put into the aquarium. The following species are kept particularly

– *C. angulata* LACÉPÈDE, 1802; Clown Labrid, Twinspot Wrasse, Red-throated Rainbow Wrasse, Orangespot Wrasse. Red Sea. To 120 c. Found on sandy surface near reefs. Lives alone. Younger fish are grey-white with 2 gleaming orange spots on the back. As it grows it soon becomes too large for domestic aquaria.

– *C. formosa* (BENNETT, 1834); Yellow-tail Wrasse; African Clown Wrasse. Indo-Pacific on sandy surfaces near reefs. To 40 cm. Juveniles dark red with 5 white, black-edged spots on the head and back. The third spot lies behind the Ps, and runs down to the underside. D is between the 3rd and 4th bands and has a black spot. The adult fish do not have these spots and are brownish-violet with black patches and 2 blue-green slanting stripes on the gill covers.

– *C. gaimardi* (QUOY and GAIMARD, 1824); Gaimard's Rainbowfish, Gaimardi, Tomato Wrasse, Red Labrid. Indo-Pacific. The young are like *C. formosa* but without the black patch on D and none of the white patches reaches the belly. The older fish are brown-violet with blue patching (fig. 363).

Fig. 363 *Coris gaimardi*

– *C. julis* (LINNAEUS, 1758); Rainbow Wrasse. Mediterranean, E Atlantic on rocky coasts. To 25 cm. Young fish and females have dark brown back and yellow white underside which are quite sharply divided. Mature males have green back, orange longitudinal band under which there is a black stripe on the front of the body (fig. 364).

Fig. 364 *Coris julis*

Corixa MÜLLER, 1764; Lesser Water Boatmen. They make chirping noises, so are sometimes called Water Cicadas (fig. 365). G of Heteroptera*. Many related and similar forms appear in large numbers in waters of very varying types, even in brackish waters. They are flat, boat-shaped, and clamp themselves firmly to water plants with the long claws of their middle leg and thus counteract the lift to which they are constantly subject through the supply of atmospheric air under the wings

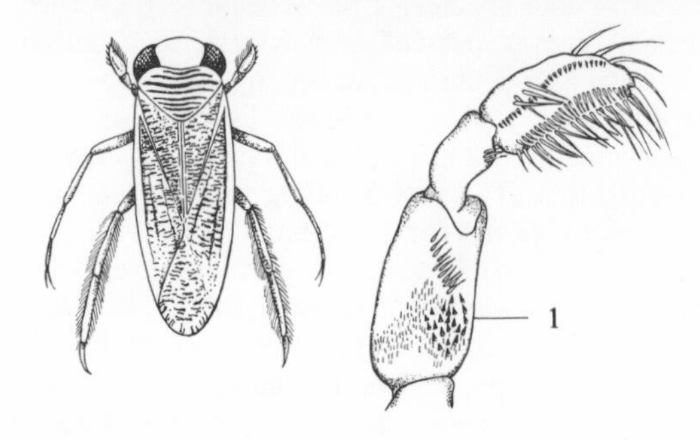

Fig. 365 Examples of the F Corixidae, *Sigara hieroglyphica*. Right, front leg with spiny-regions (1) for producing sounds

and in their stomach. The *Corixa* species rummage through the bottom for food using their short shovel-like front legs. In contrast to other Water-fleas, they eat mainly vegetable matter or dead particles of dead matter (Detritus). *Corixa* often gets into the net when one is looking through ponds or catching food and it can be kept as a study object in a pond aquarium.

Corm (fig. 366). A multi-celled vegetative body which is made up of stem axes*, leaves* and roots*. In many aquatic plants there may be no roots. A corm is characteristic of the asexual generation of pteridophytes (Pteridophyta*) and of seed plants (Spermatophyta*). Opposite to protophyte* and thallus*.

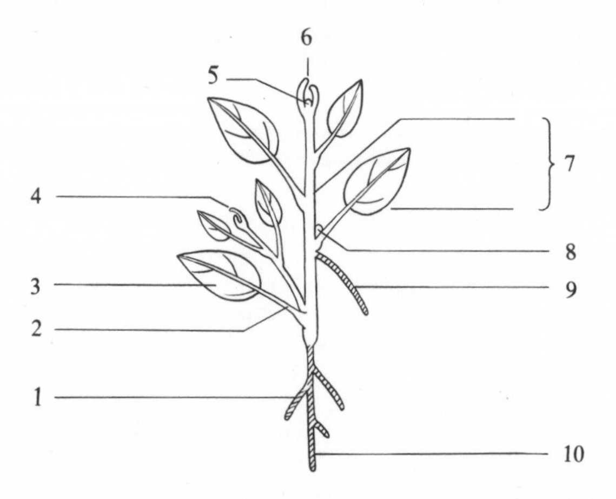

Fig. 366 Corm. 1 Lateral root 2 Leaf-stalk (petiole) 3 Leaf-blade 4 Lateral shoot 5 Vegetative cone 6 Main stem 7 Shoot with stem axis and leaf 8 Bud 9 Stem-born root 10 Main root

Cornetfishes (Fistulariidae, F) *see* Aulostomoidei

Corolla *see* Flower

Corumba Dwarf Cichlid *(Apistogramma commbrae) see Apistogramma*

Corvina CUVIER, 1829. G of Sciaenidae*. The main distribution area of this G is the tropical and subtropical Pacific. Some live in warmer parts of the Atlantic or in the Mediterranean and even penetrate freshwater areas. These peaceful fish mostly stay near cliffs, reefs and

caves. They eat crustaceans, molluscs and other small creatures. They are hardly ever any bigger than 30 cm. Some are even smaller. They are of typical perch shape and are mostly dusky coloured. Some are of economic importance. They keep well in an aquarium, eg:
– *C. nigra* (BLOCH); Corb. Mediterranean. E Atlantic. To 40 cm. Grey-brown with dark fins.

Corydoras LACÉPÈDE, 1803. G of Callichthyidae*. Catfishes found widely in tropical S America, 2.5–12 cm but mostly 5–7 cm in size, with a very uniform shape. About 95 species have been noted so far. They are bottom-dwelling fish which live in small shoals in shallow water and search the bottom for food. The end gut, which serves as an additional respiratory organ, enables them to take in and use atmospheric air and thus allows them to exist in very dirty waters. When kept in an aquarium they are contented fish which eat anything and will live for years but one must make allowance for their need of water agitation by good filtering. They need frequent changes of water and clear water which is free of infusoria. They do not need any special water composition and can also be bred successfully in medium hard water. All species are compact, fairly high-backed and laterally compressed. The line of the back is always more arched than that of the stomach. Characteristic of them is the lack of the bony plates on the swell of the back between D1 and D2 which is particular to the *Callichthys** species. The missing bony shield on the snout and the short D divide them clearly from members of the G *Brochis**. The body is covered at the side by bony plates arranged in two rows. Both D1 and D2 have a moveable spine. The C is deeply forked, the An relatively short, Ps and Vs well developed. The C has strong spines for support. 2 pairs of barbels of equal length are found on the lower and upper jaw. The countless species can be grouped according to their needs when being kept. The first group includes those which are standard species for aquarists because they are particularly easy to keep and breed. These are *Corydoras aeneus* (GILL, 1858), *Corydoras elegans* STEINDACHNER, 1877, *Corydoras eques* STEINDACHNER, 1877, *Corydoras macropterus* Regan, 1913, *Corydoras nattereri* STEINDACHNER, 1877, *Corydoras paleatus* (JENYNS, 1842), *Corydoras schultzei* HOLLY, 1940, and the like. They are unfortunately the least strikingly coloured, mostly of greenish or metallic colouration with a dark cloud-like spot or patchy markings and monotone grey fins. They are happy at temperatures of 20–24 °C and reproduce quite easily. Another group includes species which have a contrasting black spot or stripe markings (or both) on a silvery background. The fins are usually transparently colourless. The D1 has a large dark patch, which often extends to the rear section. D2, An and particularly the C are mostly covered with several dark, vertical chain-like rows of spots. These species need temperatures around 25–28 °C. They have either not been successfully bred or only erratically and mostly they are unproductive. The following species are included amongst these: *Corydoras agassizi* STEINDACHNER, 1877, *Corydoras caquetae* (FOWLER, 1943), *Corydoras caudimaculatus* RÖSSEL, 1941, *Corydoras haraldschultzi* KNAACK, 1962, *Corydoras julii* STEINDACHNER, 1906, *Corydoras melanistius* REGAN, 1912, *Corydoras punctatus* (BLOCH, 1794), *Corydoras reticulatus* FRASER-BRUNNER, 1938, *Corydoras sterbai* KNAACK, 1962, *Corydoras trilineatus* COPE, 1872, amongst others. The third group has one or several dark longitudinal bands in the upper half of the body on a silver white, yellowish or orange background and mostly they have fins which have no patterning or colour. These are *Corydoras arcuatus* ELWIN, 1939, *Corydoras axelrodi* RÖSSEL, 1962, *Corydoras boesemani*, NIESSEN and ISBRÜCKER, 1967, *Corydoras metae* EIGENMANN, 1914, *Corydoras panda* NIJSSEN and ISBRÜCKER 1917, *Corydoras rabauti*, LA MONTE, 1941, *Corydoras schwartzi* GÉRY, 1966, amongst others. Finally one can group together the very small species which partly swim freely in the central water levels: *Corydoras cochui* MYERS and WEITZMAN, 1953, *Corydoras hastatus* EIGENMANN and EIGENMANN, 1888, and *Corydoras pygmaeus* KNAACK, 1966 (previously known as *Corydoras hastatus australe*). None of these groups has, of course, any taxonomical value. The reproductive behaviour of the *Corydoras*-species is quite uniform. 2–3 females were put with 4–6 males in a large breeding tank of about 1 m in length which was filled with fresh water. After a few days, the first spawn was deposited, after violent activity, on plants, stones or sheets of glass which were leant on the side of the aquarium – the last of these can then easily be moved to the rearing tank. When mating, the male clamps the barbels of the females fast with the P and releases sperm at the same

Fig. 367
Corydoras arcuatus

time. At the same moment the female produces 3–5 eggs and puts them in the Vs which have folded to make a bag. Then she swims through the cloud of sperm to fertilise the spawn and sticks this firmly after a short search. According to KNAACK, the number of eggs varies according to the species, 30–50 *(Corydoras hastatus)*, and 800 *(Corydoras eques)*. They will spawn again after a few days and thus the parent fish should be left in a permanent position until the spawning period (of which there will be more in the course of the year) is over. The young fish hatch after 5–8 days according to the temperature, are kept in tanks without any bottom soil, and are fed with rotifers and micro-worms and later with little pot worms and minced *Tubifex*. They grow quickly. Because of their funny behaviour, the Catfishes are amongst the best-loved aquarium fish. The following species are very often kept.

– *C. aeneus* (GILL, 1858); Bronze Catfish, Bronze Corydoras. Trinidad, Venezuela and south to La Plata Basin. To 7 cm. Body sides yellow-brown with intensive green to copper-red metallic sheen. Fins grey and without patterning.

– *C. arcuatus* ELWIN, 1939; Arched Corydoras, Skunk Catfish. Amazon in area surrounding Tefé. To 5 cm. Body sides pale grey-green to whitish. A black band stretches length-ways from the tip of the mouth through the eye and follows the line of the back to the lower rays of the C. The fins are colourless. The gill-cover has a pale yellow-gold gleam. A very beautiful species (fig. 367).

– *C. axelrodi* RÖSSEL, 1962. Rio Meta in Columbia. To 5 cm. Body pale flesh coloured. Two dark bands run lengthways from the gill covers to the tail stalk. The upper follows the line of the back, the lower runs along the middle of the body to the start of the C. Underneath is a broad area with a gold-coloured shimmer. A wedge-shaped cross band stretches from the neck through the eye to the lower edge of the gill cover. The fins are translucent, the first rays of the D and the rays of the D2 are blackish (fig. 368).

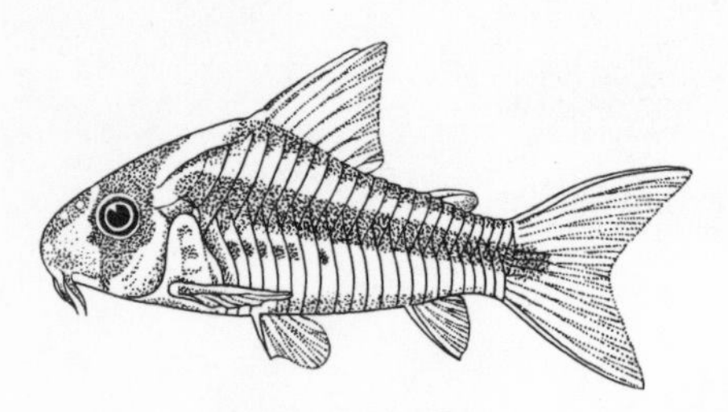

Fig. 368 *Corydoras axelrodi*

– *C. caudimaculatus* RÖSSEL, 1961; Tail-spot Corydoras. Upper Rio Guaporé. To 6 cm. The pale silver grey body sides are covered with many very small spots which also extend to the unpaired fins. The large black spot on the C gleams an intense metallic green when the light falls on it (fig. 369).

– *C. elegans* STEINDACHNER, 1877; Elegant Catfish. Central Amazon. To 6 cm. Body sides yellowish. There

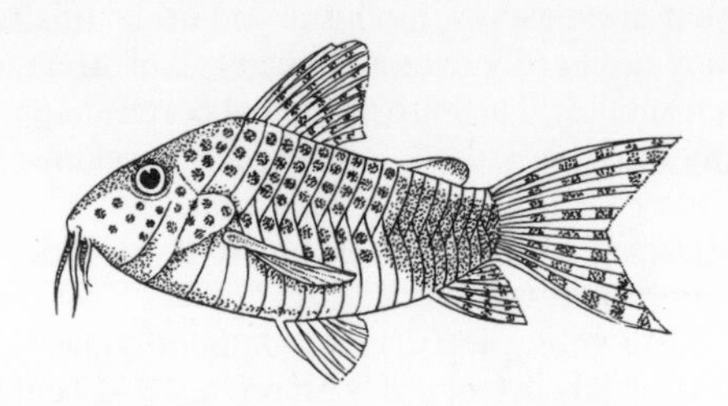

Fig. 369 *Corydoras caudimaculatus*

is a band running lengthways just under the dorsal line to the beginning of the C. It is broken up into countless spots and is black-green in the females and an intense metallic green in the male. It is bordered above and below by gold-coloured narrow zones. The long, pointed D also has green spots. The other fins are transparent (fig. 370).

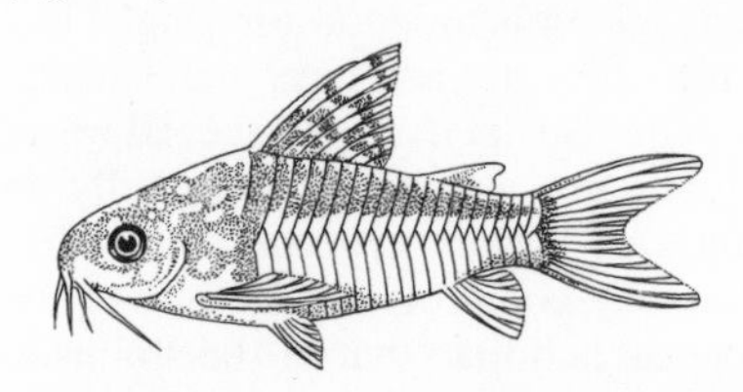

Fig. 370 *Corydoras elegans*

– *C. hastatus* EIGENMANN and EIGENMANN, 1888; Dwarf or Pigmy Corydoras. Amazon Basin near Villa Bella. To 3 cm. Body sides sand-coloured. Back brown-olive. Head, body and pale grey fins covered with small spots. A thin black band runs lengthways from the gill cover, to a black lozenge-shaped block just before the tail which is edged with yellowish white. A second black stripe runs from the Vs to the C. The fins are colourless

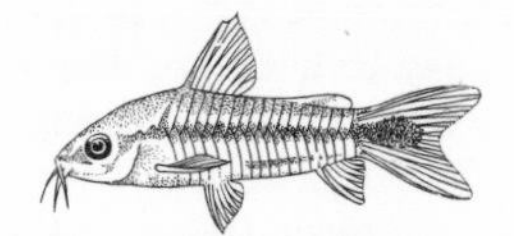

Fig. 371 *Corydoras hastatus*

and transparent. At the base of both the upper and lower caudal lobes are 3 black, brightly edged cross-bands. This free-swimming, dainty species is unfortunately not very prolific. The very small young fish are also difficult to rear (fig. 371).

– *C. julii* STEINDACHNER, 1906; Leopard Corydoras. Small tributaries of lower Amazon. To 6 cm. The silver grey body sides are studded with small black dots which run together on the head to form small worm-like lines. Between the two rows of lateral scutes runs a black longitudinal band, which is edged by a silver grey band above and below. It begins in the middle of the body and runs to the root of the C. The D1 has a large black patch in its upper third, the C has 6–7 and the An has 1–2 rows of black spots across it. *Corydoras julii* is probably the most striking species of the G. A well known synonym for this species is *C. leopardus* MYERS, 1933, (fig. 372). According to NIJSSEN, all the Leopard Corydoras species kept in Central Europe are not *C. julii*, but *C. trilineatus* COPE 1872.

Fig. 372 *Corydoras julii*

– *C. melanistius* REGAN, 1912; Black-spotted Corydoras. Northern S America. Essequibo. To 6 cm. Yellowish-grey white body sides with a pale reddish tinge. These are spotted with countless very dark dots. Two wedge-shaped cross bands run from the neck through the eye to the under edge of the gill cover and from the D1 to the edge of the rows of the scutes respectively. C and An are covered with fine dark spots and the other fins are colourless (fig. 373).

Fig. 373 *Corydoras melanistius*

– *C. paleatus* (JENYNS, 1842); Peppered Corydoras. SE Brazil and La Plata Basin. To 7 cm. The yellow-green body sides have a strong metallic sheen and are covered with irregular large dark spots. D, C and An are grey-green with dark dots which form cross bands. *Corydoras paleatus* is the best-known and most widespread species of the G. It is very hardy, not sensitive to temperature changes and long-lived (fig. 374).

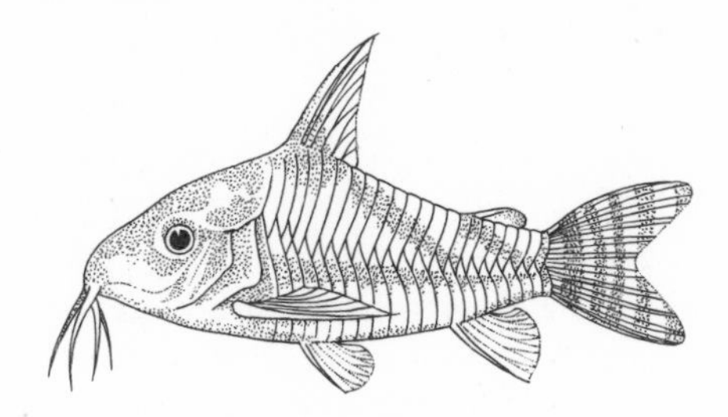

Fig. 374 *Corydoras paleatus*

– *C. rabauti* LA MONTE, 1941; Dwarf or Rabaut's Corydoras. Small tributaries of the Amazon above the estuary of the Rio Negro. To 6 cm. Body sides bright orange-red-brown. A broad black-brown band with a green metallic sheen begins at the head end and runs lengthways underneath the dorsal line to the insertion of the C. Fins light grey, often faintly orange at their base. A splendid species which is unfortunately very rarely bred (fig. 375).

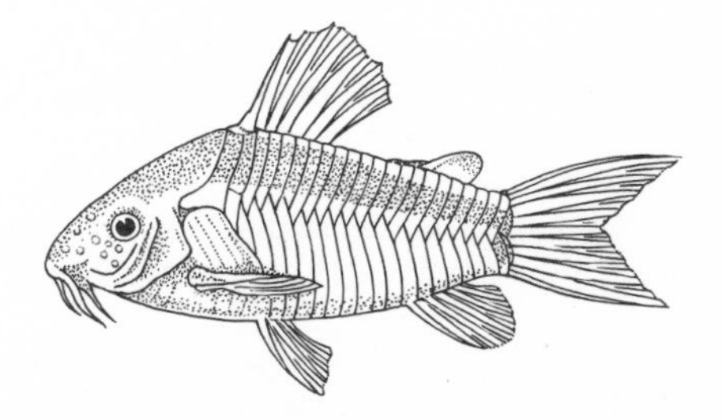

Fig. 375 *Corydoras rabauti*

– *C. schultzei* HOLLY, 1940; Gold-striped Corydoras. Small tributaries of the Amazon. To 6.5 cm. Body sides metallic green with blackish tinge, back dark-brown, stomach whitish. Parallel to the ridge of the forehead runs a short gleaming gold-yellow band up to the insertion of the D. The fins are light grey. *C. schultzei* is often kept because like *C. paleatus* it is so undemanding and easy to breed (fig. 376).

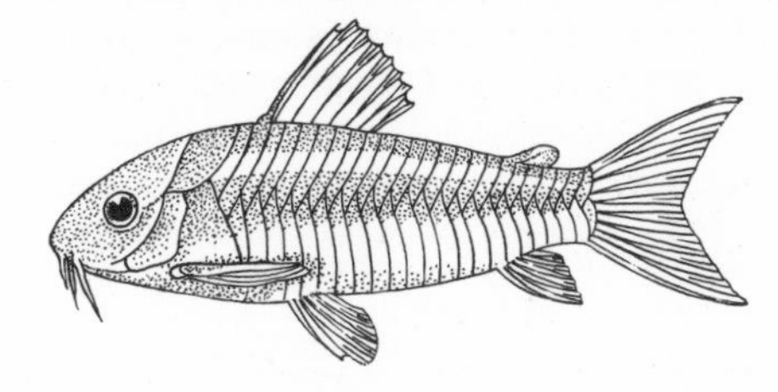

Fig. 376 *Corydoras schultzei*

Coryphaena hippurus *see* Coryphaenidae

Coryphaenidae; Dolphin fishes. F of Perciformes*. Sub-O Percoidei. The Coryphaenidae which are found world-wide in tropical seas have only two species. The one which is best known is the 1.50 m long, 30 kg Dolphin Fish or Dorado *(Coryphaena hippurus)* (fig. 377) – an elongate fish which is very strongly laterally compressed. The D begins on the head and stretches to the deeply indented slim C. The An is about half as long as the D. The scales are very small. The large males have a bulge on their forehead which makes them look rather like dolphins from the side when seen at a distance. This

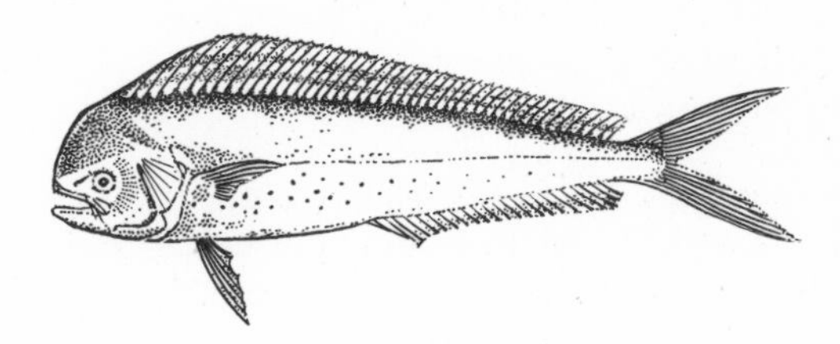

Fig. 377 *Coryphaena hippurus*

explains their name. The back is blue to green, body sides orange to cream coloured. The stomach is light, and spotted all over. Fins are yellow to orange. The splendid colouring becomes paler very soon after they are caught. The Dolphin Fishes are very fast swimmers. They feed primarily on fish and grow very quickly. They are valued foodfish in certain regions and are also good sport fish.

Coryphaenoididae, F, name not in current usage, *see* Gadiformes

Corythoichthys KAUP, 1856. G of Syngnathidae. Sub-O Syngnathoidei*. Sea fish which live on coral reefs and in the shore regions of the tropical Indo-Pacific and W Atlantic. The tail region of these fish which only measure 10–20 cm is almost twice as long as the trunk and in contrast to *Syngnathus** the snout forms a clear angle to the forward curve of the front body. The eggs are carried between open folds of skin on the underside of the front of the tail stalk. The small lively species of this G are frequently imported and keep well on a diet of small crustaceans, eg:

– *C. fasciatus* (GRAY, 1830); Messmate Pipefish. Indo-Pacific including the Red Sea. 17 cm. Light with dark spotted pattern.

Coscinasterias VERRILL, 1867 (fig. 378). G of Starfish (Asteroidea*). The *Coscinasterias* species live in warm seas and are tough predators like many starfish. They have more than 5 arms and have characteristic tufts of pedicellariae like trees on the upper side. They reproduce asexually by cross division and one hardly ever finds a symmetrical specimen amongst *Coscinasterias tenuispina*. This species is easy to keep in an aquarium.

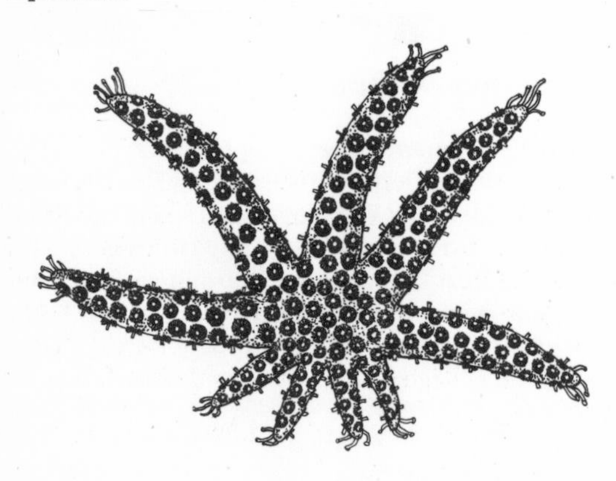

Fig. 378 *Coscinasterias* sp.

– *C. tenuispina* LAMARCK; Blue Starfish. Mediterranean, W Africa on rocky shores in shallow depths. To 18 cm. 8–10 arms. Bluish with rust coloured pedicellariae.

Cosmopolites. Species of plants and animals, or higher systematic groups, that are distributed almost worldwide in one of the three main habitats – sea, freshwater and land. Cosmopolites are therefore not restricted to particular biogeographical regions. Such organisms are usually highly adaptable, and because they are small they drift easily or are naturally transported (eg certain kinds of algae and small crustaceans).

Costia, G, *see Ichthyobodo*

Cottidae, F, *see* Cottoidei

Cottoidei. Sub-O of Scorpaeniformes*. Sup-O Acanthopterygii. The members of this Sub-O differ from other Sub-Os of the Scorpaeniformes* by a list of special characteristics which cannot be dealt with in detail here. What is more, of the total of 9 families, only 3 are described now:

– *Agonidae;* Poachers (fig. 379). The members of this family differ from the Cottidae in their rows of bony-plates which together surround the relatively slim, angular body like armour. They have thus a certain similarity to some Loricariidae* (Mailed Catfishes) although this is only external. Of the fins, the Ps are most strikingly large and the females in general have larger fins (D2, An and V) than the males. Their main habitat is the northern Pacific but some species are found in the N Atlantic and the west and east coasts of S America. They are almost all small and typical bottom-living fish of the tidal zones where they are often to be found in great masses. Some live at the base of mainland in greater depths. They feed almost exclusively on small invertebrates. Members of the G *Agonus** are often kept in marine aquaria.

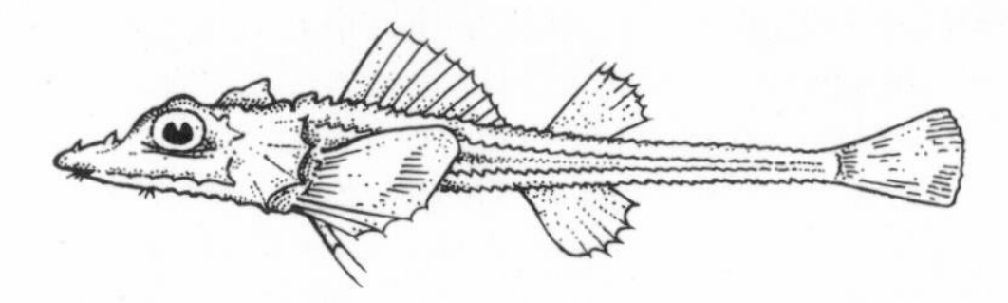

Fig. 379 Agonidae-type

– *Cottidae;* Sculpins, Sea Scorpions, Bullheads (fig. 380). Bottom-living fish with a large head which is usually somewhat flattened and armed with spines. The plump body tapers towards the back into a fairly narrow tail stalk. Mouth large, with powerful teeth. Fins large especially the Ps. D1 has spiny rays, D2 soft rayed. Scales mostly reduced or limited to the lateral line. Some examples have skin spines or thin bony plates in the skin. Most of the 300 or so species live near the coasts of the Arctic and northern temperate seas. Only the G *Cottus* is found all over in freshwater in the northern hemisphere. The oldest fossils are from the Oligocene period. Some larger species are of regional economic importance and there are aquaristic observations about various examples, *see Cottus*.

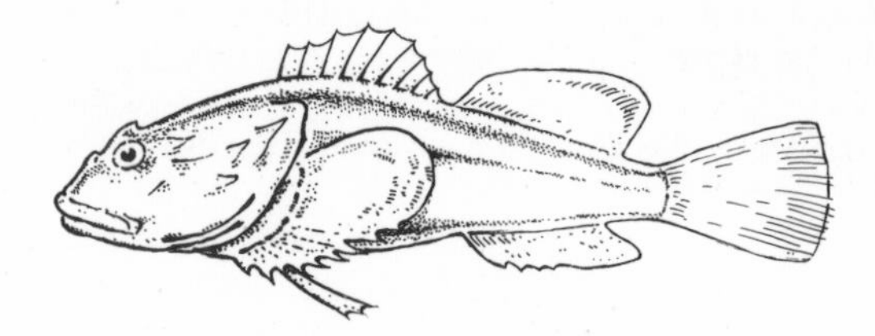

Fig. 380 Cottidae-type

– *Cyclopteridae;* Lumpfishes, Lumpsuckers and Snailfishes. The Sub-Fs of this F the Liparinae (Sea Snails) and the Cyclopterinae are occasionally dealt with as separate Fs. The members of both sub-Fs, although varying greatly in body shape, have various common characteristics amongst which the sucker on the breast (which is formed by the Vs and the scaleless skin are most typical. In the case of some Liparinae there are small prickles on the skin instead of scales. These are often arranged in rows and in some species almost enclose the fish like armour. The Liparinae are small, elongate, tadpole-like bottom-living fish with long D and An as well as relatively large pectoral fins. They are found in shallow waters near the coast but also in very great depths. Most of the 120 or so species live in the N Pacific. The Sea Snail or Unctuous Sucker *(Liparis liparis)* (fig. 381), which grows to 15 cm, is found in the north and western Baltic. It feeds primarily on shrimps and seaweed growth. In contrast to the Liparinae, the Cyclopterinae are high-backed, fairly plump fish living

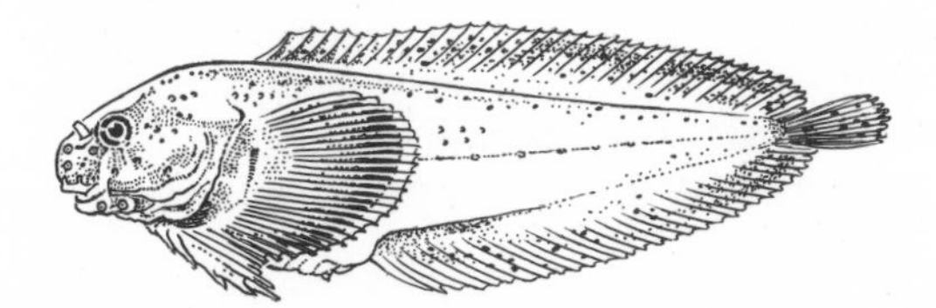

Fig. 381 *Liparis liparis*

in the Northern Atlantic and Pacific. Their skeleton only consists of cartilage and they have no swimbladder. There are various larger species and the 50 cm long grey-blue Lump Sucker *(Cyclopterus lumpus)* (fig. 382) is sold smoked. A 'caviare' substitute is made from its roe. The young fish are often kept in the aquarium, *see Cyclopterus.*

Fig. 382 *Cyclopterus lumpus*

'Cotton-wool' Disease *see* Columnaris Sickness

Cottus LINNAEUS, 1758. G of Cottidae, Sub-O Cottoidei*. To this G belong cold-stenothermic coastal fish of northern seas which lie in wait for their prey. They have a large head with many spines. Most species only grow to around 20 cm. Some of them make grunting noises. In some cases the males practise brood care. They need temperatures under 20 °C in the aquarium, a good deal of water movement and a good water-quality. Well known European species are:

– *C. bubalis* EUPHRASEN; Long-spined Sea Scorpion. E Atlantic to Baltic. To 17 cm. It has small white barbels in the corner of its mouth. Body has roughly outlined blackish-yellow patches. Male has red stomach and blue-white spots at spawning time (fig. 383).

Fig. 383 *Cottus bubalis*

– *C. gobio* LINNAEUS, 1758; Miller's Thumb, Bullhead. Europe west of the Elbe, in clear streams and lakes. To 15 cm. Has dirty brown pattern. Male practises brood care in spawning pits (fig. 384).

Fig. 384 *Cottus gobio*

– *C. scorpius* LINNAEUS, 1758 (= Myoxocephalus, *see* this also); Shorthorned Sculpin, Short-spined Sea Scorpion, Father Lasher, Bull Rout: N Polar Sea to Baltic. To 30 cm. Back dark brown. Sides marbled without barbels in the corner of the mouth. Stomach of the males red at breeding season, they probably care for the eggs. Growls if caught.

Cotyledon; Seed Leaf *see* Leaf Succession

Courtship Behaviour. Term used in ethology. Such behaviour is a prelude to mating and mostly consists of a combination of appetitive* and display* behaviour. As the disposition towards courtship* increases, the fish pass through a pre-courtship phase and later a high courtship phase, following which the mating takes place. The mating partner exhibits certain colours or movements that are typical of the species – these are known as the releasers* or key stimuli*. It is the releasers that stimulate the courtship behaviour. It is stimulated by the damming up of specific reaction energy as an internal stimulus or by the females releasing pheromones into the water as biotic stimuli. Lastly, fresh water, an increase in temperature or the introduction of rocks as a spawning substrate (especially in the case of Cichlids) can also increase courtship behaviour.

Courtship Drive, disposition towards. Term used in ethology to describe the inner state of readiness of a fish to begin courtship behaviour. Its intensity is dependent on internal stimuli, such as optical releasers*, means of communication (*see* Communication), the addition of fresh water, changes in temperature, the bringing in of spawning substrates. Courtship drive proper begins only when the stimulus threshold* is exceeded. If the sexes are separated for any length of time, it often results in a marked lowering of the stimulus threshold limit. In such cases, as soon as the sexes are reunited, the courtship behaviour of the males will immediately be accompanied by elements of the final courtship phase (leading the female to the spawning ground, immediate attempts at mating). In other words, the appetitive and display behaviour of the pre-courtship phase is largely dispensed with. Visible signs of the disposition towards courtship drive may include an intensifying of certain colour patterns, eg the courtship colouring of the males or the way in which certain sign stimuli on the females become more distinct. Another sign is the typical courtship behaviour of each species.

Courtship Swimming (fig. 385). Term used in ethology. An important element in the courtship display of male fish, in particular those species that are substrate-spawners. Part of the display involves the male swimming alongside or leading the females to the spawning-ground. Depending on the behavioural inventory* of the particular species, the male uses 3 types of display swimming—head, ventral or dorsal display swimming. As he does this, he positions himself above or below the female, synchronising the rhythms of his movements with those of the female. *Head display swimming* involves the male touching the roof of the female's cranium with his snout, eg *Nannostomus eques. Dorsal display swimming* can be observed in *Rasbora heteromorpha, Nannostomus beckfordi aripirangensis, Pantodon buchholzi*; it is often maintained even when swimming through thick clusters of plants. This type of swimming involves the male touching the female's dorsal ridge directly behind the Dl. With *ventral display swimming* the male touches the ventral surface of the female near where the An or Vs are inserted. Often the fish may bite at the same time, particularly if the female is not ready to spawn, and this may lead to the fins becoming damaged (particularly common among many species of characins and barbs). It is common for all 3 types of display swimming to be interrupted by the male swimming in an undulating fashion at a particular spot in the spawning ground. It involves the male gliding forwards over the head, back or beneath the belly of the female, sending undulating waves through his body and caudal peduncle. He follows this up by remaining briefly on the substrate and waits for the female. Subsequently, mating takes place or the male swims back to the female. With many species the undulating swimming movements of the males are followed by the fluttering dance* which is a behavioural syndrome* of display swimming.

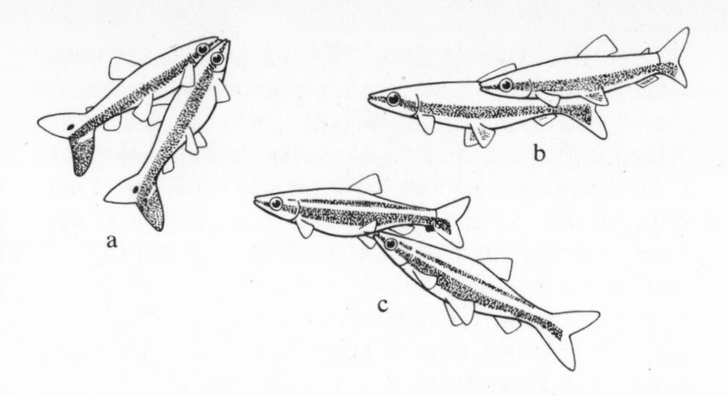

Fig. 385 Courtship Swimming (leading the female to the spawning substrate). a) Head display swimming *(Nannostomus unifasciatus ocellatus)* b) Dorsal display swimming *(N. harrisoni)* c) Ventral display swimming *(N. bifasciatus)*

Cow Shark *(Hexanchus griseus) see* Hexanchidae

Cow Sharks, F, *see* Hexanchidae

Cowbane *(Cicuta virosa) see* Apiaceae

Cowries, G, *see* Cypraea

Crabs, Sub-O, *see* Brachyura

Craneflies, F, *see* Tipulidae

Crangon WEBER, 1795; Common Shrimp. G of Shrimps (Natantia*). Various species in N Atlantic and N Pacific. They often live in dense populations on soft bottoms where, by beating their abdominal feet, they bury themselves completely in the sand but for their protruding eyes. Their body is flattened and its colour blends with its surroundings. The following species is of great economic importance in the North Sea and the Baltic and is very interesting aquarium object which is easy to keep.

– *C. crangon* (LINNAEUS, 1758); Common or Sand Shrimp. North Sea, Baltic on sand and mud in shallow water. To 5 cm. Wrongly sold as crabs.

Cranial. Situated near the head. From Latin *cranium*, the skull. Term used in anatomy* to describe the position of parts of the body. The opposite is caudal*.

Craspedacusta LANCESTER, 1880; Freshwater Jellyfish (fig. 386). G of Coelenterata*. With very few species living in freshwater. They have a small polyp phase (about 2 mm) and a medusoid phase (to over 3 cm). *Craspedacusta* prefers slowly flowing waters but is also found in clear, cool, quarry ponds and is kept in

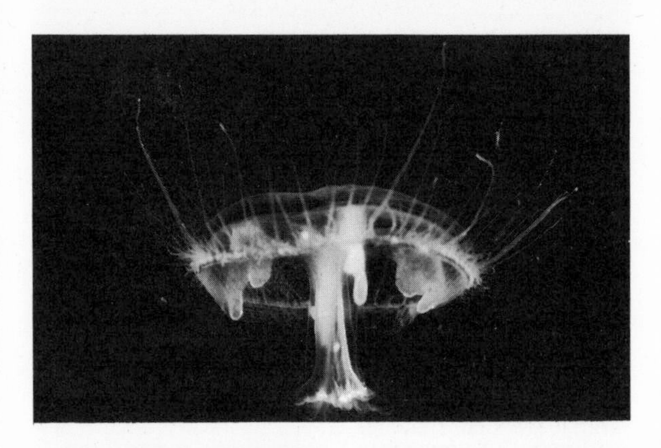

Fig. 386 *Craspedacusta*

the aquarium now and again. The only species found in Central Europe is *Craspedacusta sowerbyi.*

Crassula LINNÉ, 1753. G of Crassulaceae*, 300 species. Found almost all over the world with a great concentration in S Africa. Shrubs of medium height, polymorphic perennial plants but also annual plants. Leaf position alternate and opposite. Leaves fairly thick and they store water. Flowers are hermaphrodite and polysymmetrical. Double perianth without a separate calyx or corolla, 4 to 9 but mostly 5 part. Usually 5 stamens. Carpels usually 5, free, superior. Carpels unite to form a follicular fruit. The species of this G are predominantly plants of dry places which store water in their shoots and particularly in their leaves. There are some species amongst them which grow in places which are sometimes dry but sometimes flooded such as:

Fig. 387 *Crassula helmsii*

– *C. aquatica* (LINNÉ) SCHÖNLAND; Water Houseleek. Temperate areas of the N hemisphere but only here and there. Not yet tested by aquarists.

– *C. helmsii* (KIRK) COCKAYNE (*C. intricata*, name not in current usage, *Tillaea recurva*, name not in current usage). Australia. An undemanding aquarium plant (fig. 387).

Crassulaceae. F of Magnoliatae. 30 Ga, 1,400 species. Found almost all over the world with great concentration in S Africa and Central America. Shrubs, perennials and annual plants almost always succulent. The leaves alternate or opposite mostly with water storage tissue. Flowers polysymmetric with coloured dialypetalous corolla. The Crassulaceae are mostly plants from dry habitats. Some species grow where there is periodic flooding. One species of the G *Crassula** is occasionally kept in the aquarium.

Crawfishes, G, *see Palinurus*

Crawling Water Beetles, F, *see* Haliplidae

Crayfishes, G, *see Astacus*

Creediidae, F, *see* Trachinoidei

Creeping Jenny *(Lysimachia nummularia) see Lysimachia*

Creeping Rush *(Juncus repens) see* Juncaceae

Creeping Stem Algae, G, *see Caulerpa*

Crenal; also crenon. Source region of flowing water (Flowing Waters*). The most noticeable features of this habitat are the small fluctuations in temperature (during the year often less than 5 °C), lack of oxygen and the marked absence of higher plant forms. In accordance with these very specific ecological conditions, the animal world in this region is also typical. It consists essentially of certain kinds of turbellarians (Turbellaria*), molluscs (Mollusca*) and some groups of articulated animals such as crustaceans (Crustacea*), Water Mites (Acari*), Stoneflies (Plecoptera*) and Caddis Flies (Trichoptera*).

Crenicara STEINDACHNER, 1875. G of Cichlidae*. 3 species in tropical S America (Amazon area). There is nothing known yet about their habitat or ecological data, but from the results of breeding them one may conclude that they live in dystrophic* flowing waters. They are small, elongate, laterally compressed and elegant Cichlids which are very splendidly coloured in part. The members of *Crenicara* are also known as 'Dwarf Cichlids' together with such well known species as *Apistogramma** and *Nannacara** and some others (eg *Apistogrammoides**). The profile of the brow is strongly arched and the head therefore looks blunt. The soft-rayed sections of D and An are pointed and elongated in the males and in *Crenicara filamentosa* the C ends in two thread-like tips. The Crenicara are easy to keep as stated under *Apistogramma* and *Nannacara.* The fish are happiest in tanks with plenty of plants and hiding places and very soft, slightly acidic water. In these conditions they will reproduce. In contrast to *Apistogramma,* they are open-water breeders and the leaf of a water plant *(Echinodorus*, Cryptocoryne*)* is often chosen as the place to spawn. The mating is somewhat less excited than that of other Cichlids. The male either spreads out all his fins and quickly swims jerkily around the female or tumbles with his fins flat against his sides, at the same time becoming paler so that one has the impression that he is about to die. The few observations on breeding habits which exist agree that although the female takes over the brood care at the beginning and even chases the male away, she soon loses this drive and it is therefore advisable to attempt to rear the young 'artificially'. To this end, after one and a half days, one should transfer the leaf of the water plant together with the eggs to a bowl with a supply of running fresh water at a constant temperature of 26 °C. About 100 eggs are laid at a time and they hatch after about 60 hours. After another four and a half days, the young fish can swim freely and may be fed with the smallest *Cyclops* nauplii and rotifers. It is particularly important that they be given plenty of food and that at least 50% of the water be changed constantly (at least daily).

– *C. filamentosa* LADIGES, 1959. Details of exact habitat unknown, presumed to be central region of the Amazon basin. To 8 cm. Body yellowish with a broken black horizontal band from the nose to the caudal peduncle. On the back there is a second band next to the first, giving a chequered appearance. During the reproductive period there is a broken horizontal black band on the middle of the body. Unpaired fins which are extended to a point in the case of the male, particularly noticeable in the C. Bright red and blue markings. Adult females have red Vs (fig. 388).

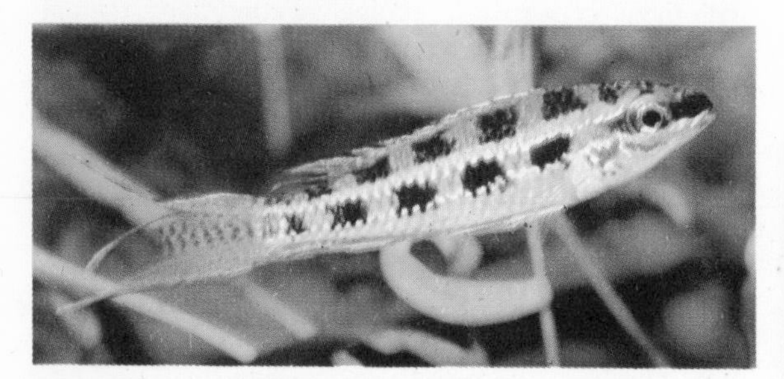

Fig. 388 *Crenicara filamentosa*

– *C. maculata* STEINDACHNER, 1875. Central Amazon. To 10 cm. Colouring similar to *Crenicara filamentosa*, and because of this, both species are also known as Checkerboard Cichlids. Females with orange-yellow Vs in the spawning period.

Crenicichla HECKEL, 1840; Pike Cichlids. G of Cichlidae* with many species. Widespread throughout S America, but predominant in tropical regions. Very elongate, medium to large sized extremely predatory fishes. Long based D. Short An, set opposite the back section of the D, both pointed in the males, and rounded in the females. C fan-or egg-shaped. Pointed head, very large mouth with a wide gape. Lower jaw usually protruding. The colouration of the various species varies greatly. They often possess a dark horizontal band, sometimes broken into spots. The *Crenicichla* species wait for their prey in a concealed place and then dart out at it as quick as a flash. They attack fairly large animals compared with themselves. They prefer fish which they overcome and then swallow headfirst. All in all, the day-time behaviour of the *Crenicichla* is similar to that of the Pikes (*see* Esocoidei). They are prized as a local delicacy.

The *Crenicichla* species are only suitable for those who are particularly fond of Cichlids, or for public aquaria. Mixing *Crenicichla* with other species is not recommended, especially not with smaller varieties. They can also be fairly intolerant of members of their own species. It is best to allow pairs to separate themselves from a shoal of fish, for the inferior partner will otherwise usually pay with his life. The Pike Cichlid aquarium should always be densely planted with sufficient hiding places, (crevices in stones and root-works). The tropical species should be kept at a temperature of 23 °C or more, or at 26–28 °C for breeding purposes. Room temperature is sufficient for the S Brazilian and Argentinian varieties. For feeding one should offer water insects (dragonflies, Ephemerata etc). The *Crenicichla* are open-water Breeders (*see* Cichlidae) and deposit their eggs in shallow depressions. It is the males which chiefly devote themselves to brood care. The following species are occasionally kept in the aquarium:

– *C. dorsocellata* HASEMAN 1911; Eye-spot or Two-spot Pike Cichlid. Brazil. To 20 cm. Greenish with a dark longitudinal band from the eyes to the start of the C. There are also narrow transverse bands mainly in young fish. Characteristic eye spot on the D. Relatively peaceable species.

– *C. lepidota* HECKEL, 1840; Pike Cichlid. From Brazil to N Argentina. To 20 cm. Yellowish-green with a dark longitudinal band which is broken up into flecks over the P, from the mouth to behind the start of the P. Caudal peduncle with a dark eye-blotch (fig. 389).

Fig. 389 *Crenicichla lepidota*

– *C. saxatilis* (LINNAEUS, 1758); Ring-tailed Pike Cichlid. Widely found in S America, including Trinidad. To 35 cm. Yellowish-green with a dark longitudinal band continuing from the mouth to the start of the C, but broken into spots over the P. Striking eye-blotch in the upper third of the C, D with a dark edge and a milky white longitudinal stripe beneath.

Crenilabrus OKEN, 1817. G of Labridae, Sub-O Labroidei*. Numerous species found on the coasts of Europe and W Africa. Sometimes also in the Baltic. The *Crenilabrus* species live in shallow depths among stones and plants where they seek out small animals and withdraw into the vegetation when danger threatens. They rarely reach more then 15–20 cm. The body is moderately high and laterally compressed. The *Crenilabrus* species are mainly mottled and are able to change colour. The males are highly coloured at breeding time. These occupy territories sometimes building plant-nests for care of the eggs, where they attract passing females by their courtship display. They do not form permanent pairs. They are very easy to keep in the aquarium. Breeding has not yet been successful. Attractive species can be caught on the South European coasts. These include, eg:

– *C. ocellatus* (FORSKÅL); Axillary Wrasse. Mediterranean, Black Sea, frequently on rocky coasts. Male to 9 cm, female to 6 cm. Dark spot edged in red on the gill-cover. Young fish and females are sociable, with brown longitudinal stripes on a light background, males mossy-green with reddish-yellow back and yellow D and An with blue edges.

– *C. tinca* (LINNAEUS, 1758); Peacock Wrasse. Mediterranean, Black Sea. To 30 cm. No eye spot on the gill cover. Young fish and females grey-white with brown longitudinal stripes. At breeding time the males are yellow with red-brown longitudinal stripes and blue-edged fins.

Crenon *see* Crenal

Crenuchidae; Sailfin Characins (fig. 390). F of Cypriniformes*, Sub-O Characoidei. These are S American Characoidei belonging to the F of Crenuchidae, comprising few species which are reminiscent of small carps in their body shape, in that they have only slight lateral

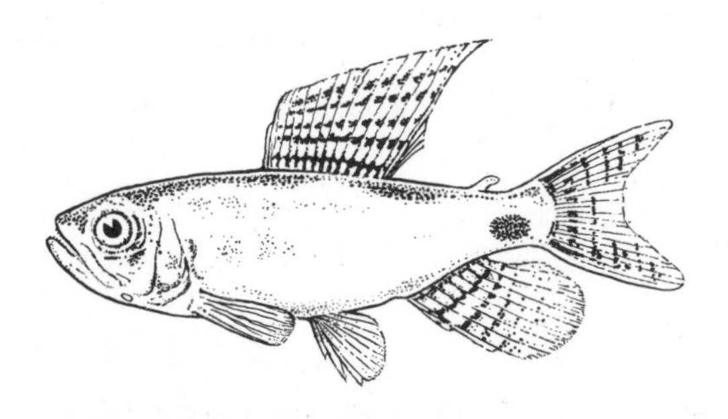

Fig. 390 Crenuchidae-type

compression with moderate elongation. Large terminal mouth. Praemaxillary bone (middle jaw) with one row of close three-pointed teeth, while the dental (lower jaw) has two rows. Maxillary bone (upper jaw) without teeth. D long, often in the shape of a flag. Adipose fin present, C indented, An shorter than the D and much elongated in *Crenuchus*. On the head, between the eyes, there is a paired frontal organ, the function of which is not known. Gill membranes not fused with the isthmus. Cycloid scales. The sexual dimorphism, that is the difference between the male and the female, is strikingly great in this family – a characteristic which is not usually common in the Characoidei*. Ga *Poecilocharax, Crenuchus*. See *Crenuchus* for tips for the hobbyist.

Crenuchus GÜNTHER, 1863. G of Crenuchidae*. Found in Guyana and the upper and central region of the Amazon. Bottom-dwelling Characins of 5–7 cm in length. Although the powerful, deeply cleft mouth gives this fish an aggressive appearance, it is in fact a peaceable fish which is sickly and short-lived if kept in hard water. They require soft, slightly acidic water at a temperature of 25 °C as well as an aquarium which is well stocked with plants. The body is fairly elongate, with much lateral compression. D and An approximately the same length, with a straight top edge; in the males both are extended and flag-like. Adipose fin present, C deeply indented. The *Crenuchus* species eat live food of all types, especially worms and mosquito larvae. According to FRITZSCHE they will spawn in a clay pipe, the red eggs being cared for by the male. Rearing has not been successful.

– *C. spilurus* GÜNTHER, 1863; Sailfin Characin. W Guyana, central Amazon and tributaries. Male to 6 cm, females to 5 cm. A dark longitudinal band, edged in yellow at the top which is not always visible, passes through the red-brown sides of the body and ends in a rectangular black spot at the end of the caudal peduncle. The upper half of the iris is blood-red. D, C, An and Vs are yellow and have splendid red-brown mosaic markings. The scales have dark edges. The less brightly coloured females have no elongated unpaired fins. These are interesting aquarium fish which have been imported more frequently of late.

Cress, F, *see* Brassicaceae

Crest-fishes (Lophotidae, F) *see* Lampridiformes

Crimson Squirrelfish *(Myripristis murdjan) see Myripristis*

Crinoidea; Feather-stars. Sea-lilies. Cl of Echinoderms (Echinodermata) with the Sea-lilies which cling by a stem and the free-living Feather-stars. They feed on floating particles, which they catch with their singularly or multi-branched delicate arms which are densely covered in lateral feathers (pinnulae). The food particles are directed to the upward-pointing mouth along furrows of cilia. The arms are attached to a cup-like base (calyx) which is extended into a stem in the case of the Sea-lilies. Feather-stars also have a stalk while they are young but they detach themselves later and can even swim by rhythmically beating their arms. Whereas the modern Sea-lilies which have a stem of up to 2 m in length, are only found in deep water, some Feather-stars are also found in shallow water. Transport and care of these delicate, fragile creatures have not been satisfactorily achieved. *Antedon** DE FREMINVILLE, 1811 is the most common G of Feather-stars, outside the tropics.

Crinoids (Sea-lilies and Feather-stars), Cl, *see* Crinoidea

Crinum LINNÉ, 1753. G of Amaryllidaceae*. 60 species. Tropics and subtropics. Bulbous perennials with lineal leaves arranged in rosettes. Hermaphroditic flowers, large, polysymmetric. Simple perianth, 6 part, white but also reddish, 6 stamens, 3 carpels, fused, superior. Capsule-fruit. The species of this G usually take root in places with a changing supply of water. Some species grow in running waters which dry up occasionally. The bulbs survive through the dry period remaining in the ground. These are decorative marsh-plants which are suited for hydro-culture, but grow very slowly in the aquarium. The best-known are:

– *C. natans* BAKER. Tropical W Africa. Leaves to 1.5 m long. Flowers white.

– *C. purpurascens* HERBERT. Tropical S America. Leaves to 40 cm long. Flowers white or reddish.

Cristiceps VALENCIENNES, 1836. G of Clinidae. Sub-O Blennioidei*. A few species of this G live in warmer seas where they hide amongst the plants in rocky coastal waters and feed on small creatures. They grow to about 10 cm and are elongate with pointed mouths. D1 small, D2 long. They move sinously with the Vs, which are near the throat, creeping forwards. The males watch

over the eggs which are laid on the plants. The one species found in Europe is very easy to keep in the aquarium but hides itself most of the time.
– *C. argentatus* Risso. Mediterranean. 10 cm. Body, D and An with black brown bands and light silver spots on its sides. The young fish are yellowish.

Croakers, F, *see* Sciaenidae

Croaking Tetra *(Glandulocauda inaequalis) see Glandulocauda*

Cromileptis Swainson, 1839. G of Serranidae* from the tropical Indo-Pacific. The *Cromileptis* species lie in wait for their prey on coral reefs and rocks. They are high-backed with a strikingly small, low set head. Only one species is known which is often imported and is easy to keep and fairly peaceful. *See* Serranidae for illustration.
– *C. altivelis* Cuvier and Valenciennes; Grace-Kelly-Fish, Panther Fish, Polka Dot Grouper, Barramundi Cod. To 50 cm. Silver grey with black spots over whole body.

Cross-breeding; Hybridisation. Mating between sex partners that have different hereditary material, the result of which is a hybrid*. Cross-breeding is important commercially to produce new combinations of features (*see* Breeding). The offspring are not genetically uniform however (ie they are not homozygous, but heterozygous). If the hybrids are allowed to reproduce, the features are once again distributed among various individuals (*see* Heredity). In practice therefore, the usual intention is only to make use of the first filial (F_1-) generation. This leads to the phenomenon known as *hybrid vigour* (Heterosis*).

When an F_1-hybrid is mated with a parent type, this is known as *back-cross breeding.*

Crossocheilus Van Hasselt, 1823. G of Cyprinidae*. Sub-F Garinae. S Asia, Indo-Australian Archipelago. These small to medium-sized fish live in small groups in flowing water. The elongated body which is slightly compressed laterally is reminiscent of the G *Botia**. The skin of the tapered protruding snout merges into the upper lip. The ventral mouth is bordered with sharp-edged jaws and has 1–2 pairs of barbels on the upper and lower lip. LL complete. These peaceable fish are solitary and are also suited to life in a mixed aquarium, but need larger tanks well-stocked with plants which provide hiding places. The water should be soft and should be slightly acidified by the addition of peat. The temperature should be 22–25 °C. The *Crossocheilus* species are omnivores which have a preference for live and dried foods which they take from the bottom. They

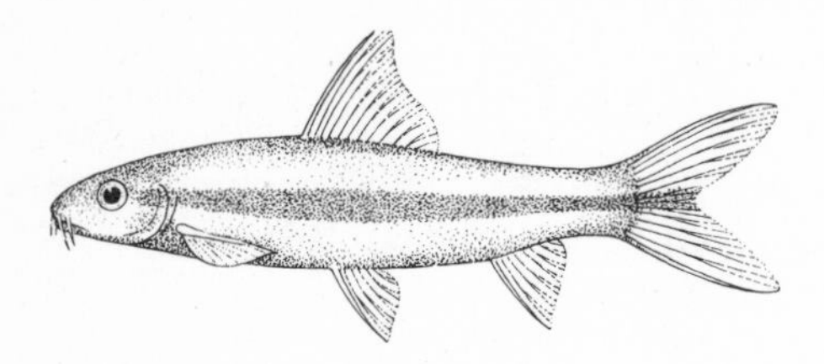

Fig. 391 *Crossocheilus oblongus*

also require additional plant food. Not yet bred in captivity. Several species:
– *C. oblongus* Van Hasselt, 1823. Thailand, Malay Peninsula, Greater Sunda Island. To 16 cm. Upper lip fringed. Two pairs of short barbels. Back grey-olive, sides of body yellowish, A black longitudinal band stretches from the snout to the caudal fin. Fins colourless to grey. Male more strongly coloured, with nuptial tubercles on the head during spawning time (fig. 391).

Crossopterygii, Sup-O, *see* Osteichthyes

Cross-pollination *see* Pollination

Crowfoot (*Ranunculus,* G) *see* Ranunculaceae

Crown of Thorns *(Acanthaster planci) see Acanthaster*

Cruciferae, F, name not in current usage, *see* Brassicaceae

Crustacea; Crustaceans. Class of segmented animals with jointed limbs (Arthropoda*) with many species and a huge variety of forms. Numerous Os, including dwarf and giant forms, and comprising plankton*, sandhole-dwellers, sessile* species, parasites, bottom-dwellers and skilled swimmers. All crustaceans are primarily aquatic and there is hardly any aquatic location which is completely devoid of them. By far the most species live in the sea, others in freshwater and the terrestial Woodlice (Isopoda*) have even succeeded in becoming almost entirely independent of water. With exception of the last-mentioned O, all crustaceans breathe through the gills or where these are reduced through the skin. A further characteristic of the Crustacea is the presence of two pairs of feelers (antennae) and three paired oral members (mandibles, maxillulae and maxillae). As a rule the body consists of three sections, the head (cephalus), the chest (thorax) and the hindquarters (abdomen). The first two are often fused and in addition it is possible for an outfolding of the skin of the last head section (carapace) to more or less completely cover the body like a shell, eg as in the water-fleas (Cladocera*) and Ostracods (Ostracoda*). In other groups of Crustaceans, the carapace is completely missing such as in Amphipods (Amphipoda*) in Copepods (Copepoda*) and Isopods (Isopoda*). In its original form, the body of the Crustacean has many sections, each segment apart from the first and the last having legs. Thus in the case of some Phyllopods (Phyllopoda*) one finds up to 40 segments. At the other extreme, the Ostracods (Ostracoda*) provide an example of Crustaceans with an unsegmented body and a greatly reduced number of limbs. Sometimes the body is modified to such an extent that one can only recognise that a specimen belongs to the Crustacea from the characteristic larval stages (Nauplius*, Zoea). Examples of this are provided by the sessile forms (Cirrepedia*) and some parasites, particularly among the Copepods. The Crustacea have achieved significance in aquatic studies as a valued form of food. As a result one can quote examples of names such as water-fleas, hoppers, *Artemia* and nauplii to have become part of the every day vocabulary of every aquarist. Furthermore, the larger colourful Decapoda* in particular represent very interesting subjects for the marine aquarium and to some extent also for the freshwater aquarium. Only various Ga of the shrimps, lobsters and prawns need be mentioned in this connection.

Crustaceans, Cl, *see* Crustacea

Cryptobia, G, *see* Blood Flagellates

Cryptocaryon Disease is caused by *Cryptocaryon irritans*, a ciliate* and an unavoidable parasite (Parasitism*) which lives on the skin and jaws of marine fish, and is closely related to *Ichthyophthirius* multifilis*. The parasite penetrates the upper layer of skin of fish and causes the formation of 0.5–1 mm long whitish-grey nodules. At the beginning the fish scrape themselves violently against stones, later symptoms are bloody, inflamed patches on the skin, and in the case of infestation of the gills a faster rate of respiration. Subsequent infections of the skin with other ciliates, Bacteriophyta* and fungus (Mycosis*). The parasite frees itself from the skin on maturity and soon forms numerous shoals by simple cell division* which infest other fish. The danger of infestion for healthy fish is very great. Baths with quinine*, copper sulphate* or zinc sulphate* are suitable methods of combating the disease, but Trypaflavin*, Methylene blue and Sulphathiazol-sodium (Sulphonamide*) can be used successfully.

Cryptocentrus VALENCIENNES, 1837. G of Gobiidae, Sub-O Gobioidei*. Different species found in the tropical Indo-Pacific and the Red Sea where they live mainly in shallow waters on sandy bottoms. Many form a close living relationship with crustaceans of the G *Alpheus**. The fish live in holes in the sand which have been dug out and maintained by the crustaceans. They warn the poorly-sighted crustaceans about enemies and are also said to bring food to them. The small *Cryptocentrus* species have the typical shape of the family and are sometimes very prettily coloured. They are easy to keep in the aquarium, eg:

– *C. lutheri* GLAUSEWITZ, 1960, Red Sea. 8 cm. Head rust-red, body olive with light bands. Lives together with *Alpheus* djiboutensis*.

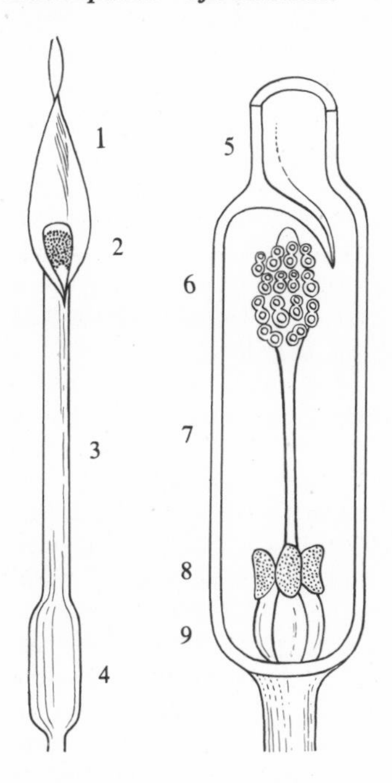

Fig. 392 *Cryptocoryne*-inflorescence. 1 Blade 2 Throat 3 Tube 4 Chamber 5 Valve 6 Stamens 7 Spadix 8 Stigmas 9 Carpels

Fig. 393 *Cryptocoryne ciliata*

Cryptocoryne SCHOTT, 1832; Water Trumpet. G of Araceae*. 50 species. S Asia, coastal regions of the mainland and islands. Small to medium-sized perennial plants with thin rhizomes, runners and intermediary formations between. Leaves in rosettes at base level, stemmed. Leaf-blades lineate, longitudinal-lanceolate, broad-lanceolate or ovate, usually pointed, the base being pointed, rounded or heart-shaped. Blade-surfaces flat, waved or bullate and with very variable colouration according to species and conditions of growth. Inflorescence enclosed in a coloured spathe* which forms a chamber at the base to which a tube of varying length is usually connected. The truncated or pointed and sometimes caudate spathe-blade forms the end of the tube. The part connecting the tube and the blade is called the throat. A characteristic coloured throat ring may be formed. At the joint between the tube and the chamber, there is a closing flap. The actual inflorescence is situated inside the chamber (fig. 392). Flowers unisexual and monoecious with the female flowers at the base of the inflorescence and males above separated by a sterile intermediate piece, all greatly simplified, without perianth. The fruit is a green berry. The species of this G are marsh and water plants, which live in the most varied places. There are species which grow emerse for most of the time and live submerged for shorter periods only. These usually flower only as terrestrial plants. Other species live permanently or predominantly under water and also flower in the submerged phase. Their spathes then grow out above the surface of the water. Finally there are also species which are subject to daily changes in the water level caused by the ebb and flood tides. Almost all species are suitable for keeping in the aquarium. The optimum temperature lies between 20° and 30°C. Contrary to earlier beliefs, they do not necessarily require soft acidic water but many species tolerate it well. Whereas most species can receive full sunlight in their natural state, subdued light is recommended when they are kept submerged. Even in this, however, there are variations from species to species. They develop best when they are left undisturbed for a long time. Certain problems result from the so-called Cryptocoryne disease*. As a result of the numerous runners, propagation presents no problems. The species can be divided into groups according to practical aspects.

1) Marsh plants which grow submerged only very slowly and unsatisfactorily.

– *C. ciliata* (ROXBURGH), SCHOTT; Mainland and archipelagos of S Asia. Most widespread and largest species also in brackish water. Leaf-blades broad-lanceolate, light to mid-green. The edge of the spathe-blade has simple or branched cilia (fig. 393).

2) Species with long, narrow leaf-blades, mostly growing well both emerse and submerged.

– *C. aponogetifolia* MERRIL (*C. usteriana*, name not in current usage). Philippines. Leaf blades lanceolate, long, dark green, embossed, spathe flat with twisted point, yellow to reddish.

– *C. balansae* GAGNEPAIN (*C. somphongsii*, name not in current usage). Indo-China. Leaf blades narrowly linear and long, light green and dentate. Spathe-blade long and spirally twisted, pale brown with violet stripes and spots.

– *C. spiralis* (RETZIUS) SCHOTT. India, Bangladesh, probably Sri Lanka. Leaf-blades narrow to lanceolate, light green. The tube of the spathe is almost absent, the blade is long and twisted, red-violet with nodular grooves across it.

Fig. 394 *Cryptocoryne costata*

Fig. 395 *Cryptocoryne lucens*

– *C. tonkinensis* GAGNEPAIN (*C. retrospiralis*, name not in current usage). India, Bangladesh and Burma. Leaves narrow-linear and long, light green, fairly flat. Spathe-blade long and twisted, greenish with violet markings.

3) Plants with relatively short leaf stalks and leaf blades, mostly growing well submerged and emerse.

– *C. affinis* N. E. BROWN ex HOOKER fil; (*C. haerteliana*, name not in current usage). South Asiatic Islands. Leaf-blades lanceolate with short base, velvety dark green. Spathe-blade long, much twisted, violet. Only grows submerged.

– *C. becketti* THWAITES ex TRIMEN. Sri Lanka. Leaf-blades broad-lanceolate with short base, olive-green to grey-green. Spathe with dark brown throat ring. Spathe-blade short, twisted and yellowish.

– *C. bullosa* BECCARI ex ENGLER. Kalimantan. Leaf-blades broad-lanceolate with short base, olive-green, bullate. Spathe-blade short, violet.

– *C. lucens* DE WIT. Sri Lanka. Leaf-blades narrow lanceolate, light green. Spathe-blade very short, brown-violet (fig. 395).

– *C. nevillii* TRIMEN ex HOOKER. Sri Lanka. Leaf-blades lanceolate, light green. Spathe-blades short, twisted, brown-violet.

– *C. petchii* ALSTON in TRIMEN. Sri Lanka. Leaf-blades broad-lanceolate with short base, olive-green to brownish. Spathe blade short, dark olive-green.

– *C. walkeri* SCHOTT. Sri Lanka. The var. *lutea* (ALSTON) RATAJ (fig. 396) and var. *legroi* (DE WIT) RATAJ belong to this. Leaf-blades broad-lanceolate, partly with short base and olive-green, light green or brown-green according to to the variety. Spathe-blade short, partly twisted, green to yellow.

– *C. wendtii* DE WIT. Sri Lanka. Several varieties. Leaf-blades broad-lanceolate, colour light green to brownish according to variety. Spathe has dark brown throat ring. Spathe-blade short, twisted, brown.

– *C. willisii* BAUM (*C. undulata*, name not in current usage). Sri Lanka. Leaf-blades lanceolate with short base, olive-green. Spathe-blade short twisted, yellow-brown.

4) Plants with relatively long leaf-stalks and short leaf-blades, mostly only growing to their best when submerged.

– *C. blassii* DE WIT. Further India. Leaf-blades ovate with short or heart-shaped base, dark green, spotted. Spathe-blade short, yellow, rough.

– *C. purpurea* GRIFFITH (*C. cordata*, name not in current usage; *C. grandis*, name not in current usage).

Southern Asian islands. Leaf-blades ovate, dark green. Spathe-blade short, red, rough.
– *C. siamensis* GAGNEPAIN. Further India. Leaf-blades ovate, dark green. Spathe-blade short, yellow, smooth.

Fig. 396 *Cryptocoryne lutea*

Cryptocoryne Disease. This is a sudden decaying of the leaves of Cryptocorynes. It is mostly caused by a change in environmental conditions such as change of water, alteration in light intensity, transfer from daylight to artificial light, addition of chemicals to the water and change of the filtration substrate etc. The opinion is occasionally expressed in specialist tests that Cryptocoryne disease is caused by bacteria and is therefore an infectious disease. In most cases the rhizomes come up again and the plants rarely die off completely. The effect on *Cryptocoryne* species may vary in intensity.

Ctenobrycon EIGENMANN, 1908. G of Characidae*. Characins of up to 8 cm in length dwelling in the coastal regions of northern S America at middle depths. This omnivorous shoaling fish, which is easy to keep, has a high back and is laterally much compressed. The pectoral region and, in adult fish, the flanks are covered in comb-edged scales. D short, pointed, C forked, An long, adipose fin present. Breeding is simple. After vigorous activity, up to 2,000 eggs are laid indiscriminately between plants preferably near the surface. At 25 °C the young fish hatch after 24 hours. They are easily reared. Soft to medium hard water is required.
– *C. spilurus* (CUVIER and VALENCIENNES, 1848); Silver Tetra (fig. 397). Found in the coastal regions of northern S America. The flanks of this fish, which grows to 8 cm long, are silver with a bluish sheen. A greenish longitu-

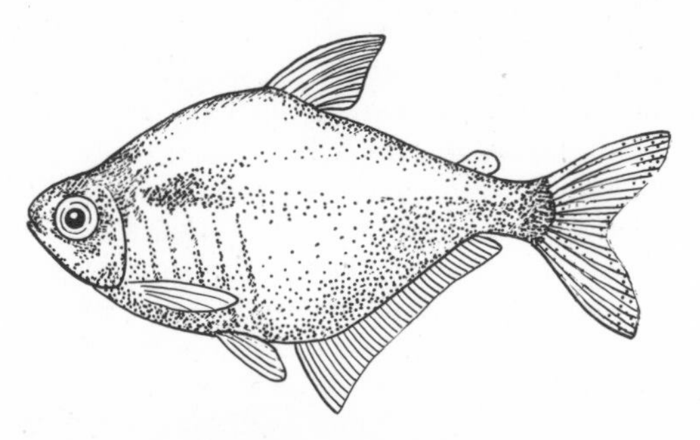

Fig. 397 *Ctenobrycon spilurus*

dinal band connects the large blue-black shoulder blotch with another black blotch on the root of the tail. The gill covers are an iridescent green. This species used to be a popular aquarium fish but is only rarely kept nowadays. According to GÉRY those fish caught in the Amazons are usually *C. hauxwellianus* (COPE, 1870).

Ctenoid Scales *see* Scales

Ctenolabrus VALENCIENNES, 1839. G of Labridae, Sub-O Labroidei*. Found in the E Atlantic and bordering seas, on rocky and pebble coasts. Features, size and behaviour are similar to that of *Crenilabrus**, but this G differs in that it possesses two rows of teeth, on the jaws. In contrast to the brooding *Crenilabrus* species, the *Ctenolabrus* has pelagic eggs and takes no care of them. The following species is well-suited to life in the aquarium, but is unfortunately difficult to obtain.
– *C. rupestris* (LINNAEUS, 1758); Goldsinny. Norway to the Black Sea, also in the western Baltic. To 15 cm. Brown with a black spot on the dorsal edge of the caudal peduncle.

Ctenoluciidae; Pike Characins (fig. 398). F of Cypriniformes*, Sub-O Characoidei*. Larger pike-like Characin with a long pointed head and a mouth with wide gape. Intermediate jaw or chin with a projection.

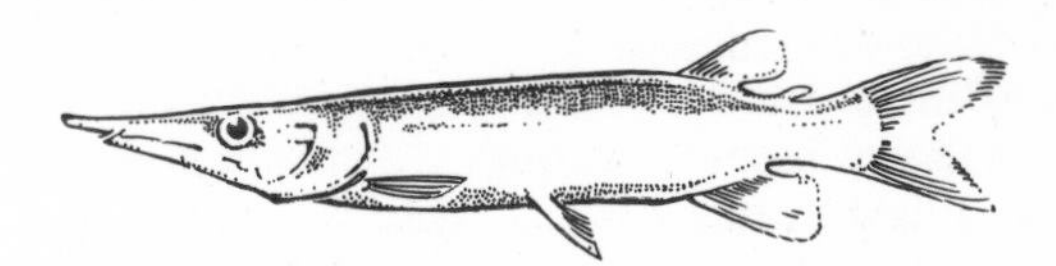

Fig. 398 Ctenoluciidae-type

Numerous very sharp conical teeth. Sometimes the gum also has teeth. Maxillary bone short. D short, set far back in front of the An, adipose fin present, C deeply indented. Lateral line sometimes incomplete. Scales numerous, cyclo-ctenoid (according to GÉRY). *Hepsetus odoe*, which grows up to 35 cm long and is found in Africa, is often grouped with the S American Pike Characins. This species with its many primitive characteristics is strongly reminiscent of our Pike. The Ctenoluciidae are predators which feed primarily on fish. Little is known about their biological characteristics. In 1980, *Ctenolucius hujeta* was bred successfully at Leipzig's Zoological Gardens. Compare G *Boulengerella**.

Ctenopharyngodon idella *see* Grass Carps

Ctenophora; Comb-Jellies, Sea Gooseberries. Only class of the Coelenterates (Coelenterata*) without stinging cells, which is why they are also treated as a sub-phylum. Acnidaria, differentiated from the stinging animals (Cnidaria*). They are glass-like, transparent and usually pelagic marine animals, with approximately eight strips of cilia, and normally two tentacles. They possess adhesive clinging cells (colloblasts). The shape of the body is melon-shaped, eg G *Pleurobrachia* FLEMING, 1821. Others are tape-like, such as the unusual Venus's Girdle (G *Cestus* CHUN, 1880). Some sessile species have irregular shapes. There are no aquarium observations.

Ctenopoma PETERS, 1844; African Climbing Perches. G of Anabantidae, Sub-O Anabantoidei*. Found predominantly in tropical W and SE Africa, in bays of larger tributaries of the the Niger, Congo and Zambesi, smaller rivers, weeded streams and ponds. Small and medium-sized fish which resemble the G *Anabas* in certain characteristics. Physically strong. Body laterally flattened, in some species more or less elongated, high in others. Edge of the gill cover toothed to a varying degree. Vs not elongated or filamentous. Brown tones predominate often with transverse bands, marbling or patterns of blotches. This outward appearance and the partly transparent fins assure these predatory fish a good camouflage. Most of them lack any distinctions between the sexes. They should be kept in roomy tanks with a lot of plant life and hiding places. They are active at dusk. Almost all species need warmth, 25–30 °C. They need substantial food, including fish. Not all species have been bred sucessfully. They de not always build a bubble-nest and some species are probably mouth-brooders. Species with elongate body are:

– *C. ansorgei* (BOULENGER, 1912); Ornate Ctenopoma. Tropical W Africa, Chiloango river systems. To 7 cm. 6–7 orange-red transverse bands on their sides which vary from greenish to bluish. D and An edged with white with fairly long tips at the rear. Fine bubble nest (fig. 399).

– *C. argentoventer* (SCHREITMÜLLER and AHL, 1922); Silver-bellied Climbing Perch. Niger Basin. To 15 cm. Underside silvery. A large eye-spot, with a paler rim on the tail.

– *C. multispinis* PETERS, 1844; Many-spined Climbing Perch. E African species. E Africa, Zambesi Basin. It has many spines on the edge of the gill-cover. Body and soft section of D clearly spotted.

– *C. nanum* GÜNTHER, 1896; Dwarf Climbing Perch. Southern Cameroons, Congo. To 7.5 cm. A few rows of light scales are found between the 6-9 dark transverse bands.

– *C. nigropannosum* REICHENOW, 1875; Two-spotted Climbing Perch. Niger Delta, Gaboon and Congo. To 17 cm. Faint dark transverse bands. A black blotch on the gill-covers and the root of the tail.

Fig. 399 *Ctenopoma ansorgei*

Species with fairly deep body are:

– *C. acutirostre* PELLEGRIN, 1899; Spotted Climbing Perch. Middle and Lower Congo. To 15 cm. Very deep body. Snout pointed. Body, P and C and soft-rayed sections of the D and An have dark spots. C as well as soft rayed section of D and An have light edge. The background colour of the fins is greenish.

– *C. congicum* BOULENGER, 1877; Congo Climbing Perch. Lower Congo, Chiloango, Ubangi. To 8.5 cm. Scales partly edged in lighter colour. Fairly complete transverse bands on body and eye-spot on root of tail. Soft D and An as well as C have lighter rim.

– *C. fasciolatum* (BOULENGER, 1899); Banded Climbing Perch. Congo Basin. To 8 cm. 8–10 dark transverse bands on the body. Blue spot on the gill-covers. Fairly pointed and elongate V, D and An. Loose bubble-nest. 1,000–1,200 eggs.

– *C. kingsleyae* GÜNTHER, 1896; Tail-spot Climbing Perch. Senegal, Gambia (to Congo). To 20 cm. Body has no distinct bands or markings. There is a black blotch on the root of the tail which in the young is rimmed with pale-yellow. C and soft D and An have light edges.

– *C. ocellatum* PELLEGRIN, 1899; Eye-spot Climbing Perch. Congo. To 25 cm. Leopard-like markings mostly in the rear half of the body due to irregular transverse bands and spots. Eye-spot on the root of tail. C, soft D and An with lighter edges.

– *C. oxyrhynchus* (BOULENGER, 1902); Sharp-nosed or Marbled Climbing Perch. Tributaries of the Lower Congo. To 10 cm. Body strongly compressed laterally, with large fairly clear eye-spot in the middle. Has dark marbling. C has sharply defined transparent middle section and dark edge. No bubble-nest. Over 1,000 eggs which float on the surface (fig. 400).

Fig. 400 *Ctenopoma oxyrhynchus*

Ctenops MCCLELLAND, 1845. G of Belontiidae, Sub-O Anabantoidei*. Found in rivers in India and Bangladesh. They are small fish which are strongly compressed laterally. They have a high back. The jaws have small, conical teeth. The lower jaw protrudes. They have a short D in the back third of the body beginning over the start of the soft section of the An. C and P are rounded. The lateral line when present is usually broken. The care of these fish is dealt with under their family entry. No information is available about breeding. The following is occasionally imported.

– *C. nobilis* McClelland, 1844. India, Bangladesh. A little over 10 cm. Brownish background colour. Whitish band from the eye to the base of the C. Under this are 2 similar stripes or rows of spots. The base of the An is often pale. There is an eye-spot on the upper base section of the C. Otherwise the fins are transparently colourless (fig. 401).

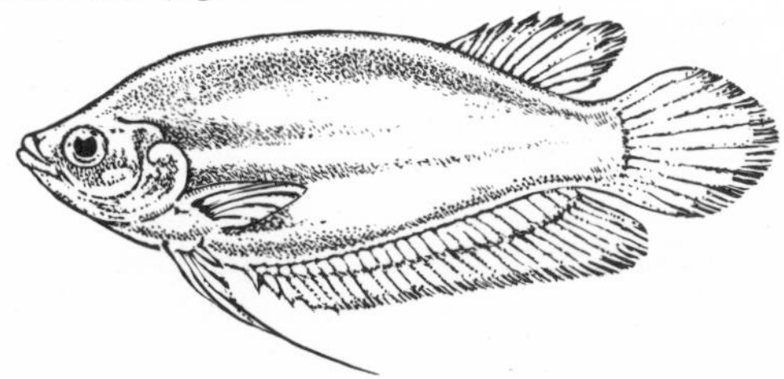

Fig. 401 *Ctenops nobilis*

Cuban Killie *(Cubanichthys cubensis) see Cubanichthys*

Cuban Limia *(Poecilia vittata) see Poecilia*

Cuban Rivulus *(Rivulus cylindraceus) see Rivulus*

Cubanichthys Hubbs, 1926. G of Cyprinodontidae*. Found in pools in Cuba and occasionally in brackish water. Small slightly elongate fish with rounded fins. D and An lying opposite to each other and similarly formed. Mouth very small and upturned. In their needs and behaviour, the *Cubanichthys* species are reminiscent of the *Aphanius** species, to which they are distantly related. Care and rearing of these peaceable fish is easy. They are most happy in tanks well stocked with plants, in sunny positions with water at a temperature of between 22 and 25 °C. In an aquarium of their species, the fish will usually reproduce without any special arrangements. They lay their eggs on fine feathery leaved plants after vigorous mating. The spawning period may extend over several weeks. The eggs are completely transparent and are, therefore, easily overlooked. The young fish hatch after 10–12 days and their rearing is rather troublesome.

– *C. cubensis* (Eigenmann, 1902); Cuban Killie. W Cuba. To 4 cm. Prettily coloured, flanks golden yellow with dark blue longitudinal stripes, the middle stripe being the broadest and most intensively coloured. Fins yellowish, edged in blue. Females paler.

Cubbyu *(Equetus acuminatus) see Equetus*

Cubomedusae; Box Jellyfish. O of Scyphozoa* with a high square bell with tentacles at each corner. The Cubomedusae live on the shores of tropical and subtropical seas. They swim fairly quickly and catch fish and other sea creatures with the help of their stinging poison. The poison of some tropical Cubomedusae, which can scarcely be seen in the water, is lethal to humans.

Cucharon *(Sorubim lima) see Sorubim*

Cuchias *(Amphipnous cuchia) see* Synbranchiformes

Cuchia (Amphipnoidae, F) *see* Synbranchiformes

Cucumaria Blainville, 1830. G of Sea Cucumbers (Holothuroidea*). They live in both warm and cold seas. The bottom-dwellers catch plankton and disturbed detritus with their crown of bushy, branched tentacles. The tentacles are put into the mouth one after the other and the particles of food which are attached are consumed. As a result of their way of life, they rarely move when in suitable environmental conditions. Various species have been looked lafter successfully in the aquarium.

Culcita Agassiz, 1836. Pincushion Starfish (fig. 402). G of predatory tropical Starfishes (Asteroidea*) with a five-pointed padded form. These attractive animals are often imported and are very hardy. Unfortunately they prefer to attack Sea-anemones, Polychaetes, Sea Urchins and other invertebrates. They should, therefore, only be kept with free-moving animals.

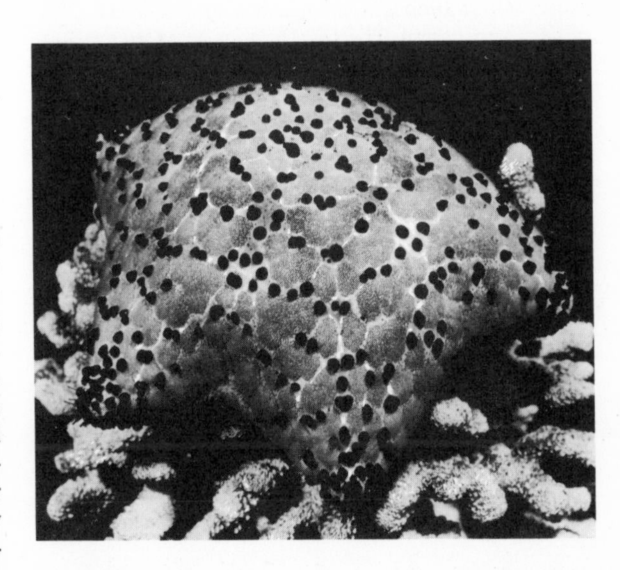

Fig. 402 *Culcita*

Culex Linnaeus, 1758 (fig. 403). G of biting Mosquitos (Culicidae, Nematocera*) with worldwide distribution. The females of the *Culex* species bite and need the blood that they suck up for the development of the eggs. The males do not bite. Larvae and pupae

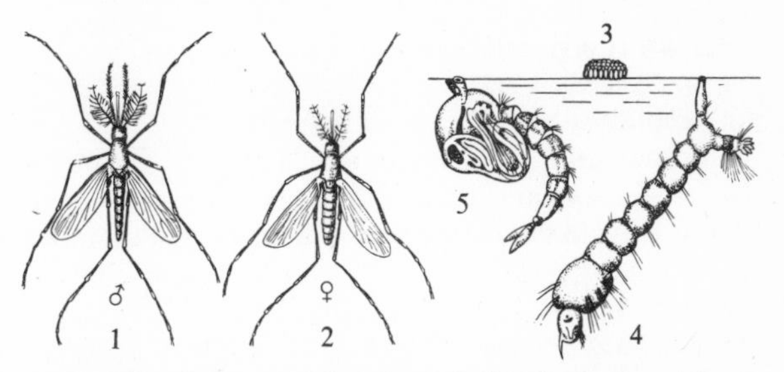

Fig. 403 *Culex.* 1, 2 Imagines 3 Clutch of eggs 4 Larva 5 Pupa

prefer to stay in still heavily polluted waters and are easily reared for feeding purposes in rain-butts (with some manure added). It is characteristic for the larvae of this G to hang diagonally pointing upwards on the surface of the water and breathe atmospheric air through a short tube. When disturbed they flee tumbling to the bottom from where they let themselves float to the surface soon afterwards. The pupae behave similarly and are easily recognised by their awkward appearance (*see also* fig. 86).

Cumings Barb *(Barbus cumingi) see Barbus*

Curimatidae (fig. 404). F of Cypriniformes* Sub-O Characoidei*. The common name Barb Characins which are found widely in Africa indicates that they are Characoidei of *Barbus*-like form, ie elongate and fairly solid. The mouth is terminal or underslung and toothless in adult fish. Various Ga have very broad, thick lips. These are toothless in the case of *Curimata* and *Curimatopsis*, and in *Prochilodus* and *Ichthyoelephas* can be turned inside out to form a circular disc which has two rows of very small teeth. The Ga *Chilodus* and *Caenotropus* have a noticeable enlargement on the 4th gill-arch which has still not been completely explained. D and An short, C deeply or very deeply indented, adipose fin usually present. Scales as a rule cyclo-ctenoid; lateral line often complete. Many Curimatidae grow to 20–25 cm and are found in great numbers in some areas. With few exceptions they feed mainly on plants and Algae which they graze on or rasp off. Some specimens are head-standers which, however, in contrast to the head-standing Anostomidae*, have a terminal mouth and can thus take food off the bottom without difficulty. The constant presence of some Curimatidae means they are useful and indispensable fish in certain areas of S America. The head-standers are those most kept by aquarists *(Chilodus*)*. Other Ga are only kept occasionally: *Curimatopsis**, *Prochilodus**.

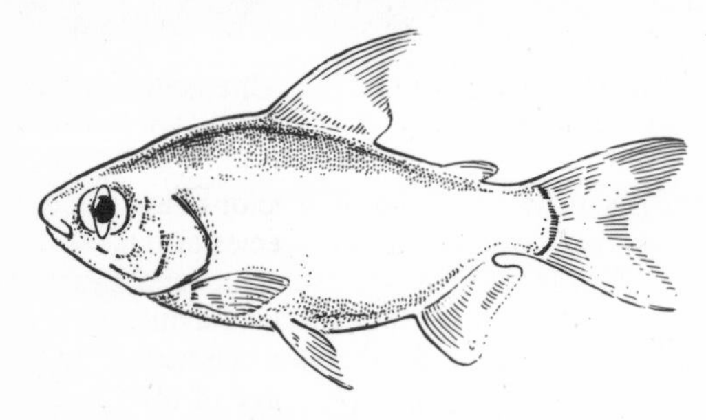

Fig. 404 Curimatidae-type

Curimatinae, Sub-F, *see* Curimatidae

Curimatopsis* STEINDACHNER, 1876. G of Curimatidae*. 4–6 cm long. Characin which lives in north Argentina. It lives in the lowest levels of water and searches for its food almost exclusively on the bottom. Body is compact, gills without teeth. Lateral line very short, D1 extended and pointed. D2 present, An short. The C of the males is only slightly forked and almost straight. The tail-stalk is broader than females of the same age. Until now only *C. saladensis* has been bred. They were bred without problems and it seems in a similar way to that of the smaller species of the G *Hemigrammus**. The young fish hatch after 26–32 hours and are easy to rear. Small live food as well as dried food is acceptable. They are peaceable and brightly coloured small Characins and well suited to life in the aquarium.

– *C. evelynae* GÉRY, 1964. Upper Rio Meta in Columbia. To 5 cm. Body sides of male pink, the tail-stalk and the base sections of the C are blood-red with a black longitudinal stripe. The females are silver with a black stripe on the tail-stalk, the fins are colourless or lightish yellow. This lovely species has been imported several times when it has been caught with other fish (Unintentional Catches*). It has not yet been bred and is susceptible to fish tuberculosis in hard water (fig. 405).

Fig. 405 *Curimatopsis evelynae*

– *C. saladensis* MEINKEN, 1933; Rose-coloured Curimatopsis. Rio Salado, Rio Paraná. To 6 cm. The upper half of the body sides of the males has grass-green or metallic green longitudinal bands, breast and belly-section pale reddish. The gill covers have a brassy gleam, the scales have a dark edge. A dark longitudinal band runs from the gill-cover to the root of the tail and finishes there in a black blotch edged with a copper colour. All the fins except the Ps are a bright yellow red to orange. They are very undemanding and colourful aquarium fish. For breeding see above. Unfortunately they have not been imported since 1945.

Curly Pondweed *(Potamogeton crispus) see Potamogeton*

Cushion Star *(Asterina gibbosa) see Asterina*

Cusk-eels *(Ophidium barbatum) see* Gadiformes

Cutis *see* Skin

Cutlassfish *(Trichiurus lepturus) see* Scombroidei

Cutter's Cichlid *(Cichlasoma cutteri) see Cichlasoma*

Cutting *see* Propagation of Aquarium Plants

Cuttlebone *see Sepia*

Cuvier George Leopold Chrétien Dagobert de (1769–1832). Naturalist and ichthyologist. Founder of Comparative Anatomy and Palaeontology. Born in Montbéliard in France. He went to the Karlsschule in Württemberg from 1784 to 1788 and subsequently became a tutor on the Normandy coast. Here, at 19, he drew almost all the fish of the English Channel, worked on Mollusc fauna and comparative anatomical studies which he later continued in Paris as Professor at the Central School of the Panthéon. Between 1801 and 1805, 5 volumes of *Leçons d'anatomie comparée* were published, which he completed in 1816 with *Memoires pour servir a l'histoire et à l'anatomie des mollusques.* Cuvier also extended his studies to cover fossil finds and his work *Recherches sur les essements fossiles* appeared in 4 volumes in 1812; by 1835 it had reached a total of

12 volumes. From 1803 to 1813 he spent time near the Mediterranean at Marseilles, Genoa and other Italian coastal towns where he became acquainted with the fish of the Mediterranean. The results are the basis of his major work *La Règne animal distribué d'après son organisation, pour servir de base à la zoologie et d'introduction à l'anatomie comparée* which was published in Paris in 1817. Cuvier inspired many ichthyological expeditions and with his pupil ACHILLE VALENCIENNES* (1794–1864) began the most comprehensive work about fish that has ever been written – *Histoire naturelle des poissons*. The first two volumes were published in Paris in 1828 and another 6 volumes followed before 1832. The 9th to 20th volumes were carried on by VALENCIENNES alone. The work remained uncompleted and includes many hundreds of species of fish which were described for the first time. Cuvier also believed in the unchanging nature of the species and explained the richness in forms of organisms with the help of a catastrophe theory.

Cuvierian Tubes *see* Holothuroidea

Cyanea, G, *see* Scyphozoa

Cyanophyta; Blue-green Algae. D of Schizobionta*. 2,000 species. Primitive, very small unicellular or unbranched thread organisms without a nucleus, ranging from blue-green to black in colour, very occasionally colourless. They contain chlorophyll which is partly masked by blue phycocyan or red phycoerythrin. The individual cells are often bound into loose colonies by gelatinous secretions. In unfavourable conditions some species form winter spores. The Cyanophyta are found all over the world. They live as gelatinous masses or fine-threaded coatings, mostly in fresh water but also on the ground, on the bark of trees and on cliffs. They are fairly resistant to variations in their environment. Some species can break up limestone, in others the lime is deposited in the gelatinous sheaths. Some forms are found in the sea and in large masses can cause Water Bloom* and be to some extent poisonous to fishes. Some live in the hollows in the tissue of other plants, such as the *Anabaena azollae* in *Azolla** leaves. The Cyanophyta can usually build up organic materials from inorganic ones. The energy needed for this is produced by photosynthesis*. Some species can also make use of organic substances. Some Cyanophyta are in the position of absorbing the nitrogen from the air. Thus in ricefields the yearly intake of nitrogen by Cyanophyta can be up to 50 kg per hectare. The colourless forms obtain the energy they need for synthesis by the oxidation of inorganic compounds (chemosynthesis). The Cyanophyta are not related to the true Algae (Phycophyta*). *See* also Algae Control.

Cyclocheilichthys BLEEKER, 1859. G of Cyprinidae*, Sub-F Cyprininae. Indo-China, Malay Archipelago. Found all over in rivers and larger areas of still waters. These small to medium-sized shoal fish are compact and fairly deep in shape and strongly compressed laterally. D longer than An, both concave at their rear end, C forked. The small underslung mouth is characterised by powerful lips which have 1–2 pairs of barbels which are also sometimes absent. The back is dark brownish colour and the sides greenish with a silver gleam. These fish need a tank which is not too small with a sandy bottom and hiding places in the form of groups of plants or roots. Larger specimens are well suited for public aquaria. They do not need any special water composition. Temperature 25–28 °C. Peaceful contented omnivores which prefer to pick their food up off the bottom. This G has numerous species.

– *C. apogon* CUVIER and VALENCIENNES, 1842; Indian River Barb. Indo-China, Malaya, Indonesia. To 50 cm, smaller in the aquarium. Back olive-brown, body sides greenish with silver gleam. Gill cover deep green. Dark spot at the base of each scale. A dark spot below the gill cover and on the root of the tail. The unpaired fins brick-red with grey edge. Only younger specimens are suited to the domestic aquarium. Sexual distinctions and breeding not known (fig. 406).

Fig. 406 *Cyclocheilichthys apogon*

Cycloid Scales *see* Scales

Cyclops MÜLLER, 1785 (fig. 407). Characteristic freshwater G of Copepods (Copepoda*). Including numerous species and found all over the world. There is hardly any still water in which some *Cyclops* have not lived for at least a little while. Some species also make their way into brackish water. Their structure and way of life corresponds to that of the Copepoda living out of the water. For aquarists and in nature, the various species of *Cyclops* form an important source of food for many fish.

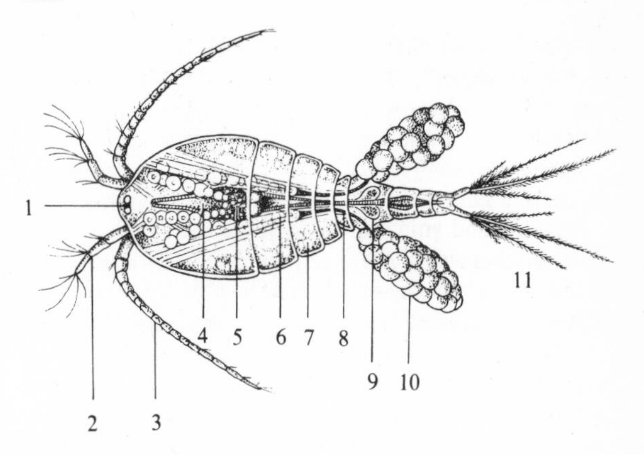

Fig. 407 Anatomy of female copepod of G *Cyclops*. 1 Eye 2 1st Antenna 3 2nd Antenna 4 Stomach 5 Ovary 6 Uterus 7 Leg muscles 8 Oviduct 9 Sperm bladder 10 Egg sacks 11 Tail fork (furca)

Cyclopteridae, F, *see* Cottoidei

Cyclopterinae, Sub-F, *see* Cottoidei

Cyclopterus LINNAEUS, 1758. G of Cyclopteridae. Sub-O Cottoidei*. *Cyclopterus* species are cold-stenothermic bottom-dwelling fish in the littoral* region of the N Atlantic and N Pacific. They attach themselves to firm substrates using the sucker on their stomach and with quick thrusts attack crabs, jellyfish and small fish amongst other things. They are thick set, high backed, plump fish with a round pug-like head. The body has rows of blunt spines. The males care for the eggs on rocks. The spawn of *Cyclopterus lumpus* is made into caviare substitute. At temperatures lower than 20 °C these fish are easy and interesting fish to keep, at least:

– *C. lumpus* LINNAEUS, 1758; Lump Sucker, Sea Hen. N Atlantic to Baltic on stony bottoms. To 60 cm. Blue-grey. Males have red stomach in the spawning season. In late summer the young are found on stones and seaweed in the waters near the shore. For illustration *see* fig. 382.

Cyclostomata; Cyclostomes. Cl of Agnatha*. The Agnatha had their heyday in primaeval times (Ordovician to Late Devonian). Only a very few Ga were flexible enough to adapt to the changing conditions of the earth's history so that they could survive into the present day. The present forms, the Lampreys (Petromyzoniformes) and Hagfish (Myxiniformes) are put together in one Cl, although it is not at all clear whether they originally came from one phylum or two. Both groups lost their bony skeleton very early in the course of the earth's history and the chance of their preservation thus became very slight. It is therefore not surprising that we have no fossils today from which we can trace their exact phylogenesis. Generally the Lampreys are traced to the Anaspida and the Hagfish from the Heterostraci. Both the Lampreys and the Hagfish are eel-like creatures, and above all the fact that they have no jaws proves that they belong to the Agnatha. Like their palaeozoic ancestors they have a head skull formed from clasps and plates but no facial skull (*see under* Agnatha). The mouth is round and shuts like an iris diaphragm. Of the two Os the Petromyzoniformes and especially their larvae have many of the primitive characteristics whereas the Hagfish are far more specialised.

– *Myxiniformes;* Hagfish. Worm-shaped sea creature with direct evolution. Head pointed; the mouth, which is somewhat underslung, is slit-shaped and surrounded by short tentacles. There is a horny-tooth in the oral cavity and the tongue is covered with several sharp horny grinding plates. They have rudimentary eyes. The organ of smell starts with an unpaired nasal passage near the oral orifice and opens out as a naso-palatal passage behind in the pharynx. During feeding, water can also be carried to the gills through this channel. The pharynx itself does not serve as a weir-basket in the Hagfish but simply as a respiratory-gut. The gills are lodged in gill-chambers and open out in pore-like individual gill-slits or as in the G *Myxine* in a collective channel. The body is laterally compressed in the tail area. A narrow edging of fin extends from the back round the tip of the tail and extends forward quite a long way on the underside. The skin is noticeably rich in glands and secretes a great deal of slimy mucus. Myxiniformes are mud-dwellers of mainland bases. They dig into the bottom and stretch the front of their body out of their conical holes. Contrary to previous ideas they

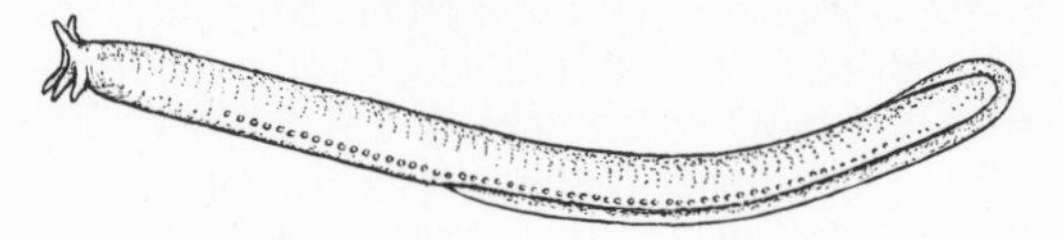

Fig. 408 *Myxine glutinosa*

are not predators of fish, but they eat carrion. Little is known about their reproduction. The bean shaped eggs which are up to 20 mm long are laid in strings or connected in meshes and are often enclosed in balls of slime. The best-known species, the 30–60 cm pale meat-coloured Hagfish *(Myxine glutinosa)* (fig. 408) is found in depths of 30–500 m on the muddy bottoms of the East Atlantic. The Myxiniformes are of no economic importance, but are however interesting subjects for zoological study.

– *Petromyzoniformes;* Lampreys (fig. 409). Eel-shaped creature but with no paired fins. Their life-cycle consists of a larval and imaginal stage. The unpaired naso-palatal passage finishes blind at the back in contrast to

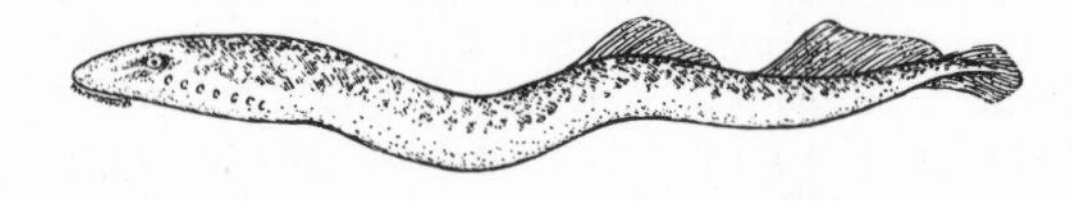

Fig. 409 Lamprey-type

that of the Myxiniformes. The larvae are mud-filterers like the primitive Agnatha; the adults are semi-parasitic, feeding on fish-blood and muscle-pulp. The larvae of Lampreys, which are also called ammocoetes or prides (fig. 410), live on the fine sand or muddy bottoms of flowing water. These worm-shaped creatures have an underslung mouth which is surrounded by lobes and finishes in a wide-meshed filter. They stretch this out from their holes against the current. Water with tiny particles of food but also with a high percentage of indigestible particles is constantly carried into the pharynx and is filtered here with the help of a mud-filterer as in the palaeozoic Agnatha*. The water flows

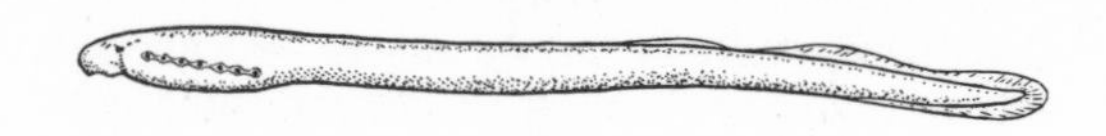

Fig. 410 Ammocoetes (Larval stage)

out through the 7 pore-like gill-slits, and the particles and the slime pass into the gut. The larvae of almost all the species grow very slowly and usually take about 4 years. They are then about 15–20 cm long depending on the species. In the metamorphosis which then follows, the eyes, which have previously been buried deep under the skin, come to the surface and the larval mouth becomes a sucking mouth with horny teeth (fig. 411), at the same time a round chisel-shaped tongue develops which is armed with teeth. 2 dorsal fins and a C develop from the undivided larval fin-border, the brownish

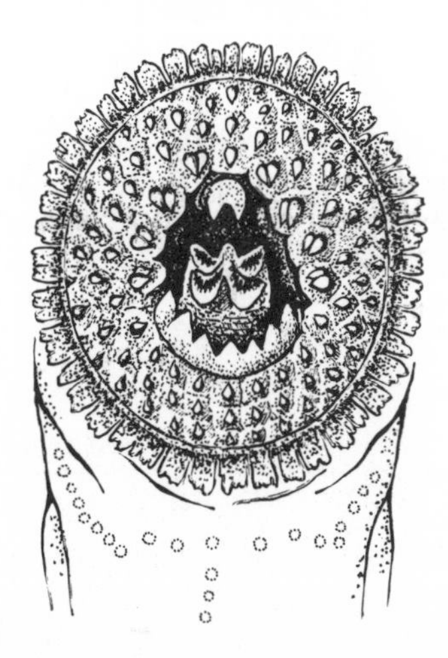

Fig. 411 Sucking mouth of the Sea Lamprey with its horny teeth

colour becomes dark on the back and has a silver gleam on the underside. In typical species eg the River Lamprey or European Lamprey *(Lampetra fluviatilis)* (fig. 412), the young imagines set off into the sea (catadromous migration) and begin their parasitic life there. They suck their prey, which consists of various fish,

Fig. 412 *Lampetra fluviatilis*

rasping holes in the musculature using their toothed-tongue and at the same time suck in blood and muscle-pulp. Thanks to this rich food they grow very quickly. When after 1–2 years they have grown to 40–50 cm long and are the thickness of a thumb, they cease taking food and go up river (anadromous migration) to their spawning grounds. When mating the male attaches itself firmly to the head of the female and encircles it with his body (fig. 413). Due to the pressure it exerts when clasping the female, the eggs are expelled from the abdominal cavity and immediately fertilised by the male. Some species fan their bodies to make little hollows in the pebbly bottoms of streams. After the mating, which takes several days, they die. The larvae develop from the yellowish eggs which are the size of millet-seeds. Some species have a shortened life-cycle and do not have the imaginal eating period or the journey to the sea. In these forms the mating and death follow a few months after the metamorphosis. A typical native example of this stationary species is the Brook Lamprey *(Lampetra planeri)*. The Petromyzoniformes are to be found only in the northern and southern temperate zones. They are never found in the tropics. A variety of up to 1.20 m long, the Sea Lamprey *(Petromyzon marinus)*, has multiplied very greatly in the large North American lakes and virtually brought the profitable fisheries to a stand-still. Lampreys are of great economic value in Europe as a delicacy. They are caught in weir-baskets during the anadromous migration. In the USSR, the chief producer, Lampreys are systematically bred and caught. In Central Europe the river pollution has long since put the original efficient Lamprey-fisheries out of production. They are smoked but usually sold in aspic. The larvae can be kept quite well in cold

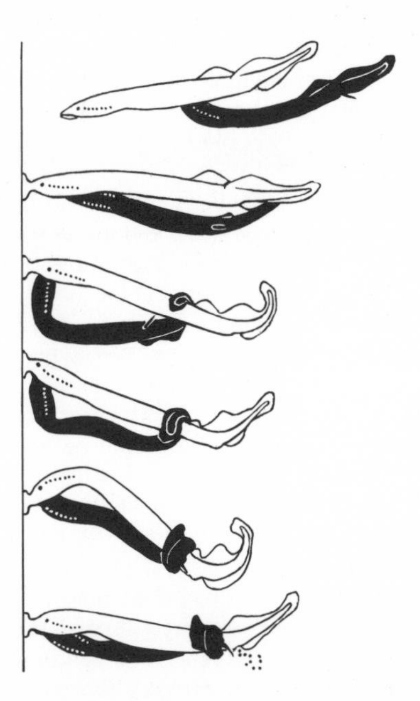

Fig. 413 Mating of the River Lamprey (male black)

water tanks for a long time and can be brought to metamorphosis. They need a deep base out of fine fairly muddy sand and good ventilation or water-circulation. Very fine stinging-nettle meal, milk-powder and fermented yeast seem to serve well as food.

Cyclostomes, Cl, *see* Cyclostomata

Cynoglossidae, F, *see* Pleuronectiformes

Cynolebias Steindachner, 1876. G of Cyprinodontidae*. Has more than 25 species. Mainly found in eastern subtropical S America (S Brazil, N Uruguay, La Plata). where they live in puddles and ponds which are liable to dry up and even in ruts which are filled with water and other such places. As befits such extreme biotopes all members of *Cynolebias* are very short-lived (annual species), ie their existence as fish is limited to the few months when their habitat has water, during and after the rainy season. They grow to maturity within a few weeks, lay their eggs in the bottom soil before the water dries up and the eggs survive the dry season without being damaged. Once it rains again, the new accumulation of water is filled with fish. The *Cynolebias* species are so well suited to this rhythm that they do not live very long in the aquarium and the eggs must be stored in a dry place in order for them to develop successfully. One finds similar behaviour in the case of the Ga *Pterolebias**, *Cynopoecilus**, *Nothobranchius** and some *Aphyosemion** species. Most of the *Cynolebias* species are small high-backed fish and they belong to the sub-genus *Cynolebias* (strictly speaking). Slimmer species are thus put in the sub-genus *Leptolebias* MYERS, 1942. *Cynopoecilus* is sometimes also thought of as a sub-genus of *Cynolebias*. In contrast to *Cynopoecilus* the 'true' *Cynolebias* have less fin-rays in the D than in the An and there are differences in the egg-shells also. The sexual dimorphism* within the G is also very striking. The males are always larger and higher-backed than the females and moreover they are more brightly coloured at spawning time. Mating males usually also take on a very dark blue to velvet black line, in contrast to the white or light blue spots on the body and fins.

Cynolebias should only be kept in tanks with their own kind. Small tanks which do not stand in too much light are suitable. They should have a good depth of bottom (boiled peat mould is suitable). As the males are keen territory-formers, there should be plenty of hiding places in the tank which can be provided using plants, stones or wood or there should only be 1–2 males to several females. They do not need any particular water-composition but the temperature should not be higher than 20 °C, 16 °C is enough for most species. They can tolerate higher temperatures but they make them even shorter lived. Before breeding the sexes should be separated for a while.

If the females show signs of spawning, mating will soon take place after the males have been put with them. The male displays himself to the female by beating his fins and swimming in circles and presses himself suddenly on his mate. The pair finally seek out the bottom and disappear in it completely to spawn. If the bottom is too hard – sand, for example, they plough the eggs in instead using the funnel-shaped anal fin. After a few weeks when all the eggs have been laid the fish should be removed and the bottom soil should be dried out so that it is still just damp. Finally the peat-mould with the spawn in it should be stored in jars or plastic bags and ventilated regularly. Whereas the embryos of some other annual species, eg *Aphyosemion**, are ready to hatch after 4 weeks, the development of *Cynolebias* and comparable species generally takes many months and can extend over years. They can be induced to hatch by soaking the peat-mould and the eggs lying in it in rain or aquarium water. As all the embryos hardly ever hatch at once the drying out procedure should be repeated several times. Otherwise the eggs can also be kept in a solution of Trypaflavin* and stored this way, free of bacteria, for up to 3 years. By changing the water, the embryos can be hatched within this space of time. As the young fish grow very quickly, they should be fed plentifully with nauplii at first and soon they will be able to cope with larger food. After 4 days the sexes can be distinguished, and after 6–8 weeks the new generation reaches sexual maturity.

– *C. bellotti* STEINDACHNER, 1881; Argentine Pearlfish. La Plata Basin. Males to 7 cm, females smaller. Males dark-blue, almost black during mating; body and fins with white to light blue spots. Female has dark cross bands or a marbly effect on a yellow-brown background. Both sexes have a dark arched band which stretches from the neck across the eyes (fig. 414).

Fig. 414 *Cynolebias bellotti*

– *C. nigripinnis* REGAN, 1912; Black-finned Pearlfish. La Plata Basin and Rio Paraná. Males to 5 cm, females smaller. Colouring similar to that of *Cynolebias bellotti* but without the dark head bands. Males rich blue-black or velvet-black when mating with lighter spots. D has longitudinal band formed by spots. Female pale grey to ochre, with darker marbly effect (fig. 415).

Fig. 415 *Cynolebias nigripinnis*

– *C. whitei* MYERS, 1942; White's Cynolebias/Pearlfish. SE Brazil. To 8 cm, female smaller. Male brownish, with pale blue-green spots; D and An base colour the same, with orange colour at the outside; point or edge black. Female yellow-brown with two dark spots on the body sides as well as two smaller dark spots in front of the caudal peduncle.

– *C. wolterstorffi* AHL, 1924; Wolterstorff's Pearlfish. SE Brazil. To 10 cm. Females smaller. Males similar to *Cynolebias bellotti*, head with deep black band, female similar to *Cynolebias nigripinnis*, fins colourless. Spawns readily at temperatures around 14 °C.

Cynolebias holmbergi *see* Cyprinodontidae

Cynopoecilus REGAN, 1912. G of Cyprinodontidae*. Found from Uruguay to Brazil; has very few species. Related most nearly to *Cynolebias** from which it can be distinguished by its slimmer body form and the fact that it has the same number of D and An rays. *Cynopoecilus* is similar in many respects to *Cynolebias* in its way of life and consequently in its requirements regarding care and rearing, and thus the *Cynolebias* should be referred to for details.

– *C. ladigesi* (FOERSCH, 1958); Ladiges' Gaucho. Brazil. To 4 cm. Body and fins of male a brilliant emerald green with dark red transverse bands. Females plain brownish.

– *C. melanotaenia* (REGAN, 1912); Fighting Gaucho. Brazil and Uruguay. To 5 cm. Colouring red-brown, pale greenish sheen, dark carmine-red longitudinal band stretches from the eye to the tail-section. Fins have red spots. Females similar colour but D and An rounded.

Cyperaceae; Reed grasses or Sedges. F of Liliatae*. 70 Ga, 37,000 species. Found all over the world particularly in temperature areas. Bushy and mottled perennial plants, grass-like in character, rarely annuals. Leaves arranged in rows of three, linear. Flowers hermaphroditic or unisexual and monoecious, inconspicuous, in inflorescences built up of little spikelets. *Carex* (Sedge) is a widespread G of Cyperaceae. There are indeed species found in mountains, dry places, and in woods, but they live mainly in marshes, on river-banks and at the edges of lakes and ponds, also in damp meadows and on moors, and on airless, poorly drained ground.

Fig. 416 *Cyperus alternifolius*

They often form larger, closed patches. The species of G *Cyperus* grow in similar places but predominantly in the tropics and subtropics. In the early days of aquarium keeping, some species were kept in the aquarium such as *C. alternifolius* (fig. 416) from Madagascar. As the plants always grow up out of the water, they are of less significance nowadays. However, one G which is useful is *Eleocharis**

Cyperus (various species) *see* Cyperaceae

Cyphotilapia REGAN, 1920. G of Cichlidae*. Only found in Lake Tanganyika*. Fish over 30 cm in length with very characteristically shaped heads. Mouth has thick lips and the profile of the head is indented in front of the eyes, and bulging out again in older fish (similar to the G *Steatocranus**). Furthermore the head is the highest part of the body thus giving it an extraordinarily bullish appearance. Fins pointed with the exception of the C, C rounded. Care and keeping as for *Lamprologus** and *Julidochromis**. Only species is *C. frontosa* (BOULENGER, 1906) (fig. 417).

Fig. 417 *Cyphotilapia frontosa*

Cypraea LINNAEUS, 1758; Cowries (fig. 418). G of Prosobranchs (Prosobranchia*). The *Cypraea* species live in warm seas mainly on coral coasts in shallow waters and feed on animal and vegetable matter. The oval, shining shells are characteristic of this G. The last spiral of the shell covers all the previous ones so that they are not externally noticeable. The toothed mouth extends along its whole length. *C. moneta* and other species were once used as a form of money in Africa and Asia. In the living animal the shell is almost always covered by the mantle. Cowries are easy to keep in the aquarium, eg *C. tigrina*, Tiger Cowrie.

Fig. 418 *Cypraea* sp.

Cyprinidae; Carps and Carp-like fishes. F of Cypriniformes*, Sub-O Cyprinoidei*. The Carps are distinguished from the members of all other Fs of the Sub-O by the presence of a millstone. This is the wart-shaped projection at the rear end of the base of the skull which is covered in a hard horny substance and which serves as a support for the pharyngeal teeth (*see* Cyprinoidei). Food is crushed or ground between the pharyngeal teeth and the millstone. All other families do have pharyngeal teeth but no millstone. Further characteristics are the very heavy pharyngeal bones as well as the very heavy, few but powerful pharyngeal teeth which can be formed as gripping, dished, chewing or grinding teeth and are often arranged in two to three rows (figs. 419 and 420). In contrast to the Cobitidae and Homalopteridae, which have at least 3 pairs of barbels, the Cyprinidae, with few exceptions, have only 2 or more often 1 pair of barbels, which are usually laid back

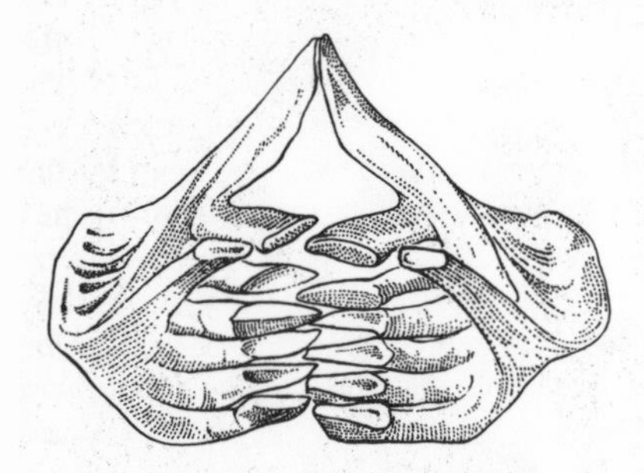

Fig. 419 Pharyngeal teeth of Chub *(Leuciscus cephalus)*

when they are swimming. Nevertheless in some species, barbels are missing altogether. With regard to many other characteristics, Carps are, however, similar to the other families (*see* Cyprinoidei). The oldest fossils are from the early Tertiary period in Great Britain. The number of modern species is extremely large and is probably about 1,250. Almost all species are purely freshwater fish which are found in Europe, Africa and N America as well as Northern Central America. They are not found in S America, on Madagascar or in Australia.

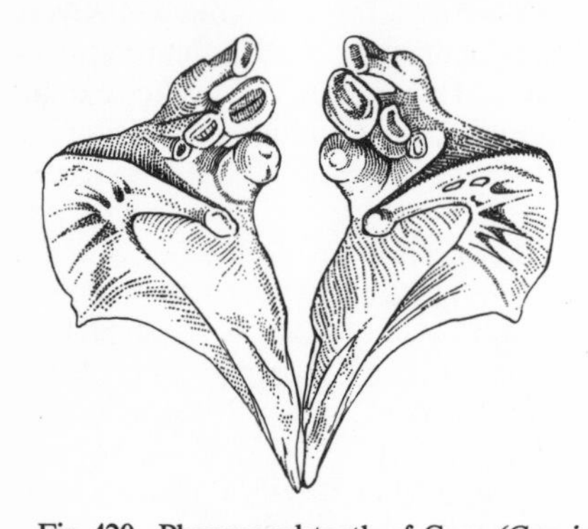

Fig. 420 Pharyngeal teeth of Carp *(Cyprinus carpio)*

The individual Ga and species live in extremely varied biotopes of all types. Nevertheless it can be shown that still or slowly flowing warmer waters form the typical environment for the Cyprinidae. Few only have accustomed themselves to conditions of cool or cold and fast-flowing waters. There are open-water species as well as those which prefer waters which are close to banks and vegetation. Many Cyprinidae are bottom-dwellers; one species *(Caecobarbus geertsi)* is blind and lives in the caves of the Ngungu (Thysville) on the lower Congo. Almost all species stay together in schools or shoals, at least as young fish. The carnivorous species which live on small animals and the purely herbivorous species are together less in number than the omnivores, a few of which however do have a specialised diet. Most Cyprinidae spawn between or above plants. The small eggs remain adhering to parts of plants where they land and in a few hours or days they are mature embryos. The young hang vertically onto various substrates until they have completely consumed the yolk-sac. Other Carps lay their eggs over the bottom, whereas still others have pelagic non-adhering eggs and the young which hatch from them have no organ of attachment (eg *Ctenopharyngodon*). Still others such as the Moderlieschen *(Leucaspius delineatus)* spawn on substrates. Spawn migration is only known in the case of forms from brackish water. Brood-care is rare in this F. Thus *Pseudogobio rivularis*, which is found in the River Amur, makes a hollow for spawning and chases other fish from the spawning ground. The males of the Moderlieschen which was mentioned above fan fresh water to the eggs. The Bitterling *(Rhodeus**)* has developed a completely unique and individual method of brood-care. These fish place their eggs in the gills of large bivalves. In 1941 BARNARD wrote of a livebearing Barb in Africa *(Barbus viviparus)*. This report was later proved to be false. Most Cyprinidae are extremely fertile. It is not unusual for eggs totalling 50,000 to 150,000 to be produced. In many regions of the earth, but above all in the northern, subtropical and temperate latitudes, Carps are important economically. More than others this applies to the Carp *(Cyprinus carpio)* and the Roach *(Rutilus rutilus)* but Bream, Tench and Goldfish are also caught in great numbers every year. The fish are brought to market freshly caught or smoked, and outside Europe they are also dried. Many species are also interesting subjects for anglers. The smaller, colourful carp-like fish from tropical and subtropical waters are kept in large numbers by aquarists. They are mainly extremely undemanding and can be accustomed to many different kinds of food. One does not require any special experience or knowledge in order to rear them. Many indigenous species can be kept in cold water aquaria without difficulty. This family supplies many subjects for scientific research, such as Goldfish, Minnows and Carps amongst others. In general, attempts are made at dividing this huge F into several Sub-Fs. However opinions vary greatly and the system which has been chosen here has like all others advantages and disadvantages. The Ga mentioned in the Sub-Fs are treated as separate entries.

– *Abraminae.* Elongate to relatively deep, laterally much compressed Carp-like fishes with completely or partly keeled belly. An very long. Pharyngeal teeth in 1–3 rows. *Abramis, Alburnus, Blicca, Chela, Leucaspius, Oxygaster, Rasborichthys.*

– *Cyprininae;* Carps in strict sense. Elongate, laterally only moderately compressed Carps, almost always with a rounded belly. An short, 5–6 and rarely 7 soft-rays. D is generally opposite the Vs. The first fin-ray of the An and D is frequently spiny and toothed on the rear edge. Pharyngeal teeth in 1–3 rows, the outer row has no more than than 7 teeth. Near the root of the tail the lateral lines almost always thus runs along the middle. *Balantiocheilus, Barbus, Barilius, Caecobarbus, Carassuis, Cyclocheilichthys, Cyprinus, Gobio, Labeo, Morulius, Osteochilus, Rhinichthys.*

– *Garrinae.* Mostly elegant and elongate Carp-like fish, frequently with relatively straight ventral profile. Upper lip fused with the skin of the conical, protruding snout and frequently granulated or set with fringe-like papillae or horny tubercles. In some Ga *(Garra, Discolabeo)* the lower lip has developed into a sucking-disc. Jaw-edges sharp, functioning as a scraper. Ps often enlarged, lying horizontal and serving as contact surfaces to the bottom. They mostly live in fast-flowing water where they feed on Algae. *Crossochilus, Epalzeorhynchus.*

– *Leuciscinae.* Elongate or fairly elongate Carp-fish. An usually has 8–11 separate soft-rays. D generally short and without spines on the front edge. Pharyngeal teeth in 1–2 rows. Barbels absent. (Exception *Tinca.*) *Chrosomus, Idus, Leuciscus, Notropis, Phoxinus, Scardinius, Squalius, Tinca.*

– *Rasborinae;* Rasboras. Small to very small, elegant, elongate Carp-fishes, which are found almost exclusively in SE Asia and on the Malayan Islands. An short with 6–7, occasionally 8, soft rays, D likewise short always beginning behind the Vs. Mouth slightly upturned, lower jaw somewhat protruding with a symphyseal knob. Lateral line generally turns down in an arch and finishes in the lower half of the caudal peduncle. It is often greatly reduced. 1 or 3 rows of pharyngeal teeth. *Brachydanio, Danio, Esomus, Hemigrammocypris, Luciosoma, Microrasbora, Rasbora, Tanichthys.*

– *Rhodeinae;* Bitterlings. Small Carp-fishes, with An situated very near the front and beginning almost on a level with the rear end of the D. Pharyngeal teeth arranged in a row. Found mainly in East Asia. *Rhodeus.*

Cypriniformes; Carp-fish types. O of Osteichthyes*. COH Teleostei. With about 3,500–4,000 modern species, this O is not only the one with most species among the Osteichthyes but also among the Chordata* altogether. The first fossil specimens are from the Upper Cretacean and since then the Cypriniformes have developed continuously. Such a strong progressive development is called radiation. To this O belong the Sub-Os Characoidei* (Characin related types) Gymnotoidei* (Knife-fish related types) and Cyprinoidei (Carp-fish related types). Together with the Siluriformes* (Cat-fish types) the Cypriniformes represent the Sup-O Ostariophysi. In other systems, eg that of BERG, the Catfishes are not placed on a par with Cypriniformes but grouped as Sub-O Siluroidei (Catfish related species), ie they are on a par with the Characoidei, Gymnotoidei and Cyprinoidei. The characteristic feature of the Carp-types is the so-called Weberian apparatus* by which the swimbladder and the inner ear are joined together. They share this characteristic only with the Siluriformes. Connections between the swimbladder and the inner-ear are quite common amongst the primitive groups of Teleostei but such connections are mostly produced by the swimbladder extending forward and making contact with the ear-region or by the swimbladder forming outfoldings which penetrate into the skull through openings in the cranial capsule and lying against certain areas of the inner-ear (Balancing Organ). The Weberian apparatus is, however, a connection of quite a different kind; here volume changes caused by pressure or by sound waves caused by vibrations of the swimbladder wall are transferred to the inner-ear and registered there by little bones joined together in a flexible chain. The apparatus consists of two parts. The first 4–5 vertebrae serve as support or hanging apparatus. They are altered to a degree; quite often there are fused vertebrae. The moveable sections which lie one under the other are anchored to this so-called sustentaculum. They form a chain on either side made up of 3, and in some case 4, small bones which make the connection with the inner ear. The moveable part of the apparatus is called *Pars auditiva.* Alterations of the Weberian apparatus specific to various groups have systematic importance. Its efficiency is shown by the fact that the Cypriniformes and also the Siluriformes react very quickly to changes in pressure and in general are good at noticing sound-waves and can differentiate them, ie they have a very primitive sense of hearing. The other characteristics of the Cypriniformes show them to be fairly primitive Teleostei. Amongst these characteristics are: the fins supported by soft-rays, even the first fin rays which are often formed as spines are reconstructed soft-rays; the Vs are on the underside far behind the Ps; the anal vent lies at the back; the swimbladder is connected to the gut. The features which distinguish them from the Siluriformes, apart from certain very specific peculiarities of the skeleton are the following: the Cypriniformes are less specialised; the parietal bones, symplecticum and subopercular bone are present in the skull; the vomer is without teeth. Bones between muscle fish-bones well-developed in general; the vertebrae are seldom fused in the Weberian apparatus and when there is any fusion then it is only between the 2nd and 3rd vertebrae; the 5th vertebra is the first to have normal ribs; the swimbladder is constricted in the middle to form a front and rear section, the body is usually covered with cycloid scales and occasionally ctenoid scales and is only naked in exceptional cases. The Cypriniformes are almost exclusively freshwater fish with large concentrations of species in S Asia, Africa and tropical S America. They are found in almost all water types and fill numerous ecological niches. Many species are carnivorous, which feed on small invertebrates or prey on other fish, some are omnivores or are purely plant or detritus eaters. In some regions, Cypriniformes are of great importance to internal fisheries and many species are favourite objects of ornamental fish care and rearing because of their small size, splendid colouring and markings, and their adaptability. For this reason the individual Sub-Os and their Fs have been dealt with as separate entries. For the classification of the Cypriniformes *see* the system of bony-fishes under Osteichthyes*.

Cyprininae, Sub-F, *see* Cyprinidae

Cyprinodon LACÉPÈDE, 1803. G of Cyprinodontidae*. There are about 12 species of these found on the coasts of temperate N and Central America including the Antilles. They live in brackish or salt water areas of the interior. Small to medium-sized, compact, similar to Carps in shape. They are a little compressed laterally, and very high-backed when older. The fins are roundish. D lies quite a long way in front of An, C fan-shaped. D and An usually smaller in the female than the male and with a black spot in each case. As befits their natural state the *Cyprinodon* species (like *Aphanius**) should be kept in brackish water with about 7% sea-salt. Their shy nature should be catered for by providing hiding-places (stones, thick groups of plants). They should not be kept with other fish as *Cyprinodon* only shows its brilliant colouring when kept in tanks with their own species. They are very greedy and are not fussy eaters but should be fed a rich variety (*Daphnia*, mosquito larvae, *Tubifex*, and additional plant food). The water temperature should be between 15°C and 20°C although it is preferable that it should be too cool rather than too warm. They are easy to breed, when they are fed well. A breeding pair spawns every 3–4 days. They are attached spawners (*see* Cyprinodontidae). The eggs should be picked out after spawning. The young hatch after 1–2 weeks depending on the water temperature. The following species is most important for aquarists.

– *C. variegatus* LACÉPÈDE, 1803; Sheepshead or Purse Minnow. Eastern N and Central America including the islands lying near. To 8 cm. Male has patches which have brilliant steel-blue and sea-green sheen; body-sides, throat and stomach orange; fin colour variable. Female has numerous indistinct dark transverse bands.

Cyprinodontidae; Egg-laying Tooth-carps (fig. 421). F of Atheriniformes*, Sub-O Cyprinodontoidei*. Up to now there are about 450 species known in the large F, which are divided into the following Sub-F.

– Rivulinae (about 250 species)	Asia, Africa and America
– Fundulinae (about 40 species)	America
– Cyprinodontinae (about 17 species)	America
– Orestiatinae (about 18 species)	America
– Aphaniinae (about 13 species)	Africa
– Procatopodinae (about 80 species)	Africa
– Pantonodontinae (about 2 species)	Africa and Madagascar
– Oryziatinae (about 10 species)	Asia

As the table shows, the Cyprinodontidae are found on all continents with the exception of Australia, but seem to prefer the tropical and subtropical areas. Almost all species are only to be found in fresh water and only a few isolated species live in brackish water. A very few can also tolerate sea water. In many respects the Egg-laying Tooth-carps correspond to the particulars given under the Sub-O Cyprinodontoidei*. Most species are very small, hardly reaching 5 cm. The largest ones *(Cynolebias holmbergi* and *Orestias cuvieri)* are found in S America and can grow to 30 cm. The smallest species live in Asia and Africa and are only 2 cm long *(Oryzias minutillus, Aplocheilichthys myersi).* The body is always elongate, the head flattened at the top, the protrusile, toothed mouth is generally somewhat upturned and particularly suited to taking food from the water-surface. Various members of the Sub-F Aphaniinae do not feed in this way nor do almost all Cyprinodontinae who in addition eat mats of Algae. In general the females have round scales and only the females of the Procatopodinae have cyclo-ctenoid scales which the males of almost all the Sub-Fs have. The scale arrangement of the upper side of the body is often specific to the genus or species and is important for determining these. The lateral line organs are absent on the body as in all members of the Sub-O Cyprinodontoidei but are, however, well developed on the head. The sense-buds (neuromasts) are situated in channels with individual openings outwards or in open grooves or even in hollows. The size and number of neuromasts varies greatly between the individual Sub-Fs and Ga. There is also great variety within species in that they appear in numerous local forms and sub-species. Sexual differences are almost always apparent; The males which are larger as a rule have longer fins and stronger markings and colouration. The Cyprinodontidae are not shoaling fish. It is only among the Cyprinodontinae, Aphaniinae and Oryziatinae that schools are formed at specific times. Breeding is simple and details about this are to be found under relevant Ga. It is sufficient to mention here that one can distinguish between attached and bottom spawners: the former lay their eggs on plants whereas

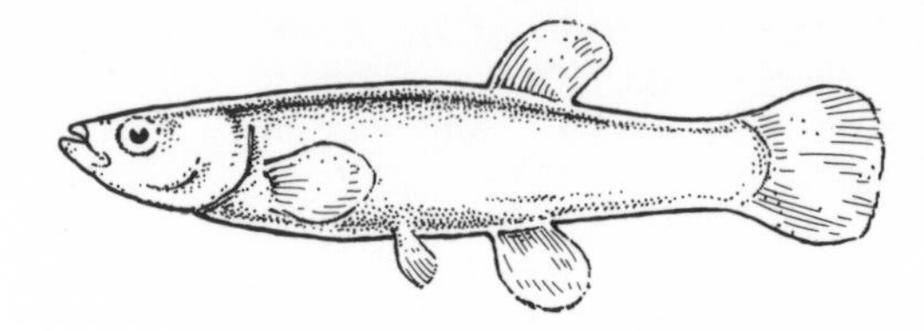

Fig. 421 Cyprinodontidae-type

the bottom spawners embed them in the ground. Some species may be either attached or bottom spawners depending on the conditions in which they live. Almost all of those which are confirmed bottom-spawners are so-called annual species, ie they live in small accumulations of water which are liable to dry up and the eggs survive the dry period stored in the damp bottom. The development of the egg is thus limited to a defined period (diapause). With the onset of rain the young fish hatch and quickly grow to sexually mature adults. The eggs themselves are unusually hard-shelled. The Cyprinodontidae are of hardly any economic importance; as they are limited to small accumulations of water they only have local significance as food for larger predatory fish. This family is of very special significance to the aquarist. There are specialists in this group of fish in almost every country. International exchange is enhanced by the ability to send eggs world-wide packed in damp substrates such as peat-mould. The activity of these specialists has given rise to numerous valuable observations in the fields of biology and classification of the Cyprinodontidae. The following Ga are treated as separate entries: *Aphanius, Aphyoplatys, Aphyosemion, Aplocheilichthys, Aplocheilus, Austrofundulus, Cubanichthys, Cynolebias, Cynopoecilus, Epiplatys,*

Fundulosoma, Fundulus, Garmanella, Jordanella, Lucania, Nothobranchius, Oryzias, Pachypanchax, Procatopus, Pterolebias, Rachovia, Rivulus, Roloffia, Valencia.

Cyprinodontinae, Sub-F, *see* Cyprinodontidae

Cyprinodontoidei; Tooth-carps, related fish. Sub-O of Atheriniformes*, Sup-O Atherinomorpha. Small to very small freshwater fish with Pike or Rasbora-like form. This fish group is represented in the REGAN-system as O Microcyprini and as O Cyprinodontiformes in the BERG-system; however, in both cases the Amblyopsidae are included which now, as a result of new information, are transferred to the O Percopsiformes* in the system of GREENWOOD and colleagues (1966) which is used here. The area of dispersion of this large and interesting Sub-O extends over the tropics and subtropics of all continents except Australia, only a few single representatives are found in the temperate latiudes. A few species have been released in different parts of the world because they eat mosquito larvae; others reached new regions through the activities of aquarists and have greatly increased their numbers there. The Cyprinodontoidei are characteristic inhabitants of small weedy stretches of water; many are almost exclusively surface-feeders, others eat predominantly insect larvae and small Crustaceans. In the family Cyprinodontidae* species of various Ga have even adapted themselves to life in temporary accumulations of water. Their annual cycle includes a phase in which the species survives only as an egg embedded in the damp bottom. The Cyprinodontoidei are characterised morphologically above all by the following: body more or less elongate, head conical, frequently with a flattened top and with a wide terminal but upturned mouth. Jaw toothed; teeth mostly long and pointed. The upper edge of the mouth is formed by the premaxilla and is protrusile (with few exceptions). All fins are supported by soft-rays only, D set relatively far back, C rounded or straight-edged, adipose fin absent, Vs on the underside. In many species parts of the uppaired fins, or certain whole fins, are elongated. The lateral line is missing from the body but the lateral line organs on the head are well developed. Swimbladder is not connected to the intestine (physoclists). Round scales usually on the head and body, a few individual cases have cyclo-ctenoid scales on some parts of the body. No inter-muscular bones (fish-bones). Sexual dimorphism very apparent as a rule, even with regard to the size of the body, the fin arrangement, markings and colouration. In live-bearing species the males also have copulatory organs which usually consist of a partly modified An. Their economic importance is slight. In many parts of the world the Cyprinodontoidei keep the plague of mosquitos in check. The oldest fossils date back to the middle Tertiary (Early Oligocene). Because of the significance of this Sub-O to the aquarist, the following Fs are treated as separate entries: Adrianichthyidae, Anablepidae, Cyprinodontidae, Goodeidae, Jenynsiidae, Poeciliidae.

Cyprinoidei; Carp related fish. Sub-O of Cypriniformes*. Sup-O Ostariophysi. About 250 Ga in total and 1,500 species most of which are found in S Asia. The area of dispersion of the Sub-O extends far beyond this, however, and includes the whole of Europe, with the exception of Northern Norway, the whole of Asia approximately to the latitude of 70° N, Indonesia, Africa, except the Sahara Belt, N America and Central America as far as Guatemala in the South. The earliest fossils date to the early Tertiary. Most Cyprinoidei represent the common idea of the shape of a fish. Some groups have a greatly elongated body, other bottom-dwelling Cyprinoidei have a much flattened body, ie a shortening of the dorso-ventral profile (eg Homalopteridae). However disc-shaped species are missing from the Sub-O. The body is densely covered with cycloid scales as a rule; the head, however, has no scales. The terminal but rarely inferior mouth which is surrounded by obvious, occasionally swollen lips can be protruded to some extent. In addition 1–3 pairs of barbels are grouped around the mouth of many Ga. The gape of the mouth is bordered above usually only by the premaxilla, the maxilla being included in the protruding mechanism of the mouth. The jaw-edges and the roof of the mouth are without teeth. The so-called pharyngeal teeth are an important characteristic for the classification of many Cyprinoidei. These consist of 1–2 rows of conical or blunt powerful teeth which are arranged on the lower bony section of the right and left 5th gill-arch in such a way that they can be moved against each other or against a horny plate at the base of the cranium. In some Ga the structure of the skull has certain peculiarities which are characteristic for the more primitive *Characoidei**. One of these is a window-like cut-out between the quadratum and metapterygoid. The Weberian apparatus (*see* Cypriniformes) is often characterised by the fusion between the 2nd and 3rd vertebrae. The rear end of the vertebral column which is so important for the classification of many groups of fish differs so much in the Cyprinoidei that it plays no systematic role. The fins are supported by soft-rays, in the D and An as well as the Ps, the 1st and 2nd soft-rays can also be formed as a hard spine. Intermuscular bones (fish-bones) are found in both the back and stomach musculature. The fin arrangement is normal, the adipose fin is absent (with the exception of some Cobitidae*). The swimbladder is constricted into two sections (and occasionally more) and connected to the gut by a channel (physostomes). This channel is, however, often closed in older fish, ie it enables no intake or expulsion of air. To regulate the amount of air other mechanisms have thus been developed. All members of the Sub-O are freshwater fish of flowing or still waters; some live for a time in brackish water. Most species expel their spawn into open water or between water-plants. The eggs adhere very well and thus stick to various substrates. Some Cyprinoidei have simple brood-care, very few have developed special forms of brood-care or ensuring the safety of the spawn. Some Cyprinoidei reach a considerable size and others grow no larger than 3 cm. Many kinds of Cyprinoidei are of regional economic importance; some are the best known of useful fish. A large number of smaller, beautifully coloured species are ideal aquarium fish. Because of their economic importance and their interest for aquarists the following

Fs are dealt with under separate entries: Catostomidae, Cobitidae, Cyprinidae, Gyrinocheilidae, Homalopteridae.

Cyprinus LINNAEUS, 1758; True Carps. G of Cyprinidae*, Sub-F Cyprininae. SE Europe, SE Asia in deep, slowly flowing, large rivers with sandy and muddy bottoms. However, the water is quite warm. The bottom-dwelling shoal fish are large or very large. The body is slightly elongate to high-backed with a broad back and slightly flattened laterally. The scales are strikingly large. D longer than An, C clearly 2 pointed. The terminal mouth is evertible and characterised by a well defined bulging lip with 2 pairs of barbels. These fish which are predominantly active at dusk feed on bottom-dwelling creatures such as molluscs, worms, small crabs, insect larvae and plant food. They should be kept in fairly large tanks with soft sand and adequate ventilation and filtration. The water should not be too fresh and should only be topped up with fresh water occasionally. Temperature should be 16–22 °C depending on where the fish come from and should be cooler in winter. The G comprises 5 species, the main area of dispersion of which is SE Asia and there is only one species from Europe.

Fig. 422 *Cyprinus carpio*

– *C. carpio* LINNAEUS, 1758. Carp. To 120 cm, generally smaller. Originally found as 3 sub-species in SE Europe and SE Asia. Since the thirteenth century as important foodfish they have been spread across the whole of Europe, N America and Australia in various hybrid forms (Mirror Carps, Leather Carps amongst others). Colouring grey-brown to grey-green and silvery, the males have a few spawning tubercles on head and Ps at spawning time. Only younger fish are suitable for keeping and they can be fed with every type of live food, oatmeal dough with fish or meat added, soaked maize and soya. The Koi from Japan has become well-known. The fish which have been specially bred are extremely variable in colour and grow to 60 cm (fig. 422).

Cypselurus (several species) *see* Exocoetoidei

Cyst. A characteristic organ for Algae* and Fungi*, in which reproductive cells are formed. The cyst develops from a solitary cell, and so in contrast to the angium* the outer boundary consists of a cell wall. If asexual reproductive cells (Spores*) are formed, the cyst is called a *sporocyst*, if sexual cells (Gametes*) are formed, it is called a *gametocyst*.

Cystoseira. G of Brown Algae (Phaeophyta*) from the upper littoral* of warmer seas. It has a tree-like growth. The Cystoseira which form large clusters on firm ground on the southern European coasts are not suitable for marine aquaria. When the water is polluted the Cystoseira are the first algae to die off and can thus serve as bio-indicators on sea coasts.

Cytology. Branch of biology* and medicine that is concerned with the structure and functions of the cells and their component parts. Cytology used to be based on examinations of cells under a light microscope, but modern research into cells has largely gone over to the electron microscope in order to explain the finest of structures. Information gained like this, together with the help of biochemistry*, has contributed considerably to the study of the molecular structure of cells and the processes of metabolism and metabolic regulation.

Cytology (cytodiagnostics) is of great practical importance in medicine for the identification and treatment of diseases. For example, it is used to determine the cellular composition of the blood.

Dab *(Limanda limanda) see* Pleuronectiformes
Dace *(Leuciscus leuciscus) see Leuciscus*
Dactylogyrus, G, *see* Trematoda
Dactylopteridae, F, *see* Dactylopteriformes

Dactylopteriformes. O of the Osteichthyes*, Coh Teleostei. Small, elongated, bottom-dwelling fish of warm seas with enormous Ps which comprise an upper and a lower sail-like part similar to those of the Triglidae (*see* Scorpaenoidei). The fish uses the lower part of the P, together with the Vs at the throat, to support itself or move along the bottom. Unlike the Ps of the Triglidae, the lower parts of the Ps in the Dactylopteriformes are not composed of single fin rays, but are simply small, normal fins. Other typical features of the O concern the structure of the cranium and the shoulder girdle. It is thought that the Dactylopteriformes are related to the Scorpaeniformes*. There are a few species, all belonging to the Dactylopteridae F (fig. 423) which seek levels of water near the surface at spawning time. They can also speed out of the water and sail for short stretches.

Fig. 423 Dactylopteridae-type

– *Dactylopterus volitans* (fig. 424), Flying Gurnard, which grows up to 50 cm long, is found on sandy and muddy bottoms in the Mediterranean. This is a brown flecked fish with dark, blue-patterned Ps and a very long spine springing from the nape (= 1st fin ray of the D).

Fig. 424 *Dactylopterus volitans*

Dactylopterus volitans *see* Dactylopteriformes

Dactyloscopidae, F, *see* Trachinoidei

Dahlia Anemone *(Tealia felina) see Tealia*

Daisy Anemone *(Cereus pedunculatus) see Cereus*

Dalatiidae, F, name not in current usage, *see* Squalidae

Dalatiinae, Sub-F, *see* Squalidae

Dallia, G, *see* Esocoidei

Dalliidae, F, name not in current usage, *see* Esocoidei

Damselfishes, F, *see* Pomacentridae

Damselfishes, G, *see Abudefduf*

Danio HAMILTON-BUCHANAN, 1822. G of the Cyprinidae*, Sub-F Rasborinae. SE Asia, India to Thailand. Small to medium-sized shoaling fishes that live in still and flowing waters. Body elongated and much flattened laterally. Mouth terminal or superior, surrounded by 2 pairs of barbels, which may be reduced or absent in some cases. This G is distinguished from the G *Brachydanio** by its larger number of divided dorsal fin rays (12–18) and by its complete LL. *Danios* are well adapted swimmers and good jumpers. They should be kept as a shoal in large, long tanks. Do not keep along with calm fish. Any vegetation may be planted loosely and must allow enough room for swimming. There are no particular requirements regarding the composition of the water, although the occasional addition of some fresh water is advisable. Temperature 20–24 °C, a little cooler in winter. As a rule, all types of live and dried food is accepted. Smaller tanks (10–30 l) with fresh water are suitable for breeding. Anchor fine-leaved plants in the substrate which is covered with coarse gravel. The hardness and pH value of the water is of little consequence. Put the females in first, then 1–2 days later add 1–2 males to every female. Spawning usually takes place in the morning sunshine. Very prone to eating the spawn. The brood hatches after 20–36 hours, nourishing themselves for a further 48–72 hours with their yolk supply. Feed the fast-growing young fish with very fine live or dried food. The parent fish can spawn again after 3–4 weeks, if kept well fed.

– *D. devario* (HAMILTON-BUCHANAN, 1822). N India to Bangladesh and Assam. Up to 10 cm. Compressed with a sharply curved ventral profile. No barbels. Dorsal surface grey to olive-green, flanks greenish-grey with a silvery iridescence. Behind the gill coverings there are indistinct, blue and yellow vertical stripes. There are 3 blue longitudinal bands, flanked by yellow lines, which unite into 1 band on the root of the tail. This band continues into the upper lobe of the C. Fins dirty yellow to reddish (fig. 425).

Fig. 425 *Danio devario*

– *D. malabaricus* (JERDON, 1849); Giant Danio. Common in SW India (Malabar coast), Sri Lanka (Ceylon). Up to 15 cm. 1 pair of barbels. Dorsal surface greenish, ventral surface reddish-yellow. Flanks have 3–4 steel-blue longitudinal bands, the middle one of which reaches into the C. These bands are separated from one another by red-gold longitudinal stripes. There are several gold-coloured oblique stripes behind the gill coverings. Large breeding tanks are essential (fig. 426).

Fig. 426 *Danio malabaricus*

– *D. regina* FOWLER, 1934. S Thailand. Up to 12 cm. Dorsal surface olive-green, flanks a green iridescent colour. A single, broad, matt-blue band reaches into the C. It is bounded on both sides in yellow. There is a black-blue blotch behind the gill slit (this is not present in *D. malabaricus*). There have been only isolated cases of successful breeding.

Danube Catfish *(Silurus glanis) see Silurus*

Daphnia MÜLLER, 1785 (fig. 427); Water-fleas. Well-known G of the Cladocera*. Distributed worldwide in the bank area of still fresh waters. The females of many species grow to over 4 mm long; the males are much

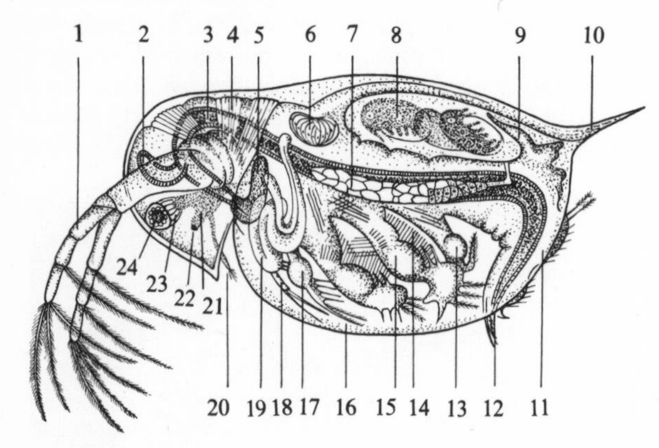

Fig. 427 *Daphnia.* Anatomy of a Water-flea. 1 2nd antenna 2 Liver cornicle 3 Intestine 4 Muscle of antenna 5 Mouth parts 6 Heart 7 Ovary 8 Embryos in brood chamber 9 Dorsal process 10 Spine (thorn) 11 Abdomen 12 Anus 13 5th breastbone with small gill sac 14 Filter bristles 15 4th breastbone 16 Shell 17 2nd breastbone 18 1st breastbone 19 Excretory organ 20 1st antenna 21 Ganglion 22 Ocellus 23 Eye muscles 24 Compound eye

smaller, besides which they are often only encountered at cold times of the year. *Daphnia* shapes vary markedly depending on the environmental conditions. *Daphnia* form a considerable portion of the diet of many fish and are used for this purpose in aquarium-keeping.

Dardanus PAULSON, 1875. G of large, predatory Hermit Crabs (Paguridea*) from warm seas. The left pincer is bigger than the right. *D. megistos* from the tropical Indo-Pacific is a very hardy species, often imported. However, it should only be kept with large crustaceans or fish.

Dark Clownfish *(Amphiprion bicinctus) see Amphiprion*

Darkling Beetles, G, *see Tenebrio*

Darter Characin *(Characidium fasciatum) see Characidium*

Darwin, Charles Robert (1809–82). Important English naturalist. Studied at the universities of Edinburgh and Cambridge. C. LYELL's book *Principles of Geology* and Darwin's own observations on a journey round the world (1831–36), which took him to S America and various oceanic islands (the Galapagos), stirred him to study the evolutionary development of life. He wrote numerous works, most importantly *On the Origin of Species by means of Natural Selection* (1859), in which he laid down his conclusions about the origin of species. His basic ideas on the subject are still valid today as descent theory*.

Dascyllus (CUVIER, 1829). G of the Pomacentridae*. Members of this G are found in tropical and subtropical parts of the Indo-Pacific, including the Red Sea. They live in pairs or in small polygamous groups on any large coral growths. Within a group all the individuals know each other 'personally' and there exists a distinct ranking order. Any aliens are attacked. When defending their territory, *Dascyllus* makes purring noises. Scarcely bigger than 10 cm, these fish are tall, snub-nosed and very lively. The rear edge of the front part of the operculum is serrated. They are substrate spawners that take care of their eggs. The larvae live amongst the plankton. In the aquarium too, they spawn regularly at brief intervals. Not yet reared successfully. All species of *Dascyllus* will accept any food, and all are extremely hardy, highly recommended aquarium fish.

– *D. aruanus* (LINNAEUS, 1758); Banded Humbug, Black and White Damsel, Humbug Damsel, White-tailed Damsel, Three-striped Damselfish. Indo-Pacific, Red Sea. Up to 8 cm. During the day pearl white with 3 black oblique bands; at night and in aquaria that are too dark, totally grey-black. C transparent.

– *D. marginatus* (RÜPPEL, 1828); Marginate Damselfish, Marginate Puller. Up to 10 cm. Dirty white, darker in the anterior part of the body. C has a blue border (fig. 428).

Fig. 428 *Dascyllus marginatus*

– *D. melanurus* (BLEEKER, 1854); Black-tailed Footballer, Black-tailed Humbug. Eastern Indo-Pacific. Up to 8 cm. Like *D. aruanus* but with a black C (fig. 429).

– *D. reticulatus* (RICHARDSON, 1845); Reticulated Damselfish, Reticulated Puller. Indo-Pacific. Yellowish-

Fig. 429 *Dascyllus melanurus*

brown, with a dark band behind the head and another in front of the caudal peduncle (fig. 430).
– *D. trimaculatus* (RÜPPEL, 1828); Domino Damsel, Three-spot Damsel, Three-spot Humbug, White-spot Puller. Indo-Pacific, Red Sea. Up to 15 cm. Velvet black with a white blotch at the forehead and a similar blotch on each of the flanks. More grey at night, when spawning and with age. When young lives as Anemonefishes*.

Fig. 430 *Dascyllus reticulatus*

Dasyatidae; Stingrays. F of the Chondrichthyes*, O Myliobatiformes. The F contains a great many species, and all possess a poison spine, serrated on both sides, which is located on the whip-like caudal peduncle. Another typical feature is the large Ps which surround the snout like a fringe (fig. 431). Ds, C and An are all absent. All Dasyatidae are livebearing. There are 6–16 young at a time, and in many species they are nourished by a cell tissue of the oviduct. Stingrays live on the bottom, fanning themselves into the sand with their Ps so that only the eyes and the large spiracles remain uncovered. Swimmers or inexperienced divers often come too close to the fish because they are so well hidden, so triggering off the fish's defence mechanism.

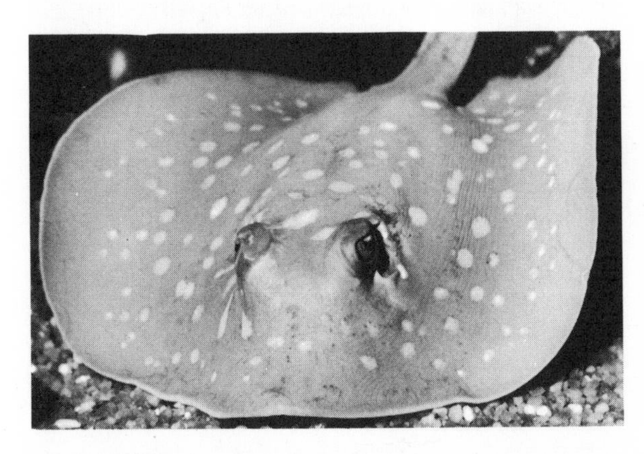

Fig. 431 *Taeniura lymma*

The Stingray strikes upwards and sideways with its whip tail, often causing wounds with its spine. This is not only unpleasant, it is also dangerous. There are reports of large stingrays which have battered through the side of wooden boats, their tail can strike so hard. The damage done by the spine itself (which often lodges in the wound) is less harmful than the poison. It is said to paralyse the heart muscles, and may also damage the nervous system, respiratory system and the kidneys. Any damage to the abdomen region can be fatal. The poison is formed in 2 poison glands which lie in grooves at the side of the spine. They often remain behind in the wound along with the spine. For this reason the wound must be cleaned thoroughly. Stingrays can replace lost spines but not the poison glands. Many species have 3–4 spines. In some places they are used as spear tips.

Stingrays occur mainly in shallow waters of tropical seas. Their food consists chiefly of molluscs, crustaceans and fish. *Dasyatis pastinaca* (fig. 432) is found in the Black Sea, the Mediterranean and the E Atlantic, and occasionally it has been observed in the North Sea in late summer. It can grow up to 1.50 m long. The Potamotrygonidae are closely related to the Dasyatidae. The freshwater forms from South America (G *Potamotrygon*) belong to this family. They occur in the Amazon Basin and are considered particularly dangerous. The Dasyatidae are of little economic importance.

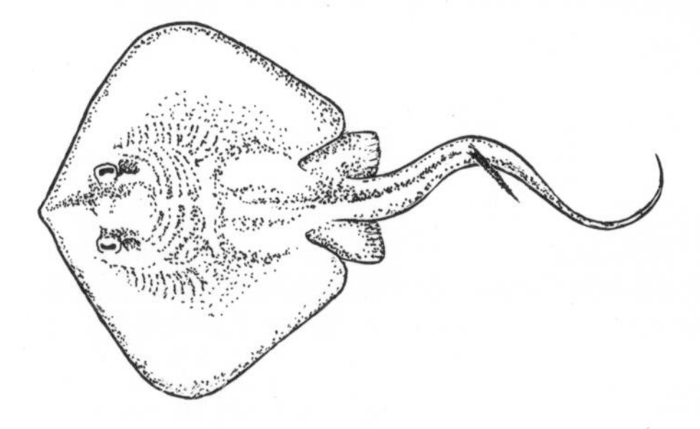

Fig. 432 *Dasyatis pastinaca*

Dasyatis pastinaca *see* Dasyatidae

Debilitating Parasites *see* Parasitism

Decabrachia, O, *see* Cephalopoda

Decapoda. O of the crustaceans (Crustacea*), belonging to the Sub-Cl Malacostraca*. Includes Crayfishes *(Astacus*)*, Common Lobsters *(Homarus*)*, Spiny Lobsters *(Palinurus*)*, Hermit Crabs (*see* Paguridea), Crabs (Brachiura) etc. In all Decapoda the first 3 pairs of legs on the breast have become jaw feet which are used exclusively to help food intake. The remaining 5 pairs of legs on the breast are remarkable for their size, which is why these animals have been given the name Decapoda. The legs allow the animal to move about, although the front pair often possesses powerful pincers which may be used to grip things or in defence. Most Decapoda live in the sea, some in fresh water, and a few on land (eg the Robber Crab). No ten-legged crustacean is truly independent of the water, however. The Decapoda also include the largest of all crustaceans, the Giant Crab *(Kaempferia)* of Japan. This species lives in deep waters off the coast of Japan, and its legs may span more than 2 m. Because of their size, bizarre appearance, interesting habits and often beautiful colouring, many Decapoda make popular specimens in the marine aquarium.

Decorative Material is used in many ways in the aquarium. Apart from its purely aesthetic function, it is also essential for many aquarium animals to live, creating hiding-places, territorial boundaries etc. Decorations should be non-toxic, free of contaminating organic matter and corrosion-proof in sea water. They must not induce undesirable changes in the water. For example, metallic compounds effect noticeable imbalances in the water, calcium and magnesium salts increase the water hardness, and organic acids lead to a lowering of the pH value. All decorative material should be cleaned thoroughly before being placed in the aquarium and possibly disinfected (*see* Disinfection).
– *Rocks.* The most suitable are those that do not release hardness-forming substances – metamorphic or volcanic rock, such as granite, basalt and gneiss. The completely inert quartz rocks can also be used in sea and fresh water. Schist and sandstone must be tested for hardness-forming substances before use in fresh water. To carry this out, drip a small amount of hydrochloric acid onto the rock. If a lot of gas bubbles up, calcium compounds are present which can make the water extremely hard (*see* Water Hardness). In marine aquaria, rocks containing calcium may be used, eg some sandstone, pure limestone and dolomite. Several types of rock gradually release small amounts of trace elements* into the water, which may create a deficiency, particularly in marine aquaria. The much-used xylolite, which is obtained by silicifying parts of plants, can be variable in quality, releasing nitrates, organic acids and hardness-forming substances under certain circumstances. Rocks containing unknown minerals and metallic compounds should not be used in the aquarium. Before placing any rocks in the aquarium, scrub them thoroughly. (It may be necessary to place them in boiling water for a time.) Finish off with a good rinse. Particularly in marine aquaria, all organic matter attached to the rocks should be meticulously cleaned off.
– *Roots and pieces of wood.* Should only be used in freshwater aquaria. The wood must be dead and have rotted in water for several years, so that the decomposition process is completed and that any remaining humic acids are released only in limited amounts into the aquarium water. Live wood or wood with rotting parts on it are as unsuitable as dead pieces of wood that have dried out in the air and contain a high proportion of organic material. It is best to use roots from elder, beech, willow or oak. Wood that has lain for a long time in bogs possesses good qualities. Before placing any wood in the aquarium, clean thoroughly with a brush, boil it in a saturated cooking salt solution, and finally rinse again. Some tropical indoor plants form aerial roots (*Monstera, Scindapsus, Philodendron* etc), which, when introduced to the aquarium, produce highly decorative, much branched aquatic roots. They are particularly suitable for aquaria with very active burrowing fish which cannot normally be put in a planted tank. At the same time, mineral break-down products are removed from the water which maintains a healthy balance in the aquarium. In a freshwater aquarium, dead reeds and bamboo which have been boiled may also be used as a decoration. It is best to seal the ends or the whole stem with a suitable resin to avoid any rot setting in.
– *Corals.* Aquaria in which coral fishes are kept can be decorated with coral of all different types. Coral consists of calcium or horn-like compounds. Living coral polyps only survive for a short while in the aquarium, even under favourable conditions. Newly acquired pieces of coral must be specially cleaned before use in order to get rid of any matter that might cause decay. This is done by placing the coral for at least 48 hours in a plastic container to be treated with potash lye or caustic soda solution (add 300 ml of 15% lye solution to 10 l fresh water – this is corrosive matter, so protect the eyes!). Then rinse under running water for several hours, clean with a brush, boil thoroughly in pure water and finally rinse once more. Corals that have not been cleaned sufficiently well will cloud the water around them and acquire a veil of bacteria after a short while in the aquarium. Remove the corals immediately and clean them thoroughly once more.

Many species of fish require places to hide. There are many things that can be used for this purpose – the rocks, corals, roots or branches found naturally in waters, as well as coconut shells that have been split open, plastic and clay pipes and flowerpots. They can be cleaned easily, and although they are of less importance than the decorative material, such hiding places are essential for the successful rearing of some types of fish. Such materials can also be used with success in biotope aquaria by skilfully disguising them as naturally occurring objects.
– *Back walls.* The simplest way of creating the illusion of infinity to the back wall of the aquarium is to paint it a dark colour or use a dark cardboard box. A flat box has endless possibilites. Stonework, corals, roots, bark, reeds and other materials can be stuck on to the box which can then be placed behind the aquarium. It is possible to create the illusion of great depth in this way. In large aquaria a rear wall is often placed directly into the tank. Stonework can be applied in such cases, which will create extra holes and crevices, but the stones must be well secured with cement (*see* Poisons) or other adhesive materials. Slabs of pressed peat (*see* Peat) or an artificial material are often used, into which depressions have been made. Such materials can be planted and can look very decorative. If peat is used, its effect on the water quality must be taken into account and its suitability for the hobbyist's purposes must be tested.

Decussate *see* Leaf Arrangement

Deep Zone *see* Abyss

Deep-sea Anglers, related species (Ceratioidei, Sub-O) *see* Lophiiformes

Deep-sea Eels (Synaphobranchidae and Serrivomeridae, Fs) *see* Anguilliformes

Deep-sea Giganturids (Giganturidae, F) *see* Cetomimiformes

Deep-sea Hatchetfishes (Sternoptychidae, F) *see* Stomiatoidei

Deep-sea Spiny Eel types, O, *see* Notacanthiformes

Defensive Broadside Position *see* Inferiority Behaviour

Deficiency Diseases occur in organisms as a result of an insufficient supply of material that is needed to sustain life. Aquatic plants will show clear signs of deficiencies if nutrients are not available in a balanced ratio or are not available in a form that can be absorbed by them. Other causes might be a lack of vital trace elements, insufficient light or light with an unsuitable spectral combination (*see* Lighting). For example, a lack of iron in plants will upset the means by which green assimilation pigments are formed, so that only white-yellow dwarf leaves (Chlorosis*) will grow. Fish will waste away if not given enough to eat. But much more common in the aquarium are deficiency diseases caused by feeding* fish too much of one thing. This will result in there being a lack of certain substances in the fish's body that are necessary for it to live. A lack of vitamins* produces very characteristic deficiency diseases. Lack of vitamin A leads to a fatty degeneration of organs (Degeneration*, Liver, Diseases of*). Lack of vitamins B1 and B2 can cause disturbances in growth and a general lowering of the body's defence mechanisms. An insufficient supply of vitamin D can disturb the hardness of the skeleton bones and so cause them to become deformed. A natural source of vitamins for fish is provided by plant food and plant-eating food animals. If vitamin deficiency diseases do occur, pure vitamin preparations can be added to the food. The reduction in vividness of colours often found in aquarium fish (as opposed to the same fish in the wild), can often be traced back to a lack of some particular substances in the food. For coral fish, the effect of light, especially controlled doses of UV light, is important in their life-styles. Other causes of deficiency diseases, of which there are many, are not yet fully understood. It is often the specialist eaters among the fish that are prone to deficiency diseases in captivity; at the very least they seem to be capable of less when kept in the aquarium and so far no adequate explanation has been found.

Definitive or Final Host *see* Parasitism

Degeneration. The cells* of an animal or plant may in part lose their characteristics, which may result in changes to tissues, organs and entire organ systems. The causes of degeneration are many and various, but may arise from the non-usage of organs, lingering infectious diseases (*see* Infection), genetic damage (*see* Genetics) etc. In fish, often as a result of unsuitable feeding, excess fatty deposits build up in the liver particularly, but also in the heart, kidneys and pancreas, so overburdening these organs. This fatty degeneration brings about severe damage to the metabolism, as the organ cells are turned into functionless fatty tissues. Regeneration* is the opposite process to degeneration.

Delesseria. G of the red algae (Rhodophyta*) containing a great many species and found in the lower littoral of cool and warm seas. The thalli are shining red and flat, with a mid-rib from which nerves branch off; they look very much like leaves. These lovely algae will usually survive well in the marine aquarium, provided they are not damaged in transit. Unfortunately, they are easily overgrown by green algae and will then die.

Dendrochirus (SWAINSON, 1876). G of the Scorpaenidae, Sub-O Scorpaenoidei*. Contains a few species from the tropical Indo-Pacific. In way of life and appearance very similar to *Pterois**. However, *Dendrochirus* only grows 15–20 cm long; its longest P rays reach only as far as the base of the C and all are united by a membrane into a point. The D spines are also scarcely longer than the body height. Poison glands open out into the D spines. In the aquarium, *Dendrochirus* are hardy, enchanting fish, but like *Pterois* they need an astonishing amount of live food fish or shrimps. It is difficult to get *Dendrochirus* to switch to dead substitute foodstuffs or earthworms. The following are often imported.

– *D. brachypterus* (CUVIER and VALENCIENNES, 1852); Turkeyfish, Scorpion. Tropical Indo-Pacific, Red Sea. Up to 17 cm. Body brick-red with indistinct oblique bands. P has greenish oblique bands (fig. 433).

Fig. 433 *Dendrochirus brachypterus*

– *D. zebra* (CUVIER and VALENCIENNES, 1849); Dwarf Lionfish, Zebra Lionfish. Tropical Indo-Pacific. Up to 20 cm. Body has very prominent brown-red and white oblique stripes with a T-shaped pattern on the root of the tail fin.

Dentalium, G, *see* Scaphopoda

Dentex *(Dentex dentex) see* Sparidae

Denticeps clupeoides *see* Clupeiformes

Denticipidae, F, *see* Clupeiformes

Dermal Bones. Bony formations of the corium (dermis). There are several features in fish that are derived from dermal bones: all bony scales and part of the cranium bones; bony plates, eg those of the Sturgeons Mailed Catfishes (Loricariidae*), the South American Catfishes or Hassars *(Hoplosternum*)*; the bony rings of the Pipe-fishes (Syngnathidae *see* Syngnathoidei) that encircle the body, and the complete coat of mail that occurs in the Trunk-fishes (Ostraciontidae*); the spines of the Porcupine-fishes (Diodontidae*) and the Pufferfishes (Tetraodontidae *see* Tetraodontiformes). (Acipenseridae), the Sticklebacks (Gasterosteidae), the A large number of extinct fish from the Earth's ancient past had a complete coat of mail (Aphetohyoidea, *see* Osteichthyes).

Dermal bones are mainly used for protection. Large-surface dermal bones severely impair the movement of the body.

Dermogenys pusillus *see* Exocoetoidei

Desalination. The removal of salts dissolved in water* either through distillation or by means of ion exchangers*. *Distillation* relies on the fact that only pure water is evaporated when it is boiled, leaving the salts behind as a residue. The steam is then condensed again by cooling it. This method of desalination is time-consuming, and demands expensive equipment and a lot of energy. Hence it is only really used for small quantities of water. With the *ion exchange* method, any salt ions* are compounded by means of certain resins and replaced with hydrogen or hydroxide ions. This will result in pure, completely neutral water (*see* Ion exchangers). In aquarium-keeping it should be noted that extremely soft water is beneficial to neither fish nor plants. Try to aim for soft water* with a hardness* of 2–3° dH. This can be arrived at by mixing the desalinated water with a corresponding amount of the original water (*see* Standard Water).

Descent Theory. Theory of evolution. The theory generally recognised today states that all the species alive at this moment have come about through a process of development known as evolution* in the course of world history. During the phylogenetic development of living organisms there has been a higher development of simple life forms up to man. An abundance of different life forms has also developed.

In essence, descent theory was established in the eighteenth and nineteenth centuries. Whereas LAMARCK* saw the origin of evolution in the inheritance of acquired characters (through use and non-use of organs), DARWIN* recognised the importance of natural selection (*see* Selection). Subsequently, the existence of evolution and phylogenesis* was repeatedly proven by a great many scientists working in all fields of biology. Modern evolutionary theories, in addition on those of Darwinism, are supported primarily by recent conclusions in the fields of genetics*, comparative anatomy*, palaeontology*, systematics* and embryology.

Desert Mouthbrooder *(Haplochromis desfontainesi) see Haplochromis*

Detritus. Fine particles of animal or plant origin (eg finely ground remains of aquatic plants) or particles of inorganic origin. In waters with a strong current these particles may be held in suspension, or they may be deposited on the bottom (sediment). Detritus feeds a whole host of animals (detritus-feeding animals), eg many annelids (Annelida) insect larvae (Insecta), small crustaceans (Crustacea) and molluscs (Mollusca).

Detritus Siphon. The simplest way of removing substrate debris in the aquarium is by using a tube to suck it up, together with a little of the water. A funnel-like attachment is fixed onto the sucking end of the tube; this reduces the suction effect so that only the light debris is sucked up, not the heavy sand particles of the substrate as well. The water that is sucked up will contain a large amount of debris, but it can be poured back into the tank, provided it is passed through a filter immediately beforehand. For smaller tanks and for day-to-day care of your tank, dip-tubes can be used. These suck up water together with the substrate debris into a sealed catchment bulb; this makes it impossible to be able to continually suck up debris from very large areas of the substrate (fig. 434). There are also technical pieces of equipment, sometimes called aquarium 'vacuum-cleaners'. These are attached to an air-pump (Pumps*) or have built-in pumps so that a continuous water suction is produced. The connecting suction pipe, which is shaped like a funnel, can be moved over the sand or pressed into it. The water is conducted over a filter bag (that can be washed out), where the pieces of debris are removed, and then it is pumped straight back into the tank.

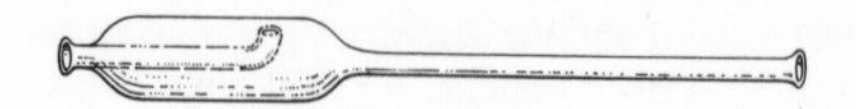

Fig. 434 Dip-tube

Devil Fish *(Geophagus jurupari) see Geophagus*

Devil Rays, F, *see* Mobulidae

Diadem Sea-urchin, G, *see Diadema*

Diadema HUMPHREY, 1797; Diadem Sea-urchin. A tropical sea-urchin (Echinoidea*) with thin, needle-sharp spines up to 30 cm long. Often found in great clusters even at shallow depths. The body can measure up to 9 cm wide and is often covered in bright blue spots. 'Sea-urchin Fishes' often live between the spines (*see Siphamia* etc). A wound caused by the spines becomes inflamed as remains of the very brittle spines always get lodged there. These sea-urchins are difficult to transport and care for.

Diagnosis. The recognition of the characteristic features of an object. A diagnosis is particularly important in the recognition of diseases, ie their type and causes. A precise knowledge of the normal condition and any deviations from it is necessary for a diagnosis. Fish diseases are extremely difficult to diagnose. Some of the difficulties are poor observation; a complicated method of examining live specimens; expensive specialised procedures which are not always possible for the layman to carry out; the large number of single-celled and multi-celled pathogens; and finally, the problem created by the fact that fish need to be in water to live. All these factors mean that fish disease diagnosis has become a complicated specialist area. For the correct diagnosis of a disease it is important to recognise the characteristic symptoms of a particular disease and distinguish them from any untypical deviations that might point to any one of a number of different diseases. When a fish dies, it may not be because of disease, as deteriorating living conditions may lead to death without any characteristic changes taking place visibly to the fish. If a fish that really *is* ill is to be treated successfully, a very exact diagnosis will be required, using a specific examination technique* and anamnesis*.

Diagnostic Feature. Used in systematics* to describe any feature of a living organism that can be used to separate one group of organisms from another (*see* Taxon). It includes all characters (colour, anatomy, behaviour etc) that do not fall within the range of variation of one taxon. Identification tables and descriptions of groups of organisms are based on such diagnostic features. However, not every diagnostic feature can be used in the reconstruction of natural (ie phylogenetic) relationships; in these cases, there has to be a character evaluation (apomorphic*, plesiomorphic*). With fish diseases diagnostic features are taken to mean any deviations from the norm. Such features are very important in the recognition of a type of disease.

Dialammus fuscus *see* Blennioidei

Diamond Tetra *(Moenkhausia pittieri) see Moenkhausia*

Diamond-back Moray *(Echidna nebulosa) see Echidna*

Diamond-flecked Pufferfish *(Canthigaster margaritatus) see Canthigaster*

Dianema COPE, 1871. G of the Callichthyidae*. Widespread armoured catfish, about 10 cm long, found in the Amazon Basin. Elongated, cylindrical body. Similar in habit to members of the G *Hoplosternum**. The dorsal profile is somewhat more arched than the ventral profile. There are 4 bony plates between the D1 and D2 which are firmly locked together with plates along the flanks. The much elongated ventral parts of the shoulder girdle cover the whole breast area and meet at the middle line. In contrast to the species of the G *Hoplosternum* the C is deeply forked. 2 pairs of long barbels are carried stretched out in front. *Dianema* is a crepuscular fish that seeks out hiding places during the day. They live in shoals. They have been bred successfully *(see D. longibarbis)*. As in *Hoplosternum* and *Callichthys* a large bubble-nest is constructed. *Dianema* is an unassuming omnivore which will survive many years in the aquarium.

– *D. longibarbis* COPE, 1872. Widely distributed throughout the Amazon. Up to 8 cm. Dorsal surface dark with a weak blue iridescence; flanks pale beige to flesh-coloured with numerous small dark dots sprinkled erratically over the body. Fins colourless. An interesting but unfortunately rarely imported species, which has often been observed building a bubble nest. Kerocczalla has been successful with rearing (fig. 435).

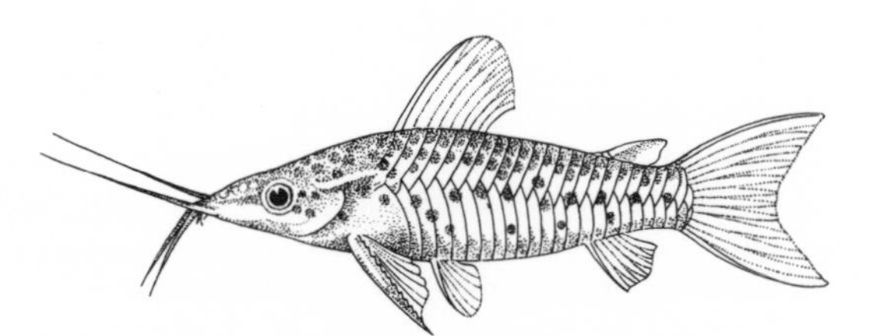

Fig. 435 *Dianema longibarbis*

Fig. 436 *Dianema urostriata*

– *D. urostriata* RIBEIRO, 1912; Stripe-tailed Catfish. Found in tributaries of the Amazon near Manaus. Up to 10 cm. Dorsal surface dark, flanks pale brown to beige-coloured with a lot of small dark dots. All fins colourless, with the exception of the C which is striped horizontally black and white. A very attractive species which has already been bred successfully in the USA, but as yet there are no detailed reports. To date only rarely imported and in small quantities (fig. 436).

Diapause. A kind of dormancy, ie a deviation from the normal even tempo of development in an individual life cycle. It is an hiatus in development with a greatly reduced metabolism* during which time an organism is adapted to unfavourable environmental conditions, thereby enabling it to survive. A diapause is common amongst invertebrates, but also occurs in fish that live in habitats with extreme conditions (eg waters that dry up periodically). For example, annual members of the Cyprinodontidae* have an egg diapause, and some Lungfishes (Lepidosirenidae, *see* Ceratodiformes) live through the dry period in a 'summer sleep' in holes in the mud.

Diaphragm or Membrane Pumps *see* Pumps

Diaptomus WESTWOOD, 1836 (fig. 437). Typical freshwater G of the copepods (Copepoda*). Unlike *Cyclops* which constantly moves about, *Diaptomus* is a true planktonic form. It can float along in open water using

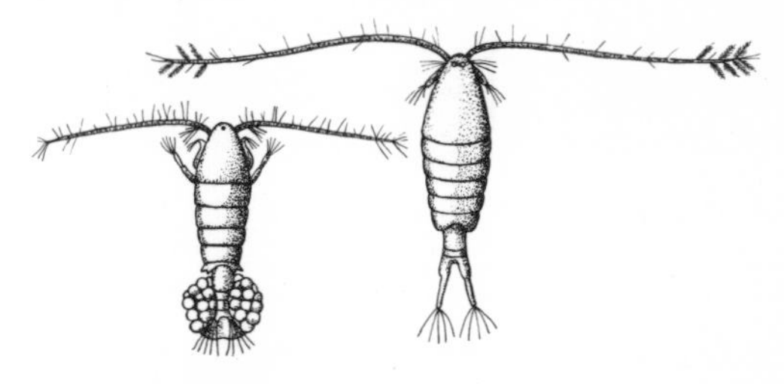

Fig. 437 *Diaptomus* sp.

its long, first pair of antennae which are held out on a horizontal plane. Only occasionally does it have to compensate for a slight sinking by making a jumping movement upwards. Taxonomically, it is important to note that many species of *Diaptomus* are today placed in separate Ga (*Eudiaptomus, Hemidiaptomus, Arctodiaptomus,* etc); as a result the G now comprises a small number of species (about 7 European ones). All the related groups of *Diaptomus* are important in lakes* and ponds*, but for the aquarist they are much less important than *Cyclops**.

Diatoms, D, *see* Bacillariophyta

Dibranchiata, Sub-Cl, *see* Cephalopoda

Dicotyledoneae, Cl, name not in current usage, *see* Magnoliatae

Dicotyledons (Dicotyledoneae, Cl, name not in current usage) *see* Magnoliatae

Diffusion. The spontaneous mixing of substances that come into direct contact with one another. For example, two salt solutions of different concentrations (such as fresh water and sea water) will mix in with each other until the concentration is even throughout the mixture. This happens even without the two solutions being stirred together. The same process can be observed with gases, but with solid objects diffusion takes place very slowly. The speed with which diffusion takes place depends on temperature and the size of the molecules that make up the substances. Diffusion also plays a vital role in living organisms. Ingestion and egestion are often carried out by means of diffusion. On the other hand, every organism has mechanisms at its disposal which can be used to maintain differences in concentration, ie work against diffusion. Such mechanisms are always associated with membranes, and their presence enables very different metabolic processes to take place side by side in one cell (*see* Osmosis).

Digestive Organs (fig. 438). An organ system found in multi-celled animals for the mechanical and chemical preparation of food, so that the components of the food can take part in the metabolism* of the body. Despite a great many peculiarities of feature, the digestive canal and digestive glands in fish correspond to the digestive organs of other vertebrates.

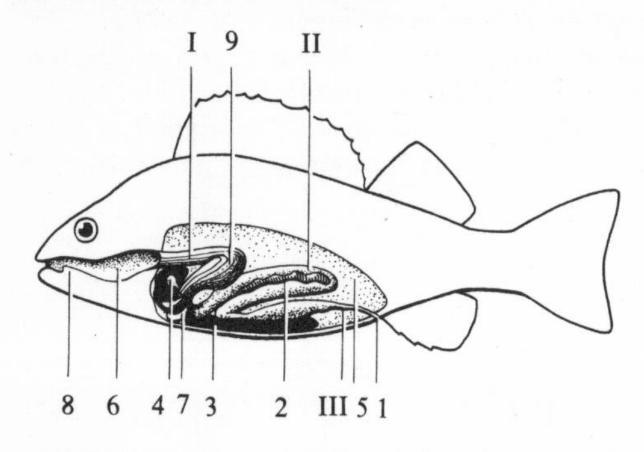

Fig. 438 Digestive organs. I Fore-gut II Mid-gut III Hind-gut 1 Anus 2 Pancreas 3 Liver 4 Gall bladder 5 Abdominal cavity 6 Gill cavity 7 Pylorus' outgrowths 8 Tongue 9 Stomach

In the front section of the digestive canal, the *mouth* and *gill cavity*, fish usually have numerous teeth* with which they acquire and partially break up their food. The tongue cannot move, and there are no salivary glands. The water that is taken in whilst eating is passed out of the body again via the gill slits*. As this happens, the food particles are retained by the gill rakers* and pass through the short *oesophagus* into the *stomach*, or where there is no stomach, into the mid-gut (eg Blennidae, Cobitidae, Cyprinidae, Cyprinodontidae, Gobiesocidae, Gobiidae, Scaridae). In the acid surroundings of the stomach the enzymatic protein digestion begins with the aid of the pepsin released from the stomach glands. These glands lie in the mucus membrane of the stomach which, like the mucus membrane of the intestine, is surrounded by a layer of circular and longitudinal muscle, and by a layer of connective tissue rich in blood vessels. A fish's stomach is often U-shaped or Y-shaped if there is a *stomach blind sac*. In rare cases, it may be elongate. At the exit from the stomach the circular musculature is often strengthened into a sphincter muscle or it is developed in a large part into a masticatory stomach (eg Mugilidae). In many cases the stomach cannot be distinguished externally from the elongate or twisting *mid-gut*. When a stomach is present, the transition is marked by tube-like blind sacs known as *pylorus outgrowths*, which enlarge the resorptive surface of the intestine. Inside the mid-gut of a fish (except bony fish) there is a spiralling mucous membrane fold. This *spiral intestine* also represents an additional resorptive surface, and so the exploitation of the food is good despite the length of the intestine being fairly small.

In the alkali milieu of the mid-gut the two large digestive glands, the liver and the pancreas, are of fundamental importance for the chemical break-down of the food pulp. Bile, the secretion of the *liver*, flows via the bile ducts (there may be no gall-bladder) and liver duct into the intestine. There, the bile, together with the bile acids, breaks up the fat into very fine drops, so that it is prepared for chemical breakdown. On the other hand, the liver receives from the intestine, via the portal vein system of the blood, the basic ingredients of the digested proteins and carbohydrates (and also of the fats in fish). These are worked upon by the liver, then as necessary, they are released into the blood system or stored. That is why a fish's liver is rich in fat. The liver has other functions to do with detoxication, the breakdown of the blood pigments into bile pigments (Spleen*) and the storing of blood. Chemically active digestive enzymes are produced by the intestine and chiefly by the *pancreas*. In bony fish, the pancreas is not a compact organ but is split up into small units all along the portal vein system to the liver. It consists of 2 components with different functions, and in fish it is often divided into separate areas. The digestive enzymes (amylase, lipase, trypsin) develop in the exocrinal parts. The endocrinal parts do not take any part in the digestion. They form the hormones insulin and glucagon with which the blood-sugar level is regulated.

The undigested parts of the food pulp are passed into the *hind-gut* and leave the body via the *anus*. Bony fish do not have a cloaca*.

Dinoflagellates, D, *see* Dinophyta

Dinophyta. Division of the Algae*. 1,000 species. Typical colouring is yellow-brown to reddish, and chlorophyll type a, and only a little type c, is present. Lentil-shaped chromatophores. In those forms that move about a lot there is a flagella for propulsion, and a second one that lies in a furrow. The Dinophyta are a major constituent of marine plankton; only a few species occur in fresh water or on land. They are single-celled organisms often with bizarre shapes, but only rarely forming colonies. Inactive forms often possess floating processes. Some have a form of external protective envelope. The species of several Ga produce a brief flash of light in the sea. Others secrete poisons, and if they congregate in large enough numbers they may poison fish. There are also species that live as parasites in marine animals.

Dinospore *see Oodinium*

Diodon LINNAEUS, 1758 (fig. 439). G of the Diodontidae*, comprising several species found in all tropical seas. *Diodon* lives alone in shallow water and feeds on molluscs and crustaceans. It is a plump, ovoid fish which grows to between 50 and 90 cm long, and is capable of inflating itself so that dermal spines stick out on all sides. In contrast to *Chilomycterus**, *Diodon*'s spines normally rest flat against the body. *Diodon* swims mainly with the aid of the Ps. Its eyes are remarkably large and highly mobile. Porcupine fish keep well in captivity if fed on molluscs, snails, shrimps etc.

Fig. 439 *Diodon holacanthus*

They become quite tame. However, they are susceptible to skin parasites. Try not to let the fish puff themselves up in the air, so if they have to be transferred to another tank, ladle them out, do not use a net. The best known species is the following.

– *D. hystrix* LINNAEUS, 1758; Common Porcupine Fish; Spotted Porcupine Fish. Found in all warm seas, with one or two in the Mediterranean. Up to 90 cm. Grey-white with brown blotches. *See* Diodontidae.

Diodontidae; Porcupine Fishes (fig. 440). F of the Tetraodontiformes*, Sub-O Tetraodontoidei. Found in coastal waters of all tropical and subtropical oceans, especially on reefs. The Diodontidae are club-shaped or extended into a long ovoid shape. Usually, there is only

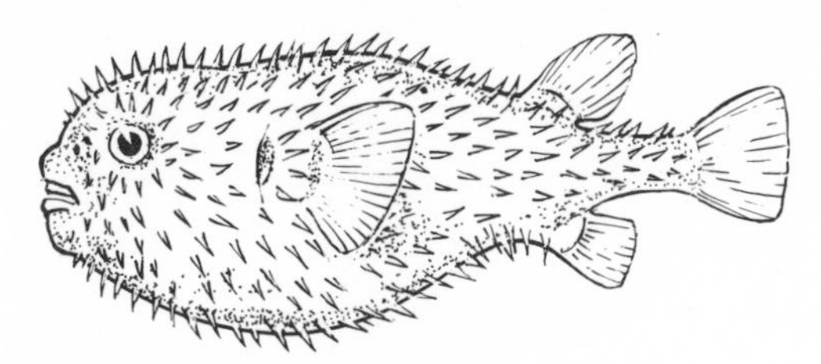

Fig. 440 Diodontidae-type

a slight flattening of the dorsal and ventral surfaces. The Diodontidae differ from the related Tetraodontidae* in one particular way: The teeth in the upper and lower jaws are fused (therefore, Diodontidae = two-toothed) to form an almost perfect bird-like beak. Another difference is that the long sharp spines are usually laid flat against the body and only rarely erected, unlike the G *Chilomycterus**, for example. The Diodontidae can inflate themselves into a spherical shape, their spines standing out absolutely rigid; it is in this form that many are caught and dried and sold as souvenirs, particularly in the South Seas. The spines are not poisonous.

There are about 8 Ga comprising 15 species. They have no Vs. *Diodon hystrix* (Common Porcupine Fish) (fig. 441) lives primarily among reefs, where it grows up to 90 cm long. The juveniles are often found among blankets of seaweed. They roam about feeding on molluscs, worms and crustaceans, like the other species. They flee at the slightest disturbance. In the aquarium the young fish and small species survive quite well.

Dioecious *see* Sex Division in Higher Plants

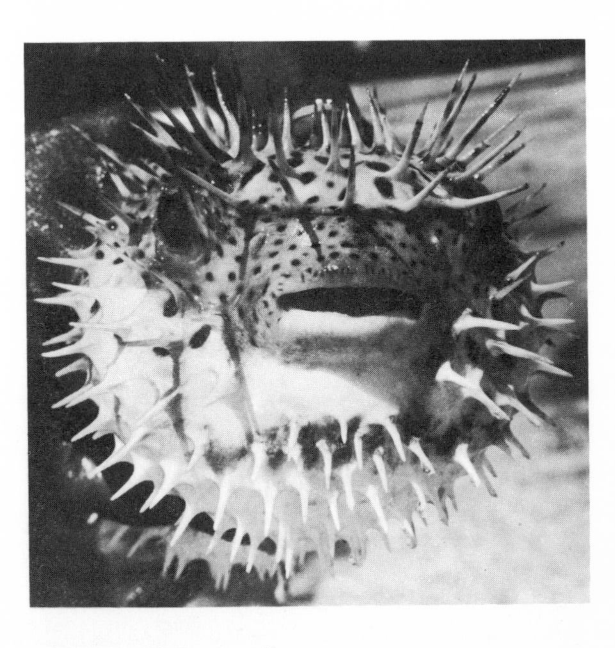

Fig. 441 *Diodon hystrix*

Diogenes DANA, 1851. G of small Hermit Crabs (*see* Paguridea) which are often found living in small groups on the shallow sandy bottom of warm seas. They often burrow into the sand as well. *Diogenes* is about 2 cm long and its left pincer is bigger than its right one. The antennae are feathery. *D. pugilator* (ROUX) lives off the coasts of Europe. It has a grey-white mercuric patterning. The genus comprises rather delicate animals which will nevertheless survive for quite some time in peaceful aquaria with no real competition for food or enemies.

Diphyllobothrium latum *see* Cestodes

Diplectrum HOLBROOK, 1855. G of the Serranidae*. *D. formosum* or Sand Perch (LINNAEUS, 1758) (fig. 442) is found in tropical parts of the W Atlantic where it lives in holes on the bottom. It is a relatively elongated fish, growing up to 30 cm long. It is also a predator. When at rest the Sand Perch has bluish to brown transverse bars on its body, but at other times it has longitudinal stripes on a grey background with a blue pattern of stripes on the head. Typical of the G is the way the C is drawn out; the tip of the upper lobe of the C is visibly elongated. Imported occasionally and well suited to life in the aquarium.

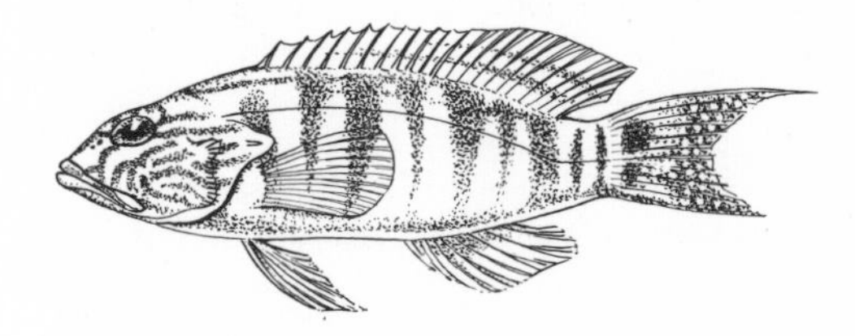

Fig. 442 *Diplectrum formosum*

Diplodus RAFINESQUE, 1810. G of the Sparidae*. Omnivores that live in the littoral. These fishes possess several rows of grinding teeth in both jaws, chisel-like front teeth but no fangs. Head profile more or less straight. Known as protandrous hermaphrodites, ie they pass first through a masculine phase and later change into females. Length varies between 20 and 50 cm. In the wild *Diplodus* often appears to be a sociable fish, yet in the aquarium they are most intolerant towards each other. Otherwise, it is simple to care for them. The following species is particularly well known.

– *D. annularis* (LINNAEUS, 1758) (= *Sargus annularis*). Found in warm parts of the E Atlantic, the Mediterranean, Black Sea. Up to 20 cm. Silvery with yellow Ps and a black ring around the root of the tail fin (fig. 443).

Fig. 443 *Diplodus annularis*

Diploid *see* Cell

Diploria, G, *see* Madreporaria

Diplozoa, G, *see* Trematoda

Diplozoans *see* Trematoda

Dipnoi, Sub-O, *see* Sarcopterygia

Diptera; True Flies. An O of the insects (Insecta*) with an abundance of species and forms found everywhere in the world. At the present time more than 100 Fs and 85,000 species of Diptera are known. There are 2 major Sub-Os within the O, the Nematocera* (Thread-horns) and the Brachycera (Short-horns), both of which contain some very well known flies indeed. A notable characteristic common to all Diptera are the reduced hind-wings known as halteres. The development cycle is complete (holometabolic), ie it consists of the egg, larva, pupa and imago stages. A large number of Diptera spend their egg, larval and pupal stages in water, from the smallest bit of water that has collected in a hole in a tree to lakes and areas of brackish water. On a purely numerical basis, the Diptera often make up a considerable part of the animal life in a stretch of water, and so they constitute an important food source for fish. In the aquarium too, certain types of Diptera are used as food, either in the form of larvae (Chironomidae* [Midges], *Corethra** [Phantom Midge], *Aedes**, *Culex** [Mosquito] etc), or in the form of the adult insect (*Drosophila** [Fruit-flies]). Some Diptera larvae and pupae found in pools of water are striking because of their odd appearance, eg the Blepharoceridae*, *Simulium** (Black-fly), the Tipulidae* (Crane-flies), *Stratiomys** (Chameleon Fly) and certain Hoverflies (*Eristalis**).

Dip-tube (fig. 444). A pipe-like piece of apparatus made of glass or see-through plastic which is used for catching fish or other aquarium specimens. It has various advantages over a net. Under water, it is barely perceptible to the fish so it is easier to catch hold of one fish without disturbing the rest of the tank. The specimen is also kept in water so damage to the skin or the fins is kept to a minimum. The device is also suitable for transferring highly sensitive creatures and fish embryos.

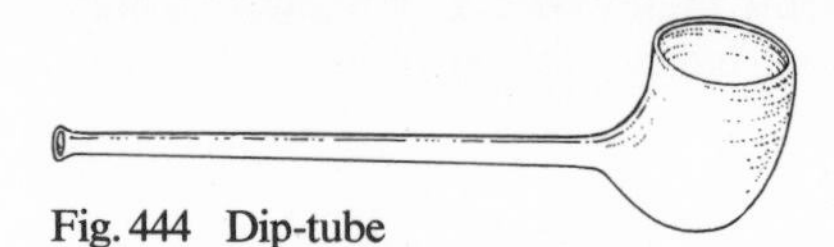

Fig. 444 Dip-tube

Discosoma, G, *see* Stoichactis

Discus (*Symphysodon discus*) *see Symphysodon*

Discus Disease (fig. 445). A greatly feared infectious disease of *Symphysodon discus* HECKEL, 1840, which is caused by the parasitic flagellate *Hexamita** *symphysodoni*. The latter is 12 μm long and has 8 flagellae* which make it highly mobile. The parasite makes its home in fish in the wild too, and is found in small numbers in the intestine and gall-bladder. If conditions in the aquarium are not good, the natural defences of a fish against an attack become weakened. As a result, the parasites increase in numbers very quickly and sometimes they are excreted into the water in great quantities along with the contents of the gut; so other

Fig. 445 Discus disease

fish become infected as they take in their food. In severe cases of intestinal infection, *Hexamita* can also get into the blood stream and infect the liver, kidneys, spleen and skin. The fish becomes apathetic, does not eat and gets darker in colour. If the biliary excretory duct and fine blood vessels become blocked, serious metabolic damage can result. In an advanced stage, blockages of fine vessels in the head region, around the LL and along the dorsal surface are visible externally. Small holes start to appear in these areas which expand dramatically like a crater, out of which pour small white threads. These are the parasites which by then are present in enormous quantities. In the final stage, these areas turn into great expanses of peeling skin with the muscle tissue below being totally destroyed. The fish becomes emaciated and will die without treatment. Specimens that are weakened through *Hexamita* are very susceptible to infection from fungi (*see* Mycosis), bacteria* (especially Fish Tuberculosis*), as well as the normally harmless saprophytes*. These kinds of germs can often be detected in the crater-like holes in the skin, and for a long time they were thought to be the cause of discus disease. Even when a fish's intestine is not severely stricken with the parasite, its presence can often be traced in freshly emitted excreta that is examined under a microscope. There are several ways of combating the disease. The temperature can be raised to 37 °C for 3–4 days or to 35 °C for a week (under certain circumstances this may have to be repeated 2 or 3 times). This action will kill off *Hexamita*, as it can no longer develop at temperatures of 31–33 °C. A chemical cure is possible using 1.3-hydroxyethyl-2-methyl-5-nitroimidazol, which are the effective ingredients of preparations (Clont®, Sanotrichom®, Vagimid®) used in the treatment of trichomonad infections in humans. Also suitable is the antibiotic tetracyclinhydrochloride (*see* Tetracyclin). Both substances are applied in a prolonged bath lasting about 4 days. The concentration level should be 4–6 mg/l water, and afterwards the chemicals should be removed from the aquarium water by passing it over activated charcoal*. Recently, effective anti-*Hexamita* preparations have become available commercially (Hexa-Ex®). Other antibiotics and sulphanilamides* are not effective against *Hexamita*, but may be suitable as a measure against secondary infections. Rivanol* is also suitable as a supplementary treatment for localised skin damage (concentration 100 mg/100 ml water). The symptoms of discus disease also appear in other Cichlidae*, especially in *Pterophyllum**. In such cases, infected fish suffer only from attacks upon the internal organs and there is no sign of the typical eruptions of the skin. It is thought that several species of the G *Hexamita* and the closely related G *Spironucleus* live as parasites on several species of fish. In marine fish too (Acanthuridae*, Pomacentridae* etc) great areas of peeling skin caused by *Hexamita* infections may be observed. However, most of the fishes continue to eat well, and there are no secondary infections, as a rule. Treatment with tetracyclinhydrochloride has proved worthwhile in these cases too.

Disease History *see* Anamnesis

Diseases that cause Swellings. An example of such a disease is found in the Cyprinidae* and is caused by *Myxobolus pfeifferi* (THELOHAN, 1870) (*see* Sporozoa). It occurs in Barbs and is greatly feared. Hard swellings form, particularly in the muscles of the main torso, as well as in the intestine, liver, kidneys and heart. In the advanced stage the swellings become jelly-like, turning the infected organ a yellow colour. The disease is incurable. Swellings may also occur because of bacterial infection* (especially of the skin), an attack of *Ichthyosporidium**, stomach dropsy*, tumour diseases*, as well as other sporozoan diseases.

Disinfection. The killing off or rendering harmless of pathogenic organisms. In aquarium-keeping, if disease breaks out, any pieces of equipment and decorative material, and the plants too in many cases, should be disinfected. Disinfection is a good idea in any case as a preventative measure. Which method of disinfection is selected depends on the type of objects that need disinfecting. Methods often used include high temperatures, prolonged boiling in water, treatment with hot steam or dry heat in an oven at 200–300 °C. Items can be disinfected chemically with strong oxidising agents (potassium permanganate*, hydrogen peroxide*) or reducing agents (hydrochloric acid). If either chemical method is used, great care must be taken when handling these concentrated solutions (protective gloves and safety goggles). Substances that cause dehydration or the precipitation of protein also make good disinfectants. These include 80% ethyl alcohol or methyl alcohol (poisonous!), 6–10% formaldehyde solutions, saturated cooking salt solutions. Copper sulphate*, Trypaflavin*, Rivanol* and bought disinfectants are also suitable on account of their toxic effects. They must be used in correspondingly high concentrations. Afterwards, the disinfected items must be rinsed thoroughly with water until the last trace of disinfectant is removed. There should be no difficulties regarding the disinfecting of glass aquaria or glass equipment, rocks or sand, as all these items are very durable. Pieces of coral, limestone or alluvial shell deposit should be disinfected with alcohol or with high temperatures, and pieces of wood need to be boiled in water for several hours. Implements that are used in several tanks (eg fishing nets) are best left standing in a saturated cooking salt solution for 20–30 mins. A short rinse in water should remove the cooking salt, thus preventing any of the disinfectant being transferred into the aquarium by the implement.

Disjunction. The division of a distribution area (*see* Areal) into two or more parts, the parts being separated from one another by zones of extinction, ie making it

Fig. 446 Disjunctive areal of the Weather-fish

impossible for individuals living in one part of the areal to reach the other part. Areal disjunctions are a common occurrence in biogeography* and can only be explained historically. For example, the Africa-South America disjunction of the Characin related species (*see* Characoidei) and the South America-Africa-Southern Asia disjunction of the Nandidae* family can be explained through continental drift. Similarly, the Western Europe-Eastern Asia disjunction of the Weather-fish (*Misgurnus**) can be explained by the Pleistocene Ice Age (fig. 446).

Displacement Activity. Term used in ethology. Actions that suddenly break forth as the result of a conflicting drive situation* without their appearing to have any connection to the simultaneously activated, mutually inhibitive drives of the two spheres of function*. Such actions can include pretending to take in food when none is there, gaping, making cleaning movements, adopting the sleep position with eyes open (among birds) when an adversary pops up. They probably help to break up the excessive action-specific energy of another sphere of function. (For further information, *see* Conflicting Drive Situation.)

Display Behaviour. Threat Postures (fig. 447). Term used in ethology. Instinctive movements, usually consisting of ritualised behaviour patterns, that are performed by one animal towards a rival or mate. Threat postures are usually reserved for rivals that happen to come into the field of view of mating partners during courtship, mating itself or during brood care. The following display behaviour patterns are particularly common among fish: broadside display behaviour describes

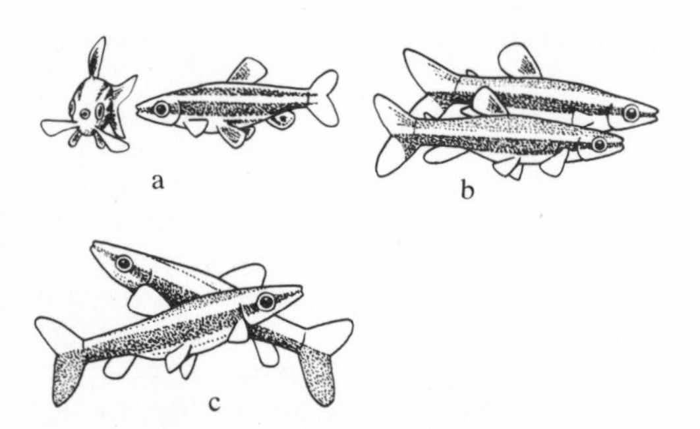

Fig. 447 Display behaviour. a) Broadside display (*Nannostomus marginatus*) b) Parallel display (*N. digrammus*) c) Anti-parallel display (*N. eques*)

the posture whereby the fish remains, broadside on, fins outstretched, in front of its rival's mouth. By this means it hopes to appear bigger than it really is (according to OHM, 1959). In the frontal position both rivals adopt a mouth-to-mouth stance, often opening the mouth wide and forcing open the gill covering. In the parallel position, both fish place themselves broadside on to one another, a little distance apart. They spread out their fins and maintain the position of head to head and tail to tail. In the antiparallel position, the head of the one fish lies next to the tail of the other. Semicircular display swimming, as the name suggests, involves one fish swimming round the head of its rival in a semicircle, after which it will often turn round and swim back again, following the same route. If the fish swims in a complete circle round its opponent, this is called circular display swimming.

Distichodontidae, F, name not in current usage, *see* Citharinidae

Distichodus MÜLLER and TROSCHEL, 1844. G of the Citharinidae*. African Characin from the Congo Basin, growing to between 8 and 60 cm long. There are numerous species in the G and the majority live in shoals near the bottom of rivers. They are very elegant swimmers that are only suited to the aquarium when young, and they will need a lot of room for swimming. The body is very flat from side to side, and is either moderately elongated or compressed. Well developed fins. In *D. affinis* GÜNTHER, 1873, the C is deeply forked. Adipose fin very strong and covered with scales at the base, like the C. Head relatively small and tapering to a point (which is typical of the G). The eyes are large and the mouth is terminal or slightly inferior. Flanks are coloured green-grey, blue-grey or bronze with a striking silvery iridescence. A number of dark transverse bars run from the dorsal ridge to the lower half of the body in the species *D. antonii* SCHILTHUIS, 1891, *D. atroventralis* BOULENGER, 1898, and *D. sexfasciatus* BOULENGER, 1897. However, these bars tend to fade with age. Reproductive behaviour still unknown. They eat vegetable food, as well as *Daphnia, Cyclops, Tubifex* and insect larvae of all kinds. Rolled oats and prepared foodstuffs will also be accepted. Many species of *Distichodus* are prized delicacies. There are also many species that are not suited to the aquarium, on account of their unstriking colouring, considerable size and their predilection for plant food. The following species are the exceptions to this rule.

– *D. noboli* BOULENGER, 1899. Congo Basin. At only 8 cm, this species is easily the smallest of the G. Silvery flanks. Scales on the lower half of the body have dark edges. A dark, round blotch is located on the caudal peduncle near the insertion of the C. All the fins, with the exception of the Ps, are orange to brick-red in colour. The first third of the D is a deep black colour. A very beautiful, but unfortunately rarely imported species (fig. 448).

– *D. sexfasciatus* BOULENGER 1897; Six-barred Distichodus. Central and lower Congo. Up to 25 cm. Flanks

Fig. 448 *Distichodus noboli*

bright orange. Ventral surface paler. 6–7 blue-black transverse bars, fins blood-red. A beautiful species which has recently begun to be imported more often. Distinguished from the species *D. lusosso* SCHILTHUIS, 1891, by its stubbier snout (fig. 449). *D. lusosso* also grows up to 40 cm long.

Fig. 449 *Distichodus sexfasciatus*

Distichous *see* Leaf Arrangement

Distillation *see* Desalination

Distribution. Species of plants and animals, as well as higher systematic units are not evenly distributed through the world. Rather, they have a specific area of distribution (Areal*) which is characteristic for the races, species, genera concerned. This distribution area is dependent on ecological factors as well as on the history of the area and the capacity of the species concerned to be able to spread. So, a distribution area is determined objectively and is therefore able to be explained scientifically (Biogeography*).

Distribution, limitations upon. There are insuperable geographical or ecological obstacles that prevent a particular species from widening its distribution area (*see* Areal). Eg many animals that live in flowing waters, such as fish and Plecoptera, are confined to particular river systems as the places where the rivers divide act as a barrier. Similarly, still-water animals are confined to lakes that have no river drainage. If these natural barriers are removed, eg through the building of canals or by artificial colonisation etc, the area of distribution can become increased – in extreme cases it may even result in a false adaptation of the flora or fauna. For example, the building of the Suez Canal has to date enabled more than 20 species of fish to reach the Mediterranean from the Indian Ocean and adjoining seas (eg the Red Sea).

Dogfish *(Scyliorhinus canicula) see* Scyliorhynidae

Dogfish Sharks, F, *see* Squalidae

Dogfishes, F, *see* Scyliorhynidae

Dohrn, Anton (1840–1909). German zoologist who studied at Jena under ERNST HAECKEL and was also a private lecturer there for a short while. In 1874 he founded the aquarium and zoological unit at Naples. He published *News from the Zoological Unit at Naples* and from 1880 the *Zoological Annual Report.* He undertook extensive surveys of the Mediterranean, as a result of which appeared the multi-volume *Fauna und Flora des Golfs von Neapel und der angrenzenden Meeresabschnitte* ('Fauna and Flora of the Bay of Naples and Neighbouring Parts of the Sea'). His son REINHARD DOHRN continued his work after his death.

Dolphin Cichlid *(Aequidens itanyi) see Aequidens*

Dolphin Fish *(Coryphaena hippurus) see* Coryphaenidae

Dolphin Fishes, F, *see* Coryphaenidae

Dolphins, O, *see* Cetacea

Domestication. The bringing of wild species under human control, including cultivated or useful plants as well as domesticated animals. The original reason for domestication was to bring animals and plants in closer proximity to humans to improve availability of food. Sheep were already domesticated 11,000 years ago in the region we now know as Iraq. And very early on, humans found great pleasure in cultivating plants indoors and keeping pets. For example, the goldfish was already popular in China 1,000 years ago, and new varieties were developed (fig. 450).

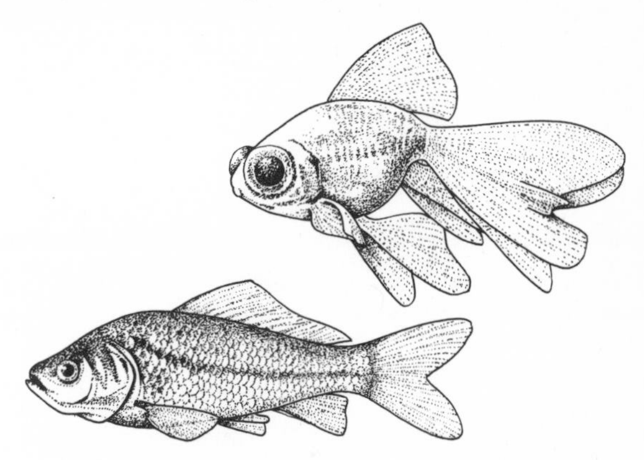

Fig. 450 Domesticated features: the Telescope Veiltail and its original form, the Gibel or Prussian Carp

The beginning of domestication is caring for wild species. By doing this, environmental conditions for the species are altered and natural selective mechanisms are replaced with the artificial selection imposed by people. In nearly all domesticated animals characters start to appear that are known as *domestication characters.* The reproductive pattern found in nature is generally interrupted, the weight of the brain is reduced and the proportions of the cranium and other parts of the body are affected. Domestication often leads to the loss of natural competitiveness and viability, particularly in ornamental species. This can be seen in aquarium fish with extreme fin developments, the 'veil-tail' varieties, for example (Veil-tail Goldfish, Veil-tail Angel, Veil-tail Guppy), in albinos or in fish with striking variations in colour.

Dominant *see* Heredity

Dominant Species. Those species of plants and animals that numerically predominate in a particular biotic community (Biocenesis*), ie are particularly common there.

Domingo Gambusia *(Gambusia dominicensis) see Gambusia*

Domino Damsel *(Dascyllus trimaculatus) see Dascyllus*

Doradidae; Thorny Catfishes (fig. 451). F of the Siluriformes*, Sup-O Ostariophysi. Small, compressed, fish from S America, often tadpole-like in shape. Head large, broad and strongly ossified. The separate dermal bony plates on the skull lie well to the upper surface, and many have a granular texture. They form a head armour stretching from the nape to the D at the very least. A very long toothed spine is often noticeable on the skull extending from the operculum backwards to the insertion of the Ps and beyond. Mouth slightly inferior, surrounded by 3 pairs of barbels, the pair on the upper lip being the longest. Jaws usually toothed. D short, inserted far forwards, and shaped like a pointed sail. Like the Ps, the D possesses a strong toothed or furrowed spine. Adipose fin small; C rounded, straight-edged or slanted. Most Ga have a row of overlapping spiny scutes along their sides (these scutes bend backwards slightly). Some species also have scutes between the D and the adipose fin; *Lithodoras dorsalis* is armoured on all sides. It is unusual for the scutes to be very small on the body or to be missing entirely (G *Trachycorystes*). It is thought that the thorny catfishes make use of supplementary intestinal respiration.

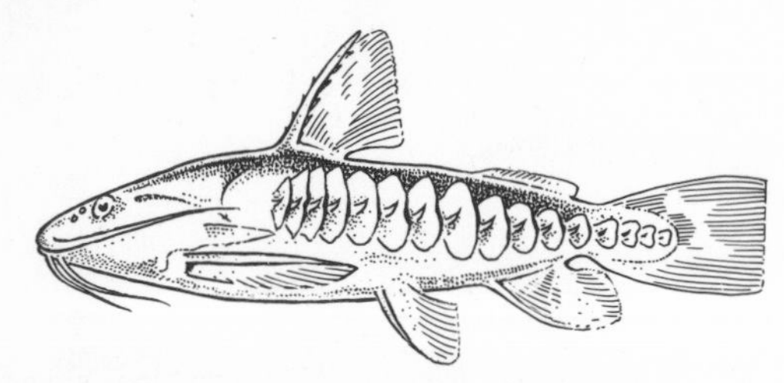

Fig. 451 Doradidae-type

The Doradidae are bottom-dwelling fish that sometimes burrow so far into the substrate that only the red iridescence of the eyes shows. Many species also produce growling or purring noises with their shoulder-girdle which may be heard whenever a fish is removed from its tank. They feed on invertebrates. Primarily active at night, these fish tend to remain hidden during the day. Many species are very attractively coloured. Little is known about their reproduction, although some are said to build nests. The following have been observed in the aquarium: *Acanthodoras**, *Agamyxis**, *Amblydoras**, *Astrodoras**.

Dorado *(Coryphaena hippurus) see* Coryphaenidae

Dormitator GILL, 1861. G of the Gobiidae, Sup-O Gobioidei*. Found in sea water and brackish water along the Atlantic coast from N Carolina (USA) to Brazil and along the Pacific coast of Mexico and Central America. Medium-sized to large fishes, 25 to over 30 cm, sexually mature at 10 cm. Body relatively compressed and covered completely in large scales. The front part of the operculum contains no spine. There are also no teeth along the vomer of the front palate. Vs separate from one another at the base. D1 has 7 spines. Ground colouring of the body brown with spots or speckles (likewise on the fins which are unpaired). *Dormitator* is a predatory species that should be kept at temperatures of 20–25 °C in spacious tanks filled with brackish water (50–100 g sea salt to 10 l water) or sea water. Feeds on live food of all kinds. Spawn is laid on rocks, and there is a high rate of production. The young fish should be reared on the finest live food. Of interest to the aquarist are the following.

– *D. latifrons* (RICHARDSON, 1844); Broad-headed Sleeper Goby. Pacific coast of Mexico and Central America, also in sea water. Up to 25 cm. Truncated head, forehead in adult specimens is very broad. Body brownish to red-brown with a greenish shimmer. Dorsal surface darker. Scales have red-brown spots. The operculum has red-brown, slighly twisting lines running across it which emanate from the eyes. D and C translucent with rows of red-brown speckles. An reddish. The young fish are grey with dark oblique bands. Female less definitely patterned (fig. 452).

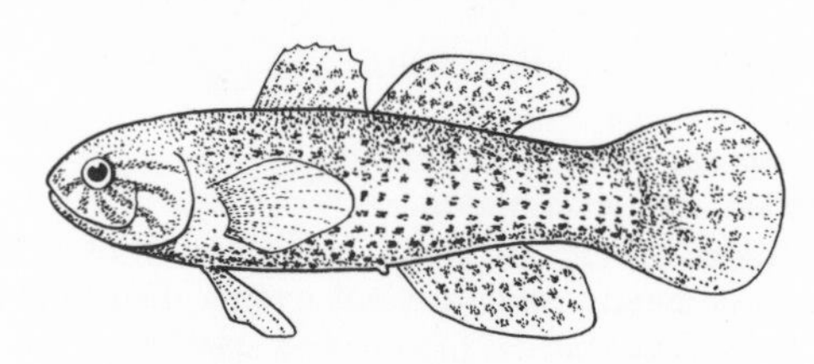

Fig. 452 *Dormitator latifrons*

– *D. maculatus* (BLOCH, 1785); Fat Sleeper, Spotted Sleeper Goby. Atlantic coast from N Carolina to Brazil, where it lives both in brackish water and sea water. Grows more than 30 cm long. Body grey-brown to dark brown with a greenish shimmer. There are dark blotches or an indistinct longitudinal band along the sides. Above the insertion of the P there is a blue shoulder spot. Dark lines spread out from the eyes over the operculum. Fins translucent. D in particular has rows of dark blotches, and there are some iridescent blue flecks on the An. Female paler and the fins less prominently patterned. In its homeland this species is consumed by humans.

Dorsal. From the Latin *dorsum* meaning back. A term used in anatomy* to describe the position of parts of the body. Opposite to ventral.

Dorsal Display Swimming *see* Courtship Swimming

Dorsal Fin *see* Fins

Dotted Barb *(Barbus stigma) see Barbus*

Dotted-scale Pyrrhulina *(Pyrrhulina brevis) see Pyrrhulina*

Dragon Fin *(Pseudocorynopoma doriae) see Pseudocorynopoma*

Dragonet *(Callionymus lyra) see Callionymus*

Dragonets (Callionymidae, F) *see* Callionymoidei

Dragonets, related species, Sub-O, *see* Callionymoidei

Dragon-finned Characin *(Pseudocorynopoma doriae) see Pseudocorynopoma*

Dragonflies, O, *see* Odonata

Dreissena BENEDEN, 1835. Zebra Mussels (fig. 453). G of the bivalves (Bivalvia*). Found in fresh and brackish water. Shells roughly triangular, up to 3 cm long, but with no shutting teeth. Colour whitish-yellow with dark zigzagging bands on top. *Dreissena* attaches itself to a hard substratum with byssus threads in a similar way to *Mytilus**. It therefore leads a sedentary (sessile*) life and is often colonial. Can be kept in unheated pond-aquaria.

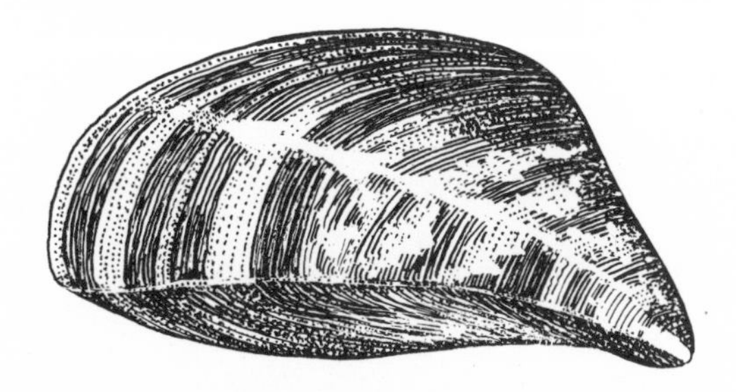

Fig. 453 *Dreissena polymorpha*

Dried Food *see* Feeding
Driftfishes (Nomeidae, F) *see* Stromateoidei
Drive *see* Instinct
Dromia, G, *see* Brachyura
Drone-flies, G, *see Eristalis*

Droseraceae; Sundew Family. F of the Magnoliatae* with 4 Ga and 90 species. Tropical to temperate regions. Small perennials with an alternate or whorled leaf arrangement. Flowers small, hermaphroditic, polysymmetrical. The Droseraceae have a special way of feeding; they use enzymes that can break down protein to digest insects etc. The prey is clasped and digested by means of glands on the leaf blade. There are also hairs on the leaf blade that can shut tight round the prey. Sundews grow in dry, sandy or marshy places. Best known is the G *Drosera*. The single species belonging to the G *Aldrovanda** grows in water.

Drosophila; Fruitflies (fig. 454). G of the Drosophilidae, O Diptera. These flies are often encountered in large numbers near fermenting substances (fruit, fruit-juices, wine, beer). Because of their high rate of multiplication and the ease with which they can be reared they have proved useful to the aquarist as a foodstuff. *D. melanogaster* is a classic object of study in genetics.

Fruitflies are best cultivated in erlenmeyer flasks or milk-bottles (or similar containers), the open end of which can be sealed with a wad of cotton-wool or gauze. Place a layer of food medium, 1–2 cm thick, into the container for the larvae, then add some cellulose or wood wool for the pupation. A sheet of paper can be put in instead of the cellulose but it must reach to the opening of the container where it has to be fixed securely to the wad of cotton-wool. The simplest kind of food medium is a mixture of meal-pap and grated apple. There are various recipes that call for media containing sugar, to which is then added some yeast so that fermentation starts up. One such example suggests that 8–10 g agar (or gelatine) be dissolved in a good 0.5 l water (by heating it). This is then cooked with 100 g meal and 90 g syrup to produce a purée. After cooling, 2–3 drops of a watery yeast froth should be added. It is simplest to compound a very thick purée of rolled oats with liquefied yeast after allowing to cool. Fermentation soon starts if left to stand in a fairly warm place (20–26 °C). After about 3 days the breeding flies can be placed in the containers. Usually, *D. melanogaster* is used for

breeding purposes – it produces forms that have stumpy wings making them incapable of flying. They are about 3 mm long. The females can be recognised by their pointed abdomens, which are blunt-ended in the males. One female can lay up to 400 eggs in a few days. The hatching larvae take up residence in the food medium first of all. 5–10 days later, pupation takes place and a further 3–11 days after that the new fly generation hatches out. Lower temperatures slow down development. To remove the flies, gently tap them out of the container into another vessel then place them on the surface of the aquarium water. Rearing media can usually only be used by one generation, so it is recommended that the cultures are renewed continually.

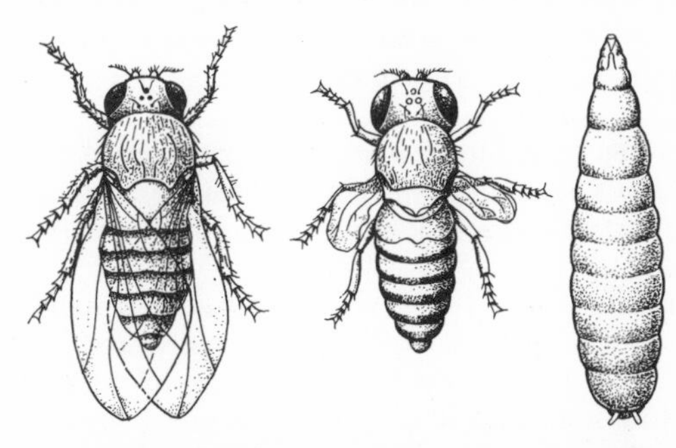

Fig. 454 *Drosophila*. From left: wild form, stubby winged 'vestigial'-form, larva

Drums, F, *see* Sciaenidae

Dryopidae (fig. 455). F of Beetles (Coleoptera*) with few species. Apomorphic* representatives of the F live in water. They are only 1–5 mm long, and unlike the true water beetles they cannot swim. When they are freed from the substratum, they climb to the surface like a bubble because of the atmospheric layer around them. Underwater, Dryopidae usually climb about on plants, wood and rocks, whilst they also like to fly around at night. Both the beetles themselves and their larvae may be found occasionally around ponds.

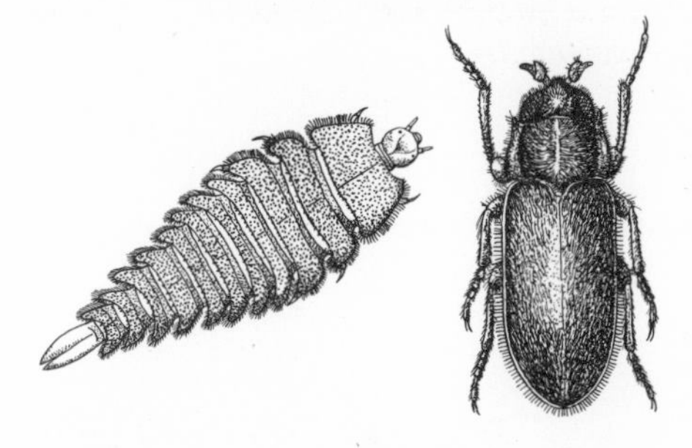

Fig. 455 Dryopidae. Left, larva; right, imago

Duckweeds, F, *see* Lemnaceae
Duckweeds, G, *see Lemna*

Dummy Fish, Experiments with. Term used in ethology. Natural sources of stimuli are replaced by artificial ones, determined by experiment. By this means both the type and quantity of stimuli can be established, by varying the shape, size and colour of the dummy and recording the reactions of the fish being experimented with. By analysing those reactions it is possible to draw conclusions about behaviour patterns. For example, when an almost round, finless dummy, its lower half painted red, is placed near the nest of a male Three-spined Stickleback, the fish will attack immediately and try to chase it away. The effect the dummy produces becomes even more dramatic when the back is painted bright blue and a brilliant blue eye is added. In contrast, a more lifelike representation of the shape of the body without the red colour, or a freshly killed fish of the same species also minus the red colour attracts little attention when placed near the nest. So, it is safe to conclude that a red ventral surface and blue eyes are prerequisites for a male stickleback to adopt aggressive behaviour; both would appear to indicate the presence of a rival. The finless dummy fish, minus the red colour, with its body shape imitating that of a female in breeding condition (swollen belly sharply marked off from the caudal peduncle), releases a male stickleback's courtship behaviour as long as he has finished building the nest.

Duncker, Georg (1870–1953). Important German ichthylogist with extensive knowledge of native and exotic fish. Director of the Zoological Museum of Hamburg. He wrote numerous systematical works of which the following are particularly of note: *Die Fische der malayischen Halbinsel*, 1904 ('The Fishes of the Malaysian Peninsula'); *Die Süßwasserfische Ceylons*, 1912 ('The Freshwater Fishes of Ceylon'); *Die Fische der Nord- und Ostsee*, 1929 ('The Fishes of the North and Baltic Seas'); *Die Fische der Nordmark*, 1935–41 ('The Fishes of the Northern Marches').

Duncker's Barb *(Barbus dunckeri) see Barbus*

Dusky Millions Fish *(Phalloceros caudimaculatus) see Phalloceros*

Dwarf Amazon Plant *(Echinodorus tenellus) see Echinodorus*

Dwarf Amazon Sword Plant *(Echinodorus latifolius) see Echinodorus*

Dwarf Amazon Sword Plant *(Echinodorus quadricostatus) see Echinodorus*

Dwarf Barb *(Barbus phutunio) see Barbus*

Dwarf Bladderwort *(Utricularia gibba) see Utricularia*

Dwarf Climbing Perch *(Ctenopoma nanum) see Ctenopoma*

Dwarf Corydoras *(Corydoras hastatus) see Corydoras*

Dwarf Corydoras *(Corydoras rabauti) see Corydoras*

Dwarf Danio *(Brachydanio nigrofasciatus) see Brachydanio*

Dwarf Duckweeds or Rootless Duckweeds, G, *see Wolffia*

Dwarf Firefishes, G, *see Dendrochirus*

Dwarf Flounder *(Achirus fasciatus) see Achirus*

Dwarf Gobies, G, *see Brachygobius*

Dwarf Gourami *(Colisa lalia) see Colisa*

Dwarf Japanese Rush *(Acorus gramineus* var. *pussillus) see Acorus*

Dwarf Limia *(Poecilia heterandria) see Poecilia*

Dwarf Lionfish *(Dendrochirus zebra) see Dendrochirus*

Dwarf Loach *(Botia sidthimunki) see Botia*

Dwarf Marbled Catfish *(Microglanis parahybae) see Microglanis*

Dwarf Panchax *(Aplocheilus blocki) see Aplocheilus*

Dwarf Pencilfish *(Nannostomus marginatus) see Nannostomus*

Dwarf Rainbowfish *(Nematocentris maccullochi) see Nematocentris*

Dwarf Seahorse *(Hippocampus zosterae) see Hippocampus*

Dwarf Sunfish *(Elassoma evergladei) see Elassoma*

Dwarf Top Minnow *(Heterandria formosa) see Heterandria*

Dystrophic Waters. Standing* or flowing* waters which have a dark brown colour, caused chiefly by a rich content of humic matter. The typical sediment on the bottom consists of material that has only slightly decomposed (peat mud). High humic acid* concentrations and extremely low amounts of calcium make dystrophic waters very acidic with a pH value of about 4.5–6. Such waters also have a low nutrient content. A negligibly small number of aquatic organisms can tolerate such extreme conditions, and a similar few actually need such conditions to live. Among aquarium

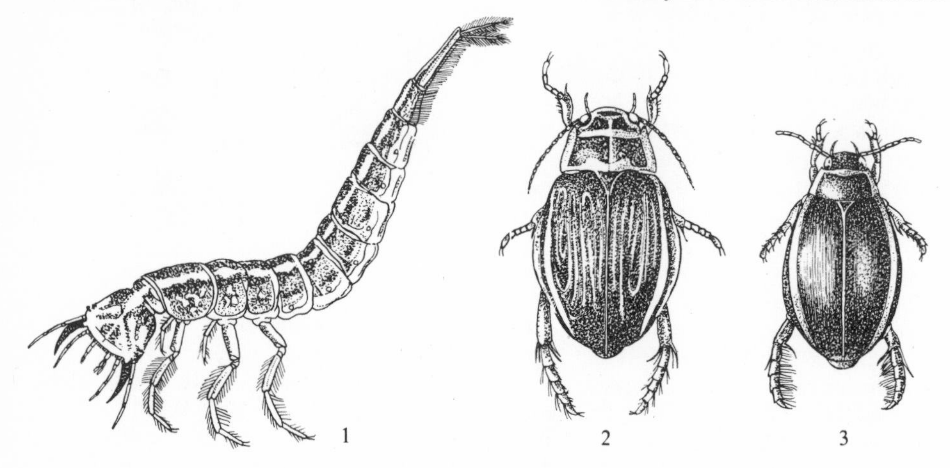

Fig. 456 Some Dytiscidae. 1 Larva 2 *Dytiscus latissimus* 3 *Cybister lateralimarginalis*

specimens there are a few 'problem fish' that belong to this small number, eg *Symphysodon**, *Uaru** and *Hyphessobrycon** *axelrodi*. Among other things, problems arise from the fact that dystrophic waters are so hostile to living forms that even bacteria and other pathogenic micro-organisms do not live there; so when the fish are put in the aquarium, they are defenceless against attack. In cold areas of the world dystrophic waters usually occur in the form of bog waters; in tropical areas they are usually black water rivers (such as the Rio Negro in South America).

Dytiscidae (figs. 456 and 457). F of Beetles (Coleoptera*) with many species and a worldwide distribution. Dytiscidae larvae live exclusively in all kinds of fresh water, whilst the pupal stage is spent on land. The adult beetles themselves return to the water, although they are still capable of lengthy overland flights. Because of this life-style, the larval and beetle stages are remarkably adapted to an aquatic existence, although they have to come to the surface now and again to breathe. The boat-shaped, flattened body of the Dytiscidae with their broad swimming legs is most easily confused with members of the F Hydrophilidae*, although the Dytiscidae always beat their legs in time towards the back. Without exception, the Dytiscidae are predators, and some of the larger species (eg members of the G *Dytiscus**) are a threat even to small fish. This is particularly true of the larvae which overpower prey bigger than themselves. First the prey is seized with the hollow, dagger-like jaws (mandibles) then it is injected with a paralysing secretion, before being killed and pre-digested. The larger species of Dytiscidae make very interesting specimens in pond-aquaria, although for the reasons already given they should only be kept on their own during the larval stage.

Dytiscus LINNAEUS, 1758 (fig. 457). Well-known G of the Dytiscidae*. Distributed throughout the holarctic region*. The various species measure more than 4 cm long, although the larvae can measure up to 8 cm long. Eggs are laid in the autumn and hatch out in late spring. Within a few weeks they reach their full size. Their food requirements are considerable; they even overpower small fishes and amphibians. So wherever fish are reared the larvae are justifiably feared. After a pupal resting phase on land, the new beetle generation hatches out in late summer and autumn. The beetles may live for 5 years! The larvae and imagines of *Dytiscus* make

Fig. 457 *Acilus sulcatus*

very interesting specimens in pond-aquaria, where it is possible to observe how they catch their prey and mate. The best known species is *Dytiscus marginalis* (Great Diving Beetle) which grows up to 4 cm long.

Eagle Ray *(Myliobatis aquila) see* Myliobatidae

Earth Eater *(Geophagus jurupari) see Geophagus*

Earthworms, F, *see* Lumbricidae

Eastern Mosquito *(Gambusia affinis holbrooki) see Gambusia*

Ecdyonurus EATON, 1868 (fig. 458). G of the Mayfly insect order (Ephemeroptera*). Consists of many species found in Europe and temperate parts of Asia

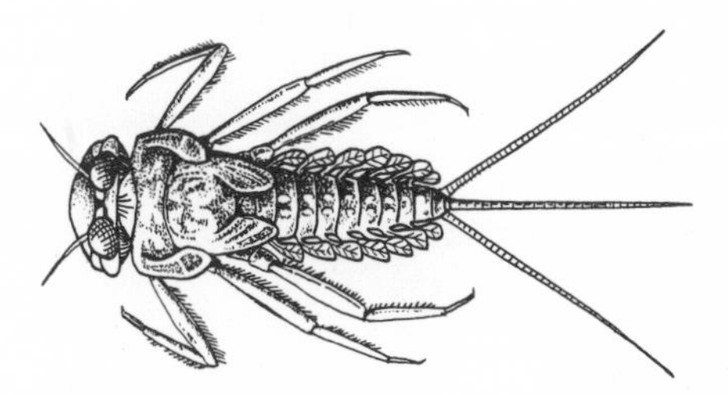

Fig. 458 *Ecdyonurus*

(Palaearctic region*), and in N America its place is taken by the closely related G *Stenonema* TRAVER. The nymphs mostly live on rocks in fast-flowing streams in the mountains and highlands. They can usually be identified by the first thoracic segment which is elongated towards the back when viewed from the side. Since they need a lot of oxygen, all the species are really only suited to pond-aquaria, although they do make a valuable foodstuff for large aquarium fish.

Echeneidae; Remoras or Shark Suckers (fig. 459). F of the Perciformes*, Sub-O Percoidei*. Often regarded

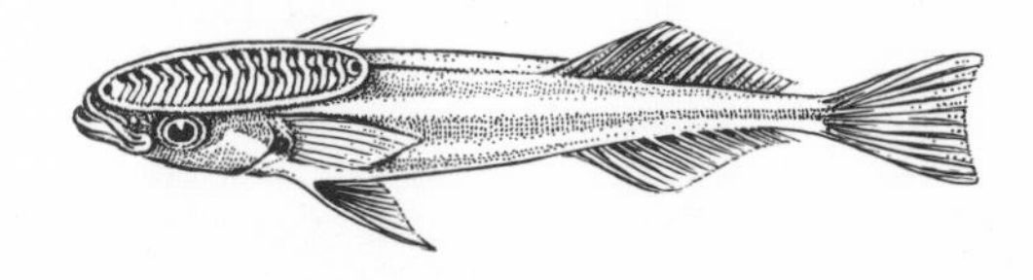

Fig. 459 Echeneidae-type

as an order unto itself in other systems. The Echeneidae are very slender fishes with a unique sucking disc on top of the head, formed from the D1. Seen from above, numerous lamellae cross the disc at an angle. It is the way in which these lamellae are arranged that the partial vacuum is created which enables the fish to stay stuck onto its host. Young fish first develop the sucking disc when they are 8–12 mm long. There are 8 known species distributed throughout tropical and subtropical seas everywhere. They can all swim well and skilfully, yet most of them move around attached to larger fish, in particular sharks. *Remilegia australis* selects whales as its means of transport. As an adaptation to this way of life, the dorsal surface (ie the surface next to the host) is often paler than the ventral surface. The Echeneidae feed in part on large skin parasites, but their chief source of food is various kinds of small fish. The largest species is the Sharksucker or Suckerfish *(Echeneis naucrates)* (fig. 460) which can reach a lengh of 1 m. It is brown with 2 yellow longitudinal bands. Remoras used to be employed by South Sea Islanders to catch turtles. They were released from boats and would attach themselves to the turtles which were then hauled aboard by a line tied round the fish's tail.

Fig. 460 *Echeneis naucrates*

Echeneis naucrates *see* Echeneidae

Echidna FORSTER, 1788. G of the Muraenidae, O Anguilliformes*. Contains several species that live on reefs and rocks in tropical seas. Predatory fish that are active at night. Very striking pattern of colours. Length between 50 and 120 cm. Various features are typical of the G: the hollow teeth, the way in which the top of the head is clearly bent, and the position of the D in front of the gill opening or just behind it. The G comprises very attractive fish that survive well in large aquaria, although some may be aggressive (*E. zebra* [SHAW, 1797] Zebra Eel or Zebra Moray) (fig. 461). The Cloudy Moray, otherwise known as Diamond-back Moray, Snowflake Moray or Starry Moray *(E. nebulosa)* (AHL, 1789) is relatively peaceful and grows up to 75 cm long. It is often imported. The species comes from the tropical Indo-Pacific, including the Red Sea.

Fig. 461 *Echidna zebra*

Echinaster MÜLLER and TROSCHEL, 1840. G of the Sea-anemones (Asteroidea*). Found in warm seas. This G has 5 (in rare instances 7) long arms, a small mouth and a reticulated upper surface. For food, *Echinaster* eats particles and small animals. Many of these sea-anemones are a bright red colour. The larvae of *E. sepositus* develop from yolk-rich eggs directly into a sea-anemone (ie with no planktonic stage). This may even happen occasionally in the aquarium. However, they are not easy to keep. The following species can be persuaded to live on finely chopped bivalve meat:

– *E. sepositus* GRAY; Purple Star. Mediterranean, E Atlantic, where it is often found in carpets of grass weed below the tidal zone. Diameter up to 20 cm. Bright red.

Echinodermata, Echinoderms. Marine-dwelling animal Ph to which the Starfishes (Asteroidea*), Sea-urchins (Echinoidea*), Sea-cucumbers (Holothuroidea*), Brittlestars (Ophiuroidea*), Sea-lilies and Feather Stars (Crinoidea*) belong. The crinoids are ontogenetically the oldest Cl of modern echinoderms. There are about 6,000 modern species of echinoderms. Fossil Echinodermata date back to the Cambrian. Characteristic features of echinoderms are: the five-rayed symmetry of their basic structure; the calcareous skeleton that lies in the layer of connective tissue of the skin; and the complicated water vascular system with the typical ambulacral feet. Sea-urchins, Starfishes and Brittlestars sit with their mouth side to the bottom; the anus is positioned more or less opposite the mouth on the upper side or it is absent altogether. They move sideways, using any one of the arms to dictate direction. The Sea-cucumbers (Holothuroidea*) are worm-like in shape with the head at the front. Feather Stars and Sea-lilies have upward-pointing mouths with the anus to the side.

The Echinodermata are oddities. Their extraordinary rayed structure does not allow them to be placed in the same group as the bilaterally symmetrical animals. Nor can the higher echinoderms be said to be related to the Coelenterata* with *their* radial structure, as the body plan of the coelenterates is completely different and simple in comparison. However, the structure of the echinoderms becomes clear when the larval development is examined. The majority of the larvae live in the plankton. They have a bilateral symmetry which only develops into a radial symmetry during the course of development. So, it is safe to assume that the echinoderms evolved from the bilaterally symmetrical forms, and indeed the larvae of some other animal Pha (eg Acorn Worms, Enteropneusta) show similarities wich the echinoderm larvae. The five-rayed basic body plan of the echinoderms is modified in various ways in the separate Cls. The symmetry is seen most clearly in the majority of the Starfishes and Brittlestars. In the crinoids the arms are more branched in places. The spherical shape of the Sea-urchin can be seen to be based on the star shape if one imagines the Sea-urchin's arms with the tips bent upwards and joined together at the sides. In contrast to the Sea-urchins, the Sea-cucumbers are very elongated in shape and lie on their sides. The head is at one end, the anus at the other. The calcareous skeleton is made up of skeletal plates that move against one another. In most Sea-urchins the skeleton has fused into a rigid sphere, while in Sea-cucumbers the plates are tiny and embedded in the leathery skin. Many Sea-urchins and some Starfishes have small stalked forceps (pedicellaria) on the skin which help in defence; they work in conjunction with poison glands. The water vascular system (ambulacral system) is used primarily in movement. It works by hydraulics and is unique in the animal kingdom. Normally it consists of 5 double rows of hollow feet arranged along the underside of a Starfish's arms or the radia of a Sea-urchin or Sea-cucumber, and ending in a small sucking disc in many cases. The tube-feet are linked to a muscular vesicle inside the body through which water can be pumped to the feet. A network of channels connects the feet to each other, the network opening out into the sea water through a perforated plate known as the madreporic plate. This is located on the upper surface of a starfish in the angle between the bases of two arms, and appears as a slight bump.

Echinoderms make interesting specimens in marine aquaria on account of their unusual appearance, way of life and attractive colouring. With the exception of the Feather Stars and Sea-lilies, many species, particularly many of the Starfishes, survive well. Some have reproduced in the aquarium *(Asterina*)*. All echinoderms require an excellent water quality. Fluctuating salinity, pH values that are too low, and water cloudiness are all especially dangerous for them. A lot of time has to be allowed for echinoderms to get used to a new environment and they must be handled extremely carefully. Specimes should not be lifted out of the water by hand, and should be transported in a small plastic case.

Echinoderms, Ph, *see* Echinodermata

Echinodorus ENGELMANN, 1848; Burhead, Amazon Sword Plants. G of the Alismataceae* containing 50 species. From tropical and temperate parts of America. Medium-sized to large perennials that form rhizomes, or smaller rosette plants that form runners. Leaves in basilar rosettes. Submerged leaves are linear, lanceolate or stalked with delicate or coarse blades of varying shape; emersed leaves are almost always divided into stalk and blade. Translucent dots and lines on the blades (but only really visible when the plant is dry) are an important feature regarding scientific classification (fig. 462). The inflorescences (called thyrsi) are of different sizes and have a complicated structure (fig. 463). Flowers hermaphroditic, polysymmetrical. Perianth parts in groups of 3, with a green calyx and usually white petals. Stamen numbers vary from 6 up to a great many; there are many carpels that are free-standing and superior. Nut formed from a multiple fruit.

Fig. 463 *Echinodorus* flower

The species of *Echinodorus* are marsh plants that are found in the vast expanses of marshy areas of America. How well each species can tolerate submersion depends on where it grows naturally. Many of them are popular aquarium plants. They grow well in tanks that are provided with a nutrient-rich substrate and temperatures of 20–30 °C. Problems only arise if the tank is not well lit and the water is too acidic. Those species that reproduce with runners can propagate without any problem. After a while there will be thick clumps of them from which plants can be taken. The larger species (which are also suited to the aquarium) will even form inflorescences underwater if left to grow undisturbed. However, the flowers do not often open out, and instead adventitious plants* develop near the flower. These may be separated from the main plant and replanted after establishing themselves sufficiently well. In the

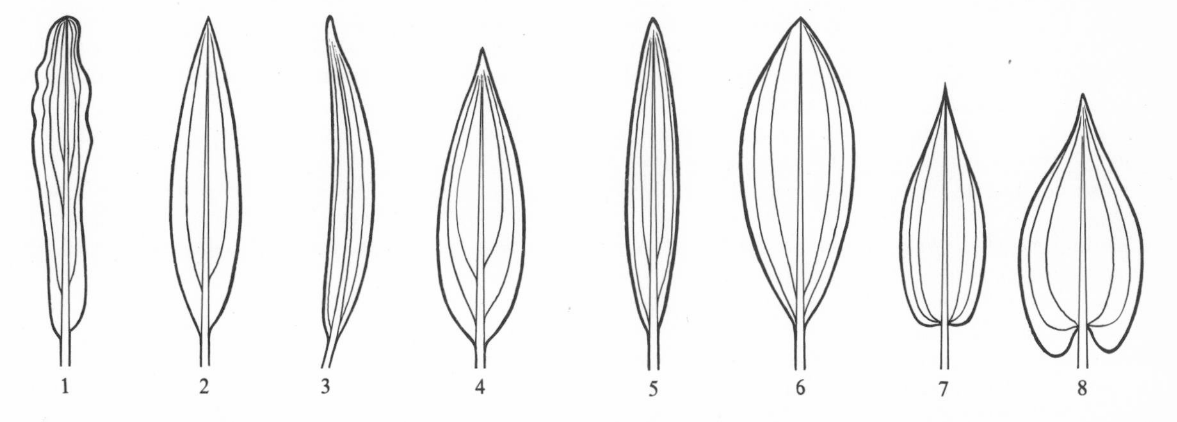

Fig. 462 Leaves of various *Echindorus species.* 1 *E. major* 2 *E. bleheri* 3 *E. amazonicus* 4 *E. parviflorus* 5 *E. horemanii* 6 *E. osiris* 7 *E. aschersonianus* 8 *E. horizontalis*

wild, the adventitious plants go free as the inflorescences decay, so permitting asexual propagation to take place. On rare occasions side-shoots develop from an old rhizome which can be separated off. There are only a few species that do not form adventitious plants, and in such cases propagation takes place through the scattering of seeds.

From a practical point of view the species can be divided into the following groups.

1. Large marsh plants that are cultivated in botanical gardens, but are unsuitable for the aquarium. Young plants can sometimes be kept in aquaria for a little while, but they will eventually start growing out of the aquarium water.

– *E. argentinensis* RATAJ; Argentinian Amazon Sword Plant. From temperate regions of S America. Leaves up to 1.2 m long, blades lanceolate to ovate.

2. Species that can be kept in the aquarium when young. Older plants always reach above the water line if the tank is illuminated for more than 10 hours at a time.

– *E. berteroi* (SPRENGEL) FASSETT (*E. rostratus*, name not in current usage). Southern parts of N America, Central America and the West Indies. There is a subspecies that grows in temperate parts of S America. Submerged leaves have variable and transparent blades (fig. 464).

Fig. 464 *Echinodorus berteroi*

– *E. cordifolius* (LINNE) GRISEBACH (*E. radicans*, name not in current usage). SE North America. Submerged leaves have coarse heart-shaped blades.

– *E. grandiflorus* (CHAMISSO et SCHLECHTENDAL) MICHELI. Central America and into the temperate areas of S America. Leaves up to 1.5 m long, blades heart-shaped.

3. Medium-sized to large species that are suited to life in the aquarium, as they do not grow beyond the surface of the water. Give them as much room as possible to be seen to their best advantage.

– *E. amazonicus* RATAJ (*E. brevipedicellatus*, name not in current usage). Small-leaved Amazon Sword Plant. Amazon Basin. Leaf blades narrowly lanceolate, in part silghtly bent on one side. Pale to medium green in colour.

– *E. aschersonianus* GRAEBNER. Subtropical and temperate regions of S America. Leaf blades lanceolate to ovate with a heart-shaped base. Variable in appearance, pale to medium green in colour.

– *E. bleheri* RATAJ (*E. paniculatus, E. rangeri*, names not now in current usage). Amazon Sword Plant. Tropical S America. Leaf blades broadly lanceolate. Colour pale to medium green.

– *E. horemanii* RATAJ. Eastern parts of temperate S America. Leaf blades lanceolate, dark green (fig. 465).

Fig. 465 *Echinodorus horemanii*

– *E. horizontalis* RATAJ. Amazon Basin. Leaf blades cordate, tapering at the apex. Pale green, reddish at first if illumination is strong. Leaves held out horizontally (fig. 466).

Fig. 466 *Echinodorus horizontalis*

– *E. major* (MICHELI) RATAJ (*E. martii, E. leopoldina*, names not in current usage). Tropical S America. Leaf blades lanceolate with rounded tips and cordate base, broadest above the middle. Edges of the leaf blades wavy. Mid-green colour.

– *E. osiris* RATAJ. Eastern part of temperate S America. Leaf blades an inverted oval shape. An intense red colour to start with, then becoming paler and finally dark green.

– *E. parviflorus* RATAJ (*E. peruensis*, name not in current usage). Tropical S America. Leaf blades broadly lanceolate, mid- to dark green. Inflorescences with adventitious plants appear only when illumination lasts less than 12 hours.

4. Small to medium-sized species that form runners and result in large clumps of vegetation.

– *E. angustifolius* RATAJ. Tropical and subtropical S America. Leaves narrowly lanceolate, up to 50 cm. Similar to *Vallisneria*.
– *E. latifolius* (SEUBERT) RATAJ (*E. magdalenensis*, name not in current usage). Central America, Northern S America, the West Indies. Leaves narrowly lanceolate, up to 25 cm long (figs. 467 and 468).
– *E. quadricostatus* FASSETT (*E. intermedius*, name not in current usage). Dwarf Amazon Sword Plant. Tropical S America. Leaves broadly lanceolate, up to 15 cm long.

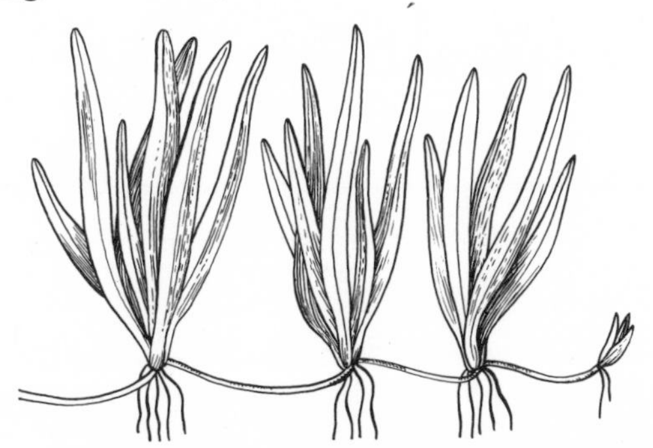

Fig. 467 Runner formation in *Echinodorus latifolius*

– *E. tenellus* (MARTIUS) BUCHENAU. Eastern N America, Central America, tropical S America. Leaves narrowly linear, up to 8 cm long. A popular plant at the front of aquaria. Smallest species of the G.

Fig. 468 *Echinodorus latifolius*

Echinoidea; Sea-urchins. Cl of the echinoderms (Echinodermata*). Various kinds of Sea-urchin are found in all seas. They are anything from apple-shaped to disc-shaped. The skeleton forms a rigid capsule except in the Leathery Sea-urchins. This capsule is covered in spines which articulate by means of ball-and-socket joints. Sea-urchins usually possess strong pedicellariae. Rows of sucking tube feet stretch along the five radia from the mouth almost as far as the end of the body, where the anus is located in the Sub-Cl Regularia. The Regularia are without exception grazing animals, tearing off plant and animal growth with their 5 small teeth, although they will also eat small animals. The teeth are fixed to a maxillary framework made of limestone–the so-called Aristotle's Lantern. This chewing apparatus is not present in the Irregularia Sub-Cl (the anus is also displaced over to the mouth side). Members of the Cl live buried in a soft substratum where they eat minute food particles. Well-known irregular echinoids are the Sand Dollars (*Clypeaster* etc) and the Heart Urchins (*Brissopsis* etc). For the hobbyist only the radially symmetrical Regularia are of interest. The spines are sometimes needle-sharp (*Paracentrotus**), very long (*Diadema**), and sometimes thick and bristly (*Cidaris**, *Eucidaris**, *Heterocentrotus**). The Leathery Sea-urchins carry poison glands in their spines. Many others have poisonous pedicellariae (eg *Toxopneustes*). Apart from defence, the spines also enable the Sea-urchin to move along as if on stilts and to anchor itself in rock crevices. Eggs and spermatozoa are mostly released freely into the water. The planktonic larva is called pluteus. In some places the gonads of certain species of Sea-urchin are eaten by humans.

Many Regularia are interesting aquarium specimens that survive well. Absolute essentials for success are a plant diet (algae, as well as lettuce leaves, raw potatoes etc) and healthy, stable conditions in the water. They seem to survive best in well-established marine aquaria rich in algae. Small pieces of meat will be eaten and even fish food pellets. Healthy Sea-urchins live attached with their spines erect and their tube feet highly active. If possible, they should not be taken out of the water. Those Sea-urchins that are suitable for the aquarium belong to the following Ga: *Arbacia**, *Cidaris**, *Eucidaris**, *Echinus**, *Echinometra**, *Paracentrotus**, *Psammechinus**, *Sphaerechinus**, *Stylocidaris**, *Tripneustes**.

Echinometra GRAY. G of the Sea-urchins (Echinoidea*). Contains various very common species found in the intertidal zone of tropical seas. Similar to *Paracentrotus**, it lives on rocks into which it bores holes. The test is oval in shape with a diameter of 5–7 cm. It is covered in strong pointed spines between 2 and 4 cm long. Colour varies. *Echinometra* is a lively species that is easy to keep in an algae-rich aquarium.

Echinus LINNAEUS, 1758. G of the Sea-urchins (Echinoidea*). *Echinus* lives as a grazing animal in the sub-littoral of cool seas. Shaped like an apple, they have relatively short, not very sharp spines. Diameter up to 17 cm. *E. esculentus* LINNAEUS, 1758, lives at depths of 10 m on hard substrata and is found from Iceland to Portugal. It is red-violet in colour and is an edible species. Keeps fairly well in cool aquaria if fed on algae and animal food.

Echiurida; Echiuroids. A marine animal Ph with few species. Their relationship to other animals is unclear. Externally they are similar to the polychaete annelids (Annelida*), but unlike them the body is not divided into segments. The Echiurida live primarily in burrows with a lobed food-collecting proboscis protruding. They eat detritus and micro-organisms. The food particles are surrounded by mucus then taken to the mouth by cilia, like on a conveyor belt. In the Mediterranean, the T-shaped black proboscis of the female *Bonellia viridis* ROLANDO is often observed. Its body is round like a balloon and often lies hidden in holes in the rocks or beneath rocks. This species has a very interesting sexual determination. The swimming larvae which hatch from the eggs are sexually indeterminate. A chance brush against the proboscis of a female determines them as males. Males remain minutely small and live in the foregut of the females. Larvae that do not come into contact with a female sink to the bottom and develop into females.

Echiuroids, Ph, *see* Echiurida

Ecological Factors. Those parts of the environment* that can have a direct effect on a living being. Because the effective environment of an organism has a lifeless (abiotic) and a living (biotic) part, the ecological factors are likewise divided into abiotic and biotic factors. Important abiotic factors for aquarium animals are temperature, light, water, salt content, hydrogen-ion concentration (pH value*); important biotic factors include food organisms, species members, enemies, competitors.

Ecological factors are normally only of benefit within a certain range, outside this range they can be harmful. This range is called the ecological valence*, and the ability of an organism to utilise a particular factor range is called ecological potence*. The reaction of a living being to the variation of an ecological factor can be represented in graph form by a so-called normal distribution curve with 3 essential points on it: the optimum, minimum and maximum (the minimal and maximal values of the factor which can be tolerated). The ranges that lie between the optimal and extreme values are known as pejus. In the wild, species are sometimes forced unwillingly into such pejus zones—perhaps because of stronger competition from other species. For example, Carps (*Carassius* carassius*) are often found in water that is less than ideal for them (low pH value, poor food supply etc). Their growth is affected, but at least they do not face much competition there, nor are there many enemies.

The success of a species in its habitat (Biotope*) reaches its peak when all the ecological factors lie in the optimal zone. Under such conditions, the species will have a very good reproductive quota, for example. So, if this principle is applied to keeping animals in an aquarium, it will mean that the optimal zones of the ecological factors will have been attained if the plants and animals reproduce well. On the other hand, the demands of a species may be so difficult to fulfil, that even managing to keep it in an aquarium at all must count as a success. This might apply to a great many marine organisms and beings that come from cool, fast-flowing waters.

Ecological Potence (fig. 469). The ability of an animal or plant to come to terms with ecological factors, eg with temperature, salt content. For example, *Maya arenaria* and the Three-spined Stickleback *(Gasterosteus* aculeatus)* have a high ecological potential in respect of fluctuating salinity levels; coral reefs, on the other hand, can only tolerate slight fluctuations in salinity level and temperature (ie they have a low ecological potential in respect of these factors). Broadly speaking, there is a distinction between eurypotent (Eury-*) and stenopotent (Sten-*) organisms—that is, organisms with broad or narrow ecological potentials in respect of the amplitudes (ranges) it is possible to encounter. So, in our drawing, *Maya arenaria* is ecologically characterised by curve c (optimum in a high salinity level),

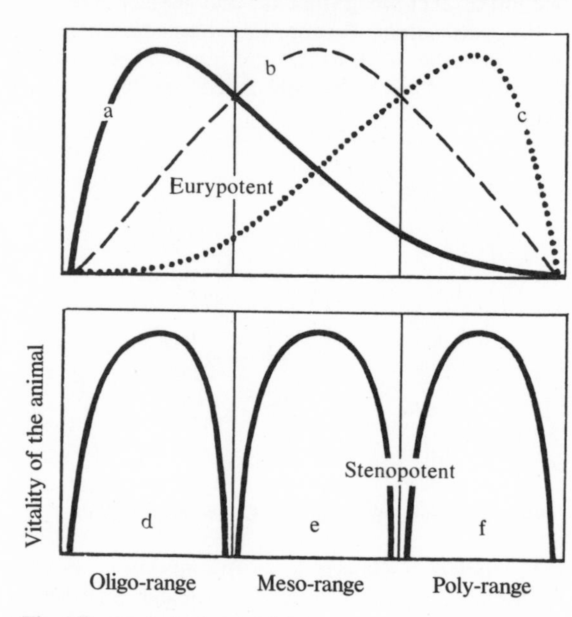

Fig. 469 Ecological potential

the Stickleback by curve a (optimum in a low salinity level), and the coral reefs, in respect of temperature and salinity level, by curve f (optimum at a high temperature and high salinity level). Compare ecological valence.

Ecological Valence (fig. 470). The value an ecological factor (Ecological Factors*) has for an organism. Usually, animals and plants can only exist within a certain range of an ecological factor (eg within a certain temperature range, pH value or salinity level). The most favourable values of these ranges are known as the optimum, the extreme values as the maximum or minimum. Between the optimum and the extreme values lies the pejus. The span between the extreme values (amplitude), together with the width of the optimum, determine the ecological valence of the factor for a particular organism. The valence of a factor, ie its value *for* the organism, as seen from the organism's point of view, corresponds to the ecological potence* the organism has to come to terms with its environment*.

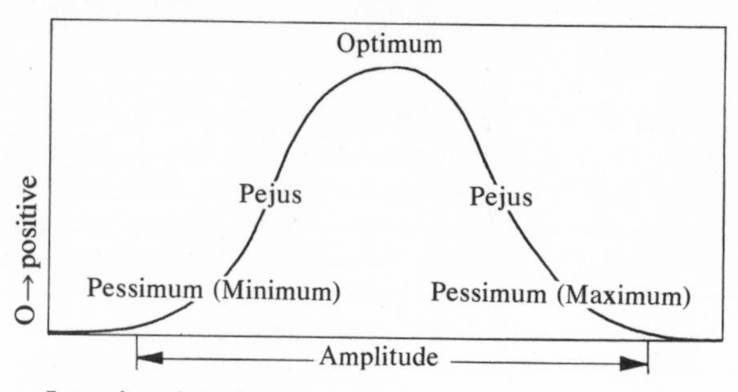

Fig. 470 Ecological Valence

Ecology. The study of the interrelationships between organisms and their environment*, or the science of the resources of nature, ie its structure, function and economy (HAECKEL). There are of course an enormous number of questions and problems with which ecologists have to concern themselves, but they can be grouped under 3 main levels – an individual organism, a population, and a collection of organisms made up of several species (all in relation to a particular biological reference point). From this, it is possible to divide ecology up into 3 disciplines: autecology*, population ecology or biodemography* and synecology*. Alongside pure natural scientific studies, ecology also has to contend with problems of a very practical nature, eg within hydrobiology, production biology, landscape husbandry, pest control and environmental pollution. In the future, what we know about ecology and how well we have implemented our findings may have a considerable bearing on how far the material requirements for *all* life on Earth, including our own, can be secured in the face of an ever increasing technological and artificial environment.

Ecosystem. The naturally defined community of a particular habitat (Biotope*) and the organisms that belong there (Biocenosis*), eg a mountain stream or a coral reef in its total complexity. How an ecosystem is made up, its function and the relationships of all its members to one another are subjects that are studied in synecology*.

Ecsenius MCCULLOCH, 1923. G of the Blennidae, Sub-O Blennioidei*, containing several species that live in the tropical Indo-Pacific. They are bottom-dwelling fish that live in shallow waters and feed on algae and small animals. Their build is typical of the family but they are distinguished by the high arch of the forehead, lack of comb on the head and lack of tentacles above the eyes. However, there are 1–3 thread-like nasal cirri. The pores of the LL form a double row, ending below the llth fin ray of the D. The following species, for example, will survive well if fed on algae and small animals and kept away from large, active specimens.

– *E. bicolor* (DAY, 1888); Two-coloured Slimefish, Bicolor Blenny. Indo-Pacific. Up to 10 cm. The males are black-brown anteriorly and yellow-orange posteriorly. The females are similarly coloured, although the pale part is restricted to a part of the caudal peduncle. C elongated.

Ectocarpus. G of the Brown Algae (Phaeophyta*) with delicate, branched strands that form clumps. Found in the upper coastal area of both cool and warm seas. Unlike most other brown algae *Ectocarpus* does occasionally flourish in a marine aquarium.

Ectoparasites *see* Parasitism

Eddy Worms, Cl, *see* Turbellaria

Edible Cockle *(Cardium edule) see Cardium*

Edible Crab *(Cancer pagurus) see Cancer*

Edible Sea-urchin *(Echinus esculentus) see Echinus*

Eel Cods (Muraenolepididae, F) *see* Gadiformes

Eel Gobies (Gobioididae, F) *see* Gobioidei

Eel Grass *(Vallisneria spiralis) see Vallisneria*

Eel-pout *(Lota lota) see* Gadiformes

Eel-related Species (Anguilloidei, Sub-O) *see* Anguilliformes

Eels *see* Anguilliformes

Egeria PLANCHON, 1849. G. of the Hydrocharitaceae* containing 2 species. Found in eastern S America from the Rio de la Plata to the source rivers of the Rio San Francisco. Has also spread into other areas. Submerged perennials with long stem axes. Leaves arranged in whorls, branching irregularly. Leaves are linear, rounded, with a slightly jagged edge. Mid-rib present. Flowers unisexual and dioecious, polysymmetrical. Perianth made up of parts in numbers of three; it is dialypetalous with a green calyx and well developed white petals. Male flowers in groups of 2–4 contained in a translucent spathe situated at the leaf axil and comprising 9 stamens. Female flowers come singly in a spathe and comprise 3 carpels which are fused and inferior. The fruit is a green berry. The various species of the G always stay submerged, rooting into the substratum or floating freely in the water. The flowers grow up to the surface where they are pollinated by insects. Female plants are thought to be very rare. So far, only one of the 2 species has been cultivated.

Fig. 471 *Egeria densa*

– *E. densa* PLANCHON (*Elodea densa*, name not in current usage). Distribution as for the G, and also introduced into Central and N America, Europe, Australia. An undemanding aquarium plant that likes a lot of light. The free-floating variety also grows well. Can be kept in the open in summer (fig. 471).

Egg *see* Gametes

Egg Development *see* Embryonic Development

Egg-laying Tooth-carps, F, *see* Cyprinodontidae

Egyptian Lotus Flower *(Nymphaea lotus) see Nymphaea*

Egyptian Mouthbreeder *(Hemihaplochromis multicolor) see Hemihaplochromis*

Eichhornia KUNTH, 1843; Water Hyacinth. G of the Pontederiaceae* comprising 6 species. Found in tropical and subtropical America, and one species in Africa. There has been some introduction to other areas. Perennials with long or short stem axes and an alternate or rosette leaf arrangement. Submerged leaves are linear, emersed leaves stalked, with round to cordate blades. The petioles are swollen up with inflated tissue, Flowers hermaphroditic and monosymmetrical. Perianth is made up of 6 fused tepals. 6 stamens, 3 carpels which are fused and superior. All the species of the G are aquatic and marsh plants which take root either in the substratum of a stretch of water or in damp ground. They may also grow floating around freely at the water surface.

– *E. azurea* (SWARTZ) KUNTH. Tropical S America. Linear aquatic leaves and stalked aerial leaves (fig. 472).

Fig. 472 *Eichhornia azurea*

– *E. crassipes* (MARTIUS) SOLMS; Common Water Hyacinth. Tropical and subtropical America. Has been introduced to many regions with similar conditions where it has become a weed in polluted waters. As a result the waters may cease to be navigable. Infested waters are treated with chemicals and the weed may also be removed by hand. No aquatic leaves. Many runners are produced (fig. 473).

Fig. 473 *Eichhornia crassipes*

– *E. diversifolia* (VAHL) URBAN. America and Africa. Linear aquatic leaves and stalked aerial leaves that float.

Eigenmann, Carl H. (1863–1927) Born in Flehingen in Swabia. He studied at Bloomington University and at Harvard. From 1889–90 he was curator of the Fish Collection at the Californian Academy of Sciences, and from 1891 he was Professor at Indiana University. He made studies of the cave-dwelling fishes of Kentucky, Arkansas, Texas and Cuba. With his pupil J. D. HASEMAN, he made several expeditions taking him to Central and South America during which he made extensive collections of fish. His book *The Freshwater Fishes of British Guyana* appeared in 1912 and in 1922 there followed *The Freshwater Fishes of Northwestern South America* and *The Freshwater Fishes of Chile.* Between 1917 and 1921 there appeared the comprehensive monograph *The American Characidae.* Eigenmann is regarded as one of the greatest experts on South American fish; he made first descriptions of 100 species.

Eimeria, G, *see* Sporozoa

Elachocharacinae, Sub-F, *see* Characidiidae

Elasmobranchii, Sub-Cl, *see* Chondrichthyes

Elasmobranchs (Elasmobranchii, Sub-Cl) *see* Chondrichthyes

Elasmoid Scales *see* Scales

Elassoma JORDAN, 1877. G of the Centrarchidae* comprising 2 species that come from south-eastern N America. Small, moderately elongated fish, only slightly compressed laterally. In contrast to other Ga of the F, the D has only 4 or 5 hard rays and the soft-rayed part is not separated off from the hard-rayed part by an incision. LL absent. *Elassoma* can 'walk' along the bottom using its Ps. Undemanding species in an unheated aquarium, they seem to be able to tolerate considerable temperature fluctuations without coming to any harm. They are best kept at cool temperatures

Fig. 474 *Elassoma evergladei*

(8–10 °C) in winter, but at up to 30 °C in the summer, provided there is ventilation. The tank should be small and with dense vegetation; reproduction will take place without further provision. After a very charming love play, about 30–60 eggs will be laid on the plants; they will hatch after about 2–3 days. After the brood has swum free, feed with finely ground powdered food. The parent fish do not need to be removed. Finally, it should also be mentioned that *Elassoma* is particularly recommended for open-air ponds. During the warm period specimens may be left alone with no special care.

– *E. evergladei* JORDAN, 1884; Everglades Pygmy Sunfish, Dwarf Sunfish. USA, from Carolina to Florida. Up to 3.5 cm. Clay-yellow in colour, silvery in parts, with black speckles. At spawning time the body and fins of the males are a beautiful velvet black with iridescent green speckles (fig. 474).

– *E. zonatum* JORDAN, 1877; Banded Pygmy or Dwarf Sunfish. USA, in the region between Illinois, Alabama and Texas. Up to 3.5 cm. Body and fins grey-green with fine black speckles and transverse bars on the C. 11 or 12 dark bands run from the back down towards the belly. A dark, sometimes black, blotch is located above the P (fig. 475).

Fig. 475 *Elassoma zonatum*

Elatinaceae; Waterworts. F of the Magnoliatae* comprising 2 Ga and 40 species. Tropical to temperate regions. Most are small perennials that either grow upright or along the ground like turf. Others are annuals. Leaf arrangement is decussate in most cases. Inconspicuous flowers, hermaphroditic, polysymmetrical. Found mainly in marshy and flooded areas, so many species are adapted to life in water. Species belonging to the G *Elatine** have already been kept in the aquarium.

Elatine LINNÉ, 1753; Waterwort. G of the Elatinaceae* containing 12 species. Tropical to temperate regions. Either small perennials that grow along the ground like turf or annuals. Leaves arranged in whorls or decussate. Inconspicuous flowers, hermaphroditic, polysymmetrical. Double perianth, sepals and petals numbering between 4 and 2, free-standing. Stamens 8–2, carpels 4–2, fused, superior. Fruit is a capsule. The various species of *Elatine* grow in damp, sometimes flooded, sandy or muddy places that are rich in nutrients. Their seeds are partly distributed by aquatic birds, so many species succeed in colonising large areals. Information regarding their suitability for the aquarium is scant.

– *E. alsinastrum* LINNÉ. Europe. Whorled leaves. Found in ponds and ditches as well as places that become dry in summer.

– *E. americana* (PURSH) ARNOLD. N America. Decussate leaves. Widespread in flooded areas.

– *E. macropoda* GUSSONE. Southern Europe, North Africa. Decussate leaves. Grows at the edge of still waters. Has been kept in aquaria for short spells.

Electric Catfish *(Malapterurus electricus) see* Malapteruridae

Electric Catfishes, F, *see* Malapteruridae

Electric Eel *(Electrophorus electricus) see* Gymnotoidei

Electric Eels (Electrophoridae, F) *see* Gymnotoidei

Electric Organs. Structures found in about 250 species of fish in which electrical currents are produced that are far in excess of the bio-electrical charge of the nervous system* or of the muscular system*. Particularly large electric organs create strong currents which can paralyse prey and at the same time protect the producer of the current (fig. 476). Examples of fish that produce such currents include the Electric Ray (*Torpedo torpedo*, Torpedinidae*) which can deliver between 60 and 80 volts, the Electric Eel *(Electrophorus electricus)* which can emit 300–600 volts from its sides, the Electric Catfish *(Malapterurus electricus)* and the Stargazer *(Uranoscopus)*. Weaker currents are produced in short bursts by other electric organs, thus creating an electric field round the body of the fish. They are obviously emitted to enable the fish to orientate itself, as well as to indicate territorial boundaries and to detect like species. The electrical potential created varies from below 10 volts up to 20. Such weaker currents are produced by the mormyrids (Mormyridae, Gymnarchidae) and the skates and rays (Rajidae), for example. Electric organs are produced from transverse

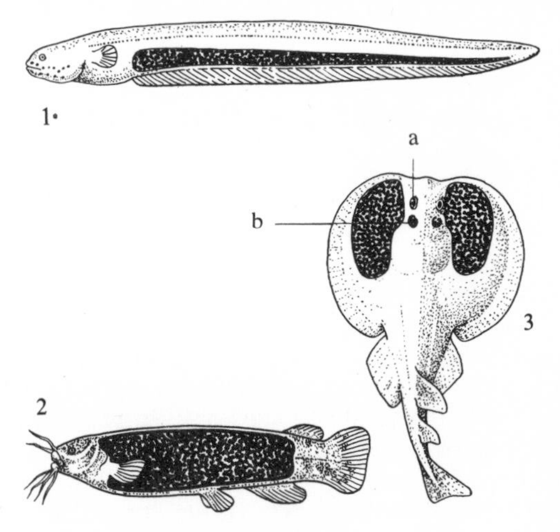

Fig. 476 Electric organs of highly electric fish (position indicated by the black areas). 1 Electric Eel 2 Electric Catfish 3 Electric Ray a Spiracle b Eye

bands of muscle, the cells of which form thin plates stacked up like pillars. The pillars lie very close to one another and each one may be composed of a great many (sometimes hundreds or even thousands) of current-producing elements. The electric organs surrounding the body of *Malapterurus* are probably derived from structures of the skin.

Electric Ray *(Torpedo nobiliana) see* Torpedinidae

Electrolyte Content (fig. 477). Electrolytes are chemical substances that break down into electrically charged particles (Ions*) when dissolved in water. Examples are acids*, bases* and salts. There is a certain amount of electrolyte content in all naturally occurring water, although the amount fluctuates enormously. For example, electrolyte content is extremely low in mountain streams and lakes that flow across igneous rock, as well as in many tropical rivers (*see* Amazon); in chalk-rich fresh water, brackish water and sea water the electrolyte content climbs much higher. Very definite living communities (*see* Biocenesis) establish themselves in relation to the electrolyte content (of course in relation to other environmental factors too). Some organisms (eg many Coral Fishes and *Symphysodon**) can tolerate only small electrolytic content and fluctuations, whereas others can tolerate higher levels and greater fluctuations (eg brackish water species and *Artemia**). The most important thing for the freshwater aquarist to know is that the electrolyte content is very low in desalinated or distilled water and in rain water,

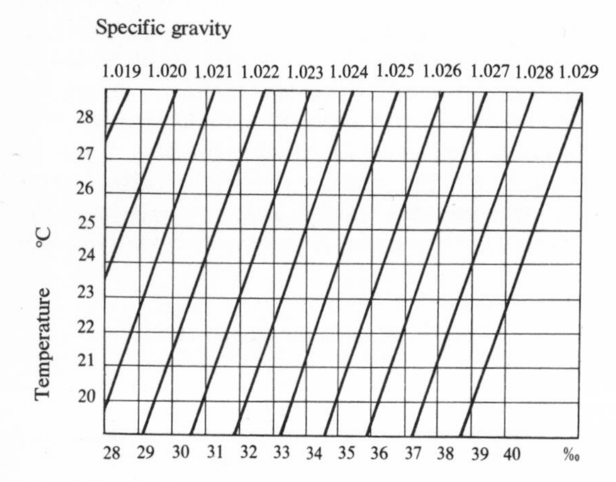

Fig. 477 Diagram to determine electrolyte content

whilst it is very high in hard water. In order not to let the electrolyte content of aquarium water climb unnecessarily high, any evaporated water should be replaced with water containing very low amounts of electrolytes. This applies to both fresh and sea water aquaria. The electrolyte content of fresh water can be established by determining its hardness* and by taking conductivity measurements*; that of sea water can be determined by measuring its density with an areometer*. The relationships of salt content, temperature and density in sea water is represented by the diagram. For example, at 22 °C and a density of 1.027, the salt content amounts to 37‰.

Electrophoridae, F, *see* Gymnotoidei

Electrophorus electricus *see* Gymnotoidei

Eledone LEACH, 1817 (fig. 478). G of the Cephalopoda*, O Octobrachia. Found in the Atlantic, North Sea and in the Mediterranean, mainly in the lower coastal waters on sandy and pebbly bottoms. The appearance and way of life of *Eledone* are like those of an *Octopus** although the suckers are in a row and there is a greater amount of skin between the arms. Members of this G remain smaller than *Octopus* so are more suitable for the aquarium – at least, as suitable as any Cephalopoda can be. One example of a well-known species is:

– *E. (Oceana) moschata* (LAMARCK). Mediterranean. Arms up to 40 cm in length. Dirty brown colour with dark blotches. Smells of musk.

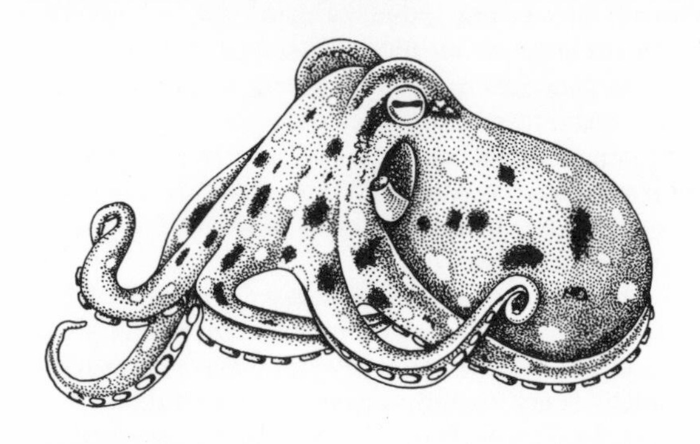

Fig. 478 *Eledone*

Eleocharis R. BROWN, 1810 (*Heleocharis*, name not in current usage). Spike Rush. G of the Cyperaceae* comprising 180 species that grow all over the world. They are small perennials, some with runners, rarely with tubers. Rosette leaf arrangement. Leaves are long and tufty; flowers small and inconspicuous arranged in spikelets. Perianth bristly. Stamens 3–2, carpels 3–2, fused and superior. The fruit is in the form of a nut. The various species of the G grow in marshes, wet meadows, on muddy terrain and sometimes on flooded river banks etc. To date only a few have acquired any significance for the aquarist, eg:

– *E. acicularis* (LINNÉ) ROEMER et SCHULTES; Hair Grass, Lesser Spike Rush. Found in certain places in almost all temperate regions, except Africa. They are small plants that form thick carpets by producing runners.

– *E. vivipara* LINK; Umbrella Plant. Southern N America. Leaves long and thin. Usually, adventitious plants are produced instead of spikelets.

Eleotridae, F, name not in current usage, *see* Gobioidei

Eleotris GRONOVIUS, 1763. G of the Gobiidae, Sub-O Gobioidei*, Found in the coastal waters of tropical parts of W and E Africa, Australia, the Malaysian Archipelago, and the Atlantic coast of America from S Carolina (USA) to Brazil. They live both in sea water and brackish water of lagoons, mangrove swamps, river mouths, as well as in inland waters near the coast. *Eleotris* is a small to medium-sized fish (10–23 cm). Its body is elongated, usually cylindrical at the front and laterally compressed at the rear. The head and mouth are large, lower jaw protruding. Sharp teeth. The middle palatal bone bears no teeth. V is never fused, and the D1 usually has 6 weak spines. C rounded. Scales small to medium-sized. Ground colour various shades of brown, often with dark speckles or blotches on the body and fins, in part producing a mottled effect. In rare cases, colours are more striking. Differences between the sexes largely unknown. Should be kept in spacious tanks with a soft sandy substrate and hiding-places. The strictly freshwater forms require 1–2 tsp sea salt to every 10 l water, the brackish water forms 50–100 g to every 10 l. Temp. 20–25 °C. These fish are mostly predatory with voracious appetites. They will feed on live food of all kinds. So far most species have not been bred in captivity. The following species are frequently imported.

– *E. africana* STEINDACHNER, 1880. Tropical W Africa where it lives in brackish water at the mouths of rivers. Up to 16 cm. Body particularly elongated. Grey-brown in colour, younger specimens with an indistinct mottled effect. There is a round black blotch on the upper part of the caudal peduncle. D and C yellowish with rows of brown dots. The lower edge of the C, the edge of the An and the tips of the Vs are whitish. The female has smaller dots on the fins.

– *E. butis* (HAMILTON-BUCHANAN, 1822). E Africa, Malaysian Archipelago and Australia, where it lives in sea and brackish water. Up to 14.5 cm. Head flat and pointed. Body yellow-brown to grey-brown with reddish to dark brown dots; irregular transverse bars may also be present. D1 grey, often with a red tip. An edged in a broad yellow band; and at the base of the P is a dark blotch edged in red.

– *E. lebretonis* STEINDACHNER, 1870 W. Africa, particularly common in fresh water. Up to 12 cm. Body yellowish to olive-brown with an irregular dark longitudinal band. Scales are dotted in brown. A blue blotch lies behind the gill covering. Unpaired fins grey to whitish with dark dots. At spawning time the upper surface becomes olive-green, the throat and belly orange-red. In the male the D2 is drawn out into a point, Vs pointed. An undemanding species that produces a lot of spawn which is laid on fine-leaved plants. The young fish are very small and should be reared on the finest live food (fig. 479).

Fig. 479 *Eleotris lebretonis*

– *E. melanosoma* BLEEKER, 1852. Found in brackish water and sea water from the Malaysian Archipelago to Central America. Up to 13.5 cm. Dorsal surface clay-yellow to red-brown, the rest of the body being dark. Some may have pale spots or pale markings in the shape of a net. Green stripes extend from the eyes across the operculum. Unpaired fins brown with dark spots.

– *E. monteiri* O'SHAUGHN, 1875. W Africa, where it lives in brackish and fresh water. Up to 23 cm. Body very elongated. On the preoperculum there is a noticeable spine. The upper surface of the body is dark brown, pale at the sides with red-brown blotches that are not clearly outlined. Young fish have a dark longitudinal band. Unpaired fins are dark brown, with a pale mottly pattern. D1 has a black edge. Quarrelsome species.

– *E. pisonis* GMELIN, 1788. N, Central and S America from S Carolina to Rio de Janeiro, including the West Indies. Found in the sea, and in brackish and fresh waters. Up to 12 cm. Upper surface of the body pale brown, otherwise dark brown; some may be mottled. Blackish, slightly twisting lines radiate from the eyes across the operculum. Unpaired fins yellowish with dark spots. V dark. Intolerant and burrows a great deal.

– *E. pleurops* BOULENGER, 1909. Tropical W Africa, particularly the Lower Niger and the mangrove swamps of the Niger Delta. Up to 10 cm. Body relatively tall. Upper surface dark brown, otherwise paler, with dark blotches in places. Unpaired fins brownish, D with white speckles.

– *E. vittata* DUMERIL, 1860. Tropical W Africa, from Senegal to the Congo. Found mainly in fresh water, sometimes in brackish water. Up to 22 cm. There is a distinct spine on the preoperculum. Upper surface of body brown to reddish, paler at the sides. Young fish have a dark longitudinal band with a pale edge at the top. Older fish have cloudy dark blotches. There are metallic blotches on the operculum. D1 has a broad brown band and a white border. Females less conspicuous, D2 lower than in the male.

Elephant-nosed Fish *(Gnathonemus petersi) see Gnathonemus*

Elephant-nosed Fish *(Mormyrus kannumae) see Mormyrus*

Elephant-nosed Fish types, O, *see* Mormyriformes

Elephant-nosed Fishes (Mormyridae, F) *see* Mormyriformes

Elfin Sharks, F, *see* Scapanorhynchidae

Elk's Antler Coral *(Acropora palmata) see Acropora*

Elodea RICHARD in MICHAUX, 1803 (*Helodea*, name not in current usage). Waterweed. G of the Hydrocharitaceae* comprising 17 species. Found in N and S America, not in Central America, and it has also been introduced into other areas. Submerged perennials with elongated stem axes. Leaves either decussate or in whorls, branched irregularly. Up to 7 leaves in a whorl, each leaf being linear, blunt or pointed at the end and slightly serrated at the edges. Mid-rib visible. Flowers hermaphroditic or unisexual and dioecious, polysymmmetrical. The perianth parts come in threes and the leaves are free-standing. The sepals are green and the inconspicuous petals are white or colourless. Stamens 3–6. Carpels 3, fused and inferior. The flowers always grow individually either in a spathe that is located in a leaf axil or in a stalked double-pointed spathe. The fruit is a green berry. The various species of the G always live submerged, taking root in the substratum or floating freely in the water. The hermaphroditic flowers (the females and the majority of the males) grow up to the water surface where they open out. In *E. nuttallii* the male flowers free themselves from the spathe and climb upwards. Drifting around on the surface, they can reach the female flowers and so transfer the pollen. Usually, the pollen is conveyed by the water. The *Elodea* species are undemanding water plants, some of which are suitable for the room aquarium, some for open-air ponds. The floating varieties also grow well, Several species have already been cultivated, but they have not all been identified for certain.

– *E. canadensis* RICHARD in MICHAUX, Canadian Waterweed. N America. Female plants have been introduced into Europe. Leaves flat and rounded.

– *E. granatensis* HUMBOLDT et BONPLAND. Northern and eastern S America. Flowers hermaphroditic.

– *E. nuttallii* (PLANCHON) St. JOHN; Nuttall's Waterweed. Eastern N America. Leaves curving in on themselves.

Elopidae, F, *see* Elopiformes

Elopiformes. O of the Osteichthyes*, Coh Teleostei. Herring-like marine fish with several primitive features, eg a bony plate in the throat region (gula), the swimbladder joined by a duct to the intestine (in addition to which the swimbladder can project outgrowths forwards nearly to the inner ear), as well as a great many peculiarities of the skeleton. Several Elopiformes also have larvae which are similar to the leptocephalus larvae of the Anguilliformes*.

– *Albulidae*; Ladyfishes. Unlike the other Fs to follow, the Ladyfishes have no gular plate. They are of interest to the scientist because of their ribbon-like leptocephalus larva, among other things. There are only 2 known species. *Albula vulpes* (fig. 480) (Bonefish, Ladyfish) grows up to 1 m long and is found in shallow waters in all warm seas. It has a slightly inferior mouth and its head is covered in a gristly substance. Not of great importance as a food, but a firm favourite with anglers.

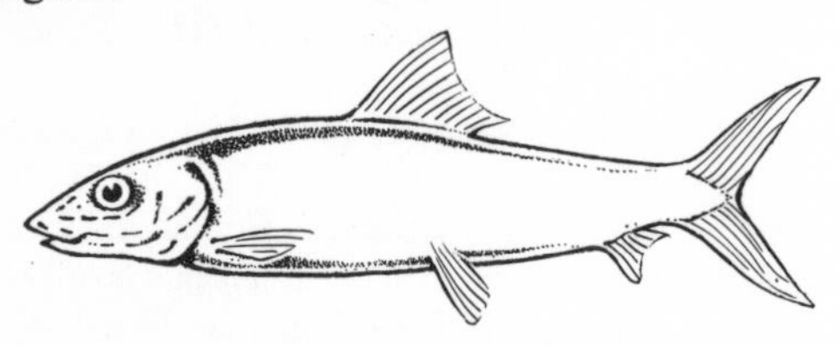

Fig. 480 *Albula vulpes*

– *Elopidae*. Gular plate present. The swimbladder is not in contact with the inner ear and the heart lacks the Conus arteriosus. For a long time 7 modern species were attributed to the Elopidae, but it is now thought that all are identical to *Elops saurus* (fig. 481) (Tenpounder, Giant Herring). However, there are several fossil Ga known since the Lower Cretaceous. *Elops* occurs in all tropical and subtropical seas, and can grow up to 1 m long. It usually swims in small shoals. The fins are relatively small, the C being deeply indented. The Vs lie directly opposite the beginning of the D. Look under the following family for information regarding its economic importance.

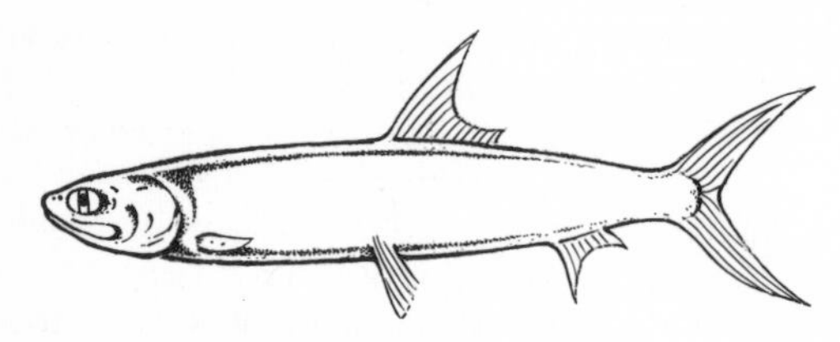

Fig. 481 *Elops saurus*

– *Megalopidae*; Tarpons. The main differences between the Megalopidae and the Elopidae are that in *Megalops* the mouth is slightly superior, the chin is swollen and protruding and the last ray of the D is elongated like a thread. The Vs also lie somewhat in front of the beginning of the D. Fossil Ga have been known since the Lower Cretaceous. The Atlantic Tarpon *(Megalops atlanticus)* grows to about 2 m long, bigger in rare cases. In the summer it remains off the American Atlantic coast from Brazil to Long Island, then it migrates northwards, occasionally entering brackish and fresh water. At cold times of year it prefers the open seas. The smaller *Megalops cyprinoides* (fig. 482) (Ox-eye or Pacific Tarpon) is widely distributed in the Pacific. The species of both *Elops* and *Megalops* contain a lot of bones so are of little economic importance. However, they make ideal sporting fish for the angler. It is said to be one of the most exciting experiences for an angler to grapple with either of the species, as the Atlantic Tarpon for example does not give up the battle till the bitter end. Each year many people go to Florida just to take part in the sport.

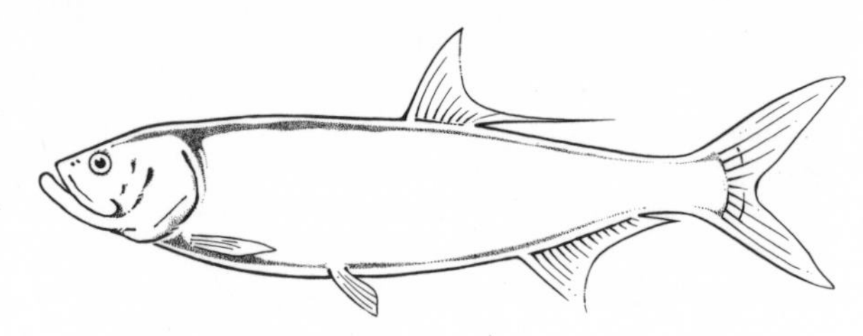

Fig. 482 *Megalops cyprinoides*

Elopomorpha, Sup-O, *see* Osteichthyes

Elops saurus *see* Elopiformes

Eluachon *(Thaleichthys pacificus) see* Salmonoidei

Emaciation in fish is the result of a deficient or incorrect diet (*see* Feeding), or of disease. In extreme cases the belly becomes hollow and the back knife-edged. Extreme emaciation may also be caused by parasitic infection of the digestive tract by Nematodes*, Acanthocephala*, Cestodes*, *Hexamita* and others, as well as Fish Tuberculosis* and *Ichthyosporidium**.

Embiotocidae; Surfperches or Seaperches (fig. 483). F of the Perciformes*, Sub-O Percoidei*. Small to medium-sized livebearing fish found in the coastal

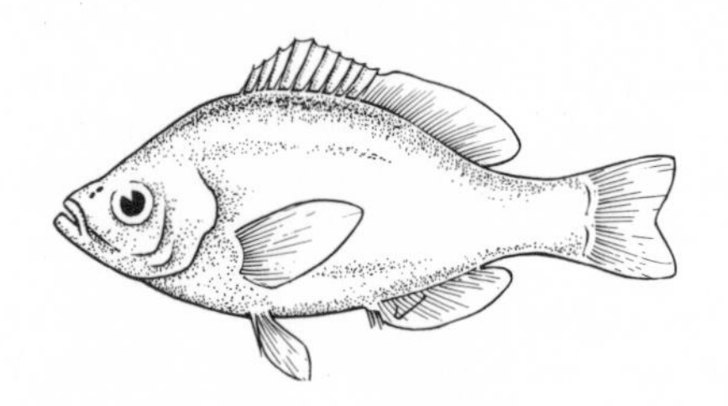

Fig. 483 Embiotocidae-type

waters of the northern Pacific. Contains 22–23 species. With their long oval bodies, the Embiotocidae are faintly reminiscent of the Barbs from the side. The mouth is relatively small but is surrounded by well defined lips. Jaws contain some teeth. D is of a uniform height and comprises at most 11 spiny rays. C indented, An comprises 3 spines and 15–35 soft rays. The head and body are covered in cycloid scales. LL complete, arched near the top end. A characteristic feature is the groove along the D separating the scales on the body from those at the base of the fin. The Embiotocidae are livebearing – a rare thing among the marine bony fish. Mating takes place without any temporal reference to fertilisation. The embryos develop in the cavity of the ovary where they are able to breathe and obtain food via outgrowths of the skin on the dorsal and anal fins

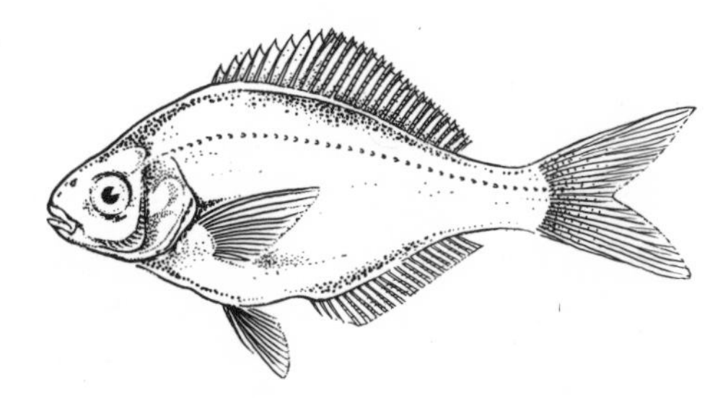

Fig. 484 *Phanerodon furcatus*

which lie close to the ovary. Each female gives birth to about 20–40 youngsters, all of which are very big at birth and almost sexually mature. The Embiotocidae live in shallow waters; one species, *Hysterocarpus trasci*, lives in the freshwater of the Sacramento Delta. *Phanerodon furcatus* (fig. 484) and some other species are used in the fishing industry. Some are popular species with anglers. For the aquarist they are of little importance.

Embryo. The early stage of development of a multi-celled organism.

Embryonic Development (fig. 485). The formation of a multi-celled organism from a fertilised egg-cell (*see* Fertilisation, Gametes). The time taken for an embryo to develop varies a great deal in fish – it may take a day or considerably longer, even months. Embryonic development is at an end when the eggs hatch out. Development is slower at lower temperatures. Embryonic development starts with cell division or *cleavage*, as daughter cells (blastomeres) are formed one after the other from the one egg-cell. With each division the daughter cells get smaller. To begin with, the yolk supply in bony fish is not used. The cleavage processes do not take place over the entire egg-cell, but only in a disc-shaped area at the animal pole that contains little yolk (the germinal disc). Gradually, this area takes on the look of a heap of mulberry-like cells (morula). Complicated cell shifting and further cell division bring about the *formation of the germ layers*. This consists first of 2, and later of 3, leaf-like layers of cells that overlap. Inside these layers are the sites where the various organs and parts of the body will form. At the same time the germinal disc begins to flatten off and envelop the mass of yolk, so forming the yolk sac*. An organising centre (the embryonic knot) also develops at a point on the fused edge from which the fundamental development processes are regulated. The yolk begins to be used up. The phase of *organogenesis* gets underway with the establishment of the central nervous system. In fish, the spinal cord and brain develop as one massive cord, in which the typical cavities form only later. The cerebral divisions, the eyes (which are colourless at first), the olfactory cavities and the auditory vesicles become visible very early on. The early stages of the body's muscular system start to develop on either side of the neural system. Later on, the primordium of the gill slits starts to form. The heart begins to beat, driving the embryonic blood through the circulatory system of the body and the yolk sac. The head and tail free themselves from the rest of the matter. The mouth and anus and the internal organs start to differentiate, and more and more pigment accumulates in the eyes. Buds forming the paired fins begin to appear alongside the unpaired fin corona of the embryo. Body movements will be observed more and more often until the egg membrane is broken through. The body stretches itself out with the aid of a rod-like support

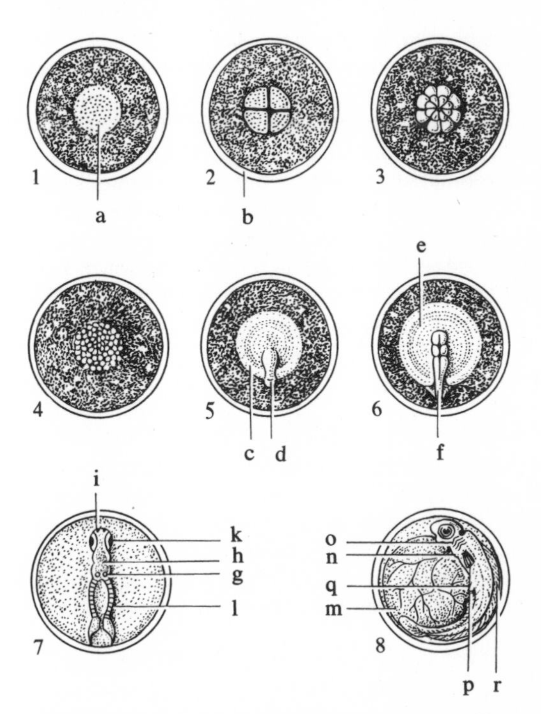

Fig. 485 Embryonic development (trout). 1 Egg with germinal disc (a) 2 4-cell stage (b Egg envelope) 3 16-32-cell stage 4 Multi-cell stage 5 Differentiation into a marginal swelling (c) from which the embryo develops (d Marginal knob), and a middle area 6 Embryonal primordium (f) enlarged (e Embryonic shield) 7 The yolk is almost completely overgrown by the germinal disc and the following can now be recognised on the embryo: g) Primordium of organ of balance, h) Brain, i) Nasal depressions, k) Primordium of the eyes, l) Primordium of the body segments 8 In this stage the yolk-sac (m) has become much reduced, and the following features are recognisable on the embryo: n) Lump that will form the heart, o) Gill slits, p) Pectoral fin, q) Pelvic fin, r) Dorsal fin fringe

called the notochord (chorda), which will later be replaced by the vertebral column.

After hatching the young fish still eats any yolk remaining in the yolk sac, usually for quite some time. Independent intake of food only begins later. In some fish a larval* period follows on from the embryonic development. During this phase, the young fish is quite different from the adult fish (European Eel, Flatfishes, Polypterids).

Emerald Perch *(Abudefduf oxyodon) see Abudefduf*

Emerald-eye Rasbora *(Rasbora dorsiocellata macrophthalma) see Rasbora*

Emersed. All plants that grow above the level of the water are described as emersed. So too are all the parts of submersed plants that rise above the surface. Opposite to submersed*.

Emperor Angelfish *(Pomacanthus [Pomacanthodes] imperator) see Pomacanthus*

Emperor Dragonflies, G, *see Anax*

Emperor Snapper *(Lutjanus sebae) see Lutjanus*

Emperor Tetra *(Nematobrycon palmeri) see Nematobrycon*

Enallagma, G, *see Coenagrion*

Enchytraeus; Whiteworms. G of the Enchytraeidae, Cl Oligochaeta*. *Enchytraeus* live in the ground in damp, sometimes wet, places rich in humus. They are found among rotting plant remains of various kinds (eg in the ground in a wood), in rotten tree-stumps, along the banks of streams and ponds or along the sea shore where the tide washes in. *Enchytraeus* is relatively easy to cultivate and is used as a foodstuff in aquaria. This is particularly true of the Common Whiteworm *(E. albidus)* which grows to over 3 cm long and is found in compost heaps. The Grindal Worm *(E. buchholtzi)* is used in a similar way; it grows up to 1 cm long. Both species are egg-laying hermaphrodites, and both are cultivated in very similar ways. *E. buchholtzi* needs to be kept at slighter higher temperatures perhaps (18–24 °C). The vessels used in cultivation and the culture medium should be adapted for the smaller sized *E. buchholtzi.*

The instructions that follow are for the rearing of *E. albidus.* You will need a wooden box measuring at least 20 cm × 30 cm at the base and 10 cm to 15 cm in height, with a lid that will shut tightly. Temperature 15–20 °C. A suitable substrate consists of loose, humus-rich earth (eg garden-mould), containing fine-grained sand. This should be mixed with leaf-mould in the ratio 1:1. The earth should be kept moist and broken up now and again. As far as possible the earth and box should be disinfected with steam. Sprinkle in the earth until it leaves room enough for some air between the soil and the lid. Make a shallow groove in the top of the soil and add a portion of whiteworms from which others will be bred. Cover the worms with a small amount of food medium then place a sheet of glass measuring about 10 cm × 15 cm over the site of the food. The food must be used up in 2–3 days and should then be renewed. Sour-smelling, rotting food remains should be removed from the box. There are various recipes for making up a good foodstuff. Usually, thick, pulpy, cooked oatflakes are the basic ingredient. These may be mixed with a little milk, egg-yolk, sugar, cooked vegetables or noodles. For vitamins, a few drops of cod-liver oil may be added. Damp bread and mashed potatoes may also be used. After about 6-8 weeks the culture will have multiplied in great numbers. They will collect at the feeding point or on the glass lid, from where they can be removed, popped into a glass and fed there. A culture is quickly exhausted, however. So, depending on requirements, it is advisable to set up several breeding boxes at once. Mites frequently appear in whiteworm cultures, and they are difficult to get rid of. If this happens, you could try spraying the lid of the box with an insecticide that kills mites, or add a further layer of soil to the culture and carefully spray that. However, usually the breeding box has to be set up anew.

Endemic Species. Animals and plants that are naturally restricted to a specific area. So, for example, one might refer to the endemic species of the Ethiopian region*, Lake Malawi* etc. All living organisms are in fact endemic species since they are not distributed worldwide and therefore cosmopolites*. Nevertheless, the term is used mainly whenever the particular distribution area (*see* Areal) is restricted to a single biogeographical region or an even smaller area.

Endoparasites, internal parasites, *see* Parasitism

Engraulis (several species) *see* Clupeiformes

Enneacanthus GILL, 1864. G of the Centrarchidae*. There are only a few species, all of which come from the eastern states of the USA where they live in clear lakes and ponds, and in slow-flowing streams and rivers with a rich submerged vegetation. The genus *Enneacanthus* is used in its broadest sense here and includes the G *Mesogonistius* GILL, 1864, which differs only in minor ways, although it is often classified separately. *Enneacanthus* species are small fish, perch-like to look at, with a tall oval shape and very compressed laterally. The D has 9 hard rays, the An 3 and the C is rounded and fan-shaped (convex). The edge of the gill covering is curved inwards and the upper corner is slightly elongated like a lobe. The V is pointed like a pennant. It is a very attractively coloured fish with some transverse bars that are an intensively dark colour.

The *Enneacanthus* species are ideal specimens for an unheated room aquarium. The tank should not be placed in too bright a spot and there should be thick clumps of vegetation to give these rather shy fish enough hiding-places. Great care should be taken when changing the water, as *Enneacanthus* is particularly sensitive to any great changes to its environment (pH value*, water hardness*). Reproduction is normally successful in a species aquarium particularly after a cool winter. As is typical for the Centrarchidae, the male makes a trough in which mating takes place with a willing female and the spawn is laid. The male will take care of the spawn and the newly hatched brood for some time. After the young fish are swimming freely they will require large amounts of powdered food (*see* Fry, Feeding of). *Enneacanthus* is also ideal for open-air ponds, and should produce a good number of offspring in the autumn. In winter, both the species *E. obesus* and

E. gloriosus should not be kept at temperatures lower than 10 °C, and for this reason should be removed from places liable to frost. In contrast, *E. chaetodon* can tolerate temperatures near freezing-point without coming to any harm.
– *E. (Mesogonistius) chaetodon* (BAIRD, 1854); Black-banded Sunfish. New Jersey to Maryland. Up to 10 cm. Dark-brown to black transverse bands on an iridescent mother-of-pearl background colour, with irregular dark blotches in between. The first two rays of the dorsal fin are deep black, the following two orange. The first two rays of the Vs are orange, the next two black. A very popular aquarium fish (fig. 486).

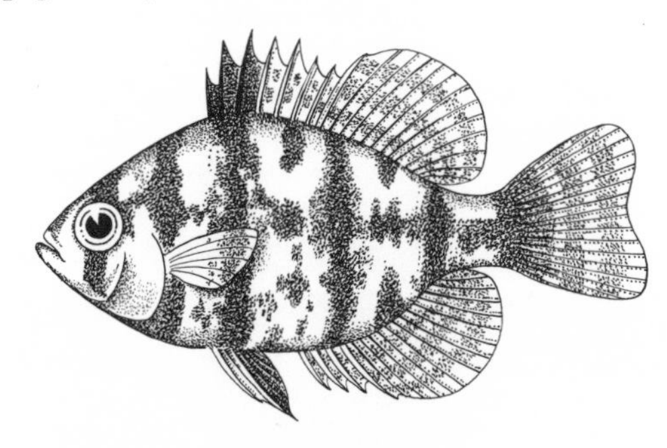

Fig. 486 *Enneacanthus chaetodon*

– *E. gloriosus* (HOLBROOK, 1855); Blue-spot Sunfish. From the state of New York south to Florida. Up to 8 cm. Olive-brown, particularly dark on the dorsal surface. There are irregular dark transverse bars in young fish. Pale-blue iridescent spots found all over the body and fins. There is a dark blotch on the operculum (but it is smaller than the eye in this case, unlike the following species!).
– *E. obesus* (GIRARD, 1854); Banded Sunfish. From Massachussetts south to Florida. Up to 10 cm. Colouring similar to that of *E. gloriosus*, except that the transverse bars are retained with age. The dark blotch on the operculum is about the size of the eye, and is outlined in a gold colour in the male. Pale spotting, mostly green-yellow in colour.

Enteromorpha; Green Seaweed. G of tube-like green algae (Chlorophyta*) found near the surface of sea water, some also in fresh water. Common along all sea coasts, and often growing more than 1 cm wide and 20–30 cm long. Cannot normally be cultivated in a marine aquarium.

Entoprocta, Ph, *see* Kamptozoa

Environment. The sum total of all the objects, happenings and energies on which a living being's existence depends. Since not all these factors are important as far as the organism is concerned, a distinction is made between the total environment and effective environment. So, the effective environment consists of only those parts of the total environment that have a direct effect on the organism (Ecological Factors*). Furthermore, the environment is divided into a non-living (abiotic) and a living (biotic) part. For example, the abiotic environment of a Carp *(Carassius* carassius)* might consist of the pond water and its temperature, light and particular chemistry, whereas the water plants, food organisms, like-species and enemies (eg Pike) make up its biotic environment.

Enzymes. Active substances produced within living organisms that have an effect on all metabolic processes. There are over 700 known enzymes. They are protein substances which are necessary either for the functions of specialised individual cells or for digestion. In the latter case they are deposited in the gastrointestinal canal in the form of secretions (digestive enzymes). Some groups of animals digest their food outside the body and so engulf the prey with enzymes.

Enzymes are biological catalysts. They trigger off biochemical processes or speed them up. Enzymes have a high reaction specificity and substrate specificity, ie each enzyme may work in conjunction with one, or only a few, substances, and only under certain conditions.

Epalzeorhynchus BLEEKER, 1855. G of the Cyprinidae*, Sub-F Garrinae. Sumatra, Kalimantan (Borneo), Thailand. Small to medium-sized fish that live near the bottom of flowing waters. The body is an elongated cylindrical shape, the D is short and the LL complete. Head narrow, conical with an inferior mouth beset with 1–2 pairs of barbels. Snout much elongated and fused with the fringed upper lip. The edges of the jaws are sharp enough for *Epalzeorhynchus* to scrape off algal growths. It makes an active specimen in the aquarium and requires a large tank with hiding-places and a soft substrate. Often intolerant towardeach other, whereas they are usually peaceful towards other fish. Temperature 23–28 °C. They will eat all kinds of live and dried foods, and a bit of plant food is advisable. Sex differences and reproduction unknown. So far 5 species have been described, but only 2 are important in aquarium-keeping.
– *E. kallopterus* (BLEEKER, 1851); Flying Fox. Sumatra, Kalimantan (Borneo). Up to 14 cm. Dorsal surface greenish-brown, flanks gold-brown, ventral surface pale, often with a reddish iridescence. A broad, black-brown longitudinal band runs from the tip of the snout to the outside edge of the C. On the dorsal side, the edge of this band is indicated by a gold-yellow stripe. Background colour of the fins is reddish. D, V and An each have a dark band. This species is a good algae-destroyer and eats planarians. Not yet bred in captivity (fig. 487).

Fig. 487 *Epalzeorhynchus kallopterus*

– *E. siamensis* SMITH, 1931. Malaysian Peninsula, Thailand. Up to 14 cm. Greenish-brown in colour, dorsal surface dark. A black longitudinal band stretches from the tip of the snout to the C and is edged with a silver-white stripe at the bottom. A good algae-destroyer. Not yet bred in captivity.

Ephemerella WALSH, 1862 (fig. 488). G of the Mayfly insect order (Ephemeroptera*). Comprises numerous species distributed across wide areas of the Northern Hemisphere (*see* Holarctic Region). *Ephemerella* larvae live in streams and rivers, sometimes in large numbers. There, they creep about somewhat clumsily on rocks, plants etc. Unlike almost all other Mayflies, the *Ephemerella* larvae possess five pairs of tracheal gills along the abdomen which makes them easy to recognise. The various members of the G make a good live food for larger kinds of fish. They also make interesting specimens in a pond aquarium if provided with the right amount of ventilation.

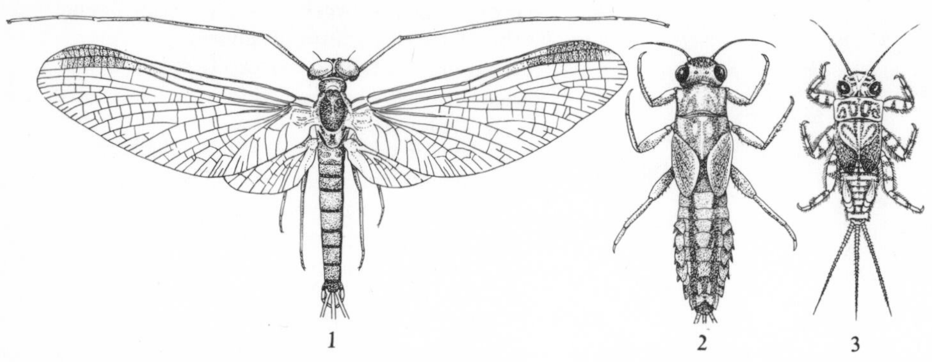

Fig. 488 *Ephemerella.* 1 Imago 2, 3 Various larvae

Ephemeroptera; Mayflies (fig. 489). Very primitive O of the Insecta*, comprising several thousand species distributed worldwide. Most closely related to the Dragonflies and Damselflies (Odonata*). Mayfly larvae have characteristics typical of those larvae that live in very varied freshwater biotopes. They can be recognised by the combination of tracheal gills on the abdomen and 2 or 3 thread-like extensions at the tail. The time the larvae take to develop varies from a few months to 3 years, but the adult flying insects may live for an extremely short period. Among the insects, Mayflies are unique in that the form that hatches from the larvae is capable of flying, yet it still has to undergo a complete moult. For this reason it is known as a subimago (the stage before the actual imago, or insect that can reproduce). The subimago can be recognised by its opaque wings. Neither the subimagines nor the imagines can take in food or liquid – their gut is full of air and their mouth parts are bent. They devote themselves entirely to reproduction and die after a few hours, or a few days at most. The Mayflies are an important part of the natural diet of fish. Most species are also very sensitive to pollution in the water and so are good indicators of biologically perfect water. If kept in a pond aquarium, the larvae afford a glimpse of a very interesting way of life; if conditions are right, they may even reach the imago stage. Common Ga include *Baetis**, *Caenis**, *Cloeon**, *Ecdyonurus**, *Ephemerella** and *Heptagenia**.

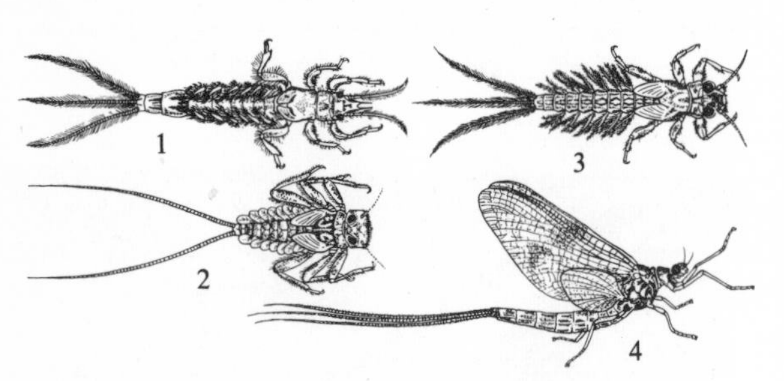

Fig. 489 Ephemeroptera. Various Mayflies. 1 Larva of the G *Ephemera* 2 Larva of the G *Epeorus* 3 Larva of the G *Potamanthus* 4 Imago of the G *Ephemera*

Ephestia. G of the Pyralidae. The various species include a number of important food pests, among them the cereal crop pest, *E. kuehniella*, which comes from India but has spread throughout the world. It is the whitish larvae that attack flour, cereal grains and flour products, dried fruit, legumes etc. When the larvae reach 2 cm long, they pupate inside a cocoon. The time taken to develop depends on the temperature. At 18–20 °C one generation takes about 3 months but in the tropics up to 6 generations can mature in one year. The adult butterflies have narrow, triangular-shaped front wings with a dark marbled patterning and paler hind wings which lie flat against the body like a roof when at rest. Wingspan about 2 cm. *E. kuehniella* is used in experiments to teach the laws of heredity. In aquarium-keeping it is the larvae that crop up in mills and warehouses that find a use as fish food. They can also be bred in cultures using the food media already mentioned.

Ephippidae; Spadefishes (fig. 490). F of the Perciformes*, Sub-O Percoidei*. Found in the tropical Indo-Pacific and in the Atlantic. All have elongated D and An fins, which stand away from the body like sails. They are particularly long in the young fish and are formed mainly from the soft-rayed part of the D and An. The spiny rays at the front of both these fins either form a continuous part of the soft-rayed section (Sub-F Platacinae, the Batfishes) or they remain as separate parts (Sub-Fs Ephippinae, Drepaninae). The best-known members of the F belong to the G *Platax*. Chief among

Fig. 490 Ephippidae-type

them is *Platax orbicularis* (fig. 491), the Common or Round Batfish, which comes from the coastal area of SE Asia. It has become a specialised omnivore near to ports. Grows up to 50 cm long. The young fish have dark transverse bands on a yellow background colour. In captivity, this species is said to be lively and hardy. They fall greedily upon any type of food and literally grow out of the tank. The attractive youthful colouring is gradually replaced with silvery grey tones. The Atlantic Spadefish *(Chaetodipterus faber)* (Ephippinae) lives

Fig. 491 *Platax orbicularis*

in the W Atlantic and grows up to 60 cm long. It is a shoaling species. Young fish are found in the mangrove region; they are black but all the fins, apart from the Vs, are transparent. This makes them deceptively similar to the fruits of the mangrove which float in the water (protective adaptation).

Ephydatia LAMOUROUX, 1816. G of the Sponges (Porifera*). Distributed throughout the Holarctic region*, with 2 representatives in Europe. Found in fresh and brackish waters, particularly lakes and slow-flowing waters. They form encrustments on rocks, dead wood, water-lily stems etc in shallow waters, sometimes forming the most remarkable shapes more than 1 m long. Colour usually yellowish-white to grey-brown, with an occasional green colour caused by the presence of algae.

Ephyra Stage *see* Scyphozoa

Epidermis *see* Skin

Epinephelus BLOCH, 1790 (fig. 492). G of the Serranidae*. Contains a great many species found in all tropical and subtropical seas. Predatory, territorial fish that live near the bottom. They feed on large fish, crusta-

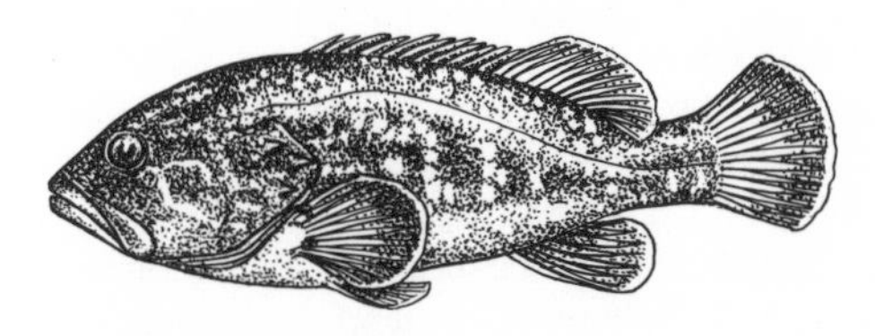

Fig. 492 *Epinephelus guaza*

ceans and molluscs. Length between 25 and 300 cm (*E. lanceolatus*, BLOCH, 1790 – Queensland Grouper or Brindle Bass). *Epinephelus* is of a typical perch shape with small scales. The Vs lie below or behind the base of the pectoral fins. The C has 15 soft rays. Teeth arranged in several rows, those of the inner rows are elongated but not bent. There is no spine at the lower edge of the preoperculum. Many *Epinephelus* species are prized sporting and food fish. Unfortunately, however, their numbers are being decimated by unsporting harpooning. Because of their size, most species cannot be considered for a room aquarium. They also need a great deal of food. However, like many predatory fish, they do survive well in aquaria.

Epiplatys GILL, 1862 (fig. 493). G of the Cyprinodontidae*. Contains more than 40 species found in tropical parts of W Africa. Almost all the species live in small expanses of water in tropical rain forests, with only a few types penetrating east into the savannah region. Fairly closely related to the Asiatic G *Aplocheilus**. Can be distinguished from the other African Ga of egg-laying Tooth-carps by the shape of the snout. Viewed from the side, it looks much more pointed than in the other Ga (eg *Aphyosemion**, *Nothobranchius**) because the front half of the upper jaw is not covered with skin from the head and it juts out. The eyes are very

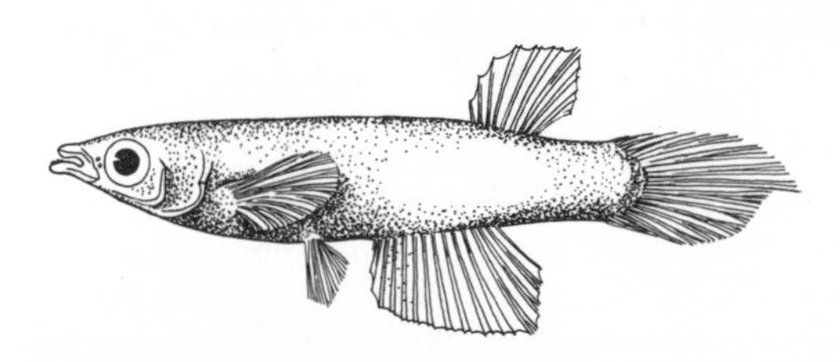

Fig. 493 *Epiplatys* sp.

big. The body is elongated like a pike's, at most 5–7 cm long, rarely more. D not inserted along an area as broad as that of the An, and placed above the rear part of the An. C in the males is well developed, usually amounting to a third the length of the body, with an elongated middle lobe. The colouring of *Epiplatys* is also similar to that of *Aplocheilus*. Particular hallmarks are the appearance of dark transverse bands (at least in the females and in rival males); and a dark longitudinal band running from the head to the insertion of the C is common in many species. The males are more luxuriantly coloured than the females and often have an intense metallic iridescence. The fins are also better developed (the D and An of many species are extended into points for example). The females do not have a round, dark blotch on the D (in contrast to *Aplocheilus* and *Pachypanchax* playfairi*). Colour is also used as a signal, particularly in the throat area. It includes the appearance of dark transverse bands or a black or red colouring. Among the Ga *Epiplatys*, *Aplocheilus* and *Roloffia** this is particularly widespread, and only occurs in some species of *Aphyosemion* apart from these.

Chromosome counts are playing an increasingly important role in the systematics of the egg-laying Tooth-carps. Originally, there were 24 (n = 24) in *Epiplatys*, although many species have reduced this number to 17–18 by fusing together some small chromosomes; a very small number of forms are distinguished by having 25 (n = 25).

The *Epiplatys* species are peaceful surface fish that often linger for hours in one spot, only flapping the elongated lobes of the C. They like to sit beneath aquatic plants whilst they lie in wait for fish broods and any insect food. Although many species are far less sensitive in respect of the composition of the water than many other African Cyprinodontidae, aquarium water should still correspond to that found in their natural habitat – ie soft and slightly acidic. *Epiplatys* is non-annual; seasonal fish occur very rarely. Within the F Cyprinodontidae, reproduction in the G *Epiplatys* (attached spawners) is very primitive. The spawning period can last for weeks, and as the parent fish may well hunt the young fish it is best to remove the spawning substrate containing the eggs into separate rearing tanks.

Even here, the young fish must continually be separated according to size, otherwise the larger ones will eat the smaller ones. If they are to be successfully reared, it is also important to stir the water round a little so that the right sort of food reaches the mouths of the growing youngsters (nauplii, rotifers). This is necessary because the young fishes do not actively search out their food. Many species are suitable for the beginner to look after.

– *E. annulatus* (Boulenger, 1915); Rocket or Clown Killie. Guinea to Nigeria. Up to 4 cm. Has typical colouring: 4 broad, very dark transverse bands on a yellow background colour. The front half of the D is red in some populations*, and the C has red or brown lyre-shaped patterning.

– *E. bifasciatus* (Steindachner, 1881). Widely distributed throughout tropical Africa where it lives in the savannahs. Up to 5 cm. Greenish-yellow with a metallic blue iridescence. 2 dark longitudinal bands run from the head to the insertion of the C. They are sometimes broken up into several pieces. The body and unpaired fins show a marked, red, net-like patterning.

– *E. chaperi* (Sauvage, 188); Red-chinned Panchax, Orange-throat Panchax. Ivory Coast and Ghana, comprising several sub-species and local species. Up to 7 cm. Colouring variable, a metallic blue iridescence on a green-yellow to orange ground colour. Scales and fins covered in red spotting. There are 5 narrow, dark transverse bands on young fish and on the females.

– *E. chevalieri* (Pelegrin, 1904); Chevalier's Epiplatys. Lower Congo Basin. Up to 6 cm. Similar in colour to *E. bifasciatus* except the dark transverse bands are less distinct. The body and fins are covered in red dots and there is no net-like pattern.

– *E. dageti* Poll, 1953. Liberia to Ghana. Up to 5 cm. Body a bronze colour with a greenish or bluish iridescence. There are also 5–6 dark, slightly oblique transverse bands (similar to *E. sexfasciatus*), but usually there are no dark bands across the Vs (fig. 494).

– *E. fasciolatus* (Günther, 1866); Banded Panchax, Striped Panchax. Guinea, Sierra Leone and Liberia. Up to 7 cm. All the features vary a great deal. The most characteristic features are: slightly oblique transverse rows of shining spots in the rear half of the body and a double red band in the lower part of the An and C.

Fig. 494 *Epiplatys dageti*

Fig. 495 *Epiplatys fasciolatus*

Body and unpaired fins are covered in red dots or have a red net-like pattern on them (fig. 495).
– *E. longiventralis* (BOULENGER, 1911); Red-spotted Panchax. Niger Delta. Up to 7 cm. Easy to confuse with *E. sexfasciatus* although the latter has at least 11 fin rays in the D, while *E. longiventralis* has only 7–8. *E. longiventralis* also has a great many dark transverse bands (especially the females).
– *E. macrostigma* (BOULENGER, 1911); Banded Panchax, Spotted Panchax. Mouth of the Congo and nearby tributaries. Up to 6 cm. A blue iridescent colour with longitudinal rows of large red spots on the scales (fig. 496).

Fig. 496 *Epiplatys macrostigma*

– *E. sexfasciatus* GILL, 1862; Six-barred Panchax. Ghana to Gaboon. Up to 10 cm. Very varied in appearance on account of its broad area of distribution. Flanks give off a metallic iridescence (in bronze colours, green or blue). The scales are covered in red dots, which are sometimes edged in black, and together the dots form several length-wise rows. Unpaired fins are blue, yellow, orange or green in colour (depending on where the fish comes from).

Epithelioma *see* Tumour Diseases

Equetus RAFINESQUE, 1815. G of the Sciaenidae*. Contains several species that live in the tropical W Atlantic near reefs and rocks in shallow waters. They feed on small animals. Length between 20 and 30 cm. Body is elongated, almost triangular in shape when seen from the side. The noticeably elongated D1 (particularly long in young fish) emerges at the highest point along the back. These are very attractive fish that should be kept in spacious tanks with plenty of swimming room. Food should consist of small crustaceans, crustacean meat etc. Usually *Equetus* survives well, although they should not be kept together with lively fish that may compete for food. Very intolerant among like species, eg:
– *E. acuminatus* (BLOCH and SCHNEIDER); Cubbyu, High Hat. 30 cm. Whitish, with 6–7 dark brown longitudinal stripes. Survives very well in the aquarium (fig. 497).
– *E. lanceolatus* (LINNAEUS, 1758); Jack-knife Fish, Ribbonfish. White with 3 broad black bands. D1 very tall. A relatively delicate species (fig. 498).

Fig. 497 *Equetus acuminatus*

Fig. 498 *Equetus lanceolatus*

Equisetaceae, F, *see* Pteridophyta

Equisetum (various species) *see* Pteridophyta

Eretmodus BOULENGER, 1898. G of the Cichlidae* comprising a single species that is confined to Lake Tanganyika* (ie is endemic there). Length up to 8 cm. Fairly elongated fish with a strikingly large head and a thick-lipped mouth containing large, spade-shaped teeth. D has a long base and consists of 23–25 hard rays; An has 3 hard rays and 5–7 soft ones; C fan-shaped. *Eretmodus* lives in a similar way to the related Ga *Spathodus** and *Tanganicodus*, ie on a rocky substratum where it grazes on the loose algal growth found there that is rich in organisms. A territorial fish that is extremely argumentative in the aquarium. They are best kept in large tanks with lots of holes in the rocks in which to hide. The water should be hard and neutral or slightly alkali, with a temp. of 22–24 °C.

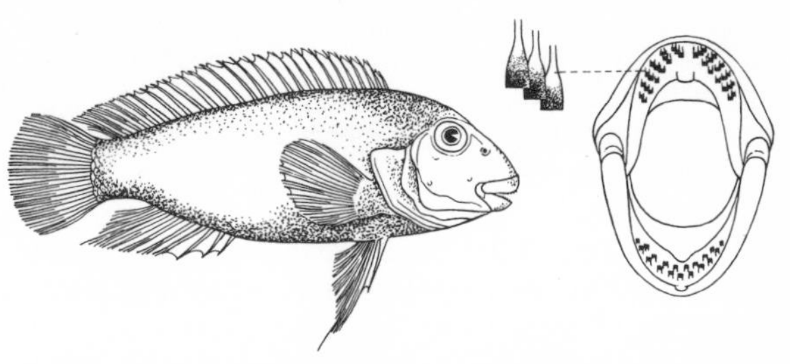

Fig. 499 *Eretmodus cyanostictus* (right, jaw with spatula-teeth)

– *E. cyanostictus* BOULENGER, 1898; Striped Goby Cichlid. Grey-brown to black-brown with brilliant blue spots (figs. 499 and 500).

Fig. 500 *Eretmodus cyanostictus*

Ergasilus, G, *see* Copepod Infestation

Eriphia LATREILLE, 1817. G of predatory Crabs (Brachyura*, F Xanthidae) found in warm seas. Has several rows of coarse teeth at the side of the forehead and the walking legs have bunches of bristles which sit on knob-like structures. The very strong species, *E. spinifrons* (HERBST), lives along the rocky coasts of southern Europe. Its carapace measures up to 10 cm broad and it has enormous pincers. Olive-brown in colour. Can only be kept alone in aquaria.

Eristalis LATREILLE, 1804; Drone-fly (fig. 501). G of the Flies (Brachycera*); belongs to the Hoverfly F (Syrphidae), many kinds of which choose to live in damp places or even in water during their larval and pupal stages. *Eristalis* larvae have a unique telescope-like breathing tube at the end of the abdomen of their bag-like bodies. This tube can be extended. The larvae live in decaying mud in particular and are often caught near ponds. They are interesting to observe if kept in suitable (flat) containers. The adult insects (imagines) are easy to recognise by the striking colour of their abdomens (yellow-red on a black background colour).

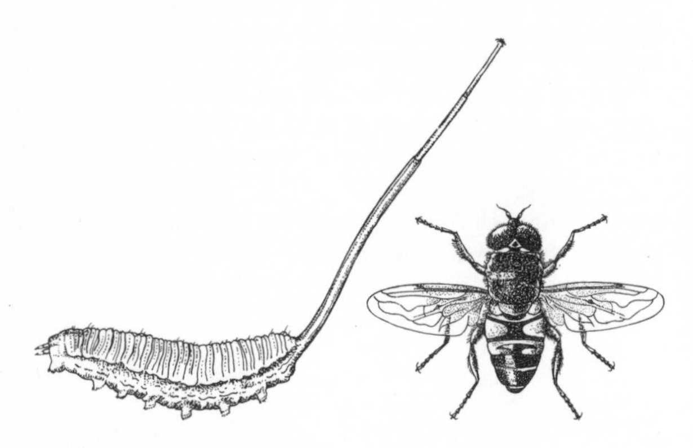

Fig. 501 *Eristalis.* Left, larva; right, imago

Errantia (collective term) *see* Polychaeta

Erythrinidae (fig. 502). F of the Cypriniformes*, Sub-O Characoidei*. Medium-sized (up to 1 m long) predatory fish from S America. The body is particularly elongated and cylindrical in shape. The mouth has a wide gape and contains a great many conical or cutting teeth. The maxillaries are long. The pterygoid is always toothed. The D is relatively short and is inserted near the middle of the body, the C is rounded and the An is short. None of the species has an adipose fin. In *Erythrinus* and *Hoplerythrinus* the front part of the swimbladder has become adapted as an aid to respiration. By such means the fish can live in water that has a poor oxygen supply by coming to the surface now and again to breathe in air. As LÜLING has reported, however, they breathe in air like this at their peril. In so doing, they trigger off turbulence in the water, so causing any lurking Electric Eels to send out electric currents that stun the offender so that it can be devoured later. All Erythrinidae species are voracious predators, sometimes lying in wait for their prey like a pike, and sometimes hunting it actively. The most voracious of all is the Tigerfish (G *Hoplias**) which is found in calm waters (also known as the Anjoemara by the Indians). The species attacks indiscriminately. LÜLING considers *Hoplias* the most dangerous of all Characins. The various members of the F are prized as a food in certain parts of their distribution zone. Large specimens are very occasionally seen in public aquaria. Young specimens are sometimes kept as unusual objects of observation by aquarists.

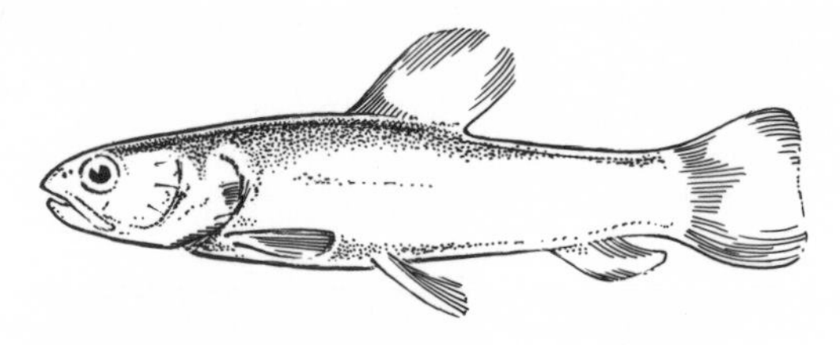

Fig. 502 Erythrinidae-type

Erythrinidoidea, Sup-F, *see* Characoidei

Escolar *(Ruvettus pretiosus) see* Scombroidei

Esocidae, F, *see* Esocoidei

Esocoidei; Pikes, related species. Sub-O of the Salmoniformes*, Sup-O Protacanthopterygii. These fish used to be classified under the O Haplomi and were first associated more closely with the Salmoniformes by BERG. The 10 cm long *Palaeoesox fritzschei* from the middle Eocene Period is one of the few known fossils belonging to this group of fish. The old description 'Haplomi' referred to peculiarities of the shoulder girdle (it lacked the mesocoracoid). All the fins are supported only by soft rays. The swimbladder is linked to the gut by a duct and also contains outgrowths that project forwards into the labyrinth. Both the upper and lower jaws are toothed; in the upper jaw only the premaxillary bone has teeth, not the maxillary bone.

– *Esocidae*; Pikes. Very elongated predatory fish with a spindle-shaped body and pointed head. Seen from above, the snout looks like that of a duck. There are pointed teeth on virtually all sides of the mouth. Unlike the fixed fang-like teeth at the edges of the jaws, the hooked teeth in the roof of the mouth can be moved

found when throttling prey. The D and An are inserted well to the rear and opposite one another. Short gut. The 5 species are from the northern hemisphere. They live in fresh water and are usually motionless. The best-known species, the Common Pike *(Esox lucius)* (fig. 503), can grow up to 1.5 m long (females) or 90 cm long (males). It has a large area of distribution stretching from Europe (not Spain or southern Italy) across vast parts of Asia. It is also not uncommon in N America. Colour varies depending on where the fish comes from and its age. But a brownish to greenish upper side with stripes emanating from it towards the front, green-yellow flanks, a pale belly and reddish-brown fins are all typical. Young fish are usually green all over. A pike lies in wait for its prey among the reeds or other vegetation. Suddenly it will pounce, swallowing its victim whole. Even members of its own species are not safe – some may not be much smaller than the hunter. In Central Europe the spawning period takes place in the months of March-April. Shallow waters with vegetation are sought out, sometimes even flooded land. Their proverbial shyness seems virtually to disappear during this time. Large females can lay up to 300,000 eggs in one spawning period. The newly hatched brood attach themselves to plants by means of an organ on the forehead; they start to take in food 8–10 days after hatching. They grow quickly, and in a cold water aquarium well-fed youngsters will literally grow too big for the tank. Pikes are very good fish to eat, and are also a favourite with anglers because they put up such a good fight. This is also true of the other species, eg the American Muskellunge *(Esox masquinongi)* or the Amur Pike *(Esox reicherti)*.

Fig. 503 *Esox lucius*

– *Umbridae*; Mudminnows. Included in this family are the Dalliidae which were treated as a separate order (Xenomi) by GILL 1885 and JORDAN 1923. Body slightly compressed and flattened laterally. Head rounded with scales both on top and at the sides. The whole of the body is covered with scales. C rounded. Of the 4 kinds of Mudminnow, only one, the Hungarian Mudminnow *(Umbra krameri)*, is found in Europe (fig. 504). It grows up to 10 cm long and lives in waters with dense vegetation and a soft substratum, into which it sometimes burrows. It is found in the Danube Basin, from Vienna downstream. Brownish or a dull red-brown in colour with a great many dark spots; In addition to the gills, this species can also use its swimbladder as a respiratory aid, so it can survive in water that has a poor oxygen content. The eggs are laid in small hollows or in simple caves made out of plants, and the females guard them. The young hatch out after 6–10 days, but they are very cannibalistic. The 3 North American species are found in similar biotopes. They are very common in many places. The common name 'Mudminnows' refers to their habit of burrowing tail-first. The *Dallia* species are very similar in their outward appearance, except that they are somewhat bigger (fig. 505). Both groups are distinguished by unusual skeletal features, eg the ossification of certain parts of the shoulder girdle in the Mudminnows and the absence of this ossification in the *Dallia* species. The latter are found mainly on the Chukotsk Peninsula (Siberia) and in Alaska. They are used mainly as dog food. They have been kept in cold-water aquaria from time to time and are said to need a great variety of live foods as they are purely meat-eaters.

Fig. 504 *Umbra krameri*

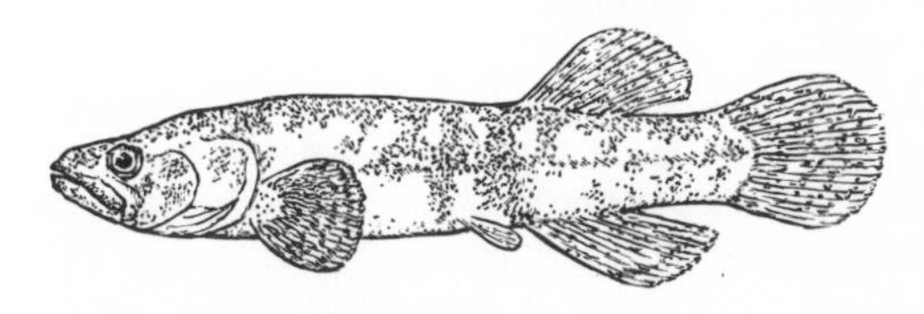

Fig. 505 *Dallia pectoralis*

Esomus SWAINSON, 1839; Flying Barbs. G of the Cyprinidae*, Sub-F Rasborinae. From SE Asia. Surface-dwelling fish that are found in waters containing a lot of vegetation. Small to medium-sized fish with an elongated body and very flattened laterally. The dorsal surface is straight and the dorsal fin is positioned right back at the tail. Mouth is superior with 2 pairs of barbels, the ones on the maxilla stretching far to the rear (a typical feature of the G). The An has 5 separate fin rays and the Ps are elongated. The *Esomus* species are highly active swimmers and like to jump. They should be kept as a shoal in long, large tanks giving a lot of room to swim in. Include a lot of plants in the tank. If kept as an individual specimen, the fish becomes shy. Temperature 18–25 °C. Live and dried food is accepted, although it is best to provide a rich variety of foods if the fish are to breed successfully (particularly insects). For breeding, prepare large, shallow tanks with fine-leaved plants that reach up to the surface. Use slightly peaty, medium-hard fresh water that has been allowed to stand. Put the female in the tank first, then add 1–2 males the following evening. They are very productive fish and will spawn in the early hours of the morning at a temperature of 24–26 °C. They are very prone to eating the spawn. The young fish hatch out after 16–20 hours, and swim freely after 2–3 days. The should be fed on finely ground powdered food. In about 15–20 weeks they should reach sexual maturity.

– *E. danrica* (HAMILTON-BUCHANAN, 1822); Flying Barb. India, Sri Lanka (Ceylon), Burma, Thailand. Up to 14 cm. Dorsal surface grey green, flanks a shiny silver with a reddish iridescence. A black-brown band runs length-wise from the snout to the root of the tail where it ends in a triangular blotch; it is bordered dorsally by a narrow gold stripe. Fins orange-yellow to brownish (fig. 506).

– *E. goddardi* FOWLER, 1937; Goddard's Flying Barb. Thailand. Up to 8 cm. Dorsal surface brownish, flanks ochre-yellow. A dark grey longitudinal stripe runs from the head to the root of the tail. LL incomplete in this species. The maxillary barbels reach to the An.

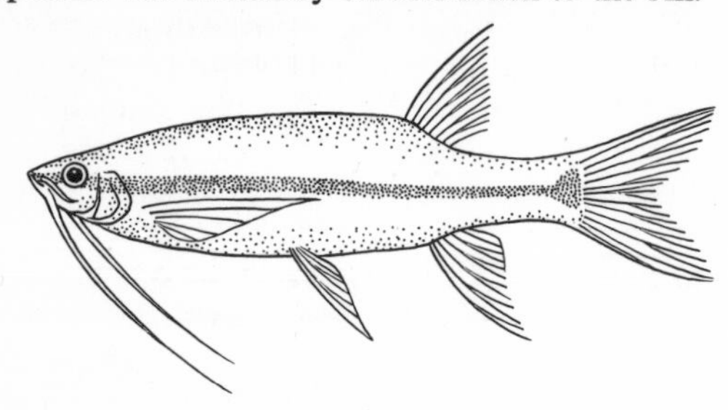

Fig. 506 *Esomus danrica*

– *E. lineatus* AHL, 1925; Striped Flying Barb. Mouth of the Ganges. Up to 6 cm. Dark dorsal surface, flanks grey with a bluish iridescence. A dark length-wise band with a silvery iridescence runs from the head to the C. Can be distinguished from *E. malayensis* by its shorter maxillary barbels and the lack of a blotch on the root of the tail.

– *E. malayensis* (MANDÉE, 1909); Malayan Flying Barb. Malay Peninsula. Up to 10 cm. Dorsal surface greenish with a steel-blue sheen, flanks silver with a blue or blue-green iridescence. An ill-defined dark longitudinal band becomes more prominent on the caudal peduncle where it ends in a black blotch edged in red-gold.

Esox (various species) *see* Esocoidei

Estuary. The mouth of a river that has been widened into a funnel shape by tidal currents. Along the bottom of an estuary, heavy salt water and substrate material push up-river like a wedge when the tide comes in, breaking the fresh water up into different layers. As the tide continues to come in and out again, the riverbank is eroded away more and more and the floor of the estuary (its sand banks and mud banks) changes shape.

The rivers Elbe, Thames and Amazon* have typical estuaries. The effect of the tidal surge can be detected up to 870 km inland from the mouth of the Amazon.

Etheostoma RAFINESQUE, 1819 (*Boleosoma*, name not in current usage). G of the Percidae*. Eastern USA and as far as the Rocky Mountains. Small to medium-sized fish that are distributed widely and live beneath rocks in oxygen-rich waters. The body is elongated and only slightly flattened laterally. Head is conical and ends in a terminal mouth. The gill-covering has scales on it. The Ps are very strong; the D is divided in two, the rear section being supported only by soft rays. C straight. All the species are peaceful and can be kept in small tanks with a coarse gravelly substrate. There should be plenty of hiding-places and ventilation must be good. Temperature 16–18 °C. Live food of all kinds is suitable. After a cool winter, spawning takes place beneath rocks. The male guards the brood. There are numerous species that are divided into various sub-species; they are widespread in eastern N America.

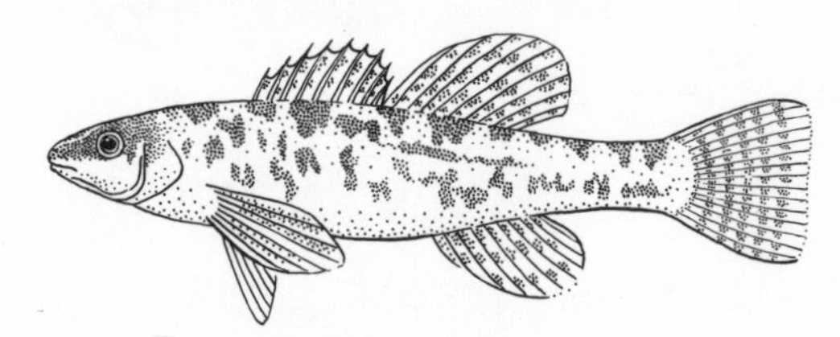

Fig. 507 *Etheostoma nigrum*

– *E. nigrum* (RAFINESQUE, 1820). SE Canada and NE USA. Up to 7 cm. Dorsal surface grey to yellow-green. Flanks clay-yellow and covered in irregular dark flecks. A black stripe runs from the tip of the tail to eyes (fig. 507).

Ethiopian Region. Zoogeographical region with an exclusively tropical or subtropical climate, encompassing the greater part of Africa in addition to Madagascar and the southernmost part of Saudi Arabia. Many animals and plants are only found in the Ethiopian region ie they are endemic to the region. They include many important aquarium fish families such as the Citharinidae*, Gymnarchidae*, Mormyridae*, Pantodontidae* and Polypteridae*. On account of the unusually large number (*c.* 260) of endemic fish Ga in the Ethiopian region they cannot be named here but chief amongst them are those that belong to the following families – Cyprinidae*, Bagridae*, Cyprinodontidae* and Cichlidae*.

Ethogram; Behavioural Inventory. Term used in ethology. It can describe either all the behaviour patterns that are typical in a species or particular spheres of function*, such as metabolically induced behaviour, aggression, display and courtship behaviour or parental care behaviour. When making such an inventory, either purely *fixed action patterns** are considered or supplementary complex patterns of behaviour (*Behavioural Syndromes**) may be taken into account instead. It is the job of ethology* to compare the ethograms of several species in order to be able to clarify any relationships in the natural system that may have been in doubt.

There follow 2 ethograms of the courtship behaviour of various *Nannostomus* species. The number of times each type of behaviour is observed in each species during a 15-minute period is noted down.

Pre-courtship phase

	Head display swimming	Dorsal display swimming	Ventral display swimming	Undulating display swimming	Pause, body vibrating	Appetitive behaviour (male)	Appetitive behaviour (female)	Attempt at mating	Mating	Total
P. eques	10	–	2	17	10	15	14	3	–	71
N. espei	5	6	9	13	12	6	7	2	–	60
N. beckfordi	11	18	3	15	18	10	11	7	–	93
N. marginatus	9	11	9	20	14	11	10	6	–	76

High courtship phase

	Head display swimming	Dorsal display swimming	Ventral display swimming	Undulating display swimming	Pause, body vibrating	Appetitive behaviour (male)	Appetitive behaviour (female)	Attempt at mating	Mating	Total
P. eques	23	–	2	28	14	2	14	6	5	94
N. espei	8	5	19	8	5	3	17	6	3	84
N. beckfordi	17	22	5	25	23	3	12	16	23	146
N. marginatus	13	11	12	22	16	8	12	10	18	106

Ethology. Study of behaviour. A young science within zoology that studies the way in which animal behaviour patterns are built up (has been extended to cover human behaviour in recent years). The zoologists VON FRISCH, LORENZ and TINBERGEN are considered the fathers of modern ethology. Very early on, ethology was recognised as a very important science. The recognition that instinctive movements (*see* Fixed Action Patterns), like organ systems, are phylogenetically reasonably stable, meant that they came to be regarded as equally important characteristics of species and higher systematical cateogries. A particular aim of ethology in general is the study and interpretation of animal behaviour. This involves an analysis of closely related disciplines like physiology and functional morphology. Specialised branches of ethology draw up monographs on behaviour in which individual types of behaviour are divided up according to their spheres of function (*see* Ethogram). In related species, the ethographical comparison of complete behavioural monographs or of certain functional systems can lead to valuable systematical knowledge. The special interest of the ethologist has always been behaviour during reproduction, as it is in this area that the richest variation of behaviour pattern is seen. In 1961 TEMBROCK suggested the following divisions when compiling a behaviour inventory: 1 General types of movement; 2 Comfort movements; 3 Metabolically induced behaviour; 3a Acquisition of food and food intake; 3b Defaecation and urination; 3c Food storage; 3d Rest and sleep; 3e Lolling movement; 4 Behaviour relating to protection and defence; 5 Territorial behaviour; 6 Reproductive behaviour; 7 Social behaviour; 8 Building activities; 9 Vocalisation; 10 Play behaviour; 11 Behavioural development (ontogenesis of behaviour). Fish, too, make interesting objects of study in ethology, eg their different ways of living together; communication within a species; methods of territorial demarcation, defence and escape behaviour; mating behaviour and the relationship between parents and offspring.

Ethusa, G, *see* Brachyura

Ethylurethane *see* Anaesthetics

Etiolation. Because of a lack of light, the stem axis of a plant grows unnaturally long and the leaves grow smaller and paler. A good example might be potatoes germinating in a cellar.

Etroplus CUVIER and VALENCIENNES, 1830. G of the Cichlidae*. Comprises 3 species which represent the only Cichlids that come from Asia. Found in India and Sri Lanka (Ceylon) where they inhabit brackish water. One species, *E. maculatus*, also enters fresh water. They are tall, oval-shaped fish with rounded, unpaired fins. Flattened into a disc shape laterally. The D has 17–20 hard rays and 8–15 soft ones; the An has 12–15 hard rays and 8–11 soft ones. In the wild, the largest species of the G reaches 40 cm in length.

The G comprises very primitive members of the Cichlidae F. The sexes are difficult to tell apart. Spawn is laid on rocks, leaves or on wood (more rarely in caves). Both parents care for the brood. Before they can swim freely, the young fish sometimes attach themselves to their parents' sides. *E. maculatus* is the most popular aquarium species; it requires slightly alkali, hard water. It is a good idea to add a little sea salt. *E. suratensis* is the second most important species as far as the aquarist is concerned. It is a typical brackish water fish, and can also be kept in pure sea water which seems to bring out more splendid colouring in the fish. In contrast, this species can tolerate pure fresh water only fleetingly. Temperature 23–28 °C. All *Etroplus* species require roomy species aquaria fitted with large rocks. The sometimes appear aggressive towards like species. Two species are kept in the aquarium.

– *E. maculatus* (BLOCH, 1795); Orange Chromide. India and Sri Lanka (Ceylon) where it lives in brackish and fresh water. Up to 8 cm. During the spawning period the flanks are yellow or orange with red dots. 3 large, round, dark blotches are also clearly visible, the middle one particularly. Dorsal surface grey-blue to brown-black. Unpaired fins yellow to orange, partially covered in red dots and with dark edges. Out of the spawning period, the colours are duller (fig. 508).

Fig. 508 *Etroplus maculatus*

– *E. suratensis* (BLOCH, 1790); Green Chromide, Silver Chromide. Sri Lanka. Lives in brackish water. Up to 40 cm, but stays considerably smaller in captivity. Blue-grey or green-grey in colour with a mother-of-pearl iridescence. 6–8 dark transverse bands on the sides. The scales have blue or green iridescent flecks and some have a black edge. DAY has reported that this species has beautiful purple-red and deep black bands during its spawning period in fresh water.

Eucidaris POMEL, 1883. G of the Sea-urchins (Echinoidea*). Found in tropical seas where it inhabits shallower water than *Cidaris**. *Eucidaris* has spines that are scarcely longer than its body and they are usually blunt-ended. The diameter of its body is about 4 cm. *E. tribuloides* (LAMARCK), which comes from the tropical W Atlantic, is frequently introduced into the aquarium. The species is a very bizarre one, but specimens will survive many years if fed on a mixed animal and plant food diet.

Eudistylia, G, *see* Polychaeta

Euglena, G, *see* Euglenophyta

Euglenophyta. Division of the algae* containing 400 species. Green algae containing chlorophyll a and b; chromatophores are shaped like lentils, rods, bands or stars. A characteristic of the Euglenophyta is a striking red 'eye-spot'. There are 2 flagella at the front end, one very long, the other very short. The Euglenophyta are single-celled organisms that can move about. They occur primarily in standing, nutrient-rich fresh water, for example in village ponds and dung-water pools. Large numbers of Euglenophyta will colour the water green. The best-known G is *Euglena*. There are also chlorophyll-free forms within the Euglenophyta.

Eukaryota; nucleated organisms. In contrast to the Prokaryota*, the Eukaryota are organisms that possess defined cell nuclei. Their cells are further divided by membranes into other reactive centres where the metabolic processes take place. The Protobionta are the lowest form of organisms to belong to the Eukaryota. They consist of a large number of independent systematical groups which display a mixture of plant and animal ways of life. The plant groups that are capable of photosynthesis* are usually described as algae*; those groups that live on organic matter that happens to be present are classified either as fungi* or as single-celled animals (Protozoa*), depending on their form and way of life. In the strictest sense, the higher eukaryotic organisms – the plants (Phytobionta) and animals (Zoobionta) – are not all that different from certain Protobionta.

Euleptorhamphus viridis *see* Exocoetoidei

Eunice, G, *see* Polychaeta

Eunicella, G, *see* Gorgonaria

Eupagurus BRANDT, 1951. G of the Hermit Crabs (Paguridea*) that are found in cool and warm seas from the upper littoral to greater depths. They live on detritus. Their body is 2–4 cm long and their right pincer is elongated. Many form a biotic community, eg the shell of *E. bernhardus* is often covered by a spiny, dark brown mass formed by a hydroid polyp colony (*Hydractinia echinata* FLEMING); the subtenant in the same shell is often the annelid *Nereis fucata* SAVIGNY. *E. prideauxi* always lives with the anemone *Adamsia**. In the aquarium, *Eupagurus* is agile and usually hardy. *E. anachoretus* from the Mediterranean has already been bred successfully.

– *E. bernhardus* (LINNAEUS, 1758); Bernhard's Hermit Crab. European Atlantic coast to the western Baltic Sea. 35 mm. Whitish-red brown in colour.

– *E. prideauxi* LEACH; Anemone Hermit Crab. Mediterranean. 25 mm. Reddish-violet. Unfortunately very sensitive.

Euphorbiaceae. F of the Magnoliatae* comprising 290 Ga and 7,500 species. Tropical to subtropical, but chiefly in America and Africa. Comparatively few species grow in temperate regions. They can be succulents, lianas, perennials and annuals with an alternate leaf arrangement and latex. The flowers are unisexual, monoecious or dioecious, and polysymmetrical. The range of the Euphorbiaceae extends from extremely dry places to swamps. For example, the F is represented by cacti-like succulents in dry parts of Africa and Asia. However, only the G *Phyllanthus** contains a typical aquatic plant.

Eupomacentrus (BLEEKER, 1877). G of the Pomacentridae*. Reef-dwelling fish that are aggressive and territorial. There are several species of *Eupomacentrus* and they live in tropical parts of the Atlantic where they represent the Indo-Pacific G *Pomacentrus**. Length 6–16 cm. In profile they look oval in shape with large scales. Unlike *Abudefduf**, the praeoperculum is serrated and in contrast to *Microspathodon** the teeth do not move and there is no furrow above the upper jaw. When young, *Eupomacentrus* exhibits a variety of bright colours, but they become duller with age. In all other respects, the information on *Pomacentrus* applies equally to *Eupomacentrus*. One species that is often imported is:

– *E. leucostictus* (MÜLLER and TROSCHEL); Beau Gregory. Tropical W Atlantic. Up to 15 cm. Relatively elongated. Upper surface blue, ventral surface yellow, the colours becoming paler with age. A hardy species.

Europe. Second smallest continent (about 10 million km^2), but together with Asia* it forms the largest connected land mass in the world. The boundary between Europe and Asia is artificial; biogeographically, Europe and Asia (with the exception of the tropical areas) are a single unit – the palaearctic realm*. The European freshwater animal kingdom contains few species in comparison to other temperate regions like N America* and Asia. This is as a result of the last Ice Ages as well as geographical peculiarities. Europe's mountains run roughly west-east, and prevented many animals from reaching warmer areas to the south. As a result, many species of animals died out.

European fish play a very modest role in aquarium-keeping. Sticklebacks hold a certain amount of interest (*Gasterosteus**, *Pungitius**). The Moderlieschen (*Leucaspius**) and the Bitterling (*Rhodeus**) are sought after because their way of life and reproductive methods are interesting. A few Egg-laying Tooth-carps (Cyprinodontidae*) are resident in Europe too.

European Anchovy *(Engraulis encrasicholus) see* Clupeiformes

European Catfish *(Silurus glanis) see Silurus*

European Frogbit *(Hydrocharis morsusranae) see Hydrocharis*

European Mudminnow *(Umbra krameri) see* Esocoidei

Eury-. A prefix that means 'broad' when it is used in a word that describes the relationship between a species and some ecological factor* (although it can be used in other contexts too). For example, species are described as 'euryhaline' if they remain largely unaffected by fluctuations in salinity (ie almost all anadromous and catadromous fishes and *Artemia**). If a species can tolerate great temperature changes it is said to be eurythermic (eg *Elassoma**). Opposite to sten-*.

Euryale ferox *see* Nymphaeaceae

Eutaeniophoridae, F, *see* Cetomimiformes

Euthyneura, Sub-Cl of the Snails (Gastropoda*) comprising almost 50,000 species. Included in this Sub-Cl are the Opistobranchiata which usually have no shell and live in the sea, and the Pulmonata (snails and slugs) which usually do have a shell. (Both are Sup-Os.) Among the Pulmonata, the Basommatophora* are interesting to the aquarist. These particular snails have retained their pulmonary respiration although they have returned from the land to live in the water. So, unlike snails that have always led aquatic existences, they must come to the surface occasionally to breathe in air.

Euthynnus pelamis *see* Scombroidei

Eutrophic Waters. The high nutrient content of such waters enables algae to multiply, sometimes in great numbers. The effects can be far-reaching. Sapropel accumulates at the bottom of eutrophic waters, causing the oxygen to be used up. It often contains a high amount of poisonous sulphides and hydrogen sulphides (H_2S) as well. As a result, a great stretch of water can become a dangerous place for the higher organisms that live there. The Black Sea* and Lake Tanganyika* are just 2 well-known places where this can happen. In inland waters*, a high pH value* (= 8) is often associated with a high nutrient content. The increase in nutrients (eutrophation) is accelerated by the introduction of waste water, especially phosphates. In the aquarium, too, similar processes go on, the only difference being that besides phosphate, nitrogen compounds also play an important role. Nutrients build up in the aquarium particularly as a result of too much feeding and too many specimens. In extreme cases, the eutrophation and the resultant oxygen destruction can take such a hold that large amounts of poisonous sulphurous and nitrogenous compounds (H_2S, nitrite) build-up, causing the tank to 'overturn' suddenly, thus killing all the occupants. To prevent eutrophation of aquarium water, make sure the water is changed regularly (*see* Water, changing of). Opposite to oligotrophic waters*.

Eutropiellus NICHOLS and LA MONTE, 1933. G of the Schilbeidae*. Catfishes of the Congo Basin, particularly around the Kasai district. Small, peaceable shoaling fish that live in the open water. They swim around tirelessly, their tail fin somewhat hanging and the caudal peduncle producing a fanning action. Up to 10 cm. Body elongated, only moderately flattened laterally. The head is short and ends in a terminal mouth surrounded by 3 pairs of barbels. This is the feature which distinguishes *Eutropiellus* from the *Eutropius* species, all of which have 4 pairs of barbels. *Eutropius* is also larger and less suitable for the aquarium. *Eutropiellus* has large eyes. The D1 is very short and inserted far forwards (almost opposite the beginning of the Ps), D2 is well developed. The C is twin-lobed and deeply forked and the An is narrow and very long. Vs very small. Diet consists of all sorts of small live food, as well as dried food. Solitary aquarium specimens seem unhappy. Not yet bred in captivity.

– *E. debauwi* (BOULENGER, 1901); Three-striped Glass Catfish. Congo Basin near Avakubi, Niapu, Poko and Lake Malebo (Stanley Pool). Up to 8 cm. Flanks tran-

Fig. 509 *Eutropiellus debauwi*

sparent, throat and ventral surface white. Three narrow, black length-wise bands become more prominent as the fish get older. The females can be recognised when they are swollen with eggs and because they have paler band markings. Very pretty shoaling fish that should be kept alongside small peaceful fish (fig. 509).

Euxiphipops Fraser-Brunner, 1934. G of the Chaetodontidae*, Sub-F Pomacanthinae. They inhabit tropical parts of the Indo-Pacific, including the Red Sea, living along the coral coasts. Their diet consists of small animals and growths of algae on the coral. All the species are territorial. Typical features are large scales, a complete LL and a rounded C. All the species are very colourful. *E. navarchus* (Cuvier and Valenciennes, 1831) (Majestic Angelfish) (fig. 510) can reach a length of 50 cm. Feeding can be a problem in the aquarium – start them off on *Mysis* and *Enchytraeus*, with a bit of

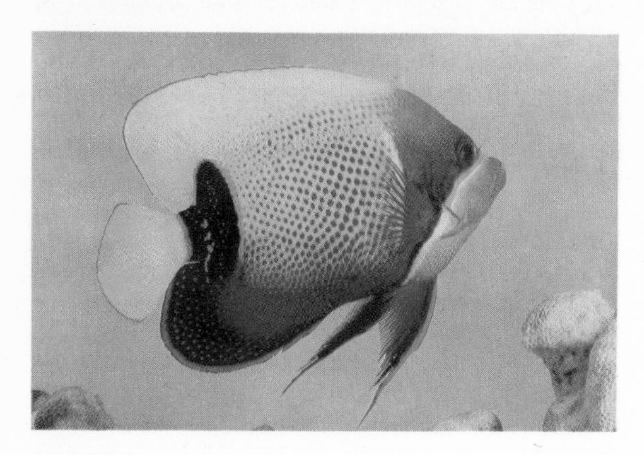

Fig. 510 *Euxiphipops navarchus*

plant food as well. They are extremely intolerant of like species and similar fish. The following is imported occasionally.

– *E. xanthometopon* Bleeker, 1858; Yellow-faced Angelfish. Malaysian Archipelago. Up to 40 cm. The body is covered in black scales edged in white. The head has a grid of blue lines and a yellow eye-mask. There is a black blotch at the end of the D. The C is yellow and all the fins have blue edges (fig. 511).

Fig. 511 *Euxiphipops xanthometopon*

Evening Primrose Family *see* Onagraceae

Evolution. The constant development of structural forms within living organisms. Evolution is the basis of phylogenesis* throughout the Earth's history and in the present time. It is through evolution that higher and higher organisms have developed from the simplest life forms right up to Man. It is also because of evolution that there is such a great variety of different species in the world. Charles Darwin*, the founder of the theory of evolution (Descent Theory*), was the first to acknowledge the existence of evolution and its basic principles.

The basis of evolution are the constantly occurring changes in hereditary material that happen quite by chance. Such *mutations** repeated time and again can gradually bring about changes in an entire group of organisms. If such changes prove not to be advantageous, the species will die out. In those cases where the mutations offer better chances of survival, however, that species will continue to live (ie the conditions in the environment will determine whether or not a mutation is an advantage, it is the environment that makes the selection). This natural *selection** is the second important factor in evolution. However, mutations and selection by themselves are not enough to cause an increase in the number of species. A new species only comes about when changes in the hereditary material in successive generations are not distributed throughout all the members of that species. For such a thing to happen, members of a species would have to be isolated from one another. For example, it may happen when the area of distribution of a species has become very large, when new habitats have been colonised or when populations have become separated as a result of geological factors.

Evolution, Theory of, *see* Descent Theory

Examination Technique. Before any kind of therapy is attempted on sick fish it is important to provide an exact diagnosis* of the type of illness. To be able to do this some knowledge about anatomy* and histology*, as well as fish physiology* and behaviour* is needed. You will also need to be skilled in the art of examining fish.

The following instruments and pieces of equipment are needed for this: a preparation dish, the bottom of which is covered with paraffin about 2 cm deep; a spatula; drop pipettes, one pointed and one truncated; pins; various preparation needles; a strong pair of sharp scissors and a small pair of tweezers; specimen mounts and glass covers; a good magnifying glass and a microscope. All the results of the examination are listed along with the information about the history of the illness (Anamnesis*) in what is known as an examination protocol. The systematic examination of sick fish is carried out in the following sequence.

1) Make a list of the ways in which the fish's behaviour differs from normal (abnormal ways of swimming, fins held in an unusual manner, rubbing of skin against objects, flight reflex action curtailed, quickened or interrupted breathing pattern, refusal to accept food etc. Now note the ways in which its appearance is altered (colour, state of skin, emaciation, deformed parts of the body or any other deviations in body structure).

2) Evaluation of the external state of the body. To carry out a more thorough examination, one specimen from a group of sick fish which has particularly striking symptoms should be caught very carefully – make sure no damage is done to the skin (use a dip-tube). If done with care, smears can be taken from the skin and fins without anaesthetic. Examinations under the magnifying glass and examinations of the gills can usually only be carried out on anaesthetised specimens (Anaesthetics*). Anaesthetised fish can be thoroughly examined in the following way for: cloudiness, inflammation and coatings on the skin, fins (Skin Diseases*) and eyes; white to dark dots or bloody ones; abscesses; scale protrusion; peeling or destroyed areas of fins; swollen or sunken-in belly area; inflamed and protruding anal region; deformations; parasites on the skin and gills (Hirudinea*, Argulidae*, Trematodes* and others). To make a skin smear, some skin mucus is scraped away using the spatula, preferably from around the insertion of the P, from the C, or from the area around the LL and the operculum (fig. 512). This is then placed onto a specimen mount along with a drop of aquarium water then covered with a piece of glass before being examined under the microscope. Parts of fins can also be cut off and examined microscopically.

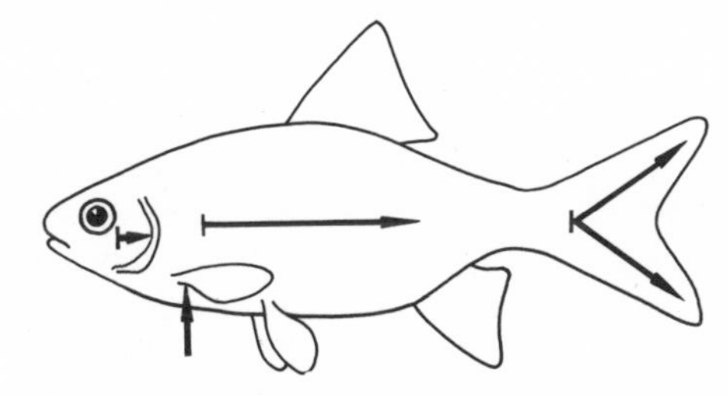

Fig. 512 Position and direction of skin smears: operculum, pectoral fin, lateral line, caudal fin

Some Indian ink (1:10) added to the preparation will colour the background dark. Parasites will remain uncoloured and so will contrast well with the background. To test for gill parasites, it is possible to spray in some water from a truncated pipette into the gill cavity of an anaesthetised fish, then suck it out again immediately. The contents of the pipette can then be examined for parasites either with a magnifying glass, or better still under a microscope.

3) To make an even more thorough examination, the fish has to be killed (Fish, Killing of*) then placed on its side in the preparation dish and held there firmly by a needle in the head and caudal peduncle. Firstly, using a sharp pair of scissors, a slit is made along the belly from the anal opening to the throat, then another slit is made from the anus along the upper edge of the body cavity to the operculum (this will make an arch shape along the side of the fish). With the tweezers, lift off the side wall of the body from the anus as far as the rear edge of the opercular cavity. Now lift up the operculum and snip it off at the front edge. Using a scalpel, make a very shallow insertion and remove the forehead skin area to reveal the brain (fig. 513). Now fill the preparation dish with water such that it covers the fish.

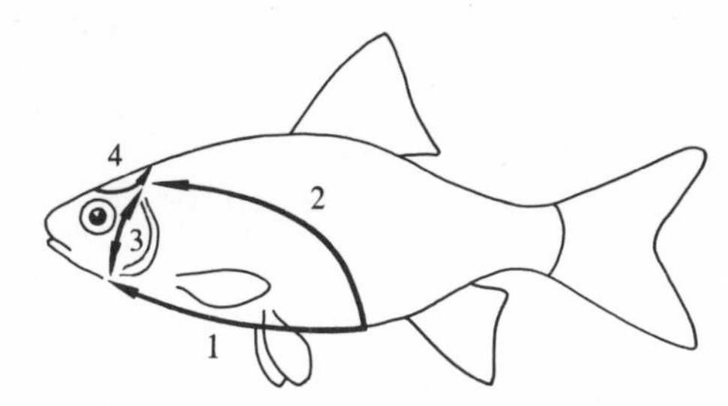

Fig. 513 Cuts to be made when making an internal examination. 1 Ventral cut 2 Lateral cut 3 Opercular cut 4 Cranial cut

4) All the internal organs should now be examined under a magnifying glass for any changes to them. Beginners are advised to prepare a healthy fish for comparison. Green-yellow discolouration of the normally red to brown liver and swellings in the gall bladder point to liver (Liver, Diseases of*) and metabolic damage (Deficiency Diseases*); other abnormalities to look for are enlargements of the spleen, pale-coloured kidneys, a transparent quality and inflamed reddishness of the intestine (Intestinal Diseases*), marked accumulation of fat between the organs (Degeneration*, Deficiency Diseases*), the appearance of nodules in the spleen, kidneys, liver, gills and other organs, a pale gill epithelium with swellings and stickiness (Gills, Diseases of*).

5) Microscopic examination of the internal organs: Small pieces of tissue are taken from the internal organs, especially the spleen, kidneys, liver and intestine, and from the musculature of the torso, then they are broken up into pieces with some cooking salt solution (0.64%) and placed on the specimen mount, or they are carefully squashed between the mount and a glass cover laid on top. In both cases, the result is that the tissue is scattered very finely, and this enables a microscopic examination to take place. For a microscopic examination of the gills, a gill arch is prepared and covered over by the piece of glass, but not much pressure is applied. If brain parasites are suspected, some pieces of brain tissue should be examined with the squashed technique. Blood parasites can be detected from blood smears (Blood, examination of*). The contents of the intestine and the gall bladder should also be tested for parasites. The substances obtained from the inside should be smeared onto a specimen mount (smear preparation) and then looked at under a microscope. Larger kinds of parasites and cysts are usually easily visible under a microscope. Changes to tissue caused by Bacteria, viruses or other factors can only be detected and analysed by using special methods. But it still cannot be emphasised enough that to arrive at an unequivocal, sound diagnosis of a fish disease is only possible after lengthy considerations, and often it takes many years of experience.

Exclamation Mark Rasbora *(Rasbora urophthalma) see Rasbora*

Excrements. Useless matter expelled from the body in animals and humans, in the form of dung. In vertebrates, excrement contains indigestible or hard-to-digest matter as well as micro-organisms and secretions of the digestive organs* such as bile and mucus. The breakdown products of the blood pigments which is contained in the bile gives the excrement its colour. Most of the metabolic end products are expelled from the blood through excretion* via the kidneys, along with the urine.

Excretion. Matter that must be expelled from the body. It comprises metabolic end products (mainly urea in fish), water, salts and poisons. Excretion is associated with osmoregulation*—in both cases, matter is largely expelled via the urinary organs*. Excess carbon dioxide is given off during respiration*. Fish have mesonephros (Wolffian bodies) for excretory organs.

Exocoetidae, F, *see* Exocoetoidei

Exocoetoidei; Flying Fish, related species. Sub-O Atheriniformes*, Sup-O Atherinomorpha. REGAN classified them as the O Synentognathi, BERG as the O Beloniformes. In the main, they are slender, marine fish that live near the surface. Only a few species live in fresh water. The oldest known fossils date from the Neogene period. In all the species, the fins are supported only by soft rays. The D and An are placed well to the rear, just in front of the indented C. The Vs are in the middle of the body and in the Exocoetidae F the Ps are often greatly enlarged. The head can be normal or pointed; the jaws are toothed and often much elongated. The upper edge of the mouth is bounded by the premaxillary and maxillary bones. The LL is sited well down towards the ventral surface. The scales are small and round, and quite loose. There is no connection between the swimbladder and the gut. A specialised, but typical feature is the way in which the lower pharyngeal bone and the third one from the top are fused. Some species are livebearing, and some are well-known flying fishes.

– *Belonidae*; Garfishes, Needlefishes or Long-toms. There are about 60 species in this F; all are reminiscent of the Garpikes (Lepisosteidae*) to look at, (although they are not related to them). The body is very elongated, the head is long and has extended jaws containing a great many teeth. The D and An are directly opposite one another and are inserted just in front of the C. Like the Halfbeaks (*see* below), young Belonidae only have an elongated lower jaw (the upper jaw develops later). Garfishes have an enormous number of vertebrae (77). They are shoaling fish that are constantly on the move. The European Garfish *(Belone belone)* (fig. 514) is found in northern parts of the E Atlantic including all

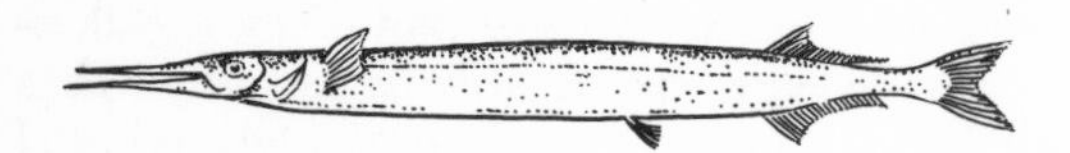

Fig. 514 *Belone belone*

neighbouring seas, and it has spread to the Mediterranean and the Black Sea. It grows up to 1 m long. Its diet consists mainly of fishes and crustaceans. In the North Sea the spawning period is from May to June. The eggs are laid among shallow clumps of seaweed. The Garfish is good to eat, but because the bones turn green when it is being prepared as a food, people shy away from it. Some Belonidae enter rivers. *Potamorrhaphis guianensis* lives in the fresh waters of the Amazon. Young *Xenentodon cancila* are occasionally kept in an aquarium. This species is widespread in the fresh waters of S and SE Asia. All Garfishes can jump exceptionally well.

– *Exocoetidae*; Flying Fishes (fig. 515). The Hemirhamphidae (Halfbeaks) are also included in this F, even although they are often considered to be a separate F. The Flying Fishes have short, virtually equal-length jaws, while the Halfbeaks have a very long, pointed lower jaw which cannot move and a short upper jaw which moves like a beak. In the middle is *Oxyporhymphus micropterus*, a species which comes from the W and E Atlantic, but which is also found in the Pacific.

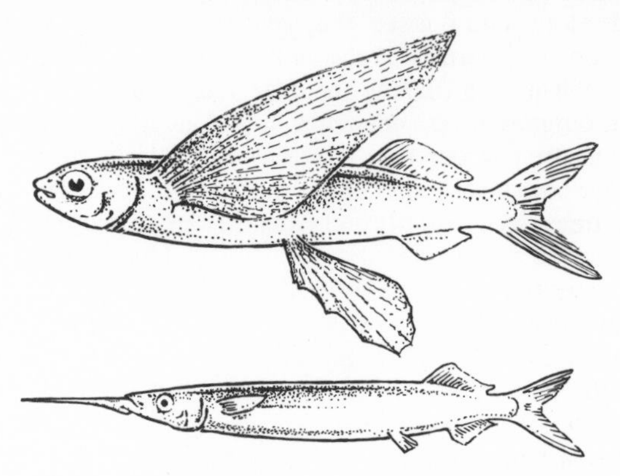

Fig. 515 Exocoetinae-type and Hemirhamphinae-type

The juveniles of the species have an elongated lower jaw which is later reduced to a normal-sized jaw. Similar changes to the jaws are known in some typical Flying Fishes, so it is clear that both groups are closely related to one another, ie their classification as Sub-Fs within one F is correct. Approximately 70 species belong to the Sub-F Hemirhamphinae*; most are marine fish. One of the largest species is *Euleptorhamphus viridis* (Oceanic Halfbeak) which comes from the Pacific. Its length may reach 45 cm. *Dermogenys pusillus* (fig. 516) from SE Asia and the Greater Sunda Islands is known to aquarists. It grows up to 7 cm long and is a typical

Fig. 516 *Dermogenys pusillus*

freshwater form. It is a surface-dwelling species, silvery in colour, often with a bluish iridescence. Typical features are its black stripe along the lower jaw (there is often a red one as well), yellowish fins and green iridescent iris. This species is livebearing. Large females give birth to 12–20 youngsters. In captivity, the females may miscarry if not fed sufficiently well with insects and if the temperature is too high (over 22 °C). The males are highly aggressive so it is best to put one male in with several females. They like to have a light covering of floating vegetation.

The Sub-F Exocoetinae contains exclusively surface-dwelling fishes from the open waters of tropical and subtropical seas. The Ps in all the species are extended into a kind of wing and they all have an extended lower lobe of the tail fin. In some species the Vs are also very large, so they are described as 'four-winged'. The Exocoetinae are gliders, shooting out of the water at a high initial speed then spreading their Ps out to glide above the water for several seconds. They can pick up speed again by dipping the elongated lobe of the tail fin into the water and waggling it very fast, so that they can glide along a bit further. In total they can glide like this for up to 12 seconds at about 55 km/h, covering 100–150 m. The juveniles of many species have very long outgrowths from the lower jaw, a bit like barbels, and in the majority of cases the colouring is very variable. Some were even described as separate species at one time. The largest two-winged species comes from California – the 45-cm-long *Cypselurus californicus*. *Exocoetus volitans* is found all over the world. It too is two-winged and grows up to 25 cm long. The four-winged *Cypselurus heterurus* lives in the mid-Atlantic. All the species live in shoals, and some have economic importance in certain areas.

– *Scomberesocidae*; Sauries (fig. 517). This F comprises 4 species, all of which look like the Garfishes described earlier. The differences are that the jaws of the Scomberesocidae are somewhat shorter and the flat,

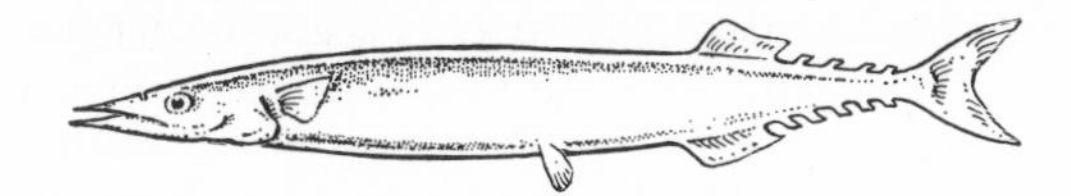

Fig. 517 Scomberesocidae-type

rear part of the D and An is broken up into a number of small finlets (a feature that occurs again in similar form among the Mackerels (*see* Scombroidei). The Scomberesocidae are not very big, but they live in enormous shoals and so are important to the economy in some places. In Japan, the Pacific species, *Cololabis saira* (Pacific Saury), is a well-liked food fish. It can grow up to 30 cm long. Another species with a tasty flesh is *Scomberesox saurus* (Skipper or Atlantic Saury). It grows up to 40 cm long and is found in the Atlantic, the Mediterranean and the Black Sea.

Exocoetus volitans *see* Ecocoetoidei

Exodon MÜLLER and TROSCHEL, 1844. G of the Characidae*. Contains only one species which is resident in the north-eastern part of S America. Up to 15 cm. Elegant and very beautifully coloured fish. Unfortunately, they are intolerant towards other fish, and because they also practise rank order* low-ranking members of their own species are also subjected to harrassment. They eat all kinds of live food, but also like to eat scales. They swim very fast up to their victim and bite the scales off with their strong, sharp set of teeth. *Exodon* has an elongated body, compressed laterally. The head is short and ends in a terminal mouth with a wide gape. The eyes are large. For information on the fins turn to the description of the species. As these fish like to swim, keep them in large containers with plenty of room for swimming. The water should be slightly acidic and soft. Temperature 24–26 °C. There should be at least 6–8 specimens in a shoal (preferably 10–15). After violent driving swimming actions, the spawn is laid on fine-leaved plants. The young hatch out after 25–30 hours, but they are difficult to rear. So far, more recent reports of successful breeding are not yet available. It is very much to be hoped that this species will be imported again soon.

– *E. paradoxus* MÜLLER and TROSCHEL, 1844. Rio Branco and Rio Rupununi. Up to 15 cm. Flanks a delicate yellow with a silvery iridescence which may look a shining pale violet in certain light. There is an intense pale yellow longitudinal band in the middle of the body. 2 large deep-black blotches in front of the insertion of the D and on the caudal peduncle. All the fins are yellow at the base, while the middle of the D, the tips of the C lobes, the front third of the An and the Vs are all brick-red or blood-red.

Exophthalmus *see* Eyes, Diseases of the

Expressive Behaviour. In fish such behaviour can include emphatic swimming movements or the uttering of noises during courtship, the way in which body shape or colour takes on a more pronounced effect during a fight or courtship, threatening or submissive gestures. The Cichlids in particular have a rich repertoire of expressive movements, such as particular ways of swimming, the spreading apart of fins or gill coverings, the way in which the fins are laid down. An intensification of certain patterns of colour, or a diminishment, both on the body and the fins, will indicate to a partner or antagonist whether the fish intends to attack, flee, become submissive or whether it is ready to court and mate.

Expressive Movement *see* Expressive Behaviour

Expulsion Reaction. A behaviour often observed in shoaling fish whereby a sick, injured or very small member of the shoal is driven out or even killed. Other reasons for such expulsion can be when a fish exhibits patterns of behaviour unusual for its species or when a non-species member penetrates the shoal.

External Filter *see* Filtration

Extremities, limbs. Movable appendages of the body, chiefly used in locomotion, although they may also serve a function during the intake of food, orientation, etc. Among the invertebrates, the arthropods (Arthropoda*) have highly specialised extremities. Besides jointed legs, the insects have wings and antennae, and oral extremities such as the mandibles and maxillae. The arms of the cephalopods (Cephalopoda*) and the ambulacrum of the echinoderms (Echinodermata*) can also be regarded as extremities according to our definition.

The 4 extremities of land-living vertebrates are developments of the pectoral and pelvic fins. They take the form of arms, legs or wings. In vertebrates that have once more returned to the water, these extremities have taken on the shape of fins again, eg the wings of penguins. Many reptiles, in particular the snakes, have much reduced extremities.

Eyes *see* Sight, Organs of

Eyes, Diseases of the. In fish, diseases of the eyes are usually an accompaniment to other diseases, rarely a disease of the eye only. Bulbous eyes (Pop-Eye) known as Exophthalmus are often brought about by fish tuberculosis*, *Ichthyosporidium**, stomach dropsy* or a bacterial eye infection. The latter is often characterised by white flecks and can lead to a clouding over of the whole eye. A whitish-grey discoloration of the eyes is a result of 'spot' diseases, *Cryptocaryon* (*see* Cryptocaryon Disease), Fish Tuberculosis and *Oodinium**. In aquarium fish, the eyes may be attacked in rare cases, by the larvae of Trematoda* and Cestoda*. If the optic nerves, the eye or the part of the brain controlling vision is destroyed by pathogenic agents, the fish will become blind.

Eye-spot Algae, D, *see* Euglenophyta

Eye-spot Cichlid *(Cichla ocellaris) see Cichla*

Eye-spot Climbing Perch *(Ctenopoma ocellatum) see Ctenopoma*

Eye-spot Hermit Crab *(Paguristes oculatus) see Paguristes*

Eye-spot Pike Cichlid *(Crenicichla dorsocellata) see Crenicichla*

Eye-spot Rasbora *(Rasbora dorsiocellata) see Rasbora*

Fairy Basslet *(Gramma loreto) see Gramma*

Fairy Shrimps, O, *see* Phyllopoda

False Catfishes, F, *see* Auchenipteridae

False Cleanerfish *(Aspidontus taeniatus) see Aspidontus*

False Rummy-nose Tetra *(Petitella georgiae) see Petitella*

False Skunk-striped Anemonefish *(Amphiprion perideraion) see Amphiprion*

False Tetra *(Hyphessobrycon heterorhabdus) see Hyphessobrycon*

False Thorny Catfishes, F, *see* Auchenipteridae

Family *see* Systematical Categories

Fan Mussels, G, *see Pinna*

Fanwort, G, *see Cabomba*

Fanwort Family *see* Cabombaceae

Farlowella EIGENMANN and EIGENMANN, 1890. G of the Loricariidae*. Armoured Catfishes from S America. Up to 20 cm. They are peaceful bottom-dwellers, and in the aquarium they will need lots of stones to which they can cling with their sucker-mouth. Their body is very long but only very slightly flattened laterally. The ventral surface of the body is covered in armour (ventral scutes). The sucking mouth is ventral and is used to graze algae from the rocks. It is completely covered by the snout which is like an elongated beak. The D and An are positioned directly opposite each other which is in contrast to the closely related G *Loricaria**. An abnormally long caudal peduncle emphasises the overall spear-like shape of the fish's body. This impression is further confirmed by the way in which the top-most ray of the unlobed C is drawn out into a threadlike extension. The bottom-most ray of the C is also extended in this G. The Vs are short, the Ps are well developed. There is no D2. So far, 2 species have been reared successfully. The eggs were laid on the front glass of the aquarium and for 10 days the male fanned and cleaned them. Unfortunately, this G is only rarely imported and in small numbers. Several species have been described, but they are nevertheless difficult to differentiate.

– *F. acus* (KNER, 1852). Central and southern Brazil, the states of the River Plate. Up to 20 cm. Body olive-green to yellow-brown, ventral surface yellowish. There are 33 bony scutes in a length-wise row (fig. 518).

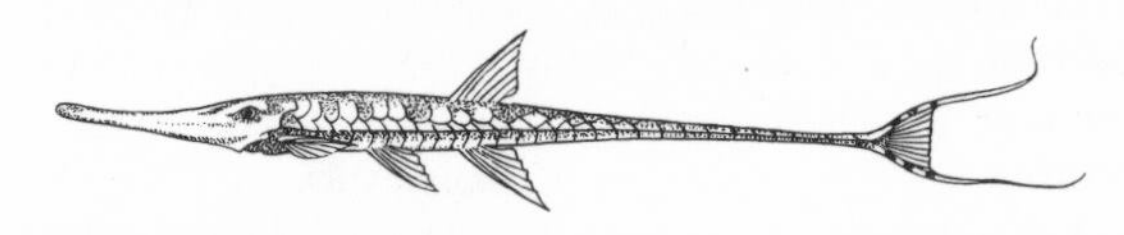

Fig. 518 *Farlowella acus*

Father Lasher *(Cottus scorpius) see Cottus*

Fatty Degeneration *see* Degeneration

Fatty Substances; lipids. Chemically disparate substances that occur naturally; they will not dissolve in water, but they are soluble in organic solvents such as benzine or ether. There are three types of fatty substances: *simple* (fats, fatty oils and waxes); *compound* (phosphatides, cerebrosides) and *fat-like compounds* (sterols, steroids, lipochrome). Fundamentally, fatty substances are a chemical combination of fatty acids and alcohols. The great many naturally-occurring fats are the result of the particular type of fatty acids and alcohols they contain, together with the way in which they combine with substances that contain phosphorus and nitrogen.

Fats are important structural elements of cellular membranes; they are also very important as reserve substances for all organisms. Plants store oils in particular in the seeds. Animals keep depot fat in special fat cells (in subcutaneous fatty tissue in vertebrates and others). One of the best-known animal fats is cod-liver-oil. As a food, fats are rich in energy.

Fauna. The animal world of a particular geographical region.

Faunal Element. Species of animal that originated from a particular zoogeographical region and bore certain characteristics as a result. Eg the Pike (*Esox lucius see* Esocoidei) is a faunal element of the holarctic region*, the Valencia Toothcarp *(Valencia* hispanica)* a faunal element of the Mediterranean countries, and the Butterfly-fish *(Pantodon* buchholzi)* a faunal element of tropical parts of Africa, the Ethiopian region*.

Faunal Falsification. The way in which species of animal spread into areas where they had not previously been endemic (autochthon*) through the activities of man. Artificial means of being able to overcome barriers that limit the distribution areas of species (eg through the building of canals) may also lead to faunal falsification. Although it is impossible today to prevent faunal falsification entirely, it should be confined to as small areas as possible. Often, species that were originally allochthonic* in a region cause obvious, and often irreparable, harm to the endemic flora and fauna.

Featherbacks (Notopteridae, F) *see* Mormyriformes

Featherfin Bullfish *(Heniochus acuminatus) see Heniochus*

Featherfoil *(Hottonia palustris) see Hottonia*

Feeding. Correct feeding is very important when keeping animals. It promotes optimum growth, good health and normal reproductive capacity. Correct feeding is also just as important whether your specimens are made up of species that have been known to hobbyists for a great number of years or whether they are specimens you have caught yourself. As far as possible, aquarium specimens should be fed on their natural diet. This means not only that the food should be of the correct type, quantity and quality, but it should also contain the right proportions of nutrients and vitamins.

Particular eating habits should be taken into account, eg how often, and at what times and places your animals are wont to eat. Another factor to bear in mind is competition for food. Only keep together those species that will not interfere with one another's food requirements. Above all, feed your specimens a varied diet, in small amounts and as far as possible, several times a day. The decaying remains of food can foul the water in an aquarium, to such a degree that the oxygen becomes used up.

There are many different types of food available. *Live food* is to be recommended most highly. You can catch it yourself (*see* Food Trapping), although it is also available through shops. Breeding food animals is also a good idea (eg. whiteworms, earthworms, brine shrimps, fruit flies). When adding food to a marine aquarium, it is as well to remember that most food animals do not live long.

Deep-frozen food is of similar value to live food, especially to see you through the winter months. You can keep Water-fleas, fly larvae, shrimps and brine shrimps, mussels, fish, meat and plant food such as algae or spinach in a deep-freeze. It is important to pack them in airtight containers to prevent them drying out. A great variety of food materials for aquarium fish is available commercially. There is *dried food* in the form of oats or grains, *food tablets* and dried animal food (water-fleas, pieces of shrimp, Amphipods). Artificial foodstuffs are of better quality and often contain a large proportion of the very best freeze-dried plant and animal parts mixed together in different amounts, eg *Tubifex* with added fish-roe, pieces of shrimp, brine shrimps, algae, yeast, egg-powder, the flesh of mussels and frogs, vegetable juices and vitamin preparations.

You will find information about the special needs and food requirements of particular fish under the descriptions of the Ga. The following list represents a survey of the types of food species used in both freshwater and marine aquaria. For information regarding the plants and animals themselves (their life-styles, how to care for them, their value as a food), turn to the relevant entry.

*Worms: Tubifex**; Whiteworms (*Enchytraeus**); Earthworms (Lumbricidae*); Marine Polychaetes (*Nereis**); Nematodes (*Turbatrix**, *Panagrillus**); Rotifers (Rotatoria*).

Crustaceous Animals: Water-fleas *(Cladocera**); *Cyclops* and their nauplii (*see* Copepoda); Brine Shrimps (*Artemia**) in various stages of development; Shrimps (Natantia*); Amphipods (*Gammarus**); Water-lice (*Asellus**); small Crabs (Brachyura*) and Crab meat.

Insects: red, white and black Midge larvae (Nematocera*); Mealworm (*Tenebrio**); Mayfly larvae (Ephemeroptera*) and other Fly maggots*; Wax Moth* and Flour Moth (*Ephestia**) larvae; Fruit Flies (*Drosophila**); Springtails (Collembola*).

Molluscs: Mussels (Bivalvia*), in particular the flesh of mussels found in rivers, ponds and along the seashore (*Mytilus**)—the meat can be given raw or it can be cooked briefly; various Snails (Gastropoda*); Cephalopod meat (Cephalopoda*).

Fishes: live young fish; Live-bearing Tooth Carps (Poeciliidae*), eg Guppies, the flesh of freshwater and marine fishes that does not contain a lot of fat, fish liver; unsalted or well-rinsed fish roe.

Amphibians: Frog meat (minus the skin); Tadpoles.

Mammals: lean meat; the heart or liver of cattle; horse meat and horses' hearts, all of which should be cut up into little pieces and fed to your specimens raw or cooked.

Plants: soft-leaved aquatic plants (*Elodea**, *Hygrophila**, *Riccia**); Green Algae (Chlorophyta*) also in dried and powdered form; marine Brown Algae (Phaeophyta*) and Red Algae (Rhodophyta*); Yellow-green Algae (Bacillariophyta*); lettuce leaves; soaked spinach and soaked rolled oats; yeast*.

For information about the feeding of a fish brood *see* Fry, Feeding of.

Feeding Equipment. You will need only a few, very simple pieces of equipment to feed your fish or invertebrate aquarium specimens. It is important to remember that metal objects should not come into contact with sea water, if you keep marine aquaria. So that you will always have a good supply of live food animals to hand, it is a good idea to make a collection of stock containers. Flat vessels (dishes, including photographic dishes) are suitable for Water-fleas and *Cyclops*. They will survive for quite some time if the water level is kept low, the temperature kept cool, and not too many specimens are kept together. Containers for fly larvae should be put in the open air, as you must always bear in mind that the larvae will hatch out eventually. Use feeding nets to remove the food from the stock containers. The width of the mesh should be appropriate for the size of the food animals you are dealing with (*see* Food Trapping). Plastic spoons are best for placing the food in the aquarium. If you need to feed specimens individually, use a feeding tube. This consists of a glass tube (diameter about 5 mm, length depending on requirements), whose broken edge is smoothed down in a gas flame. A piece of flexible tubing is then fixed into it. Small food animals (water-fleas, brine shrimps etc) are sucked up along with the water into the glass tube. Then the tube contents are blown out, which enables you to feed marine invertebrates individually. A food pipette can be used in a similar way, but instead of a flexible tube it has a rubber cap. The glass tube is pointed at the free end. Larger foods (earthworms, pieces of meat) can be handled with food tweezers made of wood or plastic. The fish become accustomed to the instrument quickly and can even be fed individually with it. In a marine tank, invertebrates such as hermit crabs, anemones, Echinoderms etc can also be fed by means of the tweezers. *Tubifex* and whiteworms are best introduced to the freshwater aquarium via a food funnel. It is usually made of glass and floats on the surface like an upturned bell. Its walls are punctuated with a great number of holes. The worms gradually crawl through the holes and are either pulled out by the fish or they are eaten as they sink to the bottom. One disadvantage of this system is that the strongest and quickest fish are always satisfied first and the worm food begins to make up too large a part of their diet. Dried food is best used in conjunction with a feeding ring, which prevents the food spreading out over the water surface. It consists of a floating glass tube shaped like a frame, and is best attached to a corner of the aquarium to stop it drifting about (*see also* Fry, Feeding of). Feeding machines have not yet been adopted by aquarists. They simply cannot fit in with the need for a varied diet and individual feeding.

Feeding Machines *see* Feeding Equipment

Feeding Ring *see* Feeding Equipment

Ferns (Pteropsida, Cl) *see* Pteridophyta

Fertilisation. The fusion of the cell nuclei of two germ cells (*see* Gametes) of different sexes, during sexual reproduction* in plants and animals. In numerous single-celled organisms 2 externally identical individuals combine. The ciliates (*see* Ciliata) exchange a part of the nucleus material when they fleetingly lie next to one another.

In multi-celled animals fertilisation takes place when a moving sperm penetrates a stationary egg cell. This happens either inside the body (mammals, birds, reptiles, live-bearing fish and amphibians, numerous invertebrates) or outside the body, primarily in water (fish, amphibians, many egg-laying invertebrates). The finding of the egg cell and the penetration of the sperm is guided by special gamones*. After the two gametic nuclei have fused, the fertilised egg cell begins to divide and the development of the embryo* begins.

Fertilisation leads to the mixing of the hereditary material of the parents, particularly that of the nucleus of the cells, ie the transference of information regarding the formation of different characteristics. The carriers of the hereditary material, the chromosomes, are reduced to a half (haploid) set as a result of a special type of cell division* (meiosis). Fertilisation unites the 2 haploid sets to a diploid chromosome set which is usually present in the body cells.

Fertilising. The placing of certain inorganic and organic matter in the ground or water to improve the nutrient supply of plants. There are numerous preparations on the market, but usually manuring is unnecessary in a normal aquarium. If aquatic plants are being intensively cultivated in a nursery, the addition of manure in correct quantities may be beneficial to growth.

Festive Cichlid *(Cichlasoma festivum) see Cichlasoma*

Fiddler Crabs, G, *see Uca*

Fierasfer acus *see* Gadiformes, also Holothuroidea

Fierasferidae, F, name not in current usage, *see* Gadiformes

Fighting Behaviour *see* Aggressive Fighting Behaviour, Symbolic Fighting Behaviour

Fighting Drive, Disposition towards. Term used in ethology. It describes the build-up of a particular amount of action-specific energy in a fish, enabling it to start up fighting behaviour or if the fish is being attacked to counteract with like fighting behaviour. For this to happen, a stimulus threshold* (which blocks the sub-threshold fighting drive) must be broken through.

The drive to begin fighting is controlled by abiotic factors (exposure to light, water temperature, air pressure etc) and biotic factors (these are the so-called *releasers** or their *key stimuli**). Its degree of intensity can often be deduced from a fish's colouring or by the appearance of species-specific colour patterns. When the fighting drive is on the wane (ie when one of the rivals in a battle is defeated), you will also see characteristic signs in the fish: the colouring fades, the fins are laid down, the body is held at an unusual angle (one sign of inferiority* behaviour), or the fish may swim away or seek a place of refuge.

Fighting Loach *(Noemacheilus kuiperi) see Noemacheilus*

Figure Eight Puffer *(Tetraodon fluviatilis) see Tetraodon*

Figwort Family *see* Scrophulariaceae

Filamentous Algae. All algae that are usually made up of branched or unbranched threads which are formed of row upon row of individual cells. They do not represent a related group, but belong to very different divisions within the algae*.

Filefishes, F, *see* Balistidae

Filefishes, G, *see Monacanthus*

File-shells, G, *see Lima*

Filling Mastic (cement) *see* Aquarium Sealants

Filter Charcoal *see* Activated Charcoal

Filtration. Aquarium and natural water differ fundamentally in their hydrochemical and biological ratios. The number of animals kept in a tank is usually much too high in relation to the amount of water. As a result, organic matter builds up to such a degree that it is no longer possible for it to be broken down naturally in sufficient quantities. It becomes difficult for the water to purify itself. This tends to result in a low redox potential* which brings with it all those consequences that are so injurious for the organisms that live there. Under such unfavourable circumstances, in order that plants and animals may continue to live, we must turn to technical aids that can maintain an artificial balance. To this end, one of the most important techniques is filtration.

With a filter, constantly occurring organic material capable of decomposition is removed from the system as rapidly as possible. This is because organic compounds, especially albuminous ones, may produce intermediate products when they are broken down that have highly toxic effects even in the smallest concentrations. Particularly dangerous are concentrations of bits of protein, amines, ammonia*, nitrite, phenols and hydrogen sulphide. When visible particles of dirt (usually in the form of mould) and their concentrates are sucked off in a filter, the end-goal is not fully achieved. Only after the dirt is removed from the filter is the amount of potentially harmful matter reduced, but even then the problem is only partially resolved. Aquarium water that looks completely clear to the eye can be biologically extremely hostile to life because of high concentrations of soluble intermediate products from proteometabolism. A filter must also remove, or render harmless, soluble substances that are subject to breakdown processes and all the intermediate products that may result. Filtration systems work on 2 main principles. They can form stable bonds between the toxic matter and the filter substrate. The bonded substances are removed from the system by cleaning, regenerating or renewing the filter material. A second method establishes a milieu in the filter that promotes the rapid oxidative break-down of the accruing substances. It is important to ensure that the breakdown is neither reductive nor incomplete and that poisonous intermediate products are not contained in high concentrations in the water that is run off from the filter. If reductive processes do take place in the filter, the water in it is left virtually without oxygen and with reductive properties. If such water is allowed to run into the tank for any length of time, the redox potential* in the aquarium water will gradually sink too, until the whole of the water in the tank will come to have reductive properties. Aquaria that contain such water cannot be enriched with a sufficient amount of oxygen even with considerable aeration, since the oxygen is immediately bonded. There are numerous kinds of technical filtration systems on the market. All filter systems suck the aquarium water up a tube by means of suitable pumps*; from there it is sent across some kind of filter substrate and finally fed back into the tank as 'purified' water. Suction tubes should always take in water from near the bottom of the tank as it has a low oxygen content. They should be constructed like sieves so that they will not suck in coarse particles, animal food stuffs or small aquarium specimens at the same time. The aquarium water is always sent through a sieve or large-meshed gauze before it reaches the actual filter. This means that large pieces of dirt are removed before passing through the filter, thus preventing the system from becoming clogged up quickly. This pre-filter must be easily accessible and should be cleaned every 1–2 days.

There are basically 2 types of filter: external and internal. The internal filter is suspended in the aquarium, whilst the external filter is only linked to the aquarium via a communicating system of tubes (fig. 519). The properties of the actual filter substrate

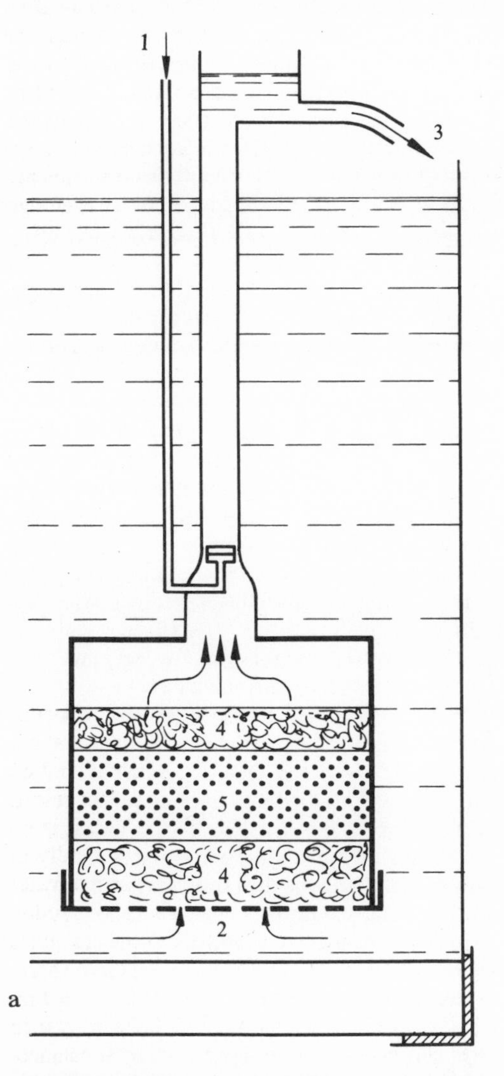

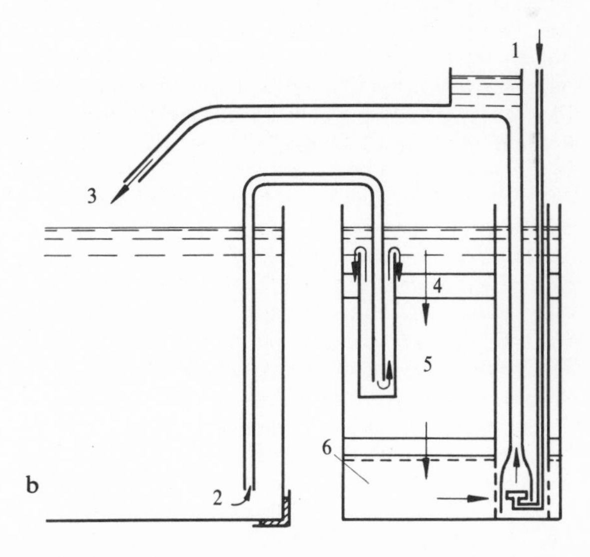

Fig. 519 Filters. a) Internal filter b) External filter 1 Air 2 Water inlet 3 Water outlet 4 Filter wadding 5 Filter substrate 6 Grate

must be known exactly. On no account can it be water-soluble or be liable to the release of any trace of toxic substances. Where the water comes in and goes out, the filter mass is sealed with a layer of filter wadding. Internal filters must be so arranged that they do not suck in water from the substrate. If this were to happen, dust particles would be introduced into another part of the substrate along with the return flow of water, thereby turning the area into a reducing system with a low oxygen content. Mechanical filters, with the aid of a fine-pored filter substrate, remove coarse-grained particles of dirt and colloidal matter from the water. Both internal and external mechanical filters are available. Suitable filter substrates include filter wadding, medium-fine quartz gravel, neutral open-pored foams etc. After a while, because of the accumulation of proteins and their properties, and independently of the substrate that has been used, such filters begin to behave as true adsorbers (*see* Adsorption). As such, they are able to bind actively mechanisms, colloids and various other substances by denaturisation, ion transfer etc. These adsorbing properties can be achieved immediately if a small amount of clay, loam, silicate or similar active adsorber is added to the filter substrate. Activated charcoal* also brings about immediate colloidal bonds; its suitability should be tested before use. With mechanical or adsorptive filters dirt from the aquarium is concentrated in the filter system. The bacterial breakdown of such matter in the filter is of secondary importance only and can be disregarded. The amount of the accumulating impurities and the size of the filter determine the point at which it will no longer be of any use. After a certain length of time, such a filter system releases poisonous fission products and ions* which will eventually find their way back into the aquarium. Spent filters contain oxidisable substances in high concentrations and remove the oxygen from the water that passes through them. They then operate as a reducing element. For most efficient operation, clean the filter regularly. The water should flow through as quickly as possible (Circulatory Pumps*), so that the water remains in the filter for as short a time as possible, thus keeping the removal of oxygen from the water to a minimum. In order to maintain the balance of gases in the water, it can be passed through a radiating tube or it can once again be brought into contact with air via aerators or suction nozzles.

Biologically active filters can likewise be employed as external or internal filters, but external filters have more advantages. They contain a substance with a large surface area and many pores (gravel, coarse sand, foams with large and open pores). The aquarium water is passed over this substance at slow speed. In order to work properly, it is essential to fit a large pre-filter that operates purely mechanically. At first such systems work purely mechanically, then they enrich organic and inorganic colloids with typical adsorber properties, and finally in the third stage they accumulate bacteria, algae and protozoa which take over the job of breaking down the ballast material in the water. Illumination and plant matter can increase efficiency still further. The filters work best only after a certain 'breaking-in' period and they must exactly suit the size of the aquarium and its animal content. Whenever the filters are cleaned, they can be inoculated again with part of the old contents of the filter so that they return to their optimal efficiency quicker. The filter mass is put in loosely, as there must always be sufficient oxygen available. Oxygen can be blown in from below through the filter substrate in a jet of air. Should there be a lack of oxygen, incomplete anaerobic* decomposition will set in, as a result of which highly toxic intermediate products will be formed. The filters are extremely sensitive. A single case of overfeeding may result in the filter becoming inundated with by-products thereof, and thus becoming poisoned. They are most suitable for those tanks that have only a moderate amount of fish in them. The system should also be modified so that the water level in the filter lies beneath the filter substrate and is only sprayed with aquarium water. Such an arrangement guarantees good aeration and a sufficient supply of oxygen. Air blown in additionally from below increases the aeration. Rapid filters that work mechanically or adsorptively as well as biological filters must not be switched off and after a while must be started up once again. If the water flow were to be interrupted, an anaerobic system would start up immediately, with its putrefactive processes and highly toxic end-products. When the filter was turned on again, these harmful products would be washed into the aquarium.

Chemical filters are able to mineralise organic substances quickly. The oxidation processes in this case, however, are not comparable with bacterial breakdown processes. Instead, the processes lead to a direct oxidation and miss out the intermediate products. The ozonisation (*see* Ozone) of aquarium water is a chemical method of purifying the water, and, if used correctly, can have very good results particularly in a sea-water aquarium. With this method, ammonia* is produced as an end-product of the proteins. This must be broken down further by nitrifying bacteria till it becomes a nitrate. For this reason, newly set up aquaria with insufficient bacterial growth may not be treated with ozone. In a similar way, treatment with the correct dosage of ultraviolet rays* can speed up the breakdown of organic substances. For those people who keep marine aquaria there is another very useful method of purifying the water – foam filters*. For the removal of organic substances there are also adsorbers. Activated charcoal, which has a large adsorbtive spectrum, must be tested thoroughly for its suitability before use. Adsorbent resins* which would adsorb definite amounts of particular substances are not yet available in a suitable form for the aquarist. Ion exchangers* are able to exchange any given ion in the water with other ions; they are of particular use in water preparation*. However, if the continually accruing metabolic products were to be removed constantly, there would be an appreciable shift in the natural ionic composition of the water, so it is not to be recommended as a filter substrate. Peat* is a very suitable filter material for freshwater aquaria with soft, slightly acidic water. Peat filters can be employed to supplement biological filters, but they must always be protected from large dust particles by a mechanical pre-filter. No filter system relieves the aquarist of his duty to change part of the water at certain intervals. This will ensure that any deficit of certain deficient material (eg trace elements) is compensated for, and on the other hand excessive concentrations of organic acids, nitrates and other metabolic products will be avoided.

It is particularly important to supply enough oxygen to the substrate, in order to prevent anaerobic breakdown processes starting up in a localised area. In many cases, an insufficiently aerated substrate is the cause of poor water conditions in the aquarium. Cancelling out any locally occurring reductive systems can only be achieved with an enormous input of filters. In some cases it may be impossible to counteract. It is a very good idea to include the substrate in the water circulation*. To this end, place a movable grate 2–3 cm above the very bottom of the aquarium. The grate should have holes or slits in it. Place a layer of perlon gauze directly on top of the grate to stop any fine sand falling through. The substrate should now be layered on top of this. The space beneath the grate receives an inflow of water from a filter via a standing pipe which reaches up to the surface. By this means, the filtered water (which also has a gaseous equilibrium) flows through the substrate from below. A movable overflow pipe is attached to the ascending pipe so that the rate of flow through the substrate can be regulated. The flow of water need only be very slight to avoid washing away the nutrients from the substrate that are required by the plants (fig. 520). A slight trickle is enough to ensure that the substrate receives sufficient fresh water and oxygen, so that only aerobic* breakdown processes take place. Plant growth is definitely improved, and multitudes of organisms with anaerobic breakdown processes and the formation of Sapropel are successfully prevented. A reversal of the water flow is not suitable. Dust particles would be drawn into the substrate, they would block up the capillary spaces which would soon lead to an insufficient oxygen supply with a gradual lowering of the redox potential throughout the aquarium. Gradual deterioration of the water in the aquarium can have various causes. Natural indicators which herald the slightest changes very early on are largely absent in the aquarium. Fish respond only when there are relatively high concentrations of poisonous metabolic products, but then their reaction is very sudden – they overturn, as it were. In order to spare yourself the trouble of taking various analyses of the water quality, and observing it constantly, it is possible to detect a deterioration in quality very early on by determining the redox potential. As the redox potential* is influenced by a great many factors, all disorders can be recognised at the right time, independently of their cause.

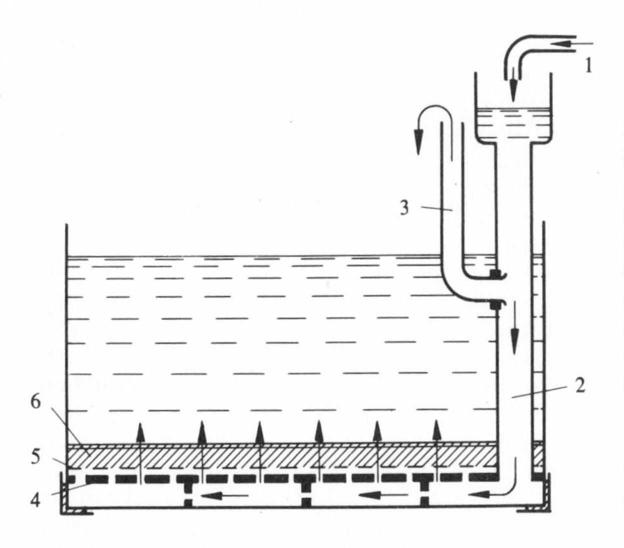

Fig. 520 Reverse flow substrate filter. 1 Water feed supply 2 Standpipe 3 Movable overflow pipe 4 Bottom grating 5 Perlon yarn 6 Substrate

Fin Formula. The description of the fins possessed by a species of fish according to the type and number of the fin rays. The fin formula, like the scale formula, is part of the identification key* and as such is an important piece of data in fish systematics. The following style of description has been adopted internationally to a large extent. The various fins* are represented by their abbreviated letters. These are followed by the number of hard rays and spines that are positioned at the front of the fin and are never articulated (written in roman numerals). An oblique stroke or comma comes next, followed by the number of soft rays (usually articulated) written in arabic numerals. For example, the fin formula for the Perch *(Perca fluviatilis)* is written thus: D1 XIII-XVII; D2 I-II/13-15; An II/8-10; P 14; V I/5; C 17. This means that the first dorsal fin has 13–17 hard rays and no soft rays; the second dorsal fin 1–2 hard rays and 13–15 soft rays; the anal fin 2 hard rays and 8–10 soft rays; each pectoral fin has 14 soft rays; each ventral or pelvic fin one hard ray and 5 soft rays; and the caudal fin has 17 soft rays. The caudal fin data is often left out because the number of its rays is often identical among different fish groups and species. The fin formula may also reveal whether the soft rays are articulated or spiny.

Fin Rays *see* Fins

Final Host; definitive host *see* Parasitism

Fingerfishes, F, *see* Monodactylidae

Fin-rot. Occurs in freshwater and marine fish. Bacteria of G *Aeromonas* and *Pseudomonas* cause inflammation* of the fins, followed by their progressive decomposition. The first sign of the disease is when the edges of the fins appear slightly turbid. Later, it may spread to the skin, causing large chunks of the epidermis to flake off. Fungi (Mycosis*) often appear as secondary parasites (*see* Parasitism). Fin-rot is a debilitating disease, which is caused by unsuitable environmental conditions – high bacterial density in the water, too low a temperature etc. Another common cause is when the fish are infected with fish tuberculosis* or with *Ichthyosporidium**. There is no easy cure; prolonged baths containing Chloromycetin*, Sulfapyrimidine, Hostacyclin, Trypaflavin* and preparations based on Nitrofurane* may be successful. At the same time, the water temperature should be raised. In every case, the prime cause of the debilitation must be identified and appropriate measures taken.

Fins (fig. 521). Flat-surfaced, movable appendages to the body, found in fish. They mainly help to propel the fish along, to steer it and to stabilise it. Some fins are paired, some are unpaired. The structure, shape and arrangement of the fins are important criteria in the systematics of fish (*see* Fin Formula).

Fins are made up of fin membrane, fin musculature, and supporting *fin rays*. In cartilaginous fishes (Sharks and Rays) and Lungfishes, the rays are mainly made of horn; in the cyclostomes they are made of cartilage. Bony fishes have embryonic horny rays which are later replaced by bony structure. There are 2 types of ray with bony structure: *soft rays* and *hard rays*. The soft rays are usually elastic, and only rarely spiny. They are paired. Usually, they are articulated; they may be unbranched or spread out like a fan to the edge of the fin. Hard rays are never articulated; they are fused from the base upwards and are usually more rigid than soft rays. Most *spines* are strong, sharp-pointed hard rays; more rarely, they are highly calcified, articulated rays. The hard rays of the fins are positioned in front of the soft rays. They are a sign of the later fish groups (phylogenetically-speaking), which are placed in the Sup-O Acanthopterygii*. The fin rays are linked by joints to the *fin ray supports* and in this way they are anchored in the body. The *fin musculature* is also located here—the means by which the fin rays are moved (fig. 521). The muscles of the paired fins have in addition large surface inserts on the shoulder girdle and on the pelvic bones (*see* Skeleton).

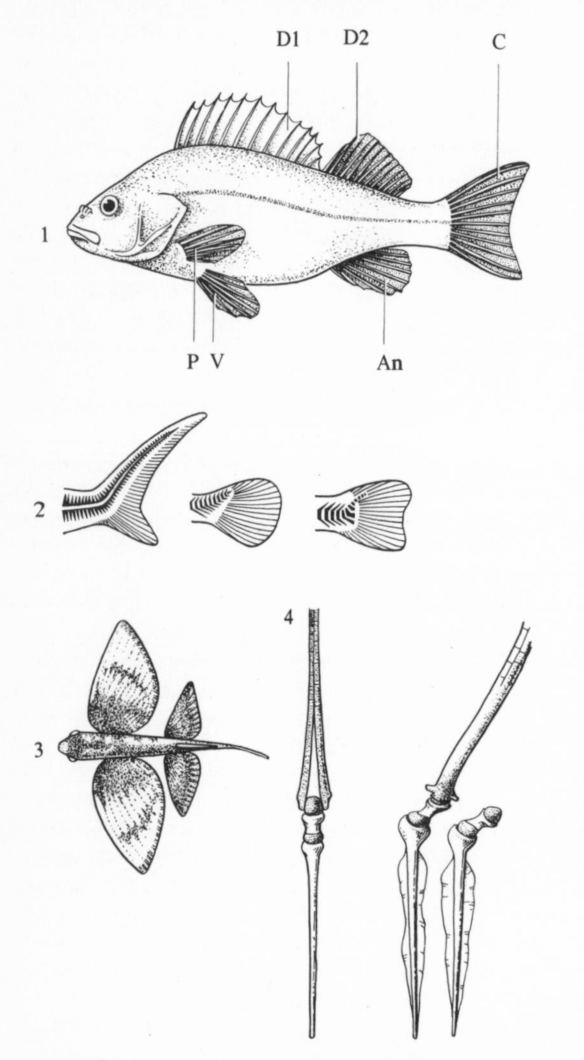

Fig. 521 Fins. 1 Indication of the fins of a bony fish (Perch). 2 Tail fin shapes and posterior end of the vertebral column; left, heterocercal; middle and right, homocercal. 3 Special fin shapes in Flying Fishes. 4 Fin ray carriers of the dorsal fin (Pike) from the front and from the side; right, with no fin ray

The dorsal fin (D), the tail or caudal fin (C) and the anal fin (An) are all *unpaired*. The ventral or pelvic fins (Vs) and the pectoral fins (Ps) are *paired*; it is these paired fins that correspond to the limbs of land vertebrates. In addition, some fish groups have something known as the adipose fin*. The *dorsal fin* works like a keel. It can be divided into several parts (eg D1, D2 and D3 in *Gadus*) or it can be broken up into a lot of small bits as in the Bichir *(Polypterus)*. How long the dorsal fin is, or where it is on the body, can vary a lot. With few exceptions *(Zoarces)*, the front part of the D is made up of hard rays; or a hard-rayed D1 is formed at the front (Perches); or the hard rays form free-standing, movable spines in front of the D2 (Sticklebacks). In many cases, hard rays and spines can be held erect by locking mechanisms without any muscle power (Sticklebacks, Pufferfishes, many Catfishes). The dorsal fin is absent only rarely *(Electrophorus)*.

The *tail fin*, in conjunction with the caudal peduncle, provides the main propulsion (fig. 521). The rays of the tail fin are closely connected with the vertebral column, so closely that the last vertebral links are often altered in shape. In rare instances, the vertebral column ends straight (diphycercal C, Lungfishes); usually, it turns off upwards in a diagonal direction. This is apparent either because the fin shape is asymmetrical (heterocercal C, sharks) or, in the case of a symmetrical C, it is only apparent from the skeleton (homocercal C, the majority of the bony fish). Very few fish have no caudal fin *(Hippocampus)*.

The *anal fin* is like the dorsal fin both in shape and function. It can be divided up in the same way, and the front sections may be supported by hard rays. The Spiny Dog-fishes *(Squalus)* are examples of fish without anal fins.

The *ventral* or *pelvic fins* are usually small. Exactly where they are on the body can vary a great deal. They may be located on the ventral surface, at the breast, and occasionally at the throat in front of the Ps. The Vs are the fins most likely to be absent in a fish, eg eels (Anguillidae), pufferfishes (Tetraodontidae) porcupine fishes (Diodontidae) and pipefishes (Syngnathidae).

The relative length of the *pectoral fins* is often typical of the species; their location varies little. Pectoral fins are used in braking, keeping the fish at the required height and they help to stabilise it. They are usually soft and transparent, although the front edge may be strengthened with hard soft rays (many catfishes). Moray eels (*see* Anguilliformes) do not have Ps.

Fish that have evolved specialised ways of life may also form specialised fins. For example, many poisonous fish* have developed poison glands that are linked to the hard rays of various fins; sharks and rays evolve copulatory organs from their Vs; and the gonopodium* of the livebearing tooth carps (Poeciliidae*) is formed from the An. The 'fishing rod' of the deep-sea angler fish (Lophiidormes*) and the suction organ* of the remoras (Echeneidae*) developed from the D. The sucking discs of the cling fish (*see* Gobiesociformes), the clinging mechanisms of the gobies (*see* Gobioidei) and the filaments of the thread-fish *(Colisa*, Trichogaster*)* are all formed from the Vs. The wing-like Ps and/or Vs enable the flying fish (*see* Exocoetoidei) to glide for long distances over the surface (fig. 521). The mudskippers *(Periophthalmus)* can move over land on account of their stalk-like Ps.

The fin-like extremities of many air-breathing vertebrates (eg whales and turtles) are indicative of a secondary adaptation* since these creatures have returned from the land to live in the sea (*see* Convergence).

Fire Clown *(Amphiprion ephippium) see Amphiprion*

Fire Corals, F, *see* Milleporidae

Fire-bodies (collective term) *see* Tunicata

Firefishes, G, *see Pterois*

Fire-mouth Cichlid *(Cichlasoma meeki) see Cichlasoma*

Fire-tailed Rivulus *(Rivulus milesi) see Rivulus*

Fish. This term was used originally as a systematic category for all vertebrates that live primarily in water. Today, however, it is used as a common name or collective term to include jawless fish (Agnatha*), cartilaginous fish (Chondrichthyes*), bony fish (Osteichthyes*) and sometimes (quite erroneously) members of the Whale family (Cetacea*) that actually belong to the mammals. The term 'fish' is usually applied to animals that live in water, have a more or less perfect fish shape, and move about by means of fins.

Fish Catch *see* Ornamental Fish, Capture of

Fish Disease Therapy. There are a great many causes of fish diseases. To stop them spreading, preventative measures are better than a therapy. Adhering to quarantine* times, taking suitable preventative measures (Prophylaxis*), and increasing and underlining the fish's natural defences by keeping it under the best conditions possible (Feeding*, Vitamins*) will usually enable the fish to ward off pathogens. If diseases do occur, an exact diagnosis* with the explanation of the causes of the illness must be obtained before treatment. Only in this way can any cure hope to be successful. Additional measures that must be adhered to are strict hygiene and a thorough disinfection* of all pieces of equipment. Simply by raising the temperature (with warm water fish up to 30–37 °C lasting 5–7 days), or by adding cooking or sea salt (Sodium Chloride Baths*), an intensification of the fish's metabolism will be effected, and at the same time its defence mechanisms will be heightened. Such conditions often make it impossible, or at least difficult, for many parasites (Parasitism*) to live. Medicaments are substances that have a poisonous effect upon pathogens, but they may also damage the fish. Good medicaments* unleash a strong poison effect, even in small concentrations; this is enough to harm the parasite, but if applied correctly will mean comparatively little damage for the fish. So, any medicaments must only be used if the instructions for use are followed exactly. The effect on different species of fish and on individual fishes can vary. The tolerance level may be decisively influenced by temperature pH value*, hardness* and other factors. So, in a lot of cases, a test has to be made to determine the effectiveness and tolerance level of individual species. A large number of medicaments have the effect of a mitotic poison (Cell Division*), so, if not applied in controlled amounts, damage may well be done to the fish's sex cells (which are actively engaged in cell division), and this may lead to sterility*. To cure skin* and gill diseases*, medicaments are applied during bathing treatments* or they are placed directly onto the affected areas. Many medicaments are received into the blood stream via the skin and gills only in small amounts; some can be introduced via the intestine, especially in the treatment of intestinal diseases*.

Fish Diseases. Aquarium fish are particularly susceptible to diseases. Three types of fish disease are distinguished: 1) hereditary diseases; 2) diseases caused by environmental factors; and 3) diseases caused by parasites (*see* Parasitism). Hereditary illnesses* are of least significance in aquarium-keeping. Cures are not possible. Any fish with a hereditary illness should be excluded from breeding, as any defects may crop up again in all subsequent generations. Much more important are the disorders caused by environmental factors as well as parasitical diseases in aquarium fish (which are often the result of the aforementioned environmentally-based disorders). They can be caused by conditions being inadequate in the aquarium, in particular poor water, overstocking, placing unsuitable animals and plants together in the same tank and inappropriate feeding. All these factors weaken the resistance of the fish, thus providing ideal circumstances for diseases to break out and spread quickly. Ornamental fish destined for the export market are often kept and transported in unsuitable conditions. Fish caught in the wild are often transported along with their parasites. Whilst being transported, conditions deteriorate further, so that when the fish arrive at the retail outlets for sale, they are weakened and starving. It is not unusual for up to 20 % of the specimens to die and the incidence of disease to be high. Containers used for transporting fish often have a small amount of anaesthetic* added to them. This, together with an enrichment of oxygen, make it possible to transport greater quantities of specimens (*see* Transportation of Aquatic Animals). However, it is very rare for medicaments to be added to counteract parasites. Imported marine fish fall ill in great numbers on account of these factors, and marine aquarists are constantly confronted by newly introduced fish diseases. Breeding establishments only rarely carry out continuous checks on health, so incurable diseases, such as fish tuberculosis*, *Ichthyosporidium** and *Pleistophora** are very much a part of the diseases that occur in the ornamental fish industry. The breeding of some fish species must even be in doubt. If the living conditions deteriorate gradually, the fish are able to adapt fairly comprehensively to such changes. Only sudden changes in the living conditions bring about shock reactions or environmentally caused diseases, which are not parasitic (poisoning, lack of oxygen, acid and alkali sickness, injuries etc). Slowly deteriorating environmental factors usually only become apparent when the fish are attacked by parasites (as a result of a lowering of their resistance). Water is a very good medium for parasites (*see* Parasitism). Large numbers of fish parasites are known to exist. Certain types, in particular skin parasites, even live on healthy fish without apparently causing them very much harm. Fishes have developed a natural defence against parasites which is reinforced when they are kept under optimal conditions. If there are noticeable shifts in the equilibrium in the tank, the natural resistance of the fish is reduced. Under such conditions, parasites begin to multiply in increasing numbers, the fish's resistance is weakened and eventually the fish will become gravely ill. In this way the natural defence of the fish is gradually reduced. Fish that have once been diseased are liable to attack by parasites that have not found the right conditions for multiplication among healthy fish. In most cases the parasites causing the disease can be determined through diagnosis*. More difficult, but of even more importance, is the recognition of the factors that caused the damage in the first place. If this cannot be determined, the success of the treatment may be in doubt. The limited space of an aquarium, and the close contact of its inhabitants, make it an ideal place for fish diseases to be transmitted. Through their excreta, diseased fish increase the microbial content of the aquarium water to such a degree that they continually reinfect themselves. Even healthy specimens eventually succumb. To prevent fish diseases, keep newly acquired fish in quarantine* for a period and make sure conditions are best suited for the particular species. If diseases do occur, make a thorough diagnosis first, and only implement a particular treatment after you are sure of the cause of the disease. If necessary, use appropriate medicaments*. In the commercial world, it is particularly important to take suitable measures against the spread of disease.

Fish Diseases, Cures for, *see* Fish Disease Therapy

Fish Food *see* Feeding

Fish Fungus *see* Saprolegniacea

Fish, Killing of. It may become necessary to kill a fish if it is suffering from an incurable disease or if a detailed internal examination is required. Small fish can be quickly and painlessly killed by cutting the back of the head with scissors. Larger specimens can first be knocked out by a sharp blow to the top of the head. A fish can also be killed by an overdose of anaesthetic* such as urethane or MS 222; this method is particularly suitable for sick fish or when a specimen is to be preserved. When an anaesthetic is used, any skin and gill parasites will quickly become detached from the host, so it will no longer be possible to tell whether they were the cause of illness or not.

Fish Leeches, G, *see Piscicola*

Fish Lice, G, *see Nerocila*

Fish Parasites *see* Parasitism

Fish Photography. Good photographs of fish are of prime importance today as an aid to species and behaviour descriptions. Taking photos of fish in the aquarium needs some experience and familiarity with photographic techniques. The meagre size of fish makes for problems in the first instance, so the magnification of the camera must be increased considerably and the depth of field must be confined to a few mm. Since the objects being photographed move around a lot, a short exposure time is needed. Fish and other aquatic animals should be photographed, where possible, in an aquarium that they are used to, and in one that shows their natural colours and types of behaviour. Only in exceptional circumstances should special tanks be used for photographing, and, similarly, separating the specimens by means of an extra pane of glass or anaesthetising them are methods that should be used only rarely. The water must be crystal-clear; the colour of the sides of the tank itself can cause a filter effect, and thus produces an unintentional colour cast. Plexiglass is absolutely colour-neutral. Reflex cameras are particularly good for fish photography, as their optics (screened lenses, adapting rings or bellows extension) are parallax-free, and can be set up for larger portraiture without difficulty. The camera has to be well screened off to achieve a usable depth of field, whereas the necessary light is produced with electronic flash lights. A pilot-light is useful (fixed into the flash) for monitoring exactly the illumination and any reflections that arise. Any opposing light makes even the smallest particles in suspension appear as bright spots. The lights must be arranged in such a way that no reflections appear in the picture. The background must suit the colour of the specimens and should not eliminate delicate contours, such as transparent fin edges. Apart from providing an over-all record of the fish species, photos also provide records of details and characteristic ways of behaviour. A series of photos of the spawning act or courtship behaviour can offer valuable clues as to phases of movement and behaviour patterns. Photographing freshwater animals is usually only successful in the aquarium, unlike photographs of animals from clear-water seas. Today, in the sea and in particularly clear fresh waters, photos of aquatic animals in their natural biotopes can be taken with the aid of watertight underwater cameras. Underwater photography involves considerable technical expense, and apart from photographic skills also demands competence as a diver.

Fish Poisons *see* Poisonous Fish

Fish Research Laboratories. Diseased fish, or fish that are suspected of having a disease, can be sent away to fishery laboratories to be diagnosed. Whenever possible, specimens should be sent live, under suitable conditions (*see* Transportation of Aquatic Animals). Diagnosis* is never certain with dead fish, as the pathogen usually vacates the dead animal very rapidly and secondary parasites like bacteria and fungi take over from them. Only those fish that would probably not survive being transported, or those that are already dead, should be packed in ice or dispatched in 4% formaldehyde* or 80% alcohol. In every case, an exact record should be enclosed, giving the necessary information about the conditions under which the fish has been kept, together with a case history (*see* Anamnesis). Without such information, correct diagnosis may be impossible. If a sample of water is to be tested at the same time, it must be sent separately in a meticulously clean bottle, containing no suspended matter and secured with a neutral cork or stopper.

Fish, Study of, *see* Ichthyology

Fish, Transportation of, *see* Transportation of Aquatic Animals

Fish Tuberculosis. One of the commonest and most feared infectious diseases in aquarium fish. It is caused by a nonmotile, rod-shaped bacterium (Schizobionta*) belonging to the G *Mycobacterium*. Numerous strains of the bacterium are known to exist. They develop best at a temperature of around 25 °C, higher in only a few cases. The bacteria are detectable in the wild in both freshwater and marine fish, but they appear to cause disease only when conditions are inadequate in the aquarium. Fish tuberculosis is a debilitating disease to which all aquarium fish are susceptible, but especially the Anabantidae*, Characidae* and Cyprinidae*. An infected fish will appear apathetic, emaciated, it will refuse its food and lose its colour. In an advanced stage, defects may occur in the scales and fins, swellings may appear in the skin, the eyes may look bulbous and the skeletal system may become deformed. The internal organs (intestine, spleen, liver, kidneys, heart, eye chambers) become impregnated with greyish-white tubercular nodules in which the fish tries to encapsulate the bacteria and render them harmless (fig. 522). Later on, these tubercles disintegrate necrotically, thereby causing great damage to the organs. The symptoms are very similar when a fish suffers an attack by *Ichthyosporidium**. The disease is often very slow to develop, and external signs of it may not show at all.

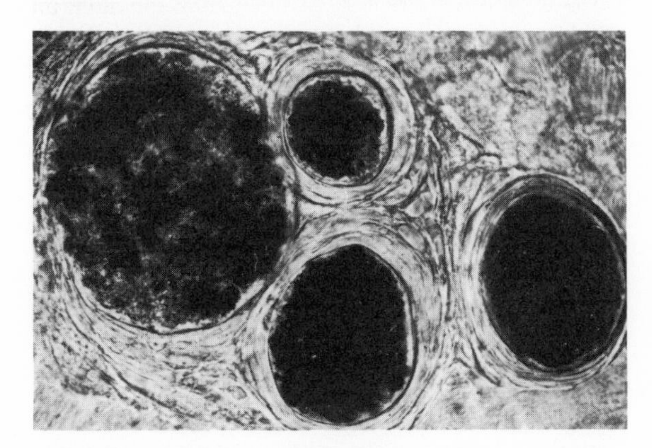

Fig. 522 Tubercular nodules in the kidney

It is transmitted directly from fish to fish and it can also be caught by amphibians* and reptiles*. With aquarium fish, 80% are likely to become infected once the disease is present. Preventative measures against fish tuberculosis include keeping the fish under the best possible conditions and never keeping too many fish in the aquarium. In the early stages medical treatment can be tried, eg tuberculostatica or Terramycin* and Streptomycin*, but success is rare. Removal of sick or dead fish is vital if further infection is to be avoided. In breeding tanks, should there be an outbreak of fish tuberculosis, the whole stock of fish that has been in contact with infected specimens should be killed.

Fisheries. A collective term for all the installations and measures relating to the use of natural and artificial waters for the improvement of fish stocks. It also includes breeding*, hatching, rearing and fattening up certain kinds of useful fish species. In recent years, techniques of intensive fish management have developed alongside traditional methods. These new techniques can give high yields in confined tanks of water. Increasingly, suitable areas of coastal waters are also being managed by the fisheries.

Fishing. Catching fish for the purposes of feeding humans and animals. Fishing is one of the oldest methods of acquiring food. For many peoples, fish is their chief and sometimes their only source of protein. The modern fishing industry includes fishing out at sea, around the coasts, and in inland waters. Deep-sea fishing is mainly the province of large concerns within the fishing industry, whereas in many places fishing in coastal and inland waters is carried on by family businesses and smaller concerns. Freshly caught fish or deep-frozen fish comes onto the market immediately; otherwise it is processed at factories. In many parts of the world, other animals and plants, besides fish, are caught (eg crustaceans, molluscs, snails, cephalopods, algae). In recent years, deep-sea fishermen have turned to catching krill to provide a new source of protein.

Fishlice, F, *see* Argulidae

Fishlouse *(Argulus foliaceus) see* Argulidae

Fistularia tabaccaria *see* Aulostomoidei

Fistulariidae, F, *see* Aulostomoidei

Five-banded Damselfish *(Abudefduf saxatilis) see Abudefduf*

Five-banded Sergeant Major *(Abudefduf saxatilis) see Abudefduf*

Fixed Action Pattern; Instinctive Movement. Term used in ethology. Fixed action patterns enable an animal to respond to a stimulus with a phylogenetically fixed type of movement which is usually preceded by an adjusting movement or taxis*. (For examples *see* Taxis.) Fixed action patterns are therefore regarded as consummatory actions*.

Flabellina, G, *see* Nudibranchia

Flag Cichlid *(Aequidens curviceps) see Aequidens*

Flag Cichlid *(Cichlasoma festivum) see Cichlasoma*

Flag Tetra *(Hyphessobrycon heterorhabdus) see Hyphessobrycon*

Flagella. Long, whip-like processes that enable a great many micro-organisms (eg Flagellata*) to be mobile. The flagellated epithelia of multi-celled animals are units of cells at the surface, and, like the ciliated epithelia (Cilia*), they carry out the functions of transport and movement in various organs.

Flagellata; flagellates. Single-celled organisms that move about by means of one or several flagella. They are ascribed to both the plant and animal kingdoms. Originally, all flagellates were plants, ie they led an autotrophic way of life, being able to change light energy into chemical energy with the aid of chlorophyll. However, in the course of evolution, nearly all the systematic groups within the Flagellata have evolved forms without chlorophyll. They derive their nourishment from organic matter, ie heterotrophically. All the forms alive today are derived from the heterotrophic group. On the other hand, all higher plants are derived from the Flagellata, so they represent the origin of both the plant and animal kingdoms.

The Flagellata inhabit a great variety of different waters, fresh and marine. Sometimes they are present in large numbers of one species, sometimes there are lots of different species. Plant-type flagellates (phytoflagellates) can multiply in such numbers when conditions are right, that they colour the water, eg certain *Chromulina* species colour the water brown, *Euglena* species green. As part of the phytoplankton*, the flagellates are an important part of the feeding inter-relationships that exist where it occurs. Various kinds of plant flagellates form cell colonies, and some animal flagellates are parasites *Ichthyobodo**, *Hexamita**, *Spironucleus (see Hexamita), Trypanosoma* and *Cryptobia* (*see* Blood Flagellates), *Oodinium**.

Flagellates, Ph, *see* Flagellata

Flagfishes, F, *see* Kuhliidae

Flagtail Surgeonfish *(Paracanthurus hepatus) see Paracanthurus*

Flagtails, F, *see* Kuhliidae

Flame Tetra *(Hyphessobrycon flammeus) see Hyphessobrycon*

Flatfishes, O, *see* Pleuronectiformes

Flat-tailed Triggerfish *(Abalistes stellaris) see Abalistes*

Flatworms, Ph, *see* Plathelminthes

Fleur-de-lis *(Iris pseudacorus) see* Iridaceae

Flier *(Centrarchus macropterus) see Centrarchus*

Floating Heart *(Nymphoides peltata) see Nymphoides*

Floating Liverwort *(Riccia fluitans) see Riccia*

Floating Moss Family (Salviniaceae, F) *see* Pteridophyta

Floating Mosses, G, *see Salvinia*

Floating Pondweed *(Potamogeton natans) see Potamogeton*

Floating Water Sprite *(Ceratopteris pteridioides) see Ceratopteris*

Flora. The plant life of a particular geographical region.

Flounder *(Platichthys flesus) see* Pleuronectiformes

Flounder-Ruffe Region *see* Potamal

Flour Moths, G, *see Ephestia*

Flower (fig. 523). Short shoot, whose leaves have a function in reproduction and are modified accordingly. The basic structural elements are the flower stalk or peduncle, the floral axis, the perianth, stamens and carpels. The great variety of flower types is caused by differences in the number, form, size and colour of the component parts of the flower; the variation in their position; and the way in which the parts grow together, multiply or revert to type. A flower is described as polysymmetrical if it can be seen to be symmetrical in many planes, and as monosymmetrical if there is only one plane in which it is symmetrical. The *peduncle* varies in length from one species to another, and if it is absent altogether it is described as sedentary. The *floral axis* can be elongated or arched, in which case all or part of the leaf organs are arranged in a spiral. Usually, however, the axis is compressed and the perianth, stamens and carpels are arranged in whorls. Some axes have a cup-like or tube-like shape, into which the carpels are set. There is a double *perianth* when it is divided into a green calyx consisting of sepals, and a colourful corona made up of petals. The perianth is simple when there is no such division, and is called

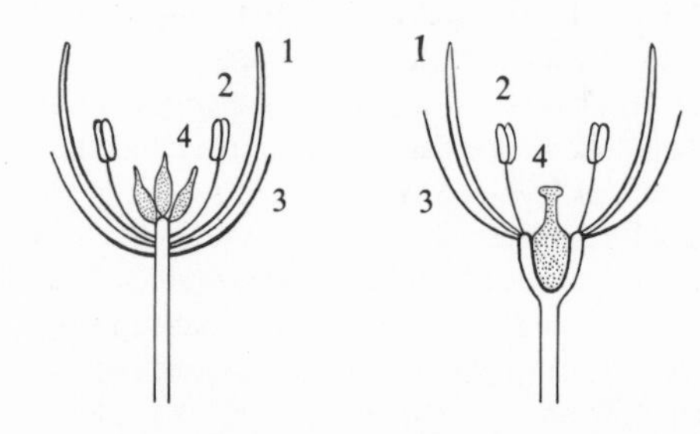

Fig. 523 Flower structure. 1 Petals 2 Stamens 3 Sepals 4 Carpels (left: free-standing, superior; right; fused, inferior)

perigone. The perigone leaves, or tepals, may be of inconspicuous shape like a calyx or they may be striking, like a corona. The sepals, petals and tepals may be separated from one another or they may grow into one another. There is rarely a complete absence of perianth. The *stamens* are concerned with pollen production. Their number can vary, but is usually the same in individual families. Stamens are made up of the *filament* and the *anther*, the latter possessing 4 pollen sacs. The *carpels* may be separate or they may combine into a uniform pistil. The separate carpels or pistil consist of the *ovary*, style and stigma. The latter receives the pollen while the style is a connecting part which is absent in some cases. Inside the ovary are the ovules, the quantity of which can vary. After successful pollination* and subsequent fertilisation the mature seeds* develop from the ovules. If the carpels lie above the perianth, they are superior. A cup-like or tube-like floral axis may unite with the carpels, in which case the carpels are inferior, ie the perianth and stamens lie above the carpels.

In some aquarium plants flowers appear only when they can grow out of the water or when they are kept as marsh plants, such as some species of *Echinodorus* and all the species of *Hygrophila.* In other aquarium plants flowers form even in submerged varieties. In *Cryptocoryne, Aponogeton, Nymphaea* or *Vallisneria* the flowers grow up to the water surface, whereas they stay underwater in *Ceratophyllum.* In species where vegetative propagation cannot take place or is carried out poorly, it is the flowers we have to turn to for seed formation.

Flowering Rush *(Butomus umbellatus) see* Butomaceae

Flowering Rushes, F, *see* Butomaceae

Flowing Waters (fig. 524). Inland waters that flow in one direction only, because of a gradient. The flow is either turbulent (eg a mountain stream) or meandering (eg a lowland river). Although the fauna in a flowing river will drift along with the current, it is usually possible for it to stop and move against the current to compensate (positive rheotaxis). Flowing waters are made up of several parts following one after the other,

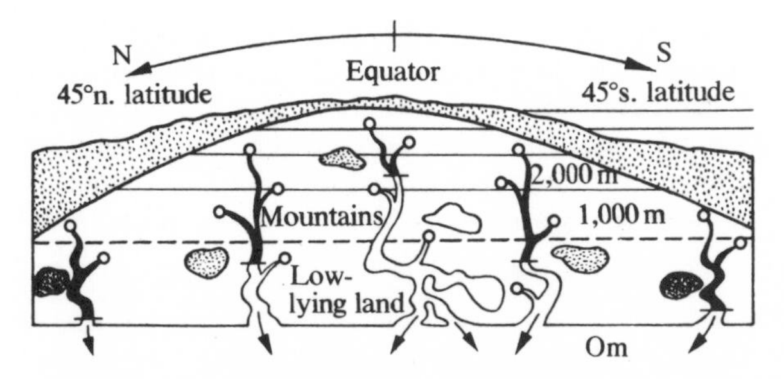

Fig. 524 The character of a stream (black) or of a river (white) in a flowing stretch of water and how it is dependent on their geographical position and altitude

each of which has its own features and therefore its own characteristic living community (Biocenesis*). Basically, three parts are distinguished: the source (crenon, crenal*), the mountain stream region (rhithron, rhithral*), and the lowland river region (potamon, potamal*). The extent of these zones is dependent upon the geographical latitude and altitude, among other things. So, the rhithral region gets longer towards the Arctic and mountain ranges, whereas the potamal region is longer towards the tropics and the lowlands.

Flutemouths (Fistulariidae, F) *see* Aulostomoidei

Fluttering Dance. Term used in ethology. It describes a typical mode of behaviour adopted by the male during intense courtship activity. It is particularly common among many species of Characin and Barb. The males swim round the females, their fins spread out, and following circular, arched or zigzag routes. Directly after the fluttering dance, the male often starts swimming towards the spawning substrate (*see* Courtship Swimming). The undulating swimming movements of the *Nannostomus** species (*see* Courtship Swimming) are primitive variations on the fluttering dance. The males of many species of *Hyphessobrycon**, *Hemigrammus** and *Megalamphodus** are particularly noted for their fluttering dances.

Fly Maggots. The larvae of the common Housefly *(Musca domestica)* or those of the large Bluebottles and Blow-flies make a very good supplementary food for surface fish, in particular. Bait (meat, fish, cheese etc) is laid out to entice the flies to lay their eggs. The eggs are then transferred to covered vessels, in which the maggots can be cultivated after they have hatched out. A broth of rolled oats with a bit of yeast added (up to 25%) will make a good nutrient substrate. It should be fed to the maggots in small amounts. In the summer, layers of fungus can be laid in the water from which fungal maggots can be obtained. Even maggots from poisonous fungi are suitable.

Flying Barbs, G, *see Esomus*

Flying Fish, related species, Sub-O, *see* Exocoetoidei

Flying Fishes (Exocoetidae, F) *see* Exocoetoidei

Flying Fox, G, *see Epalzeorhynchus*

Flying Gurnard *(Dactylopterus volitans) see* Dactylopteriformes

Flying Gurnards *(Dactylopterus,* G*) see* Dactylopteriformes

Flying Gurnards, F, *see* Dactylopteriformes

Flying Gurnards, related species, O, *see* Dactylopteriformes

Foam Filter (fig. 525). Method of separating impurities by a foaming process (not to be confused with conventional filters that use foam blocks as a filtering medium). In a liquid, matter in suspension can be separated from sinking matter by flotation. This principle is applied in marine aquaria to remove impurities in colloidal form (*see* Colloid) from the water. Particles that are heavier than water are not removed by this method, so the foam filter is not a substitute for a true filter. When fine air bubbles are introduced into sea water, protein* present in colloidal form (and other impurities in suspension) produces a foam by coating the air bubbles and thus stabilising them. This process denatures the protein and increases its ability to bind substances in solution. The protein foam will come to hold a high concentration of colloidal and other substances. As denatured protein rapidly releases ammonia and other toxic breakdown products, the foam concentrate must be removed quickly from the system. Sea water can be purified in this way to a level of 0.5 μg of protein per litre, which corresponds to conditions in depths of 5,000 metres or in waters around a coral reef. If ozone* is introduced to this very clean aquarium water the oxygen content will be increased till it becomes 20% oversaturated (*see* Redox Potential). Enriching the introduced air with measured doses of ozone increases the efficiency of the filter – the dosage should not exceed 5 mg per 100 litres water/hour.

A foam filter works in this way. Water is fed into a reaction tube where it comes into close contact with an ozone/air mixture. The foam that is produced is separated off in a separation chamber (a grating prevents it from flowing back). The purified water is pumped back into the aquarium tank once it has been passed over activated charcoal* to remove the toxic ozone. The efficiency of the method varies. The least efficient is when water and air move in the same direction. Efficiency increases when they move in opposite directions, as this gives a longer reaction time. A rotation filter works best of all. Water is pumped over an ozone/air diffuser and circulated through the separation chamber. Here the protein/air/ozone mixture is quickly separated from the purified water. Fat reduces the efficiency of the filter markedly. If food with a high fat content is fed to the fish, it is recommended that a non-toxic filterable fat emulsifier (polyvinylpyrrolidon) be added – 2–3 ml of a 3.5–4.5 % aqueous solution, Periston R, to every 500 litres of tank water. Trace elements in ionised form become bound with the proteins and are removed from the water by the filter. The addition of trace elements with complex chemical structures prevents them from becoming bound to protein. An efficient foam filter will control bad algal growth by removing necessary trace elements and nitrogen. In recent years neutral and non-toxic foam-building substances have been added to aquarium water. In marine tanks it is possible by this means to remove foam continuously and independently of the protein content. In fresh water too, the introduction of artificial foam-inducers to a foam filter facilitates the rapid and safe removal of cloudiness caused by suspended colloidal material (bacteria, algae, etc).

coarsest meshing at the top. Now sieve your collection, and you will find that different sized animals will collect at each of the different stages. *See* Food Trapping for information about the necessary mesh widths.

Food Tablets *see* Feeding

Food, Transport of. It is most important to maintain a good oxygen supply and to prevent drying out if your food animals are to be transported without loss of life. There are two ways of doing this. The most common method when transporting small quantities of food animals is to keep them in containers provided with water. Screw-top bottles, plastic containers or foil bags are most suitable. With larger catches wide tins are often used – they are provided with fine holes in the lid to admit air. Always leave about $^1/_3$ of the space inside

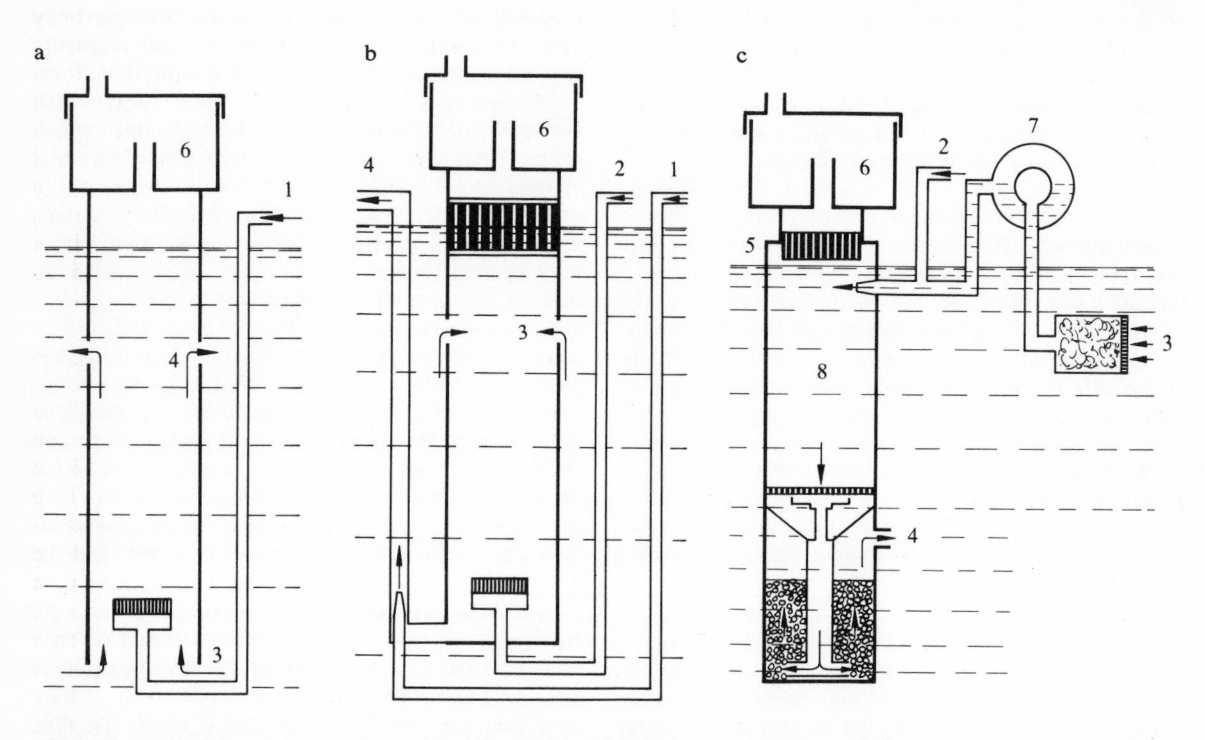

Fig. 525 Foam filter. a) Same direction principle, b) Opposite direction principle, c) Rotation filter. 1 Air 2 Air-ozone mixture 3 Water inlet 4 Water outlet 5 Foam grating 6 Foam chamber 7 Circulating pump 8 Separation chamber

Foliage Leaf *see* Leaf

Food, used for rearing purposes, *see* Fry, Feeding of

Food Animals *see* Feeding

Food Chain *see* Nutrition

Food Funnel *see* Feeding Equipment

Food Sieve. A kitchen sieve or a relatively wide-mesh net can be used to clean any food that has become highly polluted. However, it is often necessary to grade the food according to size. If very small food animals (infusorians, rotifers, nauplii) are needed to feed to young fish, they have to be caught with fine-meshed nets. Larger animals, such as *Daphnia, Cyclops,* fly larvae etc will always be caught in the net with them. You can only separate them from the smaller animals by sieving them out. For this you will need a set of sieves used for flour, stacked one on top of the other. The sieve netting should be made of metal or nylon, and the one with the finest mesh should go at the bottom, ranging upwards to the

your container free (to give sufficient air) and do not crowd your catch, especially at warm times of the year. A better way of transporting your food animals, and one that will save space and weight, is to keep them damp inside food trays. These are made of strips of hardwood with a layer of fine-mesh netting or rustless metallic gauze in the bottom. Trays measuring 14 × 20 cm can be stacked together, and then either wrapped in damp cloths or in a light wooden box. Take some of the food you have caught in your net and place it in a tray. Allow the food to spread out evenly by floating the tray briefly on the water, then lift the tray straight out again. The food layer should not be more than 3 mm thick. All food animals except infusorians and rotifers will travel in this way for long distances.

Food Trapping. Although you can buy live food animals, breeders and hobbyists regularly trap their own food. For anyone who is interested in biology, trapping food offers an ideal opportunity to study the way of life of the food animals and to get to know the countless animals and plants that live in and around the water.

There are lots of good places to look for food: village ponds, pools of water in forests and meadows, park ponds (at least those ones that do not contain fish). The food animals that you will find in these places in large quantities include water-fleas, *Cyclops*, fly larvae and rotifers. When very large numbers of them are present it is usually linked to a particular time of year and the weather. *Cyclops* and their nauplii (which are important for rearing young fish) are usually predominant in the spring, whereas water-fleas are the more common in summer.

Shrimping nets are invaluable for catching food. The rod should be made of hardwood, bamboo or light-metal segments, and by screwing on extra lengths you should be able to extend it a maximum 3–4 m. The net is usually made of nylon or terylene and is fixed firmly to a stable, non-rusting hoop with rings or clips. It should have a diameter of 20–30 cm and the length of the pouch should be 40–60 cm. Choose the width of your mesh according to what you want to catch (*see* table below).

How you catch the food animals depends on the type of water you are fishing in. In deep water, the net is moved around in the form of an 8 lying on its side so that the food animals of the lower levels of water are swirled upwards by the ensuing suction.

The food can be sorted according to size by passing it through a food sieve*. Do this while you are sitting at the side of the water where you are trapping the food.

The following is a guide to the width of the mesh needed to catch particular food animals:

Type of Animal	Width of Mesh Required (in mm)
Infusorians	0.12
Rotifers	0.18–0.2
Nauplii	0.18–0.2
Smallest Bosmina	0.2
Small *Cyclops*	0.3–0.4
Small *Daphnia*	0.3–0.4
Larger kinds of small crustaceans	0.5–0.6
Fly larvae	0.6–0.8

Food Trays *see* Food, Transport of

Fontinalis (DILLENIUS) HEDWIG, 1801; Willow Moss or Water Moss. G of the Bryophyta* comprising 30 species. Found in temperate and cold regions. *Fontinalis* is a moss with trihedral leaves on the stem which is branched in an irregular manner. Colour variable. Reproduction is through spores. All the species of *Fontinalis* are water mosses primarily, found in clear waters. The species mentioned below is commonly kept in aquaria. It attaches itself to rocks, wood etc. and can adopt different shapes according to temperature. It should not be allowed to stand in too bright a spot.
– *F. antipyretica* LINNÉ. N America, Europe, N Asia, N and S Africa.

Foraminifera. These single-celled creatures (Protozoa*) make up a class of the Rhizopoda*. The majority are marine and usually live on the sea bottom, although a few are pelagic. They are protected by a shell, that may have one or several chambers. The shell is usually made of chalk. The delicate shells of present-day foraminiferans measure on average 1–3 mm, but the nummulites of the Tertiary period reached several cm in diameter. Huge deposits of foraminiferan shells, especially during the Carboniferous and Cretaceous, created foraminiferan chalk and their presence is an important aspect of exploration for oil. Foraminiferans feed on microorganisms which they trap in their protoplasmic processes that stick out from their shells like threads. Foraminiferans undergo an alternation of generations*, ie generations that reproduce sexually and asexually alternate regularly. A wandering form which is usually flagellated develops from haploid gamonts. This form copulates and grows into diploid sporobionts which later break up into numerous haploid agametes which in turn grow into new gamonts. The sporobionts that develop from the tiny wandering forms have a tiny primary chamber (microspheric generation), whereas the gamonts that develop from the larger agametes have a larger primary chamber (megaspheric generation). Foraminifera can often be seen in the aquarium.

Foraminiferans, Cl, *see* Foraminifera

Forceps Fish *(Forcipiger longirostris) see Forcipiger*

Forcipiger JORDAN and MCGREGOR, 1928. G of the Chaetodontidae*, Sub-F Chaetodontinae. Just one species, found on coral reefs and rocky coasts of the tropical Indo-Pacific. A typical feature is the very long snout. Other characteristics include the large, crescent-shaped D which is in contrast to *Chelmon**, and the brilliant yellow colour of the body which contrasts with the dark brown of the upper half of the head and nape. Territorial fish that are thoroughly intolerant of each other. For this reason they must not be kept together in an aquarium. They are also difficult to feed, as they take a long time to get used to any diet *(Mysis, Artemia, Tubifex, Enchytraeus). Forcipiger* is also sensitive to fluctuating water quality, especially changing pH-values. Despite everything, and although it is not at all a common species in the wild, *Forcipiger* is imported:

Fig. 526 *Forcipiger longirostris*

– *F. longirostris* (BROUSSONET, 1782); Long-bill, Forceps Fish, Long-snouted Coralfish, Yellow Long-nosed Butterfly-fish. Found from the African coast to Hawaii. Up to 16 cm (fig. 526).

Formalin. An aqueous solution with about 40% formaldehyde, formalin has strong disinfectant properties. A 4–8% formalin solution will preserve animal and plant matter. It can also be used to treat ectoparasites (*see* Parasitism) in fish: freshwater fish need to be bathed for 30–45 minutes in a solution of 2–2.5 ml formalin (35–40%) to every 10 l water; marine fish for 10–15 minutes in a 10 ml formalin solution (35–40%) to every 10 l sea water.

Fossils (fig. 527). Remains or traces of plants and animals from past geological epochs. The scientific study of fossils is called palaeontology*. *Index fossils* is the name given to particular types of fossils whose presence in large numbers at a particular period in the earth's history is a sure indication of the age and history of certain parts of the Earth's crust. Fossils are usually petrified; this can happen as a result of deposits building up on riverbeds or on sea bottoms, by sand being blown over the top of the fossil, or by lava encasing it. Usually only the hard parts are preserved (eg the skeleton, individual bones, teeth, shells); organic substances are broken down. The original shape is often recorded as an impression in the stone.

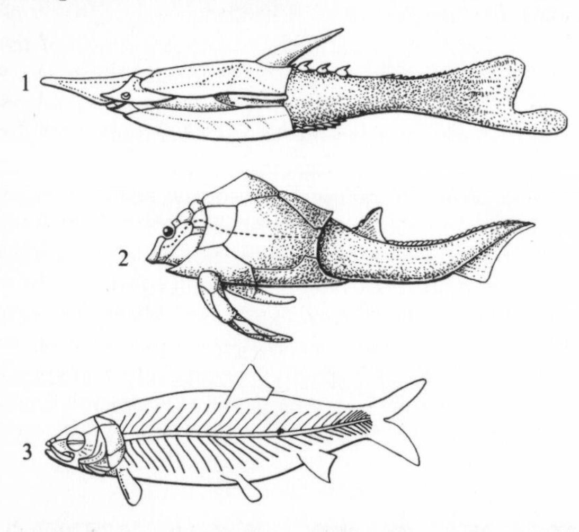

Fig. 527 Fossils. 1 Periosteal fish *(Pteraspis)* 2 Armoured fish *(Pterichthyodes)* 3 Primitive bony fish *(Leptolepis)*

Some Blue Algae are dated as being over 3,000 million years old. Remains of higher plants and animals from every invertebrate Ph have been preserved from the end of the Precambrian period (ie over 600 million years ago). The first fish-like animals – the jawless pteraspids (Ostracoderms, eg *Pteraspis**), and the armoured fishes (Placoderms, eg *Pterichthyodes**) which were the first animals to have jaws – are known to us today because their bony exoskeletons were preserved as fossils. Fossil ancestors of the cartilaginous fishes and primitive forms of the bony fishes (Osteichthyes*), in particular Coelacanths and Lungfishes, date back to the Devonian. *Palaeoniscus*, a primitive ray-fin dating from the Permian, and still with a heterocercal tail fin, is found in the German copper slate deposits. True bony fishes (Teleostei) were already widespread 150 million years ago (the Jurassic) and they were the dominant fish group in the Cretaceous, 100 million years ago. The oldest forms, eg the G *Leptolepis**, are related to the Herring-like species (Clupeiformes). They still had a prominent ganoine layer on the scales and skull bones.

Four-barred Tiger Fish *(Datnoides quadrifasciatus) see* Lobotidae

Four-eyed Fish, F, *see* Anablepidae

Four-leaved Marsilea *(Marsilea quadrifolia) see Marsilea*

Fowler, Henry W. (1878–1964). One of the best-known American ichthyologists, whose comprehensive knowledge of the entire fish fauna was condensed into various publications. *The Marine Fishes of West Africa* (1936); *The Fishes of the George Vanderbilt South Pacific Expedition* (1939); *Contributions to the Biology of the Philippine Archipelago and Adjacent Regions* (1941); and *Os peixes de aqua doce de Brasil* (1948).

Foxtail, G, *see Ceratophyllum*

Fragile Brittle-star *(Ophiothrix fragilis) see Ophiothrix*

Frame Tank *see* Aquarium

Freckled Hawkfish *(Paracirrhites forsteri) see Paracirrhites*

Freeze-dried Food *see* Feeding

Fresh Water *see* Water

Freshwater Crabs, F, *see* Brachyura

Freshwater Drum *(Aplodinotus grunnieus) see* Sciaenidae

Freshwater Herrings (Denticipitidae, F) *see* Clupeiformes

Freshwater Jellyfish, G, *see Craspedacusta*

Freshwater Mussels, G, *see Unio*

Freshwater Polyps, various Ga, *see* Hydrozoa

Freshwater Shrimps, G, *see Gammarus*

Freshwater Sponges, Ga, *see Ephydatia* and *Spongilla*

Frill Sharks, F, *see* Chlamydoselachidae

Fringe Reef *see* Coral Reef

Fringed Croaker or Drum *(Sciaena cirrhosa) see Sciaena*

Frogbit, G, *see Hydrocharis*

Frogbit Family, *see* Hydrocharitaceae

Frog-fishes (Antennariidae, F) *see* Lophiiformes

Frog-mouthed Catfish *(Chaca chaca) see* Chacidae

Frog-mouthed Catfishes, F, *see* Chacidae

Frogs (collective term) *see* Amphibia

Frontal Position *see* Display Behaviour

Frozen Food *see* Feeding

Fruit. A fruit develops from a flower, and it is an indication that the seeds are mature. Apart from the carpels, the floral axis and sepals may also form part of the structure of the fruit. To date, there has yet to appear a comprehensive and fully satisfactory account of the various types of fruit. A *simple fruit* develops from a flower with fused carpels or with one carpel; a *multiple or collective fruit* develops from a flower with free-standing carpels. The individual parts of a multiple fruit are known as fruitlets. The wall of a mature fruit may be dry (capsule), fleshy (berries) or hard like a stone (nut). Sometimes fruits are both fleshy and hard (stone fruits or drupes). Dehiscent fruits open up when mature to release the seeds one by one. Fruits that decompose form a dispersal unit made up of a seed and part of the fruit wall in each case. Juicy fruits are often eaten by animals and people, but the seeds pass through the digestive system unharmed.

Fruit-flies, G, *see Drosophila*

Fry, Feeding of. When rearing young fish (*see* Fry, Rearing of), the correct choice of food is of paramount importance. The first thing to be said is that even the best artificially manufactured food preparations can never replace live micro-organisms, particularly in the first growth phases. Described below are the important live food species that can be used as food for young fishes. The food animals can either be fished from suitable waters or grown in breeding cultures.

The nauplii* of various *Cyclops* species: the youthful stages of these small crustaceans are the most important food basis for the young of almost every kind of aquarium fish. They have a high nutritional content. Depending on the size of the brood fish, the nauplii need to be sieved thoroughly, as they grow very quickly in warm water and may then threaten the welfare of the growing fish. Their optimum development period is in the winter and spring months when they are to be found in water with a high phytoplankton content (recognisable by the greeny-grey colour of the water). They jump about in the water.

The nutritional content of *Diaptomus** nauplii is equally as high. The larval stages of these copepods are found mainly in glass-clear waters. They are not dangerous to the young fish, even while the nauplii are growing.

If no *Cyclops* nauplii are to be found, *Artemia** nauplii can be a suitable replacement. *Artemia* is the youthful stage of the small crustacean, *Artemia salina*, and it can be reared. The dried eggs can be bought from aquarium shops. They are then placed in a 3% cooking salt solution (= 30 g per litre) at a temperature of between 24 and 26 °C and with plenty of ventilation. After 24–30 hours, the eggs will hatch out. *Artemia* nauplii have certain disadvantages, however. When they hatch they are 0.5 mm long (ie bigger than *Cyclops* nauplii), and they also grow quicker in salt water. So, if they are to be supplied constantly, a fresh batch must be placed in the aquarium daily. Moreover, *Artemia* nauplii die quickly in fresh water and must be fed to the growing fish often, and in very small quantities. To this end, they must be sucked up from the bottom of the breeding glass by means of a thin piece of tubing, then transferred into a fine-mesh net. The missing water is replaced with fresh water of the same salt concentration, so that the nauplii slip through the net at regular intervals. *Artemia* nauplii move around in the water with a rowing action.

*Rotatoria** or rotifers are also very valuable as a foodstuff. Their period of development takes place in the summer months, with only a few species of the Ga *Asplanchna* and *Rotifer* having marked propagatory phases during winter. Rotifers are easy to catch in very fine-mesh perlon yarn nets. They mark out loops as they move about in the open water, almost creeping along. Sometimes they attach themselves firmly for a while to plants, rocks or panes of glass. Depending on species, in the net rotifers will appear brownish, salmon-red or greenish-grey. They cause no harm at all to growing juvenile fishes.

The use of infusorians as a food for young fish has disappeared to a large extent, since their nutritional value is scarcely comparable with that of the *Cyclops* nauplii and rotifers. If infusorians are fed to young fish, many species will starve to death. In earlier times, the Slipper animalcule, *Paramaecium caudatum*, was reared for food. *See also* Hay Infusion.

Fry, Rearing of. The successful breeding of a species of fish is the high-spot of any aquarist's aim. Once spawning has taken place and the young have hatched out, however, rearing the brood is the next most important step. This task begins the moment the embryos break through the egg-envelopes. Depending on the G or F, the length of time the young fish lies about after hatching (before it begins to swim around freely), can vary a great deal. With many Characins and Barbs it can be about 4–5 days, with Cichlids almost a week. During this period the water should be free of food animals and be well filtered constantly, in order to prevent any damage being done to the embryos. It is only after all the yolk matter has been used up that the young fish will swim freely. Most of the organs are also fully developed in this period, and just before the period comes to an end the eyes are differentiated. As soon as the swimbladder is filled with liquid gas – usually via the intestine through the Ductus pneumaticus, a connecting tube between the swimbladder and the intestine – the juveniles start to swim around, exploring their surroundings. Up until that time, they will have lain at rest on rocks, on parts of plants or on the sides of the tank, usually hanging by a secreted thread that extends from a gland on the head. Embryos that are particularly light-sensitive will seek the protection of the substrate during the day, but at night or whenever sudden darkness descends on the tank they will shoot up to the surface 'as if on command'. But once light returns, they will sink back down to the bottom again. Fry should be protected from too strong a light source by switching off the light in the rearing aquarium from time to time. In some species of fish (for example, the Egg-laying Tooth-carps, the Sand-smelts, and above all, every species of livebearing fish) this last differential phase takes place whilst the young fish is still developing in the egg. In such cases, the fry swim freely immediately after they burst out of the egg-envelope or immediately after they are born (whereby they likewise leave the egg-envelope first). They start to catch food straight away. There are extreme cases, such as the Embiotocidae*, where the young fish are already 4.4 cm long at birth and sexually differentiated, the males already being capable of reproduction. The gestation period of such species is between 5 and 6 months.

There are a whole series of important factors to bear in mind even when the young fish have reached beyond the free-swimming stage. If these factors are not heeded (through ignorance or simply through not carrying them out properly), it is almost always inevitable that the entire brood will die. The juveniles must be fed, very carefully and frequently, with the most suitable live food (*see* Feeding of). What you feed them will depend on the species you are rearing. The amounts given to the young fish must be carefully measured so that all the food animals are used up in a few hours. This will prevent the build-up of rotting matter which particularly encourages the development of harmful single-celled animals. Besides this, in warm water, some kinds of larvae (eg the larvae of *Cyclops, see* Nauplius) will grow into adults in the shortest possible period and they might kill and eat the delicate, growing fishes. As for the size of the food, it is usually safe to say that it must be sieved carefully until it is no larger than the diameter of the young fish's eye. The young of many species of fish will only eat certain types of food in the first few days (eg *Chanda nana* will only eat *Cyclops* and *Diaptomus* nauplii). In fact, they will starve to death within a few days if offered the wrong food (even if it is offered in the correct quantities). Often it helps the developing youngsters if the tank is lit up at night too (during the first few days of life). Growth will be uninterrupted if they can feed constantly and it will also prevent harmful organisms from attacking the fish if they are constantly on the move. The stomachs of the fry should always be nicely rounded – the danger of damage through over-feeding is virtually impossible when food is offered constantly. Whenever possible, avoid rearing mixed groups of fish together. It is also best to avoid rearing fry of the same species together if there are different broods present at different stages of development. This is because only the oldest will develop most satisfactorily; the younger ones will seem troubled, and despite sufficient feeding they will not grow as quickly as they should and eventually they will die. It is even worse to try to rear fish of the same age but of different species together in the same tank. The species that grows quickest in nature will very often dominate the other, causing the 'foreign' species to grow insufficiently.

Water quality is also very important when rearing young fish. Even when the filtration system is working perfectly and is in constant action, thus giving water that looks beautifully clear, the young fishes' bodies still seem to suffer damage, in particular to the fins. This will nearly always happen unless the entire tankful of water (or at least half of it), is replaced regularly (every 4–6 days) with temperature-controlled fresh water. If the brood being reared is particularly large in numbers, it is a good idea to sort them out on several occasions into different sizes (this particularly applies to Labyrinth Fishes, Characins, Barbs and Cichlids) and rear them in separate tanks. If this sorting process is not carried out, the ones that grow least well will appear troubled and either they will die or they will be eaten by their larger brothers and sisters. Dramatic temperature fluctuations should be avoided, especially a sudden large drop in temperature that lasts a long time. If this happens, the whole brood may be in jeopardy and it often causes irreversible damage to the swimbladder. So, all in all, the successful rearing of juvenile fish rests on the shoulders of the aquarist, and he must be prepared to spend a lot of time and effort checking that the living conditions of his brood are constantly at their best, during every phase of their development. The rearing stage can normally be regarded as over once the fins are fully formed and the differentiation of the sexes complete.

Fucus; Wracks. A G of marine Brown Algae (Phaeophyta*) found primarily in the upper littoral* of cool seas. The thallus is ribbon-like, branched in one plane only and it attaches itself to firm substrates by means of suckers. *Fucus* has some economic importance in certain areas (as a source of iodine, soda). All species of *Fucus*, including the well-known Bladderwrack *(F. vesiculosis)*, are unsuited to the aquarium.

Fuelleborn's Cichlid *(Labeotropheus fuelleborni) see Labeotropheus*

Fundulinae, Sub-F, *see* Cyprinodontidae

Fundulosoma AHL, 1924. G of the Cyprinodontidae* comprising a few species found in the waters of the savannah lands of tropical W Africa. In appearance they look very like the G *Aphyosemion**, but their biology is more like that of the G *Nothobranchius**. The distribution of *Fundulosoma* also serves as a link between *Aphyosemion* (tropical forest regions of W Africa) and *Nothobranchius* (the savannahs of E Africa). *Fundulosoma* is only a few cm long, elongated, and only moderately compressed from side to side. The upper jaw, like the related Ga, is remarkably short, and the mouth is superior. Large eyes. The D and An are similar in shape and are positioned almost directly opposite one another; they are pointed in the males. The top and bottom of the C are drawn out into small peaks. All the fins so far mentioned are rounded in the females. Colouring is similar to that found in some of the *Aphyosemion* and *Roloffia** species, ie flanks and unpaired fins are covered in irregular red spots and the C has a lyre-shaped patterning in red.

Like other Ga of the Cyprinodontidae that live in the savannah regions, *Fundulosoma* has adapted its development cycle to suit drought conditions. When the water dries up, *Fundulosoma* survives in an embryonic stage, the thick, watertight egg-envelope guaranteeing effective protection against desiccation. Once the savannah waters fill up again with the coming of the rainy season, the young fish hatch out and grow into sexually mature fish within a few weeks. Should the water dry up again, it will mean death for the generation current at the time, but the continuance of the *Fundulosoma* population* is assured because eggs will have been laid. Like all annual African Cyprinodontidae, *Fundulosoma* is not adapted to a seasonal life style to the same degree as its South American relatives *(Pterolebias*, Rachovia*, Cynolebias*)*, whose eggs can only develop after a dry period lasting several months. For the hobbyist, however, it is enough just to keep *Fundulosoma* spawn for several months in flat dishes filled with water, then change their surroundings in the hope that the embryos will hatch out. For example, you could add some dried food or a little fresh water. Otherwise, *Fundulosoma* can be bred and cared for as the annual

species of *Aphyosemion, Nothobranchius* and *Roloffia.*
– *F. thierryi* AHL, 1924 (fig. 528). Ghana. Up to 3 cm. Males have irregular spotting on a metallic blue background. Unpaired fins yellowish, with red spots and either red-edged (C) or red tipped (D and An). The females are grey, with a sparse number of dark spots. This species used to be known incorrectly as *Nothobranchius walkeri* or *Aphyosemion walkeri* in aquarium-keeping. Compare it with the true *Aphyosemion walkeri*!

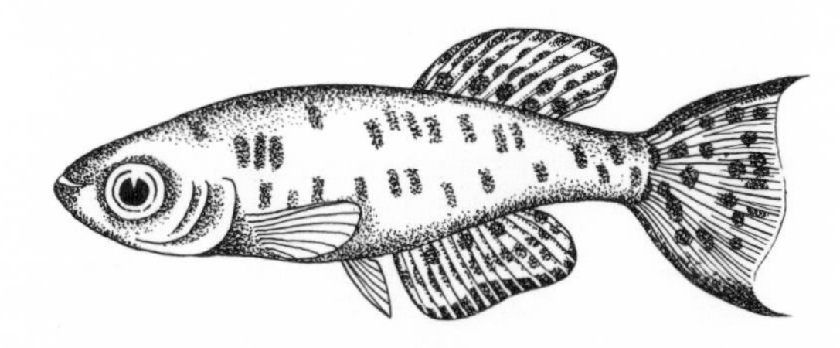

Fig. 528 *Fundulosoma thierryi*

Fundulus LACÉPÈDE, 1803. G of the Cyprinodontidae* comprising more than 25 species, distributed in the southern states of the USA primarily, but penetrating north into Canada, east as far as Bermuda and Cuba and south into Central America. Most members of the F live in fresh or brackish waters, but some species also enter sea water. Small to medium-sized fish that are elongated like a pike. *Fundulus* has unpaired fins that are rounded; D and An are similar in shape and lie opposite one another. The anal fin in the female houses a sexual sac which is a tumour-like thickening of the first anal fin ray. Many *Fundulus* species have narrow, dark transverse bands along their flanks and so look similar to the *Aphanius** species of the Old World – in fact, *Fundulus* is related to *Aphanius* in some respects. The generic term *Fundulus* points to bottom-dwelling fish, although by no means all species of *Fundulus* do live on or near the bottom. The G even contains some typical surface-dwelling fishes with broad, flattened heads and napes, and superior mouths. Many *Fundulus* species are not demanding as far as temperature, water chemistry and feeding are concerned, so even beginners can keep them and breed them. Many species do not require heated aquaria. They should be kept in large species tanks with thick clumps of fine-leaved plants (offering a spawning substrate and hiding-places) that offer sufficient room for swimming. In the warmer seasons, *Fundulus* may be kept in open-air ponds; indeed, some species will only breed there successfully. Depending on the place of origin, brackish water species require hard water or make sure the electrolyte content* is high enough by adding some sea salt (1 tablespoon to 10 l water). If you do take such measures, you may have problems with your plants, and you may have to use hardier plants like *Elodea** and certain species of *Potamogeton**. The reproductive method of *Fundulus* is quite primitive among the Cyprinodontidae; they are attached spawners. The love-play is quite violent, and the spawning period lasts several days. Since the parents prey on the eggs, it is best to transfer the spawning substrate (fine-leaved plants, Perlon-yarn), together with its attached eggs to shallow rearing tanks. The eggs take 5–15 days to mature, depending on species and temperature; the brood grows slowly, even when supplied with sufficient food.
– *F. catenatus* (STORER, 1846); Chained Top Minnow. Found in the area of the upper Mississippi. Up to 20 cm. Males have longitudinal rows of red spots on a greenish background. The females are yellow-brown with indistinct brown spotting. Suitable for unheated room aquaria and open-air ponds (fig. 529).
– *F. chrysotus* (HOLBROOK, 1860); Golden Ear. South-eastern states of the USA, found in fresh and brackish water. Up to 8 cm. Ground colour of body and fins greenish, with red spotting. The operculum has a shining blotch, and the eye is golden. This species requires warmth.
– *F. heteroclitus* (LINNAEUS, 1766); Local Killi, Bait Killi, Zebra Killi. Widespread in eastern parts of N America. Lives in fresh, brackish and sea water. Up to 12 cm. A brownish-green background colour with a great many narrow, tapering bands that are a blue iridescent colour.

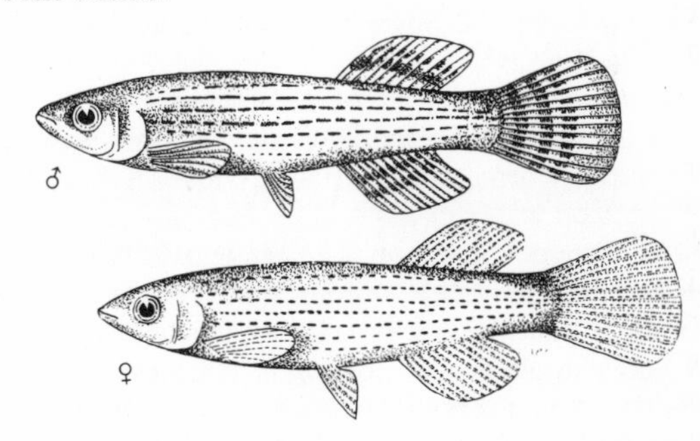

Fig. 529 *Fundulus catenatus*

– *F. notatus* (RAFINESQUE, 1820); Star-head Top Minnow. Found in south-eastern states of the USA. Up to 8 cm. Brownish with a deep-black longitudinal bar running from the snout to the insertion of the C. A particularly slender species (fig. 530).

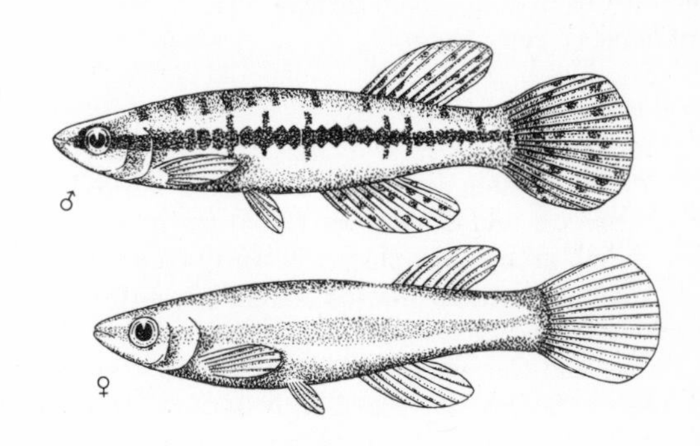

Fig. 530 *Fundulus notatus*

Fungal Attack *see* Mycoses
Fungi. Organism group of the Protobionta (Eukaryota*). Traditionally, the fungi are attributed to the plants, but because of their features they are really an independent form of organisation that should be placed

between plant and animal. They have cell nuclei and form cysts* during reproduction. Their vegetative body is rarely single-celled, usually thread-like. The threads are undivided or have cross walls. Chitin (the same substance that forms the exoskeleton of insects) is often a part of the wall structure. The highly developed forms develop tissue-like bands by the individual threads being compacted together (eg the fruiting bodies of edible fungi). Fungi have no chromatophores and no chlorophyll. They cannot photosynthesise, therefore, but have to resort to organic substances for their food. Many live on dead organic substances, breaking down dead wood, for example. Others, like many fungi of the forest floor, enter into a symbiotic relationship of mutual benefit with plants. A highly developed form of symbiosis is their co-habitation with certain algae. The vegetative bodies which result from the uniting of both components are called Lichen. Still other fungi live as parasites and are chiefly important as the stimulators of plant diseases. The fungi are primarily land-dwellers, although they do occur rarely in fresh water and even more rarely in the sea. Some have complicated alternation of generations*.

Some lower fungi only are of interest to the aquarist, and they belong to different divisions. They are single-celled or thread forms with no fruiting bodies. Some live as parasites on water plants and produce insignificant swellings. Others live on or in lower aquatic animals. Many are able to catch and suck out single-celled, mobile algae or rotifers by means of threads. Thread forms occur in water as a coating or pimples on dead plant parts, on animal bodies or on dead spawn. Parasitical fungi that attack living fish cause mycoses*.

Fungi Diseases *see* Mycoses

Fungia, G, *see* Madreporaria

Furcellaria. A G of the marine Red Algae (Rhodophyta*) which form clusters out of 2 mm thick, cylindrical branches. The black-brown coloured *F. fastigiata* (HUDS.) LAMOUR, found along the northern European coasts survives amazingly well in the aquarium and sometimes even manages to grow a little.

Fusiliers, G, *see Caesio*

Gadidae, F, *see* Gadiformes

Gadiformes; Cod-types. O of the Osteichthyes*, Coh Teleostei. The system described here is that worked out by GREENWOOD and colleagues (1966). In it, a whole series of fish groups are placed in the O Gadiformes, some of which might be ordered completely differently in other systems. For example, BERG put the Zoarcidae, Ophidiidae and Carapidae in the Perciformes, not the Gadiformes. In this he agreed with REGAN, whose O Anacanthini represents the Gadiformes in a narrower sense. The reasons given by GREENWOOD and colleagues (1966) for the wider-ranging O Gadiformes are based on highly specialised features which cannot be gone into here. In fact, the scientific appraisal of this issue is not yet concluded. So we will dispense with a general description of the O, and simply go on to a discussion of the important Fs.

– *Carapidae*; Pearlfishes (fig. 531). Also known as the Fierasferidae, this fish group is made up of small, slender Gadiformes which are elongated like an Eel. They have a fin corona of uniform height which stretches from the nape round the tip of the tail and to below the Ps on the ventral side. There are no Vs, and in a few cases no Ps either. The anus is placed well forward at the throat. There are no scales. The Carapidae are typical tail-burrowers, ie they retreat into the narrowest holes and cracks, tail-first, which they use like a feeler. Many have become specially adapted to living in the body cavities of other animals, such as Shellfishes, Tunicates and most of all Sea Cucumbers (Holothuria*).

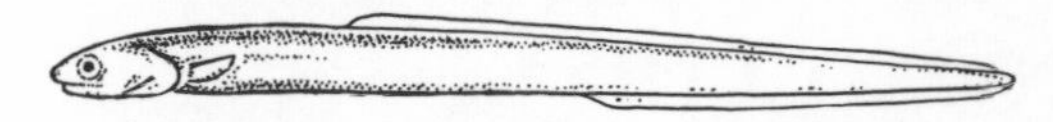

Fig. 531 Carapidae-type

Eg, *Fierasfer acus*, from the Mediterranean, lives inside the hind-gut of Sea Cucumbers belonging to the Ga *Stichopus* and *Holothuria*. It twists its way in tail-first through the anus. Young specimens enter the same way but head-first. The eggs are pelagic, and the larvae that hatch from them have a long protuberance along the back (vexillifer larvae). They seek out a Sea Cucumber, enter it and turn into another kind of larva known as tenuis. Tenuis larvae feed on the tissue of their host. Only after a second metamorphosis is the young form of the fish arrived at. The young fish leaves its host, and, head-first, seeks other hosts. It bores its way into the body and lives as a parasite in the body cavity feeding on the host's internal organs. The larger specimens only seem to supplement their food requirements from outside the host; they become, as it were, anal renters. Species that parasitise Shellfishes are sometimes encapsulated by the host and covered in mother-of-pearl.

– *Gadidae*; Cods (fig. 532). Usually spindle-shaped fish with 2 or 3 dorsal fins and 1–2 anal fins. Head conical, mouth terminal and large. The upper edge of the mouth is formed by premaxillaries. Often, there is a short barbel at the chin. None of the fins has spines. The Vs

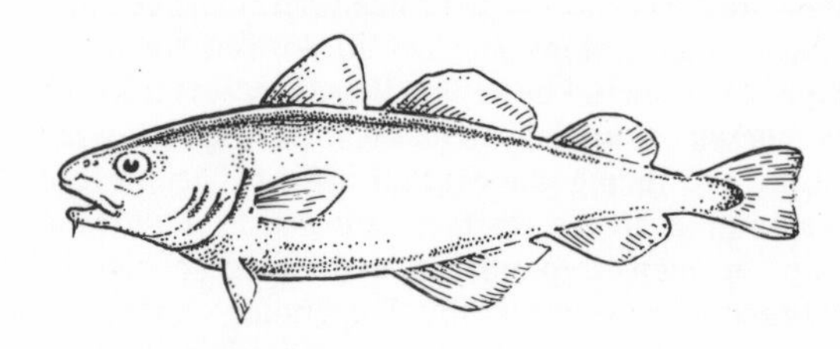

Fig. 532 Gadidae-type

are positioned below or in front of the Ps, which are relatively high up. The shoulder and pelvic girdles are joined by a ligament. The C is usually slightly indented and surrounds the tip of the tail (ie the fin extends around the caudal peduncle both top and bottom). Cods do not have any fish-bones. Their small round scales are often difficult to perceive. The olfactory lobes of the brain, placed well forward towards the olfactory organ itself, are a peculiarity of the Gadidae and other Fs belonging to the O. With one exception, Cods live in the sea. They are found mainly in temperate and cool parts, both near the sea bottom and in open waters. They eat practically any kind of marine animal they happen to encounter. There are about 150 modern species of Gadidae; the oldest fossils date back to the upper early Tertiary. Many Gadidae are very important commercially. They are caught mainly in otter trawls and seine nets. A fishmonger usually sells them fresh with the head removed, or smoked. The term 'stockfish' describes a cod that has had its head cut off and is then dried on wooden frames. A Cod's liver is rich in vitamins A and D. The largest and most important species is *Gadus morrhua* (simply called 'Cod') (fig. 533) which is found in the N Atlantic and neighbouring seas. It

Fig. 533 *Gadus morrhua*

grows up to 1.5 m long, but is usually smaller. The smaller individuals are also known as Cod–for example, the 60 cm long *Gadus morrhua* from the Baltic. It can weigh up to 3.5 kg. This species can be identified from its somewhat protruding upper jaw, its relatively long barbel at the chin, and the 3 dorsal and 2 anal fins. Geographical location and different habitats have produced a number of different forms, but the species can be distinguished from similar species by its bright LL. In the spring, at temperatures of 4–6 °C, glass-clear eggs are laid which swim about near the surface. They hatch out after about 3 weeks. Large females can produce 3–5 million eggs. The tiny fry feed on plankton and drift along with the ocean currents. They grow quickly. In the N Atlantic, roughly 1.5 million tonnes of Cod are caught each year. Another commercially useful fish is the Haddock *(Melanogrammus aeglefinus)* (fig. 534). It lives at depths of around 200 m and is found from Murmansk to Biscay. The Haddock grows up to 1 m long, and goes on extensive migrations, whilst remaining near the sea bottom. It spawns in late winter, mainly

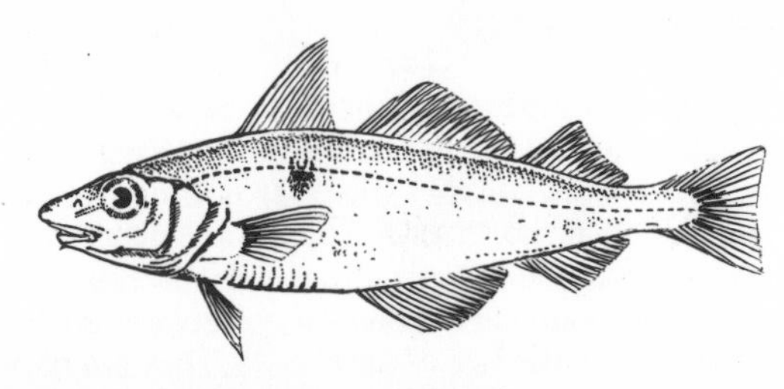

Fig. 534 *Melanogrammus aeglefinus*

just off Iceland and the Faeroe Islands. It is caught mainly in the Barents Sea and the North Sea as well as Iceland (about 250,000 tonnes each year). Several tonnes are also fished each year in the W Atlantic. Another important species on both sides of the Atlantic is the Coalfish or Saithe *(Pollachius virens)*. It is a pelagic species which feeds on Herrings mainly, and grows up to 1.2 m long. The Pollack *(Pollachius pollachius)* is very similar but comes from the eastern Atlantic. The lush carpets of seaweed of the E Atlantic, Mediterranean and Black Seas favour the Whiting *(Merlangus merlangus)*. This species reaches a maximum 50 cm in length, but usually remains smaller. Its flesh has the best taste of all the Cods. About 150,000 tonnes are caught every year. The Arctic or Polar Cod *(Boreogadus saida)* (fig. 535) is found circumpolar in the Arctic regions of both N Pacific and N Atlantic. It is rich in fat and is an important source of food for many Polar animals. A fairly untypical Cod is the Hake *(Merluccius merluccius)* (fig. 536). It grows up to 1 m long and is found in the eastern Atlantic. Because its features are not typical, it is often placed in its own F (Merlucciidae) along with its relatives.

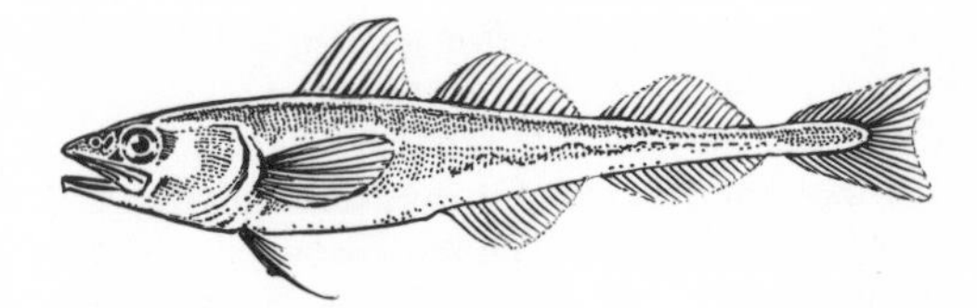

Fig. 535 *Boreogadus saida*

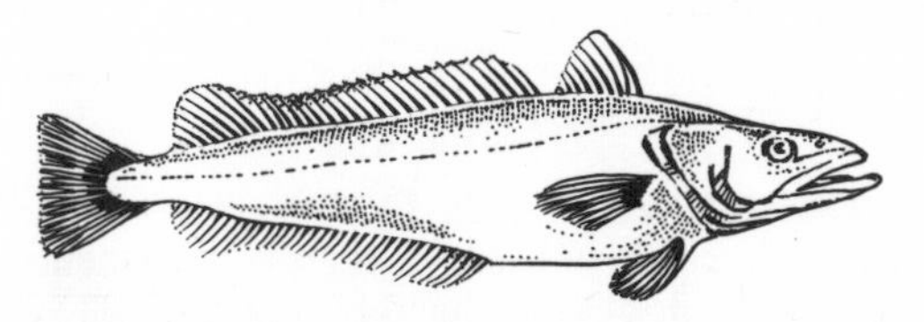

Fig. 536 *Merluccius merluccius*

Hakes have only 2 dorsal fins, D1 being short, and the long D2 being located opposite a long An. They are typical predators, hunting Mackerels and Herrings mainly. Hakes are caught in deep waters. Burbots are similarly shaped, except that they have a barbel at the chin and are even more elongate. Some Burbots have a uniform-height fin corona with the short D1 and longer D2 fusing together and extending to the rounded C (G *Brosmius*). The only freshwater species of Gadidae is also a kind of Burbot. *Lota lota* (fig. 537), called Burbot

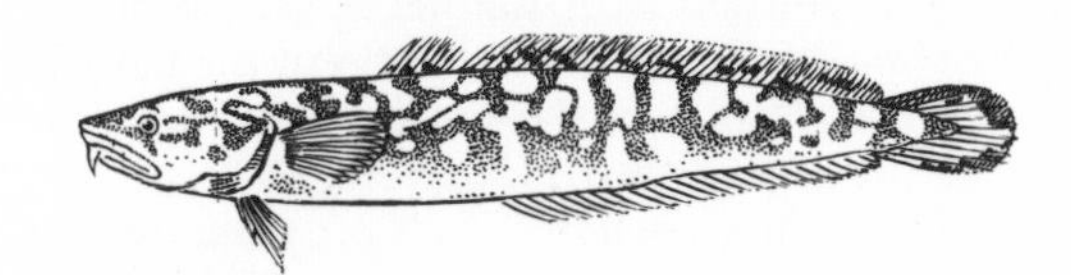

Fig. 537 *Lota lota*

for short, lives mainly in flowing waters of the Northern Hemisphere. In Europe it grows up to 50 cm long and is regarded as a good edible fish in many places. It looks a bit like a Catfish and is a well-known spawn-eater. Its breeding season is the winter months. The eggs are laid on a gravelly substrate.

– *Macrouridae*; Grenadiers. Also known as Coryphaenoididae, although this term is no longer in current usage. The Macrouridae are bottom-dwelling fish of the deep sea and occur in large numbers in some areas. They are food for many other predatory fish. Their appearance is quite unmistakable – very elongated body, tapering into a rat-like tail, and surrounded by a fin corona of uniform height. There is an additional, short in length, but tall D just behind the head. In some species, the first fin ray has developed into a serrated spine. The head itself, with its snout protruding like a nose (sometimes very long) and its inferior mouth, is reminiscent of the head found in the Chimaeriformes*, although it is obvious that they are not related. There is a strong barbel at the chin. The snout is used like a plough when searching for food in the soft mud. The scales are either cycloid or stenoid, and some Macrouridae release a luminous mucus from glands in the skin. Grenadierfishes are found worldwide, but they prefer cold-water regions. *Macrurus rupestris* (fig. 538) which can reach lengths of 1 m, is found in the northern parts of the E and W Atlantic.

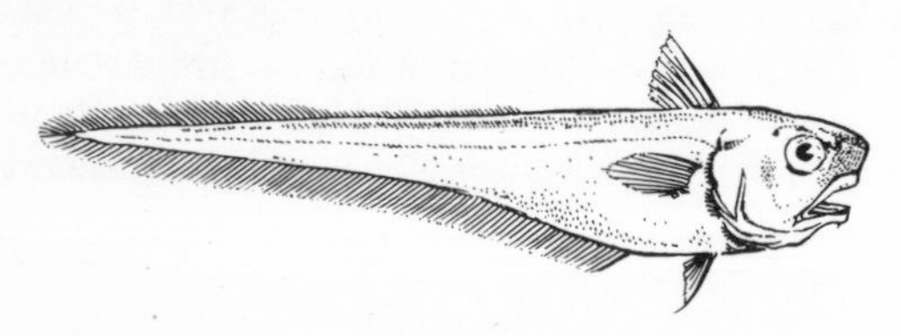

Fig. 538 *Macrurus rupestris*

– *Muraenolepididae*; Eel Cods (fig. 539). This is a small F which corresponds in many features to the Gadidae already described, despite the fact that they do not look very like each other, and that they would seem to have greater affinity with the Zoarcidae. Found only in

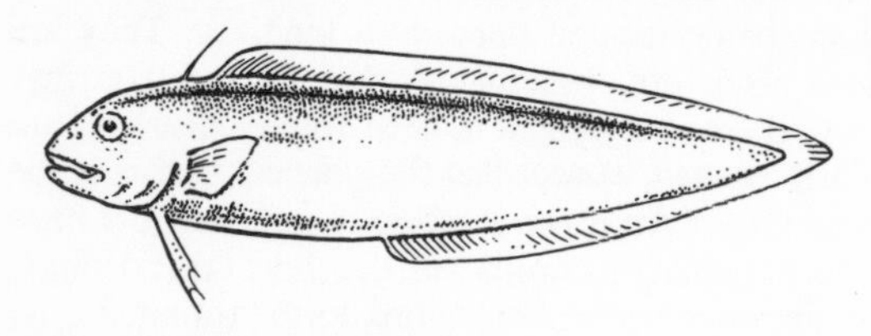

Fig. 539 Muraenolepididae-type

Antarctic and sub-Antarctic waters, the Muraenolepididae have a long fin corona of uniform height that begins near the nape, continues round the tip of the tail and along the ventral surface to the middle of the body. The corona is made up of the D, C and An. Only the first fin ray of the D has separated itself off from the rest of the corona; it looks like a bristle standing just behind the back of the head. The Vs are positioned in front of the Ps. Unlike many Gadidae, there is no barbel on the chin. Very little is known about the biology of the Muraenolepididae.

– *Ophidiidae*; Cusk-eels. Included in this F are the Botrulidae, which are often considered a family in themselves. Very elongated fish with a very long D which begins directly behind the head, continues down to the C, where it joins directly on to the long An. The D, C and An often unite to form a fin corona of uniform height. The Vs are positioned far in front of the Ps, often at the chin, in which case they resemble barbels. Sometimes they are absent. None of the fins has spines. The swimbladder is surrounded by 1–2 pairs of ribs that are spread out. The Ophidiidae live in the sea, usually near the bottom. There are several types of deep-sea Ophidiidae, and there are isolated cases of freshwater forms. The best known species is the Cusk-eel *Ophidium barbatum* (fig. 540). It lives in the Mediterranean and Black Seas and reaches lengths of 30 cm. It is an elongated species with a delicate reddish-silvery ground colour with whitish flecks. It has a fin corona of uniform height, edged in black. The Vs are located at the chin,

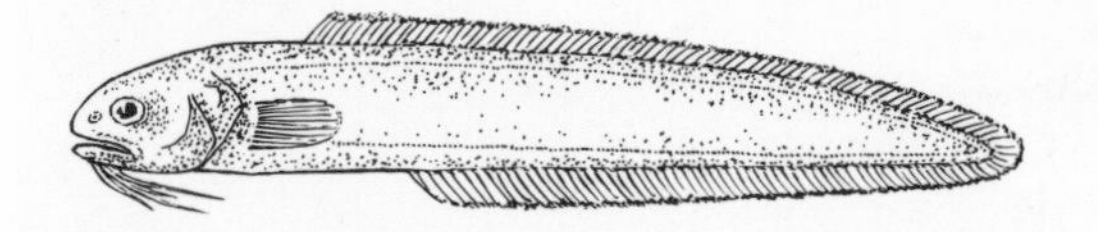

Fig. 540 *Ophidium barbatum*

each with 2 long fin rays. They sweep across the bottom like barbels, searching for food. The Kingklip *(Genypterus capensis)* is said to be an excellent edible fish. It grows up to 1.5 m long, and is found around SE Africa. It is the largest member of the F. EARL S. HARALD has reported that many Cusk-eels stand on their tails and bore with them (ie they rest vertically on their tails or twist themselves, tail-first, into cracks and fissures in the rocks). Many deep-sea varieties are livebearing. Some, such as *Grimaldichthys profundissimus*, have been caught at quite uncommon depths of 4,500–6,000 m. *Lucifuga subterranea* and *Stygicola dendatus* are both blind species, found in caves around Cuba. They measure between 2 and 12 cm long. They are glassy fish, also livebearing.

– *Zoarcidae*; Viviparous Blennies, Eel-pouts (fig. 541). The members of this family are very elongated fish, only moderately compressed from side to side. They have

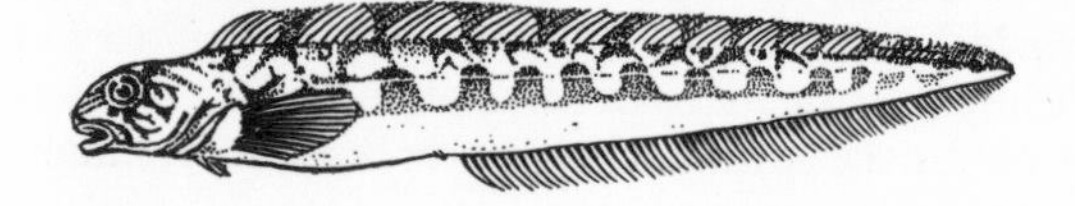

Fig. 541 Zoarcidae-type

a long fin corona made up of the D, C and An. The Vs are very small, often absent. The Eel-pout or Viviparous Blenny *(Zoarces viviparus)* (fig. 542) is found in the seaweed belt of the northern E Atlantic and neighbouring seas. It reaches lengths of 45 cm and is livebearing. The number of young in a brood varies between 20 and 400, depending on the size of the female. They are born in winter. It is a predator, eating mainly crustaceans and fish. It is sold in shops in two forms: fresh or smoked. Because the bones tend to turn green when the fish is cooked, many people will not eat it. Other species of this F lay eggs.

Fig. 542 *Zoarces viviparus*

Gadus morrhua *see* Gadiformes

Galathea, G, *see* Galatheidae

Galatheidae; Squat Lobsters. F of the Anomura (*see* Reptantia) with a segmented, symmetrical abdomen, which is held flexed beneath the body, but not against the breast. The Galatheidae have well-developed tail sections and long pincers. The 5th pair of walking legs have developed into small cleaning feet. Squat Lobsters live in the sea. They can move backwards very quickly by making beating movements with their abdomen. Active at night, Squat Lobsters feed on matter from the sea-bed. They pick the food up with their long, bristled third pair of jaw-feet. To date, these interesting animals have been rarely observed in the aquarium. *Galathea* FABRICIUS, 1793, and *Munida* LEACH, 1820, are two well-known Ga.

Galaxiads (Galaxidae, F) *see* Galaxioidei

Galaxiads, related species, Sub-O, *see* Galaxioidei

Galaxidae, F, *see* Galaxioidei

Galaxioidei; Galaxiads (fig. 543). Sub-O of the Salmoniformes*, Sup-O Protacanthopterygii. The most important F Galaxidae is found only in the southern hemisphere in S America, S Africa, Australia, Tasmania and New Zealand. They are small, scaleless fish which look like very slender Pikes. The D and An are positioned well to the rear and opposite each other. Adipose fin absent, and in some species there are no Vs either. *Neochanna apoda* (New Zealand Mudfish) can live through dry periods buried in mud. It grows to a length of 15 cm. A brackish water species with the same distribution area lays eggs on plants that can also survive a dry period.

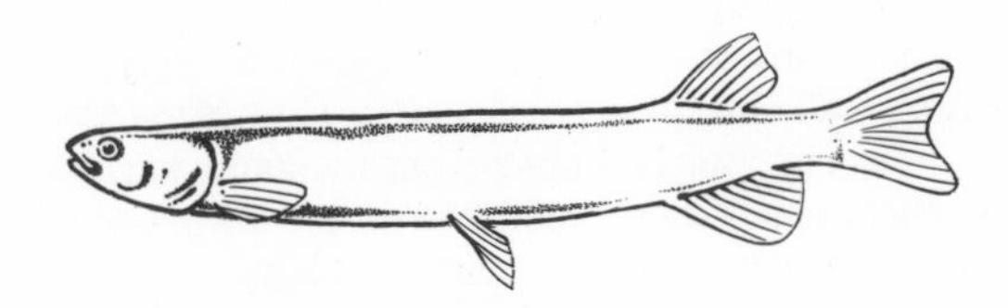

Fig. 543 Galaxioidei-type

Galba, G, *see Lymnaea*

Galeocerdo cuvieri *see* Carcharhinidae

Galeorhinus galeus *see* Triakidae

Galilee Cichlid *(Tilapia galilaea) see Tilapia*

Gall Bladder *see* Digestive Organs

Gambusia POEY, 1854; Mosquitofishes. G of the Poeciliidae*, Sub-F Poeciliinae. The richest G of the F, comprising 34 species. Found in south-eastern parts of the USA, the Atlantic sides of Mexico and Central America as far as Columbia, most of the islands of the Bahamas and Greater Antilles. They inhabit calm waters: clear streams, weedy areas near the banks of rivers and lakes, ditches and marshes, brackish water of lagoons and coastal waters. *Gambusia* are small, sometimes very small, fish. Males 2.5–5.5 cm, females 3–9 cm. The front part of the body is moderately compressed from side to side. Forehead profile horizontal. Lower jaw protrudes. Teeth are conical and sharp, arranged in a band. The females are more robust. Colouring, particularly in the females, is usually unremarkable. Upper part of the body is often brown to olive-green, paler on the belly. Both sexes usually have a dark eye-streak stretching from the forehead to the operculum. The flanks in many species have a dark length-wise band and dots or flecks. Scales often dark-edged. Fins generally colourless or very pale. D and C often have rows of dark dots. *Gambusia* can be intolerant, and they are best kept in well-planted tanks. If they are a brackish-water species, add 5% salt. Temperature 20–22 °C, southern species 24–28 °C. Most are omnivores. Live food, particularly fly larvae, supplemented with algae, plus dried and artificial foods. For rearing purposes, the pregnant females are best kept alone in well-planted tanks, as they tend to chase after the youngsters. A brood numbers 10–30 usually, but over 80 in the case of *G. affinis*. For the aquarist, only some members of the G have any importance. They are well-known, nevertheless, for their value as biological pest controllers. *G. affinis holbrooki* was introduced successfully in malarial areas to destroy the malarial mosquito larvae. The following are regularly kept in aquaria.

– *G. affinis* (BAIRD and GIRARD, 1853); Silver Gambusia, Spotted Gambusia, Western Mosquito. There exist two very similar subspecies: *G. affinis* (BAIRD and GIRARD, 1853), also known by the names given above. Found only in Texas. Male up to 4 cm, female up to 6.5 cm. D has 6–7 rays. The other species is *G. affinis*

Fig. 544 *Gambusia affinis*

holbrooki GIRARD, 1859, also known as the Eastern Mosquito. Found in SE parts of the USA, and northern Mexico. Has been introduced worldwide into warmer areas since it destroys mosquito larvae. Male up to 3.5 cm, female up to 8 cm. D has 8–9 rays. Males are covered in black flecks; the male of *G. affinis holbrooki* can also be completely black or yellow. D and C have rows of dark spots. Best kept at a temperature of 18–20 °C, and suitable for open-air ponds. Will over-winter at temperatures of 10–12 °C (fig. 544).
– *G. dominicensis* REGAN, 1913; Domingo Gambusia. Haiti, Cuba. Male up to 3.5 cm, female up to 6 cm. They have dark spotting on the upper flanks, towards the rear. Male has a dark length-wise band. D and C reddish with spotting. 24–28 °C.
– *G. nicaraguensis* GÜNTHER, 1866; Nicaraguan Gambusia. S Mexico to Panama. Also found in brackish water. Male up to 3 cm, female up to 6.5 cm. Grey-greenish, hardly any pigmentation. Upper half of body, plus D and C, have rows of dots. The female displays a pregnancy blotch which is orange-red in colour. 24–28 °C.
– *G. puncticulata* POEY, 1854; Blue Gambusia Blue Mosquitofish. Cuba. Male up to 4.5 cm, female up to 7 cm. Blue-grey with a pale blue iridescence across the body and fins. Dorsal surface light brown. Body, D and C have rows of red-brown dots. The D and C in the male have red edges. Female paler. Undemanding species, relatively tolerant. Easy to breed.
– *G. rachowi* (REGAN, 1914); Rachow's Gambusia. S Mexico. Male up to 3.5 cm, female up to 5 cm. Greenish-grey, with a silvery iridescence. There is a dark length-wise band on the body that ends in a fleck in front of the C.
– *G. wrayi* REGAN, 1913; Wray's Gambusia. Jamaica. Male up to 4.5 cm, female up to 6 cm. Greenish-grey, with a bluish iridescence. Flanks, and the D and C, have rows of dark spots. Fins yellowish.
– *G. yucatana* REGAN, 1914; Yucatan Gambusia. Found in Mexico, the Yucatan Peninsula, to Panama. Male up to 5.5 cm, female up to 8 cm. Grey-green iridescence. D and C, and the body (apart from the breast and belly region) have rows of dark spots. Prominent eye-streak. 22–28 °C.

Gametangium *see* Angium

Gametes; Sexual Cells; Germ Cells (fig. 545). Gametes are the basis of reproduction* and heredity*. In lower plants and single-celled animals, the gametes can move by means of flagella and have the same size and shape. The smaller mobile male gamete is called the spermatozoid: the larger non-mobile female gamete is called the egg-cell. Gametes unite to form zygotes.

In multi-celled plants and animals, besides structural details, and in comparison with the body cells, the gametes are characterised by having half the number of chromosomes in the nucleus (*see* Cell, Cell Fission). During fertilisation in invertebrate and vertebrate animals, the chromosomes fuse (ie the hereditary material of a male sperm unites with that of a female egg-cell). At this point, embryonic development* begins. *Sperm* are microscopically small. They are made up of a head section, which contains the nucleus, a mid-section and a flagellum-like tail section. In most fish, the sperm becomes mobile only as it is deposited in the water. There, it stays alive for only a few seconds or minutes. In that time, the sperm must reach the *egg-cells*. The sperm's head and middle sections pierce the *egg-envelope* via a small opening called the *microphyle*. Some eggs have several microphyles. Fish eggs *(spawn)*, besides the primary egg membrane, also have a secondary egg-envelope. The eggs of Sharks and Rays which are fertilised before release have a further horny *egg-capsule*. Apart from the livebearing species, fish egg-cells are equipped with a large *yolk-supply* which is used as a food source as the fertilised egg-cell develops (*see* Yolk Sac). The yolk is concentrated at one pole (vegetative pole) whilst the cell nucleus and cytoplasm lie at the opposite pole (animal pole). Fish eggs can vary in size from fractions of a mm up to several cm. The egg capsules of many Sharks may reach lengths of over 20 cm. Usually, eggs sink to the bottom in water (the *demersal eggs* of most freshwater fish and some marine

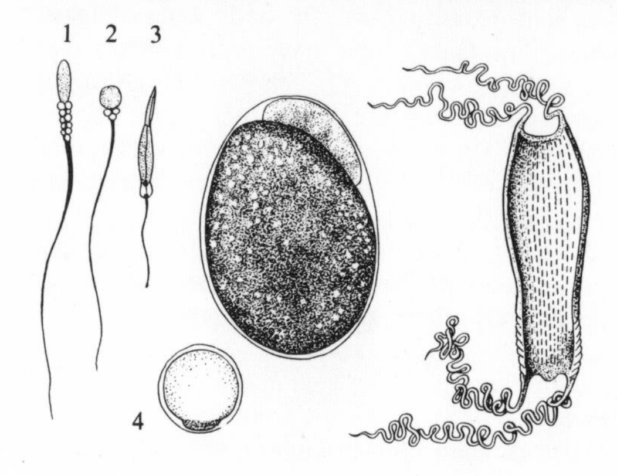

Fig. 545 Gametes. Sperm from 1 Guppy 2 Herring 3 Eel. 4 Pelagic egg of the Flounder (germinal disc at the bottom). Above, middle: Cichlid egg (germinal disc at the top). Above, right: Egg-capsule of *Scyliorhinus* (not drawn to scale).

fish, eg Herrings). In many marine species, eggs can be made to float at certain depths or climb to the surface by means of drops of oil (*pelagic eggs*). Usually, the eggs are small and often transparent. They often have some means of attachment—perhaps a sticky surface, sticky threads or other thread-like outgrowths – which enable the eggs to remain attached to plants, rocks etc. In this way, the eggs are able to group together into clumps, or they can be fixed to a substrate. The number of eggs laid is greatest (up to several million) among marine species because they have a small chance of survival, and relatively small among species that practise parental care.

Gametocyst *see* Cyst

Gametophyte (sexual generation) *see* Alternation of Generations

Gammarus FABRICIUS, 1775 (fig. 546). G of the Amphipods (Amphipoda*). The best-known species is *G. pulex* (Fresh-water Shrimp). It is 1–2 cm in length. The way in which the amphipods are systematised currently has meant that the *Gammarus* genus has been divided up differently, but it is of little consequence to the aquarist.

Most species of *Gammarus* choose to live in the sea, but inland waters* have been colonised by several different forms. Amphipods often occur in very large numbers near the bank in plant-rich areas (*see* Littoral). That is why they are an important source of food for fishes, water birds etc. Increasingly, Amphipods are used for food by aquarists – either as live food, dried or frozen.

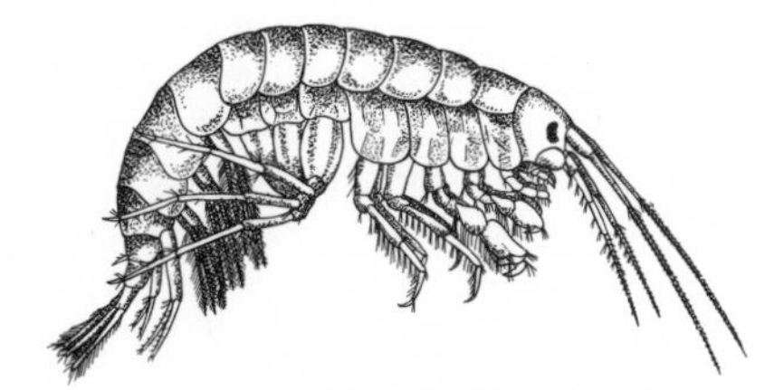

Fig. 546 *Gammarus* sp.

At the same time, keeping Amphipods for food purposes also enables the aquarist to study their way of life (at least, those forms that live in still waters* and are kept in pond-aquaria). In particular, look at the varied ways in which Amphipods move (swimming, climbing, crawling about on their sides), and their mating patterns.

Ganges Shark *(Carcharhinus gangeticus) see* Carcharhinidae

Ganoid Scales *see* Scales

Ganoin. A hard, enamel-like substance which forms the top layer of ganoid scales. Phylogenetically, these are a very old form of scales*; but they were very common among the ancestors of today's bony fish. Among modern-day fish, you will only find ganoid scales in relict species ('living fossils'), eg the Garpikes (*see* Lepisosteiformes) and the polypterids (Polypteriformes*).

Garden Pond. An artificially created, shallow, still stretch of water, usually with an inflow and an outflow; it can also be drained. The living conditions in a garden pond largely correspond to those found in a natural pond* – in the course of time, analogous animal and plant communities settle there (so long as man does not take steps to prevent this happening, such as letting the pond stay dry at times, removing mud, overstocking with fish).

Garfish *(Belone belone) see* Exocoetoidei

Garfishes (Belonidae, F) *see* Exocoetoidei

Garmanella HUBBS, 1936. G of the Cyprinodontidae*. One species, found in S Mexico (Yucatan Peninsula). They are very tall fish, that grow up to 4 cm long. Their shape is very like that of *Cyprinodon** and *Jordanella**, to both of which *Garmanella* is closely related. The D is considerably longer than the An, and is particularly well-developed in the male. The C is broad and shaped like a fan. *Garmanella* lives in brackish waters, near the coasts. So, when keeping specimens in the aquarium, you should add some sea salt (5‰, ie 5 g to every litre of water). Older specimens can also be kept in hard fresh water, but they are more delicate as a result and the spawn has a much greater tendency to go rotten. Temperature 20–23 °C; brief fluctuations by a few degrees will not do any great harm. Sometimes, *Garmanella* can even be kept in unheated aquaria, for short spells. They should be fed on live food and artificial foods (all varieties), with some vegetable matter besides (in the form of algae). *Garmanella* will reproduce in species aquaria without any special preparations being necessary. They are attached spawners (compare Cyprinodontidae), although there does seem to be some tendency towards laying the spawn in a substrate. Since the males drive rather vigorously, it is best to keep the females in the majority and to provide sufficient hiding-places among the vegetation. The plants also provide somewhere for the spawn to be laid, although perlon yarn can also be used. The male sets about his courtship by circling round his chosen partner and touching her with his fins. Vigorous driving follows this phase, during which time the male continually tries to nestle alongside the female. Finally, both partners form an S-shaped curve near the bottom and deposit their sexual products. The spawn is only slightly sticky, and many eggs will sink to the bottom. As the parents will eat the spawn, it is best to rear the youngsters separately. Flat dishes, with a lot of algae, are best suited for this. Use a pipette to transfer the eggs, or shake them free of the substrate. The embryos look white, so eggs that are developing normally are difficult to distinguish from fungal growths. At 24 °C, the eggs hatch after 10–12 days; the youngsters continue to feed from the yolk supply, however, for several days afterwards. After that time, feed them on *Artemia* nauplii and *Turbatrix**. Suitable food will develop of its own accord in the algae dishes (especially particular kinds of protozoans [Protozoa*]). Even the young sometimes live as vegetarians.

– *G. pulchra* HUBBS, 1936. Yucatan, in areas of brackish water near the coasts. Up to 4 cm. Ground colour of the males orange-red, silver-grey in the females. Both sexes have an irregular pattern of dark transverse stripes and reticulated markings. There is a large black blotch in the middle of the body. A black band, bent forwards, runs through the eyes.

Garnet Minnow *(Hemigrammocypris lini) see Hemigrammocypris*

Garnet Tetra *(Hemigrammus pulcher) see Hemigrammus*

Garrinae, Sub-F, *see* Cyprinidae

Gars or Garpike-types, O, *see* Lepisosteiformes

Gars or Garpikes (Lepisosteidae, F) *see* Lepisosteiformes

Gas Bubble Disease. At a given temperature and pressure, a definite amount of gases will dissolve in water. At high pressure, more will be dissolved than at lower pressure, and conversely less gas is dissolved at higher water temperatures than at lower ones. If the pressure suddenly drops, or there is a rapid increase in temperature, part of the dissolved gases will be released in the form of little bubbles. If you open a bottle of soda water, for example, carbon dioxide, which is dissolved under high pressure, will escape as bubbles. Similarly, aquatic animals have gases dissolved in their blood and tissue fluids. The amount of these gases is dependent on the surrounding water. In the aquarium, the dissolved gas content in the specimens can be reduced rapidly; this can be caused by a number of different factors, such as a great quantity of oxygen being used up suddenly, a rapid temperature rise, the introduction of water with a low level of gas saturation etc. Under such conditions, the gas pressure in the water falls much more quickly than the pressure inside the tissue fluids and blood of the aquarium specimens. The pressure inside the organisms is therefore greater than that of the surrounding water. This is compensated for by the gases in the tissue fluids giving off bubbles, and it is this which brings about gas bubble disease inside the organism. In fishes, the bubbles of gas are particularly visible beneath the skin and around the LL and the eyes. Fine blood vessels can become blocked by gas bubbles and the organisms will die (this is called gas embolism). Gas bubble disease frequently occurs when fish are transferred to fresh tap water. Tap water has a lowered pressure and it is also usually heated up, both of which cause bubbles of gas to be released in any case. Fish that are introduced to it too early on will suffer from gas bubble disease. Any specimens with the disease must be transferred to normal water immediately. Excess gases can be expelled by vigorously aerating the water and increasing the water movement. This will prevent the aquarium water from becoming oversaturated with gases.

Gaseous Exchange *see* Respiration

Gasteropelecidae; Hatchetfishes (fig. 547). F of the Cypriniformes*, Sub-O Characoidei*. There are about 9 modern species of this F, all of which are characterised by an almost straight profile along the head and dorsal surface, and a pronounced curvature of the throat and ventral surface (like a steep keel on a boat). This deep body is largely composed of enlarged pectoral fin musculature and the accompanying muscle attachments of the bony shoulder girdle. The hypocoracoid, in particular, is especially long and broad. The Gasteropelecidae are flying fish. Unlike all other flying fish, which glide over the water, the Gasteropelecidae can fly properly; ie they can beat their pectoral fins rapidly.

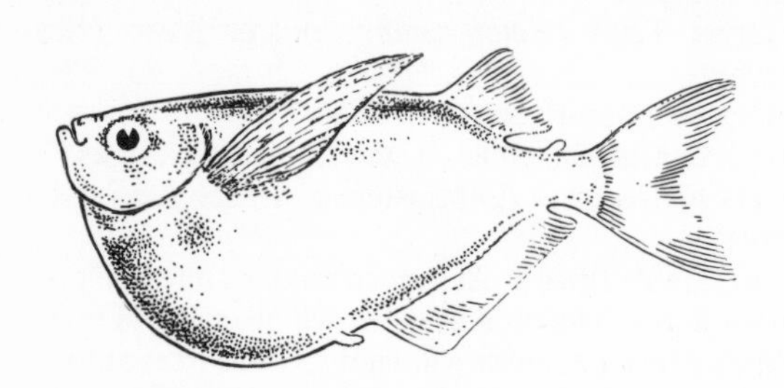

Fig. 547 Gasteropelecidae-type

The fin beat produces a characteristic noise – a buzz followed by a slapping on the water. Some species carve a furrow in the water surface with their belly, others fly along completely free of the water. Flights do not last long, the distance covered being at most a few metres. They are able to fly because the pectoral fin musculature is so powerfully developed (similar to the Birds). Flight is a means of escape for these typical surface-dwelling fish. Other morphological features include an upward-pointing mouth slit and a LL that extends in an arch-shape from the operculum downwards, ending directly before the front edge of the anal fin (or it may be reduced to 0–3 scale lengths). In the G *Carnegiella*, the small adipose fin is absent. The Vs are very small, standing just in front of the long An. The Gasteropelecidae are shoaling fish that either stand among the plants making rocking movements or undulate gently towards them. Their diet consists almost entirely of food that falls onto the water surface. Length a maximum 10 cm. Distribution area stretches from the River Plate to Central America. They are found mainly in the plant zone near the banks of calm flowing waters. They often live alongside other small Characoidei. All the species are interesting aquarium fish, but success with breeding has been scant. *See Carnegiella, Gasteropelecus, Thoracocharax.*

Gasteropelecinae, Sub-F, *see* Gasteropelecidae

Gasteropelecus Scopoli, 1777. G of the Gasteropelecidae*. These fish come from the lower and middle Amazon Basin and Guyana. Length 6–9 cm. They live in shoals beneath the water surface. They are good jumpers and are only fully active after twilight. For information on body shape and fins *see* Gasteropelecidae. Adipose fin present. Long LL that reaches down to where the anal fin is inserted. In contrast to the G *Carnegiella**, which has 5 undivided fin rays in the Vs, the G *Gasteropelecus* has just one undivided ray and four divided ones. They also have a bony sensory canal in the pteroticum which is absent in the G *Carnegiella*. *Gasteropelecus* should be kept in long tanks with plenty of room for swimming near the water surface. With good care and attention, and plenty of food, these specimens will survive many years and will rarely suffer from illnesses. Slightly acidic, soft water suits them best. They are greedy eaters that like to grab food that happens to land on the surface – eg fruit-flies, very small cockroaches, small flies. They will also feed on *Daphnia*, *Cyclops* and fly larvae if they sink into the water. Viewed from the front, the flanks of the females bulge outwards, which is one way of telling them apart from the smaller males. *G. sternicla* has already been bred successfully. According to W. Rönitz, the eggs were hanging from *Myriophyllum* tendrils in the morning, or they lay on the bottom of a metre-long tank that had a large shoal of the species in it. Spawning was not observed. It is possible that they lay the spawn at twilight, or in the moonlight. More than 200 fry hatched and they were relatively easy to rear on rotifers and *Cyclops* nauplii, changing later on to full-grown *Cyclops*. These are harmless and charming fish, and a a small shoal will enhance any community aquarium, but unfortunately the demand cannot be met from the meagre imports.

– *G. maculatus* Steindachner, 1879; Spotted Hatchetfish. West Columbia to Panama. Up to 9 cm. Flanks grey-green with a strong silvery iridescence. Brownish flecks arranged in oblique bands. A black band runs along the body; above it is another band which has a metallic silver sheen, and above that is a third band made up of speckles. The fins are colourless except for the dark edge of the D. Unfortunately, this lovely species has not been imported alive since 1945.

– *G. sternicla* (Linnaeus, 1758); Common Hatchetfish. Central Amazon, Guyana. Up to 6.5 cm. Flanks yellowish to delicate grey with a strong silver iridescence. A black longitudinal band runs from the rear edge of the operculum to the insertion of the C. It is bounded on either side by a yellow band. Fins colourless. This species is very hardy and long-lived (fig. 548).

Fig. 548 *Gasteropelecus sternicla*

Gasterosteidae, F, *see* Gasterosteoidei

Gasterosteiformes; Stickleback types. O of the Osteichthyes*, Coh Teleostei. Small fish, primarily marine, they are characterised by having dermal bones that are either separate or united into a body armour. GREENWOOD and colleagues (1966) place relatives of the Sticklebacks, Trumpetfishes and Pipefishes (ie Sub-Os Gasterosteoidei*, Aulostomoidei* and Syngnathoidei) together in the O. Other people regard these groups of fish as Os in their own right or as Sub-Os with different names (eg Thoracostei, Aulostomi, Lophobranchii, Solenichthyes, Syngnathiformes). Apart from the bony plates already mentioned, members of the O can be recognised by various skeletal features, the tube-like mouth in many cases, the fin rays and arrangement of the fins, the sealed swimbladder and, in some cases, peculiarities of reproduction. For further details turn to the relevant Sub-Os.

Gasterosteoidei; Sticklebacks, related species. A Sub-O of the Gasterosteiformes*. Sup-O Acanthopterygii. To begin with, REGAN regarded this group of fish as an O in itself (Thoracostei), but later he included it as a Sub-O in the O Scleroparei (Scorpaeniformes*). BERG classified the Gasterosteoidei as the O Gasterosteiformes. They are small or very small fish, shaped like a spindle or needle. The snout is pointed and the caudal peduncle is long and thin. The mouth is small, the upper part comprising pre-maxillaries only. Teeth small but mostly well developed. The skin has bony plates—either in large numbers or in small numbers. In front of the soft-rayed part of the D there are either a number of small spines that can be erected, or there are a few larger ones. These spines are separate from one another. The Vs are located at the breast, and often have a single, long spine that is very strong. C relatively small. The Gasterosteoidei are found in inland and coastal waters of northern temperate regions, as well as in the sea. Some species can tolerate high salt concentrations. The Indostomidae are somewhat of an oddity. They were discovered in Lake Indawgyi in Upper Burma. They grow up to 3 cm long and have bony plates covering the

entire body. The oldest fossils date back to the middle Tertiary. Many Gasterosteoidei are the best-known fish of their particular distribution area. The Sub-O is not important commercially.

– *Aulorhyhchidae*; Tubenoses or Tubesnouts. This F comprises one species—the 16 cm long *Aulorhynchus flavidus* (fig. 549). It occurs in the eastern Pacific from Alaska to Florida and lives in enormous shoals. It has

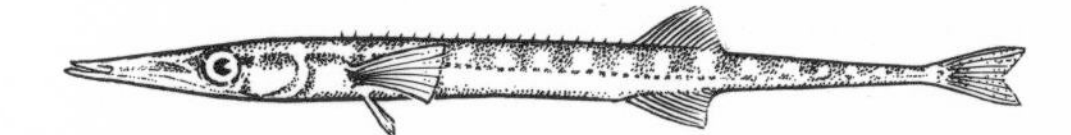

Fig. 549 *Aulorhynchus flavidus*

a very elongated shape. There are 25 spines in front of the small, soft-rayed D which is situated opposite the An. It is a timid species that prefers depths of 10–15 m. The orange-coloured eggs are laid amongst seaweed. The male builds a nest out of plant parts cemented together with fine mucous threads from the male's urogenital region, then he guards the eggs. Because the fish occurs in such vast numbers, it forms an important food for larger marine fish, and so is indirectly important from the commercial point of view.

– *Gasterosteidae*; Sticklebacks (fig. 550). The members of this F can be distinguished from the Aulorhynchidae described above, chiefly because the ribs are not joined

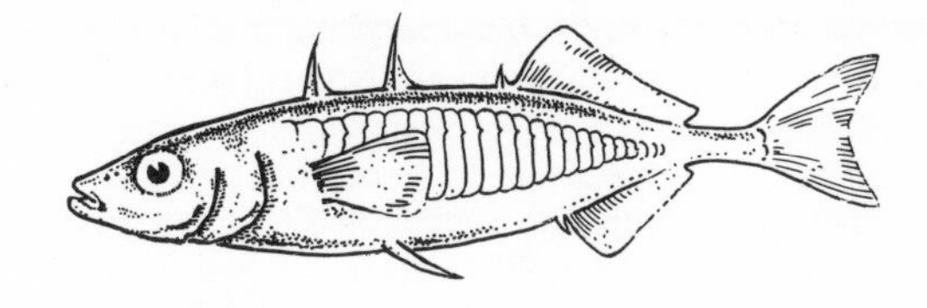

Fig. 550 Gasterosteidae-type

with the bony plates on the flanks. There are between 3 and 16 separate spines in front of the soft-rayed D. The Vs have one large spine which can be locked in an upright position, and 0–3 soft rays. Found in inland and coastal waters of northern temperate regions. 12 known species, several of which are regularly kept in aquaria. They exhibit an interesting courtship and parental care behaviour, and have often been the subject of scientific study. For further details turn to the relevant Ga *Gasterosteus**, *Pungitius**, *Spinachia**.

Gasterosteus LINNAEUS, 1758. G of the Gasterosteidae*, Sub-O Gasterosteoidei*. Found in fresh and sea water in the northern hemisphere. *Gasterosteus* eats a wide variety of small animals, and will also raid spawn and newly hatched fish. The body is shaped like a spindle, compressed from side to side, and has a stout body armour made up of lateral bony plates. The caudal peduncle is narrow and there are 3 (2–5) spines in front of the D. Sticklebacks often live in shoals and adult males are often territorial at spawning time. *G. aculeatus* has a very wide distribution area. Within this one species, 3 morphological variants are distinguished (on account of variations in the bony plates): *G. aculeatus* *trachurus* has 29–35 lateral plates which form a continuous armour stretching from the operculum to the base of the tail fin; *G. aculeatus leiurus* has only 4–10 lateral plates in the breast region; *G. aculeatus semiarmatus* is an in-between form with plates at the breast region and on the caudal peduncle. Pure *trachurus* populations live, for example, along the northern European Atlantic coasts as well as in the sea and in the fresh waters of E Europe. Pure *leiurus* populations are found in mid-European inland waters, roughly from the Elbe westwards. Mixed populations containing all 3 variants occur in the regions in between–*semiarmatus* proved to be a hybrid of the two other forms. In the ecological sense, only 2 types are distinguished: a stationary form that lives in inland waters and a marine form that is migratory, journeying back into fresh water to spawn. The reproductive behaviour of this species has been the subject of extensive examination, and can easily be observed in the aquarium. The behavioural scientist TINBERGEN has conducted many such observations. The males occupy territories in the spring, which they defend against all comers. They react particularly violently towards other males who also exhibit courtship colours. It was TINBERGEN who carried out what are now regarded as classic experiments using a dummy fish, its ventral surface being painted red, which released aggressive behaviour on the part of the Stickleback. An unusual form of threatening behaviour can be observed at the territorial boundary: the male positions himself vertically, head-down, and begins burrowing ito the bottom. This mode of behaviour is true displacement activity* and can be traced back to the conflict between the inclination towards attack and flight. It has come to have the significance of a ritualised threat gesture. The nest is built by the male on the bottom. It is made out of parts of plants that are stuck together by a secretion from the kidneys. Females that are ready to spawn react to the males' display, the so-called zigzag dance, by raising the front part of their bodies in a characteristic way. The male leads her to the nest where he lies on his side, his snout pointing towards the nest entrance. The female penetrates inside the elastic-like nest. Before she will lay her eggs, the male must touch the female's abdomen with his snout. After that, the female leaves the nest via the other side, the male takes her place and fertilises the eggs. One male will normally mate with several (up to 9) females, shooing them away from the nest after they have spawned. The spawning behaviour of the Stickleback is a classic case of a series of actions, one mode of behaviour precipitating the next, and so on; it breaks down if one link in the chain is left out. The male guards the spawn and diligently fans fresh water towards it. The young hatch out after 8–12 days, but the male watches over them for a few more days. Sticklebacks make extremely attractive specimens in the aquarium and they are to be highly recommended. You can switch them between fresh water and sea water, just as you like.

– *G. aculeatus* LINNAEUS, 1758; Three-spined Stickleback. Europe, N Asia, N America, Algeria. Found in fresh water and in the sea. Up to 9 cm. Females, juveniles and the males outside the breeding season have a black patterning with a silver belly. When ready to spawn, and during the period that they care for the young, the males have a green iridescence along the dorsal surface and in the eyes; they also have a red belly.

Gastropoda; Snails (fig. 551). Cl of the Molluscs (Mollusca*), with a great many species and different forms. There are marine Snails as well as freshwater and land-dwelling forms. A typical feature of most species is the coiled calcareous shell; in a very few species the shell is bowl-shaped or absent altogether. The Class

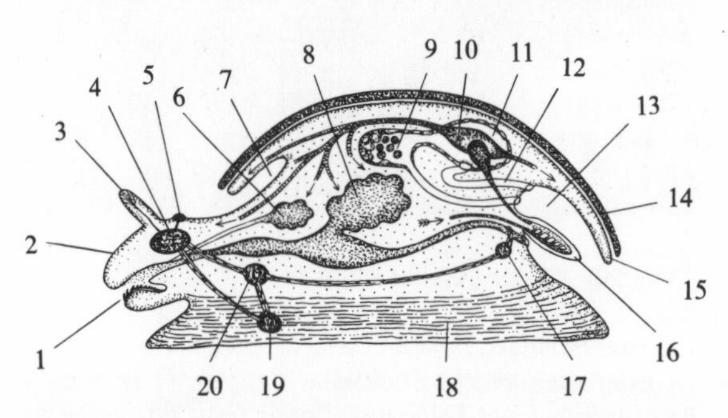

Fig. 551 Anatomy of a Snail. 1 Radula 2 Head 3 Antenna 4 Cephalic ganglion 5 Eye 6 Salivary gland 7 Mantle cavity 8 Mid-gut gland 9 Genital gland 10 Heart 11 Pericardium 12 Excretory organ 13 Gill cavity 14 Shell 15 Mantle 16 Gill 17 Visceral ganglion 18 Foot 19 Foot ganglion 20 Pectoral ganglion

Gastropoda is sub-divided into the Prosobranchia* and the Euthyneura*. The former live mainly in shallow waters of the sea (Littoral*) and often have bizarrely shaped shells that are also oddly coloured. Two freshwater Ga belonging to this group (*Ampullaria** and *Viviparus**) are of particular interest to the aquarist. Many Prosobranchia can be recognised by the 'lid' that is placed towards the rear of the foot that shuts off the shell. The majority of the Euthyneura are land-Snails, but they also include the Sub-Os Nudibranchia* (Sea Slugs)–the naked Snails of the sea which have their gills hanging freely in the water–and the Basommatophora* which contains some very interesting Ga for the aquarist *(Helisoma**, *Planorbis**, *Planorbarius** and *Physa*).*

Gaterin, name not in current usage, *see* Plectorhynchus

Gecarcinidae, F, *see* Brachyura

Geisha Girl Medaka *(Oryzias latipes) see Oryzias*

Gem Anemone *(Bunodactis verrucosa) see Bunodactis*

Gempylidae, F, *see* Scombroidei

Gene *see* Heredity

Genetics; Theory of Heredity. Branch of biology that is concerned with the structural elements and mechanics of how hereditary material (genes) is passed on. It also involves the study of the causes and possibilities of altering that material by natural or artificial means. Traditionally, genetics has been studied by the observation of the natural process of heredity and by crossbreeding experiments, but modern genetics has been making great strides with the aid of biochemistry* and cytology*. In this way, molecular geneticists have been able to analyse the biochemical principles of genetic structure as components of nucleic acids. They have also been able to explain the process of gene duplication which is the basis of heredity*, as well as the way in which genetic information is released (Cell*, Cell Division*). The knowledge acquired by theoretical geneticists has been applied in the field of breeding*, in particular. Plants, animals and micro-organisms can all be subjected to genetically controlled influences so that their use to humans is improved. The main concern of human genetics is the study of hereditary diseases* in humans.

Genital Papilla *see* Sex Organs

Genotype *see* Heredity

Genus *see* Systematic Categories

Genypterus capensis *see* Gadiformes

Geobotany (Plant Geography) is the study of how plants are distributed across the Earth's surface. Its aims include recognising the essential features and regularity of plant zones and explaining their origins.

Geophagus HECKEL, 1840 (fig. 552). G of the Cichlidae*. Comprises several species distributed throughout the eastern parts of S America from tropical Guyana to temperate parts of Argentina. Almost exclusively typical freshwater fish, although *G. brasiliensis* does enter brackish water. *Geophagus* species are large fish (about 30 cm long). The local populations catch them for

Fig. 552 *Geophagus surinamensis*

human consumption. *Geophagus* are fairly tall fish, flattened from side to side. Their upper profile is often bent more obviously than the lower profile. The eyes are often placed high on the head and the snout is long *(G. jurupari).* The fins are similar to those of the G *Aequidens** – the Vs are elongated like swords and the An has 3 hard rays. The C can have 3 differently shaped edges: rounded like a fan, straight or slightly indented. Most species have mother-of-pearl iridescent flecks on the head, body and fins; they show up more intensively during the breeding season. *Geophagus* species form territories and so are intolerant. They also chew their way through the substrate looking for something to eat, which can cause cloudiness in the water and harm any plants that are growing there. The needs of the fish can be met by providing a spacious aquarium with a thick layer of clean sand, suitable conditions and hiding-places. This will help keep the unpleasant aspects of the fish's nature in check to a certain extent. The temperature your specimens should be kept at will vary from species to species (depending on their origin). Tropical species require a constant 23 °C and above, whereas the Argentinian varieties can survive a drop in temperature down to about 12 °C without suffering any ill-effects. Most species are typical open-water spawners (*see* Cichlidae). *G. jurupari* and others show signs of being mouth-brooders.

– *G. brasiliensis* (QUOY and GAIMARD, 1824); Pearl Cichlid. Found in regions near the coast of E Brazil, in both fresh and brackish water. Up to 28 cm. Body and fins dotted with mother-of-pearl colours. Juveniles have several dark oblique bands and the adults have a black blotch in the middle of the body and an arched, black eye band.

– *G. gymnogenys* HENSEL, 1870; From S Brazil to N Argentina. Up to 21 cm. Dark olive-green to coffee-brown, front part paler. Head, body and fins a lovely pale blue or flecked in mother-of-pearl. There is a dark blotch on the caudal peduncle. Very hardy, can survive low temperatures.

– *G. jurupari* HECKEL, 1840; Earth Eater, Devil Fish. Tropical America. Up to 25 cm. Body relatively elongated. Ventral profile almost straight. Ground colour very changeable, mostly fairly pale. Scales have fine speckles with a mother-of-pearl sheen. Peaceable Cichlids.

Georgette's Tetra *(Hyphessobrycon georgettae) see Hyphessobrycon*

Gephyrocharax EIGENMANN, 1912. G of the Characidae*, Sub-F Glandulocaudinae. 4–6 cm long. The G is made up of about 7 species and they are found in Panama and Venezuela. They are harmless Characins that are highly active swimmers. Their body is moderately long and very compressed from side to side. Caudal peduncle relatively broad, LL complete, D beginning well to the rear. The lowest ray of the C is free-standing – in the male it has developed into a kind of spine. The first 10 rays of the male's An are serrated and the middle section is arched forwards like a lobe. Seed capsules are deposited by the male in the female's oviduct where they are stored. There are enough seed capsules for several egg-laying sessions, sometimes enough for an entire life-span. Without the presence of the male, the female lays the spawn on leaves, releasing some sperm from storage each time to fertilise the egg-cells. The eggs hatch out after 24–36 hours and the young fish are easy to rear on *Cyclops* nauplii or rotifers. Despite having an interesting reproductive method, *Gephyrocharax* species are not popular with aquarists on account of their somewhat dull colouring.

– *G. atracaudatus* (MEEK and HILDEBRAND, 1912); Platinum Tetra, Silver Tetra. Found in the rivers of Panama, except the Rio Chane. Up to 6 cm. Dorsal surface olive-green, flanks pale bluish with a white, length-wise band bounded by pale blue on its lower side. There are 2 luminous blotches on the deep black caudal peduncle. The caudal peduncle extends like a fork-tail into the lobes of the C. Fins colourless or a faint reddish colour. The females are more compressed with a straight outer edge to the An. Neither do they have the spine underneath the C which is typical of the males (fig. 553).

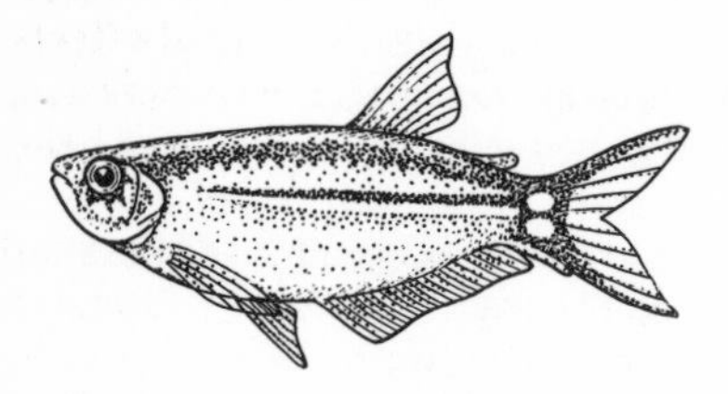

Fig. 553 *Gephyrocharax atracaudatus*

– *G. valencia* EIGENMANN, 1920. Venezuela, Lake Valencia. Up to 5 cm. Dorsal surface olive-green or blue-green, flanks an iridescent silver colour. The fins are colourless or a delicate yellow colour, apart from the free edge of the D which is tipped in white. A faintly visible dark longitudinal band broadens out into a dark blotch on the caudal peduncle. Sex differences as described in *G. atracaudatus.*

Gerlach's Rasbora *(Rasbora gerlachi) see Rasbora*

Germ Cells *see* Gametes

Germination is the further development of a dormant seed*. It begins with the seed absorbing water thus causing the seed envelope to burst open. The reserve substances stored within the seed are now activated and the embryo begins to grow. First to appear is the radicle, followed by the embryo stem containing the plumule. The seed leaves are either the last to be withdrawn from the seed envelope and to turn green, or they remain in the seed and only the primary leaves begin to assimilate.

Gerreidae; Mojarras (fig. 554). F of the Perciformes*, Sub-O Percoidei*. Small, mostly silver Percoidei, similar to the Leiognathidae*. Found in the coastal regions

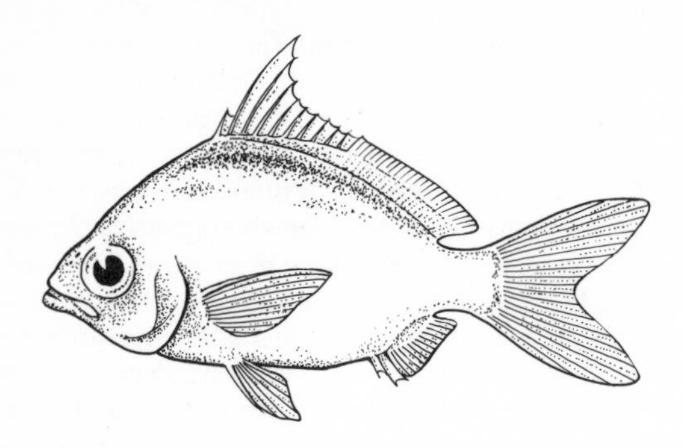

Fig. 554 Gerreidae-type

of tropical and subtropical seas, occasionally entering brackish and fresh water. The D and An may also be sunken in the Gerreidae but the scaly sheaths at the corresponding fin bases do not have spines as do members of the Leicognathidae. Some of the 35 species are important economically.

Gerris FABRICIUS, 1794; Pondskater (fig. 555). G of the insect order Bugs (Heteroptera*). Despite the fact that they live an aquatic life for the most part, they belong to the land bugs. *Gerris* species attract a lot of attention because of the peculiar jerky skating action with which they move across the water surface. They are known by various popular names, but mainly 'Pondskater'. Their middle and hind legs are thin and covered with hairs, thus enabling the bugs to make use of the surface tension of the water to avoid sinking through. The much shorter front legs are largely used for gathering food. *Gerris* eat small insects that fall onto the surface, be they dead or alive. The males of this G are noticeably smaller than the females. During mating, they are often carried on the backs of the females for days.

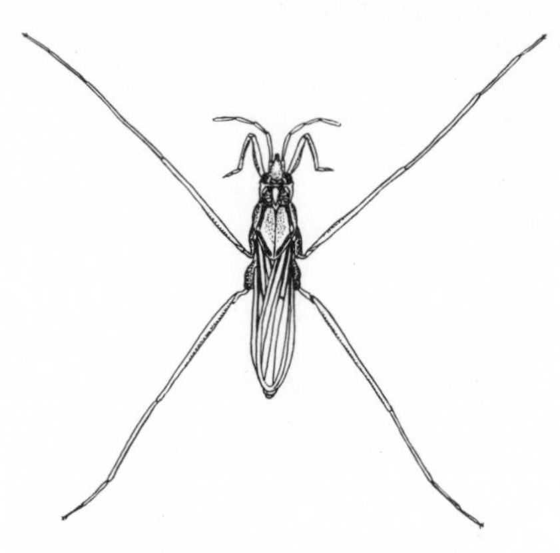

Fig. 555 *Gerris* sp.

Ghost Crabs (*Ocypode*, G) *see Uca*

Ghost Pipefishes (Solenostomidae, F) *see* Syngnathoidei

Giant Clams, G, *see Tridacna*

Giant Coolie *(Acanthophthalmus kuhli kuhli) see Acanthophthalmus*

Giant Danio *(Danio malabaricus) see Danio*

Giant Devil Ray *(Manta birostris) see* Mobulidae

Giant Gourami *(Colisa fasciata) see Colisa*

Giant Rivulus *(Rivulus harti) see Rivulus*

Giant Sailfin Molly *(Poecilia velifera) see Poecilia*

Giant Scissortail *(Rasbora caudimaculata) see Rasbora*

Giant Sea Bass *(Stereolepis gigas) see* Serranidae

Giant Squids (*Architheutis*, G) *see* Cephalopoda

Giant Vallisneria *(Vallisneria gigantea) see Vallisneria*

Giant Water-lilies, G, *see Stoichactis*

Giant Water-lily *(Nymphaea gigantea) see Nymphaea*

Gibbons Starlets, G, *see Asterina*

Gibbula RISSO, 1826. G of small Topshells (F Trochidae) that belong to the Prosobranchia*. At one time, *Gibbula* and other Ga were placed in the large G *Trochus. Gibbula* is a common species in the upper littoral of warm seas. It feeds on algae. Its shell is 1–2 cm high, cone-shaped and coiled, and it is usually prettily coloured. Several species live along the southern European coasts, and they are most suitable for the aquarium.

Gibel Carp *(Carassius auratus gibelio) see Carassius*

Giganturidae, F, *see* Cetomimiformes

Giganturiformes, O, name not in current usage, *see* Cetomimiformes

Gill Arches *see* Skeleton

Gill Rakers (fig. 556). A gill-filter system in fish formed from sieve or weir-basket processes (branchictenia) of the gill arches. The rakers prevent bits of food escaping through the gill slits*. They also protect the gills themselves from damage. The shape, length and arrangement of the gill rakers depend on the way the fish feeds. In predatory fish you will often find knob-shaped rakers that interlock with those of the neighbouring gill arch. Plankton-eaters usually have long sieve processes, often toothed, that cover the gill slits on the pharynx side like combs. The type of gill rakers is often taken into account when determining species.

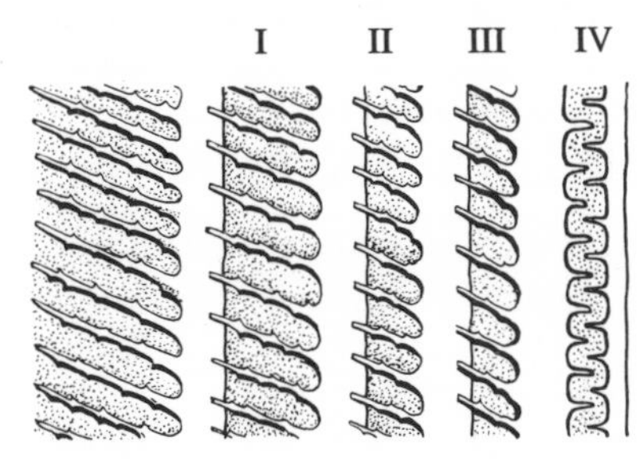

Fig. 556 Gill rakers with comb-like processes (I–IV gill arches)

Gill Rot. A disease caused by algal fungi of the G *Branchiomyces*. It is confined to only a few species of fish – *Cyprinus* carpio*, *Tinca* tinca*, *Carassius* auratus gibelio*, various species of the G *Gasterosteus** and *Esox lucius* (Esocoidei*). Signs of the disease are the epithelium of the gills turning yellow-brown and eventually leading to necrotic decay. Gill rot can be avoided if the fish are kept under good conditions. The disease occurs very rarely in aquarium-keeping.

Gill Slits. Lateral or ventral openings in a fish's pharynx which are separated by gill arches. In sharks and rays the slits are arranged in pairs along the body surface. In bony fishes the slits open out into the outer gill cavity which is sealed off by the operculum. The chief function of the gill slits is to expel the respiratory water. Sharks and rays usually have 5 gill slits on each side, although some species have 6 or 7 (eg the Cow Sharks *[Hexanchus* and *Heptranchias]*). The Rat or Rabbitfishes (Chimaeridae) have 4 gill slits. Most bony fish have 5, although some groups have only 4 (Synbranchiformes*, Tetraodontiformes*).

Gill-flukes *(Dactylogyrus) see* Trematodes

Gills *see* Respiratory Organs

Gills, Diseases of the. Gills are the most important respiratory organs in fish. Like the skin, the gills are extremely sensitive to changes in the environment. Quickened breathing, which is apparent from the increased frequency with which the operculum moves, usually means the gills have suffered some damage. The causes are many – fluctuations in pH value (Alkali Sickness*, Acid Sickness*), lack of oxygen (Oxygen Deficiency*), excess carbon dioxide (Carbon Dioxide Poisoning*), non-combined chlorine, oils, phenols, iron compounds and other poisons such as ammonia*. All of these things can effect a very quick alteration or destruction of the gill epithelium, thus causing respiratory problems. Gills are also prone to skin parasites such as *Ichthyophthirius**, *Trichodina**, *Chilodonella**, *Cryptocaryon* (Cryptocaryon Disease*), *Ichthyobodo**, *Oodinium**, as well as some Sporozoa*. Following on from the initial damage, there is often a fungal attack, such as *Branchiomyces* (Gill Rot*) or *Saprolegnia see* Mycoses*), and sometimes there may also be a bacterial attack (Bacterial Infections*). Other typical gill parasites include the crustaceans of the G *Ergasilus* (*see* Copepod Attack) and the trematodes* of the Ga *Dactylogyrus* and *Sanguinicola*.

Ginglymostoma cirratum *see* Ginglymostomatidae

Ginglymostomatidae; Nurse Sharks. F of the Chondrichthyes*, O Orectolobiformes. Nurse Sharks are closely connected with the Orectolobidae*. In appearance they look very like the better known Cat Sharks or Dogfish, but they can be distinguished by the barbels at the end of the naso-labial depressions. The head is truncated, the mouth very slightly inferior. The 4th and 5th gill slits lie much closer together than any of the others. Blow-holes small. Typical of the F is the Common Nurse Shark *(Ginglymostoma cirratum)* (fig. 557) which grows up to 3 m in length, longer in rare cases. It comes from tropical and subtropical parts of the W and E Atlantic. They are very sluggish animals and usually rest motionless beneath the surface. They feed mainly on squids and crustaceans. As far as humans are concerned, they are not dangerous. Their skin makes a good leather, and their meat is suitable for human consumption.

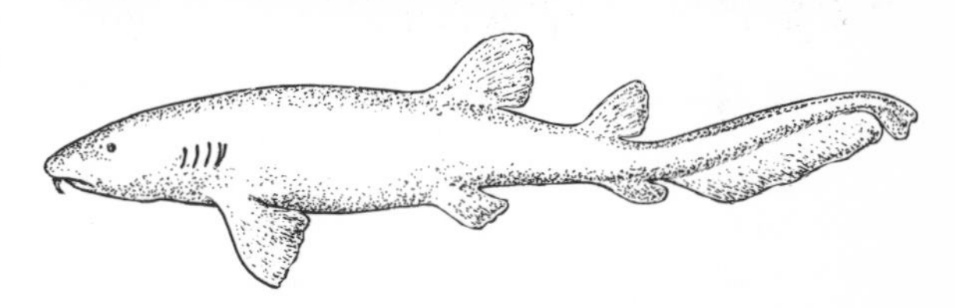

Fig. 557 *Ginglymostoma cirratum*

Girardinus Poey, 1854. G of the Poeciliidae*, Sub-F Poeciliinae. Found in Cuba, one species found in Costa Rica. They inhabit small flowing waters, but also ponds, lakes and marshes. Small fish, some may be tiny. Males 2.5–5 cm, females 4–9 cm. Most of the species are inconspicuous, the ground colour being yellowish to brown-olive. The fins are often colourless or they may have a slight tinge of colour. The female is always somewhat paler. Macroscopically recognisable generic features are not present. *Girardinus* needs to be kept in a tank that is in part densely planted. Temperature 22–25 °C. Clear, filtered and aerated water is most suitable. It should be partially replaced at intervals. Besides live food, *Girardinus* will also accept artificial and dried foods. As a supplement to this diet, plant food is also recommended, particularly algae. For the most part, *Girardinus* species are not very important to the aquarist. The following are often kept in aquaria.

– *G. falcatus* (EIGENMANN, 1903); Yellow Belly. Male up to 5 cm, female up to 7 cm. Body is very flat from side to side and translucent. The gonopodium is very long. Body colouring brownish with distinct silvery longitudinal stripes. Eyes a shining green colour. Fins yellowish. Up to 60 fry in a brood.

– *G. metallicus* POEY, 1854; Girardinus. Found in Cuba and Costa Rica. Male up to 5 cm, female up to 9 cm. Body moderately flattened from side to side and relatively tall, particularly in the caudal peduncle of the male. Gonopodium very long. Body yellow-brown, with a metallic sheen. There is a row of alternate dark and silver transverse bands along the flanks, often less prominent in the female. Beneath the eyes and on the operculum there are some shining green dots. The upper edge of the eyes are an iridescent yellow. There is a dark blotch at the base of the D. Up to 100 fry in a brood (fig. 558).

Fig. 558 *Girardinus metallicus*

Girardinus *(Girardinus metallicus) see Girardinus*

Glandulocauda EIGENMANN, 1911. G of the Characidae*, Sub-F Glandulocaudinae. Up to 7 cm. Found in northern America, Venezuela, especially in the Rio Grande do Sul. Moderately elongate, very flat, small Characins with a stout, short head and a broad caudal peduncle. L. P. SCHULTZ's suggestion that the G *Glandulocauda* could be regarded as a synonym for *Mimagoniates** was opposed by GÉRY and NELSON. They found that there were differences in the teeth as well as in the number of anal hooks and the form of the gland beneath the caudal fin (each ray of the An in the species of the G *Glandulocauda* possesses several little hooks). *Glandulocauda* is an active, peaceable genus, well suited to the community aquarium. In its reproductive behaviour it resembles that of the species of the G *Gephyrocharax**. Although a hardy genus, *Glandulocauda* is not very popular with aquarists. Only *G. inaequalis*, which is an altogether atypical species of the G (as only the first 7 rays of the An have 2 hooks each), is occasionally found in aquaria.

– *G. inaequalis* EIGENMANN, 1912; Blue or Croaking Tetra, Broad-tail Characin. Found in Uruguay, in the Rio Grande do Sul. Male up to 6 cm, female up to 4.5 cm. This species has an accessory respiratory organ with the aid of which it can produce distinctly audible chirping noises. Flanks bluish-white. When light strikes it side-on, the flanks take on a lovely blue-green iridescent colour. Belly silver-white. A broad, pale, longitudinal band begins behind the droplet-shaped, blue shoulder fleck. This band joins up with a second, pale blue zone below, which gets narrower towards the rear and continues into the caudal fin. D positioned well to the rear: it is colourless near its base whereas further out it is green-blue with a narrow, fawn-brown band. C yellowish with a dark edge above and below. An colourless with a clay-yellow band. A long, flat adipose fin is present. The males have a gland beneath the C which is absent in the females. This gland has a complicated structure and is covered with scales. The males are also different from the females in that they have a broader lower caudal fin lobe, and the front part of the An is also broader. The overall appearance of the females is paler in colour. This species used to be a popular aquarium fish but it is not known at all today. Aquarists would surely welcome their becoming available again (fig. 559).

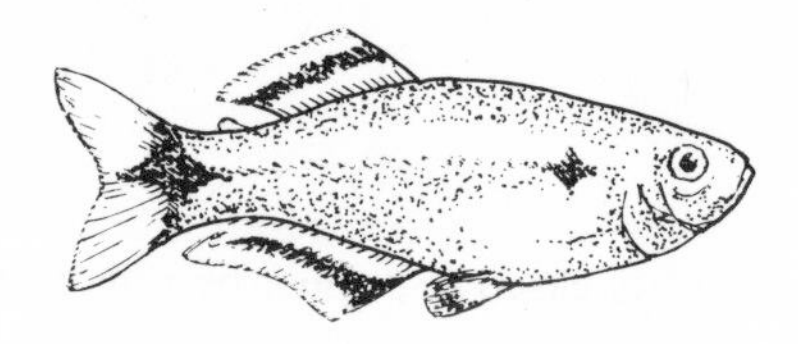

Fig. 559 *Glandulocauda inaequalis*

Glandulocaudinae, Sub-F, *see* Characoidei

Glass Barb *(Oxygaster oxygastroides) see Oxygaster*

Glass Bloodfin *(Prionobrama filigera) see Prionobrama*

Glass Catfish *(Kryptopterus bicirrhis) see Kryptopterus*

Glass Knifefish *(Eigenmannia virescens) see* Gymnotoidei

Glass Tetra *(Moenkhausia oligolepis) see Moenkhausia*

Glassfishes, F, *see* Centropomidae

Glochidia (fig. 560). The name given to the larval stage of the freshwater Mussel family Unionidae, to which well-known Ga such as *Unio** and *Anodonta** belong. Glochidia live as parasites on fish. They have a bivalved shell with sharp hooks, and whenever they come into contact with a fish the shell snaps shut and remains attached to the fish's skin. If a large number of glochidia

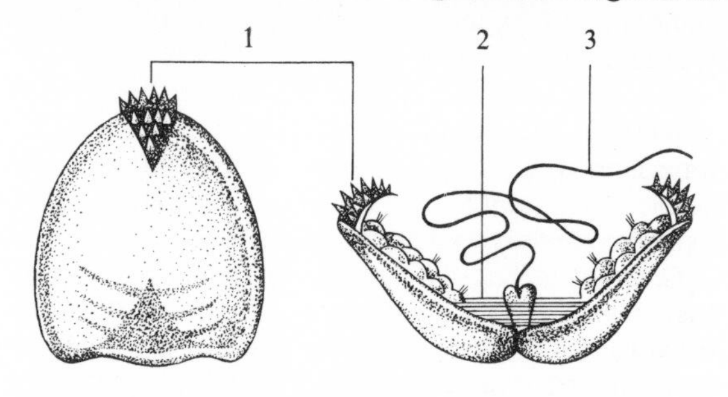

Fig. 560 Glochidia. 1 Shell hook 2 Shell shutting mechanism 3 Thread of attachment

attack the fish, it could impair the fish's health, so it is strongly advised that you do not keep river and pond Mussels alongside fish in the aquarium. (An exception is *Rhodeus* which needs Mussels for its brood to develop.) It is astonishing to think that a single Mussel can produce up to 400,000 glochidia per reproductive period.

Glowlight Rasbora *(Rasbora pauciperforata) see Rasbora*

Glowlight Tetra *(Hemigrammus erythrozonus) see Hemigrammus*

Glugea. This G comprises a large number of species that all belong to the Sporozoa*. They are without exception single-celled parasites, all of which undergo a characteristic spore stage (Parasitism*) during their development cycle. In the early stages, *Glugea* produces very small nodules in the connective tissue of fish which later become visible to the naked eye. These nodules are filled with a large number of infectious spores. The symptoms are similar to those observed during an *Ichthyosporidium** attack, but *Glugea* attacks are much rarer. *G. anomala* MONIEZ is often found in the Stickleback, *G. pseudotumefaciens* PFLUGFELDER, 1952, has repeatedly been observed among aquarium fish in their ovaries, livers, spleens, kidneys, eyes and nervous systems. Sea-horse disease is caused by a marine form of *Glugea*; the small white nodules in the skin spread out later into large, glassy areas. A cure for such an attack is not known. Diseased fish must be removed immediately, and the tank should be thoroughly disinfected (*see* Disinfection).

Gnathonemus GILL, 1862. G of the Mormyridae, O Mormyriformes*. Found in Central Africa in slow-flowing, muddy waters. These are medium-sized to large fish that form territories and live alone. The body is elongated and flattened laterally. The D and An are of equal length and are placed virtually directly opposite one another. The mouth is small and terminal and it is characterised by an extension to the chin which juts forwards a considerable distance. *Gnathonemus* is active at dusk and during the night. Specimens require large tanks with a soft substrate and places of darkness where they can hide (such as hiding-places formed from stones and roots). They are intolerant of like species but peaceful towards other species. Good quality 'old' water is recommended in the aquarium, with the occasional addition of some fresh water. Temperature 20–28 °C. Diet consists mainly of live food, particularly worms, but dried food will also be taken up from the bottom. The various species of *Gnathonemus* have interesting modes of behaviour. There is no well-substantiated information regarding sex differences and reproduction.

– *G. macrolepidotus* (PETERS, 1852); Shortnosed Elephant Fish. E Africa (Rovuma) up to Shaba. Up to 30 cm long. Dorsal surface dark grey-brown, flanks pale silver, belly silvery-white. Dark flecks are distributed in an irregular fashion over the fins and body.

– *G. moori* (GÜNTHER, 1867); Roundnosed Elephant Fish. Cameroun, Gaboon, Congo. Up to 20 cm. Flanks brown, paler towards the belly. The An and D are linked by a dark transverse band.

– *G. petersi* (GÜNTHER, 1862); Elephant-nosed Fish, Longnosed Elephant Fish. Congo, Cameroun up to Niger. Up to 23 cm. The lower jaw has an extension of the chin shaped like a finger which is highly mobile. The body is brown-black with a violet-coloured sheen. There are two yellowish-white, irregular transverse bands at the level of the D and An. The D, An and C have paler markings (fig. 561).

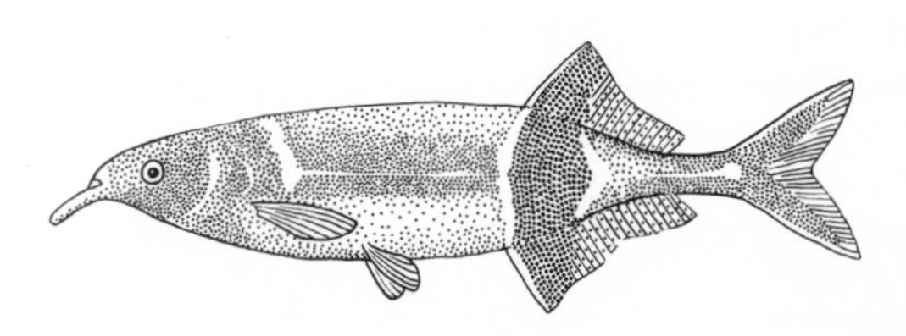

Fig. 561 *Gnathonemus petersi*

– *G. schilthuisiae* BOULENGER, 1899. Central Congo. Up to 10 cm. Flanks brown with a violet-blue sheen. There is one dark band at the level of the An and D. This species is relatively tolerant of like members.

– *G. stanleyanus* (BOULENGER, 1897). Congo, River Gambia. Up to 40 cm. Body grey-brown, belly silver-grey. The flanks, D and An are covered in irregular dark flecks.

Gnathostomata. Sub-Ph of the Chordata*. In contrast to the Sub-Pha Tunicata*, Acrania* and Agnatha*, none of which has jaws and which feed primarily on micro-organisms filtered from the water, all Gnathostomata have both an upper and a lower jaw which can be moved against one another like a pair of pincers (fig. 562). The Gnathostomata are made up of a number of different classes of animals: Aphetohyoidae*, Chondrichthyes* (cartilaginous fishes), Osteichthyes* (bony fishes), Amphibia* (amphibians), Reptilia* (reptiles), Aves* (birds) and Mammalia* (mammals). The upper and lower jaw (palatoquadratum and mandible) developed from skeletal elements of the first gill arch found in the jawless ancestors of the Gnathostomata dating back to

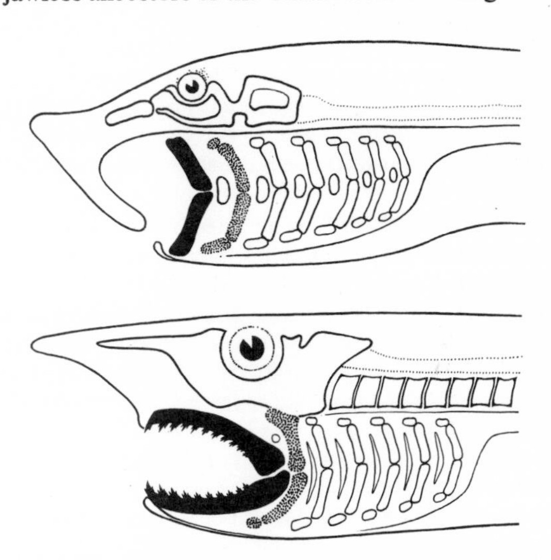

Fig. 562 The development of the upper and lower jaws, as well as the hyoid bone and jaw process from the skeletal elements of the 1st (black) and 2nd (dotted) gill arches

the Cambrian Period. From then on the gill arch is known as the mandibular arch. With the exception of the primitive fish (Aphetohyoidea), this primitive jaw mechanism has been improved upon in all higher classes by the addition of other skeletal elements. The second gill arch now forms the hyomandibular arch and the hyoid bone, the former having the function of fixing the mandibular arch to the neuro-cranium, the hyoid bone having the function of supporting the bottom of the mouth. In cartilaginous and bony fishes, the mandibular arch and the second gill arch (hyoid arch), together with the unaltered gill arches (usually numbering 4) make up the viscerocranium. In land-dwelling Gnathostomata (amphibians, reptiles, birds and mammals), only parts of the mandibular and hyoid arches are present in the viscerocranium. The other gill arches degenerated. It is also worth mentioning that in the Gnathostomata both the viscerocranium (development as described above) and the primitive neurocranium had skeletal elements added to them, primarily in the form of dermal bones. These bones joined together to form bony plates which united with the original cranium (primordial cranium) to make up a single-unit skull. In the higher classes, such skeletal elements that have evolved from dermal ossification have become the dominant parts of the skull. So, in the mammals, for example, they have taken over completely the function of the primitive jaw articulation. However, here too there are remains of the mandible, palatoquadratum and hyomandible. They make up the hammer, anvil and stirrup – the auditory bones of the middle ear. The commonly used term 'vertebrate animals' is not synonymous with the term 'jawed animals' (Gnathostomata). Nor must vertebrates be used to denote an exact systematic category, as many primitive Gnathostomata do not have true vertebral columns.

Gobies (Gobiidae, F) *see* Gobioidei

Gobies, related species, Sub-O, *see* Gobioidei

Gobiesocidae, F, *see* Gobiesociformes

Gobiesociformes; Cling-fishes (Xenopterygii is another name but it is no longer in current usage). O of the Osteichthyes*, Coh Teleostei. Small or very small Goby-like fish which occur in the coastal regions of tropical and subtropical seas, sometimes penetrating brackish water areas. There are about 90 modern species, and there are no known fossils. The most characteristic feature of the Gobiesociformes is the ventral sucking disc located almost at the throat. It is formed from both ventral fins and enables the fish to attach themselves firmly to things. The head is large and often drawn into a point towards the front. There is only one long D, located directly in front of the C and opposite the equally long An. The D, An and Ps do not have spines. The skin is naked and contains a great many glands, making it appear smooth and slimy. No swim-bladder. Many species have opalescent eyes. There is only one family (Gobiesocidae) (fig. 563). Most Gobiesociformes have camouflage colours, some can change colour to suit their background. The females as a rule are considerably larger. One of the best-known species is *Diademichthys lineatus*, because of its association with the long-spined Sea-urchin *Diadema*. The fish lives among the Sea-urchin's spines, thereby gaining a very good protection. The largest species, the South African Rock-sucker *(Chorisochismus dentex)*, grows to a length of 30 cm. Several species have been kept in marine aquaria.

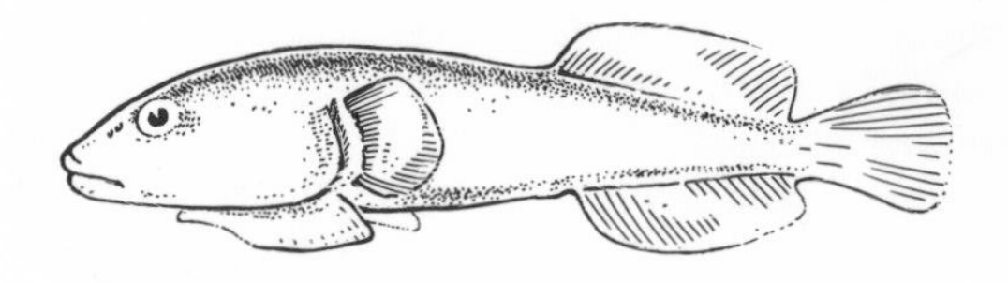

Fig. 563 Gobiesocidae-type

Gobiidae, F, *see* Gobioidei

Gobio Cuvier, 1817; Gudgeons. G of the Cyprinidae*, Sub-F Cyprininae. Found in Asia and Europe. Small to medium-sized bottom-dwelling fish that live a sociable existence in clear, fast-flowing waters, more rarely in shallow parts of still waters and in brackish waters. The body is elongate, shaped like a spindle. The head is large, the mouth inferior with 1–2 pairs of barbels. Large scales. LL complete. *Gobio* species spawn in shallow water. The males have spawning tubercles. This G makes hardy coldwater aquarium specimens. The tank should not be too small or too shallow. Coarse gravel makes a good substrate. Clear, clean water is needed, and up to a quarter of it should be replaced with fresh water on a regular basis. Temperature not above 18 °C. Gudgeons search for their food primarily on top of the substrate and in it. Their diet consists mainly of small crustaceans, worms and insect larvae, but they will also accept plant food and dried food. *Gobio* can be bred successfully in the aquarium if provided with shallow water and if allowed to overwinter at a cool temperature. The fry grow quickly.

– *G. gobio* (Linnaeus, 1758); Gudgeon. Up to 20 cm, but most remain smaller. Dorsal surface black-green or grey-green, flanks silver-grey with dark, violet-blue, shining blotches. The scales have a dark edge. There is an indistinct longitudinal band which breaks up into a row of blotches in many cases (fig. 564).

Fig. 564 *Gobio gobio*

There are 3 more species found in the drainage area of the Danube: *G. albipinnatus* Lukasch, 1933, in which the D and An do not have flecks across them; *G. kessleri* Dybowski, 1862, in which there are 2 pairs of barbels and *G. uranoscopus* (Agassiz, 1828) in which there is one pair of barbels that stretch far behind the eyes.

Gobiodon BLEEKER, 1856; Coral Gobies (fig. 565). G of the Gobiidae, Sub-O Gobioidei*. Beautifully coloured fishes from tropical parts of the Indo-Pacific. Length up to 5 cm. They live among coral branches, feeding on small organisms and fish spawn. In contrast to most other types of goby, this genus is tall and later-

Fig. 565 *Gobiodon citrinellus*

ally compressed. The strikingly large head almost forms a semicircle when seen in profile. The D1 has 6 hard rays. The males take care of the eggs. Specimens have been known to spawn several times in the aquarium. *Gobiodon* species seem unable to stand up to food competitors. The information given here also applies to the similar G *Paragobiodon* BLEEKER, 1873.

Gobioidei; Gobies, related species. A Sub-O of the Perciformes*, Sup-O Acanthopterygii, comprising a great many species. Small bottom-dwelling fishes found in warm and temperate seas. There are several freshwater species too. Body elongate, in rare cases eel-like, only slightly compressed from side to side. The head is pointed or truncated with no protruberances from the forehead. The mouth may be small or have a wide gape. There are certain peculiarities of the cranium, among them the lack of parietal bones. In most cases the D is divided up into a shorter D1 and a longer D2. The C is rounded, whereas the An is similar in shape to the D2 and is positioned opposite it. Sometimes, the D1 and D2 are fused together, and so too might be the D, An and C. The Vs are located at the breast and part or all of them may be transformed into a sucking disc. The skin is usually covered with scales, but some species are naked. There is no LL, but there are a large number of dermal sense organs on the head and body. Only a few Ga have a swimbladder. The oldest known fossils belonging to this Sub-O date from the early Tertiary Period. GREENWOOD and colleagues (1966) distinguish 6 Fs, but we need only look at 2 here.

– *Gobiidae*; Gobies (fig. 566). Small fish of the kind just described. The 3 most important Sub-Fs can be distinguished from one another by the fin rays in the D and Vs. The Gobiinae (Gobies) have 5–6 thin spines in the D1 and the Vs are joined up completely; the Eleotrinae (Sleeper Gobies) have 6 thin spines in the D1, but the Vs are completely separate; in the Periophthalminae (Mudskippers) the D1 can be of varying length, ie it may have a few or very many spines, but the Vs may be only partially or completely fused. In addition, the part of the Ps nearest to the body in the Mudskippers is muscular. The Gobiidae eat meat and feed on small invertebrates. They occur chiefly in the shallower waters of warm seas, but they do also reach far into the more temperate zones. Several species have become adapted to life in the intertidal zones where they live hidden beneath rocks or in crevices. Some species live for a period, or exclusively, in the brackish water of estuaries, and several species have conquered fresh water. Many Gobiidae are covered with a thick layer of mucus. Their colouring may be inconspicuous or quite striking, and is subject to great variation within the one species. Usually, the males have longer fins and brighter colour-

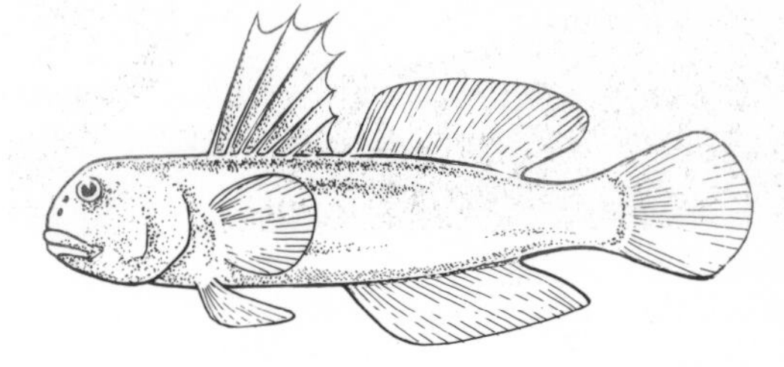

Fig. 566 Gobiidae-type

ing. Most interesting of all are the Periophthalminae. They inhabit shallow pools, ditches and water-holes all of which may at times almost dry up completely. They can support themselves off the ground on their muscular, arm-like Ps, thus raising their heads out of the water, and they may even scuttle across the damp ground.

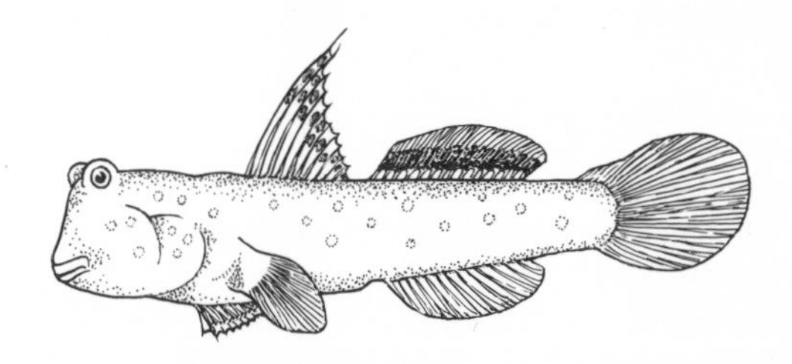

Fig. 567 *Periophthalmus* sp.

When danger threatens, they can dart over the ground on their caudal peduncles – ie escape by taking great leaps through the air. Their eyes are located high on the head like a frog's. The wall of the gill cavity is also enlarged and is particularly well supplied with blood vessels, thus enabling the fish to breathe in the air. Mudskippers spawn in deep mud-holes and the females practise brood-care. They are not easy to keep in the aquarium (fig. 567). There are several well-known species, for example, *Periophthalmus barbarus* (Mudskipper) which is found along many coastlines of E Africa, SE Asia and the Malaysian Archipelago. It grows to a length of 15 cm. Another example is *Periophthalmus chrysospilos* (fig. 568) which has the same distribution area apart from E Africa. Several Ga belonging to the Gobiidae family (Sub-Fs Gobiinae and Eleotrinae) are often found as specimens in both marine and freshwater aquaria. For information about keeping them in aquaria, plus additional biological details, look under the generic entries *Bathygobius, Brachygobius, Carassiops, Dormi-nator, Eleotris, Gobiodon, Gobiosoma, Gobius, Mogurnda, Stigmatogobius.*

Fig. 568 *Periophthalmus chrysospilos*

– *Gobioididae*; Eel-like Gobies (fig. 569). Small, eel-like fish from the tropical Indo-Pacific. They have a long D and An which border directly onto the pointed C, thus forming a fin corona of almost uniform height. The head is usually characterised by a protruding lower jaw. The mouth is vertical and filled with strong teeth and there

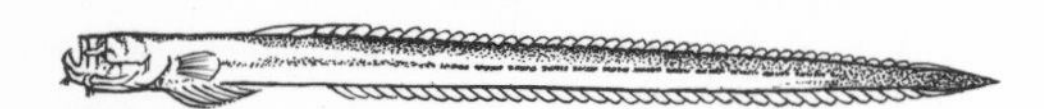

Fig. 569 Gobioididae-type

are tiny barbels at the chin. Small eyes. The Vs form a sucking disc. Scaleless. Eel-gobies like to live hidden away in deeper water layers. They are predators. They have been mentioned here only to demonstrate that some of the fish that belong to this Sub-O do not look at all like the standard Goby type.

Gobioididae, F, *see* Gobioidei

Gobiosoma GIRARD, 1858. G of the Gobiidae, Sub-O Gobioidei*. There are a large number of species in this genus, and their main distribution area is the tropical W Atlantic. Their diet consists of small animals found among reefs. Length 2–5 cm, colouring very striking. All the species are either only partially covered in scales or they are completely lacking in scales. The D1 has 7 hard rays and the Vs form a kind of funnel. The body is almost as broad as it is high, the caudal peduncle usually being the same width as the rest of the body. The body shape is elongated and cylindrical. *G. multi-fasciatum* STEINDACHNER likes to lurk between the spines of Sea-urchins and Sponges. It is a green fish with yellow transverse stripes. *G. genie* BÖHLKE and ROBINS (fig. 570) and *G. oceanops* JORDAN are both cleaner fishes. *G. oceanops* is also known to form firm partner-ships, and both sexes are said to practise brood care. There have also been some successes with breeding in the aquarium. The fry find their first food in algae-rich aquaria, then they will turn to soaked egg-yolk that is added in. After a month they will feed on *Artemia* nauplii. *Gobiosoma* species do not stand up well to food competitors, but otherwise they make delightful, hardy aquarium specimens. The only species to be imported on any regular basis actually belongs to the Sub-G *Elacatinus*:

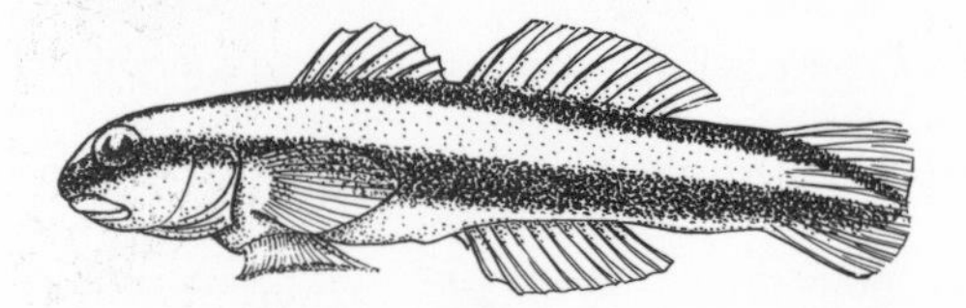

Fig. 570 *Gobiosoma genie*

– *G. oceanops* JORDAN; Neon Goby. Found in tropical parts of the W Atlantic. Up to 5 cm. It has black-blue longitudinal stripes and is similar to the Indo-Pacific Blue Streak or Bridled Beauty *(see Labroides).*

Gobius LINNAEUS, 1758. G of the Gobiidae, Sub-O Gobioidei*. This genus comprises numerous species found in the sea, as well as some that are found in brackish and fresh water. They are bottom-dwelling fish of shallow waters that feed on small animals – there are both typical sand-dwellers as well as species that live along rocky and gravelly coasts. The G was originally a very large one that has since been divided up into various Sub-Ga or Ga, although there is always some dispute as to how exactly the G should be divided up. *Gobius* species range in length from 5–30 cm. They have a body shape that is typical of the family – more or less cylindrical. The D1 has 6 hard rays. Determining which species is which can be a difficult task at times – among other things, it usually depends on the sense papillae on the head. Those species that live in the sand have camouflage colouring: rock-dwelling species are nor-mally very prettily coloured and many species exhibit a marked sexual dimorphism*. Some types of *Gobius* behave in a fashion that denotes an order of rank *(G. niger)*, others are specialised in interesting ways – the Mediterranean species *G. bucchichi* is a facultative Anemonefish*, for example. *G. fluviatilis* has adopted a completely freshwater existence. Spawn is laid beneath rocks and mollusc shells and protected by the males – they are helped in this by their ability to emit rasping noises *(G. niger)*. All the species are recom-mended for the aquarium.

– *G. minutus* (PALLAS) (= *Pomatoschistus m.*); Found along European sea coasts. Up to 11 cm. Sandy colour-ing with 5 dark blotches along the LL.

– *G. niger* LINNAEUS, 1758; Black Goby. Found along European sea coasts. Up to 18 cm. Brown with dark fleck markings. The male adopts territories and may be almost black in colour *(G. niger jozo)*, from southern European coasts (fig. 571).

Goblin Sharks, F, *see* Scapanorhynchidae

Goddard's Flying Barb *(Esomus goddardi) see Eso-mus*

Goggle-eye *(Ambloplites rupestris) see Ambloplites*

Gold Dust Disease *see Oodinium*

Gold Foil Barb *(Barbus schwanenfeldi) see Barbus*

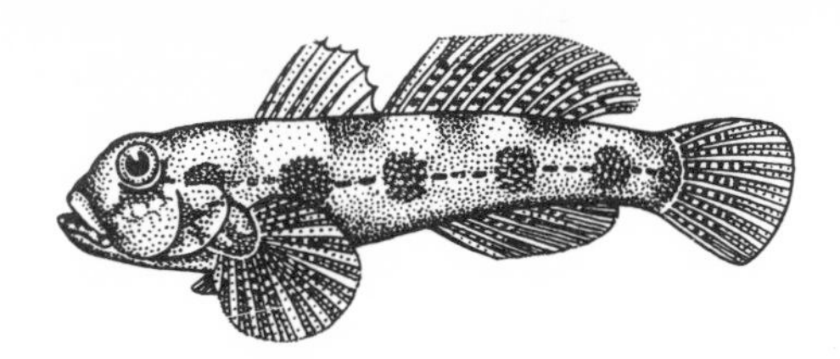

Fig. 571 *Gobius niger*

Golden Crown Aphyocharax *(Aphyocharax alburnus) see Aphyocharax*

Golden Danio *(Brachydanio albolineatus) see Brachydanio*

Golden Dwarf Barb *(Barbus gelius) see Barbus*

Golden Ear *(Fundulus chrysotus) see Fundulus*

Golden Lake Nyasa Cichlid *(Pseudotropheus auratus) see Pseudotropheus*

Golden One-spot *(Phalloceros caudimaculatus auratus) see Phalloceros*

Golden Orfe *(Idus idus) see Idus*

Golden Panchax *(Pachypanchax playfairi) see Pachypanchax*

Golden Pheasant *(Roloffia occidentalis) see Roloffia*

Gilden Rivulus *(Rivulus urophthalmus) see Rivulus*

Golden Sea-rose *(Condylactis aurantiaca) see Condylactis*

Golden Trevally *(Caranx speciosus) see Caranx*

Golden Trout *(Salmo aquabonita) see Salmonoidei*

Golden-eyed Dwarf Cichlid *(Nannacara anomala) see Nannacara*

Golden-striped Grouper *(Grammistes sexlineatus) see Grammistes*

Goldfinch Tetra *(Pristella riddlei) see Pristella*

Goldfish *(Carassius auratus auratus) see Carassius*

Gold-lipped Barb *(Barilius christyi) see Barilius*

Gold-rimmed Surgeonfish *(Acanthurus glaucopareius) see Acanthurus*

Gold-striped Corydoras *(Corydoras schultzei) see Corydoras*

Gomphosus (LACÉPÈDE, 1802); Beakfishes or Bird Wrasses (fig. 572). G of the Labridae, Sub-O Labroidei*. Comprises several species found in the tropical parts of the Indo-Pacific. Primarily inhabitants of reefs

Fig. 572 *Gomphosus varius*

where they can make use of their elongated snouts to prize small animals from rock crevices. The body is elongate and covered in large scales. Male and female *Gomphosus* may have completely different colouring, which can lead to their being classed as different species. They need careful looking after in the aquarium and a very varied diet. Any lively species that are liable to compete with *Gomphosus* for food should never be put in the tank with them.

Gonads *see* Sex Organs

Gonodactylus, G, *see* Stomatopoda

Gonopodium *see* Sex Organs

Gonorhynchidae, F, name not in current usage, *see* Gonorynchiformes

Gonorynchidae, F, *see* Gonorynchiformes

Gonorynchiformes. O of the Osteichthyes*, Coh Teleostei. The families described here, the Chanidae, Gonorynchidae, Kneriidae and Phractolaemidae, were often classified differently in older systems. REGAN and BERG placed the first-named families near the Mormyridae* and the ones after that with the Herring-types. GREENWOOD and colleagues (1966) give a large number of features which indicate that they should be grouped in one O. The most important unique feature of all representatives of the Gonorynchiformes is the formation of 1–2 head ribs on the exoccipital and the specialisation of the first 3 vertebral bodies behind the head. It seems that the Gonorynchiformes are closely related to the Salmoniformes* on the one hand but have features of the Cypriniformes on the other. The exceptional position of the Gonorynchidae has been acknowledged by the formation of its own Sub-O Gonorynchoidei. All the other Fs make up the Sub-O Chanoidei. – *Chanidae*; Milk-fishes. There is only one modern species, but fossil forms date back to the Lower Cretaceous. The well-known Milk-fish *(Chanos chanos)* (fig. 573) is an elongated fish of typical proportions with

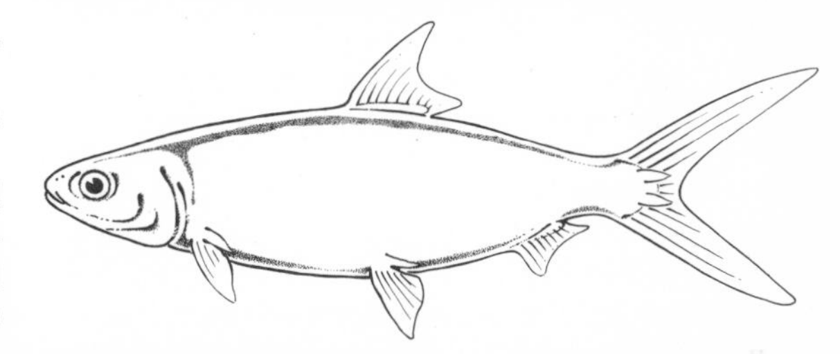

Fig. 573 *Chanos chanos*

a very deeply indented C. It can grow up to 3.6 m long, but most are much smaller. The D is shaped like a pennant and is located in the middle of the body opposite the Vs. The mouth is terminal, fairly small and not protrusible. It is only bounded by premaxillary bones on the upper edge. The jaws and palate lack teeth. The eyes are covered with skin. Cycloid scales. The swimbladder, as in the Cypriniformes*, is divided, it has an open connection with the intestine (physostomi) and is in contact with the balancing organ. Also present is a suprabranchial organ of unknown function. Without doubt the Chanidae are the most primitive fishes within the O. The silvery Milk-fish is found in tropical parts of the Indo-Pacific from S America to Australia and from S Asia to E Africa. Milk-fishes remain at sea in open waters except during the breeding season. For reproduction purposes they journey into coastal waters, usually spawning in the brackish water of large river-mouths. A single female can lay up to 9,000,000 eggs. The fry grow very quickly. Fry have been put into breeding pools, in the Philippines for example, where they have been raised to a marketable size. Their diet consists of various types of algae. Large *Chanos* are good fish to eat despite the fact that they contain a lot of bones.

– *Gonorynchidae*; Sand-eels. There exists only one marine species, but there are several fossil forms dating from the Upper Cretaceous. The Ratfish, Mousefish or Sand Eel *(Gonorynchus gonorynchus)* (fig. 574) is a very elongate cylindrical fish with a Sturgeon-like, pointed snout that juts out. It grows to about 40 cm long.

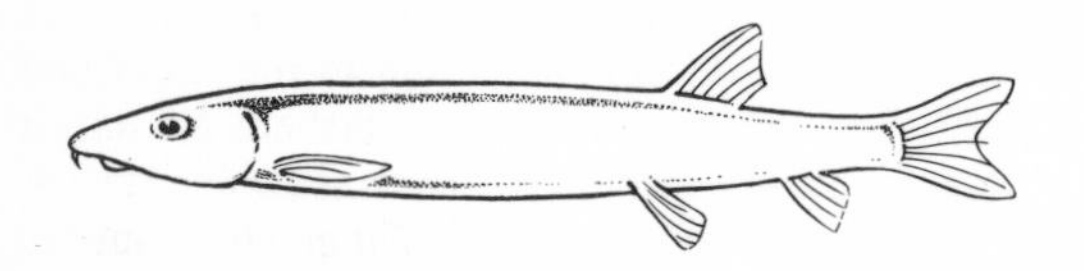

Fig. 574 *Gonorynchus gonorynchus*

It has a barbel just in front of the inferior, protrusible mouth. The jaws are toothless and the upper edge of the mouth is formed only from maxillary bones. Premaxillaries only rudimentary. Skin covers the eyes. The D is located well to the rear opposite the Vs. The C is indented and there is no adipose fin. The head and body are covered in small ctenoid scales. No swimbladder. There is a suprabranchial organ of unknown function. Sand-eels are very attractively coloured – grey-green to black on the upper surface and a meat-red colour underneath. The black C has shining red tips. Its distribution area includes the coasts around S Japan, Australia, New Zealand and E Africa. It seems to prefer sandy substrates and depths of up to 1.50 m. Sand-eels often bury themselves in the sand. They are considered a delicacy in New Zealand.

– *Kneriidae* (fig. 575). Slender fish found exclusively in African mountain streams. Mouth slightly inferior, very protrusible, the upper edge being bounded only by premaxillary bones. Jaws and palate have no teeth. The

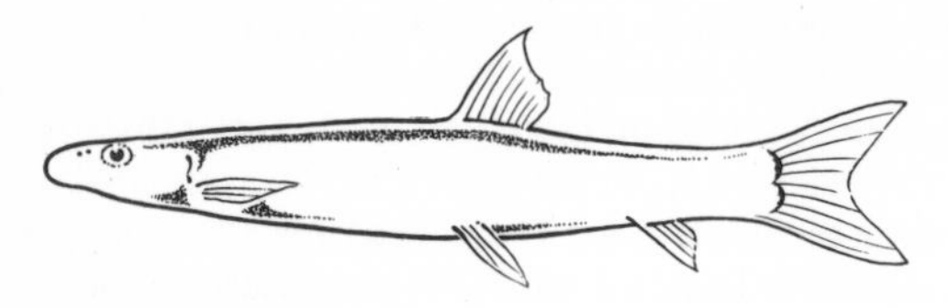

Fig. 575 Kneriidae-type

D is short and in the middle of the body, opposite the Vs. There is an open connection between the swimbladder and the intestine (physostomi*). In *Cromeria* the swimbladder is a single unit, in *Kneria* it is divided. The intestine is very long. A suprabranchial organ is present. Very little is known about the individual species. *Kneria* species have been kept in the aquarium, but they number very few. The males have an unusual, bean-shaped lamellar organ which lies in front of and above the insertion of the Ps. It is usually called the occipital organ. From his observations in the aquarium, PETERS (1967) suspects that the male clings onto the female with it during mating. Other Ga belonging to this F are *Parakneria*, *Cromeria* and *Grasseichthyes*.

– *Phractolaemidae*; African Mudfishes. Only one modern species *(Phractolaemus ansorgei)* (fig. 576) which grows up to 15 cm long. Found in muddy, weedy waters

Fig. 576 *Phractolaemus ansorgei*

in the Niger delta, Ethiopian rivers and the Congo. Body elongate, almost round when seen in cross-section, except that is slightly flatter towards the rear. Head small and broad, mouth small and extendable like a trunk with hardly any teeth. The upper edge of the mouth is bounded by premaxillary and maxillary bones. The D is short, inserted just beyond the middle of the body. Cycloid scales. There is an open canal between the swimbladder and the intestine, and the swimbladder serves as an accessory organ of respiration. The relationship between the Chanidae and the Phractolaemidae can be seen most clearly (according to BERG) by the lack of a Fossa temporalis posterior and the structure of the supraoccipital bone. *Phractolaemus* is occasionally kept in aquaria. It is a rather dingily coloured fish, but it is an interesting one. Sexually mature males have spawning tubercles on the head in the form of white nodules and 2 rows of sharp outgrowths on the caudal peduncle. The substrate must be soft and muddy to enable the fish to root for food there (mainly worms of all types). They also require a number of hiding-places. There are no records of any successful breeding attempts.

Gonorynchus gonorynchus *see* Gonorynchiformes

Gonostoma, G, *see* Stomiatoidei

Gonostomatidae, F, *see* Stomiatoidei

Goodea JORDAN, 1879. G of the Goodeidae*, Sub-F Goodeinae. Found in the rivers and lakes of Mexico. Small to medium-sized fish; males 5–8.5 cm, females 8–13 cm. Body usually elongate, laterally compressed. Fins relatively small, not pointed. The front 5–6 rays of the An are stiffened in the male, forming a slightly shorter section of fin set somewhat apart. It is used as a copulatory organ. The An of the female is rounded. Ground colouring of the body greenish-brown or yellowish, often with a dark longitudinal band or several bands. Fins colourless, yellowish or brown/black with varied markings. *Goodea* species should be kept in spacious tanks with a moderate amount of plants. Temperature 18–24 °C. The regular addition of fresh water is recommended. Feed on live food of all kinds

with a supplementary diet of plant food. Gestation period of the livebearing females is about 6–8 weeks at temperatures of 20–24 °C. Copulation is necessary for every litter. The following species is regularly kept and bred in the aquarium.
– *G. atripinnis* JORDAN, 1879; Black-fin Goodea. Found in Central Mexico, especially in the upper Rio Panuco. Males up to 8.5 cm, females up to 13 cm. Body olive-green with a bluish sheen. There is a distinct longitudinal band along the flanks which gets increasingly darker towards the rear. Ventral surface yellowish. Fins yellow with dark edges and fine dots, alternately yellow to black in the female. Large females give birth to 20–40 fry, each about 2 cm long. The fry grow quickly (fig. 577).

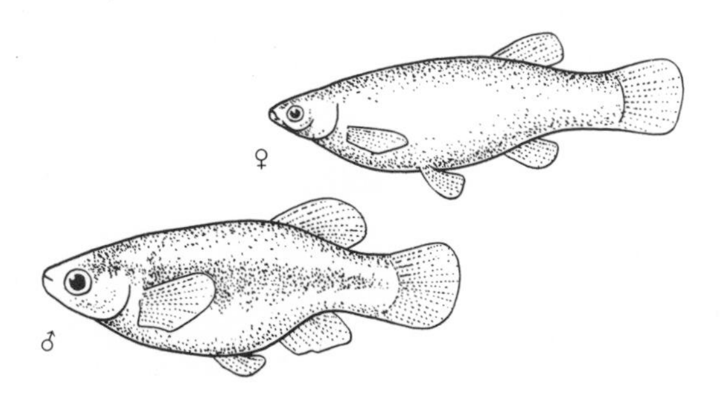

Fig. 577 *Goodea atripinnis*

Goodeidae; Goodeiids (fig. 578). F of the Atheriniformes*, Sub-O Cyprinodontoidei*. This F consists of small, livebearing fish that are confined to the high land of Mexico and Central America. Here, they are found in small streams and lakes. Altogether there are some 2 dozen species, of which the majority remain under 10 cm in length. The body is somewhat taller than that of the well-known Livebearing Tooth-carps (Poeciliidae*), and there are also differences between them in reproductive details. The male copulatory organ, for example, is of much simpler structure and is made up of the first 6 (–8) rays of the An which are shortened and compressed together. The females are also unable to store sperm, in fact copulation is necessary for every newly developed egg batch. If this does not take place,

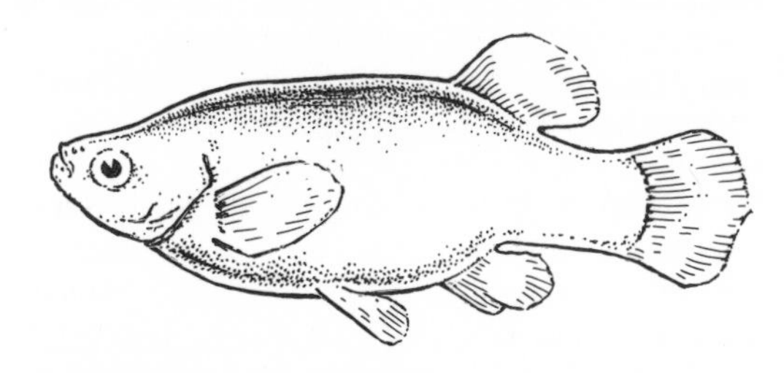

Fig. 578 Goodeidae-type

the eggs are ejected unfertilised. The eggs themselves do not contain much yolk and they develop in the ovarial cavity. Food is transported from the mother to the embryo via nutrient tubing (trophotaenia). These grow out of the anal region of the embryos and can be laid next to the wall of the ovarial cavity. The fry are born without protective covering, ie the Goodeidae are true viviparous fish. The species of the F are not commonly found anywhere and have no economic importance. In the aquarium they are only of interest to the specialist: *Characodon, Goodea*, Ilyodon, Lermichthys, Limnurgus, Neotoca.*

Goodeids, F, *see* Goodeidae

Goose Barnacles (*Lepas*, G) *see* Cirripedia

Gorgonaria; Horny Corals. O of the Anthozoans (Anthozoa*), Sub-Cl Octocorallia. Comprises about 1,200 species found mainly in warm seas. The central axial skeleton is often pliant in the same sense that horn is pliant, and it is covered by a rind out of which stem the individual polyps. The Gorgonaria are very attractive animals, with finely branched shapes and pretty colouring. They survive quite well in the aquarium if fed daily with small crustaceans (eg *Artemia*) and mash. However, they are liable to become overgrown with algae, particularly if the lighting is too strong. The colonies must be able to stand freely in the water and the rind must on no account be squashed. Damaged parts should be cut off. Species of the following Ga have been kept with success in aquaria: *Eunicella* VERRILL, 1869, *Gorgonia* LINNAEUS, 1758, *Paramuricea* KÖLLIKER, 1865 (fig. 579) and *Rhipidogorgia* VALENCIENNES, 1855.

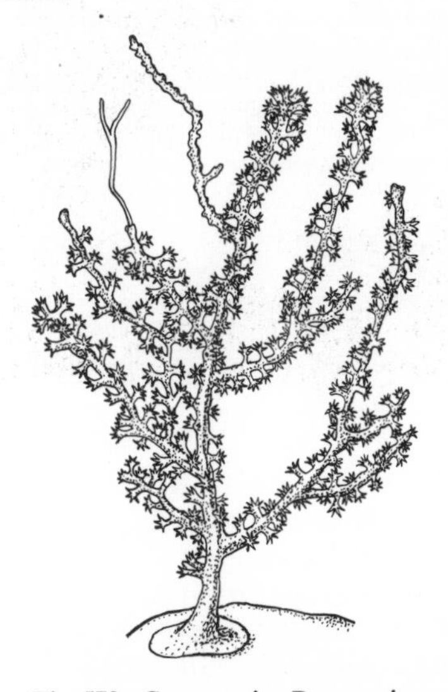

Fig. 579 Gorgonaria: *Paramuricea chamaeleon*

Gorgonocephalidae, F, *see* Ophiuroidea

Gorgonis, G, *see* Gorgonaria

Gourami *(Osphronemus goramy) see Osphronemus*

Gourami, G, *see Colisa, Trichogaster*

Gouramis (Osphronemidae, F) *see* Anabantoidei

Government Bream *(Lutjanus sebae) see Lutjanus*

Grace-Kelly-Fish *(Cromileptis altivelis) see Cromileptis*

Gramineae, F, name not in current usage, *see* Poaceae

Gramma POEY, 1868. G of the Serranidae* (sometimes also placed in its own F Grammidae). Consists of several species found in tropical parts of the W Atlantic. *Gramma* species live in shady spots along coral reefs and rocks as well as in the lower littoral. They are sociable fish that feed on small animals. Length 7.5–10 cm. Large-eyed, slim fish with extraordinarily beautiful colouring. The LL is in two parts: a front section lying towards the dorsal surface and a rear section lying in the middle of the caudal peduncle. The D at the front has 11–13 spiny soft rays and the V has one strong spine and 5 soft rays. The C is indented. When keeping *Gramma* in the aquarium, it is important to have a peaceful tank with plenty of hiding-places and it should be as free as possible of food competitors. Provide a good lid as these fish can jump well. Only the following species has been imported so far, and it has proved to be well suited to the aquarium. It will accept any type of food.

– *G. loreto* (= *hemichrysos*) POEY, 1868; Fairy Basslet Royal Gramma. Found in the Caribbean Sea. Up to 7.5 cm long. The front two-thirds of the body are a shining violet colour, the rear third a beaming yellow. The front portion of the D has a dark blotch (fig. 580).

Fig. 580 *Gramma loreto*

Grammistes BLOCH and SCHNEIDER, 1801. G of the Serranidae*. There is only one species, *G. sexlineatus* (THUNBERG, 1792), commonly called the Golden-striped Grouper or Six-lined Grouper. It can grow up to 25 cm long. It lives in tropical parts of the Indo-Pacific, including the Red Sea. *Grammistes* has a compressed shape and has a dark-brown or blackish ground colour with 3–9 yellow longitudinal stripes running across it. As the fish gets older so the number of these stripes increases. It is a greedy predator which normally likes to remain in open water above the corals. When distressed, or when it is fished out of the water, *Grammistes* secretes a strong poison (Grammistin) which quickly kills other fish – for which reason it is best to keep *Grammistes* alone in a tank. It is a very hardy species in the aquarium (fig. 581).

Grapsus, G, *see Pachygrapsus*

Grass Carps. This term is used to describe various large, plant-eating Cyprinids from East Asia which have been introduced in increasingly large numbers as specimens for pond aquaria in E, S and Central Europe. They grow almost as quickly as the Carps themselves and in

Fig. 581 *Grammistes sexlineatus*

some places they are very popular in the trade. The Grass Carp *(Ctenopharyngodon idella)* eats higher aquatic plants mainly, thus preventing the water from becoming too overgrown with weed. The Silver Carps *Hypophthalmichthys molitrix* and *Aristichthys nobilis* are filter-feeders, sifting out phytoplankton from the water with their specialised pharynxes. Grass Carps can also be reared artificially at higher temperatures, feeding at first on zooplankton but going over very soon to a vegetarian diet. The intestine is long, as in all plant-eaters. Grass Carps are hardy through the winter even in Europe.

Grass Wracks, G, *see Zostera*

Gravel *see* Substrate, Filtration

Grayling *(Thymallus thymallus) see* Salmonoidei

Graylings (Thymallidae, F, name not in current usage) *see* Salmonoidei

Great Barracuda *(Sphyraena barracuda) see* Sphyraenoidei

Great Barrier Reef *see* Coral Reefs

Great Duckweed, G, *see Spirodela*

Great Pipefish *(Syngnathus acus) see Syngnathus*

Great Scallops, G, *see Pecten*

Great Silver Beetle, F, *see* Hydrophilidae

Greater Scissortail *(Rasbora caudimaculata) see Rasbora*

Greater Weever *(Trachinus draco) see* Trachinoidei

Green Acara *(Aequidens portalegrensis) see Aequidens*

Green Algae, D, *see* Chlorophyta

Green Barb *(Barbus semifasciolatus) see Barbus*

Green Characin *(Alestes chaperi) see Alestes*

Green Chromide *(Etroplus suratensis) see Etroplus*

Green Damsel *(Chromis coeruleus) see Chromis*

Green Discus *(Symphysodon aequifasciata) see Symphysodon*

Green Fringe-lipped Labeo *(Labeo frenatus) see Labeo*

Green Knifefish *(Eigenmannia virescens) see* Gymnotoidei

Green Neon *(Hemigrammus hyanuary) see Hemigrammus*

Green Panchax *(Aplocheilus blocki) see Aplocheilus*

Green Parrot Wrasse *(Thalassoma lunare) see Thalassoma*

Green Rivulus *(Rivulus cylindraceus) see Rivulus*

Green Rivulus *(Rivulus urophthalmus) see Rivulus*

Green Seaweed *(Acetabularia mediterranea) see Acetabularia*

Green Seaweed, G, *see* Enteromorpha

Greenland Shark *(Somniosus microcephalus) see* Squalidae

Grenadiers (Macrouridae, F) *see* Gadiformes

Grey Angelfish *(Pomacanthus arcuatus) see Pomacanthus*

Grey Gurnard *(Trigla gurnardus) see* Scorpaenoidei

Grey Mullets (Mugilidae, F) *see* Mugiloidei

Grey Mullets, related species, Sub-O, *see* Mugiloidei

Grey Triggerfish *(Balistes capriscus) see Balistes*

Grimaldichthys profundissimus *see* Gadiformes

Griseofulvin. An antibiotic* which is obtained as a metabolic product of *Penicillium griseofulvum.* It has a marked inhibitive effect on parasitical moulds and it can be used with much success against Saprolegniacea* and other mould parasites (*see* Mycoses) in fish. Griseofulvin is only very slightly soluble in water and must be dissolved in alcohol before it is introduced into the aquarium water. For prolonged bathing treatments*, 25 mg/l should be added to a marine aquarium and 10 mg/l to a freshwater aquarium. The treatment should last for 1–3 days. Plants take in Griseofulvin very quickly so it is best to treat them in separate tanks.

Grooming; Body Care. Term used in ethology. In fishes, it usually describes the behaviour adopted by them to rid themselves of skin parasites. To do this, the fish will swim along against plant stems, pieces of wood or rocks. For the same purpose, smaller fish will often rub against the dorsal ridges of large, quiet fish. Various species have specialised in providing body care for other kinds of fish, eg the *Labroides** species, the so-called Cleaner fishes of the coral reefs. In this way, they are also able to acquire food for themselves. Freshwater fish, too, eg *Gyrinocheilus* (Gyrinocheilidae*), often perform comfort movements towards foreign species. However, no exact tests have been carried out to see whether this behaviour is intended to remove foreign bodies from the fish's skin or to remove pieces of skin (epidermal layer) as food. Impressive displays of body care are shown by the birds as they clean and oil their feathers, as they bathe in grit or in shallow water, or as mating pairs clean each other's head. Mammals, too, have similar behaviour patterns. They lick their coats clean, cleanse their anal regions after defaecating, and lick their newly-born offspring and growing youngsters.

Grooming or Cleaning Behaviour. Term used in ethology. An interspecific relationship of value to the participants. Grooming behaviour can be observed between species of fish, between shrimps and fish, birds and mammals and other groupings. The cleaner rids his 'client' of parasites, usually, and so from his point of view he acquires a continual supply of food. Cleaner and client make themselves understood by signals. The best-known Cleanerfish, for example, *Labroides dimidiatus (Labroides*)*, makes a rocking movement as a sign of its grooming behaviour. Client fish normally react by adopting a particular cleaning position. The Cleanerfish just mentioned is imitated by the predatory *Aspidontus**. Other examples of Cleanerfishes include *Symphodus melarnocercus* from the Mediterranean or the *Anisotremus** species.

Groundling *(Noemacheilus barbatulus) see Noemacheilus*

Growth. The enlargement of cells, parts of the body or the whole body by the processing of structure-forming building materials. The rate of growth is greatest in the youthful phase, as a general rule (*see* Ontogenesis). In animals, with the onset of sexual maturity, the rate of growth is greatly reduced. Hormones* are the main regulating factors. For a normal development in vertebrates, the secretion of the growth hormone somatotrophin by the pituitary body and the activity of the thyroid gland and the thymus gland are of great importance. The check on growth results from sex glands.

When rearing fry (Fry, Feeding of*), various factors play an important role in the pattern of growth: nutrition*, spatial conditions and temperature. A clear case of this is provided by the periodical growth of fish from temperate zones that is caused by the seasons. As this periodical growth can be read off from the annual rings of the scales, bones or otoliths, it raises the possibility of being able to tell a fish's age in this way.

Growth Form; form of growth. With reference to a plant, growth form are the essential features of what the plant looks like (phenotype) as determined by the structure of its stems and the way they are arranged spatially, its type of root-system, its growth rhythm and its life-span. The flowers and inflorescences play a subordinate role only.

Growth of Young Fish *see* Fry, Rearing of

Growth Substances. Active substances* found in plants, eg β-indole acetic acid, which largely direct elongate growth, but which also have an effect on organ-building. With many plant stems, it is possible to dab on some growth substance paste or to emerse it in a solution of the growth substance, to bring about the development of stem-born roots in large numbers.

Grunts, Ga, *see Anisotremus, Haemulon*

Guanin. Organic base with a high nitrogen content which was first discovered in guano – the excrement of sea birds. Guanin makes up a greater proportion of the excretion* in proteometabolism but only in some invertebrates. In fish guanin and other chemically similar substances are deposited in crystalline form in certain cells, the guanophores or iridocytes (Chromatophores*). The way in which light is refracted and reflected in the crystals produces the silver sheen of the skin, the eye-ball, the iris and the swimbladder. It also produces the reflective effect of the luminescent organs*. Guanin is also an important constituent of nucleic acids*.

Guatemalan Glass Characin *(Roeboides guatemalensis) see Roeboides*

Gudgeons, G, *see Gobio*

Guianas Cichlid *(Tilapia guinasana) see Tilapia*

Guineas Cichlid *(Tilapia guineensis) see Tilapia*

Guitarfish *(Rhinobatos rhinobatos) see* Rhinobatidae

Guitarfishes, F, *see* Rhinobatidae

Günther, Albert C. L. (1830–1914). Well-known ichthyologist, born in Esslingen. Studied theology in Tübingen and later medicine at Berlin and Bonn. His book *Die Fische des Neckars* ('The Fishes of the Neckar') appeared in 1855. In 1856 he went to the British Museum in London where he studied snakes at first, but later he was to become the leading ichthyologist of his day. From 1875 to 1895 he was Director of the Zoological Department of the British Museum. Between 1859 and 1870 there appeared the 8 volumes entitled *Catalogue of Fishes* in which he gave a comprehensive survey of the whole field of ichthyology, describing some 8,525 species of fish. This work appeared in German translation in 1880 under the title *Handbuch der Ichthyologie*. Günther worked on a very wide range of subjects within the field of ichthyology and wrote a large number of one-off publications. His comprehensive work *Die Fische der Südsee* ('The Fish of Oceania') appeared in 1873–1910.

Guppy (*Poecilia reticulata*, standard fish) *see Poecilia*

Gurnards (Triglidae, F) *see* Scorpaenoidei

Guttation. In contrast to transpiration*, guttation describes the process whereby water is actively pressed out of certain openings on leaves. That is why after warm damp nights you might see water droplets hanging from the edges and tips of leaves, giving the impression of dew. Some aquatic plants are also able to guttate. The water is squeezed out through slits known as hydathodes or through apical openings. Hydathodes are stomata which have lost their mobility. Apical openings occur at the tips or edges of leaves when the epidermis and part of the tissue underneath decays, thus revealing the ends of vascular tissue.

Gymnarchidae, F, *see* Mormyriformes

Gymnarchus niloticus *see* Mormyriformes

Gymnochanda FRASER-BRUNNER, 1955. G of the Centropomidae*. Comprises one species, thought to be native to the still waters of Malaysia. Small fish with a glassy transparent look. They have a moderately elongated shape with no scales and large eyes. The D2 and An have particularly long fin rays (2–7), at least in the males. These elongated rays are preceded by hard rays – one in the case of the D2, and 3 in the case of the An. The C is deeply indented. Very little is known about the biology of *Gymnochanda*. They have been kept with success in the aquarium in weak brackish water at a temperature of around 25 °C. The tank needs to offer enough room for swimming about in and the fishes should be fed on live food.

– *G. filamentosa* FRASER-BRUNNER, 1955. Malaysia. Up to 5 cm. Of transparent appearance, like glass, but when light shines upon it the fish has a delicate golden shimmer. The iris and operculum are an iridescent grass-green colour. 7–9 fine, vertical lines. The male has very long fin rays, the ends of which are thickened and white in colour. Mouth reddish (fig. 582).

Gymnocorymbus EIGENMANN, 1908. G of the Characidae*, Sub-F Tetragonopterinae. Up to 6 cm. This G consists of small Characins that live in the Amazon Basin. They are almost round in shape but very compressed from side to side. The D and An are relatively large. The bottons of the C and An are scaly and there is an adipose fin. The lack of scales at the nape is typical of the G. LL complete. The various species of *Gymnocorymbus* need similar care to those of *Hemigrammus**. Their breeding requirements are also similar. According to STALLKNECHT (1965), the male performs a very typical fluttering dance* to court the female before mating takes place. To date, only one species, the Black Widow *(G. ternetzi)* has been of importance to the aquarist. In the USA, another species, *G. sokolofi* GÉRY, 1964, has also been introduced into the aquarium. It is similar in colouring and has a similar habitat to *G. ternetzi*, but it has not yet been imported alive to Europe.

Fig. 582 *Gymnochanda filamentosa*

– *G. ternetzi* (BOULENGER, 1895); Black Widow, Black Tetra, Blackamoor, Petticoat Fish, Butterfly Tetra. Up to 6 cm. Dorsal surface olive-green, ventral surface silver-white. In young fish the flanks are deep black, especially towards the rear end, in adults they are smoky-grey. There are two very large, black shoulder flecks, almost in the shape of bands, just behind the operculum and beneath the insertion of the D. The adipose fin, D and An are all deep black; the C (at least as far as the white tips), Vs and Ps are colourless. This is a very attractive species, particularly when young, and it is popular in the aquarium. In recent years, a mutant species with veil-like elongated fins has become more widespread (Mutation*) (fig. 583).

Gymnophionia, O, *see* Amphibia

Gymnospermae, Sub-D, *see* Spermatophyta

Gymnothorax BLOCH, 1795. G of the Muraenidae, O Anguilliformes*. Comprises numerous species found in tropical seas among rocks and coral reefs. They are predators, active at night. Length 50–150 cm. Some are prettily patterned fish, and they all resemble *Muraena**, except that the rear nasal opening does not jut forwards like a tube. *Gymnothorax* species are aggressive fish, and because of this and their size they are only suitable for large aquaria. The Leopard Moray, *G. favagineus* BLOCH and SCHNEIDER, 1801, has proved to be extremely hardy in the aquarium. It grows up to 150 cm long and comes from the Indo-Pacific. It is white in colour with black blotches.

Fig. 583 *Gymnocorymbus ternetzi* (Veil-tail mutants)

Gymnotidae, F, *see* Gymnotoidei

Gymnotoidei; Electric Eels and Knifefishes. Sub-O of the Cypriniformes*, Sup-O Ostariophysi. In body shape reminiscent of a knife-blade or an eel. Their distribution area stretches from Guatemala in the north to the Rio de la Plata in central S America. The skeleton forming the cranium has much in common with that of the Characoidei*, the most primitive Cypriniformes. Similarly, the Weberian apparatus (*see* Cypriniformes) is simple, and it is not fused with the vertebral bodies. However, the Gymnotoidei must not simply be regarded as a relatively primitive group of fish belonging to the Cypriniformes; rather, they are types of fish that are highly specialised in various respects. They are easiest to recognise by these specialised features. The most striking one is the very long An which in some species stretches from the throat region to the tip of the long tail like an undulating fin corona. In other species the An begins in front of where the Ps are inserted. The anal fin is the method of locomotion. Swimming backwards or forwards is made possible by the direction of the wave movement in the anal fin corona. This is similar to the method used by the Featherbacks of Africa (F Notopteridae, O Mormyriformes*), with which the Gymnotoidei share some common body-shape features, but with which they are not related. When Gymnotids remain in one spot, the waves in the front and rear parts of the fin corona work opposed to one another. Moving along with the help of an undulating fin corona causes much less turbulence in the water than normal swimming methods, so Gymnotids give the impression of gliding or stealing through the water. With the exception of the Ps and the highly developed An, all the other fins are much reduced or they have regressed completely. A true D is absent altogether, although some species may have a thread-like adipose fin (without fin rays) in its place. The C is very small or it has regressed completely, the tail ending in a long point with no fins. Not one species has ventral fins. The head is truncated or it may be elongated at the front like a snout. All Gymnotoidei have very small, button-like eyes. The anus always lies in front of the An, ie very far forwards. In the most extreme case the anus is on the underside of the head, and it is never further back than the Ps. The scales are very small and deeply embedded in the thick skin. Many Gymnotids have electric organs which help them to orientate themselves in the water and to catch their prey. They have also developed various forms of accessory respiration. Only very little is known about their reproduction. The 4 Fs described below belong to the Sub-O and in total they include about 35 species.

– *Apteronotidae*; American Knifefishes. This F is occasionally known as the Sternarchidae. They are easiest to distinguish from the Rhamphichthyidae by the long thread-like or whip-like adipose fin on the back. Another feature is the tip of the tail which ends in a small C. Some species have an elongated trunk-like mouth, reminiscent of a Tapir's trunk. The best known species is the 50-cm-long Tovira Cavallo *(Apteronotus albifrons)* (fig. 584). It can emit weak surges of electrical current in very rapid succession (up to 1,000 per second). It is thought that this electric organ is used in

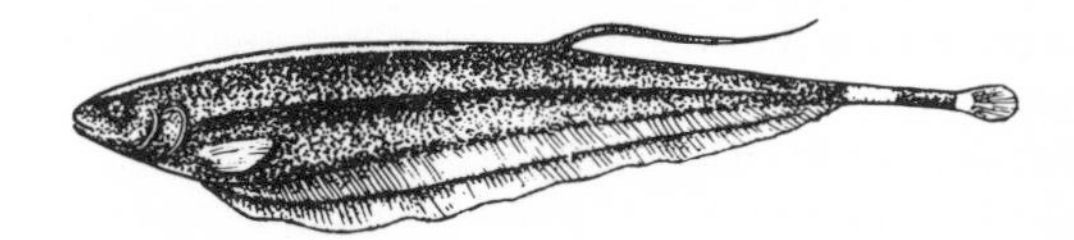

Fig. 584 *Apteronotus albifrons*

orientation. The Apteronotidae are black-brown in colour with a shining white streak across the forehead and along the dorsal surface. There is also a broad white band forming a ring around the caudal peduncle. These fish come from the Amazon region and NE South America. Occasionally, they are kept as specimens in large public aquaria. They apparently require similar care to the Rhamphichthyidae *(see below).*

– *Gymnotidae*; Knifefishes. Unlike other Fs, the Gymnotidae do not have an electric organ. The most characteristic member of the F, the Banded Knifefish *(Gymnotus carapo)* (fig. 585), is found from Guatemala to La Plata. It is an eel-like fish growing up to 60 cm long. In

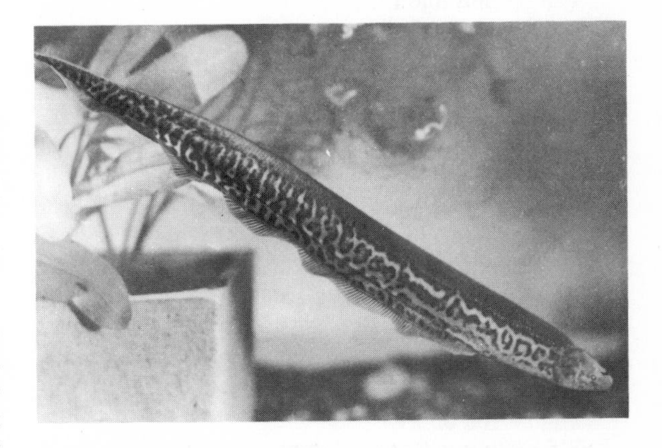

Fig. 585 *Gymnotus carapo*

cross-section it looks almost cylindrical at the front but towards the rear it becomes increasingly flatter from side to side. The tail ends in a long pointed tip. The head is truncated and the mouth is broad and set at an angle. A single row of conical teeth are fixed into the jaws. The D, C and Vs do not exist, but there are a great many, very small scales. A means of accessory respiration is present. The fish swim to the surface to breathe in air. Juveniles are fleshy colours or a dirty ochre with a number of irregular, transverse bands, whereas adults have pale slanting bands on a dark background colour. Gymnotidae are active at night, when they feed on small animals and other fish. Young *Gymnotus* specimens are imported from time to time. They are very greedy fish which also settle down quickly in the aquarium. If several specimens are kept together, there will always be fights connected with order of rank, and severe wounds may result. Generally speaking, Knifefish pay no attention to larger fish.

– *Electrophoridae*; Electric Eels. The only member of the F is the 2 m long Electric Eel *(Electrophorus electricus)* (fig. 586). It lives in the north-eastern parts of S America (Guyana) and in the central and lower Amazon region. Its body shape is like an eel's and it has a broad, heavy head with a wide mouth. The eyes are small, usually sea-green in colour, and the nasal openings are tubular and elongated towards the outside. The

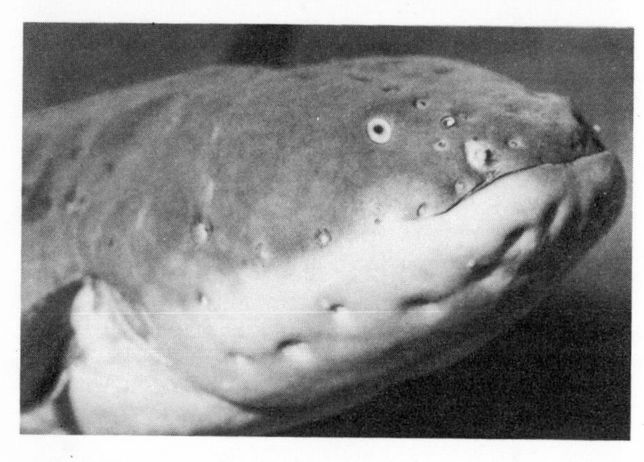

Fig. 586 *Electrophorus electricus*

head is covered with a number of bowl-shaped sense depressions. The upper side is dark, the lower side (particularly the throat region) is orange-coloured. Like all Gymnotoidei, this species has a very long An; the D, C and Vs are absent and the Ps are small. The head and body cavity take up only one-fifth of the total length, the remaining four-fifths consisting of the tail region with the electric organs. The Electric Eel has several types of electric organs. There is a main organ which is made up of a large number of wafer-like electroplates arranged in columns (20–50 of them). In addition, there are 2 smaller organs (the Sachs organ and the Hunter organ). The main organ is used in defence and in hunting for food. It discharges up to 600 volts in a series of strong separate pulses. Any prey or enemies within range will be killed or stunned. The shocks are not fatal for humans, but if stunned by them a person may drown. Mammals, such as horses, do not normally survive strong shocks. The main battery is supplemented by the Hunter organ whilst the Sachs organ sends out a continuous charge (20–50 weak pulses per second, which are registered by the sense depressions on the electric eel's head. In effect the whole system is like a radar network enabling the fish to orientate itself and find its prey. There is a distinct polarity between the front and rear ends – the front being the positive pole, the rear being the negative pole. The voltage, which can be measured by electrodes placed on the eel, depends on the distance between the electrodes (ie it depends on the length of the eel). It can measure 300–600 volts. Voltage no longer increases in old specimens but the strength of the current does increase.

Another peculiarity of the electric eel is the way it respires with its mouth cavity. It swims to the surface every 15 minutes or so and fills its mouth with air. Gaseous exchange takes place in the mucous membrane of the mouth which contains a great number of capillaries. If deprived of air, electric eels suffocate. Little is known about their reproductive methods. The fry are cared for by their parents until they are about 15 cm long. Electric eels are very interesting specimens in public aquaria, but they are perhaps a little boring to watch. Usually, they hide beneath rocks or in crevices, and can only be encouraged to show any sign of movement when being fed. Specimens should be kept in individual tanks because several of the eels together cause a lot of problems with their electrical activities, and this in effect constitutes animal torture. The largest specimen ever caught was almost 3 m long.

– *Rhamphichthyidae*; American Knifefishes. All Rhamphichthyidae are shaped like a very long knife-blade. The long An begins beneath the Ps and reaches not quite to the end of the very slender body. The tip of the body always lacks a C, and looks like a flexible rod. It can move a bit like a snake and it is thought to be covered in large numbers of sense organs. When these fish are swimming slowly backwards, it is easy to observe that they feel around with this tail, manoeuvring themselves into crevices and hiding-holes with it. There are no Vs and no D. Several species are able to make use of their branchial apparatus as an accessory respiratory organ. The Rhamphichthyidae are distributed across S America, as well as Central America and some West Indian Islands. They are found in calm, weedy waters. Young specimens of several species have been imported for aquarium-keeping purposes. One particularly popular species is the Green or Glass Knifefish *(Eigenmannia virescens)* (fig. 587) which can

Fig. 587 *Eigenmannia virescens*

reach lengths of 45 cm. It is mainly flesh-coloured, but when light shines on it it shimmers a delicate green or blue colour. Over-all, however, the Green Knifefish is very translucent. In the aquarium, specimens are fairly hardy. They adapt well once they have got over their initial timidity. Avoid too bright lighting and make sure there are enough places of refuge. All types of food can be given to them.

Gymnotus carapo *see* Gymnotoidei

Gyrinidae; Whirligig Beetles (figs. 588 and 589). F of the Beetles (Coleoptera*). There are about 800 species, of which about 17 are found in Europe. The Gyrinidae are of medium size (usually 5–7 mm long), and have an elongated oval shape and dull colouring. They are found in ponds where they attract a great deal of attention on account of the fact that they swim around very fast on the water surface marking out curves and spirals. They are completely adapted to their way of life. Because the two halves of their eyes are constructed differently,

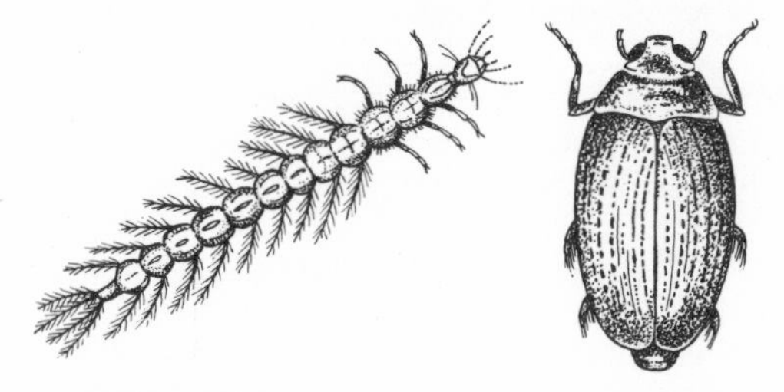

Fig. 588 *Gyrinus notator*

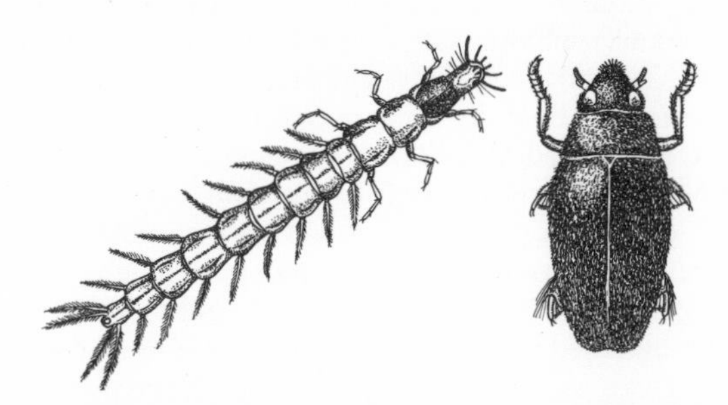

Fig. 589 *Orectochilus villosus.* Left, larva; right, imago

they can see above and below the water at the same time, for example. Their middle and hind legs also show a high degree of adaptation – they have developed into very efficient 'rudders' of complicated structure. The front legs, on the other hand, have remained in a much more primitive state and are used to attach the beetle firmly under water and to catch prey. Gyrinidae larvae live completely submerged. They are very long in shape, have hairy tracheal gills and live as predators.

Gyrinocheilidae; Algae-eaters. F of the Cypriniformes*, Sub-O Cyprinoidei*. Although it is not quite certain, there are probably 3 species belonging to the F, and they live in the mountain streams of Thailand and Kalimantan. They are small fish, like Loaches, measuring between 15 and 20 cm long. Their chief distinguishing feature is the special form of their gill slits. They do not take their respiratory water in by the mouth, which is the usual method in fish, but suck it in via two narrow openings located at the sides in the nape region. So the upper parts of both opercular slits are completely separated from one another. The water reaches the gills via the pharynx, then it leaves the body again via the lower part of the opercular slits. Every minute water is sucked in and expelled about 230 to 240 times. It is this special mechanism that enables the fish to pursue its chief activity of browsing on algae, without having to break off to breathe. Because the mouth is not used for respiration, the fish can browse anywhere for as long as it likes. The mouth is inferior and is surrounded by thick lips, which act as a sucking organ. There are rasp-like folds on the lips. In contrast to all the other Fs of the Cyprinoidei*, the Gyrinocheilidae have no pharyngeal teeth. The main colouring of the body is yellow-brown but it is interrupted by numerous blotches or by a dark longitudinal band. The females are bigger and stronger and, just like the males, have spawning tubercles when sexually mature. These look like little knots around the snout. In aquarium-keeping, *Gyrinocheilus aymonieri* (fig. 590) is very important as an algae-eater. However, only individual specimens should be kept in the tanks, because when there is more than one specimen they will constantly be fighting over ranking order. As a supplement to their diet, soaked spinach can be included.

Fig. 590 *Gyrinocheilus aymonieri*

Gyrinocheilus aymonieri *see* Gyrinocheilidae

Gyrodactylus, G, *see* Trematodes

Habitat. 1) The ecological character of the place where a plant or plant community grows. It consists of everything in the environment that can have an effect upon the plant, such as substrate, climate or other organisms. 2) The geographical location a plant or animal hails from.

Haddock *(Melanogrammus aeglefinus) see* Gadiformes

Haeckel, Ernst (1834–1919). Zoologist and nature philosopher. Born in Potsdam. From 1852 he studied natural history and medicine in Berlin and Würzburg. In 1862 he became Professor of general and specialised Zoology at Jena University. He undertook the scientific examination of countless lower marine animals. His work *Die Radiolarien* ('The Radiolarians') appeared in 1862, in 1869 there followed *Zur Entwicklungsgeschichte der Siphonophoren* ('The Ontogenetic History of the Siphonophora'), and in 1872 *Die Kalkschwämme* ('The Chalk Sponges'), etc. Haeckel was an important force in spreading Darwin's descent theory. From his own ontogenetic comparable studies, Haeckel was able to formulate biogenetic principles and descent theory (he included human evolution in the latter). In a great many of his works he tried to draw philosophical generalisations from his studies, eg in *Natürliche Schöpfungsgeschichte* ('Natural History of Creation') (1868), *Generelle Morphologie der Organismen* ('General Morphology of Organisms') (1866), *Anthropogenie* ('Anthropogeny') (1874). Besides all this, he also made a living out of popular (*Die Welträtsel* ['The Riddles of the Universe']) and aesthetic natural history books (*Kunstformen der Natur* ['Nature's Works of Art'] 1899–1903).

Haemoglobin *see* Blood

Haemopis SAVIGNY, 1822; Horse Leech (fig. 591). G of the Leeches (Hirudinea*). Distributed throughout the holarctic region*. There are about 7 species, one of which is European *(H. sanguisuga)*. They grow to about

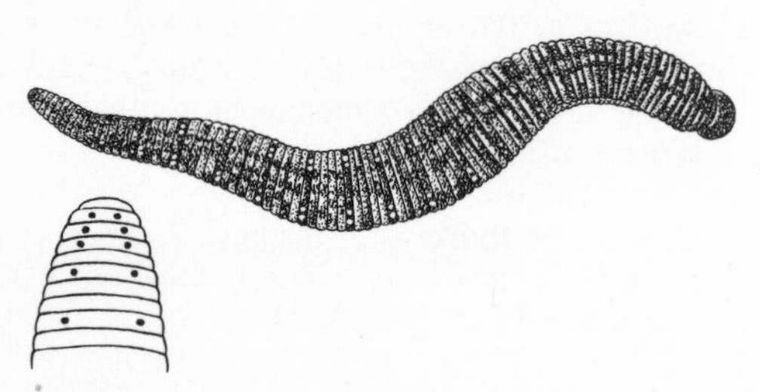

Fig. 591 *Haemopis sanguisuga*. Below, head with eyes (considerably enlarged)

15 cm long. Dorsal surface a uniform olive-green to dark brown colour (in contrast to *Hirudo**). Found mainly in plant-rich ponds where they live as predators. Their diet consists of snails, insects and their larvae, worms etc, all of which are eaten whole. *Haemopis* do not suck blood. They are suitable specimens for a pond aquarium.

Haemulon CUVIER, 1829; Grunt (fig. 592). G of the Pomadasyidae* comprising numerous species found in tropical parts of the W Atlantic. The smaller species live in shoals, the larger ones live alone in coastal waters,

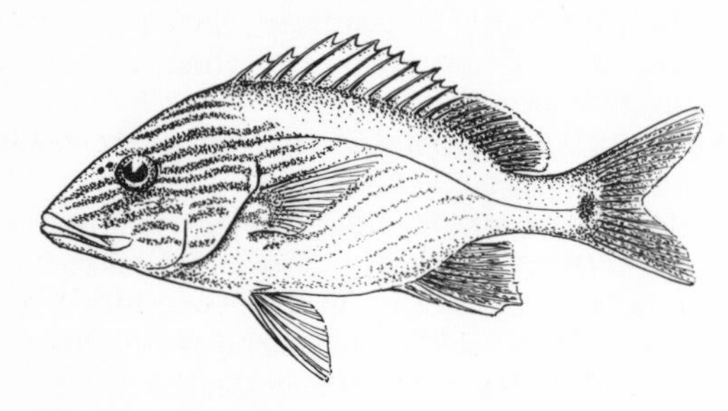

Fig. 592 *Haemulon plumieri*

often among coral reefs. Their food consists of small animals. *Haemulon* has a typical Perch shape and usually grows to between 20 and 30 cm long. It is a fairly colourful fish. In contrast to *Anisotremus** the soft D and An are scaled. *Haemulon* can produce noises with its pharyngeal teeth; the noises are amplified by the swimbladder which acts as a resonator. A typical feature of these fishes is their ritualised aggressive behaviour; they position themselves face to face, their mouths open wide to reveal their colourful, shining interiors. Grunts survive well in the aquarium but they are rarely imported.

Hagfish types (Myxiniformes, O) *see* Cyclostomata

Hagfishes (Myxinidae, G) *see* Cyclostomata

Hagfishes (Petromyzoniformes, F) *see* Cyclostomata

Hairtails (Trichiuridae, F) *see* Scombroidei

Hairy Crab *(Pilumnus hirtellus) see Pilumnus*

Hairy Fish (Mirapinnidae, F) *see* Cetomimiformes

Haiti Limia *(Poecilia heterandria) see Poecilia*

Haiti Molly *(Poecilia elegans) see Poecilia*

Hake *(Merluccius merluccius) see* Gadiformes

Halacarines, F, *see* Acari

Half-banded Coolie Loach *(Acanthophthalmus semicinctus) see Acanthophthalmus*

Halfbeak *(Dermogenys pusillus) see* Exocoetoidei

Halfbeaks (Hemirhamphidae, F, name not in current usage) *see* Exocoetoidei

Half-striped Barb *(Barbus semifasciolatus) see Barbus*

Halibut *(Hippoglossus hippoglossus) see* Pleuronectiformes

Halichoeres RÜPPEL, 1861. G of the Labridae, Sub-O Labroidei*. Labroidei*. Comprises numerous species found in tropical parts of the Indo-Pacific and Atlantic, among coral reefs and along rocky coastlines. Food consists of small animals. The majority are only up to 15 cm long and are attractively coloured. They are elongated fishes, laterally compressed, with large scales. The C is rounded. Only a few species are imported even although they survive well in the aquarium. As they grow older, the specimens often become intolerant of one another. Young specimens are sociable, eg:

– *H. gymnocephalus* (BLOCH and SCHNEIDER, 1801); Black-blotched Wrasse. Indo-Pacific. Up to 17 cm. A grass-green colour with a short, dark, vertical stripe behind the eyes.

– *H. marginatus* (RÜPPEL, 1835); Black-blotched Wrasse, Splendid Rainbowfish. Indo-Pacific, Red Sea. Up to 16 cm. Head and front part of the body yellow with blue stripes, rear portion of the body covered in emerald-green longitudinal stripes that broaden out and an inky-blue honeycomb patterning. All the fins have a yellow and green pattern with blue edges.

Halichondria FLEMING, 1828. G of marine-dwelling sponges (Porifera*) with a horny skeleton and siliceous spicules. They grow like a crust on shellfish, seaweeds and rocks. Their outline consists of tower-like peaks on which are found the pores through which water flows out. The Bread-crumb Sponge *H. panicea* (PALLAS) is relatively common from the Baltic to the Black Sea. It can be yellow, reddish, grey or greenish in colour. As an experiment small specimens can be kept in the aquarium.

Halimeda. G of the Green Algae (Chlorophyta*) found in warmer seas. They live at depths of a few metres or in shady places. The thallus is limy. *H. tuna* (ELLIS and SOL.) LAM., is found in the Mediterranean. Its shape is like that of a Prickly Pear Cactus

(Opuntia). *Halimeda* can be kept for a short time in the aquarium.

Halimochirurgus triacanthus *see* Triacanthidae

Haliotis LINNAEUS, 1758; Ormer, 'Sea-ear'. Primitive G of the Molluscs (Gastropoda*, Prosobranchia*). Found in warmer seas. The Ormer is crepuscular and eats algae and dead flesh. It lives in the upper littoral. Its shell is flat and shaped like an ear punctuated with a row of small holes through which sense tentacles are stretched. The upper surface of the shell is usually thickly encrusted so it is difficult to recognise the creature. The inside of the shell is a shimmering mother-of-pearl. *Haliotis* can attach itself very firmly by means of its foot. Can possibly be kept in the aquarium.

Haliplidae; Crawling Water Beetles (fig. 593). F of the Beetles (Coleoptera*) comprising relatively few species. Represented in central Europe by about 20 species. The Haliplidae are mostly only about 2 mm long and they live in standing or slow-flowing waters

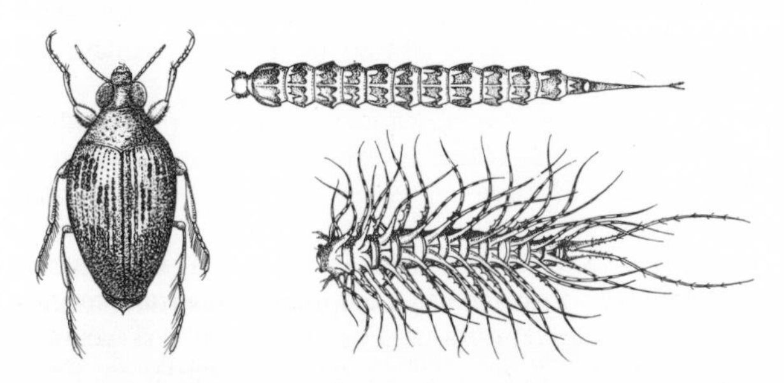

Fig. 593 Haliplidae. Left, imago; right, various larvae

both in their larval and imaginal stages. In contrast, the pupal stage is spent on land. This Water Beetle can only swim feebly (with alternate leg action as in the Hydrophilidae*) as its legs are not so modified for swimming as are some members of the Water Beetle group. The middle and rear feet are noticeably hairy, and the hind coxae form very broad plates.

Halocynthia LINNAEUS, 1758. G of the solitary Sea-squirts (Ascidiacea*). The generic characteristics are not visible from the outside. A very beautiful, unmistakable species comes from the Mediterranean: *H. papillosa* LINNAEUS, 1758. This species grows to between 6 and 10 cm long and at least half of its body is a shining red colour. They have bristles on the siphons and the coarse tunic is very solid. *Halocynthia* survives well in the aquarium.

Haloragaceae; Water Milfoil or Marestail Family. F of the Magnoliatae*, comprising 8 ga and 160 species. Tropical to subtropical, with only a few species in temperate regions. The majority of them are herbs with an alternate or whorled leaf arrangement. Flowers hermaphroditic or unisexual and monoecious, small, polysymmetrical. In the main, the Haloragaceae are marsh and aquatic plants, with only a few species growing in drier places. The extraordinary rhubarb-like plant, *Gunnera*, is a member of the F (it is often seen in botanical gardens). Two Ga, *Myriophyllum** and *Proserpinaca**, are of importance to the aquarist.

Hammerhead Sharks, F, *see* Sphyrnidae

Hancock's Amblydoras *(Amblydoras hancocki) see Amblydoras*

Haplochromis HILGENDORF, 1888 (fig. 594). A G of the Cichlidae* with a large number of species. Widely distributed in tropical parts of Africa, a few species also penetrate into the subtropical regions, such as N Africa, Asia Minor, S Africa. *Haplochromis* prefers standing waters, so in Lake Malawi* alone there are over 100 different species. They are medium-sized, fairly tall fish with a typical Perch shape. Because they have a relatively large head and thick lips, *Haplochromis* species look robust. The D usually has 15 hard rays, the An always has 3. The soft-rayed parts of these fins are only moderately elongated, rounded or slightly pointed. The C is shaped like a fan. In its reproductive method, this G is highly specialised (apomorphic*). As far as we know, they are (almost?) exclusively mouth-brooders (but compare below!). The female takes the eggs into her mouth as soon as they are released, even before they are fertilised, thus cutting short the period in which the spawn can come to harm. In the course of their evolution, the *Haplochromis* species and related Ga (eg *Pseudotropheus**, *Labeotropheus**) have developed a way of guaranteeing egg-fertilisation. The males have a number of well-developed flecks on their anal fin which look very similar to the real eggs. The female snaps at what she thinks are eggs, and in doing so receives sperm into her mouth where fertilisation occurs. The spawn takes a long time (2–3 weeks) to hatch out.

Fig. 594 *Haplochromis* sp.

During this period, the female does not eat anything, but she does chew the eggs round in her throat pouch every so often to make sure they get enough oxygen. During the first few days that the youngsters are swimming about freely, they will still return to the mother's mouth when danger threatens and at night. However, it is not certain that all *Haplochromis* species are mouth-brooders; amongst other Ga too (eg *Tilapia**), primitive (plesiomorphic*) methods of reproduction have been retained alongside newer methods. For example, the so-called *'Pelmatochromis' thomasi* BOULENGER is regarded by some authors as *Haplochromis* on account of its morphological features, although it is a typical openwater spawner.

Haplochromis should be kept in large aquaria with plenty of rocks and roots. Unlike many other African Cichlid Ga, *Haplochromis* will not normally seize upon water plants, so you will not have to dispense with them—in fact, *Haplochromis* can even be kept in a community tank. At spawning time, however, you will have to remember that the males will form territories and territorial battles may follow. The water temperature should be about 24–26 °C, but there are no special requirements regarding the water chemistry. Feed with live food of all kinds; many species also like some supplementary vegetarian food (cooked rolled oats, well-soaked spinach etc). Only a few of the numerous species have so far proved important to the aquarist. They include:

– *H. burtoni* (GÜNTHER, 1893); Burton's Nigerian Mouthbrooder. Distributed throughout 10 cm. Yellowish to greenish-grey with a pale blue iridescence and dark oblique stripes. The head has a distinct eye-band running from the nape to the corners of the mouth, plus two bands on the forehead. The oval flecks on the An are particularly large and intensive on the male (fig. 595).

Fig. 595 *Haplochromis burtoni*

– *H. desfontainesi* (LACÉPÈDE, 1802); Desert Mouthbrooder. Widespread in N and E Africa. Up to 15 cm. Brownish or greenish with a dark longitudinal band and similar oblique stripes. Unpaired fins are dotted in orange. Males are a steel blue colour during the spawning season.

– *H. polystigma* (REGAN, 1921); Polystigma, Poly, Leopard Cichlid. Lake Malawi. Body and fins a pale ground colour with irregular dark flecks. Oval flecks small, less well defined than in the species mentioned above.

– *H. thomasi* (BOULENGER, 1915); African Butterfly Cichlid, Central Africa (the Congo). Up to 7 cm. A yellowish-grey background colour with shining pale blue speckles. About 6 dark oblique bands. It is questionable whether this species belongs to the G (see above!) (fig. 596).

– *H. wingati* (BOULENGER, 1902); Nigerian Mouthbrooder. Central and E Africa. Up to 12 cm. Very similar in colouring to *H. burtoni*, but the oval flecks are small in this case and only begin from the 8th anal ray.

Haplodoci, O, name not in current usage, *see* Batrachoidiformes

Fig. 596 *Haplochromis thomasi*

Haploid *see* Cell

Hard Water. Water that contains a lot of calcium and magnesium salts—ie the water comes originally from chalky or limestone regions or chalk-rich sandstone. Standing waters* often contain hard water too. This is as a result of the tributary waters continually feeding in hardness-forming substances, whilst only the water itself evaporates. Water is termed hard when its hardness (Water Hardness*) is more than 20° dH, water with a degree of hardness higher than 30 is very hard.

Many organisms, and their embryonic stages in particular, are highly sensitive towards high calcium and magnesium concentrations. For this reason, it is recommended that soft water* be used in the aquarium; this is true even for hardy specimens such as representatives of the Ga *Tilapia** and *Haplochromis** and other African Cichlids (Cichlidae*) which may live in very salty lakes. Many livebearing Tooth-carps also live in hard water.

Hard-bellied Characin *(Mylossoma duriventre) see Mylossoma*

Harlequin Bass *(Serranus tigrinus) see Serranus*

Harlequin Fish *(Rasbora heteromorpha) see Rasbora*

Harlequin Shrimp *(Hymenocera picta) see Hymenocera*

Harpodontidae, F, *see* Myctophoidei

Harrison's Pencilfish *(Nannostomus harrisoni) see Nannostomus*

Hart's Rivulus *(Rivulus harti) see Rivulus*

Hasemania marginata, name not in current usage = *Hemigrammus nanus, see Hemigrammus*

Hassar *(Callichthys callichthys) see Callichthys*

Hasselt's Bonylip *(Osteochilus hasselti) see Osteochilus*

Hatchetfish *(Carnegiella myersi) see Carnegiella*

Hatchetfishes, F, *see* Gasteropelecidae

Hawkfishes, F, *see* Cirrhitidae

Hawkfishes, G, *see Cirrhitichthys*

Hay Infusion. Tried and tested method of culture for various single-celled, water-dwelling creatures. Hay infusion is suitable for the production of infusorians as a food for fish broods. It is also a valuable material for studying minute forms of aquatic life under a microscope. For this you will need glass containers measuring about 25 cm high and with a 15 cm diameter. Pack them loosely with hay or straw cut short, then fill them with tap water. After a few days, provided the containers are sealed and left to stand in a bright and reasonably warm spot, large numbers of micro-organisms will appear. The peak of development is usually reached after 2–3 weeks, after which time the density of the infusorians will decrease again. If a constant supply is needed, new cultures are best started off by using $^1/_3$ of the old culture water. Should infusorians fail to appear, add some pond water or rain water that has stood for a while. However, you must take care not to pour small crustaceans (*Daphnia, Cyclops* etc) in at the same time. Instead of hay, various other kinds of material can be used for your culture medium: 4–5 lumps of air-dried turnip, banana skins (about 1 cm^3 in size), or dried lettuce leaves and some mud from a pond.

The first bacteria that appear and cloud the water are the food of a whole string of ciliates (Ciliata*) which multiply very quickly. It is highly likely that among them will be the sessile bell-shaped protozoan *Vorticella* and *Paramecium**, which can even be picked out with the naked eye as moving white dots. In older cultures you may often find quite a large number of amoebae in the surface layer or on plant parts.

Head and Tail Light Tetra *(Hemigrammus ocellifer) see Hemigrammus*

Headband Pyrrhulina *(Pyrrhulina vittata) see Pyrrhulina*

Headstander *(Chilodus punctatus) see Chilodus*

Healing of Wounds *see* Regeneration, Injuries

Hearing. Among vertebrate animals, the sense of hearing is developed least in fish. The frequency range they can hear varies from species to species but lies somewhere between 100 and 1,500 Hz. Among those species that are capable of uttering sounds, the sense of hearing is probably a means of communication between individuals, too. Fish do not have an ear drum or a middle ear, and they only have a very simple equivalent of the stato-acoustic organ (inner ear, balancing organ*). In many cases the swimbladder* acts as an organ of resonance, taking over the function of sound perception. In a number of species the sound waves can be transmitted directly to the auditory organ via an opening in the bony capsule of the inner ear. This occurs in the Cypriniformes* by means of the bony Weberian apparatus.

Heart *see* Blood Vascular System

Heat Damage *see* Temperature Damage

Heating *see* Temperature Control

Hectocotylus *see* Cephalopoda

Heleocharis, G, name not in current usage, *see Eleocharis*

Heliaster GRAY, 1840. G of the predatory Starfishes (Asteroidea*) found in warm seas. It has a broad disc and 20–44 short, slender arms when fully grown. (Young specimens have only 5 arms to start with.) The members of the G are voracious eaters, devouring sea-urchins, starfish, molluscs and other invertebrates. Easy to keep in the aquarium.

Heliozoa, Cl, *see* Rhizopoda

Helisoma SWAINSON, 1840. G of American freshwater Snails belonging to the Basommatophora*, closely related to the European *Planorbis**. Because of its shell shape, *Helisoma* like other Ga (Planorbis*, Planorbarius*) is called a Ramshorn Snail. The most important species as far as the aquarist is concerned are those that come from tropical S America (they have little pigment [red]). They are usually kept in heated freshwater aquaria where they survive well with little attention.

Helminthiasis. Diseases caused by parasitical 'worms' or their larvae. As far as fish are concerned, these worms include Trematodes*, Cestodes*, Nematodes* and Acanthocephala*. Helminthiasis is commonly noticed in fish caught in the wild. Aquarium fish do not suffer from it so commonly, because the parasite's host or intermediate host (*see* Parasitism) may not be present in the aquarium. The parasites and their various stages appear on the skin and on the gills, but they are also present in the body cavity, intestine, liver and in the musculature. A cure is difficult, particularly during the larval stages.

Helodea, G, name not in current usage, *see Elodea*

Helogenes marmoratus *see* Helogenidae

Helogenidae (fig. 597). F of the Siluriformes*, Sup-O Ostariophysi. There are only 2 Ga in this F, each containing one species. Both are found in central and northern South America. The body is naked, moderately elongated; the D is small and stands well to the rear with no spine. Adipose fin small or absent, C deeply indented, An very long and broad and covered in skin.

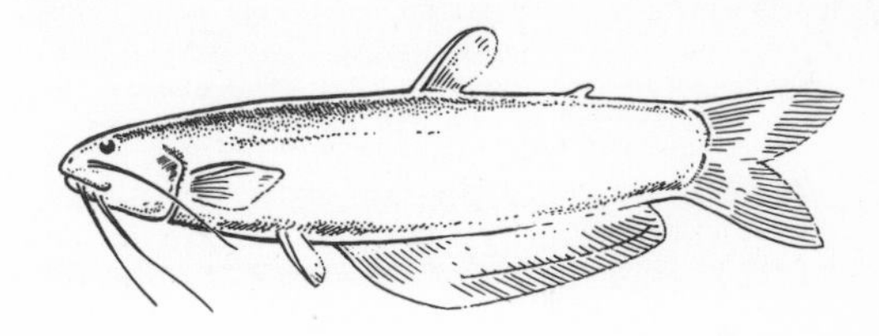

Fig. 597 Helogenidae-type

Helogenes marmoratus from the eastern part of the Amazon river complex and Guyana has occasionally been kept in the aquarium. It grows up to 10 cm long and has a fawn-brown marbled effect on a dark ground colour. Active at night and an omnivore.

Helostoma CUVIER, 1829; Kissing Gourami. G of the Helostomatidae, Sub-O Anabantoidei*. Found in SE Asia: Greater Sunda Islands, Thailand, Malaysian Peninsula, where it lives in rivers, ponds and swamps. Large fish, up to 30 cm long. Body ovate and very flat from side to side. The mouth has strong lips which can be stretched forwards and which are set with small teeth. The D and An stretch from the front third of the body to the root of the tail. The soft parts of these two fins are noticeably higher. The Vs are pointed but not elongated like threads. Background colour of the upper surface olive-green to grey, belly pale. There are some distinct dark, length-wise stripes running along the rows of scales on the flanks. On the operculum there are 2 vertical elongated blotches. A dark band extends from the edge of the hard-rayed D and An in an arch over their soft parts to the caudal peduncle. The fins are otherwise greenish or grey-yellow. The colouring of the male is more intense as it gets older. There is a slightly pigmented variety with a reddish body colour. These warmth-loving fish should be kept in roomy tanks at a temperature of at least 24°C. They will need to be fed on a vegetarian diet of algae and lettuce etc, supplemented with fine live food. The eggs are laid without a bubble nest. Highly productive Gouramis, often laying over 1,000 eggs.

– *H. temmincki* (CUVIER and VALENCIENNES, 1831); Kissing Gourami. The popular name refers to the habit these fish have of grasping or touching one another with their protruding lips. These kind of lips are particularly good for grazing algae. A valuable edible fish in its native countries (fig. 598).

Fig. 598 *Helostoma temmincki*

Helostomatidae, F, *see* Anabantoidei

Hemichromis PETERS, 1857. G of the Cichlidae*, comprising 2 species widely distributed in tropical parts of Africa. Medium-sized to large Cichlids with very pretty colouring. Moderately elongated shape, compressed laterally. D has 13–15 hard rays and 9–13 soft rays; An 3 hard rays and 7–10 soft rays. The rear parts of both these fins are only slightly elongated; C shaped like a fan. In contrast to related Ga (eg *Haplochromis**), *Hemichromis* species are typical open-water spawners, forming definite partnerships (*see also* Cichlidae). Brood-caring activities of both sexes are strongly marked – the young fish are cared for even when they are several cm long. Although *Hemichromis* is characterised by an intense, brilliant colouring, particularly during the spawning season, it is not often kept in aquaria. The main reasons for this are that these Cichlids are intolerant and have a tendency to stir up the substrate. In fact, within a short time, *Hemichromis* can completely upset the aquarium, uprooting plants, digging holes and causing stonework to tumble down. A Cichlid lover will get great pleasure out of *Hemichromis*, especially when a well-matched pair devote themselves unreservedly to caring for their brood, guiding their offspring round the tank showing off their splendid colouring. Caring for your specimens should present no problems. They are not choosy about their live food, but one thing that must be provided is an efficient filter system (to ensure clear water at all times). Also make sure that the temperature is high enough (25–28°C).

– *H. bimaculatus* GILL, 1862; Jewel Cichlid. Found in Africa, except in the south, widely distributed. Up to 15 cm. Already sexually mature when half grown. Sky-blue speckles on a red ground colour with two blue-black flecks on the gill covering and shoulder.

– *H. fasciatus* PETERS, 1857; Banded Jewel Fish. Tropical W Africa. Up to 30 cm. Body slimmer than the above-mentioned species. Head, mouth and eyes very large. A brassy iridescent colour with five black, oval flecks or transverse bands on the flanks (fig. 599).

Fig. 599 *Hemichromis fasciatus*

Hemigrammocypris FOWLER, 1910. G of the Cyprinidae*, Sub-F Rasborinae. Small, shoaling fish found in streams and standing waters in the region of Hong Kong. Body elongate, laterally compressed. LL absent. Mouth terminal. Unlike the G *Tanichthys* the front and rear nasal openings are not connected. Undemanding fish that will live happily alongside other fish. Temperature 18–22°C. A cool overwintering at 16–18°C will make breeding more successful. Omnivorous, will accept any kind of small live and dried food. For rearing purposes, fresh water at a temperature of 20–22°C is used. The bottom is decked with fine-leaved plants, then, first the male is introduced, followed by the fuller female. Specimens that have spent the winter at cooler temperatures are highly productive. The brood hatch out after about 48 hours and are easy to rear.

– *H. lini* WEITZMAN and CHAN, 1966; Garnet Minnow. S China. Up to 3.5 cm. Dorsal surface dark brown. A narrow longitudinal band begins at the height of the lower jaw and ends in a black fleck on the C. It is bounded on the top by a blue-green iridescent band. There is a golden-yellow fleck behind the eye. Fins colourless, fin rays dark. During the spawning season, the males have a length-wise band made up of dots.

Hemigrammus GILL, 1858. G of the Characidae*, Sub-F Tetragonopterinae. There is a large number of species in the G, and they consist of small American Characins, 4–8 cm long. Most are attractively coloured. They are peaceable shoaling fish that inhabit small and very small stretches of water (both standing and flowing). Very fond of swimming. Moderately elongated body, laterally much compressed, although the body height can vary. D short and pointed, small adipose fin present. An usually long, drawn out into a point at the front. C twin-lobed, Vs and Ps well developed. *Hemigrammus* can be distinguished from the closely related Ga *Cheirodon**, *Gymnocorymbus**, *Pristella** and *Hyphessobrycon** by looking at the teeth. It can further be told apart from *Hyphessobrycon* because it is only in *Hemigrammus* that the base of the caudal fin is scaly. *Hemigrammus* feeds mainly on small live food and insects that alight on the surface, although some species particularly like plants, especially *H. caudovittatus*. Nearly all the species are popular aquarium fish that will accept artificial foods. They are usually bred in small glass aquaria filled with soft, slightly acidic natural water of about 2–5 °DH and a pH value of 6.3–6.8. Or softened tap water can be used, but a small amount of alder cone or peat extract should be added. The plants, *Vesicularia* and *Myriophyllum*, and if necessary, green Perlon tissue, are all suitable spawning media. If the latter is used, the males sometimes get caught on it by the tiny anal hook. Spawning behaviour may begin the morning after the pair are introduced into the tank. Stallknecht has described this behaviour as being very typical of the genus, and certainly modified in relation to other Tetragonopterinae: the male drives at the female usually with a ventral swimming action (*see* Courtship Swimming). Just before mating, he either swims directly underneath the female or slightly behind her. A circular movement at the moment of pairing does not occur or it is only hinted at in a flat arch movement. The male does not perform a fluttering dance* either. Both sexes in *H. erythrozonus* complete a full body roll next to one another in the plant substrate as pairing takes place. Some of the eggs do not stick together very well. Most of them fall to the bottom where they must be made safe from the marauding parents by fixing in a spawning grate (*see* Breeding Aquarium). Remove the parents from the tank immediately after spawning takes place. The brood will hatch out after about 20–24 hours, and they can be reared on very fine *Cyclops* nauplii or rotifers. Growth is uninterrupted. A frequent change of water (Fry, Rearing of*) is advisable, using slightly harder water each time to get the growing youngsters used to it gradually. All the species can be kept happily in community tanks with like species or with species of related Ga. A group division within the G, as in the closely related G *Hyphessobrycon*, does not seem to make much sense here.

– *H. caudovittatus* Ahl, 1923; Buenos Aires Tetra, Red-tailed Tetra. Found in the River Plate system. Up to 7 cm. A relatively elongated body, with an intense silver sheen. Dorsal surface brown-olive, upper half of iris red. There is a comma-shaped shoulder blotch behind the operculum (although it is not always clearly visible). The horizontal tips of a rhomboid, black blotch on the caudaul peduncle and the base of the C stretch as far forwards as the insertion of the adipose fin and as far to the rear as the tips of the middle caudal fin rays. The bases of the D, the adipose fin and the Vs, and the whole of the An and the lobes of the C are all brick-red or yellow-red (lemon-yellow in the case of a colour mutant). There is a yellowish-white blotch in the corners of the rhomboid blotch on each of the C lobes. The females have less intense colours and are easy to recognise by the spawn-filled curve of their bodies. This is a very robust species, easy to rear, and it was known

very early on in the history of aquarium-keeping. It is a well-established aquarium favourite (fig. 600).

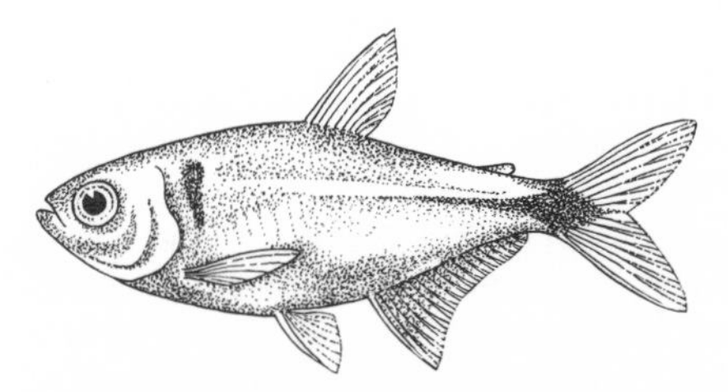

Fig. 600 *Hemigrammus caudovittatus*

– *H. erythrozonus* Durbin, 1909; Glowlight Tetra. Found in the Guyana lands. Up to 4 cm. The shape of the body is similar to that of the well-known Neon-Tetra. Body flesh-coloured, dorsal surface brown-olive, ventral surface silver. A broad, ruby-red shining band, beginning at the upper lip, stretches through the upper half of the iris, across the sides of the body, then broadens out into a shining spot which continues over the caudal peduncle into the base of the C. The first dorsal fin rays shine just as brightly red. The remaining fins are all a delicate pink or colourless, although the first rays of the An are yellow-red at the bottom. The tips of the An, D and Vs are pastel-white. This is one of the most popular and attractive Characin species. It readily spawns and grows very quickly (fig. 601).

Fig. 601 *Hemigrammus erythrozonus*

– *H. hyanuaryi* Durbin in Eigenmann, 1918; Green Neon, January Tetra. Found in Lake Hyanuary near Manãos in the Upper Amazon from Iquitos to São Paulo de Olivenca. Up to 4 cm. The shape of the body, like that of the Glowlight Tetra, is elongated. The upper half of the body is yellow-green to olive-green, the lower half greenish silver. A grass-green or bluish-green iridescent band which gets slowly wider stretches from the operculum to the caudal peduncle. In the last third of the body it is accompanied by a very broad, black zone which stretches to the middle rays of the C. There is a red-gold shining blotch in the upper half of the caudal peduncle. Iris a shining grass-green. Fins colourless, adipose fin present. It is a very attractive Characin when kept in a well-planted tank with a dark substrate. Not difficult to breed. Lays its spawn in the twilight hours (fig. 602).

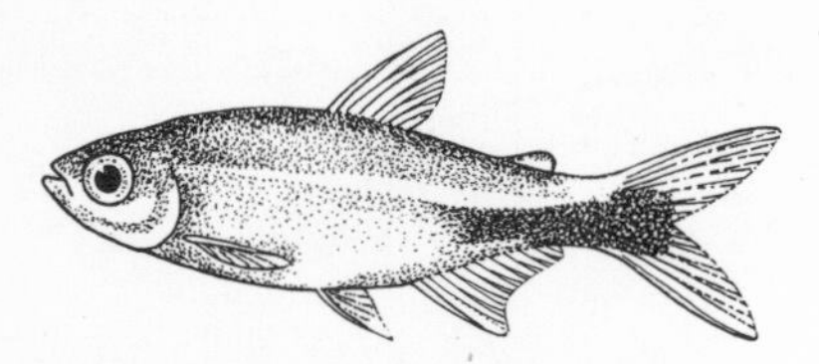

Fig. 602 *Hemigrammus hyanuaryi*

– *H. marginatus* ELLIS, 1911; Bassalm Tetra. Found in Venezuela to Argentina. Up to 6 cm. Flanks silvery, dorsal surface olive-green. A narrow, black, longitudinal band extends from the operculum to the caudal peduncle where it broadens out into a dark blotch. Above the band is another paler one. The D and An are yellow-green with shining pastel-white tips. The bottom of the C is lemon-yellow whereas the middle of both caudal fin lobes have a broad, black band and the tips are yellowish-white. The females can be recognised from their deeper bodied ventral areas. This is certainly a prettily coloured Characin, but it will always stand in the shadows of both the species *Petitella* georgiae* and *Hemigrammus rhodostomus*, as, unlike them, *H. marginatus* does not have a head and mouth area that is a brilliant red colour. Not difficult to breed, *see* genus description (fig. 603).

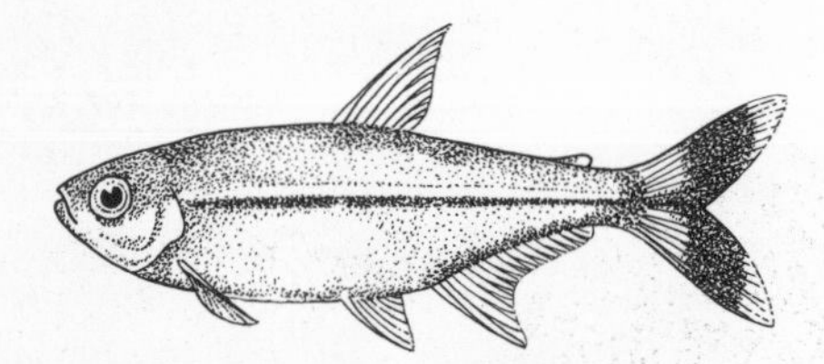

Fig. 603 *Hemigrammus marginatus*

– *H. nanus* (LÜTKEN, 1874); Silver-tipped Tetra, Silver Tip. There is another name for this species, *Hasemania marginata*, but it is no longer used. Found in SE Brazil, the San Francisco Basin. Up to 4 cm. Flanks silver with a bluish sheen and dark edges to the scales. A silvery longitudinal band extends from the operculum to the middle of the caudal peduncle. D, An and C are all a strong fawn-brown colour with shining white tips. Vs and Ps a delicate yellow, the iris silvery. Sexually excited males emit a lovely copper-brown colour along their flanks. The gold-coloured base of the C is divided in two by a black blotch which extends from the caudal peduncle to the middle caudal fin rays. There is no set time of the day for spawning. The eggs are very small and olive-brown in colour (similarly, the young fish), making them difficult to see among the plants. Easy to rear if fed on rotifers and *Cyclops* nauplii (fig. 604).

– *H. ocellifer falsus* (MEINKEN, 1958); False Beacon Fish. Found in the coastal areas of Guyana. Up to 4.5 cm. Flanks brownish to greenish-yellow with a silvery iridescence. A narrow, black, length-wise band commences at the level of the D, broadening out into a black blotch on the caudal peduncle, which is crossed by a transverse bar. The cross that this forms lies on top of a shining bright, golden-yellow spot. There is the hint of a dark shoulder fleck in a green iridescent zone behind the operculum. Fins colourless with white tips. All the specimens kept by aquarists in Europe prior to 1960 belonged to this sub-species. It was only in 1960 that the primitive form *H. ocellifer ocellifer* (STEINDACHNER, 1833) was introduced and bred. Unfortunately, it has died out once more in the meantime. It is a much more attractive species than the false variety because of its large, deep black shoulder fleck bounded front and back in gold, and because of the flash of red at the base of both caudal fin lobes. Both subspecies are relatively easy to breed and rear (fig. 605).

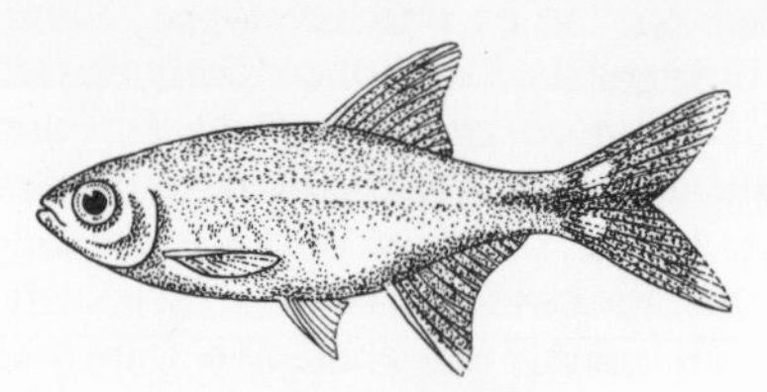

Fig. 604 *Hemigrammus nanus*

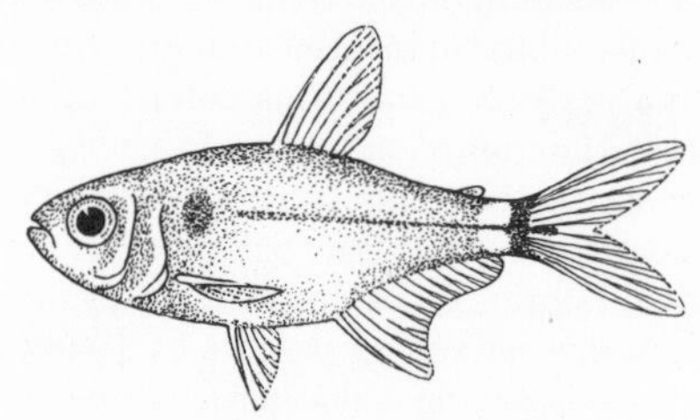

Fig. 605 *Hemigrammus ocellifer*

– *H. pulcher pulcher* LADIGES, 1938; Pretty Tetra, Black Wedge Tetra, Garnet Tetra. Peruvian part of the Amazon on the upper side of Iquitos. Up to 5 cm. The short, tall body shape is untypical of the genus. Depending on the angle of light, the flanks are a lovely pale grey-green or copper. There is a striking shoulder fleck with a copper-red sheen and a shining golden-coloured blotch in the upper half of the caudal peduncle. Below that there is a broad, wedge-shaped, black zone which begins beneath the rear end of the D. D, C and An are an intense copper-red to violet colour. The upper half of the iris is a brilliant purple-red, the lower half blue-green. The breeding of this beautiful species is not without its problems, as not all the pairs will readily spawn. Specimens that have spawned once are more willing to do so again. If you are trying to get them to breed, use an aquarium that is at least 40 cm long (fig. 606).

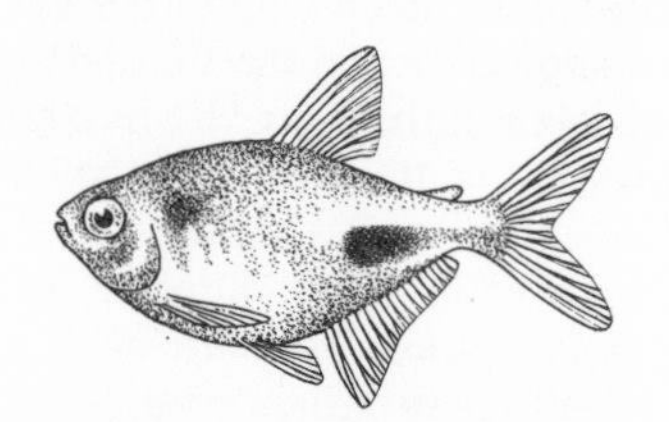

Fig. 606 *Hemigrammus pulcher*

– *H. rhodostomus* AHL. 1924; Rummy-nosed Tetra, Red-nosed Tetra. Lower Amazon. Up to 5.5 cm. Flanks a silvery iridescent colour. A narrow longitudinal band starts in the middle of the body and broadens out as it reaches towards the caudal peduncle. Here it ends in a black, rhomboid blotch covering the base of the C and the middle caudal fin rays. A greenish band runs parallel to the main band (on its top side). When content, the mouth area and the iris are blood-red. There is a black blotch in the centre of each of the yellowish-white caudal fin lobes. The remaining fins are colourless with white tips. A somewhat larger, more intensively coloured variety used to be regarded as a local species, but it has now been identified by GÉRY as a species belonging to the newly established G *Petitella**. It is quite possible that all the descriptions of successfully bred *H. rhodostomus* species were in fact *Petitella georgiae*. The species makes beautifully coloured specimens but they are shy and not necessarily easy to keep (fig. 607).

Fig. 607 *Hemigrammus rhodostomus* (below) and *Petitella georgiae*

– *H. ulreyi* (BOULENGER, 1895); Ulrey's Tetra, Tetra Ulreyi. Found in the Mato-Grosso region and in the River Paraguay. Up to 5 cm. Body fairly tall, very transparent, otherwise the colouring is very similar to that of the well-known Belgian Flag Tetra, *Hyphessobrycon* heterorhabdus* (although slightly less intense). Because of this similarity, the *Hyphessobrycon* species was for a long time mistaken for *H. ulreyi*. According to FRANKE, specimens of this species have been imported as an unintentional catch* on various occasions. Not yet bred in captivity (fig. 608).

– *H. unilineatus* (GILL, 1858); Feather Fin, Lone Line Tetra. Found in Trinidad and throughout the northern part of S America. Up to 5 cm. The shape of the body and the colouring are reminiscent of a pale *Pristella riddlei*. Flanks transparent which will look a gleaming bluish-silver in reflected light. A narrow, golden longitudinal band runs along the vertebral column which is visible through the fish. The blackish shoulder fleck is not always clearly visible. The D, C and An are reddish or a definite red. The first anal fin rays and the tips of the dorsal fin are pastel-white. There is a dark, triangular spot on the D and a dark, oblique blotch in the front third of the An. This is a lively, undemanding Characin which used to be popular with aquarists and was often bred. Today, however, it has been ousted from the aquarium by more strikingly coloured species (fig. 609).

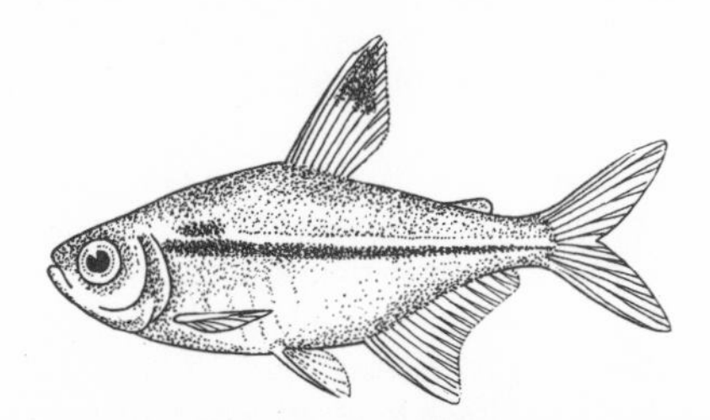

Fig. 608 *Hemigrammus ulreyi*

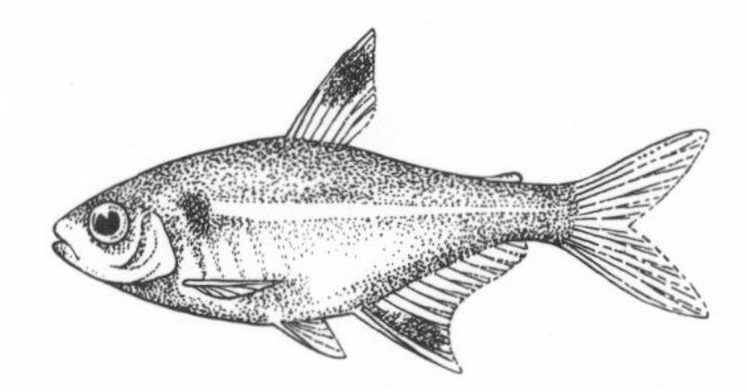

Fig. 609 *Hemigrammus unilineatus*

Hemigymnus GÜNTHER, 1861. G of the Labridae, Sub-O Labroidei*. Consists of only a few species found in the Indo-Pacific realm. Length up to 75 cm. Relatively compressed body and with a strong build. The body is covered with large scales, the breast and nape region have small imbedded scales. The lips are very thick. The following hardy species is often imported to be kept in aquaria.

– *H. melapterus* (BLOCH, 1791); Black-eyed Thicklip, Half and Half Wrasse, Thick-lipped Wrasse. Found in the Indo-Pacific and the Red Sea. Up to 38 cm. There is a sharp contrast between the pale anterior part of the body and the dark brown posterior part.

Hemihaplochromis WICKLER, 1963. G of the Cichlidae*. Comprises two species native to Africa, which were formerly ascribed to the G *Haplochromis**. In recent times the G has also become known as *Pseudocrenilabrus* FOWLER. These are fish of typical Cichlid build and shape. The head is strikingly large and solid-looking. The D has 13–15 hard rays and 8–11 soft rays; the An has 3 hard rays and 6–10 soft ones. The rear portions of these fins are rounded, the C is shaped like a fan. Like many related Ga (eg *Haplochromis* and many *Tilapia** species), *Hemihaplochromis* is a mouthbrooder. It is always the female that looks after the brood in this particular case. True partnerships are not formed. A male that is ready to mate will adopt a territory in the middle of which he will hollow out a spawning area. He now approaches any passing females with characteristic courtship movements, and if they are willing to spawn he entices them to the hollow he has made. With circling movements, the female now lays her eggs in batches, storing them immediately in her throat sac. In all, about 30–80 eggs are laid. It is only at this point that the eggs are fertilised. The male folds down his anal fin which has the effect of making the yellow edge of the fin look like an egg. This triggers the female into snapping at the fin, with the result that she scoops sperm into her mouth. The eggs take about 10 days to hatch, although more time will elapse before the brood is allowed out of the mother's mouth. During the first few days that they can swim around freely, the mother still takes very good care of her offspring, giving them shelter in her mouth when danger threatens or during the night. Ethologists have made a thorough examination of this type of brood care, and they have established that the youngsters are born with the ability to seek out the mouth and that they orientate themselves towards it purely by visual means. This species can be kept at a temperature of 24–26 °C, and they should give little problem. In all other respects they can be cared for as instructed for the G *Haplochromis*.

– *H. multicolor* (HILGENDORF, 1903); Egyptian Mouthbreeder. Found in Eastern Africa, especially in the Nile region. Up to 8 cm. Males have a multi-coloured iridescent sheen on a clay or russet-coloured background. The operculum has a striking blotch on it. The unpaired fins are rust-red with a lovely blue and green patterning. The females are less attractively coloured.

– *H. philander dispersus* (TREWAVAS, 1936). Widely distributed in the south of Africa. Up to 11 cm. Colouring very variable, but even more attractive than *H. multicolor*. The flanks have a strong golden sheen and the fins are red with shiny cobalt blue patternings. The lower lip is also blue. Females are not so strikingly coloured. Intolerant species.

Hemiodidae (fig. 610). F of the Cypriniformes*, Sub-O Characoidei*. Elegant, slender fish found in S American rivers. Widely distributed. They have a large, deeply indented C. The mouth is turned downwards slightly or inferior. There is a reduced number of teeth in the lower jaw in particular – it may have none at all or only have teeth at the sides. The upper jaw has incisor teeth which are often compressed into a uniform cutting edge. There is usually an adipose fin. Scales cycloid or cyclo-ctenoid. In many Ga the scales in the lower part of the body are larger. The Hemiodidae are speedy and agile shoaling fish found in open water

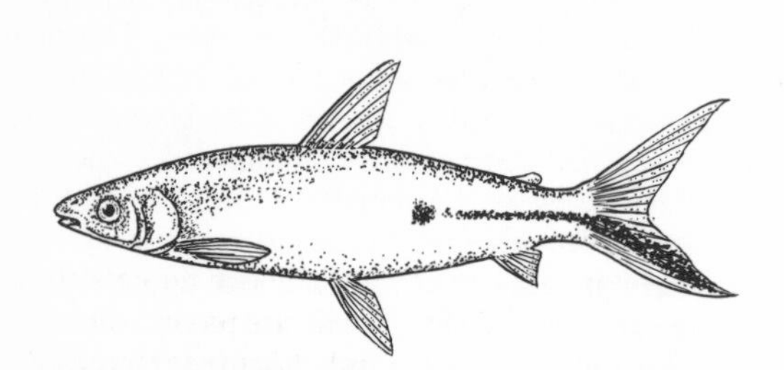

Fig. 610 Hemiodidae-type

zones. They swim along normally – there are no headstanders in this F. Hardly anything is known about their reproductive method. In their homelands, the larger species are considered very tasty. As far as the aquarist is concerned, the Hemiodidae make interesting specimens, although they must be kept in large tanks offering plenty of swimming room, since they always like to be on the move. They are a timid family. For further details, *see Hemiodopsis* and *Hemiodus*.

Hemiodopsis FOWLER, 1906. G of the Hemiodidae*. Widely distributed throughout S America. Up to 20 cm. Very similar to the G *Hemiodus**, and demanding the same care and feeding. Most of the species in this G were once placed under the G *Hemiodus*. The reason for this change relates to the scales on the body. In the *Hemiodopsis* species the scales in the upper and lower halves of the body are the same; in *Hemiodus* the scales are smaller in the dorsal region of the flanks, and larger in the ventral region. The 'semitaeniatus-group' usually have a black blotch on the flanks at the level of the Vs, or a black length-wise band begins there and extends into the lower lobes of the C. The ground colouring of the body is a shining silver. In addition, the upper lobes of the C may be crossed by black stripes. Besides *H. semitaeniatus* (KNER, 1859), *H. goeldii* and *H. gracilis* also belong to this group. In *H. gracilis*, there is an additional blood-red longitudinal streak on the lower lobe of the C. The 'quadrimaculatus-group' which was worked out by GÉRY, is recognised by the four, broad, black transverse bands on the flanks. The various species that belong to this group can only be distinguished by their different scale count – *H. huraulti* GÉRY, 1964, *H. quadrimaculatus* (PELLEGRIN, 1908), *H. sterni* GÉRY, and *H. vorderwinkleri* GÉRY, 1964. These species have not yet been bred in captivity.

– *H. quadrimaculatus* (PELLEGRIN, 1908). Found in the Camopi River in French Guyana. Up to 18 cm. The flanks are a strong silvery iridescent colour with a slight touch of pale red or violet. Four broad saddle-like bands runf from the dorsal surface to the belly edge, getting narrower as they approach the ventral surface: the first one extends across the head and the eyes, the second in front of the insertion of the D, the third and fourth between the D and the adipose fin. There is an oval black blotch on the caudal peduncle. This continues into the lower lobes of the C as a broad black band. There is also a narrower black and on the upper lobes. Scale count: $6^1/_2$–$7^1/_2$/42–44/4–$4^1/_2$ (fig. 611).

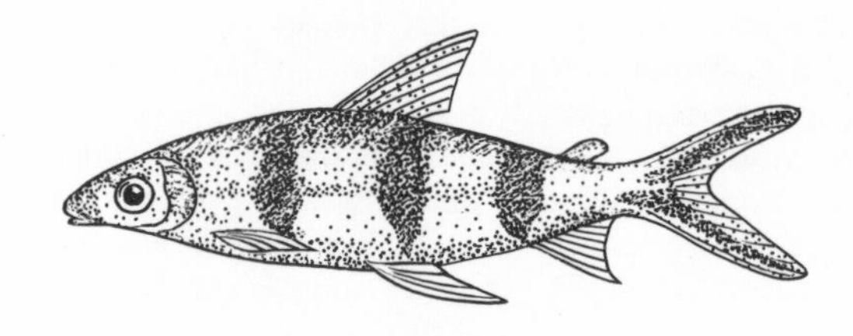

Fig. 611 *Hemiodopsis quadrimaculatus*

– *H. semitaeniatus* (KNER, 1859). Found in central and lower Amazon. Up to 20 cm. Flanks a very strong gleaming silver with a green or steel-blue shimmer. Dorsal surface brown-olive, ventral surface silver-white. There is a large, round, black blotch on the flanks located between the Vs and the An, which continues as a fine dark streak onto the caudal peduncle. From there it broadens out considerably and extends into the lower half of the C. The tips of both lobes are often tinged a delicate red colour (fig. 612).

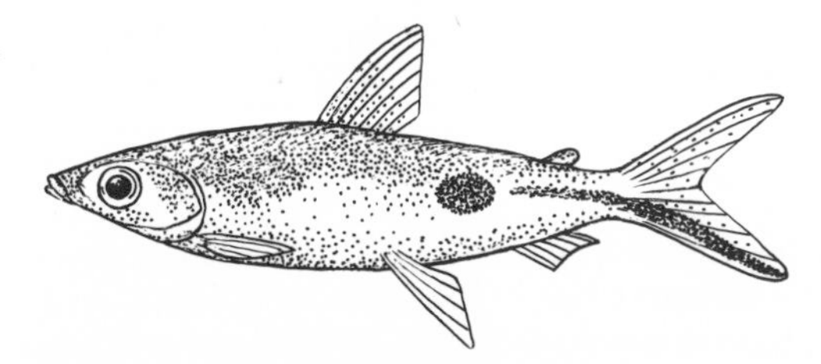

Fig. 612 *Hemiodopsis semitaeniatus*

– *H. sterni* GÉRY, 1964. Found in the Mato Grosso in Brazil, the Rio Juruena. Up to 18 cm. In habits just like *H. quadrimaculatus*, colouring and band markings very similar, except that the silvery colouring of the body is intermingled more strongly with flesh colours. Within the group, this species also has the highest scale total: 12/64–67/7$^{1}/_{2}$–8. The other two species, *H. huraulti* GÉRY 1964 (scale count: 9$^{1}/_{2}$–10/50–52/5–6) and *H. vorderwinkleri* GÉRY, 1964 (scale count: 7–7$^{1}/_{2}$/44–45/5) are barely distinguishable from *H. quadrimaculatus* and *H. sterni* in body shape, colouring or band markings.

Hemiodus MÜLLER, 1842. G of the Hemiodidae*. Widely distributed in S America. Length up to 20 cm. Peaceable shoaling fish, well skilled at swimming, and found in the middle and upper levels of the water. In the aquarium they need plenty of swimming room. As they can also jump spectacularly well, make sure you have a good lid on the aquarium. Besides robust forms of live food, they also require plenty of plant food. Most of the species of the G were transferred to the newly established G *Hemiodopsis**. According to GÉRY, *Anisitia* EIGENMANN, 1903, is another name for this G, but it is no longer used. The body is shaped like an elongated torpedo of uniform height and it is only slightly flattened laterally. The C has broad lobes and is deeply indented; the lower lobe is a little bit longer than the upper lobe. *Hemiodus* can be distinguished from the closely related G *Hemiodopsis* because it has larger scales in the ventral region and somewhat smaller scales in the dorsal region the flanks. The eyes are large and the mouth is terminal or slightly inferior as well as protrusible. Typically of the family, the lower jaw only has teeth along the sides, or it has no teeth at all. The D is short, triangular. Adipose fin present. The An is very small. The swimming position usually adopted is at a slight angle, head-up in the water. Nothing is known about sex differences or reproductive behaviour.

– *H. unimaculatus* (BLOCH, 1786) MÜLLER, 1842. Found in central and lower Amazon. Up to 20 cm. The body is silvery with a black dot at the level of the Vs. There is a dark, lengthwise stripe on each caudal fin lobe.

Hemipteronotus (LACÉPÈDE, 1802). G of the Labridae, Sub-O Labroidei*. Members of this G are found in tropical parts of the Atlantic and Pacific. They live in sandy-bottomed areas and like to burrow into the sand. They are permanently resident in the habitat and some dig out hollows to which they will return regularly. Typical features are the steep forehead profile and the high position of the eyes. The cheek areas are also scaly. The body is elongated and laterally compressed. Length 20–30 cm. Diet consists of small animals that live in the sand. Reports so far indicate that they survive well in the aquarium.

Hemirhamphidae, F, name not in current usage, *see* Exocoetoidei

Hengel's Rasbora *(Rasbora hengeli) see Rasbora*

Heniochus CUVIER, 1817. G of the Chaetodontidae*, Sub-F Chaetodontinae. Comprises several species found in the tropical Indo-Pacific where they live in shoals among reefs and rocks. They feed on animal and plant growth. The 4th ray of the D is very long, in *H. acuminatus* it is as long as the body. *Heniochus* species are deep-bodied fish. In the aquarium they are fairly hardy (in contrast to other Chaetodontids) if given a very varied diet. The species most often imported is:

– *H. acuminatus* (LINNAEUS, 1758); Featherfin Bullfish, Pennant Butterfly-fish, Pennant Coralfish, Poor Man's Moorish Idol Wimplefish. Found in the Indo-Pacific and the Red Sea. Length up to 20 cm. White with 2 broad black transverse bands. The 4th ray of the D forms a long white pennant. The D is soft and the C yellow.

Henricia, G, *see* Asteroidea

Hepaticae, Cl, *see* Bryophyta

Heptagenia WALSH, 1863 (fig. 613). G of the insect order Mayflies (Ephemeroptera*). Comprises a great number of species distributed over wide areas of the northern hemisphere (Holarctic*). *Heptagenia* larvae live mainly in rivers warmed by the summer sun. They have a very characteristic shape and are several cm long, including the tail threads. They are ideal as a live food for large aquarium fish and as specimens in pond aquaria.

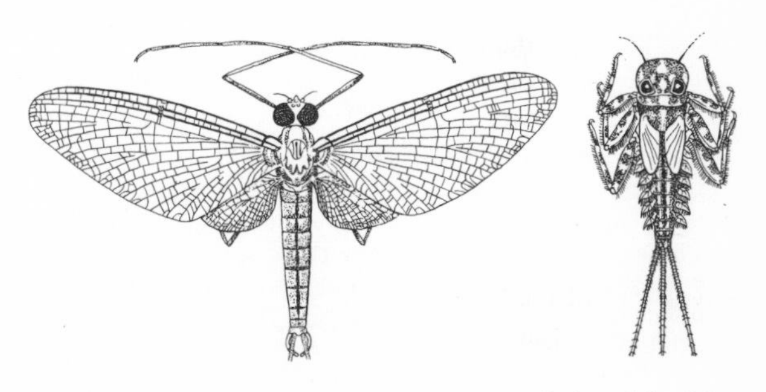

Fig. 613 *Heptagenia*. Left, imago (no tail threads); right, larva

Herbicides *see* Poisoning

Herb-two-pence *(Lysimachia nummularia) see Lysimachia*

Hereditary Character *see* Heredity

Hereditary Illnesses are brought about by defective genetic material (*see* Genetics) and are passed on from the parents to all subsequent generations (*see* Heredity). Chief amongst the causes of genetic defects in aquarium fish must be in-breeding, domestication and the formation of hybrids. Lethal hereditary factors lead to life becoming inviable even in the early stages of development. Particular susceptibility towards certain infectious diseases, yolk sac dropsy*, sterility*, swimbladder defects and some tumour diseases* can all be caused through heredity. Malformations of the fins, scales and the skeletal system are always associated with hereditary illnesses too. Certain deviations determined through heredity (eg fin shape, body shape and colouration [melanism, xanthochroism, albinism* etc.]) are made use of in fish breeding to produce desired effects.

Heredity. The passing on of genetic information (hereditary material) to offspring. The biochemical basis of heredity is deoxyribose nucleic acid (DNA), which is capable of identical duplication (Nucleic Acids*). The DNA, by a specific linear sequence of its 4 basic components, determines the specificity of the numerous hereditary factors *(genes)* in the chromosomes of the cell nuclei. The cytological basis of heredity are the processes during which the gametes mature (Cell Division*) and the processes of fertilisation*. In diploid organisms (animals and higher plants) during fertilisation, the simple (haploid) chromosome set of male gametes are joined with the haploid chromosome set of the egg-cell to become a diploid set. The zygotes and the body cells that develop from these therefore contain pairs of chromosomes with corresponding hereditary factors (known as alleles). If both homologous factors have the same effect on the formation of feature, the individual in this pair of alleles is *homozygous*, in other instances the individual is *heterozygous*. At first, the composition of the *genotype* is not apparent from the *phenotype* of an individual, as one allele from a pair may determine the feature (ie is *dominant*). In such cases, the other allele is suppressed *(recessive)*. In the Guppy, for example, the allele for the grey wild colour is dominant, while the allele for the gold colour is recessive. This means, even when both hereditary factors (symbols W and g) are present in the one animal, the phenotype is the wild colour. If the alleles from a pair are different but both determine feature, then an *intermediary* phenotype develops and this is apparent externally. Because of the mechanics of maturation division (meiosis), heterozygous individuals form genetically different sex cells and therefore they produce, in a particular ratio, offspring that are differentiated genetically and externally. The possibilities with the passing on of hereditary matter was analysed by GREGOR MENDEL* in 1865, and summed up in the 3 *Mendelian laws* (fig. 614):

– *1st Law* (Law of Uniformity). If you cross pure-bred parents (P generation) which differ in one or several features, the offspring (F_1 generation) are the same as one another.

The F_1 generation is always heterozygous in the genotype, but in the phenotype it can resemble one of the parents or stand somewhere in between the parents. It is then immaterial whether dominant hereditary factors stem from the father or the mother.

– *2nd Law* (Law of Segregation). If the hybrids of the F_1 generation are crossed with one another, the offspring (F_2 generation) are no longer the same, but split into the ratio: 25 %: 50 %: 25 %. This 2nd law is only valid if the parents (P) of the F_1 generation differ in one feature only. The ratio is only apparent genotypically if dominant features appear. For example, if Guppy hybrids resulting from the mating of a pure-bred wild

WW = wild colours (dominant)
gg = gold colours

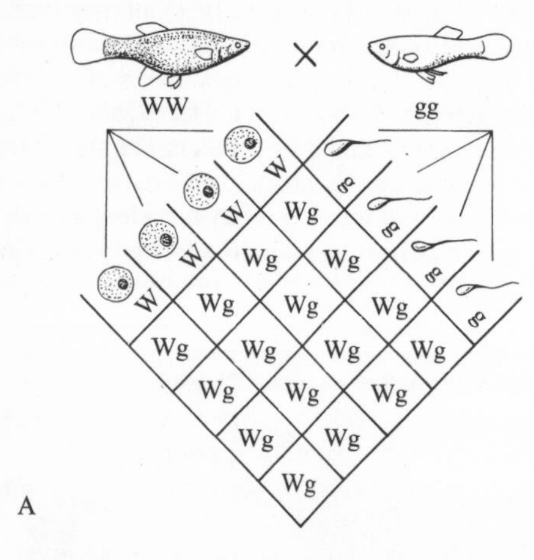

Fig. 614 A: 1st Mendelian law (Law of Uniformity). The two Guppies differ in one colour feature. The letters indicate the genes. The female has 2 dominant genes that produce the wild colour (WW); the male has 2 recessive genes that cause the gold colour (gg). As the diagram shows, all the offspring has the same combination of genes (100 % Wg), ie they are heterozygous. The dominance of W at the same time causes all the animals to have a uniformly wild-coloured phenotype.

WW = wild colours
gg = gold colours

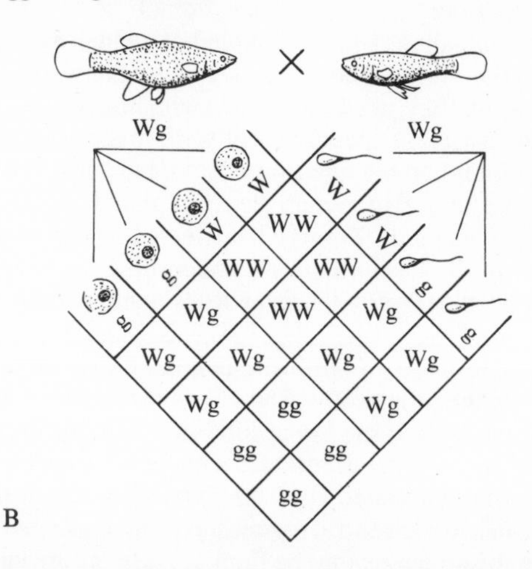

B: 2nd Mendelian law (Law of Segregation). If the hybrids of the 1st generation (F1) are crossed with one another, there will be a division in the F2-generation in the ratio 25:50:25. 25 % will have the same gene combination of one partner in the grandparents (WW) and 25 % that of the other (gg); the remaining 50 % are heterozygous like their parents. But because of the dominance of W, the phenotype of the 25 % WW and 50 % Wg is a uniform wild colour.

form and a pure-bred gold form are themselves cross-bred, 25 % of the offspring will be a pure-bred gold colour and 75 % the wild colour. But of those only 25 % are pure-bred wild colour forms.

– *3rd Law* (Law of the Free Combination of Genes). If individuals are crossed that differ in more than one feature, new combinations will occur in the F_2 generation. For example, if a pure-bred, wild colour-pale red

wild colour.
WW = wild colours
ww = gene absent for wild colours
RR = red
rr = gene absent for red colour

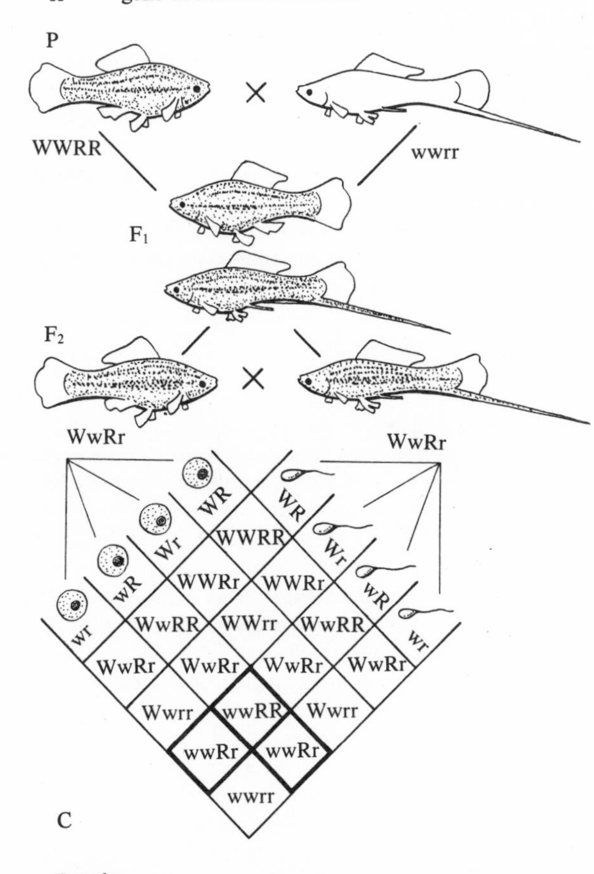

Result:

Genotype	Phenotype	
6.25 % WWRR	homozygous	
12.5 % WwRR	heterozygous	pale-red 9
12.5 % WWRr	heterozygous	
25 % WwRr	heterozygous	
6.25 % wwRR	homozygous	
12.5 % wwRr	heterozygous	blood-red 3
6.25 % WWrr	homozygous	
12.5 % Wwrr	heterozygous	wild colours 3
6.25 % wwrr	homozygous	yellow 1

C: 3rd Mendelian law (Law of the Free Combination of Genes). The two parents (P) differ in 2 features. The combinations heavily outlined in the bottom part of the diagram are particularly interesting to the breeder. These fishes are a strong blood-red, although only the combination wwRR is homozygous.

Swordtail *(Xiphophorus helleri)* (genetic formula WWRR) is mated with a yellow Swordtail that does not have these hereditary factors (wwrr), among the offspring of the uniform F_1 generation (WwRr) there are a possible 9 different combinations. Of these, 4 are homozygous. In this case, the pure-bred deep-red specimens (wwRR) are of particular interest to the breeder.

In many cases, individual features cannot be inherited in isolation from other features, since several features are bound to a locus (point on the chromosome) as *groups of features*. A special example of this is *sex-linked* heredity. So, in the Guppy (which in contrast to many other fish has particularly differentiated sex chromosomes), the hereditary material for numerous colour features and for elongated fin rays are bound to the Y chromosome, and therefore to the male sex (Cell*, Cross-breeding*, Sex Determination*).

Hermaphrodite *see* Hermaphroditism

Hermaphroditism. Bisexual individuals, or hermaphrodites, within a species are widespread among the plants. It is not uncommon among the invertebrates (eg some Sponges and Cnidaria, the Turbellaria, Tapeworms and Gastropods), but among the vertebrates hermaphroditism is the exception. In many cases the germ cells mature at different times in hermaphrodites (protandrous or protogynous hermaphroditism), thus preventing self-fertilisation. Among the fish, natural hermaphroditism occurs, for example, in the Serranids (Serranidae*), sometimes it is protogynous, and sometimes self-fertilisation takes place. In species where the sexes are separate, abnormal individuals may occur that are neither clearly masculine nor feminine. Such an individual is known as an *intersex*.

Hermit Crabs, Sub-F, *see* Paguridea

Herring *(Clupea harengus) see* Clupeiformes

Herring types O, *see* Clupeiformes

Herrings (Clupeidae, F) *see* Clupeiformes

Herrings, similar types (Clupeomorpha, Sup-O) *see* Osteichthyes

Heterandria Agassiz, 1853. G of the Poeciliidae*, Sub-F Poeciliinae. Found in N America: SE the USA; and in Central America: the Atlantic regions of S Mexico, Guatemala, Honduras. They live in fast-flowing waters up in the mountains and in weedy areas near the bank in flowing and still waters on the plains. *H. bimaculata* is also found in brackish water. Males 2–5 cm, females 3–9 cm. The body is elongate and moderately flat from side to side. The ground colour is brownish and the markings in both sexes are similar. *Heterandria* species are predators, sometimes even cannibals. They should be kept in tanks that have areas of thick vegetation and some floating plants. Their diet consists mainly of live food of appropriate size. The young are born over fairly lengthy periods. Fertilisation of the eggs does not take place, as in other members of the F, after the birth of one generation, but internally, as they develop. Therefore, inside pregnant females there are always embryos of different ages which are born one after the other. The two species at present attributed to the G are:

– *H. bimaculata* (HECKEL, 1848); Pseudo Helleri. Found in S Mexico, Guatemala, Honduras. Male up to 5 cm, female up to 9 cm. There are various races within the species, the highland and lowland forms in particular having a different body shape and colouring. The highland forms are brown and have a robust build. Most of the different races are attractively coloured: The body is an iridescent green and the scales are dark-edged. The operculum is green with an orange-red blotch. There is a black blotch on the root of the tail and at the insertion point of the P. The D is yellowish-green with brown dots and streaks. The C in the male has a red edge at the bottom. This is a very predatory species. Temperature 22–26 °C. The birth period lasts for 4–6 weeks, during which time more than 100 young may be born. The young fish grow quickly.
– *H. formosa* AGASSIZ, 1853; Mosquitofish, Dwarf Top Minnow. Found in the USA: N Carolina and Florida. Male up to 2 cm, female up to 3.5 cm. The body is brown, shining, and has a dark brown longitudinal band running along it which is crossed by 8–15 transverse bands. The belly region is pale. The fins are brownish and at the bases of the D and An there is a black blotch outlined in orange-yellow. It is a relatively tolerant species. 20–24 °C. Besides small food animals, algae are also necessary as food. Artificial foods are also possible. The birth period lasts for 1–2 weeks and more than 50 young may be born (fig. 615).

Fig. 615 *Heterandria formosa*

Heteranthera RUIZ et PAVON, 1794: Mud Plantain. G of the Pontederiaceae* comprising 10 species. Found in tropical to northern temperate parts of America and tropical parts of Africa. Small perennials with elongated stem axes and alternate leaf arrangement. The leaves are linear or divided into a stalk and leaf-blade of varying shapes. The flowers are hermaphroditic and either polysymmetrical or monosymmetrical. The perianth consists of a six-part perigonium that is slightly fused together. There are 3 stamens, in rare cases just one. The carpels number 3 and they are superior and fused. Usually the fruit is in the form of a capsule.

Heteranthera species are usually found in water, but some also live as creepers in marshes. Many of them make decorative and undemanding aquarium plants, but so far only a few have been kept in aquaria. Propagation is possible with cuttings.
– *H. limosa* (SWARTZ) WILLDENOW. Found in America. There are both floating and aerial leaves which have stalks and ovoid leaf-blades. The flowers are simple and are blue or white. This species has not yet been transferred to the aquarium.
– *H. reniformis* RUIZ et PAVON. Found in America. Stalked floating and aquatic leaves with kidney-shaped leaf-blades. The inflorescence consists of a few, small, blue flowers. Not suitable for the aquarium.
– *H. zosteraefolia* MARTIUS. Found in S America. Linear aquatic and aerial leaves. The inflorescences are made up of a few, small, blue flowers. This species makes a very decorative aquarium plant.

Heterocentrotus BRANDT, 1834; Slate-pencil Urchin. G of the tropical Sea Urchins (Echinoidea*). *H. mammilatus* (LINNAEUS, 1758) comes from the tropical Indo-Pacific where it lives beneath rocks or in holes that it bores out for itself. The body is about 8 cm across and it is covered with styloid red spines that can be up to 1.3 cm thick. The ends of the spines are blunted. Small specimens can survive for a considerable length of time in the aquarium if fed on a rich diet of plant food.

Heterocercal C *see* Fins

Heterodontidae; Bull-headed Sharks. F of the Chondrichthyes*, O Heterodontiformes. These are small sharks with a strange shape of head which looks very much like a pig's from the front. There is no rostrum and because of this the mouth is only slightly inferior. The teeth are grouped in an unusual fashion in the jaws. At the front there are several rows of small, sharp teeth which are very close together and point backwards. Farther back these pass into teeth which have pads on them. At the front edge of both Ds there is a spine. The C is relatively small and asymmetrical. Bull-headed Sharks are found in all tropical and temperate seas with the exception of the Atlantic and the Mediterranean. The females lay their large eggs in characteristic horny egg-cases (which are brown in colour). In shape they look like a spindle with two flat flanges spiralling down the outside which form at the ends of the spindle tips. The best-known species is the Japanese Bullhead Shark *(Heterodontus japonicus)* (fig. 616), which can grow up to 1.5 m long. None of the Heterodontidae is dangerous to man.

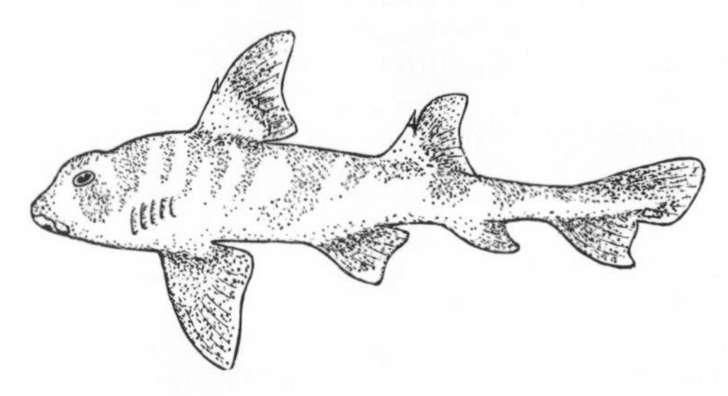

Fig. 616 *Heterodontus japonicus*

Heteromi, O name not in current usage, *see* Nothocanthiformes

Heterophylly. Term used to describe the occurrence of different types of foliage leaf on successive nodes of a stem axis. Heterophylly is normally caused by different environmental conditions in aquatic and marsh plants. In *Sagittaria*, for example, the part of the plant that is submerged always has narrow, ribbon-like leaves, whereas the aerial leaves are usually divided into a stalk and leaf-blade. Many *Limnophila* species have pinnate leaves on the submerged parts of their stems whereas those leaves above the water are entire and undivided. When a plant has both aquatic and floating leaves, as in *Nuphar*, this is also an example of heterophylly.

Heteropneustes MÜLLER, 1839. G of the Clariidae*. Asiatic Catfish. Length up to 70 cm. The body is shaped like a torpedo and it is naked. The head is small and is covered with strong plates, and the mouth is broad and lies at an angle. There are 4 pairs of barbels which project forwards. For information on the structure and function of the gill sac, turn to the F Clariidae. The D1 is small and short, beginning in front of the middle of the body at the same level as the small Vs. The C is moderately large and rounded whereas the An is very long but narrow. Ps well developed, but there is no D2. Adult specimens are really only suitable for large public aquaria. They have been bred successfully on many occasions. According to FRÄNKEL the female sucks at the sexual orifice of the male. The eggs are about the size of millet seed and they are laid in balls in depressions that have been previously hollowed out by the fish. Both parents guard the eggs and the growing youngsters. *Heteropneustes* species are inactive Catfish that live as predators. In the aquarium they require a soft substrate and stonework and roots that contain hiding-holes. They can survive on land for several hours but you should still guard against sudden changes in temperature. Feed with fish and mammal flesh, worms, potato and rolled oats. Should be kept as pairs or as individuals only.

– *H. fossilis* (BLOCH, 1792). Found in Sri Lanka (Ceylon), E India, Burma, S Vietnam. Length up to 70 cm. Colouring of the body varies between a pinky-brown, olive-brown or blackish with dark dots. Two white or yellowish longitudinal bands stretch from the end of the operculum to about the middle of the body then they gradually disappear. The fins are a fawn-brown colour although the An is often covered with dark flecks. The eyes are small and have a yellow iris. Has been bred in captivity (fig. 617).

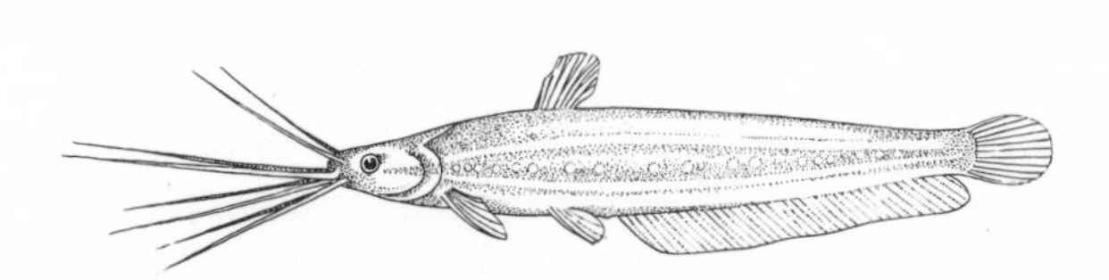

Fig. 617 *Heteropneustes fossilis*

Heteroptera; True Bugs. O of the Insecta* with an extremely large number of species and forms. They live partly on land and partly in the water. Their metamorphosis is incomplete (no pupal stage) and they have piercing mouth-parts adapted for sucking. The base of the front wings is hardened to some extent and clearly set apart from the membranous wing tips. The true water bugs with which we are concerned here are not necessarily closely related but all are amazingly well adapted to their habitat – still and flowing waters of every kind. The Ga *Gerris**, *Hydrometra** and *Velia** live on the water surface and so can be attributed to the pleuston*; other Ga, such as *Corixa**, *Naucoris**, *Nepa**, *Notonecta** and *Ranatra** are found primarily in the water itself. Although the biology of the individual Ga can be very different, the life-styles of the various stages of a species (larva, imago) are very similar. This is an important point to make, for it does not apply to other aquatic insect orders (compare Ephemeroptera*, Plecoptera*, Goleoptera*, Diptera*). When pond-collecting or food-collecting, water bugs often find their way into the net. They are very interesting creatures to watch because their shape is so unusual and because they move in such a strange way. The species *Velia* at least can be a match for the lovely colouring of many of the land bugs. Water bugs make interesting specimens in a pond aquarium where you will be able to observe how they move, how they catch their prey and how they mate.

Heterosis; Hybrid vigour. It is sometimes observed that the offspring of cross-mating between different races or species often grow quicker in the first generation and ultimately grow larger than the parents. This effect has been made use of in the breeding of Carps, for example, and it is also important in the cultivation of plants that are useful to humans. More luxurious types of hybrid have also been achieved in this way among aquarium fish (particularly among the Livebearing Tooth-carps).

Heterosomata, O, name not in current usage, *see* Pleuronectiformes

Heterostichus rostratus *see* Blennioidei

Heterotis niloticus *see* Osteoglossiformes

Heterozygous *see* Heredity

Hexacorallia, Sub-Cl, *see* Anthozoa

Hexamita (formerly known as *Octomitus*) (fig. 618). G of the parasitic flagellates* which often take up residence in the hind-gut and gall-bladder of several species of fish. Length 7–13 μm, *Hexamita* has 8 flagella and is highly mobile. Their presence in fish can only be determined by examining a fresh smear preparation (*see* Examination Technique) under a microscope. They are debilitating parasites which occur as a result of unsuitable feeding, fish tuberculosis* and *Ichthyosporidium** attacks. A fish infected with *Hexamita* loses

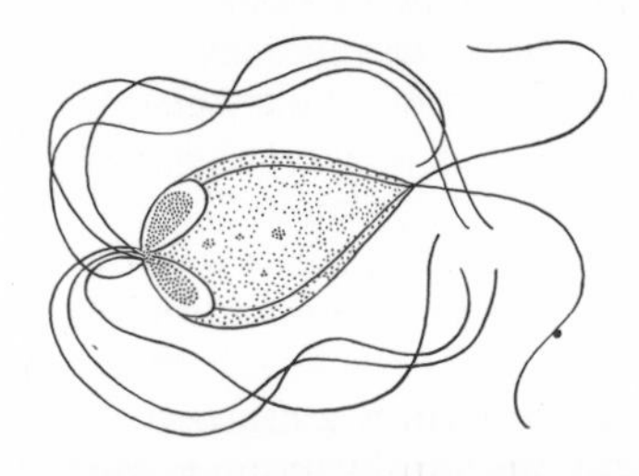

Fig. 618 *Hexamita*-flagellates

weight and does not eat well. *Hexamita* is commonly found in *Cichlasoma* severum*, *Heterandria* formosa* and in Cyprinids: when similar symptoms occur in *Pterophyllum**, it is caused by the closely related G *Spironucleus*. A species of *Hexamita* can also cause illness in *Symphysodon*–in this case, the symptoms are so characteristic that the disease is known as discus disease* or 'hole-in-the-head disease'. One cure for fish infected with *Hexamita* is to mix Calomel, Carbason* and Trypaflavin* in with the food. Prolonged baths may also meet with success. For this method of treatment you can use trichomonad medicaments (Clont, Sanotrichom, Vagimid) and the antibiotic* tetracyclinhydrochloride. Make up a concentration of the substance to be used in the proportion of 4–6 mg for every 1 litre of water, and bathe the fish in it for 4 days.

Hexanchidae; Comb-toothed Sharks. F of the Chondrichthyes*, O Hexanchiformes. Relatively primitive sharks with 6 or 7 gill slits, all of which lie in front of the Ps. The body is elongated and the head pointed with a broad, inferior mouth. The teeth point backwards and are arranged like a comb. The blow-holes are small. The upper lobe of the C is very long. Comb-toothed Sharks live in all warm seas and prefer depths below 200 m. All the species give birth to live young and there is a large number of baby sharks in each litter. The best-known species, the Six-gilled Shark or Cow Shark *(Hexanchus griseus)* (fig. 619) is grey or brownish in colour. As its name suggests, it has 6 gill slits and is on average 4–5 m long and weighs about 750 kg. It feeds mainly on fish. Its area of distribution stretches from the Atlantic and the Mediterranean to the southernmost parts of the Indian and Pacific Oceans. The people of Sicily like to eat its flesh. The 3-m-long Perlon *(Heptranchias perlo)* has 7 gill slits. Comb-toothed Sharks are not dangerous to people.

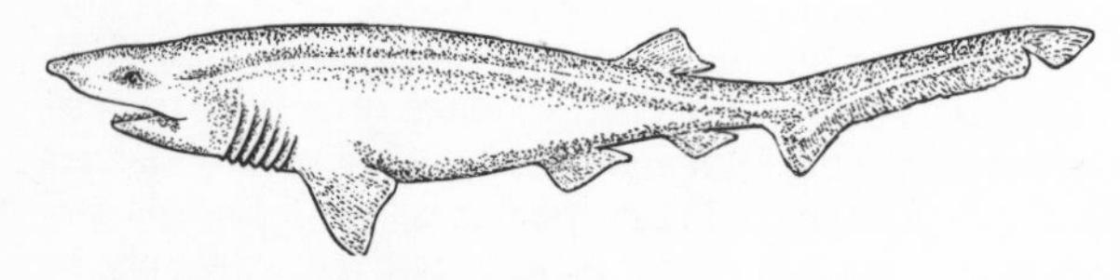

Fig. 619 *Hexanchus griseus*

Hexanchus griseus *see* Hexanchidae

Hexapoda, Cl, *see* Insecta

Hierarchic Division of Behaviour. Term used in ethology. After a sphere of function* has been activated, the individual elements of behaviour unfold in a particular sequence, ie hierarchic division. Courtship behaviour, for example, is introduced by elements of a pre-courtship phase, such as seeking a mate (Appetence Behaviour*) and display behaviour*. Once the high courtship phase is reached, the characteristics of swimming towards the spawning ground (*see* Courtship Swimming) gradually start appearing, and this eventually leads to mating taking place. Aggressive behaviour too increases in a similar hierarchic manner from display and threatening gestures to pure aggression such as ramming*, ramming circles or mouth-to-mouth fighting*. This will continue until the defeated rival is chased away. Several hierarchic phases can be left out only under certain circumstances, such as when specific-action energy has been pent up for some considerable time (eg if the sexes are separated for a long time during the spawning period). In such a case, the males would dispense with certain elements of the courtship and start swimming towards the spawning ground immediately or even start trying to mate.

High Courtship Behaviour *see* Courtship Behaviour

High Hat *(Equetus acuminatus) see Equetus*

High-backed Headstander *(Abramites hypselonotus) see Abramites*

Higher Bony Fish (Teleostei, Sub-Cl) *see* Osteichthyes

Higher Crustaceans, Sub-Cl, *see* Malacostraca

Hillstream Fishes, F, *see* Homalopteridae

Hind-gut *see* Digestive Organs

Hiodon tergisus *see* Mormyriformes

Hiodontidae, F, *see* Mormyriformes

Hippocampidae, F, name not in current usage, *see* Syngnathoidei

Hippocampus (RAFINESQUE, 1810); Sea-horse. G of the Syngnathidae, Sub-O Syngnathoidei*. Comprises several species found in all seas of the tropical and temperate zone. They live in calm water in clumps of vegetation where they feed mainly on small crustaceans which they suck in through their tube-shaped, toothless mouth. Their distinctive shape – the horse-like head, the belly arched forwards and the finless, grasping tail – makes *Hippocampus* unmistakeable. Most of the species have camouflage colours of brown or greenish so that they can scarcely be seen among the plants. They are also protected by a hard, ringed, bony armour running the entire length of the 10 to 20-cm-long body. The main means of locomotion is the propeller-like, winding D with which the fish can slowly work its way through the water in an upright position. Usually, however, they are to be found anchored in amongst the plants, looking around for suitable prey. Mating is introduced by a very obvious and long-lasting courtship phase during which time the pair may grip each other by the tail for a while *(H. guttulatus)* and twist and turn up to the surface ('carousel swimming'). Once there, they place their ventral surfaces towards each other and the female transfers her eggs with the aid of her genital papilla to the male's brood sac which is located on his ventral surface. Even outside the spawning period, the male has this brood sac, and it is the easiest way to tell the sexes apart. The sac remains shut while the young develop; this may take from 10 days to 6 weeks. There may be 50–600 young, depending on species. When fully developed, they are released by the male making pumping movements with his tail. At first, the young remain beneath the surface. In the aquarium, they can be reared on newly-hatched *Artemia* nauplii. All Sea-horses make good aquarium specimens. Normally, they only eat small, live crustaceans (for example, small portions of *Daphnia*), but they may also eat small fish. After a while, *H. guttulatus* will also learn to take dead waterfleas from the bottom. Because sea-horses are relatively defenceless and slow, they should not be put in community tanks with lively fish, anemones or larger crusta-

ceans. It is best to keep them in a species tank which has been planted with *Caulerpa**. Their life-span is about 2–3 years. The following are some of the better-known species:
– *H. guttulatus* (CUVIER); Mediterranean Sea-horse. Found in the Black Sea and in the Mediterranean. Lives among seaweed and blankets of algae. Length up to 16 cm. There are several geographical races. Colouring usually a dark olive-brown, but pale yellow when excited or during courtship.
– *H. kuda* (BLEEKER, 1852); Oceanic Sea-horse, Spotted Sea-horse, Yellow Sea-horse. Found in tropical parts of the Indo-Pacific. 30 cm long. When content a yellow colour, otherwise, blackish (fig. 620).

Fig. 620 *Hippocampus kuda*

– *H. zosterae* (JORDAN and GILBERT); Dwarf Sea-horse, Pygmy Sea-horse. From the tropical W Atlantic. Length 4 cm. Colouring unremarkable. Readily reproduces in the aquarium. The brood develops over a period of about 10 days, with up to 50 young in a brood. Only lives to be one or two years old.

Hippoglossus hippoglossus *see* Pleuronectiformes

Hippolysmata STIMPSON, 1860. G of small Shrimps (Natantia*). Sociable creatures that live in the littoral of tropical seas where they feed on small animals. Some types perform the function of cleaning fish. *Hippolysmata* species have attractive colouring and have begun to be imported more frequently in recent times, particularly *H. grabhami* (fig. 621). This species has a sharply delineated shining red back with a white lengthwise streak running along the edge of the back. *Hippolysmata* species are examples of protandrous hermaphrodites, ie they start off as males that later turn into females. The females have a sperm pouch in which they can store sperm for several months. This then means that even isolated females can lay fertilised eggs for a certain period. The larvae have been reared in the aquarium. Specimens have survived in the aquarium for up to three years.

Fig. 621 *Hippolysmata grabhami*

Hippolyte, G, *see* Natantia

Hippuridaceae; Mare's-tail. F of the Magnoliatae*. The only G is *Hippuris**.

Hippuris LINNÉ, 1753; Mare's-tail. The only G of the Hippuridaceae* and containing a single species. Has a widespread distribution in temperate and cold places of the Earth. A perennial plant with creeping rhizomes located in the ground and upright, unbranched shoots. The leaves form six-part or multi-part whorls and they are linear. There are only slight differences between the aquatic and aerial leaves. The flowers are hermaphroditic, in rare cases unisexual, monosymmetrical and small. The sepal is inconspicuous and there are no petals. There is one stamen and one carpel which is inferior. Stonefruit. The Mare's-tail grows in still waters or slowly flowing waters. The water will be shallow and rich in nutrients and calcium. Very occasionally also found in marshes. Mare's-tails are highly recommended for planting in shallow parts of open-air ponds.
– *H. vulgaris* LINNÉ; Mare's-tail.

Hirudinea; Leeches (fig. 622). Cl of the annelids (Annelida*). Contains about 300 species, most of which are freshwater leeches, but there are also some marine and brackish water forms, and some that live on the land.

The Hirudinea are highly specialised annelids with a constant number of segments (33). The outer ring system, however, does not correspond with the inner structure. With the exception of one species, that also differs in other respects, the Hirudinea do not have any bristles. At the rear end of the body there is a sucker, and in some cases there is another sucker at the front end. There are usually several eyes which are visible as dark spots of pigment on the front part of the body. The secondary body cavity, which is typical of the annelids, is largely filled with fibrous tissue in the Hirudinea. All Hirudinea are hermaphrodites and lay their eggs in cocoons; some Ga crawl onto land to lay their eggs *(Hirudo*, Haemopis*)*, others practise brood care, carrying the eggs and the young Leeches around with them. The Hirudinea have been divided into 3 categories according to the structure of the pharynx: pharyngeal leeches (with a muscular pharynx and no specialised features), jawed leeches (with three toothed jaw plates), and proboscis leeches (pharynx protrusible). Leeches feed in one of two ways; they are either predators (living on small animals) or they are parasites (blood-suckers). They also move along in one of two ways, either looping along by stretching forward and attaching the front end to an object by suction, then pulling the body up after it, or, by swimming along in a snake-like fashion. Leeches make interesting specimens in a pond aquarium (eg *Hirudo* and *Haemopis*). Some species may also crop up as fish parasites (eg *Piscicola*).

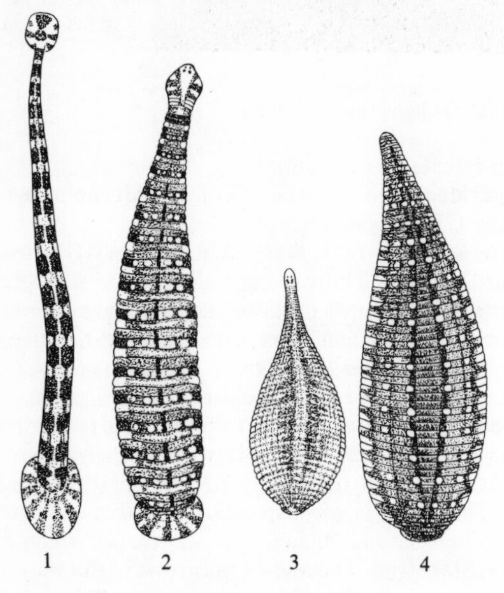

Fig. 622 Various leeches. 1 *Pisciola geometra* 2 *Hemiclepis marginata* 3 *Glossiphonia heteroclita* 4 *Glossiphonia complanata*

Hirudo LINNAEUS, 1758. G of the Hirudinea*. There is one European species (*H. medicinalis*, the Medicinal Leech) (fig. 623), but there are other species in the Palaearctic Realm* and in the Oriental realm*. *Hirudo* species are elongated creatures, up to 20 cm long, and they live in waters warmed by the summer sun and that

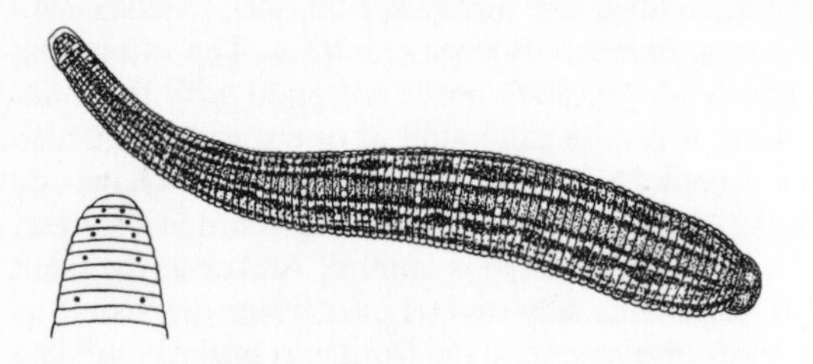

Fig. 623 *Hirudo medicinalis*. Below, head with eyes (considerably enlarged)

are rich in vegetation. In contrast to the similarly sized *Haemopis** (the Horse Leech), *Hirudo* always has red markings along its back. The various *Hirudo* species all suck the blood of warm-blooded animals, including man. They pierce the skin with their three toothed jaws (which can be quite painful). The blood is prevented from clotting by a protein substance present in the Leech and known as hirudin. This also means that the wound the Leech has made will often continue bleeding for some time after sucking has stopped. *Hirudo* was formerly used in medieval times in healing the sick through the practice of blood-letting. Modern medical practice is also taking fresh interest in the use of *Hirudo*. Artificially obtained hirudin is also dispensed to patients suffering from thrombosis.

Hirudo is well worth while keeping in a pond aquarium. Leeches do not suffer even in high temperatures or when there is little oxygen in the water (ie they are very tenacious creatures). They only need to be allowed to suck blood once or twice a year. If breeding Leeches, the container must have a suitable piece of dry land as *Hirudo* lays its egg cocoons out of the water.

Histology. A branch of biology* and medicine that is concerned with the structure and function of tissues. It is chiefly carried out by studying tissue under a microscope. In modern histology, electron microscopy and histochemistry are of particular importance. Pathological histology is the examination of changes to cells and tissues brought about by disease. The greatest practical use of this knowledge is in the diagnosis of certain diseases (both in animals and in humans), especially cancer.

History of Aquarium-keeping and Ichthyology *see* Aquarium-keeping; Aquarium, Care of; Ichthyology

Histrio FISCHER and WALDHEIM, 1813. G of the Antennariidae, O Lophiiformes*. Includes the species *H. histrio* (LINNAEUS, 1758), which is known as the Sargassofish. This fish lives in warm parts of the Atlantic and the Indo-Pacific, particularly among floating clumps of seaweed. It is well adapted to life here through its ability to change colour and through outgrowths along the body which break up the fish's outline. *Histrio* grows up to a length of 15 cm. It crawls through the weedy clumps by means of its limb-like paired fins, feeding on fish, spawn, crustaceans, medusae etc. Spawning has been observed in the aquarium. The larvae are pelagic and are surrounded by a cellophane-like covering.

Hogfish *(Lachnoleimus maximus) see Lachnoleimus*

Holacanthus LACÉPÈDE, 1803. G of the Chaetodontidae*, Sub-F Pomacanthinae. In 1933, FRASER-BRUNNER divided the originally very large G *Holacanthus* into several Ga (*Centropyge**, *Chaetodontoplus**, *Euxiphipops**, *Pomacanthus**, *Pygoplites** etc). *Holacanthus* species are found in all tropical seas, where they live territorially along coral and rocky coastlines. They feed on animal and plant growths and many of them are food specialists. Side on, *Holacanthus* is oval in shape with a very compressed body from side to side. It is very beautifully coloured and is covered in large ctenoid scales. The LL ends beneath the soft D. The C is more or less straight-edged, occasionally with a short lobe on the dorsal edge. Young specimens are often very different in colour from older specimens. In the aquarium, *Holacanthus* requires an extremely good water quality and a very varied diet, including plant food (algae, lettuce leaves). Particularly susceptible to infectious diseases *(Ichthyosporidium*)* and highly intolerant of like species and similar fish.

– *H. (Angelichthys) ciliaris* (LINNAEUS, 1758); Queen Angel, Queen Angelfish. From the tropical parts of the W Atlantic. Length up to 60 cm. Green-golden in colour with blue edges to the D and An, a yellow C and P and a blue-edged blotch at the nape. Young fish have blue, arched, transverse stripes and an eye band (fig. 624).

Fig. 624 *Holacanthus ciliaris*

– *H. tricolor* (BLOCH); Rock Beauty. From tropical parts of the W Atlantic. Length up to 60 cm. Head, nape, ventral surface and C a shining yellow, rear portion of the body black. Young fish are yellow with a dark blotch on the rear part of the body (fig. 625).

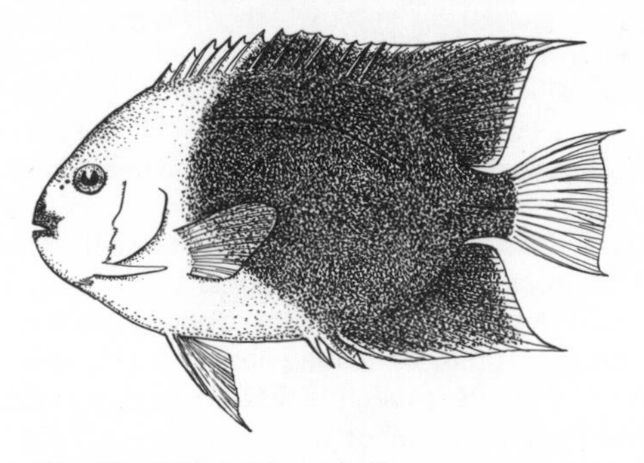

Fig. 625 *Holacanthus tricolor*

Holarctic. The largest zoogeographical realm* on earth. It includes Europe*, the northern part of Africa*, Asia* (with the exception of its tropical regions) and N America. Fish families distributed throughout the Holarctic realm include the Salmonidae (Salmonoidei*), the Gasterosteidae (Gasterosteoidei*), the Percidae* and the Esocidae (Esocoidei*). For a long time in the Holarctic realm no unimpeded exchange between the freshwater and land animal kingdoms has been able to take place (the Bering Strait acts as a natural barrier (*see* Distribution, limitations upon) between Eurasia and N America). As a result, the fauna of Eurasia and N America has developed in an increasingly isolated state, especially as far as species and genera are concerned. For this reason, the Holarctic is divided into the Palaearctic* and the Nearctic*.

Holocentridae, F, *see* Beryciformes

Holocentrus SCOPOLI, 1777 (fig. 626). G of the Holocentridae, O Beryciformes*. Comprises several species found in coastal waters of all tropical seas. Predatory fish that are active at dusk and at night. During the day they lie hidden in crevices. Most species are solitary. Length varies between 20 and 60 cm. The majority of species are red in colour and have strikingly large eyes, a large mouth and, in contrast to the G *Myripristis*,

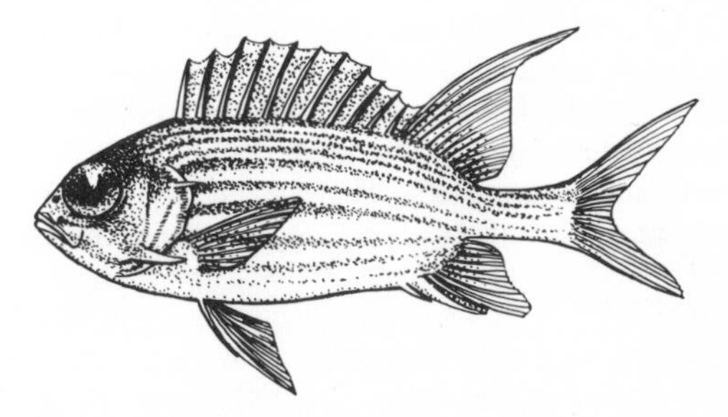

Fig. 626 *Holocentrus rufus*

they have a long spine protruding from the lower corner of the operculum. Open-water spawners. *Holocentrus* survives well in the aquarium if fed on live fish to start with, then, after a settling in period, fed on dead animal food. Unfortunately, *Holocentrus* tends to grow too large too quickly. Damage is often done to the eyes when the fish is caught and during transportation. The following species is often imported:

– *H. (Adioryx) rubra* (FORSKÅL, 1775); Common Squirrelfish, Red Soldierfish. Found in tropical parts of the Indo-Pacific, including the Red Sea. Length 23 cm. Body crossed by red longitudinal stripes.

Holocephali, Sub-Cl, *see* Chondrichthyes

Holostean Fishes (Holostei, Coh) *see* Osteichthyes

Holostei, Coh, *see* Osteichthyes

Holothuria LINNAEUS, 1758 (fig. 627). G of the Sea Cucumbers (Holothuroidea*). Contains a number of different species found in the littoral of warm seas and of the tropics. They eat from the substrate, conveying the food material to the mouth opening with their tenta-

Fig. 627 *Holothuria*

cles. *Holothuria* can be up to 30 cm long and 5–6 cm thick. On its underside it has 3 rows of sucker-feet, whereas the upper side usually has papillae only, with no suckers. As a defence mechanism many species push out very sticky threads known as the Cuvierian Organs; others release a haemolytic poison through the skin which will kill fish living in the same aquarium but which does no harm to molluscs or sea-anemones. In all other respects, *Holothuria* makes a hardy aquarium specimen—including the species that are often found along the European Mediterranean and Atlantic coasts. They need only be provided with sufficient food, which is usually to be found in longer established aquaria.

Holothuroidea; Sea Cucumbers. Cl of the Echinoderms (Echinodermata*). There are a large number of different Sea Cucumbers and they are found in seas everywhere. The body consists of a long, mobile muscular tube of skin which can allow the animal to move along with a worm-like creeping action. However, in contrast to the other classes of Echinoderms, Sea Cucumbers do not creep along on their oral surface but on three rows of tube feet (trivium). Some Holothuroidea have adopted a free-swimming way of life. The skeleton consists only of tiny, isolated pockets of calcium. The mouth is surrounded by a crown of tentacles made up of specialised ambulacral feet, which can be pulled inwards. Those species that feed on tissue have very fine tentacles to enable the animal to catch plankton and food particles (eg *Cucumaria**). Sea Cucumbers that feed from the substrate have strong mouth tentacles to probe into it to find detritus (eg *Holothuria**). The Holothuridea respire by the use of branched outgrowths of the intestine, usually referred to as respiratory trees. Some species also have cuvierian organs which are used in defence and are extremely sticky. If sorely threatened, many Sea Cucumbers will eviscerate themselves with some violence *(Holothuria*)* while others will break up into little pieces *(Synapta)*. The fish *Fierasfer (Carapus) acus* lives inside the body of many Sea Cucumbers. Sea Cucumbers are not difficult to keep in the aquarium. In well-established marine aquaria, they will usually find enough food without any special feeding.

Holothuroids, Cl, *see* Holothuroidea

Homalopteridae; Hillstream Fishes (fig. 628). F of the Cypriniformes*, Sub-O Cyprinoidei*. There are about 50 species in the F and they live in the fast-flowing mountain streams of SE Asia and the Indo-Australian archipelago. They are very well adapted to these biotopes and it is these adaptations that distinguish the Homalopteridae from all other Fs of the Sub-O. Dorsoventrally, the body is very flat (from a distance similar to many Armoured Catfishes *(Hypostomus*)*. The Homalopteridae also have very large, paired fins which project out horizontally from the body and which bend downwards slightly at the edges. Together with the underside of the body, these fins form a bowl- or dish-shaped sucking disc with which the fish can attach itself to firm surfaces. The grip is so strong that it stops the fish from being washed away even in rushing streams. Both Vs in the G *Gastromyzon* are fused—ie the sucking disc is particularly complete. Other features characteristic of the Homalopteridae include the very high position of the gill slits (almost at the nape), the short D and An, the small cycloid scales and the large number of barbels (3–5 pairs). As in all Cyprinoidei*, pharyngeal teeth are present but there is no grinding mill. The swimbladder is surrounded by a bony capsule similar to that found in the Cobitidae*. Their food consists mainly of algae which they tear off with their small mouths. The respiratory technique is also peculiar to the Homalopteridae: the gill cavity can be enlarged to become a water reservoir and the fish is able to live for a while on the oxygen in the stored water. During this period there is no outward sign that the fish is breathing. Such highly specialised fish can only be kept in the aquarium for a limited time. Little is known about their reproduction.

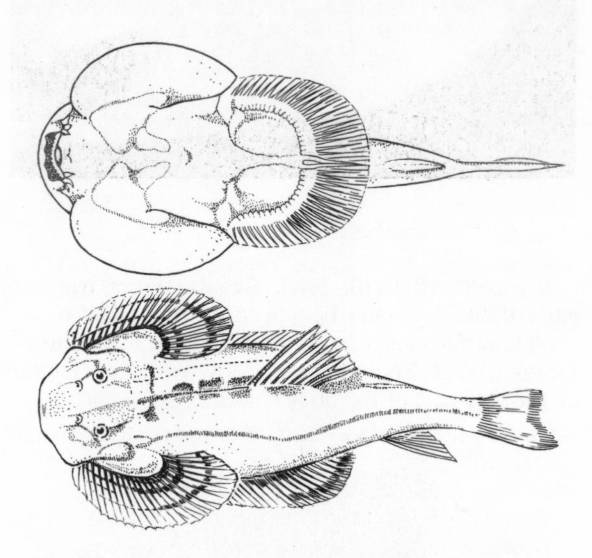

Fig. 628 Homalopteridae-type, underside and upperside

Homarus WEBER, 1795; Lobster. G of the Decapoda*, Sub-O Reptantia, F Homaridae. Consists of several species found in cool and warm seas. Solitary animals, lobsters can grow up to 903m long and have extremely broad pincers. They are predators that live hidden away. Their anatomy is very similar to that of the crayfishes and sometimes they are even placed in the same G *Astacus**. Lobsters have been fished intensively and as a result their numbers have declined dramatically. In the aquarium they are relatively hardy but they tend to grow very quickly. They should be kept on their own or with a few sea-anemones. *H. gammarus* (LINNAEUS, 1758) lives on rocky substrates and is found from N Europe to the Black Sea.

Homocercal C, *see* Fins

Homoiothermal; warm-blooded. Birds and mammals are described as homoiothermal because their body temperatures are kept constant by their producing sufficient warmth or through temperature regulation. The brain is the control source. Opposite to poikilothermic*.

Homologous Organs. Parts of the body in various plants or animals that originated from the same embryonal hereditary factor independently of their structure and function in the adult organism. A good example of this are the five-rayed extremities of air-breathing vertebrates. The front legs of four-footed animals correspond to the wings of birds, the excavating limbs of moles or the pectoral fins of whales. That there is a homologous relationship with the pectoral fins of fish is less easy to see, but it can easily be deduced from looking at Coelacanths and Lungfishes. *See also* Analogous Organs.

Homonym. 1) In the generic sense. A name which is (unwittingly) given to different taxa (*see* Taxon) by different authors. For example, the generic name *Acaropsis* was proposed by STEINDACHNER in 1875 for a genus of Cichlids, when, quite independently, MOQUIN-TANDON had suggested it for a genus of Mites in 1863. As a generic name may only be conferred once in the animal kingdom (Zoological nomenclature*), a replacement name had to be found for STEINDACHNER's name. MYERS was to confer the new name in 1940 (Acaronia*).

2) In the specific sense. The same specific names can only be termed homonyms when they both belong to the same G. So, for example, in the G *Pseudotropheus* the specific name '*auratus*' may not be given to any other *(Pseudotropheus)* species. Homonyms can also occur because Ga have to be grouped together on account of their natural relationships, but in such a case one of the homonymous names must still be altered – the one that had its name conferred on it most recently (providing that the rules do not suggest an alternative plan).

3) In the family sense. Here too, homonyms are not allowable.

4) In categories above the family. In the animal kingdom, no set rules apply here. For example, the name Agnatha is used for both the Cyclostomata* as well as the insect order Ephemeroptera*.

Honey Gourami *(Colisa chuna) see Colisa*

Hooknose *(Agonus cataphractus) see Agonus*

Hoplias GILL, 1903. G of the Erythrinidae*. These fish are predatory Characins with a wide distribution in northern and central parts of S America. Length up to 50 cm. Elongate body, cylindrical, and covered in large cycloid scales. The head is short with a relatively broad gape and very strong teeth. The D1 and An are fairly short and are not drawn out into a point, although in fully-grown specimens the D1 rays extend as filamentarous outgrowths. The free end of the C is rounded. *Hoplias* species are greedy predators which live on fish exclusively, lying in wait for them on the substrate or among reeds (rather as Pikes do). Only young specimens should be considered for the aquarium as adult specimens are not attractive enough even in the show aquaria of zoos – the youthful colouring largely fades away as the fish gets older. Nothing is known about its reproduction. There is only one species.

– *H. malabaricus* (BLOCH, 1794); Tiger Fish, the Trehira. Found in northern and central S America. Length up to 50 cm. *Macrodon trahira* is another name for this fish although it is no longer in current usage. Colouring is very variable. Youthful colouring: dorsal surface olive-brown to red-brown, flanks a brighter yellow-brown, this colour continually extending into the delicate reddish yellow colouring of the ventral surface. Several irregular bands extend from the mouth across the operculum. A broad, blackish-green longitudinal band begins just behind the operculum and ends in a dark blotch at the base of the C. With age, this longitudinal band is increasingly covered over with blotches that almost form transverse bands. All the fins are brown and covered in rows of dark blotches (fig. 629).

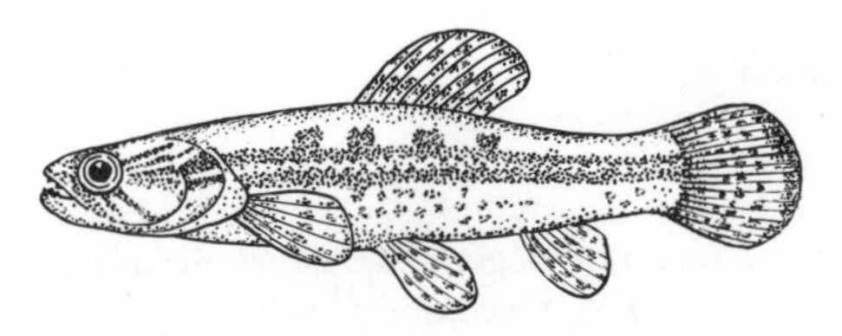

Fig. 629 *Hoplias malabaricus*

Hoplosternum GILL, 1858. G of the Callichthyidae*. Found in Brazil, Trinidad and the Guyana lands. Length up to 20 cm. The body is a plump cylindrical shape and, in contrast to the G *Corydoras*, the bony plates between the D1 and D2 are firmly linked with the side plates. The C is either slightly indented or straight-edged. There are two rows of bony plates along the flanks which interlock like roof-tiles. Two pairs of long barbels project forwards from the upper jaw of the short, plated head. *Hoplosternum* is an omnivore. The males build bubble-nests at the water surface among large plant leaves (they will also build such nests in an inverted ceramic bowl). After vigorous driving, the male presses firmly on the barbels of the female with his pectoral fins. The sperm and several large, orange-yellow eggs are released at the same time. The female catches the eggs in a pouch (formed from the Vs) and swims with them up to the nest. Here, she presses the eggs into the nest by blowing bubbles of air. Up to 800 eggs can be kept together in this way. The male guards them till they hatch out (after about 4 days). The female goes off after mating is finished. Artificially breeding these fish does not always work, even when there is a nice current present on account of the level of aeration in the tank. The eggs will often be attacked by fungal growth or the young fish will hatch out too early and die. Juveniles are beige or dark-brown in colour with transverse bands. In the aquarium they will grow at an uninterrupted rate if fed on finely chopped *Tubifex. Hoplosternum* species are attractive armoured catfish which are particularly interesting on account of their mating and brood care behaviour. They are not aggressive towards any smaller fish.

– *H. thoracatum* (CUVIER and VALENCIENNES, 1840); Port Hoplo, Atipa. Panama to Paraguay. Length up to 18 cm. With such a large distribution area, colouring tends to be very variable: Body and fins dark-brown to red-brown with large numbers of dark spots. The upper surface is often blackish, the belly grey-violet or dirty white and covered in a great many black blotches. Adult males can easily be distinguished from the females because the first ray of the Ps is very flat and broad, and is a striking fawn-brown colour. In both sexes, the C has a pale transverse band at the base. Omnivorous. After TEICHFISCHER carried out the first successful breeding of this species, it became popular very quickly (fig. 630).

Hora's Loach *(Botia horae) see Botia*

Fig. 630 *Hoplosternum thoracatum*

Horizontal Fighting Behaviour *see* Symbolic Fighting Behaviour

Hormones. Active substances formed both in plant and animal organisms that regulate large numbers of life processes, eg growth, maturation, reproduction, certain types of behaviour and different metabolic processes. Depending on their chemical composition, hormones can be proteins, steroids and other combinations. In vertebrates, most hormones are produced in glands with internal secretion* and transported along with the blood to have an effect upon certain cells or organs, or upon the whole body. It is in this way, for example, that thyroxin from the thyroid gland influences the general metabolism of proteins and fats, or that insulin and glucagon from the pancreas influences the level of sugar in the blood, or that the hormones from the adrenal cortex controls the mineral and glycogen content. In fish, the hormone in the pituitary that stimulates melanophores causes the body to turn dark (*see* Chromatophores). The lack of certain hormones, or their overproduction, usually causes behavioural imbalances or changes to the body. The occurrence of sex reversal* in fish is something that is well known in aquarium-keeping. It is caused by an imbalance of the sex hormones.

Horn Sharks, F, *see* Heterodontidae

Horned Pondweed (*Zannichellia*, G) *see* Zannichelliaceae

Horned Pondweeds, F, *see* Zannichelliaceae

Hornwort *(Ceratophyllum demersum) see Ceratophyllum*

Hornwort *(Ceratophyllum submersum) see Ceratophyllum*

Hornwort Family *see* Ceratophyllaceae

Horny Corals, O, *see* Gorgonaria

Horny Seaweed, G, *see Ceramium*

Horse Leeches, G, *see Haemopis*

Horse Mackerel *(Trachurus trachurus) see Trachurus*

Horsefaced Loach *(Acanthopsis choiorhynchus) see Acanthopsis*

Horseshoe Crabs, Cl, see Xiphosura

Horseshoe Worms (Phoronida, Cl) *see* Tentaculata

Horse-tail (*Equisetum*, G) *see* Pteridophyta

Horse-tails (Sphenopsida, Cl) *see* Pteridophyta

Hostility Factor. Term used in ethology. The hostility factor is an important biological factor in the regulation of the population density of a species within its living-space. A lack of hostility factor can lead to a population explosion, particularly among animal pests (eg migratory locusts, fieldmice). This can cause catastrophic damage to the environment. Particularly in countries where the land is cultivated, such disproportions in the animal world occur because certain enemy organisms are decimated or die out. In agriculture and in forestry in particular, biological pest control is to a large extent based on the promotion of the hostility factor. In aquarium-keeping the hostility factor is especially important among those species of fish that care for their brood. Cichlids often eat their own offspring when there is a lack of hostility factor; it is thought that when the Cichlids are not sufficiently active in defending their brood there is a premature lowering of the specific reaction energy related to caring for their offspring (ie, it falls below its stimulus threshold*).

Hottonia LINNÉ, 1753; Featherfoil. G of the Primulaceae*, comprising 2 species. Found in N America, Europe, W Asia. Featherfoils are perennial plants with creeping rhizomes and upright shoots. The leaves are arranged in alternate pattern, and they are simply pinnate. The flowers are hermaphroditic and polysymmetrical. There is a double perianth. The sepals and petals are in threes and fused. 5 stamens, 5 carpels. The carpels are fused and superior. Capsule fruit. The various species of the G grow in standing waters such as ponds, ditches or bogs. They prefer water that does not contain much chalk. None of the plants is suitable for the aquarium and even in open-air installations it is rare for *Hottonia* to survive for very long. Propagation is possible with cuttings.

– *H. inflata* ELLIOTT. Found in south-eastern parts of N America. The inflorescence axes are thickened.

– *H. palustris* LINNÉ; Common Featherfoil. Found in Europe and W Asia. The inflorescence axes are normal.

Houttuynia cordata *see* Saururaceae

Humbug Damsel *(Dascyllus aruanus) see Dascyllus*

Humic Acids. Substances found in all types of humus (and therefore also found in peat*). They have the qualities of a weak acid and tannin. Humic acids can easily be washed out and in enough concentration they will colour the water brown (*see* Dystrophic Waters). In aquarium-keeping, humic acids have become important chiefly because they are used to simulate tropical black waters by filtering the water through peat or by adding peat extracts to the water. Humic acids have both astringent (ability to contract vessels and tissues) and antiseptic (ability to impede the development of bacteria and fungal growth) properties. Like all tannic substances, humic acids become poisonous in too high concentrations. So, an aquarium must never be allowed to have brown water – at most, it should only appear amber-yellow. If too much humic acid is present, it can be removed by filtering the water through activated charcoal.

Humpbacked Headstander *(Charax gibbosus) see Charax*

Humpbacked Limia *(Poecilia nigrofasciata) see Poecilia*

Humuhumu-nukunuku-a-puaa *(Rhinecanthus aculeatus) see Rhinecanthus*

Hyas LEACH, 1830. G of Spider Crabs (Brachyura*, F Majidae) that live in northern seas. They live on molluscs, sea-urchins, crustaceans and algae. Their carapace is ovate and grainy. The pseudorostrum consists of two teeth which are located very close to one another.
– *H. araneus* (LINNAEUS, 1758) is found from Spitzbergen to the western Baltic and the Channel. It is a stenohaline species. The carapace is up to 11 cm long. Young specimens (which in contrast to older specimens like to camouflage themselves) are suitable for the aquarium, but bear in mind that they are predators.

Hybrid. The offspring of two animals or plants whose hereditary material differs from each other in at least one respect. Therefore, hybrids join together different hereditary material, so they are not homozygous (*see* Heredity). The hybrid is described as a monohybrid, dihybrid or polyhybrid, depending on the number of dissimilar characteristics and their genetic structures (pairs of allelomorphs). If hybrids are cross-bred, further dissimilar features will result. This was the prerequisite for the compilation of the laws of heredity or Mendelian laws. The production of hybrids is the basis of breeding*. Besides the normal cross-breeding of races (eg the breed races of the Guppy) it is also possible to achieve species hybrids from time to time. Such hybrids are often infertile but may possess the particular combination of characteristics the breeder wished to achieve. In the breeding of ornamental fish, the crossbreeding of species has produced some notable forms, particularly in the Swordtails *(Xiphophorus*).*

Hybridisation *see* Cross-breeding

Hybridisation Zone. The parts of an area of distribution of geographically adjacent races (compare Species, Concept of) that overlap and in which mixed populations occur. The presence of a hybridisation zone proves that there are no restrictions on reproduction between the forms found there and so they belong to a common species (Sympatry*).

Hybrid-Ludwigia *(Ludwigia repens palustris) see Ludwigia*

Hybrid-Mackerel *(Trachurus trachurus) see Trachurus*

Hybrids, Growth in, *see* Heterosis

Hydra LINNAEUS, 1758 (fig. 631). G of the Hydrozoa*. About 6 different species are represented in Europe where they live in contrasting types of inland waters (streams, rivers, lakes, fishponds, etc). *Hydra* and related Ga (eg *Chlorochydra*) are sessile creatures – they attach themselves to a suitable substrate by means of a sucker. When in a stretched position, *Hydra* look like a slender bottle and in this state measure up to 15 mm long. The mouth area is surrounded by a crown of tentacles which can measure as long as the body. The moment the tentacles touch a suitable prey animal (such as a small crustacean) it is harpooned with stinging cells, or nematocysts, and the poison they contain paralyses the prey. The *Hydra* then forces the prey into its stomach.

Hydra usually propagate themselves asexually, forming buds on the body that grow into small polyps. In this way, an enormous number of *Hydra* can develop within

Fig. 631 *Chlorohydra*

a short period. If living conditions are not very good, in autumn for example, eggs and spermatozoa may be formed, ie sexual reproduction takes place. The fertilised eggs survive through the winter.

Certain Ga, eg *Chlorohydra*, play host to single-celled green algae inside their bodies (hence the green colouring) and live symbiotically in conjunction with them (*see* Symbiosis). The algal cells are deposited in the egg-cells too, so they are transferred to the next generation.

Hydra are often introduced into the aquarium along with live food. If they manage to feed well, they can sometimes develop in such large numbers that steps must be taken to cut down those numbers. Copper ions, ammonium ions and temperature increase all help in this regard. The last mentioned is the simplest method. Simply remove the fish and raise the temperature to 42 °C. Plants will not be harmed. If adding ammonium ions in the form of ammonium nitrate (1–3 g for every 10 litres of water), the pH value* must be determined *exactly* beforehand, and if necessary corrected. Remember that the ammonia-ammonium balance in the water is dependent on the pH. Under alkali conditions, ammonium (which is largely non-poisonous as far as fish and plants are concerned) will be converted into highly poisonous ammonia. When combating the presence of *Hydra* with ammonium, the pH value must not be higher than 7.2! If in doubt, the fish should be removed from the tank before adding the ammonium nitrate, and afterwards the water should be changed.

If instead copper ions are to be used, either add copper sulphate (0.03 g to every 100 litres of water), or hang copper plates in the tank (copper ions will be formed by electrochemical increase processes). To speed up the process and increase its strength, apply a weak current with the aid of a torch battery, thus turning the plates into electrodes. As copper ions are highly toxic for fish, the whole tank must be emptied before beginning treatment. Once the treatment has taken effect, the dissolved copper can be turned into a hard-to-dissolve deposit with tannin (0.1 g for every 10 litres of water). This in turn can be filtered out using activated charcoal.

Hydracarina (taxonomic collective group) *see* Acari

Hydractinia echinata *see Eupagurus*

Hydrilla L. C. RICHARD, 1814. G of the Hydrocharitaceae*, comprising just one species. Found in E Europe, Asia, Africa, Australia. *Hydrilla* is a small perennial with long stem axes that are branched at irregular intervals. The leaves come in 3-part to 5-part whorls, and they are linear, finely serrated. The flowers are unisexual, monoecious or dioecious, small, and develop singly within a spathe. There is a double perianth, the sepals and petals being inconspicuous and in threes. Male flower: 3 stamens, plus 3 staminodes. Female flower: 3 carpels, fused, inferior, with usually 3 staminodes. The fruit is like a berry. *Hydrilla* always lives submerged. In the tropics and subtropics, the plant keeps its vegetation at all times, but in more temperate regions winter buds are formed. *Hydrilla* makes an undemanding plant for cold and heated aquaria. It can either be planted in the aquarium or allowed to float about freely.

– *H. verticillata* (BESSER) DANDY (*H. lithuanica* – name not in current usage).

Hydrobiology. The study of organisms that live their entire lives in water or those that have to live in water for a certain period of their development. It includes the study of the relationship of these organisms to their environment.

Hydrocharis LINNÉ, 1753; Frogbit. G of the Hydrocharitaceae*, comprising 3 species. Found in Europe, Asia, Africa, Australia. Small rosette perennials with runners. The leaves are stalked with round leaf blades. The flowers are unisexual and dioecious, several developing together within a spathe. There is a double perianth, the sepals and petals being free-standing and in threes. The petals are white. Male flowers: up to 12 stamens; Female flowers: 3 carpels, fused, inferior, with usually 3 staminodes. The fruit is like a berry. The various species of the G are aquatic plants, their rosettes floating on the surface in wind-protected places of standing waters. More rarely, they take root in marshy substrates. In autumn, winter buds form in the leaf axils of the rosette leaves and at the tips of the runners. These buds sink to the bottom. *Hydrocharis* is suitable for open-air ponds. The following species is well known:

– *H. morsus-ranae* LINNÉ; European Frogbit. Europe, W Asia.

Hydrocharitaceae; Frogbit Family. F of the Liliatae* comprising 15 Ga and 100 species. Mainly found in tropical-subtropical areas, but also in temperate zones. Frogbits are perennial plants from very varied habitats. Their stem axes are compressed or elongate with an alternate or whorled leaf arrangement. In rare cases the flowers are hermaphroditic, but usually they are unisexual and dioecious, polysymmetrical, and in some cases, inconspicuous. The Hydrocharitaceae are very definitely aquatic plants and usually they propagate vegetatively in very large numbers. Only a few species are able to live in marshes, too. They inhabit very varied types of fresh water, and some species also live in marine areas close to the coast. Many species have an interesting flower biology because of the fact that they are pollinated by water. Important Ga include *Elodea**, *Egeria**, *Lagarosiphon**, *Hydrilla**, *Vallisneria**, *Ottelia**, *Blyxa**, *Stratiotes**, *Hydrocharis**, and *Limnobium**.

Hydrocleis nymphoides *see* Butomaceae

Hydrocotyle LINNÉ, 1753; Water Pennywort. G of the Apiaceae* comprising 160 species. Found in almost every part of the Earth, but chiefly in the southern hemisphere. Small perennial plants with long stem axes that usually grow horizontally and are much branched. Alternate leaf arrangement. The leaves are stalked with variously formed round or kidney-shaped blades. Many species have peltate leaves. The flowers are hermaphroditic, small, arranged in umbels. There is a double perianth. The sepals are inconspicuous and the petals come in fives. 5 stamens. 2 carpels which are fused and inferior. Schizocarp. The various species of the G inhabit marshes, either floating with their leaves at the surface or staying submerged for a lengthy or short period. Numerous species are said to be suitable for the aquarium, but to date only a few have been tried out. They like light but otherwise are undemanding and very decorative. The best known species are:

– *H. leucocephala* CHAMISSO et SCHLECHTENDAL. From tropical S America. The submerged stems reach upwards and float at the surface. The leaf blades are round or kidney-shaped.

– *H. verticillata* (A. RICHARD) FERNALD. From south and east N America, Central America. The stem axes grow along the ground. Peltate leaves.

Hydrogen Ion Concentration *see* pH value

Hydrogen Peroxide (H_2O_2). A strong oxidising agent with good disinfecting properties (Disinfection*). The non-poisonous nature of hydrogen peroxide and its dissociated products (water and oxygen) is beneficial. If there is acute oxygen deficiency*, 1 ml of a 15% hydrogen peroxide solution can be added to every 20 litres of aquarium water. To combat many skin diseases*, short-term hydrogen peroxide bathing treatments* lasting 10–15 minutes can be carried out in separate tanks (10 ml of a 3% hydrogen peroxide solution to every 1 litre of aquarium water). Localised treatments of skin defects take place out of the water with a 3% solution.

Hydrogen Sulphide Poisoning *see* Poisoning

Hydrology. The study of waters and water cycles. It is divided into two fields – limnology* and oceanology*.

Hydrometra LATREILLE, 1796; Water Measurer (fig. 632). G of the Bugs (Heteroptera*). There are two European species and like the well-known Pond Skater *(Gerris*)*, it belongs to a family of water-loving land

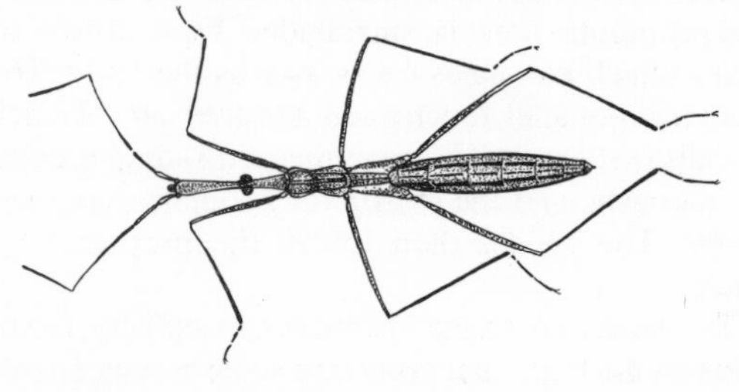

Fig. 632 *Hydrometra stagnorum*

bugs. In contrast to *Gerris*, *Hydrometra* does not move forwards in a jerky fashion, but rather as other land insects, with alternating leg movements. The body is as thin as a needle and an unremarkable dark grey colour. For this reason they may often be overlooked, especially as they normally rest comfortably on the surface or move about on the bank.

Hydrophilidae; Water Beetles (fig. 633). F of the Beetles (Coleoptera*). In Central Europe there are about 150 different species. At most, only two-thirds of all species of Hydrophilidae lead on aquatic existence, others preferring the zone where water meets land, others living completely terrestrially, eg in dunghills.

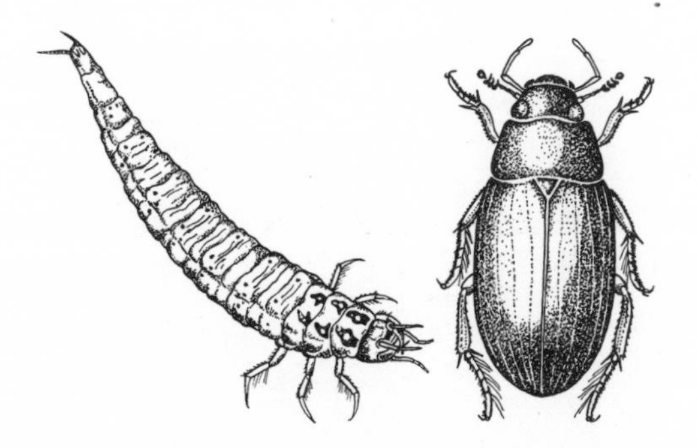

Fig. 633 Hydrophilidae. Left, larva; right, imago

Water-dwelling members of the Hydrophilidae can be distinguished from the often similar-looking Dytiscidae* by their alternate leg swimming action. In addition, the Hydrophilidae carry their air supply around with them, usually on the underside of the body. Whereas the larvae are notable predators, the sexually mature adults (imagines) live mainly on plant food. As with all other aquatic beetles, the pupal stage is spent on land. Some species can grow to a considerable size, such as the well known Great Silver Beetle *(Hydrous*)*. Hydrophilids often get caught in fishing nets and they can make very interesting specimens, particularly their methods of swimming, feeding, air-breathing, mating and egg-laying.

Hydrophily *see* Pollination

Hydropsyche Pictet, 1834. G of the Caddis Flies (Trichoptera*). In Europe there are more than 20 species. Body length up to 2 cm. The larvae are typical of these that live in fast-flowing waters. They build the familiar caddis cases from mouth saliva. There are various features by which you may recognise *Hydropsyche* larvae – the horny plates on the thorax and the bushy tracheal gills on the abdomen. *Hydropsyche* species make good food animals and specimens in their own right in the aquarium.

Hydrostachyaceae. F of the Magnoliatae*. There is only one G *Hydrostachys*, and it consists of 30 species. Found in Central and S Africa, Madagascar. Rosette perennials with long, pinnate leaves. The flowers are small, much simplified, unisexual and usually dioecious. The Hydrostachyaceae grow on rocks in rushing rivers and streams, in rapids and in waterfalls. As yet they have not been cultivated.

Hydrotriche Zuccarini, 1832. G of the Scrophulariaceae* and consisting of only one species. From Mada-

gascar. It is a perennial plant with long stem axes, irregularly branched. The leaves are arranged in whorls. The leaves are like Broom, pointed, pale green. The flowers are hermaphroditic and monosymmetrical. There is a double perianth. The sepals and petals come in fives and they are fused. The petals can just be seen to be divided into two lips, and they are white and yellow. 2 stamens. 2 carpels which are fused and superior. Fruit is a capsule. *Hydrotriche* grows in sunny spots in flowing and standing waters. In the aquarium, it needs a lot of light and so far success in caring for it has been intermittent.
– *H. hottoniiflora* Zuccarini (fig. 634).

Fig. 634 *Hydrotriche hottoniiflora*

Hydrous; Silver Beetles. G of the Water Beetles (Hydrophilidae*). The best-known species, the Great Silver Beetle *(H. piceus)* (fig. 635), can grow to over 70 mm as a larva, and over 50 mm as an imago. The various members of *Hydrous* live in standing waters* rich in vegetation. As larvae, they are predatory, as Beetles they are mainly vegetarian. Because of their size and their lack of specific needs in pond aquaria, they make very interesting objects of study. *H. piceus* is a protected species almost everywhere. Features to look for are the peculiar swimming action first of all, the way in which the feelers are used to help draw breath, and the long air-tube fixed to the egg cocoon.

Fig. 635 *Hydrous piceus*

Hydrozoa (fig. 636). Cl of the coelenterates (Coelenterata*) comprising large numbers of species and forms. There are about 2,700 species and with few exceptions (eg *Hydra**), they live in the sea. Most live a sessile* existence. The medusoid phase that is characteristic of many coelenterate groups is considerably played down in the Hydrozoa (an exception is *Craspedacusta**, for example); instead, the polyp phase is prominent. The polyp either lives a solitary existence, in which case it is microscopically small or at most a few cm long, or many polyps combine to form a colony. Among the highly specialised hydrozoan forms there is a complicated division of labour within the polyp colony. For example, among the Siphonophora the individual animals are practically reduced to the function of organs whilst the colony as a whole represents one organism.

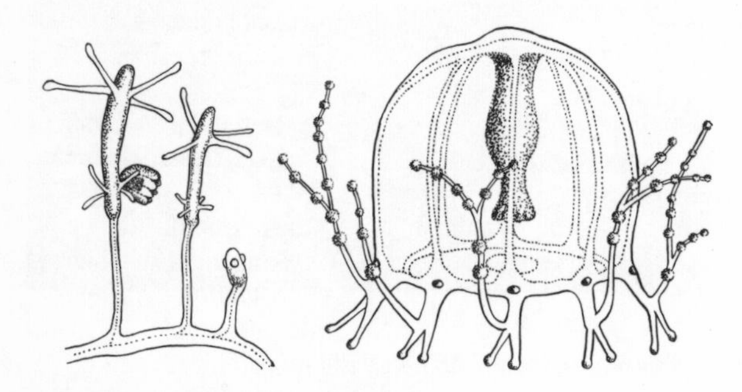

Fig. 636 *Cladonema radiatum*

Hydrozoan Polyps, Cl, *see* Hydrozoa

Hygrophila R. Brown, 1810. G of the Acanthaceae* comprising 60 species. (Many botanists include the Ga *Nomaphila** and *Synnema** along with *Hygrophila.*) Found in all tropical regions. *Hygrophila* species are small to medium-sized perennials with erect, ascending stems. Decussate leaf arrangement. The aerial leaves are undivided whereas the aquatic leaves are undivided or pinnate. The flowers are hermaphroditic, monosymmetrical, and appear as thick inflorescences at the leaf axils. There is a double perianth with a five-part calyx fused to the middle. The corolla is two-lipped, the lower lip being flat and the corolla tube therefore open. 4 stamens, rarely 2. 2 carpels, fused, superior. Capsule fruit. *Hygrophila* species are marsh plants, some of which also grow well when submerged. So far, only a few species have been tried out in the aquarium. They like a bright spot but otherwise have no special requirements. Propagation is usually possible with cuttings. However, adventitious plants will also be produced if separated leaves are allowed to float on the water surface. Widespread species include:

– *H. polysperma* T. Anderson. From SE Asia. Aquatic leaves up to 5 cm long (fig. 637).

Fig. 637 *Hygrophila polysperma*

– *H. salicifolia* (Vahl) Nees (*H. angustifolia,* name not in current usage). From SE Asia. Aquatic leaves up to 12 cm long.

Hymenocera Latreille, 1819. G of small Shrimps (Natantia*) from the coastal areas of the tropical Indo-Pacific. Rather bizarre animals, they are fairly compressed, with a hunch-backed rear portion of the body. The second pair of pincers are like lobes and the first set of antennae look like a couple of flags. *H. picta* is noted as a food specialist: it eats starfish almost exclusively, although it occasionally touches brittle-stars and some types of sea-urchin. Alone, *Hymenocera* can kill even large starfish by cutting a hole in the body wall of its prey and tearing out the insides. If *Hymenocera* does not meet with success dorsally, it will turn the

Fig. 638 *Hymenocera picta*

starfish round. In the aquarium, if you are able to feed *Hymenocera* on starfish, your specimen will live for a long time. Copulation and the deposition of eggs has also been observed, but to date the larvae have not been reared successfully. The following species is often imported from E Africa:
– *H. picta* DANA; Harlequin Shrimp, Clown Shrimp. From the tropical Indo-Pacific. 5 cm. White with large reddish-brown blotches edged in blue (fig. 638).

Hypertonic. Developing a higher osmotic pressure (tonus, tone). The blood plasma of freshwater fish, for example, is hypertonic in relation to the surrounding water. It is caused by the blood plasma containing a larger propertion of dissolved, osmotically effective substances. Opposite to hypotonic*. The term is also used to describe blood pressure that is above normal.

Hyphessobrycon DURBIN, 1908. G of the Characidae*, Sub-F Tetragonopterinae. Native to northern and central S America, *Hyphessobrycon* species are usually very attractively coloured. They are small shoaling fish from the middle and lower water levels. Length up to 7 cm. Since they were first introduced to the aquarium large numbers of the species included in this G have become very popular. Unfortunately, to date, there has been no comprehensive modern revision of this very important Characin G (as far as aquarists are concerned). So for the moment, it is extremely difficult and sometimes impossible, even for scientists, to make categorical statements about the classification of some imported species, just as it is difficult for them to be sure about the validity of different scientific names for species that are currently used among aquarists. The systematical groupings and species descriptions given here are therefore only an indication of current views. The conditions this G requires for successful breeding and care are largely the same as those for them closely related G *Hemigrammus**. *Hyphessobrycon* species differ from *Hemigrammus* in respect of the teeth, and secondly in respect of the lack of scales at the base of the C. For further particulars *see* the individual species. STALLKNECHT has indicated that the spawning behaviour has been modified in a way that is typical of the G: courtship behaviour is introduced by the male beckoning the female by swimming round her in a series of fluttering dances*. Later, the male swims in front of the female, leading her towards the spawning substrate with his fluttering dances. The spawning ground is located on or above some matted vegetation (depending on species) and is selected by the female. There, and only directly before mating takes place, the male appears alongside the female for the first time, usually swimming towards her from a position above and to the side of her. In the moment of mating, they twist around each other as if around an imaginary vertical axis. The G can be split up as follows:

1) The *Bentosi group* (also known as the Ornatus group): *H. bentosi*, *H. erythrostigma*, *H. ornatus*, *H. rosaceus*, *H.* sp. (= Sickle Characin). These species all exhibit a marked sexual dimorphism because of the pennant-like elongated D in the males. With the exception of *H. erythrostigma*, there is no shoulder fleck and the C, An and Vs all have zones that are mainly coloured a strong red. Our notions about *H. rosaceus* and *H. bentosi* remain largely unclear. It could be that the former is identical to the Sickle Characin which has been kept in tanks in recent years and has been known by the scientifically incorrect term '*H. robertsi*'. Recently, the scientific species name for the Ornate Tetra has been questioned, since GÉRY has informed STERBA that he suspects that *H. ornatus* is another name for *H. bentosi*. If GÉRY's opinion is confirmed, the Ornate Tetra will in future be known as *H. bentosi*. That is why we have not included a separate entry for *H. bentosi* here.

2) The *Bifasciatus group*: *H. bifasciatus*, *H. flammeus*, *H. griemi*. These species are characterised by the 2 large comma-shaped flecks at the shoulder.

3) The *Callistus group*: according to GÉRY, the following species belong to this group: *H. callistus*, *H. hasemani*, *H. minor*, *H. serpae*, *H. heraldschultzi*. The Coffee Bean Tetra, *H. takasei*, should probably be included here too. Most of these species have a comma-shaped shoulder fleck and are coloured yellowish or reddish/blood red. With the absence of a modern revision of the G *Hyphessobrycon*, it is difficult to correctly classify these species that are kept by modern aquarists. We cannot be sure, for example, that *H. serpae* has ever been imported since 1945. This gives rise to the possibility that all the 'Serpae' species that have been kept in tanks since that time really belong to the species *H. callistus*, the Jewel Tetra. For further information *see* the specific descriptions. Thus, the most popular fish of all from this group, the blood red 'Minor' which has no shoulder fleck or only a very small one, would then be a variety of *H. callistus* and not *H. serpae*, unless it should turn out to be a species in its own right. According to GÉRY the description of the species *H. minor* does not relate to this fish. The two species, *H. hasemani* and *H. haraldschultzi* have never been kept in aquaria in Europe and so they are not described here. Some authors also place the tiny Georgette's Tetra *(H. georgettae)* in this group.

4) The *Metae group* (also known as the Heterorhabdus group): *H. herbertaxelrodi*, *H. heterorhabdus*, *H. loretoensis*, *H. metae*, *H. peruvianus*, *H. scholzei*. A typical feature of this group is a dark longitudinal band that may also be accompanied by other colours. Only the validity of the species *H. peruvianus* is open to doubt. It may be another name for *H. metae*. For this reason there is no separate entry for *H. peruvianus*.
– *H. bifasciatus* ELLIS, 1911; Yellow Tetra (fig. 639). From the coastal areas of SE Brazil. Up to 5 cm long.

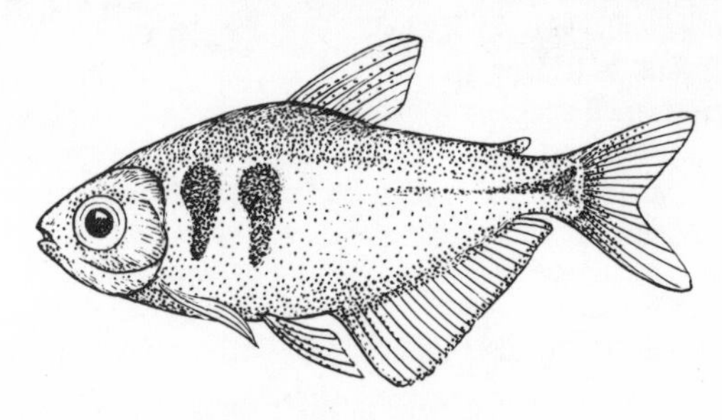

Fig. 639 *Hyphessobrycon bifasciatus*

Body relatively tall, similar to the better known species, *H. flammeus*. Flanks greenish-yellow with a silvery sheen. In front of two large, fairly long, dark shoulder flecks there is a shining green blotch which lies just behind the operculum. In the male the fins are yellow. The An has a narrow, brown-red edge to it. The first ray of the Vs is shining white – the same colour as the tip of the D and An. Adipose fin present. The female's fins are grey-yellow. Particularly attractively coloured specimens have a red base to the C, even as they get older. (This stems from the blood red fin colouring of the juvenile fish.) Males do not retain this red colouring. Young specimens of between 1 and 1.5 cm in length can barely be distinguished in colouring from young *H. flammeus*. Not difficult to rear – the young fish grow very quickly. For further information turn to the G *Hemigrammus*.

– *H. callistus* (BOULENGER, 1900); Jewel Tetra, Callistus Tetra. From the Rio Paraguay Basin. Up to 4 cm long. Taller than the other species in the Callistus group. Often wrongly named as a 'Serpae', although it is not identical with that species. Body colouring brown-red, with a silvery iridescence when light falls directly on it. Ventral surface yellow-brown, blood red in males that are ready to mate. The base of the D is a transparent yellowish-brown. The remaining surface of the fin is a black patent colour and there is a colourless adipose fin present. The C is blood red and the free edges of the fin lobes are a black colour. The base of the An is brown-red or red, and in the males there is a broad black edge, in the females a narrower one. The front third of the fin is edged in a pastel white. The Vs are red with a black-white front edge, the Ps are reddish or red. There is a comma-shaped, vertical, black blotch at the shoulder behind the operculum. The iris is blood red with a black transverse streak. The ventral surface in sexually mature females is very rounded. The popular blood red species normally called 'Minor' is probably a different colour variety of the Callistus species (although it might prove to be a species in its own right). It is most definitely *not* a variety of *H. serpae* as was thought till recently. All the fins, except for the deep black middle bit of the D edged in white, are blood red like the body. This fish has largely supplanted *H. callistus* in the aquarium. Not particularly difficult to breed. The mating pairs spawn usually in the morning after vigorous driving. The spawn is laid on finely leaved plants. The eggs are a brown colour and because they are not very sticky they sink to the floor. A spawning grate is recommended because this species produces young that will feed on the spawn. The fry can be reared on live food (*see* Fry, Feeding of). It is a good idea to add some extract of peat or alder cone to the water. As soon as they hatch and whilst they grow, the pale red fry of the blood red variety are easy to distinguish from the darker pigmented fry of *H. callistus*. Unfortunately, adult specimens (particularly those of the blood red variety) do not live very long in the aquarium. They seem to die prematurely of fish tuberculosis*. This is a very attractive species, and the blood red variety is one of the most popular Characins among aquarists. (figs. 640 and 641).

Fig. 640 *Hyphessobrycon callistus*

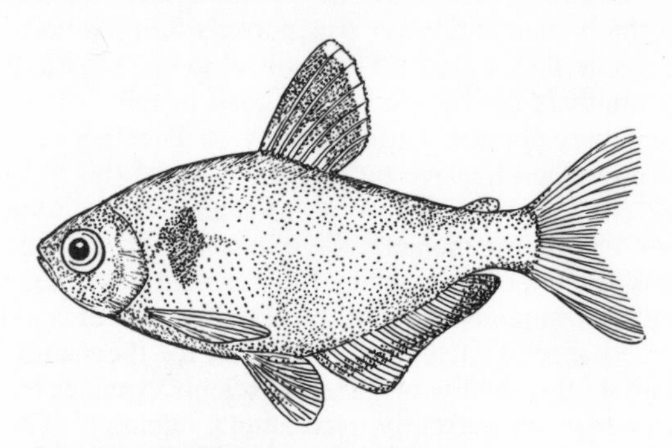

Fig. 641 *Hyphessobrycon callistus*

– *H. erythrostigma* (FOWLER, 1943); Bleeding Heart Tetra. From Colombia. Up to 7 cm long. According to GÉRY, *H. rubrostigma* HOEDEMAN, 1956, is another name for this species. The shape of the body is reminiscent of the well known Ornate Tetra *(H. ornatus*)*, although it grows much bigger. Flanks silver with a mother-of-pearl sheen, dorsal surface grey-green to grey-brown, often with a reddish tinge. The throat and ventral regions are orange-red. A red longitudinal band extends from the middle of the body to the insertion of the C. There is a large, cherry-red shoulder fleck in the front third of the body. The red iris is traversed by a vertical black streak. The D is red, drawn out into long pennant shapes in the male. The first D rays are black with white edges. In the female, the D is of normal length with a large black blotch in the middle of the fin surrounded by a reddish-white zone. In the males, the An is extended at the front into a sickle shape. Its base is bluish-white whilst its free edge is black. The An in the females is reddish with a small bluish-white area in the front third of the fin. The C is red with blue-white speckles. Vs reddish, adipose fin bluish-white. When excited, the whole body has a reddish tinge. In recent years, specimens have often been introduced that only reach a length of 5 cm and male specimens, although they have had the same colouring, have not had the characteristically elongated D and An. This would seem to indicate a geographical race. So far, there has been no rationalised approach to breeding these delightful fishes, and there have only been isolated cases of success (REED in *Tropical Fish Hobbyist* 6/65). Occasionally the males are observed driving at the females, shaking their fins in an enticing manner as they climb up onto finely leaved plants. Often, however, the ripe females do not react at all to this. A very attractive species that lives for a long time (fig. 642).

Fig. 642 *Hyphessobrycon erythrostigma*

– *H. flammeus* MYERS, 1924; Flame Tetra, Red Tetra from Rio. From the area around Rio de Janeiro. Flanks a brassy yellow, dorsal surface grey-olive, ventral surface silver. The rear section of the body is a lovely red colour, as are all the fins, except for the colourless Ps. The An and the Vs in the male are edged in a deep black. There are two comma-shaped, dark shoulder flecks behind the operculum. Has been popular with aquarists ever since it was first introduced. Lovely colouring, undemanding and easy to rear; for further information turn to *Hemigrammus** (fig. 643).

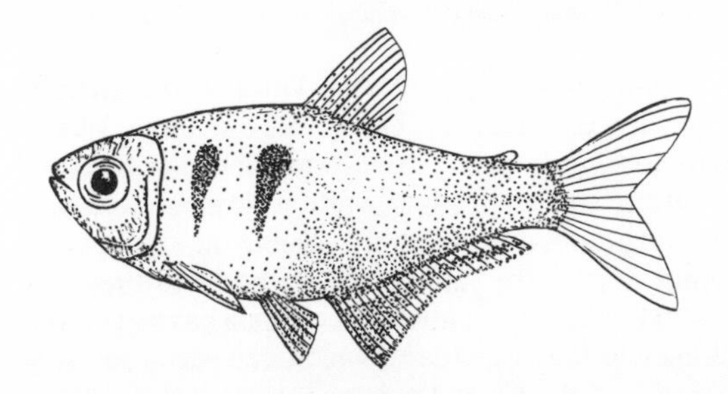

Fig. 643 *Hyphessobrycon flammeus*

– *H. georgettae* GÉRY, 1961; Georgette's Tetra. From Surinam, near to the Brazilian border. Males measure less than 2 cm, females just over 2 cm. The colour of the body and the fins in the male is strawberry-red to blood red, yellowish-red in the female, and there is no shoulder fleck. The D in the male has a large, wedge-shaped black blotch, whereas in the female it is small and round. The front edge of the D is milky white, as are the first rays of the An. If you include *Rasbora maculata*, it is one of the smallest species of fish. It is a pretty species, undemanding and easy to rear, although not very productive. The fry will grow very quickly if fed on *Cyclops* nauplii (fig. 644).

– *H. griemi* HOEDEMAN, 1957; Griem's Tetra. From Brazil, around Goyaz. Up to 3 cm long. Very similar to

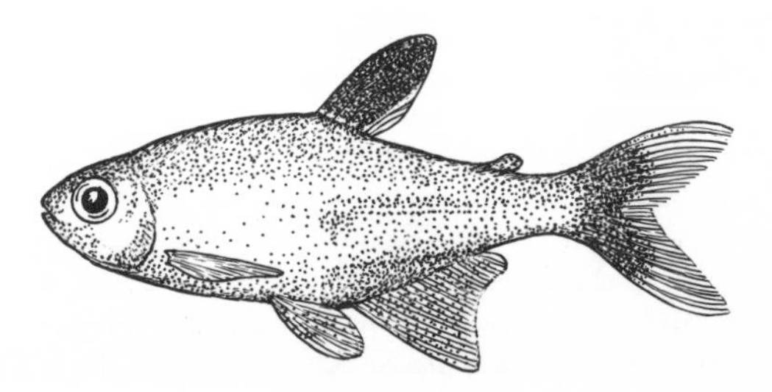

Fig. 644 *Hyphessobrycon georgettae*

the closely related *H. flammeus* in habitat and colouring, but more vermilion red. In this species the two comma-shaped shoulder flecks are located in a yellow zone, and the first one is sometimes a bit indistinct. The D and An have white tips. An undemanding, pretty species which spawns readily. Easy to rear. For further information *see Hemigrammus**.

– *H. herbertaxelrodi* GÉRY, 1961; Black Neon Tetra. From the Rio Taquari near Coxim in the Brazilian state of Mato Grosso. Up to 3.5 cm long. Very similar in shape to the well known Belgian Flag Tetra *(H. heterorhabdus)*, and closely related to *H. loretoensis* and *H. metae*. Dorsal surface a transparent brownish-olive with fine reticulated markings. Ventral surface silver. The flanks are crossed by a broad, deep black longitudinal band that runs from the operculum to the insertion of the C. It is bounded on the dorsal side by a grass-green or yellow-green shining band. The upper part of the iris is a shining blood red, the lower part a green iridescent colour. The common name 'Black Neon' is not a very apt description. According to FRANCKE, it is not too difficult to rear this species in soft, slightly acidic water. Mating takes place in open water, often as the fish are swimming along the sides of the tank. Will eat spawn, therefore remove the parents after spawning. Best fed on rotifers to start with, then move on to *Cyclops* nauplii. Frequent changing of water is recommended. When rearing this attractive species, it has been found that males tend to predominate (fig. 645).

– *H. heterorhabdus* (ULREY, 1895); Belgian Flag Tetra, Flag Tetra, False Tetra. From the lower Amazon, Rio

Fig. 645 *Hyphessobrycon herbertaxelrodi*

Tokantins. Up to 4.5 cm long. Flanks yellowish-brown, dorsal surface red-brown, ventral surface silvery-white. Three longitudinal bands run along the flanks: a narrow blood red one on top, a yellowish-white one below that, and a somewhat broader, deep black one below that. The fins are colourless or a faint yellow. The upper half of the iris is shining red. A very pretty species, but unfortunately it is very prone to diseases. For information about rearing this species turn to *H. herbertaxelrodi* (fig. 646).

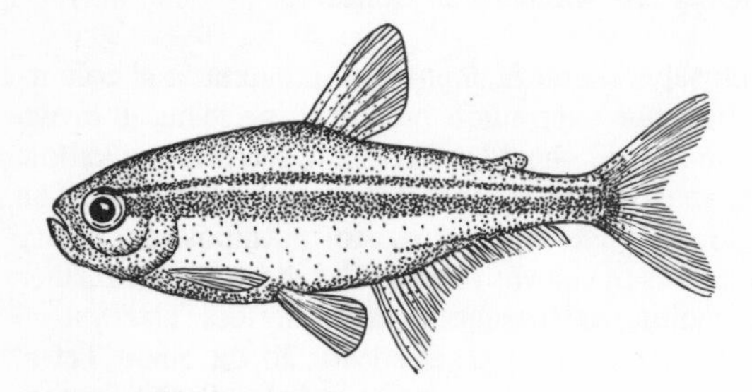

Fig. 646 *Hyphessobrycon heterorhabdus*

– *H. loretoensis* LADIGES, 1938; Loreto Tetra. From the Loreto region of the Peruvian Amazon area, Rio Meta. Up to 4 cm long. The body is elongate, the dorsal surface brownish-olive, the ventral surface silvery-white. A broad, blue-black band runs along the flanks as far as the C insertion. The operculum is a metallic iridescent colour. The C is a strong blood red colour. The D in the males has a yellow centre and a white tip. Adipose fin yellowish. Ps and Vs colourless, as is the An except for the first rays which are brown-red with white tips. A very attractive species, but unfortunately very prone to diseases. A large proportion of the imported fish die after a short spell from fish tuberculosis. Few known successes with breeding, only KLUGE has reported any success. Often falsely named as *H. metae*, although there are only closely related features in both species (fig. 647).
– *H. metae* EIGENMANN and HENN, 1914; Loreto Tetra. From the river basins of the Rio Meta and Ucayali. Up to 4.5 cm long. Body more compressed than *H. loretoensis**. Dorsal surface dark brown-olive, ventral surface a gleaming silver-white. A very wide, blue-black or violet-black longitudinal band runs across the flanks. It begins at the tip of the mouth, continues through the eyes, over the operculum to the lower rays of the C. A copper-coloured shining band nestles alongside the dark dorsal colouring. It begins as a narrow band at the edge of the operculum, slowly broadens out, and ends at the upper rays of the C. Iris coppery colour. D, adipose fin, An and Vs a delicate copper brown with pastel white tips. The males are much slimmer than the ripe females. Examples of this attractive species have been imported on many occasions in past years but they have never been bred successfully. A new import would be very welcome. *H. peruvianus* LADIGES, 1938, is probably another name for this species and so it has not been given a separate entry. The name '*Hemigrammus nigro*' was used at first in the commercial world, but it has no scientific validity. At first glance, *H. metae* looks like a slim Emperor, *Nematobrycon palmeri* (fig. 648).

Fig. 647 *Hyphessobrycon loretoensis*

Fig. 648 *Hyphessobrycon metae*

– *H. ornatus* AHL, 1934; Ornate Tetra. From Guyana and the lower Amazon. Up to 4.5 cm long. Flanks brown-olive, if in prime condition, tinged with blood red. Ventral surface a silver white iridescent colour. A silver longitudinal band, only visible at odd points. stretches across the middle of the body to the C insertion. Shoulder fleck absent. The male has a very long D, elongated like a pennant, which can reach as far as the middle of the C. At the base it is coloured red, the rest of the fin being black. In the female, the D is shaped normally and has a round black blotch in the middle of the fin. This blotch is bounded on the upper edge first by a white zone, then by a red zone. The C is a sooty colour with pastel white, long tips and a blood red blotch on each fin lobe. The An is also sooty with a white elongated tip and a red zone in the front part of the fin. Vs red with white tips, Ps colourless, adipose fin small. Iris silver with a vertical black streak. Even in large breeding tanks, getting these fish to mate is not easy, as only young specimens that are not quite adult will readily spawn. During courtship, the males swim up to the spawning substrate with enticing twisting movements and shaking of the fins. Usually, they adopt an oblique or vertical position. The eggs and the newly hatched fish are brownish in colour. Rearing the fry is not completely without its problems. They must receive good food (rotifers to start with, followed by *Cyclops* or *Artemia* nauplii), good filtration or frequent change of water. A very popular species since it is decorative and lives for a long time. Best kept in a well-planted aquarium (fig. 649).

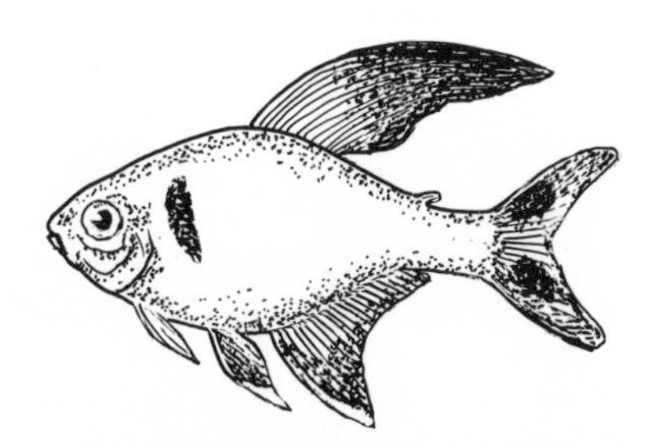

Fig. 649 *Hyphessobrycon ornatus*

– *H. peruvianus* LADIGES, 1938; Loreto Tetra. *See H. metae* EIGENMANN and HENN, 1914.
– *H. pulchripinnis* AHL, 1937; Lemon Tetra. From S America, any more precise location not recorded. Length up to 4 cm. Body relatively tall, a glassy translucent colour with yellow tones. There is the indistinct outline of a shoulder fleck that appears to lie deep beneath the epidermis. Large eyes. The upper half of the iris is a shiny blood red. All the fins are yellowish, the upper third of the D being black and its tip a lemony yellow. The front fin rays of the An are also lemon-yellow and they are drawn out into a point. In the females, the free edge of the An has a narrow black area, whereas in the males there is a broad, deep-black area. Fairly easy to rear, although specimens will not always be willing to spawn. The eggs are a pale yellow colour. The fry grow quickly and they are not difficult to rear if fed on *Cyclops* nauplii (fig. 650).

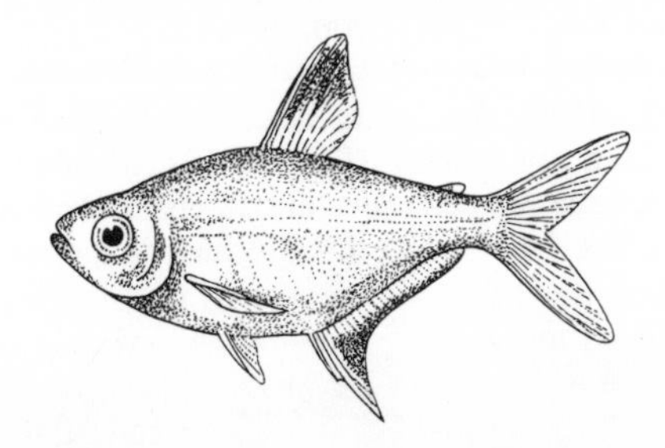

Fig. 650 *Hyphessobrycon pulchripinnis*

– *H. scholzei* AHL, 1936; Scholze's Tetra, Black Line Tetra. Found in the area around Para. Length up to 4 cm Flanks bluish with a silvery iridescence. Dorsal surface brownish-olive, ventral surface silver, iris yellow. A narrow, black, longitudinal band starts at the operculum and ends as a rhomboid, black blotch on the middle rays of the C. The tip of the An is pastel white, the other fins being colourless or a faint reddish-brown. Easy to rear and breed (for further information *see H. ornatus*). Compared with the other species of the G, *H. scholzei* is rather colourless and so it is not often kept by aquarists, even although it is very hardy (fig. 651).
– *H. serpae* DURBAN, 1908; Serpae Tetra. From the Amazon, Rio Guapore. Length up to 4 cm. Flanks and dorsal surface very transparent, pale yellow in colour

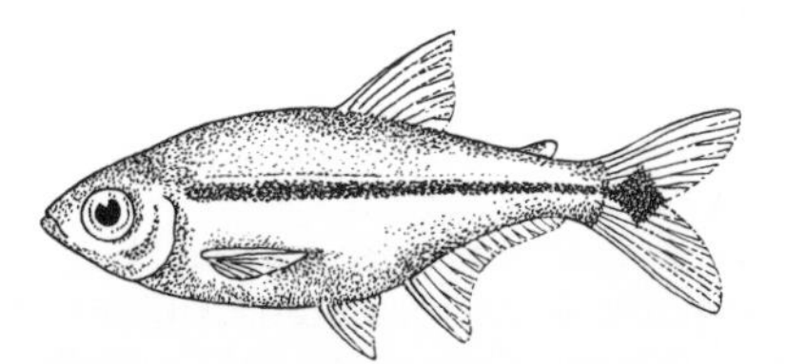

Fig. 651 *Hyphessobrycon scholzei*

or colourless with a silvery iridescence (sometimes with just a tinge of pink). Colouring of the D very striking: the base is a delicate pinky white, next to which is a black blotch in a longish oval shape. This is in turn bounded by a pinky white zone at the free, upper edge of the fin and in this area there is a reddish blotch. The rear rays of the D are colourless. Adipose fin pink, C a uniform brick red. The base of the An is also a brick red colour, the free edge being colourless. The first anal fin ray is a pastel white colour and the second ray is black. Vs blackish in colour with a pastel white first ray. Ps colourless. Behind the operculum is a comma-shaped, black shoulder blotch. The iris is silver with a vertical black streak. The mating behaviour and method of rearing the young are very similar to *H. callistus**. However, the eggs and newly hatched fry are almost colourless.
– *H. simulans* GÉRY, 1963; Schwartz's Neon. From the Rio Iufaris, a tributary of the Rio Negro. Length up to 2.5 cm. Extremely similar to the well known Neon Tetra (*Paracheirodon* innesi*), but smaller and with less intensive colouring. As a result, they look like paler specimens of the Neon Tetras and not quite fully grown. Body transparent, apart from the metallic green-blue or pure blue shiny band that runs from the tip of the mouth, through the eye and as far as the insertion of the C. Because of this, the red zone that begins at the level of the P insertion and stretches to the base of the C looks pale in comparison with the red zone in *Lamprocheirodon* axelrodi* and *Paracheirodon* innesi*. Dorsal surface brownish-olive, iris green-golden and fins colourless. Has been bred successfully on many occasions (fig. 652).

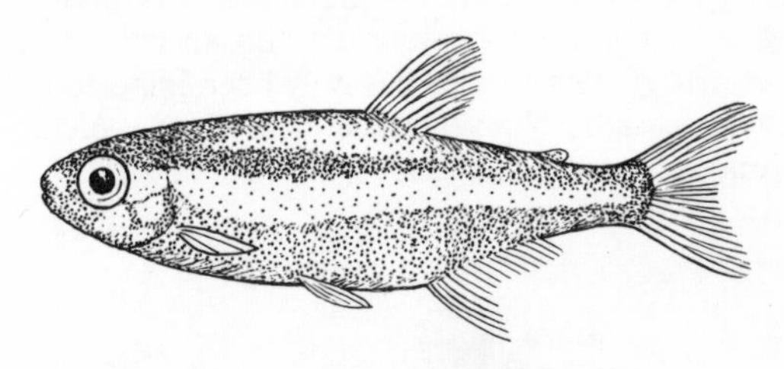

Fig. 652 *Hyphessobrycon simulans*

– *H.* sp.; Sickle Characin. Origin unknown. Length up to 5 cm. In body shape and colouring very similar to the closely related Ornate Tetra (*H. ornatus**), except that the flanks have stronger blood red or violet tones. The very long D in the males is a deep black colour up to the reddish base. The front third of the An, coloured black, swings outwards in a sickle shape. Unlike *H. ornatus*, the tips of the An and D are not pastel white, The C is a strong red colour. Female *H.* sp. are very easy to mistake for female *H. ornatus*. One distinguishing feature is that the round, black blotch in the middle of the D (which is not elongated as in the male) is bounded above and below by a white zone in the Sickle Characin, but only above in *H. ornatus*. *H.* sp. has been bred successfully on numerous occasions, but no exact details are available. For hints about breeding, turn to *H. ornatus*. '*H. robertsi*' is another name often used for the Sickle Characin, but it has no scientific validity. *H.* sp. may be identical with *H. rosaceus*. It is a very attractive species, and could well offer serious competition to the better known Ornate Tetra if it were to be made more generally available (fig. 653).

Fig. 653 *Hyphessobrycon* sp.

– *H. takasei* GÉRY, 1964; Coffee Bean Tetra. From Serra do Navio near Macapa (Amapa region) in the lower Amazon Basin. Length up to 4 cm. Body shape similar to that of *H. callistus*, but more elongate. The An begins at the level of the last rays of the D. The body is transparent with faint pinky tones. There is a large, black, coffee-bean-shaped, shoulder fleck behind the operculum. The base of the D is orange-yellow with a black blotch in the centre. The adipose fin is orange. The bases of the An and C are vermilion red. It is an interesting specimen for the aquarium and survives well. Unfortunately, this species has only been imported on isolated occasions. Not yet bred in captivity (fig. 654).

Hyphessobrycon stigmatias = *Axelrodia stigmatias see Axelrodia*

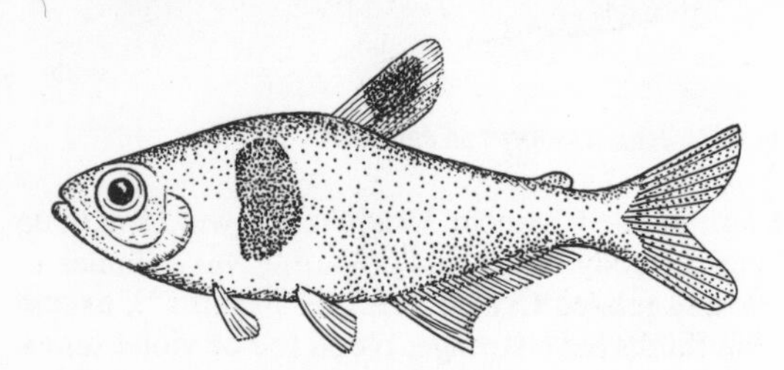

Fig. 654 *Hyphessobrycon takasei*

Hypoplectrus GILL, 1861; Butter Hamlet (fig. 655). G of the Serranidae*, species of which are found in tropical parts of the W Atlantic. 10–20 cm in length. Attractively coloured, predatory fish that live in fairly shallow waters. Very tall fishes with a 10-rayed D. The lower edge of the preoperculum is toothed. The species are difficult to tell apart. Occasionally imported for the aquarist, *Hypoplectrus* has proved to be a hardy aquarium specimen.

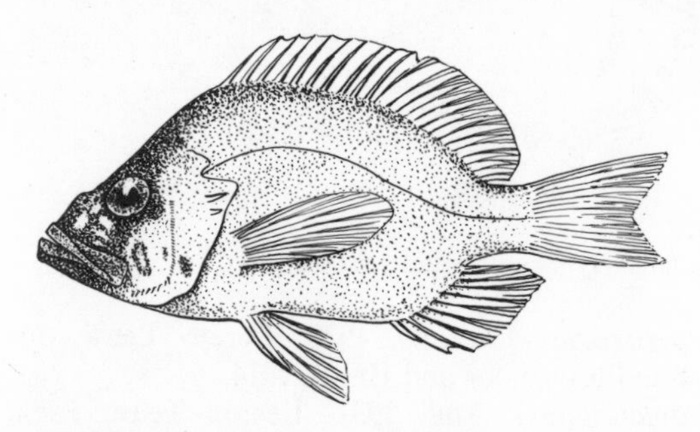

Fig. 655 *Hypoplectrus gummigutta*

Hypostomides, O, name not in current usage, *see* Pegasiformes

Hypotonic. Term used to describe a solution that has developed a lower osmotic pressure (tonus) than another. The blood plasma of marine fish, for example, is hypotonic in relation to sea-water (*see* Osmoregulation). The term is also used to describe below-normal blood pressure.

Hypsophyll *see* Leaf Succession

Hysterocarpus traski *see* Embiotocidae

Icefishes (Channichthyidae, F) *see* Notothenioidei

Ichthyboridae, F, name not in current usage, *see* Citharinidae

Ichthyo Disease *see* Cryptocaryon Disease

Ichthyobodo PINTO, 1928; *(Costia)*. G of the Flagellata*. The best-known species in aquarium-keeping. *I. necatrix* (HENNEGUY, 1884) (fig. 656), is a small (10–15 μm long), bean-shaped flagellate. It has 2 flagella and reproduces itself by simple longitudinal division.

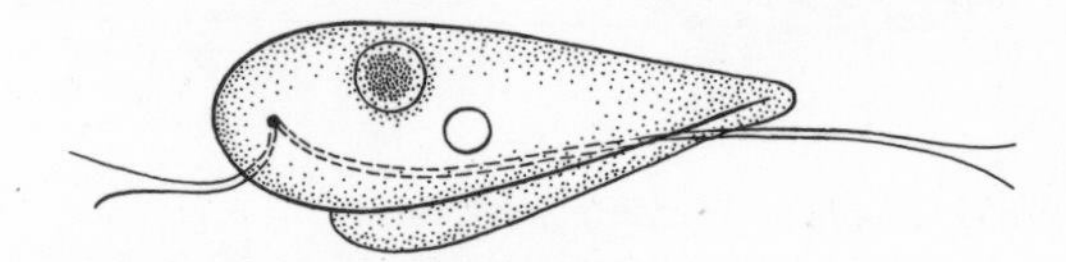

Fig. 656 *Ichthyobodo necatrix*

It is a typical debilitating parasite (Parasitism*), living on the skin of fish. Without a host, it will die in 30–60 mins. Delicate, cloudy areas will appear on a stricken fish's skin, and, if the attack is severe, the skin will start to look red and inflamed. Eventually, the skin will flake off and the fish will start swimming in a very unsteady fashion. To detect the presence of *Ichthyobodo*, smears from the skin and gills of live fish have to be taken and studied under a microscope at great magnification. Warm-water fish are attacked more rarely, as a temperature of around 25 °C is not suitable for the flagellates and at 30 °C they die. Prolonged bathing treatments with Trypaflavin* or Malachite

Green* are recommended. The temperature should also be raised to about 30 °C. Short bathing treatments with formaldehyde* are also good for removing the parasites.

Ichthyology. Branch of zoology* concerned with the study of fish. Depending on the particular zoological aspect in question, ichthyology is divided into disciplines which are concerned with very different topics. These include questions about the structure of the body and organs (Anatomy*), their functions (Physiology*), fish life-styles (Ecology*), heredity and breeding (Genetics*) or fish relationships (Systematics*, Palaeontology*). Because fish are a very important food source for humans, ichthyology has developed a practical field of research to deal with it – fishery biology. It deals with the study and analysis of fish stocks, how they grow, how to catch them, how much they yield and their profitability, how to breed them artificially and how to combat fish diseases. As it is a vast subject, scientists all over the world are actively engaged on various aspects of fishery biology. Research is carried out in scientific institutes and museums, in marine biology stations and aboard research ships.

Fish have long been of great interest to researchers. The Greek philosopher and natural historian ARISTOTELES (384–322 B.C.) knew of more than 150 different species and was able to describe details of their anatomy and biology. The Swede, PETER ARTEDI*, is usually considered the father of scientific ichthyology. His chief work, published by KARL VON LINNÉ in 1738, lays down a systematic division of fish into orders, genera, species and varieties, Later, VON LINNÉ was to perfect this system. Since them, various scientists have developed ichthyology into a very important branch of biology. They include BLOCH*, LACÉPÈDE*, CUVIER*, VALENCIENNES*, AGASSIZ*, REGAN*, BERG*.

Ichthyophonus, name not in current usage, *see Ichthyosporidium*

Ichthyophonus Disease *see Ichthyosporidium*

Ichthyophthiriasis *see Ichthyophthirius*

Ichthyophthirius FOUQUET, 1876 (fig. 657). G of the Ciliata*. One of the commonest parasitical ciliates found in fish is *I. multifilis* FOUQUET, 1876. It attacks almost all freshwater fish and causes the disease known as ichthyophthiriasis (White-Spot disease). The highly mobile parasite itself is spherical, 0.2–1 mm in size, and is covered with numerous cilia (*see* Ciliata*). Fish suffering from the disease press their fins against rocks or stones, trying to rub them. They behave in an apathetic manner, unable even to swim away when danger threatens. The parasites show up as white spots or off-white blotches on the skin (particularly on the fins and gills). Their presence can be detected in smears taken from the body, up to 3 hours after death. The life-cycle of *I. multifilis* has very definite characteristics (fig. 658).

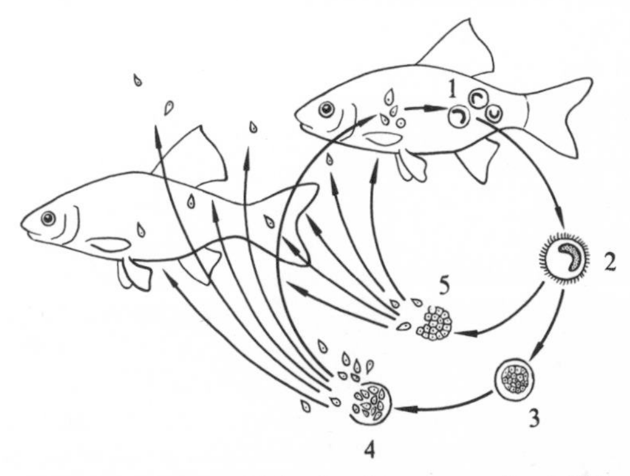

Fig. 658 *Ichthyopthirius* development-cycle. 1 Skin stage 2 Free-living stage 3 Cyst (bottom-dwelling) stage 4, 5 Daughter-cell stage as they leave the cyst

The parasite is located between the epidermis and the subcutaneous tissue of the fish's skin (*see* Skin*) skin stage – where it feeds on tissue fluid and fragments of cells. Depending on temperature, it will grow to its full size in 1–3 weeks. After boring its way through the epidermis and into the water, the parasite sinks to the bottom and develops a cyst around itself – bottom stage. Through simple cell division, 20–300 (1,000) ciliated daughter cells develop. They leave the cyst after 12–20 hours (daughter cell stage) and once more penetrate the fish's skin. Without a host, the parasites will only live for 33–48 hours. Exchanges of sexual nuclear material between the daughter cells (conjugation) can increase the period of survival to several weeks. If the parasite frees itself from the skin before reaching its full maturity, it will form a smaller number of daughter cells, but without forming a cyst. Badly infected fish are very

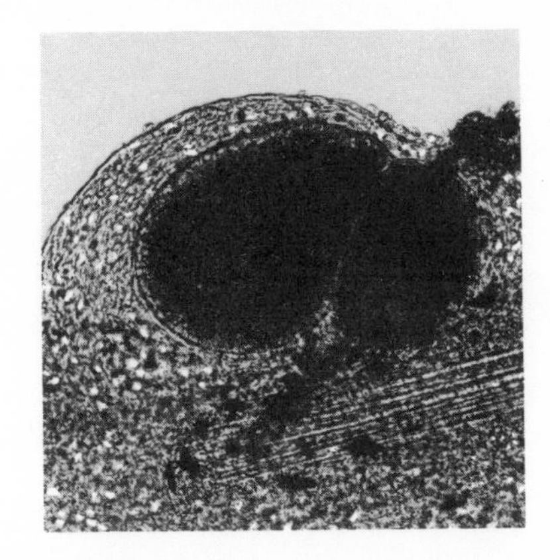
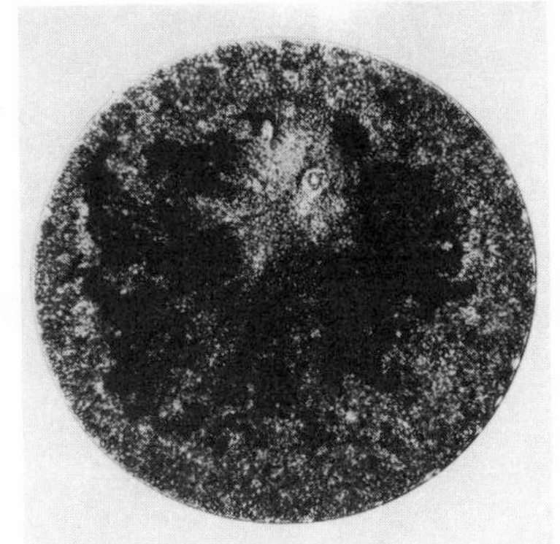
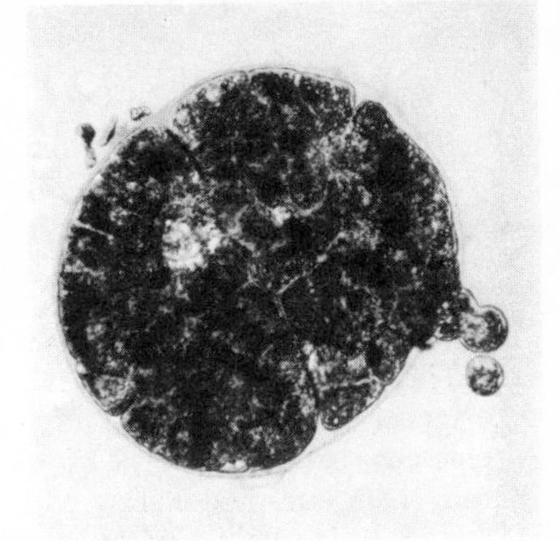

Fig. 657 *Ichthyophthirius.* From left: skin stage, cyst (bottom-dwelling) stage, daughter-cell stage

difficult to cure as the skin stage is very hard to treat successfully. Daughter cells can be killed off by prolonged bathing techniques, using Trypaflavin* or Malachite Green*. Occasionally there are some very tenacious cells that seem to have a certain amount of resistance* to both these substances. Raising the temperature to 30 °C speeds up the development of the parasites, so treatment should normally last for 8–10 days (ie enough time for all the skin stages to have developed into daughter cells). If the disease has reached the chronic stage, the treatment will have to be applied for several weeks. There is another method of treatment which is less upsetting for the fish: every 12 hours the fish are transferred into a tank that is free of parasites. At a temperature of 28 °C, the parasite is unable to produce daughter cells within this time, so a new infection becomes impossible. Furthermore, since daughter cells are the only means by which a fish can become newly infected, and because it appears that these daughter cells can only be formed from the non-mobile stages of the parasite's life-cycle, there may be yet another method of treatment. Turbulence could be set up in the aquarium water (through a diffuser or a nozzle) to stop the parasite from finding somewhere to come to rest as it leaves the fish's body. (This could only be tried in tanks that are not sandy and do not have any vegetation.) This would result in the fish gradually becoming freed of the parasites. New imports of fish must be kept in quarantine* for at least 4 weeks as part of health control.

Ichthyosporidium CAULLERY and MESNIL, 1905 (Ichthyophonus, name no longer in current usage) (fig. 659). G of the Algal Fungi, the species *I. hoferi* (PLEHN and

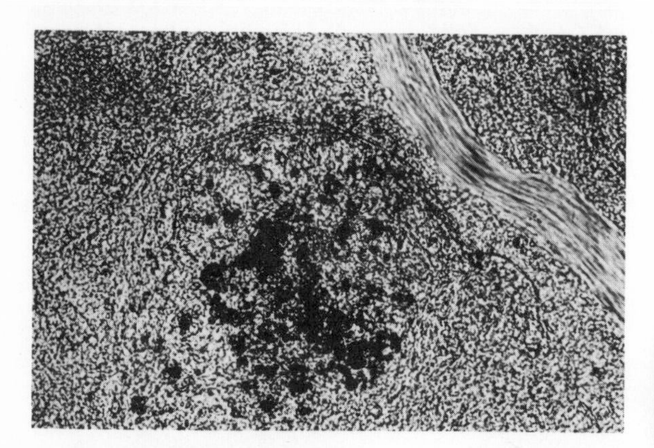

Fig. 659 *Ichthyosporidium*

MULSOW, 1911) for a long time being regarded as the cause of one of the commonest contagious diseases in fish. This was because some of the symptoms of fish tuberculosis* and several virus infections* bore a strong resemblance to one another. *I. hoferi* is a parasite found in both freshwater and marine fish, although it occurs fairly rarely in the aquarium. External signs of the disease in fish include emaciation, swelling of the abdomen, flaking off of the fins, skin defects and abnormal swimming movements. The fungus will inflame the tissues of all internal organs, especially the liver, kidneys and intestine. The fish tries to encapsulate the parasite in connective tissue. This causes white, necrotic nodules to appear (in an advanced stage, they will also appear in the skin itself). As a result, the organs shrivel up, thus causing disturbances to the fish's metabolism. The life-cycle of the fungus is fairly complicated and includes resting periods during which the parasite still remains infectious although outside a host. Infection can probably follow from faeces dropped by fish that are already infected, from dead fish and possibly from Copepods* too. (Copepods are sometimes given as food and could well carry the parasites.) Usually, a cure is not possible. Phenoxyaethanol* can be tried, and raising the temperature to about 38 °C has proved successful in a few cases. As a rule, however, infected fish stocks have to be killed. The aquarium must then be thoroughly disinfected (*see* Disinfection). To prevent the disease taking hold in the first place, keep your aquarium in the best possible condition, and if necessary, add extra vitamins A and B_{12} (*see* Vitamins). In addition, make sure to follow the quarantine rules for any newly imported fish (*see* Quarantine).

ICM *see* Innate Character-forming Mechanism

Ictaluridae; Catfishes. F of the Siluriformes*, Sup-O Ostariophysi. Small to medium-sized Catfishes with an elongated body, almost round at the front when viewed in cross-section but getting increasingly flattened from the D to the C. The head is large, broad, and has 4 pairs of barbels. The first pair grow from the rear nasal openings, the longest pair from the upper lip and 2 pairs from the chin. The jaws and palatal bone are filled with small teeth. The D is fairly far forward and is short and rounded. Like the Ps, it has a strong, toothed spine. The An is very long and broad, the C rounded or slightly indented. Adipose fin like a small flag, although sometimes flat or very low. Naked skin. First known in the Neogene Period, there are nowadays about 25 species in the Ictaluridae. They inhabit the temperate and subtropical rivers and lakes of N America. *Ictalurus nebulosus* (fig. 660), which grows to a length of 50 cm, was introduced into Europe around the turn of the century. It has since become very widespread. Food consists mainly of invertebrates. It is a dark brown fish, more bronzy when light falls directly on it, with a greenish or violet iridescence. It is crepuscular, only leaving its

Fig. 660 *Ictalurus nebulosus*

hiding-place when the sun goes down. In Europe, the spawning season is between March and May, and it takes place in flat, sunny spots near the bank. The male and the female make a shallow hollow into which they deposit the eggs in balls (which are very sticky). The fry hatch after 8 days and resemble tadpoles. The male looks after them for some time. Catfishes are easy to keep in the aquarium. As experimental fish, they have proved of great value to scientists. K. v. FRISCH, for example, was able to show that Catfishes detect sounds and can even be trained to respond to names. Catfishes have often been used in experiments that study the fish senses of taste and smell. In Europe, Catfishes are of little economic importance, but in N America, other kinds of Ictaluridae are very important edible fish. Chief among them is the 1.2 m long Spotted Channel Catfish *(Ictalurus punctatus)*. The smallest representatives of the F, the Stone Cats, are interesting specimens for the coldwater aquarium. One species, the 9-cm-long Mad Tom *(Noturus gyrinus)* (fig. 661), is very attractively coloured. It comes from eastern N America.

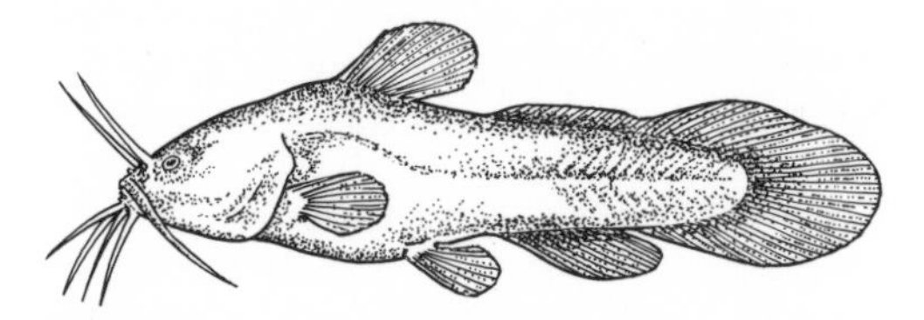

Fig. 661 *Noturus gyrinus*

Ictalurus, G, *see* Ictaluridae

Icterus (Jaundice) *see* Liver, Diseases of

Ictiobus cyprinella *see* Catostomidae

Identification Key. Aid to taxonomy*, with which the names of plants and animals can be determined. It works as follows. A series of questions must be answered relating to typical anatomical features, where the animal is found, way of life etc, so that the specimen can be classified according to a larger systematic unit such as phylum, class, order, family. Using the same principle the latin generic term and, lastly, the name of the species can be arrived at. The key given in identification books mostly refers to the plants and animals in geographically defined regions, eg BROHMER, EHRMANN, ULMER: *Die Tierwelt Mitteleuropas* ('The Animals of Central Europe') or HEGI: *Illustrierte Flora von Mitteleuropa* ('Illustrated Flora of Central Europe'). Many ichthyological works also contain identification keys, eg BÖHLKE and CHAPLIN: *Fishes of the Bahamas*; SMITH: *The Freshwater Fishes of Siam or Thailand*; or WEBER and BEAUFORT: *The Fishes of the Indo-Australian Archipelago*.

Idiacathidae, F, *see* Stomiatoidei

Idotea WEBER, 1795. G of marine-dwelling Lice (Isopoda*), comprising several species. It lives mainly in seaweed, feeding on plant matter. In places, *Idotea* is extremely common. Individuals within a species can vary enormously in colouring. They are well worthwhile keeping in a marine aquarium. Some species can grow up to 40 mm long, *Mesidotea entomon* (LINNAEUS, 1758) up to 8 cm.

Idus HECKEL, 1858. G of the Cyprinidae, Sub-F Leuciscinae. From Europe, north of the Alps, as far as Siberia. Large fish that live a sociable existence near the surface in rivers and lakes. Elongate body, only slightly flattened laterally. Mouth terminal and small. LL complete. Can be distinguished from the G *Leuciscus** by the concave shape of the An. *Idus* enjoys swimming and if kept in the aquarium it will need plenty of room. Keep as a shoal in large tanks with a good lid. Temperature not above 22 °C. Feed with nourishing live food and plant food. Successful breeding only possible in open-air installations. There is only one species:

– *I. idus* (LINNAEUS, 1758); Orfe. Up to 60 cm. Dorsal surface grey-green, flanks a gleaming silver with a blue-green or reddish sheen, ventral surface silvery. D and C grey, An, Vs and Ps reddish. A xanthic variety of Orfe is known – the Golden Orfe. It occurs under natural conditions too and is reared on a large scale. The upper half of the body is red-gold in colour, whereas the flanks are orange and the ventral surface a white-yellow with a silver sheen. Very hardy fish, well suited to ponds in the garden or in parks (fig. 662).

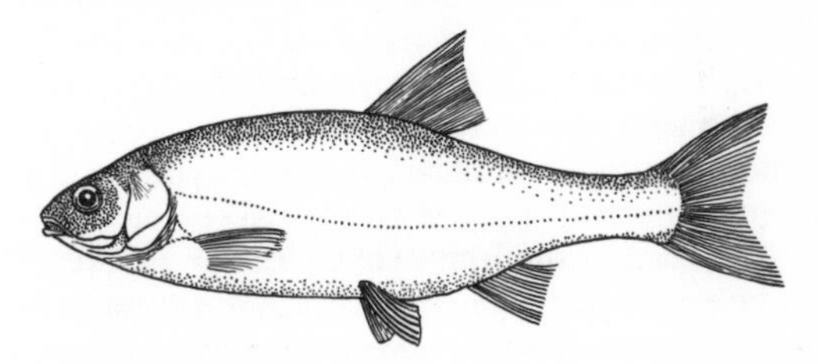

Fig. 662 *Idus idus*

Iguanodectinae, Sub-F, *see* Characidae

Imago (plural 'imagines'). General term for the last stage in the development of an insect. In many insects, the imago has the ability to fly unlike the eggs, larvae and pupae. In addition, only the imago has the ability to reproduce.

Immersion Baths *see* Bathing Treatments

Immunity. The state in which the body can successfully defend itself against certain pathogens and harmful foreign bodies. In contrast to hereditary resistance*, immunity is something that is acquired through various causes in the course of an animal's life. The antigens (eg various strains of bacteria, viruses or poisons) that have to be resisted possess chemical properties that enable the organism to produce substances that work effectively against them. In vertebrates, as soon as the antigens enter the body, these substances, or *antibodies*, are produced in the plasma cells of the spleen and lymphatic node. They then circulate in the blood as special protein substances (immunoglobulin). Through a highly specific *antigene-antibody-reaction*, these protein substances will only combine with those antigens for which they were specifically formed. With bacteria, they settle on the surface, thus killing off the intruder. If the same antigen should enter the body again, immunity is increased and the defence reactions improved.

To hasten immunity against commonly occurring and dangerous pathogens, preventive immunisation programmes are carried out.

Immunobiology. The science of the body's defence reactions. Includes the study of the reaction chain and its specificity, the production and chemical structure of the defensive substances as well as genetic problems that touch on immunity* and resistance*.

Imperial Angelfish *(Pomacanthus [Pomacanthodes] imperator) see Pomacanthus*

Imported Fish. Fish that are imported from foreign countries (as far as aquarium-keeping is concerned, this normally means fish from tropical countries) that are often shorter-lived and more prone to disease than offspring that are born in captivity and reared there. This is because imported fish may find it difficult to adapt to new surroundings when taken away from their homelands, and their bodily constitution may also be poor. A distinction should be made between imported fish that are truly caught in the wild and those that are the offspring of fish kept in tropical and subtropical, open-air breeding installations. (The latter are often sold in Europe as 'imported fish'.) Getting true imported fish accustomed to new surroundings is very problematical and large numbers may die, normally because the fish often arrive in the hands of the aquarist already riddled with organisms that may cause disease. Also, information regarding required water conditions and other environmental factors is not passed on to the aquarist in many cases. This means that the aquarist is forced to experiment, thereby causing even more problems for the already weakened specimens (unless, by chance, the ideal conditions are implemented straight away). The correct food combination is also a matter of finely tuned consideration. Nevertheless, the care of imported fish is particularly absorbing for many aquarists. It is extremely satisfying when, after many unsuccessful attempts at rearing certain fish, success is achieved. It means that a gap has been closed in the behavioural inventory* of a new species of fish and this will ensure that that species will become more widespread in captivity.

Inachus, G, *see* Macropodia

In-breeding. The mating of closely related individuals within one species. Particularly important for breeders (*see* Breeding), as it quickly leads to homozygous breeds, if carried out in conjunction with continual selection*. Essential for line-breeding.

Fears of damage caused by inbreeding (malformations, occurrences of degeneration etc) are largely over-emphasised, and in a lot of cases are caused by poor selection. Sudden alterations to the genetic make-up are of course transmitted very quickly by in-breeding and any defects in the hereditary material (although they may be suppressed [recessive]) will be revealed more readily by in-breeding for particular characters. It is up to the breeder to exclude less perfect specimens from further reproduction and so keep the in-bred breeds healthy.

Indian Fern *(Ceratopteris thalictroides) see Ceratopteris*

Indian Glass Barb *(Chela laubuca) see Chela*

Indian Glassfish *(Chanda ranga) see Chanda*

Indian Lotus Flower *(Nelumbo nucifera) see* Nymphaeaceae

Indian Ocean. Measuring about 75 million km^2, including neighbouring seas. This makes it the smallest of the three oceans. It is bounded on the west by Africa*, in the north by Asia*, in the east by the Sunda Islands and Australia*, and in the south it merges with the South Polar Sea. Average depth about 3,900 m, greatest depth 7,450 m in the Sunda trench. Salinity on average 36‰, lower in the south, and in the north-west (Persian Gulf, Red Sea*) 40‰. Whilst the fish fauna of the pelagic zone* belong, for the most part, to the F Clupeidae (*see* Clupeiformes) and a great variety of different Sharks, corals (*see* Anthozoa) and Coral Fishes (*see* Coral Reefs) predominate in the littoral* of tropical parts of the ocean.

Indian River Barb *(Cyclocheilichthys apogon) see Cyclocheilichthys*

Indicator. Either a chemical substance that changes colour to indicate particular chemical conditions, or a living thing that indicates certain environmental conditions by its presence (or absence). For example, the CZENSNY-indicator changes colour in relation to the pH value*, and the presence of the crustacean *Artemia** indicates that there is a high salt content in inland waters.

Indo-Pacific. The largest ocean in the world, split into two unequal parts by the Malaysian Archipelago and Australia—the Pacific* and the Indian Ocean*. Connected to the Atlantic via the Polar Seas.

Infection. The transfer of germs into an organism where they grow and multiply. Germs come in the form of viruses, bacteria, fungi and protozoans, for example. They produce either a localised infection (when the germs are confined to a very narrow area), or a general infection (when the germs spread throughout the organism). Pathogenic germs cause diseases. Depending on the type of germ, characteristic disease symptoms will develop after a specific incubation period. Should these fail to appear, the infection is described as latent. Germs can be transmitted either directly through contact with infected organisms or indirectly through infected objects, faeces, etc.

Inferiority Behaviour; Submissive Behaviour. Term used in ethology. During a fight, defeated or low-ranking fishes behave in such a way that they indicate their inferiority to the highest-ranking fish or their opponent (such inferiority behaviour is typical of the particular species). This inhibits the aggression of the victor and prevents the loser from being killed. A common indication of inferiority includes a *defensive broadside posture*, at the same time turning the body along its length-wise axis. In this way, the defeated fish offers the victor his dorsal or ventral side. Other forms of inferiority display include *offering of the throat, laying down of the fins, bending in of the body.* At the same time, a gradual colour change is brought into effect that is species-typical (*see* Inferiority Colouring).

Inferiority Colouring; Submissive Colouring. Term used in ethology. In low-ranking fishes, or in fishes that have been defeated in battle, the intensive body colouring and patterning on the fins fades away to become less striking, more like that of the females. In this state, the fish is no longer chased away during courtship by its opponent or higher ranking rival, since at the same time the defeated fish displays inferiority behaviour* typical of its species. This inhibits the aggression of the stronger fish and prevents the fight continuing to the death. Inferiority colouring is often identical to the night colouring* of the species concerned (eg as in the *Nannostomus** species). Low-ranking males can retain their inferiority colouring (especially in a cramped space) for as long as it takes for the highest ranking fish to swim away, eg as in *Pseudotropheus auratus*. During such a period, these low-ranking males are indistinguishable in colour from the females. The females also adopt inferiority colouring between matings or whenever they are not ready to spawn. At the same time they will seek to remove themselves from the view of their mate.

Infertility *see* Sterility

Inflammation. A localised, pathological change in tissue brought about by an irritant. The source of the irritant might be mechanical, thermal and chemical reactions, but it is parasites (*see* Parasitism) and their poisons which are more likely to be the source. In fish, inflammation starts with an increased flow of blood to the tissue and red corpuscles spilling out into it. The infected area then turns a vivid red colour. As blood plasma starts to pour out of the blood vessels, the tissue begins to swell up. The inflamed area will degenerate and the cells will die off, but regeneration will take place later. Inflammation occurs in fish as a result of spot diseases*, stomach dropsy*, intestinal diseases*. It may also be caused by infection from ectoparasites and other things.

Inflorescence (fig. 663). The flower-bearing part of a stem or a flower-bearing stem system. Typical inflorescences, such as spikes, racemes, spadices, panicles etc are clearly set apart from the vegetative region and either have hyposyllary leaves *(Echinodorus)* or much more rarely no leaves at all *(Sagittaria)*. There are also flowering stems and stem systems which possess normal foliage leaves and which are not clearly set apart from the vegetative region *(Nomaphila stricta)*. These are difficult to characterise.

Infusorians (Ph and collective term) *see* Ciliata

Iniomi, O, name not in current usage, *see* Myctophoidei

Initial Action. Term used in ethology. Initial actions precede the concluding consummatory act*. They occur either singly or in multiples, as a chain of fixed action patterns* combining into a behavioural syndrome*. For examples *see* Behavioural Syndrome.

Injection. Large fish can be injected with medicaments*, anaesthetics*, vitamins*, immunoreagents (Immunity*) and other substances. This guarantees an exact dosage, specifically geared for the individual fish, as well as a speedy effectiveness. Syringes and hypodermic needles used for injecting should be dismantled and sterilised by boiling in water for 20 mins, then, still sterilised, put back together again. Now suck up a little bit more of the injection fluid than is necessary into the syringe, hold the needle pointing upwards, then squirt out some of the fluid until all the air is expelled. Keep squirting until the required quantity of fluid is left in the syringe. In all cases, only the smallest amount of fluid should be injected, and this will, once again, depend on the size of the fish and the solubility of the medicament. Injections into the body cavity (intraperitoneal) have a very quick effect and can be given in large quantities. The needle should be inserted above the V and to the side. Hold the needle flat and inject from the back forwards. This will prevent any damage to the internal organs (fig. 664). Unpractised people are able to give injections into the muscle (intramuscular). In this case

Fig. 664 Injection, intraperitoneal

however, only small quantities of the injection fluid can be used. Insert the needle just down from the dorsal region opposite the level of the D. The needle should be held at an angle and pointing forwards. It needs to be stuck deep into the fish. Before injecting, pull the syringe back out a little way (fig. 665). Fish are taken

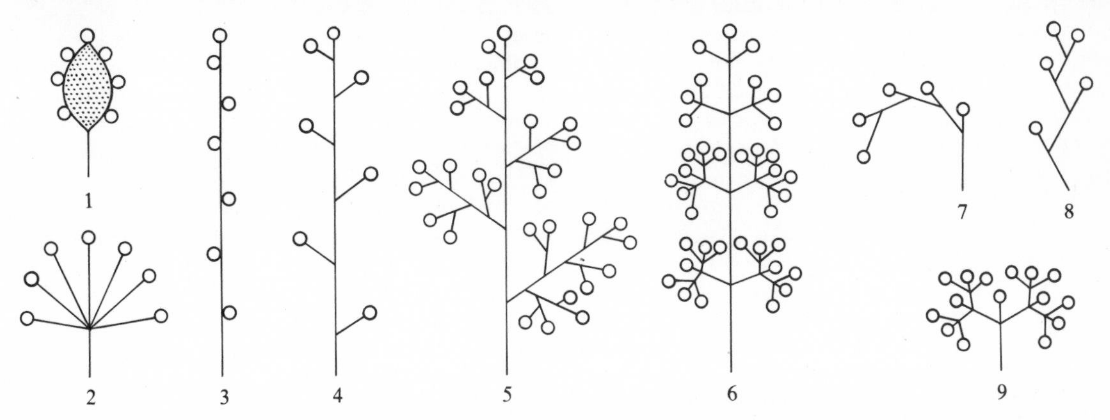

Fig. 663 Inflorescences. 1 Small head 2 Umbel 3 Spike (ear) 4 Raceme 5 Panicle 6 Thyrsus 7 Cincinnus 8 Rhipidium 9 Dichasium

Fig. 665 Injection, intramuscular

out of the water for an injection and held in a damp cloth. When the injection is complete, replace the fish in the water immediately.

Injuries heal very quickly in fish as a rule, and without complication. Not only do scales and fins with fin rays regenerate (Regeneration*), but even quite deep muscle wounds quickly heal over. With fairly large wounds, infections* should be avoided where possible by treating the area concerned with Rivanol*, tincture of iodide* or Mercurochrome*.

Inland Waters. Any standing or flowing waters into which it is impossible for the sea to flow. Included here, notwithstanding the historical origin of the inland water (perhaps as part of the sea) is its salt content and the character of its flora and fauna. The Caspian Sea is one example. Typical characteristics of inland waters are that they are small in area in comparison to the sea and the land masses, they are short-lived and they are isolated from one another. So the flora and fauna found there are subject to special conditions, eg a high speed of evolution but also of extinction. Phylogenetically and biogeographically inland water flora and fauna have much closer links to the land than to the sea, ie the amount of species that are descended from land-dwelling forms is very high (cormophytes, arachnids, insects, air-breathing snails).

Innate Character-forming Mechanism (ICM). Term used in ethology. A system assumed to exist that coordinates the characteristic behaviour fed to it by the sense organs over and above obligatory learning. In this way the individual time-place relationships necessary for the maintenance of the species is transmitted to the fish. For example, eels and salmon reach their far-distant spawning grounds, following precise routes established centuries ago, after their ICM has relayed the necessary information to them. Purely functional term with no reference to its place in the central nervous system.

Innate Release Mechanism (IRM). Term used in ethology. A species-specific principle that has come into being during phylogenesis that receives a particular stimulus or series of stimuli (key-stimuli*) from a stimulus releaser* into reactions that are characteristic of the species. For example, a low-ranking *Danio malabaricus* male starts swimming with its body pointing obliquely upwards when it meets a higher-ranking animal. This is a mark of its inferiority. The sight of the high-ranking animal is the key-stimulus in this case which acts upon the IRM of the lower-ranking animal and induces the inferior behaviour. IRM is a purely functional term, and expresses nothing about the spatial relationships in the central nervous system.

Insecta; insects (fig. 666). Also given the scientific name of Hexapoda. Cl of the arthropods (Arthropoda*). With over a million known species, the insects make up by far the largest part of the animal kingdom. Basically, the insect body is split into the head, a three-part thorax, and an eleven-part abdomen. A sub-class of the winged insects (Pterygota), however, also has two pairs of wings which developed from outfoldings of the skin. Nevertheless, in many groups, the wings have disappeared again in the course of evolution. Some groups of insects develop directly from the larval stage into the sexually mature imago, eg Mayflies (Ephemeroptera*), Dragonflies (Odonata*) and Stoneflies (Plecoptera*). Others, such as the Caddis Flies (Trichoptera*) and the Alder Flies and Snake Flies (Megaloptera*), insert a pupal stage between the larva and imago. The examples given also make up the most important insect orders that live at some time or other in water. A few other Fs ought to be mentioned here, too: some of the Bug Fs (Heteroptera*), Beetle Fs (Coleoptera*) and True Fly Fs (Diptera*).

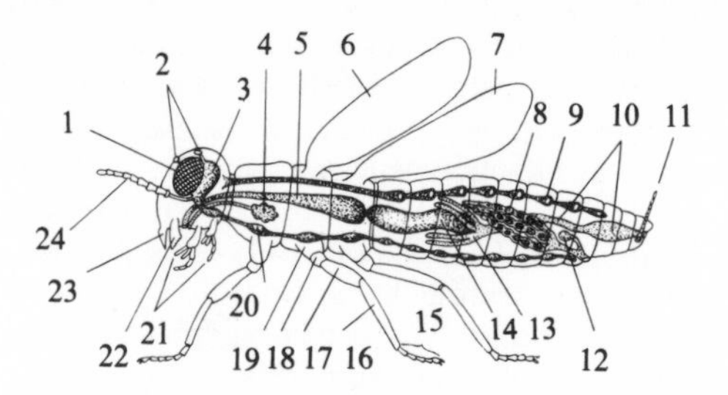

Fig. 666 Anatomy of an insect. 1 Compound eye 2 Ocellus 3 Brain 4 Salivary gland 5 Foregut 6 Front wing 7 Rear wing 8 Ovary 9 Heart 10 Hind-gut 11 Tail bristle (cercus) 12 Seminal receptacle 13 Excretory organ (Malpighian tubules) 14 Mid-gut 15 Jointed feet (tarsi) 16 Tibia 17 Femur 18 Trochanter 19 Coxa 20 Ventral cord 21 1st and 2nd maxilla 22 Jaw (mandible) 23 Upper lip 24 Antenna

All insects descend from land-dwelling forms; the period of their lives spent in water is a secondary development. That is why aquatic insects do not breathe through gills but through newly acquired or modified accessory breathing organs, such as tracheal gills, abdominal outgrowths or hind-gut respiration. Without any of these aids, the insect has to come to the surface to breathe. Small forms are also able to carry a cloak of atmospheric air around with them, rather like aquatic Ticks and Mites (Acari*). It is interesting to note that no insect order has been able to conquer the sea, the aquatic groups being confined to fresh water and brackish water. For the aquarist, insects are only indirectly of interest, mainly as fish food (Midge larvae, *Drosophila*). Other than that, some insect forms are interesting to observe in pond aquaria.

Insecticides *see* Poisoning

Insects, Cl, *see* Insecta

Instinct. Term used in ethology. Instinct means the ability of an animal to convert internal or external stimuli into patterns of behaviour typical of the species. In the physiological sense, instinct is the co-ordination of certain reflexes, usually called a reflex chain, which is not imposed by any nervous system. Instincts enable an animal to make astonishing achievements within its rhythm of life, something which is particularly apparent among insects, fish and birds. The multiplicity of instincts surrounding migration and brood care are perhaps the most striking examples. Like anatomical features, instincts are characteristic of a species, since they are determined by heredity, and not subject to any learning process. An instinct is usually made up of three different components—drive, appetitive behaviour, consummatory act. The state of readiness within a fish to react to internal or external stimuli is therefore the *drive*; the search for spawning water, for a sufficiently stimulated partner, or for a rival etc are all examples of *appetitive behaviour**; whereas the *consummatory act** or instinctive movement can take the form of mating, attack, capture of prey etc. *See also* Fixed Action Pattern.

Integument *see* Skin

Intention Movement. Term used in ethology. Intention movements are movements that are interrupted almost as soon as they began (ie remain incomplete). They occur because two spheres of function become activated at the same time, but neither function seems to predominate. When two fish swim alternately towards one another in a frontal threatening posture (in a dispute over territories), the urge to attack is uppermost to start with, as each fish takes it in turn to adopt the threatening posture. The attacking fish will prepare to ram (*see* Ramming), but as soon as it swims across into the other's territory, the urge to flee takes over, the fish turns and swims away. In the same moment, the attacked fish will become the aggressor and vice versa, with the result that true aggressive behaviour does not take place. In other words, we are talking about ritualised intention movements that have developed in the course of phylogenesis. The actions carried out have come to have a different meaning and now they have a threat function. Often, such ritualised intention movements are 'mimically exaggerated'. For example, a courting male Sparrow that indicates it will fly up next to the female shows that it has lost its original tendency to fly away, and so, secondarily, this movement has become an element of the courtship behaviour*.

Interference Actions. Term used in ethology. Interference actions are the consequent actions of a *conflicting drive situation** which occurs because two spheres of function* are activated at the same time. Such actions may be released, for example, when a rival fish swims into the view of a courting fish, thereby causing the courtship behaviour to be interrupted. However, should the rival disappear from view very suddenly, no fight will break out. At the same time, the ripe female will once again approach her mate. However, at first, the male will attack the female as if she were a rival, because there is still a strong urge within him to satisfy his action-specific energy for a battle. He is quite unable to continue with the courtship activity straight away.

Interfertility. The ability of hybrids that have developed from different species to reproduce. Under natural conditions in the animal world, interfertility occurs extremely rarely. The sexual isolation* that is caused in this way has been particularly important for defining individual species—it is the only generally recognised criterion for defining a species. One of the few natural hybrids known is the Amazon Molly *(Poecilia* formosa)*. However, under artificial conditions, such as keeping fish in an aquarium, species often cross-breed and produce fertile offspring, especially among the Fs Cyprinodontidae* and Poeciliidae*.

Intermediate Host *see* Parasitism

Internal Filter *see* Filtration

Internal Parasites (Endoparasites) *see* Parasitism

Internal Secretion. The release of active substances (Hormones*) into the blood in order to regulate bodily functions. For this purpose vertebrates have various endocrine organs (eg the thyroid gland, the adrenal glands, parts of the pancreas and sex glands), whose activity is controlled by the pituitary gland (hypophysis).

Internode *see* Stem Axis

Intersex *see* Hermaphroditism

Inter-territorial Fighting. Term used in ethology. Many species of fish occupy territories either as individuals or as pairs. Sometimes the territory is held permanently, sometimes only during the breeding season. The boundaries of the territory are defended against all rivals within the species even to the point of self-sacrifice (*see* Territorial Behaviour). Inter-territorial fighting takes place along the boundaries between neighbouring territories. At such times these boundaries are only crossed for short periods and only for short distances. The fish swim alternately at each other, front on, in a threatening fashion (*see* Intention Movement). Only on rare occasions does inter-territorial fighting progress to aggressive fighting behaviour or, even further, to fights to the death (Aggressive Fighting Behaviour*): However, inter-territorial fighting does involve all the elements of display behaviour* and symbolic fighting behaviour*. Usually, inter-territorial fighting ends inconclusively because one of the combatants is overcome with fatigue.

Intestinal Breathing *see* Respiration

Intestinal Diseases in fish may have a number of different causes. In many cases, inflammation of the intestine manifests itself with a swollen abdomen, a protruding, inflamed anus and white, sometimes bloody, thread-like excrement. One of the commonest causes of this is a deficiency in the diet, usually brought about by giving one type of food which lacks roughage. The same symptoms may also occur as a result of feeding sewage-infested foodstuffs (as is often the case with Chironomidae* and *Tubifex**) or as a result of harmful substances (*see* Poisons) dissolved in the water. Diseases of the intestine can also be caused by bacteria* or viruses* in which cases there is the danger of infection (unlike intestinal diseases caused by diet deficiencies). The intestine, along with other organs, is often infected when a fish has stomach dropsy* and fish tuberculosis*. With inflammation of the intestine, the fish should not be fed for 2–8 days, thereafter it should be given a rich variety of food containing lots of roughage. If the infection has not definitely been eliminated, mix up antibiotics* such as chloromycetin* and terramycin* with the food and raise the temperature at the same time. The intestine also plays host to various parasites and their larvae, such as the protozoa* *Eimeria* (*see* Sporozoa), *Hexamita** and *Oodinium**. Larger parasites such as Trematoda*, Cestoda*, Nematoda (*see* Nematode Diseases) and Acanthocephala* often live in fish intestines too. They are combated according to their particular cycle of development. If too many large parasites congregate in one area, the intestine may be blocked off and the fish will die. A decision about the state of the intestine can be made by making a smear preparation of the excrement and examining it under a microscope; the evidence to look for is bacteria, as well as the egg-cells and larvae of worm parasites. An examination of the intestine itself can only be carried out on fish that have been killed (*see* Examination Technique).

Intestine *see* Digestive Organs

Invertebrates (collective term) *see* Vertebrata

Ions. Electrically charged atoms or groups of atoms; the charge occurs because electrons are released or taken up. When electrons are taken up, the charge is negative (ie an anion results); when electrons are released, the charge is positive (ie a cation results). For example,

$$Cl + 1e^- \rightarrow Cl^-$$
$$Mg \rightarrow Mg^{++} + 2e^-$$

Ions have fundamentally different properties to the corresponding uncharged atoms. There are a great many chemical substances whose individual components would not be able to combine without an electrical charge. When these substances are dissolved in water the strength of this bond is much reduced and the substance breaks up partly or completely into ions, ie it dissociates.

For example, when cooking salt is dissolved in water, sodium and chloride ions are formed; when sulphuric acid is diluted, hydrogen and sulphate ions are formed; and when potassium hydroxide is dissolved, potassium and hydroxyl ions are formed:

$$NaCl \rightarrow Na^+ + Cl^-$$
$$H_2SO_4 \rightarrow 2H^+ + SO_4^{--}$$
$$KOH \rightarrow K^+ + OH^-$$

The number and type of ions dissolved in the water decisively determine its qualities, eg its conductivity, its hardness (*see* Water Hardness) and its pH value*.

Ion Exchange *see* Desalination, Ion Exchangers

Ion Exchangers. Artificial resins with the property of removing certain ions* from watery solution by exchanging them for other ions. Such an exchange is reversible and depends solely on the concentration ratios. So, ion exchangers can be easily regenerated and in theory may be used for an unlimited time. It is these properties of ion exchangers that are used in aquarium-keeping. It rids the aquarium of unwanted ions, particularly calcium and magnesium ions (*see* Water Hardness), and so helps the aquarist achieve the quality of water he requires.

There are basically two types of ion exchangers: cation and anion. If cation exchangers only are used, a neutral exchange* comes about, if a combination of cation and anion exchangers are used (note the sequence!), the water will become desalinated (*see* Desalination).

Neutral exchange is in most cases the desired effect in aquarium-keeping. With this process a reverse distribution of the cations takes place, so that the total salt content remains the same, eg the unwanted calcium and magnesium ions are replaced with sodium ions. The regeneration can be carried out with a concentrated cooking salt solution (sodium chloride), and so, in contrast to the procedure explained later on, it is not dangerous. Suitable exchange resins are: Permutit RS, Permutit AG, Wofatit KPS, Wofatit F, Lewatit S 100, Lewatit G 1, Amberlite IR-120.

Original Water
H_2O
$Ca(HCO_3)_2$
$Mg(HCO_3)_2$
$CaSO_4$
$MgSO_4$
↓

Na^+ Na^+ Na^+ — Neutral Cation Exchanger
Na^+ Na^+ Na^+
Na^+ Na^+ Na^+ — before
Na^+ Na^+ Na^+ — ↓
Ca^{++} Na^+ — passing through water
Na^+ Mg^{++} Mg^{++} — ↓
Ca^{++} Na^+ Na^+ — after

↓
H_2O
$NaHCO_3$
Na_2SO_4

If these exchange resins are treated with hydrochloric acid instead of sodium chloride, the metal ions dissolved in the water (Ca^{++}, Mg^{++}, Na^+ etc) are exchanged for

Original Water
H_2O
$Ca(HCO_3)_2$
$Mg(HCO_3)_2$
$CaSO_4$
$MgSO_4$
↓

H^+ H^+ H^+ H^+ — Cation Exchanger before
H^+ H^+ H^+ — ↓
H^+ H^+ H^+ — passing through water
Ca^{++} Mg^{++} H^+ — ↓
H^+ Ca^{++} Mg^{++} — after

↓
H_2O
H_2CO_3 (breaks up partially into H_2O and CO_2)
H_2SO_4
Partly desalinated water

hydrogen ions (H^+): This then means that acids develop with the remaining anions (SO_4^{--}, CO_3^{--}, Cl^-, NO_3^- etc). The result is a considerable lowering of the pH value*, which will only partially be raised again by vigorously aerating the water (forcing out the carbonic acid).

This process, also known as partial desalination*, is therefore only of any point if the pH value can be raised again to 6.5–7 by expelling the carbonic acid and admixing some of the original water. Every neutralisation of the acids with bases* or with salts that have a basic reaction (eg soda), amounts to a neutral exchange which can be more easily achieved by treating the ion exchangers with sodium chloride. There are also ion exchangers that are only receptive to the calcium and magnesium that is bound to hydrogen carbonate, for example. When the carbonic acid is expelled through vigorous aeration, a pH value of between 6.5 and 7 will be established. Permutit C-65 works on this basis, for example. For the successful rearing of certain 'problem' fish, a complete desalination of the water can sometimes be essential. This produces extremely pure water, recognisable by its very slight conductivity (about 3–5 microsiemens). The original water is first of all sent across a cation exchanger charged with hydrogen ions *(see above)*, then across an anion exchanger charged with hydroxide ions, eg Permutit EM 13, Amberlite IRA-400, 401, 402 etc, Dowex 2, Wofatit SBW. For regeneration purposes, usually 8 % caustic soda solution is used. The fully desalinated water must be prepared for aquarium purposes, as there needs to be a certain concentration of salt ions to guarantee the necessary buffer* effects. To this end, mix in some of the original water to achieve a degree of hardness of between 2 and 3.

Original Water
H_2O
$Ca(HCO_3)_2$
$Mg(HCO_3)_2$
$CaSO_4$
$MgSO_4$
↓
H^+ H^+ H^+ H^+ — Cation Exchanger
H^+ H^+ H^+ — before
H^+ H^+ H^+
↓
passing through water
↓
Ca^{++} Mg^{++}
Mg^{++} H^+
H^+ Ca^{++} — after
↓
OH^- OH^- — Anion Exchanger
OH^- OH^- — before
OH^- OH^-
↓
passing through water
↓
OH^- SO_4
OH^- OH^-
OH^- SO_4 — after
↓
Fully desalinated water

If a particular operation requires fully desalinated water, so-called mixed bed exchangers are used, eg Amberlite MB-1 or MB-3. MB-3 contains an indicator that shows when the ion exchangers are exhausted. Ion exchangers are also able to remove other substances from the water, such as high-molecule organic compounds like tannins*, humic acids*, proteins* etc. It is best to find out about these types of exchangers from the relevant catalogues provided by the manufacturers.

Technically, setting up an ion exchanger can be arranged quite simply. Basically, funnels are used filled with the particular resin. For desalination purposes, allow the water to run through, for regeneration purposes, the necessary acids and bases. The only important thing is that the fluid is allowed to remain in the exchange resin for some time – ie do not let the fluid run through too quickly. A technically more perfect type of ion exchanger consists of several cylindrical columns inserted one after the other. Information regarding the resins (pouring them out and loading them into the containers) is usually contained in the manufacturer's leaflet. The need for regeneration is easy to recognise by determining the water hardness (Water Hardness, Determination of*) or measuring the conductivity.

Iridaceae. F of the Liliatae* comprising 70 Ga and 1,500 species. Tropical or subtropical plants, mainly from America and S Africa. There are also a few species in northern temperate regions. Primarily rhizomed and tuberous perennials with a distichous leaf arrangement. Flowers hermaphroditic, usually strikingly coloured, and polysymmetrical or monosymmetrical. The Iridaceae grow in dry or wet places. A large number of ornamental plants belong to the F, such as species of the Ga *Crocus, Gladiolus* and *Iris*. There are over 200 species of the G *Iris*, one of which, the Yellow Flag *(Iris pseudacorus)* with its pale yellow flowers, is found along European and West Asian waters. This one is particularly good for planting around open-air ponds.

Iridescent Rasbora *(Rasbora kalochroma) see Rasbora*

Iris (*Iris*, G) *see* Iridaceae

Iris (part of the eye) *see* Sight, Organs of

Irises, F, *see* Iridaceae

IRM *see* Innate Release Mechanism

Iron Poisoning *see* Poisoning

Irregularia, Sub-Cl, *see* Echinoidea

Isoetaceae, F, *see* Pteridophyta

Isoetes LINNÉ, 1753; Quillwort. G of the Pteridophyta* comprising 75 species. Distributed throughout the world. Smallish perennials with truncated, tuberous stems, leaves arranged in rosettes and roots branching like a fork. Leaves awl-shaped, with three edges or linear. Has a complicated method of reproduction using spores. Sporangia located at the base of the leaves, encased in a sheath. The majority of the different species of this G grow in marshes that are flooded at times or they grow submerged all the time. Rarely kept in aquaria. Examples are:

– *I. lacustris* LINNÉ; Quillwort. From N America, Europe, N Asia. Leaves awl-shaped, stiff, bent, and dark-green. A coldwater plant.

– *I. malinverniana* CESATI et DE NOTARIS. From S Europe, where it grows in rice fields. Leaves awl-shaped, slightly wavy, pale green. Suitable for heated aquaria (fig. 667).

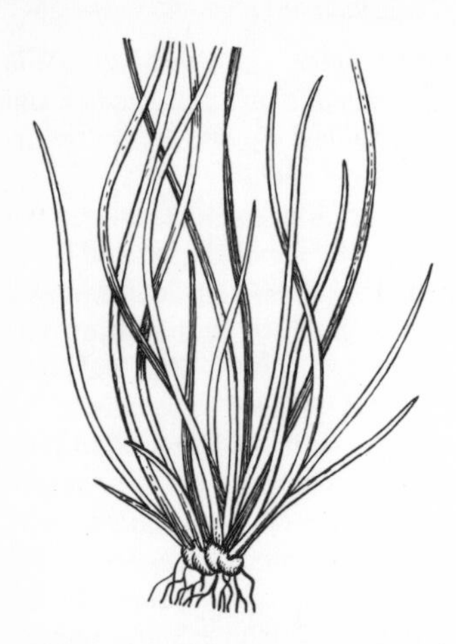

Fig. 667 *Isoetes malinverniana*

Isoperla Banks, 1906. G of the Stone Flies (Plecoptera*). Found throughout the holarctic region* and represented in Europe by more than 30 species. The various species are yellow or greeny-black in colour, usually with an X-shaped or U-shaped marking on the head. 7–15 mm long. The larvae are basically yellow or olive-green in colour with pale flecks on top. In some places, very common in different kinds of flowing waters*. *Isoperla* are suitable as food animals. Those species that live in slow-flowing waters can be kept in small aquaria.

Isopoda; Isopods, Woodlice. O of the crustaceans (Crustacea*), Sub-Cl Malacostraca*. A crustacean group that contains a large number of species, and, at the same time, a unique group. The Isopods have successfully conquered land. However, many Isopod families have not left their original habitat, the sea, and one family lives in fresh water. Isopods vary in size from 1 mm to 8 cm, and usually, the body is flattened. Aquatic Isopods often get trapped in the net when fishing, but to the aquarist they are of little interest. Occasionally, the freshwater genus *Asellus** is kept in pond aquaria for purposes of observation or given to larger species of fish as food.

Isotonic; iso-osmotic. Developing the same osmotic pressure (tonus).

Istiompax marlina *see* Scombroidei

Istiophoridae, F, *see* Scombroidei

Istiophorus orientalis *see* Scombroidei

Isuridae, F, name not in current usage, *see* Lamnidae

Isurus oxyrinchus *see* Lamnidae

Ivy-leaved Duckweed *(Lemna trisulca) see Lemna*

Jack Dempsey *(Cichlasoma biocellatum) see Cichlasoma*

Jack-knife Fish *(Equetus lanceolatus) see Equetus*

Jacks, F, *see* Carangidae

Jacobson's Rasbora *(Rasbora jacobsoni) see Rasbora*

January Tetra *(Hemigrammus hyanuary) see Hemigrammus*

Japanese Barb *(Barilius neglectus) see Barilius*

Japanese Giant Crab *(Kaempfferia kaempfferi) see* Brachyura

Japanese Medaka *(Oryzias latipes) see Oryzias*

Japanese Pond Lily *(Nuphar japonica) see Nuphar*

Japanese Rush *(Acorus gramineus) see Acorus*

Japanese Tang *(Naso lituratus) see Naso*

Japanese Weatherfish *(Misgurnus anguillicaudatus) see Misgurnus*

Jaundice (Icterus) *see* Liver, Diseases of

Java Fern *(Microsorium pteropus) see Microsorium*

Java Moss, G, *see Vesicularia*

Javan Mouthbrooding Fighting Fish *(Betta brederi) see Betta*

Javanese Fighting Fish *(Betta picta) see Betta*

Javanese Ricefish *(Oryzias javanicus) see Oryzias*

Javanese Starfish *(Nardoa variolata) see Nardoa*

Jawed Animals, Sub-Ph, *see* Gnathostomata

Jawfishes (Opisthognathidae, F) *see* Trachinoidei

Jawless Fishes, Sub-Ph, *see* Agnatha

Jellyfish, True, Cl, *see* Scyphozoa

Jenynsia lineata *see* Jenynsiidae

Jenynsiidae; One-sided Livebearers (fig. 668). F of the Atheriniformes, Sub-O Cyprinodontoidei*. Distributed from S Brazil to N Argentina. The few species of the F are all small, very reminiscent of the Livebearing Tooth-carps (Poeciliidae*), although they can be distinguished from them by peculiarities of reproduction. The copulatory organ (gonopodium) is not formed directly from the An, but develops from a ring-shaped fold of skin (genital papilla) which grows out into a tube and thereby envelops the An. As in the Anablepidae*, this gonopodium can only be moved to the right or to the left, ie there are 2 types of male. The oviduct opening

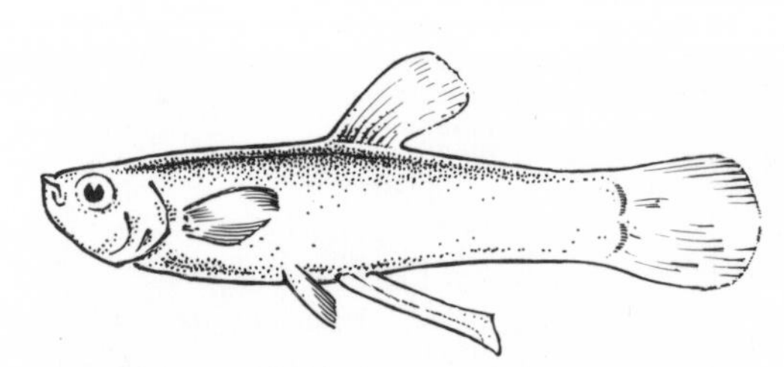

Fig. 668 Jenynsiidae-type

in the female is covered over by a large scale in such a way that the female can only be entered from the right or left. So, this leads to left-females being able to mate with right-males only, and vice versa. The embryos are fed in the ovarial cavity via outgrowths that grow from the cavity wall into the mouths of the developing youngsters. *Jenynsia lineata* is occasionally kept in the aquarium. Females up to 12 cm, males only 4 cm long. Body colouring basically a delicate green-silver, often with a cobalt-blue shimmer, and covered with a number of streaky longitudinal lines. Temperature not more than 23 °C. Other than that, the species is very undemanding.

Jewel Cichlid *(Hemichromis bimaculatus) see Hemichromis*

Jewel Grouper *(Cephalopholis argus) see Cephalopholis*

Jewel Tetra *(Hyphessobrycon callistus) see Hyphessobrycon*

John Dories (Zeidae, F) *see* Zeiformes

John Dory *(Zeus faber) see* Zeiformes

John Dory types, O, *see* Zeiformes

Johnius hololepidotus *see* Sciaenidae

Jordan, David Starr (1851–1931). American zoologist and ichthyologist. Pupil of AGASSIZ* and an important expert on American fish. Professor at the Bloomington University, Indiana, and later President of Stanford University. It was in 1876 that his greatest book was published, *Handbook of the Vertebrates of the North American United States.* In 1896–1900, together with B. W. EVERMANN, he published the 4-volume work *The Fishes of North and Middle America,* and *Checklist of the Fishes of North and Middle America* (along with B. W. EVERMANN and H. W. CLARK).

Jordanella GOODE and BEAN, 1879. G of the Cyprinodontidae*. One species, found in Florida and Mexico, where it lives in marshes and ponds that contain a lot of aquatic plants. Related most closely to *Cyprinodon** and *Garmanella**. A very compressed fish, tall, and measuring up to 6 cm long. Body flattened from side to side, caudal peduncle very tall. Unpaired fins rounded, D attached to the body over a much longer length than the An. C a broad fan shape.

It is best to keep *Jordanella* in a species aquarium and rear it here. Very small tanks are suitable for just one mating pair. If you prefer keeping several specimens, you will need to have some areas of dense vegetation for the females to hide in as the males drive at them very hard. Tanks with algal growths and a dark substrate seem to suit *Jordanella* best and it is then that their colouring will be seen at its best. *Jordanella* has no special requirements regarding water quality, temperature or food. It will even live happily in unheated room aquaria, and during the warm part of the year you can even transfer your specimens to open-air installations. These are greedy fish, and they will also need additional plant food to keep them in prime condition (algae, delicate aquatic plant shoots). *Jordanella* likes to stay near the bottom. The males are very intolerant in the aquarium as they form territories. They prepare a spawning area at spawning time which consists of a shallow hollow in the substrate. The love-play is violent and the spawning itself may take place over a period of several days. The male practises brood care just as we know many cichlids (Cichlidae*) do. Depending on temperature, the eggs hatch out after 5–10 days. 2 weeks after the fry have swum free, they should be transferred into a separate rearing tank with plenty of algae.

– *J. floridae* GOODE and BEAN, 1879; American Flag Fish. Found from Florida to S Mexico (Yucatan). Length up to 6 cm. Colour brownish to greenish. The scales are covered with large, pale speckles that have a lovely iridescent sheen. The flanks give the impression of having dark, broad, transverse bands across them. The D has a round, black blotch on it (fig. 669), much more prominent in the female.

Fig. 669 *Jordanella floridae*

Joy-weeds, G, *see Alternanthera*

Julidochromis BOULENGER, 1898. G of the Cichlidae*. Only found in Lake Tanganyika. Inhabit areas of rocky bed. Body very elongate, the D being inserted for a long way along the dorsal ridge and comprising 20 or more hard rays. The soft-rayed part is slightly pointed. C roundish or shaped like a fan. Dorsal profile more strongly curved than the ventral profile. Mouth slightly inferior. Head relatively small in comparison with the body. In some species *(J. marlieri),* the area of forehead above the eyes is clearly hunched up. Colouring—yellow background colour with dark longitudinal stripes going across it or dark lattice-work marking. *Julidochromis* requires a large tank with plenty of rock work and room for swimming. Since they are very intolerant of one another, either keep them as a pair or create enough hiding-holes and obvious places of retreat so that the specimens will be able to set up their own individual small territories. Another way of reducing their aggressiveness is to introduce other similar-sized Cichlids from Lake Tanganyika. Depending on conditions in their natural habitat, have the aquarium water at a temperature of about 24 °C. The water should be hard and slightly alkali (due to the presence of hydrogen carbonate). *Julidochromis* species are typical cave-spawners. The male practises intensive brood care.

– *J. marlieri* POLL, 1956; Marlier's Julie. From Lake Tanganyika. Length up to 10 cm. A white ground colouring with broad, black-brown bands running across it in both length-wise and cross-wise directions. Unpaired fins are covered in semi-dark speckles. The edges of the fins are white with a black border (fig. 670).

Fig. 670 *Julidochromis marlieri*

– *J. ornatus* BOULENGER, 1898; Julie. From Lake Tanganyika. Length up to 7 cm. The body, D and An are a yellow ground colour with dark-brown longitudinal bands running across it. The unpaired fins have a white-blue edge bounded on both sides by brown (fig. 671).

Fig. 671 *Julidochromis ornatus*

Three other species are imported:
– *J. dickfeldi* STAECK; Dickfeld's Julie.
– *J. regani* POLL, 1942; Regan's Julie.
– *J. transcriptus* MATTHES, 1959; Masked Julidochromis.

Juncaceae; Rushes. F of the Liliatae* with 8 Ga and 300 species. Primarily from temperate and cold regions. A few are annual plants, but most are perennials with long leaves, many with runners. Flowers small and inconspicuous. Rushes grow mainly in wet places. In recent years, a species of the G *Juncus, Juncus repens*, Creeping Rush, which comes from south-eastern parts of N America, has occasionally been kept in an aquarium.

Juncus repens *see* Juncaceae

Jussiacea, G, name not in current usage, *see Ludwigia*

Juvenile. The juvenile phase of an animal's life lasts until it becomes sexually mature. In many animals, this period is characterised by a larval (Larva*) stage. Opposite to adult*. *See also* Ontogenesis.

Kaempfferia kaempfferi *see* Brachyura

Kamptozoa (Entoprocta). A Ph of mainly marine animals. Only a few mm long, Kamptozoa either live alone or in colonies. Their body is divided into a stalk, and a mouth and anus surrounded by a circlet of tentacles. Filter-feeders. In marine aquaria, Kamptozoa sometimes develop too abundantly (eg *Loxosoma* spp.), since they are often confused with polyps or bryozoans.

Kerr's Danio *(Brachydanio kerri) see Brachydanio*

Key Stimulus; Sign Stimulus. Term used in ethology. Every species-specific instinctive movement (Fixed Action Pattern*) is put in motion by an innate release mechanism (IRM)*, which in its turn reacts to certain stimuli, or key stimuli. An example can be taken from a male stickleback guarding a nest–for him, a fish with a red ventral surface acts as a key stimulus. Immediately, species-specific instinctive movements relating to the 'fighting' sphere of function* are activated by an IRM, and so the adversary is chased away.

Keyhole Cichlid *(Aequidens maronii) see Aequidens*

Kidney *see* Urinary Organs

King Crabs, Cl, *see* Xiphosura

Kingfish *(Caranx speciosus) see Caranx*

Kingklip *(Genypterus capensis) see* Gadiformes

Kissing Gourami *(Helostoma temmincki) see Helostoma*

Kneria, G, *see* Gonorynchiformes

Kneriidae, F, *see* Gonorynchiformes

Knife Livebearer *(Alfaro cultratus) see Alfaro*

Knifefishes (Gymnotidae, F) *see* Gymnotoidei

Knifefishes, related species, Sub-O, *see* Gymnotoidei

Knight Goby *(Stigmatogobius sadanundio) see Stigmatogobius*

Komplexon *see* Chelaplex

Kribensis *(Pelvicachromis pulcher) see Pelvicachromis*

Kribensis *(Pelvicachromis* sp.) *see Pelvicachromis*

Krill. Plankton that occurs in vast quantities, usually in giant shoals. It is mainly made up of crustaceans that are about 3–8 cm long. Krill forms the basic diet of many fish and whalebone whales. It is found mainly in cool seas. Arctic krill contains large amounts of Copepods from the G *Calanus**, whereas in the Antarctic krill consists mainly of *Euphausia* species (Euphausiacea). Because of the virtual extermination of whalebone whales, krill stocks have become much larger. The great fishing nations (USSR, USA, Chile etc) lay great store by the location of shoals of krill to make their catches. According to recent surveys, krill represents the world's greatest store of protein.

Kryptopterus BLEEKER, 1858. G of the Siluridae*. From Indo-China and the Sunda Islands. Glass Catfishes that belong to the true Catfishes. Length up to 12 cm. They are peaceable, shoaling fish from the middle and lower levels of water, and they are active during the day. They will float freely in among the plants, constantly beating their caudal peduncles to and fro to keep the rear part of their bodies in balance. Their body is scaleless, elongated, and very flat from side to side. There are a few plates on top of the head and only one pair of very long barbels on the upper jaw. The D1 is much reduced and consists of the first fin ray only; D2 missing. The C is deeply forked, the upper lobe being longer. The An is very long and may be joined to the C *(K. macrocephalus)*. Vs very small, Ps strong. There is a great similarity to the G *Physailia** from Africa, although they are not closely related. Provide a large, well planted aquarium when keeping specimens, with enough free-swimming area. The lighting should be subdued and the temperature 25–27 °C. Keeping one specimen only is not advisable, since it will not settle down. Feed with live food only. *Kryptopterus* species are very sensitive fish.

– *K. bicirrhis* (Cuvier and Valenciennes, 1839); Glass Catfish. From Indo-China and the Greater Sunda Islands. Length up to 10 cm. Body and fins colourless, transparent like glass, with the vertebral columns and swimbladder visible. When light falls directly on the fish, it shimmers in all colours of the rainbow. There is a violet shoulder fleck behind the operculum. This species has so far resisted all attempts to breed it, and as a result it is still only a rarity in the aquarium (fig. 672).

Fig. 672 *Kryptopterus bicirrhis*

– *K. macrocephalus* (Bleeker, 1858); Poor Man's Glass Catfish. From the Greater Sunda Islands. Length up to 11 cm. Similar in build to *K. bicirrhis* but not quite so transparent on account of a number of melanophores creating dark spots in the upper half of the body. Body colouring a delicate yellow with a blue shimmer. 2 narrow, brownish, longitudinal bands run alongside either side of the vertebral column starting from the operculum and continuing to the C insertion, where they unite to form a large dark blotch. A third narrow band lies above the long An. Fins transparent. The C has a semi-circle of dark dots in the middle (fig. 673).

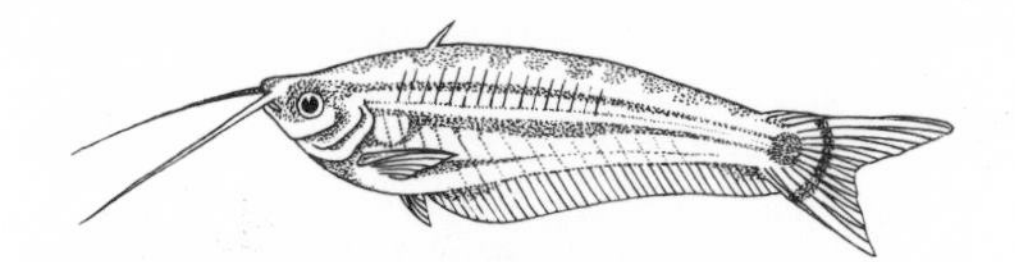

Fig. 673 *Kryptopterus macrocephalus*

Kuhliidae; Flagfishes or Flagtails (fig. 674). F of the Perciformes*, Sub-O Percoidei*. Silvery marine fish from the coastal waters of the tropical Indo-Pacific. Similar to the Sun-fish (Centrarchidae*). Body elongated, flattened laterally. D of uniform height, C indented. There are about 20 different members of the F, all of which belong to the G *Kuhlia*. Some grow to around 40–45 cm long. Some species are common in particular regions where they have become important edible fish.

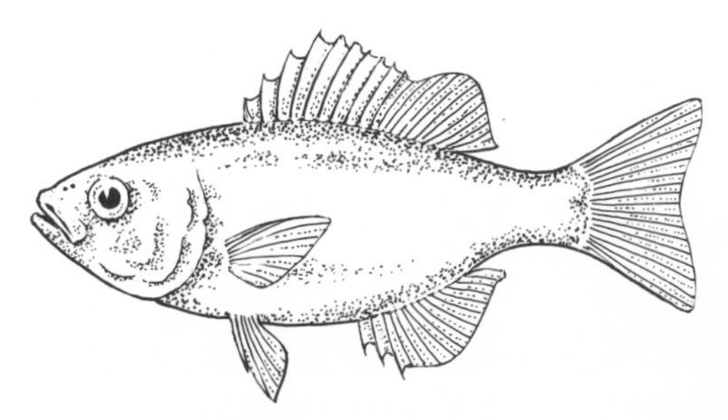

Fig. 674 Kuhliidae-type

Kyphosidae; Rudderfishes, Sea Chubs (fig. 675). F of the Perciformes*, Sub-O Percoidei*. Small to medium-sized Percoidei, in a lot of cases almost Barb-like in shape. Mouth small. On the outer edge of the jaws there are cutting teeth with long roots, and behind them are

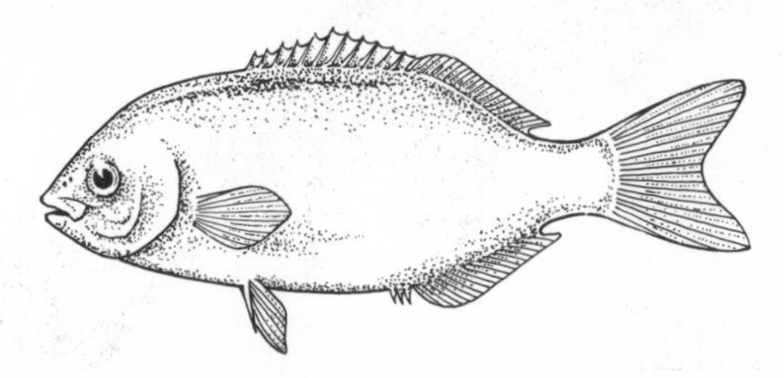

Fig. 675 Kyphosidae-type

fine crown teeth. The D is of uniform height. Scales very small and extending onto the bases of the unpaired fins. The swimbladder is divided at the rear and reaches far into the caudal peduncle. The Kyphosidae are shoaling fish that live mainly in the coastal areas of shallow waters. They feed on algae mainly. The meat from these fish is palatable but of little value.

La Plata *see* Rio de la Plata

Labeo Cuvier, 1817; 'Sharks'. G. of the Cyprinidae*, Sub-F Cyprininae. Widespread in Africa, S and SE Asia. The G comprises fairly small fish as well as species measuring as much as 150 cm in length. The ventral surface is more or less straight while the dorsal surface is arched. The mouth is inferior with 1–2 pairs of barbels. On the inner edges of the thick lips there are villi, sharp ridges and warty knobs which are used for grazing on algae and animal growth. Large scales. The fins are well developed, and the D especially is held in a splayed-out position. Only the smaller species or young specimens should be kept in normal-sized room aquaria, but larger specimens are suitable for public aquaria. *Labeo* species are good swimmers and need large tanks, positioned in relatively dark spots. There should also be plenty of places for the fish to hide. The G comprises robust, hardy fish that are territorial by nature. Thus, they can often be aggressive towards like species but it is rare for them to attack other fish. The water should be soft and slightly acidic. Temperature 22–26 °C. *Labeo* is an omnivore and will eat all forms of live and dried food. An additional supply of plant

food should also be provided. There are only a very few particulars available about reproductive behaviour and breeding. In most species, the males practise brood care.
– *L. bicolor* SCHMITH, 1931; Red-tailed Black Shark. From Thailand, where it lives in small rivers and lakes. Length up to 12 cm. Body, including the D, Vs and An are velvety black, the C is red and the Ps orangey-red. Specimens that are not well attended to will lose their colouring and become grey-brown. There are also local forms with different colouring. Has been bred successfully in a few isolated cases (fig. 676).

Fig. 676 *Labeo bicolor*

– *L. erythrurus* FOWLER, 1937; Red-finned Shark, Ruby Shark. From Thailand. Length up to 12 cm. Colouring alternates between pale brown to dark black-blue. Ventral surface is flecked in pale colouring. A dark band runs from the tip of the snout, through the eye to the level of the P. There is a dark transverse streak on the root of the tail fin. Vs colourless or grey, C orange or red.
– *L. forskali* (RUPPEL, 1853). Found in the Nile and in tributaries of the Blue Nile. Length up to 36 cm. Body grey-green with a silvery iridescence. Ventral surface pale, fins yellowish-grey. Will also tolerate lower temperatures.
– *L. frenatus* FOWLER, 1934; Green Fringe-lipped Labeo. From N Thailand. Length up to 8 cm, which makes it the smallest species. Body grey-brown with a greenish-violet shimmer, ventral surface pale. A black band extends from the tip of the snout, across the eye to the operculum. There is a triangular blotch on the caudal peduncle. Fins reddish-yellow to red, Ps, Vs and An with black edges. Has been bred successfully in a few isolated cases. If kept in large tanks with enough hiding-places, this species is tolerant of like species.
– *L. wecksi* BOULENGER, 1909. From the upper and middle Congo. Length up to 23 cm. Flanks yellowish-green with a gold sheen and narrow, brown longitudinal striping. Scales have dark edges. The fins are greenish-grey, the D and C with a wine-red edge. Often found to be tolerant of like species.
Labeotropheus AHL, 1927. G of the Cichlidae*. Only found in Lake Malawi* (ie it is endemic there). *Labeotropheus* lives over rocky ground and it is adapted for grazing on algal growths containing small animals. This can be seen in the inferior position of the mouth and its spatulate teeth, as well as the enlarged, overhanging, upper lip. Because of this, this fish looks a little like a Tapir. The D has 16–19 hard rays and 7–9 soft ones. The An in the males features typical ovate blotches, a pointer to the fact that this G comprises highly specialised mouth-brooders. The body colouring is similar to some species of the G *Pseudotropheus** – dark transverse bands crossing a blue background colour, the bands on the head producing a mask-like effect. The intensity of the colours is to a large extent dependent on mood, and furthermore it is species-specific. There are more colour variants among the females, and alongside similarly coloured (homeochromous) male forms you may also find orange-coloured female specimens with dark, dappled markings. *Labeotropheus* species are not difficult to keep, especially as they are not nearly so intolerant as other Lake Malawi Cichlids. The nearest duplication of their natural surroundings can be provided with large tanks containing medium-hard water, a number of algae-infested rocks and a temperature of around 24 °C. If desired, water plants can also be added.
– *L. fuelleborni* AHL, 1927; Fuelleborn's Cichlid. From Lake Malawi. Up to 12 cm long. Relatively tall, bluish with dark transverse bands. The female dappled variety is orange-black (fig. 677).

Fig. 677 *Labeotropheus fuelleborni*

Fig. 678 *Labeotropheus trewavasae*

– *L. trewavasae* FRYER, 1956; Red Top Zebra. From Lake Malawi. Up to 10 cm long. A slim species with intense colouring. Brilliant blue with dark transverse bands. The female dappled form is orange with black, red and blue flecks (fig. 678).

Labridae, F, *see* Labroidei

Labroidei; Wrasses and Parrotfishes. Sub-O of the Perciformes*, Sup-O Acanthopterygii. Comprises a great many different species. The Labroidei come in all sorts of shapes but they are all found in shallow coastal waters of warm and temperature seas. The earliest known fossil forms date back to the Eocene period. The body is elongate, moderately tall and then either compressed from side to side, or a low, spindle shape. The head is usually shaped like a truncated cone with the mouth slightly protruding. Particularly noticeable are the thick lips. There are never any spines on the operculum or praeoperculum. For arrangement of teeth, turn to the individual Fs. The lower pharyngeal bones form a plate containing grinding teeth. The D is always uniform in height, the front, spiny-rayed part being longer in most cases (the spines are sometimes elongated towards the outside edge). The C is rounded, straight, or more rarely with elongated rays at the edges. The Vs are always located below the Ps and in some Ga they are shaped like a sabre. The skin is usually thick and contains a great many mucous glands. Cycloid scales that are small or sturdy. LL complete, although divided in two in some Ga. The Labroidei live among coral reefs, along rocky coasts and among carpets of seaweed. Their area of distribution is centred on the tropics, although a few individuals penetrate deep into temperate waters. Many species have delightful colouring, whereas others have a marked tendency to change colour. The colouring of the young fish can be entirely different from that of the sexually mature adults, and even the adults can vary so much in colour that male and female are often described as separate species. All Labroidei have a very typical way of swimming that makes them easily recognisable: The two Ps are pushed simultaneously backwards like oars, so that the fish seem to push themselves through the water.

– *Labridae*; Wrasses (fig. 679). Roughly 600 species. These fish have all the features detailed above, but their denture is characteristic: the jaws are beset with conical

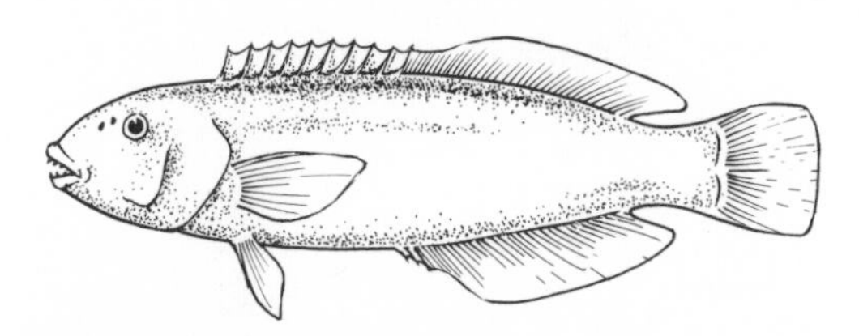

Fig. 679 Labridae-type

or chisel-shaped teeth. They feed on hard-shelled animals as well as on small crustaceans and algae. Most Wrasses are small and live alone, and some are cleaner fishes (fig. 680). The larger species, including all members of the G *Cheilinus*, can grow more than 2 m long. Commercially, Wrasses are only of regional importance, although many are considered good sporting fish. Occasionally, fish poisoning has occurred after people have eaten Wrasse meat. To the aquarist, this F is very interesting (*see under* the Ga *Bodianus, Coris, Crenilabrus, Ctenolabrus, Halichoeres, Hemigymnus, Gomphosus, Hemipteronotus, Labroides, Labrus, Lachnoleimus, Xyrichthys*).

Fig. 680 Wrasse *(Anampses rubrocaudatus)*

– *Scaridae*; Parrotfishes (fig. 681). Roughly 80 species. In contrast to the Labridae, in the Parrotfishes the outer teeth of the jaws have been fused into cutting edges or plates. Parrotfishes are mainly distributed among the reefs and rocky coasts of the tropics. They far surpass most Wrasses in the beauty of their colouring. Many species have a variety of different colouring at different

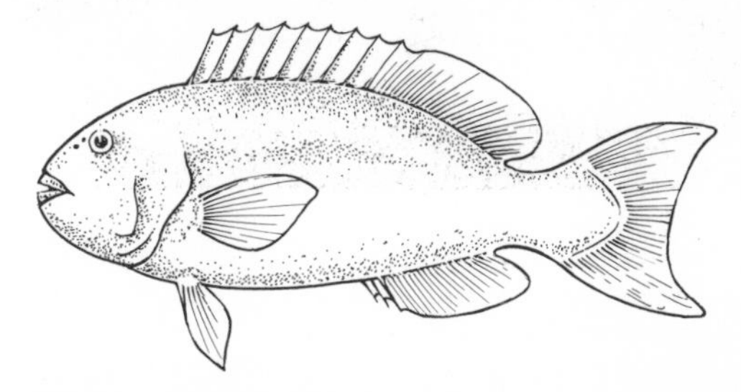

Fig. 681 Scaridae-type

periods in their growth cycle. At night, some species build large mucous cocoons. Parrotfishes can be small or medium-sized, whereas some may be giants, 2–3 m in length and at the same time very tall. With age, the males often form a tumour on the forehead. Most Scaridae feed on algae and crustaceans, and many break off bits of coral with their beakteeth and crush them up. Parrotfish meat is not highly regarded, especially as it is widely thought to be poisonous. However, they are very important to the aquarist (*see under* the Ga *Callyodon, Scarus, Sparisoma*).

Labroides (BLEEKER, 1851). G of the Labridae, Sub-O Labroidei*. There are several species that come from the Indo-Pacific. They are found only among coral reefs and many other fish seek them out to rid them of skin parasites. *Labroides* species are slender fishes, around 10 cm long, with pointed mouths. They advertise their services as cleaner fish by their characteristic see-sawing swimming action and by their striking colouring. Their 'clients' stand motionless in front of the cleaner, even allowing it to swim into their open mouths. One cleaner fish can treat about 300 fish daily. The shape, colour and behaviour of the cleaner fish has been imitated to an astonishing degree by the False Cleanerfish *(Aspidontus* taeniatus)* although, in contrast to the true Cleanerfish, its mouth is inferior. At night, *Labroides* hides among the coral and secretes a thin layer of mucous (which will often be found the morning after if the fish is being kept in an aquarium). Cleanerfishes are easy to care for; they will readily accept animal food, particularly small crustaceans. The following species is often imported:

– *L. dimidiatus* (CUVIER and VALENCIENNES, 1839); Blue Streak, Bridled Beauty. From the Indo-Pacific up to the Red Sea. 10 cm. Sky-blue with a black length-wise band that gets broader towards the rear of the fish. Juveniles are black with a blue stripe on the dorsal surface (fig. 682).

Fig. 682 *Labroides dimidiatus* and *Hemipteronotus taeniurus* in typical position; the Cleanerfish is thereby advertising its cleaning services

Labrus (LINNAEUS, 1758). G of the Labridae, Sub-O Labroidei*. Comprises several species found in the E Atlantic, as far as Norway, and also in the Mediterranean. They are shy, solitary fish that live hidden among vegetation and in holes in the rocks. They feed on small animals. Length up to 60m. *Labrus* species are slim with an elongated snout. In young fish, the praeoperculum is toothed. The males look after the eggs in nest hollows. Fish belonging to this G are only rarely caught and kept in aquaria, eg:

– *L. bimaculatus* (LINNAEUS, 1758). From the Mediterranean, E Atlantic and the North Sea. Up to 45 cm. At spawning time the male is yellow-green with blue stripes on the head and anterior part of the body; the female is red-orange with 3–4 black flecks on the upper side of the caudal peduncle. Adult males are very aggressive.

Labyrinth Fishes (Labyrinthici, Sub-O, name no longer in current usage) *see* Anabantoidei

Labyrinth Organ *see* Respiratory Organs

Labyrinth Organ (respiratory) *see* Anabantoidei

Labyrinthici, Sub-O, name no longer in current usage, *see* Anabantoidei

Lace Gourami *(Trichogaster leeri) see Trichogaster*

Lacépède, Bernhard-Germain Etiènne de la Ville, Count of (1756–1825). French natural historian who studied in Paris. In 1785 he gained a position in the natural history department of the museum and by 1795 he was Professor of Natural History. He conducted various studies in different natural history fields. His 5-volume work *Histoire des Poissons* ('History of Fishes') appeared from 1798 to 1809.

Laceration (fig. 683). An asexual means of reproduction found in some sea-anemones (Actiniaria*), eg *Aiptasia**, *Metridium**. Small pieces of the anemone's foot break loose and develop into new specimens.

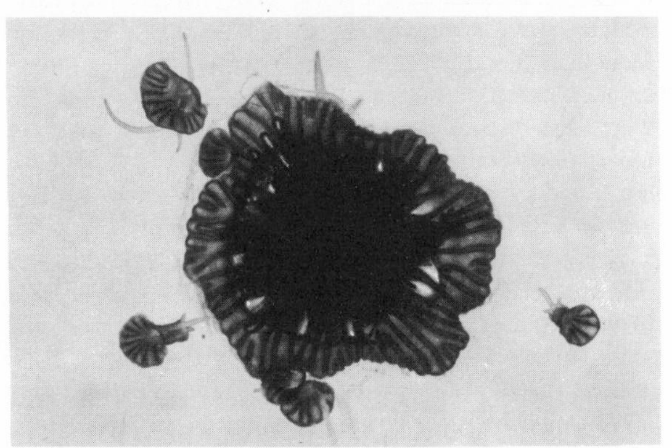

Fig. 683 Laceration in *Aiptasia*

Lachnoleimus (CUVIER, 1829). G of the Labridae, Sub-O Labroidei*. Comprises one species found in tropical parts of the W Atlantic – *L. maximus* (WALBAUM), the Hogfish or Hogsnapper. It is a tall fish with an elongated snout and very long first 3 dorsal fin rays. Length up to 90 cm. An imposing fish that lives as a predator. Not very suitable for a hobbyist's aquarium (fig. 684).

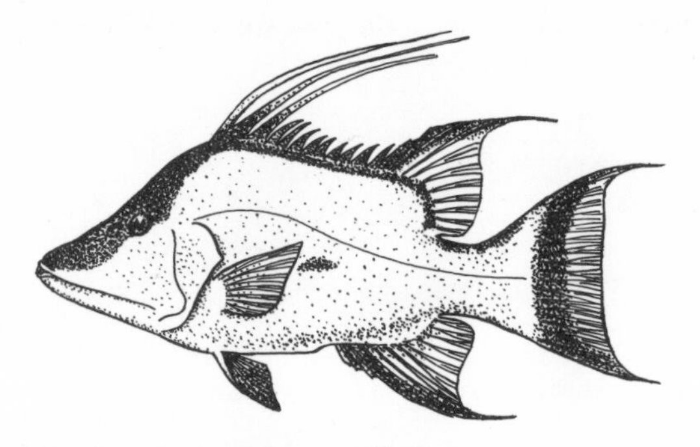

Fig. 684 *Lachnoleimus maximus*

Lactoria JORDAN and FOWLER, 1902. G of the Ostraciontidae* comprising a few species from the tropical Indo-Pacific. All the species have a typical family build and they feed on small animals. Length varies between 12 and 45 cm. In appearance, it is mainly distinguished by having a forward-pointing spine above each eye and another at each hind corner of the belly. *Lactoria*, like other Boxfishes, is sensitive towards food-competitors in the aquarium, is particularly prone to skin parasites and to skin damage. For this reason, the tank should not be decorated with sharp-edged coral skeletons. Food, to start with usually small shrimps only, is readily accepted from the sandy substrate. Specimens that are sick or dying may be a danger for other fish, as *Lactoria* secretes a poison. The following species will usually become accustomed to a great variety of foods:

– *L. cornuta* (LINNAEUS, 1758); Horned Boxfish, Longhorned Cowfish. Indo-Pacific to the Red Sea and S Africa. Up to 50 cm. In older fish, the C is very long. Long spines above the eyes and on the ventral surface. Sandy colouring with pale blue flecking (fig. 685).

Fig. 685 *Lactoria cornuta*

Lacustrine. Term used to describe organisms that prefer still, peaceful waters, ie the typical inhabitants of ponds* and pools*. They are not very sensitive towards fluctuations in oxygen content and temperature.

Ladyfishes (Albulidae, F) *see* Elopiformes

Lagarosiphon HARVEY, 1841; G of the Hydrocharitaceae* comprising 15 species. Found in Africa and in Madagascar. Small perennials with long stem axes and a great many branches. In contrast to *Elodea** and *Egeria**, the leaves are alternate, although sometimes almost whorled. They are also linear, pointed, and with a single vein. Flowers hermaphroditic and dioecious, very small. They develop in a spathe located at the leaf axil. Double perianth, calyx and corolla in threes, inconspicuous. Male flower: 3 stamens, with 3 long, thread-like, coloured staminodes, many in a spathe. Female flower: has a long calyx tube and 3 carpels that are fused and inferior. The fruit is a capsule. *Lagarosiphon* species are all aquatic plants that cannot live in an emersed state. They have an interesting flower biology, similar to that found in *Vallisneria**. The calyx tubes of the female flowers stretch, enabling the flowers to reach up to the atmosphere; at the same time the male flowers work free, climb up, and swim on the surface by means of their sepals. Nearly all the species are undemanding aquarium plants that can either be planted or left to swim around freely. The following species are kept in aquaria:

– *L. madagascariensis* CASPARY. One of the smaller species. Leaves only slightly curving. Temperature 20–30 °C (fig. 686).

Fig. 686 *Lagarosiphon madagascariensis*

– *L. major* (RIDLEY) MOSS. A strong plant with recurving leaves. Temperature not more than 25 °C. Grows well in summer in the open air.

Lagena *see* Balancing Organ

Lagenandra, G, *see* Araceae

Lagoon *see* Coral Reefs

Lake Baikal, USSR. Oldest, and with depths of almost 2,000 m the deepest, freshwater lake on earth. Extremely clear water and in contrast to comparable waters (eg Lake Tanganyika*) there are life forms at great depths (Abyss*). As a consequence of the fact that this lake has long been isolated (since the early Tertiary), it provides a home for an extraordinarily large number of endemic species*. Eg, of the 240 freshwater shrimps that live in Lake Baikal, 239 are exclusively found there; of the 50 species of fish, 23 are exclusively found there; Because of its special features of age, structure, fauna and flora, the lake is regarded as one of the most interesting natural areas of Earth, and therefore warrants a great deal of protection.

Lake Limpets, G, *see Acroloxus*

Lake Limpets, G, *see Ancylus*

Lake Malawi. Earlier known as Lake Nyasa. Located in the East African Rift Valley, surrounded by steep mountains, it is 550 km long and up to 50 km wide. Its deepest point measures 786 m. The whole deep zone of the lake is azoic, ie has no animal life. This is because there is a build-up of hydrogen sulphide. As far as the water chemistry is concerned, Lake Malawi is similar to Lake Tanganyika*: the water has an alkali reaction because of its rich content of hydrocarbons and its electrolyte* content is high for an inland stretch of water (Inland Waters*). Lake Malawi, along with Lake Tanganyika and Lake Baikal*, is one of the oldest lakes* on Earth. It has been isolated since the Tertiary period which is why it has a highly unusual fauna. The systematic differentiation in comparison to Lake Tanganyika may be less advanced, but 50 % of the species of animal found in Lake Malawi are *only* found there (*see* Endemic Species). As far as the fish fauna is concerned, Lake Malawi contains a lot of Cichlids (Cichlidae*). Apart from the few *Tilapia** species, the 175 or so endemic Cichlid species all stem from the G *Haplochromis**. In the course of time, some have developed into their own Ga (in total more than 20 Ga) (eg

*Pseudotropheus**, *Labeotropheus**); *Haplochromis* itself is represented by more than 100 species in Lake Malawi. It is also interesting to note that in each case several species are very closely related to one another (so-called species flocks), a fact that confirms the species development in Lake Malawi.

Lake Nyasa *see* Lake Malawi

Lake Tanganyika. This is a biologically very interesting lake situated in the south-east of the central African rift valley. Its coastal length is approaching 2,100 km and its maximum depth is 1,470 m. Like Lake Baikal*, Lake Tanganyika (and many other East African lakes like Lake Malawi* and Lake Victoria), has a highly unique fauna, caused by the fact that the lake has been isolated from other stretches of water for some time. Especially characteristic is the large number of endemic* species. For example, of the 44 Mollusc Ga found in Lake Tanganyika (Mollusca*), 25 are *only* found there; 36 Cichlid Ga are endemic in Lake Tanganyika (Cichlidae*), among them *Julidochromis** and *Tropheus**. Also very interesting is the occurrence of groups of extremely closely related species (so-called 'species flocks'); this is a demonstration that a species has broken up and developed further.

The water in Lake Tanganyika is surprisingly rich in potassium and magnesium bicarbonate, the pH value* is 8.5 and the water temperature around 27.5 °C. Because of the salt concentration, only certain, salt-tolerant animal groups can exist in the lake; for example, the fish fauna consists essentially of Cichlids (Cichlidae*, approx. 140 species), Glass Barbs (Centropomidae*, 5 species) and Herrings (*see* Clupeiformes, 2 species). On the other hand, fish groups that are typical of the rest of Africa (Catfishes, Characins and Barbs) are hardly found at all in Lake Tanganyika.

Life in Lake Tanganyika is restricted to the top water levels; at depths below 125 m the accumulation of poisonous hydrogen sulphide is just too strong.

Lake Trout *(Salmo trutta lacustris) see* Salmonoidei

Lake Trout *(Salvelinus namaycush) see* Salmonoidei

Lake Whitefishes (Coregonidae, F, name not in current usage, *see* Salmonoidei

Lakes. An inland body of water (Inland Waters*) of middling depths but with no flow of water in any particular direction. Along with the bank zone (Littoral*), a lake must have a bottom zone (Profundal*) where there is no light and no plants; on the other hand, the size of the water surface plays no role. The open body of water (Pelagial*) of a lake is also known as limnion—this can have a very characteristic stratification of temperature (fig. 687) or it can circulate fully. In temperate latitudes full circulation is restricted to spring and autumn, whereas in the tropics it can occur simply as a result of tremendous temperature differences between day and night. Depending on the content of nutrient and humus particles, lakes are described as oligotrophic*, dystrophic* and eutrophic* waters.

Lamarck, Jean (1744–1829). French natural historian. He studied both medicine and natural sciences. His botanical studies led him to suggest an analytical method for the classification of plants (1778 *Flore française*). Later on, he worked on zoological subjects, and in 1792 became Professor of Natural History at the Jardin des Plantes. Lamarck presented the first coherent theory on the evolution of species. According to him, species came about through the active adaptation of organisms to their environment (Lamarckism). His most important works are: *Histoire naturelle des animaux sans vertèbres* ('Natural History of Invertebrate Animals') (1815–1822), and *Philosophie zoologique* ('Zoological Philosophy') (1809).

Laminaria. G of giant brown algae (Phaeophyta*) found in cold seas, the southernmost boundary being areas with a summer temperature of 20 °C-isotherms. Cannot be considered for the aquarium.

Lamna nasus *see* Lamnidae

Lamnidae; Mackerel Sharks. F of the Chondrichthyes*, O Lamniformes. Comprises predatory Sharks found in most seas near the surface. They are extremely good, fast swimmers, and include the most dangerous Sharks of all—the 'Man-eaters'. The Lamnidae have a strong, torpedo-shaped body and a half-moon-shaped C that is almost symmetrical (ie the upper tail-fin lobe is scarcely enlarged). There are folds of skin along the sides of the tail which act as stabilisers, along with the very large, crescent-shaped Ps. The D1 is considerably larger than the D2 which lies opposite the An. Blowholes either very small or absent altogether; 5 large gill slits. The large mouth contains several rows of very strong, triple-pointed teeth, the sharp edges of which can be used in a sawing action. All species of Lamnidae are livebearing. They feed on fish, including other sharks, turtles and marine mammals. Usually, the prey is swallowed whole—inside their stomachs, scientists have found seals that have been swallowed whole and fish measuring between 1 and 2 metres. Mackerel Sharks often follow in the wake of ships devouring anything edible that happens to fall in their path. All the species have a similar colouring, with a characteristic

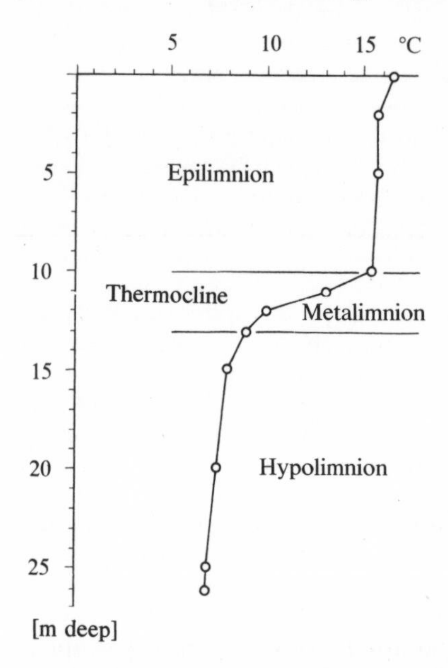

Fig. 687 Temperature layers of a lake in summer

black-blue or lead-grey upper surface, which gradually gives way to a pure white or yellowish ventral surface. An especially typical member of the F is *Carcharodon carcharias* (the White Shark or Man-eater) (fig. 688), which is on average 6–7 m long, but may grow as long as 10 m or more. It is found in all warmer seas, although it may also wander into cooler areas. Mako Sharks, which also belong to the same F, are a little smaller and are usually found in a group. *Lamna nasus* (the Porbeagle or Mackerel Shark) (fig. 689) is found in the Mediterranean and temperate regions of the N Atlantic; *Isurus oxyrinchus* (Shortfin Mako or Sharp-nosed Mackerel Shark), on the other hand, is distributed worldwide in tropical and subtropical regions. They will also attack boats and people, but fortunately the large adult sharks will not normally venture in the coastal waters. Commercially, the Lamnidae are of little importance, although many species are a favourite with sporting fishermen. Although there are innumerable horror stories about the sharks belonging to this F, it should be mentioned that three times as many people are killed by lightning every year than by sharks.

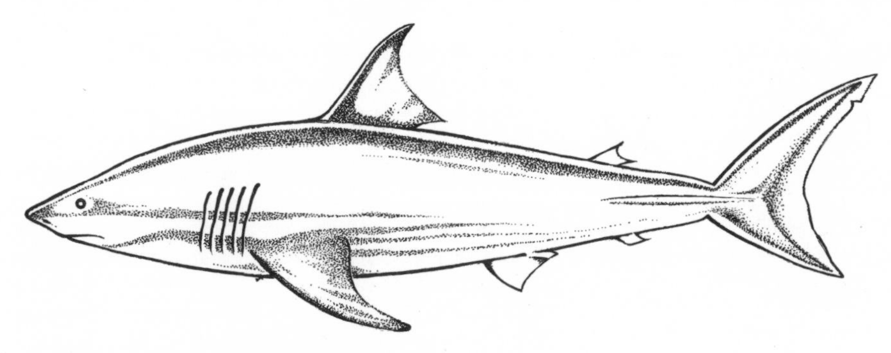

Fig. 688 *Carcharodon carcharias*

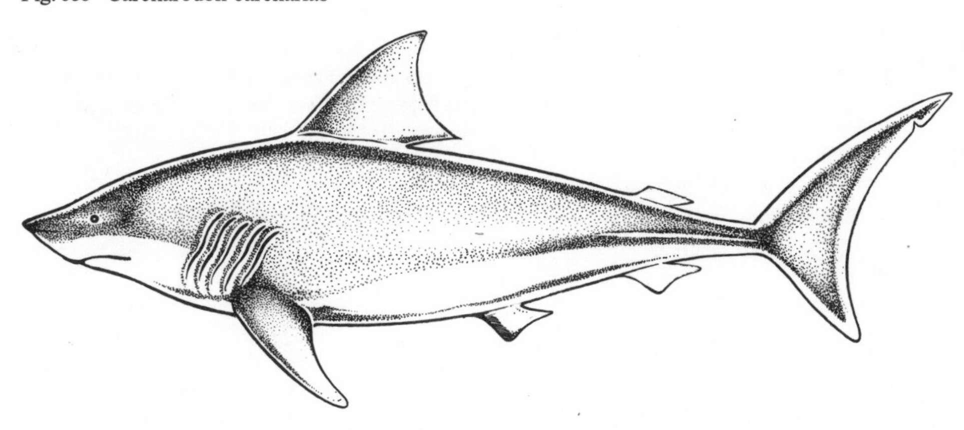

Fig. 689 *Lamna nasus*

Lampern *(Lampetra fluviatilis) see* Cyclostomata

Lamp-eyed Panchax *(Aplocheilichthys macrophthalmus) see Aplocheilichthys*

Lampreta, G, *see* Cyclostomata

Lampreys (Petromyzoniformes, F) *see* Cyclostomata

Lampridae, F, *see* Lampridiformes

Lampridiformes. O of the Osteichthyes*, Coh Teleostei. The O comprises fish of great diversity of appearance. In the REGAN-system the Lampridiformes are known as the Allotriognathi. There are skull and shoulder girdle features that all the fish belonging to the order have in common. The upper side of the mouth is made up of the premaxillary bone and the maxilla. A unique pushing mechanism of the lower jaw makes the mouth protrusible. Teeth are either weak or absent altogether. The fins contain only soft rays. The D is very long, and the Vs are either placed just behind the Ps or are absent altogether. Physoclists. Small, smooth scales or scaleless. The Lampridiformes are probably related to the Beryciformes*. The oldest fossils date from the middle Tertiary period. The Lampridiformes are found in all oceans, usually at fairly great depths.

– *Lampridae*; Opahs or Moonfishes. The single representative of the F, *Lampris regius* (fig. 690), is found

Fig. 690 *Lampris regius*

mainly in warm seas. Viewed from the side, it is oval and very flat from side to side, with a long D, that has a tall pennant shape towards the front. An long and narrow. The Opah can measure 1.80 m long and can weigh between 150 and 200 kilograms. They are rare fish, but are beautifully coloured. The upper surface is steel-blue or green, flanks with a purply or golden sheen, lower surface pink. The jaws and fins are scarlet red. There are also large numbers of silver dots over the whole body. The mouth is small and toothless, so the Opah can only prey on small Squids and crustaceans. Its red meat is considered a great delicacy. Unfortunately, the Opah is rare.

– *Lophotidae*; Crest-fishes (fig. 691). Elongated, ribbon-like, scaleless fish from cold and warm seas. Forehead profile almost vertical or even jutting out in front of the mouth. The very long, first fin ray of the D is located on the crown of the head, and the fin extends right back to the small C. An tiny, Vs absent. *Lophotes cepedianus* can grow up to 2 m long.

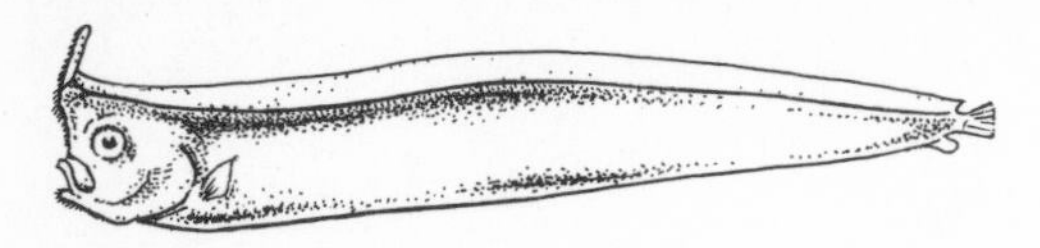

Fig. 691 Lophotidae-type

– *Regalescidae*; Oar-fishes, Ribbon-fishes (fig. 692). Monster Oar-fishes can be 10 m long and weigh 250 kg. They have no scales, although they do have 4–6 hard, length-wise ridges and a large number of wart-like tubercles. The D begins at the nape with a Kiwi-style, erectile crest and extends right back to the tip of the tail. No C. The Vs are located beneath the small Ps and consist only of a very long fin ray. Ribbon-fishes swim in a snake-like fashion at the surface or at middling depths. The delightful horror stories about 'sea snakes' that we have all read about are based on these fish.

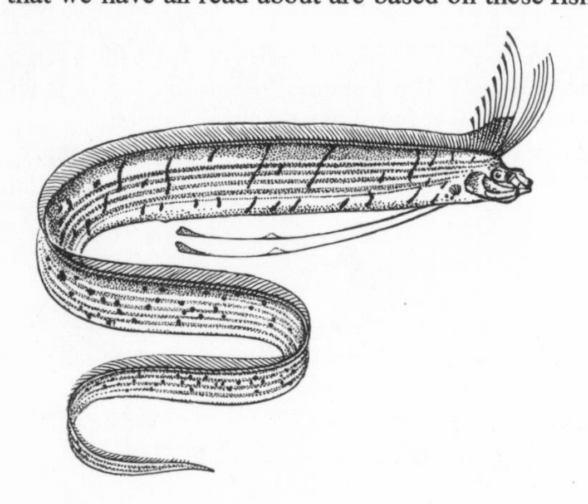

Fig. 692 *Regalecus glesne*

– *Trachipteridae*; Ribbon-fish (fig. 693). There are approximately 10 species in this F, and all are elongated in shape, like a ribbon. At a height of 25–30 cm, several species reach lengths of 2 m or more, but are only 5–6 cm in width. The D begins as a crest at the nape and stretches back to the C, of which only the upper lobe is formed. There is no An. One species, *Trachipterus trachipterus*, is occasionally trapped in fishing nets in the Mediterranean, as well as in the mid-Atlantic and Pacific. The embryos of this species have telescopic eyes – they lack the very long fin rays so characteristic of the larval stage.

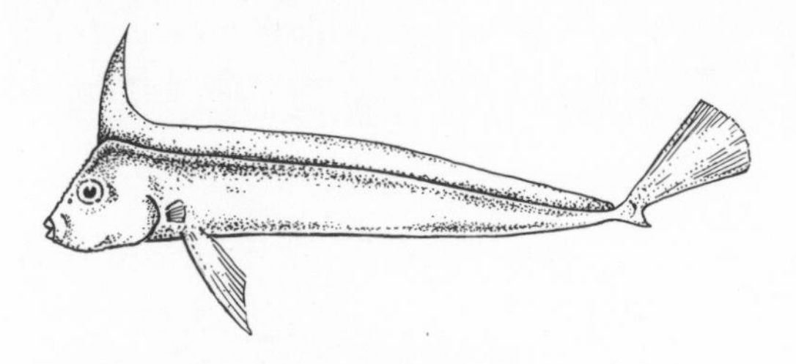

Fig. 693 Trachipteridae-type

Lampris regius *see* Lampridiformes

Lamprocheirodon axelrodi, name not in current usage, *see Cheirodon*

Lamprologus SCHILTHUIS, 1891 (fig. 694). G of the Cichlidae*. Includes more than 35 species found in tropical Africa, of which the majority are endemic* to the east African Lake Tanganyika*. They are elongated fish, to a greater or lesser degree, and compressed laterally. They vary in size from about 4 to 31 cm. In many species, the forehead has a characteristic padding of fat and the dorsal profile is arched more noticeably than

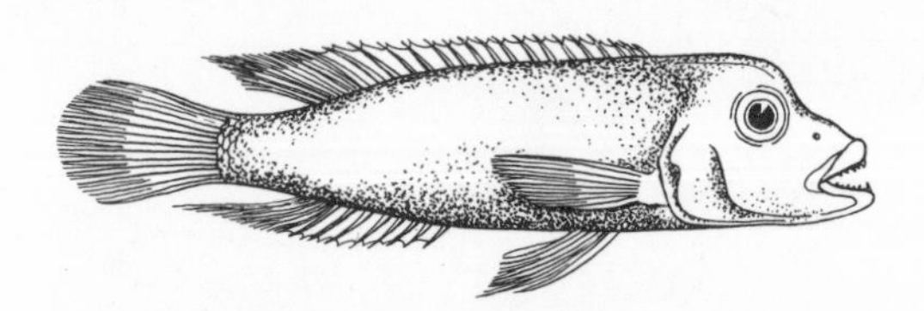

Fig. 694 Representative of the G *Lamprologus*

the ventral profile. A lot of species also have a large, thick-lipped mouth (eg *L. compressiceps, L. congicus, L. congolensis*) which lends them the appearance of a true predator. Fins, in part, very well developed: D has 14–21 hard rays, An 6–7 hard rays, the soft rayed parts being to some extent drawn out into long, pennant-like points (eg *L. brichardi*). In some cases, the D is very tall *(L. compressiceps).* The colours of most species are remarkably inconspicuous (exceptions *L. brichardi, L. leleupi*). *Lamprologus* species can be distinguished from closely related Ga such as *Julidochromis** chiefly by the type of pharyngeal bony teeth present and the way in which they are used. There are also other cranial differences. In other words, without expert ichthyological knowledge, it is difficult to tell them apart.

Lamprologus has a very interesting biology. Without exception, the species appear to be territorial. It is the males that form territories, in which they will not tolerate any like-species of their own sex. Transferring this behaviour to the aquarium, it normally means that a male's territory will coincide with the entire tank, so that it is only possible to keep one pair together at a time.

However, with a few 'artificial ploys', such as the provision of enough hiding-places (which act as spawning grounds) and enough rock structures (positioned in such a way that they shield the fish's view), it is possible to keep several pairs in the same tank. The reproductive behaviour of *L. brichardi* is particularly well known: both partners clean the spawning area, usually the darkest corner of a hole in the rock. Eggs are laid and insemination takes place at the same time. The female then devotes herself to the care of the spawn, and the male is chased away, or occasionally allowed to defend the territory. Some species dig holes for themselves under stones (eg *L. werneri*). None of the species contains any mouthbrooders. If you are keeping *Lamprologus* in the aquarium, it should be simple to care for them and rear them, provided you give due consideration to their territorial behaviour patterns. Specimens from Lake Tanganyika should be kept in very hard, slightly alkali water, at a temperature of around 27 °C.
– *L. brichardi* POLL, 1974; Lyretail Lamprologus. Also known by the name *L. savoryi elongatus* TREWAVAS and POLL, 1952. From Lake Tanganyika. Up to 10 cm. The soft-rayed parts of the fins are very elongate, edges white, body cream colours, edges of scales orange. A dark band runs through the eyes, and there is also a dark blotch on the operculum with a pale edge (fig. 695).

Fig. 695 *Lamprologus brichardi*

– *L. compressiceps* BOULENGER, 1898; Compressed Cichlid. From Lake Tanganyika. Up to 13 cm. Body relatively tall and flattened laterally. The D is well developed, dorsal surface hump-backed. Colouring dark olive-brown, Ps orange-red.
– *L. congolensis* SCHILTHUIS, 1891. From the Congo area. Up to 10 cm. Head massive, male with a fatty padding on the forehead. A very elongate species. Colouring grey, deep black at spawning time.
– *L. leleupi* POLL, 1948; Lemon Cichlid. From Lake Tanganyika. Up to 10 cm. An elegant and beautifully coloured species, yellow to red-golden. At spawning time, the iris is an intense blue.
– *L. werneri* POLL, 1959. From the Congo area. Male up to 12 cm, female much smaller. Ground colour yellowish-grey to greenish-grey, with shining dots.

Lancelet *(Branchiostoma lanceolatum) see Acrania*

Lancet Fishes (Alepisauridae, F) *see* Myctophoidei

Land Hermit Crabs, G, *see Coenobita*

Lanice, F, *see* Polychaeta

Lantern Eyes (Anomalopteridae, F) *see* Beryciformes

Lantern-fishes (Iniomi, name no longer in current usage) *see Myctophoidei*

Lantern-fishes (Myctophidae, F) *see* Myctophoidei

Lantern-fishes, related species, Sub-O, *see* Myctophoidei

Laomedea LAMOUROUX, 1812. G of the Hydrozoa* comprising numerous species found in the sea and deep into brackish water, especially on plants. The millimetre-sized individual polyps of the branch-like colonies sit in small, toothed cups (hydrotheca). Some species form free-swimming medusae (eg *Obelia*); in others, the sexual forms stay linked to the colony, and are known as gonophores. *L. flexuosa* ALDER is found well into the eastern Baltic. *Laomedea* species have to be kept in special aquaria, but offer the hobbyist very interesting observations. Feed with fine rotifers and single-celled creatures.

Largemouth Bass *(Micropterus salmoides) see Micropterus*

Large-scaled Rasbora *(Rasbora paucisquamis) see Rasbora*

Large-scaled Scorpionfish *(Scorpaena scrofa) see Scorpaena*

Larva. A juvenile stage found in animals. It can live an independent existence, but its body structure and life-style differs markedly from that of the adult animal. The larval stage is brought to an end with a metamorphic process (Metamorphosis*). A larva often has special larval organs for locomotion or attachment, for breathing and feeding itself.

Larval forms are very common among the invertebrates. Among the vertebrates, only amphibians and some fish exhibit true larvae with metamorphosis. In many cases, the larva is so different from the adult that it has been described as a separate species. This was true, for example, for the larva of the Eel – the glassy, transparent *Leptocephalus* with its long teeth, and anus located well towards the caudal region. Lamprey larvae *(Ammocoetes)* live in the substrate, filtering out their food. They do not have the sucking disc found in the adults, and their eyes lie beneath the skin at this stage. Flatfish larvae are symmetrical to begin with and live in the surface waters. The larvae of the Sun-fishes have a normal tail and are provided with large spines. Bichirs and Lungfishes have larvae with feathery external gills and larval organs of attachment.

In a wider sense, all young fish that retain a yolk sac* and a more or less continuous fin corona after hatching out are often called larvae.

Lateral. From the Latin word *latus*, meaning side. A term used in anatomy* to indicate the position of parts of the body.

Lateral Line System; Lateral Organ (fig. 696). An organ found in fish and many amphibians that can sense current. It enables objects to be detected too, without the need for seeing them with the eyes or touching them directly with touch organs. Sometimes called a long-distance sense of touch, the lateral line detects the slightest alterations to the conditions of the current in the surrounding water – perhaps reflective waves from firm objects or vibrations set up by the movements of other animals. There are sensitive sense cells in the lateral line that are arranged in groups as a line along the body. Often, there are several branching lines in the head which usually join up on both sides with a line

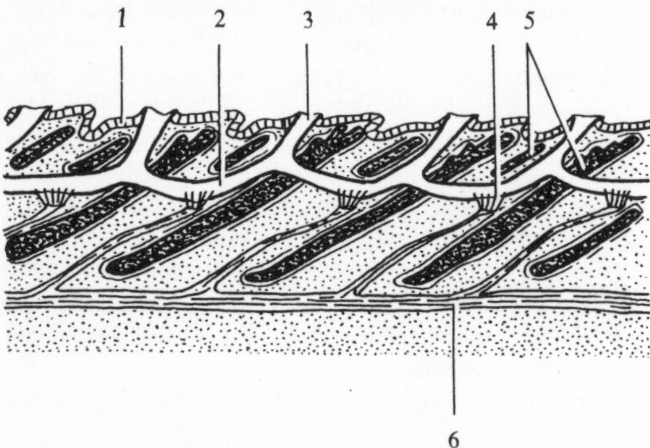

Fig. 696 Lateral line system of a bony fish. 1 Epidermis 2 Lateral line canal with sense cells (4) and pores (3) 5 Lateral line scale 6 Lateral line nerve

running along the flanks. This line along the body sides may be short or it may reach to the tail region. In many bottom-dwelling fish and in those that lead a peaceful, calm life-style, the sense cells lie in depressions in the skin. With fast swimmers and river fish, the cells will be found at intervals in the wall of a connected canal system of the skin that comes to the body surface at numerous pore-points. These longitudinal canals are filled with mucus and run their course through the scales that stick into the skin. Lateral line scales are therefore broken up and usually bend outwards.

The sense cells of the lateral line system are covered with small rod-like projections that become deformed in the canals by the movements of a mass that covers them (cupula). This stimulates the little rods and the message is picked up by the sense cells and directed to the brain via the lateral line nerves. The formation of the lateral line system is important in systematics. The lateral lines on the head (fig. 697) are also a valuable identification feature when trying to order fossil forms, as they can easily be picked out as furrows on the skull bones.

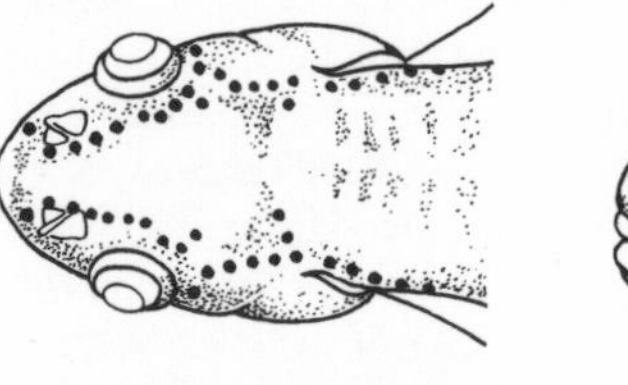
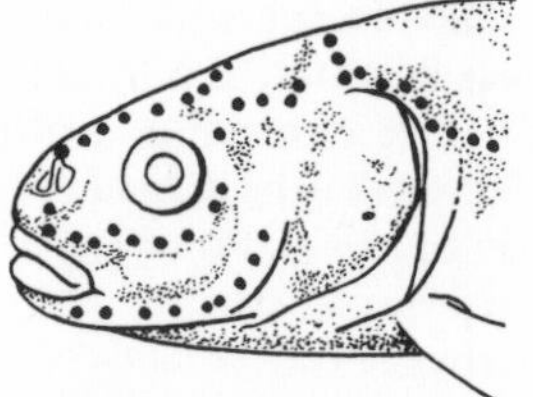

Fig. 697 Lateral line organs in the head of the Minnow

Lateral Organ *see* Lateral Line System
Lates niloticus *see* Centropomidae
Latimeria chalumnae *see* Coelacanthiformes
Lattice Dwarf Cichlid *(Nannacara taenia) see Nannacara*
Lead Poisoning *see* Poisoning
Leaf. The leaf, along with the stem axis* and the root*, forms part of the cormus* of the pteridophytes

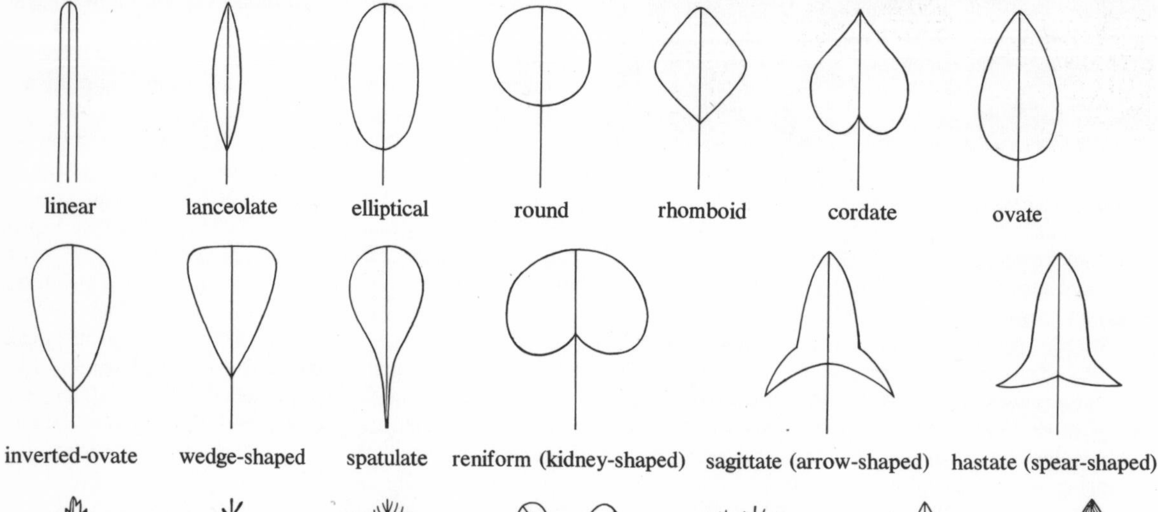

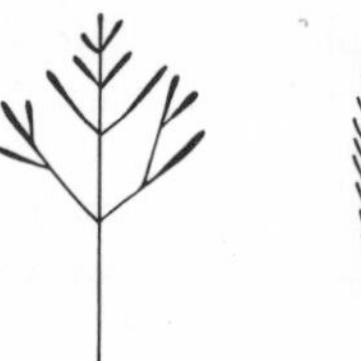

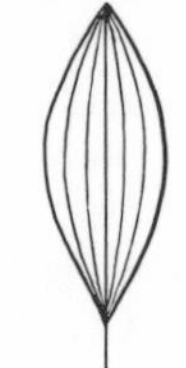

Fig. 698 Leaf shapes

(Pteridophyta*) and the seed plants (Spermatophyta*). Its position is as an axillary organ to the stem axis, but it lacks the root. Normally the leaf is flat and its growth is well-defined. A leaf is developed in the atmosphere or in water, but it can occur underground. It is the chief organ of photosynthesis* and transpiration*, but can also fulfil other functions such as storage of reserve substances, storage of water or the formation of reproductive cells. Leaf and stem axis emerge from a common vegetative cone*.

The typical leaf is also described as foliage leaf. There is a great variety of leaf shapes (fig. 698). Each leaf can consist of 3 main parts, the leaf blade, the leaf stalk or petiole, and the leaf base. The *leaf blade* is the true leaf surface. It may be undivided and can have different outlines, such as lineal, lanceolate, ovate, spear-shaped. Also significant in the description of leaves are the shape of the tip of the blades, the base of the blades, the edge of the blades as well as the state of the leaf surface itself. The edge of the blades of submerged leaves is usually complete. The surface is flat, wavy or bulbous, ie the upper or lower surface is thick with protrusions. Veins containing the vascular bundles* traverse the leaf surface. In the monocotyledons (*see* Liliatae) well-defined, longitudinal parallel veins are often present alongside the mid-rib. In the dicotyledons (*see* Magnoliatae) axillary veins branch off the mid-rib. Leaves with lobed and feathery (ie not indented to the middle) blades are on the way to becoming true pinnate leaves, whose blades are made up of separate leaflets or pinnae. The pinnae are either arranged along an axis-like spindle or they are arranged in a hand-like shape. If the pinnae are themselves further divided, the leaf is then described as multiply pinnate. A special type of foliage leaf is the *paltate leaf*, in which the leaf stalk is placed near the middle of the underside of the leaf (eg as in many species of the G *Hydrocotyle**). The *leaf stalk* can be of varied length. When it is absent, we speak of a sedentary leaf. The *leaf base* is no longer visible but is represented by a slight broadening out of the stalk base, through which the leaf is attached to the stem axis. Occasionally, 2 small outgrowths (the stipules) protrude from it. In some marsh and aquatic plants, especially from the Sub-Cl of the monocotyledons, a *leaf sheath* of varied length is formed in place of the leaf base. The leaf sheath lies very close to an elongated stem axis and in more compressed stem axes it envelops the developing young leaves. Thus, in the area of the tip of the stem there is good protection for the vegetative cone. The leaves of flowers* and those of the typical foliage leaf differ through reduction of cataphyllary and hypsophyllary leaves as well as through different modifications of the leaves.

The internal structure of the leaf is closely related to the plant's habitat (fig. 699). In cross-section, the leaf of a plant growing in a habitat with a balanced water economy reveals the palisade layer below the upper epidermis. This palisade layer contains densely packed cells that stretch vertically up to the upper surface of the leaf blade. The lower half of the blade is taken up with spongy parenchyma, a loose tissue with much hollow space. Lastly, the lower epidermis with the stomata* forms the bottom surface. In marsh plants, and particularly in aquatic plants, the hollow tissue system inside the leaf is enlarged. In submerged plants, the stomata are usually absent. Often, their leaves have very narrow blades. In an extreme case, only the two epidermal layers remain, eg as in the G *Elodea**. *See also* Leaf Succession, Leaf Arrangement, Anisophylly, Heterophylly.

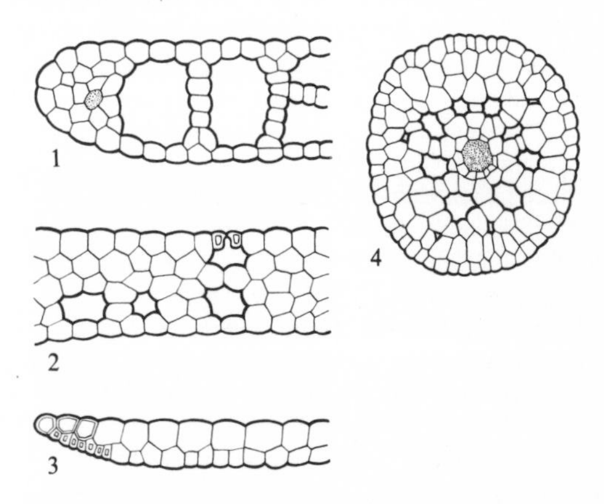

Fig. 699 Leaf anatomy. 1 *Sagittaria subulata* 2 *Ludwigia repens* 3 *Elodea canadensis* 4 *Myriophyllum spicatum*

Leaf Arrangement (fig. 700). This refers to the way in which the leaves are organised on the stem axis. In *alternate* leaf arrangement there is a leaf on every node. Each consecutive leaf is placed at a constant angle. In *distichous* leaf arrangement too, there is only one leaf per node. The angle between 2 consecutive leaves in this case is 180°. A *decussate* leaf arrangement is characterised by 2 leaves on each node, which are situated opposite each other. The median planes of the leaves of 2 consecutive nodes are displaced by 90°. In a *verticillate* or *whorled* leaf arrangement there are 3 (up to many) leaves per node. In a *rosette* leaf arrangement there is usually one leaf per node, but as a result of a much compressed stem axis, the leaves are extremely close together. Rosettes are primarily *basilar*, ie their leaves emerge right above the ground.

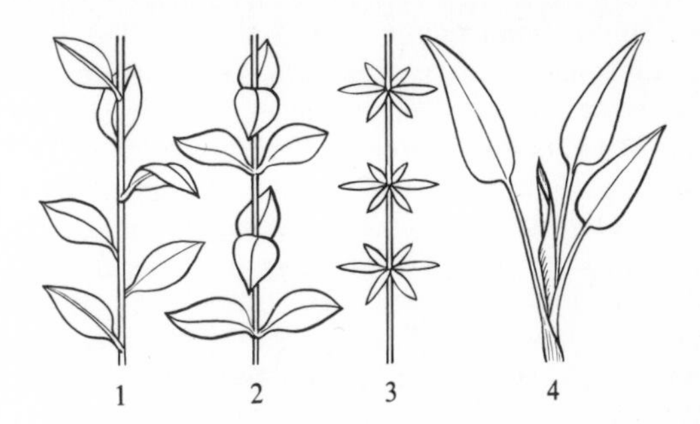

Fig. 700 Leaf arrangements. 1 alternate 2 decussate 3 whorled 4 basilar

Leaf Axil. The corner between the stem axis and the insertion of the leaf. Buds form in the leaf axils and out of them axillary buds, inflorescences or flowers may grow.

Leaf Succession is the sequence of differently formed leaves on the stem axis. The first leaves of a plant are the *seed leaves* (cotyledons*) which are already contained in the seeds*. Next come the *foliage leaves*, ie the leaves that are typical of the plant (*see* Leaf). The lower foliage leaves can have a simpler form, in which case they are *primary leaves* as distinct from *nature leaves*. Smaller, simplified or modified leaves above the foliage leaves of a shoot are known as *hypsophyllary leaves*. Examples of this are many flower bracts* or the spathe of the Araceae*. *Prophyllary leaves** and the leaves on underground stems can also be simplified. They are called *cataphyllary leaves* and they are located below the foliage leaf zone of a stem. When very different leaves occur within the foliage leaf zone, it may be a case of anisophylly* or heterophylly*.

Leaf-fish, South American *(Monocirrhus polycanthus) see Monocirrhus*

Leafy Liverworts (Hepaticae, Cl) *see* Bryophyta

Leander, G, *see Palaemon*

Leathery Corals, O, *see* Alcyonaria

Lebiasinidae (fig. 701). F of the Cypriniformes*, Sub-O Characoidei*. Fairly large, small, or tiny S American Characoidei of an elongated or slender, spindle-shaped build. Flanks often only slightly flattened, belly rounded. In many species, the head is pointed and has a small, narrow mouth, which in some species, eg *Copeina*, is upward-pointing. The arrangement of the teeth is not always the same. There is only ever one row

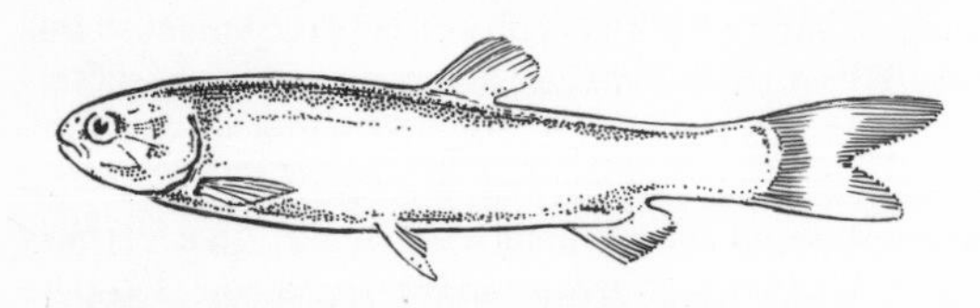

Fig. 701 Lebiasinidae-type

of teeth on the premaxilla, consisting of conical, cutting teeth with many tips. The maxillary bone is long or very small, and usually lacking in teeth. The dentary bone has two rows of teeth (eg *Lebiasina, Pyrrhulina, Copeina*) or one row *(Nannostomus, Poecilobrycon)*. The D is short, in some Ga extending into a point *(Copeina, Pyrrhulina, Copella)*. Adipose fin may be present or absent. The C is deeply forked, the lobes often of dissimilar size. The An is fairly short, the males often having specialised fin rays. LL incomplete or absent. Cycloid scales, which are very large in *Lebiasina* and which extend into the part of the C which lies next to the body. Sexual dimorphism exists – in the males of some Ga (eg *Copeina, Copella*), the fins are elongated, in other Ga the dimorphism is slight and only apparent through the intensity of colouring. Members of the G *Lebiasina* can be up to 20 cm long; they are predators that are also distinguished by having an accessory respiratory capability with the aid of their swimbladder. The species of the other Ga remain fairly small and are usually prettily coloured and patterned. In their natural habitats, the Lebiasinidae are found mainly in calm, weedy waters that lie half in the shade. They live in loosely knit groups, but they are not shoaling fish. Because they have only a small mouth, they can only eat the smallest of food animals. Some species of the Ga *Copeina, Copella* and *Pyrrhulina* have developed primitive forms of brood care. One species, the Spraying Characin *(Copella arnoldi)*, has developed a most unusual form of brood care that is unique in the animal kingdom. It places its eggs on the undersides of leaves and branches which hang directly above the water surface. The mating fish must spring out of the water together to be able to spawn in this way, the male spraying the eggs that have been laid (G *Copella**). Nearly all the Ga belonging to this F are very popular in the aquarium. H. J. FRANKE is particularly well-known for his comparative studies of behaviour between the Ga *Nannostomus* and *Poecilobrycon*. The following Ga are described under separate entries: *Copeina, Copella, Nannostomus, Pyrrhulina*.

Lebiasininae, Sub-F, *see* Characoidei

Lebistes, G, name not in current usage, *see Poecilia*

Leconte's Loach *(Botia lecontei) see Botia*

Leeches, Cl, *see* Hirudinea

Left-eye Flounders (Bothidae, F) *see* Pleuronectiformes

Leiocassis BLEEKER, 1858. G of the Bagridae*. Spiny Catfishes from the Asiatic part of the USSR, Thailand, Kampuchea (Cambodia), Java and Kalimantan. Length to 25 cm. These fish are active at twilight or at night, and rest in hiding-places during the day. The body is elongated into a torpedo shape, slightly flattened laterally, with a snub head. There are two pairs of barbels on both the upper and lower jaws. D1 short, rounded towards the outside edge, D2 very long. C deeply forked with long tips, An rounded and short. Vs and Ps well developed. The latter, like the D1, has a spiny first fin ray. According to KIRCHSHOFER, the *Leiocassis* species exhibit a marked territorial behaviour. They can be kept in large aquaria with plenty of vegetation and hiding-places. Subdued lighting. Some kinds of *Leiocassis* often rest on their backs *(L. siamensis)*, others emit single tones *(L. brashnikowi)* or series of tones *(L. siamensis)*. They are all omnivores that will accustom themselves both to lower and higher temperatures. *L. brashnikowi* lays its spawn between water plants and roots. The male tends the spawn and the young fish (which hatch after 3 days and swim freely after 7). MACHLIN reports that the males are larger and have a longer caudal peduncle. They are attractive and colourful Catfishes that must only be kept with others of their species or with other larger fish.

– *L. brashnikowi* BERG, 1907; Russian Catfish. From the lower and middle reaches of the Amur, Ussuri, Sungari and Lake Chauka. Length to 22 cm. Dorsal surface grey or brown, flanks and fins pale ochre colours, ventral surface white. The head is blue-grey as far as the insertion of the D1. Contrasting violet-black longitudinal and transverse bands run across the main body colouring.

– *L. siamensis* REGAN, 1913; Bumble-bee Catfish. Ground colour black-violet, dark coffee brown or blue-grey with a yellowy ventral surface. There are 4 irregular, broad transverse bands, pale beige or white in colour, one behind the operculum, another behind the

D1, a third at the same level as the D2 and the fourth at the base of the C. The unpaired fins are delicate yellow with brown blotches, the Vs and Ps being a uniform brown. An attractive and hardy species for the aquarium (fig. 702).

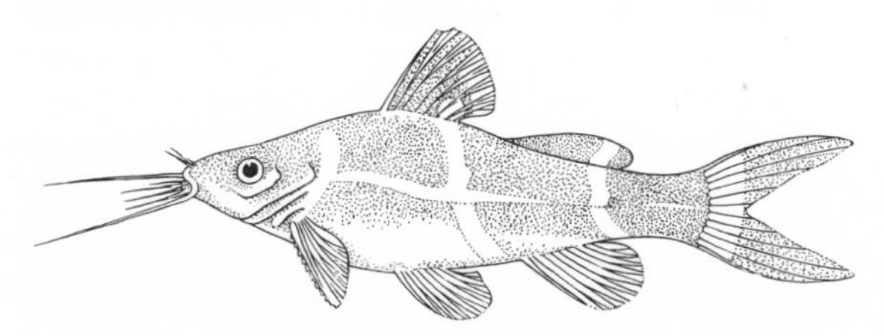

Fig. 702 *Leiocassis siamensis*

Leiognathidae; Slimies, Slipmouths or Soapies (fig. 703). F of the Perciformes*, Sub-O Percoidei*. Smallish fish found in the tropical Indo-Pacific. Whenever they are caught, they are very slimy to the touch. The body is fairly tall, compressed laterally, mouth protrusible like a tube. The long D and An have a lengthwise swelling at their bases, beset with spines. C indented. The scales are small, embedded deep in the skin, head lacking in scales. The Leiognathidae are found mainly in shallow coastal waters and estuaries where they occasionally occur in very large numbers. Many species have a luminescent organ in the oesophagus.

Fig. 703 Leiognathidae-type

Lemna LINNÉ, 1753; Duckweed. G of the Lemnaceae* comprising 10 species. Found almost everywhere in the world. Duckweeds are small perennial plants with very much simplified vegetative bodies. Daughter plants develop from the two lateral pockets of the individual members of the plant, each member plant having only one root. The flowers are unisexual and monoecious, very small. There are 2 male flowers and 1 female flower in a spathe. No perianth. Male flower: 1 stamen. Female flower: 1 carpel, superior. Fruit like a berry.

The various species of the G are found in still, nutrient-rich waters, where they either float on the surface forming a thick covering or grow submerged. Many species will therefore become a tiresome weed if sufficient quantities are not removed continually. This can happen in aquaria and in open-air ponds, as well as in aquatic plant nurseries and botanical gardens. The following two species will often develop quickly in the aquarium.

– *L. minor* LINNÉ: Lesser Duckweed. Found almost everywhere on earth. The individual members of the plant are ovate, flat, quick to isolate themselves and float on the water surface.

– *L. trisulca* LINNÉ; Ivy-leaved Duckweed, Single-rooted Duckweed. Found almost everywhere on Earth. Individual members of the plant are long-stalked, will clump together for long periods and live submerged.

Lemnaceae; Duckweed Family. F of the Liliatae* comprising 4 Ga and 25 species. Found almost everywhere on earth. Very small plants with much simplified vegetative bodies, so much so that it is difficult to distinguish between the stem axis and the leaves. The individual members of the plant are made up of a complexus of leaf and stem axis. In part, there are no roots. The flowers are unisexual and monoecious, extremely small and simplified. They also appear only rarely. All Lemnaceae are hardy aquatic plants with rapid vegetative reproduction by means of daughter plants isolating themselves from the parent plant. They live floating on the water surface or submerged. Some types that live in temperate zones sink to the bottom of the water to over-winter. The fact that Duckweeds sometimes cover broad expanses of water, and that they sometimes appear in a new site very quickly, can be explained by migrating waterfowl helping to disperse new plants. Duckweeds are the smallest flowering plants. *Lemna**, *Spirodela** and *Wolffia** are well-known Ga.

Lemon Cichlid *(Lamprologus leleupi) see Lamprologus*

Lemon Tetra *(Hyphessobrycon pulchripinnis) see Hyphessobrycon*

Lentibulariaceae; Bladderwort Family. F of the Magnoliatae* comprising 5 Ga and 300 species. Tropical to temperate, with a few from the Arctic. Small perennials with elongate or stumpy stem axes, an alternate or rosette leaf arrangement, usually without roots. The flowers are hermaphroditic, usually monosymmetrical and two-lipped. Most Bladderworts grow in water, in marshes and in other wet and damp sites. All of them are food specialists with apparatus for catching animals, through which they obtain an additional source of nitrogen. The G *Utricularia** is of importance to the aquarist.

Leopard Corydoras *(Corydoras julii) see Corydoras*

Leopard Danio *(Brachydanio frankei) see Brachydanio*

Leopard Eel *(Acanthophthalmus kuhli kuhli) see Acanthophthalmus*

Leopard Lungfish *(Protopterus aethiopicus) see* Ceratodiformes

Leopard Puffer *(Tetraodon schoutendeni) see Tetraodon*

Leopard Shark *(Triakis semifasciata) see* Triakidae

Leopard Sharks, F, *see* Triakidae

Lepidocephalus BLEEKER, 1858. G of the Cobitidae*, From the Near East to India, Sri Lanka (Ceylon), Thailand, Java, Sumatra. Small to medium-sized fish that live in clear, flowing and still waters with a sandy substrate. The body is elongate, cylindrical at the front, flattened at the rear and covered with tiny scales. Head small, partially covered in scales, mouth inferior. The eyes are covered by a transparent skin. The G *Lepidocephalus* is distinguishable from the closely related G *Noemacheilus** HASSELT, 1823, by 4 pairs of barbels. Aquarium tanks should not be too small and they should be placed in a shady spot. Add some thick vegetation and provide hiding-places in the form of roots and rocks. As *Lepidocephalus* likes to burrow into the substrate, fine sand must be provided. The water should be clear and clean, of medium hardness and slightly acidic. A regular addition of fresh water seems to help members of the G a great deal. Temperature about 24 °C. Food consists mainly of worms and insect larvae and is taken from the bottom. Sex differences and reproduction unknown.

– *L. guntea* (HAMILTON–BUCHANAN, 1822). From the Near East to India where it lives in flowing waters. Length up to 20 cm. Flanks yellowish-grey, belly pale. A yellow-brown stripe runs from the head to the root of the tail, flanked above and below by yellow-brown flecks. There is a black blotch on the root of the tail. D and C have rows of dark dots.

– *L. octocirrhus* (VAN HASSELT, 1823). From Indo-China, Java, Sumatra. Length up to 12 cm. Flanks yellowish-white to grey, dorsal surface dark, ventral surface pale. Irregular dark flecks gleam steel-blue, and there is a dark blotch on the caudal peduncle.

– *L. thermalis* (CUVIER and VALENCIENNES, 1846). From India, Sri Lanka (Ceylon). Length up to 8 cm. Flanks grey-green with irregular dark flecks. A paler zone extends from the head to the root of the tail (fig. 704).

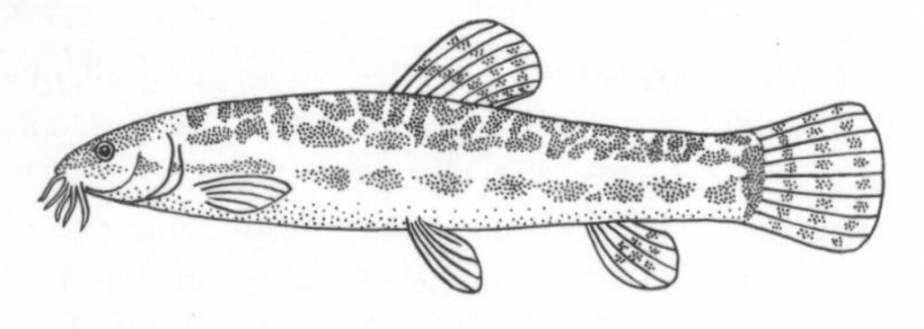

Fig. 704 *Lepidocephalus thermalis*

Lepidosiren paradoxa *see* Ceratodiformes

Lepidosirenidae, F, *see* Ceratodiformes

Lepidurus, G, *see* Triopsidae

Lepisosteidae, F, *see* Lepisosteiformes

Lepisosteiformes; Garpike types. O of the Osteichthyes*, Coh Holostei. There is only one F in this O (Lepisosteidae) which is made up of about 8 species. They are elongate, Pike-like fish with a very long snout and jaws that contain a lot of teeth. The Lepisosteiformes first appeared in the upper Cretaceous period of Europe and N America, ie 100 million years ago. The modern species have still retained some of their primitive features. Included among them is the covering of ganoid scales*, the small teeth on the scales, the angled scales (fulcra) at the front edges of the fins, the numerous bones around the eye-socket and the suggestion of a spiral fold in the intestine. The swimbladder is used as an aid to respiration. The Garpikes are the only group of fish to have vertebrae that are concave behind and convex in front (opisthocoelian vertebrae). The vertebrae are also linked with one another like joints. Garpikes are found in western parts of N America from S Canada to Cuba. They seem to prefer shallow, weedy waters where they lie in wait for their prey in the pike fashion, dashing out to seize any likely fish at lightning speed. Spawn is laid between March and May. The eggs are around 5 mm in size and are very sticky. The young fish attach themselves by means of an organ on the forehead in order to eat up their yolk sac. The largest species, the 3-m-long Caiman Fish or Alligator Gar *(Lepisosteus tristoechus)* (fig. 705), is found in southern parts of the USA, in Cuba and in Central America. It is eaten by people in Mexico. The Common Garpike or Longnose Gar *(Lepisosteus osseus)* (fig. 706) can grow up to 150 cm long. It has the biggest distribution range stretching from the Great Lakes to the Rio Grande del Norte. Body olive-grey to silver-grey and covered in large flecks just like the greenish or orange-coloured fins. Often found as an exhibit in large aquaria. Young Garpikes can also be kept in smaller tanks, but they grow so quickly that they will have to be removed after a short time. Garpikes are not highly prized among the fishermen and anglers of N America.

Fig. 705 *Lepisosteus tristoechus*

Fig. 706 *Lepisosteus osseus*

Lepisosteus, G, *see* Lepisosteiformes

Lepomis RAFINESQUE, 1818. G of the Centrarchidae*. Consists of several species distributed throughout the eastern and central part of N America. Some have also been successfully introduced in central and S Europe. *Lepomis* species are typical fish from clear, still-water zones containing much vegetation. Their body is compressed into an egg-shape, very flat from side to side, and can reach lengths of more than 20 cm. They can be distinguished from similarly built Ga of the Centrarchidae, eg *Pomoxis**, because they have a larger number of hard rays in the dorsal fin (10–12); the difference between *Lepomis* and another similar G *Ambloplites** lies in the smaller number of hard rays in the An (3). In addition, all members of *Lepomis* species have an outgrowth on the operculum made of skin. Normally called the fish's 'ear', it is strikingly coloured and varies in shape depending on species. None of the species has any great economic emportance in their natural habitats, and are only of any interest to anglers and aquarists. They are also suitable for garden ponds, where they can even spend the winter if it is mild enough.

Lepomis is well suited to the unheated room aquarium, especially as they will not then reach their maximum length. They like well planted tanks placed in a sunny spot and filled with clear, hard, neutral or slightly alkali old water. Take care with any sudden change in their environment—*Lepomis* does not happily tolerate fresh water, acidic water or impurities. Specimens will reproduce in a species tank without any special preparations on the aquarist's part. It has been noticed, however, that a cool over-wintering heightens the zest for reproduction. The male makes a nesting hollow and also takes care of the spawn and newly hatched fry.

– *L. auritus* (LINNAEUS, 1758); Redbreast Sunfish. Up to 18 cm. Operculum outgrowth narrow, long and deep black. Body grey-green or yellow-green, ventral surface and Ps a lovely orange colour. Head patterned in a pretty pale blue. Becomes increasingly more brown with age (fig. 707).

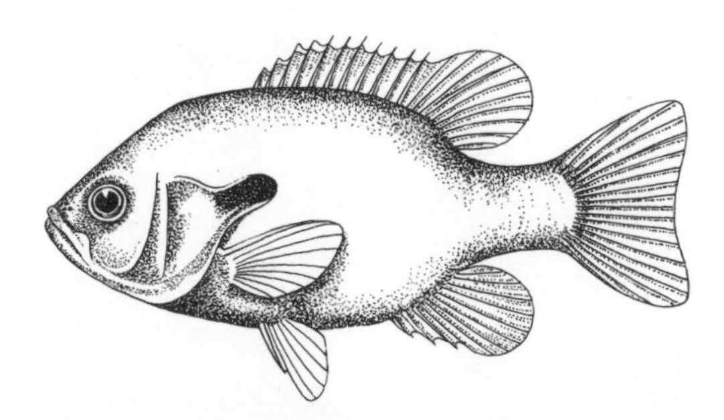

Fig. 707 *Lepomis auritus*

– *L. gibbosus* (LINNAEUS, 1758); Pumpkinseed Sunfish. Up to 24 cm. Young specimens are particularly splendidly coloured. Body brownish-yellow with shimmering, mother-of-pearl, green-blue transverse bands and large numbers of red dots. Operculum a lovely gleaming green. The outgrowth is deep black with an orange-red blotch on the end. Throat and belly an intense orange colour (fig. 708).

– *L. megalotis* (RAFINESQUE, 1820); Long-ear Sunfish. Up to 16 cm. Body very compressed and tall. Operculum outgrowth very large and pointing upwards at an oblique angle. It is blue-black and edged in a golden colour. The flanks are olive with violet-coloured tones, and green, blue or reddish spotting. Iris an intense red. Territorial, therefore aggressive in the aquarium.

Leporellinae, Sub-F, *see* Characoidei

Fig. 708 *Lepomis gibbosus*

Leporinus AGASSIZ, 1829 (fig. 709). G of the Anostomidae*. The G comprises narrow-mouthed Characins from S America, east of the Andes. Length to 40 cm.

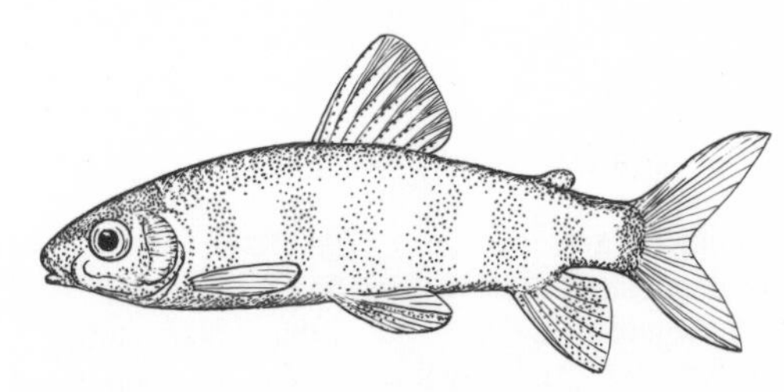

Fig. 709 *Leporinus pellegrini*

Leporinus species are peaceable fish, and very colourful, at least in their youthful phase. Body torpedo-shaped and only slightly flattened laterally. Head pointed like a cone. The mouth is fairly small, terminal, or slightly superior or inferior, and it looks a bit like a hare's snout, which is where the generic name *Leporinus* ('little hare') comes from. This hare-like impression is strengthened by the two middle teeth of the upper jaw jutting out considerably (reminiscent of the middle cutting teeth of the hare). Eyes relatively large. LL present. The D is usually drawn into a triangular point and short. No adipose fin. C moderately deeply forked, the upper lobe being longer. The An is short and triangular. Vs and Ps well developed. *Leporinus* requires a roomy aquarium not less than 1 m long. Hiding-places can be provided with hard-leaved vegetation, bamboo shoots and pieces of root. The water should be changed regularly. Since most of the species reach lengths of 30 cm and because most of them lose their pretty colouring with age, you will usually only find young specimens being kept in an aquarium. Adult specimens, because of their elegant body shape and methods of swimming, as well as their longevity, are very suitable for public aquaria in zoos and the like. They are omnivores that will also accept plant food such as algae, cabbage and rolled oats. Unfortunately, they also graze on the plants in the tank. *Daphnia* and *Cyclops*, as well as worms of all kinds, are skilfully extracted from the sand. So far, only *L. maculatus* has been bred successfully. The systematical classification of the G *Leporinus* according to GÉRY leaves many points unclear, and so far there does not exist a modern revision of his system. That is why it is sometimes difficult (particularly with younger specimens) to give an exact species diagnosis.

– *L. arcus* EIGENMANN, 1912; Lipstick Leporinus. From Surinam, the Cottica river. Up to 32 cm long. Flanks yellow-brown with a violet shimmer. 4 longitudinal bands, of which the broadest runs across the middle of the body and begins at the rear edge of the eyes. The Vs and An are blood-red, the adipose fin with a black blotch. The other fins are colourless or slightly reddish. A very attractive species (fig. 710).

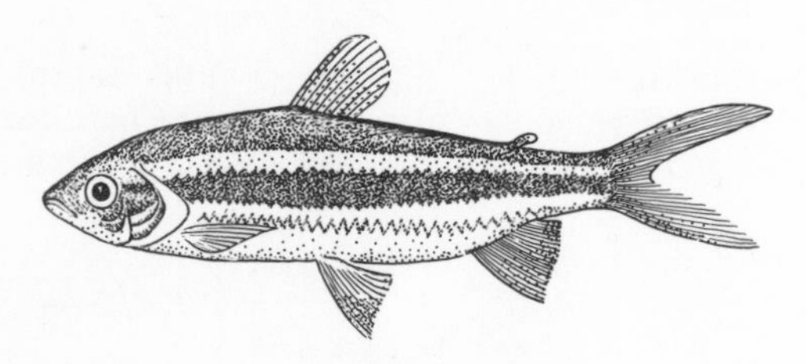

Fig. 710 *Leporinus arcus*

– *L. fasciatus fasciatus* (BLOCH, 1794); Striped or Banded Leporinus. Widely distributed in northern and central S America. Up to 30 cm long. Flanks a lovely lemon-yellow or golden-yellow with 10 deep black transverse bands. A dark band running around the lower

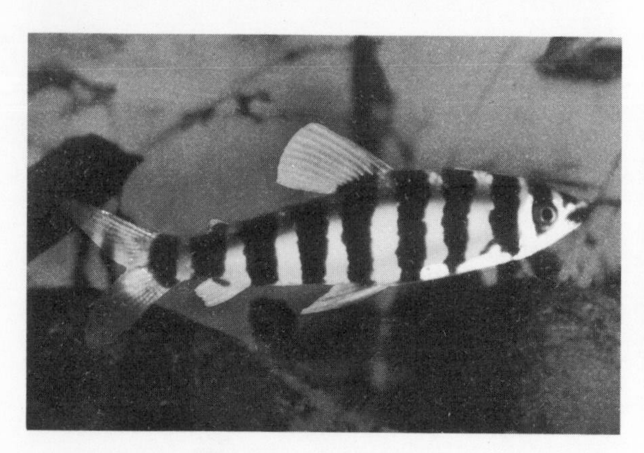

Fig. 711 *Leporinus fasciatus fasciatus*

jaw links the two eyes. Fins a delicate grey or colourless. The adult males are slimmer and often have an orange-red or blood-red throat area (fig. 711).

– *L. fasciatus holostictus* COPE, 1878. From the Amazon Basin, any more precise location doubtful. Up to 25 cm long. Very similar in colouring to *L. fasciatus fasciatus*. A juvenile has 9 transverse bands, of which the third last and the next to last caudal band split down their lengths to give 11 bands in an adult specimen. Adult males are said to have no red throat area.

– *L. frederici* (BLOCH, 1794); Spotted Leporinus, Frederici. From Guyana to the Amazon Basin. Up to 35 cm long. Upper surface grey, flanks yellow-grey with a faint silver sheen, caudal peduncle a strong yellow colour. A black band runs from the front edge of the eye to the tip of the snout. Several irregular dark, transverse bands run along the dorsal ridge, but they fade away with age to a large extent. There are usually 3 black dots on the LL, the largest beneath the D, another in front of the adipose fin and the third at the end of the caudal peduncle. Occasionally, there is a fourth dot behind the operculum. Fins colourless. The An can have black streaks on it (fig. 712).

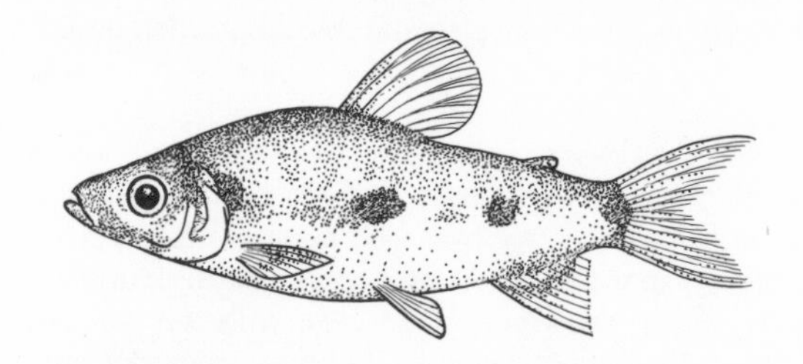

Fig. 712 *Leporinus frederici*

– *L. maculatus* MÜLLER and TROSCHEL, 1845; Spotted Leporinus. From the Amazon Basin and the Guyana lands. Up to 25 cm long. Flanks golden-yellow with large, grey-black dots which give the impression that they lie deep beneath the skin. There are also 3 deep black, oval flecks, one beneath the D, one in front of the adipose fin, and one on the caudal peduncle. Another black fleck is located on the operculum. The fins are yellowy transparent. A very attractive species which is always striking to the eye because of its unusual fleck markings. AZUMA tells us that the species is

Fig. 713 *Leporinus maculatus*

a free-spawner, the young hatching out after 3–4 days (fig. 713).

– *L. melanopleura* GÜNTHER, 1864. From W Brazil, between the Amazon Basin and Rio de Janeiro. Up to 20 cm long. Flanks yellowish-brown with a silvery iridescence, getting paler towards the belly. Dorsal surface olive-brown, the caudal peduncle having a reddish-violet shimmer and fin brown dots. A broad, shimmering brown longitudinal band extends from the operculum to the insertion of the C. Fins colourless or a delicate grey-brown. In larger males, the D and An may have dark edges.

– *L. striatus* KNER, 1859; Striped Leporinus. From the Mato Grosso, Marañon, the river Magdalena. Up to 25 cm long. Flanks pale yellow or ochre with 4 black-brown longitudinal bands, the broadest of which runs through the middle of the body and stretches from the tip of the snout to the base of the C. This species seems to have the most contrasting colouring. Fins colourless. The adipose fin and the base of the An may be dark brown in colour (fig. 714).

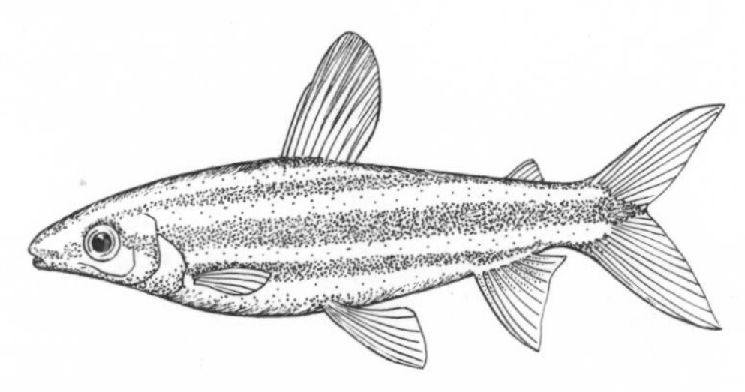

Fig. 714 *Leporinus striatus*

Leptocephalus Larva *see* Anguilliformes

Leptotilapia PELLEGRIN, 1899. G of the Cichlidae* (fig. 715). Comprises 3 species found in tropical W Africa (Ghana, Congo Basin). They live in fast-flowing waters and are similar to *Teleogramma** and *Steatocranus**. *Leptotilapia* grows up to 15 cm long and is extremely well adapted to its habitat. In shape it is very

Fig. 715 Representative of the G *Leptotilapia*

elongate, about 4.5–5.5 times as long as it is tall. It does not possess a swimbladder that functions, and to a large extent it has lost the ability to swim – it glides along jerkily. The head and mouth are very large, which also make up the tallest part of the body. The body simply tapers away towards the rear. Above the eyes the forehead juts out, but not so much as in *Steatocranus*. The D is inserted along a wide area and has 19–21 hard rays; the soft-rayed section is pointed. C rounded. Colour very monotone, grey predominantly. *Leptotilapia* should be kept in large, shallow tanks with plenty of ventilation or even a centrifugal pump, giving a strong current. Coarse-grained sand and gravel make a good substrate, and plenty of hiding-holes should also be provided (rock structures, flower pots). *Leptotilapia* species are territorially orientated bottom-dwelling fish which will also defend their territory against larger fish. The cavity where they live forms the centre of the territory and is usually occupied by an individual or by a pair. Like *Teleogramma*, members of the G *Leptotilapia* are extremely quarrelsome among themselves when kept in the narrow confines of an aquarium. Weaker specimens are gradually bitten to death by the strongest pair. It is advisable only to keep larger Cichlids of other Ga in the same tank with them. Spawning and brood care take place in the same cavity, ie *Leptotilapia* is a typical secret spawner (compare Cichlidae). Often, it is only obvious that spawning has taken place once the fry appear. Special rearing food is not necessary in the first few weeks because the parent fish chew up the food in the brood hole and blow it out through the gill slits. Feed with live food of a suitable size, eg fly larvae, large *Daphnia*, mealworms, Mayfly larvae etc. Of the 3 species, *L. irvinei* TREWAVAS, *L. rouxi* PELLEGRIN and *L. tinanti* (POLL), only the last one has so far been introduced and reared (fig. 716).

Fig. 716 *Leptotilapia tinanti*

Lernaea, G, *see* Copepod Infestation

Lesser Duckweed *(Lemna minor) see Lemna*

Lesser Spike Rush *(Eleocharis acicularis) see Eleocharis*

Lestes LEACH, 1830. G of the insect order Dragonflies (Odonata*). Represented in Europe by 6 species. All of them have a metallic green body, and when at rest the wings are held pointing upwards at an oblique angle. *Lestes* larvae are commonly found in standing waters and often get trapped in the net when people fish in ponds. These larvae can be used as food for larger kinds of predatory fish, but because they are extremely lively specimens in their own right they are most suitable for keeping in pond-aquaria. They can be observed catching their prey, shedding their skin and metamorphosing into an adult capable of flight. Since the eggs over-winter in most *Lestes* species and since they can already fly when the summer comes (July, August), the life of the larva is restricted to a few months.

Leucaspius HECKEL and KNER, 1858. G of the Cyprinidae*, Sub-F Abraminae. Found in Central and E Europe, as well as Central Asia. Small, sociable fish that live in the top water levels of ditches, ponds and shallow lakes, where they sometimes occur in large numbers. The body is elongate and very flat from side to side. Between the Vs and An the belly is shaped like a keel. The silver shining scales easily fall out. LL incomplete. The slit of the superior mouth points steeply upwards. Very active swimmers, peaceable. Provide large, long tanks in which a small shoal can be kept. Vegetation should not be too thick and should leave enough room for swimming. Temperature 18–20 °C; higher temperatures will also be tolerated well for short spells. *Leucaspius* species are all contented omnivores. They spawn between April and June. The females have a short laying tube with which they can stick rings or spirals of eggs to the water plants. The male looks after the brood,

smearing the eggs with a dermal mucous that inhibits bacteria. For breeding, place 3–4 pairs in the tank together. Success usually follows if the fish have been allowed to over-winter in cool conditions.
– *L. delineatus* (Heckel, 1843); Moderlieschen. From Central and E Europe. Up to 9 cm. Dorsal surface brownish-grey, flanks and ventral surface silvery. A steel-blue, metallic longitudinal band reaches as far as the root of the tail. Fins colourless and translucent or faintly yellowish (fig. 717).

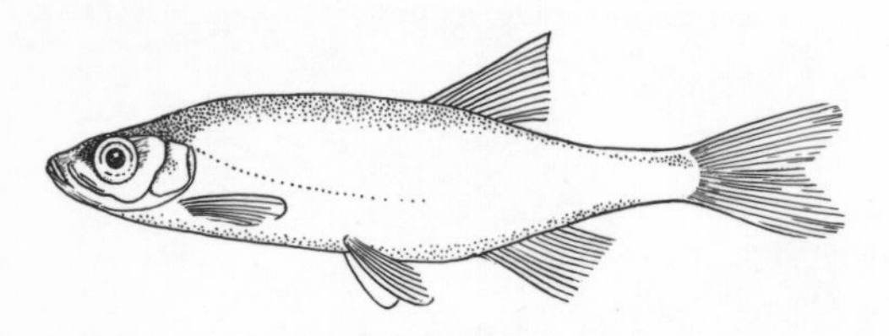

Fig. 717 *Leucaspius delineatus*

– *L. irideus* Ladiges, 1960 is found in Turkey and *L. marathonicus* Vinciguerra, 1920 in Greece.

Leuciscinae, Sub-F, *see* Cyprinidae

Leuciscus Cuvier, 1817. G of the Cyprinidae*, Sub-F Leuciscinae. Found in N America, Europe, Central and E Asia. Large, sociable fish that live in the surface waters of cool, fast-flowing rivers and clear lakes. Body elongate and only slightly flattened laterally. Mouth either terminal or inferior with no barbels. LL complete. The D is inserted opposite the Vs, D and An short. At spawning time the males have spawning tubercles*. As juveniles, *Leuciscus* species can be kept in unheated room aquaria that are well venlated. Some fresh water should be added occasionally. They need sufficient room aquaria that are well ventilated. Some fresh water Every sort of live food is eaten, as well as plant food. Adult specimens should only be kept in large public aquaria.

The various species of the G divide up into several sub-species.
– *L. cephalus* (Linnaeus, 1758); Chub. From Europe, but not found in some northern and north-eastern regions. Up to 60 cm. Dorsal surface grey-brown with a greenish sheen. Flanks silvery, belly white. Scales with dark edges. D and C dark grey-green. Fin rays of the V and An are red. Can be distinguished from *L. leuciscus* by its deeply slit mouth and by its An which is arched towards the outside edge.
– *L. leuciscus* (Linnaeus, 1758); Dace. From Central Europe, Siberia. Up to 25 cm. Dorsal surface black-grey with a blue shimmer. Flanks a silvery iridescent colour, often with yellowish tones. D and C grey-yellow, all the other fins weakly yellow. Distinguishable from *L. cephalus* by its small mouth and indented An.
– *L. souffia* Risso, 1826; Stroemer. With 3 sub-species found in central S Europe. Up to 25 cm. Dorsal surface grey-black with a metallic blue sheen. Flanks silvery, belly white. A broad, violet-black longitudinal band extends from the eye to the root of the tail (fig. 718).

Lids *see* Aquarium Lids

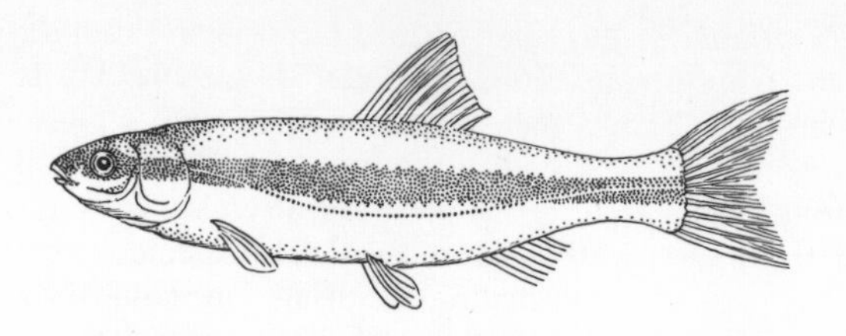

Fig. 718 *Leuciscus souffia*

Life Span. The length of time between the formation and death of an organism. The average life span of a species is fixed genetically. Single-celled organisms that reproduce through simple division have, potentially, an unlimited length of life. Among the plants, apart from annuals and biennials, there are great numbers of perennials, the oldest of all being the 4,000-year-old Bristlecones of N America. Among the animals, life-spans range from years to decades. Giant Tortoises can live to be over 150 years old. Among the fish, the lifespans of different species vary enormously. In the wild the South American *Cynolebias* species survive for only 6–8 months, the Herring for about 20 years, and the Flounders and Carps for more than 50 years. The age of temperate zone fish can normally be determined, because they have an annual, rhythmical growth pattern that is visible from the scales, bones or otoliths (rather like the annual rings of trees).

Lighting. Light affects vital processes in the aquarium in many ways. It is responsible along with numerous other factors for the continued stimulation of life processes and its function can only be understood in relationship with them. Light works through its different spectral compositions, ie its quality, through its intensity (measured in lux) and through its length of exposure, ie the particular rhythm between day and night. Living beings place different demands on the various light factors, something which must be borne in mind when constructing a community tank with different plants and animals. In aquarium-keeping it is rare for natural sunlight alone to be sufficient. With artificial lighting, which today presents no technical problems, natural colour should be created as near as possible to that found in fresh water – good plant growth with a moderate amount of algae. In healthy sea water a good algal growth should be achieved. Normal light bulbs with a relatively high red content and good colour reproduction nevertheless give out a lot of heat and a small amount of light. Increasingly they are being replaced with fluorescent tubes. These do not heat up nearly as much, so have a higher production of light. Various kinds are sold, offering whatever spectral composition is required, and some have UV content. The light source should be positioned above the water surface of the aquarium. When installed, it must be protected from damp as well as from the direct effects of water. A reflector provides optimal light efficiency as well as an even-tone illumination, without glare. In many cases complete light boxes are used, covering the whole aquarium, but they must contain vents for warm air to escape. To avoid accidental deaths of fish through fright at the light being suddenly switched on or off, additional light-bulbs can be installed above a regulating device to give corresponding dimmed lighting, or fluorescent tubes with gradually alterable glare shields can be used. The most suitable spectral composition and light intensity must be determined separately for each aquarium. This depends on the needs of the plants and animals being kept, the size and location of the tank and, moreover, the quality of the water. Light is absorbed (*see* Absorption) according to the depth, cloudiness and colour of the water and its spectral composition is altered. A tropical aquarium should receive 12–16 hours of light daily. Prolonged lighting is harmful, as along with sufficient lighting a rest period of darkness or minimal lighting is necessary for a biological balance. Sudden alterations in lighting should be avoided in the aquarium, as it can cause the whole balance to be upset. pH values and the redox potential may change, plants may be damaged, the water may cloud etc. To grow well, plants need light for assimilation* of carbon dioxide as well as sufficient nutrient content and a corresponding water quality. Whereas algae and higher plants from great water depths or very shaded tropical forest waters need less light, the light requirements of shallow water plants, marsh plants and floating plants from open water systems may be extremely high. Aquatic plants, in relation to the light factor, can decisively influence water quality in the aquarium (assimilation, dissimilation, biogenic, decalcification, redox potential). The fluorescent tubes with a high red and blue light content, often sold to increase plant growth, have the disadvantage of unnatural colour reproduction. Fish spawn and embryos are not helped in their development by a high blue content. Fish and other animals kept in the aquarium are adapted to different light conditions, depending on where the specimens come from. Fish that live in deep water, cloudy water or tropical forest waters, as well as cave-dwelling fish and nocturnal animals, all prefer subdued light or complete darkness. In the marine aquarium little consideration is given to the fact that many of its specimens will have come from depths of up to 40 m, where light loses a great deal of its intensity and changes its spectral composition. Attempts are made to imitate those light conditions with lamp filters or with dyes added to the water. Animals from savannah waters and surface-water fish from lakes, ponds and streams, as well as coral reef creatures, are much more likely to need a high intensity of light. A particular rhythm of lighting is especially important. In the case of tropical fish, light remains constant throughout the year and consists of 14–16 hours per day. Creatures that live in more temperate zones are subject to temperature changes in winter and an altered day-night rhythm besides. Light plays a significant role in certain fish life-cycles; many species, for example, only spawn in the morning sunlight or at dusk, others at night. Fish spawn is usually sensitive to light and can be hindered in its development by too strong an illumination. Under natural conditions in water, light is very much influenced by other seasonal factors, such as water cloudiness in the rainy seasons, hydrophily in the spring and autumn months, ice, etc. In aquarium-keeping, little notice can be taken of these factors, although they may be significant as stimuli for reproduction, particularly in temperate zone creatures. There is insufficient information about the needs of aquatic animals in relation to a particular spectral composition of light. In particular, the usually absent UV light content and its effect are of interest. Its ability to inhibit the growth of or kill micro-organisms is well-known. For example, with regular dosages of UV radiation, the susceptibility of coral fishes to skin parasites (*see* Parasitism) is considerably reduced. The UV content of light also works directly upon the metabolic processes in the aquarium and it can alter the redox potential of the water to a marked degree. With optimal lighting, even in a marine aquarium, poisonous breakdown products may be compensated for to a certain extent and a satisfactory growth of algae may be achieved. Light and lighting are essentially biological problems to the aquarist and can only be regarded as such.

Ligula intestinalis *see* Cestodes

Lilaeopsis, G, *see* Apiaceae

Liliatae (Monocotyledoneae, name not in current usage). Cl of the Spermatophyta*, comprising 50 Fs and 72,000 species. The embryo of the Liliatae has only one cotyledon, in contrast to the Magnoliatae*. The leaves are arranged alternately and in whorls. The leaf-blades are usually simply built with parallel veining. The vascular bundles* are sealed and scattered on the plant's stem structure. A normal, secondary growth in thickness is absent. The radicle is short-lived and is always replaced by roots that have developed from shoots. In the flower structure, 3-part circles predominate. The Liliatae are primarily herbaceous plants.

Lima CHEMNITZ, 1784; Gaping File-shells (fig. 719). G of the molluscs (Bivalvia*, O Anisomyaria) found pri-

Fig. 719 *Lima*

marily in warm seas. *Lima* has a long oval-shaped shell covered with fine spines. The two sides of the shell cannot be closed completely. Another key feature is the long, often colourful tentacles. *Lima* can swim in a similar way to *Pecten**. Some species build themselves 'nests' with the aid of byssus threads and little stones, lying safely within them. Although interesting specimens in the aquarium, *Lima* species are a little sensitive; their tentacles are often bitten off by fish.

Limanda limanda *see* Pleuronectiformes

Limia, name not in current usage, *see Poecilia*

Limnephilus BURMEISTER, 1839 (fig. 720). G of the Caddis Flies (Trichoptera*). Distribution confined to the holarctic realm and represented in Europe by more than 50 species. The larvae like to live in still or slowly flowing waters with a great deal of vegetation, and build characteristic caddis cases. They are excellent subjects for observing in pond aquaria and also make good fish food.

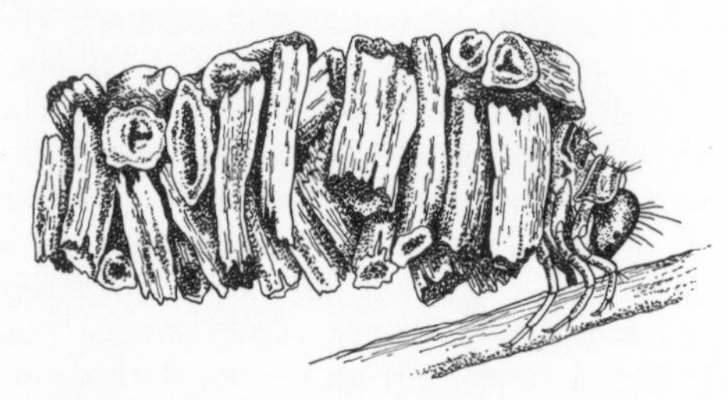

Fig. 720 *Limnephilus.* Climbing larva with caddis case

Limnichthyidae, F, *see* Trachinoidei

Limnion *see* Lakes

Limnobium RICHARD, 1812 (Hydromystria, name not in current usage); American Frogbit. G of the Hydrocharitaceae* comprising 2 species. From America, some have dispersed into other regions. *Limnobium* species are small rosette-like perennials with runners. The leaves are stalked and the leaf-blades are roundish with a heart-shaped base in some cases; the undersides have spongy air tissue. The flowers are unisexual and dioecious, and develop within a translucent spathe. Double perianth, calyx and corolla inconspicuous, in threes. Male flower: stamens 12–6; female flower: carpels 3, fused, inferior. The fruit is like a berry. The various species of the G are water plants and marsh plants. Their rosettes float on the water surface or grow on the marshy bottom, where the leaves lie flat or stand upright. Because Frogbits make water very shady, they are not used very often in today's aquarium, but they are cultivated regularly in botanical gardens. In the summer they can be kept in the open-air. The two species are:

– *L. laevigatum* (HUMBOLDT et BONPLAND ex WILLDENOW) HEINE (*L. stoloniferum,* name not in current usage); South American Frogbit. From northern S America, Central America and the West Indies. Leaf-blades clearly arching (fig. 721).

– *L. spongia* (BOSC) RICHARD: North-American Frogbit. From central N America. Leaf-blades relatively flat.

Fig. 721 *Limnobium laevigatum*

Limnochromis REGAN, 1920. G of the Cichlidae* comprising around 11 species, all of which are endemic* to Lake Tanganyika*. They are fairly elongate fish of typical perch shape. Their large head is particularly striking and characteristic, as are their large eyes and deeply slit, superior mouth. The soft sections of the D, Vs and An are pointed. The C curves inwards slightly. *Limnochromis* can be cared for in a similar way to other Cichlids from Lake Tanganyika *(Lamprologus*, Julidochromis*, Eretmodus*, Tropheus*).*

Limnology. The science of inland waters*, ie still waters, flowing waters and ground water. Limnologists study the physical-chemical structure of such waters, as well as their hydrographical and biological make-up, in order to perceive the inland waters as ecosystems*. So, limnology is both a branch of general hydrology as well as ecology*. The science is mainly concerned with the material economy of the water, which, amongst other things, is dependent on climatic and physical-chemical factors as well as the colonisation of organisms. With increasing industrialisation, limnology has become more and more important. The theories about inland waters that abounded without any practical examples being performed have been strengthened through the work of limnologists. Their studies have largely brought about such important realisations as the need to care for and protect our waters, as well as information about water preparation and the purification of waste water. Another of their tasks which is of immense interest to humans is the opening up of natural and artificial waters and waste waters as production sites for human food (fish, algae), thus helping to cut down the protein deficiency that exists in the world's food supplies.

Limnophila R. BROWN, 1810 (*Ambulia,* name not in current usage); G of the Scrophulariaceae* comprising 35 species. Found in tropical-subtropical parts of Asia and Africa, and also introduced into some parts of N America. They are perennial plants with creeping and upright elongated stem axes, but some annuals do exist. Leaves arranged in whorls, the leaves being either undivided or pinnate. The flowers are hermaphroditic, monosymmetrical. Double perianth, their parts coming in fives and fused. The corolla is tube-like, two-lipped. Stamens 4. Carpels 2, fused and superior. Capsule fruit. All the species of the G are marsh plants, and many are able to tolerate long periods of flooding very well. They

grow in marshes, on the edges of lakes and slowly flowing waters, and also occur as weeds in rice fields. Some species make decorative aquarium plants. They are easy to propagate from cuttings. The most appealing species are those that are heterophyllous (*see* Heterophylly) and those that form pinnate aquatic leaves.
– *L. aquatica* (ROXBURGH) ALSTON. From India, Ceylon. Aerial leaves undivided, aquatic leaves finely pinnate, up to 5 cm long (fig. 722).
– *L. aromatica* (LAMARCK) MERRILL. From E and S Asia. Aerial and aquatic leaves undivided.
– *L. indica* (LINNÉ) DRUCE. From S Asia up to central Africa, Australia. Flowers stalked, otherwise like the previously mentioned species. There is a hybrid form which has been introduced in N America (Louisiana).
– *L. sessiliflora* BLUME. From E and S Asia, introduced into N America (Florida). Aerial leaves almost undivided, aquatic leaves finely pinnate. Flowers sedentary.

Fig. 722 *Limnophila aquatica*

Limosella LINNÉ, 1753; Mudwort. G of the Scrophulariaceae* comprising 10 species. Distributed throughout the world. Either small rosette perennials with runners or annual plants. Leaves awl-shaped or stalked with small blades. Flowers hermaphroditic, small. Double perianth. Calyx and corolla 5-part, fused. Stamens 4. Carpels 2, fused, superior. Capsule fruit. The species of the G are small marsh plants that can also tolerate lengthy periods of flooding. Because they have runners, many of the species form thick carpets of vegetation and are well suited as foreground plants in the aquarium, although they are seldom used. One example is the following.
– *L. aquatica* LINNÉ; Mudwort. Distributed throughout the world. Aquatic leaves awl-shaped, aerial leaves with spatula-shaped blades. Suitable for cold and heated aquaria.

Limpets, G, *see Patella*

Limulus polyphemus *see* Xiphosura

Linckia FORBES, 1839 (fig. 723). G of the Starfishes (Asteroidea*) found in shallow waters of tropical and subtropical seas. Eats small animals and detritus. Has a small disc and 5 slender, cylindrical arms. They can also reproduce themselves asexually by one arm tearing itself free; a new disc will form at the end of the arm and new arms will follow. Easy to keep in the aquarium. *Linckia* will even eat fish food tablets and they are not predatory. Shining blue species are often imported.

Fig. 723 *Linckia laevigata*

Lindane (y-Hexa-chlorocyclohexane). An insecticide that works very quickly and one that is poisonous for higher organisms, including Man. Fish that are infected with Argulidae* and parasitical Copepods (Copepod Infestation*) can be treated in a container with 15–20 litres of water. Take enough Lindane to cover the point of a knife and scatter it on the water surface. Leave the fish for 15 mins in the bath, observing them the whole time, then rinse them with clean water and put them back in the aquarium. Not every kind of fish will tolerate Lindane bathing treatments.

Line Breeding is not really artificial breeding in the narrowest sense, but a controlled reproduction of existing races or sorts according to preconceived standard conditions. To ensure uniformity, any deviant forms must be continually weeded out. Continual in-breeding is essential.

Lineus, G, *see* Nemertinea

Linné, Carl von (1707–78). Known as Linnaeus. Swedish natural historian, doctor and botanist. He studied medicine and natural sciences at Lund University. Made a research trip through Lapland (1737 *Flora Lapponica*) and later devoted his time to botanical studies in Holland. In 1735 *Systema Naturae* appeared, which in the tenth edition of 1758 forwarded for the first time a scientifically uniform specification for the animal kingdom. It still forms the basis of today's binary nomenclature. Each species was given generic and specific names, the genera were placed in classes and these were grouped under orders. In 1739 he founded the Stockholm Academy of Sciences, in 1741 he became

Professor at Uppsala, and in 1762 he bacame a member of the Paris Academy of Sciences. His comprehensive botanical studies are reflected in his numerous publications (1751 *Philosophia Botanica*, 1753 *Species Plantarum*, 1774 *Systema Vegetabilium*).

Liparidae, F, name not in current usage, *see* Cottoidei

Liparinae, Sub-F, *see* Cottoidei

Liparis liparis *see* Cottoidei

Lipids *see* Fatty Substances

Lipoidosis *see* Liver, Diseases of

Lipstick Surgeon *(Acanthurus glaucopareius) see Acanthurus*

Lithodidae, F, *see* Paguridea

Lithophaga lithophaga *see* Boring Molluscs

Littoral (fig. 724). The bank or shore area of a stretch of water through which light can shine, including the upper section of the benthal* zone. As it falls away into the depths it joins up with the profundal* zone where no light can penetrate and no plants can grow. Depending on how clear the water is, the boundary between the littoral and the profundal lies between 3 and 30 m in inland waters* and around 200 m in the sea. It follows, therefore, that shallow waters can sometimes consist of a littoral zone only, since light can penetrate everywhere (eg Ponds*). The littoral zone in lakes is normally divided into different plant communities following one after the other.

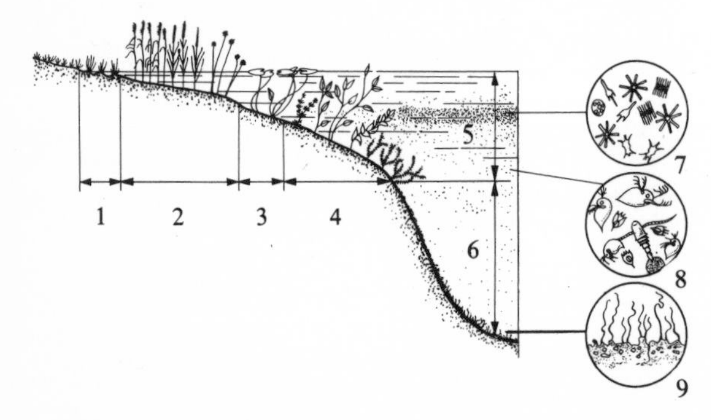

Fig. 724 Cross-section through a still stretch of water with typical shore area. 1–4 Bank zone (littoral): 1 Great sedges 2 Reed and shelf zone 3 Water-lily zone 4 Area of Pondweeds and Charophyta 5 and 6 Open-water zone (pelagial): 5 Illuminated pelagial 6 Non-illuminated pelagial 7 Vegetal plankton 8 Animal plankton 9 Living community of the depths (profundal) with *Tubifex* worms and non-biting Midge larvae.

The habitats of the littoral can be successfully imitated in the aquarium, ranging from a pool tank to a coral fish tank. The other aquatic regions, the profundal and pelagial*, can scarcely come into consideration for the aquarium. Compare Still Waters.

Littorina FERRUSSAC, 1822; Periwinkles. G of marine-dwelling molluscs (Gastropoda*, Prosobranchia*). They live in very specific zones in the top part of the shoreline, some of them often leaving the water altogether or drying up at low tide. All of them are grazing herbivores. Their shells are spherical and very firm with stumpy, cone-shaped tips, measuring 5–30 mm in diameter. They can seal themselves firmly by means of a lid. Not really suitable for the aquarium because they usually creep out of the water.

Livebearing Barb *(Barbus viviparus) see Barbus*

Livebearing Tooth-carps, F, *see* Poeciliidae

Liver *see* Digestive Organs

Liver, Diseases of. The liver (*see* Digestive Organs) is one of the most important glands of the vertebrate body. In healthy fish, it is coloured red-brown to ochre. If it looks more yellowish-brown in colour, this often indicates one of the commonest liver diseases, fatty liver degeneration (lipoidosis). It is normally caused by disturbances to the metabolism which is in turn caused by errors in feeding (*see* Feeding) and lack of vitamin A (Deficiency Diseases*), thus leading to the fatty degeneration of the liver. But the liver is also very prone to other pathogens because it is an organ of the body with a lot of blood flowing through it. The formation of little nodes is characteristic of infections from Sporozoa*, fish tuberculosis* or *Ichthyosporidium**, which, just like stomach dropsy*, cause the build-up of organisms of decay in the liver tissue which is then repaired by connective tissue. But just like lipoidosis, as the connective tissue multiplies (cirrhosis), the liver's function is impeded. A common parasite of the liver bile ducts is *Hexamita** (*see* Discus Disease). It seals off the exits to the bile ducts, causing the bile fluid to build up and overflow into the blood and tissues. In such a case, the bile colouring matter, bilirubin, turns the liver a yellowy-green colour (jaundice). If attacked by trematodes*, their eggs will often be found in the blood vessels of the liver. Poisoning* from various substances will also cause great damage to the liver.

Living Community *see* Biocenesis

Living Rocks. In marine aquarium-keeping, living rocks mean any rocks or stones with animal or plant growth on them that can be transported quickly and carefully from the sea and placed directly into the aquarium. The first result of this will be that marine bacteria are encouraged to colonise the tank. A great variety of different organisms will also find a home on the stones: Foraminiferans (Foraminifera*), small sponges (Porifera*), hydroid polyps (Hydrozoa*), calcareous corals (Madreporaria*), Anemones (Zoantharia*), Sea-anemones (Actiniaria*), Tube Worms (*see* Polychaeta), small Crabs (Brachyura*), Snails and Molluscs, Brittle-stars (Ophiuroidea*), Sea-squirts (Ascidiacea*) etc. Of course, many of these things only start to colonise the stones in the course of time. This is also true for numerous forms of algae. You can collect living rocks from any coastline, but they must be transported like living animals. Avoid large sponges and algae, because they usually die. The most suitable stones or rocks come from calm, shady areas, where you will usually find that they have an abundant animal growth on them. Living rocks make very interesting objects for the aquarium.

Livoneca symmetrica *see Anilocra*

Lizardfishes (Synodontidae, F) *see* Myctophoidei

Loaches (collective term) *see Botia*

Loaches, F, *see* Cobitidae

Lobelia LINNÉ, 1753; Lobelia. G of the Lobeliaceae* comprising 360 species. Found almost everywhere on earth. Small rosette perennials with elongated stem axes or annual plants. Alternate leaf arrangement. Flowers hermaphroditic, monosymmetrical. Double perianth. Calyx and corolla 5-part, the corolla being fused like a lipped flower. Stamens 5. Carpels 2, fused and inferior. Capsule fruit. The various species of the G colonise very different habitats. Many are important as ornamental plants. Numerous species occur in wet places, even places that are occasionally flooded, so several of them should be suitable for the aquarium. However, so far, only the following undemanding species is regularly seen in the aquarium.

– *L. cardinalis* LINNÉ; Scarlet Lobelia, Cardinal Flower. From N America. Elongated stem axes with alternate leaf arrangement. An emersed plant with shining red flowers.

Lobeliaceae; Lobelia Family. Often grouped with the Campanulaceae (Bell flowers). F of the Magnoliatae* comprising 30 Ga and 900 species. Temperate to subtropical, with only a few species in the tropics. Perennial plants with an alternate leaf arrangement, more rarely annuals. Some produce latex. Hermaphroditic, monosymmetrical flowers with a fused corolla. Lobelias colonise very varied habitats that range from the plains to high up in the mountains. Some are important as ornamental flowers. Although several species belonging to various Ga are aquatic or marsh plants, only the G *Lobelia** has so far proved of any significance to the aquarist.

Lobelias, F, *see* Lobeliaceae

Lobochilotes BOULENGER, 1915. G of the Cichlidae*. There is only one species and it is found in Lake Tanganyika*. Body moderately elongate, dorsal ridge quite arched. The head is large and the mouth has a very characteristic shape. The lips are extremely thick and have long, membranous outgrowths. The D has 18 hard rays and about 10 soft ones. The rear edge of the D is cut off straight. For care in the aquarium, see details as for other Lake Tanganyika Cichlids *(Lamprologus*, Julidochromis*).*

– *L. labiatus* BOULENGER, 1915, Lake Tanganyika. Up to 37 cm long (fig. 725).

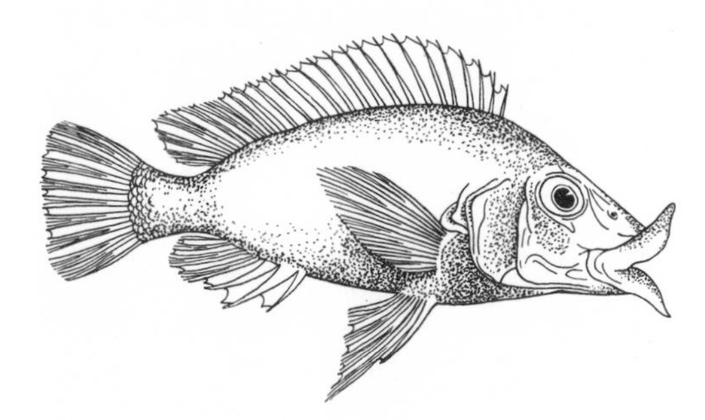

Fig. 725 *Lobochilotes labiatus*

Lobotes surinamensis *see* Lobotidae

Lobotidae (fig. 726). F of the Perciformes*, Sub-O Percoidei*. Easy to recognise from the soft-rayed parts of the C and An which broaden out into lobe-like extensions. From a distance, these extensions, together with the rounded C, make the fish look as if it has 3 tails. The Lobotidae comprise a small number of robust predators that live in the tropical Indo-Pacific and Atlantic. Some prefer brackish water zones or even live in fresh water mainly. The Black Triple-tail *(Lobotes surinamensis)* can grow up to 1 m long and occurs in almost all warm seas. The G *Datnoides* is represented in the brackish water and freshwater realm of SE Asia,

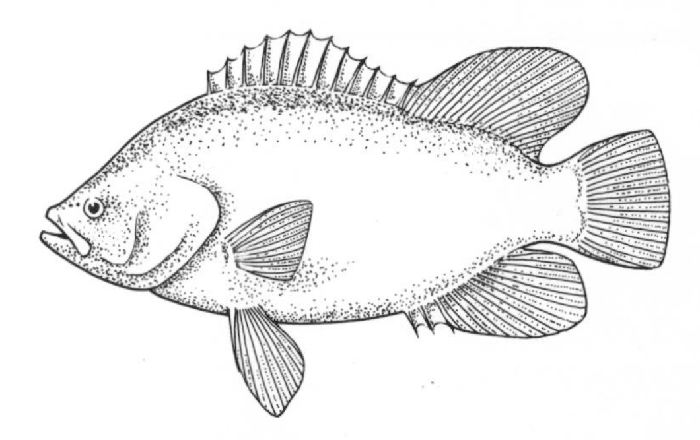

Fig. 726 Lobotidae-type

including Indonesia. *Datnoides quadrifasciatus,* which grows up to 30 cm long, exhibits a pronounced colour change from a uniform dusky olive-brown to a silvery iridescent colour with 8–10 dark transverse bands. In SE Asia it is considered a tasty delicacy. It has also been observed in the aquarium by LÜLING.

Lobsters, G, *see Homarus*

Local Killi *(Fundulus heteroclitus) see Fundulus*

Loennberg's Spiny Eel *(Mastacembelus loennbergi) see Mastacembelus*

Loligo, G, *see* Cephalopoda

Long-bill *(Forcipiger longirostris) see Forcipiger*

Long-day Plants *see* Photoperiodism

Long-ear Sunfish *(Lepomis megalotis) see Lepomis*

Longfin Barb *(Barbus arulius) see Barbus*

Long-finned Characin *(Alestes longipinnis) see Alestes*

Long-horned Cowfish *(Lactoria cornuta) see Lactoria*

Longnose Gar *(Lepisosteus osseus) see* Lepisosteiformes

Long-nosed Filefish *(Oxymonacanthus longirostris) see Oxymonacanthus*

Long-nosed Loach *(Acanthopsis choiorhynchus) see Acanthopsis*

Long-nosed Triple Spine *(Halimochirurgus triacanthus) see* Triacanthidae

Long-snouted Coralfish *(Forcipiger longirostris) see Forcipiger*

Loosestrife, F, *see* Lythraceae

Loosestrife, G, *see Lysimachia*

Lophiidae, F, *see* Lophiiformes

Lophiiformes; Anglers. O of the Osteichthyes*, Coh Teleostei. Called the Pediculati in the Regan-system, the Lophiiformes are bulky, bottom-dwelling marine fish, although in rare cases they can be free-swimming deep-sea types. Found in warm and temperate seas. They have a bizarre, often monster-like shape underlined by the fact that they have an overlarge head and a giant mouth that is usually full of teeth. The Vs are located at the throat, ie well in front of the Ps, and because they have muscular stumps they tend to look like arms (in fact, they are moved rather like arms). The hole-shaped gill slits are also found near the P insertion. The swimbladder is not connected with the intestine (physoclists). A few bones are missing in the cranial skeleton, eg the parietal bones. The most striking feature of most Lophiiformes is the rod-like projection near the snout. It is actually the first dorsal spine which has been displaced towards the front of the body and has become highly specialised. It is usually thickened at its free end and it can be oscillated back and forth to act as a lure for prey. As soon as the spine approaches any possible food source, the giant mouth flies open very suddenly and the prey is sucked in. The oldest known fossils belonging to this O date from the Neogene period. There are approximately 200 modern species and they are grouped under the Sub-Os Lophioidei, Antennarioidei and Ceratioidei.

– *Antennariidae;* Frog-fishes (fig. 727). Sub-O Antennarioidei. Bulky fish whose shape is reminiscent of marine piglets. Approximately 75 species, many of which have a very odd appearance on account of the

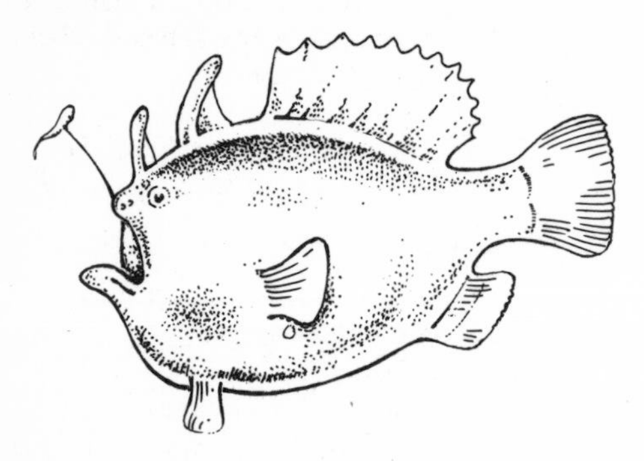

Fig. 727 Antennariidae-type

wart-like or lobe-like outgrowths that serve to camouflage the fish in its natural habitat. The body colouring can also become slowly adapted to the surroundings. The rod is well formed, occasionally with no process.

Fig. 728 *Histrio histrio*

The eggs are laid in strings of jelly. Distribution: shallow water areas of tropical and temperate seas. Particularly well known is the 18-cm-long Sargasso Fish *(Histrio histrio)* (fig. 728). It uses its pectoral fins as clasping organs when climbing among seaweeds. *Histrio* is also the only species that does not live on the sea bed. Frog-fishes survive well in the aquarium. Several species have already been cared for in aquaria and a few have also spawned in captivity, but the fry did not hatch out. Important Ga include *Antennarius** and *Histrio**.

– *Ceratioidei;* Deep-sea Anglers (figs. 729 and 730). This Sub-O consists of 10 Fs. All the members of the Fs live a bathypelagic existence in depths of 400–4,000 m. They have some extremely interesting adaptations, apart from the obvious grotesque body

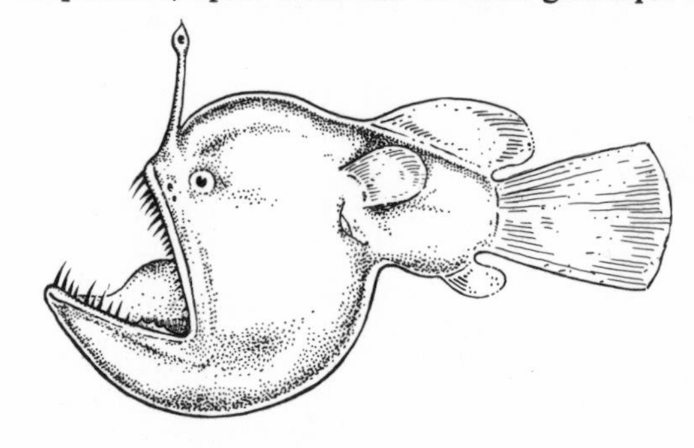

Fig. 729 Ceratioidei-type, F Melanocetidae

shapes and lack of ventral fins. Only the females have a rod, for instance, which is often supplied with a luminescent organ* on the end. The males in some Fs are very small. These dwarf-males grip firmly onto the

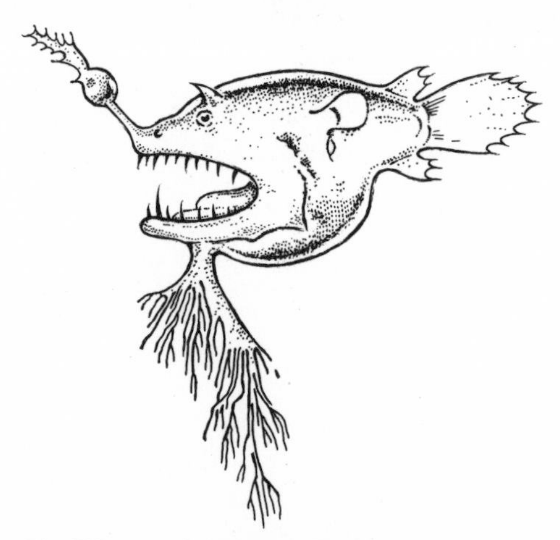

Fig. 730 Ceratioidei-type, F Linophrynidae

female when they are still young. They virtually fuse with the female and in the end they seem to become little more than a sperm-forming outgrowth. Sometimes, 2 or 3 dwarf males have been found on the one female (fig. 731).

– *Lophiidae;* Anglers or Goosefishes. Sub-O Lophioidei. Flattened fish with an enormous head and a particularly ugly appearance. The rods have a worm-shaped appendage. These are very greedy fish; even diving birds have been found in their stomachs, alongside sharks, rays, bony fish and all sorts of crustaceans. The

Fig. 731 *Edriolynchus schmidti,* female with dwarf male

best known species is probably the 2-m-long Anglerfish *(Lophius piscatorius)* (fig. 732), which is found at the foot of the European mainland as well as in the Mediterranean and the Black Sea. It is a very agile species despite its bulky shape. It spawns at great depths, the

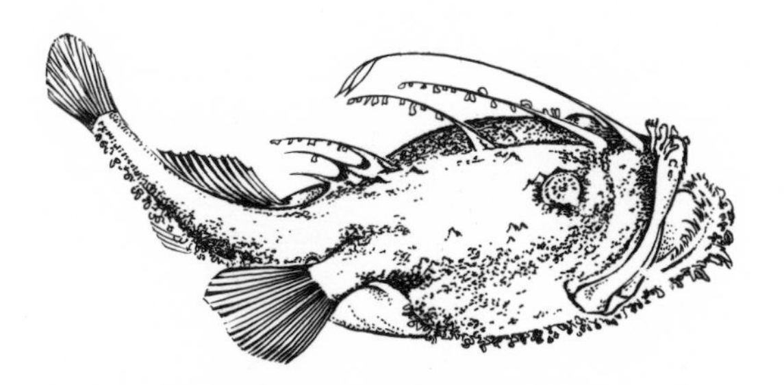

Fig. 732 *Lophius piscatorius*

jelly-like strings of eggs climbing to the surface. The young fish only change into typical Anglerfishes when they reach lengths of 10–12 cm. The Anglerfish, like other members of the F, is a prized edible fish.

– *Ogcocephalidae;* Batfishes (fig. 733). Sub-O Antennarioidei. Peculiar, flat, bottom-dwelling fish, very broad at the front end. They shuffle about over the bottom with their stalked pectoral and ventral fins. Occasionally, there is a pointed process on the head that

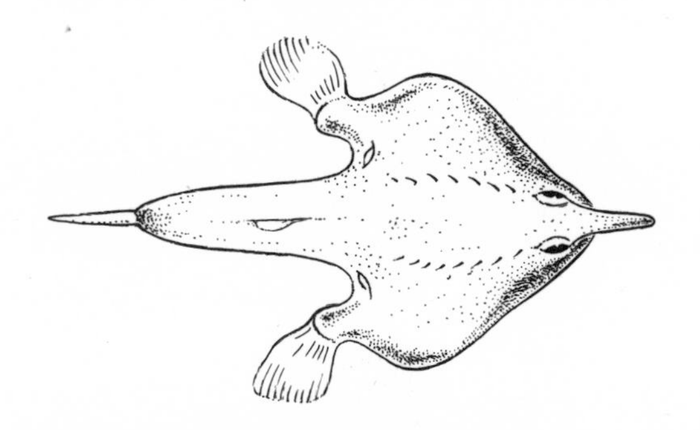

Fig. 733 Ogcocephalidae-type

is directed towards the front. The rod is short and lies in a groove above the mouth and is withdrawn as necessary. In use, the rod rotates across the mouth, the worm-like outgrowth vibrating to the right in some species, to the left in others. There are approximately 30 species, 5–35 cm long. Most live in deep water in tropical and subtropical regions of the Atlantic, the Pacific and the Indian oceans. There have already been some aquarium observations.

Lophius piscatorius *see* Lophiiformes

Lophobranchii, O, name not in current usage, *see* Gasterosteiformes

Lophotes cepedianus *see* Lampridiformes

Lophotidae, F, *see* Lampridiformes

Lophotocarpus, name not in current usage, *see Sagittaria*

Loreto Tetra *(Hyphessobrycon loretoensis) see Hyphessobrycon*

Loreto Tetra *(Hyphessobrycon peruvianus) see Hyphessobrycon*

Loricaria Linnaeus, 1758. G of the Loricariidae*. Mailed Catfishes from flowing waters in northern and central S America. Length to 25 cm. Peaceful, very contented bottom-dwelling fish that are active at dusk. In the aquarium, they require moderate light, a rich supply of plants, rocks and stones (with algal growth if possible) providing hiding-places, as well as good ventilation. The caudal peduncle is very elongate which in turn makes the body very long. Dorso-ventrally very flat. Its resistance against the force of the current in the water is therefore slight. *Loricaria* has a sucking mouth which is hidden by the much flattened head. It is surrounded by broad lips and contains teeth. The mouth is used to graze on algae as well as to attach the fish to rocks and plants. Dl, An and C are small, the latter always having a very long, upper fin ray. In some species, the lowest ray is also drawn out like a filamentous thread. The Vs and Ps are well developed, the latter having bristle-like dermal processes on the upper surface (at least in the males). The males have a distinct beard on the cheeks and a thickened first fin ray in the Ps. For the G *Loricaria,* as for the closely related Ga *Farlowella** and *Otocinclus**, the absence of a D2 is just as typical a feature as the armoured undersurface of the body. It is not easy to tell the various species apart. Besides *Tubifex, Enchytraeus, Cyclops, Daphnia* and dried food, *Loricaria* must always be given plant food as well (such as algae, lettuce leaves etc). When seeking food, these fish will usually flit around on the bottom making zig-zag movements. If disturbed, however, they will swim through the water very fast in order to attach themselves once more. Franke was the first person to describe the breeding of *L. filamentosa* (at that time known wrongly as *L. parva*). The fish clean the spot they have chosen to spawn in (it usually consists of small channels, flat rocks or, failing that, plant leaves). During the spawning act, both fish lie next to one another in parallel positions, so that the anal region of the male rests alongside the head of the female. She now sucks on one of the male's ventral fins, at the same time attaching the spawn; the male meanwhile raises his caudal peduncle and releases his sperm. Between 100 and 200 eggs are laid; they are about 2 mm long and are amber-coloured. The male guards them for about 9 days, resting all the while on top of the spawn and using his fins to fan fresh, oxygen-rich water over them. When the fry hatch out they are already 10 mm long. According to Foersch, the male shakes his fins and sucks at the spawn to help free the brood. Rear the young fish in fairly shallow water with good filtration and frequent changes of water. Feed on well-soaked powdered micro-organisms (which will quickly sink to the bottom), well chopped-up *Tubifex* and soaked lettuce leaves. Since *L. filamentosa* was successfully reared, this species has become one of the most popular kinds of Catfishes in the aquarium.

– *L. filamentosa* STEINDACHNER, 1878; Whiptail Catfish. From the River Magdalena. Length to 15 cm, although thought to reach lengths of 25 cm in the wild. There are 31–2 bony plates in a long row. Only the uppermost ray of the C is elongated. Upper surface yellowish-brown or grey-brown with a large number of irregular, dark brown dots and streaks that unite to form dark brown transverse bands on the caudal peduncle. Fins glassy with dark spots (fig. 734).

Fig. 734 *Loricaria filamentosa*

A few individual specimens of the following species have also been imported, although they are scarcely distinguishable from their colouring.

– *L. lanceolata* GÜNTHER, 1868. The upper reaches of the Amazon, Xerebros and Canelos. Length to 13 cm. 29–30 bony plates. The uppermost and lowermost fin ray of the C is elongated.

– *L. microlepidogaster* REGAN, 1904. The Rio Grande do Sul. Length to 10 cm. 29 bony plates and no elongated C rays (fig. 735).

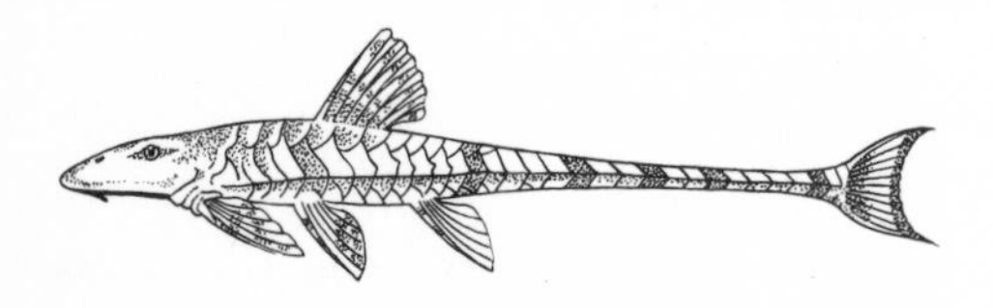

Fig. 735 *Loricaria microlepidogaster*

– *L. parva* BOULENGER, 1895. Paraguay and La Plata. Length to 12 cm. 28 bony plates. The uppermost and lowermost fin ray of the C is elongated.

Loricariidae; Mailed Catfishes (fig. 736). F of the Siluriformes*, Sup-O Ostariophysi. Found in central and northern S America, where they are well adapted to life in gushing streams and rivers (some also inhabit larger rivers as far as the rivermouth). The oldest fossils belonging to this F that can be proved as such date back to the Tertiary period in Brazil. Mailed Catfishes have the typical flattened underside of the body common to bottom-dwelling fish. The head and anterior part of the body are broad and arched and the caudal peduncle is long, sometimes very long. There are marked deviations from this pattern (eg *Farlowella**). The mouth is inferior and surrounded by lips that have broadened out into a sucking organ. It is by this means that the fish can attach itself firmly to substrates or rocks to prevent it from being washed away with the current. In contrast to the similarly armoured Callichthyidae*, that have only 2 rows of bony plates along their flanks, the Loricariidae

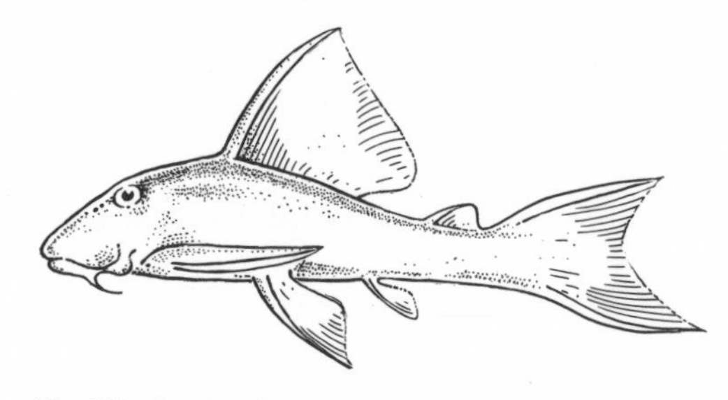

Fig. 736 Loricariidae-type

have 3–4 rows that continue right up to the similarly armoured head. In many Ga, the underside of the body is also armoured. These Ga are also characterised by having a small adipose fin. With the exception of the C all the fins begin with strong spines. In some Ga, the D is very large, and the outside edge rays of the C can also be very long. However, there are 2 features that are particularly characteristic of this F: a) the presence of small placoid scales and b) the unusual construction of the eye. Unlike all other vertebrates, in the Loricariidae the amount of light is controlled not by enlarging and contracting the pupil, but by distending and relaxing a lobe that rises from the upper edge of the iris into the pupil. All Mailed Catfishes' jaws have small, rasping teeth, some of which are spoon-shaped. They are used to graze among carpets of algae. For this reason, many species are valued highly as a means of killing off algae in the aquarium. Their sucking-disc mouths enable these fish to settle in zones in their natural habitats that could not be considered by other fish on account of the speed of the river. They flit from rock to rock, safe in the knowledge that they have their sucking mouth to attach themselves if necessary. For reproductive purposes, they seek out channels and crevices in the rock. In many Ga, sexually mature males are easy to distinguish from the females because they have a number of different processes on their bodies: eg branched or unbranched tentacle-like outgrowths near the snout (G *Ancistrus**); bearded cheeks made up of bristles (G *Loricaria**); or a long spiny border on the front edge of the Ps (G *Pseudocanthicus*). Many of the species, except *Otocinclus,* practise brood care. It should also be mentioned that the gill slits lie well down towards the ventral area, and when the fish is attached to something they are used as entry and exit points for water. Larger species are prized by many South American Indians as delicacies. They are prepared either by roasting or by drying. In some places, dried armoured Catfishes are turned into souvenirs. The Loricariidae are very popular with aquarists. For further details see *Ancistrus, Farlowella, Loricaria, Otocinclus, Plecostomus.*

Lota lota *see* Gadiformes

Lotus Flower (*Nelumbo,* G) *see* Nymphaeaceae

Loxosoma sp. *see* Kamptozoa

Lualaba *see* Congo

Lucania GIRARD, 1859. G of the Cyprinodontidae*. 2 species found in the lowlands of eastern and southeastern N America. The members of the G are up to 6 cm large and very reminiscent of *Fundulus**, but more graceful to look at. In contrast to *Fundulus,* the D is clearly located in front of the An; both the fins are of similar build and colouring, and roundish. C fan-shaped. The body is elongate, barely flattened laterally. The ventral profile is more arched than the dorsal profile, particularly in the throat and breast areas. In the *Lucania* G, the scaling on the head is very primitive (a feature that is very important in the systematics* of the Egg-laying Tooth-carps). Apart from 3 small, uncovered scales from the first row behind the mouth, all the other scales point backwards and overlap each other like roof-tiles.

Lucania species are shoaling fish that like to live in waters with plenty of vegetation, near the surface. So, in captivity, these fish should be kept in tanks that are not too tall and contain finely pinnate clusters of plants. Make sure that enough room is left for them to swim about in. *L. parva* only requires room temperature; *L. goodei* needs to be kept at a temperature of 20–24 °C. There are no special requirements as far as water preparation is concerned, but a frequent top-up with fresh water will keep the fishes happy and will encourage them to spawn. The reproductive period lasts for several weeks. The male will drive at the female with great vigour, then the mating pair will seek out a plant cluster near the surface and lay the spawn there. The female only lays a few eggs each day. Since *Lucania* has a strong tendency to eat its spawn, it is a good idea to transfer it to separate rearing tanks. The fry hatch out on the 10th–12th day and start eating immediately (very fine nauplii, rotifers, Infusoria). Older specimens should be fed with black fly larvae *(Aedes*, Culex*)*, larger types of *Artemia**, Enchytraeids etc.

– *L. goodei* JORDAN, 1879; Blue-fin Top Minnow, Killifish. From Florida. Length up to 6 cm. Body a gleaming brassy colour, edges of the scales paler, which makes the fish look as if it has reticulated markings. A deep black, zig-zagging band extends from the snout to the insertion of the red C. There is also a thin, dark line between the P and An. The D and An have an orange-red base with a bow-shaped blue zone above and blue-white tips. The female's fins are less colourful.

– *L. parva* (BAIRD and GIRARD, 1955). From the State of New York south to Florida and Texas. Length up to 5 cm. Less attractively coloured, the middle scales on the flanks have black borders.

Lucernaria, G, *see* Scyphozoa

Lucifuga subterranea *see* Gadiformes

Luciocephalidae, F, *see* Luciocephaloidei

Luciocephaloidei; Pike-heads. Sub-O of the Perciformes*, Sup-O Acanthopterygii. This Sub-O is only represented by one F (Luciocephalidae) containing one species, *Luciocephalus pulcher,* the Pike-head (fig. 737). It grows up to a length of 18 cm and is like a Pike in shape. It is found in the Malaysian Peninsula, in Sumatra, Kalimantan, as well as Banka and Biliton. The head is long, the mouth superior, with a wide gape. There are folds embedded in the mouth, which, when protruding, unite with the protruding mouth to form a funnel. An accessory air-breathing organ is located in the gill cavity, which is rather like a labyrinth organ (*see* Anabantoidei). The D is short, located well to the rear. The C is rounded and the An lies opposite the D. The Vs are very long and extended like a sword. The pelvic bones are directly joined to the shoulder girdle. There are scales on the skin and there is no swimbladder. Colouring yellow-brown with a broad, dark longitudinal band extending from the lower jaw to the root of the tail. Either side of this band is a paler area. There are also lots of spots all over the body and fins. Information about reproduction is not reliable; although some think that it is livebearing, whereas others say that the Pike-head is a mouth-brooder. So far, specimens have only been kept in the aquarium for short periods. They die from unknown causes.

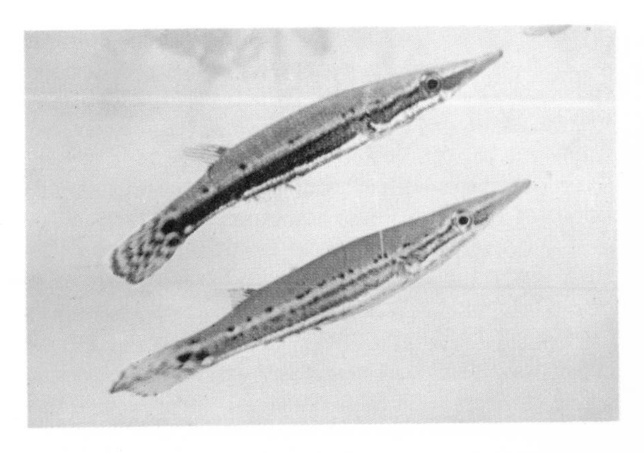

Fig. 737 *Luciocephalus pulcher*

Luciocephalus pulcher *see* Luciocephaloidei

Lucioperca, G, name not in current usage, *see Stizostedion*

Luciosoma BLEEKER, 1855. G of the Cyprinidae*, Sub-F Rasborinae. From SE Asia. Medium-sized to large shoaling fish of elongate, Pike-like shape. Only slightly flattened laterally, the dorsal profile is more or less straight. The head is pointed and is characterised by a deeply cleft mouth. 2 pairs of barbels, of which the ones from the maxillary bone extend as far as the eyes. LL complete. The D and An are located well to the rear, C deeply forked. *Luciosoma* prefers the upper levels of water and should be kept in large aquaria with good lids. The water must be soft and slightly acidic. Temperature 24–26 °C. Food consists almost exclusively of insects and they are taken from the surface. Other live food should be given only as a substitute. No information available regarding reproductive behaviour and breeding.

– *L. spilopleura* BLEEKER, 1855; Apollo Shark. From Indonesia, Thailand, Sumatra, Malaysian Peninsula. Up to 25 cm long. Dorsal surface olive or yellowish, flanks a light ochre-yellow. There is a blue-black fleck edged in gold behind the operculum which continues as a matt-blue longitudinal band strewn with dark-violet and orange-coloured flecks. Fins yellowish or reddish. Young specimens only are suitable for room aquaria (fig. 738).

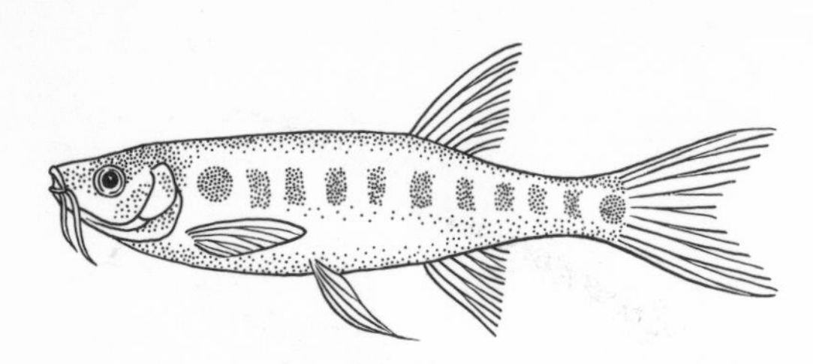

Fig. 738 *Luciosoma spilopleura*

– *L. trinema* BLEEKER, 1855. From SE Asia. Up to 30 cm long. Flanks yellow-grey with a blue gleaming band extending from the eyes to the upper lobe of the C. In the anterior part of the body, this band breaks up into flecks with bronze-coloured dots. There is a gold gleaming fleck in front of the eye and a silver gleaming fleck on the upper edge of the gill-covering. Fins colourless, C edged in a dark colour. Vs have long, white fin rays. Young specimens only are suitable for room aquaria.

Ludwigia LINNÉ, 1753 (including the G *Jussiaea*); False Loosestrife. G of the Onagraceae* comprising 75 species. Found almost everywhere on earth, except in cold regions. Shrubs and perennial plants with upright or horizontal stem axes. Alternate or decussate leaf arrangement. Several species have breathing roots*. Flowers hermaphroditic, polysymmetrical. Perianth usually double. Calyx and corolla usually 4-part, but also 3-part to 7-part. In rare cases, there is no corolla. There are as many stamens as sepals, or twice as many. There are also equal numbers of carpels and sepals, the carpel being fused and inferior. Capsule fruit. The various species of the G grow in wet and marshy places and form floating mats on the water surface. However, some live submerged. The larger species are often cultivated in botanical gardens. The small species (which usually have a decussate leaf arrangement) like light, but apart from that they are popular, undemanding and widespread aquarium plants with a variety of leaf shapes and colours. All the species can be propagated with cuttings.

Fig. 739 *Ludwigia arcuata*

– *L. arcuata* WALTER. From eastern N America. Leaves narrowly linear, olive-green to reddish in colour (fig. 739).

– *L. palustris* (LINNÉ) ELLIOTT. From W Europe, south-eastern N America. Leaves stalked with roundish blades, pale-green. A cold water plant.

– *L. repens* FORSTER. From south-eastern N America. This species has several forms. The leaves are lanceolate or stalked with roundish blades. Olive-green or reddish or pale-green.

There is a hybrid form – a cross between *L. repens* and *L. palustris* – which is widely used in aquarium-keeping.

Lugworm *(Arenicola marina) see Arenicola*

Lumbricidae; Earthworms. The F of the earthworms comprises about 170 species and they are found in all the continents. Cl Oligochaeta*. These worms live in the ground in fairly damp or damp places. Some of them help improve the soil (by loosening it up, improving ventilation, or forming humus), and so they are important in agriculture. After heavy rainfall, earthworms are forced to come to the surface because there is a lack of air in the soil. They eat organic matter in the soil, especially the dead parts of plants. They are hermaphrodites, laying their eggs individually or several at a time in firm mucous cocoons.

For aquarists, earthworms are a valuable food animal; for anglers, they make good bait. The Common Earthworm *(Lumbricus terrestris)* is useful for both purposes. It lives in clay soil and grows up to 30 cm long. It is a brownish colour. Other species are also suitable, including the smaller Dung Worm *(Eisenia foetida)* which often occurs in large numbers in dung heaps and compost piles. It is a dark-red colour. Other usable species belong to the G *Allolobophora*. They are found in wet foliage and moss. Earthworms are readily eaten by predatory fish and a whole series of invertebrates. If the weather is too dry or too cold, earthworms tend to withdraw deep into the ground and then they are not easy to find. So, it is worthwhile keeping an earthworm box where they will live for years under the right conditions and propagate. Large wooden boxes are best, provided they have a lid that closes well. Put a layer of autumnal leaves into the bottom then cover it with a layer of earth. Make sure the earth stays moist. At lengthy intervals, feed your worms with waste vegetable matter from the kitchen (the amount will depend on how many worms you are keeping). Vegetables, fruit, potatoes and grated carrots are all suitable and should be mixed into the earth. Before feeding the worms to the fish, rinse them in a sieve and leave them in a container for a while to empty their guts. They can also be shredded up to be fed to your specimens.

Luminescent Organs. Organs that are capable of producing light. Some Sharks have them *(Etmopterus, Isistius)*, as do marine bony fish from the Os Anguilliformes*, Clupeiformes*, Gadiformes*, Lophiiformes* and Perciformes*. It is not just deep-sea species that have luminescent organs, but some coastal fish as well (eg *Acrosoma, Anomalops, Porichthys, Sephamia*).

In fish, luminescent organs usually consist of special gland cells in the epidermis that either contain symbiotic luminescent bacteria or produce light themselves by means of fermentative oxidation of a luminescent substance (luciferin). The gland cells are often distributed over the whole body. Often, they are concentrated in the lower region of the body and appear as flecks or luminous patches. Sometimes, they appear near the eyes and on the forehead. In other fish, the luminescent organs are built like spot-lights. Beneath the gland cells, the subcutaneous tissue, together with pigment cells, forms a reflector. The luminous effect can also be strengthened by lens-shaped groups of cells and iris-like folds of epidermal skin. Usually, the light that is beamed out is yellowy-green, but if pigments are added red and violet-coloured tones may appear. We can only be sure of the purpose of luminescent organs in a few cases. In many Deep-sea Angler-fishes (Ceratioidei), for example, the luminescent rod acts as a lure. The way the organs are arranged according to species may also help individuals of a species to identify one another. Luminescent organs that light up for brief periods are thought to have some form of defensive function.

Lumpenus lampretiformis *see* Blennioidei

Lumpfish *(Cyclopterus lumpus) see* Cottoidei, *Cyclopterus*

Lumphead *(Steatocranus casuarius) see Steatocranus*

Lumpsucker *(Cyclopterus lumpus) see* Cottoidei, *Cyclopterus*

Lumpsuckers (Cyclopteridae, F) *see* Cottoidei

Lumpsuckers (Cyclopterinae, Sub-F) *see* Cottoidei

Lung *see* Respiratory Organs

Lungfishes, Sup-O, *see* Osteichthyes, Ceratodiformes

Lutjanidae; Snappers (fig. 740). F of the Perciformes*, Sub-O Percoidei*. Robust predatory fish mainly from the tropical Indo-Pacific. Some are very difficult to tell apart from the Serranidae*. Body moderately or very elongate, compressed laterally. The head is scaled. The mouth has a deep gape with strong, conical teeth on the jaws and small teeth on the palatal bone. In some Ga the praeoperculum is serrated. D long with 10–12 spines; the soft-rayed part, which is sometimes covered with scales at the base, is relatively short and stands opposite the short An. C slightly indented or straight. Medium-sized or small comb-scales. Snappers live mainly in shoals or as individuals on reefs or rocky coastlines, as well as in estuaries. They feed on fish mainly which they snap up very quickly (hence the name!). Many species are attractively coloured and some grow to a fair size. They are considered very edible, although in particular regions the meat is poisonous (it probably depends on the fish's diet). The Snappers are closely related to the Nemipteridae, also from the Indo-Pacific. The latter have elongated Vs and often an elongated upper caudal fin lobe. They are usually a shining pink colour or yellow. Snappers are commonly kept in aquaria. For further details see under the Ga *Caesio* and *Lutjanus*.

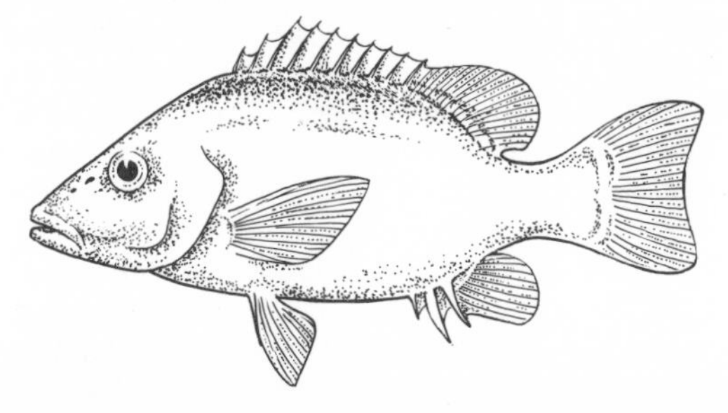

Fig. 740 Lutjanidae-type

Lutjanus BLOCH, 1790; Snapper. G of the Lutjanidae*. Contains a large number of species with very varied forms and life-styles. They live in warm seas. They are predators and either live alone or in shoals. Some eat small snimals. The name 'Snapper' refers to the greedy way in which they eat. Length 20–100 cm.

Fig. 741 *Lutjanus sebae*

Lutjanus species are colourful fish with a typical perch build, the majority of which are fairly tall with a straight ventral profile. They also have a large, protrusible mouth. The D and An are covered in scales to a greater or lesser degree, and in contrast to *Caesio** the C is at most only slightly indented. There is also a groove in the praeoperculum which accommodates an outgrowth of the interoperculum. Strong palatal teeth both on the palatinum and vomer. These fish survive well in the aquarium; they grow very quickly if fed on all type of animal food and they are often aggressive. Provide plenty of room for swimming and plenty of hiding-places. Frequently imported species are:

– *L. decussatus* (CUVIER and VALENCIENNES, 1828); Checked Snapper. From the Indo-Pacific. 30 cm. Red-brown longitudinal stripes, together with transverse bands form a check pattern in the upper half of the body. Ground colouring white.

– *L. sebae* (CUVIER and VALENCIENNES, 1828); Blue-banded Sea Perch; Emperer Snapper; Government Bream; Red Emperor; Red Snapper. From the Indo-Pacific. 1 m long. White with 3 broad, dark-red bands (fig. 741).

Lux. A unit of illumination in respect of area illuminated.

Lybia, G, *see* Brachyura

Lyes *see* Bases

Lymnaea LAMARCK, 1799. G of the pulmonate Snails (Basommatophora*). Includes the taxa *Radix* and *Galba*, which are occasionally regarded as independent Ga. *Lymnaea* comprises a large number of species that live in very varied types of fresh water and brackish water throughout the holarctic* realm. The broad, triangular feelers are typical of the G. The larger species are well suited to the pond aquarium and open-air installations (fig. 742).

Lymphatic System. A thin-walled discharge system for tissue fluid (lymph) that opens out into the venal system. White blood cells (leucocytes and lymphocytes) only are present in the colourless or yellowish lymphatic fluid.

In the lower vertebrates, the lymph flow is helped along by contractile lymph hearts. Many bony fish have a chambered lymph heart at the point where the caudal lymph vessels join up with the caudal vein. Fish do not have lymph nodes. These are aggregations of cells found in mammals that both filter the lymph as well as form lymphocytes and and defensive matter. Fish have lymphatic tissue in several organs. The lymphatic vessels in the intestine take over the transportation of digested fats.

Lymphocystis. A virus* that is capable of developing in a fish's skin cells. The connective tissue cells that are attacked are stimulated into excessive growth and become giant cells. At first, small white nodules appear on the fins and in the skin, gradually developing into pearl-like or raspberry-like tumours. In this initial stage it is easy to confuse with *Ichthyophthirius**. Lymphocystis occurs in marine fish, particularly coral fish, but is rare in freshwater fish. There is no easy cure. Diseased fish should be killed as quickly as possible and any fish that have been in contact (but without showing any signs of the disease) should be kept in quarantine for 2–4 months. With valuable fish, you can try cutting out the bits of fin that are diseased. Dab the wounds immediately afterwards with tincture of iodide*, Mercinochrome®* or Rivanol®*. Placing these solutions directly onto the tumour itself does not appear to be successful. In a few instances, diseased coral fish have been treated with fluor-hydrocortisone. This substance is contained in Volon®-A-ointment. It should bring about a complete cure if smeared once or twice directly onto the tumours.

Lyomeri, O, name not in current usage, *see* Anguilliformes

Lyre Urchin (*Brissopsis* sp.) *see* Echinoidea

Lyretail *(Poecilia reticulata* standard) *see* Poecilia

Lyretail Coralfish *(Anthias squamipinnis) see Anthias*

Lyretail from Togo *(Aphyosemion filamentosum) see Aphyosemion*

Lyretail Wrasse *(Thalassoma lunare) see Thalassoma*

Lyretailed Panchax *(Aphyosemion australe) see Aphyosemion*

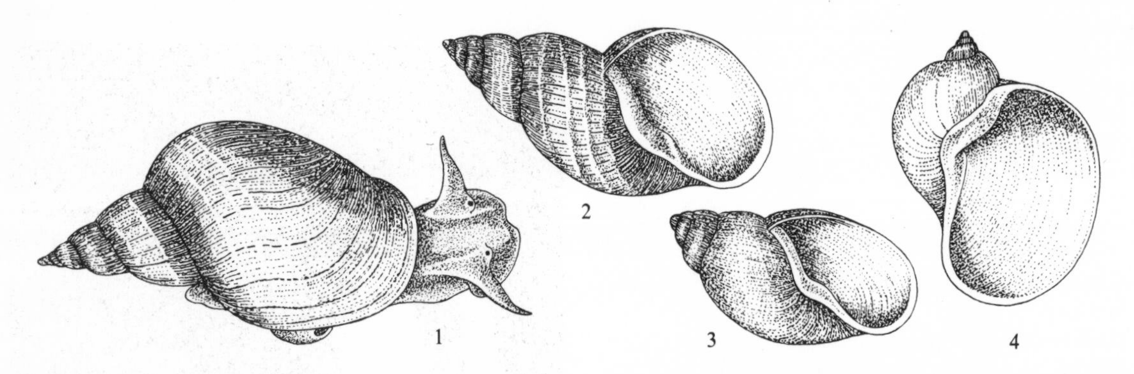

Fig. 742 Various members of the G *Lymnaea*. 1 *L. stagnalis* 2 *L. palustris* 3 *L. peregra* 4 *L. auricularia*

Lysimachia LINNÉ, 1753; Loosestrife. G of the Primulaceae*. 200 species, that are distributed throughout the world. They are perennial plants with rhizomes or thin stem axes that grow horizontally. The various species can be of very varied sizes. Decussate leaf arrangement, in rare cases whorled. Flowers hermaphroditic, polysymmetrical. Double perianth. Calyx and corolla 5-part, the corolla being fused at the base. Stamens 5. Carpels 5, fused, superior. Capsule fruit. The various species of the G grow mainly in marshes and in shallow water. *L. nummularia* is a very adaptable species that was widely used in cold water aquaria in the early days of aquarium-keeping.

– *L. nummularia* LINNÉ; Moneywort, Creeping Jenny, Herb-two-pence. Found in Europe. The stem axes creep along the ground but grow upright in water.

Lysmata RISSO, 1816. G of the Shrimps (Natantia*). These Shrimps are peaceable, sociable creatures that live in the rocky littoral of warm seas. They feed on a

variety of animal food. Typical of the G is the elongated, penultimate joint (carpus) of the second walking leg with its 20–30 joints. The following species is one of the few hermaphroditic crustaceans and makes a delightful aquarium specimen.
– *L. seticaudata* RISSO. From tropical and subtropical seas and the Mediterranean. Found in small groups on the rocky littoral. 5 cm long. Has lovely red longitudinal stripes.

Lysol. A watery solution of cresols, which, as derivatives of phenol (aqueous solution = carbolic acid), have a highly antiseptic effect. 1–2 ml lysol for every 5 litres of water is used to disinfect (Disinfection*) aquarium containers and equipment. Large fish suffering from an attack of Argulidae* or Hirudinea* can be treated with lysol. Make up a solution of 1 ml lysol to 5 litres of water and bathe the fish in it for 5–15 secs.

Lythraceae; Loosestrife Family. F of the Magnoliatae*, comprising 25 Ga and 550 species. Tropical-subtropical, especially in America. Only a few species are found in temperate regions. In rare instances Loosestrifes can be trees or shrubs, but most are perennial plants. Some annuals also exist. The leaves are opposite or whorled, in rare cases alternate. The flowers are hermaphroditic, some are small, and polysymmetrical or monosymmetrical.

Loosestrifes colonise both dry and wet habitats. Several species can therefore be considered for the aquarium – *Ammannia**, *Didiplis* and *Rotala**. The G *Lythrum*, which is found all over the world, contains about 25 species. One, the Purple Loosestrife *(Lythrum salicaria)*, is native along the waters of Europe and Asia, and it has also been introduced in eastern N America. It can be used along the edges of open-air ponds. It is possible that many *Lythrum* species are suitable for the aquarium.

Lythrum, G, *see* Lythraceae

Mackerel *(Scomber scombrus) see* Scombroidei

Mackerel Shark *(Lamna nasus) see* Lamnidae

Mackerel Sharks, F, *see* Lamnidae

Mackerels (Scombridae, F) *see* Scombroidei

Mackerels, related species, Sub-O, *see* Scombroidei

Macrocheira, G, name not in current usage, *see* Brachyura

Macrodon tralura, name not in current usage = *Hoplias malabaricus, see Hoplias*

Macrognathus LACÉPÈDE, 1800. G of the Mastacembelidae, Sub-O Mastacembeloidei*. Found in India and as far as Burma, Thailand, Malaya, Indonesia. Brackish and freshwater species. These are large fish, elongated like an eel. Their head is elongated by having a mobile extension on the snout. Can be distinguished from the G *Mastacembelus** because it has a distinct corrugated effect beneath the snout extension. The mouth and gill openings are small. The D and An are inserted along a long stretch of the body and they are located well to the rear, clearly distinct from the C. The D and An are both used for swimming and they enable the fish to swim forwards and backwards. *Macrognathus* species are typical crepuscular fish that lie hidden among plants or cocooned in the substratum during the day. They are best kept in species tanks with soft substrates and hid-

ing-places made up of rocks and roots. A bit of sea salt is beneficial, as is a regular topping up with fresh water. Temperature 22–28 °C. Feed with sturdy live food, especially worms, insect larvae and small crustaceans. There is no information about breeding.
– *M. aculeatus* (BLOCH, 1787); Spiny Eel. From India, the Malaysian Archipelago, Sumatra, Sri Lanka (Ceylon), Thailand. Up to 38 cm. Body dark brown with a distinct marbly effect and transverse band markings. There are 3–10 peacock-eye blotches on the light brown D (fig. 743).

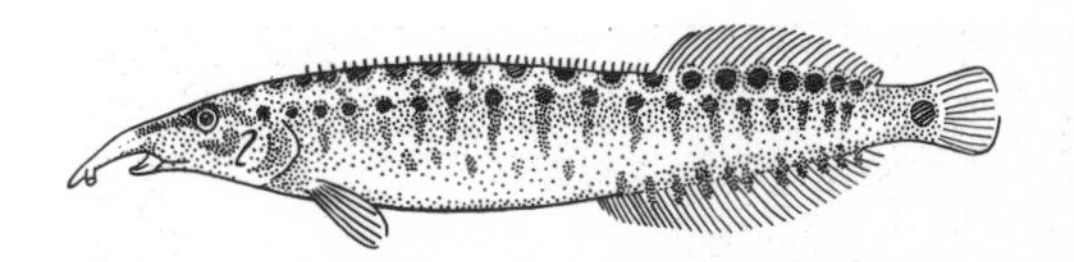

Fig. 743 *Macrognathus aculeatus*

Macrones, G, name not in current usage, *see Mystus*

Macropodia LEACH, 1830; Spider Crabs. G of the Brachyura*, F Majidae. *Macropodia* lives among vegetation in cool and warm seas. They are delicate crabs that feed on growths and small animals. Their back has no spines, the pseudorostrum is pointed and made up of two spines located very close to one another, and the eye-stalks are long and irretractible. The body is shaped like a pointed pear from which sprout very long, thin legs. The body is about 1–2 cm in length. *Macropodia* is fragile-looking and can completely camouflage itself. They are peaceful, slow creatures that can be kept in calm, algae-rich aquaria. Predatory fish and crustaceans can be a danger for them. These details also hold for the similar G *Inachus* WEBER, 1795, which has a spiny back, and for *Stenorhynchus**.

Macropodus SHERBORN, 1928; Paradise-fish. G of the Belontiidae, Sub-O Anabantoidei*. From SE Asia; India to Korea, where it lives in calm waters in the lowlands, as well as in brackish rivermouths. Small fish, 7–12 cm in length. Body elongate, flattened laterally. Fins large. The D and An begin almost in the same region, in the males, they are pointed at the rear. The first ray of the V is elongated a bit like a thread. LL reduced. There are small conical teeth on the jaws. Ground colour of the body varying shades of brown; red, green and blue-black tones predominating in the fins. In some species there is a metallic coloured blotch on the gill-covering. The colouring in the male is more intense. They require temperatures of 20–24 °C. They seem to do best in tanks with plenty of vegetation. Apart from live food, artificial and dried foods are also accepted. All the species build a bubble-nest, sometimes only a small one. Interesting species that have proved successful in the aquarium include the following.
– *M. chinensis* (BLOCH, 1790); Round-tailed Paradise-fish. S and E China, Korea. Up to 7 cm. Body colouring yellow-brown with dark transverse bands. The C is rounded, blue-black in colour with a yellow edge. Reared fish are not productive.

– *M. cupanus cupanus* (CUVIER and VALENCIENNES, 1831); Spike-tailed Paradise-fish. From the Near East and Further India, the Malaysian Peninsula, Sri Lanka. Up to 7.5 cm. Ground colouring various shades of brown. There is a clear greenish shimmer along the longitudinal band markings that are visible. Throat and breast areas dark. The middle of the C extends into a point. The D, An and C are bluish-grey with reddish speckles. This species will often change colour rapidly. When excited or during the spawning period, the male is pale red, the female almost black (fig. 744).

Fig. 744 *Macropodus cupanus cupanus*

– *M. cupanus dayi* (KÖHLER, 1909); Striped or Red Spike-tailed Paradise-fish. From the Near East and Further India. Up to 7.5 cm. Particularly slim. Ground colour brown with 2 dark longitudinal bands on the body. Throat, breast, belly and fins all red, the fins also having green borders. The middle rays of the C, particularly in the male, are elongated and black. This species needs warmth, for breeding purposes up to 30 °C (fig. 745).

Fig. 745 *Macropodus cupanus dayi*

– *M. opercularis opercularis* (LINNAEUS, 1758); Paradise-fish. Korea, China, Vietnam, Taiwan. Up to 9 cm. The sides of the body have green and red transverse bands. The fins are red with blue markings. The D and An have white edges. Both the upper and lower parts of the C extend into points. The operculum is held ajar and has a blue-green fleck on it, edged in red. This species is somewhat intolerant, particularly in the spawning season. Kept as an aquarium fish in Europe since 1869 (fig. 746).

Fig. 746 *Macropodus opercularis*

– *M. opercularis concolor* AHL, 1935; Black Paradise-fish. Probably native to S China. Up to 12 cm. Dark colour tones predominate in this sub-species. The body is grey-brown or reddish-grey. The scales have black edges. The D, An and C are yellowish with dark rays and flecks. The edge of the D and the elongated rays of the C are white-blue. In the spawning period, the male is almost black in colour, the female with irregular brown transverse bands. Very attractive forms result from a cross between *M. opercularis opercularis* and *M. opercularis concolor*. A relatively tolerant fish.

Macrorhamphosidae, F, *see* Aulostomoidei

Macrouridae, F, *see* Gadiformes

Macruriformes, O, name not in current usage, *see* Gadiformes

Macrurus rupestris *see* Gadiformes

Madagascar Lace Plant *(Aponogeton madagascariensis) see Aponogeton*

Madagascar Rainbow *(Bedotia geayi) see Bedotia*

Madreporaria; Calcareous Corals (fig. 747). O of the Anthozoans (Anthozoa*) comprising about 2,500 species found in cold seas, but mainly in warm seas, where they play a considerable part in the forming of

Fig. 747 Madreporaria

coral reefs*. The polyps are mainly colonial, although some are solitary. They have a hard, calcareous skeleton which is in contrast to the otherwise very similar Sea-anemones (Actiniaria*). No medusae are formed. The shape of the coral skeleton is dependent on species. There are bush-like, branched forms, for example, such as *Acropora** and *Cladocora** and massive block-types such as *Diploria*. The solitary species usually have fungi-type skeletons, eg the *Caryophyllia* species or the tropical corals of the G *Fungia* (fig. 748) (which can reach a diameter of 25 cm). These solitary species will also survive well in the aquarium. Coral skeletons are highly prized decorative material for a marine aquarium. But the gleaming white colour of the cleaned skeletons is not the true colour of the living animals. In nature, the calcium is covered by a layer of living tissue that often contains zooxanthella*, which gives the coral its greenish-brown colouring. The polyps are fully retractable and on many reefs they only open out at night. They feed on plankton.

Fig. 748 *Fungia*

In the past, it was not possible to keep living coral in the aquarium, but modern techniques and the understanding of the chemistry of sea water means that it is possible today. With good protein filtration (*see* Foam Filters) and a high redox potential*, Madrepore corals with large polyps will survive particularly well. Many shallow-water forms need very bright light, and those that live in shady areas or at great depths must be looked after so that the colonies do not become overrun with algae.

Madrepore Plate *see* Echinodermata

Magnoliatae (Dicotyledoneae, name not in current usage). Cl of the Spermatophyta* comprising 250 Fs and 172,000 species. In contrast to the Liliatae*, the Magnoliatae embryo has two seed-leaves or cotyledons. Every kind of leaf arrangement is found among the Magnoliatae. The leaves come in a variety of shapes, simple or pinnate and net-veined. If the stem were to be seen in cross-section, the vascular bundles* are usually arranged in a circle at the beginning and they are usually open, which means that a secondary growth of thickness is possible. The radicle is usually long-lived and in some cases forms the whole root-system of the plant by branching out; but roots that have developed from shoots also exist. In flower structure, 5-part circles predominate. Woody stemmed plants and shrubs belong to the Magnoliatae.

Mailed Catfishes, F, *see* Loricariidae

Maja LAMARCK, 1801 (fig. 749). G of the Brachyura*, F Majidae. These crabs live in warm seas where they

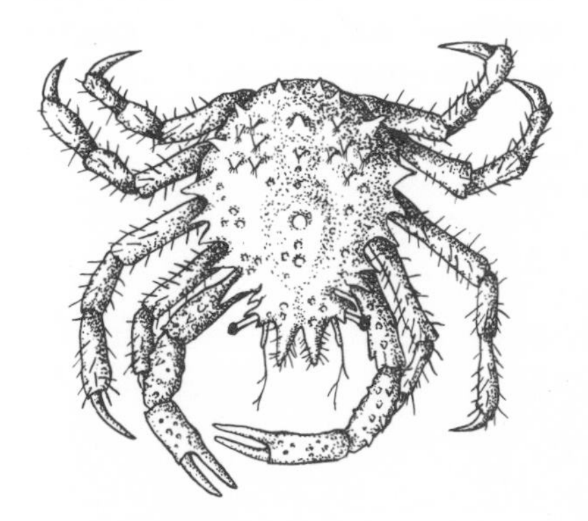

Fig. 749 *Maja verrucosa*

live on growths of all kinds. The ovate carapace, which can grow up to 20 cm long, has strong spines along the edges and shorter ones on the top. The rostrum (pseudorostrum) is forked. Young specimens especially camouflage themselves with any kind of growth they can find in their surroundings. Sexually mature members of the species *M. squinado* (HERBST) collect together great quantities of material in early summer in order to mate in safety with the newly moulted, ripe females. Despite their size, *Maja* species make peaceful, unusual specimens for the aquarium (unlike *Hyas**). Some members of the G *Pisa* LEACH, 1830, which in other respects are similar to *Maja*, can be more sensitive in the aquarium.

Makaira audax *see* Scombroidei

Makos, G, *see* Lamnidae

Malachite Green. A triphenylmethane colouring substance which can be bought as double salt zinc chloride (brassy yellow crystals) and as oxalate (green crystals with a metallic sheen). The crystals dissolve easily in water to give a blue-green colouring. Solutions of malachite green-oxalate (0.03 mg/l litre water) or malachite green-chloride (0.06 mg/l litre water) are used in the treatment of pathogenic ectoparasites in fish (*Ichthyophthirius**, *Ichthyobodo**, *Chilodonella**, *Trichodina**, *Dactylogyrus*, *Gyrodactylus* [Trematodes*]). Treatment is in the form of prolonged baths (*see* Bathing Treatments). For an exact dosage, dissolve 1.5 g malachite green-oxalate in 1 litre of water, then take 2 ml of this solution for every 100 litres of aquarium water. Since malachite green is broken down into ineffective compounds, further doses must be applied after 3, 5, and 7 days (a further 1 ml/100 litres water every time). Malachite green is an active component of many preparations designed to combat fish ectoparasites that are available through the trade. It can be removed from the aquarium water by filtering it through activated charcoal (Adsorption*). Be careful when handling Malachite green; it has cancer-inducing properties. It is best to deal with it in solution form only and wear rubber gloves.

Malacosteidae, F, *see* Stomiatoidei

Malacostraca (fig. 750). Sub-Cl of the Crustaceans (Crustacea*). Known as 'higher crustaceans', a term which is phylogenetically incorrect, since the Malacostraca still have some very primitive features, such as extremities on the abdomen. Anatomical features of note include the constant number of 8 thorax segments and 7 abdomen segments, the frequent presence of a dermal duplication of the last head segment (carapace) and stalked eyes. Well known groups of crustaceans that belong to the Malacostraca include the Lice (Isopoda*), ten-legged crustaceans such as Crabs, Lobsters and Shrimps (Decapoda*), Amphipods (Amphipoda*), Opossum Shrimps (Mysidacea*) etc. Larger forms that are particularly attractive in shape and colour play a considerable role in marine aquarium-keeping, and are often among the attractions in public aquaria.

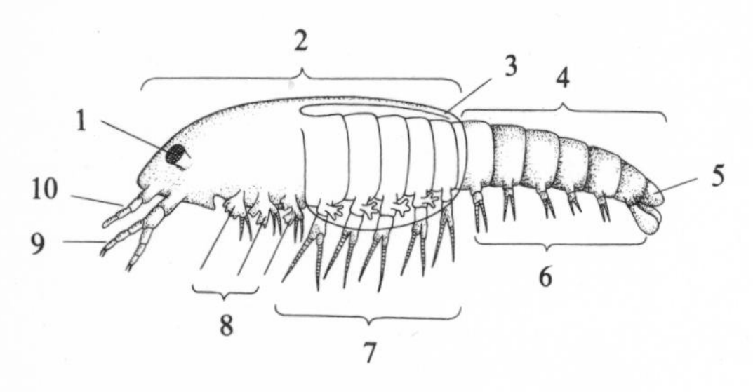

Fig. 750 Malacostraca. Anatomy of a 'higher crustacean'. 1 Stalked complex eye 2 Cephalothorax 3 Carapace 4 Abdomen 5 Telson 6 Abdominal legs 7 Walking legs 8 Mouth parts 9 2nd antenna 10 1st antenna

Malania anjouanae *see* Coelacanthiformes

Malapteruridae; Electric Catfishes. F of the Siluriformes*, Sup-O Ostariophysi. There is only one species in the F, the 1 m-long Electric Catfish *(Malapterurus electricus)* (fig. 751), which is a bulky fish with a thick head and swollen lips. It lives in the Nile region and further afield in tropical Africa. The eyes are small, phosphorescent in the dark, although older specimens are usually blind. There are 3 pairs of fleshy barbels. There is no D and the adipose fin is large, located just in front of the fan-shaped C. The An is fairly long and broad. Ps with no spine. The electric organs are paired and they are positioned in the thick skin in such a way that they envelop the body like a cloak. Unlike all the other fish that are known to have electric organs, those of the Electric Catfish have not developed from the musculature, but from gland cells in the skin. The negative pole is at the front, the positive behind. A strong current is released during a discharge – 100–300 volts. Several weaker currents follow. The electric organs helps the fish to obtain its food and to defend itself, and probably to orientate itself (radar system?). Large numbers of black speckles are strewn among the dirty flesh colours or dark grey-yellow colours that form the body colouring. The Electric Catfish was known in ancient Egypt; it was part of their food and also crops up in their drawings. Its meat is still consumed in many parts of Africa even today. Its electric organs are occa-

Fig. 751 Head of *Malapterurus electricus*

sionally used for healing purposes. There are still some uncertainties about reproduction. Their diet consists mainly of fish. During the day, *Malapterurus* hides away on the bottom, only coming out as dusk falls. Young specimens, up to about 6 cm, are suitable for normal aquaria. Larger examples are very liable to attack and will damage other fish when they discharge their electricity, so they are best kept on their own. They will grow very quickly when fed on live food of all types.

Malapterurus electricus *see* Malapteruridae

Malarial Mosquitoes, G, *see Anopheles*

Malayan Angel *(Monodactylus argenteus) see* Monodactylidae

Malayan Flying Barb *(Esomus malayensis) see Esomus*

Malpulutta DERANIYAGALA, 1937. G of the Belontiidae, Sub-O Anabantoidei*. From Sri Lanka. Found in small pools within river systems and in very weedy, flowing waters. Length to 6.5 cm. Body elongate, flattened laterally. LL only above 6 scales. The D and An, especially in the male, are pointed towards the back, and the two middle rays of the C are extended like threads. Vs pointed, the first ray elongated. The body is red-brown with dark marbling, sometimes appearing like indistinct longitudinal bands. In the male, the D, An and C may have dark dots on them and they may have blue-green edges. The elongated fin rays are also blue. *Malpulutta* should be kept in soft, slightly acidic water at a temperature of 24–26 °C. Feed with all kinds of small live food. A small bubble-nest is made for the spawn. The only known species is:

– *M. kretseri* DERANIYAGALA, 1937.

Mammalia; Mammals. Cl of the Vertebrates (Vertebrata*) made up of around 6,000 species. It is usually divided into 2 Sub-Cls, the monotremes (Prototheria) and the true mammals (Theria). The egg-laying mammals (Monotremata, O) of Australia, Tasmania and New Guinea belong to the first group – the Duck-bill Platypus and the Spiny Ant-eater or Echidna. The true mammals are divided into 17 Os. Among these, the pouched mammals (Marsupialia) make up a fairly independent systematic unit; they have a pouch on the breast for rearing their young. They live in the Australian realm (eg Kangaroos, Koala Bears, Possums) and in America (eg the Oppossums). The other Os are made up of placental mammals; the embryos are joined to the mother's blood circulatory system via a placenta.

The anatomical features mammals have in common concern the milk glands (mammae) and a covering of hair on the body. Only in some water-dwelling mammals are the hairs reduced to a large extent, although they are present in the embryonal stage (Whales, Sea-cows, Hippos). The blood circulatory system and the heart are divided up completely separately into an arterial and a venal section. The body temperature is constantly warm (35–40 °C) and can be lowered for periods during hibernation. Typical skeletal features include the 7 cervical vertebrae, the topmost pair of which (the atlas and epistropheus) enable the head to move in all directions. In the cranium, the formation of a secondary gnathic joint is of importance. The efficient nervous system with the highly developed sense organs has reached a peak of achievement in the mammals that is unsurpassed by any other living beings. The digestive organs have become adapted to different ways of feeding. Some mammals are true predators, others are vegetarians, but there are also a great many omnivores. Mammals are found from the polar regions to the tropics, and so they live in a huge variety of biotopes. The Chiroptera, which include the Bats and Flying Phalangers, have conquered the air for example. Other mammals have colonised the sea and have undergone the necessary adaptations of the body (eg the extremities or tail turning into kinds of fins). These include the Whales, and Dolphins (Cetacea*), Sea-cows (Sirenia) and Seals (Pinnipedia). The largest mammal in the world is the 30-m-long, 150-tonne Blue Whale *(Balaeoptera musculus)*.

Mammals, Cl, *see* Mammalia

Mandarin Fish *(Synchiropus splendidus) see Synchiropus*

Man-eating Sharks, F, *see* Carcharhinidae, *see also* Lamnidae

Mangrove. Plant communities* that are typical of the flat, calm coasts and rivermouths of the tropics and subtropics. Such areas are very much influenced by the tides. Mangroves are largely made up of trees that can tolerate sea water. To do this, they have developed particular adaptations, such as breathing roots*, buttress roots, or seeds that germinate on the parent plant. When the area is flooded, often it is only the tree-tops that rise above the water.

Manta birostris *see* Mobulidae

Mantas, F, *see* Mobulidae

Mantis Shrimps, O, *see* Stomatopoda

Many-rooted Duckweed *(Spirodela polyrrhiza) see Spirodela*

Many-spined Climbing Perch *(Ctenopoma multispinnis) see Ctenopoma*

Marbled Cichlid *(Astronotus ocellatus) see Astronotus*

Marbled Climbing Perch *(Ctenopoma oxyrhynchus) see Ctenopoma*

Marbled Hatchetfish *(Carnegiella strigata) see Carnegiella*

Marcusenius GILL, 1862. G of the Mormyridae, O Mormyriformes*. From central and NW Africa. Small to medium-sized fish that inhabit muddy waters with thick vegetation. Like other members of the F, most of them are territorial, solitary fish. The body is short, flattened laterally and narrows at the back into a slim caudal peduncle. The head is large and compressed and ends in a snub, rounded snout that only juts out slightly above the small, inferior mouth. No outgrowths on the snout. Distinguishable from other Ga of the Mormyridae* by having just one row of teeth confined to the middle of the upper and lower jaw. The rear nasal opening is displaced well to the rear by the oral cavity. Keep specimens in tanks provided with plenty of hiding-places made up of rocks and roots, and stand the tank in a dark spot. The substrate should be soft. Good-quality old water is best, and fresh water should be added to it at intervals. Temperature 20–28 °C. Food consists of small live food, especially worms, and dried food. *Marcusenius* only becomes active at dusk and at night. Most are intolerant of like species by peaceful towards other species. Not yet bred in captivity.

– *M. isidori* (CUVIER and VALENCIENNES, 1846). From the Nile, upper Zambesi. Up to 10 cm long. Dorsal surface grey-brown, flanks paler with a weak silvery-violet sheen. Fins not transparent, grey-brown. There is occasionally a dark longitudinal streak and indistinct flecks. A sociable species (fig. 752).

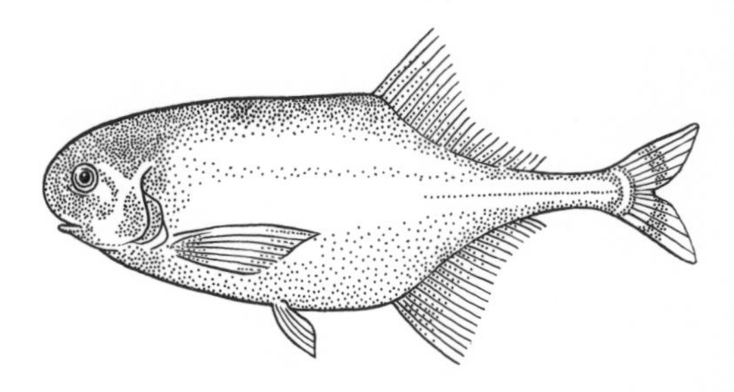

Fig. 752 *Marcusenius isidori*

– *M. longianalis* BOULENGER, 1901. From the lower Niger, SW Cameroun. Up to 15 cm long. This species is particularly slim. An very long, D much shorter. The dorsal surface is black-brown, flanks and fins brown with irregular black flecks strewn across them.

Mare's-tail *(Hippuris vulgaris) see* Hippuris

Mare's-tail Family *see* Hippuridaceae

Marestails, F, *see* Haloragaceae

Margaritana margaritifera *see* Pearls

Marginate Damselfish *(Dascyllus marginatus) see Dascyllus*

Marginate Puller *(Dascyllus marginatus) see Dascyllus*

Marine Eels (Congridae, F) *see* Anguilliformes

Marine Jewelfish *(Microspathodon chrysurus) see Microspathodon*

Marine Water *see* Sea Water

Marlins (Istiphoridae, F) *see* Scombroidei

Marsh Marigold *(Caltha palustris) see* Ranunculaceae

Marsh Plant. A marsh plant grows in habitats that are permanently wet or in shallow water so that the stems or leaves can unfurl over the water surface. Most marsh plants can tolerate being completely flooded over for a short or fairly long time.

Marsilea LINNÉ, 1753; Pepperwort. G of the Pteridophyta* comprising 65 species. Found in the tropics and subtropics and radiating out in to temperate zones. Pepperworts are small perennials with horizontal stem axes. The leaves have pinnate, clover-leaf-like blades consisting of between one and four leaflets (pinnae). Reproduction is through a complicated spore system. The sporangia are located in special containers at the base of the leaves. The various species of the G grow in marshes and in water, and the leaves either float or are fully submerged. Some types can also grow on land. The smaller species are suitable for the aquarium. They are undemanding and form very decorative blankets of foliage. The best known species is:
– *M. quadrifolia* LINNÉ; Four-leaved Marsilea. From Europe. A cold water plant.

Marsilea Family (Marsileaceae, F) *see* Pteridophyta

Marsileaceae, F, *see* Pteridophyta

Marsipobranchi, Sub-Ph, name not in current usage, *see* Agnatha

Marthasterias JULLIEN, 1878. G of the Starfishes (Asteroidea*) found in cold and warm seas. It has strong spines on the upper surface of the 5 arms. As in the closely related *Coscinasterias** the spines are surrounded by a thick cushion of stalked pedicellaria. *Marthasterias* survives very well in the aquarium. It is extremely greedy, and can even open up molluscs. They should only be kept in the same tank as larger, mobile animals for this reason. *M. glacialis* is found from Iceland to the coast of W Africa and in the Mediterranean. It lives at depths of 0–50 m.
– *M. glacialis* (LINNAEUS, 1758); Spiny Starfish. Diameter up to 70 cm. Colouring brownish to olive-green (fig. 753).

Fig. 753 *Marthasterias glacialis*

Mastacembelidae, F, *see* Mastacembeloidei

Mastacembeloidei; Spiny Eel types (fig. 754). Sub-O of the Perciformes*, Sup-O Acanthopterygii. Small to medium-sized, Eel-shaped or ribbon-shaped fish from the fresh waters of Africa, S and E Asia and Indonesia. They have a pointed head which is chiefly noted for its elongated rostrum that can be moved like a trunk. During the post-embryonal formation of the rostrum the front nasal openings are also displaced towards the

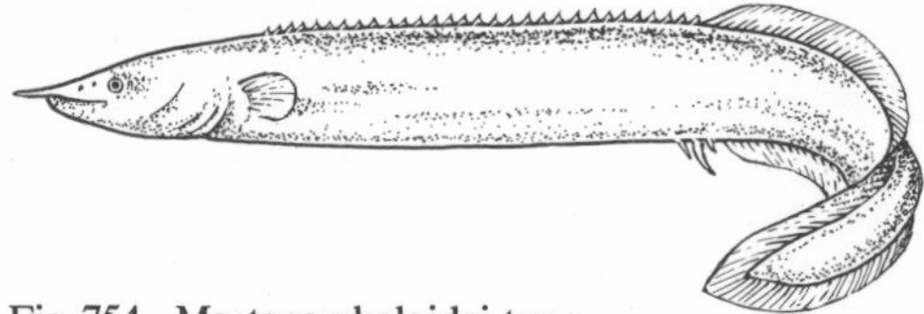

Fig. 754 Mastacembeloidei-type

front, opening to the outside via little tubes at the sides of the rostrum. The rear nasal openings remain in front of the eyes. The mouth is inferior and the jaws bear teeth. There are several peculiarities of the skull, but the only things worth mentioning are that the premaxillary bone and maxillary bone are fused along their entire lengths. The shoulder girdle is joined to the vertebral column, not the skull. The D begins behind the head; the front part consists of individual, erectile spines, whilst the rear part is normal and soft-rayed, beginning in the middle of the body. The An has 3 spines, about as long as the soft-rayed D. The D, C and An form a fin corona of uniform height, with the exception of the G *Macrognathus**. No Vs. Swimbladder present, physoclists. Many Spiny Eels are prettily coloured and patterned. They are bottom-dwelling fish that live in weedy waters that can be muddy or sandy. They can burrow into the substrate very quickly to escape from danger. Many species remain like this during the day and only come out at night to hunt for food. 2 Ga are known and they comprise a total of 50 species; both Ga belong to the one F (Mastacembelidae). Larger species are fished in some areas. For aquaristic information see under the Ga *Macrognathus** and *Mastacembelus**.

Mastacembelus GRONOVIUS, 1763; Spiny Eel. G of the Mastacembelidae, Sub-O Mastacembeloidei*. From S and SE Asia, tropical Africa. Spiny Eels live in brackish and fresh water and they are medium-sized or large fish with an elongated, Eel-like shape. The head is pointed and has a mobile extension to the snout. The gill openings are narrow and placed well towards the underneath. The D and An are inserted along a lengthy stretch of the body, and usually join up with the C in one continuous fin. The fin corona thus produced provides the thrust to move the fish along; it can swim both forwards and backwards. *Mastacembelus* like overgrown waters with muddy substrata, in which they can hide. They are active at night, only leaving their hiding-places as the evening draws in. If possible, specimens should be kept in species aquaria. They like the tank to be in a darkened spot, with a soft substrate and plenty of places to hide. The composition of the water presents no problems provided some of it is exchanged regularly for fresh water and a little sea salt is added. Temperature 22–28 °C. Live food, especially worms and insect larvae. Has only been bred successfully in a few isolated cases.

– *M. argus* GÜNTHER, 1861; Peacock-eyed Spiny Eel. From Burma, Thailand. To 25 cm. Flanks yellowish or dark brown with irregular, whitish-green transverse stripes. Brown-red eye-blotches, edged in lighter colours are found on the dorsal ridge.

– *M. armatus* GÜNTHER, 1861 (fig. 755); Spiny Eel. From India, Sri Lanka (Ceylon), Thailand, Sumatra. A widely distributed species. Up to 75 cm. Dorsal surface

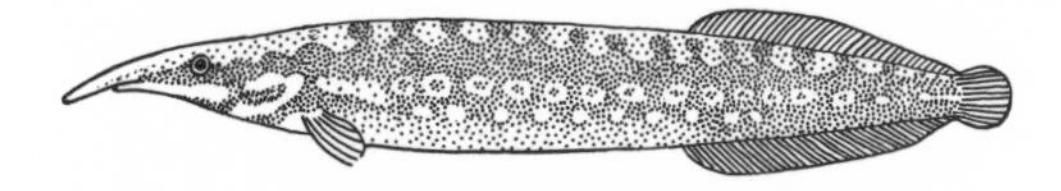

Fig. 755 *Mastacembelus armatus*

dark brown, flanks ochre-yellow. A broad, black-brown longitudinal band extends from the tip of the snout to the root of the tail. Its boundary is irregularly marked, also has a longitudinal band but it is broken up into irregular flecks.

– *M. loennbergi* BOULENGER, 1898; Loennberg's Spiny Eel. From tropical W Africa, Liberia, Cameroun to Lake Chad. Up to 20 cm. A very slim species with a dark olive-brown markings.

– *M. pancalus* (HAMILTON-BUCHANAN, 1822); Spotted Spiny Eel. From the Near East where it is found in large rivers and in brackish water. Up to 20 cm. Dorsal surface dark olive, flanks brown or grey-brown with small, pale yellow speckles. There is a pale longitudinal stripe at the level of the LL and narrow, dark, transverse bands in the rear half of the body. Fins yellow, with dark dots. This species has been bred repeatedly. The mating pair lay their spawn near the surface between floating plants.

Materials. Technical equipment that comes into direct contact with aquarium water must be made out of completely neutral and stable materials. Traces of poisonous substances must not be released into the water, nor must there take place any gradual disturbance of the materials (which occurs particularly in sea water). Often, materials are also unsuitable because they are gradually attacked by bacteria, algae or plant roots, for example. Metals are poisonous in every form (Poisoning*), since even when the smallest amounts are dissolved they can release uncontrollable catalytic processes which are often the cause of insufficient plant growth or sudden deaths among fish. Even the synthetic products that are widely used today in a variety of forms must be thoroughly tested before use. Often they contain poisonous fillings, softening agents and remains of monomer building blocks, eg phenols, which slowly dissolve in the water and cause an insidious kind of poisoning*. Plastics which smell after washing in hot water are not suitable for use in the aquarium, as they contain volatile substances. Materials that have proved usable, since they behave perfectly neutral, are glass or quartz, kaolin, cork, asbestos cement (Aquarium*), as well as synthetic substances like silica rubber, PVC, Piacryl, epoxy resins, and others. Unknown substances must always be tested before use to make sure there can be no objections to them; once in the aquarium it is difficult to isolate the source of any damage. The clearest tests are those that examine unknown substances under aquarium conditions for their effect on fish or sometimes also on small crustaceans (*Daphnia**). A further test can be made for phenols, metals etc, and the redox potential* can also be ascertained, depending on whether or not such tests are possible.

Mating *see* Spawning Behaviour

Mating Behaviour *see* Spawning Behaviour

Mayaca AUBLET, 1775. The only G of the Mayacaceae* and comprising 7 species. S America as far as south-eastern N America. Small perennials with creeping and upright stem axes. Leaves arranged alternately. They are narrowly linear, pointed or double-pointed. The flowers are hermaphroditic, polysymmetrical. Double perianth, calyx and corolla free-standing, three-part. Petals white, pink or violet. Stamens 3. Carpels 3, fused, superior. Capsule fruit. The various species of the G grow in marshes and in water. Although *Mayaca* is often mentioned in books about aquarium-keeping, there are very few reports about how to look after them. Some of them probably need low pH water. The best known species is:

– *M. fluviatilis* AUBLET (*M. vandellii,* name not in current usage). From south-east N America and the West Indies.

Mayaca-like Rotala (*Rotala* sp.) *see Rotala*

Mayacaceae. F of the Liliatae*. The only G is *Mayaca**.

Mayflies, O, *see* Ephemeroptera

Mealworms (larvae of the Darkling Beetles) *see Tenebrio*

Medicaments. Substances that are used to protect against, contain, or cure diseases. Those medicaments used to treat fish diseases are mentioned under the entry of the disease itself or reference is made to the trade name, also under the disease entry. The normal preparations you can buy usually contain a main effective ingredient as well as other substances, which may not always make the preparation suitable for use in the aquarium. Special preparations designed for the treatment of fish diseases without there being an indication of the ingredients should only be used with reservation. Usually, fish are given medicaments in the form of baths, admixtures to food, and more rarely as injections. The instructions for use must be followed exactly, in particular the times of treatment, exact dosages and fully dissolving the medicament, if necessary. Compounds that will not dissolve in water can be dissolved in solvents such as ethylalcohol and acetone and then turned into aqueous solutions. Very small amounts of any substance must be weighed in an analytical balance. If you do not have access to such a set of scales, make up a concentrated solution from larger amounts of the substance that you *are* able to weigh (use a scale for weighing letters, for example), then you can dilute it by the required amount. Concentrated solutions made up like this must contain exact amounts of substances and solvents (measuring cylinder). The effect of medicaments on animals will vary from one to the next and it can also depend on external factors as well, such as water hardness and pH values (*see* Fish Disease Therapy).

Medicinal Leech *(Hirudo medicinalis) see Hirudo*

Mediterranean. A neighbouring sea of the Atlantic Ocean, linked to it via the Straits of Gibraltar (a stretch of sea at times only 14 km wide); joined at the other end via the Dardanelle Straits and the Bosphorus to the Black Sea* and via the Suez Canal (since 1869) to the Red Sea and the Indian Ocean*. The Mediterranean covers a surface area of 2.5 million km^2, and measures N–S, on average, a distance of 1,500 km, and W–E 4,000 km. Its deepest point is 5,120 m. Its water temperature is about 12 °C at the lowest. The Apennine Peninsula divides the sea into a western basin made up of the Iberian, Ligurian and Tyrrhenian seas, and an eastern basin made up of the Ionian, Adriatic*, Aegean and Levantinian seas. Because vast areas of the Mediterranean are arid and receive little rainfall, and because water does not filter to and from the Atlantic to any marked degree, the salinity level in the Mediterranean is above that of the world's oceans. It increases steadily going from west to east and in the Levantinian Basin it reaches 39 ‰. In recent years, various measures have been taken to regulate the Nile (eg the building of the Aswan Dam), and this has raised the salinity level even higher. As a result, yields in the fishing industry have been drastically reduced.

In the Tertiary Period, the Mediterranean had a very rich variety of tropical fauna. However, it was virtually wiped out by the coming of the Ice Age, since, in contrast to the Caribbean Sea, there was no escape route south. The fauna today is made up of relatively few autochthonic* Mediterranean species such as the Coral *Corallium* rubrum* and the crustaceans *Pachygrapsus* marmoratus* and *Palinurus* vulgaris*. Various migratory species also enter the Mediterranean from the Atlantic – eg members of the Ga *Balistes**, *Coris**, *Hippocampus**, *Julis, Mugil* (Mugiloidei*), *Thunnus* (Scombroidei*) and *Xiphias* (Scombroidei*) – as well as, since 1880, from the Indian Ocean and the Red Sea, via the Suez Canal.

Mediterranean Brown Forktail Damselfish *(Chromis chromis) see Chromis*

Mediterranean Sea-horse *(Hippocampus guttulatus) see Hippocampus*

Medium Rainbow Cichlid *(Pelvicachromis pulcher) see Pelvicachromis*

Megalamphodus EIGENMANN, 1915. G of the Characidae*, Sub-F Tetragonopterinae. Small S American Characins, 4–5 cm long, that have a typical *Hyphessobrycon** shape. The front halves of the D and An are extended into sickle-like shapes, especially in the males. This makes them look a lot like members of the 'Ornatus-group'. They are distinguished from the G *Hyphessobrycon* by differences in the teeth as well as typical generic differences in the construction of the cranium. The number of anal rays is also a distinctive feature. *Megalamphodus* species are lively, very colourful shoaling fish and they will live very happily with other Characins of similar size which makes them ideal for a community tank. Unfortunately, *M. megalopterus* is extremely prone to fish tuberculosis, possibly because this species is unable to adapt very well to water that is different in composition from that found in its natural habitat. That apart, both species of Phantom Tetras like subdued lighting and well-planted tanks. Under such conditions, their colouring will be at its best, and it will be easy to see why these fish have become some of the best-loved aquarium species. They are not easy to breed, only very young specimens appear willing to spawn; older females that are obviously ripe are difficult to get to lay. Very soft, slightly acidic water is suitable for *Megalamphodus*, and absolutely necessary for breeding purposes. The reproductive behaviour is similar to that of *Hyphessobrycon callistus*. Adding peat to a breeding tank* is ill-advised, particularly with *M. megalopterus*, as the spawn will develop but the fry will not be able to break out of the egg-cases (because of the tannins caused by the peat). This means that the fry will die eventually. The fry grow quickly to start with, and should be fed on finely sieved rotifers or *Cyclops* nauplii.

– *M. megalopterus* EIGENMANN, 1915; Black Phantom Tetra. From the Rio Guapore. Length up to 4.5 cm. In the males the body is coloured a smoky grey, with a blackish dorsal surface, paler on the ventral surface. There is a very large, black, wedge-shaped shoulder blotch that is encased by a gleaming, turquoise-green zone. All the fins, with the exception of the colourless Ps, are deep black. The females are over-all much more colourful. The flanks are brownish-red, only smoky-coloured on the dorsal surface. The D and C are smoky grey or blackish; the An, adipose fin and Vs are a strong red colour. Ps also reddish. Since *M. megalopterus* was bred successfully, it has become one of the most popular Characins in aquarium-keeping (fig. 756).

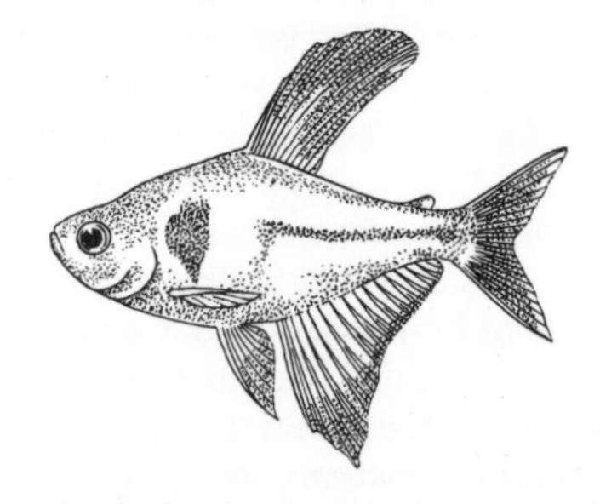

Fig. 756 *Megalamphodus megalopterus*

– *M. sweglesi* GÉRY, 1961; Red Phantom Tetra, Swegles' Tetra. From the Rio Muco and the upper Rio Meta. Length up to 4.5 cm. The body colouring in both sexes is a lovely yellow-red, as if shining from within. There is a large, drop-shaped, deep black blotch on the shoulder. All the unpaired fins and the Vs are just as intense a yellow-red. In the male, the D is drawn out into a long tip and there is no black blotch; in the female, the D has two varieties of colour from the base to the tip, either pale-pink/black-white or lemon-yellow/black-white. FRANKE maintains that the lemon-yellow variety of female is the more willing to spawn. Like its black cousin, the Red Phantom Tetra is a popular aquarium fish. It is also less prone to disease and so usually lives longer in the aquarium (fig. 757).

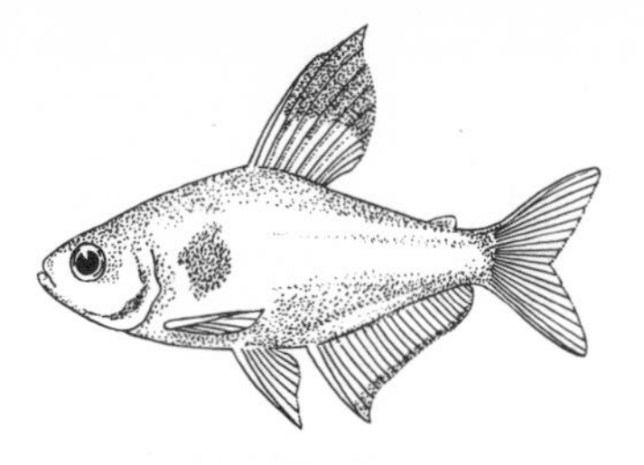

Fig. 757 *Megalamphodus sweglesi*

Megalopidae, F, *see* Elopiformes

Megalops atlanticus *see* Elopiformes

Megaloptera; Alder Flies and Snake Flies (fig. 758). O of the Insecta*. There are about 100 known species. Their larvae live in water. The Megaloptera are particularly closely related to the Lacewings, an almost exclusively terrestrial insect order. The few species of the only European G *Sialis* have large, brown wings, folded back like a roof over the insect. They are easy to confuse with Caddis Flies (Trichoptera*) and with certain Lacewings. In early summer, they sit lethargically on plants growing in the bank zone. They seem unwilling to fly, and only the sunshine and the mating drive seems to make them any more mobile. The eggs are attached above the water, perhaps to reeds, and appear as a brown, angular mass. After the larvae hatch out, they fall into or make their way to the water. They live here as predators for some months. People net-fishing in ponds will often find the larvae among their catch. You can recognise them from their 7 pairs of kidney-shaped, feathery tracheal gills and from the similarly constructed tail-thread. The larvae crawl back onto land to pupate, digging themselves in near the water. The imago emerges about a week later.

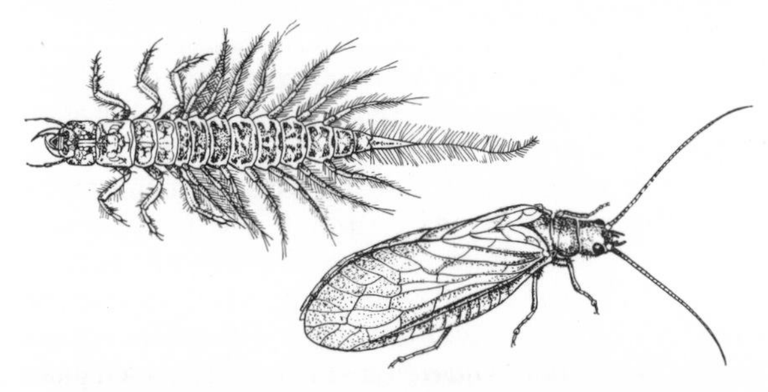

Fig. 758 Megaloptera. Top left, larva; bottom right, imago

Meinken's Rasbora *(Rasbora meinkeni) see Rasbora*

Meiosis *see* Cell Division

Melanogrammus aeglefinus *see* Gadiformes

Melanosarcoma *see* Tumour Diseases

Melanotaeniidae, F, *see* Atherinoidei

Melichthys SWAINSON, 1839. G of the Balistidae*, Sub-F Balistinae. From tropical parts of the Indo-Pacific and Atlantic. Shape typical of the F, teeth incisor-types, scales rough, and the head profile very convex. Length up to 60 cm. Colouring murky. Survives well in the aquarium and peaceful towards other, similar-sized fish. Eg:

– *M. ringens* (OSBECK, 1765); Black-finned Triggerfish, Black Triggerfish. Tropical Indo-Pacific and Atlantic. Length to 60 cm. Black, base of the D and An white.

Membranipora BLAINVILLE, 1830. A common G of marine-dwelling bryozoans (Bryozoa*), that penetrate deep into brackish water. Colonies of them form hard crusts on various objects (these crusts are made up of the small chambers of individual animals [zooids]). They will also develop from time to time in marine aquaria.

Mendel, Gregor Johann (1822–84). The discoverer of the laws of heredity. Born in Heinzendorf near Odrau. After studying in Vienna, by 1854 he was lecturer in natural history and physics at the upper secondary school in Brno. From 1856 he was engaged in botanical studies as a monk in the Augustine brotherhood in Brno. He concentrated on trying to cross-breed peas and beans. It was from these studies that he was able to make his claims for the universal laws governing heredity (Mendelian laws).

Mendelian Laws *see* Heredity

Menhaden *(Brevoortia tyrannus) see* Clupeiformes

Menyanthaceae; Buckbeans. F of the Magnoliatae* comprising 5 Ga and 40 species. Distributed throughout the world. Small perennials with an alternate leaf arrangement. The leaves are simple or pinnate. Flowers hermaphroditic, polysymmetrical, with vividly coloured corollas. The tissues often contain bitter principles. All Buckbeans are marsh or aquatic plants. Widespread in the northern hemisphere is *Menyanthes trifoliata* (Marsh Trefoil or Bogbean) which has three-part pinnate leaves. The species of the G *Nymphoides** are occasionally kept in the aquarium.

Mercurochrome®. A 2% aqueous solution of hydroxymercuridibromfluoresceinsodium which has an intense red colour. It can be used instead of tincture of iodide* to dab skin wounds in fish.

Merluccius merluccius *see* Gadiformes

Merry Widow *(Phallichthys amates amates) see Phallichthys*

Mesidothea entomon *see Idotea*

Meso-. In compound words, it means middling or between. For example, mesotrophic waters = waters that contain a medium amount of nutrients.

Mesogonistius chaetodon, name not in current usage, *see Enneacanthus*

Metabolism. The acceptance and surrender *(external metabolism)* of substances as well as their processing *(intermediary metabolism)* by the living organism; it is a necessary process for the organism to be able to maintain its structure and all its life functions. Metabolism consists of a large number of building-up and breaking-down processes, all of which are interdependent. The basic materials implicated in metabolism are proteins*, carbohydrates*, fats*, minerals and water. Among green plants, the energy needed for metabolism is obtained from the sun (Photosynthesis*). Animals get their energy from food substances.

Metacercaria *see* Trematodes

Metallic Salt Poisoning *see* Poisoning

Metamorphosis. Among animals, metamorphosis is associated with a larval form (Larva*) which is usually very different in body structure and life-style from the adult form. In many species the metamorphosis is a one-time-only, short-lived affair *(holometabolism).* In many insects this takes the form of a pupal stage, during which the organs of the grubs, maggots etc (with the exception of the nervous system) go into retrograde development and the organs of the adult insect (imago) develop from remaining embryonal tissues. Far-reaching changes also occur during the metamorphosis of frogs and toads (*see* Amphibia) and of some fish. Metamorphosis can also take place step by step; in insects, for example, the eyes and wings will gradually get bigger with each larval moulting (hemimetabolism). Special larval organs, for example the tracheal gills of the Dragonfly larva, will be retained until the very end of the larval period.

The timing of metamorphosis, and the way in which it actually unfolds, is largely controlled by hormones, the production of which is in turn controlled by the nervous system. Insects and crustaceans have moulting hormones in special glands. The thyroid hormone is particularly important in the metamorphosis of lower vertebrate animals. Experiments have been carried out to show that tadpoles injected with this hormone (thyroxin) will start to metamorphose and if injected with substances that will inhibit thyroxin, metamorphosis will not take place.

Metazoa. Multi-celled animals or animals that live in co-operative cell groups, as distinct from single-celled animals (Protozoa*).

Methylene-Blue-B (Tetramethylthionine chloride). A basic thianine dyestuff, normally in the form of double salt zine chloride which is highly soluble in water. Methylene blue has antiseptic properties and can be used for prolonged bathing treatments (Bathing Treatments*) in fish infected with *Chilodonella**, *Ichthyophthirius**, *Ichthyobodo**, *Gyrodactylus* and *Dactylogyrus* (Trematodes*). Treatment should last for 3–5 days. A concentrated mixture is made up (1 g methylene blue to 100 ml water), of which 3 ml is placed in 10 litres aquarium water. The dye can be removed from the water by filtering through activated charcoal (Adsorption*). The strong blue colour of the water is not good for plants, since light is absorbed and damages them.

Metridium OKEN, 1815; Plumose Anemones. G of the Sea-anemones (Actiniaria*). The species *M. senile* (LINNAEUS, 1758) occurs in the cold seas of the northern hemisphere. It can measure up to 30 cm tall and its mouth may be 20 cm wide and lobed. Eats small animals and tissues. At the mouth end there is a thick covering of fine tentacles (as many as 1,000). They are very pretty anemones and may be white, salmon-pink or brown. They reproduce by means of eggs as well as asexually through laceration*. Unfortunately, plumose anemones are scarcely able to tolerate temperatures above 20 °C and they also lose their colouring if the right plankton food is not available. As a substitute, *Artemia* nauplii and finely grated food can be given to them in the aquarium.

Metynnis COPE, 1878. G of the Serrasalmidae*, distributed in S America, from the Orinoco to La Plata. 12–14 cm in length. In 1951, GOSLINE revised this G and reduced the original large number of species to six. The round, disc-shaped body is very compressed laterally. The edge of its belly is pointed because of a number of toothed plates (which vary with species) which are known as serrae. Typical features of the G include the long, flat D2 and a small spine located just in front of the fairly short D1. The D1 is pointed at the front and falls away sharply at the back. The splayed, strong C is barely indented, and this indicates that *Metynnis* species swim in a flat trajectory. In the males, the long An is broadened like a lobe for the first third of the fin, in the females it is straight at the free edge, the first fin rays being drawn out into a point. The Vs and Ps are relatively small. *Metynnis* species are pretty shoaling fish, but unfortunately they are a bit shy and timid. They need plenty of room to swim about in. Despite having a similar body structure to the Piranhas* and despite being closely related to them, *Metynnis* is in fact a peaceful fish. It can even be put in a community tank with very small species. However, they do eat plants a great deal, so they must not be placed in tanks that have a lot of vegetation. Apart from plant food, they also eat all the normal kinds of live food. So far, the species, *M. maculatus*, has been bred successfully only once. This was achieved by FRANKE. The fish, 3 males and 2 females, spawned once late in the evening under artificial light, and again during the day. Both times, this was preceded by violent driving, and then the spawn was laid on *Myriophyllum* stems in the middle and upper levels of the water. Some was also laid on the back wall of the tank which was covered in algae. The tank measured 140 cm in length. During mating, the male gripped the female with the caudal peduncle behind the D1, then he caught the spawn with the lobe-like extension of his An (which was wrapped round the sexual orifice of the female). About 150 eggs were laid, more than 2 mm ∅, on each occasion. They were not very sticky and sank to the bottom. At a temperature of around 25 °C, and after 4 days, the fry hatched out – they were about 8 mm long. After a further 4–5 days, they were already eating larger kinds of nauplii and were quick to grow. The breeding of *M. hypsauchen*, first observed in the Wuppertal Aquarium by SCHMIDT, has been achieved on several occasions. In contrast to *M. maculatus*, mating takes place directly above the bottom and the species considerably more productive, laying around 2,000 eggs. At a temperature of 28 °C, the fry hatched out after about 70 hours and are then reared in the same way as *M. maculatus*. They also grow as quickly. STERBA has commented that for him the sight of a shoal of 200 or so *M. hypsauchen* is one of the most fascinating he has come across; they swim with the most extraordinary degree of unison and harmony.

– *M. hypsauchen* (MÜLLER and TROSCHEL, 1845). From Guyana, across the Amazon Basin to the Paraguay river system. Length up to 14 cm. *M. schreitmülleri* and *M. calichromus* are both other names for this species that are no longer used. The ventral edge is formed by 27–31 toothed plates. The flanks are an intense gleaming silvery colour with a bluish sheen. The dorsal surface is greenish, brownish, or blue-grey. Depending on origin, there may be visible some indistinct, narrow, dark, transverse bands, that extend from the back to the

LL. Dark fleck markings may also be visible. The D1 has dark dots, D2 yellow or reddish. The C has a broad dark edge and the An is reddish, particularly at the front, with blood-red elongated tips in the female, and with a dark edge in the male. Eyes silver with a vertical, black streak. In sexually excited males, the body has a lovely golden-red colour and two large black shoulder blotches are visible; the edges of the C and An have broad black edges at this time. This is a very attractive species, in shape and in colour, and is ideal for large public aquaria (fig. 759).

Fig. 759 *Metynnis hypsauchen*

– *M. maculatus* (Kner, 1859); Spotted Metynnis. Catchment area of the Rio Madeira. Length up to 12 cm. *M. roosevelti* is another name for this species, but it is no longer used. The flanks have similar colouring to *M. hypsauchen*, although large, dark flecks, distributed irregularly over the body, are always clearly visible (even in young fish). The D1 is colourless or yellowish, but has no dots. It often has a dark tip. The free edge of the C is black, with a red border in front of it. An reddish, the outside edge of which is brick-red. Sexually excited males are among the prettiest of all the Characins. The body is tinged red at such times and a vivid red,

Fig. 760 *Metynnis hypsauchen* and *Metynnis maculatus* (top)

sometimes very large shoulder blotch appears that also shines with an intense metallic blue colour. The free edge of the C has first a red band, followed by a black band. The widened front section of the An is also coloured red. Unfortunately, this species seems to have disappeared from aquaria. A new import would be most welcome (fig. 760).

Mexican Tetra *(Astyanax mexicanus) see Astyanax*

Micralestes Boulenger, 1899 (fig. 761). G of the Characidae*. These are African Characins, mainly from the river systems of the Congo, Nile, Niger and Zambesi. Up to 10 cm. Shy shoaling fish that swim very elegantly. They live in the open water zone, orientating themselves very well by sight. Like the closely related *Alestes** species, they require soft, slightly acidic water with some peat added. A frequent topping-up with fresh water suits them particularly well, and under these conditions they will produce long, pennant-like filaments in the D and in the middle of the C. The body is long and spindle-like, compressed laterally. The D is inserted either above or just behind the Vs. Scales relatively large. Sexually mature males often have elon-

Fig. 761 *Micralestes humilis*

gated rays in the D and C, and more rarely in the An. Apart from sturdy live food, *Micralestes* also likes insect food (it eats things that drop onto the water surface). They spawn in the shoal (Selection of Breeding Fish*), the act itself taking place in the morning usually. After vigorous driving, the spawn is laid on Java Moss or on *Myriophyllum* plants, near the bottom. There are roughly 300 eggs in all, and they swell up considerably but are not sticky. The fry hatch out after 5–6 days and are able to swim immediately, seeking out rotifers or *Cyclops* nauplii. Slow growth rate. For breeding purposes, tanks with sloping bottom panes are most suitable (*see* Breeding Aquarium). Very rarely *M. hilgendorfi* (Boulenger, 1899) used to be imported, but recently *M. acutidens* (Peters, 1852) has been more common. The latter is only 6.5 cm long.

– *M. interruptus* Boulenger, 1899; Congo Tetra, Congo Salmon. From the Congo area. Male up to 8 cm, female up to 6 cm. The flanks of the male have a green-blue shimmer in direct light, with violet, red-brown and yellow tones. Dorsal surface brownish. The unpaired fins, elongated like filamentous threads, are a smoky-grey colour. The elongated middle rays of the C are black with white borders. The females are a uniform soft green colour and the fins are not elongated. This is rightly one of the standard Characin species in any Characin tank (fig. 762). *Micralestes humilis* is another name no longer used, see *Alestes*.

Fig. 762 *Micralestes interruptus* and *Alestes longipinnis* (2nd and 3rd from top)

Microbrycon cochui *(Boehlkea fredcochui) see Boehlkea*

Microcosmus, G, *see* Tunicata

Microcyprini, O, name not in current usage, *see* Atheriniformes

Microfood *see* Fry, Feeding of

Microglanis EIGENMANN, 1912. G of the Pimelodidae*. Small Antenna-catfishes from S America. Length up to 10 cm. Peaceable fish active at dusk and at night, lying hidden beneath pieces of root or rocks or in the sand during the day. Untypical of the F, the body shape is compressed, torpedo-like, only flattened laterally from behind the P insertion. There is one pair of barbels on the upper jaw and 2 pairs on the lower. Naked body. D1 short, D2 well developed, C either deeply forked or hardly at all. An well developed, Vs small, Ps large with a toothed first fin ray. During the day, they only come out when being fed, guided by their good 'nose' for food. They love eating *Tubifex* and *Enchytraeus*, as well as *Daphnia*, and even dried food. In the tank they need a soft substrate, and it should be positioned in not too bright a spot. Nothing is known about their breeding habits. They make ideal aquarium fish on account of their pretty colouring, small size and peaceful nature. They are also suitable for community tanks.

– *M. parahybae* (STEINDACHNER, 1880); Dwarf Marbled Catfish. From the Rio Parahyba and the Rio Doce. Length up to 7 cm. Colouring very variable. Yellow-brown to brown-orange and usually with 3 irregular, dark brown transverse bands and black spots on the flanks. Fins colourless with dark-brown bands or rows of dots (fig. 763).

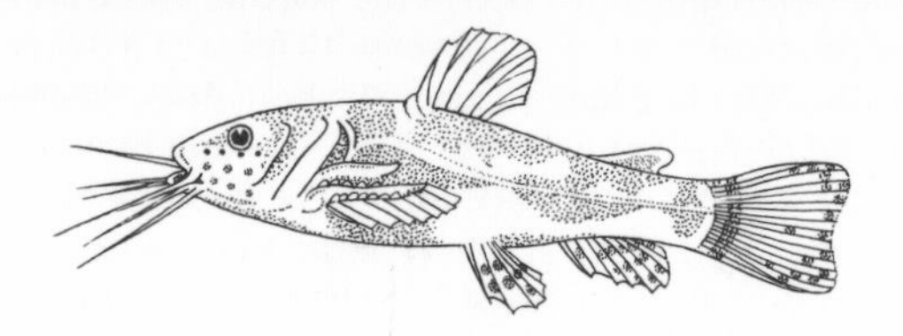

Fig. 763 *Microglanis parahybae*

Micropoecilia, G, name not in current usage, *see Poecilia*

Micropterus LACÉPÈDE 1802. G of the Centrarchidae*. Comprises two species in eastern and central N America, from Florida to S Canada. Also introduced into some areas of western N America and Europe (Portugal, Soviet Union). Large, elongate, predatory-looking fish, flattened laterally. The mouth has a wide gape; in *M. salmoides* the upper jaw reaches to behind the eyes. The D is flat, almost cut in two by an indentation in the middle. An has 3 hard rays, C gently curving. *Micropterus* species are especially important as sporting and edible fish. In normal room aquaria, only young specimens are kept usually. The biology of the G is very interesting. They live in clear lakes and rivers, lying in wait for prey like Pikes. In early summer, the male finds a large spawning depression which he cleans thoroughly (in *M. salmoides*, it is also decorated with plant material). The female lays several thousand eggs inside it, and they take 1–2 weeks to develop. Both parents practise brood care.

– *M. dolomieu* LACÉPÈDE
1802; Black or Smallmouth Bass. Up to 50 cm long. Greenish, the flanks of younger fish have transverse rows of dark flecks. Becomes a fairly murky olive-green with age.

– *M. salmoides* (LACÉPÈDE, 1802); Largemouth Bass or Sun-fish. Up to 90 cm long and weighing over 10 kg. Olive-green with a longitudinal band broken up into blackish flecks and extending from the mouth to the root of the tail. The dorsal surface is also covered in dark flecks. Monotone in colour as it gets older.

Microrasbora ANNANDALE, 1918. G of the Cyprinidae*, Sub-F Rasborinae. From Burma. Very small fish that inhabit thickly vegetated edges of small and large lakes. They live in large shoals. Body shape elongated, laterally compressed, similar to the G *Brachydanio**. It is differentiated from this and another G, *Rasbora**, because it has no LL and no barbels. The scales are large and transparent. Specimens should be kept in small, well planted tanks. The water should be clear and relatively hard (12–15° total hardness). Temperature 24–26 °C. Every kind of live food and dried food is eaten. If breeding, a lowering of the water hardness will stimulate the desire to spawn.

– *M. rubescens* ANNANDALE, 1918. From Burma, where it lives in the waters of the He-Ho lake basin and in Lake Inlé in the Shan mountains. Length up to 3 cm. Flanks silver-blue. Beginning at the level of the D, a dark longitudinal band extends to the root of the C. There is a small dark fleck in front of the An insertion. The underneath of the head, the D, An and C are all orange or red. Courting males have a red shimmer along the whole body.

Microsorium LINK; Java Fern. G of the Pteridophyta* containing 40 species. From Asia. Small to medium-sized ferns with elongated rhizomes and either simple leaves or, at most, indented ones. Reproduction by spores. The species of the G grow in shaded spots, normally, where humidity is high. They are mostly found on rocks or on trees in the forests, usually near water. The species described below flourishes on river-banks that are flooded in the rainy season. They are ideal for submerged growth in the aquarium; simply attach them to wood or stones.

– *M. pteropus* (BLUME) CHING. From SE Asia. Has a marked tendency to develop adventitious plants on leaves and roots.

Microspathodon GÜNTHER, 1862. G of the Pomacentridae* from tropical parts of the Atlantic. Similar to *Pomacentrus** and *Eupomacentrus**, except that both sides of the preorbital bone, which borders onto the upper jaw, have a pronounced curve. These are aggressive fish that live a solitary existence among reefs. They will often attack like species and unlike species quite ferociously.

– *M. chrysurus* (CUVIER); Yellowtail Damselfish, Marine Jewelfish. From both sides of the tropical Atlantic. Up to 15 cm. Juveniles blue-black with pale-blue gleaming flecks across the whole body. Older fish are more soberly coloured with a yellow C and only a few small flecks on the dorsal surface. In the aquarium this species survives well, but needs additional plant food.

Midges, Sub-O, *see* Nematocera

Midget Sucker Catfish *(Otocinclus affinis) see Otocinclus*

Migration. Can mean either the movement of animals to or from certain areas on account of changes in environment (eg the edulcoration of seas, change in climate), or movements of animals caused by a species-specific biology governing behaviour and development. Anadromous* and catadromous* fish, for example, migrate from the sea into freshwater to reproduce, and vice versa. It is also known that a great many other fish go back again and again to reproduce in the same place (eg herring and mackerel). The fishing industry centred on the oceans and inland waters decides when and where to catch its fish primarily through fish migration.

Migratory Area *see* Areal

Migratory Fish *see* Anadromous Fish, Catadromous Fish

Milfoil, G, *see Myriophyllum*

Milk-fish *(Chanos chanos) see* Gonorynchiformes

Milk-fishes (Chanidae, F) *see* Gonorynchiformes

Milleporidae; Fire Corals. Stinging, chalky, colonial polyps found among tropical coral reefs and belonging to the Hydrozoa*. Externally similar to the Madreporaria*. In the Milleporidae, however, the small individual animals sit in tiny pores, not in large cups as in madrepore corals. Some Milleporidae are formed like shrubs, others are fan-shaped. Their skeletons are used for decoration in the marine aquarium.

Miller's Thumb *(Cottus gobio) see Cottus*

Millions Fish *(Poecilia reticulata) see Poecilia*

Mimagoniates REGAN, 1907. G of the Characidae*, Sub-F Glandulocaudinae. From Brazil and Paraguay. Up to 7 cm long. They are small shoaling fish that like swimming and jumping, and are found in the upper and middle water levels. Since they are seldom imported, they have warranted little regard from aquarists. The systematic relationship to the closely related G *Glandulocauda** has not been fully explained, even today. The body is elongate and flattened laterally. The D1 is located well to the rear and towers above the D2. The An is long and has no little hooks. The C is twin-lobed, its middle rays being thickened and called a gland. LL incomplete. *Mimagoniates* species have internal fertilisation, the spawn being stored here and laid alone by the females. The sperm is transferred during a highly unusual mating dance on the part of the male; with lightning speed, he wraps himself round the female in a ring-shape, releasing his sperm as he does so (for more information see *Glandulocauda*). It is not clear whether or not the species described here are the same as those that have been described as such in aquarium literature to date:

– *M. barberi* REGAN, 1907; Barber's Tetra. From the black waters of SE Brazil and Paraguay. Up to 4.5 cm long. Flanks brownish-yellow. A deep-blue longitudinal band extends from the gill-covering to the middle rays of the C. It is bounded on the upper edge by a narrower, greenish or copper-red stripe. The unpaired fins are red-brown with a blue longitudinal band on the free edges of the D1 and An (fig. 764).

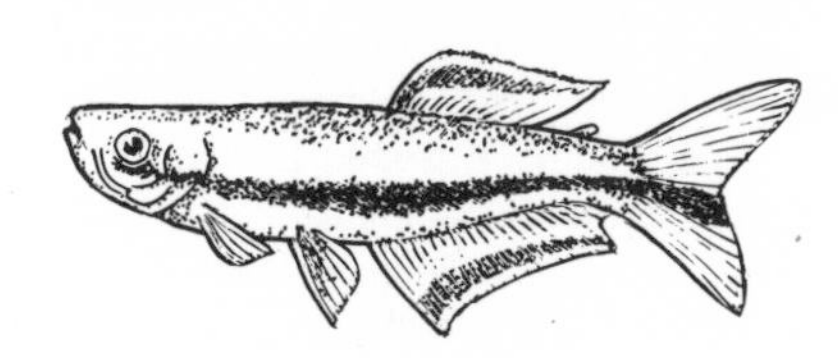

Fig. 764 *Mimagoniates barberi*

– *M. (Coelurichthys) microlepis* (STEINDACHNER, 1876); Blue Tetra. Flanks silvery with a greenish-violet sheen. There is a very broad, gleaming blue longitudinal band that gets narrower towards the back and ends in the middle C-rays (where the colouring is also more intense). The D1 and An are a green iridescent colour with a blue longitudinal band and an orange-yellow edge. Sometimes, a dark shoulder blotch is also visible.

Mimicry *see* Adaption

Mineral Acid Hardness *see* Permanent Hardness

Minerals *see* Plant Nutrition

Miniature Rasbora *(Rasbora urophthalma) see Rasbora*

Minnow *(Phoxinus phoxinus) see Phoxinus*

Minstrel Pufferfish *(Canthigaster valentini) see Canthigaster*

Minytrema melanops *see* Catostomidae

Miracidia *see* Trematodes

Mirapinnidae, F, *see* Cetomimiformes

Misgurnus LACÉPÈDE, 1803. G of the Cobitidae*. From central and E Europe, Asia. Medium-sized to large bottom-dwelling fish that inhabit warm and shallow still waters with muddy substrata. Body elongate, cylindrical in cross-section, caudal peduncle flattened laterally. Scales very small. The mouth is inferior with thick lips, and surrounded by 3 pairs of barbels on the upper jaw and 2 pairs on the lower. Eyes and gill openings small. The spines beneath the eyes are covered by a muscular layer. Fins not very large. C rounded, LL incomplete. In oxygen-poor waters, *Misgurnus* swallows air from the atmosphere and breathes via the hind-gut, which contains a large number of blood vessels. During the winter and during dry periods, *Misgurnus* burrows into the mud and remains there for long stretches of time. Active at night, but also said to swim around restlessly during the day before storms. Diet consists mainly of smaller bottom-dwelling animals, insect larvae, snails and worms. These are very hardy, long-lasting fish that are among the most interesting of European aquarium fish. Tanks should be large and standing in a dark spot. Provide places to hide and a thick layer of humus as a substrate. Ventilation not necessary. Temperature 18–22 °C. After a cool overwintering period, *Misgurnus* will normally spawn in the aquarium too. One species is native to Europe, 6 species to Asia.

– *M. anguillicaudatus* (CANTOR, 1842); Japanese Weatherfish, Dojo. From NE Asia as far as central China. Length 25 cm. Body clay-brown with greenish-grey-brown marble markings. Ventral surface pale silver.

– *M. fossilis* (LINNAEUS, 1758); Weatherfish, Pond Loach. From Central and E Europe. Length 30 cm. Dorsal surface dark brown, flanks ochre-yellow to grey-yellow, ventral surface orange-yellow. Several dark, black-brown longitudinal bands run across the body with speckles in between. Fins a dirty yellow-brown with dark fleck markings. The larvae have external gills (fig. 765).

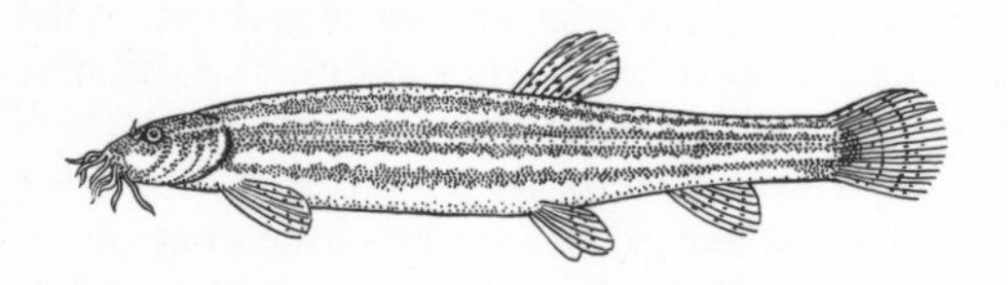

Fig. 765 *Misgurnus fossilis*

Mites, O, *see* Acari

Mitosis *see* Cell Division

Mobulidae; Devil Rays (fig. 766). F of the Chondrichthyes*, O Myliobatiformes. These rays, unlike all the other Ray, Fs, are plankton-eaters. In body shape, the Devil Rays or Mantas are reminiscent of the Eagle Rays (Myliobatidae*). Like them, they have huge, wing-like Ps and a relatively short, whip-like tail. The poisonous spine is normally missing too. Devil Rays are immediately recognisable from their two long, mobile head-lobes, which look a bit like horns. They beat continuously with them, snaffling the plankton-rich water into their slightly inferior mouths. The tiny food particles gather in a weir-basket area just in front of the pharynx, and they are swallowed from there from time to time. Teeth are reduced; besides species that have small teeth in both the upper and lower jaws, there are also species with teeth in one jaw only or with no teeth at all. The skin is rough. Devil Rays are livebearing. The number of offspring produced is small, and like the Myliobatidae it is said that they are born when the parent fish leaps out of the water. Devil Rays occur in all tropical and subtropical seas. They often swim in pairs near the surface, the tips of the Ps sometimes

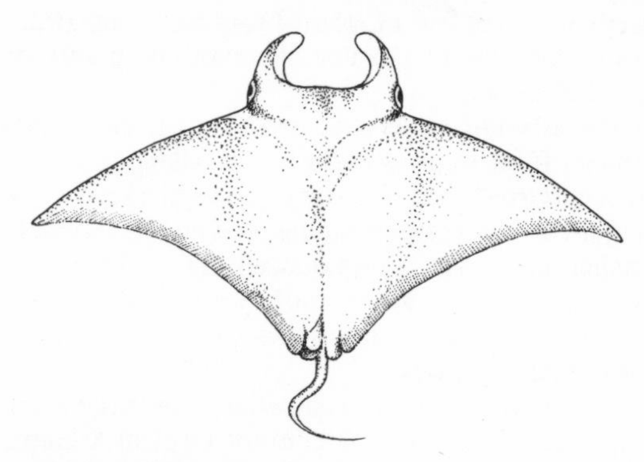

Fig. 766 Mobulidae-type

breaking through the water with each 'wing beat'. On rarer occasions, small shoals have been observed. Most species are very large. The Giant Devil Ray or Atlantic Manta *(Manta birostris)* which is found in the Atlantic, Indian Ocean and W Pacific has a 'wingspan' of 6.5 m, and weighs 1,500 kg. A dwarf form found in the waters around Australia *(Mobula diabolis)*, with a wingspan of 60 cm, is only about half as big as the new-born offspring of *Manta birostris*. Devil Rays have a tasty flesh. Their liver is also rich in oil and their skin, when prepared in a certain way, is used as sandpaper. Mantas are not dangerous to Man.

Mochokidae; Scaleless Catfishes (fig. 767). F of the Siluriformes*, Sup-O Ostariophysi. Distributed south of the Sahara, in tropical parts of Africa. This is a very uniform group of fish, reminiscent in body shape of large Mailed Catfishes *(Corydoras*)*, although the Mochokidae are not bottom-dwelling and they do not have any body armour in the form of bony plates. Characteristic features are the large, capsular head and nape armour made up of strong bones, out of which extends a pointed process that reaches to the insertion

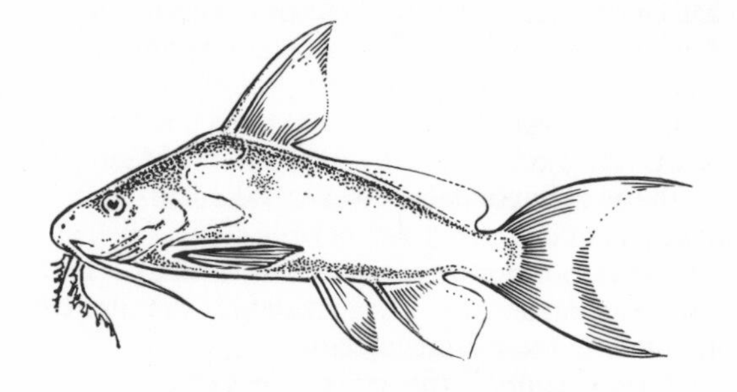

Fig. 767 Mochokidae-type

of the Ps; also typical is the slightly inferior mouth equipped with 3 pairs of barbels. The ones on the upper lip are longest, the gill barbels are shorter and usually feathery. The D stands well to the front end and is shaped like a pointed sail; often it has a strong, toothed spine, like the Ps. The adipose fin is very large, often reaching from the D to the indented C. An relatively small. The Mochokidae are crepuscular fish that live in small shoals in slowly flowing or still waters, and are particularly numerous in lagoons. During the day they seek out places of refuge near the bank or they rest vertically on posts or tree-trunks. Their food consists mainly of small invertebrate animals which they find by feeling for them constantly with their barbels on the undersides of pieces of plants and other substrata. When seeking food, they will often turn over on their backs, and for some species this is in fact the preferred way of swimming. Two species that swim in this way are *Synodontis batensoda* and *nigriventris*. The last-mentioned has even adapted to this way of life by having a darker ventral surface than dorsal surface. During courtship, these fish shoot towards one another, using their heads as ramming posts. The juveniles of those species that swim on their backs swim normally at first and only learn to turn over at a later date. Some species grow quite large and are prized as food locally. An example is the 50 cm long *Synodontis schall*. Many species, perhaps all of them, utter squeaking or purring noises, probably caused by the movement of the D and P spines in their sockets. The smaller species, at least the juvenile forms, are often prettily coloured. Because of this, and because they have an interesting appearance and life-style, the Mochokidae have found great favour among aquarists. But they do upset other fish with their constant feeling around for food. Observations in the aquarium could still teach a lot about their reproductive habits. See also the G *Synodontis*.

Mock Fights *see* Intention Movement

Moderlieschen *(Leucaspius delineatus) see Leucaspius*

Modification. Changes in the formation of bodily features during an individual's development, brought about by external influences (eg nutrition, temperature, use or lack of use of organs). For example, temperature influences the colour of insects or the body shape of Water Fleas. Such features can differ from the norm within certain parameters only. The hereditary material is not altered structurally; in other words, modifications, in contrast to mutations*, are not hereditary.

Moenkhausia EIGENMANN, 1903. G of the Characidae*, Sub-F Tetragonopterinae. From tropical S America. Up to 12 cm long. This G comprises a large number of species, and all are colourful shoaling fish that enjoy swimming. They inhabit open water zones. Together, the species make up a whole series of popular aquarium fishes. The body shape can vary a great deal, being spindle-shaped to tall. The LL, with the exception of *M. sanctaefilomenae*, is complete and straight, or slightly arched downwards. The D and An in the male may be elongated. The base of the C, as in the G *Hemigrammus**, is scaled. Adipose fin present. Many species will form territories at certain times and if they are kept in tanks with very hard water or where insufficient fresh water is used as a top-up, they may prove susceptible to fish tuberculosis, particularly *M. sanctaefilomenae*. For details relating to reproduction and care turn to *Hemigrammus*. Many species are open-water spawners*, eg *M. sanctaefilomenae*. They will spawn just below the water surface in small breeding tanks without any plant substrate. Most are very productive and are not prone to eating their own spawn. The fast-growing fry are easy to rear, although they do eat a great deal.

– *M. oligolepis* (GÜNTHER, 1864); Glass Tetra. From small water-courses of the Amazon region and Guyana, in still or slow-flowing water. Length up to 12 cm. The body is elongate and tall with a lovely silvery sheen. Scales large and often with dark edges. A broad, deep black zone covers the root of the tail, against which two, golden-yellow flecks contrast. Eyes large with a two-tone iris – blood-red above and golden-yellow below. Unpaired fins yellowish; Vs, Ps, and adipose fin colourless.

– *M. pittieri* EIGENMANN, 1920; Diamond Tetra. From Lake Valencia in Venezuela. Length up to 6 cm. Body and fin colouring in the males a milky-violet, with a gold iridescence in direct light and a smattering of small splinters of green iridescent colours. The D and Ps are elongated like pennants, the tips of which are pastel-white, like the C and An. The females are yellowish with a weaker iridescence and their fins are not elongated. This is an attractive species suitable for large community tanks (fig. 768).

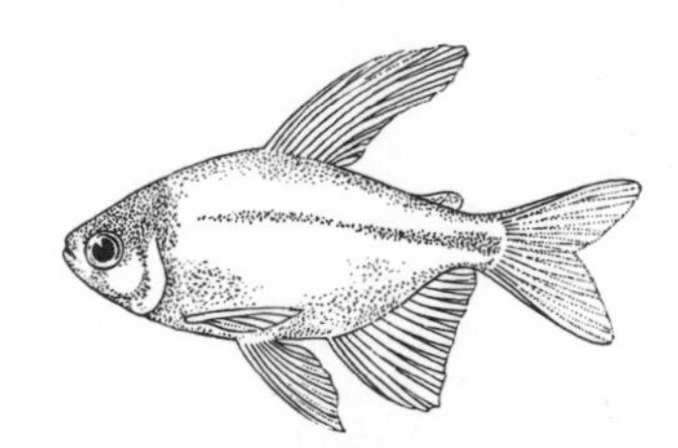

Fig. 768 *Moenkhausia pittieri*

– *M. sanctaefilomenae* (STEINDACHNER, 1907); Red-eyed Tetra, Yellow-banded Tetra. From the rivers of the Rio Paraguay and Paranahyba basins. Length up to 4 cm. Flanks a brilliant silvery gleaming colour with strong, dark-edged scales. Dorsal surface brown-olive. A broad, wedge-shaped, patent black zone covers the base of the C, and in front of it, on the caudal peduncle, is a broad band that gleams a silvery-golden colour. The upper half of the iris is a shining blood-red. Apart from the white tips of the D and An, the fins are colourless or a pale grey. This species is really a smaller but more colourful version of *M. oligolepis*, and, as such, it is a popular type of Characin in the aquarium. Adult females of both species are easy to spot from the curvature of their spawn-filled ventral areas (fig. 769).

Mogurnda GILL, 1863. G of the Gobiidae, Sub-O Gobioidei*. From Australia and New Guinea. Medium-sized fish that live in smaller stretches of fresh water, in rivers and on the coast. Body elongate, cylindrical at the front, flattened laterally at the rear. Large head with

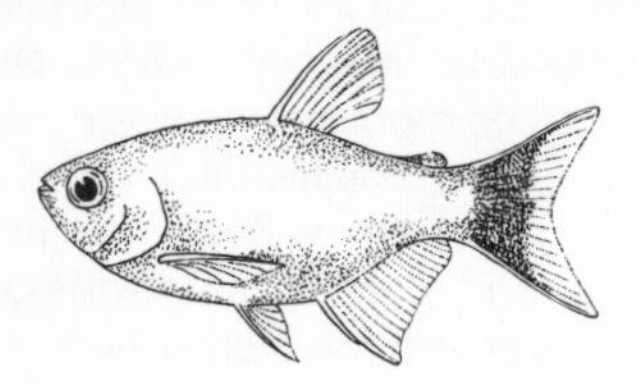

Fig. 769 *Moenkhausia sanctaefilomenae*

protruding lower jaw. Vs not fused with one another. Bodily features in general very similar to the G *Eleotris*, although the D1 does not have 6–7 hard rays, but 7–10. The D2 and An, with a hard ray at the front and over 10 soft rays, are both longer than those of related Ga. A very prettily coloured species is:

– *M. mogurnda* (RICHARDSON, 1844); Purple-striped Gudgeon, Australian Gudgeon, Sleeper Trout. From Central, N and E Australia, as well as New Guinea. Particularly common in fresh water. Length up to 17 cm. Upper surface usually olive-brown, clay-brown on the flanks with an irregular row of dark flecks and scattered, red dots. Ventral surface paler. Two green or dark streaks extend from the eyes to the insertion of the Vs. A single blue blotch, with a pale edge, lies behind the gill-covering. The bases of the D and C are covered in dark or red speckles. The female is usually larger and less conspicuously coloured. Fresh water, 18–25 °C. Feed with sturdy live food of every kind. Spawn is laid on rocks, and the male looks after the brood. The young fish should be reared on the very finest live food.

Mojarras, F, *see* Gerreidae

Mola mola *see* Molidae

Molecule. The smallest unit of a chemical compound. The chemical and physical properties of a molecule are determined by the type, number and arrangements of its atoms*. A water molecule, for example, consists of 2 atoms of hydrogen (H) and one atom of oxygen (O). The molecules of organic compounds are usually much bigger. Protein molecules can contain thousands of atoms of carbon (C), hydrogen, oxygen, nitrogen (N) and sulphur (S) in various combinations. The molecular weight, which is of importance in the recognition of a substance, is the sum of the atomic weights of a molecule.

Molecular Genetics *see* Genetics

Molidae; Sun-fishes. F of the Tetraodontiformes*. Sub-O Tetraodontoidei. Large, very peculiar-looking fish found in warm and temperate seas. Their relationship to the Porcupine-fishes (Diodontidae*) and others is apparent from the very spiny larvae produced in each case. Adult Sun-fishes are tall, a long oval shape in side-view, and very flat laterally. The fish is, as it were, chopped behind the D and An (its organs of propulsion). The mouth is small with a veritable tooth-beak, formed by the fusion of the teeth. The gill-slit is shaped like a hole and is positioned in front of the Ps. The D and An are short but very tall and with no spines. The strip of fin at the rear end is often called the C. A large part of the skeleton is made of cartilage and there is no swim-bladder. Naked skin. The Sun-fish *(Mola mola)* (fig. 770) which grows up to 3 m long and weighs about 130 kg, is primarily a pelagic fish that lives on small fishes and squids. Often it moves through the water lying on its side. In a lot of areas in the South Seas it is considered a harbinger of catastrophe. Occasionally, smaller specimens have been known to penetrate into the North Sea and the Baltic. Its flesh is not highly prized. They are very difficult to keep in captivity, and are rare even in oceanaria.

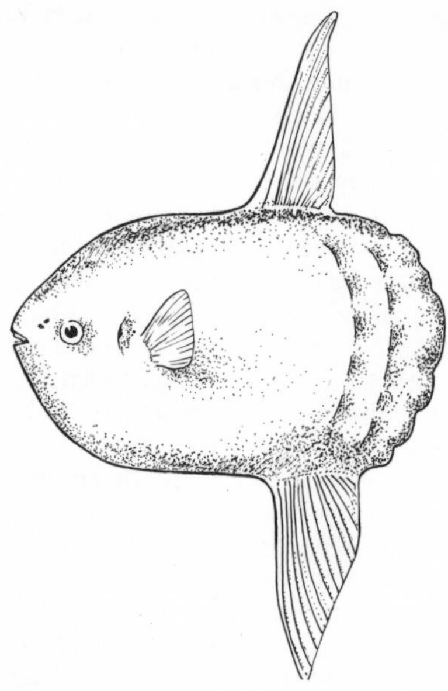

Fig. 770 *Mola mola*

Mollienisia, G, name not in current usage, *see Poecilia*

Molluscs. An animal Ph comprising roughly 130,000 species, which makes it the second largest Ph in the animal kingdom, after the Arthropoda*. Well-known classes of the Molluscs include the Snails (Gastropoda*), the Bivalves (Bivalvia*) and the Cuttle-fish or Squids (Cephalopoda*).

The variety of animal life within the molluscs is enormous; alongside small animals only a millimetre in length there are also the largest invertebrate animals in the world, such as the Giant Squids (belonging to the Cephalopod group) which can measure more than 20 m in length. Physiologically and anatomically, the complexity and degree of organisation within the molluscs is very much comparable with that of the vertebrates. In general, molluscs are of compressed build and, from the outside at least, they do not appear to be articulated. Areas that are recognised are the ‘head’, ‘foot’, visceral sac and mantle. The mantle secretes a protective calcareous shell which we known as the familiar snail shell, the two-part shell of the Bivalves or the cuttlebone of the Squids. Another characteristic feature of the molluscs is the radula, a palatal plate used like a grater to rasp food particles. This basic mollusc structure is adapted in an endless variety within the individual mollusc groups.

Molluscs originally lived in the sea. Today, all the classes of molluscs are either still confined to the sea (eg the Placophora*, the Scaphopoda* and the Cephalopoda*) or, like the Snails and Bivalves, they have many members that are still marine species. It is only the latter two groups that have penetrated fresh water, and only the Snails that have conquered life on land.

Many kinds of molluscs are important in aquarium-keeping. They make specimens in their own right, but some are also used in the control of algae (eg some aquatic Snails), as a host for the spawn of Bitterlings *(Rhodeus*)*, such as *Unio** species, or as splendid display specimens in large public aquaria, such as many kinds of Octopuses (eg *Octopus**, *Sepia**).

Monacanthidae, F, name not in current usage, *see* Balistidae

Monacanthus CUVIER and OKEN, 1817; Filefishes. G of the Balistidae*, Sub-F Monacanthinae. From tropical and subtropical regions of the Atlantic and Indo-Pacific. They eat growths and small animals from the littoral, preferring things like algae or coral polyps. Found among reefs and among clumps of plants. This G originally contained a large number of species, but is now often divided up into different Ga (*Amanses, Stephanolepis*, etc). They are laterally flattened fish, shaped like leaves, and measuring 5–40 cm long. The first spine of the D is usually long and checked against the second, small spine. There is also an erectile ventral spine. Filefishes have various camouflage colours and have a striking ability to change colour. Unlike most Triggerfishes (Sub-F Balistinae), Filefishes are for the most part peaceful in the aquarium, although they are much more sensitive. They must always receive enough supplies of plant food and the aquarium must always be calm. They are often extremely timorous in the face of food-competitors. It is best not to keep them in a community tank with delicate invertebrate animals. The following species are occasionally imported.
– *M. pardalis* RÜPPEL, 1835; Leopard Filefish. Found in tropical parts of the Indo-Pacific, the Red Sea, and tropical Atlantic. Length up to 38 cm. Usually with brown flecks and paler reticulated markings. There is a white fleck behind the last D ray. Colouring very variable. Forehead slightly concave.
– *M. spinosissimus* (QUOI and GAIMARD, 1824); Ragged Filefish. From the Indo-Australian Archipelago. Length up to 18 cm. Dirty yellow with irregular dark longitudinal stripes and flecks. There are lots of feathery outgrowths fringing the edge of the body (fig. 771).

Fig. 771 *Monacanthus spinosissimus*

Monascidea, O, *see* Tunicata

Moneywort *(Lysimachia nummularia) see Lysimachia*

Monkfish *(Squatina squatina) see* Squatinidae

Mono *(Monodactylus argenteus) see* Monodactylidae

Monocentridae, F, *see* Beryciformes

Monocentrus japonicus *see* Beryciformes

Monocirrhus HECKEL, 1840. G of the Nandidae*. One species only that is native to the black water rivers of tropical S America (Brazil, Guyana). Large, deeply ovate fish, very flat laterally, and with a characteristic worm-like process on the bottom-lip. Head pointed, forehead compressed, mouth superior and with a wide gape. The mouth can also be extended forwards quite some way. The hard-rayed parts of the D and An are incorporated in the body colouring, the soft-rayed parts and the C are transparent, like glass. Colouring very variable, yellow-brown with a dark marbling effect. A dark band runs from each eye to the snout, to the nape and to the throat. In shape and colouring, *Monocirrhus* is easily mistaken for withered leaves; it is usually motionless, its head inclined slightly downwards, hiding among thick vegetation. It waits there for prey to come along. When something suitable comes into view, usually smaller kinds of fish, it swims in slow-motion up to its victim, suddenly extends its mouth, and sucks it in. In the aquarium, it is important to provide for the natural needs of *Monocirrhus*; keep them in a species tank filled with very soft, acidic water, occasionally filtered through peat. Temperature about 25 °C. Thick vegetation. Add pieces of root and keep the lighting subdued. It is also essential to provide large quantities of food fish (Guppies perhaps). If *Monocirrhus* feels contented, it will reproduce in captivity now and again. The mating pair will clean the chosen spawning site (aquatic plant leaf, rock, or even the tank's pane of glass), and then, without any apparent sign of excitement, the eggs and sperm will be laid. The male guards the spawn, fanning it with fresh water (ie he practises a simple form of brood care). After 3–4 days, the fry hatch out, and as soon as they are swimming freely they can be fed with small crustaceans *(Cyclops)*. Within a few weeks they are as successful as their parents as predators, and even attempt to devour some of their own species. This is the latest stage at which they should be sorted according to size or even kept in isolation from one another.
– *M. polycanthus* HECKEL, 1840; Leaf-fish. From the Amazon Basin. The Rio Negro and in Guyana. Length up to 8 cm. Colouring similar to that of a dead leaf, clay-yellow or dark-brown, depending on environment. Has very specific requirements in the aquarium, and only recommended for very experienced aquarists (fig. 772).

Monocoelium, G, *see* Trematodes

Monocotyledoneae, Cl, name not in current usage, *see* Liliatae

Monocotyledons (Monocotyledoneae, Cl, name not in current usage) *see* Liliatae

Fig. 772 *Monocirrhus polyacanthus*

Monodactylidae; Fingerfishes (fig. 773). F of the Perciformes*, Sub-O Percoidei*, comprising only a small number of species. In side-view, these fish are shaped like a disc and very compressed laterally. They

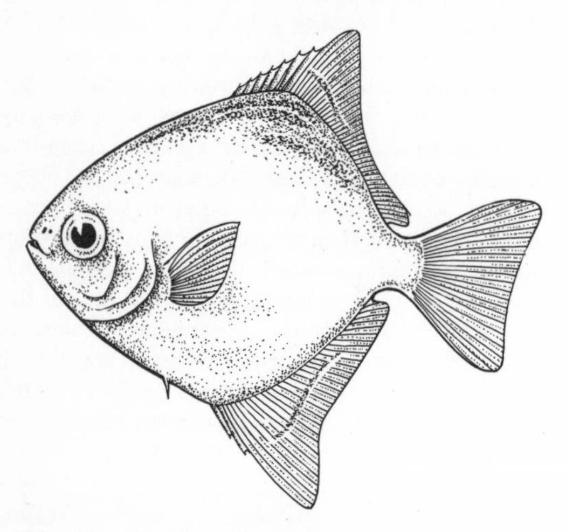

Fig. 773 Monodactylidae-type

are particularly agile, and live in tropical parts of the Atlantic and Indo-Pacific. Eyes large, mouth relatively small with fine teeth. The D and An are almost of equal length and stand opposite one another. The anterior sections of these two fins have elongated fin rays. The spines of the D are short and usually stand on their own in front of the fin. The comb-edged scales are very small and also cover the bases of the D, C and An. LL is arched upwards and complete. The Monodactylidae live mainly in coastal waters, but also penetrate into brackish and fresh waters. Almost all the species are silver, with narrow, dark, transverse bands running through the eyes and over the body. A species that is well-known and frequently imported is the Silver-leaf Fin *(Monodactylus argenteus)* (fig. 774), also known as Malayan Angel, Silver Dollar or Singapore Angel. Its distribution area extends from the Red Sea to Australia. It can reach 23 cm in length but will be smaller in the aquarium. Juveniles can also be kept in fresh water, while older specimens should be transferred to brackish water or sea water, at least from time to time. These are very lively shoaling fish with lovely yellow-coloured unpaired fins. In large aquaria, they will usually remain

Fig. 774 *Monodactylus argenteus*

timid and easily frightened. Feed with live food (all types).

Monodactylus argenteus *see* Monodactylidae

Monodonta Lamarck, 1799. G of the Topshells (F Trochidae) which belong to the Prosobranchia*. These molluscs live as algae-eaters in the upper littoral of warm seas. Their shell is a round-cone shape and very firm. The species *M. turbinata* (Born) is common along southern European coasts. Its shell can be up to 3 cm large and is flecked in grey. Occasionally, it will clamber out of the water.

Monoecious *see* Sex Division in Higher Plants

Monophyletic Group (fig. 775). A systematic group of any rank (eg a phylum, an order, a family, a genus), the members of which are all descended from the same primitive form. In a natural system* of organisms based on phylogenetic principles, only monophyletic groups are allowable. Compare also polyphyletic and paraphyletic groups.

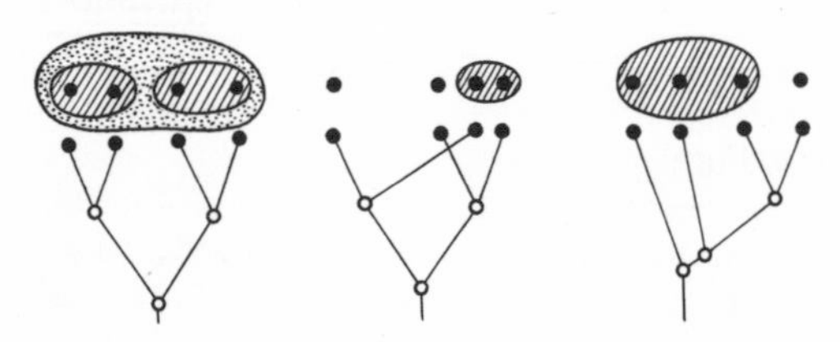

Fig. 775 Left to right: mono-, poly-, paraphyletic group

Monopodium *see* Stem System

Monopterus albus *see* Synbranchiformes

Montezuma Swordtail *(Xiphophorus montezumae) see Xiphophorus*

Moon Wrasse *(Thalassoma lunare) see Thalassoma*

Moon-eye *(Hiodon tergisus) see* Mormyriformes

Moon-eyes (Hiodontidae, F) *see* Mormyriformes

Moonfish *(Lampris regius) see* Lampridiformes

Moonfishes, F, *see* Citharinidae

Moonfishes (Lampridae, F) *see* Lampridiformes

Moonfishes, related species, O, *see* Lampridiformes

Moorish Idol Fishes, G, *see Zanclus*

Moorish Idols (Zanclidae, F, name not in current usage) *see* Acanthuroidei

Morays (Muraenidae, F) *see* Anguilliformes

Moringuidae, F, *see* Anguilliformes

Mormyridae, F, *see* Mormyriformes

Mormyriformes; Mormyrids. O of the Osteichthyes*, Coh Teleostei. The Mormyriformes, together with the Osteoglossiformes*, form the Sup-O Osteoglossomorpha. Despite their very varied shapes, the Mormyriformes also have the following characteristic features in common: A primitive form of dentition made up of 2 toothed surfaces for biting, located between the base of the cranium (parasphenoid bone) and the tongue; and a paired tendinous bone springing out from the second gill arch, on the throat side. According to GREENWOOD (1973), the Notopteridae, the Mormyridae and the Gymnarchidae are the most closely related of today's species. It is only in these Fs that the lateral line channels do not have any pores on the head. In addition, it is only in these groups that the balancing organ and auditory organ have a very unusual structure. For anyone interested in comparable anatomy, it should also be mentioned that the semi-circular canals and the utriculus are completely separate from the sacculus. The Hiodontidae and probably the Lycopteridae from the lower Cretaceous period, have a whole series of specialised features that would seem to indicate that they are related to the Notopteridae.

– *Gymnarchidae. Gymnarchus niloticus* (fig. 776) grows up to 2 m long, and has many features in common with the Mormyridae. In fact, it is often included in that F. Typically, it has an eel-like shape, with a fairly pointed head, and there are no Vs, An or C. Apart from the small Ps, only a dorsal fin is formed – a fringe of fin that is very long and begins at the nape. *Gymnarchus* builds floating nests out of parts of plants, in which about 1,000 eggs are laid, perhaps measuring 1 cm each. The eggs and the hatched fish brood are carefully guarded. The juveniles have external gills. The Gymnarchidae are eaten in many areas, and their flesh is said to taste very good.

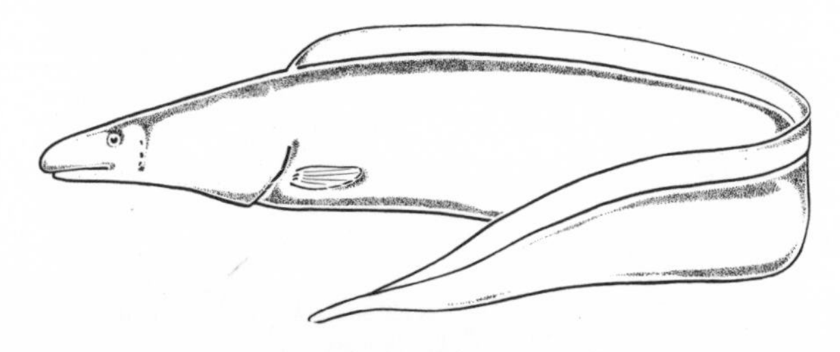

Fig. 776 *Gymnarchus niloticus*

– *Hiodontidae;* Moon-eyes. This F has various features that place it somewhat apart from the other Fs; it is probably most closely related to the Notopteridae within the O. There are 3 modern species, all of which are found in fresh waters of central and eastern N America. They prefer clear, oxygen-rich water. The best-known species, the Moon-eye *(Hiodon tergisus)* (fig. 777), can grow up to 30 cm long, sometimes even longer. It has a lovely mother-of-pearl sheen and very large eyes. It is only rarely eaten, but is very popular as a sporting fish because it is particularly tenacious.

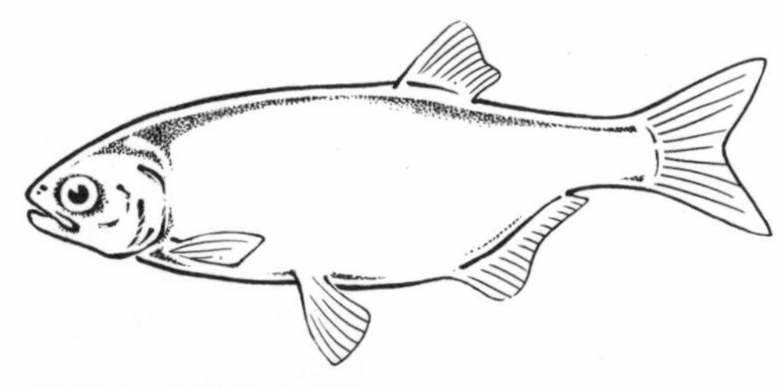

Fig. 777 *Hiodon tergisus*

– *Mormyridae;* Elephant-nosed Fish (fig. 778). Exclusively native to Africa. Most species are a uniformly dark colour, but there is an enormous variety of body shapes within the F, particularly as far as the head is concerned. Characteristic features of the body include the usually slim caudal peduncle and the deeply indented C. The D and An are often located directly opposite. The shape of the head can range from stumpy, long-snouted types, through species with a very thick or elongated chin, to those with a trunk-like extension to the head (fig. 779). Those species with elongated heads use them to hunt for food in the muddy bottom, the trunk being used like a plough. The more unusual forms of Mormyrid fish had an important role to play in the religion and art of the ancient Egyptians – preserved for ever in numerous paintings and in the plastic art-forms. The mouth is always very small, making it possible to eat small food animals only, especially worms. There are about 100 species, most of which are less than 15 cm long, and only a very few of which grow to more than 1 m long. Our knowledge of their reproduction is still very patchy. Some species are said to build nest of plant material, others spawn in sandy hollows. Two unusual anatomical features of particular interest are the large brain and the electric organs. In relation to total body mass, the mass of the brain is almost as large as a human brain (about 1/50), and

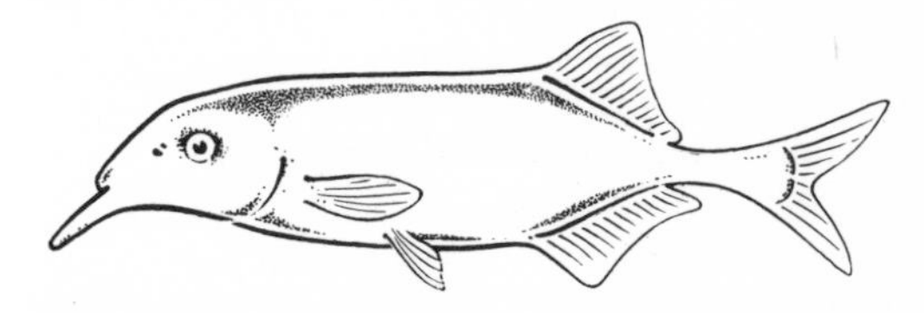

Fig. 778 Mormyridae-type

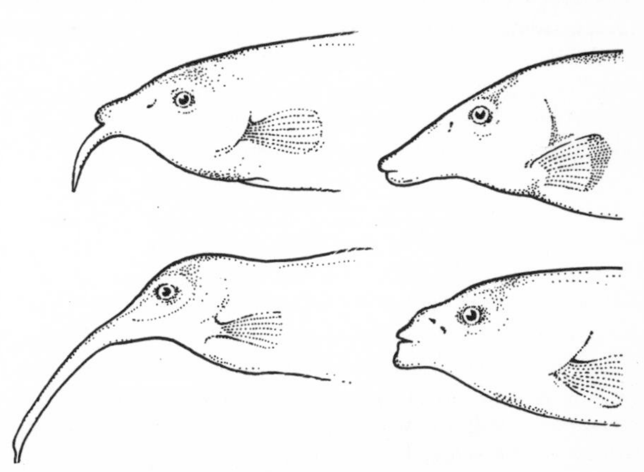

Fig. 779 Head shapes found in Mormyrid fish

sometimes even larger. This enlargement of the brain can be traced back mainly to the very strong development of the cerebellum. Since it is this part of the brain that co-ordinates movement, it is easy to see why the Mormyridae are such excellent swimmers. They can move equally well backwards or forwards, and swim belly-up or at an angle. Anyone who has tried catching them in a net will testify to how agile they are. Having such a large brain also enables the mormyrids to behave in certain ways that other fish cannot. They like to play, for example (ie they make use of objects in a playful manner). Their ability to learn is also very good. The electric organs develop in the caudal peduncle region and are composed of certain muscles. They are stimulated by special nerves, and are used mainly to help the fish to orientate itself, to recognise like species and to define territories. The discharges of electricity follow rapidly one after the other (about 200/minute). The electrical field thus created will enable any disturbances in its range to be detected by sense organs on the fish's head. By this means of orientation, Mormyrid fishes can also move around in particular directions in the murky waters where they usually live. It is probable that the eyes have become small because these fish have become more used to using electric signals to find their way about. Many species have been kept in aquaria, but none has so far been bred successfully. For details regarding care turn to the various Ga – *Gnathonemus, Marcusenius, Mormyrops, Mormyrus.*

– *Notopteridae;* Featherbacks. Freshwater fish, like a knife-blade in appearance, and native to Africa and SE Asia. The anus is very far forwards, and directly behind it begins the An, which fuses with the C to produce a long fringe of fin of uniform height. It is by means of this fin fringe that Featherbacks propel themselves along. They make rhythmical wave movements, running from front to back or vice versa, and this enables the fish to swim backwards or forwards. To stand still, the fish makes opposed wave movements in the front and rear sections of its fin fringe. Featherbacks are similar in shape and methods of movement to the American Knife-fishes, although they are not closely related. The D is very small or absent altogether. The Vs are also small, whereas the Ps are quite large and are used for steering. The swimbladder is used as an accessory breathing organ – it is divided into chambers and is often branched, too. In their natural habitats, Featherbacks prefer the weedy bank zones of slowly flowing waters. During the day, they usually stay in small groups, head slightly inclined, hidden among large tree stumps or overhangs from the bank. They only come out at dusk to search for food. The smaller species eat insect larvae, worms and other small animals, whereas the larger species prey on fish. WICKLER reports that they spawn on firm substrata (rocks, wood), and the spawn is guarded by the male and fanned by him. The young hatch out after 6–7 days and are also guarded by the male for a few more days. The Clown Knifefish *(Notopterus chitala)* (fig. 780) can measure as long as 80 cm, sometimes up to 100 cm. It is widely

Fig. 780 *Notopterus chitala*

distributed in the Indo-Australian realm and is economically important in Thailand as an edible fish. Up until they are 6 cm long, *Notopterus* species have worm-like transverse bands, that later become a more uniform velvety brown, and in adult specimens large eye-spots appear on the flanks. A common species in the aquarium is the African Knife-fish *(Xenomystus nigri)* (fig. 781). It grows up to 20 cm long and is coloured a mousy-grey or dark brown. It has a distribution area that stretches from the source rivers of the Nile, west to Liberia. It has no D at all and the Vs are very small. Sometimes, several longitudinal stripes appear on the body. During the spawning period, MEINKEN reports that the ground colouring of this species takes on red-brown or purply tones. They utter short sounds caused by air transferring from the swimbladder duct to the oesophagus. This is a shy, peaceable species that is not difficult to keep. Simply provide enough kinds of substrata and a varied diet of live food, especially worms. A rare species in the aquarium is *Notopterus afer* which can grow up to 60 cm long. However, it is a rather intolerant species.

Fig. 781 *Xenomystus nigri*

Mormyrops MÜLLER, 1843. G of the Mormyridae, O Mormyriformes*. From central Africa. Medium-sized to very large fish from muddy, thickly vegetated waters. The body is usually elongate and only slightly flattened latterally. The mouth is terminal or slightly inferior, and can be very elongated. A characteristic feature of the G is the way in which the teeth are arranged in a simple row over the whole jaw. The An and D are inserted well to the rear and their lengths differ. The C is very small. *Mormyrops* species are solitary fish, active at dusk and at night. They are usually very intolerant of like-species. Only young specimens can be kept in room aquaria. Place the tank in a dark spot and provide plenty of places to hide and a soft substrate. Large specimens are very attractive in public aquaria. Fill the tank with old water, and add some fresh water at regular intervals. Temperature 24–28 °C. Diet consists of larger kinds of live food, especially worms and snails. The larger species are partially predatory. No information available on sex differences or breeding.

– *M. nigricans* BOULENGER, 1899. From the lower Congo. Length up to 35 cm. A dark black-brown fish, sometimes with indistinct transverse bands and fleck markings. Adult fish are pronounced predators.

Mormyrus LINNAEUS, 1758. G of the Mormyridae, O Mormyriformes*. From central and NE Africa. Large fish that live in the rivers of the Nile Basin in thickly vegetated places with very muddy bottoms. The body is elongate and only moderately flattened laterally. The D is usually twice as long as the An. Mouth terminal with several rows of fine teeth on the jaws (a characteristic of the G). *Mormyrus* is a solitary fish, active at night. It searches for food along the bottom. Large aquaria are needed to keep members of this G. Stand the tank in a darkened spot and make sure there are plenty of hiding-places in the tank as well as a soft substrate. The fish are usually intolerant of one another, and they can only be kept in the same tank as larger fish of other species. Fill with old water and add fresh water regularly. Temperature 24–28 °C. Live food, particularly worms and snails. No information regarding sex differences and reproduction.

– *M. kannumae* FORSKÅL, 1776; Elephant-nosed Fish, Tapir-snout Fish. From the rivers of the Nile Basin. Length up to 50 cm. The mouth opening is small and is located at the end of a small, mobile trunk. Body grey or black-brown with a reddish shimmer. Young specimens only are suitable for room aquaria (fig. 782).

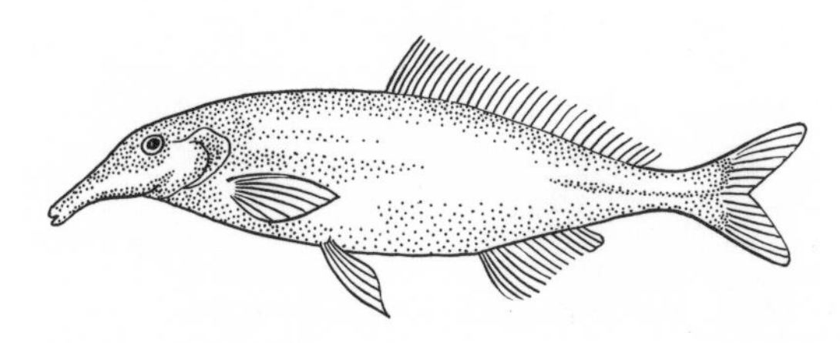

Fig. 782 *Mormyrus kannumae*

Morphology. A branch of biology and medicine that is initially concerned with the external forms of plants, animals and humans, including the form of the body's organs and their locations in respect of one another. In a wider sense, morphology is the science of the structure of organisms. Morphological disciplines include anatomy*, histology* and cytology*.

Morulius HAMILTON-BUCHANAN, 1822. G of the Cyprinidae*, Sub-F Cyprininae. From SE Asia, Thailand, Indonesia, Sumatra Large fish, similar in form to the G *Labeo**, with elongated bodies, only slightly flattened laterally. The ventral profile, particularly among older specimens, is straight. Snout covered in a large number of pores, also with a lateral lobe and one pair of rostral barbels and one pair of maxillary barbels. The lips are fringed. The fins are strikingly large, the strong D being inserted well to the fore of the Vs. Since *Morulius* species are large and active, they can only be kept in a room aquarium when young. The tank should provide sufficient hiding-places and enough swimming-room, and it should be stood in not too bright a position. The substrate should be dark and the water should be soft and slightly acidic. Temperature 24–27 °C. These fish are very hardy and will eat all kinds of live food and dried food, although plant food is also advisable. No information available regarding sex differences and reproduction.

– *M. chrysophekadion* (BLEEKER, 1850); Black Shark. From Thailand and the Greater Sunda Islands. Length up to 60 cm. Body and fins a uniform blue-black colour. Part of the flanks may have small yellow or orange-red flecks. Very suitable for public aquaria.

Mosaic Gourami *(Trichogaster leeri) see Trichogaster*

Mosquitofish *(Heterandria formosa) see Heterandria*

Mosquitos, F and Ga, *see Aedes, Anopheles* and *Culex*

Moss Animals, Cl, *see* Bryozoa

Mossbunker *(Brevoortia tyrannus) see* Clupeiformes

Mosses (Musci, Cl) *see* Bryophyta

Mosses and Liverworts, D, *see* Bryophyta

Motorist Fish *(Hemigrammus ocellifer) see Hemigrammus*

Mousefish *(Gonorynchus gonorynchus) see* Gonorynchiformes

Mouth; oral cavity. The mouth in vertebrates is bounded by the jaws, but in fish it can come in a variety of shapes. 3 basic forms are distinguished. The mouth is described as *terminal* when the upper and lower jaws are of the same length. It is called *superior* when the lower jaw reaches beyond the upper jaw, and *inferior* when the upper jaw or the tip of the head extends beyond the lower jaw. An inferior mouth is typical of sharks and rays, the mouth actually lying on the underside of the head. Fishes' mouths often have lips. Depending on what type of food is eaten, the mouth can be narrow or it can enlarge itself to quite a marked degree, and sometimes it even has a special shape (eg like a beak in the Snipe Eel [*Nemichthys*], like pincers in the Long-snout Butterfly Fish [*Chelmon*] or like a trunk as in many mormyrid fish [Mormyridae*]). The mouth can also be used as an organ of attachment* in some species.

Mouth-brooders *see* Cichlidae

Mouth-brooding Fighting Fish *(Betta pugnax) see Betta*

Mouth-to-mouth Fighting. Term used in ethology. This describes the high phase of aggressive fighting behaviour* as observed in Cichlids, and sometimes also among male Egg-laying Tooth-carps. Mouth-to-mouth fighting normally follows on from when the fish swim at one another, their gill-coverings opened out; sometimes it may begin after they have begun to ram each other (Ramming*). The opponents ram each other from the front, grasping each other's lips. There then follows a 'tug-o'-war' competition during which the lips may be torn away. The defeated party takes to flight, often showing the signs of inferiority behaviour* and inferiority colouring*, thus preventing the outbreak of a fight to the death (Aggressive Fighting Behaviour*).

Moxostoma, G, *see* Catostomidae

Mozambique Cichlid *(Tilapia mossambica) see Tilapia*

Mucous Gland *see* Skin

Mud Plantains, Ga, *see Heteranthera, Zosterella*

Mudfishes, African (Phractolaemidae, F) *see* Gonorynchiformes

Mudfishes (Amiidae, F) *see* Amiiformes

Mudfish-types, O, *see* Amiiformes

Mudminnows (Umbridae, F) *see* Esocoidei

Mudskippers (Periophthalmidae, F, name not in current usage) *see* Gobioidei

Mudwort *(Limosella aquatica) see Limosella*

Mugil cephalus *see* Mugiloidei

Mugilidae, F, *see* Mugiloidei

Mugiloidei; Grey Mullet types. Sub-O of the Perciformes*, Sup-O Acanthopterygii. In most of the older classification systems, the Fs of the Mugilidae, Sphyraenidae and Atherinidae were placed together in this fish group which was recognised as a Sub-O in its own right, or even as an O (Percesoces or Mugiliformes). The system described here was devised by GREENWOOD and colleagues (1966) and it reduces the Sub-O Mugiloidei to the F Mugilidae only, which then becomes a Sub-O of the Perciformes*. The Sphyraenoidei with their solitary F Sphyraenidae are attributed the same rank in the same O. The Atherinidae, on the other hand, have been removed to a completely different place in the system (*see* Atheriniformes). It is not possible to go into the scientific reasons for such a swap-around here, but it can be looked up in the relevant literature. See also the system applied to the Osteichthyes.

The Mugiloidei have been known since the Neogene period and are related to the Apogonidae*. They are distinguishable from most of the other representatives of the O Perciformes through the position of the Vs. They are not located well forwards, as is typical for the O, but they have been moved, as a secondary development, further back again. Nevertheless, the pelvic girdle has remained joined to the shoulder girdle via a ligament. The body is elongate, only moderately flattened laterally. The mouth is small, its upper edge being bounded by premaxillary bones only. Small, conical teeth that are not inserted in deep alveoli (sockets). They eyes are deep-set, easier to see from below than from above. The D1 has 4 spines and stands well in front of the D2 which is short. The D2 is located opposite a similar-looking An. The C is indented and the Ps are high-up. No LL. Scales large and distinct. The pharynx contains a filter apparatus made of horny papillae. Food is broken up into very small pieces by a particularly strong, muscular stomach (similar to a bird's), and following on from there is an exceptionally long intestine. Colouring usually plain silver. The Mugiloidei live in warm and temperate seas, some species penetrating brackish water also, and some being freshwater species only. Most of them are shoaling fish that feed on tiny living organisms and algae; some are typical grazers, others chew their way through the substrate. Several species have regional importance economically. They are unusual specimens to find in an aquarium.

– *Mugilidae;* Grey Mullets (fig. 783). For characteristics, see above. The best-known species is probably the Striped Mullet *(Mugil cephalus)*, otherwise known as the Flathead Mullet or Haarder. It grows up to 80 cm long and weighs 10–12 kg. It occurs in warmer seas, as well as the Mediterranean. Because it grows quickly, this species is cultivated in many parts of the world in brackish-water areals. *Chelon oligolepis,* a freshwater species from SE Asia, grows up to 15 cm long. There are about 100 species in all.

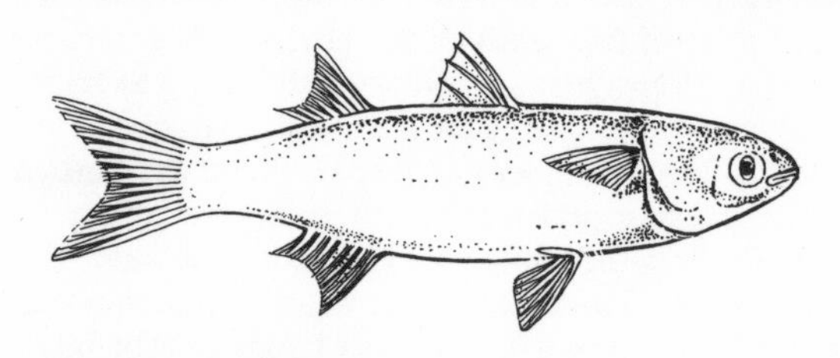

Fig. 783 Mugilidae-type

Mullets, F, *see* Mullidae

Mullidae; Red Mullets (fig. 784). F of the Perciformes*, Sub-O Percoidei*. Small to medium-sized bottom-dwelling fish from warm and temperate seas. Body elongate, the head in some species being snub like a bulldog's. The eyes are positioned high-up. D1 and D2 about the same length and sail-like. The C is deeply indented, the An short and opposite the D2. There are about 60 species in all, and typical features include 2 long barbels at the chin (retractable into a groove).

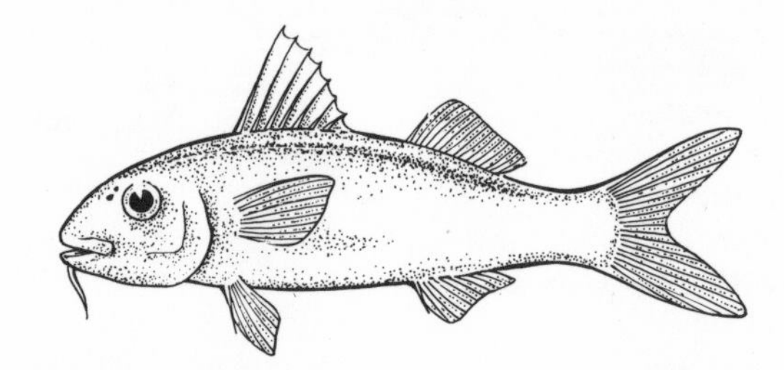

Fig. 784 Mullidae-type

These barbels are used both in touch and in smell, and help the Mullets to find food either on or in the substrate. Most Red Mullets are shoaling fish continually searching for food in small bands. The majority of them are reddish in colour. Red Mullet eggs, like the young fish, are pelagic. Their flesh is tasty, and so many species are important in the fishing industry, and are caught in large numbers. For aquaristic details see under the Ga *Mullus**, *Pseudupeneus**, *Upeneus**.

Mullus LINNAEUS, 1758. G of the Mullidae*. Comprises some species from southern European coasts and the Atlantic. *Mullus* species will be found on sandy and muddy expanses. They are sociable fish, with a typical family shape, and they feed on small animals. *Mullus* grows to about 40 cm long, and unlike the other Ga does not have teeth on the upper jaw. They are lively fish, able to change colour rapidly. They are valuable edible fish that are rarely kept in aquaria. If they are, they prove to be very sensitive. The best-known species is:
– *M. barbatus* LINNAEUS, 1758. Dorsal surface brown-red, ventral surface white, dark stripes run along the LL. The D is colourless. At night, and when excited, has large, purply flecks (fig. 785).

Fig. 785 *Mullus barbatus*

Munidia, G, *see* Galatheidae

Muraena LINNAEUS, 1758. G of the Muraenidae, O Anguilliformes*. Members of this G are found in tropical and temperate parts of the Indo-Pacific, the Atlantic and the Mediterranean. They are predators that are active at night, and they live in shallow water. Length 60–100 cm. The various species have a typical family shape, and both nasal openings protrude like tubes. Body has dull patterns only on it. *Muraena* is also prone to biting a lot. Only suitable for large aquaria. The following Mediterranean species has been prized as an edible fish since ancient times:
– *M. helena* LINNAEUS, 1758. From the Mediterranean and E Atlantic. Found on rocky substrates. Up to 100 cm long. Marbly patterns, dark-brown/yellowish.

Muraenas (Muraenidae, F) *see* Anguilliformes

Muraenolepididae, F, *see* Gadiformes

Murex LINNAEUS, 1758. G of predatory and carcass-eating Prosobranchs (Gastropoda*, Prosobranchia*) from warm and tropical seas. Their shell is knotty or spiny, they have a long siphonal canal, and they have a mobile, protrusible proboscis. The shell can grow up to 10 cm long. *Murex* attacks bivalves, in particular, boring through their shells in part; but they will also seize upon calcareous tubed Polychaetes, Sea-urchins and other organisms that move along slowly. *Murex* often carries parasitical Sea-anemones on its shell *(Calliactis*)*, Sponges and other sessile animals. The egg-cocoons are laid in large balls, often by several of the Snails together. All the *Murex* species that have been kept in aquaria so far have proved very hardy, but enough regard must be given to their predatory way of life. They will accept various types of animal food. They can be transported for 2–3 days at a time without water; just keep them damp. There are two common species in the Mediterranean:
– *M. brandaris* LINNAEUS, 1758. 10 cm. Shell is spiny and this species has a very long siphonal canal. Bores into molluscs (bivalves).
– *M. trunculus* LINNAEUS, 1758. 7 cm. Shell knotty and has a shorter siphonal canal.

Musci, Cl, *see* Bryophyta

Musculature (fig. 786). The musculature of invertebrates and vertebrates consists of muscle cells that contain proteinous fibrils that are capable of contraction. According to the structure and arrangement of these fibrils, three types of musculature are distinguished: smooth muscle (eg stomach, intestine, blood vessels), skeletal muscle and cardiac muscle. The impulses that produce the contraction of the muscles are supplied by the nervous system. In humans, skeletal muscles are subject to the will.

The musculature in bony fish is whitish, in some species also red (eg Tunnies) or transparent like glass (eg Glass Barbs, Glass Catfishes). The muscles attached to the extremities are relatively weak, whereas the torso musculature is strongly developed. Depending on how the vertebral column breaks down into vertebrae, the torso muscles will be made up of numerous segments (myomeres), separated by connective tissue (myosepta). Myomeres do not lie one after the other like a row of books, but are packed into one another like pointed teats. So, in cross-section, a fish will be seen to have several, often concentric layers of muscle, arranged in a way that is typical of the species. A horizontal layer of connective tissue divides the torso musculature of a bony fish into a smaller, upper band and a stronger, lower band. On the outside of the segmented torso musculature is an unsegmented lateral muscle (also known as red muscle), which is striking because of its red colour. It is usually rich in fatty tissue and blood vessels. A paired strand of muscle also runs across the dorsal and ventral edges, to be replaced around the D and An by the fin musculature.

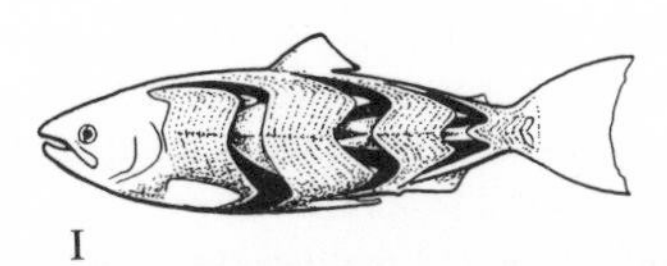

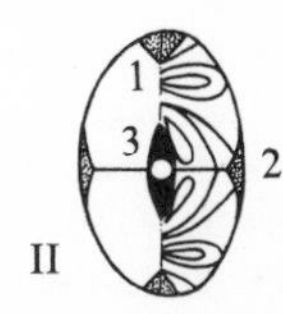

Fig. 786 Musculature. I Torso musculature of a salmon. At 3 points a row of myomeres has been removed to show the folding of the muscle segments. II Cross-section through the middle of a salmon's body. 1 Myomeres 2 Lateral muscle 3 Vertebral column

Unpaired fins have muscles to splay the fins out and fold them back. The muscles in this case are located at the fin rays and fin-ray sockets. These muscles have lateral control over the fin rays, and when they are contracted alternately they produce a wave-like motion along the fins or fin corona. The musculature of the C is influenced by the form of the tail. Usually, muscular threads are present between the bottom parts of the fin rays next to flexor muscles above and below. The paired fins have opposed groups of muscles which either hold the fins against the body or keep them away from the body. Particularly complicated muscular systems are found in the head region. These include the muscles of the jaws, the gill arches and the eyes. In some fish, particular groups of muscles have developed into electric organs*.

Muskellunge *(Esox masquinongi) see* Esocoidei

Mussels, G, *see Mytilus*

Mustelus mustelus *see* Triakidae

Mutagenes *see* Mutation

Mutation. A structural alteration in the genetic material which, to a varying degree, results in changes to the characters of an organism. A mutation can take place in each cell. If body cells are concerned *(somatic mutation)*, the mutation is confined to those particular cells and their daughter cells. A mutation in germ cells or their precursors *(generative mutation)* can be inherited, however. Generative mutation is one of the central processes which has made the evolution* of organisms possible. The frequency with which a mutation occurs *(mutation rate)*, in the germ-cells of the Vinegar-fly for example, is in the region of 5%.

The order of magnitude of a mutation is very variable. *Point or gene mutation* takes place in the molecules. In this case, individual building blocks of the genetic information-carriers, the deoxyribonucleic acid (DNA), are either restructured chemically or their position in the DNA-chain is altered. A mutation of this sort is not always perceptible from outside. *A chromosome mutation,* on the other hand, can be detected under a microscope during nuclear division. When chromosomes break up (*see* Cell) and the pieces come together again, but wrongly, this will lead to a new configuration of the hereditary factors (genes), which often causes great changes to an organism's features. If the number of chromosomes is changed, this is called ploidy mutation. The lack of individual chromosomes or the lack of pieces of chromosomes leads to death usually, whereas too many chromosomes usually cause severe damage (deformities, idiocy etc). It commonly occurs in plants that the whole set of chromosomes is polyploid, ie doubled, quadrupled etc, which means that features are multiplied. (This rarely occurs in animals.)

The mutation-releasing factors *(mutagenes)* consist of various physical and chemical influences; in nature they are mainly cosmic radiation or temperature changes and food components. During experiments to produce mutations, which are an important part of breeding*, these influences are replaced by, among other things, X-rays, temperature shocks or chemicals (eg nitrites or colchicin).

Mycoses. Diseases that are caused by parasitical fungi. Of those that cause damage to fish, the algal fungi (Phycomyceta) are particularly important. Fungi occur primarily as pathogens only rarely. Most are typical debilitating parasites (Parasitism*) and they occur in conjunction with or as the result of other fish pathogens. Fungi of the Ga *Saprolegnia** and *Achlya** occur as ectoparasites in aquarium-fishes. *Ichthyosporidium** and *Branchiomyces* (Fin-rot*) occur relatively infrequently as endoparasites. Endoparasitical fungi are much more difficult to treat than ectoparasitical ones.

Myctophidae, F, *see* Myctophoidei

Myctophoidei; Lanternfish-types. Sub-O of the Salmoniformes*, Sup-O Protacanthopterygii. In REGAN's system (1911) considered as the O of the Iniomi. These are very variable, usually small fish that live primarily in the depths of the ocean, although some are also pelagic. Many species have light organs on various parts of their body. The upper edge of the mouth is bounded only by the premaxillary bone, and the mesocoracoid bone is missing from the shoulder girdle. The mouth is large and often contains large numbers of teeth, which are long and pointed. The fins are supported by soft rays only, some of which may be very elongated. An adipose fin is always formed behind the D. There is no swimbladder in many species, but when there *is* one it is always connected openly to the intestine (physostomi). The eyes are often greatly modified and there are none at all in some deep-sea types. Fossil forms date from the Upper Cretaceous.

– *Bathypteroidae* (fig. 787). Elongate deep-sea fish with an elongated stilt-like first fin-ray in the Vs. The right and left stilts and the lower, similarly elongated tail-fin lobe make a kind of three-legged stool which supports the fish off the bottom. There are about 13 known species.

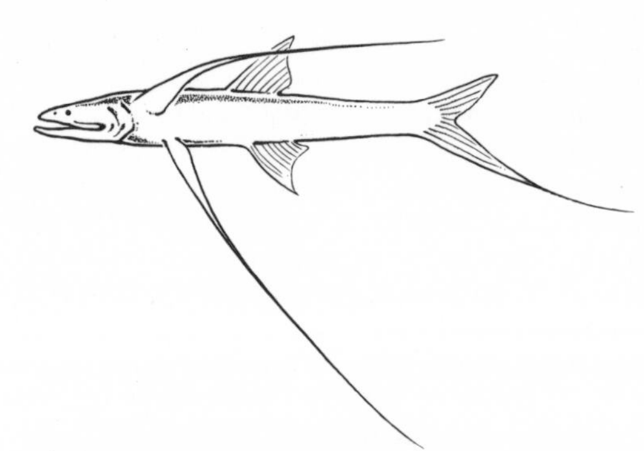

Fig. 787 Bathypteroidae-type

– *Myctophidae;* Lanternfishes (fig. 788). Small or very small deep-sea fish of elongated shape. Very large eyes; mouth deeply slit, reaching to behind the eyes. They also possess an array of different luminescent organs. The Lanternfishes filter plankton food which they sieve from the water with the aid of their pharynx. They have certain biological specialities including migrations from the depths to the surface and vice versa, which take place according to the daily rhythm. At night, they can often be observed from ships, because they are lured in close by the strong light coming from them. As the day dawns, they disappear again, seeking depths of 500–800 m. This daily act of surfacing and diving again is an amazing feat, particularly for the smallest species

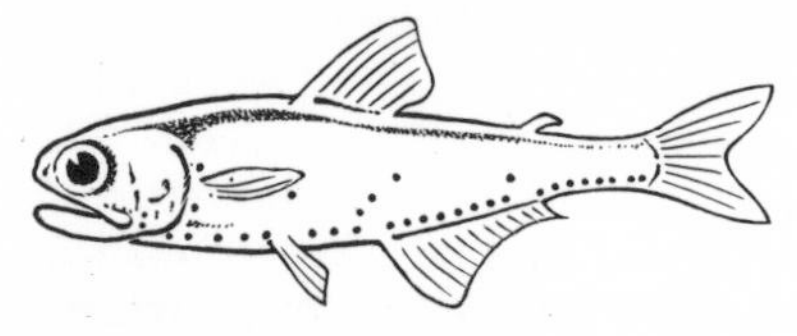

Fig. 788 Myctophidae-type

that are no more than 2–3 cm in length. The luminescent organs of the Lanternfishes are organised in a variety of ways and often have means of switching themselves off. This protects them from being chased and also enables rhythmical signals to be emitted, possibly as an aid to species recognition. The sexes are often equipped with different luminescent organs. In *Myctophum punctatum*, patches of light are sited on the caudal peduncle just in front of the C; in the female they are on the underside, and in the male they are on the upperside. The number of light patches and their intensity can also vary. From this, it is safe to assume that the luminescent organs, at least in this F, act as a kind of signal between partners. Unfortunately, we know very little about their reproduction. *Myctophum punctatum* spawns in the Mediterranean in late winter. The number of eggs is small, about 200–400. The larvae are quite often found near the surface. A total of about 150 species are known, some of which are indirectly important economically.

– *Synodontidae;* Lizard Fishes. Predatory fish, widely distributed in tropical and subtropical seas, and usually about 30 cm long. They live in sandy, shallow-water zones. The head is like a lizard's, the mouth slit reaching well to the rear. Body elongate, cylindrical, usually with dark flecks. Adipose fin present. They lie in wait for prey, resting on the sand with the front part of their body raised up. When something suitable appears, they ram it, gripping it in their long, sharp teeth. Lizard Fishes are also able to burrow into the sand with their strong Ps, so that only the forehead and eyes remain free. In many species, the embryos that hatch from the eggs grow first into a slender larva, changing into the definitive fish only when they reach a length of 4–5 cm. There are about 35 species, of which 2 are important to mention here: *Synodus synodus* (fig. 789) from the tropical Atlantic and the 50-cm-long *Saurida endosquamis* from the Indo-Pacific. The latter is reputed to be a particularly tasty table fish. Lizard Fishes occasionally crop up in marine aquaria. But as they will only accept live food animals, such as shrimps and small fish, they are not easy to care for. *Bathysaurus ferox* (fig. 790) is also placed in this F. It can reach a length of 60 cm and lives at depths of 900–3,000 m in all seas. It has a very long snout and is armed with strong teeth. No adipose fin. Apart from the Fs described here, there

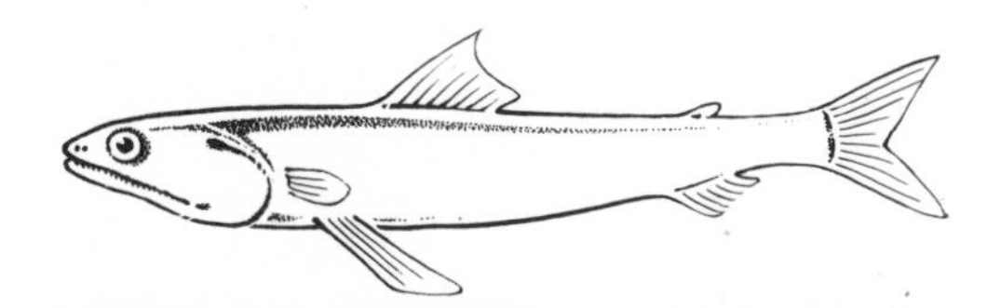

Fig. 789 *Synodus synodus*

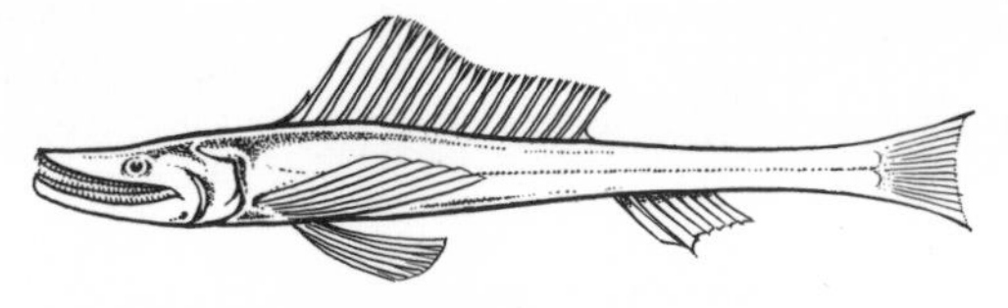

Fig. 790 *Bathysaurus ferox*

are a further 15 Fs that belong to the Sub-O Myctophoidei. Most of them are deep-sea fish, eg the Aulopodidae *(Aulopus)* (fig. 791), which live in the middle depths of the ocean. They can be recognised from their very long, broad, dorsal fin. The 20-cm-long *Aulopus japonicus* is caught at depths of around 100 m off the coast of Japan, and is said to be very tasty. The Alepisauridae (Lancet Fishes) have an even larger, sail-like D. *Alepisaurus ferox* (fig. 792) can grow up to 1.8 m long, has a very slim body, and a voracious set of teeth set in a pike-like mouth. Occasionally caught at middling depths. Members of the F Paralepididae are known as the Barracudinas. They are elongated deep-sea fish with a pointed head and a set of hard teeth inside a deep mouth (fig. 793). Lastly, the Harpodontidae (fig. 794) includes 5 species belonging to the G *Harpodon*. One, *Harpodon nehereus*, is usually about 30 cm long and is found off the northern coast of India, particularly be-

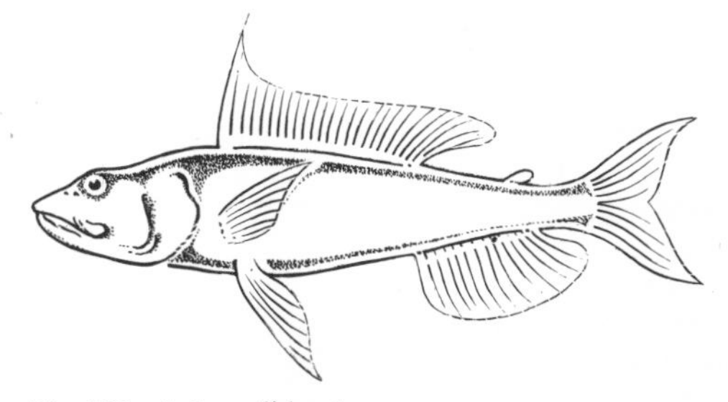

Fig. 791 Aulopodidae-type

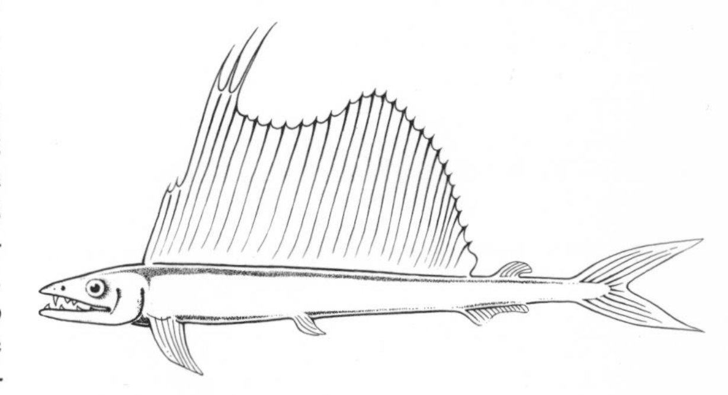

Fig. 792 *Alepisaurus ferox*

Fig. 793 Paralepididae-type

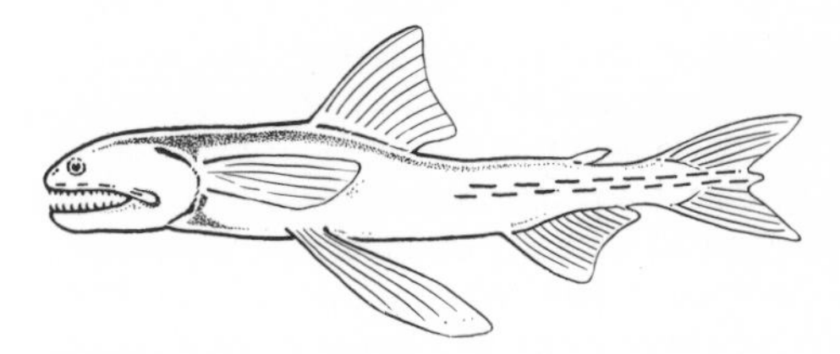

Fig. 794 Harpodontidae-type

fore the mouth of the Ganges. It is caught and eaten in large quantities. Its dried flesh, prepared in various ways, is known as 'Bombay Duck'.

Myctophum punctatum *see* Myctophoidei

Myer's Lampeye *(Aplocheilichthys myersi) see Aplocheilichthys*

Myer's Loach *(Acanthophthalmus myersi) see Acanthophthalmus*

Myleinae, Sub-F, *see* Characoidei

Myletes, G, name not in current usage, *see Alestes*

Myliobatidae; Eagle Rays (fig. 795). F of the Chondrichthyes*, O Myliobatiformes. Large Rays with huge, pointed Ps, spreading out far to the sides (like wings). The tail is thin and like a whip; on it is a small D near to where the tail joins onto the body. Like the Sting Rays (Dasyatidae*), the Myliobatidae also have

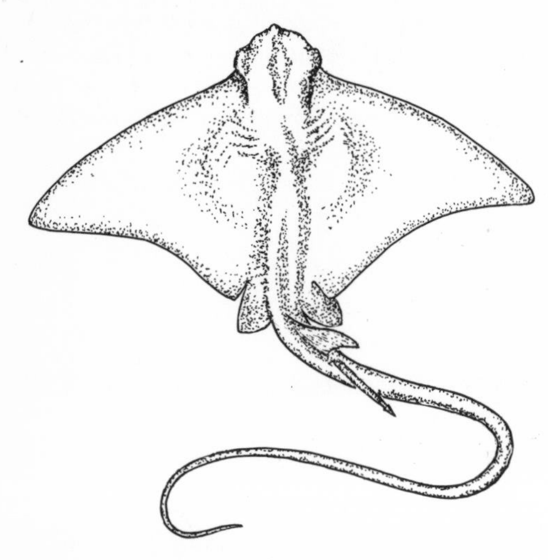

Fig. 795 Myliobatidae-type

a poison spine, although it is set further forward and is usually smaller. The front edges of the Ps are inserted at eye-level in the side of the head, then they peter out for a short stretch, and finally form a lip-like tumour on the snout. Spray-holes large. There is either one long row, or several long rows of teeth, arranged like paving-stones. The middle row is very broad. Such dentition is well suited to cracking open bivalves and snails, although Eagle Rays also eat crustaceans, worms and fishes. No placoid scales; only in the males will a few isolated ones appear on the upperside. The females are dissimilarly large and are livebearing. The number of offspring is small, and they are said to be born as the female makes leaps out of the water. Eagle Rays are known to date from the Eocene period. They float gently through the water for the most part, flapping their 'wings', but they are also able to shoot quickly through the water and leap out of it. Like the Sting Rays, they use their whip-tail as a defence weapon and try to spear their opponent with their spine. A poison gland opens out into the spine groove, and injuries from it are said to be very painful. Eagle Rays prefer the warmer seas and very rarely occur in colder regions. The Mediterranean and the central area of the E Atlantic are the main areas of distribution of the Common Eagle Ray *(Myliobatis aquila)*. It can reach a total length of nearly 4 m. A particularly pretty species is the Spotted Eagle Ray *(Aetobatis narinari)* from tropical and subtropical seas. It has whitish or yellowish dots distributed very evenly across its brown or grey upperside. Its 'wingspan' can be about 2 m wide. Various kinds of Eagle Ray can be bought at the fish markets of tropical countries.

Myliobatis aquila *see* Myliobatidae

Mylossoma EIGENMANN and KENNEDY, 1903. G of the Serrasalmidae*, Sub-F Myleinae. From the southern Amazon Basin, Paraguay. Length up to 25 cm. The D1 is short; the C outspreading, but not indented, or only slightly so. An long, and in contrast to the G *Metynnis**, it broadens out in the last third to give it a triangular shape. Thus is not a sex feature as is the lobe-like extension of the An in the *Metynnis* male, which has an important function to fulfil during mating. The relatively short D2 also serves to distinguish this G from the G *Metynnis* with its broad, flatter D2. *Mylossoma* species are disc-shaped elegant Characins with an outspreading ventral profile. They make decorative specimens in large aquaria, although unfortunately they graze on water plants, so this must be borne in mind when setting up the aquarium. Their requirements in the aquarium are similar to those of the G *Metynnis*. Sex differences and breeding unknown.

– *M. duriventre* (CUVIER, 1818); Hard-bellied Characin. From the southern Amazon Basin, Paraguay, Paraña, La Plata. Length up to 23 cm. *M. argenteum* AHL, 1929, and *M. aureum* AGASSIZ, 1829, are both alternative names no longer in use. The flanks are silvery with an occasional golden gleam. A right-angled black blotch on the operculum is immediately striking. The D1, D2 and C are a weak reddish colour, the base of the An being a lovely red, with a strong golden iridescence, and a black outside edge. The Ps and Vs are colourless. Juveniles have dark, wedge-shaped transverse bands, ending in a point in the middle of the body. They also have a black blotch, edged in a paler colour, in the middle of the body (fig. 796).

Fig. 796 *Mylossoma duriventre*

Myoxocephalus, G, name not in current usage, *see Cottus*

Myriophyllum LINNÉ, 1753; Water Milfoil. G of the Haloragaceae* comprising 30 species. Distributed throughout the world. Perennials with elongated stem axes and irregular branching. Leaf arrangement usually whorled, but sometimes alternate. Aquatic leaves pinnate, aerial leaves pinnate or undivided. Flowers hermaphroditic or unisexual, usually monoecious, but also dioecious, small and inconspicuous. Perianth can have a varied shape. Calyx 4-part, sometimes almost absent. Corolla is hermaphroditic and male flowers well developed and 4-part; in female flowers often absent. Stamens 8 or 4. Carpels 4, fused, superior. Fruit a schizo-carp. The various species of the G are aquatic plants, their inflorescences rising up above the surface. Some species can also form land-dwelling plants. Their finely pinnate leaves and varied colours (pale green to dark green, olive-green to reddish) make these plants very attractive, many of which are suitable for the aquarium. Depending on their origin, *Myriophyllum* species have specific temperature requirements, but are otherwise relatively undemanding. Some are sensitive to algal growths and humus being churned up. The species that are common in the aquarium are not always identified for sure. Examples are:

– *M. brasiliense* CAMBESSEDES; Brazilian Milfoil. From S America and also introduced into southern N America and SE Asia. Stem axes pale green, leaves pale green, often with reddish tips. Suitable for cold and heated aquaria.

– *M. heterophyllum* MICHAUX. N America. Leaf whorl 4 or 5-part. Stem axes and leaves dark green. A cold water plant.

– *M. hippuroides* NUTTALL. N America. Stem axes red. Leaves olive-green to reddish. Suitable for heated aquaria (fig. 797).

– *M. spicatum* LINNÉ; Spiked Milfoil. From Europe, N Asia, and some isolated occurrences in SE Asia and Africa. Leaf whorl usually 4-part. Stem axes and leaves dark green. A cold water plant.

– *M. ussuriense* (REGEL) MAXIMOWITZ; Japanese Milfoil. From SE Asia. Stem axes and leaves dark green, leaves very stiff. Suitable for heated aquaria.

– *M. verticillatum* LINNÉ; Whorled Milfoil. From the northern hemisphere. Leaf whorl 5 or 6-part. Stem axes and leaves dark green. A cold water plant.

Fig. 797 *Myriophyllum hippuroides*

Myripristis CUVIER, 1829. G of the Holocentridae, O Beryciformes*. Representatives of the G live in tropical parts of the Indo-Pacific and the Atlantic. They are predatory fish, active at dusk and during the night. During the day, they remain hidden. Similar to *Holocentrus**, except they are sociable and do not have a spine on the lower edge of the gill-covering. 20–30 cm long. The majority are coloured red. Open-water spawners (at least as observed in the aquarium). Survives well in larger aquaria.

– *M. murdjan* (FORSKÅL, 1775); Big-eye Squirrelfish, Blotcheye, Crimson Squirrelfish, Red Soldierfish, Red-striped Soldierfish. From tropical parts of the Indo-Pacific, and the Red Sea. Length up to 30 cm. A shining red colour, and large eyes with a black band (fig. 798).

Fig. 798 *Myripristis murdjan*

Mysidacea; Opossum Shrimps. O of the Crustacea*, Sub-Cl Malacostraca, which are similar to the Shrimps (Natantia*) but which do not belong to the Decapoda*. Most of them live in the sea, often in large numbers, but some also live in brackish and fresh water. In contrast to the Shrimps, they swim with the aid of extremities located on the head and breast region (pleopoda), whilst their body stands up vertically in the water. Length 5–25 mm. Opossum Shrimps are filter-feeders mainly *(Neomysis)*, although some live a predatory existence in part *(Praunus)*. The eggs develop in a brood sac on the underside of the females. Reproduction can also be observed in the aquarium. Opossum Shrimps play an important role as food animals; they are a particularly ideal food for Sea-horses and Pipefishes. Among the Ga represented along European coasts are *Mysis* LATREILLE, 1802, *Neomysis* CZERNYAVSKY, 1882, and *Praunus* LEACH, 1830.

Mysis, G, *see* Mysidacea

Mystus GRONOVIUS, 1854. G of the Bagridae*. Asiatic Catfishes. Length up to 20 cm. *Mystus* is a very active swimmer that likes warmth (24–28 °C). It is active during the day. In the aquarium it will need enough room for swimming and a soft substrate; also provide clumps of plants and pieces of root. Its body is naked and elongated in a torpedo shape, only slightly flattened laterally. The dorsal profile is more strongly arched than the ventral profile. The rounded, short, colourless D1 is inserted at the highest point of the front third of the body. D2 very strong, C deeply forked. An moderately long, Vs broad. There are 4 pairs of fairly long barbels, the longest pair being on the upper jaw, one pair on the nose and 2 pairs on the lower jaw. They are held out from the body, pointing forwards. Eyes relatively large. Colouring usually variable, with longitudinal band markings. Omnivores that prefer worm food most of all. ARNOLD-AHL reports that *M. tengara* spawns between plants in the humus; the fish conduct an elegant, lively form of love-play, circling each other and uttering 'chirping' noises at the same time. The eggs are fairly large, yellowish-white in colour. However, they have not yet been reared successfully. The report on spawning technique did not say whether they were observed in the open or in an aquarium. *Macrones* is another name for this G but it is no longer used.

– *M. tengara* (HAMILTON-BUCHANAN, 1822). From N India and Burma. Length up to 18 cm. Flanks a delicate green or a strong golden-yellow colour with 4–5 curving, broad longitudinal bands, in a dark cobalt-blue colour. These bands get narrower towards the back of the fish. Dorsal surface a delicate pale brown, ventral surface shining white. There is a large, round, black-blue blotch, edged in gold-yellow, on the shoulder and at the insertion of the C. The top part of the barbels are white and the bottom part blackish. Fins colourless and transparent-looking or a delicate shade of blue.

– *M. vittatus* (BLOCH, 1794). From India, Burma, Thailand. Length up to 21 cm. Flanks very variable in colour, a delicate silver-grey or a shining gold colour. Depending on origin, several broad, pale-blue or pale-violet (sometimes even dark brown or deep black) longitudinal bands contrast against the colours of the flanks. Towards the caudal peduncle, these bands taper away. The barbels are white with dark undersides. The fins are colourless and transparent, often with dark tips. A dark shoulder blotch lies above the strong Ps. Unfortunately, this species has only been introduced on rare occasions so far (fig. 799).

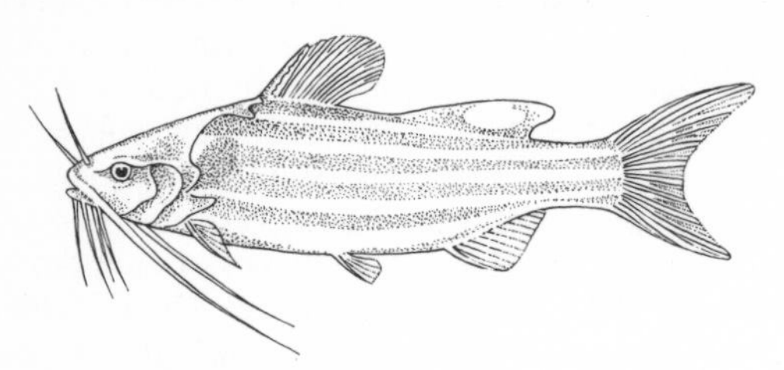

Fig. 799 *Mystus vittatus*

Mytilus LINNAEUS, 1758; Common Mussels. A widely distributed G of marine molluscs (Bivalvia*, O Anisomyaria) with a steeply triangular black shell. They sit in vast numbers below the tidal zone on rocks and on breakwater posts, attached there by their byssus threads. Length up to 10 cm. Mussels are bred in large quantities to be eaten (particularly in France and the Netherlands). For marine aquarists, they are especially important as food animals – they are best frozen in large quantities for this purpose. Smaller specimens will survive well in aquaria. They are often brown-striped, and climb about by means of their mobile foot. *M. edulis* LINNAEUS, 1758 (fig. 800) lives along northern European coasts; a more hardy species is *M. galloprovincialis* LAMARCK which is found in the Mediterranean and in the Black Sea.

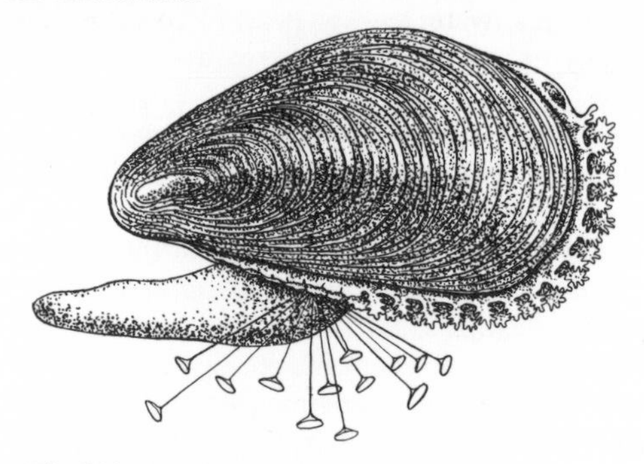

Fig. 800 *Mytilus edulis*

Myxine glutinosa *see* Cyclostomata
Myxiniformes, O, *see* Cyclostomata
Myxobolus pfeifferi *see* Sporozoa
Myxopterygium *see* Chondrichthyes
Myxosoma cerebralis *see* Sporozoa

Naias or Naiad Plants, G, *see Najas*

Najadaceae; Pondweeds. F of the Liliatae*. The only G is *Najas**.

Najas LINNÉ. The only G of the Najadaceae*. 35 species, that are found almost everywhere on Earth. Small perennials with elongated, often very brittle stem axes that are much branched. There are some annual plants as well. Leaf arrangement opposite. Leaves narrowly linear. Flowers unisexual and usually monoecious, small and inconspicuous. Perianth simple or usually absent. Male flower: 1 stamen. Female flower: 1 carpel, superior. Nut-fruit. The species of the G are plants that grow in fresh and brackish water. Some would probably make ideal aquarium plants, but only a few have been tried out so far. They are undemanding and they can either be planted in the aquarium or allowed to float around freely. They form thick skeins of vegetation very quickly and are excellent for tanks holding Livebearing Tooth-carps. The various species are difficult to determine. The following species is often encountered in the aquarium.

– *N. indica* (WILLDENOW) CHAMISSO. From tropical SE Asia (fig. 801).

Najas Plant Family *see* Najadaceae

Fig. 801 *Najas indica*

Naked Catfishes, F, *see* Mochokidae

'Naked Seeds' (Gymnospermae, Sub-D) *see* Spermatophyta

Name-giving *see* Botanical Nomenclature, Zoological Nomenclature

Nandidae; Nandids or Leaf-fishes. F of the Perciformes*, Sub-O Percoidei*, comprising only a few species. The Nandidae are certainly the most primitive F among the freshwater groups of the Percoidei. This view is apparent from their regionally limited distribution in S America, W Africa and SE Asia, which is the result of the land-bridge that used to connect these continents. The Nandids are relatively small fish, robust, with a leaf-shaped body, rarely elongated. The head is large and has a deeply slit mouth which can usually be protruded like a tube. 6 gill arches on each side. The D is large, the spiny section being long, and the spines strong. The soft-rayed part is short, like a pennant and usually colourless. C rounded. An usually shorter than the D, but in principle similarly shaped. Strong comb-edged scales cover the body, and the LL is incomplete or absent. Colouring usually murky; many of the species imitate leaves and float around in the water like them too. The more predatory forms devour fish that can measure $^3/_4$ of their own size. Not important in fish economy. For details regarding reproduction and aquarium experiences turn to the Ga *Afronandus, Badis, Monocirrhus, Nandus, Polycentropsis, Polycentrus.*

Nandids or Leaf-fishes, F, *see* Nandidae

Nandus Cuvier and Valenciennes, 1831. G of the Nandidae* comprising only a few species. Native to S and SE Asia, including the Malaysian Archipelago. They live mainly in fresh water, but also in brackish water. Medium-sized fish of predatory appearance, body fairly elongated, eyes and mouth very large. The D is inserted along a greater stretch of the body than the An. The An, as in the related G *Pristolepis**, has only 3 hard rays. The soft-rayed parts of the D and An, as well as the C, are either not included in the body colouring or only insignificantly so. The gill-covering has a flat spine. Colouring dark, usually with dark patterns on a brown or greenish background. *Nandus* species are highly predatory and can only be kept in species-aquaria. If keeping *Nandus nandus,* fill the tank with slightly brackish water (about 0.2–0.5% sea salt content). Depending on their natural habitat conditions, specimens will require large, well-planted tanks with some hiding-places and a large quantity of live food, especially food fish. Aquatic insect larvae, amphipods, shrimps and other food of similar proportions will also suit. Water temperature about 24°C. No information available regarding any successful breeding.

– *N. nandus* (Hamilton-Buchana, 1822); Nandus. From the Near East and Further India. Length up to 20 cm. 7–9 soft rays in the An. Colouring marbly because of the presence of large blotches, but overall a murky green or brown.

– *N. nebulosus* (Gray, 1930). From Further India and the Malaysian Archipelago. Length up to 12 cm. Brownish with dark, irregular transverse bands. A distinct, dark band runs from the mouth, arches through the eye and continues to the insertion of the dorsal fin. 5–6 soft rays in the An (fig. 802).

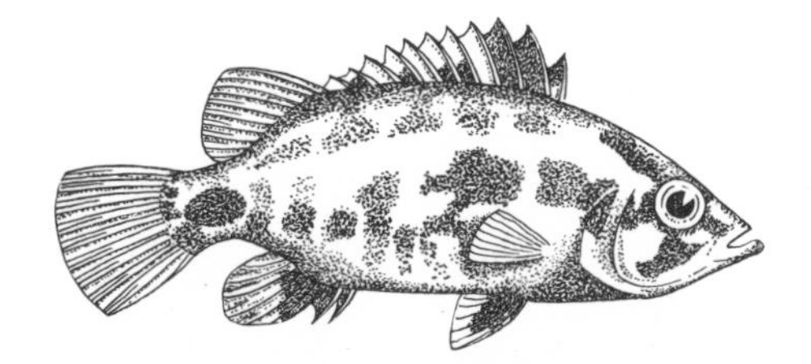

Fig. 802 *Nandus nebulosus*

Nandus *(Nandus nandus) see Nandus*

Nannacara Regan, 1905. G of the Cichlidae* with 2 species native to tropical S America. Fairly small, relatively elongate fish, moderately flattened laterally. D inserted along a long stretch of the body, the soft-rayed parts of this fin and of the An being elongated into points (at least in the males). C roundish, elongated like a dagger. The males are considerably larger than the females. The females often have lattice-work markings made up of longitudinal and transverse stripes. These are peaceful Cichlids that do not bother other fish or plants to any great extent which makes them very suitable for community tanks. Make sure the tank contains some hiding-places though, such as holes in the rocks, flowerpots cut in two, pieces of roots etc. Such places are also needed for spawning, since the spawn is laid hidden in the substrate. Care and breeding is the same as for the related G *Apistogramma**. Like them too, the *Nannacara* species form a similar social structure.

– *N. anomala* Regan, 1905; Golden-eyed Dwarf Cichlid. From Guyana. Length up to 8 cm. Colouring very delightful. Upper surface olive-brown, flanks a metallic green or bronze, and every scale bears a triangular dark fleck. The cheeks and operculum are dotted with an iridescent green and the iris is reddish. D dark, its edge being a shining red and bluish. Females over-all more clay-yellow, but when excited covered in intensely dark longitudinal and transverse bands (fig. 803).

Fig. 803 *Nannacara anomala*

– *N. taenia* Regan, 1912; Lattice Dwarf Cichlid. From the Amazon Basin. Length up to 5 cm. Colouring very variable – a brownish ground colour with a broad, deep black longitudinal band running from the eye to the root of the tail fin. Female more plainly coloured.

Nannaethiops GÜNTHER, 1871. G of the Citharinidae*. Widespread in the whole of equatorial Africa. Length up to 6.5 cm. Only one species. It is a small Characin from the bottom zone. It makes jerky swimming actions, and is reminiscent of the species of the G *Nannostomus**. Body shape elongated and only moderately flattened laterally. Can be distinguished from the closely related G *Neolebias** because of the covering of scales on the base of the C. The shape and position of the fins is also reminiscent of members of the G *Nannostomus*; adipose fin present and LL complete. The gill membranes are not joined with the isthmus. Feed with small live food as well as artificial food. Has been bred successfully on several occasions. The fish usually spawn in the morning sunshine indiscriminately, between finely pinnate plants. They will eat their own spawn. The fry hatch out in about 24 hours at a temperature of 25 °C, and from the fifth day they will accept fine live micro-organisms as food. There should be no problems with rearing the fry. This is a very productive species.

– *N. unitaeniatus* GÜNTHER, 1871; African One-striped Characin, African Tetra. Dorsal surface brownish-olive, ventral surface and throat silvery-white. A brown-black or blue-black band extends from the mouth, through the eye and body to the root of the tail, where it ends in a broader, black patch. The band is bounded above by a gold or coppery coloured band. The fins are a pale yellowish colour or greeny-white. During the breeding season, the front half of the D and the upper tail-fin lobe in the male are a shining blood-red colour (fig. 804).

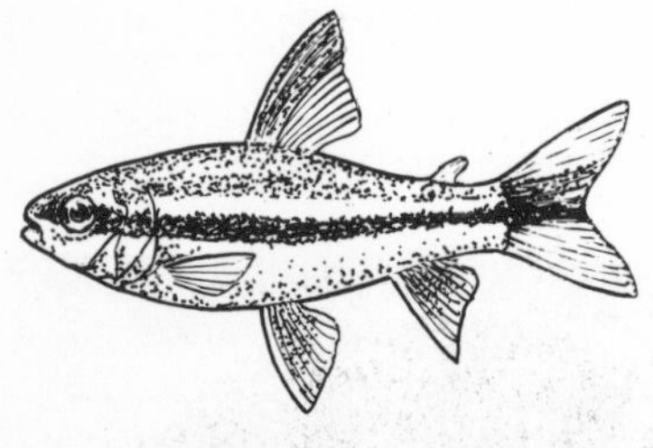

Fig. 804 *Nannaethiops unitaeniatus*

Nannaethiops tritaeniatus, name not in current usage, *see Neolebias*

Nannobrycon, G, name not in current usage, *see Nannostomus*

Nannochromis PELLEGRIN, 1904. G of the Cichlidae* comprising several species all confined to central Africa (Congo Basin). Small, elongate Cichlids with a characteristic LL. The anterior section of the LL runs in an arch shape near the dorsal ridge to the soft-rayed part of the D, and the posterior section is confined to the caudal peduncle region. The D has 17–18 hard rays and 8 soft ones, whereas the An has 3 hard rays and 6–7 soft ones. The naso-forehead profile is much more strongly arched than the chin, similar to that found in many *Pelvicachromis** species. It is similar to this G in other respects too.

In line with their tropical origin, *Nannochromis* species should be kept at a temperature of 24–27 °C. The water should be soft and occasionally filtered through peat, ie so that it contains humic acids* and tannic substances*. *Nannochromis* does not attack plants, and they particularly like murky places to hide, such as root formations or upturned flowerpots with an entrance at the side. They also like to spawn in such pots, and they are typical of those fish that hide their spawn in this way (*see* Cichlidae). The female practises brood care. The eggs take about 3 days to mature, and after about a week the fry are swimming freely.

– *N. dimidiatus* (PELLEGRIN, 1900); Dimidiatus. From the Congo. Length up to 7 cm. Body a gleaming pink-violet, head and dorsal surface particularly strongly coloured.

– *N. nudiceps* (BOULENGER, 1899); Congo Dwarf Cichlid. From the Congo. Length up to 7 cm. Ochre colours. Flanks a gleaming blue, ventral surface a shining emerald green.

Nannostominae, Sub-F, *see* Characoidei

Nannostomus GÜNTHER, 1872; Pencilfishes. G of the Lebiasinidae*, Sub-F Nannostominae. From the Amazon Basin and Guyana lands. Up to 6.5 cm long. Small, peaceful shoaling fish from the middle and top water levels. According to FRANKE the sexes of all the species can be told apart by the shape of the An – in the male it is rounded at the front like a lobe, in the female it is straight-edged and pointed at the front. The elongated body is only slightly flattened laterally. The head is short and the mouth terminal. The D is short, inserted around the middle of the body, and the adipose fin has very much regressed. In the species *N. beckfordi, N. bifasciatus, N. espei, N. marginatus,* the adipose fin is missing altogether, in *N. trifasciatus, N. unifasciatus* and *N. harrisoni* it is usually still present, and in *N. eques* some members of one brood will have an adipose fin, others will not. The C is fairly deeply indented, the lobes being round. The An and Vs are small. The Ps, like the upper C lobe, is constantly being beaten in a wave-like motion. Pencilfishes prefer soft, lightly acidic and clear water, at a temperature of 23–25 °C. The lighting in the tank should be subdued and it should be well planted. Because the mouth is small, Pencilfishes can only devour *Cyclops*, small *Daphnia*, chopped up *Tubifex* and Enchytraeids. *N. eques* and *N. unifasciatus* adopt a slanting position in the water, head upwards. A similar position is also suggested by *N. espei.* Formerly, the species *N. eques, N. harrisoni* and *N. unifasciatus* were attributed to the G *Poecilobrycon* EIGENMANN, 1909. The only feature that can still be said to differentiate the 2 Ga is a bony canal which leads to the second infraorbital arch in *Poecilobrycon*, and which is absent in the G *Nannostomus*. In his revision of the Nannostominae (1966), WEITZMANN considers it of little value to maintain the G *Poecilobrycon* unless other generic characteristics can be established (eg through comparable studies of behaviour). In 1972, FRANKE was able to study and describe in a comprehensive ethological report the courtship, aggression and mating behaviour of all 9 species, and he was not able to conclude from them any new generic characteristics. He therefore suggests that the G *Poecilobrycon* be demoted to a Sub-G, and at the same time included in the G *Nannobrycon* HOEDEMAN, 1950. As a result, all 9 species are incorporated in the G *Nannostomus*. Most of the species are not difficult to breed. The male drives at the female, usually involving a dorsal or ventral courtship display (*see* Courtship Swimming), and often interrupted by a horizontal or vertical fighting display (*see* Symbolic Fighting Behaviour). In the species *N. beckfordi aripirangensis, N. digrammus, N. marginatus*, and *N. trifasciatus*, the eggs are usually laid among finely pinnate plants, or more rarely beneath leaves; in the other species they are more commonly laid beneath leaves. For each spawning act, each of the species mentioned produces 1–2 eggs, *N. eques* 4–5. FRANKE maintains on the other hand that *N. espei*, with a smaller number of matings, lasting 60 seconds and more each, lays 10 or more eggs in several instalments. The spawn is caught by the male in his arched An, then it is stuck onto the leaf or wrapped in the plants. All the species have a marked tendency to eat their own spawn, and *N. unifasciatus* has not been able to be reared in the aquarium so far. The juveniles have characteristic colouring and are not difficult to rear with live food suitable for them (*see* Fry, Feeding of). None of the species is especially productive, but will often spawn in rhythmical cycles for a few days one after the other. FRANKE was able to show that the young fish of the horizontal swimming species, *N. harrisoni*, which until then had been reared only rarely, also swim in a slanting position during the first few weeks of life, just like young *N. eques*. So there exists a transition stage between those species that swim horizontally and those that swim in a slanting position. These are the species that were originally included in the G *Poecilobrycon* and are now included in the G *Annobrycon*. Most of the *Nannostomus* species are among the most popular standard aquarium fish. Nearly all of them have species-specific night colouring*, usually in the form of transverse bands. FRANKE reports that these colours are the same as the inferiority colours.

– *N. beckfordi aripirangensis* (MEINKEN, 1931); Brown or Red Pencilfish. From the Island of Aripiranga in the lower Amazon Basin. Length to 6.5 cm. In the male, the flanks and the base of the Dl, C, An and Vs are blood-red when excited. The Vs also have shining pale blue tips. A broad, blue-black longitudinal band runs through the body, bounded above by a golden-yellow band. The females are more plainly coloured, only the base of the fins being reddish. They are pretty, hardy species and are therefore very popular. *N. beckfordi anomalus* (STEINDACHNER, 1876), a not so prettily coloured species, has not been imported since 1945 (fig. 805).

Fig. 805 *Nannostomus beckfordi*

– *N. bifasciatus* HOEDEMAN, 1954; Two-banded Pencilfish. From the Guyana lands. Length to 6.5 cm. 2 black

longitudinal bands run across the silver-white flanks, the lower one of which is by far the stronger, reaching from the tip of the snout, through the eyes and body to the C insertion. The top, narrower band, made up of fine dots, extends from the upper edge of the eyes to the upper edge of the caudal peduncle. It divides the darker toned ventral area from the silver-white of the body. The base of the C and An is a weak brick-red. The other fins are colourless, apart from the tips of the Vs in the male. The females can be recognised by their sturdier ventral profile and the lack of pale-blue tips to the Vs. A new import of this attractive species would be worthwhile (fig. 806).

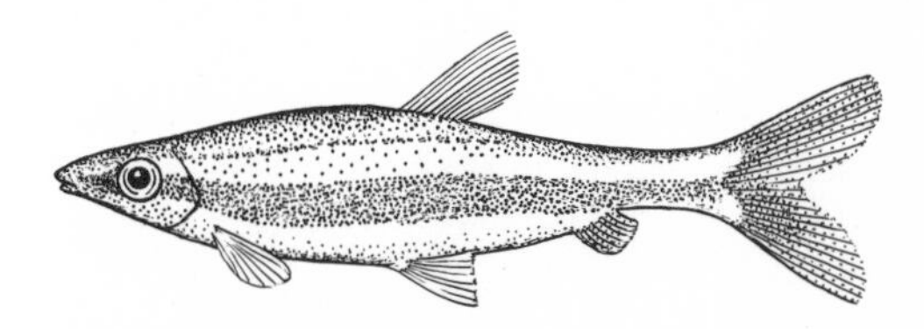

Fig. 806 *Nannostomus bifasciatus*

– *N. digrammus* (FOWLER, 1913); Two-striped Pencilfish. From the Amazon Basin, the Rio Madeira, the Rio Branco, the Rio Negro and the Rio Rupununi. Length up to 3 cm. Dorsal surface grey-olive. A narrow, black-brown longitudinal band runs along the length of the body (in the lower third of it), starting from the tip of the snout and ending at the C insertion, where it continues as a red-brown wedge shape on the middle rays of the C. Above it there is a broad, gold-yellow or green-gold band that also begins at the tip of the snout. The ventral surface is silvery. The middle of the D is a faint reddish colour, with a dark edge in the male. Adipose fin dark grey, the other fins colourless. The males are easy to recognise by their lobe-like An, although it is usually held in the collapsed position. This is a delightful small species, but so far it has only been a rare import, as part of a make-weight catch. It spawns between finely leaved plants, according to FRANKE.

– *N. (Poecilobrycon) eques* (STEINDACHNER, 1876); Tube-mouthed Pencilfish. From the central Amazon Basin. Length to 5 cm. Dorsal surface pale brown or silvery-grey with 5 dark rows of dots. Beneath this lie a beige-coloured and a broad black-brown longitudinal band. The latter extends from the tip of the snout to the lower lobe of the C, the upper lobe of which is colourless. The An is red and black, its outside edge (like the tips of the Vs in the males) being a lovely blue-white. The D is colourless and a small adipose fin may be present or absent altogether. This species swims at an angle, its head pointing upwards, forming an angle of 45° between the body and the water surface. This is a very popular, peaceful, surface fish that must only be kept in the same tank as small peaceful fish (fig. 807).

– *N. espei* (MEINKEN, 1956); Barred Pencilfish. From the Rio Mazaruni and the Guyana lands. Length to 3.5 cm. Dorsal surface a delicate olive, ventral surface silver. A golden-coloured longitudinal band runs along

Fig. 807 *Nannostomus eques*

the flanks in the middle of the body. There is the suggestion of a dark longitudinal band beneath it, running from the tip of the snout to the insertion of the C, but it is covered over by 5 very striking, deep black transverse bands that are short and slanting. All the fins are colourless. The slimmer males are easy to tell apart from the ripe females by their rounded An, which is typical of the G. For spawning behaviour see generic description. This is a pretty species, but unfortunately one that is very sensitive. It has only recently been imported, but as more and more are bred they are bound to become more popular, on account of their striking, untypical markings (fig. 808).

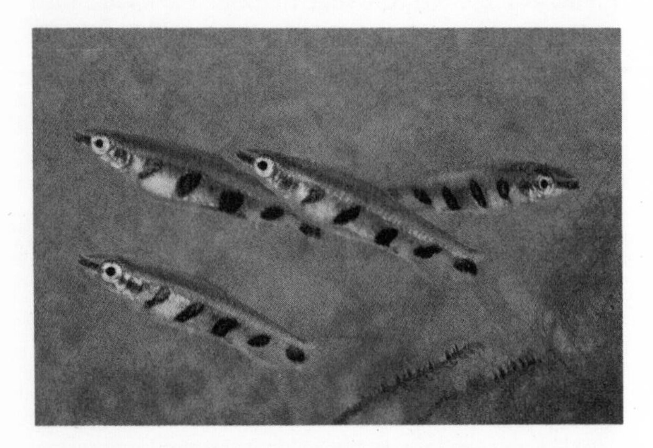

Fig. 808 *Nannostomus espei*

– *N. (Poecilobrycon) harrisoni* EIGENMANN, 1909; Harrision's Pencilfish. From the Guyana lands, especially Guyana itself. Length to 6 cm. Body very elongate, head and mouth extending into a point. Swims horizontally as it gets older, although as a juvenile is swims at an angle, just like the young of *N. (P.) eques* (this according to a study carried out by FRANKE). Older juveniles of this species were mistakenly described as *Archicheir minutus* by EIGENMANN, 1910. It represents a transitional phase between the slanting swimming position of *N. (P.) eques* and *N. (P.) unifasciatus* and the horizontal swimming position of the other *Nannostomus* species. Dorsal surface yellow or chocolate-brown. A straw-yellow longitudinal band extends from the tip of the snout to the upper lobe of the C, bounded below by a black longitudinal band that ends at the middle rays of the C. Ventral surface silver-white. D colourless, adipose fin very small. The end of the black longitudinal band on the C is red in the females, which is edged in blood-red in the males. The An and Vs in the females are reddish, blood-red in the males. The tips of the Vs are pale blue in the males. The An in the males is rounded as is typical of the G. A very attractive species, but not very widespread (fig. 809).

Fig. 809 *Nannostomus harrisoni*

– *N. marginatus* EIGENMANN, 1909; Dwarf Pencilfish. From the Guyana lands. Length to 3 cm. The smallest species, although it is an extraordinarily colourful one. It swims horizontally. The ventral surface is silver-white, yellowish in excited males. 3 black longitudinal bands run along the flanks, the upper and middle one being divided by a gold-yellow or green-golden band. In the middle of the body, on the middle black longitudinal band there is a blood-red streak. The front edge of the D is black with a bright red blotch. The male anal fin is rounded (typical of the G), lemon-yellow in colour with a blood-red blotch and a black edge. In the females, there is no yellow middle area to the fin. The Vs also have a blood-red blotch. Eggs lying in the oviduct of ripe females shimmer in cross-light, producing a golden-yellow colour near the edge of the ventral area that shines through the ventral wall. If well fed, spawn will be laid every 4–5 days. However, they are not very productive, 30 fry per brood being a good result. The eggs swell up considerably in the acidic water and are readily eaten by the parents, so use a thick layer of Java Moss! The juveniles are covered in dark-brown bands and are not difficult to rear. This is certainly one of the most attractive and popular Characins in the aquarium, where they will survive for a long time. They are seen to best effect if kept as a small shoal (fig. 810).

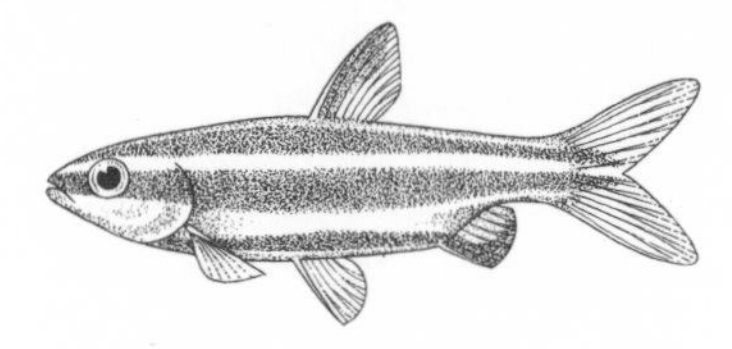

Fig. 810 *Nannostomus marginatus*

– *N. trifasciatus* STEINDACHNER, 1876; Three-banded Pencilfish. From Guyana, the central Amazon Basin and the Rio Negro. Length to 6 cm. Dorsal surface olive, flanks and ventral surface silver-white. A broad, black longitudinal band extends from the tip of the snout through the eyes to the beginning of the C; a second, narrower band extends from the upper eye edge to the C insertion, and a third begins on the ventral surface in front of the Vs and continues to the insertion of the An. Adipose fin very small or absent. There is a blood-red fleck on each of the D, An, and Vs at the base, whilst there are two on the C. The first rays of the An and the tips of the Vs in the males have shining pale blue tones. The An in the male is rounded which is typical of the G. There is a gold-coloured row of scales between the first and second longitudinal bands. A blood-red blotch appears on the shoulder. This species has a very characteristic night colouring – instead of 3 longitudinal bands, there appear 2 broad, black transverse bands. According to FRANKE there is a smaller form of this species with a particularly broad middle longitudinal band and very bright red fin markings. This is possibly a local race, or it might be a sub-species. These delightfully beautiful Characins have, unfortunately, only been bred on rare occasions and in not very large numbers, either. FRANKE reports that they spawn in a finely leaved substrate, such as Java Moss etc. In appearance, *N. trifasciatus* is a bigger version of *N. marginatus*. *N. trilineatus* LADIGES, 1948, *Poecilobrycon erythrurus* EIGENMANN, 1910 and AHL, 1933, *Poecilobrycon vittatus* are all alternative names for this species, but they are no longer in use (fig. 811).

Fig. 811 *Nannostomus trifasciatus*

– *N. (P.) unifasciatus ocellatus* (EIGENMANN, 1909); One-lined Pencilfish. From central and lower parts of the Amazon Basin and tributaries, the Guyana lands. Length to 7 cm. This species swims only at a slight angle. Dorsal surface a delicate translucent beigey colour, ventral surface silver-white. A broad, black longitudinal band extends from the tip of the snout through the eye and body, across the caudal peduncle to the lower caudal lobe, which it covers completely by broadening out. From time to time it is accompanied above by a narrow, golden streak. The upper lobe of the C is glassily transparent. There is a black eye-spot in the lower part of the upper lobe, outlined in white on the top side and in red on the bottom side. There is also a white and red zone in the middle of the dark lower lobe. The remaining fins are colourless; the front edge of the An is pastel-white in both sexes (and rounded in the male [generic-typical]). The Vs have white tips in the male. Occasionally, the One-striped Pencilfish (*N. (P.) unifasciatus unifasciatus* STEINDACHNER, 1876) is also imported. It has the same colouring and band markings but lacks the eye-spot. Instead, the lower, dark, caudal

lobe has 2 white transverse bands. Unfortunately, neither of the two sub-species has been bred successfully so far. FRANKE reports that the males of this species, just like *N. beckfordi aripirangensis*, occasionally adopt symbolic vertical fighting behaviour. This species is highly valued by aquarists, but unfortunately it is rarely imported (fig. 812).

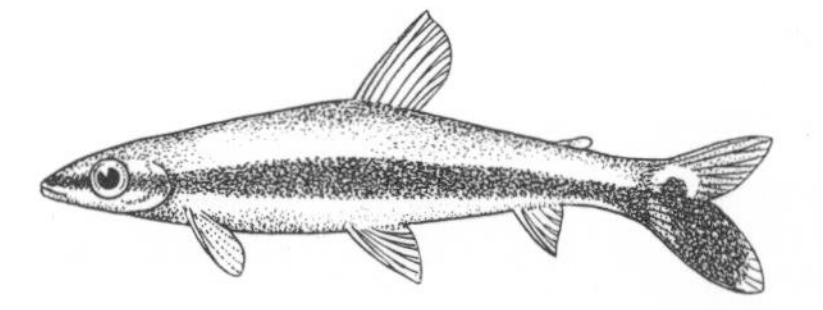

Fig. 812 *Nannostomus unifasciatus ocellatus*

Nardoa GRAY, 1840. G of tropical Starfishes (Asteroidea*). They feed on small animals and particles. The arms of these Starfishes are slim and round and their surface is knotty. *N. variolata* is often imported. It comes from tropical parts of the Indo-Pacific. In the aquarium it survives well, feeding on pieces of Bivalve meat, fish food tablets and other things. Diameter 10–15 cm, colour brown or red, often with a dark net-pattern on the upper surface.

Naso LACÉPÈDE, 1802; Unicornfishes (fig. 813). G of the Acanthuridae, Sub-O Acanthuroidei*. Found in tropical parts of the Indo-Pacific, including the Red Sea. These are sociable fish, wandering along the coast, feeding on plant growth mainly. Length 30–80 cm. Unicornfishes are tall and flattened laterally, with immobile spines on the root of the tail (but not like stiletto blades as in *Acanthurus** or *Zebrasoma**).

Fig. 813 Representative of the G *Naso*

Many Unicornfishes have a horn or some kind of hump on the forehead as they get older. Colouring usually unremarkable, but rapid changes of colour are possible. Courting specimens within a shoal will suddenly adopt shining colours that disappear just as quickly (*N. tapeiposoma*). Very hardy in large aquaria. Requires a great deal of supplementary plant food.

– *N. lituratus* BLOCH and SCHNEIDER, 1801; Japanese Tang, Smooth-head Unicornfish, Striped-faced Unicornfish. From tropical parts of the Indo-Pacific, the Red Sea. Length up to 50 cm. No process on the forehead. Blue-green in colour, D and An yellow.

Nassa LAMARCK, 1799. G of marine Prosobranchs (Gastropoda*, Prosobranchia*). *Nassa* eats carcasses. Most of its life is spent digging, usually on soft substrata. The shell is 10–25 mm long and is a pointed egg-shape. There is a large siphon which is allowed to extend above the bottom. *Nassa* can detect food even from a considerable distance. As lively snails, *Nassa* species are well suited to aquarium life, and are particularly useful for clearing up any remains. An example is *N. reticulata* LINNAEUS, 1758 (fig. 814) which occurs from Norway to the Black Sea, sometimes in vast numbers. Its dark-brown shell is sculpted in net-like markings. This species can turn 'somersaults' to evade Starfishes.

Fig. 814 *Nassa reticulata*

Nastic Movements. The movement of parts of sessile plants that shows no relation to the source of stimulation. It includes the opening and shutting of flowers in respect of the brightness, eg in the G *Nymphaea**. In the sleep movements of certain aquatic plants, eg the folding up of the youngest leaves of *Hygrophila* or *Limnophila* species, an inner rhythm may form that admits of no relationship to the external stimulus source (eg a 12-hour rhythm). *See also* Taxis and Tropism.

Natantia; Shrimps. Sub-O of the Ten-legged Crustaceans (Decapoda*) with several Fs and about 2,000 species. Most are marine, but some live in brackish and fresh water. In contrast to the Sub-O Reptantia*, a shrimp's body is compressed laterally, it has feet for swimming on the rear section of the body, and its walking legs are slim. The first and second antennae are also large. The calcareous covering is usually delicate – many species are transparent like glass, the warm water forms often being fantastically coloured. Many Natantia are economically important *(Crangon*)*, and many are ideal for the aquarium. They will usually accept any kind of animal food and they are also useful in the aquarium for eating up remains. Shrimps can be kept together with most types of animals, but larger fish are dangerous for them. *Hymenocera* picta* will eat Starfishes, however. Some tropical shrimps are cleaners *(Hippolysmata*, Stenopus*)*. Tropical *Periclimenes* species live in Sea-anemones (Actiniaria*). Apart from the Ga mentioned above, species of the Gs *Hippolyte*, *Lysmata**, *Macrobrachium* and *Palaemon** are also found in the aquarium.

Natterer's Piranha *(Serrasalmus nattereri) see Serrasalmus*

Natural Selection *see* Selection

Naucoris FABRICIUS, 1775; Water Bugs (fig. 815). G of the Bugs (Heteroptera*) comprising only a few species. One species is common and widespread in Europe *(N. cimicoides)*. Water Bugs are similar in shape to the Water Beetles, and among the aquatic bugs they are certainly the most skilful swimmers. The middle and rear pairs of legs are used in locomotion; the front legs are adapted into a kind of jack-knife (as in *Ranatra**) for catching hold of prey (eg fish broods and other water

Fig. 815 *Naucoris* sp.

insects). Being pricked with the short, strong, sucking tube is painful, even for humans. *Naucoris* gets trapped in nets sometimes when fishing for food animals in ponds. It makes an interesting specimen for the pond-aquarium. In a room aquarium, *Naucoris* will attack fish and kill them.

Naucrates ductor *see* Carangidae

Nauplius (fig. 816). A typical larval phase found in many lower crustaceans (Crustacea*), characterised by an unpaired eye and 3 pairs of extremities. Nauplii are ideal as a food for rearing fish broods (compare Feeding* and Fry, Feeding of*).

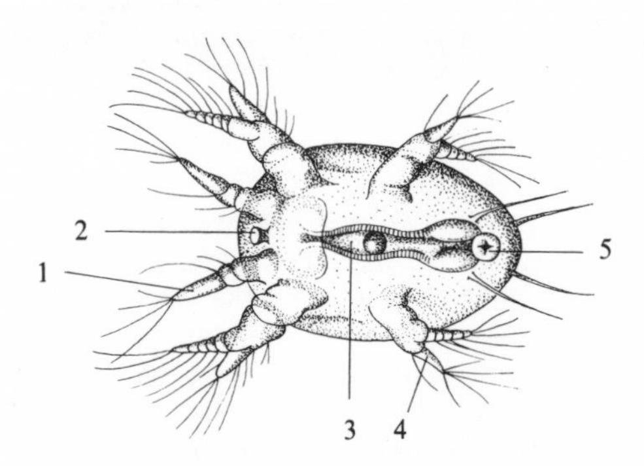

Fig. 816 Nauplius. 1 Rudimentary 1st antenna 2 Eye 3 Rudimentary intestine (with oil droplets) 4 Rudimentary mouth parts 5 Anus

Nautilus, G, *see* Cephalopoda

Nearctic Realm. Zoogeographical region, comprising the whole of N America* and reaching south to California–N Mexico–Florida. Although climatically comparable to the palaearctic* realm, the animal kingdom of the nearctic is dissimilarly richer than that of the palaearctic. This does not only apply to the number of species, but also to the higher taxonomic groups. This abundance can be explained by the N–S orientation of the mountain ranges in the nearctic realm, which afforded the animals in the last ice ages an unimpeded migration route south. The river basin of the Mississippi was particularly important in this respect for the migration of aquatic animals. Fish Fs confined to the nearctic include the Amiidae (Amiiformes*), the Lepisosteidae (Lepisosteiformes*), the Percopsidae (Percopsiformes*) and the Centrarchidae*; the last named have also a certain significance in aquarium-keeping. Because of the similarities in relation to the flora* and fauna* in the nearctic and palaearctic relams, the two are grouped together in the holarctic realm*.

Necton; Nekton. A characteristic community of animal organisms of the open-water zone (Pelagial*); in contrast to the plankton*, necton can actively move along, independently of the direction of current. Examples of organisms that belong to the necton are the fish of the open waters that have no relationship at all to the bottom (eg open-water spawners such as Herrings and Muraenas). Whales are an example from the mammals.

Nemachilus, G, *see Noemacheilus*

Nemathelminthes; Aschelminthes. This Ph contains a very large number of species, and includes harmless, free-living, external and internal parasites, as well as dangerous ones. They populate all the major habitats (sea, fresh water, land), sometimes in large numbers. In aquarium-keeping, they are important not only as fish parasites, but also as a highly valued fish food in some cases (especially members of the Cl Rotatoria*).

Nematobrycon EIGENMANN, 1911. G of the Characidae*, Sub-F Tetragonopterinae. From Colombia. Length to 7 cm. Small, very colourful Characins that love moving about. In the aquarium they will need sufficient free swimming areas, a good amount of vegetation and a dark substrate. Subdued lighting, provided by the presence of floating plants, will ensure *Nematobrycon* is content, and this will mean that it will be seen in its most beautiful colours. The body is relatively short and tall. The D has a small base and is elongated. No adipose fin. The An is very long, elongated into a point at the front. The C is triple-pointed (3-lobed), and the LL is incomplete. Not entirely easy to breed and not very productive. FRANKE and FRANKE, 1964, report that the males drive very hard and that the eggs are laid between Java Moss or *Myriophyllum*. To begin with, the fry remain hidden and they should be fed with fine micro-organisms (Fry, Feeding of*). Breeding pairs that have been tried out should not be separated.

– *N. palmeri* EIGENMANN, 1911; Emperor Tetra. From the Rio San Juan and tributaries in Colombia. Length to 5.5 cm. Dorsal area olive-brown with a broad, lovely grass-green or blue-green longitudinal band below, extending from the operculum to the root of the tail. This is in turn bounded below by a broad, blue-black band that extends from the operculum to the tip of the middle lobe of the C. Ventral surface whitish, the iris of the eyes being a lovely blue-green iridescent colour, yellow-green at the base, and brown-red on the outside. The bottom of the An is blue-green, brown-violet in the middle, then there follows a narrow, blue-black longitudinal band and a narrow border area at the free edge (which is a pale yellow-green). Vs yellow-green, Ps lemon-yellow. The females are smaller, and less intensely coloured over-all; the middle caudal lobe is suggested only (fig. 817).

Fig. 817 *Nematobrycon palmeri*

Nematocentris PETERS, 1886. G of the Melanotaeniidae, Sub-O Atherinoidei*. From Australia. These fish inhabit the middle water levels of rivers and lakes, although many also penetrate into brackish water. They are small and live in large shoals. Body elongate, sometimes quite tall, flattened laterally. The mouth is small and superior. No LL. The D is divided, D2 and An are inserted along a good stretch of the body. Caudal peduncle shorter than the An. The D1 is inserted either just in front of the Vs, or, more usually, behind them. This is a very peaceful and undemanding genus, the species of which should be kept in a shoal in large tanks with plenty of room for swimming. The water should not be too soft, 12° hardness and above. A regular part-exchange with fresh water and the addition of sea salt will ensure the fish remain content. Temperature 22–25 °C. They are omnivores, and will accept live and dried food. Fairly easy to breed in medium-hard, neutral fresh water. The male has a pointed D, the female a stumpy one. Specimens will often spawn over several weeks (Breeding Season*), laying the eggs between fine-leaved plants. The eggs are quite large and they attach themselves to the plants by means of short threads. The fry hatch out after 8–10 days, at a temperature of 23–25 °C. After 1–2 days they can swim freely, and to start with they should be fed on micro-organisms.

– *N. fluviatilis* (CASTELNAU, 1878); Rainbowfish. E Australia, south to Sydney. Length to 10 cm. Dorsal surface yellow or grey-yellow. Flanks green or blue-green (iridescent), root of the tail green. Scales dark at the front, with a red border at the back. There are several dark longitudinal bands with iridescent zones in between. The gill-covering has a red blotch on it with a golden-green and white edge. The D and An are yellow-grey with red streak markings and a black zone at the edge. Vs yellow (fig. 818).

Melanotaenia nigrans (RICHARDSON, 1843) is not a synonym for this species, as was wrongly assumed before.

– *N. maccullochi* OGILBY, 1015; Dwarf Rainbowfish. N Australia. Length to 7 cm. Body silver-grey with a bluish shimmer and 7 red or brown longitudinal stripes. Operculum turquoise with a red blotch, edged in golden-

Fig. 818 *Nematocentris fluviatilis*

green. During the breeding season the body takes on a reddish shimmer all over. The D and An are red, green at the base, yellow at the edges. C brown-red.

Nematocera; Midges. A Sub-O of the Two-winged or True Flies (Diptera*) containing a great many species and forms. They are universally known on account of the Mosquitos of the Ga *Aedes**, *Culex**, *Anopheles**. Other aquatic Fs and Ga include the Non-biting Midges (Chironomidae*), the Black Flies *(Simulium*)*, the Blepharoceridae*, the Crane-flies or Daddy-long-legs (Tipulidae*) and *Corethra**. Many of the Fs and Ga are very much linked to certain biotopes; as a result, they are often referred to when describing the typical characteristics of stretches of water. The larval and pupal stages of the midges are extremely important in the diet of many fish, and they are also used as fish food in the aquarium. The larger forms are very suitable for observing in pond aquaria – interesting studies can be made of their anatomy and biology.

Nematocysts *see* Stinging Capsules

Nematoda; nematodes, threadworms. Cl of the Nemathelminthes*, estimated number of species between 20,000 and 100,000. Nematodes are found in every kind of life zone. They are either land, freshwater or marine creatures that live an independent existence, or they live as parasites in plants and animals. Nematodes are unsegmented, unpigmented, and unlike many other worm groups they have a mouth and an anus, and they are heteroecious. In length they range from fractions of a mm to almost 9 m, depending on species and life-style. Included among them are well-known parasites of the human intestine such as maw-worms. Some members of the Nematoda are also used as food animals for aquarium fish, especially *Anquilla* and *Panagrillus silusiae**. Compare Feeding*.

Nematode Diseases. Parasitical nematodes* are extraordinarily common among wild forms of freshwater and marine fish. Fish may be intermediate hosts (Parasitism*), the larvae being carried in the fish's skin, in its muscles or in almost every internal organ. The larvae are 0.5–1 mm in diameter and are encapsulated in the fish's connective tissue. If the fish is the definitive or final host, the adult parasites are almost always found in the intestinal lumen. Most nematodes that cause disease in fish reproduce by means of one or several intermediary hosts, so they cannot reproduce in the aquarium. A mild parasitical attack is not dangerous for fish and normally it is not even noticed; a stronger attack will lead to the slow but steady emaciation* of the fish. In aquarium-keeping, only a few nematodes are of any importance. The G *Capillaria* ZEDER, 1800, has several species and is described as a parasite of the gastro-intestinal canal; it is common in the Cichlidae* and the Siluridae*. Some species of this G do not undergo a change of host during their development cycle; others require Copepods (Copepoda*) and Gammaridae as intermediate hosts. The eggs that are released can be detected microscopically (Examination Technique*) in excreta smears. *Camallanus* RAILLIET and HENRY, 1915, attacks the Poeciliidae*, in particular, and other kinds of aquarium fish as well. It anchors itself by the head in the hind-gut, where it attracts attention through its red colouring. This parasite is livebearing – its larvae can be detected in fresh excreta smears. Nematode diseases can sometimes be cured with an antihelminth substance, but success is not guaranteed. Trichhorphone*, Thiabendazol, Concurat* or Parachlorometaxylenol are effective against intestinal parasites, but the larvae encapsulated in connective tissue cannot be effectively treated.

Nematognathi, O, name not in current usage, *see* Siluriformes

Nemertinea (Rhynchocephala); Nemertines, Proboscis Worms. This is a predominately marine phylum of worms. They are parenchymatous and colourful with a ciliated epidermis and protrusible proboscis. They come in an extreme variety of forms, Earthworm-types can often reach several metres in length, yet be as thin as a piece of string. Large species can reach lengths of 25–30 m. Nemertines are predators, and some also eat dead organisms. Sometimes nemertines of the G *Lineus* SOWERBY, 1806, will be inadvertently introduced into marine aquaria, where they will reproduce. They can grow up to 20–30 cm long. Food remains or dead organisms will often tempt them out of their hiding-places in the substrate.

Nemichthyidae, F, *see* Anguilliformes

Nemipteridae, F, *see* Lutjanidae

Nemoura PICTET, 1841. G of the Stone Flies (Plecoptera*), comprising large numbers of species. Represented in Europe by more than 40 species. They are dull-coloured insects, up to 10 mm long. The larvae are typical inhabitants of flowing waters, and are sometimes very common. They are a favourite food among larger kinds of aquarium fish.

Neoceratodus forsteri *see* Ceratodiformes

Neochanna apoda *see* Galaxioidei

Neoheterandria HENN, 1916. G of the Poeciliidae*, Sub-F Poeciliinae. From central America: Costa Rica, Panama, Colombia. Small or very small fish that live in river areas. Males 1.5–5 cm, females 2–6.5 cm. Macroscopic generic features absent. Many of the species are very difficult to tell apart from each other or from members of other Ga. Brown colour tones predominate, often with dark transverse bands or streaks in the rear part of the body. Keep at a temperature of 22–26 °C, in a partially well-planted tank. Feed with live food mainly, of appropriate size. This G is not very important in aquarium-keeping. The smallest species, and one of the smallest vertebrates in the world is:

– *N. elegans* HENN, 1916; Teddy. From Colombia. Male about 1.5 cm, female up to 2 cm. This species is similar to *Heterandria formosa*. Features: 6–9 dark transverse bands in the rear part of the body, the second one appearing like an oval blotch. The D and An have blackish bands. The gonopodium can be up to $^1/_3$ of the length of the body.

Neolebias STEINDACHNER, 1894. G of the Citharinidae*. Small Characins from central Africa. Length to 4 cm. They are peaceful fish that live near the bottom. Only suitable for small biotope aquaria, since they will withdraw into the vegetation in a community tank, out of fear, and lose their colouring. Preferred temperature around 20–22 °C and a sandy substrate. The body is elongate, the dorsal and ventral areas being almost equally arched. Unlike the G *Nannaethiops**, the base of the C is not covered with scales and the gill membranes are not joined to the isthmus. The D, An and Vs are relatively short, the C moderately forked, and there is no adipose fin. Breeding should be attempted in small, all-glass tanks containing plenty of vegetation. About 300 eggs will be laid and they will hatch out after 24–30 hours. The fry are very small and sensitive. Only a small number will survive, even when they are fed on the smallest of rotifers. Unfortunately, *Neolebias* species are only kept rarely in the aquarium; *N. trilineatus* appears to have disappeared from the aquarium altogether.

– *N. (Micraethiops) ansorgei* BOULENGER, 1912; Ansorge's Characin. From central Africa. Length to 3.5 cm. Body colouring of sexually excited males: dorsal surface brownish-olive, flanks covered by a broad, longitudinal band, grass-green or mossy-green in colour. The ventral area is yellowish to brassy-yellow. Throat, Ps, Vs and An a deep blood-red, the latter also narrowly outlined in black. There is a narrow, black longitudinal band on the caudal peduncle. The D and C are reddish. The females have a stronger ventral area. Just before mating begins, the spawn is visible in the body cavity when light strikes it from the side. A female's normal colouring includes a dark longitudinal band. STALLKNECHT reports that the spawning act is not typical of the Characins; the male performs a clearly defined fluttering dance*, interrupted by ramming*, which lures the females to the spawning ground. During mating, the female stands a little behind the male, thus leaving her mouth about level with the sexual opening of the male. The male now beats his anal fin beneath the female. The two bodies entwine one another in a loose S-shape, and the eggs and sperm are released. The male then beats his tail fin to bring the spawn in among some plants. There is no turning movement of the body during this process, which is characteristic of many types of small Characins. This is an extremely prettily coloured Characin. *N. landgrafi* AHL, 1928, is another name for the species that is no longer in use fig. 819).

– *N. (Rhabdaethiops) trilineatus* BOULENGER, 1899; African Three-banded Characin. From the upper Congo. Length to 4 cm. Flanks a pale beige colour, dorsal surface brownish, ventral surface silver-white, often with reddish tones. 3 deep black longitudinal bands run across the flanks; the middle one is the broadest and ends in a black blotch, edged in gold, on the caudal peduncle. Golden-coloured zones lie in between the bands. The D is reddish-yellow, the C and An reddish to reddish-brown. The females are somewhat larger and more rounded in the ventral region. During the mating period, the males have a blood-red longitudinal band and similarly coloured fins. In aquarium-keeping, this species was also known by the name *Nannaethiops tritaeniatus* BOULENGER, 1913, but it is no longer in use. Care and breeding as for *N. ansorgei* (fig. 819).

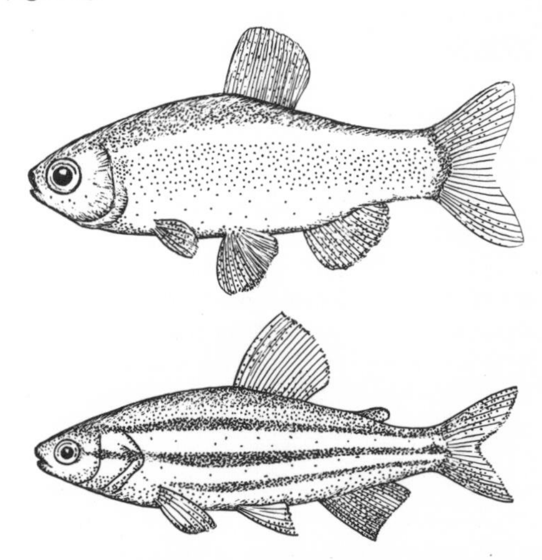

Fig. 819 *Neolebias ansorgei* and *Neolebias trilineatus*

Neomysis, G, *see* Mysidacea

Neon Disease *see Pleistophora*

Neon Goby *(Gobiosoma oceanops) see Gobiosoma*

Neon Reef Perch *(Abudefduf oxyodon) see Abudefduf*

Neon Tetra *(Paracheirodon innesi) see Paracheirodon*

Neopetrolisthes oshimae *see* Porcellanidae

Neopterygii, Sub-Cl, name not in current usage, *see* Osteichthyes

Neotropic Realm. Zoogeographical region*, including the whole of S and Central America, bordering onto the nearctic realm* in the north. Important areas from the aquaristic point of view include the giant tropical rain forests near the equator, and the grassland areas that border on to these forests, to the north and south. These grasslands are the home of the biologically very interesting season fish from the F Cyprinodontidae*. The most important river basins in the neotropic realm are, from north to south, the Orinoco, the Amazon* and the Rio de la Plata* (River Plate). All of them are very important as the homes of a great many well-known aquarium fish. (The Fs and Ga which now follow are described under the relevant entry.) In this connection, we ought to mention here such neotropic Ga as *Hemigrammus, Hyphessobrycon* and *Megalamphodus* from the Characins (Characidae), *Corydoras* from the Mailed Catfishes (Callichthyidae), and *Aequidens, Cichlasoma, Pterophyllum* and *Symphysodon* from the South American Cichlids (Cichlidae). Because the neotropic realm is largely isolated from the other zoogeographical regions, it has a large number of endemic fish families, such as the Characin families Gasteropelecidae, Characidiidae, Lebiasinidae, Hemiodidae and Anostomidae; also the Electric Eels (Electrophoridae, *see* Gymnotoidei), the Knifefish families Gymnotidae (*see* Gymnotoidei), Rhamphichthyidae (*see* Gymnotoidei), and Apteronotidae (*see* Gymnotoidei), the Catfish families Doradidae, Pimelodidae, Aspredinidae and Callichthyidae, plus many others. It is also worthwhile mentioning that the otherwise widely distributed F of the Carps (Cyprinidae) does not have a single representative in the neotropic realm. Over-all, the neotropic realm bears a closer resemblance in terms of fish fauna to the Ethiopian realm* than it does to the nearctic realm on its northern border. This is clearly demonstrated by the distribution of the Characin-type species (Characoidea) and the Cichlids (Cichlidae).

Nepa LINNAEUS, 1758; Scorpion Bugs (fig. 820). G of the Bugs (Heteroptera*) belonging to the F Nepidae. There are about 150 species, of which only 2 – the Water Scorpion *(N. cinerea)* and *Ranatra** – are found in central Europe. *Nepa* is flat, about 3 cm long (including

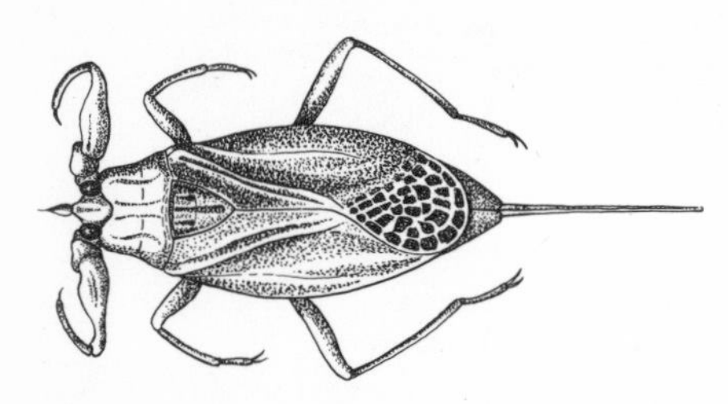

Fig. 820 *Nepa rubra*

the breathing tube at the rear end of the body), and lives near the bank in still waters. It is usually found between plants or in the muddy substrate. The front legs are used only to catch food and are shaped like a jack-knife. The prey is gripped and stabbed with the small proboscis to paralyse it, then its innards are sucked out. A prick from the proboscis can even be painful for a man. *Nepa* will often find its way into a net when fishing in ponds, although it is often overlooked on account of its unremarkable colouring and slow movements. A fascinating glimpse of the Scorpion Bug's life-style is possible if specimens are kept in a suitably arrranged (shallow!) pond-aquarium along with other aquatic animals.

Nephrops LEACH, 1830. G of the Decapoda*, Sub-O Reptantia. It includes the Norway Lobster *(N. norvegicus)*, LINNAEUS, 1758) which is found from Iceland to the Mediterranean on soft substrates. In the north it lives at a depth of 10 m and less, in the south at the depth of 50 m and less. They grow to over 20 cm long and have slim, angular pincers. Their upperside is a brick-red colour, and the underside pinky-white. They also have dark eyes that are strikingly large. The Norway Lobster is a lovely aquarium animal that survives well; unfortunately, it is hardly ever imported.

Neptune Grass *(Posidonia oceanica) see Posidonia*

Nereis LINNAEUS, 1758. G of the Bristle Worms (Polychaeta*). *Nereis* is widely distributed on sea bottoms near the coast, but is also found in brackish water, and some in the moist strip along beaches. They eat small animals, plants and detritus. They measure 10–20 cm long and are oval in cross-section. They are evenly segmented worms that burrow out channels for themselves with their extendable proboscides; the channels are then lined with mucus. *N. diversicolor* O. F. MÜLLER, is common along European coasts. It is an omnivore. In the initial part of its U-shaped tube, it spins a net; then, with the movements, it sweeps water through the net, food particles become trapped, which the worm then eats (net and all). *Nereis* worms are often introduced into the marine aquarium – they are very useful for getting rid of any remains, and they are a favourite food of fish and crustaceans. *N. fucata* SAVIGNY lives in the shell of Hermit Crabs (*Eupagurus**).

Nerocila LEACH, 1818; Fish Lice. G of parasitical Lice (Isopoda*, F Aegidae). *Nerocila* is particularly prevalent along southern European coasts and is prone to attacking members of the G *Crenilabrus**. The parasites can grow up to 25 mm long, and they are oval in shape. They attach themselves firmly to the front dorsal section of the fish in a way that is hard to miss. They suck the tissue fluid up through the wound they make, just like other Aegidae (eg *Anilocra*, fig. 821). Scissors have to be used to cut them off. Tearing them from the fish's skin will damage the fish.

Fig. 821 Fish Louse *Anilocra* on a Three-fin Blenny *(Tripterygion)*

Nerophis RAFINESQUE, 1810; Worm Pipefishes. G of the Syngnathidae, Sub-O Syngnathoidei*. Comprises several species from the algal and seaweed zones of the Atlantic and neighbouring seas. The P and C have regressed. It is a needle-shaped fish, cylindrical, and is prone to wrapping its tail round plants. The females are larger and more colourful than the males, and they take on the active role during courtship. The eggs are stuck onto the belly of the males in a double row, and the fry hatch out after about 3 weeks. They are easy to keep in calm aquaria and should be fed on small crustaceans. Particularly rewarding is:

– *N. ophidion* (LINNAEUS, 1758). Found in the Atlantic, along the European coast from the Baltic to the Black Sea. Female 25 cm, male 15–18 cm. Greenish to brownish in colour. Ripe females have shining blue stripes on the front part of the body.

Nervous System. Highly specialised bands of cells in multi-celled animals for regulating and co-ordinating the body's functions; the controlling mechanism is bioelectrical/biochemical. In its simplest form in invertebrate animals, the nervous system consists of a diffuse network of nerve cells. But, even here, there are usually central interneural complexes (ganglia) and paths of conduction (nerve cords). Accordingly, in the vertebrates the brain and the spinal cord are called the *central nervous system*, and the nerves that depart from there, plus the sense organs and other regions of the body where nerves enter, are collectively known as the *peripheral nervous system*.

In vertebrates, the central nervous system has a fluid-filled hollow cavity running through it, the cerebral ventricles and the central canal of the spinal cord. The *brain* is divided into 5 parts, to which 10–12 cerebral nerves belong. Many regions of the brain have only altered slightly in the course of phylogenesis, whilst in others they have undergone a significant development. This is true of the fore-brain (prosencephalon) which is fairly small in fish and represents a centre for the sense of smell. In mammals it is a voluminous cerebrum with many functions, superior indeed to the brain as a whole. The inter-brain (diencephalon) which contains the pituitary body (hypophysis) is involved (among other things) in the hormone regulation of the body. At the foot of this part of the brain, the two optic nerves enter. The mid-brain (mesencephalon) is the most important optic centre in lower vertebrate animals. Another part of the brain, the cerebellum, controls and co-ordinates the body's movements. In species that live on the substrate, this part of the brain is usually smaller than in species that move around freely in space. In the after-brain (myelencephalon) there are important regulating centres, eg the respiratory centre. The *spinal cord* connected to the brain is a reflex centre, particularly involved in muscular movements; its function is influenced by the brain (fig. 822).

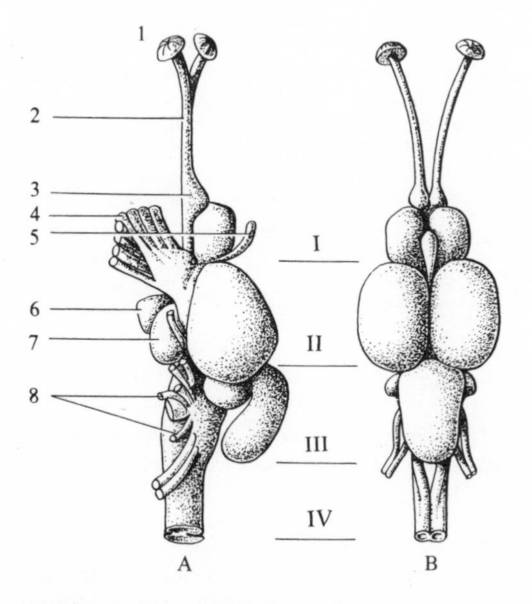

Fig. 822 Brain of *Salmo trutta* (Trout). A In side-view, B From above. I Fore-brain II Mid-brain III Cerebellum IV After-brain. 1 Olfactory nodule 2 Olfactory nerves 3 Olfactory lobe 4 Optic nerves 5 Pineal gland (epiphysis) 6 Hypophysis 7 Inter-brain 8 Cranial nerves

The structural elements of the nervous system are the nerve cells with branching processes like a tree (dendrites), the contact points (synapses) of which receive microelectrical impulses from sense cells or other nerve cells. These impulses are added together and then transmitted via a (usually) long process (axon) to other nerve cells or to an effector organ in the body (muscle, gland etc). The specific circuit of nerve cells with very varied biochemical and structural specialities is the reason why the nervous system is able to perform such a large number of very different functions, from muscular movement to memory.

Nets are essential for catching fish and other aquarium animals. For use in the aquarium, they should not be too big so that they can be handled easily even in a tank that has been fully set up. Nets with cornered frames enable the aquarist to reach into the corners of the tank, whereas round-framed nets are less useful. The net frame should be made of plastic or inert metal wire, whereas the netting itself should be made of a strong, unrippable perlon yarn. The size of the frame, the width of the meshing and the depth of the net should suit whatever purpose it is being used for (eg the size of the fish). Very fine-meshed, thick nets stand up very well to the rigours of the water; wide-meshed nets make it easy for the animals to become entangled (caught in the mesh with their spines, fin rays or other outgrowths) and this may damage them. A deep net is difficult to hold open in an installed tank, but fish will be able to swim out of a shallow net all too easily. Very sensitive and delicate animals are better fished out with a dip-tube*. Fish diseases* are most commonly transmitted by nets, so if a net is being used in several different tanks make sure it is disinfected (Disinfection*) before use.

Neuston. A living community of micro-organisms (Bacteriophyta*, Fungi*, Algae*) on the water surface. They in turn are eaten by many small crustaceans (Crustacea*), snails (Gastropoda*) and tadpoles (Amphibia*). A visible layer of neuston which hangs together is called scum and it consists essentially of fungi and bacteria; it is found mainly in water that has a very high nutrient content and is putrefying. Scum is undesirable in the aquarium, since it impedes the gaseous exchange between the water and the atmosphere and also hinders fish from taking food at the water surface. The situation can be remedied by having a group of snails in the aquarium, creating more water movement, changing the water, and lifting off the scum with old newspaper. In an extreme case, the tank will have to be set up afresh and the source of the problem removed (eg perhaps overfeeding!).

Neutral Exchange. The exchange of hardness-forming salts (calcium and magnesium salts) for salts that are less damaging to certain fish and plants that prefer soft water* (eg sodium chloride [cooking salt]). Such an exchange is brought about by certain artificial resins (Ion Exchangers*). Many aquarists do not bother with neutral exchange, but replacement is possible if the original water has a hardness degree of no more than about 8–10°dH (Water Hardness*). In this case, the calcium and magnesium salts are largely removed as desired, while the sodium chloride content thus formed takes effect without disturbance. When preparing hard water*, salt should be removed (Desalination*) in any case.

New Zealand Mudfish *see* Galaxioidei

Nicaragua Shark *(Carcharhinus nicaraguensis) see* Carcharhinidae

Nicaraguan Gambusia *(Gambusia nicaraguensis) see Gambusia*

Nichol's Limia *(Poecilia nicholsi) see Poecilia*

Nicotine Poisoning *see* Poisoning

Niger. The largest river in W Africa, more than 4,100 km long. It has its source in Guinea. From there it flows to the Sahara (with only a slight incline), fans out into a flood area, then winds down south-east to enter the Gulf of Guinea in a large delta. The Niger is linked to the Lake Chad area via its most important tributary, the Benue. The Lake Chad fauna has close relationships with the Nile*, whilst there are considerable differences in this respect with the Congo*. Aquarium fishes of the Niger river system are, eg, *Distichodus**, *Phago** and many *Barbus** species, the Catfishes *Auchenoglanis occidentalis, Eutropius niloticus* as well as *Synodontis** species, and, among the Cichlids, various *Pelmatochromis** and *Tilapia** species.

Night Tetra *(Prochilodus theraponura) see Prochilodus*

Nile. The longest river in the world, the Nile rises in SE Africa, flows through Lake Victoria and Lake Albert, forms a huge marsh and flood area immediately following on from this, then unites with the Blue Nile near Khartoum to form the White Nile. For the last 2,700 km there is no significant inflow of water, then the Nile finally enters the east Mediterranean* via a large delta. The river's total length is 6,671 km. Although the Nile and the Congo* come quite close in E Africa, each of the rivers has retained a characteristic fauna. On the other hand, the Lake Chad area of the Niger* has closer faunal relationships with the Nile. Aquarium fish from the Nile include *Marcusenius** *isidori* and *Mormyrus** *kannume*, and among the Catfishes, some *Synodontis** and *Clarias** species. The countries through which the Nile flows are provided with a great many edible fish.

Nile Mouthbrooder *(Tilapia nilotica) see Tilapia*

Nile Perch *(Lates niloticus) see* Centropomidae

Nile Snakehead *(Polypterus bichir) see Polypterus*

Nine-spined Stickleback *(Pungitius pungitius) see Pungitius*

Nitella, G, *see* Charophyta

Nitrate Content. Nitrates are the salts of nitric acid and the end-product from the oxidation of nitrogen compounds. In the aquarium they are produced mainly through the breakdown of animal protein and ammonium compounds (urine, excreta, foodstuff remains). Whereas most freshwater organisms are very insensitive towards nitrates, in the marine aquarium (Sea Water*) great care must be taken. Precautionary measures against too high a build-up of nitrates include feeding sparingly and only having a small animal population in the tank – again, both these measures are especially important for marine tanks.

A reduction of the nitrate content is effected most cheaply by a routine part change of water. In a freshwater tank, if the aquarist wishes to undertake something more sophisticated, he can learn to comprehend the rising nitrate content to a certain extent by the increase in conductivity, and only when it reaches a certain level must he change the water or prepare it anew by using ion exchangers*. In a marine tank, a partial change of water is still the surest method, assuming that the presence of a lot of algae does not keep the nitrate content low in any case. Foam filters* can also be fixed up in a marine tank.

Nitrite Content. Nitrites are the salts of nitrous acids. They occur in the aquarium either through the partial oxidation of ammonium compounds (Ammonia*) or through the reduction of nitrates. Unlike the compounds just mentioned (with the exception of ammonia), nitrites are poisonous even in very weak concentrations, which is why they are dreaded by the aquarist. The danger of nitrite formation occurs especially with too low a redox potential*, ie in tanks where too much oxygen is used up as a result of overpopulation and decaying processes. The simplest way to prevent nitrite build-up is to feed sparingly, make sure there are not too many animals in the one tank and regularly carry out a partial water change or periodically prepare the water again with ion exchangers* (*see* Water Preparation).

Nitrofurane. A derivative of furane. It comes in a crystallised yellow prismatic form and is insoluble in water. It inhibits the growth of bacteria and fungi and is used as an effective ingredient in medicaments. Recently, preparations based on nitrofurane have become available for treating bacterial infections* in fish.

Nocturnal Colouration (fig. 823). Term used in ethology. After taking up their sleeping position, many species of fish adopt a special night colouring, which is often very close to the inferiority* colouring. In many

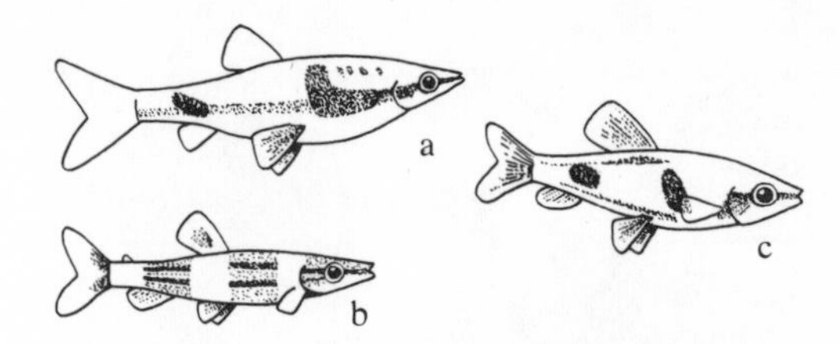

Fig. 823 Nocturnal colouration a) *Nannostomus bifasciatus* b) *N. trifasciatus* c) *N. beckfordi aripirangensis*

*Nannostomus** species, for example, the longitudinal bands that are visible by day disappear and give way to transverse bands or fleck markings. In the well-known Neon-Tetra the lovely green-blue/red colouring of the daytime gives way to inconspicuous fleshy tones. It is only when a light source has been switched on for about 20–30 mins that the full intensity of the daytime colouring is achieved once again. The colour change is brought about in this case by a displacement of the pigment granules in the chromatophores into the neck of the cells (daytime colouring) and by their return to the bottom of the cell (night-time colouring).

Nodes *see* Stem Axis

Noemacheilus VAN HASSELT, 1823. G of the Cobitidae*. Consists of numerous species and sub-species distributed in Europe, and S, E and central Asia. They are small to medium-sized fish that live in the bottom zones of clear waters (which might at times be fast-flowing). The body is elongate, often slender, and cylindrical in cross-section. The caudal peduncle is slightly flattened laterally. The skin is naked or covered with tiny scales. LL complete or incomplete. The head has no spines beneath the eyes. 3 pairs of barbels are inserted in the upper jaw. The D is inserted at the level of the Vs or just in front of them, the C is straight, rounded or inward-curving. The various members of this G are usually hardy fish that should be kept in tanks that are not too small and which contain a sandy or gravelly substrate. The water should be clear and medium-hard. Good ventilation and the regular addition of fresh water is beneficial. Temperature 20–24 °C, or 18–20 °C for temperate zone species. Do not stand the tank in too bright a spot and make sure it has enough hiding-places, since *Noemacheilus* species are often crepuscular. Many species will only show their natural behaviour if they are kept as small groups; others are typical territorial fish. As a rule, *Noemacheilus* is a hardy fish that will eat all kinds of live and dried food, but especially food animals that can be taken up from the bottom. Some species require supplementary plant food. Only isolated details are available about reproductive behaviour.

– *N. barbatulus* (LINNAEUS, 1758); Stone Loach, Groundling. From Europe, and Central to E Asia. There are also many sub-species that live in fast-flowing streams and in the bank zone of clear lakes. Length to 15 cm. Colouring very variable. Dorsal surface greenish, flanks pale yellow-grey with irregular, black-brown to greenish, fleck Fins fins yellowish-grey with dark spotting. This species can sometimes require a lot of oxygen. Temperature not more than 20°C. If allowed a cool overwintering, it may breed in the aquarium. The male practises brood care.

– *N. botia* (HAMILTON-BUCHHNAN, 1822); Mottled Loach. NW India to Assam. Length to 12 cm. Varying colouring, body pale yellowish-white, grey-greenish or grey-brown with narrow, irregular transverse bands or flecks. Additional plant food required. Temperature 20–24°C.

– *N. fasciatus* VAN HASSELT, 1823; Barred Loach. From Sumatra, Java, Kalimantan (Borneo). Length to 9 cm. C inward-curving. Dorsal surface olive-green or brown, flanks ochre-yellow or clay-yellow with about 15–20 brown-red transverse bands. Fins colourless or a delicate yellow with rows of dark spots. Temperature 20–24 °C.

– *N. kuiperi* DE BEAUFORT, 1939. From the island of Billiton. Length to 7 cm. C inward-curving. Flanks ochre yellow, ventral surface pale. There are 12–14 brown transverse bands running across the flanks over the back; they are distributed unevenly and some are forked. Fins colourless, except for a black blotch, red at the front, on the first fin ray of the D. Temperature 22–24°C. This species is active during the day too. If several specimens are kept together, they show some interesting types of behaviour.

Nomaphila BLUME, 1826. G of the Acanthaceae* comprising 8 species. (It is attributed to the G *Hygrophila** by many botanists.) From tropical Africa and Asia. Small to medium-sized perennials with elongated stem axes that grow upright. Leaf arrangement decussate. Aerial and aquatic leaves undivided. Flowers located in loose inflorescences placed at the leaf axils; they are hermaphroditic, monosymmetrical. Double perianth. Calyx 5-part, fused up to the middle. Corolla 2-lipped, the lower one being arched upwards and sealing off the corolla tube. Stamens 4, carpels 2, fused, superior. Capsule-fruit. The various species of the G grow in damp and wet habitats, and some can tolerate long periods of flooding. Many of them are therefore suitable for the aquarium. They like a lot of light, but otherwise are undemanding. Propagation normally by means of cuttings, but adventitious plants can also be obtained if leaves that have become separated are allowed to float on the surface. The best-known species is:

– *N. stricta* (VAHL) NEES. (*Hygrophila corymbosa*, name not in current usage.) From SE Asia. The large pale green aquatic leaves are clearly distinguished from the grey-green aerial leaves (figs. 824 and 825).

Fig. 824 *Nomaphila stricta*

Nomeidae, F, *see* Stromateoidei

Nomenclature. The scientific conferring of names, administered according to rules and advice laid down internationally, which provides a uniform method of naming animals, plants and micro-organisms. People working in the fields of taxonomy* and systematics* must bear in mind the nomenclature when making known the finds of their particular studies. *See also* Botanical Nomenclature and Zoological Nomenclature.

Fig. 825 *Nomaphila stricta,* in flower

Nomeus albula *see* Stromateoidei

Non-biting Midges, F, *see* Chironomidae

Non-carbonate Hardness (of water) *see* Permanent Hardness

Non-parasitical Diseases *see* Fish Diseases

Norm of Reaction *see* Taxis Components

Norman, John Roxbrough (1898–1944). English ichthyologist. By 1921 he was working at the British Museum in London, where he was a pupil and colleague of C. T. REGAN. Together with G. S. MYERS, he continued work on GÜNTHER's *Catalogue of Fishes.* From 1930–39 Norman put together the ichthyological finds of the Discovery II expedition conducted by the British, Australian and New Zealand Antarctic Expedition, as well as the finds of the MURRAY expedition. Norman also had published *A History of Fishes* (1936) and, together with F. C. FRASER, in 1937 *Giant Fishes, Whales and Dolphins,* which reached a wide reading public.

North America. The third largest land mass in the world, measuring 23.5 million km^2. North to south it measures 8,000 km, west to east 6,500 km. A considerable amount of the landmasses of N America lie in the temperate and Arctic zone, so the fauna that requires warmth is concentrated in the south (California, Mexico, Florida). It is here also that the border of the two zoogeographical regions, the nearctic* and neotropic*, lies. Climatically very comparable with large parts of Eurasia, N America nevertheless has a far greater number of freshwater animals both in respect of the higher systematic taxa (Taxon*) as well as the number of different species. During the Ice Age in N America, the warmth-loving organisms found more favourable conditions by withdrawing to refuges* in the south. For the rest, N America, like the palaearctic realm*, has a very marked west-east contrast, the eastern regions having a greater abundance of species and forms than the west. Alongside the course of the thermal continentality, the possibilities for migrating south also play a role here.

The fish fauna of N America has close ties with that of the palaearctic. This is apparent from the occurrence of fish families in N America that are also found in the holarctic realm*, eg Salmonidae (Salmonoidei*), Umbridae (Esocoidei*), Esocidae (Esocoidei*), Percidae*, Gasterosteidae (Gasterosteoidei*) and the Sub-F Leuciscinae (Cyprinidae*). 5 endemic freshwater families are particularly typical of N America – the Amiidae (Amiiformes*), Hiodontidae (Mormyriformes*), Aphredoderidae (Percopsiformes*), Percopsidae (Percopsiformes*) and Centrarchidae*. In the south of North America the holarctic and nearctic kinds of fauna become rarer, while the neotropic fauna begins to win through. This can be seen when species of the Cyprinodontidae, Poeciliidae and Cichlidae begin to occur. (The Fs that have been mentioned here are described under the relevant entries.)

North American Cave-dwelling Fishes (Amblyopsidae, F) *see* Percopsiformes

Northern Platy *(Xiphophorus couchianus couchianus) see Xiphophorus*

Norway Lobster *(Nephrops norvegicus) see Nephrops*

Notacanthiformes; Deep-sea Spiny Eel types. O of the Osteichthyes*, Coh Teleostei. Smallish, eel-shaped fish, on average about 60 cm long. Like the Anguilliformes*, they have a very long anal fin fringe, but unlike them they have a short D that is not part of the fringe and the Vs are always present (fig. 826). In one F, the Notacanthidae, the D consists of a row of fin rays only; the skin in between is missing (fig. 827). The tail region is usually very long and pointed at the end. The chorda (Skeleton*) is retained throughout life and is surrounded by the vertebral bodies which is shaped like a half-cylinder. The swimbladder is not connected to the intestine. The head and body are covered with cycloid scales (Scales*). Some species have luminescent organs. The Notacanthiformes are found in every ocean, although they seem to avoid regions that are very definitely cold. So far they have been caught in depths ranging from 350–2,600 m.

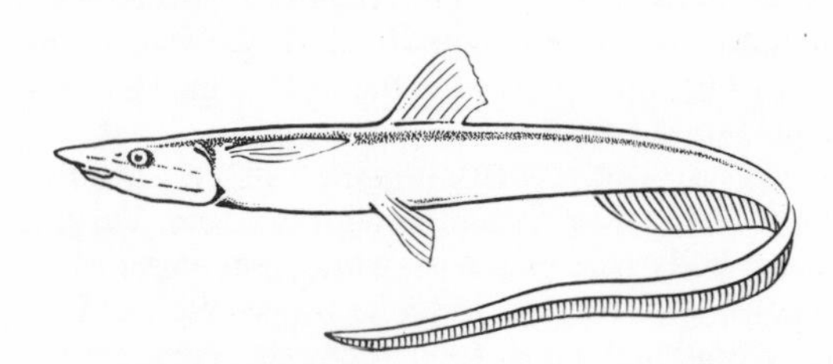

Fig. 826 *Halosaurus oveni*

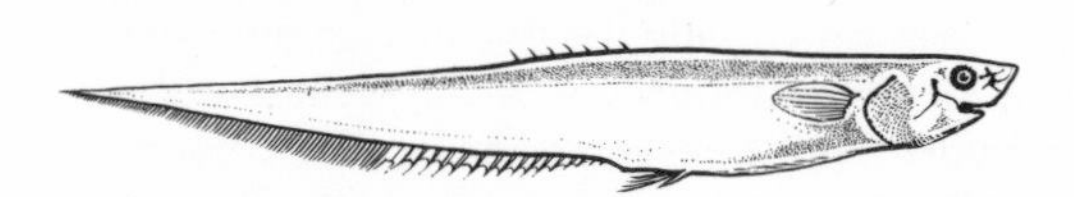

Fig. 827 *Notacanthus sexpinnis*

Nothobranchius PETERS, 1868. G of the Cyprinodontidae*. There are about 10 species and their distribution area is centred around tropical E Africa. They live in still waters of the savannas, although some species penetrate north-west to Lake Chad, and others to the islands off the East African coast (Zanzibar, the Seychelles). Their closest relatives are members of the West African G *Fundulosoma**; the scale patterning on the head and the thickening of the first anal fin rays in the females are reminiscent of American Ga *(Cynolebias*, Fundulus*). Nothobranchius* is 5–10 cm long and fairly tall with lovely colouring. Scales usually with intense red edges, thus forming a net-like pattern overall. The D and An are rounded, virtually equal in size and located opposite each other. C fan-shaped. Females are smaller and less attractively coloured. The systematic situation within this G is still very much unclear. The individual populations are inclined to develop local forms that are mainly differentiated by certain colour features. For this reason, it is best from the aquarist's point of view to assign the individual forms to a few species-groups to begin with, without committing himself to a particular species. These *Nothobranchius* forms are annual Cyprinodonts, similar ones being found in the Ga *Roloffia**, *Aphyosemion** and *Fundulosoma*, and in an extreme form among certain South American Ga *(Cynolebias, Rachovia*, Pterolebias*)*. This seasonal life-style is a form of adaptation to conditions in the stretches of water that cross the grasslands – usually, they only contain water during the rainy season and after a few weeks will dry up again for several months. For the fish, this means that they have to reach sexual maturity in a very short time and be able to spawn. The eggs are encased in a very robust envelope, thus protecting the developing embryos from any moisture loss. With each new rainy season, a new generation of fish hatches out. However, the *Nothobranchius* forms are not nearly so specifically geared to such conditions as the S American Ga already mentioned. For example, *Nothobranchius* species are also found in stretches of water that have a continuous flow, nor do the eggs have to experience a dry phase for them to hatch. This makes it easier to keep *Nothobranchius* in the aquarium, and to breed them. Small tanks are required, thickly vegetated, so that the specimens can hide from each other as necessary. Rivalry is strong among the males and weaker individuals are often bitten to death. The substrate must be soft enough to allow the fishes to 'plough in' their eggs – boiled, well rinsed peat appears to be best. Water temperature around 22–24 °C, any higher and the lifespan becomes shorter. A slight problem is that *Nothobranchius* is susceptible to fish tuberculosis and other diseases (eg *Oodinium**). So to prevent this as far as possible make sure that the water is changed regularly or prepared anew (Water Preparation*) to stop waste matter accumulating and ion relationships from becoming disturbed. The water hardness should be about 6–8° and the pH value around 6.5–7, although in nature these fish usually live in extremely soft, slightly acidic water with a pH of 6. Since *Nothobranchius* spawn can tolerate drought, but does not need to experience it, it is best kept in shallow dishes, filled with water. The embryos will hatch out after a few weeks if something is done to change the environment (perhaps adding fresh water, or blowing in air bubbles by mouth, or adding a little organic matter [sugar, dried food]). Just how far the decline in oxygen content plays a decisive role here is still uncertain (despite frequent debates on the subject). For an overall appraisal of annual species, turn to the Ga *Aphyosemion* and *Cynolebias*.

We have already mentioned that there are still problems with the systematics* of *Nothobranchius*, so here we have decided to combine the important aquarium forms into groups.

– *N. taeniopygus*-group. Includes *N. taeniopygus* HILGENDORF, 1888, *N. brieni* POLL, 1938 (fig. 828), and *N. rachovi* AHL, 1926 (fig. 829). From the area around Lake Victoria, Shaba, Mozambique. Length to 5 cm. None of these forms has a purely red C, although its base is sometimes patterned in red. The outside edge of the C is black, with a blue or orange band just before this. Flanks blue to green, with well-developed red net-markings.

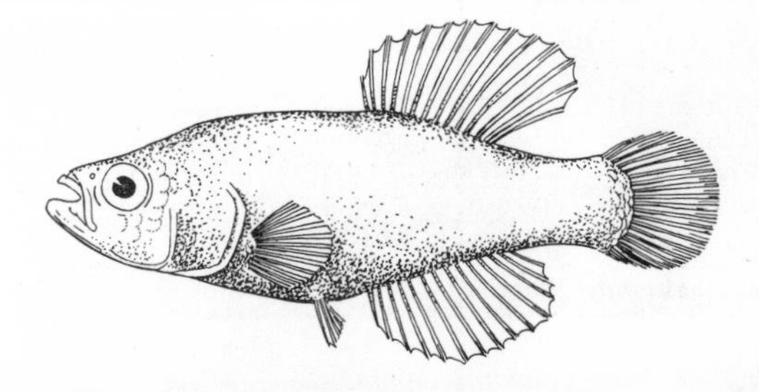

Fig. 828 *Nothobranchius brieni*

Fig. 829 *Nothobranchius rachovi*

– *N. palmquisti*-group. Includes *N. palmquisti* (LÖNNBERG, 1907) (fig. 830) and the 'false' *N. guentheri* (fig. 831). From Kenya and Tanzania. Length to 5 cm. Red net-patterns complete over the whole body. C a uniform red or red with a black edge, but no other colours.

– *N. orthonotus*-group. Includes *N. orthonotus* PETERS, 1844, *N. guentheri* (PFEFFER, 1895), *N. mayeri* AHL, 1935, *N. melanospilus* (PFEFFER, 1895) and *N. kuhntae* AHL, 1926. Has a similar distribution and is similarly coloured to the *N. palmquisti*-group, except that the *N. orthonotus*-group is more robust and up to 10 cm long. The red net-markings are usually incomplete, the C is a uniform red and usually has a broad black edge. This group is further distinguished from the *N. palmquisti*-group by anatomical features.

The females of all the groups are smaller than the males and an inconspicuous grey colour.

The following species is very much an isolated case within the G and its fin colouring is reminiscent of the South American G *Cynopoecilus**:

– *N. korthausae* MEINKEN, 1973. From the island of Mafia (Tanzania). Length to 6 cm. Steel blue with red net-markings. Caudal peduncle and unpaired fins are crossed by brown transverse bands, and the D and An have blue-white edges.

Fig. 830 *Nothobranchius palmquisti*

Fig. 831 *Nothobranchius guentheri*

Nothobranches, G, *see Nothobranchius*

Notonecta LINNAEUS, 1758; Backswimmers or Water Boatmen. G of the Bugs (Heteroptera*). There are 6 species in Europe. *Notonecta* species are among the better-known water insects because they frequently get trapped in nets when fishing for live food in ponds. They have a very unusual way of swimming which makes them particularly noticeable. Water boatmen are bubble-breathers, the air usually being taken with them under the water on their ventral sides. This makes them drift belly-side up, so that they swim on their backs. The combination of bubble air-supply and strong beating of the hind legs results in a method of swimming under water that is typical of *Notonecta* species – they shoot down through the water at an angle, then drift passively up to the surface again. Undisturbed, water boatmen like to stay directly at the surface, lying in wait for prey – especially other insects and fish broods. The prey is detected mainly through the sense of vibration, then seized with the front legs, stabbed to death and the innards sucked out. Stings can even be painful for Man. It is not generally known that adult water boatmen are good fliers; they can perform a quick take-off from the water surface or from land. They can perceive gleaming surfaces of water with their eyes and steer towards them. *Notonecta* species are ideal for observations in pond-aquaria.

Notopteridae, F, *see* Mormyriformes

Notopterus, G, *see* Mormyriformes

Notothenioidei; Antarctic Cod types. Sub-O of the Perciformes*, Sup-O Acanthopterygii. Small to medium-sized fish that live almost exclusively in the cold waters of the Antarctic. There are 6 Ga in all, with 55–60 species, divided into 4 Fs. The most interesting F is the Chaenichthyidae (Icefishes) (fig. 832); it has 18 species. This is the only fish group that does not have any red blood, ie it contains no red blood corpuscles.

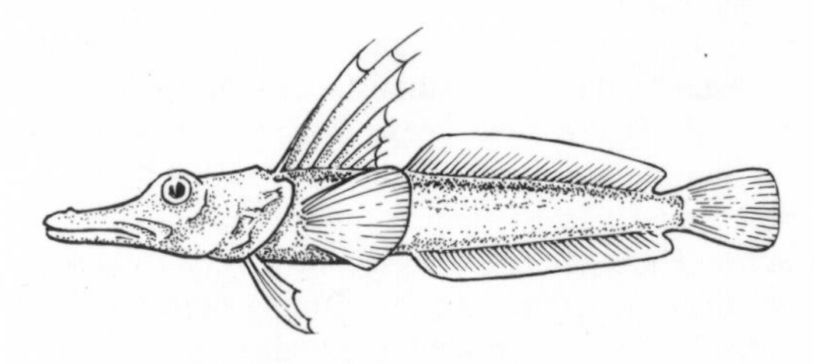

Fig. 832 Chaenichthyidae-type

As a result, there are no cells either for transporting oxygen through the body with the aid of haemoglobin. White blood cells are present. The lack of red blood is a biological curiosity, since it still leaves unexplained how the animal's tissues receive the necessary oxygen supply. The blood plasma has a maximum 1 % dissolved oxygen in it, and this would appear not to be sufficient for the animal's needs. Much research is still going on to solve this puzzle. Icefishes are elongated, with long, pointed snouts and strong teeth in the jaws. There is only one nasal opening per side. The D1 is short and often has long rays. The D2 is long and stands opposite a similar-looking An. The C is rounded and the Ps large; the Vs are located at the throat. No scales. The largest Icefish is *Chaenocephalus aceratus* at 60 cm long and weighing 1 kg. Another species, *Champsocephalus esox*, is also found around the southern tip of S America.

Notropis RAFINESQUE, 1818. G of the Cyprinidae*, Sub-F Leuciscinae. Several species are distributed in N America. They are small shoaling fish that live in clear still waters and flowing waters. The body is elongate and very flat from side to side. There are no barbels around the mouth, which is terminal. Scales fairly large, LL complete. Specimens should always be kept as a shoal in sufficiently large tanks with plenty of room for swimming. They need a lot of oxygen, so provide clear, clean water. Temperature not above 22 °C – if kept too warm, *Notropis* will not live long. The species are all omnivores, and will accept all types of live food; they will also become adapted to dried food. Allow to over-winter at cool temperatures (14–16 °C) then chances will be good that they will spawn in the tank come the spring months. The spawn is laid between plants. The fry are not entirely easy to rear.

– *N. hypselopterus* (GÜNTHER, 1868); Sailfin Shiner. From southern parts of the USA, Florida, Alabama, Georgia. Length to 6 cm. Dorsal surface greenish-brown, flanks brown with a greenish shimmer, ventral surface pale. A blue-black longitudinal band extends from the tip of the snout to the root of the tail. It is bounded dorsally by a red stripe. The D, An and C are yellowish with a brick-red base.

– *N. lutrensis* (BAIRD and GIRARD, 1853); Red Shiner. From central and southern parts of the USA as far as Mexico. Length to 8 cm. Dorsal surface grey-blue, flanks silvery with a reddish, at times blue-green shimmer. There is a dark longitudinal band at the level of the An that reaches to the root of the tail. Behind the operculum is a blue-violet blotch. The D is colourless, and all the other fins are yellowish or reddish. During the spawning season, the belly and flanks, together with the An and C, are a bright crimson red. The males have spawning tubercles. Their breeding has not yet been described.

Noturus gyrinus *see* Ictaluridae

Nucleic Acids. These are substances found in all cells. They are usually macromolecular, and are used in the storage, transmittance and realisation of hereditary information. Nucleic acids are shaped like strings and are made up of a large number of only 4 different structural elements (mononucleotides). The majority are bound to proteinous substances as nucleo-proteins.

There are basically two types of nucleic acid. *Deoxyribose nucleic acid* (DNA) is the main constituent of the chromosomes in the cell nucleus (Cell*). It forms the chemical basis of the genes. A smaller amount lies outside the nucleus in the mitochondria and, among the plants, in the plasts. DNA consists of a double string, wound in a spiral; in conjunction with the cell division, it splits up and duplicates itself identically. The bottom parts of the mononucleotides of both DNA strings lie opposite one another in a characteristic paired arrangement; the sequence they follow is species-specific. This arrangement is the genetic key for the large number of protein synthesis processes starting from the DNA. Enzyme* proteins are formed mainly, which make possible other synthesis processes and which regulate cell metabolism. *Ribonucleic acid* (RNA) actually carries out the protein synthesis processes that are regulated by the DNA. RNA occurs in various forms. A form with high molecular content (messenger-RNA) transmits the programme (rooted genetically in the DNA) to the protein synthesis and transfers this into the cytoplasm. Another RNA is located in the synthesis areas of the cytoplasm, the ribosomes, and in the nucleolus of the cell nucleus. A form of RNA with a lower molecular content is used in the transport of amino acids to the protein synthesis areas.

Nudibranchia; Nudibranchs. O of the Opisthobranchiata, *see* Euthyneura. Nudibranchia are shell-less, marine gastropods with gill processes on the dorsal surface which can have a variety of shapes. Most of them are very colourful, bizarrely shaped animals from warm seas. Unfortunately, the majority of them are also food specialists, some living on particular Sponges *(Peltodoris)*, others on Hydroid Polyps *(Flabellina)* (fig. 833), Sea-anemones *(Aeolidia)*, 'Moss-animals' or Sea Squirts. They will only survive for a short time in the aquarium if not given the right food. *Peltodoris atromaculata* BERGH, a species from the Mediterranean,

Fig. 833 *Flabellina*

is said to accept fish food tablets (Tetramin). This Nudibranch is a porcelain-white colour with brown flecks.

Nudibranchs, O, *see* Nudibranchia

Nummulites (collective term) *see* Foraminifera

Nuphar SMITH, 1809; Yellow Pond Lily, Spatterdock. G of the Nymphaeaceae* comprising 15 species. Found in subtropical and northern temperate regions. They are perennials with strong rhizomes, and very wavy aquatic and floating leaves in basilar rosettes. Flowers hermaphroditic, polysymmetrical. Perianth made up of 5–9 yellow sepals and a large number of yellow petals. There are many stamens with crossing points between them and the petals. There are also many carpels; the floral axis is fused, and the carpels are superior. The fruit is a berry.

Nuphar species grow in still or slow-flowing waters that are rich in nutrients. As their flowers cannot compete with those of the Water-lilies *(Nymphaea*)*, they are found only rarely in open-air, managed ponds. They are cultivated in the same way as the Water-lilies. With many species, young plants that have been grown from seed or shoots that have developed from weak rhizome sections can be kept in the aquarium for a limited period.

– *N. advena* AITON. From eastern N America. The blades of the aquatic leaves are roundish. 6 sepals. Suitable for open-air ponds.

– *N. japonica* DE CANDOLLE. From Japan. the blades of the aquatic leaves are a compressed sagittate shape. There is also a variety that has red-brown leaves. Suitable for warm-water aquaria.

– *N. lutea* (LINNÉ) SMITH: Yellow Water-lily. From Europe, N Africa, N Asia. The blades of the aquatic leaves are roundish. 4 sepals. Can be used in open-air ponds.

– *N. pumila* (TIMM) DE CANDOLLE. From Europe. Smaller than the previous species. A decorative plant for cold water and warm water aquaria.
– *N. sagittifolia* PURSH. Carolina Yellow Pond-lily. From south-eastern N America. The blades of the aquatic leaves are an elongate sagittate shape. Suitable for warm water aquaria.

Nurse Hound *(Scyliorhinus stellaris) see* Scyliorhynidae

Nurse Shark *(Ginglymostoma cirratum) see* Ginglymostomatidae

Nurse Sharks *see* Ginglymostomatidae

Nutrient-yeast. An important means of nutrition and food that is obtained by the culture of yeast-fungi in the sugary lye of the wood pulp industry. It contains 40–50% protein and is rich in vitamins, particularly vitamins B_1, B_2, C, E and provitamin D. In dried form, it is also a constituent of various kinds of common fish foods available on the trade. In aquarium-keeping, it is also used in the rearing of food animals (Whiteworms, Vinegar-fly larvae), ie it is added to the nutrient medium.

Nutrition. In animals, all the substances that are needed for growth and the maintenance of the bodily functions (with the exception of oxygen) are introduced into the body through nutrition. One part of the food serves to build up and renew cells and tissues, another part provides the energy for metabolism*. In nature, all animals have a particular place in the biological balance* between food supply and food consumers. As one species always forms the food of another species, a great many *food chains* become established. At one end of the chain there are always the plants, which can build up organic substances from inorganic material. The various kinds of plant-eaters are eventually eaten by animals that will become the food of other animals in their turn. This is demonstrated in the following chain: marine diatoms–small crustaceans–herring–cod–spurdog–blue shark. The method of nutrition is reflected in the anatomy* – particularly in the shape of the body, the form of the mouth*, the type of teeth* and the digestive organs*. Among fish, species that only eat plant food are relatively rare (eg Acanthuridae, Myleinae), although plants often represent the main food or are eaten at least occasionally (eg Cyprinidae, Characidae, Loricariidae). Many fish feed on the minute organisms that live in open waters (*see* Plankton), others eat mainly bottom-dwelling invertebrates; some fish living at the surface feed mainly on insects (eg *Dermogenys, Pantodon*). Many predatory fish only eat other fish. Because each type of food has its own particular composition, the ratio of different food substances (such as proteins*, fats*, carbohydrates*, vitamins* and minerals*) contained in each food will vary. The protein content is usually very high, while the fat content is usually less than 2%. There is also little carbohydrate (such as sugar and starch) in the natural diet of a fish, except among the plant-eaters. These ratios must be adhered to when feeding fish or invertebrates in the aquarium. As the following table shows, pork or herring meat are not suitable as fish food on account of their high fat content.

Food requirements are particularly high when a fish is fed on foodstuffs containing little nutritive value. Every day, young fish and many adult plankton-eaters can eat several times their own body weight in food. Many species feed continuously, some also at night. Generally speaking, food requirements in poikilothermal animals are dependent on the surrounding temperature; when the temperature falls, requirements are considerably less (eg that of our native fish during the winter months). All these factors are important to take note of when keeping animals in captivity.

	Protein	Fat	Water
	(in % of total weight)		
Pork	17	10–40	45–75
Beef	20	10–20	65
Cow's (Ox) liver	20	5	70
Horse meat	22	3	74
Carp	16	5	76
Halibut	19	5	75
Cod	20	0.5	75
Herring	20	15	60

Nymphaea LINNÉ, 1753; Water-lily. G of the Nymphaeaceae*. 40 species, that are distributed in almost every part of the world. They are perennial plants with elongated or tuberous rhizomes, rarely with short runners. In the youthful phase, or in the early phase of growth after rest periods, the aquatic leaves are fragile and the floating leaves are located in basilar rosettes. Flowers hermaphroditic, large, polysymmetrical. The perianth consists of 4 sepals and many white, yellow, red or blue petals. There are a great many stamens, and there are often crossing points between them and the petals. The carpels are numerous too, sunk into the floral axis and fused with it. The fruit is a berry.

Nymphaea species are found in still or slow-flowing waters, in a sheltered spot. The water may dry up for certain periods. Water-lilies have reached their peak of development in the tropics, where they occur in a whole array of colours. White and dark red are typical of those that flower at night, while those that open their flowers by day are blue, pale red or yellow (although some are white, too). Many of the tropical species are cultivated in the aquatic plant houses of botanical gardens. Most of the northern temperate species have white flowers. The brilliantly coloured open-air Water-lilies found in garden and park ponds are derivatives of a red variety of *N. alba* which comes from N Europe. Some are also derivatives of other colour deviants and hybrids. Soil mixed with plenty of nutrients is recommended for their cultivation. They can be planted in pots, boxes, baskets or in the pond bottom to which the substrate has already been added. Frost must not be allowed to get to the rhizomes. If there is to be no water in the pond during winter, fill it with foliage. If cultivating in containers, these must also be kept well away from frost, but make sure they stay reasonably moist all the time. In the aquarium, even the smaller species tend to cast too

much shade with their floating leaves. Propagation is possible by rhizome division, whilst rearing plants from seed is a lengthy operation. In some places, Water-lily rhizomes are eaten for food.
– *N. alba* LINNÉ; White Water-lily. From Europe. Flower white. A large number of colour deviants and hybrids are cultivated in open-air ponds (fig. 834).

Fig. 834 *Nymphaea alba,* artificially developed form

– *N. daubenyana.* A hybrid form of *N. micrantha* and *N. caerulea.* Flower blue. Adventitious plants form easily from separated leaves, and these can be used for propagation. Often kept in aquaria.
– *N. gigantea* HOOKER; Giant Water-lily. From Australia. Flower blue; it has a long stalk which sends it way above the water surface. Found in botanical gardens.
– *N. helvola* (*N. pygmaea,* name not in current usage); Dwarf Water-lily. A hybrid form of *N. tetragona* and *N. mexicana.* Flower yellow or white. A very small plant that is often recommended in aquarium-keeping. Also suitable for open-air ponds.
– *N. lotus* LINNÉ: Egyptian Lotus. From Africa. There is also a variety from the hot springs of SE Europe. Flower white, aquatic leaves flecked. As young specimens produce only aquatic plants for a long period, they are suitable for the aquarium.
– *N. odorata* AITON. From eastern N America. Flower white. Suitable for keeping in open-air ponds.
– *N. rubra* ROXBURGH: Red Water-lily. From S Asia. Flower and leaves dark red. Found in botanical gardens.

Nymphaeaceae; Water-lilies. F of the Magnoliatae*, comprising 6 Ga and 60 species. Found in all parts of the world. They are perennial plants with rhizomes and tubers. The leaves are arranged in rosettes; they are stalked and develop into aquatic, floating or, more rarely, aerial leaves. There are large air canals in the leaf and flower stalks. The solitary flowers are large, hermaphroditic and polysymmetrical. Water-lilies are true water plants, with only a few species having any emersed forms (which grow on marshy substrata). A focal point in any botanical garden is provided by species of the G *Victoria** which are often cultivated in aquatic plant houses. Less well known is *Euryale ferox* which comes from tropical parts of E Asia. Its circular floating leaves are nowhere near as big as those of *Victoria,* their edges are not curved upwards and they are a dark violet colour on the underside. Its flowers are

fairly small and have violet petals. Some of them remain submerged. The species of the G *Nelumbo* are characterised by floating, hoary blue, peltate leaves. They can be cultivated in temperate regions in sheltered spots, even in open-air ponds. The yellow-flowered *Nelumbo lutea* (American Lotus) is native to eastern parts of N America, whereas *Nelumbo nucifera* (Indian Lotus) comes from E Asia and has pale red flowers. Important Ga in aquarium-keeping include *Nymphaea**, *Nuphar** and *Barclaya**. So far, there have been no reports about the newly discovered Australian G *Ondinea* and how they adapt to the aquarium.

Nymphoides SEGUIER, 1754 (*Limnanthemum,* name not in current usage); Floating Heart. G of the Menyanthaceae*, comprising 25 species. Found in tropical and subtropical parts throughout the world, and radiating out into temperate regions. They are perennial plants with rhizomes, runners and floating leaves. Blossoming shoots are elongated, with one or several leaves. Flowers hermaphroditic, polysymmetrical. Double perianth, clayx and corolla fused, 5-part. Some of the free-standing parts of the petals have tiny hairs on them, and the petals are orange, yellow or white. 5 stamens. 2 carpels, fused, superior. Capsule-fruit. The various species grow in still waters, and, like the Water-lilies *(Nymphaea*),* they belong to the floating-leaf-plant zone. Some can also survive in the marsh zone. Since their leaves cast a lot of shade, *Nymphoides* species are not much use in the aquarium, although a few species can be kept in open-air ponds either all the year round, or at least in summer. Propagation is possible sometimes from leaves that have become separated; they should be pressed down into the damp substrate. These will then form adventitious plants. The best known species are:
– *N. aquatica* (WALTER) O. KUNTZE; Underwater Banana Plant. From the coastal area of eastern and southern parts of N America. Leaf-blades kidney-shaped, flowers white. It is the thick roots that are compared with bananas. Often kept in aquaria.
– *N. peltata* (GMELEN) O. KUNTZE; Common Floating Heart. From N America, Europe, N Asia. Leaf-blades heart-shaped. Flowers yellow. Hardy in winter.

Oar fishes (Regalecidae, F) *see* Lampridiformes

Oceanic Sea-horse *(Hippocampus kuda) see Hippocampus*

Oceanic Whitetip Shark *(Carcharhinus longimanus) see* Carcharhinidae

Oceanology. The science of marine ecosystems*.

Ocellated Pufferfish *(Canthigaster margaritatus) see Canthigaster*

Ocellatum Disease *see Oodinium*

Octobrachia, O, *see* Cephalopoda

Octocorallia, Sub-Cl, *see* Anthozoa

Octomitus, G, *see* Hexamita

Octopus LAMARCK, 1799. G of the eight-armed Cephalopods (Cephalopoda*, O Octobrachia). There are several *Octopus* species and they live in the shelf area of the Atlantic and Pacific. They are bottom-dwelling animals that live hidden during the day in caves and self-made rock castles. Their diet consists of crustaceans, molluscs and fish. *O. vulgaris* has an armspan of up to 3 m and weighs 25 kg; other species can weigh up to 100 kg. The body is shaped like a bag; the arms have a double row of suckers on them and are joined together at the base by a membrane. Octopuses can adapt themselves amazingly well to their environment, both in colour and by altering the surface structure. When excited, vivid colours will appear. The Octopus is a territorial animal. It catches its prey with its muscular arms and wraps it between the membrane surrounding the mouth. A secretion from the salivary glands then paralyses the prey, and it is bitten open by the Octopus' strong jaws, before being dissolved by digestive juices and sucked in. During mating, the male transfers a spermatophore to the female (who is sitting close by) with his highly specialised hectocotylus tentacle. A female *O. vulgaris* lays up to 150,000 eggs in jelly-like threads, which she then guards and constantly supplies with fresh water. The young Octopuses live in the plankton for one or two months, then they become bottom-dwelling. Even large Octopuses are not very dangerous to humans. Many people eat them. They are intelligent, very interesting aquarium specimens, that can even be tamed. However, they grow quickly and need enormous amounts of food. The inky fluids released by Octopuses can be dangerous, even for them. When they are kept in an aquarium, make sure the tank is large enough, that the filtration is extremely efficient, that part of the water is changed frequently, and that there is a constant supply of food animals, especially Crabs. When setting up the aquarium, it is important to remember that an Octopus is strong enough to move even large rocks effortlessly. The tank must also have a secure lid, otherwise the Octopus will crawl out. A species that lives along the European coast is:

– *O. vulgaris* LINNAEUS, 1758; Common Octopus. Found in the Atlantic, the North Sea and the Mediterranean. In some places they are common. Usually brownish camouflage colours.

Ocypode, G, *see Uca*

Odonata; Dragonflies (fig. 835). O of the insects (Insecta*). Next to the Mayflies (Ephemeroptera*), this is the oldest group of winged insects. They were already known in the Carboniferous, and today there are 4,000 species distributed worldwide. In Europe there are about 120 species. Dragonflies are among the best-known insect groups because of their considerable size, their charming colouring and the elegant flight of some Ga *(Aeschna, Anax)*. They do not bite and they are not poisonous. Both the larvae and the imagines (sexually mature adults) are true predators, and will even attack their own relatives. Dragonfly larvae are found in very varied stretches of water, from the smallest of streams to the largest of rivers, from the brackish water of lagoons to fens and marshes. They are cylindrical in shape and have a very characteristic mask with modified jaws for catching prey. This mask can be extended very quickly (fig. 836). The larvae breathe either by means of the leaf-shaped outgrowths on the abdomen (eg *Lestes*) or through the hind-gut (eg *Aeschna*). The metamorphosis of the larva into the adult capable of flight takes place out of the water. There is no pupal stage.

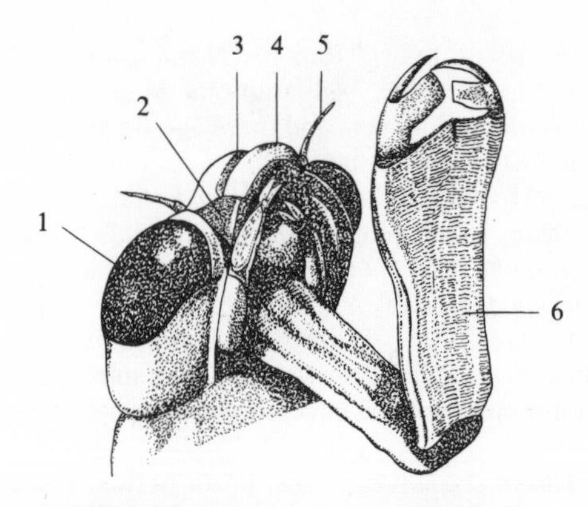

Fig. 836 Odonata. Head of a Dragonfly larva, mask half-extended. 1 Eye 2 Jaw 3 Forehead 4 Upper lip 5 Feeler 6 Mask

Dragonfly larvae make very interesting objects of observation. Types that live in still waters usually present no problem when kept in a pond-aquarium.

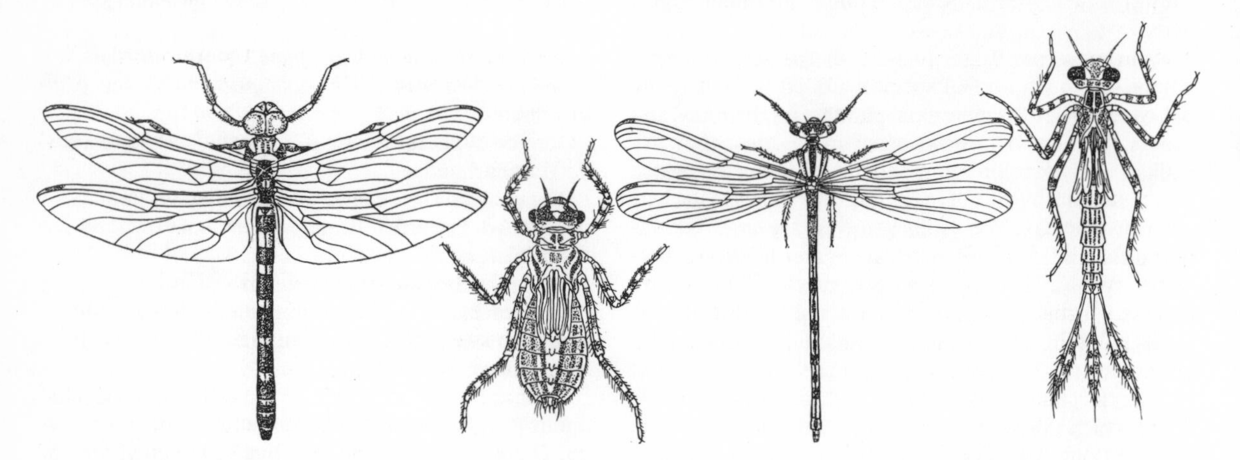

Fig. 835 Odonata. Form of the Dragonflies (Anisoptera), left, and the Damselflies (Zygoptera), right; imago and larva in each case

Because many larvae are cannibals, however, it is best to keep them on their own in very small aquaria. Particularly striking or common Ga are, eg, *Aeschna**, *Anax**, *Calopteryx**, *Coenagrion** and *Lestes**.

Odontaspididae; Sand Sharks. F of the Chondrichthyes*, O Lamniformes. Medium-sized fish with a typical shark shape; head flattened, snout pointed, teeth large and long, 5 gill slits either side in front of the Ps, and 2 dorsal fins, roughly the same size. The upper tail fin lobe is much bigger than the lower. Sand Sharks live in shallow water regions and swim restlessly. They feed on fish, crustaceans and cephalopods. The females give birth to two pups only. Of the 6 species, *Odontaspis taurus* (fig. 837) [the Sand Shark] is probably best known. It can grow up to 3 m long and is found in the E and W Atlantic and in the Mediterranean. Despite its grim appearance, it is not a very dangerous shark. Its upper surface is grey, and its skin is made into leather. Otherwise, its value is not very great.

Odontaspis taurus *see* Odontaspididae

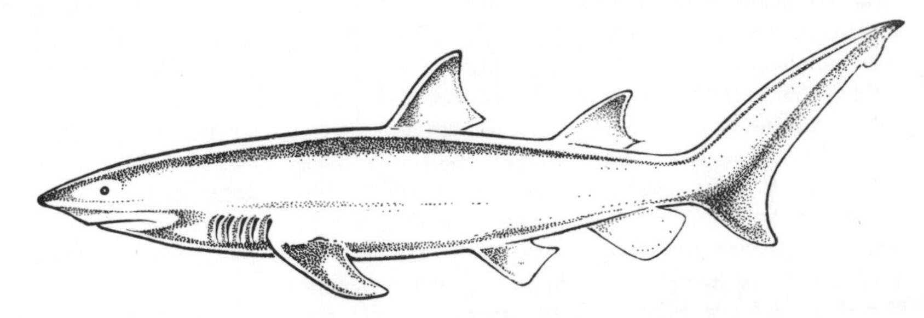

Fig. 837 *Odontaspis taurus*

Odontodactylus, G, *see* Stomatopoda

Odonus GISTEL, 1848. G of the Balistidae*, Sub-F Balistinae. There is only one species, found in tropical parts of the Indo-Pacific. It is quite unmistakable:

– *O. niger* (RÜPPEL, 1835); Black Triggerfish, Redfang Trigger, Red-toothed Triggerfish, Royal-blue Triggerfish. Length up to 50 cm. It is a sociable fish in the wild, and is also fairly peaceable in the aquarium. It is blue, with red teeth, two of which jut out sideways.

Oenotheraceae, F, name not in current usage, *see* Onagraceae

Oesophagus *see* Digestive Organs

Ogcocephalidae, F, *see* Lophiiformes

Oilfish *(Ruvettus pretiosus) see* Scombroidei

Old Wench *(Balistes vetula) see Balistes*

Old Wife *(Balistes vetula) see Balistes*

Olfactory Organs (fig. 838). Sense organs which are receptive to different chemical substances in the surrounding medium. In fish, the olfactory organs are used for finding food, in migration, in shoal formation and when a fish is trying to escape.

In vertebrates, the sensitive part of the olfactory organs is the olfactory mucous membrane. It is here that the branches of the olfactory nerves begin, which lead to the front of the brain. Fish (apart from Dipnoi) do not have any connecting nasal-mouth passage; their olfactory organs are located in depressions on the surface. Sharks and rays have nasal openings on the lower surface of the head, bony fish have them on the upper surface between the mouth and the eyes. The nasal depressions are paired and either have a common opening (Cyclostomata), or a single opening each (eg Cichlidae, Labridae), or more commonly one entry opening at the front and an exit opening at the back. In many cases, swimming movements send water to the olfactory epithelium of the nasal depressions. This can create a funnel effect because of projecting folds of skin at the front of the nasal openings. Many kinds of fish have large internal nasal sacs, and in such cases water is sucked in by the rhythmic movements of the mouth, then forced out again (Perciformes, Mugiliformes). In rarer cases, water is transported by means of a ciliated epithelium *(Anguilla)*. The sensory olfactory epithelium often has the shape of a conical rosette, or it may be arranged like a two-sided chamber. Its size is proportional to the efficiency of the fish's olfactory system.

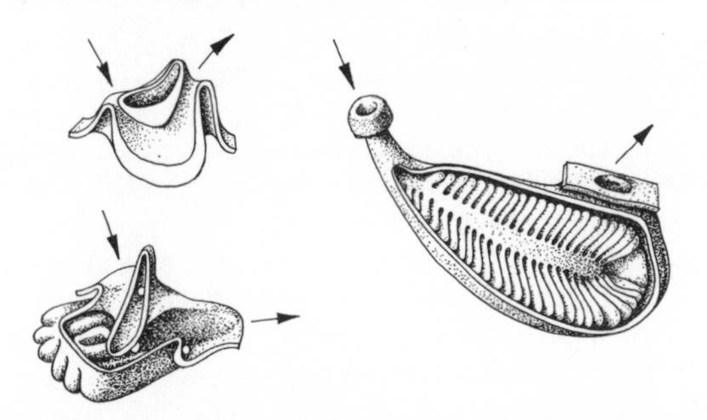

Fig. 838 Olfactory organs. Top left, Pipefish; right, Goldfish; below, Eel

Oligo-. In word compounds, meaning 'little' or 'few', eg, an oligotrophic stream = a stream that contains little nutrient.

Oligochaeta; Oligochaetes. Cl of the Annelida*. There are more than 3,000 species, most of which live in fresh water. Many members of the O have adapted to life on land; nevertheless, they still remain closely bound to biotopes with moist atmospheres. Many species also live in brackish and sea water. Oligochaetes are largely uniformly segmented organisms, like a worm in shape. They range in size from a few mm to about 3 m, and the segments are equipped with individual little bristles or groups of bristles. Most types crawl along, although some can swim freely. They feed mainly on dead plant parts, although some are predatory and others parasitical.

The best-known members of the O are the terrestrial Earthworms (eg *Lumbricus, Eisenia*). They help improve the soil in agriculture, by loosening it, ventilating it and forming humus. In aquarium-keeping, oligochaetes are used mainly for fish food, especially the Ga *Lumbricus, Eisenia, Enchytraeus**. The aquatic G *Tubifex** is also used for the same purpose.

Oligochaetes, Cl, *see* Oligochaeta

Oligotrophic Waters. Waters with a low nutrient content (especially phosphate and nitrate). As a result, only a small amount of algae develops and the water is very clear. However, the depth of visibility can be reduced by humus substances (Dystrophic Waters*) or by mineral particles in suspension, which is what happens in the tropical white water rivers, for example (Rio Madeira, lower Amazon*, White Nile). Oligotrophic waters are completely free of sapropel sedimentation, so that poisonous hydrogen sulphide never arises, and the oxygen content appears to be quite high even in the deeper zones. Many aquarium animals can be seen from their origin to be bound to water with a low nutrient content, eg some American Characins. Opposite to eutrophic waters*.

Olive Limia *(Poecilia versicolor) see Poecilia*

Onagraceae (Oenotheraceae, name not in current usage); Evening Primrose Family. F of the Magnoliatae*. 20 Ga, 650 species. In rare cases tropical, but mainly subtropical and temperate, especially in America. The majority are perennials or annuals with alternate or opposite simple leaves. Flowers hermaphroditic, polysymmetrical, with either a conspicuous or inconspicuous perianth. Evening Primroses colonise dry or wet habitats. They include various ornamental plants, as, for example, Fuchsias. For aquarium purposes, some species of the G *Ludwigia** have proved suitable.

Oncorhynchus, G, *see* Salmonoidei

Ondinea, G, *see* Nymphaeaceae

One-lined Pencilfish *(Nannostomus unifasciatus ocellatus) see Nannostomus*

One-sided Livebearer *(Jenynsia lineata) see* Jenynsiidae

One-sided Livebearers, F, *see* Jenynsiidae

One-spot Barb *(Barbus terio) see Barbus*

One-spot Livebearer *(Phalloceros caudimaculatus) see Phalloceros*

One-spot Livebearer *(Poecilia vivipara) see Poecilia*

Ontogenesis. The history of an individual's development, as distinct from phylogenesis*. Whereas in the higher plants, a youthful phase can be distinguished from a gerontic phase, in multi-celled animals life is divided into 4 phases. After *embryonic development** the *progressive phase* (juvenile period) begins. It is marked by tremendous growth of the body. With a great many animals, as well as with some fish, this period is characterised by a larval form (Larva*) that develops into a juvenile form after a metamorphic phase (Metamorphosis*). The juvenile phase is often marked by a juvenile form of colouring. Following on from here is the *stage of maturity* (reproductive period). It begins with the maturing of the sex glands and the completion of the external sex features, especially typical colouring and the final body forms. In most animals, this adult stage is the longest period of life. However, there are some invertebrates that die after reproducing, so this period is relatively short (eg Butterflies, Mayflies). The fourth phase of life, the *regressive phase* (gerontic phase) is characterised by the absence of sexual functions, by signs of age and breakdown in various organs, particularly in the nervous system, the circulatory organs, the glands and the skeleton. The causes of death in an organism are not simply connected with the collapse of certain functions of the organs or with the reduced ability to defend itself, they also have a lot to do with the defence systems against diseases breaking down.

In many fish, eg some salmon and eels, the regressive phase is absent altogether, and they die shortly after spawning.

Oodinium CHATTON, 1912 (fig. 839). G of the Flagellata*. Many of the species of the G are ectoparasites (Parasitism*), that cause similar disease symptoms in fish. *O. ocellatum* BROWN, 1931, causes Coral Fish Disease or Ocellatum Disease, and it can occur in all marine fish. Catches from Singapore and Hawaii are commonly infected. *O. pillularis* SCHAPERCLAUS, 1951 causes Pillularis Disease (sometimes known as Colisa disease, since it was first discovered on *Colisa lalia* [HAMILTON-BUCHANAN, 1822]). However, this parasite occurs on many other freshwater fish besides. *O. limneticum* JACOBS, 1946, has so far only cropped up in freshwater aquaria in N America. The parasites are 20–150 μm long and are round, oval or pear-shaped.

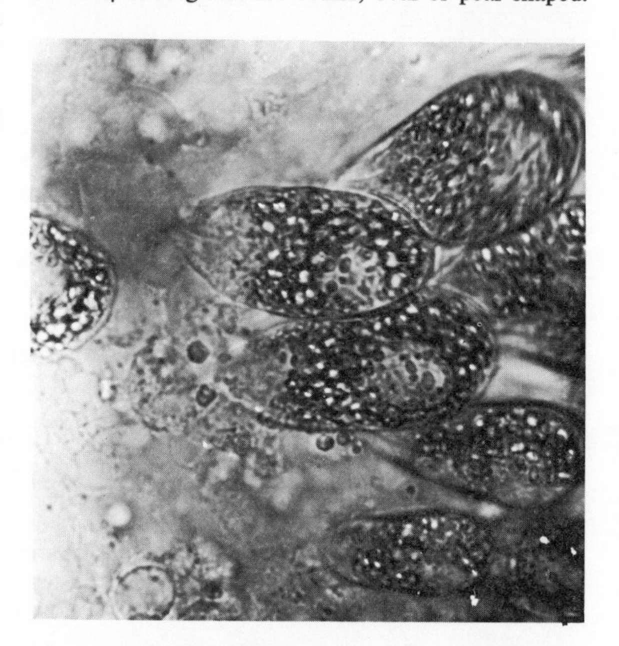

Fig. 839 *Oodinium.* Skin stage

They are encased in a protective cellulose envelope and are able to anchor themselves firmly in the fish's gills and skin by means of root-like outgrowths from the cell (parasitic stage). When fully mature, the parasite falls from the fish and forms a cyst (non-mobile cyst stage). The cell now divides repeatedly, and, depending on temperature, after 3–4 days a large number of daughter cells (dinospores) will develop. They are roughly 9–15 m in length. It is these dinospores that infect fish, seeking them out actively with the aid of their two cilia (fig. 840). If they do not find a host, the dinospores die after 12–24 hours *(O. pillularis)*, or after 18–72 hours in the case of *O. ocellatum*. They attach themselves to gills, skin and fins where they change into the parasitical stage. Grey-white or yellow dots will appear on an

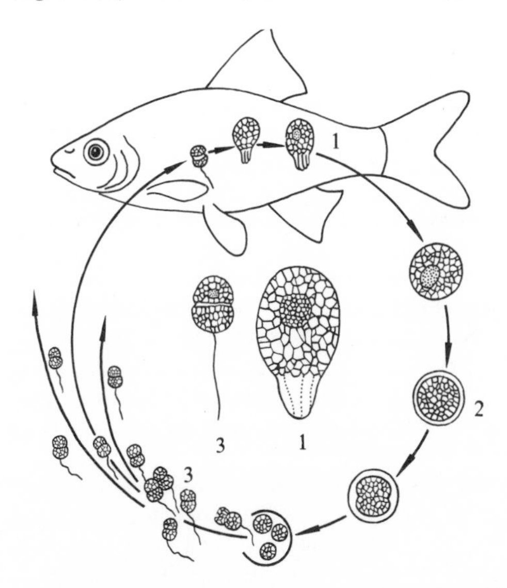

Fig. 840 *Oodinium* development-cycle. 1 Parasitical stage 2 Immobile cyst stage 3 Dinospores

infected fish's skin (easily confused with the symptoms caused by *Ichthyophthirius** in the initial stages); in the advanced state, these dots spread out to form an unbroken layer. The fish rub against stones and leaves and often have ragged skin; emaciation* sets in. Parts of the skin look inflamed and bloody, and if the gills are attacked the breathing frequency increases. Only the mobile dinospores can be effectively treated with medicaments. Prolonged baths (Bathing Treatments*) of up to 12 hours are best; use Trypaflavin*, Aureomycin*, copper sulphate* or quinine*. Quite frequently strains occur that are resistant to individual medicaments. Increasing the temperature for 3 days to 38–40 °C will also kill off the cyst stages. With *Oodinium* infections there is the strong likelihood of a relapse. On a very few occasions, the parasite has also been detected in the mouth and in the intestine. It is difficult to deal with here and it can herald the start of a mass infection. If the cyst stage does not divide, it can survive for months and still be capable of reproduction. So it is advisable to repeat the treatment every 10–14 days.

Opah *(Lampris regius) see* Lampridiformes

Opelet Anemone *(Anemonia sulcata) see Anemonia*

Open-water Spawners. Species of fish that release their sperm and egg-cells into open water, so that fertilisation as well as the protection of the developing egg-cells become largely a matter of chance. Such species usually produce very large amounts of spawn (eg many Characins, Barb related species) which is in contrast to the substrate spawners*. This is to ensure that the species survives despite the high loss rate which results from insufficient protection (*see also* Cichlidae).

Open-water Zone *see* Pelagic

Ophiactis, G, *see* Ophiuroidea

Ophicephalidae, F, name not in current usage, *see* Channiformes

Ophichthyidae, F, *see* Anguilliformes

Ophidiaster AGASSIZ, 1835. G of the Starfishes (Asteroidea*). Found in warm seas, *Ophidiaster* species have a small disc and very long cylindrical arms. They eat small animals and particles. Their requirements in the aquarium are the same as for *Linckia**. *O. ophidianus* LAMARCK comes from the Mediterranean and the coast of W Africa. It can grow up to 25 cm long. It is a uniform violet-red colour or it has irregular fleck markings.

Ophidiidae, F, *see* Gadiformes

Ophidium barbatum *see* Gadiformes

Ophiocara GILL, 1863. G of the Gobiidae, Sub-O Gobioidei*. Found from Madagascar and the Seychelles, across S and SE Asia, including the Malaysian Archipelago as far as New Guinea and Australia. *Ophiocara* lives along the coastal regions in sea water, brackish water and fresh water. They are medium-sized fish, sometimes longer than 30 cm. The body is elongate, only flattened at the rear. The lower jaw is slightly protruding. There are many rows of teeth, the outside ones being bigger. No spine on the front part of the gill-covering (praeoperculum). The scales are medium-sized; 30–40 in the middle longitudinal row. Vs not fused with one another, C rounded. The basic colouring is brownish or olive-green tones, with varying, very attractive colour patterns. *Ophiocara* should be kept in a roomy tank with a soft sandy substrate. Temperature 18–25 °C. Depending on origin, freshwater forms need 1–2 tsp sea salt, brackish water forms 50–100 g to every 10 litres of water. Feed on live food of any kind. Species that are frequently imported include:

– *O. apores* (BLEEKER, 1854). SE Asia, including the Malaysian Archipelago of the Philippines, Madagascar. Found in sea water, brackish water and in fresh water. Length up to 30 cm. Colouring variable. Upper side of the body is dark brown to olive-green, ventral side organge or yellowish. Flanks dotted in black, with some green flecks. 3 red bands radiate out from the eye across the operculum. Fins dark, fin rays yellowish. The D, Vs and An have a red edge, and the C has yellow dots. There is a black transverse band edged in red at the base of the P. The female is more inconspicuous, with rather more brown colour tones. An undemanding species that will also reproduce in captivity (fig. 841).

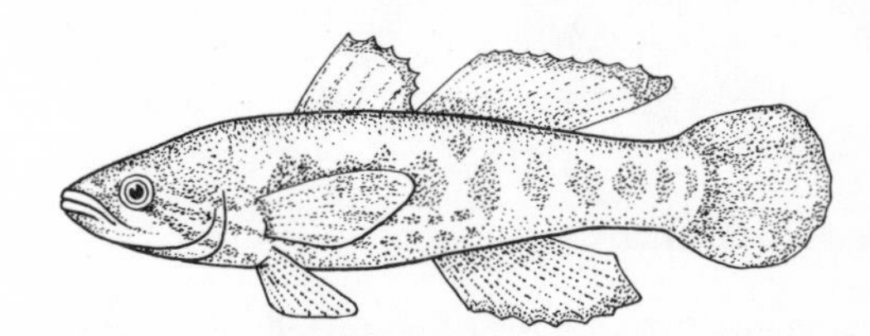

Fig. 841 *Ophiocara apores*

– *O. porocephala* (CUVIER and VALENCIENNES, 1837); Snake-head Goby. SE Asia, Malaysian Archipelago, China, Australia, Madagascar. Found in brackish and fresh water. Length up to 32 cm. Colouring variable. Upper side of the body is dark green or olive-green, also red-brown; paler below. There are dark longitudinal stripes along the flanks, that become less distinct with age. Fins brownish to violet, rays orange. In many specimens, the body, D2 and An have orange dots. Other than that, the D2 and C have brown dots and red edges. Young specimens have silver transverse bands on the body.

Ophiocephaliformes, O, name not in current usage, *see* Channiformes

Ophiocoma GASSIZ, 1835. G of the Brittle-stars (Ophiuroidea*). Contains numerous species found in shallow tropical coastal waters, where they live beneath stones. They sweep up food particles and also eat small animals. To catch their food, they extend 2 or 3 of their arms from their hiding-place.

– *O. scolopendrina* grazes among the water surface at low tide. The mouth disc is grainy, no plates being visible, and has a diameter of up to 4 cm. The arms can measure up to 14 cm long and have strong spines on their sides. This species is fairly hardy in a marine aquarium if fed on *Tubifex* and pieces of bivalve flesh.

Ophioderma MÜLLER and TROSCHEL, 1840. G of the Brittle-stars (Ophiuroidea*). The various species live in the littoral* of warm seas where they feed on small animals and dead animals. The leathery body disc is grainy on the upper surface, and the arms have short, close-set spines which make them appear cylindrical. *O. longicauda* RETZIUS comes from the Mediterranean. This species is pale or dark brown and its arms can measure up to 15 cm long. In the aquarium it is fairly easy to care for.

Ophiothrix MÜLLER and TROSCHEL, 1840 (fig. 842). G of small Brittle-stars (Ophiuroidea*). Often found in large numbers in cool and warm seas from the shore area to depths of several 100 m deep. They eat particles and small animals, catching the plankton and detritus

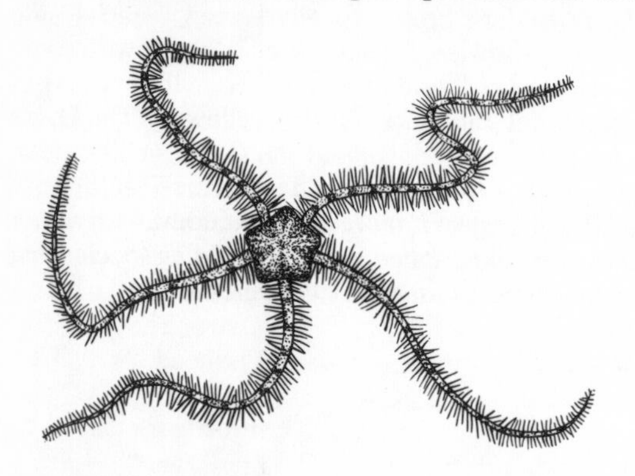

Fig. 842 *Ophiothrix*

that contain them with their fine-spined arms and then guiding them to the mouth. The body disc is spiny on the upper surface and juts forward between the bases of the arms (which makes it look star-shaped). *Ophiothrix* allows its arms to drop off easily, but they will regenerate quite quickly if conditions are good. The commonest European species is:

– *O. fragilis* ABILGD. This species is very variable in colour and is found in shallow water beneath stones. The disc measures about 1 cm in diameter and the arms are 5 cm long. It survives well in old-established aquaria. The method of acquiring food is easy to observe: several of the arms are extended from their hiding-place, and food particles become trapped between the spines. The arms are then entwined to grip the food and it is transported to the mouth.

Ophiura, G, *see* Ophiuroidea

Ophiuroidea; Brittle-stars. Cl of the echinoderms (Echinodermata*) containing the greatest number of species. They live on the bottom and eat particles and small animals. Members of the Cl are found in all seas. The arms are a characteristic feature of the Brittle-stars; they are distinctly set apart from the body, are round, and divided up like vertebrae. In the Gorgonocephalidae the arms are much branched. The mouth side is turned towards the bottom and there is no anus. The ambulacral feet have no suckers – Brittle-stars push themselves forwards in a jerky fashion by means of their mobile arms. Reproduction is either sexual or through splitting in two. A number of species practise brood care. Brittle-stars live hidden in or on the bottom, and the majority either graze or actively catch their food. To do this, they extend some of their arms from their hiding-place, firmly grasp small animals or particles in them, and transport them to the mouth. The monster Gorgonocephalidae catch their prey in 'nets', spreading out their finely branched arms at night. In the aquarium, many Ophiuroidea make long-lasting, interesting specimens. Well known species include *Amphiura**, *Ophiactis*, *Ophiocoma**, *Ophioderma**, *Ophiothrix**, *Ophiura*.

Opisthognathidae, F, *see* Trachinoidei

Opistobranchiata, Sup-O, *see* Euthyneura

Opossum Shrimps, O, *see* Mysidacea

Optical Sense Organs *see* Sight, Organs of

Optimum *see* Ecological Potence

Oral Cavity *see* Mouth

Orange Chromide *(Etroplus maculatus) see Etroplus*

Orange Dorsal Livebearer *(Phallichthys amates pittieri) see Phallichthys*

Orange Filefish *(Oxymonacanthus longirostris) see Oxymonacanthus*

Orange Sea Perch *(Anthias squamipinnis) see Anthias*

Orange-cheek Barb *(Barbus orphoides) see Barbus*

Orange-emerald Filefish *(Oxymonacanthus longirostris) see Oxymonacanthus*

Orange-fin Minnow *(Barbus barilioides) see Barbus*

Orange-finned Loach *(Botia modesta) see Botia*

Orange-green Filefish *(Oxymonacanthus longirostris) see Oxymonacanthus*

Orange-green Trigger *(Balistapus undulatus) see Balistapus*

Orange-spot Wrasse *(Coris angulata) see Coris*

Orange-spotted Filefish *(Oxymonacanthus longirostris) see Oxymonacanthus*

Orbicular Velvetfishes (Caracanthidae, F) *see* Scorpaenoidei

Orchid Family *see* Orchidaceae

Orchidaceae; Orchid Family. F of the Liliatae* comprising 700 Ga and 20,000 species. Found mainly in the tropics and subtropics, but also in all other regions.

They are perennials that come in a variety of shapes, some with stem tubers, more rarely with root tubers. Leaves alternate with simple shapes. Flowers hermaphroditic, monosymmetrical, usually with a strikingly coloured perianth.

The Orchidaceae are ground plants that grow in a variety of habitats; in warm places they also colonise rocks and some grow in large numbers as epiphytes on trees. They have complicated mechanisms to ensure pollination. Orchids live in association with certain fungi that settle in the Orchid's roots (symbiosis). Their powder-fine seeds lack nutritive tissue and can only develop in nature if, from an early stage, they are associated with these fungi (since the fungi provide the necessary nutrients). When Orchids are reared artificially, the fungi can be dispensed with in the early development stages. The seeds are allowed to germinate on nutrient substrata under sterile conditions, these substrata containing the necessary organic and inorganic substances. Many species of Orchid are ornamental plants and are bred specifically. *Spiranthes cernua* (LINNÉ) RICHARD, the Water Orchid, comes from tropical S America and grows on marshy substrates, where it can also tolerate flooding. It has been kept in aquaria with varying degrees of success.

Orconectes limosus *see Cambarus*

Order *see* Systematic Categories

Order of Rank; Social Hierarchy. Term used in ethology. Nearly all species of animals that live partly or completely as social creatures develop a social hierarchy within their community (among poultry, especially, this is also known as the pecking order). Similarly, within a shoal of fish, an order of rank also appears, usually split up according to male and female individuals. All are subject to very strict rules of conformity. If, for example, there are 4 males in a fish shoal, then the highest-ranking one is dominant – the strongest fish A dominates the other fish B, C and D, all of whom must show signs of inferiority behaviour* when in his presence. If A is not around, then fish B dominates C and D. If both A and B are absent, C still forces D into a position of inferiority, the position of the lowest-ranking. If the reproductive drive of a lower-ranking male is suddenly activated, then the order of rank can just as suddenly switch in his favour, as in this moment fish A is forced from his top position and the lower-ranking fish takes over. A strict order of rank also develops frequently among males around favourite feeding places; here too, both sexes can be subject to a mixed ranking order, which both a strong male and female can lead.

Oreaster MÜLLER and TROSCHEL, 1842 (fig. 843). G of predatory tropical Starfishes (Asteroidea*). They all have a broad base to their strong arms that also have long rows of large, stumpy spines on them. Diameter up to 50 cm. *Oreaster* species have some delightful colours and they survive in the aquarium for lengthy periods. However, remember that they are highly predatory, and will attack all slow-moving or sessile animals as well as smaller kindred species. These details also apply to the related G *Pentaceraster* DÖDERLEIN, 1922 (fig. 844).

Orectolobidae; Carpet Sharks. F of the Chondrichthyes*, O Orectolobiformes. Small to medium-sized

Fig. 843 *Oreaster nodosus*

Fig. 844 *Oreaster (Pentaceraster) mammilatus*

sharks that occur mainly in the coastal waters, near the bottom. Characteristic features are the 2 deep grooves that connect both nasal holes with the mouth (nasolabial grooves) and, most of all, the short, strong barbel on the front edge of each nasal opening. The head is often very compressed, the snout very stunted and short, which leads to the mouth lying almost at the front end of the head. There is no nictitating membrane, and the slit-like spray-holes are located behind and beneath the eyes. 2 Ds that are positioned well to the rear. The top lobe of the C forms an almost straight continuation of the body. The best-known species, the Carpet Shark or Wobbegong *(Orectolobus barbatus)* (fig. 845), has tufts of skin along its flanks as well as on the chin of its broad head; these may be organs of touch and taste. This species is found in the E Pacific. Its diet consists of bottom-dwelling animals. Many species are livebearing; others lay eggs. A large number of pups are given birth to in each litter. Carpet Sharks are usually harmless, although they will attack if molested. Since several species, including the Wobbegong, rest on the bottom during the day, swimmers are occasionally bitten when they tread on them. A fine leather is made out of the skin.

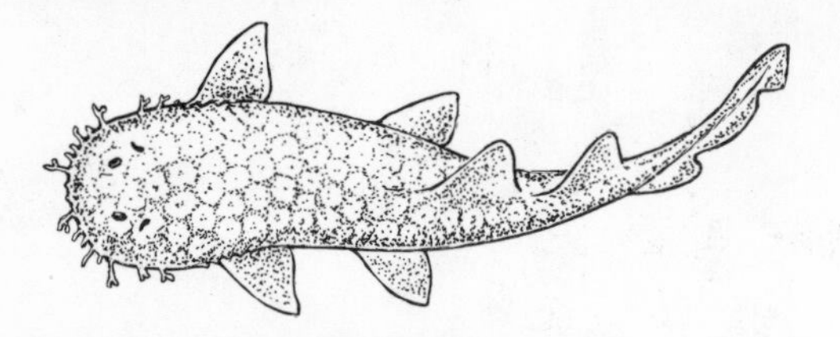

Fig. 845 *Orectolobus barbatus*

Orectolobus barbatus *see* Orectolobidae
Orestias cuvieri *see* Cyprinodontidae
Orestiatinae, Sub-F, *see* Cyprinodontidae
Orfe *(Idus idus) see Idus*
Organ of Attachment (fig. 846). A number of fish that live in turbulent waters or where there is a strong current have developed special anatomical features that enable them to attach their bodies to the substrate. By attaching themselves in this way, some species actually move about (passively, as it were) or obtain their food. Most organs of attachment look like a movable sucking disc. They can be formed from the mouth or the lips,

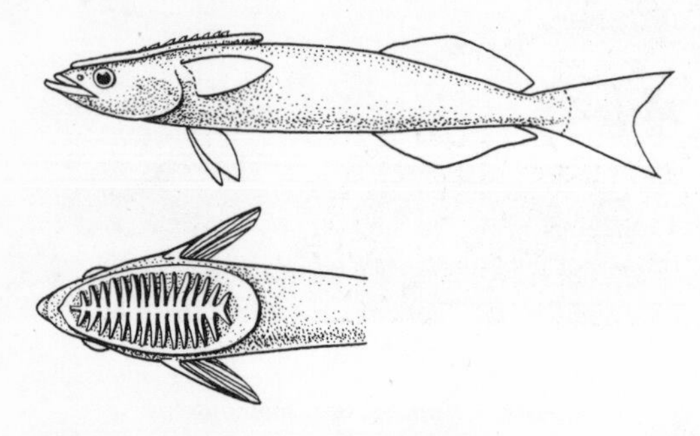

Fig. 846 Organs of attachment on the Sharksucker *(Remora remora).* Below, as seen from above.

eg as in the Cyclostomes (Cyclostomata), the Loaches (Cobitidae) and some Catfishes *(Glyptosternum, Exostoma).* The ventral fins are often used for attachment, eg as in the Gobies (Gobiidae), the Clingfishes (Gobiesocidae) and the Lump-sucker *(Cyclopterus).* The Remoras (Echeneidae) have a sucking disc on top of the head which has developed from the front dorsal fin. By changing the position of the modified fins, the lumen of the sucking chambers can be altered, thus creating the necessary vacuum for attachment. Other kinds of organs of attachment include the sticky threads found in some fish eggs, and the sticky glands, body folds and embryonic sucking discs found in some larvae.
Organ Pipe Corals, G, *see* Tubipora
Oriental Realm. Zoogeographical region made up of the Near East and East India south of the Himalayas, S China and the Greater Sunda Islands (Sumatra, Java, Kalimantan), including the Philippines. The oriental realm has features in common with the palaearctic* realm that have been caused geographically, as well as features in common with the Ethiopian realm* that have been caused historically. Characteristic freshwater fish of the Ethiopian-oriental region include the Notopteridae*, the Mastacembelidae (Mastacembeloidei*), the Channidae (Channiformes*) and the Anabantoidei*. There are relatively few endemic* species in the oriental realm, but those that do exist belong to the Fs Luciocephalidae (Luciocephaloidei*) and the Phallostethidae (Atherinoidei*).
The freshwater fauna of the oriental realm is extrememly rich in forms and species, and many well-known aquarium fish are native to the region, eg *Barbus** species, plus the Ga *Rasbora*, Brachydanio*, Danio*, Botia*, Colisa*, Trichogaster**.
Origin of Species *see* Descent Theory
Oriole Angelfish *(Centropyge bicolor) see Centropyge*
Ornamental Fish, Breeding of. For many aquarists, successfully breeding the fish they are caring for is the high-point of their hobby. That is why, since very early on in aquarium-keeping, successful attempts have been made to mate aquarium specimens, particularly tropical ones. The first fish to be reared expanded the supply of imported fish* brought in by seamen (which was rather sporadic in the beginning). Their owners swapped them or gave them away as presents, and so the number of hobbyists grew rapidly. The first specialist shops selling everything for the aquarium and terrarium then began to appear, and this now made it pssible to buy fish, plants and everything else the aquarist needed. Soon, however the supply of specimens reared by hobbyists was not enough to meet demand, and so the first commercial breeding stations for ornamental fish came into being. (Some of them, besides their own successfully bred species, also offered wild catches for sale that they had imported themselves.) Since 1945 especially, the ornamental fish trade* and ornamental fish breeding have developed greatly. In Florida, enormous breeding stations sprang up which (favoured by the climate) managed to produce in *open-air breeding stations* quantities of ornamental fish quite unheard of before. Because these could be run at very little energy costs (no heating or lighting needed for the breeding* and rearing* aquarium), these fish could be offered on the world market at very low prices. Nevertheless, breeding by hobbyists is still very important, as many aquarists publish their knowledge about the care and first breeding of new species, and thus are able to give useful hints to the breeding stations. At the same time valuable information is made available to ethologists (Ethology*) and systematicians.
Ornamental Fish, Catching of. There are various different ways of catching ornamental fish – the method chosen depends largely on the biotope where the catch is to be made and on the habits of the particular species concerned. So, if catching tropical coral fish, the techniques are different from those applied when catching tropical freshwater fishes. Often, the latter are caught only during the dry season, as at this time the fish congregate naturally in their particular habitats because of the lower amount of water available. Large

drag-nets, adjustable nets or hand-nets are used mainly, although narcotics are occasionally used particularly in very weedy waters or if a stretch of water contains dangerous fish. In the early days of aquarium-keeping, large firms handling ornamental fish undertook their own fishing expeditions, but today export companies supply the European and North American markets. The wild catches are sorted according to species and size then packed in large, well ventilated container tanks for dispatch elsewhere. Fish sent by plane are always put in pouches made of synthetic materials which are pumped up with oxygen and packed in isolated card-board-boxes or in boxes made of foam. To keep the transport costs low, the fish are often anaesthetised or they are kept calm by some other means. In this way, a great many animals can be transported in a small amount of water. However, not all species will tolerate such conditions of transport. Large adjustable or drag-nets cannot be used when catching fish on the coral reefs. Usually, coral fish are caught by local divers in hand-nets or the fish are taken by hand from their hiding-holes among the coral formations. Often pieces of coral have to be broken off and brought up to the fishing vessel. Only then will a lot of the fish hidden inside emerge. Some species are caught with a rod to which small hooks are attached that do not have the usual barbs.

Ornamental Fish Export Trade *see* Ornamental Fish Trade

Ornamental Fish Trade. The ornamental fish trade is an important element in the home and export markets of many industrial countries. Whereas specialist zoological business concerns satisfy the home market demand for ornamental fish, water plants, live and artificial foods, plus all aquarium-keeping accessories, large companies are concerned with the export of a considerable part of the inland production of ornamental fish and water plants. In various socialist countries the export of ornamental fish is part of a state institution created expressly for this purpose. Even before the First World War, to give an example of the export activity of the time, Germany's annual cotton imports were covered by the export of ornamental fish. After World War II the export rate of various European states and American monopolistic companies increased by leaps and bounds. Recently the Peoples Republic of China has developed into an important partner in the ornamental fish trade having built efficient inland breeding stations. For a short time now, some South American countries have been realising the value of their native species of aquarium fish, regarding them as natural reserves for increasing the efficacity of their export trade. In earlier times, these South American fishing grounds were visited by expeditions from American and European companies for the purposes of catching fish. (Ornamental Fish, Catching of*.)

Ornate Ctenopoma *(Ctenopoma ansorgei) see Ctenopoma*

Ornate Dwarf Cichlid *(Apistogramma ornatipinnis) see Apistogramma*

Ornate Limia *(Poecilia ornata) see Poecilia*

Ornate Tetra *(Hyphessobrycon ornatus) see Hyphessobrycon*

Orontium aquaticum (Golden Club) *see* Araceae

Ortmann's Dwarf Cichlid *(Apistogramma ortmanni) see Apistogramma*

Oryza LINNÉ, 1753; Rice. G of the Poaceae* comprising 25 species. Found in tropical and subtropical parts of the earth. Biennial or annual grasses. Flowers hermaphroditic enclosed in glumes. Perianth consists of a glume. Stamens 6; carpels 3, fused, superior. The fruit wall is fused fast with the seed-shell. The various species of rice grow in marshes and in water, although many also grow in dry habitats. The most important species is:

– *O. sativa* LINNÉ: Rice. Rice is the most important cereal in the tropics and subtropics, but it is also cultivated in a few temperate regions, such as S Europe and N America. The main areas of cultivation are in SE Asia. Whether rice can be cultivated or not depends on an area having an average temperature of 20 °C for at least 3 months of the year. The thousand or so cultivated species can all be traced back to several wild species. Water rice is the most important; during its vegetative period, its bottom parts must stand in water, so the seeds are sown in well irrigated places. In Asia the rice fields are formed in terraces on hilly land, separated by embankments. The seeds are sown in irrigated seed-beds, and they are transplanted after 30–50 days. In America, the seeds are sown directly into the rice fields. Depending on variety, the rice will be harvested 3–7 months later. Rice fields harbour large numbers of species of fish and aquatic plants. Mountain rice, which consists of less productive forms of rice does not require as much moisture as water rice.

Oryzias JORDAN and SNYDER, 1906; Ricefishes. G of the Cyprinodontidae*. There are roughly 10 species widely distributed in tropical and subtropical Asia and closely related with the African Lamp-eyes *(Aplocheilichthys*, Procatopus*), Oryzias* species live in flooded rice fields, in the lower courses of rivers and in river-mouth areas (Estuaries*). They are very small fish, only a few cm long, and with a characteristic shape. The dorsal profile is virtually straight, a little bit concave at the nape; the ventral profile on the other hand is much more arched. The head is very pointed, the mouth broad, and the eyes large. The D is smaller than the An and is positioned above the rear part of it. Vs often tiny. C fan-shaped, very well developed, sometimes slightly concave at the rear edge. Unlike its African relatives, the sexes are about equal in size or the females are bigger. *Oryzias* likes to swim and is always found in the surface waters. Easy to care for in the aquarium: provide medium hard water at a temperature of 19–24 °C. They are shoaling fish, so it is best to buy several specimens; solitary specimens will not develop well. The reproduction of Ricefishes is especially interesting. Mating normally takes place in the morning, as the day dawns. The male lays his strikingly large anal fin around his partner's belly, forming a kind of bag. The female releases her eggs immediately afterwards; they hang like a bunch of grapes in the genital region of the female, stuck there by the long, sticky threads of the eggs. Oxygen is supplied to the eggs inside the cluster via very fine, stiff processes that extend from the egg surface. The spawn is fertilised in the anal fin bag-shape, although there is some discussion as to whether some species have internal fertilisation as in the Livebearing Tooth-carps (Poeciliidae*). For breeding, soft water is needed (which is in contrast to the normal requirements in captivity). Perhaps this is because in nature something similar happens to stimulate reproduction (maybe rain accumulates and causes flooding). The newly hatched fry have very strong pigments (ie they are dark-coloured), and they swim restlessly around on the surface in an Eel-like fashion. The African relatives already mentioned (*Procatopus* and *Aplocheilichthys*) have the same features. Because the fry expend a lot of energy on swimming, they need a lot of food and only grow slowly to start with.

– *O. javanicus* BLEEKER, 1854); Javanese Ricefish, Blue Eyes. From the Malaccan Peninsula across Java to Lombok. Length to 4 cm. Body a translucent metallic pale blue that shines. Iris and edge of the anal fin similarly blue; otherwise, fins are yellowish (fig. 847).

Fig. 847 *Oryzias javanicus*

– *O. latipes* (SCHLEGEL, 1850); Geisha Girl Medaka, Medaka, Japanese Medaka, Ricefish. From the islands of Japan. Length to 4 cm. Similar to *O. javanicus*, but body more elongate and the unpaired fins are usually covered with dark dots and have orange-coloured edges.

Oryzias minutillus *see* Cyprinodontidae

Oryziatinae, Sub-F, *see* Cyprinodontidae

Oscar *(Astronotus ocellatus) see Astronotus*

Osmeridae, F, *see* Salmonoidei

Osmerus (various species) *see* Salmonoidei

Osmoregulation. The control of the osmotic ratios in the body, especially water and salt, which must be kept constantly within certain limits. Freshwater fish have, in their blood and tissues, a higher concentration of salts and other osmotically effective substances than the surrounding water. Because of this difference in concentration, water forces its way through the body surface and has to be actively removed via the urinary organs*. So freshwater fish produce large amounts of urine, although they do not drink.

In marine bony fish, the blood concentration is less than the concentration of the surrounding sea water. Water is constantly being removed from the body and the fish have to make good this loss by drinking. They are therefore forced to take in salts too, which must be removed once again by special salt-secreting cells in the gills. Very little urine is produced. Sensitivity towards fluctuating salinity levels in the water is at its lowest among species that live in brackish water and tidal zones, as well as among fish that migrate between freshwater and sea water.

In marine sharks and rays, the osmotic level of the blood is somewhat higher than that of the sea water because of a high urea concentration.

Osmosis. The diffusion of substances through a membrane. If the membrane is sufficiently permeable, the solvent and dissolved particles (molecules or ions) of the two adjacent solutions will move in the direction of the lower concentration, until the two concentrations become equal. Biological membranes, eg the cell membrane, are usually only osmotically permeable for a particular size of particle. They are semipermeable. Water always flows in the direction of the higher concentration of dissolved particles. Thus, in relation to the number of dissolved particles in the cytoplasm (ie its *osmotic value*), *osmotic pressure* will be produced in the cells.

Osmotic processes occur in all living organisms. It is of fundamental importance in the water economy of the cells and tissues, in fact for the whole organism Osmoregulation*). Some species, eg Ribbonworms (Cestodes*), feed osmotically through the surface of the body. Most plants support their leaves and stems by means of the osmotic pressure of their cells (Turgor*).

Osphronemidae, F, *see* Anabantoidei

Osphronemus LACÉPÈDE, 1802; Gourami. G of the Osphronemidae, Sub-O Anabantoidei*. Originally found only in the Greater Sunda Islands. They live in rivers, lakes, weedy ponds and in brackish water. Have been introduced into a great many countries in SE Asia and Australia. Length up to 60 cm. Body more or less oval, flattened laterally. Head relatively small. The lower jaw forms a definite chin as the fish gets older. The D and the An especially are elongated, pointed at the rear in the males. The rear edge of the C is straight or slightly concave. Vs elongated like a thread. Large scales and LL distinct. In older specimens, the colouring is brown, yellow on the ventral surface or silver. Juveniles have dark transverse bands on the body. There is a dark blotch, edged in yellow or silver, above the soft part of the An. Fins bluish, only the V is orange.

Juveniles only are suitable for keeping in a room aquarium. Temperature 20–25 °C. *Osphronemus* is mainly a plant-eater. Salad and rolled oats are accepted, with a supplementary supply of live food, as well as insects. Gouramis build bubble-nests, mostly threaded through with plant parts. The only species is:

– *O. goramy* LACÉPÈDE, 1802; Gourami (fig. 848).

Ostariophysi, Sub-O, *see* Osteichthyes

Osteichthyes; Bony Fish. Primitive Cl of the Gnathostomata*. Fish whose skeleton is made entirely of a bony substance. In the embryo the individual bones are laid

Fig. 848 *Osphronemus goramy*

down as cartilaginous moulds, and in the course of development replaced by bones, although in some cases, the individual bones develop directly in the connective tissue. In this way, the skeleton is a uniform piece of apparatus composed of replacement bone and connective tissue bone. In individual groups, particularly primitive ones, the process of ossification is not yet complete or, as a secondary adaptation in the course of phylogenesis, it has been suppressed. In such cases, many parts of the skeleton remain cartilaginous. It is also important to mention here that the replacement bones are usually older in terms of phylogenesis than the connective tissue bones, and so often provide very valuable clues for the systematician. At the same time, it is essential to rid oneself of the commonly held view that cartilage was first 'invented' during the evolution of fish, to be followed later by bones. A great number of fossils brove the exact opposite. The skeleton of the very primitive Agnatha* is largely made of bone, for example. In the epochs that followed, various processes took place within the individual groups, so that sometimes the bony elements were more strongly developed and the cartilaginous ones suppressed, whereas at other times the bony skeletal parts were reduced and the cartilaginous parts strengthened. So, we should be able to conclude from this that it is not right to regard the cartilaginous fish (Chondrichthyes*) simply as the more primitive fish, and the bony fish as the more highly developed.

The most important parts of the skeleton are the cranium or skull, the vertebral column, the shoulder girdle and the pelvic girdle and the skeletal elements of the paired and unpaired fins. The cranium is made up of the neurocranium and the viscerocranium. Both sections have numerous additional skeletal elements (connective tissue bones) that make them much more complicated than in the cartilaginous fish. The jaws are bounded by several, usually toothed, connective tissue bones, the link between the jaws being made of the replacement bones, quadratum and articular bone. These correspond to the upper jaw (palatoquadratum) and lower jaw (mandibular bone) of the bony fish. The viscerocranium is joined at the back to the neurocranium by an arch of bone which is in part formed by the jaw process (hyomandibula). They are connected at the front mainly by ligaments. A new development is the group of bones behind the jaw process that form the operculum (fig. 849). The top section of the shoulder girdle is usually joined to the cranium. Only in primitive fish does the rear end of the vertebral column continue into the upper part of the C. In the higher Osteichthyes, the vertebral column ends with a slight bend upwards in the root of the tail, and the C in this case is symmetrical. The fins are supported by bony fin rays. Two types are distinguished: unarticulated, spine-like hard rays and articulated soft rays that are usually divided like a fan on the outside edge. 'Fish-bones' are y-shaped bony rods that lie between the muscle segments.

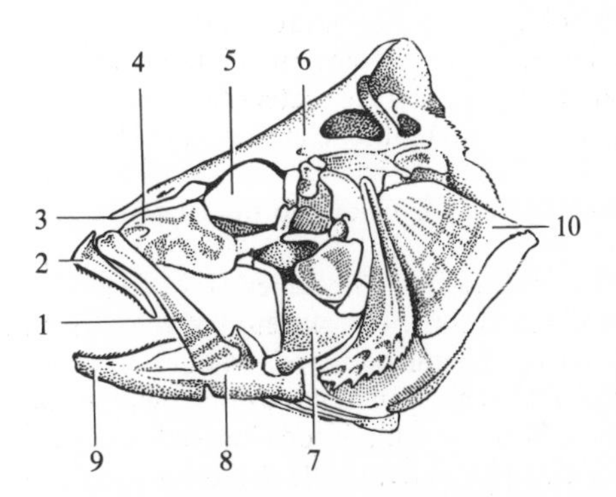

Fig. 849 Skull of a higher bony fish 1 Maxilla (upper jaw bone) 2 Premaxilla 3 Nasal bone 4 Ethmoid bone (lacrimal bone) 5 Orbit 6 Parietal bone 7 Quadrat bone 8 Articular bone 9 Dental bone (lower jaw) 10 Opercular bone

The body is usually spindle-shaped, more rarely snake-like, or it may be laterally or dorso-ventrally flattened. It is divided into a head, torso and tail region. The head is characterised by the mouth (which is usually located at the front end), the depressions of the olfactory organs, the lidless eyes and the operculum. Barbels may also be formed. The tail region, together with the C, is the main organ of propulsion. The D and An help to some extent in propulsion, but their main function is stability. The paired fins are the most important means of steering. In many bony fish, the D is divided into 2 fins, and behind the D in some fish groups there is a small adipose fin, usually without rays. The skin is full of glands. Typical bony fish are covered in bony scales, which are arranged in rows that overlap from front to back like roof-tiles. They are formed from the subcutaneous tissue and may be round-edged or comb-edged (Scales*). In many groups, the covering of scales is reduced. Some primitive Osteichthyes have ganoid scales (Scales*) instead of bony scales. The intestinal canal consists of a short oesophagus, a stomach (may be absent), a small intestine usually in coils, and a hind-gut. In the intestinal canal of some primitive Actinopterygia and Sarcopterygia there is a spiral valve. There may be some protrusions like blind sacs between the stomach and small intestine. The swimbladder, the hydrostatic organ in the bony fish, is a protrusion of the oesophagus. Some fish retain the link with the oesophagus for life in the form of a canal (physostomatous fish);

in others the link has regressed (physoclistous fish). Many Osteichthyes have no swimbladder at all, and in some primitive groups it can function as a lung, in which case it will also lie on the ventral side of the intestine. Apart from the typical sense organs, the majority of bony fish also have a touch sense organ that operates from a distance. It consists of branched canals in the head region and a length-wise canal on each flank. These canals are linked to the outside by pores which traverse the scales along the flanks. Sense buds lie in the canals. When determining species, important points to look at are the shape of the body, the fin arrangement, the type of scales, skeletal features and dentition. Increasingly in recent years, patterns of behaviour and genetic and immunological characters have been taken into account when attempting clarification of relationships.

The following Sub-Cls belong to the Cl of the bony fish.

– *Actiopterygia*: Ray-fins. The paired fins in particular are typical of this Sub-Cl. The skeleton of these fins is made up of a row of small bony rods (radialia), roughly equal in size and with a parallel arrangement in one plane. In the typical case, these rods insert directly into the shoulder girdle or pelvic girdle. On the outside, the bony fin rays sit directly on top of them. The olfactory organ is always in the form of a depression on the top of the head. As a general rule, this depression has an entrance and a separate exit. There is no connection between the olfactory organ and the mouth. In some primitive groups there are ganoid scales, otherwise cycloid or ctenoid ones. The earliest Ray-fins known date back to the middle Devonian period. More than 1,000 fossil Ga are known. The number of modern species lies somewhere between 20,000 and 30,000. This makes the Ray-fins the group of jawed animals (Gnathostomata) with by far the most species. How the Sub-Cl should be divided up further is still uncertain. REGAN sub-divides them into the Palaeopterygii (old fins) and the Neopterygii (new fins). Others have adopted 3 groups: the Chondrostei (cartilaginous-ganoid), the Holostei (bony ganoid) and the Teleostei (higher bony fish). STENSIÖ and BERG are of the opinion that the Chondrostei gradually merge into the Holostei. Even more recent researches (eg PATTERSON, 1973; H. L. JESSEN, 1973) have not been able to clear up the uncertainties. Here, the Chondrostei, Holostei and Teleostei are treated as separate groups with the rank of a cohort. As the following system shows, the Teleostei in this dictionary are divided into 8 Sup-Os (this concurs with the suggestions published in 1966 by P. H. GREENWOOD, D. E. ROSEN, St. H. WEITZMAN and G. S. MYERS): Elopomorpha, Clupeomorpha, Osteoglossomorpha, Protacanthopterygii, Ostariophysi, Paracanthopterygii, Atherinomorpha, Acanthopterygii. In the meantime, D. E. ROSEN has put forward a completely new phylogenetic system (Zool. Journ. of the Linnean Soc. 53, Suppl. 1, 1973), but researches into this are still very much going on, so that it is impossible to conclude any generalization from them.

– *Sarcopterygia*; Muscle-fins. The skeleton of the paired fins consists mainly of an articulated axis with skeletal rods arranged on both sides (archipterygium-type). The base of the fins may take the form of muscular stumps. The olfactory organ is connected to the mouth (with the exception of the Coelacanthiformes). The swimbladders are formed like lungs and enter the pharynx on the ventral side via an air canal. The intestine has a spiral valve. The two Sup-Os Dipnoi (Lungfishes) and Crossopterygii (tuft fins) reached their peak in ancient times, with only 3 Ga of Dipnoi and only one species of Crossopterygii surviving into modern times. The Crossopterygii are more closely related to the land vertebrate animals than to the fishes. For further details see under Ceratodiformes and Coelacanthiformes.

CL Osteichthyes (Bony fish)
- SUB-CL Actinopterygia (Ray-fins)
  - COH Chondrostei (Cartilaginous-ganoid)
    - O Polypteriformes (Bichir-types)
      - F Polypteridae (Bichirs)
    - O Acipenseriformes (Sturgeon-types)
      - F Acipenseridae (Sturgeons)
      - F Polyodontidae (Paddle-fishes)
  - COH Holostei (Bony ganoid)
    - O Amiiformes (Bowfin-types)
      - F Amiidae (Bowfins)
    - O Lepisosteiformes (Gar Pike-types)
      - F Lepisosteidae (Gar Pikes)
  - COH Teleostei (Higher bony fish)
    - SUP-O Elopomorpha (Tarpon-types)
      - O Elopiformes (Tarpon-types)
        - F Elopidae (Tenpounders)
        - F Megalopidae (Tarpons)
        - F Albulidae (Bone Fish)
      - O Anguilliformes (Eel-types)
        - SUB-O Anguilloidei (Eel-related species)
          - F Anguillidae (True Eels)
          - F Moringuidae (Worm or Spaghetti Eels)
          - F Muraenidae (Muraenas, Moray Eels)
          - F Congridae (Conger Eels)
          - F Ophichthidae (Serpent Eels)
          - F Synaphobranchidae (Deep-sea Eels)
          - F Serrivomeridae (Deep-sea Eels)
          - F Nemichthyidae (Snipe Eels)
        - SUB-O Saccopharyngoidei (Swallowers)
      - O Notacanthiformes (Deep-sea Spiny Eel types)
    - SUP-O Clupeomorpha (Herring-like Fishes)
      - O Clupeiformes (Herring-types)
        - F Denticipitidae (Freshwater Herrings)
        - F Clupeidae (Herrings)
        - F Engraulidae (Anchovies)
        - F Chirocentridae (Wolf-herrings)
    - SUP-O Osteoglossomorpha (Bony Tongue-types)
      - O Osteoglossiformes (Bony Tongue-types)
        - F Osteoglossidae (Bony Tongues)
        - F Pantodontidae (Chisel-jaws)
        - F Notopteridae (Featherbacks)
        - F Mormyridae (Elephant-nosed Fishes)
        - F Gymnarchidae
        - F Hiodontidae (Moon-eyes)
    - SUP-O Protacanthopterygii
      - O Salmoniformes (Salmon-types)
        - SUB-O Salmonoidei (Salmon-types, related species)
          - F Salmonidae (Salmons)
          - F Osmeridae (Smelts)
        - SUB-O Galaxioidei

F Galaxidae (Galaxids)
SUB-O Esocoidei (Pikes and Mudfishes, related species)
F Esocidae (Pikes)
F Umbridae (Mud Minnows)
SUB-O Stomiatoidei
F Gonostomatidae (Bristlemouths)
F Sternoptychidae (Deep-sea Hatchetfishes)
SUB-O Myctophoidei (Lantern-fishes, related species)
F Synodontidae (Lizard Fishes)
F Bathypteroidae
F Myctophidae (Lantern-fishes)
O Cetomimiformes (Whalefish-types)
F Cetomimidae (Whalefishes)
F Giganturidae (Telescope-eyed Fishes)
O Gonorynchiformes (Sand Eel types)
F Gonorynchidae (Sand Eels)
F Chanidae (Milk-fishes)
F Kneriidae
F Phractolaemidae (African Mudfishes)
SUP-O Ostariophysi
O Cypriniformes (Carp-types)
SUB-O Characoidei (Characins, related species)
F Characidae (True Characins)
F Serrasalmidae (Piranhas)
F Gasteropelecidae (Flying Hatchetfishes)
F Erythrinidae
F Ctenoluciidae (Pike-characins)
F Chalceidae
F Crenuchidae (Sail-fin Characins)
F Characidiidae (Characins)
F Lebiasinidae
F Anostomidae
F Hemiodidae
F Curimatidae
F Citharinidae
SUB-O Gymnotoidei (Knifefishes, related species)
F Gymnotidae (Knifefishes)
F Electrophoridae (Electric-Eels)
F Apteronotidae American Knifefishes)
F Rhamphichthyidae (American Knifefishes)
SUB-O Cyprinoidei (Carps, related species)
F Cyprinidae (Carps)
F Catostomidae (Suckers)
F Gyrinocheilidae (Malay Cyprinids)
F Homalopteridae (Hillstream Loaches)
F Cobitidae (Loaches)
O Siluriformes (Catfish-types)
F Ictaluridae (Catfishes)
F Bagridae
F Siluridae (True Catfishes)
F Schilbeidae
F Pangasiidae
F Amphiliidae
F Clariidae (Airbreathing Catfishes)
F Chacidae (Frogmouth Catfishes)
F Malapteruridae (Electric Catfishes)
F Mochokidae (Scaleless Catfishes)
F Ariidae (Sea Catfishes)
F Doradidae (Thorny Catfishes)
F Auchenipteridae
F Aspredinidae (Frying-pan or Banjo Catfishes)
F Plotosidae (Catfish Eels)
F Pimelodidae
F Helogenidae
F Trichomycteridae (Parasitic Catfishes)
F Callichthyidae (Armoured Catfishes)
F Loricariidae (Mailed Catfishes)
SUP-O Paracanthopterygii
O Percopsiformes (Cave-dwelling Amblyopsids)
F Amblyopsidae (Blindfishes, North American Cave-dwelling Fishes)
F Aphredoderidae (Pirate Perches)
F Percopsidae (Trout Perches)
O Batrachoidiformes (Toadfish-types)
F Batrachoididae (Toadfishes)
O Gobiesociformes (Coastal Cling-fish-types)
F Gobiesocidae (Cling-fishes)
O Lophiiformes (Angler-fish types)
F Lophiidae (Anglers or Goosefishes)
F Antennariidae (Frogfishes)
F Ogcocephalidae (Batfishes)
SUB-O Ceratioidei (Deep-sea Angler-fishes, related species)
O Gadiformes (Cod-types)
F Muraenolepididae
F Gadidae (Cods)
F Ophidiidae (Brotulids, Cusk-eels)
F Carapidae (Pearlfishes)
F Zoarcidae (Eelpouts)
F Macrouridae (Grenadiers)
SUP-O Atherinomorpha (Silverside types)
O Atheriniformes (Silversides, Sand-smelt-types)
SUB-O Exocoetoidei (Flying Fishes, related species)
F Exocoetidae (Flying Fishes)
F Belonidae (Gardfishes, Needlefishes or Long-toms)
F Scomberesocidae (Sauries)
SUB-O Cyprinodontoidei (Tooth-carps, related species)
F Adrianichthyidae
F Cyprinodontidae (Egg-laying Tooth-carps)
F Goodeidae (Goodeids)
F Anablepidae (Four-eyes, Four-eyed Fishes)
F Jenynsiidae (One-sided Livebearers)
F Poeciliidae (Livebearing Tooth-carps)
SUB-O Atherinoidei (Silversides, related species)
F Melanotaeniidae (Rainbowfishes)
F Atherinidae (Silversides, Sand-smelts)
F Phallostethidae
SUP-O Acanthopterygii
O Beryciformes
F Berycidae (Alfonsinos)
F Monocentridae (Pine-cone Fishes)
F Anomalopidae (Lanterneye Fishes)
F Holocentridae (Soldierfishes)
O Zeiformes
F Zeidae (Dories)
F Caproidae (Boarfishes)
O Lampridiformes (Opah-types)
F Lampridae (Opahs, Moonfishes)
F Lophotidae (Crestfishes)
F Trachipteridae (Deal-fishes)

F Regalecidae (Oarfishes, Ribbonfishes)
O Gasterosteiformes (Stickleback-types)
SUB-O Gasterosteoidei (Sticklebacks, related species)
F Gasterosteidae (Sticklebacks)
F Aulorhynchidae (Pacific Sticklebacks)
SUB-O Aulostomoidei (Trumpetfishes, related species)
F Aulostomidae (Trumpetfishes)
F Fistulariidae (Flutemouths)
F Macrorhamphosidae (Snipe-fishes)
F Centriscidae (Shrimpfishes)
SUB-O Syngnathoidei (Pipefishes, related species)
F Solenostomidae (Ghost Pipefishes)
F Syngnathidae (Pipefishes, Sea-horses)
O Channiformes (Snake-head types)
F Channidae (Snake-heads)
O Synbranchiformes
F Synbranchidae (Swamp Eels)
F Amphipnoidae (Cuchia)
O Scorpaeniformes (Scorpion-fish types)
SUB-O Scorpaenoidei (Scorpion-fishes, related species)
F Scorpaenidae (Scorpionfishes)
F Triglidae (Gurnards, Sea Robins)
F Caracanthidae (Orbicular Velvetfishes)
F Synancejidae (Stonefishes)
SUB-O Cottoidei (Bull-heads, related species)
F Cottidae (Bull-heads)
F Agonidae (Poachers)
F Cyclopteridae (Lumpsuckers)
O Dactylopteriformes (Flying Gurnard types)
F Dactylopteridae (Flying Gurnards)
O Pegasiformes (Sea Moth types)
F Pegasidae (Sea Moths)
O Perciformes (Perch-types)
SUB-O Percoidei (Perches, related species)
F Centropomidae (Snooks)
F Theraponidae (Tigerfishes)
F Kuhliidae (Flagfishes, Flagtails)
F Centrarchidae (Sunfishes)
F Priacanthidae (Big-eyes)
F Apogonidae (Cardinalfishes)
F Percidae (True Perches)
F Pomatomidae (Bluefishes)
F Echeneidae (Remoras, Shark Suckers)
F Carangidae (Jacks, Scads, Pompanos)
F Coryphaenidae (Dolphinfishes)
F Leiognathidae (Slimies, Soapies)
F Lutjanidae (Snappers)
F Lobotidae (Triple-tails)
F Gerriedae (Mojarras)
F Pomadasyidae (Grunts)
F Sparidae (Sea Breams, Porgies)
F Sciaenidae (Croakers, Drums)
F Mullidae (Goatfishes, Red Mullets)
F Monodactylidae (Fingerfishes)
F Toxotidae (Archerfishes)
F Kyphosidae (Rudderfishes)
F Ephippidae (Spadefishes)
F Scatophagidae (Scats)
F Chaetodontidae (Butterflyfishes)
F Nandidae (Leaf-fishes, Nanders)
F Embiotocidae (Surfperches, Sea Perches)
F Cichlidae (Cichlids)
F Pomacentridae (Damselfishes)
F Cirrhitidae (Hawkfishes)
SUB-O Mugiloidei (Grey Mullets, related species)
F Mugilidae (Grey Mullets)
SUB-O Sphyraenoidei (Barracudas, related species)
F Sphyraenidae (Barracudas)
SUB-O Polynemoidei (Thread-fins, related species)
SUB-O Labroidei (Wrasses and Parrotfishes, related species)
F Labridae (Wrasses)
F Scaridae (Parrotfishes)
SUB-O Trachinoidei (Weaverfishes, related species)
F Trichodontidae (Sandfishes)
F Opistognathidae (Sawfishes)
F Trachinidae (Weaverfishes)
F Trichonotidae (Sand Divers)
F Dactyloscopidae (Stargazers)
F Uranoscopidae (Stargazers)
SUB-O Notothenioidei (Icefishes, related species)
F Channichthyidae (Icefishes)
SUB-O Blennioidei (Blennies, related species)
F Blennidae (Blennies)
F Anarhichadidae (Sea Catfishes or Sea Wolffishes)
F Clinidae (Scaled Blennies)
F Stichaeidae (Pricklebacks)
F Pholididae (Gunnels)
SUB-O Ammodytoidei (Sand Eels, related species)
F Ammodytidae (Sand Eels, Sand Lances)
SUB-O Callionymoidei (Dragonets, related species)
F Callionymidae (Dragonets)
SUB-O Gobioidei (Gobies, related species)
F Gobiidae (Gobies)
F Gobioididae (Eel-like Gobies)
SUB-O Acanthuroidei (Surgeonfishes, related species)
F Acanthuridae (Surgeonfishes)
F Siganidae (Rabbitfishes, Spinefeet)
SUB-O Scombroidei (Mackerels, related species)
F Gempylidae (Snake Mackerels)
F Trichiuridae (Cutlass-fishes, Hair-tails)
F Scombridae (Mackerels)
F Xiphiidae (Swordfishes)
F Istiophoridae (Billfishes)
SUB-O Stromateoidei (Butterfishes, related species)
SUB-O Anabantoidei (Climbing Perches, related species)
F Anabantidae (Climbing Perches)
F Belontiidae (Siamese Fighting Fishes)
F Helostomatidae (Kissing Gouramis)
F Osphronemidae (Gouramis)
SUB-O Luciocephaloidei (Pike-heads, related species)
F Luciocephalidae (Pike-heads)
SUB-O Mastacembeloidei (Freshwater Spiny Eels, related species)
F Mastacembelidae (Spiny Eels)
O Pleuronectiformes (Flat-fish types)
F Bothidae (Left-eye Flounders)

F Pleuronectidae (Flatfishes)
F Soleidae (True Soles)
F Cynoglossidae (Tongue-soles)
O Tetraodontiformes (Pufferfish-types)
F Triacanthidae (Triple-spines)
F Balistidae (Triggerfishes, Filefishes)
F Ostraciontidae (Trunkfishes, Boxfishes)
F Tetraodontidae (Pufferfishes)
F Triodontidae (Three-toothed Puffer)
F Diodontidae (Porcupine-fishes)
F Molidae (Marine Sunfishes)
SUB-CL Sarcopterygia (Muscle-fins)
SUP-O Dipnoi (Lungfishes)
O Ceratodiformes
F Ceratodidae (Australian Lungfishes)
F Protopteridae (African Lungfishes)
F Lepidosirenidae (South American Lungfishes)
SUP-O Crossopterygii (Tuft-fins)
O Coelacanthiformes (Coelacanth-types)
F Coelacanthidae (Coelacanths)

Osteochilus GÜNTHER, 1868. G of the Cyprinidae*, Sub-F Cyprininae. These fish are widely distributed throughout Indonesia where they live in rivers, lakes and small stretches of water. There are large numbers of different species, and they are shoaling fishes. The G includes small, medium-sized and large species, among which there are some important edible fish. The body is elongate and flattened laterally. Older specimens look more compressed and become tall. The mouth is terminal and has protrusible lips beset with papillae, and together with the sharp-edged lower jaw, they form a kind of scraper for grazing on plant and animal growths. Barbels present and LL complete. The D is well formed and is inserted in front of the Vs; it is longer than the An. *Osteochilus* species are very active swimmers and need roomy tanks with loose-knit groups of plants and suspended root-work. The substrate should be made up of clean, not too fine sand, since the fish like to probe. Temperature 22–24 °C. Feed with live and dried food, plus additional plant food. There are no details available regarding breeding.
– *O. hasselti* (CUVIER and VALENCIENNES, 1842); Hasselt's Bonylip. Thailand, Indonesia, very common. Length up to 30 cm. Dorsal surface olive-green, flanks brassy-yellow with a silver sheen. Ventral surface pale. Young fish have a dark fleck at the bottom of each scale. Fins orange to red, D and C yellowish. Becomes increasingly paler with age.
– *O. vittatus* (CUVIER and VALENCIENNES, 1842); Bony-lipped Barb. Greater Sunda Islands, especially in rivers. Length to 25 cm. Dorsal surface greenish, flanks a silvery iridescent colour with a broad, black longitudinal band running across, extending from the tip of the snout to the C. Fins colourless, D spotted.

Osteoglossidae, F, *see* Osteoglossiformes

Osteoglossiformes; Bony Tongue-types. O of the Osteichthyes*, Coh Teleostei. Freshwater fish of the tropical regions of S America, SE Asia, Africa, Australia and New Guinea. The number of modern species is small, and fossil forms are known from the Eocene period. 2 very different Fs belong to the O, although a very primitive denture is common to both. It consists of the parasphenoid bone beset with teeth and the toothed tongue; these are positioned opposite one another and act as the biting surfaces. They also have paired tendon bones in common that develop from the 2nd gill arch on the throat side. Other important features include the musculature of the pharynx, the anatomy of the inner ear and the way in which the swimbladder is connected to the inner ear. In 1973, GREENWOOD was able to show that the Notopteridae and Hiodontidae, which used to be grouped under the Osteoglossiformes, are in fact closely related to the Mormyridae and Gymnarchidae. So they are now placed in the Mormyriformes*. The Osteoglossiformes and the Mormyriformes together form the Sup-O Osteoglossomorpha.
– *Osteoglossidae*; Bony Tongues. The 5 modern species are all elongated fish with large, distinct scales, composed of many parts. The bones on the upper side of the cranium are also sculptured. The largest species, *Arapaima gigas* (fig. 850), is usually about 2 m long, sometimes up to 3 m. It lives in the rivers of tropical S America (Amazon Basin and Orinoco). The D and An are long, and are so close to the C that when the fish is swimming along it looks as if it has an almost uniform fin fringe. In the front part of the body, the scales are greenish and outlined in red, but towards the tail they become increasingly red in tone. Specimens brought onto land shimmer in all colours of the rainbow. *Arapaima* eats fish and shrimps. It spawns near the bank in self-made troughs. The male takes care of the spawn and guides the brood around, a secretion from the skin on the male's head providing the source of contact (ie

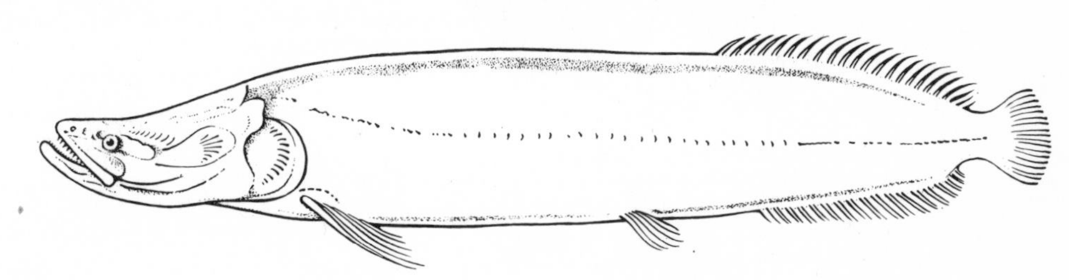

Fig. 850 Arapaima gigas

even in deep water the fry can detect the close proximity of the father, thus ensuring all of them keep in contact). *Arapaima* is important economically; its reddish, fat-rich flesh is immediately of value, or it can be cut up into strips, salted and dried.

Closely related to *Arapaima* is the more compressed, smaller African member of the F *Heterotis niloticus.* Another well-known species is the Green Arowana *(Osteoglossum bicirrhosum)* (fig. 851), which grows up to 60 cm long. Its typical features include the two

Fig. 851 *Osteoglossum bicirrhosum*

strong chin barbels, the upward-slanting mouth slit, the deep red dots on the scales, and the lovely gleaming colouring that shimmers in all the rainbow colours. Its distribution area extends from the Amazon Basin to Guyana. The Green Arowana, like its relatives from the Indonesian-Australian realm (G *Scleropages*), is a mouth-brooder.

MAUPIN has observed that the male lets the first 8–10 cm long fry out of his mouth after 55 days. For a while afterwards, the fry still sought refuge in his mouth, and from the 64th day onwards they were eating *Daphnia* and *Artemia.* Young specimens survive well in normal warm water aquaria, and will accept live food of every sort. Large specimens are showpieces in public aquaria.

– *Pantodontidae*; Butterfly-fishes. The only species of this F, the 10 cm long Butterfly-fish *(Pantodon buchholzi)* (fig. 852) lives in still waters in tropical W Africa.

Fig. 852 *Pantodon buchholzi*

The body is shaped like a boat, ie flattened on top, narrowing like a keel on the ventral side. Ps elongated like wings, Vs small with 4 very long fin rays with black-and-white circles all the way up them. The An and C are broadened out like flags. The upper surface of the Butterfly-fish is brownish, the remainder being flecked in various ways. It is a surface-water fish and eats insects. It can jump out of the water in a flat trajectory, spread its 'wings' and glide through the air for a short stretch. During courtship, the male rides on top of the female, and during mating they both entwine round each other. The oil-rich eggs float at the surface, and the fry hatch out after about 3 days. The parents do not take care of either the eggs or the fry. *Pantodon* is often cared for in the aquarium, but so far it has not been bred successfully very often. Slightly acidic, peaty water is needed and a high temperature. Provide a covering of floating plants with plenty of gaps in between, and feed on insect food. This should promote the well-being of your specimens. If breeding, it seems to be very difficult to feed the fry properly. Success has been recorded with Springtails (Collembola*) and small leaf-lice.

Osteoglossomorpha, Sup-O, *see* Osteichthyes

Osteoglossomorphs (Osteoglossomorpha, Sup-O) *see* Osteichthyes

Osteoglossum bicirrhosum *see* Osteoglossiformes

Ostracion LINNAEUS, 1758 (fig. 853). G of the Ostraciontidae*. Several species found in tropical parts of the

Fig. 853 *Ostracion meleagris*

Indo-Pacific. *Ostracion* feeds on small animals which it can blast from the bottom with a jet of water. Length 15–45 cm. The build is typical of the family. Unlike other Ga, there are no spines on the body and no ridge to the back (although a ridge may be suggested by a row of flat knobs in front of the D). Like all Trunkfishes, *Ostracion* makes very great demands on the aquarist. Calm tanks are needed with soft, sandy bottoms and the water quality must be excellent. There must not exist any stiff competition for food, and a good variety of food must be offered. *Ostracion* is particularly prone to damaging the delicate skin on top of its bony armour and to being attacked by skin parasites. Do not keep *Ostracion* species in the same tank as Cleaner Wrasses *(Labroides*).* Sick and dying Trunkfishes sometimes release a dangerous fish poison. Sunken-in flanks are a sign of a bad state of health – in healthy specimens the sides of the belly are straight. Most commonly imported is:

– *O. cubicus* LINNAEUS, 1758; Blue-spotted Boxfish; Coffre; Black-spotted Boxfish. Tropical Indo-Pacific. Length up to 45 cm. Yellow with black dots that become outlined in blue as the fish gets older.

Ostraciontidae; Trunkfishes, Boxfishes (fig. 854). F of the Tetraodontiformes*, Sub-O Balistoidei. Small, armoured fish from tropical seas, mainly from the reefs. Body and head encapsulated in a stiff, hard, bony armour which may be 3, 4 or 5-cornered in cross-section. It is made up of polygonal dermal bones that are bound very firmly to one another. At the edges of the armour there are very often spiny processes pointing forwards

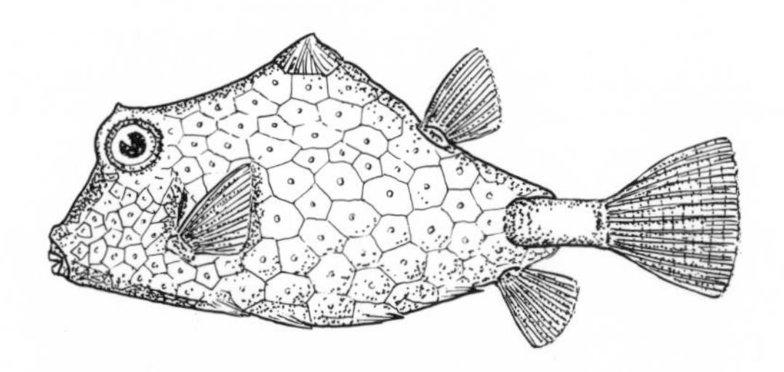

Fig. 854 Ostraciontidae-type

or backwards. The small mouth, the eyes, the anus, the short D and An, the Ps and the caudal peduncle are all still able to move in the armour. In the Sub-F Aracaninae, the armour is shorter, ie the whole tail region together with the D and An can be moved freely. The fish is able to move by beating its Ps, and the D and An, in a propeller-like action. The C is only used for steering and there are no Vs. Trunkfishes glide gently through the water; they are highly manoeuvrable and often drift along for considerable stretches. In their dentition there are about 10 teeth, very close together, in each of the jaws. For further information about this interesting F turn to the Ga *Acanthostracion**, *Lactoria**, *Ostracion**, *Tetrosomus**.

Ostracoda; Ostracods (fig. 855). O of the Crustacea* with many more than a 1,000 species distributed world-

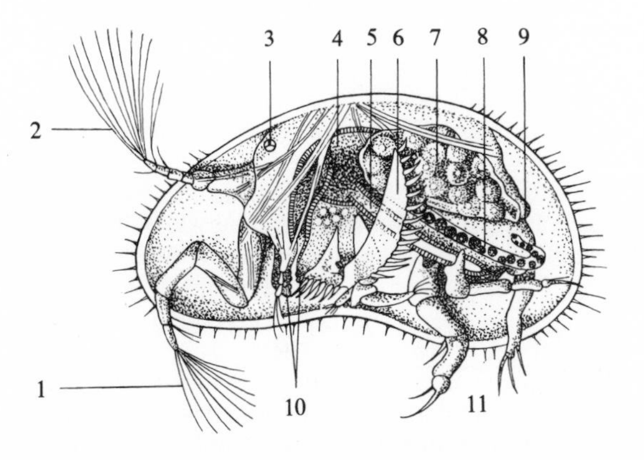

Fig. 855 Ostracoda. Organisation of an Ostracod. 1 2nd antenna 2 1st antenna 3 Eye 4 Stomach 5 Liver 6 Gill 7 Oviduct 8 Ovary 9 Anus 10 Mouth parts 11 Extremities

wide. Most live in the sea, but some live in fresh water. Despite being found almost everywhere, Ostracods are usually overlooked, as they only have a body size of 1–2 mm. They are completely encased in a hinged shell, which makes them look, from the outside, like small Bivalves. During locomotion, normally only the antennae are extended from the shell. Most species crawl along the bottom and on water plants, whereas others are skilled swimmers. Pelagic forms (Pelagial*) occur in the sea (open-water forms). Ostracod eggs in particular, but also the larvae and the adult animals, have an astonishing resistance to environmental influences. The water where they live can dry up or freeze completely with no damage being done to them. As a fish food, Ostracods mostly play a minor role. Larger specimens are suitable for observation purposes in a pond-aquarium (use a magnifying glass!).

Ostracods, O, *see* Ostracoda

Ostrea LINNAEUS, 1758; Oyster. G of the marine Bivalves (Bivalvia*, O Anisomyaria). The shell is unsymmetrical and the valves coarse. The deeper, left half of the shell is cemented to rocks below the tidal zone. The Common Oyster (*O. edulis* LINNAEUS, 1758) has a shell diameter of 10 cm and is cultivated as a delicacy. Oysters are very bizarre animals, and survive well in the aquarium.

Otocinclus COPE, 1871. G of the Loricariidae*. Small, sucking Catfishes native to central and northern S America. Length up to 6 cm. Peaceful and harmless crepuscular fish that rest during the day, usually against rocks, plant stems or leaves. The body is not as flattened dorso-ventrally as in the members of the G *Loricaria**, but they do have features in common, such as the plating on the ventral surface, the form of the sucking mouth and the arrangement of the fins. However, *Otocinclus* does not have any filamentous elongations of the top and bottom-most caudal rays, and the caudal peduncle is only slightly elongated. Here too, the different number of bony plates in a longitudinal row helps determine species. Preferred temperature around 25 °C. Like the closely related Ga *Farlowella** and *Loricaria**, *Otocinclus* grazes among carpets of algae, but will also eat whiteworms, chopped up *Tubifex, Cyclops, Daphnia*, dried food and rolled oats. So far, 2 species have been bred with success. SCHMIDT made his report of the breeding of *O. affinis* in *Tropicarium Frankfurt.* After a form of love-play which resembled that of the *Corydoras** species, the eggs were laid indiscriminately on panes of glass and parts of plants. The fry hatched out after 2–3 days and were reared on microworms, fine nauplii and grated egg-yolk. The report of a successful rearing of *O. maculipinnis* was made by FRANKE. Here too, the eggs were laid indiscriminately on plants. No care was taken of the brood. The fry hatched out after 2 days and were fed on rotifers and chopped up *Tubifex.* Growth was uninterrupted if filtration was good and the water was changed regularly. Unfortunately, these have been the only successes with breeding, so that *Otocinclus* species are still comparatively rare and sought-after.

– *O. affinis* STEINDACHNER, 1877; Midget Sucker Catfish. From SE Brazil, around Rio de Janeiro. Length up to 4 cm. 23–24 bony plates. Flanks silver-white, and a broad, black longitudinal band extends from the operculum to the insertion of the C. Underside whitish, fins colourless, with no spots in the vertical fins (fig. 856).

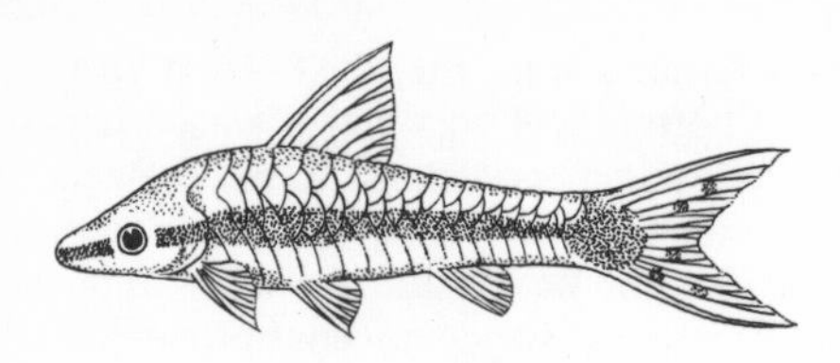

Fig. 856 *Otocinclus affinis*

– *O. maculipinnis* REGAN, 1912. From La Plata region. Length up to 4 cm. Body yellow-olive to yellow-brown with a large number of dark olive-brown flecks and spots distributed irregularly and extending into the glassily transparent fins. Underside a weak yellow (fig. 857). Other species that are occasionally imported include:

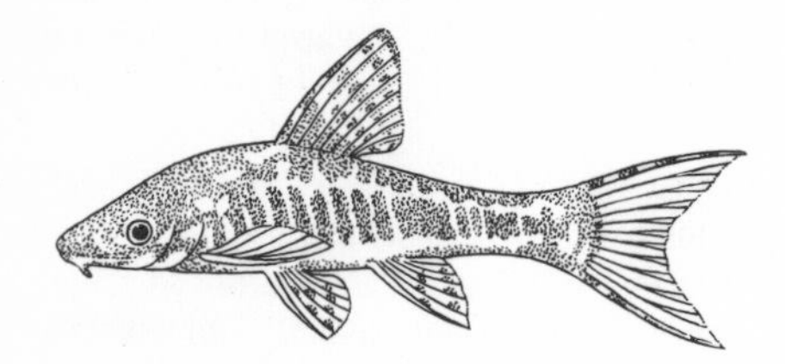

Fig. 857 *Otocinclus maculipinnis*

– *O. flexilis* COPE, 1894. From Rio Grande do Sul. Length up to 6 cm. 25 bony plates. The only colour difference between it and *O. affinis* is that the vertical fins are covered in spots.
– *O. maculicauda* STEINDACHNER, 1876. From SE Brazil. Length up to 6 cm. 24 bony plates. Colouring similar to *O. maculipinnis*, but with a large, dark fleck at the base of the C.
– *O. vittatus* REGAN, 1904. In the region of the Rio Paraguay, Mato Grosso. Length to 5.5 cm. 21–22 bony plates. Very similar in colour to *O. affinis* and *O. flexilis*, except that the base of the C is dark and near the outside edge there are dark spots, often uniting to form transverse bands.

Otoliths. Calcium deposits found in various parts of the inner ear of fish which are important for hearing* and for the function of the balancing organ*. The form and size of the otoliths depends on their position in the inner ear, although they are often species-specific and so have a certain importance in the determination of species. Three types of otoliths are distinguished – the *lapillus*, the *sagitta* (often arrow-shaped) and the *asteriscus* (often star-shaped). Otoliths grow with annual layers, so rather like tree-rings they can be studied to try and ascertain the age of a fish.

Ottelia PERSOON, 1805. G of the Hydrocharitaceae* comprising 40 species. From tropical and subtropical Africa and Asia, and has also spread to other regions in part. One species is said to be native to S America. Ottelias are perennial and annual plants with rosettes. The aquatic leaves are very varied in shape, and some are a floating variety. Flowers hermaphroditic or unisexual, in the latter case dioecious. A varying number of flowers are located in a spathe. Double perianth, calyx and corolla 3-part, petals white or yellow. Hermaphrodite flower: stamens 6–15, carpels 3, fused, inferior. Berry-fruit.

The various species colonise a wide variety of waters, even ones that dry up periodically. The G certainly contains a larger number of useful aquarium plants, but only the following species is regularly kept in botanical gardens and occasionally in aquaria.
– *O. alismoides* (LINNÉ) PERSOON. NE Africa to E Asia, and it has spread into southern European rice fields. Flowers hermaphroditic. Needs to be illuminated for not less than 12 hours (fig. 858).

Fig. 858 *Ottelia alismoides*

Outdoor Breeding Installations *see* Ornamental Fish, Breeding of

Ovary *see* Sex Organs

Ovary (part of the carpel) *see* Flower

Oviduct *see* Sex Organs

Oviparity *see* Reproduction

Ovoviviparity *see* Reproduction

Ovule (part of the carpel *see* Flower

Oxyeleotris BLEEKER, 1874. G of the Gobiidae, Sub-O Gobioidei*. Malaysian Archipelago, Thailand. Lives in fresh water. Its bodily features correspond very much to the G *Eleotris*. The following species is of importance to the aquarist.
– *O. marmorate* (BLEEKER, 1853); Marbled Sleeper Goby. Length to 40 cm. Body elongate, only flattened laterally in the caudal peduncle region. Head and mouth large, lower jaw protruding. Eyes positioned high-up and arching forwards. Colouring variable, very dependent on surroundings. Body usually grey-brown with blurred dark marbling effect. All the fins are brownish with dark flecks. Female less conspicuous. A greedy species that is extremely predatory. Omnivore. Crepuscular fish that burrow into the substrate. 22–28 °C. Not yet bred in captivity.

Oxygaster VAN HASSELT, 1823. G of the Cyprinidae*, Sub-F Abraminae. S and SE Asia. Small, sociable surface-dwelling fish from still and flowing water. The body is longish, very flat from side to side and with an almost straight dorsal profile. The ventral side is very arched with a sharp keel. Mouth superior. The P reaches to the insertion of the poorly developed V, the second ray of which is elongated. In the closely related G *Chela** the forehead is scaled between the eyes, in

Oxygaster there are no scales here. *Oxygaster* can jump well and should be kept as a shoal in broad tanks with plenty of vegetation. There are no particular requirements regarding water quality, but the occasional addition of fresh water does prove advantageous. Temperature 23–26 °C. Food, consisting of live and dried foodstuffs, will only be taken from the top-most water levels. To produce well-developed specimens capable of reproduction it is necessary to feed with insects. For breeding purposes, provide roomy tanks with fine-leaved plants and medium-hard water. The male wraps himself round the female and spawning takes place during the evening and at night. The fry hang by a sticky thread at the surface and must be reared with the finest micro-organisms.

– *O. atpar* (HAMILTON-BUCHANAN, 1822); Blue Keelbelly. The Near East, Bengal. Length up to 6 cm. Dorsal surface pale olive, flanks translucent with a silver sheen, ventral surface white. A green shimmering longitudinal band begins at the level of the V, becoming more prominent towards the rear. Fins a weak yellow.

– *O. bacaila* (HAMILTON-BUCHANAN, 1822). Found almost everywhere in India. To 18 cm. Dorsal surface grey-green, flanks a silvery iridescent colour. A white-green, gleaming longitudinal band begins at the edge of the operculum and continues into the root of the tail. Fins colourless. Not yet bred in captivity.

– *O. oxygastroides* BLEEKER, 1852); Glass Barb. From the Indo-Australian Archipelago. To 20 cm. Dorsal surface yellow-brown, flanks glassily transparent, silver-white with a blue-green shimmer. A silver shining stripe extends from the edge of the gill-covering to the root of the tail. Fins colourless with fine dots.

Oxygen Content. Oxygen is a chemical element, symbol O, which is essential in the respiration of all higher living organisms. In the atmosphere, oxygen accounts for a little less than 21 % by volume; on the other hand, in water, the amount of oxygen varies between 1 and 0 % in volume, depending on pressure, temperature, salinity, redox potential*. Under normal pressure, assuming maximum air saturation, the following amounts of oxygen are contained in 1 litre of water:

	Fresh water	Sea water (36 ‰)
0 °C	10.3 ml	8.0 ml
15 °C	7.2 ml	5.8 ml
30 °C	5.6 ml	4.6 ml

As a general rule, oxygen content in fresh water* is given in mg, in sea water* in ml. The conversion factor from mg to ml is 0.7 in this case. In natural stretches of water, the oxygen content figures differ from those given above. Biotopes like a mountain stream and a coral reef are extremely rich in oxygen – oversaturation is the rule here. Conversely, a very polluted river can have an oxygen content of nil. Provided the individual water-dwelling live forms do not breathe atmospheric air, they are remarkably adapted to a particular oxygen content. In the aquarium, it is normal to strive for a higher oxygen content. It is a common mistake in aquarium-keeping to believe that vigorous ventilation will raise the oxygen content of the aquarium and that it will help maintain it at a level of saturation; in fact, oxygen content is determined much more by other factors. On the one hand, plants constantly produce oxygen when illuminated as a metabolic by-product of photosynthesis, and this can result in oversaturation; on the other hand, oxygen is constantly being used up during animal and plant respiration, as well as through the oxidation of organic and inorganic substances in the aquarium such as excretory products, food remains, ammonia, nitrite and so on. In this case, oxygen can be exhausted to such a degree that all animal life in the aquarium is endangered. Marine and flowing water organisms react particularly badly to such conditions. A low oxygen content leads at the same time to a lowering of the redox potential*, and then there is the danger that non-poisonous substances will be transposed into poisonous ones. To prevent any of these undesirable situations from arising, make sure that the aquarium is well-planted, that only a few animals are placed in the tank, and that the water is continually circulating gently by means of ventilation. Also make sure that food remains and excretory products are not allowed to build up. If there is an acute lack of oxygen in the tank, HÜCKSTEDT advises that an exact dosage of hydrogen peroxide* will provide relief for a short spell.

Many an aquarist may wish to determine exactly the oxygen content. There are 2 methods of doing this: electrical and chemical. The first method is very expensive, as precision equipment is needed, and the majority of the pieces of equipment will only give sufficiently accurate results in sea water, in any case. Chemical determination is based on the tried and tested WINKLER method: Here, manganese salts are oxidised by the dissolved oxygen. These salts release iodine from calcium iodide, and this can be determined volumetrically. For aquarium-keeping purposes, the simplified drop bottle method gives a good enough result – the complete set-up of equipment and chemicals can be bought through the trade.

Oxygen Deficiency. Different organisms have different oxygen requirements. Individual species of fish, even individuals within one species, may have very different oxygen requirements from the water (much will depend on their place of origin and metabolic situation). If the oxygen content sinks below a minimum level, the fish will die from suffocation. The commonest cause of oxygen deficiency in the aquarium is too many specimens (both plant and animal), as well as simply piling in too much food. The organic food remains that stay in the tank and the excretory products of the fish will break down and lead to a rapidly rising oxygen requirement, and this, together with the imbalance in the aquarium (usually onto the reductive side) (*see* Oxygen Content, Redox Potential), can lead to a severe lack of oxygen. Other causes that can bring about an acute oxygen deficiency include defects in ventilation*, filtration* and heating*. When there is a lack of oxygen in the tank, the fish will lie right underneath the surface and snap after air; they will eventually die with operculum held open and the mouth open wide. The gill epithelium (Respiratory Organs*) of fish that have suffocated is clearly bright. To relieve acute oxygen deficiency HÜCKSTEDT recommends adding 1 ml of 15% hydrogen peroxide* to 20 litres of water. But the first priority is to tackle what was causing the oxygen deficiency in the first place. With fish embryos, a lack of oxygen in the water leads to disturbances in development – deformations may result or development may stop altogether, and the embryos will die.

Oxymonacanthus BLEEKER, 1865. G of the Balistidae*, Sub-F Monacanthinae. Contains a single, unmistakable species found in tropical parts of the Indo-Pacific:

– *O. longirostris* (BLOCH and SCHNEIDER, 1801); Beaked Leatherjacket, Black-saddled Leatherjacket, Long-nosed Filefish, Orange-emerald Filefish, Orange Filefish, Orange-green Filefish, Orange-spotted Filefish. Length up to 10 cm. This species lives a sociable existence among coral reefs. It has an elongated shape with a very elongate snout for preying on coral polyps and small animals. The shining blue-green body is crossed by longitudinal rows of oval, orange-coloured blotches. Unfortunately, in the aquarium, these wonderful fish are very delicate. At least 2 specimens should be kept together. They eat small crustaceans, and the flesh of bivalves and crustaceans (fig. 859).

Fig. 859 *Oxymonacanthus longirostris*

Oysters, G, *see* Ostrea

Ozone is a 3-atom modification of oxygen which has strong oxidising properties. This effect can be made useful in aquarium-keeping and brings with it valuable benefits. Ozone is formed from normal atmospheric oxygen by electric discharges at a high-tension electrode. Ozonisers work on this principle and can be regulated. They can be switched into the air-stream of a pump and the air can then be enriched with the necessary amount of ozone. Rubber and some synthetic products are destroyed very quickly by ozone. Nor must ozone be introduced directly into the aquarium, as even small quantities are poisonous to fish and plants. The ozonisation must take place outside the tank in an external filter (Filtration*) or in an ozone reaction tube (fig. 860). For safety's sake, the water output should be put through a layer of activated charcoal*, which will

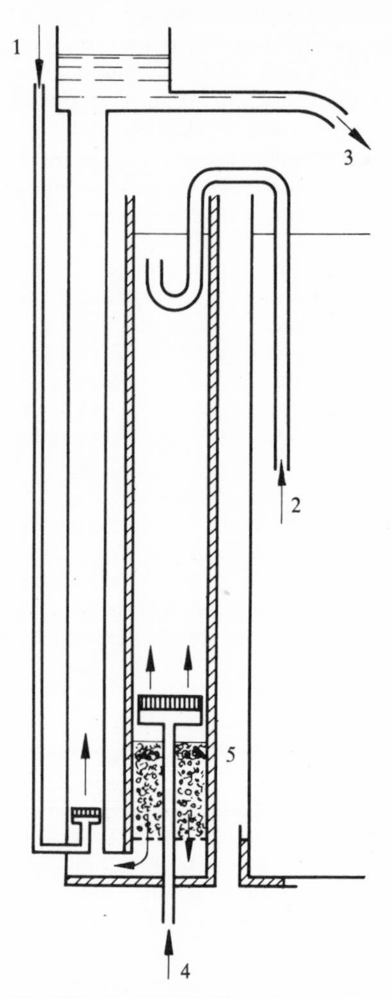

Fig. 860 Ozone reaction tube. 1 Air 2 Water inlet 3 Water outlet 4 Air-ozone mixture 5 Activated charcoal

remove even the smallest traces of ozone immediately. The chief benefit of ozone treatment is the great increase in redox potential*, so that an optimal oxygen saturation of the water becomes possible. All reducing compounds, especially proteins and corresponding breakdown stages, are oxidised without poisonous intermediary products. The oxidation of proteinous bodies will only take place up to ammonia*, however. So, an ozone treatment only makes sense if there are sufficient nitrifying bacteria in the filter systems that follow after it or in the aquarium itself, since they will break down the ammonia into nitrate. Too strong ozonisation will lead, after feeding, to a sudden increase in ammonia*, that will not be able to be broken down fast enough by the bacteria, and poisonous concentrations will build up. In addition, because of the strong oxidation effect, bacteria, viruses, algae, fungi and other single-celled organisms will be killed. An undesirable effect of over-using the ozoniser is the formation of nitrous gases, which are present in the water as nitrates. Water with insufficient alkali buffer capacity will experience a rapid drop in the pH value during ozonisation – something which can sometimes be traced back to the formation of salpetre and nitrous acid. Fresh water should be treated with ozone for short spells only, because longer treatments start to produce the undesirable effects more and more. For sea water, careful doses of ozone can be effective. The optimal dose must be determined for each tank separately and it will depend on its contents. If a good amount of green algae develops in a sea water tank (which is perhaps a slow-developing indicator but at least a reliable one), then the amount of ozone being introduced can be at a minimum. Ozone is also used in foam filters* and increases their effect dramatically.

Pachygrapsus RANDALL 1840. G of the Crabs (Brachyura*, F Grapsidae) that live amphibiously. Found in warm seas. PACHYGRAPSUS is an omnivore and has a four-cornered carapace. Its forehead edge is straight and broad, and its back is flat and smooth. Like the species of the related tropical G *Grapsus*, *Pachygrapsus* remains out of water much of the time. They are highly mobile, alert creatures that live along rocky coasts, sometimes also along muddy banks. They survive well in the aquarium, best of all, if they are provided with a stretch of land too. Relatively peaceful. Definitely cover the aquarium, for the specimens will manage to escape from the tiniest of cracks. A species commonly found along the southern European coast is:

– *P. marmoratus* FABRICIUS. Carapace up to 4 cm wide. Fine dark-brown markings. At 10°C can survive for up to 5 days out of water.

Pachypanchax MYERS, 1933. G of the Cyprinodontidae*. 2 species that occur mainly on some islands off the east coast of Africa (Madagascar, Zanzibar, Seychelles). The G could also be represented on the African continent itself. It is related chiefly to *Aplocheilus** and *Epiplatys**, and in terms of distribution *Pachypanchax* is in the middle in relation to these 2 Ga. Length up to about 10 cm. Their body structure is strong and elongated like a pike's. Dorsal profile only slightly arched, forehead and nape almost straight. Mouth superior, lower jaw clearly protruding. The D is inserted along a shorter length of the body than the An and is located opposite its rear section. Both fins are more strongly developed in the male than in the female and in *P. homalonotus* they are clearly pointed. The C in both sexes is fan-shaped. Patterning is very primitive, so in contrast to the related Ga there are no distinct dark transverse stripes or a dark longitudinal band. During the spawning season, *P. playfairi* has an unusual feature – the scales stand up from the body to some extent (but this does not indicate any disease symptom). *Pachypanchax* species are typical surface fish that lurk like a pike in the protection of plants while waiting for prey to come along (eg fish broods, food that alights on the surface, as well as crustaceans of suitable size). *Pachypanchax* is very tolerant of increases in salinity levels, so they are also able to live in brackish water*. When kept in the aquarium, the tank must not be too tall and it should be well planted. Floating vegetation provides the necessary cover and also dims the light. The composition of the water plays a subordinate role. *Pachypanchax* species are attached spawners (compare Cyprinodontidae), that lay their eggs on plants near the surface. Depending on species and temperature, the incubation period lasts about 10–16 days. The fry are easy to rear and if sufficient hiding-places are provided, they can even be reared in community tanks.

– *P. homalonotus* (DUMERIL, 1861); Green Panchax, Homalonotus Killi-fish. From Madagascar. To 9 cm. Flanks emerald-green, nape and dorsal surface cinnamon-brown, ventral surface pale yellow or white. Iris yellow. A peaceful species.

– *P. playfairi* (GÜNTHER, 1866); Playfair's Panchax, Golden Panchax. From Zanzibar, Seychelles, possibly also E Africa. To 10 cm. Flanks greenish with rows of red dots. C, D and An yellow, likewise covered in red dots. The D in the female has a dark blotch at the base. Great rivalry within this species, and prone to biting (fig. 861).

Fig. 861 *Pachypanchax playfairi*

Pacific Herring *(Clupea pallasi) see* Clupeiformes

Pacific Ocean. The greater part of the Indo-Pacific* and at the same time the largest ocean of all, bounded in the west by Asia and Australia and in the east by America. The whole eastern area contains very few islands. Including the neighbouring seas, the Pacific measures 180 million km^2 and so covers over a third of the earth's surface. From N to S it measures about 15,000 km, from W to E about 20,000 km. On average, the Pacific is 4,300 m deep, but in the Marianas Trench, it drops to 11,034 m. Near the equator, the average temperature at the surface is around 29 °C, in the Bering Sea around 1 C. Salt content* varies between 32‰ in the Sea of Okhotsk and 36.9‰ near the southern tropics. The Pacific does not contain a great deal of living organisms; only where cold and warm sea currents meet and mix is the plankton production high enough to sustain large numbers of fish. That is why the fishing industry is restricted almost exclusively to the continental shelf areas.

Pacific Pompano *(Peprilus simillimus) see* Stromateoidei

Pacific Sailfish *(Istiophorus orientalis) see* Scombroidei

Pacific Salmon (*Oncorhynchus*, G) *see* Salmonoidei

Pacific Stickleback *(Aulorhynchus flavidum) see* Gasterosteoidei

Pacific Sticklebacks (Aulorhynchidae, F) *see* Gasterosteoidei

Paddlefishes (Polyodontidae, F) *see* Acipenseriformes

Padina. G of the Brown Algae Phaeophyta*) from warm seas. Peacock Weed (*P. pavonina* (L.) GAIL) is a characteristic member of the G. It occurs in the upper rocky littoral of the Mediterranean coast and is 5–10 cm tall. It is whitish in colour because of calcium deposits and has funnel-shaped thalli. Not suitable for the aquarium.

Paguridea; Hermit Crabs. Sup-F of the Decapoda*, Sub-O Reptantia. Included in this Sup-F are the marine Hermit Crabs (F Paguridae), the land Hermit Crabs (F Coenobitidae, *see* Coenobita) and Stone Crabs from cold seas (F Lithodidae). A feature they all have in common is the usually worm-shaped, soft-skinned, asymmetrical abdomen that is often hidden inside empty Snail-shells. Forms that differ markedly from this, such as the well-known Robber-crab *(Birgus latro)* LINNAEUS, 1758) and the crab-like Lithodidae clearly belong in this group when one considers their development and structure. For the aquarist, it is the typical Hermit Crabs of the F Paguridae that are of most interest. They are lively creatures that accept a variety of animal food as well as detritus and growths; the larger forms are predators. It is particularly charming to watch the different kinds of living communities that the Hermit Crabs enter into with Sponges *(Suberites, see Paguristes)*, Hydrozoans *(Hydractinia, Podocoryne)*, Sea-aneomones *(Adamsia*, Calliactis*)* and Zoantharia*. Hermit Crabs commonly kept in marine aquaria belong to the Ga *Clibanarius**, *Dardanus**, *Diogenes**, *Eupagurus**, *Paguristes** and *Pagurus*.

Paguristes DANA, 1851 (fig. 862). G of the Hermit Crabs (Paguridea*) with one common species in the Mediterranean – the very pretty Eye-spot Hermit Crab (*P. oculatus* FABRICIUS). Its body length is only 4 cm and it is coloured brick-red. On the inside of the pincer legs there is a striking dark violet blotch, and the left pincer is somewhat larger than the right. It lives at depths of 4–40 m, and is common in places. On its shell it carries parasitical anemones *(Calliactis)*, the Sponge *Suberites* or hydroid polyp colonies *(Podocoryne)*. In the aquarium, *Paguristes* survives well and it can be fed easily on a diet of different animal foods.

Pagurus, G, *see* Paguridea

Pakistani Loach *(Botia lohachata) see Botia*

Palaearctic Realm. Part of the holarctic realm*, made up of the whole of Europe as well as the northern part of Africa and the whole of Asia, with the exception of the tropical south. In comparison with the second zoogeographical sub-region of the holarctic, the nearctic*, the palaearctic contains significantly less freshwater animals (especially so in Europe). There are few fish of interest to the aquarist, but those that are interesting include the Bitterling *(Rhodeus*)*, the Mudminnow (*Umbra krameri, see* Esocoidei) and the Sticklebacks *(Gasterosteus*, Purgitius*)*, plus certain kinds of Catfishes from the Amur*.

Fig. 862 *Suberites* on top of *Paguristes*

Palaemon WEBER, 1795. G of the Natantia*. Contains numerous species in the littoral of cool to tropical seas, with some also in brackish water. *Palaemon* lives on small animals, carcasses, plants and detritus. The very similar G *Leander* DESMAREST, 1849, lives in the fresh water of tropical countries. The first pair of walking legs have small pincers, the second pair larger ones. The forehead process (rostrum) is large and denticulated – the number of teeth is important in determining species. They are graceful animals, between 5 and 10 cm long. They are glassily transparent and usually very attractively coloured. Several species of *Palaemon* live along the European coasts. The Baltic Shrimp is caught in shallow waters in late spring, and when cooked is a delicacy. *Palaemon* species will last years in an aquarium, where they are useful because they destroy any remains there may be. Easy to feed on a varied diet of animal food.

– *P. elegans* RATHKE. From the North Sea to the Black Sea, W Africa. 6 cm. Upper edge of the rostrum has 7–10 small teeth. The body is crossed by thin, reddish-brown transverse stripes, and the joints of the walking legs are orange. Often used as anglers' bait (fig. 863).

– *P. squilla* (LINNAEUS, 1758); Baltic Shrimp or Prawn. From the Baltic, North Sea. 6 cm. Upper edge of rostrum has 5–6 small teeth. Coloured brownish or greenish.

Palaeontology. The science of the living beings from the Earth's past epochs. *Palaeobotany* is the study of fossil* plant remains, while *palaeozoology* is concerned with fossil animals. It is very interesting to try to work out the relationships of these fossil organisms to the species alive today, as well as to compare anatomy, life-styles and living conditions. Another major task of palaeontologists is the ontogenesis of the formation of the first forms of life up to the plants and animals of the present day. Palaeontology is particularly tied up with geology. For information on the palaeontology of fish, *see* under Agnatha, Chondrichthyes, Osteichthyes, as well as under the Sup-Os, Sub-Os and sometimes the Fs of the named systematic categories.

Palaeopterygia, Sub-Cl, name not in current usage, *see* Osteichthyes

Fig. 863 *Palaemon elegans*

Palinura, Sup-F, *see* Reptantia

Palinurus FABRICIUS, 1798; Spiny Lobster; Crawfish; Rock Lobster (fig. 864). G of the Decapoda*, Sub-O Reptantia*. F Palinuridae. There are several species found in the littoral of warmer seas. They live mainly on molluscs, sea-urchins and dead animals; larger or more mobile animals cannot be caught with their pincerless extremities. Body length 30–55 cm, and covered in a large number of spines. The antennae are very long, often reaching out from wherever the Lobster is hiding during the day (Spiny Lobsters are active at night). Purring noises are produced by rubbing the bottom sections of the antennae against the head armour; they are probably meant to help defend territories. *Palinurus* species are very important in the economy. In the aquarium they make attractive specimens that survive well and are tolerably peaceful (but usually only small specimens are made available, and then only rarely).

Fig. 864 *Palinurus ornatus*

P. vulgaris LATREILLE is found in the deep parts of the rocky littoral along the steep coastlines of the E Atlantic and Mediterranean. It is red-coloured and may weigh up to 8 kg. The beautiful butterfly-coloured Lobsters of the G *Panulirus* WHITE, 1847, come from the tropical Indo-Pacific. They get too large for the aquarium very quickly and are often difficult to adapt to conditions in captivity.

Palm Crab *(Birgus latro) see* Paguridea

Palythoa, G, *see* Zoantharia

Pamphorichthys, G, name not in current usage, *see Poecilia*

Panagrillus. G of the Nematodes. The tiny *P. silusiae* is used in particular as fish food for juveniles (Fry, Feeding of*). Cultivate and feed as instructed for *Turbatrix**.

Pancreas *see* Digestive Organs

Pangasianodon gigas *see* Pangasiidae

Pangasiidae (fig. 865). F of the Siluriformes*, Sup-O Ostariophysi. Fairly large to large Catfishes of elongated build, laterally compressed. In many respects similar to the Schilbeidae*, along with which they are often grouped as one F. In the majority of features, the 2 Fs do indeed agree, although the Pangasiidae usually have only 2, sometimes even 1, pairs of barbels. The spiny first fin rays of the D and P, in rare cases of the V too, may be very elongate, protruding like a stiletto blade above the fin skin (eg *Pangasius sanitwongsei*). A giant Catfish, *Pangasianodon gigas*, also belongs to this F; it is found throughout SE Asia. Large specimens may reach lengths of 2 m or more. Great numbers of them are fished in the Mekong river system, for example, as they swim upriver during their spawning migration. They are aiming to reach Lake Talin in the Jünnan Province (Peoples Republic of China). In Kampuchea (Cambodia) it is called 'trey reach' (King Fish) and in Laos 'pla bük' (Giant Fish). Many other species are also valuable commercial fish.

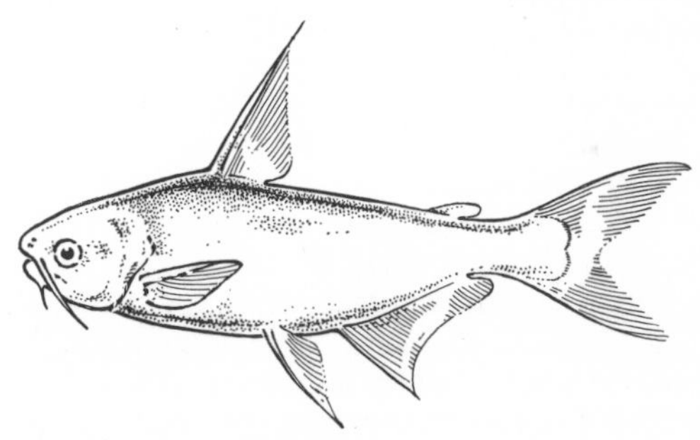

Fig. 865 Pangasiidae-type

Panther Fish *(Cromileptis altivelis) see Cromileptis*

Pantodon buchholzi *see* Osteoglossiformes

Pantodontidae, F, *see* Osteoglossiformes

Pantonodontinae, Sub-F, *see* Cyprinodontidae

Pantopoda. Marine-dwelling O of the Spiders (Arachnida*). There are 4–6 overlong pairs of legs sprouting from a tiny, rod-like torso. Pantopoda bore into Sea-anemones, Hydrozoans and other soft-skinned marine animals.

Panulirus, G, *see Palinurus*

Paper Nautilus *(Argonauta argo) see* Cephalopoda

Paracanthopterygii, Sup-O, *see* Osteichthyes

Paracanthurus BLEEKER, 1863. G of the Acanthuridae, Sub-O Acanthuroidei*. One species only found in

the tropical Indo-Pacific. Similar to *Acanthurus**, but with 3 soft rays in the V:
– *P. hepatus* LINNAEUS, (1758) (= *P. theutis* LACÉPÈDE, 1802); Blue Surgeon; Blue Tang; Flagtail Surgeonfish; Wedgetailed Blue Tang. Length 26 cm. A shining blue colour with a black, palette-shaped blotch on the dorsal surface, C yellow. Younger fish relatively long-lasting. Intolerant towards one another.

Paracentrotus MORTENSEN, 1903. G of the Sea-urchins (Echinoidea*). Found in the rocky littoral of temperate zones, often in vast numbers. Some excavate rock. *Paracentrotus* is a grazing animal and one species found along the European coast of England and into the Mediterranean is:
– *P. lividus* LAMARCK. Pale or dark violet-brown in colour. Has pointed spines and a greenish shell. The ripe gonads of this species are a delicacy in the Mediterranean region. It survives a long time in alga-rich aquaria. The urchin masks itself well with bivalve shells or algae.

Parachaetodon BLEEKER, 1874. G of the Chaetodontidae*. Sub-F Chaetodontinae. Contains a few species in the tropical Indo-Pacific. Almost like a disc in shape, *Parachaetodon* has fairly large scales but no protruding snout area that is particularly striking.
– *P. ocellatus* (CUVIER and VALENCIENNES, 1831); Six-spined Butterfly. Indo-Pacific. Length to 15 cm. Yellowish-white with 4 brown transverse bands. In the dorsal area of the fourth one there is a black eye-spot. This species adapts fairly well to life in captivity (fig. 866).

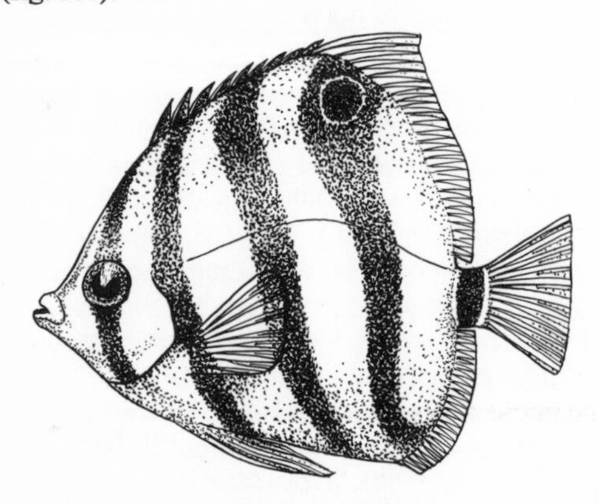

Fig. 866 *Parachaetodon ocellatus*

Paracheirodon GÉRY, 1960. G of the Characidae*, Sub-F Cheirodontinae. Found in the upper reaches of the Amazon from São Paulo de Olivenca to Iquitos and upriver, especially the Rio Putumayo. Lives in small water channels. Length to 3.5 cm. Small, extremely colourful shoaling fish that love swimming and hug the middle and lower levels of water. They also like a dark substrate and subdued lighting in the aquarium. Provide soft, at most medium-hard, water with a small amount of peat added. *Paracheirodon* species are content to live in a community tank with similar-sized species. The best temperature to keep them at is 20–22 °C. They readily accept live and artificial foods (fairly small kinds). The species is very susceptible to the dangerous Neon Tetra disease, *Pleistophora** – whole broods fall victim to it time and time again. It might be that the disease is encouraged to spread by keeping the fish at too warm a temperature and not allowing enough fresh water to flow through. The body is elongate, moderately compressed laterally. The C is not covered in scale at the bottom and an adipose fin is present. Breeding them is not without its problems. Youngish sex partners should be used, and the females should not be too full of spawn, otherwise they will usually release the egg-cells overnight, stimulated by the fresh water of the breeding aquarium. Small containers measuring 18 × 15 × 15 cm are plenty big enough for a pair. Set the tank up with fresh spring water (1–2 °dH and 5.5–6.5 dH) or with pure rainwater and a small clump of plants above a spawning grid. The fish will even spawn in tanks where there are no plants. If they are put in the tank in the evening, the spawning act will usually take place on the following morning. At this point the shining band of the male is an intense violet tone. The glass-clear spawn does not stick very well and mostly lies on the bottom. The parent fish will eat their own spawn. It is a good idea to use a pipette to take out any egg-cells that have died during the evening on the day that spawning took place. The fry hatch out after about 20 hours; subdue the lighting and filtrate the tank water. About 5 days later, the fry will be swimming freely. They grow quickly and need a lot of care with rearing. There is only one species:
– *P. innesi* (MYERS, 1936); Neon Tetra. Dorsal surface a dark olive-green, ventral surface yellowish-white. A lovely greeny-blue iridescent shining band runs along the flanks, extending from the front edge of the eyes, across the upper eye-arch to the height of the adipose fin. The iris is a shining blue-green and has small gold dots in the upper section, that sparkle like diamonds. All the fins, with the exception of the pastel-white tips of the An, are colourless. The exquisiteness of colouring in this species makes it one of the most popular aquarium fish of all, along with the Scalare Angelfish and its cultivated varieties. The sight of a shoal of these fish in the aquarium is always most delightful (fig. 867).

Parachlorometaxylenol. A firm, white mass that dissolves well in alcohol and in a small amount in water.

It has strong disinfecting properties and is used as an admixture to disinfectants and conserving substances. In aquarium-keeping, it can be used in prolonged baths (5–10 mg to 1 litre of water), during the treatment of nematodes (Nematode Diseases*) over a period of some hours up to several days. (Alternatively, it can be mixed in with food.)

Paracirrhites BLEEKER, 1875. G of the Cirrhitidae* containing species found in tropical and subtropical parts of the Indo-Pacific and in the Red Sea. Length around 20 cm. Similar in life-style to *Cirrhitichthys**. Will survive well in the aquarium if fed on shrimps, small fish etc. Eg:

– *P. forsteri* BLOCH and SCHNEIDER, 1801; Freckled Hawkfish. 30 cm. From the Red Sea to the Pacific. Head bluish with black flecks (outlined in red on the operculum), dorsal surface blackish, ventral surface reddish.

Paradisefish *(Macropodus opercularis) see Macropodus*

Paradisefishes, G, *see Macropodus*

Paradontinae, Sub-F, *see* Characoidei

Paragobiodon, G, *see Gobiodon*

Paragoniatinae, Sub-F, *see* Characoidei

Paralichthys californicus *see* Pleuronectiformes

Parallel Display Swimming *see* Display Behaviour

Paraluteres BLEEKER, 1865. G of the Balistidae, Sub-F Monacanthinae. Contains a single species found in the tropical Indo-Pacific and it is very reminiscent of the Pufferfish *Canthigaster* valentini* in shape and colouring:

– *P. prionurus* (BLEEKER, 1851); Black-finned Triggerfish, Valentini Mimic, False Pufferfish. 11 cm. The first D-spine is not fully erectile. Towards the tail, the scales have longer spines on them. At night, the species sticks itself firmly into the coral.

Paramecium MÜLLER, 1773 (fig. 868). Well-known G of the Ciliata* that lives mainly in water that contains a lot of decaying matter. Paramecia are occasionally bred in aquarium-keeping to become 'Infusoria' for juvenile fish (*see* Feeding). These ciliates often colonise

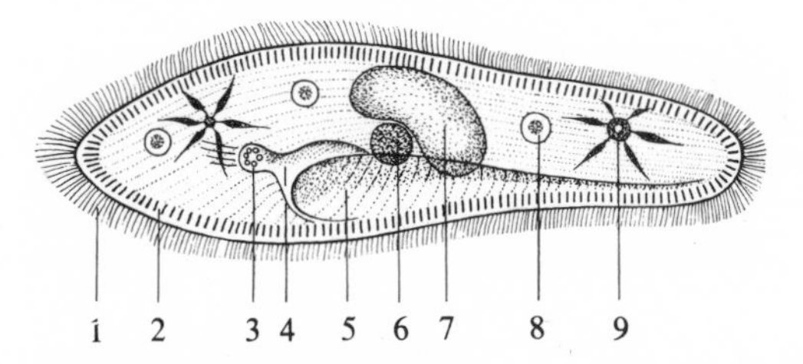

Fig. 868 Anatomy of *Paramecium*. 1 Cilia 2 Defensive organelles (trichocysts) 3 Formation of a food vacuole 4 Cell-mouth 5 Mouth area 6 Micronucleus 7 Macronucleus 8 Food vacuole 9 Pulsating vacuole

aquaria and are excellent for studying under a microscope. Magnify 50–100 × for observing locomotion, feeding and basic structure. Things to note are the 2 cell nuclei (large and small) and the way in which the Paramecium changes its position through a co-ordinated beating of the cilia. Paramecia move along in an elongated spiral path, rotating around their own axis at the same time.

Paramuricea, G, *see* Gorgonaria

Paraphyletic Group. A systematic group of whatever rank (eg phylum, order, genus) that encompasses progeny that do not necessarily stem from one and the same basic form, and so the relationships of the organisms are not reflected exactly. The term 'pisces' (fishes) for example, represents a paraphyletic group of organisms, even although both the fish as well as all the land vertebrates are descended from one original form. Consequently, there are fish that are more closely related to the land vertebrates phylogenetically than they are to any other fish—the well-known Stump-fins (Crossopterygii, *see* Osteichthyes), for example. For illustration turn to Monophyletic Group.

Parasitic Catfishes, F, *see* Trichomycteridae

Parasitic Sea-anemone *(Calliactis parasitica) see Calliactis*

Parasitism. Parasites (Greek, meaning 'table companions') are organisms that can exist only in a one-sided dependence on another organism, the host. In close contact with the host-organism, the parasite receives protection and food, and thus becomes absolutely dependent on it. For the parasite, the host is the determining habitat without which it would die. Pathogenic parasites, which are of great importance among aquarium fish, damage their host in various ways: purely mechanically by their size and form; by drawing food away; by destroying tissues and organs; by creating poisonous metabolic products; by effecting abnormal developments and over-sensitivity etc. Specimens that are attacked succumb to a parasitical disease, each type of disease having characteristic symptoms that provide clues as to the species of parasite causing it. The host is not directly killed by a parasite, as is the case with predators killing their prey. There may, in fact, be no undesirable effects at all upon the host (apathogenic parasites).

Parasitism is known among many plants and in nearly all animal groups. The numerous fish parasites belong to the fungi (Mycophyta), viruses, bacteria (Schizomycetes*), Protozoa*, worm-like animals (Worm Parasites*), Arthropoda* and a few vertebrate animals (Cyclostomata*). On the other hand, there are fish too that lead a parasitical way of life: the larvae of the Bitterling (*Rhodeus sericeus* PALLAS, 1776) live parasitically in Bivalves (Bivalvia*); some species of Pearlfishes (Fierasferidae, *see* Gadiformes) are parasites on Mollusca*, Tunicata* and Holothuriodea*. Fish are particularly often attacked by parasites, which can attach themselves as external or ectoparasites to the outside of the body and in easily accessible body cavities (gill cavity, mouth cavity) (eg *Ichthyobodo**, *Trichodina**, *Ichthyophthirius**, *Oodinium**, *Gyrodactylus* and *Dactylogyrus* [Trematodes*], Argulidae*, Copepoda* etc). Internal or endoparasites penetrate the internal body organs via the intestinal tract, the skin and blood system (Blood Parasites*, Sporozoa*, *Ichthyosporidium**, Discus Disease*, Stomach Dropsy*, Fish Tuberculosis*). Because of their specific life-style, parasites from very different groups of animals form very similar adaptations (Analogy, Convergence*). Among them are defence mechanisms against the effects of the host organism, organs of attachment (such as claws, hooks and suckers), sucking tubes and piercing organs, flattened, worm-like or spherical bodies, the reduction of sense organs and the organs of motion, the lack of pigmentation, the regression of the digestive tract, highly developed, often dominating sex organs and hermaphroditism*. Permanent parasitism, in which an organism lives as a parasite in all its stages of development, is rarer than periodical parasitism, in which only certain stages of development live parasitically. Highly specialised parasites often form one or several larval stages during an indirect course of development. This involves a change of host via specific intermediary hosts which are the only places where the next stage of development can take place. The adult form then develops in the definitive or final host. The probability of finding the right host or intermediary host can be slight under certain circumstances. During the evolution* of the parasites, a great number of mechanisms have been developed for making their complicated life-cycles more secure. Usually, for example, the number of offspring is very large, self-fertilisation can take place through hermaphroditism, and even egg-development without fertilisation (Parthenogenesis*) and the ability of the larvae to reproduce (Paedogenesis) occur quite often among parasites. If more than one host animal is necessary for the development-cycle, the natural contacts between host animals or the members of a food chain (Nutrition*) are used in order to increase the chances of reaching the required next host (Cestodes, Trematodes, Sporozoa). Unfavourable periods (temperature, drought) can be survived by developing hardy resting forms (cysts, spores). Many parasites can exist in numerous hosts, eg in all the species of a family; others are specialised towards a particular host or towards certain organ systems within a host. Often, a kind of main host is preferred, and only in exceptional cases will such parasites resort to secondary host. A host animal may be colonised by various different kinds of parasites at the same time. Depending on their particular adaptations, they will be confined to certain organs or organ systems. Ectoparasites are subject not only to the influences of their host, but also to the effects of the environment. With a change in habitat, host animals often change their typical parasites. Migrating fish (Migration*) are often attacked by different parasites as they transfer from the sea to fresh water, or vice versa. Individual parasites can be found regularly on every fish (Fish Diseases*). The host organism usually mobilises his defences against attack and this usually prevents a mass-accumulation of the parasites, and thus no acute signs of disease appear. Only unfavourable environmental factors or a mass infection of the host will lead to its becoming weakened, which in turn will encourage the development of the parasites, and under certain circumstances the fish will become gravely ill. So-called debilitating parasites can only develop in great number in fish that are already damaged. Despite all advances and technical achievements, the aquarium is still obviously an inadequate imitation of natural living conditions for aquatic animals. The specimens are therefore particularly endangered and susceptible to parasites.

Parasphromenus BLEEKER, 1879. G of the Belontiidae, Sub-O Anabantoidei*. From Sumatra, the island of Bangka. Lives in river systems. Length up to 3.5 cm. Body elongated, slim, flattened laterally, head pointed. The jaws contain small, conical teeth. Large ctenoid scales arranged regularly. LL indistinct. C shaped like a tongue. Vs elongated into points, but not like long threads. The D is very long with 13 hard rays in the front section (this is in contrast to the similarly shaped members of the G *Trichopsis* or *Betta*). Ground colouring yellowish to reddish, paler below. There are 2 dark longitudinal bands running along the flanks, beginning at the snout. Another longitudinal streak extends from the P insertion to the lower edge of the caudal peduncle. More rarely, there are dark transverse bands in the rear portion of the body and along the edges of the D and An. Fins yellowish otherwise, the base of the C being reddish. The male is more intensively coloured, and the D and An are pointed. Tanks should be well planted and the temperature 23–26 °C. Feed with small live food of every sort. The only species is:
– *P. deissneri* BLEEKER, 1859 (fig. 869). Widespread in aquarium-keeping.

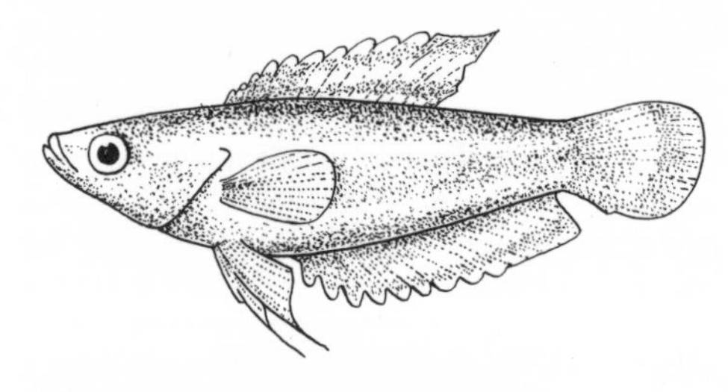

Fig. 869 *Parosphromenus deissneri*

Parazoanthus axinellae *see* Zoantharia

Parental Care Behaviour. Term used in ethology to describe the continuation of spawning behaviour* in order to protect and help the development of the spawn and young fish. It is important to distinguish between behaviour related to care of spawn and parental care behaviour.

The parent fish take care of the spawn by fanning it to ensure a constant oxygen supply; by removing any unfertilised egg-cells or those egg-cells that cannot develop; and by defending the spawn etc. Parental care behaviour begins as the young fishes hatch – parents often use their mouths to free the brood from their egg-shells. Other care behaviour includes lifting their babies out of holes, removing them to new quarters, defending the brood, chewing up large pieces of food after the babies can swim freely catching any of the brood that breaks away from the shoal, returning the brood to a hole where they spend the night etc. Tropical Cichlids, in particular, exhibit impressive displays of parental care behaviour. The females of many species of *Haplochromis** and *Pseudotropheus** take the eggs into the mouth after spawning where they remain in a throat pouch until hatching. Even afterwards, the young fishes can return to the throat pouch at lightning speed when danger threatens. Parental care can be practised by both parents, in which case the partners often take it in turns, or by the male, or by the female. In the aquarium parental care may not proceed entirely satisfactorily due to the presence of a hostile factor*.

Parkeriaceae, F, *see* Pteridophyta

Parrot Pufferfish *(Colomesus psittacus) see Colomesus*

Parrotfishes (Scaridae, F) *see* Labroidei

Parsley Family *see* Apiaceae

Parthenogenesis. A monosexual form of reproduction*, in which the offspring develop from unfertilised egg-cells. Parthenogenesis is not uncommon among plants and invertebrate animals, but only occurs exceptionally among vertebrates (some lizards, among the fish it occurs in certain populations of the Silver Carps). A distinction is made between *haploid* and *diploid* parthenogenesis, eg in the summer generations of the Water Fleas. This means that the developing organism is haploid on account of the lack of fertilisation, but it can still be diploid through special regulations. In many species, parthenogenesis is coupled with sexual determination. For example, among the Hymenoptera (eg bees), males (drones) always develop from unfertilised eggs. Parthenogenesis in the larval stage (eg among the larvae of parasitical Worms) is called *paedogenesis.*

Partial Desalination. The reducing of the salts contained in water either by boiling, since heat changes the dissolved calcium and magnesium carbonates into insoluble carbonates that precipitate (Temporary Hardness*), or by means of ion exchangers*. The most commonly used is the exchange of metallic cations for hydrogen ions by means of highly acidic cation exchangers. In this case the salts are turned into the corresponding acids*, eg carbonates and hydrocarbonates into carbonic acid, chlorides into hydrochloric acid, nitrates into nitric acid, and so on. Water treated in this way will react highly acid therefore, so for aquarium purposes it has to be prepared. Attempts can be made to raise the pH value* to 6.5–7 by mixing the initial water and strong ventilation (to drive out the carbonic acid), but this is not possible in every case. The acids can be neutralised by carrying out a neutral exchange*, but this is not desirable in every case. So for this reason aquarists are turning increasingly to those ion exchangers that work in the slightly acidic to slightly basic range. Certainly, this method does not achieve desalination*, but water prepared in this way is suitable for most aquarium purposes and requires only slight corrections to the pH. Exchange resins that work in this way include Permutit C-65 (cation exchanger) and Amberlite IR-45 (anion exchanger).

Patella LINNAEUS, 1758; Limpets. G of marine-dwelling Molluscs (Prosobranchia*). Lives on the rocks of the tidal zone where it feeds on growths. Has a dish-shaped, flatly conical shell with no coils. Limpets are attached amazingly firmly to the rock. They will return always to a firm resting place, the uneven texture of which corresponds exactly to the shape of the edge of the shell. No recorded observations in the aquarium.

Pathogenic; producing disease

Pathology. The study of diseases and abnormalities. It includes not only infectious diseases but also various kinds of organic damages and abnormal developments. The diseases of plants are the object of study in phytopathology.

Peacock Blenny *(Blennius pavo) see Blennius*

Peacock Demoiselle *(Pomacentrus pavo) see Pomacentrus*

Peacock Lionfish *(Pterois volitans) see Pterois*

Peacock Wrasse *(Crenilabrus tinca) see Crenilabrus*

Peacock Wrasse *(Thalassoma pavo) see Thalassoma*

Peacock-eyed Spiny Eel *(Mastacembelus argus) see Mastacembelus*

Peacock-eyed Sunfish *(Centrarchus macropterus) see Centrarchus*

Peanut Worms, Ph, *see* Sipunculida

Pearl Cichlid *(Geophagus brasiliensis) see Geophagus*

Pearl Danio *(Brachydanio albolineatus) see Brachydanio*

Pearl Essence. A silvery preparation extracted with ammonium solutions from the skin of Whitefishes. It is used in the making of glass pearls. The essence consists of tiny crystals (guanine). To produce imitation pearls, hollow glass spheres are coated with pearl essence on the inside. Today, glass pearls coated with pearl essence have almost been completely supplanted by products that contain aluminium powder instead of essence.

Pearl Gourami *(Trichogaster leeri) see Trichogaster*

Pearl Mussel *(Margaritana margaritifera) see* Pearls

Pearl Tilapia *(Tilapia lepidura) see Tilapia*

Pearlfishes (Carapidae, F) *see* Gadiformes

Pearls. Concretions of nacre in some Bivalves, more rarely in marine snails or other shellfish. Pearls are formed from the epithelium of the mantle which is also responsible for the secretion of the calcium shell in molluscs. If a small foreign body happens to come between the mantle epithelium and the shell, it is covered with a layer of mother-of-pearl. When the mantle epithelium lies all the way round the foreign body, that is when the typical, round, mantle pearls develop with numerous concentric layers of mother-of-pearl. Natural pearls are found primarily in the Indo-Pacific Bivalves *(Meleagrina margaritifera)* and *Trigona* (various species) as well as in the European Pearl Mussel *(Margaritana margaritifera).* In Japan and China cultivated pearls are produced industrially by implanting foreign bodies. Their value climbs with the number of mother-of-pearl layers around the artificial nucleus. Apart from the white-silver pearls, there are also ones with a pink or bluish lustre and dark (so-called black) pearls.

Pearly Rasbora *(Rasbora vaterifloris) see Rasbora*

Peat. A break-down product made of plant material that develops in a state completely sealed off from air (incarbonisation). Peat can have a highly variable character determined by the original nature of its plant material, its special way of developing and its age. It contains complicated mixtures of numerous compounds, such as organic acids, resins, waxes, hormones, salts etc, which among other things determine the slightly acidic to neutral reaction that peat has. In aquarium-keeping, a suitable peat as a substrate* or filter material (Filtration*) can have a very good effect on the composition of the water. Treating the water with peat is particularly advantageous for fish that come from slightly acidic black water rivers with extremely small amounts of minerals and bacteria. Peat reduces the total hardness* of the water. By the release of humic acids* which are an excellent buffer system, the pH value* falls and becomes stable in the slightly acidic range. The substances released by the peat have a beneficial effect on the growth of plants, especially the development of their root system; on the other hand, they stop fungi and bacteria from growing too proliferously. For aquarium-keeping, good-quality peat from high moors is normally best, but before use any peat must be tested for suitability – above all, it must contain no fertiliser additives. The peat should be crumbly, as stranded peat contains a large proportion of incompletely broken down plant remains and releases inhibitive substances that prevent good plant growth. To test the content of volatile organic acids and humic acids, a little of the peat should be soaked in some distilled water and after 10–20 hours the water's pH should then be measured. The humic acids that leach out usually colour the water a clear yellow-brown and result in a clearly acidic reaction that remains such even when the test water is ventilated. If the pH climbs with ventilation, the acidic reaction can be traced primarily to volatile organic acids, and this means that the peat is unsuitable for use in the aquarium. It is also important to make a test for ammonia* in the probe-water. Peat with a high content of ammonium salts is not suitable in aquarium-keeping either, as the salts that are released would be present as poisonous ammonia at a pH value over 7.0. In order not to destroy the good properties of a suitable peat, it must never have boiling water poured over it before use, nor must it be boiled itself.

Peat Moss, G, *see Sphagnum*

Pecten MÜLLER, 1776; Great Scallops. Marine-dwelling G of the Bivalves (Bivalvia*, O Anisomyaria). The two valves of the shell are dissimilar with ribs running across them like rays. On each side of the hinge joining the valves is a wing-shaped flap (ears). Scallops are attached either by byssus threads or they lie freely on the bottom. The edge of the mantle bears tentacles and between them are numerous eyes with which to perceive approaching enemies, eg Starfishes. Scallops can swim jerkily by opening and shutting their shell, which is how they reach safety. The diameter of a Scallop may be 15 cm, and in the aquarium it makes an interesting, fairly hardy specimen.

Pectorals *see* Fins

Pedicellaria *see* Echinodermata

Pediculata, O, name not in current usage, *see* Lophiiformes

Pegasidae, F, *see* Pegasiformes

Pegasiformes; Sea Moth types. Small O of the Osteichthyes*, Coh Teleostei. Also known as Hypostomides, the Pegasiformes comprise 5 modern species all of which occur in the coastal waters of the Indo-Pacific, and all of which belong to the F Pegasidae (fig. 870). The body is elongate and completely encased in regularly arranged bony plates, forming a stiff armour around the head-torso region and a movable armour in the tail section. The mouth is inferior and toothless, and towering over it is a long, pointed rostrum. The large Ps are splayed out like aircraft wings. whereas the other fins are small. The Vs, which are located on the ventral surface, usually consist of only one spine. The Pegasiformes feed mainly on small animals. Very little is known about their biology.

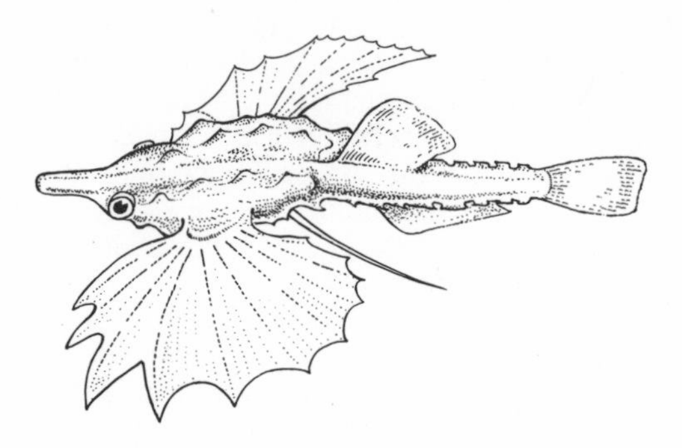

Fig. 870 Pegasidae-type

Peje Puerco *(Balistes vetula) see Balistes*

Pejus *see* Ecological Valence

Pelagial. A zone of the open water in natural and artificial lakes and in the sea. Divided into 2 layers defined by the boundary of where the light falls – the upper pelagial or epipelagial and the lower pelagial or bathypelagial. In the layer that is illuminated, the plant plankton* produces organic nutrients by means of photosynthesis. Not only are all the other inhabitants of the pelagial directly or indirectly reliant on this plankton, but also all the organisms that live in the bottom zone (Profundal*, Abyss*).

The pelagial zone contains 2 very characteristic groups of organisms, the plankton just mentioned and the necton*. It is common to both that they have no or, at most, insignificant relationships to the bottom besides their plant growth. Typical fish of the pelagial zone, such as herrings, tuna or whitefish, are important in the fishing industry, but in the aquarium they are not important. For illustration see under Littoral.

Pelmatochromis STEINDACHNER, 1894 (fig. 871). G of the Cichlidae* containing a large number of species. Native to tropical W Africa where they live mainly in the lower reaches of larger rivers (eg the Niger*) and in brackish-water pools near the coast. They are typical small animal eaters that prefer shallow-water zones with thick vegetation and the possibility of hiding-places (sunken wood, holes in the bank etc). They are medium-sized Cichlids of moderately elongated shape, the dorsal profile being somewhat more arched than the ventral profile. The mouth is often sunken between the nose and eyes. The D has about 15–17 hard rays and 9–12 soft rays; the An with 3 hard rays and 7–8 soft ones. The rear sections of these fins are drawn out into points, and the C is usually fan-shaped.

The various members of *Pelmatochromis* are typical

hidden-brooders, usually laying their spawn in caves. In some species *(P. guntheri, P. kingsleyae),* the male even keeps the brood in his mouth. As is typical of hidden-brooders, there are clear differences between the sexes; the males usually have more strongly developed fins. In the aquarium, *Pelmatochromis* is mainly peaceful towards other fish, but amongst themselves fights relating to rank may break out. So, it is best to keep a pair of specimens only, or, at most, keep them in the same tank as different fishes. *Pelmatochromis* does not stir up the substrate normally and also leaves the plants in peace. Conditions in the tank ought to approximate the natural conditions as far as possible, ie a temperature of between 25 and 27 °C, hiding-places in the form of clumps of plants, roots and rock-caves (also flowerpots cut in two!), and the water should not have too low an electrolyte* content. Too soft fresh water can have some sea salt added to it therefore (1 tsp to 10 litres water), or use can be made of 'biologically treated' water that inevitably has a really high ion* content. The composition of the water is in fact less important for older specimens than it is for breeding purposes, since in fresh water the spawn tends to have fungus growing on it. Where possible, allow the pairs to select themselves from amongst a shoal of young fish, If the right kind of spawning grounds are provided (caves), they will even spawn in community tanks. On

Fig. 871 Representative of the G *Pelmatochromis*

average, the eggs number 100–250, although many species are much more productive – the number of eggs they lay is comparable only with the Cichlids that brood in the open waters. Usually, one of the parents devotes itself particularly intensively to taking care of the brood, in *P. ansorgei*, for example, it is the female, in *P. guntheri* the male. The systematics of the genus has been much revised of late and so it cannot escape the notice of the aquarist either. Numerous species that are important in aquarium-keeping have been removed from the *Pelmatochromis* group and now belong to separate Ga *(Pelvicachromis*, Haplochromis*)* However, despite this re-grouping, the taxonomic status of some forms is still not explained completely satisfactorily.

The G *Pelmatochromis*, in its new narrow sense, comprises the following aquaristically important species.

– *P. ansorgei* BOULENGER, 1901. From tropical W Africa. Length to 13 cm. Body a gleaming sea-green. In the female the belly is pink, in the male red. There is a blue-green blotch on the operculum, edged in red, and a row of dark flecks on the flanks. Synonyms for this species are *P. arnoldi* BOULENGER, 1912, and *P. annectens* BOULENGER, 1913.

– *P. guentheri* (SAUVAGE, 1912); Guenther's Mouthbrooder. From tropical W Africa. Length to 20 cm. Mouth very noticeably pointed. Olive-green with 2 irregular, dark. lorgitudinal bands, the top one of which is much thinner. D edged in red, the An and the iris also red. A mouth-brooder.

– *P. kingsleyae* (BOULENGER, 1898). From Gabon. Length to 25 cm. Similar to *P. guntheri*, Also a mouthbrooder.

Peltandra, G, *see* Araceae

Peltodoris, G, *see* Nudibranchia

Peltate Leaf *see* Leaf

Peltogaster paguri *see* Rhizocephala

Pelvicachromis THYS VAN DEN AUDENAERDE, 1968. G of the Cichlidae*. Distributed in tropical W Africa where the G is represented by several species. They are smallish, fairly elongate Cichlids that were formerly attributed to the G *Pelmatochromis**, although they clearly differ from them in the shape of the head, among other things. The nasal area is strongly arched, whereas the chin profile is almost straight. The *Pelvicachromis* species are also even more colourful than any members of the G *Pelmatochromis*; the females especially are reminiscent of Coral fish in their intensity of colour. Care and breeding as for *Pelmatochromis.*

Despite modern revisions, the taxonomic situation of *Pelvicachromis* is still still very unclear. For some time, we have not been sure if we are talking about very variable species or if the individual colour varieties are species in their own right. The following are important in aquarium-keeping.

– *P. pulcher* (BOULENGER, 1901); Kribensis; Medium Rainbow Cichlid. From S Nigeria. Length to 10 cm. Upperside brownish, lower side ivory colours, belly and V a deep wine-red or violet. In the rear section of the D there are 1–5 dark, round blotches with pale yellow edges. Younger specimens have the suggestion of a dark longitudinal band as well. *P. aureocephalus* (MEINKEN, 1960) is regarded as a synonym for this species (fig. 872).

– *P. subocellatus* (GÜNTHER, 1871). From tropical W Africa. Length to 10 cm. Dorsal surface blackish-

Fig. 872 *Pelvicachromis* 'Kribensis'

green, flanks olive to ochre, belly in the male reddish, in the female red-violet. Fins rust-red with lovely pale blue markings, C with a characteristic eye-spot. During spawning, the female is a velvet-black with a chalky white dorsal surface. The so-called 'Pelmatochromis klugei II' probably belongs here, too.

– *P.* sp., so-called 'Camerunensis'. From S Nigeria. Length to 10 cm. Similar in colour to *P. pulcher*, but the longitudinal band running from the upperside of the mouth to the C is still very dark in the older specimens too (fig. 873).

Fig. 873 *Pelvicachromis* sp.

– *P. taeniatus* (BOULENGER, 1901). From Nigeria, Cameroun. Length to 9 cm. Olive-yellow to golden-yellow, with a black longitudinal band bounded by gold. Ventral surface blue-violet to green, never red (which is in contrast to the above-mentioned species). The D and C have several dark, round blotches on them. *P. klugei* (MEINKEN, 1960) is regarded as a synonym for this species.

Pencilfishes, G, *see Nannostomus*

Penguinfish *(Thayeria boehlkei) see Thayeria*

Penguins (Sphenisciformes, O) *see* Aves

Penicillin. One of the most commonly used antibiotics* that is formed from several species of mould fungus of the G *Penicillium*. Its effect is to stop the growth of bacteria. Pure penicillin forms colourless, bitter crystals that dissolve easily in water. In aquarium-keeping it is used against bacterial infections*, especially against spot diseases*, swimbladder inflammation*, bacterial intestinal infections (Intestinal Diseases*), scale protrusion*, fin-rot* and others. For prolonged baths (Bathing Treatments*), 10,000–40,000 international units (IU) are dissolved in 1 litre of aquarium water; the bath should be renewed every 2 days. Often, penicillin is combined with streptomycin* when being used.

Pennant Butterfly-fish *(Heniochus acuminatus) see Heniochus*

Pennant Cardinalfish *(Apogon nematopterus) see Apogon*

Pennant Coralfish *(Heniotus acuminatus) see Heniochus*

Pennatularia; Sea Pens. O of the Anthozoa*, Sub-Cl Octocorallia. Pennatularia live on soft substrata in the sea at depths of about 20 m. They feed on small animals. The club-shaped end to the elongated primary polyp is stuck into the bottom; the secondary polyps sit directly on the upper part or on side-leaves, and perform the function, among other things, of catching food. Many Pennatularia shine. In the aquarium, they are very sensitive and require very fine sand, at least 10 cm deep. Feed on *Artemia* nauplii.

Penny Algae *(Halimeda tuna) see Halimeda*

Pentaceraster, G, *see Oreaster*

Peplis LINNÉ, 1753; Water Purslane. G of the Lythraceae* comprising 3 species. With the exception of Australia, they grow in places on all the other continents. They are small perennial or annual plants with decussate leaves. Flowers hermaphroditic, small and inconspicuous. Perianth made up of 3-part circles, doubled; often, the corolla is missing in part or completely. Stamens 6–12. Carpels 2, fused, superior. Capsule-fruit. The species of the G are small marsh plants that can also tolerate flooding. The following has been kept in an aquarium with varying degrees of success.

– *P. diandra* DE CANDOLLE (*Didiplis diandra*, name not in current usage); American Water Purslane, American Marsh Wild Thyme. N America. Leaves linear to narrowly lanceolate (fig. 874).

Fig. 874 *Peplis diandra*

Peppered Corydoras *(Corydoras paleatus) see Corydoras*

Pepperwort, G, *see Marsilea*

Peprilus simillimus *see* Stromateoidei

Perca LINNAEUS, 1758. G of the Percidae*. From Europe, Asia, N America, where they often live in still and flowing waters, but also in brackish water. Large fish that often form a shoal when young, but in later life they are pronounced solitary animals from very specific habitats, and they are predators. Body moderately elongate, tall and only slightly flattened laterally. Covered in small comb-scales. The head is snub and the terminal mouth is large and protrusible. The jaws contain uniformly shaped teeth. The back of the operculum extends into a strong spine. LL complete. The D is separated, the Dl with hard rays; C curving inwards slightly. Provided they do not stem from oxygen-rich waters, young specimens will adapt well to life in a room aquarium. Largish tanks with good ventilation, a gravelly substrate and groups of plants all help the fish to settle. Temperature not more than 22°C. All types of live food are eaten. Since older specimens are definite predators, they should also be fed some fish. In the spring, the females lay the spawn in long jellied bands on flat parts of the bank. After a cool overwintering, *Perca* will also spawn in large aquaria. Has some economic importance locally.

– *P. flavescens* (MITCHILL, 1814); Yellow Perch. NE USA, from the area of the Great Lakes to the SE coast. Length to 30 cm. Similar to *P. fluviatilis*. Body clay-yellow with a golden sheen. Operculum covered with dark streaks. D greenish-yellow, D1 with a black blotch. V and An orange to red.

– *P. fluviatilis* LINNAEUS, 1758; Perch. Found almost throughout Europe and NE Asia, also in N America. Length to 50 cm. Dorsal surface black-grey or grey-blue, flanks green-yellow, in part with a bluish iridescence and 6–9 dark transverse bands. Ventral surface pale. The rear part of the Dl has a black blotch on it. C yellowish-green to orange-red. V and An orange to brick-red, P clay-yellow. There are often ecological forms with colours that deviate from those given (fig. 875).

Fig. 875 *Perca fluviatilis*

Perch *(Perca fluviatilis) see Perca*

Perches, O, *see* Perciformes

Perches, related species, Sub-O *see* Percoidei

Perches, True, F, *see* Percidae

Percidae; True Perches (fig. 876). F of the Perciformes*, Sub-O Percoidei*. Confined to the inland waters of the northern temperate zone, they are elongated predatory fish, usually flattened laterally. Typical of the Perches are the separated dorsal fins (although sometimes they are not fully separated, only a large indentation being present), and the anal fin, which begins with 1–2 spines only. Often the D2 and An stand opposite each other. The head is large, the mouth terminal or subterminal, large or relatively narrow. The premaxillary bones are often, although not always, protrusible. The jaws, vomer and palatal bone have rows of teeth on them, the teeth being strong or fairly fine. LL usually complete. The Perches are closely related to the Serranidae*. There are 2 Sub-Fs: Percinae and Etheostomatinae. The centre of distribution is N America. The oldest fossils date from the Neogene. Perches are found in warmer and cold waters that may be still or flowing. Many species, at least when they are young, seek out weedy bank zones. Various species, eg the Zander, are very important in the fisheries industry. Some species adapt well to a habitat aquarium. For further details see under the Ga *Acerina**, *Etheostoma**, *Lucioperca**, *Perca**, *Stizostedion**.

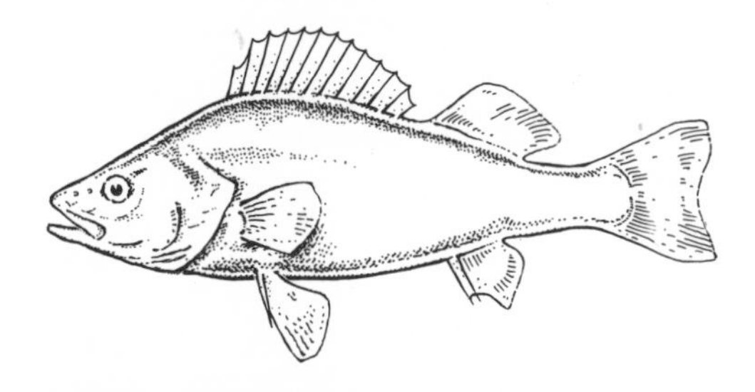

Fig. 876 Percidae-type

Perciformes; Perch-types. O of the Osteichthyes*, Coh Teleostei. This O of bony fishes has the largest number of forms and species, and in the system devised by GREENWOOD, and colleagues (1966), it is divided into 20 Sub-Os. In older systems, some of the Sub-Os given here were regarded as Os in their own right, and, vice versa, some of the Os that appear here used to be considered as Sub-Os, Sup-Fs or Fs. So, from this fact alone, it is clear that problems still exist with regard to the natural relationship of the individual fish groups in this O and of fish groups closely related to this O. Such circumstances are in fact apparent from the 1966 study of the Sup-O Acanthopterygii conducted by GREENWOOD and colleagues. The oldest known Perciformes fossils date from the upper Cretaceous. In the Eocene (Tertiary), the number of Ga and species increased in leaps and bounds. Perches are mainly marine fish, with only a few Fs penetrating the fresh water in the course of evolution. Many members of the O are reminiscent in their habits of the well-known River Perch, others bear only a slight resemblance or none at all. The front spiny-rayed part of the dorsal fin and the soft-rayed part to the rear are divided into a D1 and D2, or they are split down a large indentation; then again sometimes the fin is uniform. The C has at most 17 fin rays, the An often a few spiny rays at the front. The Ps are positioned high up on the body, and the Vs are located at the breast or at the throat. There is a connection between the pelvic and shoulder girdles. The swimbladder is usually present but is not connected to the intestine. As a rule, the scales are comb-edged; no fish-bones. Other important features concern the skeleton, especially the cranium and the vertebral column. The following 17 Sub-Os are treated under separate entries: Acanthuroidei, Ammodytoidei, Anabantoidei, Blennioidei, Callionymoidei, Gobioidei, Labroidei, Luciocephaloidei, Mastacembeloidei, Mugiloidei, Notothenioidei, Percoidei, Polynemoidei, Scombroidei, Sphyraenoidei, Stromateoidei, Trachinoidei.

Percoidei; Perches, related species. Sub-O of the Perciformes*, Sup-O Acanthopterygii. This Sub-O contains the largest number of species within the Perciformes, and GREENWOOD and colleagues (1966) attribute about 70 Fs to it. The Percoidei are the most primitive members of the O – various skull bones are not yet fused firmly with one another, as is characteristic of many specialised Sub-Os. In habits, the members of this Sub-O resemble the typical Perch-type, as a rule. The Vs are always beneath the Ps or even at the throat, and they are usually supported by 1 spine and 5 soft rays. The number of vertebrae is nearly always 24–25; physoclists; ctenoid scales, in rare cases, cycloid scales. Most of the Fs are represented only in the sea, but some also in fresh water. Many Fs are very important in the aquarium, and numerous species of the Sub-O have specific uses. 31 Fs are treated as separate entries in this dictionary, and their systematic position is detailed under the Osteichthyes*.

Percopsidae, F, *see* Percopsiformes

Percopsiformes; Trout-Perch types. O of the Osteichthyes*. Coh Teleostei. In other systems also known as the Salmopercae or Xenarchi. This is a small O usually represented by the Fs Aphredoderidae and Percopsidae. In the system give here (GREENWOOD and colleagues 1966), the Amblyopsidae are included too. Physoclists. In body shape, and with the 0–4 spines at the beginning of the otherwise soft-rayed D, An and Vs, the Percopsiformes look like Perch-types (Perciformes*); but with the small adipose fin (which is not formed in every case), they are reminiscent of the Salmon-types (Salmoniformes*). They can be distinguished from the former by the position of the Vs. These lie always behind, never below, the Ps, as is typical of the Perch-types. The oldest known fossils of this O date from the Neogene. All the members live in the fresh waters of N America.

– *Amblyopsidae;* Blindfishes, North American Cave-dwelling Fishes. Small, spindle-shaped, elongated fish from central and south-eastern N America. Of the 5 species, only one, the 15 cm long *Chologaster cornutus* (fig. 877) occurs in waters above ground. It is characterised by 3 narrow, dark longitudinal bands. All the other members of the F are cave-dwellers. *Amblyopsis spelaeus* (fig. 877) has the most extreme adaptations to this life underground – its eyes have almost regressed

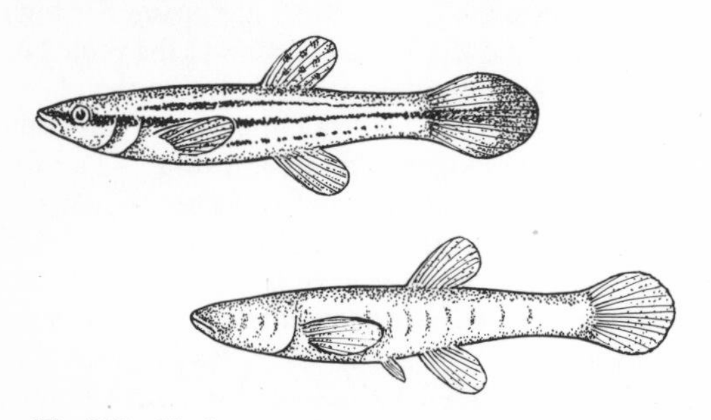

Fig. 877 *Chologaster cornutus* and *Amblyopsis spelaeus*

completely. The species grows up to 13 cm long and it comes from the limestone region of the eastern Mississippi Basin. Without eyes, *Amblyopsis* uses highly developed long-distance sensory organs to orientate itself. These are embedded in bulging folds on the head, body and tail. Young members of the species still retain small eyes. The old belief that the Amblyopsidae were livebearing does not hold true. Rather, they seem to be gill-cavity brooders – in fact, females can provide a home for up to 70 eggs in the gill cavity. They take about 8 weeks to mature. However, this still begs the question of how the eggs get in there in the first place. As the anal papilla in the Amblyopsidae is located far forwards beneath the throat, it is possible that the eggs transfer directly into the gill cavity from the oviduct. Individual specimens of the F are occasionally kept in the aquarium.

– *Aphredoderidae;* Pirate Perches. This F is represented only by the 13 cm long Pirate Perch *(Aphredoderus sayanus)* (fig. 878), which is found in the eastern states of the USA. It is like an elongated Sunfish

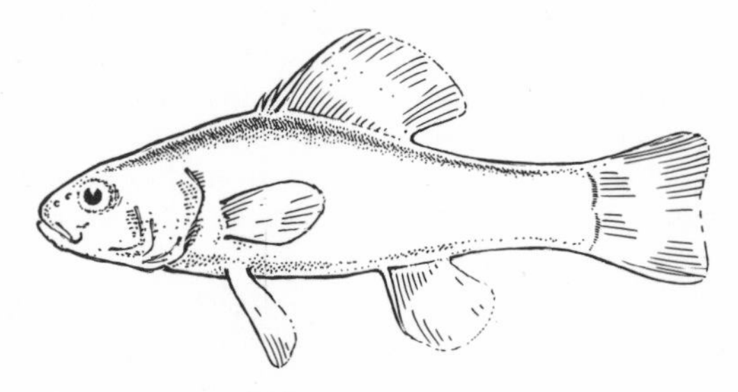

Fig. 878 *Aphredoderus sayanus*

(Centrarchidae*), and is dark olive-green to brownish in colour with spotting. The fish sticks very much to habitat and lives in still or slow-flowing waters. The position of the anal opening is particularly characteristic. In young specimens, it is located in front of the An; later on, it moves further and further forward and ends up in the throat region. No adipose fin. The D has 3 spines, the An 2, and the Vs one. The eggs are laid in simple nests and guarded by the female.

– *Percopsidae;* Trout-perches. The two species of this F live in lakes and rivers. They are silvery and covered in dark spots. The well-known species, *Percopsis omiscomaycus* (fig. 879) is distributed from Alaska to the NE states of the USA and grows up to 15 cm long. In a lot

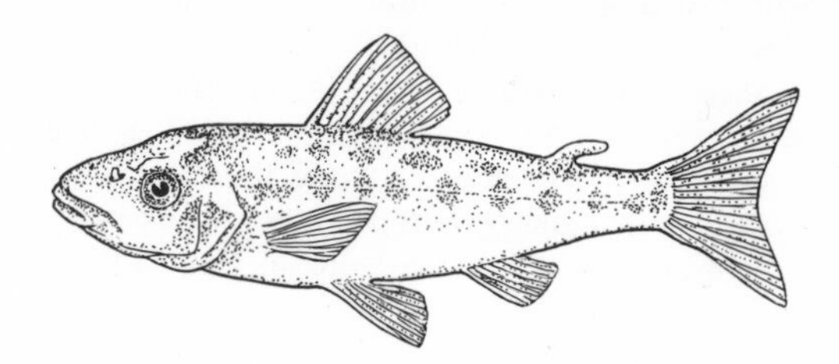

Fig. 879 *Percopsis omiscomaycus*

of places, it is one of the most important foods of larger kinds of predators. The Percopsidae have a small adipose fin, the D and An have 1–2 spines, and the V one spine. They lay their spawn in open water in shallow zones, and most species die after spawning. Since the species seeks out deep water during the day and only comes up to shallower water near the bank at night, it is very rare to see one.

Percopsis omiscomaycus *see* Percopsiformes

Perennial Plant. In contrast to an annual plant*, a perennial grows for several or many years. In order to flower, it must have reached a certain strength. The structure of the vegetative body and the pattern of growth are many and various among perennial plants. Almost all aquarium plants are perennials.

Pergonium *see* Flower

Periclimenes, G, *see Natantia*

Perigone *see* Flower

Periophthalmidae, F, name not in current usage, *see* Gobioidei

Periophthalmus, G, *see* Gobioidei

Peripheral Nervous System *see* Nervous System

Periphyton. Microscopically small algae and protozoans which live attached to aquatic plants, particularly their leaves, as well as to rocks, wood and other substrate material. Snails and many fish, such as Livebearing Tooth-carps, feed on periphyton.

Periwinkles, G, *see Littorina*

Perla GEOFFROY, 1764 (fig. 880). G of the Stone Flies (Plecoptera*). Represented in Europe by about 7 species. Yellowish-brown in colour, these insects are 2–3 cm long. Their larvae live primarily in streams and the upper reaches of rivers, and are especially striking because of their size and very contrasty yellow-brown patterning. They lead a predatory life-style, lying beneath rocks. The larval phase last 3 years.

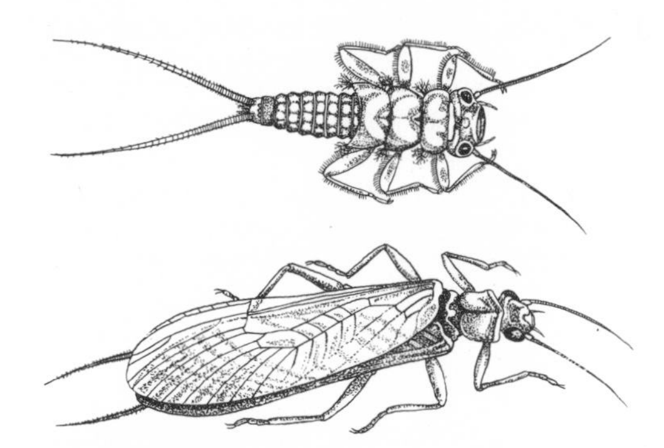

Fig. 880 *Perla.* Top, larva; below, imago

Permanent Hardness. That part of the water hardness* which will not be removed by boiling, ie all the calcium and magnesium compounds with the exception of the hydrogen carbonates and carbonates. Hydrous calcium sulphate ($CaSO_4 \cdot 2H_2O$), otherwise known as gypsite (gypsum), constitutes the main part of permanent hardness, which is why it is sometimes called gypsum or sulphate hardness. Other substances that contribute to the hardness are magnesium sulphate, calcium and magnesium chloride, -nitrate, -phosphate and -silicate; the term mineral acid hardness refers to these things. Lastly, as organic calcium and magnesium salts such as humates (Humic Acids*) must be taken into account, in addition to permanent hardness, the term non-carbonate hardness seems the most exact.

Permanent hardness can be calculated from the difference between the total hardness and the temporary hardness (Water Hardness, Determination of*). To remove it, ion exchangers* are increasingly being used in aquarium-keeping.

Permanganate Consumption. The contamination of water* with organic oxidisable substances can be determined roughly with potassium permanganate ($KMnO_4$) as a means of oxidation; in addition, permanganate solutions will be needed in exactly titrated amounts. Although this method is constantly used to judge the purity of drinking water, industrial-use water and sewage water, it is not important in aquarium-keeping. From the aquarist's point of view, the most important thing is to be able to interpret correctly the published values of the fishing waters that interest him. Usually, permanganate consumption is given in terms of potassium permanganate consumption – so many mg to one litre of test water. In very pure water, the values lie between 0 and 10 mg, in very impure water they may climb to 1,000 mg and more.

Petal *see* Flower

Petiole *see* Leaf

Petitella GÉRY and BOUTIERE, 1964. G of the Characidae*, Sub-F Tetragonopterinae. From the lower Rio Huallaga in the Loreto district of Peru, the Rio Maranon near Iquitos. Length to 6 cm. They are peaceful shoaling fish from the middle and lower levels of water. Clear, soft, and slightly acidic water of 2–4 °dH and 6.4–6.8 pH is preferred with thick vegetation in parts of the tank, leaving a free swimming zone. Temperature 25–27 °C. The body is elongate, compressed laterally. The D and An are short, there is no adipose fin, and the C is deeply forked. The only species, *P. georgiae*, is distinguished from the deceptively similarly shaped and coloured *Hemigrammus* * *rhodostomus* (Red-nosed Tetra) because is has only one (not two) rows of teeth in the premaxillary jaw bone and there are also differences in the structure of the cranium. The upper jaw and premaxilla are also somewhat longer. (The *Hemigrammus* species is also a bit smaller.) *Petitella* has been bred successfully on more than one occasion, but the number that hatched out was small and the losses at the rearing stage have always been large. Breeding should be attempted in roomy breeding tanks*, otherwise as instructed for *Hemigrammus*.

– *P. georgiae* GÉRY and BOUTIERE, 1961; False Rummy-nose Tetra. Flanks silvery with a lovely, blue-green

Fig. 881 *Petitella georgiae* (top) and *Hemigrammus rhodostomus*

iridescence, dorsal surface a delicate brown-olive. The D in the female is colourless, in the male it is a delicate reddish shade. The An is colourless or whitish with a black longitudinal band. The adipose fin and Vs are milky-white, and the C is basically pastel-white with 3 broad, black, longitudinal bands running across it (the middle one of which continues into the caudal peduncle). Unlike *Hemigrammus rhodostomus*, this species does not have the 2 black blotches on the caudal peduncle. The shining blood-red colouring extends across the whole head and the top half of the iris and, unlike the Red-nosed Tetra, the colour always maintains its intensity. It is only because of the few successes with breeding that this species has not become more widespread in the aquarium (fig. 881).

Petrolisthes oshimae *see* Porcellanidae

Petromyzon marinus *see* Cyclostomata

Petromyzoniformes, O, *see* Cyclostomata

Petroscirtes RÜPPEL, 1828. G of the Blennidae, Sub-O Blennioidei*. Contains numerous 5–12 cm long species in tropical and subtropical parts of the Indo-Pacific. *Petroscirtes* swims around freely in the water, near the bottom, but flees into caves when danger threatens. Feeds on small animals. Body an elongated cylinder shape, snout pointed. The long D begins at the head, the C is rounded. On the lower jaw, there are 2 long sabre-like teeth with which they can give a painful bite. Several species are covered in yellow-black longitudinal stripes. In the aquarium, *Petroscirtes* seem to be peaceful among themselves and towards other fish. They eat small crustaceans, earthworms and other things. *P. temmincki* BLEEKER, 1851 (the Striped Slimefish) is frequently imported. It comes from the Indo-Pacific and is 12 cm long. It has 3 black longitudinal stripes and the caudal peduncle and C are flecked.

Petrotilapia TREWAVAS, 1935. G of the Cichlidae*. There is only one species and it is endemic to Lake Malawi*. It grows up to 20 cm long, is very tall and looks bulky, although it is a skilled swimmer. Like the other Lake Malawi Cichlids (*Pseudotropheus**, *Labeotropheus**), *Petrotilapia* is highly specialised in respect of the way it acquires food and reproduces. Its mouth is thick-lipped, extremely mobile and inside there are dense rows of spatula-shaped teeth. Examinations of the stomach reveal that in nature the fish grazes only on algal growth (and the small animals contained inside) which it finds on the rocks. In front of the eyes, the forehead is sunken, above the eyes it juts forward in a bulge. The eyes themselves are remarkably small. The D is inserted along a good stretch of the body, as is also the case in *Pseudotropheus* and *Labeotropheus*. The C curves gently inwards. Colouring very variable; apart from uniformly brown specimens, there are also some with different shades of blue and orange-coloured bellies, some that are red-golden all over etc. This phenomenon also occurs among other Lake Malawi Cichlids (eg *Pseudotropheus* zebra*). For care and breeding, see under *Pseudotropheus* (ie very large tanks with rock constructions and hard, neutral water). In captivity, *Petrotilapia* is fond of live food, so feeding is no problem. Additional plant food (algae) should also be offered, however. The female is a mouth-brooder. After the fry can swim freely, they are already about 15 mm long and can tackle small *Cyclops*. With plenty of food, the fry grow very quickly. At a length of 10–12 cm, the males achieve the full intensity of their colouring.

– *P. tridentiger* TREWAVAS, 1935. Lake Malawi. Length to 20 cm. Colouring extremely variable, adult males usually a shining blue colour, the throat and belly may be brown, or yellow. There are egg-shaped blotches on the An. The females are usually brown, some with orange-coloured bellies or golden all over. Juveniles have dark transverse stripes (fig. 882).

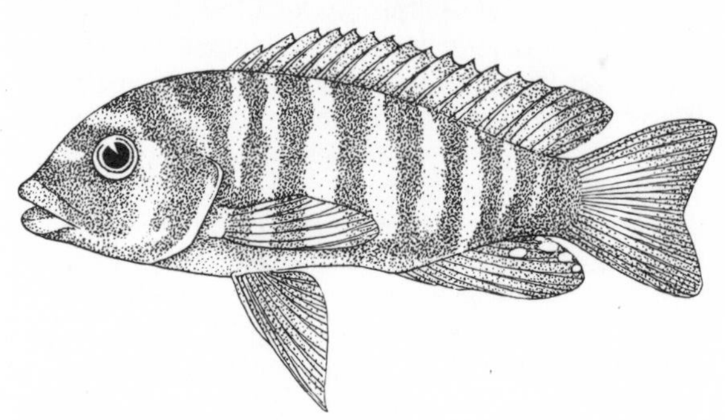

Fig. 882 *Petrotilapia tridentiger*

Petticoat Fish *(Gymnocorymbus ternetzi) see Gymnocorymbus*

Peyssonellia. G of the marine Red Algae (Rhodophyta*) that is found in shady spots and in greater depths. It has dark red, round-leaved thalli that attach themselves to the substrate with rhizoids. The common Mediterranean species *P. squamaria* (GMEL.) DEC. survives quite well in a marine aquarium. Sometimes, it even grows a little, but it is easily killed off by green algae.

pH value. The pH value is a measure of how strongly acid or alkali water reacts. The acidic reaction is caused by an excess of hydrogen ions (H^+), the alkali reaction by an excess of hydroxide ions (OH^-). Chemically pure water will react neutral, since it contains equal amounts of these types of ions*, but in natural waters the pH value can differ enormously because of the acid (Acids*) and salt content.

The product of hydrogen ion and hydroxide ion concentrations in aqueous solutions is constant. The chemist expresses this by the following equation:

$$C_{H_+} \cdot C_{OH_-} = K$$

From this can be seen the dependence of the one ion concentration on the other, and so it suffices to give just one of them. To be consistent, the H^+-ion concentration is the one always referred to. The constant mentioned above is 10^{-14}. If both types of ion are present in the same numbers, it amounts to $C_{H^+} = 10^{-7}$ therefore (which is the case in totally pure water). The pH (*pondus hydrogenii*, weight of the hydrogen ions) is now defined as a negative decadic logarithm of the C_{H^+}, and thus the pH of this water is 7. The smaller the negative exponents of the C_{H^+} become, the more strongly acid will the water react. Greater values than 7 signify a predominant OH-ion concentration, and as a result an alkali reaction. The connections between the pH value scale and reaction can be represented thus:

0 1 2 3 4 5 6 7 8 9 10 11 12 13 14

increasingly acid increasingly alkali
neutral

Depending on water type, the pH value of naturally occurring waters lies between 4 and 9. Particularly acid are the high moors and other dystrophic waters*; most tropical waters have values of between 6.5 and 6.8; Lakes Tanganyika* and Malawi* have a weak alkali reaction (7.5), and finally, sea water has a pH value of 8.3–8.4. Interesting in this respect is the significance of the salts dissolved in the water for the constancy of the pH (chemists refer to a buffer* effect). This gives rise to one of the basic rules when setting up an aquarium: never go below a certain level of salt content (Water, Preparation of*). In aquarium-keeping, the Czensny-indicator used in conjunction with a comparison scale is particularly suitable for measuring pH values. In this way, the pH value can be calculated to within 0.2 of a unit. 4 drops of the indicator solution should be placed in 5 ml of the test water, and the colour that results should be compared against the 14 grades of the scale. (If testing sea water, read instruction under Sea Water* first.) To correct a pH value, basically only weak acids and bases, or their salts can be used. Hardish, well buffered water is scarcely usable at all, and must be prepared in advance (Ion Exchangers*, Water, Preparation of*). To increase the acidity of soft water, humic acids* or tannic substances* are used, so as to imitate as closely as possible natural conditions. To raise the pH value, however, the alkali carbonates or alkali hydrocarbonates (Na_2CO_3, $NaHCO_3$) are especially suitable. If strong acids and bases are used, once the buffer capacity is exhausted the pH will alter dramatically and then it will be at a level outside the one wanted!

Phaeophyta; Brown Algae. D of the Algae* with 2,000 species. Brown algae with chlorophyll a and a little b, which are covered by brown fucoxanthin. Chromatophores spherical or lens-shaped. The actively mobile cells have 2 flagella on the side, one of which points forwards and the other backwards.

The vast majority of the species belong to the marine benthos* that is closest to the surface. They live in the sea-spray zones of the coasts, in the tidal zones or completely submerged; many live on top of other algae. In particular, Brown Algae occur in temperate and cold seas. Only a few species live in tropical sea areas or in fresh water.

There are no single-celled forms of Brown Algae. Even the simplest kinds have branching threads, such as *Ectocarpus**. Broad-surfaced or ribbon-shaped vegetative bodies join onto these threads. The most highly differentiated forms (the seaweeds, eg species of the Ga *Fucus** or *Laminaria**) have multi-layered thalli made up of various kinds of tissue. Some have gas-filled bladders to increase buoyancy. Many of the species look as if they are divided into stems, leaves and roots, and have a much greater extent than even the largest land plants. In some, the change in season is expressed by phases of growth and resting periods. During the resting period, certain parts of the thallus are allowed to drop off, rather like the leaf-fall of our summer-green trees. A unique form of Brown Algae is *Sargassum** which is found from the West Indies to the Azores, and forms a floating mass of vegetation. Iodine is obtained from many kinds of Brown Algae, and others yield mucins for the cosmetic and food industries; some are even eaten in E Asia.

Phago GÜNTHER, 1865. G of the Citharinidae*. From tropical W Africa. Length to 15 cm. These Characins have strong teeth and a wide-gape mouth, caused by the beak-like elongation of the jaws. *Phago* species are predators that lie in wait for prey between reed stems. ARMBRUST reports that they feed almost exclusively on the fins of larger species of fish, which is different from what used to be thought. The fins are bitten off as *Phago* comes towards the victim at break-neck speed. Unfortunately, specimens arrive in Europe so hungry that nothing can usually be done to save them. The body and caudal peduncle are very elongated and not very compressed laterally; the forehead is flattened. Each jaw has 2 rows of teeth. The strong ctenoid scales (Scales*) have bristles and edges on them, and the nape too is covered with scales. The fins are fairly small, and they have hard rays in part. The C is deeply forked and the adipose fin very small. LL in the middle of the body. Provide well planted tanks as the fish are shy. Nothing is known about their reproduction.

– *P. loricatus* GÜNTHER, 1865; Armoured Phago. From the Niger river system. Length to 15 cm. Flanks reddish-brown with 2–3 dark longitudinal bands, the middle one of which is the broadest; in between lie narrow, golden-yellow bands. Dorsal surface dark brown, ventral surface yellowish-white. The D has 2 black longitudinal bands, the lobes of the C have 3–4. The middle fin rays of the C covered in black wedge-shapes (fig. 883).

Fig. 883 *Phago loricatus*

– *P. maculatus* AHL, 1922; Pike Characin, Spotted Phago, African Pike Characin. Flanks yellow-brown with numerous, downturned transverse streaks along the LL. The dorsal surface is a cloudy fawn-brown or black-brown, the ventral surface is whitish-yellow and the upper jaw brown. Fins a weak yellowish tone. The D and C have the same patterning as *P. loricatus*. The body of this species is more compact than that of *P. loricatus*.

Phallichthys HUBBS, 1924. G of the Poeciliidae*, Sub-F Poeciliinae. Central America: Guatemala to Panama. Lives in slow-flowing and still waters: streams, calm river bays, ponds. Small or very small fish, males 2–6 cm, females 3.5–10 cm. The body in most species

is tall and looks compressed. Ground colouring of the body yellowish to brownish, otherwise very attractively coloured and patterned. Keep in tanks that have patches of dense foliage, at a temperature of 20–26 °C. Usually peaceful in the aquarium, *Phallichthys* is an omnivore that also requires plant food as a supplement. The 2 sub-species are often kept by aquarists:
– *P. amates* (MILLER, 1907). *P. amates amates* (MILLER, 1907); Merry Widow. From Guatemala, Honduras, Costa Rica, Panama. Male up to 4 cm, female up to 7 cm. The sides of the body have a clear longitudinal band running across them, which in the male is crossed by 10–12 dark, narrow, transverse lines. In direct light, the body has a blue shimmer. Fins yellowish. The D has a black band and a black edge, with a white border. The C has a faintly dark edge. A black streak runs through the eye and continues at an angle towards the rear. Pregnant females are best kept on their own for breeding purposes. Thick vegetation is essential. Up to 40 young fish develop, but they grow slowly (fig. 884).

Fig. 884 *Phallichthys amates*

– *P. amates pittieri* (MEEK, 1912); Orange Dorsal Livebearer. From Panama, Costa Rica. Male up to 6 cm, female up to 10 cm. Colouring similar to the previous sub-species. However, the male has 6–12 dark-blue transverse stripes. The C has a distinct blue-black border. The gonopodium is very long. More than 100 young fish may develop, and they grow quickly.

Phalloceros EIGENMANN, 1907. G of the Poeciliidae*, Sub-F Poeciliinae. S America: S Brazil, Paraguay, Argentina, Uruguay, where they live mainly in smallish, flowing waters of the lowlands and mountains, but also in other small stretches of water and in brackish water. Small fish, males 3–4 cm, females 5–7 cm. Body elongate, flattened laterally very little at the front, much more so at the back. Gonopodium forked like a set of antlers. Ground colouring brownish to yellow, usually flecked. The aquarium should contain thick clusters of vegetation in part, otherwise the parents will eat their spawn. 20–25 °C. Repeated partial water exchange and filtration are to be recommended. Pregnant females are best kept on their own; they produce 10–80 offspring. Omnivores; besides live food, dried and artificial foods, as well as plant food, are accepted. To date, the only species known to aquarists are:
– *P. caudimaculatus* (HENSEL, 1868); Caudo, One-spot Livebearer, Dusky Millions Fish. Male to 3 cm, female to 6 cm. Dorsal Dorsal surface olive-greenish, otherwise the body is yellow. There is a black, oval fleck beneath the D, and it has a silver, also golden, gleaming edge area. Fins yellowish; in the male the D has a black border. The standard form, in contrast to the following varieties or sub-species, can be kept at temperatures below 20 °C.
– *P. caudimaculatus auratus* (HENSEL, 1868); Golden One-spot. A colour variation. From Brazil, Argentina. Male to 3.5 cm, female to 5 cm. Usually there is no flank blotch. In the rear body section of the male there are 7–8 dark transverse bands.

Fig. 885 *Phalloceros caudimaculatus reticulatus*

– *P. caudimaculatus reticulatus* (HENSEL, 1868 ; Spotted Livebearer. From Brazil, Paraguay, Uruguay. Male to 4 cm, female to 7 cm. There are numerous black flecks of irregular shape on the body and at least at the bottoms of the fins. There is another variety of the sub-species, *auratus*, which has, in addition, golden and pale blotches (fig. 885).

Phalloptychus EIGENMANN, 1907. G of the Poeciliidae*, Sub-F Poeciliinae. S America: Brazil, Uruguay, Argentina. Lives mainly in smallish flowing waters. Small or very small fish; males 2.5–3 cm, females 3.5–4.5 cm. Body only moderately flattened laterally, elongated. Long gonopodium. There is information about care and breeding in the following species only:
– *P. januarius* (HENSEL, 1868); Barred Millions Fish, Striped Millions Fish. From SE Brazil, Uruguay, Argentina. Male to 3 cm, female to 4.5 cm. Body olive-brown on the dorsal surface, greenish-yellow on the flanks and with a brownish iridescence. Ventral surface pale. 8–12 dark, distinct transverse stripes. Fins colourless or slightly yellowish. Female paler, body translucent. Tanks should be well planted and the temperature 20–25 °C. A little addition of cooking salt or, even better, sea salt (1–2 tsp to 10 litres of water) is recommended. Live food is eaten in the main, but also dried and artificial foods, as well as algae. 10–25 fry are given birth to over a period of about one week. They are difficult to rear.

There is a black-piebald variety of this species.

Phallostethidae, F, *see* Atherinoidei

Phallostethids (Phallostethidae, F) *see* Atherinoidei

Phallostethiformes, O, name not in current usage, *see* Atheriniformes

Phallusis, G, *see* Ascidia

Phanerodon furcatus *see* Embiotocidae

Phantom Midges, G, *see Corethra*

Pharyngeal Teeth *see* Teeth

Phases of Life *see* Ontogenesis

Phenol Damage *see* Poisoning

Phenotype *see* Heredity

Phenoxetol *see* Phenoxyaethylene Alcohol

Phenoxyaethylene Alcohol. Known as Phenoxetol in the trade, this is a 2.2% aqueous solution, and is used to disinfect wounds, on account of its known effect of killing off bacteria and fungi. In aquarium-keeping, it is used in the treatment of Mycoses*, especially to combat *Ichthyosporidium**. Can be applied in the form of prolonged baths (Bathing Treatments*), using 10–20 ml of an original solution (1 ml to 10 litres of water) to every 1 litre of aquarium water; at the same time, it can be mixed in with the food.

Pheromones *see* Communication

-phil(ic). In compounds with other words means liking; eg thermophil(ic) = liking warmth; psammophil(ic) = liking sand; halophil(ic) = liking salt.

Philippine Rasbora *(Rasbora philippinica) see Rasbora*

Philippine Surgeonfish *(Acanthurus glaucopareius) see Acanthurus*

-phob(ic). In word compounds means fearing or avoiding, eg: halophobic = avoiding salt; acidophobic = avoiding acid.

Pholas, G, *see* Boring Molluscs

Pholididae, F, *see* Blennioidei

Pholis gunellus *see* Blennioidei

Phoronida, Cl, *see* Tentaculata

Photoblepharon, G, *see* Beryciformes

Photoperiodism. Term used to describe the fact that the formation of flowers in a plant is dependent on how long it is illuminated each day. In *short-day plants*, which are distributed primarily in the tropics and subtropics, flower formation takes place over a daily illumination period of around 12 hours *(Echinodorus parviflorus)*. If illumination lasts longer, they grow only vegetatively. Many short-day plants, especially floating plants, are put in severe danger if they receive considerably less than 12-hours-a-day illumination. This is why so many species die in northern botanical gardens during the winter, unless some additional light is made available. *Long-day plants*, which are distributed mainly in temperate zones, will only flower if they are lit daily for more than 14 hours; any less, and they too grow vegetatively only *(Echinodorus berteroi)* (fig. 886). *Day-neutral plants* do not show any dependency on the daily amount of illumination for the formation of their flowers *(Echinodorus paniculatus)*. Among marsh plants, the formation of aerial and aquatic leaves may also be controlled by how long illumination lasts.

Photosynthesis is the formation of organic substances (Carbohydrates*) from carbon dioxide and the hydrogen of water, under the catalystic influence of chlorophyll and the utilisation of light-energy. Light is trapped and stored in the form of chemical energy.

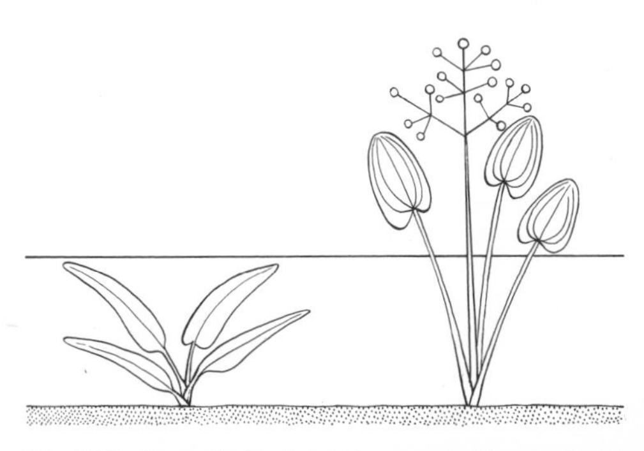

Fig. 886 Growth rhythm. Photoperiodism in *Echinodorus berteroi*; left, the plant in a short-day growth; right, the plant in a long-day growth

Oxygen is released with this process which is in turn used in respiration*. The main organs of photosynthesis are the leaves. Photosynthesis takes place in several partial reactions, such as light absorption, water dissociation, carbon dioxide combining and reduction. The physical-chemical details of these reactions are complicated and only partly explained. The intensity of photosynthesis is influenced by various factors, those of importance including the general condition of the plant, the provision of water, the width of the stomata*, the illumination, the temperature and the carbon dioxide content of the surrounding medium. (Carbon Dioxide Assimilation*.)

Phoxinus RAFINESQUE, 1820. G of the Cyprinidae*, Sub-F Leuciscinae. Large numbers of species across Europe, N and Central Asia, Japan and N America. Small, very lively shoaling fish that prefer the surface regions of oxygen-rich flowing waters and clear lakes. Body elongated and only slightly flattened laterally. Mouth terminal or slightly inferior, no barbels. Scales very small, LL complete or interrupted. D and An of equal length, C strongly inward-curving. Ventral surface between the V and An not keel-shaped. Peaceable fish that should be kept as a shoal in largish tanks with sufficient room for swimming; also provide groups of plants and clean gravel for a substrate. Add fresh water regularly and make sure the ventilation is good as *Phoxinus* requires plenty of oxygen. Temperature not above 18–20 °C. Some species have a good sense of hearing and smell. Diet consists mainly of insects and insect larvae, and some other live food is also accepted; difficult to adapt to dried food. After a cool overwintering, *Phoxinus* will also breed in the aquarium. Several females and males should be placed in a large tank with clear fresh water, good ventilation and some fairly large rocks and clumps of plants. Open-water spawners: the eggs attach themselves to rocks and plants. The fry hatch after 6 days and are easy to rear.

– *P. neogaeus* (COPE, 1866); North American Minnow. N America, the river system of the Mississippi and Missouri. Length to 8 cm. Dorsal surface black-olive, flanks dark greenish-brown with a broad, black, longitudinal band bounded above by white, and extending over the whole length of the body. Fins colourless to brownish.

– *P. phoxinus* (LINNAEUS, 1758); Minnow. Europe, Asia. Length to 14 cm. Colouring very variable. Dorsal surface brown-green, flanks paler with a silver sheen, ventral surface pale with a reddish shimmer. About 15 dark-brown transverse bands extend from the dorsal surface to the level of the LL, and these may unite into a longitudinal band on the flanks. A narrow, gleaming gold longitudinal stripe begins behind the eyes and reaches to the root of the tail. Fins yellowish-grey. The males have a red belly at spawning time, and both the males and females have spawning tubercles (fig. 887).

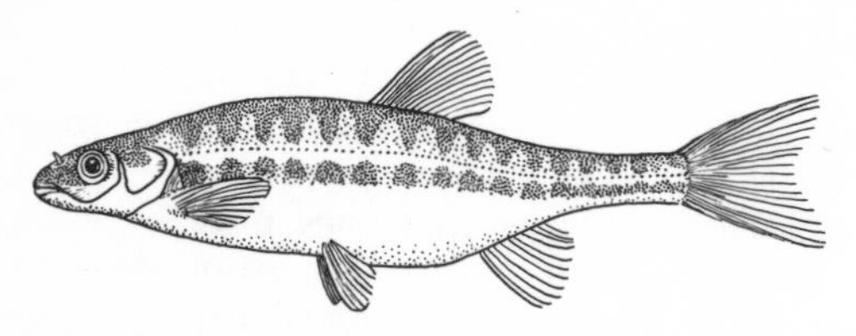

Fig. 887 *Phoxinus phoxinus*

Phractolaemidae, F, *see* Gonorynchiformes

Phractolaemus ansorgei *see* Gonorynchiformes

Phragmites communis *see* Poaceae

Phryganea LINNAEUS, 1758 (fig. 888). G of the Caddis Flies (Trichoptera*). 4 species in Europe, which are particularly noticeable because of their size (length of body 4 cm, wingspan up to 6 cm). The larvae prefer to live in still waters and build ice-cream-cone-shaped caddis cases. The plant material used for this is bitten off to the right length and arranged in a spiral. The larvae are agile creatures and live as predators. They are highly recommended as objects of observation in the pond-aquarium.

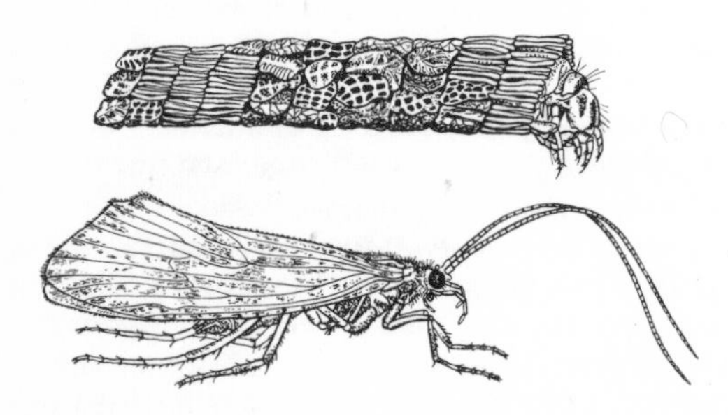

Fig. 888 *Phryganea.* Top, larva with caddis case; below, imago

Phycophyta, name not in current usage. Scientific name for the algae*, but one which is no longer valid, since it is made up of several independent divisions.

Phyllophora. G of the Red Algae (Rhodophyta*) from both cool and warmish seas. They have a variously shaped, usually ribbon-like, gristly thallus that is branched. Phyllophora will survive well in the marine aquarium and often grows slowly.

Phyllopoda; Fairy Shrimps. O of the Crustaceans (Crustacea*). Well-known groups such as the Water Fleas *(Cladocera*)* and the Triopsidae* are included here, as well as the largely overlooked Conchostraca*. All Phyllopoda are remarkable for their shell formations (bivalved or shield-like), and their size ranges from 1 mm to 10 cm. Almost exclusively confined to fresh water, and most species are very strictly bound to certain types of water. Some Phyllopoda are very important as food animals for fish.

Phylogenesis. The formation of all the living organisms of today in a long process of higher development (Evolution*), in the course of which more and more new forms have come into being and are still doing so. That is to say, a part of the offspring of an originally homogeneous group of organisms develops by alteration of the hereditary material* (*see* Mutation) in a different direction. Gradually, the degree of relationship with the original form becomes less and less, as each generation develops further, and so on. Early in the Earth's past, the division between plant and animal organisms took place, and later on the development of today's plant and animal phyla, which for their part are divided into numerous related groups (systematics). For example, over 350 million years ago the first ancestors of today's amphibians developed from certain Stump-fins. Later, the first reptiles developed from these, and also the mammals and the birds.

We gain our knowledge of phylogenesis and the system of the natural relationships of the organisms from the remains of prehistoric living beings (Fossils*) and from the results of comparison studies of today's animal and plant kingdoms. The explanation of phylogenesis and the mechanics of evolution are the province of descent theory*, combined with various fields of research, eg comparable anatomy*, physiology* and biochemistry*, genetics*, palaeontology* and systematics*.

Phylogenetic Tree *see* System

Phylum *see* Systematic Categories

Physa DRAPARNAUD, 1801; Bladder Snail (fig. 889). G of the Basommatophora*. 7 species in Europe, some distributed in the holarctic realm*, and some introduced from N America. Typical of the Bladder Snails and other Ga of the F Physidae (eg *Aplexa*) the thin shell with a sinistral coil. The species are well suited to unheated aquaria and for open-air installations.

Fig. 889 *Physa fontinalis*

Physailia BOULENGER, 1901. G of the Schilbeidae*. Glass Catfishes native to Africa. To 10 cm. Shoaling fish fond of swimming that live in middle and lower water levels, and do not do well when kept alone. Keep in large, well-planted aquaria with a free swimming zone and subdued lighting. Temperature about 25 °C. Feed only with live food such as *Daphnia*, Midge larvae, *Tubifex* and Whiteworms; they will not accept dried food. Body elongated, much flattened laterally. Head slightly flattened with 4 pairs of long barbels, held away from the body. D1 missing, D2 very small. C has two lobes and is moderately forked. An very long, Vs small and Ps strong. Reproduction of Glass Catfishes is unknown.

– *P. pellucida* BOULENGER, 1901; African Glass Catfish. Upper Nile region. Length to 10 cm. Body almost colourless and transparent like glass. In the light, it has a blue iridescent shimmer. The vertebral column and swimbladder are clearly visible. Isolated dark blotches of pigment occur on the dorsal surface. *P. pellucida* is a very interesting, utterly peaceful Catfish species (fig. 890).

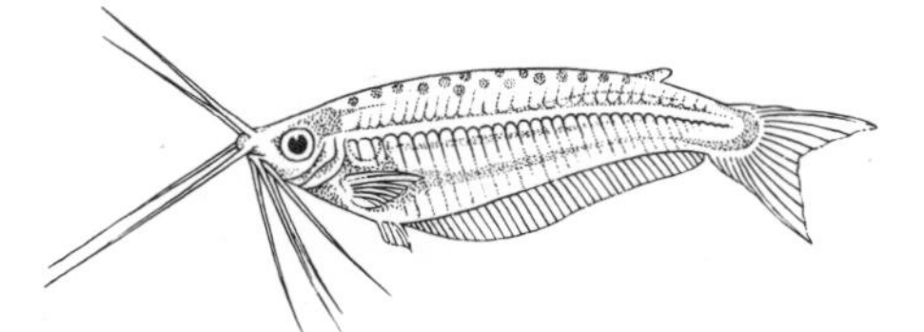

Fig. 890 *Physailia pellucida*

Physalia, G, *see* Stinging Capsules

Physiology. The study of the life processes of organisms. A branch of biology* and medicine. The whole of physiology is subdivided into the physiology of microorganisms, plants, animals and humans, which in their turn are made up of numerous specialist fields. Important disciplines of animal physiology deal with, eg, the nerves, senses, locomotion, digestion, metabolism, as well as development. Cell physiology is concerned with the functions of the cell. Modern physiology mainly employs biochemical and biophysical methods of working.

Physoclists *see* Swimbladder

Physostomes *see* Swimbladder

Phytoplankton *see* Plankton

Picasso Trigger *(Rhinecanthus aculeatus) see Rhinecanthus*

Pigment, Lack of *see* Albinism

Pigment Cells *see* Chromatophores

Pike *(Esox lucius) see* Esocoidei

Pike Characids, F, *see* Ctenoluciidae

Pike Characin *(Acestrorhynchus microlepis) see Acestrorhynchus*

Pike Characin *(Phago maculatus) see Phago*

Pike Cichlid *(Crenicichla lepidota) see Crenicichla*

Pike Cichlids, G, *see Crenicichla*

Pike Top Minnow *(Belonesox belizanus) see Belonesox*

Pike Top Minnows, G, *see Belonesox*

Pike-head *(Luciocephalus pulcher) see* Luciocephaloidei

Pike-heads, related species, Sub-O, *see* Luciocephaloidei

Pike-perch *(Stizostedion lucioperca) see Stizostedion*

Pikes (Esocidae, F) *see* Esocoidei

Pikes, related species, Sub-O, *see* Esocoidei

Pilchard *(Sardina pilchardus) see* Clupeiformes

Pillularis Disease *see Oodinium*

Pillwort, G, *see Pilularia*

Pilotfish *(Naucrates ductor) see* Carangidae

Pilularia LINNÉ, 1753; Pillwort. G of the Pteridophyta*, with 6 species. Found in all temperate regions. Small perennials with stem axes growing horizontally, similar to the G *Marsilea**, but with kidney-shaped leaves. Complicated reproduction through spores. Sporangia in special containers at the base of the leaves. The species of the G colonise ditches and the edges of ponds and rivers. They are often recommended as carpet-forming foreground plants in aquaria, but are only kept rarely. The best-known is:

– *P. globulifera* LINNÉ; Pillwort. Europe. Rare.

Pilumnus LEACH, 1815; Hairy Crabs. G of small marine Crabs (Brachyura*, F Xanthidae), that live on stony ground on banks of Bivalves and in sponge-holes. They eat mainly molluscs and other bottom-dwelling animals, the shells of which they crush between their strong pincers. Legs and back thickly covered with bristles; the edge of the forehead is narrowly denticulated. Because of their small size (carapace diameter 2–3 cm), they are very good for the aquarium and survive well. However, it is advisable not to keep them in the same tank as sessile invertebrate animals or slowly moving ones. The following species is frequently caught along southern European coasts:

– *P. hirtellus* (LINNAEUS, 1758); Bristle or Hairy Crab. N Sea to the Black Sea. Carapace width about 2 cm. Brown or shining brick-red, pincers asymmetrical.

Pimelodella EIGENMANN and EIGENMANN, 1888. G of the Pimelodidae*. Scaleless Antenna-catfishes from central and northern S America. Length to 25 cm. Crepuscular fish from the middle and lower levels of water. The torpedo-shaped body is slightly compressed laterally and the head is broad and a little flattened. A pair of very long upper-jaw barbels and 2 pairs of shorter lower-jaw barbels surround the inferior mouth. Eyes relatively large. D1 pointed in a triangular shape or rounded, D2 long, and C deeply forked. The An is fairly long and rounded, the Vs are small and the Ps well developed. There is a close relationship with the members of the G *Pimelodus** – in fact, externally, they can scarcely be distinguished. Differences do exist however, in that *Pimelodella* has not teeth on the vomer and does possess a process on the parietal bone. Also, the process on the occipital bone does not extend into a point, as in *Pimelodus*, but remains equally broad all the way along, nor is it connected to the dorsal plate. *Pimelodella* should not be put in the same tank with fish that are too small. They require large aquaria standing in a shady spot with a good covering of plants around the aquarium-edge and free swimming areas. *Pimelodella* species are greedy omnivores, that will eat the flesh of fish and the lean meat of mammals, as well as soaked or boiled rolled oats. Nothing is known about their reproduction. Interesting aquarium fish, but unfortunately only *P. gracilis* has so far been imported on any regular basis.

– *P. gracilis* (CUVIER and VALENCIENNES, 1840); Slender Pimelodella (fig. 891). From the Orinoco, Amazon and La Plata regions. Length to 17 cm. Juveniles dark

Fig. 891 *Pimelodella gracilis*

grey on the upper side, silver-white to silver-green on the sides. A black longitudinal band extends from the operculum to the base of the C, along the entire length of the body. Adult females are similarly coloured, but the males becomes a monotone blue-black with a strong metallic sheen. Fins colourless. In the past few years, a very attractive form of Antenna-catfish *(Pimelodella pictus)* has become available through the trade. Its shape and fins are like a shark's, its basic colour is greeny silver in the upper half of the body and the fins are transparent, and it is covered with patent-black dots. These Catfishes are active during the day and they might possibly be young examples of *Pimelodus* clarias*. They swim about in small schools, never seeming to rest.

Pimelodidae; Antenna Catfishes (fig. 892). F of the Siluriformes*, Sup-O Ostariophysi. The largest F of the Siluriformes in the New World, the distribution area of which ranges from S Mexico across the whole of central

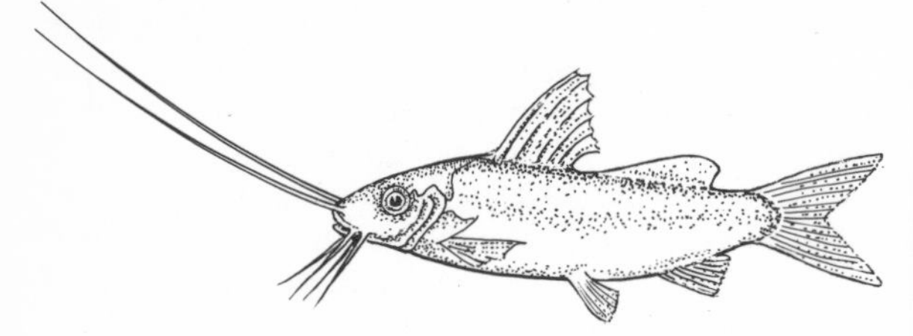

Fig. 892 Pimelodidae-type

and S America to the 40° south latitude. The F is closely related to the Bagridae* of the Old World. Elongate, or very elongate, completely naked Catfishes with 3 pairs of long barbels, pointing stiffly towards the front. In many Ga, the head is flattened in a spatula shape (eg the Ga *Sorubim, Pseudoplatystoma*). Mouth terminal, not very large, jaws and dermopalatine bone beset with comb-like teeth. D positioned well forwards, relatively short and, like the Ps, it has a strong spine. Adipose fin present, very long in many Ga and fused with the C which is almost always twin-lobed and often asymmetrical. The Pimelodidae are typical bottom-dwelling fish, active mainly at night. They constantly search their surroundings with their barbels for any sign of food. Many species are found only in murky waters (white water rivers). There is a blind species *(Typhlobagrus kronëi)* which lives in the Areias Caves near São Paulo. It is probably a descendant of the above-ground species *Pimelodella transitoria*. The larger kinds of Pimelodidae are regionally important as a food for various races of Indians, at least if they are fairly common in a region. In aquarium-keeping, isolated specimens of the smaller species or young specimens seem to be the only ones that are imported, and then only rarely. *See* also under the Ga *Acentronichthys, Microglanis, Pimelodella, Pimelodus, Sorubim.*

Pimelodus BLEEKER, 1864. G of the Pimelodidae*. Widespread in Central America, the northern and central parts of S America, as well as the West Indies. Length to 40 cm. Active at dusk and during the night, with only a few species active by day. Body torpedo-shaped, moderately flattened laterally. Head broad and fairly flat with 3 pairs of barbels, the upper jaw pair being very long, and the lower jaw pairs being much shorter. Eyes large. Closely related to the G *Pimelodella**; in fact, only distinguished by the teeth on the vomer, the lack of process on the parietal bone, the presence of an occipital bone process that extends into a triangular point and which is fused with the dorsal plate, and the fact that the An is shorter. Because of their size, most species are only suitable for the aquarium when young. The tank should not be placed in too bright a spot. Omnivores that like worm food in particular. Reproduction unknown. Of the numerous species, only *P. clarias* with its striking colouring is of any significance in the aquarium.

– *P. clarias* (BLOCH, 1795); Spotted Pimelodus, Pintado. Widespread from Central America and the West Indies to the northern and central parts of S America. Length to 30 cm. With such a large distribution area, the body of younger specimens is very variable. Dorsal surface

Fig. 893 *Pimelodus clarias*

dark grey to olive, flanks grey or greenish-grey to steel-blue with large, medium-sized or small black blotches (paler in the steel-blue coloured varieties). Fins pale grey with black blotches or speckles. Barbels white, underneath dark. In recent years, a variety called *Pimelodella pictus* has been imported frequently. It is very attractively coloured and swims around very elegantly the whole day; it is probably only a young specimen of *P. clarias* (fig. 893).

Pincushion Starfishes, G, *see Culcita*

Pine-cone Fishes (Monocentridae, F) *see* Beryciformes

Pink Rainbowfish *(Nematocentris fluviatilis) see Nematocentris*

Pink Skunk Clown *(Amphiprion perideraion) see Amphiprion*

Pink-tailed Characin *(Chalceus macrolepidotus) see Chalceus*

Pink-tailed Characins, F, *see* Chalceidae

Pinna LINNAEUS, 1758; Fan Mussels. G of marine Bivalves (Bivalvia*, O Anisomyaria). Has a pointed, triangular shell which it sticks into the substratum, then it anchors itself with byssus threads. The shell may reach a length of 80 cm, and only smaller specimens should be kept in an aquarium. Crabs belonging to the G *Pinnotheres* (Pea Crab) often live in the shell.

Pinnate Leaf *see* Leaf

Pinnipedia, O, *see* Mammalia

Pinnotheres, G, *see* Brachyura, *Pinna*

Pipefish, related species, Sub-O, *see* Syngnathoidei

Pipefishes (Syngnathidae, F) *see* Syngnathoidei

Piranha *(Serrasalmus piraya) see Serrasalmus*

Piranhas, F, *see* Serrasalmidae

Pirate Perch *(Aphredoderus sayanus) see* Percopsiformes

Pirate Perches (Aphredoderidae, F) *see* Percopsiformes

Piraya *(Serrasalmus piraya) see Serrasalmus*

Pisa, G, *see Maja*

Pisces *see* Fish

Piscicola DE BLAINVILLE, 1818. G of the Leeches (Hirudinea*). There are about 5 species distributed throughout the holarctic realm*; they are external parasites on fish that live in fresh water, brackish water and in the sea. Body virtually cylindrical, with sucking discs at front and rear ends. Only the species *P. geometra* is found in central Europe. It is up to 5 cm long, its body is very thin, and it is coloured greenish-grey with lengthwise and crosswise rows of black or red-brown dots. *Piscicola* lies in wait for prey on water plants, motionless; then, when a fish swims by, it makes oscillating search movements and attaches itself firmly to the fish's body. It can cause severe damage to the fish simply by sucking its blood and also because it transmits the highly infectious stomach dropsy*. Well suited to aquarium observations (in a pond, at least). For illustration see under Hirudinea.

Pisciculture *see* Breeding

Pistia LINNÉ, 1753; Water Lettuce. G of the Araceae*. Only one species. Found in all tropical and subtropical regions. Small rosette perennial with runners. Leaves sedentary, wedge-shaped, spongy. Flowers unisexual and monoecious, very small and simplified, contained in a small number in a delicate spathe. No perianth. Male flower: 2 stamens. Female flower: 1 carpel, superior. The fruit is a berry. The only species of the G floats on the water surface, but it can also root itself in a marshy substrate. In many places, especially where the water is impure, *Pistia* has become a weed. When this happens, it has to be controlled mechanically and chemically in order to keep the water free. It is cultivated in all botanical gardens, but in aquarium-keeping it is rarely used, because it casts a shadow over everything.

– *P. stratiotes* LINNÉ (fig. 894).

Fig. 894 *Pistia stratiotes*

Place of Retreat *see* Refugium

Placodermi, Sub-Cl, *see* Aphetohyoidea

Placoderms (Placodermi, Sub-Cl) *see* Aphetohyoidea

Placoid Scales *see* Scales

Placophora *(Polyplacophora)*; Chitons. Cl of the Mollusca*, Sub-Ph Amphineura. Has a broad creeping foot and a shell made up of 8 plates. Chitons are typical of animals that live in the intertidal zone and along rocky sea coasts. They feed on algae and other growths. The various species of the G *Chiton** can most certainly be kept in the aquarium; usually they live hidden.

Plaice *(Pleuronectes platessa) see* Pleuronectiformes

Plaice-types, O, *see* Pleuronectiformes

Plainfin Midshipman *(Porichthys notatus) see* Batrachoidiformes

Plankton. A characteristic group of organisms from the open water (Pelagial*) of still inland waters and of the sea. In contrast to the mobility of the necton*, plankton can only float passively or at most only compensate for the tendency to sink by its own movement. Currents carry away plankton, which is why none of the species contained in the plankton is associated with flowing water* or prefers it.

Included in plankton are bacteria, single-celled and multi-celled free-swimming algae and small invertebrate animals, particularly rotifers (Rotatoria*) and lower crustaceans (Crustacea*). Many of the plankton elements have processes for floating; taken all together, this creates a parachute effect and so the rate at which the plankton would otherwise sink is reduced. The disc-like, star-like, and ribbon-like forms in the plankton also have this same effect. Animal plankton in particular is inclined to migrate vertically at times; the decisive factor for this strange up and down movement (usually following a daily rhythm) is light. Plant plankton (phytoplankton) has a very significant role within the food relationships of the pelagial zone – it is quite categorically in the number one position. Phytoplankton produces organic material by means of photosynthesis*. And all other living things in the same living area are reliant on this, either directly like the animal plankton (zooplankton) or indirectly like peaceable and predatory fish.

Planorbarius DUMÉRIL, 1806. G of freshwater pulmonate Snails (Basommatophora*). Together with other Ga such as *Helisoma** and *Planorbis** it is referred to as Ramshorn Snails (on account of the characteristic shape of their shells). In Europe, the G is represented by the Great Ramshorn Snail *(P. corneus)* which is found throughout the palaearctic realm. Its shell may have a diameter of more than 3 cm (fig. 895). It is often found in thickly weeded small stretches of water and is suitable for keeping in an unheated room aquarium or in open-air ponds.

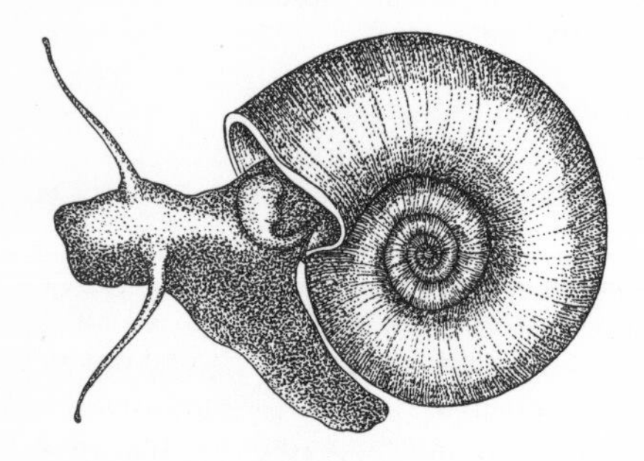

Fig. 895 *Planorbarius corneus*

Planorbis (MÜLLER, 1774) (fig. 896). G of freshwater pulmonate Snails (Basommatophora*). Distributed in the holarctic realm. Together with members of other Ga *(Planorbarius*, Helisoma*)*, this G comprises the Ramshorn Snails. The shells of *Planorbis* can measure up to 17 mm wide and they are keeled at the periphery. The various species are ideal for unheated room aquaria and for open-air ponds.

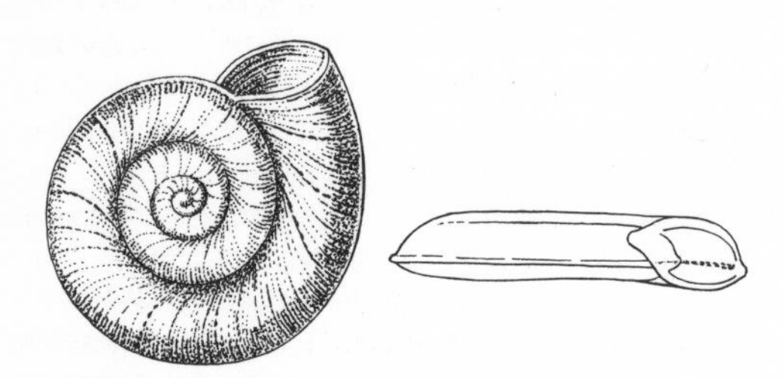

Fig. 896 *Planorbis* sp.

Plant Colouring Matter. There are 2 types of colouring matter: those that dissolve in fat and those that dissolve in water. The former are located in chromatophores in the protoplasm of the plant cell. They include chlorophyll which is responsible for giving all green plants and green plant parts their colour; chlorophyll is also important in photosynthesis*. Today, we know of several forms of chlorophyll with very slight differences in chemical structure; to distinguish them they are given letters from a to d. In the Brown Algae (Phaeophyta*), the chlorophyll is covered with a brown colouring matter, fucoxanthin, and in the Red Algae (Rhodophyta*) it is covered with red phycoerythrin. In green chromatophores or in other chromatophores, there are also found other fat-soluble colouring matter – red or orange-red carotenoid and yellow xanthophyll. They are inactive during photosynthesis and cause the colouring of many perianth leaves. After the breakdown of the chlorophyll they are also responsible for the colour of fruits and autumn foliage. Water-soluble colouring matter (blue, violet or red violet anthocyanin or pale yellow anthoxanthin) is found in the sap-filled vacuoles of the plant cells.

Plant Community. The association of a different number of plant species with a particular habitat*. A plant community can be typical for a large geographical habitat, or it may be confined to a very narrow area. For example, there are aquatic plant communities, marsh plant communities, forest plant communities etc.

Plant Dishes. Shallow containers made of plastic or glass (or other materials), in which plants with special requirements relating to the composition of the substrate* can be placed. They should always have a few holes in the bottom to allow sufficient water circulation. The use of plant dishes also enables you to put plants in a tank even when the fish like to stir up the substrate violently, which usually causes the plants to be uprooted all the time. Put the plant dishes directly onto the substrate in order to avoid getting stagnating sand layers trapped underneath (Substrate*).

Plant Food *see* Feeding

Plant Geography *see* Geobotany

Plant Names *see* Botanical Nomenclature

Plant Nutrition. All living things need organic and inorganic substances for their metabolism*, both as construction matter and energy matter. Green plants, unlike all other organisms, have the ability to create organic matter from inorganic matter. Carbon dioxide plays an important role here. Its component parts, carbon (C) and oxygen (O), by means of photosynthesis*, form simple carbohydrates* to start with. Land plants obtain carbon dioxide from the air, aquatic plants from the water. Another necessary ingredient, hydrogen (H) is obtained by water dissociation. For submerged plants, water is all around and can be absorbed through the entire body surface. In land plants, absorption takes places via the roots mainly, from the ground.

Also important for metabolic processes are inorganic *minerals*, which are in the main locked into organic compounds. As weathering products, minerals are a part of the earth and are absorbed by plants in dissolved form. Important ones are the cations K^+, Ca^{++}, Mg^{++}, Fe^{++} and the anions NO_3^-, SO_4^{--} and PO_4^{---}. Potassium does not occur in an organic compound; it promotes the imbibition of the plasma and activates the enzymes of protein formation. Calcium has the opposite effect on the plasma and is also engaged in the structure of the cell walls. Magnesium is a component part of chlorophyll and activates enzymes concerned with energy metabolism. Iron is contained in the basic substance of the green chromatophores and also takes part in the structure of enzymes. Nitrate is the starting substance for the entire nitrogen and protein metabolisms. It can normally be replaced by the ammonium ion (NH_4^+). Sulphate is a part of the structure of protein compounds containing sulphur, as well as some enzymes. Phosphate is needed for the cell nucleus proteins and numerous other organic compounds.

Extremely small amounts of *trace elements** (boron, manganese, copper, molybdenum and chlorine) also play a part in a plant's life processes. An absence of them will lead to severe damage. The amounts required vary considerably from species to species. *See also* Manuring.

Plant Pins. Glass or plastic rods bent into a hairpin shape, which are used in the aquarium to anchor particularly large plants in the substrate (*see* Aquarium, Installation of).

Plant Single-celled Organisms *see* Protophyte

Plantaginaceae; Plantains. F of the Magnoliatae*. 3 Ga and 270 species. Distributed worldwide. In rare instances they are small shrubs, but most are perennials or annuals. Leaves usually alternate, often in rosettes. Flowers small and inconspicuous. Plantains colonise very varied habitats, and some can tolerate a high salt content in the substratum. The species of the G *Litorella* are small marsh plants that can also grow submerged.

Plantain Family *see* Plantaginaceae

Plant-eating Fish *see* Nutrition

Plasma Circulation is the movement of the protoplasm within a cell. It can be observed under a microscope, for example in the cells of the leaves in aquatic plants such as *Egeria* or *Vallisneria.*

Platacidae, F, name not in current usage, *see* Ephippidae

Platax orbicularis *see* Ephippidae

Plataxoides, G, name not in current usage, *see Pterophyllum*

Plat(y)helminthes; Flatworms. An animal Ph that contains a great many forms and species. It includes the free-living Eddyworms (Turbellaria*) and the parasitical Sucking and Ribbon-worms (Trematodes*, Cestodes*).

Platichthys (various species) *see* Pleuronectiformes

Platinum Tetra *(Gephyrocharax atracaudatus) see Gephyrocharax*

Platy *(Xiphophorus maculatus) see Xiphophorus*

Playfair's Panchax *(Pachypanchax playfairi) see Pachypanchax*

Plecoptera; Stone Flies. A very primitive O of the Insecta* comprising about 2,000 species. Most closely related to the Mayflies (Ephemeroptera*) and the Dragonflies (Odonata*). In contrast to these, however, the Stone Flies can fold their wings on their abdomen. Other recognition features include 2 thread-like outgrowths from the tail, although it is true that some Mayflies also have them. Like the related Os, the Plecoptera also spend their larval stage in water, and there is no pupal stage. The sexually mature adults (imagines) normally sit lethargically very close to the water where they live, perhaps on a clump of vegetation or beneath rocks. They are most reluctant to fly and their colouring is unremarkable. Perhaps these are the reasons why few people know of the Stone Flies and why only a few specialists show much interest in them. The larvae are found mainly in flowing waters and they can take up to 3 years to develop. Whereas the larvae of smaller species *(Nemoura*)* live mainly on what grows on rocks, and the medium-sized forms, such as *Isoperla**, are omnivores, the larvae of the large Ga such as *Perla** and *Dinocras* are among the greediest predators of all the invertebrates. In contrast to the larvae, the imagines live for only a few weeks and hardly eat at all during this period. Stone Flies are of no significance in aquarium-keeping, although very occasionally the larvae are used for feeding purposes. Observations in the pond-aquarium meet with difficulties in so far as most species are adapted extremely well to a high oxygen content and a low temperature, thus making it almost impossible to take care of the larvae successfully.

Plecostomus GRONOVIUS, 1754. G of the Loricariidae*. Native to northern and central S America where they live in flowing waters and the brackish-water river-mouths. Length up to 50 cm. *Plecostomus* species are hardy, contented, peaceable Mailed Catfishes that are not sensitive to temperature. They are active at dusk. The body is elongated, only slightly compressed laterally. The ventral area is flattened to better resist the water current and to be able to lie flat on the bottom. Caudal peduncle somewhat elongated. The flanks are covered in 3–4 rows of bony plates and the belly is naked (although in older specimens it may be covered in tiny bony plates). Head very broad, flattened; the inferior sucking mouth is shaped as in *Loricaria** and is supplied with fine teeth for rasping algae. The pennant-shaped D1 is only raised rarely, the D2 is small, and the C is only slightly concave, the lower tip being longer than the upper. The An is very small, the Vs and Ps are well developed (the latter also has a strong supporting ray). It is very difficult to tell the species apart. None of the species imported so far has been positively identified. *Plecostomus* lives mainly on algae which they graze from rocks and plants, without damaging the plants. So, particularly when young, they are very useful in the community tank. Apart from algal food, they also accept worms of every sort, dead *Daphnia* and *Cyclops*, rolled oats, lettuce leaves etc, as well as dried food (all of which is taken up from the bottom). To encourage settling in, add some sea salt (about a teaspoon to 10 litres of water). *Plecostomus* likes places to hide in dark corners of the tank; also provide plenty of vegetation and a moderate amount of light. No successes so far with breeding. Interesting specimens for large community tanks. Recent researches suggest the generic name should be *Hypostomus* (LACÉPÈDE, 1803).

– *P. commersoni* (CUVIER and VALENCIENNES, 1840); Sucker Catfish. La Plata region, Rio Grande do Sul. Length to 40 cm. Dorsal surface and flanks grey-olive, grey-brown or dark-brown with thinly or thickly dispersed dark dots. Underside pale grey-green with many small dark speckles. Fins brownish, with rows of dark blotches along the fin rays in many specimens. Can be distinguished from *P. punctatus* by the length of the D1 base, which in this case is clearly longer than the distance between the D1 and D2 (fig. 897).

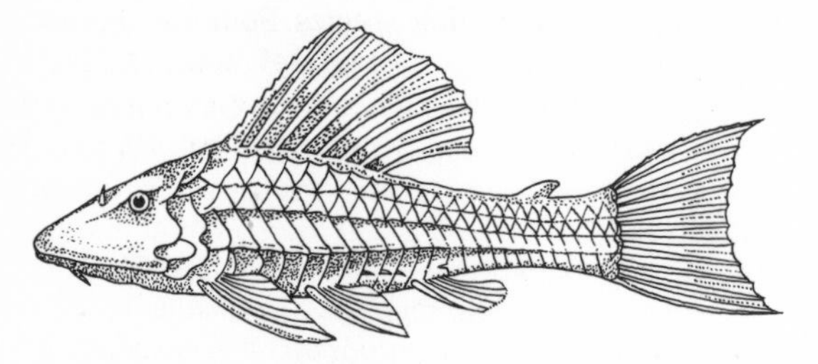

Fig. 897 *Plecostomus commersoni*

– *P. punctatus* (CUVIER and VALENCIENNES, 1840). S and SE Brazil. Length to 30 cm. Dorsal surface and flanks brown-grey or brown with dark dots and flecks and 5 oblique transverse bands. The nasal openings are joined by a dark streak. Underside pale brown or whitish. Fins brownish with large, round speckles arranged in rows. The length of the D1 base is the same as the distance between the D1 and D2 (fig. 898).

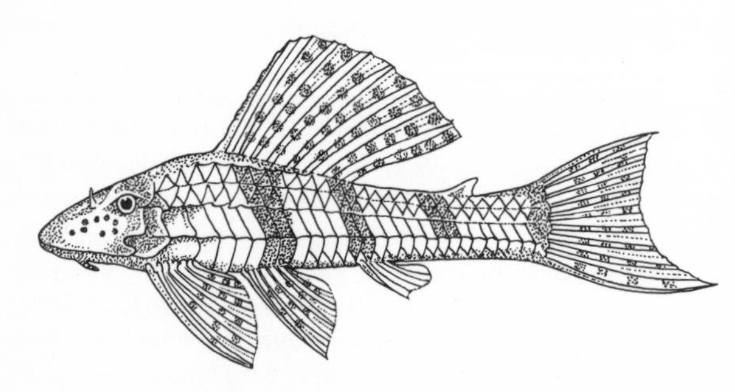

Fig. 898 *Plecostomus punctatus*

Plectognathi, O, name not in current usage, *see* Tetraodontiformes

Plectorhynchidae, F, name not in current usage, *see* Pomadasyidae

Plectorhynchus LACÉPÈDE, 1802. G of the Pomadasyidae*. There is a large number of species in the tropical Indo-Pacific. They are sociable fish when young, but later become territorial. Omnivores that seek out shallow, sunny waters. Length 20–60 cm. Perch-shaped fish with small scales, the most striking features being the steep, convex forehead profile and the relatively large eyes. Along the path of the LL there are 53–100 scales, and a row of 6 pores on the chin. The operculum has one stumpy spine; the D has 9–14 spines and 15–26 soft rays. In young fish, the C is rounded and only slightly concave in older specimens. The colouring is lively, and many species have a characteristic youthful set of colours that is completely different from that of the sexually mature fish. With a varied diet of live food, many species survive well in large, calm tanks with no strong competition for food. Others are extremely sensitive *(P. chaetodontoides, P. orientalis).* A hardy species is:

– *P. albovittatus* RÜPPEL, 1837; Sweetlips, Yellow-lined Sweetlips. From the Indo-Pacific and the Red Sea. 23 cm. 3 narrow white longitudinal stripes running across a dark background (fig. 899).

Fig. 899 *Plectorhynchus albovittatus*

Pleistophora GURLEY, 1893 (fig. 900) (Plistophora, name not in current usage). G of the Sporozoa*. Many species of the G live as parasites on freshwater and marine fish. *P. hyphessobryconis* (SCHÄPERCLAUS, 1941) was first described in *Paracheirodon* innesi* as the cause of 'Neon disease'. Apart from this species, *Hyphessobrycon gracilis* (REINHARDT, 1874), *Hyphessobrycon* flammeus, Hemigrammus* ocellifer* (STEINDACHNER, 1883) and *Brachydanio* rerio* are often attacked too, so the disease is more aptly described as Pleistophora disease. The pathogens can be detected under a microscope as pansporoblasts (28–30 μm) in the torso musculature. These contain numerous spores which can be released into the muscles, where they turn into amoeboid germs. These undergo a propagatory phase, then form new pansporoblasts again. So any fish that is attacked keeps infecting itself again and again. Spores that get into the open water via the kidneys, skin

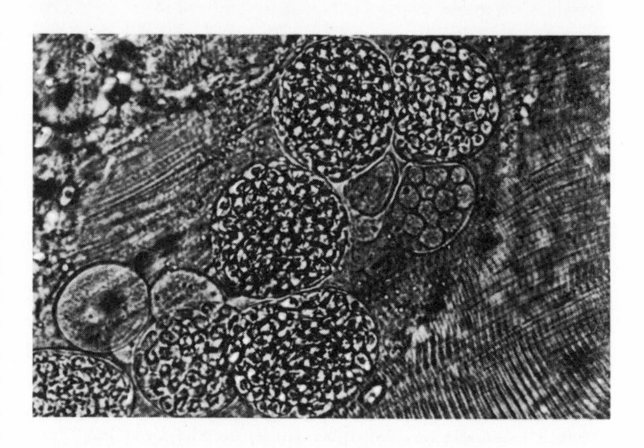

Fig. 900 *Pleistophora.* Spore cysts

grazes or from dead fish can also be taken in by healthy fish. They form an amoeboid germ in the intestine which reaches the musculature via the blood stream, where it once again attaches itself to become a pansporoblast. The area of muscle affected will break up in part or degenerate. The external signs of the disease are very varied: blotchy discolouration of the skin and a fading of the colour bands, disturbances of the equilibrium and emaciation are all common, but none of them can be sure signs. There is no known cure. All the aquarist can do is take preventative measures: keep a close check on the state of health of the fish and apply the strictest hygiene, especially in breeding tanks. This way, at least the disease will be stopped from spreading (Prophylaxis*).

Plesiomorphic. Word used to describe a primitive feature development in systematics*, as opposed to a feature that has subsequently developed (apomorphic*). An example might be a fish fin made up of soft rays only. This is a plesiomorphic feature, whilst the occurrence of a fin with hard rays is apomorphic. Plesiomorphic features in common between groups of organisms are no proof whether or not natural or phylogenetic relationships exist between them.

Pleurobrachia, G, *see* Ctenophora

Pleuronectes platessa *see* Pleuronectiformes

Pleuronectidae, F, *see* Pleuronectiformes

Pleuronectiformes, Flatfish-types. O of the Osteichthyes*, Coh Teleostei. Marine fish, distributed worldwide, with many species also found in brackish water and some in pure fresh water. Also known as the O Heterosomata, this group of fish is known from fossils that date back to the late Tertiary. The chief characteristic of Flatfishes is that one flank really functions as the underside of the fish and the other as the upperside (ie the fish are not flattened along the dorsal surface-ventral surface axis, but practically lie on one side of the body). Usually it is determined genetically which side will function as the underside. Only in the primitive Psettodidae can both the left and right sides become the underside. The juvenile fish are built perfectly normally and have to go through a kind of metamorphosis before the function of the flanks is determined. At the same time that this is happening, other forms of morphological asymmetries take place. The eye of the future underside, for example, migrates to the other side which will now be functioning as the upperside. The mouth becomes oblique, the nasal and gill openings are removed to a different position, various skull bones develop asymmetrically, and on the upperside a dark pigmentation develops. With the exception of the Psettodidae, the fins have no spines. The long D and An are the main organs of propulsion and make undulating movements. C rounded. Vs located at the breast or at the throat. Ps relatively small, often reduced on the underside. Scales mainly well developed, adult specimens have no swimbladder. Colour change and adaptation is highly developed and controlled by the eyes. Most Pleuronectiformes live at the foot of the continental land masses, the young fish often in very shallow water. The eggs are small, pelagial and are produced in large numbers. Many species are very important economically and are usually caught in drag nets. Only the 4 most important Fs (out of 7) are treated here.

– *Bothidae*; Left-eye Flounders. Long oval-shaped fish, the left side of which always forms the upperside. The D is very long and begins at the head an reaches to the C; the An is similarly long and goes from the C to the breast region. Vs very asymmetrical, usually with 6 fin rays. Most Bothidae are covered in scales, ctenoid scales on the upperside and cycloid on the lower side. There are a great many species in shallow water, and only a few deep-sea forms. The best-known species is probably the 90-cm-long Turbot *(Scophthalmus maximus)* (fig. 901). It is a grey fish with greenish flecks and is highly prized as an edible species. Found along almost the whole coast of Europe. Instead of scales, the Turbot has bony tubercles which can give quite a thick covering on the head. The somewhat smaller, but equally important Brill *(Scophthalmus rhombeus)* does have scales and no bony tubercles. It is covered in brown flecks and the first rays of the D are elongated like paintbrushes. It too is distributed from the Mediterranean to Norway. The 40-cm-long *Citharichthys sordidus* (fig. 902) comes from the W Pacific, and it is regarded rather highly in N America especially. The fairly common California Halibut *(Paralichthys californicus)* can grow up to 1·5 m long. Some prettily coloured and patterned species live in the warm seas; some are quite elongate. Nearly all Left-eye Flounders are valuable useful fish.

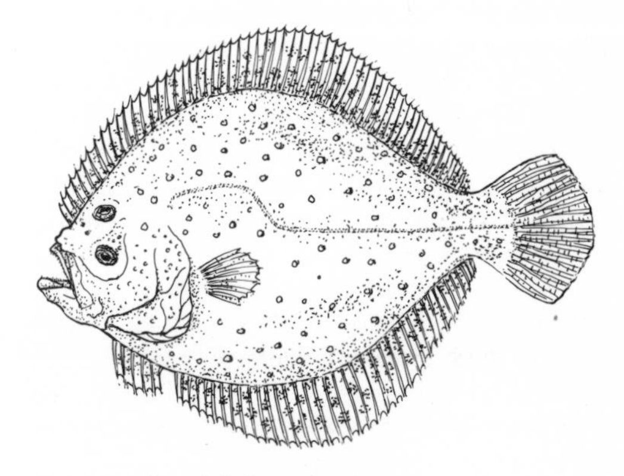

Fig. 901 *Scophthalmus maximus*

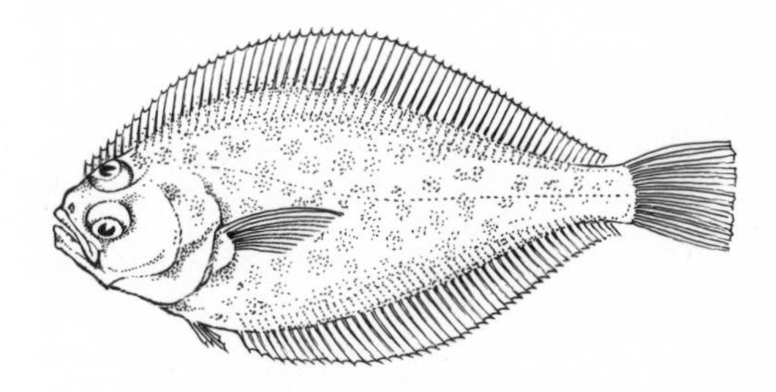

Fig. 902 *Citharichthys sordidus*

– *Cynoglossidae;* Tongue-soles (fig. 903). Small Flatfishes of the type described below for Soleidae, except that the upperside is always the left side. Also, the body is usually pointed at the back. The D, C and An form a fin corona of uniform height that is pointed at the rear end. No Ps. LL present, formed in several strings or absent. Tongue-soles normally live burrowed into the substrate in warm seas. There is no species around Europe.

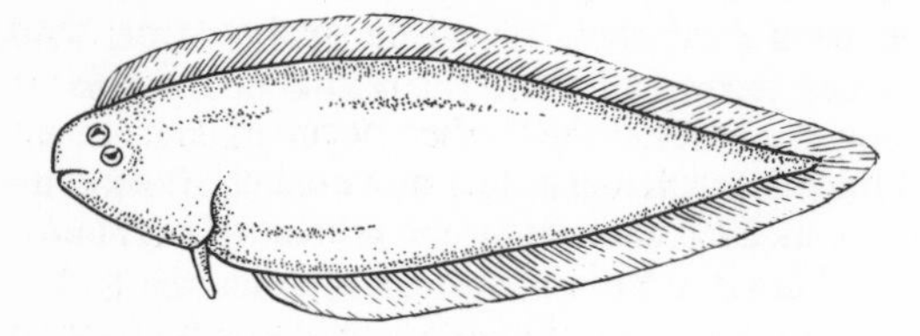

Fig. 903 Cynoglossidae-type

– *Pleuronectidae;* Right-eyed Flounders. An elongated oval shape, and it is always the right side that forms the upperside. The long D begins far forwards on the head, whereas the An begins in the breast or belly region; neither quite extends to the C. The front fin rays of the An are often fused and hidden beneath the skin. The Vs usually have 5 fin rays. Many very important useful fish belong to the Pleuronectidae. Perhaps most important of all is the 60 cm long Plaice *(Pleuronectes platessa)* (figs. 904 and 905) which occurs on sandy substrata along the European Atlantic coast. It is covered with orange-coloured flecks. Its very smooth skin easily distinguishes it from the rough, brown-yellow Flounder *(Platichthys flesus)* (fig. 906) which is also not quite so large. It is found in the Black Sea and right across even as far as the northern E Atlantic, although not around Iceland. Also occurs in brackish water. About 20% of all Flounders are reverse-sided, ie the upperside is not the right, but the left side. A spawning ground of the Flounder lies in the Baltic, north of Rügen. The small eggs float in water and will be found from January to April. As every holidaymaker to the Baltic will know, you can be knee-deep in fry when you stand in the water. The most common Flatfish of the Baltic, the 40 cm long Dab *(Limanda limanda)*, is easy to recognise from its rough upperside and arched LL as it skirts the Ps. The well-known Halibut *(Hippoglossus hippoglossus)* (fig. 907) belongs to this F too. It is 2–3 m long and will then weigh over 100 kg, although the males remain smaller. It is found mainly in the N Atlantic around Iceland and Greenland at depths of 100–600 m. Its main diet consists of largish crustaceans and fish. Its spawning grounds lie at greater depths. One female can lay up to 3.5 million eggs, which float in the water. A sub-species of the Halibut lives in the N Pacific. A few individual species live in fresh water or at least penetrate far upstream. An example is the Starry Flounder *(Platichthys stellatus)* which is found in the inland lakes of Japan, and also in the rivers of other countries that border into the N Pacific. Young Flounders and Plaice are very suitable for keeping in cool, well ventilated aquaria. Use fine sand for the substrate. They are easy to catch in knee-deep water using a sliding-shut net.

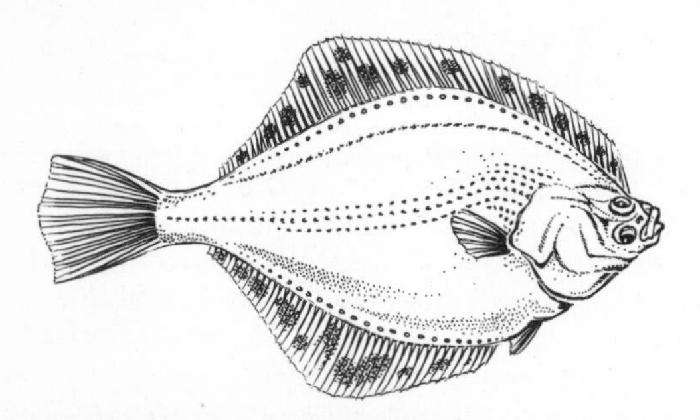

Fig. 904 *Pleuronectes platessa*

Fig. 905 *Pleuronectes platessa*

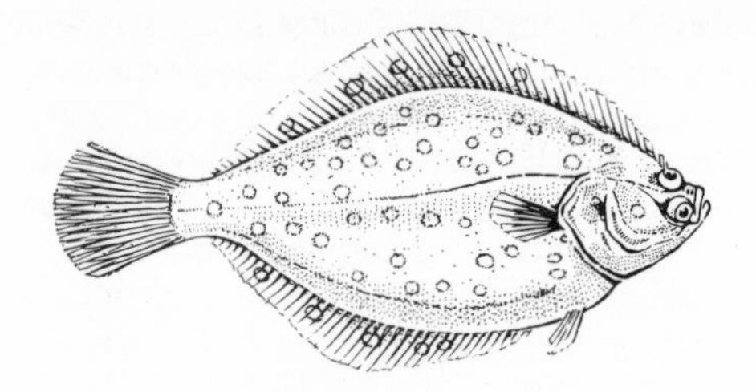

Fig. 906 *Platichthys flesus*

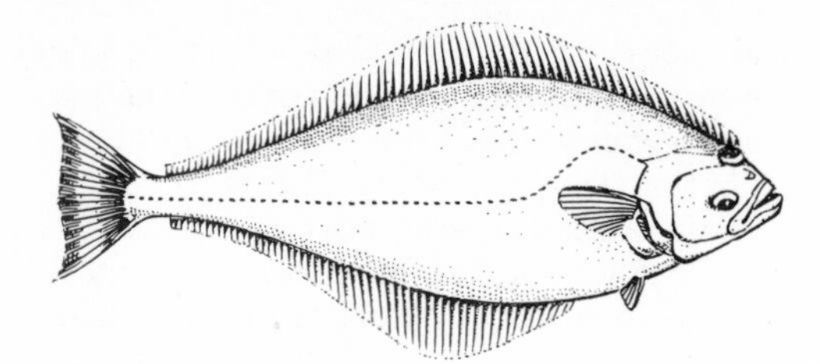

Fig. 907 *Hippoglossus hippoglossus*

– *Soleidae;* Soles. More or less elongate to look at, either droplet-shaped or a long-oval shape. The head is rounded at the front. The upperside is usually the right side. The edge of the operculum is covered over with skin. The long D begins in front of the eyes. Often, the D, C and An form a fin corona of uniform height. Most species are covered with scales, although some species are naked. Many Soles have very striking colours. They are found worldwide, except for very cold seas. Many of them are among the tastiest of all fish. The Dover

Sole *(Solea solea)* (fig. 908) is perhaps the most highly regarded. It grows up to 50 cm long, and is found in the Mediterranean, the North Sea and in western parts of the Baltic over sandy and muddy substrata. The species is characterised by a brown-black marbly colouring and often burrows into the substrate. Some individuals also penetrate brackish water. Its diet consists of various kinds of invertebrates as well as small fish. Some tropical and subtropical species live all the time, or some of the time in fresh water. For information about members of this F in the aquarium, turn to the Ga *Achirus* and *Solea.*

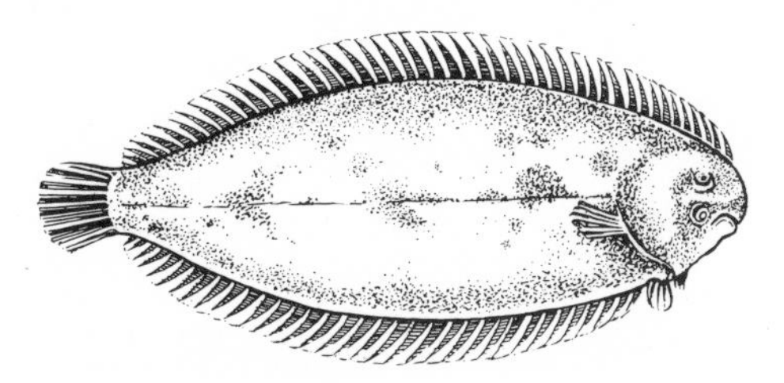

Fig. 908 *Solea solea*

Pleuston. A living community of fairly large animals and plants on the surface of the water – in contrast to the neuston*, they are clearly visible. Well-known member-species of the pleuston are the Duckweeds (*Lemna**) and the Water Bugs *Gerris** and *Velia**. In still and slow-flowing waters in the tropics the plant pleuston is often extremely well developed, especially when there is a mass proliferation of the Water Hyacinths *Eichhornia**. Floating vegetation like *Eichhornia*, or the floating Ferns *Salvinia** and *Azolla**, are used in the aquarium. Apart from providing a mirror-image of the natural biotope or simply providing decoration, such plants also provide somewhere for surface fish to lay their spawn. For many fish, the natural protection from the light that they provide is essential to their well-being.

Plistophora, G, name not in current usage, *see Pleistophora*

Plotosidae; Coral Catfishes (fig. 909). F of the Siluriformes*, Sup-O Ostariophysi. Almost all Siluriformes are freshwater fish; only in some Fs (eg the Ariidae*, Pangasiidae* and Aspredinidae*) are there also some brackish water and marine forms. The Plotosidae, on the other hand, are primarily sea fish that only enter brackish water occasionally; there are also a few that

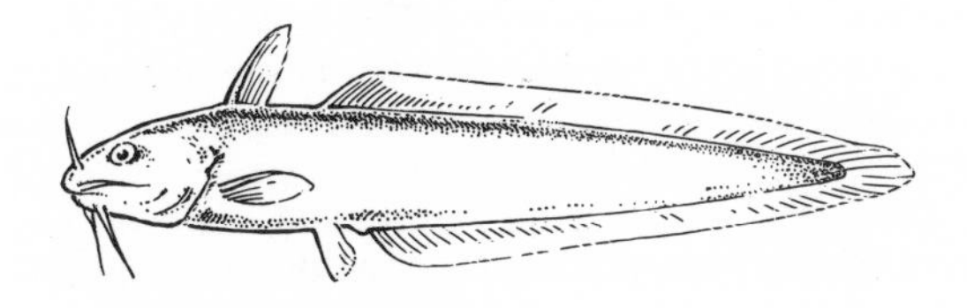

Fig. 909 Plotosidae-type

live permanently in estuaries. In shape, the Plotosidae look like a fat, foreshortened Eel, and their main feature is the long fin fringe along their body. This begins on the back, well in front of the mid-point of the body, and forms a point at the rear end, finally coming to a stop on the ventral side almost directly opposite where it began.

There are 3–4 pairs of barbels around a broad mouth, and the jaws have teeth in them. The D is very short and is inserted directly behind the head. Like the Ps, it too has a strong, very sharp, denticulated spine.

Poison glands emerge at all three spines, and they can cause painful injuries. Behind the anal opening, there is a small tree-shaped organ, the function of which is unknown. The Plotosidae often occur in large shoals, their bodies undulating as they move calmly through the water or as they stand under the protection of rocks or corals. They lay their spawn in crevices in the rock (concealed-spawners). Their distribution area stretches along the tropical and subtropical coasts of the Indian Ocean and corresponding area in the western Pacific. In Africa, their flesh is highly prized, but in Japan much less so. An occasional specimen in the marine aquarium is the 30 cm long *Plotosus anguillaris* (fig. 910), a species that may in fact be identical to *Plotosus lineatus.* It has 8 barbels, and young specimens have 2 lovely yellow stripes running through the body over a brownish background colour. Keep as a shoal; isolated specimens will not thrive. They are very greedy predators that grow quickly, and they will also become adapted to brackish water.

Fig. 910 *Plotosus anguillaris*

Plotosus anguillaris *see* Plotosidae

Plumed Lyretail *(Aphyosemion filamentosum) see Aphyosemion*

Plumose Anemones, G, *see Metridium*

Pluteus-larva *see* Echinoidea

Poaceae (Gramineae, name not in current usage); F of the Liliatae*, with 700 Ga and 8,000 species. Distributed worldwide. The Poaceae are perennial plants of varying size that may form tufts or mats; however, there are also some annual plants. Leaf arrangement distichous, leaves clearly divided into sheath and linear blade. Flowers usually hermaphroditic, inconspicuous and enveloped in hypsophyllary leaves (glumes) and located in inflorescences made up of spikelets. The Poaceae colonise a wide variety of habitats from the plains to the mountains, and in many biotopes, such as savannah, steppes or prairies, they are the dominant plant growth. Our most important strength-giving plants belong to this F (we know them as the cereal crops) as well as the G *Oryza**, or rice. Related to this is the G *Zizania* (Water Rice), the species of which are native to N America and E Asia. *Z. aquatica* is occasionally sown on fish ponds, since its fruits are suitable as fish food. There are numerous marsh and water grasses, all of which are probably not suitable for keeping in an aquarium, but one G that should be mentioned here is *Glyceria* (Sweet Grass). It contains about 30 species and the only place they are not found is Africa. The most important grass along the land border area of stretches of water is the Common Reed *(Phragmites communis).* It is found worldwide. It can spread rapidly on account of its strong runners. There is a group of plants within the Poaceae, containing about 100 Ga, that is particularly characteristic of the tropics and subtropics. These plants usually grow very tall and have woody stems, to some extent. The G *Bambusa* (Bamboo) belongs to this group.

Poachers (Agonidae, F) *see* Cottoidei

Podostemaceae. F of the Magnoliatae*. 40 Ga, 200 species. Tropical, particularly centred around S America. Small perennials or annuals, the shape of their vegetative body being very varied. It differs markedly from that of the normal higher plants, and some even look more like algae or mosses. The roots are flattened and are adapted into organs of attachment and assimilation. Flowers very small and simplified. The Podostemaceae are aquatic plants that, like the Hydrostachyaceae*, live primarily on rocks and stones, in rarer cases attaching themselves to the trunks of trees, in waterfalls and rapids that are subject to periodical fluctuations in water-level. There is no information available about the cultivation of the Podostemaceae.

Poecilia BLOCH and SCHNEIDER, 1801 (fig. 911). G of the Poeciliidae*, Sub-F Poeciliinae. America: SE of the USA, Central America, northern and eastern S America as far as Uruguay as well as neighbouring island groups, especially the Greater and Lesser Antilles. Found in calm waters of every description: streams, river lagoons, ponds, lakes, often also in the brackish water of rivermouths and lagoons. The Guppy, *P. reticulata*, was introduced in various tropical places to combat mosquitoes. They are very small to medium-sized fish. Males 2–15 cm, females 2–18 cm. The various species do not have a uniformly shaped body. The members of the Sub-G *Limia* have a relatively compressed shape with a tall back. The male and female in *Limia* and in the earlier G *Mollienisia* are built similarly. *P. latipinna, P. petenensis* and *P. velifera* are characterised by a very large D. The dominant tones of the body colouring are olive-yellow, brownish or greenish. Patterning is often very lively and the females are usually paler. In the Sub-G *Limia* and the former *Mollienisia*-species, the colouring of the sexes is similar. In the last-named, all

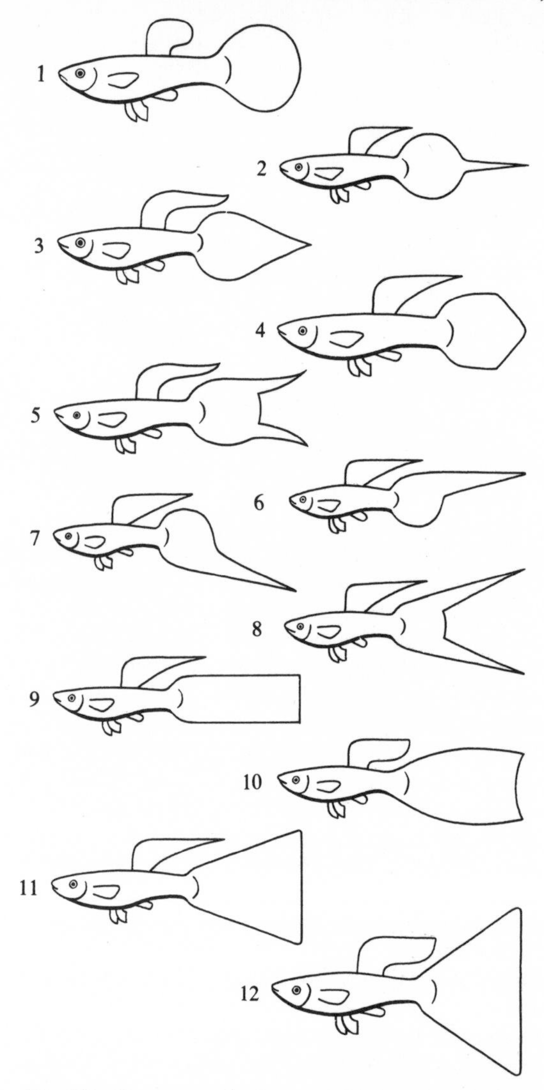

Fig. 911 Guppy-standards. 1 Round-tail 2 Pin-tail 3 Pointed tail, Spear-tail 4 Spade-tail, Cofer tail 5 Lyre-tail 6 Top Sword 7 Bottom Sword 8 Double Sword 9 Flag-tail, Scarf tail 10 Veil-tail 11 Fan-tail 12 Triangular tail, Delta tail

the members of the G have a marked tendency to be black. There are various cultivated forms of *P. reticulata*, which clearly differ in colour and fin type. Keep *Poecilia* species in well planted tanks (large tanks may be necessary if keeping the larger species), and leave enough room for swimming freely. Add some sea salt (up to 50 g for every 10 litres of water), and regularly change part of the water. (At least, this is to be recommended for many of the species.) Usual temperature required 22–26 °C. For breeding purposes, gently raise the temperature and preferably keep the females on their own. In general, *Poecilia* species are peaceful. Most of them are omnivores. Apart from live food, feed also some plant matter (which in some cases may be the main part of the diet). With many species, experiences of care and breeding are inadequately recorded, and the task of rearing fry is fraught with problems. As a result, the different species of the G vary greatly in importance as far as the aquarist is concerned.

In 1963, ROSEN and BAILEY placed the following Ga in the G *Poecilia: Lebistes, Limia, Micropoecilia, Mollienisia, Pamphorichthys, Poecilia.* Many well-loved and highly recommended aquarium fish belong to the G:

– *P. branneri* EIGENMANN, 1894; Branner's Livebearer. Brazil, Guyana. Male to 3 cm, female to 3.5 cm. The sides of the body are crossed by 7–8 green-yellow transverse bands. There is a black blotch on the caudal peduncle that is yellow-green in colour with a red border area at the back. The D in the male is elongated at the back and ends in a point, orange, with a black border at the top. Brackish water. Difficulties with breeding.

– *P. caucana* (STEINDACHNER, 1880); South American Molly. Colombia. Male to 3.5 cm, female to 5 cm. The body in the male has 8–15 dark-blue transverse bands on it. D orange with a black edge. C yellow with a dark edge. Add sea salt to the aquarium water. Feeds mainly on plant food.

– *P. caudofasciata* (REGAN, 1913); Blue Poecilia, Steel Blue Limia. Sub-G *Limia.* Jamaica, Haiti, Cuba. Male to 5 cm, female to 6.5 cm. There is a clear, dark longitudinal band on the body that is crossed by 6–10 transverse bands in the tail region. In the male, this band is covered by green dots, particularly in the front. In the female, the base of the D has a dark fleck on it; in the male, there is a dark band here.

– *P. elegans* (TREWAVAS, 1948); Haiti Molly. Haiti, Dominican Republic. Male to 4 cm, female to 6 cm. An olive-yellow body with a black longitudinal band stretching from the P to the root of the tail. Iris blue. Fins colourless; C and D with fairly distinct dark dots. 20–22 °C.

– *P. formosa* (GIRARD, 1859); Amazon Molly. From the Atlantic region of Mexico to Panama. Length to 8 cm. Body olive-green. Scales with dark edges. D covered with black dots. So far, only females have been identified as belonging to this species and they must reproduce parthenogenetically (Parthenogenesis*). Obviously, the eggs are stimulated to develop by the sperm from males of other species, without fertilisation taking place. The offspring are therefore never hybrids, but always duplicate females. Depending on the natural communities formed, male *P. latipinna* or *P. sphenops* are used for breeding purposes. 22–26 °C.

– *P. heterandria* (REGAN, 1913); Dwarf Limia, Haiti Limia. Sub-G *Limia.* Haiti, Barbados. Male to 3 cm, female to 5 cm. There is a dark longitudinal band between the eye and caudal peduncle. In the middle of the male's body there are 3 dark transverse bands. In the male, the D has an orange-coloured middle area, bounded by 2 dark bands; in the female, there is only one black blotch.

– *P. latipinna* (LESUEUR, 1821); Sailfin Molly. SE of the USA, Mexico. Male to 10 cm, female to 12 cm. D large; in the male, from the second year, it is elongated like a sail. Scales have a mother-of-pearl sheen. Colouring variable. Up to 9 longitudinal bands on the sides that may be made up of fine, red, blue or green dots. In the rear section there are 6–7 dark transverse bands. D blue with black streaks and dots, with an orange-red edge. Dappled and completely black forms are also known. Add some sea salt to the aquarium water. Temperature 20–24 °C. Without plant matter for the main part of the diet, the sail-like elongation of the D will not take place in the males. Apart from algae, cooked spinach or lettuce, live food and dried food can also be given, however. If kept in algae-rich open-air tanks, especially beautiful forms will be reared in the summer.

– *P. latipunctata* MEEK, 1904. Mexico. Male to 5 cm, female to 6.5 cm. The body has a bluish shimmer. A longitudinal band made up of dark dots is crossed by distinct dark transverse bands in the rear part of the body. D and C with dark flecks. Mainly eats plant matter.

– *P. melanogaster* GÜNTHER, 1866; Blackbellied Limia, Blue Limia. Sub-G *Limia.* Jamaica, Haiti. Male to 4 cm, female to 6 cm. Body a shimmering steel-blue with 6–8 dark transverse bands in the rear section. As the males get older, their throat and belly become orange. A black blotch on the root of the tail. Females paler, but during pregnancy they always have a very large, black-blue blotch.

– *P. nicholsi* (MYERS, 1931); Yellow Limia, Nichol's Limia. Sub-G *Limia.* Dominican Republic. Male to 4.5 cm, female to 5.5 cm. Across the sides of the body run dark transverse bands, fused with one another. C yellowish. In the male the D is red-brown with a black edge and a black blotch on the front border. Gonopodium sometimes black.

– *P. nigrofasciata* (REGAN, 1913); Humpbacked Limia, Black-barred Limia. Sub-G *Limia.* Haiti. Male to 4.5 cm, female to 7 cm. There are 8–12 distinct black transverse streaks on the body. Older males have a deep body and a fan-shaped D. Lots of fine dots produce a metallic yellow-green sheen. Throat and belly black. D covered in black dots and streaks. Both the sexes when young are coloured the same.

– *P. ornata* (REGAN, 1913); Ornate Limia. Sub-G *Limia.* Haiti. Male to 4 cm, female to 6 cm. Body covered with numerous black dots and flecks. Edges of the scales metallic green. D and C with dark flecks. 24–28 °C.

– *P. parae* EIGENMANN, 1894; Two-spot Livebearer. Brazil, Guyana. Male to 2.5 cm, female to 3 cm. There is a dark shoulder-blotch, with a pale outline, in the male. The C has a dark eye-spot; the upper side of it, like the D, is yellow to red in colour, covered in dark dots. The D has streaks as well, besides dots. Females inconspicuous. Temperature higher than 26 °C. Add sea salt to aquarium water. Difficult to rear.

– *P. petenensis* (GÜNTHER, 1866); Spike-tail Molly. Guatemala. Male to 13 cm, female to 11 cm. In the male, the D is broad, like a sail. C pointed in the lower half. Body a gleaming blue colour with rows of black and green iridescent speckles. D and C a similar colour to *P. latipinna.* Add sea salt to the aquarium water. Mainly eats plant food.

– *P. reticulata* (PETERS, 1859); Guppy Millions Fish, Spade-tail, Pointed-tail, Bottom Sword. Sub-G *Lebistes.* N Brazil, Guyana, Venezuela, Barbados,

Trinidad. Male to 3 cm, female to 6 cm. Males gaily patterned in lots of different colours. Female inconspicuous. There are numerous natural form and colour varieties, from which a large number of races have been developed over the last few decades by means of systematic line breeding (Heredity*). With regard to the fin type (C and D) in the males, there are at the moment 12 internationally recognised Standards (fig. 911). Alongside these standard shapes, or in combination with them, there also exist some very prominent standard colour varieties, eg 'Carpet-pattern', where the patterning on the C is like a mosaic; 'Emerald', where the basic colour of the body is emerald green; 'Half-Black', where the rear section of the body is almost black; the Golden Guppy, Bronze Guppy etc. The females are usually less striking, but they too can be altered through breeding (fig. 912).

Fig. 912 *Poecilia reticulata* (Viennese Guppy)

– *P. sphenops* Cuvier and Valenciennes, 1846; Black Molly; Pointed Mouth Molly. Southern USA (Texas) to Colombia and Venezuela. Common in brackish water. Male to 7 cm, female to 10 cm. Numerous colour varieties but with some features that occur regularly: flanks bluish with 4–6 longitudinal rows of orange-red dots, between which are distributed bluish or greenish iridescent dots. The D is large and angular and has a red edge surrounded by black, as well as black speckles between the fin rays. C similarly coloured. Gonopodium orange-red. Female paler in colour. The dappled and black forms are especially popular (Black Mollies). They come in a variety of cultivated forms, that differ mainly in the shape of the fins. Sea salt should be added to the tank water. Feed with plant food, apart from the usual foodstuffs normally recommended.

– *P. velifera* (Regan, 1914); Giant Sailfin Molly. Mexico, especially Yucatan. Male to 15 cm, female to 18 cm. Similar in colour to *P. latipinna*, but also with similar variations in colour and cultivated forms. The D in the males is very large, and with its 18–19 rays it is a species-specific recognition feature. 25–28 °C. Care for in the same way as *P. latipinna* (fig. 913).

– *P. versicolor* (Günther, 1866); Olive Limia. Sub-G *Limia*. Haiti. Male to 4.5 cm, female to 6.5 cm. Brown to olive-green with a blue shimmer in direct light. Large

Fig. 913 *Poecilia velifera*

numbers of iridescent dots. Usually 12 dark transverse bands on the body. Female paler, with a clear, dark blotch on the bottom of the D. 25–28 °C.

– *P. vittata* Guichenot, 1853; Cuban Limia. Sub-G *Limia*. Cuba. Male to 6 cm, female to 10 cm. In direct light, the body has a blue sheen. There is a distinct dark longitudinal band crossed by some black transverse bands. D and C yellow or orange-red with dark flecks and a dark edge. Females paler. Add sea salt to water (fig. 914).

Fig. 914 *Poecilia vittata*

– *P. vivipara* Bloch and Schneider, 1801; One-spot Livebearer. Eastern S America, Lesser Antilles. Male to 6 cm, female to 9 cm. Young specimens have a black shoulder blotch enclosed by a golden-gleaming zone. There are narrow dark transverse bands in the rear part of the body. The D is large, orange at the base with a black band above it, as well as a yellow and black edge. Female less striking.

Poeciliidae; Livebearing Tooth-carps (fig. 915). F of the Atheriniformes*, Sub-O Cyprinodontoidei*. The original distribution area of this large F stretched from the southern states of the USA across the whole of Central America, including the West Indian Islands, as far as N Argentina. Today, the distribution especially of this F is almost complete throughout the tropical and subtropical areas, where some kind of species will be found locally. This situation has been brought about partly because certain forms with a wide ecological distribution and which eat mosquito larvae, have been introduced in various parts of the world specifically to combat mosquitoes, because mosquitoes can carry malaria. In addition, aquarists have also ensured (intentionally or unintentionally) that they have spread. The situation has been compounded even further since the large breeding stations discovered that the warm springs of temperate climates make ideal natural spots for breeding Livebearing Tooth-carps. In respect of

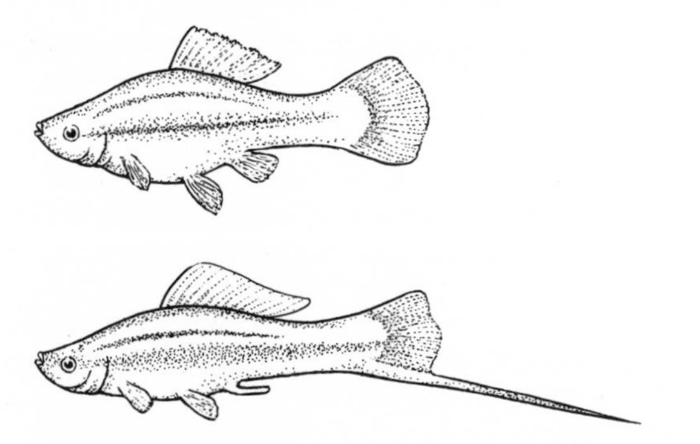

Fig. 915 Poeciliidae-type of the G *Xiphophorus*. Top, female; below, male

many features, the Poeciliidae correspond to those given for the Sub-O Cyprinodontoidei*. Only the sexual dimorphism and the reproduction needs to be discussed here. In contrast to the Egg-laying Tooth-carps (Cyprinodontidae*), the females are usually bigger, whereas the males are more strikingly patterned and coloured. Often, the males also have longer fins, or special fin processes, eg the well-known Swordtail. The copulatory organ (gonopodium) of the males is highly specialised: it serves to transfer packets of spermatozoa and is developed from the An. This anal development takes place at the onset of sexual maturity and is linked to the displacement forwards of the anal opening and the Vs. The gonopodium itself develops from the 3rd, 4th and 5th fin rays of the An. These become elongated and sometimes adopt very complicated generic-specific forms. At the same time, the necessary musculature for the movement of the gonopodium is formed. The remaining fin rays of the An are reduced. When the packets of sperm are being transferred, the tip of the gonopodium is placed in the female's genital opening. During this, the gonopodium is directed forwards, its fin ray forming a channel which is sealed off into a tube by the Vs. It is down this channel that the sperm packets travel into the oviduct. There, the sperm is released, some of it fertilising the eggs that are ripe at the time, the rest being stored in folds in the oviduct. As a rule, one copulation is enough for several pregnancies. The eggs contain plenty of yolk, so no special mechanisms are required in order to nourish the embryos; they can also breathe via the follicle epithelium. During birth, eggs that are about to hatch out are often released – the egg-case is split open while in the oviduct or immediately after the eggs reach the outside world. In the species that have eggs with little yolk, mechanisms have developed whereby nutrients are transported from the mother to the embryo (eg *Heterandria**). Superfetation is also known. This is the state in which eggs at different stages of advanced development occur in the mother simultaneously. In such cases, the offspring are given birth to one after the other during a birth period that lasts several days *(Heterandria*, Poeciliopsis*)*. It is the exception, rather than the rule, for the Poeciliidae to be important in the fisheries industry. They are of importance wherever they occur in mass quantities, and are used for pig food. Many species have proved of great value in scientific researches, leading to new knowledge about the heredity of features of colour and shape, sex, certain tumours and immunological properties. In aquarium-keeping, the various members of the F have played a significant role for a very long time. Many of them are ideal ornamental fish, because of their meagre size, often lively colouring, interesting reproduction and easy propagation. For some years now, there has been a big up-swing in the controlled breeding of new forms and colour varieties. For details see under the Ga *Alfaro, Belonesox, Brachyrhaphis, Cnesterodon, Gambusia, Girardinus, Heterandria, Neoheterandria, Phallichthys, Phalloceros, Phalloptychus, Poecilia, Poeciliopsis, Priapella, Priapichthys, Quintana, Xenodexia, Xiphophorus.*

Poeciliopsis REGAN, 1913. G of the Poeciliidae*, Sub-F Poeciliinae. Found mainly in the Pacific regions of Central America, from Arizona (USA) in the north to Colombia. Lives mainly in calm tributary rivers, but also in ponds and other still waters, as well as in brackish water. Small to medium-sized fish. Males 2–6 cm, females 3–15 cm. The majority of the species are inconspicuous with a brownish ground colour. Some have no patterning, others have dark-black speckles, flecks, stripes or transverse bands. There is a considerable lack of experience regarding care and breeding. So far, only the following have importance in aquarium-keeping.

– *P. gracilis* (HECKEL, 1848); Porthole Livebearer. S Mexico, Guatemala. Male to 3 cm, female to 7 cm. Grey-green, both sexes with a long row of black flecks. Male has a very long gonopodium, female has no blotch during pregnancy. 22–26 °C. Omnivore (fig. 916).

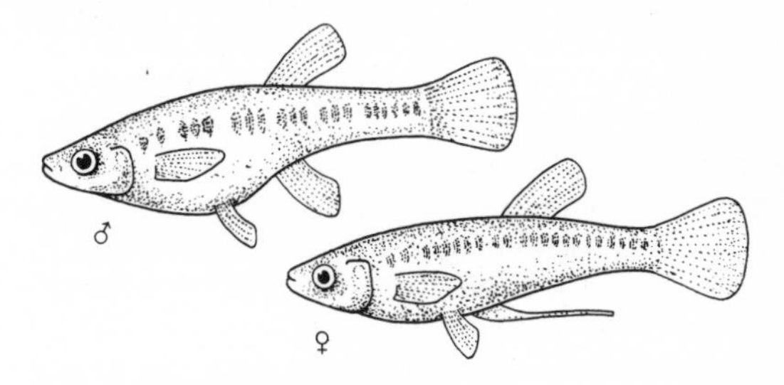

Fig. 916 *Poeciliopsis gracilis*

– *P. turrubarensis* (MEEK, 1907). Pacific regions of Guatemala to Colombia, also in brackish water. Male to 4 cm, female to 8 cm. Pale brown, with about 5 dark-brown transverse stripes on the flanks. 23–25 °C.

Poecilobrycon erythrurus, name not in current usage, *see Nannostomus*

Poecilobrycon vittatus, name not in current usage, *see Nannostomus*

Poecilurichthys, Sub-G, *see Astyanax*

Pogge *(Agonus cataphractus) see Agonus*

Pogges (Agonidae, F) *see* Cotteidei

Pogy *(Brevoortia tyrannus) see* Clupeiformes

Poikilothermic; cold-blooded. Describes all animals whose body temperature is clearly altered by changing temperatures in the environment. Fish, for example, are poikilothermic. Birds and mammals, on the other hand, are homoiothermic*.

Pointed Mouth Molly *(Poecilia sphenops) see Poecilia*

Pointed-head Pufferfishes (Canthigasteridae, F, name not in current usage) *see Tetraodontidae*

Pointed-tail (*Poecilia reticulata,* standard fish) *see Poecilia*

Poisoning. The reaction of individual species of fish to poisons varies. Young fish are usually more sensitive than adult fish. The poisonous effect of many substances is very much dependent on the quality of the water and its hardness, its pH value and its temperature. In hard water the efficacity is often less than in soft water, and with rising temperature the poison effect of most substances increases. Ammonia* develops in the aquarium chiefly as a result of the bacterial breakdown of proteins and even in concentrations of 0.2–0.5 mg/litre it is very poisonous to the blood. Non-combined ammonia is only present when there is an alkali pH value, under pH 7 it is present entirely as non-poisonous ammonium ions. Nitrites (Nitrite Content*) and nitrates (Nitrate Content*) likewise develop as breakdown products (containing nitrogen) of proteins, but they only become poisonous in concentrations of 10–20 mg/litre or 100–300 mg/litre. If the protein breakdown process is disturbed, there is a build-up of intermediate products with a high poison effect. Phenols which on the one hand can come about through protein breakdown and on the other may be introduced into the aquarium by using food-animals from polluted waters are poisonous even if present as traces only. They can gradually build up in the fatty and muscular tissue of fish. Phenols increase the poison effect of other substances sometimes manyfold. For example, if chlorine is present, stable chlorophenols develop and these cause severe damage even in concentrations of 0.1 mg/litre. Chlorine is present in tap water in concentrations of 0.1–0.2 mg/litre. If fish are subjected to such concentrations for a few days, the effect is fatal. Chlorine can be removed from water by strong ventilation, increased temperature, by filtration through activated charcoal* or sodium hyposulphite ($Na_2S_2O_3$). In different ways, metals and their salts can directly or indirectly damage fish. The composition of the water is very decisive in this case – fish kept in soft and slightly acidic water are particularly in danger. Metallic ions can upset the balance (Redox Potential*) in water and have a negative effect upon metabolic processes in the aquarium; they can also block the enzymes of respiration. Iron, zinc, lead and copper can get into the water if the aquarium is defective or if any implements are poorly isolated. Iron salts, which are often present in dissolved form in ground and spring waters, precipitate under the effect of oxygen as ferric hydroxide ($Fe(OH)_3$). It very easily flakes off onto the alkali fish gills and onto spawn and this hinders gaseous exchange (Respiration*). Zinc, which is quite often found in fairly high concentrations in tap water or as an admixture in cooking salt, is precipitated only very slowly as non-poisonous zinc sulphide in the presence of proteins. In contrast, lead and copper are quickly transformed into insoluble carbonates. Harmful substances break free from all kinds of materials that are used in aquarium-keeping, such as isolation materials, putties (Aquarium Sealants*), rubber and synthetic products, and coloured coats of paint. Before any materials are used, they should be tested thoroughly to make sure there can be no objection to them. Synthetic substances that give off odours in hot water (60 °C) are not suitable. Aquaria built with the aid of cement, concrete or eternite should be soaked thoroughly for about 20 days before use (change the water a lot!). Cases of poisoning can also occur occasionally if tobacco smoke (nicotine), or gases from the motor or heating system are sucked in via the airpump. Materials used in biological pest control (eg insecticides), herbicides (materials for getting rid of weeds), algicides (for combatting algae), disinfectants and detergents can sometimes be very powerful fish poisons, and even the smallest traces of them should be kept well away from the aquarium. Unlike parasitical diseases, the symptoms of poisoning in fish are not very characteristic. Often the blood picture is altered, the colours fade, the gills corrode, the skin looks cloudy and the mucus coating is disturbed; there may also be dot-like haemorrhaging. In many cases, if fish or plants suddenly die en masse, this is a sign of poisoning. If transferred into water with perfect conditions, even badly damaged fish will usually recover quickly.

Poisonous Fish (fig. 917). Species that have poison mechanisms or species that have poisonous substances in their organs and tissues that will poison anything or anybody that eats the fish. Poison mechanisms are found in several kinds of fish and are used chiefly for defence: eg, some Sharks and Rays *(Squalus*, Heterodontus, Myliobatis, Trygon)*; Ratfishes (Chimaeridae); various Catfishes *(Galeichthys, Heteropneustes*, Ictalurus*, Noturus, Plotosus, Schilbeodes)*; Scorpion-

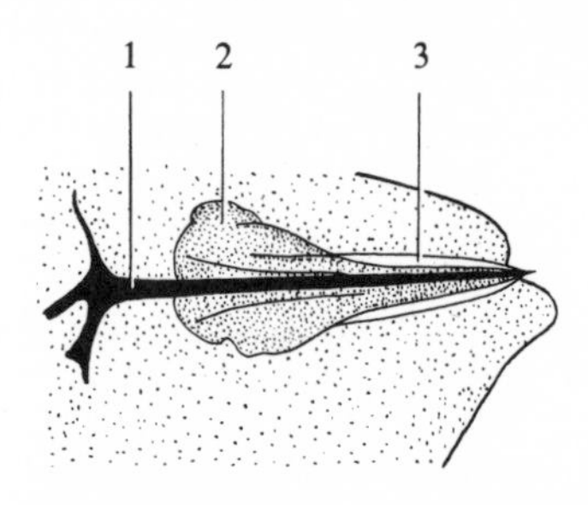

Fig. 917 Poisonous fish. Opercular spine on the Greater Weever Fish *(Trachinus draco)* with poison gland. 1 Spine 2 Gland 3 Sheath

fishes (Scorpaenidae); Stonefishes (Synancejidae) and Weeverfishes (*Trachinus, see* Trachinoidei). In the sharks and rays, the poison mechanisms consist of the front rays of the dorsal fins or of dermal spines in the tail region; in the bony fish they consist of the spines of various fins or spines along the operculum. The poison glands are located in the epidermal layer (*see* Skin) of the spines which are often furrowed or serrated. Whenever they pierce the skin of an enemy, the secretion from the glands flows into the wound. This can cause severe pain, localised inflammation, paralysis or it may interfere with the blood circulation and respiration, leading to death in certain cases. Fish that are poisonous to humans produce toxins that affect the nervous system, causing very severe poisonings indeed. Off tropical coasts there are about 500 species of fish that can produce poisons (they are known by the term *Ciguatera*). They include various Snappers (Lutjanidae*), Surgeonfishes (Acanthuridae, *see* Acanthuroidei) and Serranid fishes (Serranidae*). There are a number of different symptoms of poisoning, which may sometimes prove fatal. They include confusion of the ability to sense heat and cold, muscular pain and headaches as well as long-lasting bouts of paralysis. A person's nails may also fall out. Tetraodon poisoning can have even worse consequences. Even today, more than 50 % of known Tetraodon poisonings prove fatal. It is the poison tetrodotoxin which is the cause – it is found mainly in Pufferfish-types (Tetraodontiformes*). Tetrodotoxin is concentrated mainly in the ovaries at spawning time, and in the liver, but it may also be present in the muscles, in the testes and in the blood. Symptoms of such poisoning include a lowering of the blood pressure, interruption of the heart's rhythm, and in severe cases, total muscular paralysis. The number of Tetraodon fatalities is highest in Japan, where Pufferfish flesh is prepared as a delicacy in special restaurants.

Polka Dot Grouper *(Cromileptis altivelis) see Cromileptis*

Polka-dot African Catfish *(Synodontis angelicus) see Synodontis*

Pollachius (various species) *see* Gadiformes

Pollack *(Pollachius pollachius) see* Gadiformes

Pollen contains simplified male reproductive cells (Gametes*). It is characteristic of seed-plants (Spermatophyta*) and corresponds to the spores of many Pteridophytes (Pteridophyta*). Pollen is formed in the stamens and is transferred to the female plant organs during pollination*.

Pollen Sac (part of the stamen) *see* Flower

Pollination is the transfer of pollen from the stamens to parts of the carpels, via the bract scales onto the stigmas. In *self-pollination* (autogamy) pollen is transferred within the same flower or between different flowers on the one plant. In *cross-pollination* (allogamy) seeds will only form after the transfer of pollen between two different plants of the same species. If the pollen is transferred by animals (mostly insects) or, as in the tropics, also by birds and bats, it is called *zoogamy*. In these cases the pollen is sticky and the flowers attract the animals visually and chemically (eg the colours of the perianth or odorous substances), as well as offer food such as nectar and excess pollen. In *wind-pollination* (anemogamy) the pollen is easily separated and floats well on the wind. The flowers do not attract, and the perianth is inconspicuous or completely reduced. *Water-pollination* (hydrophily or hydrogamy) may take place in aquatic plants. In such cases the male flowers float to the surface *(Vallisneria)*, or the pollen floats to the surface *(Callitriche)* or remains in the water *(Ceratophyllum)*. The flowers are inconspicuous.

Poly-. In word compounds means 'many', 'excessive'. For example, polysaccharide = multi-sugared; polysaprobiotic = excessively decaying (perhaps a polluted stretch of river).

Polycentropsis BOULENGER, 1901. G of the Nandidae*. Only one species, native to tropical W Africa, where it lives in the region of larger rivers (Niger, Ogowe). Body short and tall, very flat laterally. Head pointed, mouth inferior, deeply slit. D and An shaped similarly, hard-rayed parts included in the colouration of the body, soft-rayed parts of both fins, plus the C, glassily transparent. Caudal peduncle extremely short. LL visible only above a few scales. Colouring variable, but always very murky, with dark, irregular flecks on an ochre, brownish or grey-green background. A dark band extends from the eye to the snout, another to the throat, and a third to near the insertion of the dorsal fin. Care for *Polycentropsis* in a species tank filled with soft, slightly acidic water, occasionally filtered through peat. Provide thick vegetation and root-work. Temperature about 24 °C. Floating plants and subdued lighting increase their well-being. It is typical of the F for *Polycentropsis* to lie in wait for prey, hidden among the vegetation. Prey may be smaller fish and water insect larvae. Food requirements are high. With appropriate care, specimens will also reproduce. A small bubblenest is made beneath the floating plants, and inside the female lays about 100 eggs, one by one. The male cares for the brood (like Cichlid species), busying himself round the nest and spawn. When the fry hatch out after a few days, he carries them to shallow hollows and looks after the brood for a while, even after they can swim freely. Like the adults, the fry need a lot of live food and grow quickly.

– *P. abbreviata* BOULENGER, 1901; African Leaf-fish. Tropical W Africa. To 8 cm. Dark patterning on a murky yellow-brown or grey-green background. Only suitable for keeping in a species aquarium (fig. 918).

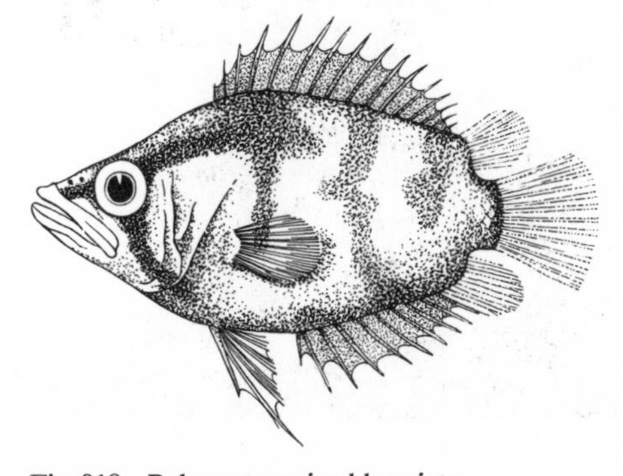

Fig. 918 *Polycentropsis abbreviata*

Polycentrus MÜLLER and TROSCHEL, 1848. G of the Nandidae*. One species native to north-eastern S America. Largish, roughly ovate fish, very compressed laterally. D and An similarly shaped, made up of a lower hard-rayed part a higher, lobe-like, soft-rayed part (the soft-rayed part being transparent). C fan-shaped, likewise transparent. Tail noticeably short. LL absent. Mouth deeply slit, protrusible. Colouring very changeable, dependent on surroundings, the water temperature and the level of excitement. Usually, a grey or brown background colour with bright and dark, often gleaming silvery flecks. A dark band extends from the eye to the snout, another to the nape and a third to the lower edge of the operculum. Hard-rayed parts of the D and An have large pale blotches at their bases: Vs elongated with a yellow-green front edge. Sexes easy to distinguish during the spawning season: male dotted in gleaming silver and blue-green on a velvet-black background; female very brightly coloured. Life-style and behaviour typical of the family. *Polycentrus* needs richly planted tanks with additional places of refuge in the form of root-work and soft, slightly acidic water, occasionally filtered through peat. Temperature about 25–26 °C, subdued lighting. The fish lie in wait for prey in amongst the vegetation; they then rush out and grab hold of the victim in their protrusible mouths. Diet consists of small fish, worms, insect larvae and other similar things; many species will also accept pieces of horse flesh and beef. As *Polycentrus* is only suitable for a species aquarium, breeding means no particular preparations are necessary. Like most Nandidae, the species likes to lay its spawn in caves or on undergrowths. The male conducts a simple form of broodcare.

– *P. schomburgki* MÜLLER and TROSCHEL, 1848; Schomburgk's Leaf-fish. North-eastern S America, and the island of Trinidad. Up to 10 cm long. Colouration as described above (fig. 919).

Polychaeta; Bristle Worms. Cl of the annelids (Annelida*) with about 4,000 species. Widely distributed marine-dwellers. They crawl about or burrow on or in the bottom, or they live in self-built tubes; a very few are free-swimming. The method of feeding varies: predatory species devour small animals *(Aphrodite*, Eunice, Nereis*)*, others eat nutrient-rich mud *(Arenicola*)*, or graze from the bottom with their mobile feelers *(Lanice, Polymnia)*; yet others catch prey with nets (eg *Nereis* once again). Many tube-dwelling species which are particularly attractive for the aquarium filter fine particles by means of their fine-feathered, often very colourful crowns of tentacles *(see Sabella)*. Characteristic for the Polychaeta are the numerous long bristles on the limb-like paropodia, and the head has so-called prostomial outgrowths. In primitive species, there is a ciliated plankton larva stage of development (trochophora). The systematic division of the Polychaetes is unclear; the earlier device of splitting them up into equally segmented, usually free-moving, predatory Errantia and the overwhelmingly tube-dwelling, often sessile Sedentaria does not reflect the natural conditions of relationship. Errantia-types are usually only introduced into the aquarium by chance, and their presence only discovered some time later (*Nereis, Eunice* etc). The Tube-worms on the other hand have an important role to play in aquarium-keeping. Some build chalky tubes *(Protula*, Serpula*, Spirorbis*, Pomatoceros*)*, others produce leathery tubes made of mud and secreted particles *(Bispira*, Sabella*, Sabellastarte*, Spirographis*)*. The tubes are formed with the aid of glands that lie at a collar-like extension of the front end. Many species are extremely hardy in the aquarium (*Eudistylia, Pomatoceros, Sabellastarte, Serpula* and others); others are more sensitive, eg *Protula* and *Spirographis*. Finely chopped artificial food, soaked oats or specially bred plankton can be blown by a feeding-tube into the range of their crown of tentacles. (Suitable 'microalgal plankton' is cultivated, according to WILKENS, in glasses filled with sea water and a thin layer of old, unmanured garden soil. The glass is then stood in a bright spot and some bivalve flesh or *Tubifex* is added.) In well-established, algae-rich marine aquaria, separate feeding is often unnecessary. Various animals are dangerous to Polychaetes, such as Crustaceans, predatory Starfishes and very stinging Sea-anemones. However, they can be put in a community tank with many other invertebrates and peaceful fish.

Polychaetes (collective term) *see* Polychaeta

Polycladida (fig. 920). O of the Eddy Worms (Turbellaria*). Leaf-shaped, often strikingly coloured animals from the warmer seas, that grow up to 16 cm long. Their mid-gut is often much branched. There are only a very few reports of observations in the aquarium.

Fig. 919 *Polycentrus schomburgki*

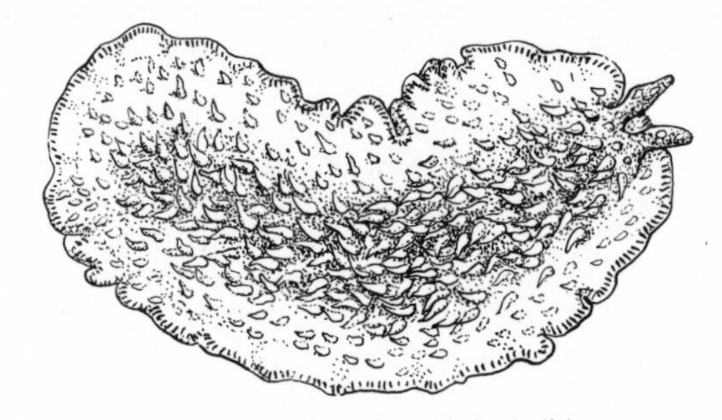

Fig. 920 Representative of the Polycladida

Polygonaceae. F of the Magnoliatae*, comprising 40 Ga and 800 species. Found on all the continents, but mainly in northern temperate regions. In rare cases, they are trees or shrubs, but most are perennials and annuals. Leaf arrangement usually alternate. Leaves often have a sheath at the base enclosing a stem. Flowers hermaphroditic or unisexual, fairly small, with either a striking or inconspicuous perianth. The Polygonaceae colonise the most varied of habitats, even very wet biotopes. A useful plant that belongs to the Polygonaceae is Rhubarb. *Polygonum amphibium* is occasionally kept in open-air ponds. It is a member of the genus *Polygonum,* which has about 200 different species. *Polygonum amphibium* forms broadly lanceolate floating leaves on the pond's surface, but the plant also grows well on land.

Polygonum, G, *see* Polygonaceae

Polymnia, G, *see* Polychaeta

Polynemoidei; Thread-fins, related species (fig. 921). Sub-O of the Perciformes*, Sub-O Acanthopterygii. Elongated fish, slightly flattened laterally, with a rounded snout and an inferior mouth. Found in warm seas.

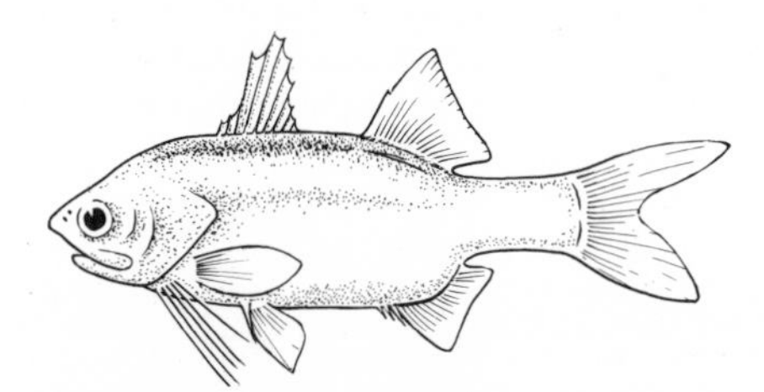

Fig. 921 Polynemoidei-type

They look particularly like the Mugilidae (Mugiloidei*), and indeed there is a close relationship. Especially characteristic are the Ps divided in two, the lower part being located almost at the throat. It consists of isolated, very long fin rays which serve as organs of touch and taste. In young specimens, they reach as far as the C if folded back; in adults, only as far as the belly. The D1 and D2 are short, and separated by a wide distance. C deeply forked. The Polynemoidei prefer shallow waters near the coast and also enter brackish water. They feed mainly on small animals. All Polynemoidei are small, apart from one species, *Eleutheronema tetradactylus,* which comes from south of India and can grow to over 150 cm long. The Sub-O includes some of the tastiest fish of anywhere and so large numbers are fished in certain regions.

Polyodontidae, F, *see* Acipenseriformes

Polyphyletic Group. A systematic group of any rank (eg a phylum, an order, a genus) with members that stem from different primitive forms, and so are not closely related, to one another. Species that belong to a polyphyletic group are often grouped together because of primitive features that they have in common (Plesiomorphic*). Such features give no clues as to phylogenetic relationships. The demonstration of polyphyletic groups and their elimination is of essential importance in modern systematics*. Compare the entry, and illustration, under Monophyletic Group.

Polyploid *see* Cell

Polypodiaceae, F, *see* Pteridophyta

Polyprion americanus *see* Serranidae

Polypteridae, F, *see* Polypteriformes

Polypteriformes. O of the Osteichthyes*, Coh Chondrostei. Only one family (Polypteridae, Bichirs) (fig. 922). These fish are exclusively native to tropical Africa. They are elongated, only slightly flattened laterally, and in rare cases like a snake in shape. They have an unusual D that is composed of numerous, small, pennant-like fins, arranged one after the other. At the

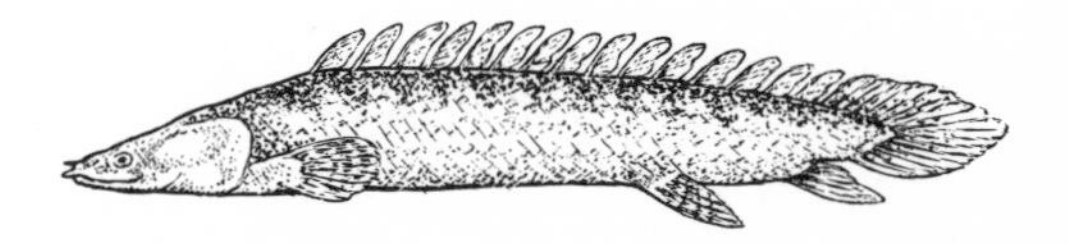

Fig. 922 Polypteridae-type

front edge of each one there is a spine. These small finlets can be erected or laid down. Another very primitive feature are the ganoid scales arranged in thick transverse rows – this type of scale is often found among the Palaeoniscids, very ancient fish from very early on in the Earth's history. A thick layer of ganoin* causes the upper surface of the scales to be hard and gleaming (ganoin is a substance similar to the enamel on our teeth). Beneath the ganoin lies a thin honeycombed layer (cosmine) and beneath that a bony layer once again (isopedine). Many parts of the cranium remain unbony, and the upper jaw is firmly joined to the neurocranium. The head is broad, the mouth large, and the nasal openings are elongated on the outside like tubes. Another peculiarity of the Polypteriformes are the stalked, fan-like Ps, which are used like oars when swimming and like props for resting on the substrate. There are sometimes no Vs (G *Calamoichthys**), the An is short, and the C symmetrical on the outside, although the vertebral column runs through the top part. The swimbladder is on the ventral side of the intestine and consists of a small left sac, and a large right sac, both of which are joined to the pharynx via a common passageway. The swimbladder forms a primitive kind of lung and it is indispensable as a respiratory organ. If the Polypteriformes are prevented from breathing air, they will suffocate. Like many other primitive groups of fish, the intestine has a spiral valve. Polypteriformes larvae look like the larvae of newts with their tree-like external gills (fig. 923). In their natural habitats, they are found in the shallower regions near the bank. They only leave their hiding places as dusk falls to steal along after their prey, usually padding slowly forwards, then pausing and finally snapping the victim up in one gulp, unchewed. The prey consists mainly of worms, insect larvae, crus-

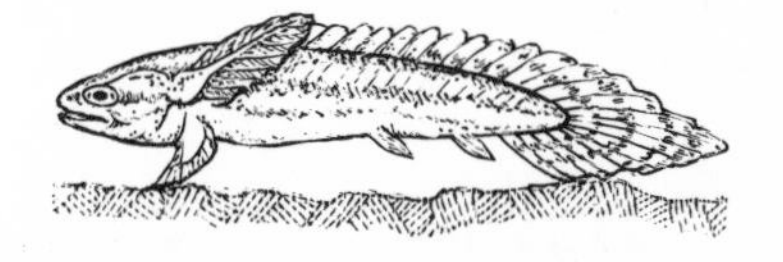

Fig. 923 Larva of the Polypteridae

taceans and smaller fish. To breathe in air, the fish meander up to the surface then shoot down to the bottom again. The reproductive period begins with the approaching end of the rainy season. The spawn is laid in areas that will be flooded. The larger species, such as the 70-cm-long *Polypterus bichir* with its 14–18 little finlets, are eaten by native Africans. The meat is roasted in hot ash or on a spit over a fire. It is said to taste very good. Souvenirs are made from the hard scaly armour. In recent years, various species have been kept in aquaria. For further details see under the Ga *Calamoichthys** and *Polypterus**.

Polypterus GEOFFROY, 1802. G of the Polypteridae, O Polypteriformes*. From tropical Central and W Africa in muddy waters with dense vegetation, in flood areas, and also in brackish water. Medium-sized to large fish, elongated like eels and flattened laterally. The head is broad, flattened dorsoventrally and also has a wide-gape mouth. There are tube-like elongations to the front nasal openings. The entire body is covered with ganoid scales. LL complete. The fin rays of the long D form separate little finlets and join on directly to the C. In contrast to the G *Calamoichthys**, the *Polypterus* species have well developed Vs. The Ps are broad and sit on top of a muscly stump covered with scale. There is an open passageway from the mouth cavity to the paired swimbladder; the latter contains a great many blood vessels and serves as an accessory respiratory organ. *Polypterus* is active at night and may be intolerant of like species. Keep specimens in large tanks stood in a dark spot. Supply medium-hard water, not too coarse a sandy substrate and plenty of places of refuge. Once accustomed to their new surroundings, *Polypterus* species are very hardy. Temperature 23–28 °C. Feed them with substantial worm food, insects and their larvae, fish, bivalve and fish meat, as well as cow's heart. Large specimens are only suitable for public aquaria. There have been isolated successes with breeding. The eggs are laid near the bottom in between plants, during which the male pushes his An beneath the genital opening of the female. The larvae that hatch out have feathery external gills and are not easy to rear.

– *P. bichir* GEOFFROY: 1802; Nile Snakehead. Africa, the Nile, Lake Rudolph, Lake Chad. Length to 70 cm. Dorsal surface olive, flanks grey, ventral surface pale yellow. Young specimens have 10–12 transverse stripes that end in a wedge-shape at the bottom; in addition, there are often 2–3 longitudinal stripes. Fins a delicate grey, Ps and Vs with dark patterning. Not yet bred in captivity.

– *P. delhezi* BOULENGER, 1899. Upper and middle Congo. Length to 35 cm. Body brown-yellow with 6–7 irregular, black-brown transverse bands, often edged in a paler colour. These bands extend from the back to the level of the LL. Scattered in between are individual dark flecks, with some on the C, too. Fins yellowish with dark transverse markings. Not yet bred in captivity (fig. 924).

– *P. ornatipinnis* BOULENGER, 1902. Upper and middle Congo. Length to 37 cm. Body grey-yellow with dark-brown reticulated markings, that stretch right across the body, and across the head and fins. Fins yellow; the D with black-white speckles, and there are also black transverse bands on the Ps and Vs. This species has been bred in the aquarium.

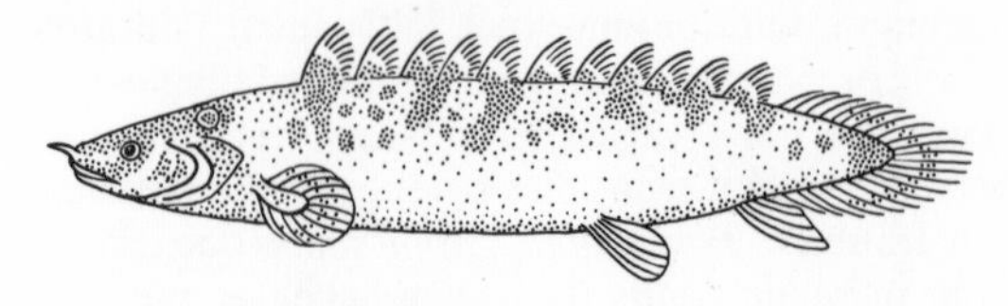

Fig. 924 *Polypterus delhezi*

– *P. palmas* AYRES, 1850. Sierra Leone, Liberia, Zaïre. Length to 30 cm. Dorsal surface grey-green, flanks paler, ventral surface yellow. Young specimens have irregular dark flecks and bands. Fins covered in dark markings. The muscular stump of the Ps has a black blotch on it. Not yet bred in captivity.

– *P. weeksi* BOULENGER, 1898. Upper Congo, Shaba. Length to 40 cm. Dorsal surface grey to olive, flanks paler with irregular, dark transverse bands which branch out. Belly yellow. Fins covered in black dots and bands. Each of the finlets has a black blotch on it.

Pomacanthidae, F, name not in current usage, *see* Chaetodontidae

Pomacanthodes, Sub-G, *see Pomacanthus*

Pomacanthus LACÉPÈDE, 1803 (figs. 925, 926 and 927). G of the Chaetodontidae*, Sub-F Pomacanthinae. There are numerous species found in all tropical seas, where they live on coral reefs and rocky coasts. They live alone (as they get older, in pairs), and form very definite territories: they feed on growths. The main area of distribution is the Indo-Pacific realm. The Indo-Pacific Sub-G *Pomacanthodes* is distinguished from the Atlantic Sub-G *Pomacanthus*. Length 30–60 cm. They are tall fish, but more strongly built than other Chaetodontids. The scales are small, LL complete, and in older fish the D is elongated at the back. Youthful colouring dark with pale bands arched towards the rear. The adult colouring is completely different and the change normally takes place when the fish is about 10 cm long. The striking pattern of colours helps define territories – like species and similarly coloured fish are attacked, whereas fish of different appearance are largely unmolested (even in the aquarium). *Pomacanthus* species

Fig. 925 *Pomacanthus annularis*

survive better than other Chaetodontidae. Apart from a varied diet of animal food, they also require plant food. The aquarium must offer many places of refuge as well as free-swimming zones. Well-known species are:

Fig. 926 *Pomacanthus semicircularis* (juvenile form)

– *P. (Pomacanthus) arcuatus* (LINNAEUS, 1758); Black Angelfish, Grey Angelfish. Tropical Atlantic. Length to 60 cm. Young fish black with 4 yellow transverse bands. Adults black-brown with a black dot on each scale.
– *P. (Pomacanthodes) imperator* (BLOCH, 1787); Emperor or Imperial Angelfish. Indo-Pacific, including the Red Sea. Length to 40 cm. Young fish have ring-shaped markings at the rear. Adults are blue with about 20 yellow longitudinal lines, a yellow head and black eye-band.

Fig. 927 *Pomacanthus semicircularis* (adult form)

Pomacentridae; Damselfishes (fig. 928). F of the Perciformes*, Sub-O Percoidei*. They are small fish, closely related to the Cichlidae*, and they live almost exclusively on reefs. Body tall to elongate, flattened laterally. Mouth relatively small, surrounded by clear lips; there are one or several rows of conical or wedge-shaped teeth in the jaws. There is only one nasal opening per side. The D is uniform, with a fairly long, hard-rayed section. C rounded or indented, in some species, deeply forked. An short and almost the same shape as the soft-rayed part of the D. Distinct comb-edged scales; physoclists. Most species are delightfully coloured. The youthful colouring may be very different from the adult colouring. It provides camouflage, enables species-recognition, and also acts as warning colours when defending territories. When danger threatens, many Pomacentridae will retreat into crevices in the coral. Some species live in pairs, and the young fish form small shoals, in many cases. Several species swim in an undulating manner, as if balancing out the waves of the surf. They eat small invertebrate animals mainly, although many species feed on growths. The way in which many species live together with certain giant anemones or *Actinia* is of great interest. Some live in such a symbiotic relationship with a particular kind of anemone only, others will be found in several kinds of anemone. The anemone forms the centre of the fish's territory, and the eggs are often laid next to the anemone's foot.

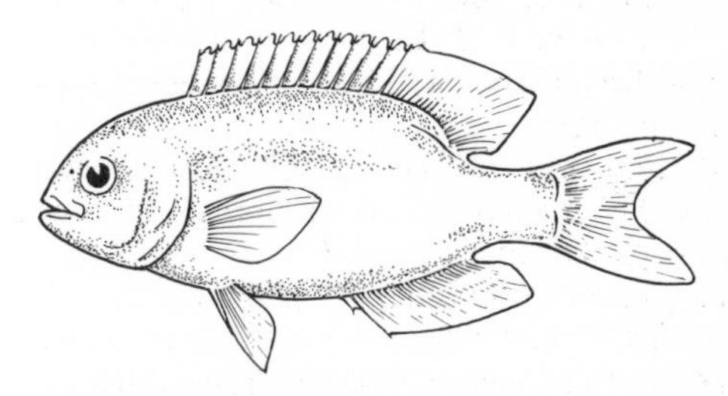

Fig. 928 Pomacentridae-type

To be able to colonise an anemone, the fish has to become accustomed to being in contact with it, so that, in the end, the 'guest' fish is no longer stung. The details of the mechanisms involved are not yet fully understood, but it is probable that chemical recognition substances play an important role. For further information on this particular kind of co-habitation and on aquaristic experience with Damselfishes, turn to the Ga *Abudefduf**, *Amphiprion**, *Chromis**, *Dascyllus**, *Eupomacentrus**, *Pomacentrus**, *Premnas**.

Pomacentrus LACÉPÈDE, 1803 (fig. 929). G of the Pomacentridae*. Numerous species in the tropical and subtropical realm of the Indo-Pacific, including the Red Sea. They live mainly on coral reefs as territorial, aggressive, solitary fish. Maximum length scarcely more than 10–15 cm. *Pomacentrus* has an oval body, large scales and compressed teeth; the rear edge of the praeoperculum is denticulated. Especially when young, these

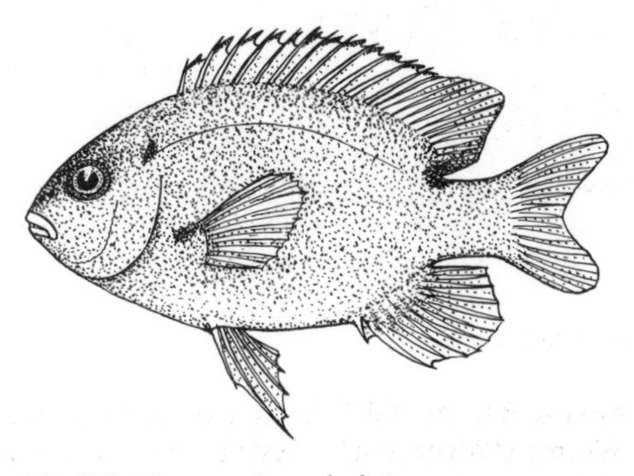

Fig. 929 *Pomacentrus nigricans*

fish are attractively coloured: as they get older, they are often very murky in colour. This then means it is difficult to determine species by inspection of colour. As far as is known, the males take care of the eggs. In the aquarium these pretty fish are hardy; just remember to take their extraordinary aggression into account. In reality, it is only possible to keep at most two specimens in the one domestic aquarium. Other fish will also be in danger from them if they are similarly coloured and not much bigger in size. *Pomacentrus* species need plenty of places to hide, but feeding presents no problem. Among the well-known species are:

– *P. tripunctatus* (CUVIER and VALENCIENNES, 1830); Three-spot Damselfish. From vast areas of the Indo-Pacific. Length to 13 cm. Brown; the young fish has a pale blue band on the dorsal surface and a black blotch edged in blue on each of the caudal peduncle, dorsal surface and behind the Ps; These features gradually disappear with age.

– *P. pavo* (BLOCH, 1787); Peacock Demoiselle. Indo-Pacific, Red Sea. Length to 12 cm. Shining blue; caudal peduncle, belly; C and An yellow.

Pomadasyidae; Grunts (fig. 930). F of the Perciformes*, Sub-O Percoidei*. Small to medium-sized, Perch-like fish from tropical seas, reminiscent from the outside of the Lutjanidae* (indeed, they are sometimes placed with them in one F). However, they are distinguished from the Lutjanidae because, amongst other things, they have a fairly small mouth, the dentition in the jaws is weak, and the pharyngeal teeth are well developed. When these pharyngeal teeth are rubbed together, a grunting noise is produced, which is amplified by the swimbladder. This sound is often heard when the fish is pulled out of the water, and this is where the common name comes from. The Indo-Australian Pomadasyidae, which are united in the Sub-F Plectorhynchinae, have very strong colour patterns that can vary a great deal in young fish and sexually mature adults. Many species live on coral reefs, others along rocky coasts, and some penetrate brackish water. The larger species have excellent tasting flesh. For information about keeping the Pomadasyidae in the aquarium turn to the G *Plectorhynchus**.

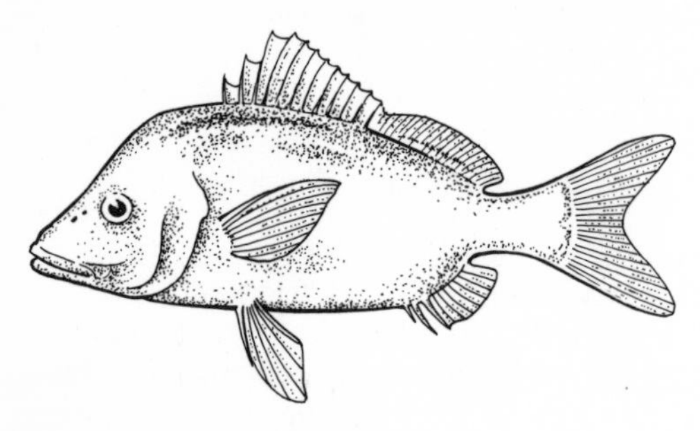

Fig. 930 Pomadasyidae-type

Pomatoceros BRAUN, 1883. G of the small, sessile Bristle Worms (Polychaeta*). Distributed worldwide and found from the littoral* to very great marine depths. They eat small floating matter. *Pomatoceros* has a keeled, chalky tube, a variable kind of crown of tentacles, and a stalked lid. In contrast to the similar G *Serpula**, the tube in cross-section is triangular, and the whole length of the tube lies on the substrate. A very hardy specimen in the aquarium is:

– *P. triqueter* (LINNAEUS, 1758). A cosmopolitan. 5 cm. Tentacles grey, blue, red or yellow.

Pomatomidae; Bluefishes. F of the Perciformes*, Sub-O Percoidei*. This F is known mainly through the Bluefish *(Pomatomus saltatrix)* (fig. 931). Blue-grey to greenish in colour, this fish grows up to 1 m long and weighs 12 kg. It lives in all warmer seas, with the exception of eastern and central parts of the Pacific. It is easy to recognise by its spindle shape, tiny scales, and most of all the shape of the D. The D1 is much lower

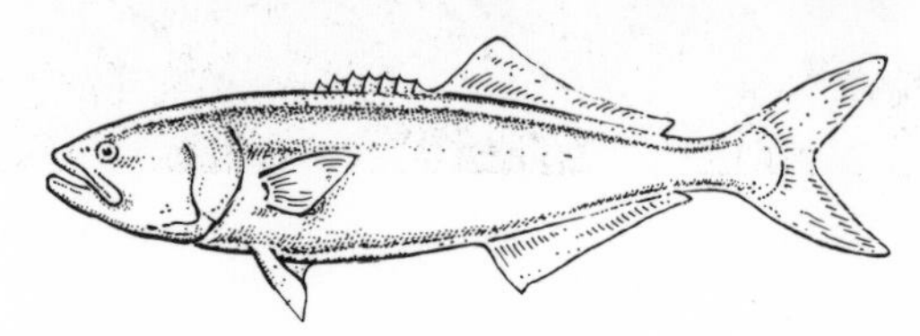

Fig. 931 *Pomatomus saltatrix*

and shorter than the D2; the D2 also lies more or less opposite a very similarly shaped An. The Bluefish has razor-sharp teeth and is one of the most predatory and aggressive of all the Percoidei. Usually, it will kill more than it needs. The Bluefish tastes delicious and is popular with sport fishermen.

Pomatomus saltatrix *see* Pomatomidae

Pomatoschistus, G, *see Gobius*

Pomoxis RAFINESQUE, 1818. G of the Centrarchidae* containing few species. Widely distributed in lakes and rivers of central and eastern N America, reaching S Canada in the north; on rare occasions also found in brackish water. Medium-sized to large fish, tall-backed and flattened laterally, like a Perch to look at. The eyes are large, the mouth deeply slit, the lower jaw protrudes. In contrast to the other Ga of the Centrarchidae, there are only 6–8 hard rays in the D; the G *Elassoma** has even fewer (4–5). The D and An are similarly formed, the C is weakly curved. In the early summer, the male adopts a territory, in which a spawning hollow is made. *Pomoxis* nests of this type may lie on their own, or there may be colonies of them, up to 50 at a time. The females are not tied to a particular partner, and lay spawn with several males. The father performs a simple type of brood-care – he builds the nest, guards it, repairs it, and even watches over the hatched brood for the first few days.

As *Pomoxis* species are not particularly attractive to look at, and as they grow very large in any case, only young specimens are suitable for the aquarium. On the other hand, *Pomoxis* is a popular, tasty sporting fish.

– *P. annularis* RAFINESQUE, 1818; White Crappie. Central and eastern N America, also introduced in many places (eg California) that are outside its natural areal. Length over 50 cm. A bright yellow-grey ground colour with an irregular covering of olive-green to olive-brown flecks. 6 hard rays in the D.

– *P. nigromaculatus* (LESUEUR, 1829); Black Crappie. Central and eastern N America. Found in fresh water and more rarely in brackish water. Length to 49 cm. Dorsal surface dark olive or metallic green or brown with paler flanks and a greenish-silver iridescence. Also covered irregularly with dark flecks. 7–8 hard rays in the D (fig. 932).

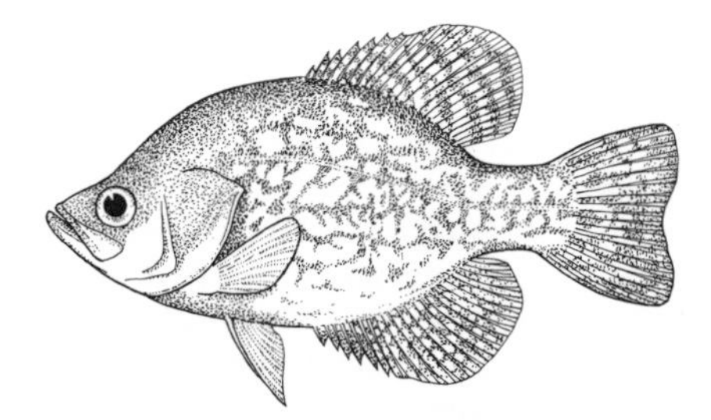

Fig. 932 *Pomoxis nigromaculatus*

Pompadour *(Symphysodon discus) see Symphysodon*

Pompanos, F, *see* Carangidae

Pond Horse-tail *(Equisetum fluviatile) see* Pteridophyta

Pond Lily, G, *see Nuphar*

Pond Loach *(Misgurnus fossilis) see Misgurnus*

Pond Snails, G, *see Lymnaea*

Ponds. A still stretch of shallow water of any size that has developed naturally. In contrast to pools, ponds permanently contain water and normally they also have extensive stocks of water plants. Another characteristic of ponds is the rich variety of species from a great many different groups of organisms. Most are typical still-water species that can tolerate simultaneously great fluctuations in water temperature as well as oxygen content.

Pondskaters, G, *see Gerris*

Pondweed, G, *see Potamogeton*

Pondweeds, F, *see* Potamogetonaceae

Pondweeds, Ga, *see Egeria, Elodea, Lagarosiphon*

Pontederiaceae; Pickerelweed Family. F of the Liliatae*, comprising 7 Ga and 30 species. Tropical-subtropical, also radiating out into temperate regions in America. Perennials, some with rhizomes or runners, as well as annual plants. Leaf arrangement rarely alternate, usually distichous, but sometimes in rosettes. The leaves come in a variety of shapes. Flowers hermaphroditic, usually monosymmetrical, large, but of varying sizes. The perianth is usually a striking colour. Pickerelweeds are marsh or aquatic plants.

The G *Pontederia* is distributed from S to N America. One eastern N American species, *Pontederia cordata*, can also be used as a pond plant in Europe, although it should be lightly covered over in winter. 2 Ga important in aquarium-keeping are *Eichhornia** and *Heteranthera**.

Pool. A still stretch of water of any size which only has water in it temporarily, perhaps after heavy rainfall, after flooding or after snow has melted. Organisms that truly live in pools have a life-cycle that is adapted to the periodical nature of the water in which they live. As a rule, they survive the dry period as an egg or embryonal stage, and usually there are only a few weeks between hatching out and reaching sexual maturity. Characteristic inhabitants of pools include certain groups of lower Crustaceans (Anostraca*, Ostracoda*, Conchostraca*) and from the fish certain Ga of Egg-laying Tooth-carps (Cyprinodontidae*), such as *Cynolebias**, *Pterolebias** and *Roloffia**.

Poor Man's Glass Catfish *(Kryptopterus macrocephalus) see Kryptopterus*

Poor Man's Moorish Idol *(Heniochus acuminatus) see Heniochus*

Pope *(Acerina cernua) see Acerina*

Poptella EIGENMANN, 1908. G of the Characidae*. Widely distributed in S America. Up to 12 cm. Hardy shoaling fish from the mid water levels, and requiring plenty of free swimming area (young specimens, in particular, are skilled swimmers). Omnivores. Body disc-shaped, LL complete, scales fairly large. A blunt spine, shaped like a hook and bent forwards, stands in front of the D. Open-water spawners*, that require a breeding tank with minimum dimensions of $50 \times 20 \times 20$ cm. *Poptella* is extremely productive, and will be encouraged to breed if the tank is filled with fresh water of 2–4° dH and a pH of 6.2–6.8. Temperature 25–27 °C. Fine live micro-organisms should be fed to the fry, and success should be assured.

– *P. orbicularis* (CUVIER and VALENCIENNES, 1848); Salmon Discus. Previously known as *Ephippicharax orbicularis*. Guyana, Amazon Basin, Paraguay. Length to 12 cm. Flanks silvery, with a bluish, greenish, violet or yellow sheen (depending on illumination). The dorsal surface is blackish-olive. Occasionally, a fine, dark longitudinal streak and 2 long, comma-shaped shoulder flecks make their appearance. The unpaired fins are colourless, with fine fleck markings at their base. The first rays of the An are brown and end in a point. Because of its rather drab colouring, the Salmon Discus is rarely kept in an aquarium (fig. 933).

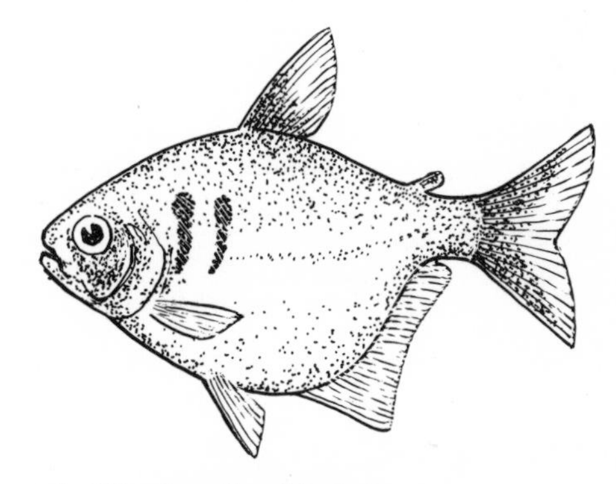

Fig. 933 *Poptella orbicularis*

Population. An organisation collective of the same species within a defined locality – perhaps some Trout (*Salmo trutta, see* Salmonoidei) in a particular mountain stream, or Striped Aphyosemion *(Aphyosemion* bivattatum)* in a particular savannah pool. All the members of a population have at least the potential of producing fertile offspring by mating amongst themselves; genetic material can be exchanged between them without restriction. The population as a whole is inter-related with its environment*, but the individual members of the population also exert a mutual influence upon one another in a variety of ways. For example, the state of health of the individual members, their features (eg colour), sex distribution, behaviour and death-rate can all be dependent on the population density. Furthermore, one of the most fundamental phenomena in biology, evolution* and the subsequent formation of species is based on the grouping of living beings into individual, sexually isolated populations.

Population Ecology *see* Biodemography

Porbeagle *(Lamna nasus) see* Lamnidae

Porcelain Crabs, F, *see* Porcellanidae

Porcellana, G, *see* Porcellanidae

Porcellanidae; Porcelain Crabs. F of the Crab-like Anomura (Reptantia*). Has a thin abdomen flapped beneath the cephalothorax. The last segment of the abdomen has uropoda and the fifth pair of walking legs are small and lie in the branchial cavity. The Percellanidae live in the littoral of tropical to temperate seas. They fish suspended matter from the water using their finely bristled 3rd jaw-feet. The tiny crabs of the G *Porcellana* LAMARCK, 1801, have a carapace that measures a mere 0.5 to 1 cm in width. They are found in the upper littoral of southern European seas, and are extremely common there. The so-called Anemone Crabs, *Petrolisthes (Neopetrolisthes) oshimae* MIYAKE from the tropical Indo-Pacific, also belong to this F (fig. 934). As their name suggests, they live in the tentacles of Sea-anemones (Actiniaria*).

Fig. 934 *Petrolisthes oshimae*

Porcupine Fish *(Diodon hystrix) see Diodon*

Porcupine Fishes, F, *see* Diodontidae

Porichthys notatus *see* Batrachoidiformes

Porifera; Sponges (fig. 935). A primitive phylum (with about 5,000 species) that belongs to the multi-celled animals (Metazoa) and which is descended from certain forms of Flagellates (Flagellata*). Sponges lead a sessile* life-style and they are found in a great variety of species and forms in the sea; a few species have also penetrated fresh waters, via brackish water *(Ephydatia*, Spongilla*)*. Sponges have a characteristic internal skeleton made of a chalky or silica substance, together with a horny substance. A well-known one is the skeleton of the Bath Sponge *(Euspongia)* as used in many a household. Various kinds of Sponges can be kept successfully for a long time in a marine aquarium.

Porkfish *(Anisotremus virginicus) see Anisotremus*

Port Acara *(Aequidens portalegrensis) see Aequidens*

Port Hoplo *(Hoplosternum thoracatum) see Hoplosternum*

Port Jackson Sharks, F, *see* Heterodontidae

Porthole Livebearer *(Poeciliopsis gracilis) see Poeciliopsis*

Porthole Rasbora *(Rasbora cephalotaenia) see Rasbora*

Portuguese Man o' War *(Physalia) see* Stinging Capsules

Portunus WEBER, 1795. G of the Swimming Crabs (Brachyura*, F Portunidae). Found in cool and warm seas. Predatory omnivores from sandy expanses that never leave the water. The carapace is heart-shaped and toothed on the front side. Unlike the similar G *Portumnus* from the same habitat, *Portunus* does not have flattened 2nd-4th walking legs. It swims by beating its paddle-like 5th walking legs very fast. *Portunus* is sandy-coloured and its carapace 1–4 cm in width. Like

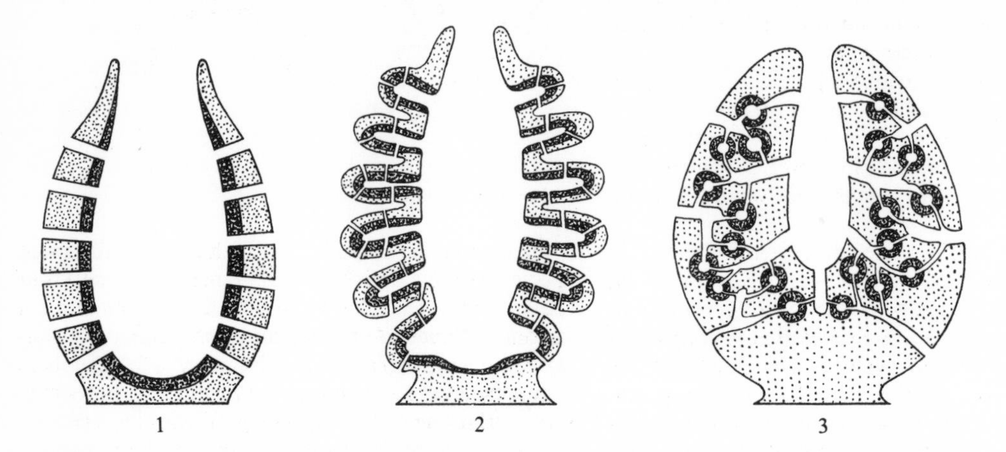

Fig. 935 Porifera. Various types of organisation within the Sponges (black portions indicate the position of the choanocytes). 1 Ascone-type 2 Sycone-type 3 Leucone-type

many creatures that inhabit sandy places, they are not as robust as, say, the related Beach Crabs *(Carcinides*)*; so there are few observations in the aquarium. There are many species along the European coasts, all of which look very similar.

Posidonia. A marine-dwelling G of the Pondweed Family (Potamogetonaceae*). *P. oceanica* (L.) DEL. is a magnificent plant with dark-green, linear leaves up to 80 cm long and fibrous root-work. In the Mediterranean, it is found at depths of 3–40 m and forms large mats. Usually, a rich variety of fauna lives in amongst these mats. *Posidonia* has so far resisted cultivation in the aquarium.

Potamal, also potamon. A flowing water region with the character of a river of the lowland area (ie the water flows relatively slowly, its oxygen level is medium, and the (summer) temperature is considerable). Ecologically, the potamal can be divided into far more numerous sections than the other flowing water regions (Crenal*, Rhithral*), and it therefore harbours a corresponding variety of species and life-forms. A considerable number of aquariumfish also stem from waters that are attributed to the potamal, namely Characins and Barbs, as well as Catfishes, Cichlids and many others.

Formerly, the potamal was also named after its characteristic fish species, and was thus divided into the Barb-region, the Bream-region and the Flounder-Ruffe region. In contrast to the characteristic fish named, the corresponding river sections are found worldwide, so the tendency is (quite rightly) to use more generalised terms and speak of the epipotamal, the metapotamal and the hypopotamal.

Potamogeton LINNÉ, 1753; Pondweed. G of the Potamogetonaceae* with 100 species, distributed worldwide. Small perennials with thin rhizomes or runners that grow in or on the ground. Upright shoots, variously branched. Leaf arrangement distichous. The leaves come in a variety of shapes, sometimes there are aquatic leaves only, sometimes there are floating leaves as well. Flowers hermaphroditic, inconspicuous. There is no perianth; instead, there are outgrowths of the stamens. Stamens usually number 4. Carpels 4–1, free, superior. The fruit is a collective nut. The various species live in still and flowing fresh waters, more rarely in brackish water. As yet, we do not know a great deal about their suitability for keeping in the aquarium or open-air pond. Almost certainly, however, other plants from the G will prove useful. Some examples are:

Fig. 936 *Potamogeton crispus*

– *P. crispus* LINNÉ; Curly Pondweed. From temperate regions of the northern hemisphere. Aquatic leaves lanceolate with curly edges and red-brown in colour. No floating leaves. Found in still and flowing water, rich in nutrients (fig. 936).

– *P. gayi* A. BENNET; Gay's Pondweed. S America. Aquatic leaves linear, green to reddish; no floating leaves. Suitable for heated aquaria.

– *P. natans* LINNÉ; Floating Pondweed. Temperate regions of the northern hemisphere. Aquatic leaves linear, floating leaves ovate. Often found in the floating leaf zone of of the water-land junction. Suitable for open-air ponds.

– *P. octandrus* POIRET. Tropical Africa and Asia. Aquatic leaves lanceolate, green, translucent; floating leaves coarser. Suitable for heated aquaria.

Potamogetonaceae; Pondweed Family. F of the Liliatae*. 5 Ga and 110 species, distributed worldwide. Small perennials with elongated stem axes. Leaf arrangement distichous, in some species the leaves are differentiated into aquatic and floating leaves. Flowers hermaphroditic, rarely unisexual, inconspicuous. Perianth replaced by outgrowths of the stamens. Pondweeds are aquatic plants from fresh and salty water. The G *Ruppia* is found all over the world in salty and brackish water. The 2 Ga *Zostera** and *Posidonia** are found along the sea coasts of temperate regions, but for the marine aquarist they are both of no value. In freshwater aquaria, the G *Potamogeton** is of some importance.

Potamon *see* Potamal

Potamonidae, F, *see* Brachyura

Potamorrhaphis guianensis *see* Exocoetoidei

Potamotrygon, G, *see* Dasyatidae

Potassium Iodide. In fish, excessive growth of the thyroid gland (*see* Tumour Diseases) caused by lack of iodine can be cured by giving extra amounts of iodine. Iodine is insoluble so must be given in the form of soluble potassium iodide in prolonged baths (2–3 weeks). To make up some potassium iodide solution, dissolve 10 g potassium iodide (KI) + 0.1 g iodine (I) in 100 ml distilled water. Use 0.5 ml of this solution for every 1 litre of fresh water or every 2 litres of sea water. If necessary, repeat after 10 days. Filters and UV lighting should not be used during treatment.

Potassium Permanganate ($KMnO_4$) is a strong oxidising agent. Its crystals are a purply-red with a metallic sheen, and they dissolve in water easily, producing a vived violet-red colour. Because of these properties, potassium permanganate is used by aquarists as a disinfectant (Disinfection*) and as a treatment for skin parasites. Separate tanks for bathing the specimens *must* be used. Short-term baths (*see* Bathing Treatments) lasting from 5 to 30 mins are suitable for treating bacterial infections* of the skin, *Saprolegnia* (Saprolegniacea*), *Ichthyobodo**, *Chilodonella**, *Trichodina** and parasitical Copepods (Copepod Infestation*). Use a concentration of 0.1–0.5 g potassium permanganate to every 10 litres of water. Immersion baths lasting 30–50 secs will kill any Argulidae* on large fish (concentration 1 g potassium permanganate to every 10 litres water). Small fish with fungi or Argulidae on them can be treated by carefully dabbing the infected areas with a potassium permanganate solution made up of 1 g $KMnO_4$ to 1 litre of water. Aquatic plants can be disinfected by placing them in a solution of 1 g potassium permanganate to 20 litres of water and leaving them there for 5 mins. Tools and containers can be disinfected in highly concentrated solutions (1–3 g to 1 litre of water).

Potassium Permanganate, Consumption of, *see* Permanganate Consumption

Powder-blue Surgeonfish *(Acanthurus leucosternon) see Acanthurus*

Praunus, G, *see* Mysidacea

Precious Coral *(Corallium rubrum) see Corallium*

Pre-courtship Behaviour *see* Courtship Behaviour

Premnas (CUVIER, 1817). G of the Pomacentridae*. Very similar to *Amphiprion**, but with a spine beneath the eyes. There is only one species:

– *P. biaculeatus* (BLOCH, 1790); Spine-cheeked Anemonefish, Maroon Clownfish, Tomato Clownfish. Tropical Indo-Pacific. Length to 15 cm. Brown-red with 3 white transverse bands. *See* Anemonefish. Often sensitive in the aquarium.

Preparation *see* Examination Technique

Preparation of Swabs *see* Examination Technique

Preparations, Equipment used in, *see* Examination Technique

Pretty Tetra *(Hemigrammus pulcher) see Hemigrammus*

Preventative Measures *see* Prophylaxis

Preventative Treatment (disease) *see* Prophylaxis

Priacanthidae; Big-eyes (fig. 937). F of the Perciformes*, Sub-O Percoidei*. Marine fish, usually a monotone deep-red colour, and like a perch in shape. The head has a large mouth set at an angle and very large eyes. The edge of the operculum is saw-toothed and there are spines in the corner. D uniform with 10 spines in the front part and a high rear section. The An is long and broad, the Vs fairly large. Rough ctenoid scales. Big-eyes are found both in shallow and deep areas of

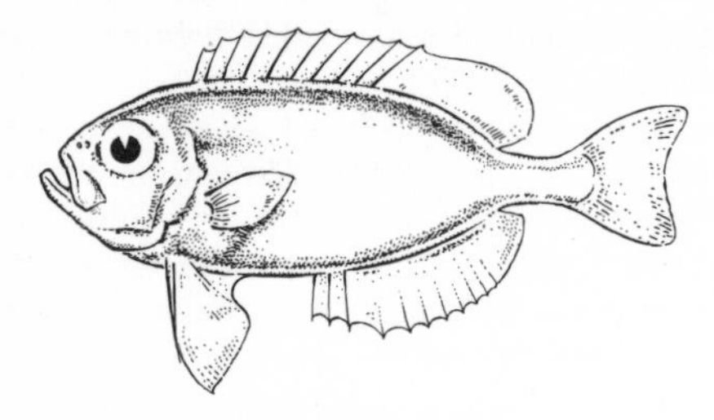

Fig. 937 Priacanthidae-type

tropical and temperate seas, and some live on reefs. Length up to 50 cm. The Priacanthidae have some economic importance regionally.

Priapella REGAN, 1913. G of the Poeciliidae*, Sub-F Poeciliinae. Mexico. Lives in calm parts of smallish rivers and various tributary waters. Small fish, males 3–5 cm, females 4.5–8 cm. Body elongated, only slightly flattened laterally. There are no macroscopically recognisable generic-specific features. Keep specimens in tanks offering sufficient room for swimming and parts of which should be densely vegetated. 22–25 °C. Omnivores, live food preferred. To date, the G has been of little importance in the aquarium. The following are kept:

– *P. bonita* (MEEK, 1904). Male to 5 cm, female to 6 cm. Greenish-grey with a dark longitudinal band. Fins yellowish; in the male, the D and C are dark-edged. A peaceful, undemanding species.

– *P. intermedia* ALVAREZ, 1952. Blue-eyed Livebearer. Male to 5 cm, female to 7 cm. Body yellowish with a dark longitudinal band; towards the dorsal surface there are the suggested outlines of longitudinal stripes. A green blotch is located on the operculum. Iris blue. Fins have a translucent bright quality, with a white edge in part. A shoaling fish.

Priapichthys REGAN, 1913. G of the Poeciliidae*, Sub-F Poeciliinae. Central America and northern S America: Costa Rica, Panama, Colombia, Ecuador. Lives in smaller and medium-sized flowing waters, as well as brackish water. Very small to small fish, males 2.5–4.5 cm, females 3–6 cm. Ground colouring usually a brownish-olive, also grey-green. Several species have dark transverse bands on the body. So far, the G has not been important in aquarium-keeping. The conditions required for care and breeding are largely unknown.

Pricklebacks (Stichaeidae, F) *see* Blennioidei

Primary Leaf *see* Leaf Succession

Primitive Fish, Cl, *see* Aphetohyoidea

Primitive Mammals (Prototheria, Sub-Cl) *see* Mammalia

Primrose Family *see* Primulaceae

Primulaceae; Primrose Family. F of the Magnoliatae*, comprising 40 Ga and 800 species. They grow mainly in the northern temperate regions, with only a few representatives in the tropics and subtropics. Most are perennials or annuals. Leaves arranged alternately, in rare cases opposite or in whorls, with some as rosettes. Flowers hermaphroditic, polysymmetrical, with a strikingly coloured corolla, the leaves of which are fused. Various ornamental plants belong to the Primulaceae, such as Primroses and Alpine Violets. Many of them are able to tolerate low temperatures very well and are able to live in Arctic regions and high up in the mountains. Only a few colonise wet places. In aquarium-keeping, only the Ga *Hottonia** and *Lysimachia** are of interest.

Prionace glauca *see* Carcharhinidae

Prionobrama FOWLER, 1913. G of the Characidae*, Sub-F Paragoniatinae. From the Amazon Basin. Length to 6 cm. Small shoaling fish, skilled swimmers, that resemble the Bloodfin, *Aphyocharax anisitsi,* in colour, care and breeding requirements (*see under* relevant entry). Body elongate, very flat from side to side. They fairly long An has elongated first fin rays that stand on their own, and this feature distinguishes *Prionobrama* from the closely related G *Aphyocharax**. Temperature around 25–27 °C. Feed with small live food, and make sure the tank offers plenty of room for swimming. Finally, only small species should be kept in the same tank together.

– *P. filigera* (COPE, 1870); Translucent Bloodfin, Glass Bloodfin. From the river area of the Rio Madeira in the Amazon Basin. Length to 6 cm. Body glassily transparent, a delicate pale grey-yellow in colour. When illuminated side-on, there is a bluish or greenish shimmer; dorsal surface somewhat darker. In the female, the root of the tail and the base of the C are reddish, in the male they are blood-red; the other fins are colourless. The tips of the An are pastel-white, in the male, it joins on to a black streak. This species is a very pretty shoaling fish, and a new import would be most welcome (fig. 938).

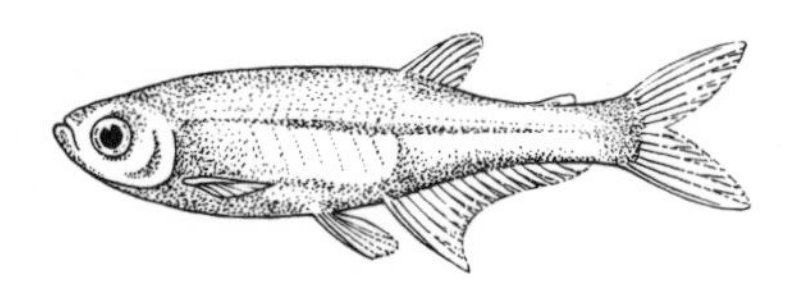

Fig. 938 *Prionobrama filigera*

Pristella EIGENMANN, 1908. G of the Characidae*, Sub-F Tetragonopterinae. Native to South America, these are colourful Characins, 4–6 cm in length. They are peaceful shoaling fish, very fond of swimming, and will survive well in a community tank if provided with enough free-swimming area. They will feed on smaller kinds of live food as well as artificial foods. The body of the males and younger specimens is fairly elongated and flattened laterally. Females full of spawn look much more robust. *Pristella* is closely related to the G *Megalamphodus**. The D1 is short, triangular, and a D2 is present. C two-lobed, deeply indented and covered in scale at the bottom. The An is fairly long and drawn out into a point at the front end. LL incomplete. Breeding is easy, once a suitable pair has been got together (the pair must not be separated). The procedure for breeding is the same as for the species of the G *Hemigrammus**. STALLKNECHT reports that bred fish, if looked after in the correct manner, are capable of reproduction up to the fifth year of life. The young fish hatch out after 24 hours, and are very small. Feed them on very fine juvenile food (Fry, Feeding of*). While rearing them, a frequent change of water is recommended. To date, only the following species has been imported alive.

– *P. maxillaris* (ULREY, 1894); X-ray Fish, Water Goldfinch, Goldfinch Tetra, Riddle's Pristella. From the lower Amazon, the Guyana lands, northern S America.

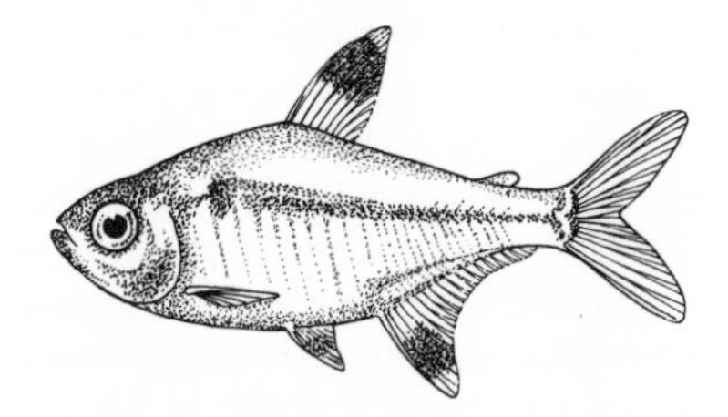

Fig. 939 *Pristella riddlei*

Length to 4.5 cm. Body transparent, yellow-green with a silvery iridescence. The front parts of the D1, An and Vs are lemon-yellow with a deep-black blotch and white tips. The males are a bit smaller and slimmer, and the colouring of the Vs and An is less intense. Ca a delicate red. A very popular aquarium fish. *P. riddlei* (MEEK, 1907) is a widely used synonym for this species (fig. 939).

Pristidae; Sawfishes (fig. 940). F of the Chondrichthyes*, O Pristiformes. The Sawfishes have their gill slits on the ventral side and so are identified as true rays. At the same time, this feature is a sure way of distinguishing between the Pristidae and what at first glance look similar – the Saw Sharks (Pristiophoridae*). In body shape, the Sawfishes look roughly like the Guitarfishes (Rhinobatidae*) – ie Sawfishes do not have the typical disc-shaped ray body, but look much more like ventrally flattened sharks. The way they move is also rather reminiscent of the sharks – the C is used as an organ of propulsion and the relatively small Ps as steering and stability organs. But the most striking feature of all is the long, sword-like rostrum, beset with even-sized teeth all along the sides – the obligatory wall decoration to be bought as a present at almost every port. The teeth sit in sockets and are in fact enlarged placoid scales. The sole natural function of this saw-like rostrum is to acquire food. The fish stirs up the substrate with it to find suitable shelled prey which it then cracks open with its cuboidal dentition. Alternatively, Sawfishes chase through fish shoals, brandishing their 'sword'; the fish they manage to kill or wound are then devoured. Sawfishes are livebearing and give birth to up to 25 offspring at a time, whose saws and teeth remain soft and pliable up to birth, so that the mother suffers no damage. The main distribution area of the Sawfishes is the coastal waters of tropical and subtropical seas. They enter brackish water and the lower reaches of rivers quite often. Of the 6 known species, some grow very large. The largest ever caught were said to measure over 10 m, with over a third of that being attributed to the saw. The Common Sawfish *(Pristis pristis)* is found in the E Atlantic, including the Mediterranean; it could be identical to the West Atlantic Sawfish *(Pristis pectinatus)*. Both grow to, at most, 5–6 m long. Their meat is quite coarse and fibrous, but is eaten in many places all the same. The liver-oil is of great value, and the skin makes a good leather. In India, the saw is of religious import in some places. Attacks on people have not been definitely proved.

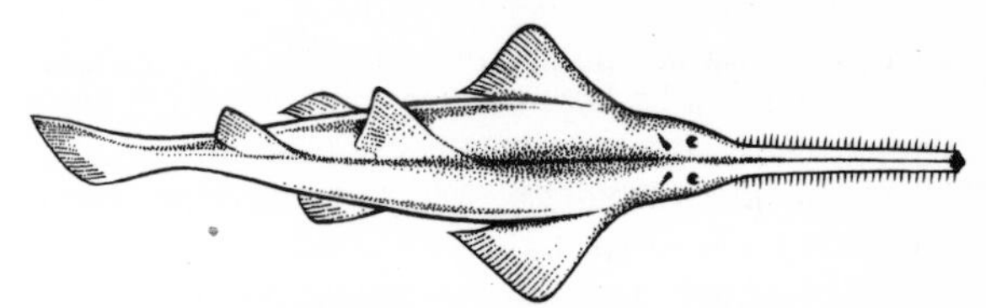

Fig. 940 Pristidae-type

Pristiophoridae; Saw Sharks (fig. 941). F of the Chondrichthyes*, O Pristiophoriformes. Small, livebearing Sharks with an upper jaw elongated into a kind of sword-blade. Along the sides of this blade are small and large dermal teeth, and there are 2 long barbels on the underside. There are 5–6 gill slits in front of the small Ps (a distinguishing feature in respect of the

Sawfishes, F Pristidae*). There are 2 Ds, and no An; C small. The distribution area stretches from S Africa across the whole of the Indo-Pacific realm. Saw Sharks live at deep water levels, and use their 'swords' to plough through the soft substrate in search of prey. They eat fish and molluscs. The Saw Sharks have a very tasty flesh and so are valuable edible fish.

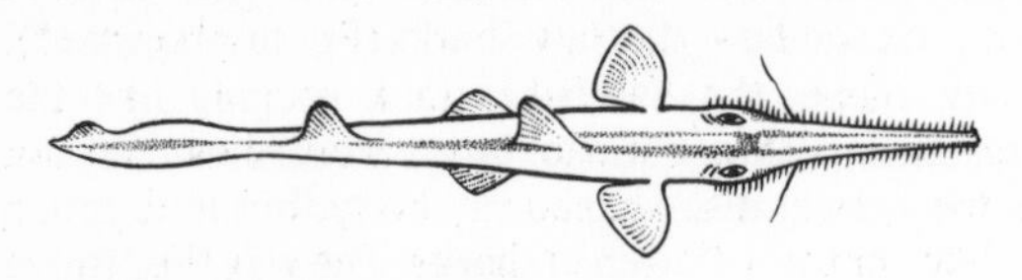

Fig. 941 Pristiophoridae-type

Pristis pristis *see* Pristidae

Pristolepis JERDON, 1849. G of the Nandidae*. One species, found in SE Asia, where it live in rivers, lakes and ponds. It is a large, ovate fish, very flat laterally and with a steep forehead and small mouth typical of the F; the An is inserted along a small stretch of the body, and like the related G *Nandus** it has 3 hard rays. But in contrast to *Nandus* there are 2 flat spines on the gill-covering. Colour: a yellowish-green background with 8–12 dark transverse bands. Little is known about keeping *Pristolepis* in captivity, but if typical of the F it will require to be kept in a separate tank with dense plant growth and undergrowths of root-work.

– *P. fasciatus* (BLEEKER, 1851). From East India, Kalimantan (Borneo). Length to 21 cm. Colouring as described above, although in young specimens the transverse bands are especially distinct.

Proalaria, G, *see* Trematodes

Procatopodinae, Sub-F, *see* Cyprinodontidae

Procatopus BOULENGER, 1904. G of the Cyprinodontidae*. Mainly found in primeval forest rivers and streams in Cameroun and Nigeria; some forms also found in savannah waters. There are about 5–10 species, but the individual populations* vary tremendously and form so-called local varieties, the interrelationships of which are not fully clear. *Procatopus* species are small, elongated fish, moderately compressed laterally, with a well defined sex dimorphism*. The males are very tall and have better developed fins. The top of the C is clearly elongated into a tip. The females are smaller, less tall and have rounded fins. Typical for *Procatopus* and some related Ga is a spine-like elongation of the lower corner of the operculum. They are attached spawners (compare Cyprinodontidae). The eggs are very sticky because they have long, sticky threads. The spawn is not laid on plants, as with other attached spawners, but sunk into crevices and cracks. This peculiarity is explained as an adaptation to the strong current of the water in their natural habitats. *Procatopus* species are agile swimmers that readily come together in shoals and seek out food at the surface that may have landed there. Younger specimens stay in the calmer inlets of the rivers, whereas older specimens prefer more turbulent waters, free of vegetation. If the aquarist is to be successful with keeping *Procatopus* in captivity, these life-styles must be taken into account. Specimens must always be kept as a shoal in shallow, roomy tanks. The necessary water turbulence (and the continual drawing in of oxygen that this causes) is achieved by ventilating well or with a circulating pump. The temperature should be 23–24 °C; higher temperatures have an adverse effect upon the capacity for dissolving oxygen. Oxygen content and the colour of many species are directly related. It must further be taken into account that *Procatopus* reacts with great senstivity to sudden changes in electrolyte content (eg the gills may be damaged). Care is also advised when transferring specimens to a new tank or when changing the water. Try hard to obtain soft water with weak humic acids; nor should it contain much waste matter. These conditions are more or less essential if *Procatopus* is to breed successfully. On the other hand, in soft water, specimens often succumb to fish tuberculosis.

– *P. gracilis* CLAUSEN, 1959. SW Nigeria. Length to 5.5 cm. Colour a uniform bluish-green with a metallic shimmer. The base of C, D and An is covered with brown, red or orange-coloured spots.

– *P. nototaenis* BOULENGER, 1904. Southern Cameroun. Length to 5 cm. Body a shimmering matt blue. The outer parts of the D, C and An are yellow or orange; the base of these same fins are dotted in the same colours. An orange-coloured streak stretches from the upper edge of the eye to the dorsal ridge (fig. 942).

Fig. 942 *Procatopus nototaenis*

Prochilodinae, Sub-F, *see* Characoidei

Prochilodus AGASSIZ, 1829. G of the Curimatidae*. Widely distributed in S America. Length to 35 cm. Highly prized as an edible fish. Body elongated, flattened laterally and usually tall. D1 short, beginning in the middle of the body; D2 strongly developed. The An is short, and the C is deeply forked. Eyes strikingly large. Mouth protrusible like a sucking-disc, and the thick lips are beset with small teeth. Juvenile *Prochilodus* species are very colourful. Roomy aquaria are needed for these species since they like to swim about. They are peaceful fish, but unfortunately they have a marked tendency to eat water plants. Provide plenty of hiding-places and a good lid, as *Prochilodus* can jump well. If possible, keep specimens as a shoal because only then do they show their true agility. Their diet consists mainly of garden-lettuce, soaked spinach, algae and rolled oats; however, *Daphnia*, *Tubifex* and midge larvae are also readily eaten. Nothing is known about the reproduction of any of the species.

– *P. theraponura* FOWLER, 1906; Night Tetra. From the Amazon Basin. Length to 35 cm. The young specimens in the various areas of distribution are very varied in colour. Flanks silver with a weak blue or green sheen. Ventral surface a delicate reddish colour or violet. There are a great many longitudinal streaks on the body, the middle ones of which broaden out on the caudal peduncle. Unpaired fins a delicate yellow-green, Ps and Vs reddish. The D1, An and C are covered with dark-blue longitudinal stripes. A fine, dark middle line ends in a dark blotch at the insertion of the C. Adult specimens are coloured more plainly. Up till now, this species has been confused with *P. insignis* SCHOMBURGK, 1841, (which has not yet been imported). This latter species is found only in Guyana and the Rio Branco. *P. amazonensis* is an out-of-date term for *P. theraponura*. H. SCHULTZ caught a lovely coloured variety of this fish in the Rio Urubu: its flanks were dark and the ventral surface orange. The D2, C and An are yellow, the first-mentioned with a dark edge. The C has 9 black longitudinal bands, the An four. D1, Ps and Vs orange, the D1 with a black front edge. A beautiful fish, but unfortunately only young specimens are suitable for the aquarium (fig. 943).

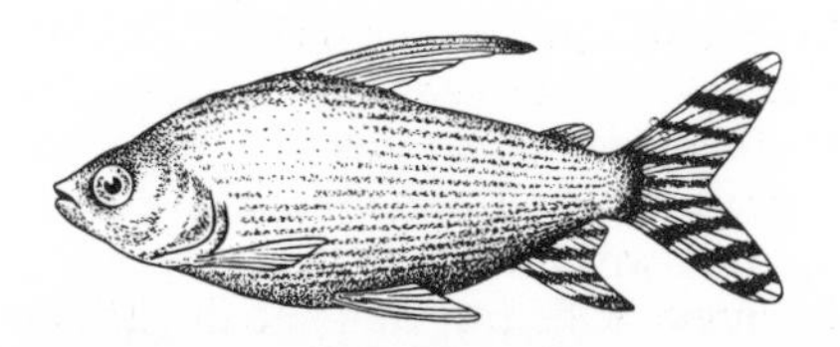

Fig. 943 *Prochilodus theraponura*

Profundal. The bottom zone of a stretch of water that receives no light and therefore has no plants. Together with the bank zone (Littoral*), it forms the benthal*. In inland waters, the profundal begins at depths of 3–30 m (depending on light absorption); in the sea, it begins at about 200 m deep. The profundal region below 1,000 m is also known as the abyssal (Abyss*). For illustration see under Littoral.

Prokaryota. Organisms without a cell nucleus, but with a nucleus equivalent. In contrast to the Eukaryota*, all the organisms included in the Prokaryota have no cell nucleus, nor are their cells divided up by membranes into reaction spaces in which the metabolic processes take place. Organisms known as the Schizobionta* belong to the Prokaryota. Other examples are the much-simplified viruses which are very important as pathogens (Virus Diseases*) and other similar groups.

Prolonged Baths *see* Bathing Treatments

Propagation of Aquarium Plants (fig. 944). In aquarium-keeping, sexual propagation of plants, as opposed to asexual propagation, plays a subordinate role.

The commonest form of *asexual propagation* is through cuttings which can be taken from all aquarium plants with elongated stem axes, eg *Ludwigia, Hygrophila* or *Limnophila.* Tips of shoots, lateral from shoots and even rearward sections of the stem can be used, from which dormant buds will then sprout. It is enough when the cutting has 3–4 completely formed internodes. The snip should be made just below a node. When the cutting is placed in the substrate, 1–2 nodes must be in the ground, as the new roots will develop from there.

Propagation from runners is the simplest form of asexual reproduction. It occurs of itself in all species that form underground runners or runners that grow on the top of the ground, eg as in the species of the G *Vallisneria*, many *Echinodorus* and *Cryptocoryne* species. At a little distance from the mother-plant, daughter plants develop from the runners, which can be separated off once the new plant is strong enough. Sooner or later the old runners disintegrate by themselves. In plants with rhizomes, eg larger *Echinodorus* species, fairly elderly rhizome sections can be removed without damage to the mother-plant. These sections can then be cut into pieces 2–3 cm long and fixed onto the substrate, eg with glass pins. After a little time, dormant buds will push forth and so new plants will develop.

Another important method of propagation is through adventitious plants*, which once again after gaining enough strength can be used further at their places of development. They occur on the inflorescences of many *Echinodorus* species, on the leaves of *Ceratopteris*, and on the leaves and roots of *Microsorium*. With *Hygrophila, Synnema, Rorippa* and other plants, adventitious plants can be obtained if you allow leaves that have been cut off to float on the water surface.

Fig. 944 Propagation of aquarium plants. Left to right: Cuttings, runners, adventitious plants, *Aponogeton* seedlings

Sexual propagation is necessary, however, if the asexual method is wasteful or not possible. It is dependent on flowers forming. The prerequisites for flower formation are very different from species to species. Many aquarium plants flower when cultivated in a submerged state but others must first be transferred to land. Once flowers appear, many species form seeds without any intervention from the aquarist, eg *Ludwigia* species, *Echinodorus berteroi.* With others, artificial pollination is advisable, as with most *Aponogeton* species or with *Sagittaria platyphylla.* To do this, pollen must be transferred on a paintbrush (artist's variety) to the stigmas of other flowers. With *Aponogeton ulvaceus* it is vital that cross-pollination takes place, ie 2 plants will have to flower at the same time. With Hydrocharitaceae* and Aponogetonaceae*, the rescuing of the seeds needs particular attention, as the fruits disintegrate very quickly on reaching maturity. Most seeds can be kept dry, but some must lie in water so as not to lose their germination capacity. The seeds of *Aponogeton* species cannot be stored at all as they start to germinate immediately after dropping to the ground. The temperature needed for germination depends on where the species come from. With tropical and subtropical plants the best results are obtained with a fairly high temperature, around 30 °C. With most species, the seeds are best sown in small flowerpots. The filling in the pot must be wet when the seeds are scattered. Then some wet sand is added so that there will be no whirling disturbance when they come to be placed in the water. Contrary to the statements found in most aquarium literature, the level of the water is not important. But small, germinating plants are threatened by being overgrown with algae. With some species, such as *Vallisneria, Aponogeton, Barclaya,* plantlets develop better if they can float freely for a time at the beginning, before being planted properly.

Prophylaxis; preventative treatment. This term describes all the measures taken to prevent diseases effectively. In aquarium-keeping, preventative medicine is extremely important. If an aquarium contains too many specimens, if the conditions of the water are poor (pH Value*, Water Hardness*, temperature, salinity), or if feeding* is inappropriate, then the aquarium animals will be weakened, their resistance* will be lowered and thus parasites will be encouraged to proliferate (Fish Diseases*, Parasitism*). Effective prophylactic measures to prevent diseases include having a thorough knowledge of the living conditions of the chosen specimens, keeping an appropriate number of specimens, adhering to quarantine regulations (Quarantine*), constantly checking conditions in the tank and the state of health of the specimens, carefully selecting food animals (so that none may inadvertently introduce parasites into the tank), and disinfecting (Disinfection*) any pieces of equipment and tools.

Prophyll *see* Leaf Succession

Proserpinaca LINNÉ, 1753. G of the Haloragaceae* with 2 species. From N and Central America. Small perennials with upright, elongated stems. Leaf arrangement alternate. Aquatic leaves pinnate, aerial leaves pinnate or undivided. Flowers hermaphroditic, small and inconspicuous. Perianth simple, calyx-like, 3-part. Stamens 3. Carpels 3, fused, inferior. Fruit like a nut. The various species of the G are aquatic plants but some also grow emersed on marshy ground. In the aquarium *Proserpinaca* is undemanding, and will survive at low temperatures; nevertheless, the species are rarely kept. The two species are:

– *P. palustris* LINNÉ. Eastern to central N America. Aerial leaves undivided, serrated. Aquatic leaves pinnate. Several varieties.

– *P. pectinata* LAMARCK. The eastern coast of N America. Aerial leaves and aquatic leaves pinnate.

Prosobranchia; Prosobranchs. Sub-Cl of the Gasteropoda*, with more than 57,000 species. The overwhelming majority are aquatic Snails breathing through gills (sea, brackish and fresh water, although in the tropics some species have adapted to life on land. Most are heteroecious. The shell is usually thick-walled and like porcelain to a greater or lesser degree; most have a lid for sealing off the shell. Well-known Ga that are also important in the aquarium are *Ampullaria**, *Viviparus** and *Nassa**.

Prosobranchs, Sub-Cl, *see* Prosobranchia

Protacanthopterygii, Sup-O, *see* Osteichthyes

Protein Foam Filter *see* Foam Filter

Proteins. Organic compounds which consist of amino acids primarily, 20 of which occur in living organisms. Because the 20 amino acids can combine in different ways and because the number of amino acids in a protein varies, there is practically an infinite number of possible proteins. Many proteins can also combine with other chemical substances (eg carbohydrates, phosphoric acids, lipids). The vast number of specific protein bodies found in organisms is made up of various combinations of these 20 amino acids. Proteins form those parts of cells and tissues that make up the structure of a body; they also form the basis of all life processes. Their many functions include: contractile muscular movement (eg myosin); enzyme* and hormone* action that regulates metabolism; the formation of blood pigments (eg haemoglobin) involved in respiration*; the formation of antibodies to protect the body from infection (*see* Immunity). Proteins are constantly renewed through the body's protein metabolism and are in part used to produce energy. Humans and animals must acquire proteins in their food to be able to build up protein substances, for unlike plants they cannot manufacture some of the amino acids for themselves out of other compounds.

Prothallium *see* Pteridophyta

Protistans *see* Protozoa

Protophyte; protophyton. A protophyte is a single-celled vegetative body. The individual cells can be combined into loose bands of cells, but each cell will still retain its independence. Protophytes occur in the Schizobionta*, the Algae* and the Fungi*. Opposites are Thallus* and Corm*.

Protoplasm *see* Cell

Protopteridae, F, *see* Ceratodiformes

Protopterus (various species) *see* Ceratodiformes

Protoreaster DÖDERLEIN, 1920. G of predatory, tropical Starfishes (Asteroidea*). The body disc is large and the arms have a broad base. On the upper side are strong spines or tubercles. One of the most frequently imported tropical Starfishes belongs to this G, but always remember it will attack almost every animal that cannot escape to safety, including many stinging animals and sea-urchins as well as smaller examples of its own species:

– *P. lincki;* Green Spiny Sea Star, Red Spiny Starfish. Indo-Pacific. Diameter around 20 cm. Yellowish in colour with rows of shining red spines. Very hardy in the aquarium (fig. 945).

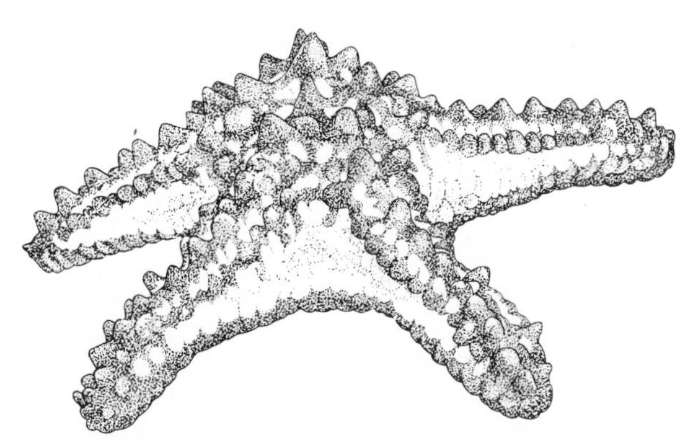

Fig. 945 *Protoreaster lincki*

Prototheria, Sub-Cl, *see* Mammalia

Protozoa (single-celled animals). Animal organisms that consist of only one cell. All the life functions take place in this cell (feeding, respiration, reception of stimuli, reproduction etc). There are 4 phyla that make up the Protozoa, and each of them contains a large number of species: animal Flagellates (Flagellata*), Rhizopoda*, Spore animals (Sporozoa*) and Ciliates (Ciliata*). The Flagellates are the original group of all higher animals and plants, whereas the other 3 groups represent side-shoots of evolution and are distinguished by specialised features of varying kinds. Although the Sporozoa have become specialised to a parasitical way of life and in doing so have become morphologically very simplified, many Ciliates show a remarkable degree of differentiation.

Unlike multi-celled animals, single-celled organisms are, potentially at least, immortal. This is because they propagate themselves through simple or multi-division, so that each parent individual merges completely with the daughter individual. If incomplete division takes place, then cell colonies will be formed in certain organisms. Among the Protozoa, alternation between sexual and asexual reproduction is the norm. With sexual processes, the hereditary material is newly combined, without a 'multiplication' of the individuals taking place. Protozoans feed on detritus* and micro-organisms such as bacteria and tiny algae, or they adopt a parasitical way of life. Certain of the parasitical Flagellata* and Sporozoa* are very important because they cause dangerous fish diseases.

Protozoans, Ph, *see* Protozoa

Protula RISSO, 1826. G of the Bristle Worms (Polychaeta*). These are sessile worms that live in a chalky shell, round in cross-section, and with a diameter of about 1 cm. They eat tissue. The crown of tentacles is several cm wide, usually a shining red colour and distichous; the lid is stalked. *Protula* is native to warm and tropical seas where it lives on rocks and between corals. The species are extremely attractive, but unfortunately they will only survive a short while in the aquarium. *P. intestinum,* RISSO, is an example of a species that is found in the Mediterranean.

Prussian Carp *(Carassius auratus gibelio) see Carassius*

Psammechinus AGASSIZ, 1846. G of smallish Sea-urchins (Echinoidea*). Found among clumps of plants in cool to warmish seas. The shell is flat and about 3.5 cm wide, with short, pointed spines on it. *P. miliaris,* GMELIN, is found from the Baltic to Morocco, and the very similar *P. microtuberculatus* BLAINVILLE is found in the Mediterranean. Both are pale green and if fed sufficient algal growth they make very hardy, worthwhile aquarium specimens.

Pseudo Helleri *(Heterandria bimaculata) see Heterandria*

Pseudobalistes BLEEKER, 1866. G of the Balistidae*, Sub-F Balistinae. From the tropical Indo-Pacific. Lifestyle and shape typical of the family. There is no furrow in front of the eyes in this G, the cheeks are naked at the front and covered with scales at the back that are smaller than those on the body. Very intolerant in the aquarium. Feeds on hard-shelled Echinoderms and molluscs.

– *P. fuscus* (BLOCH and SCHNEIDER, 1801); Yellow-spotted Triggerfish, Brown Triggerfish, Blue-banded Triggerfish. Indo-Pacific and Red Sea, where it lives on sandy substrata that are fairly deep. Maximum length 50 cm. Young fish have blue stripes on a yellow, later on green or brown, background colour and dark saddle markings that get larger when excited (fig. 946).

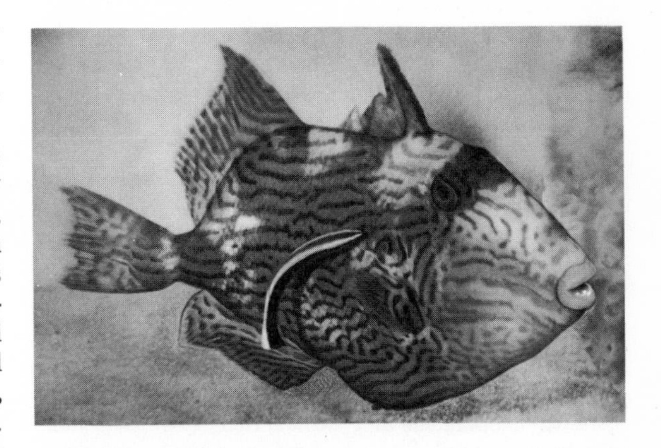

Fig. 946 *Pseudobalistes fuscus*

Pseudocorynopoma PERUGIA, 1891. G of the Characidae*, Sub-F Glandulocaudinae. S America. Length to 9 cm. These Characins live in shoals in the middle and upper water levels. The body is fairly tall and bulky, and it is flattened laterally; the ventral edge is sharp. The throat and breast region is arched forwards and downwards. D1 and An elongated; in the male *P. doriae,* the first rays of both these fins are drawn out like filaments. D2 present, C two-lobed and deeply forked. The mouth is slightly superior, the eyes large, and the LL complete. There is a gland on the caudal peduncle. In the aquarium, *Pseudocorynopoma* survives a long time and will tolerate an occasional drop in temperature to 1 °C. The various species are omnivores that will also readily accept artificial foods. They also get keen enjoyment out of swimming. Easy to care for and to breed; for further details see under *Stevardia**. The males of the 9-cm-long species, *P. heterandria* EIGENMANN, 1914, do not have the elongated rays in the D1 and An. So far, they have not been imported alive, in any case.

– *P. doriae* PERUGIA, 1891; Dragon-finned Characin, Dragon Fin. S Brazil and the River Plate states. Length to 8 cm. Flanks a transparent olive-green or brownish-green with a silvery iridescence and a steel-blue shimmer. Dorsal surface darker, ventral surface paler. The fins are grey, the C with black tips. The females do not have the filamentous elongations of the fin rays in the D1 and An. This used to be a very popular species but at the moment it is no longer kept by aquarists. A new import would be welcomed (fig. 947).

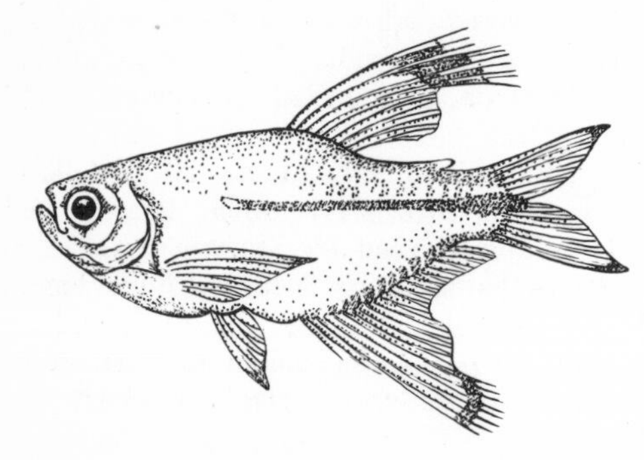

Fig. 947 *Pseudocorynopoma doriae*

Pseudocrenilabrus, G, *see Hemihaplochromis*

Pseudomonas, G, *see* Stomach Dropsy, Spot Diseases

Pseudomugil KNER, 1866. G of the Atherinidae, Sub-O Atherinoidei*. Australia: N and E Queensland. These are small shoaling fish that prefer the middle water levels of oxygen-rich, clear stretches of water. Body elongated, slim with moderately arched ventral and dorsal profiles, and flattened laterally. The mouth is small and superior. D divided; D2, together with the An, inserted well to the rear. Ps elongated and inserted high on the body. In the male, the first fin rays of the D2 and An are elongated. Distinguished from the G *Telmatherina** because the An emerges a little in front of the rear edge of the D1 and the latter begins exactly at the level of the V or a little behind it. Scales large. No LL. *Pseudomugil* species are very active swimmers that seek out oxygen-rich, clear water and require large tanks with plenty of room for swimming. The total hardness should be 12 °dH and higher, pH value neutral. A frequent part-exchange of water is very beneficial. Temperature 23–28 °C. Omnivores, that will accept any kind of live food from the surface, as well as dried food. When breeding, provide medium-hard, neutral fresh water with a small amount of salt added. Finely pinnate plants are suitable as a spawning substrate. Have 1 male for every 2 females. The fish are very productive (the spawning period lasts a long time, *see* Spawning Period). At 25 °C, the fry hatch out after 24–43 days and can immediately swim freely. Rear on micro-organisms.

– *P. signifer* KNER, 1865; Southern Blue Eye. N and E Queensland. Length to 4.5 cm. Body a translucent yellow-green, the lower half with a blue shimmer, ventral surface silver-white. A dark band extends from the P to the C, and is bounded above by a gleaming line. D1 glassy, the first fin ray in the male black. D2, C and An yellowish to orange-red.

Pseudotropheus REGAN, 1921. G of the Cichlidae*. Confined to Lake Malawi, ie endemic* there. In most species, the body is elongate (exception *P. zebra* and closely related forms), low, and only slightly flattened laterally. D inserted along a great stretch of the body, and has 17–18 hard rays and 8–9 soft ones. The An has 3 hard rays and 7–8 soft ones. The rear edge of the C curves slightly. *Pseudotropheus* stems from *Haplochromis** and shows a particularly close relationship to *Labeotropheus**, as is evidenced by type of reproduction, colouring and behaviour. Very characteristic of these Ga are the face masks made up of dark transverse stripes and the fact that in some species the females come in a variety of colours. Apart from homoeochromous forms, similar to the males, there are dappled varieties, for example, with irregular dark flecks on a paler background. Not even the flanks of one fish will necessarily correspond in markings. An obvious preference for a particular colour variety is not in evidence among the males, otherwise the less favoured forms would have been eliminated long ago through selection.

The G is highly specialised with regard to its method of reproduction. The species are most definitely mouthbrooders, and it is the females who take on this role. In contrast to open-water spawners (compare Cichlidae), the number of eggs is small, but their size is considerable – up to 7 mm. In *Pseudotropheus* the danger-period for the spawn when it is unprotected is extremely short. If some eggs fall away from the female, she turns round immediately and deposits them in her throat-sac. It is here that fertilisation takes place. In the course of evolution, certain hereditary colour and behaviour patterns have developed that help ensure that fertilisation occurs; on the An of the male there are blotches which look for all the world like the eggs, and so the female snaps at them, believing them to be her eggs (the male encourages her by presenting his anal fin to her, vibrating it, during the courtship ceremony). In this way, then, sperm from the male goes into the female's mouth along with her respiratory water.

Pseudotropheus species live above rocky ground in Lake Malawi and are specialised eaters of algal growth. In such growths, there is a whole host of small animals (especially Chironomidae*), so *Pseudotropheus* cannot be described as a vegetarian. Although not so strikingly adapted to to this way of feeding as *Labeotropheus, Pseudotropheus* can nevertheless be considered as a specialist growth-eater with its thick-lipped mouth and spatula-like teeth. In the aquarium too, *Pseudotropheus* will work its way across rocks and panes of glass with its protruding mouth, even although there may be no algal growth there.

The various species are very fast, agile swimmers with a marked tendency to adopt territories. So, although the fish are smallish, the tanks provided can never really be big enough. To concur with the natural living conditions of Lake Malawi Cichlids, in the aquarium one needs to build an underwater rocky landscape offering plenty of gaps and crevices. Make sure that your creation provides enough places of refuge as well as enough free-swimming zones, that are nevertheless isolated from one another by natural barriers. There is no need to add plants, although *Pseudotropheus* will not disturb them. Choose neutral, medium-hard water and a temperature of around 27 °C. The intensity of light should still guarantee the growth of green algae, for in subdued lighting *Pseudotropheus* is very vital and displays its best colours. Because of this delightful colouring, which is reminiscent of the Coral Fishes, there has been a real upswing in the number of Lake Malawi Cichlids kept. Due to their aggressiveness, however, the interest in them has largely subsided again. Males that are ready to spawn adopt a slanting pose and flutter their body and fins at the female, following her relentlessly, even if she is unwilling to mate. The rivalry between the males is extremely bitter, at the same time. In the tank, the only way round these problems is to use as large a tank as possible and populate it with as many fish as feasible, made up of different species. Like this at least, the aggression is not concentrated on just a few tormented specimens, as the occupier of the territory is constantly being distracted from his would-be victim by other fish, and so he becomes involved in other scraps. As a whole, the *Pseudotropheus* species behave according to an inherited pattern which in nature the individual members distribute evenly through the biotope*; it is therefore of significance. Under the unnaturally narrow confines of the aquarium, the terrible biting goes on and on because the fish on the receiving end cannot escape from the territory of the aggressor. When observed in open-air habitats, *Pseudotropheus* and other Cichlid Ga of the African lakes (Lake Tanganyika*, Lake Malawi*) behave much more peaceably, and even form large families or small shoals—ie they exhibit a certain amount of social behaviour. So, to keep *Pseudotropheus* successfully in the aquarium, the aquarist must know a great deal about their conditions in nature and he must be prepared to spend quite a lot on getting a big enough tank and sufficient numbers of fishes.

Fig. 948 *Pseudotropheus auratus*

– *P. auratus* (BOULENGER, 1897); Golden Lake Nyasa Cichlid. Lake Malawi. Length to 11 cm. Adult males brown-black with 2 pale turquoise-coloured longitudinal bands in the upper half of the body. Females and young males golden-yellow, with 2 dark longitudinal bands (fig. 948).

– *P. elongatus* FRYER, 1956; Slender Cichlid. Lake Malawi. Length to 9 cm. Head and breast black, rear half of the body a shining dark blue, with black transverse bands. Intensity of colouring very much dependent on mood (fig. 949).

Fig. 949 *Pseudotropheus elongatus*

– *P. novemfasciatus* REGAN, 1921. Lake Malawi. Length to 10 cm. Males a shining dark blue, with 9 blackish transverse bands. Females golden-yellow (fig. 950).

Fig. 950 *Pseudotropheus novemfasciatus*

– *P. zebra* (BOULENGER, 1899); Cobalt Blue Cichlid, Zebra Nyasa Cichlid. A tall species. There are various colour varieties, some of which are still difficult to place systematically; they are probably a 'collective species' of biologically isolated small-species. Normal form blue, with dark transverse bands. Other colour variations have no bands, and have a blue or bluish-white ground colour. Females either homoeochromous (similar to the males) or irregularly dappled. Other species of *Pseudotropheus*, some of which have yet to be identified for sure, have already been introduced and reared successfully (figs. 951 and 952).

Fig. 951 *Pseudotropheus zebra*

Fig. 952 *Pseudotropheus zebra*

Pseudoxiphophorus, G, name not in current usage, *see Heterandria*

Pseudupeneus BLEEKER, 1862. G of the Mullidae*. From tropical and subtropical seas, where they live on soft substrata and on shallow reefs. They have a typical family build, and feed on small animals. Sociable fish. Both jaws have fairly strong teeth, and the D and An are scaleless. These are fairly colourful fish, 20–50 cm long, and they survive well in the aquarium. *P. indicus* (SHAW, 1803) is occasionally imported.

Pteridophyta; Ferns. D of the plants, with 12,000 species. The vegetative body of the Pteridophyta is a corm*; it consists therefore of a stem axis, leaves and roots (as in the Seed-plants [Spermatophyta*]). Distribution through spores.

The Ferns are characterised by an alternation of generations*, in which, in contrast to the Mosses and Liverworts (Bryophyta*), the asexual generation is brought about. Initially, an inconspicuous proembryo (prothallium) develops from a germ-spore; in structure, this prothallium is a thallus*. This is the sexual generation and it forms egg-cells and spermatozoa in special containers (gametangia). The actual Fern plant then develops from a fertilised egg-cell; this is the asexual generation. At the same time, the prothallium dies down. The asexual generation is therefore an independent plant and forms spores once again. Unlike the majority of Pteridophyta, the spores in the Water Ferns (Hydropterids) are differentiated sexually. Fairly large megaspores (female) and small microspores (male) are distinguished, from which correspondingly different proembryos develop. These are simplified to a greater or lesser degree and develop largely inside the spore wall. So, in these water Pteridophyta, the processes of generation alternation are very complicated.

Most Ferns colonise places where the provision of water is good; only a few can settle in dry biotopes. They are distributed in all kinds of climatic zones, but are at their most abundant in the tropics, which is where the largest species, the Tree-ferns, live. Ontogenetically, the Pteridophyta have developed from the Green Algae. The Fern species alive today are remains of related groups that reached their peak in the Mesozoic and in particular in the Palaeozoic. They are divided into 3 very distinct Classes:

– *Lycopsida;* Club Mosses. 1,000 species. Stems branching like forks. Leaves alternate, some in rosettes, simple and single-veined. The length of the leaves can vary. The sporangia are isolated and located at the bottom of the upper side of the leaf. In aquarium-keeping, the F Isoetaceae (Quillworts) containing the G *Isoetes** is of interest.

– *Pteropsida;* Fens. 11,000 species. Stems branching laterally. Leaves alternate, often in rosettes, and of varying size; also simple or pinnate, with several veins. There is a large number of sporangia located on the leaf underside. There are numerous Fs, of which only some are of interest to the aquarist. They include the Polypodiaceae (Polypody Family), to which the Ga *Bolbitis** and *Microsorium** belong, and the Parkeriaceae (Ceratopteris Family) which includes the G *Ceratopteris**. The Hydropterids (Water Ferns) form their own Sub-Cl. They include the Marsileaceae (Marsilea Family) containing the Ga *Marsilea** and *Pilularia**, and the Salviniaceae (Salvinia Family) with the Ga *Azolla** and *Salvinia**.

– *Sphenopsida;* Horse-tails. 25 species. Stem branching laterally. Leaves in whorls, small and scaly, with a single vein. Several sporangia located on the underside of shield-shaped leaves. The only F still living today is the Equisetaceae (Horse-tails). It contains just one G *Equisetum* that is only absent in Australia. Some Horse-tail species colonise wet habitats. In the northern hemisphere, for example, you will find *Equisetum fluviatile* growing in ponds, ditches and among reeds.

Pteridophytes, D, *see* Pteridophyta

Pterois (CUVIER and VALENCIENNES, 1876); Fire-fishes, Scorpionfishes (fig. 953). G of the Scorpaenidae, Sub-O Scorpaenoidei*. There are several species in the tropical Indo-Pacific where they live as predators near the bottom below the surf zone. Length around

20–30 cm. Because of their bizarre shape, *Pterois* species can scarcely be confused with any other—except perhaps *Dendrochirus**. Characteristic are the giant, fan-like Ps with free-standing rays; the long, erectile D-rays; and the high tentacles (especially so in younger specimens) above the eyes, as well as other dermal outgrowths on the head. Red tones predominate in the body colouring—in all the species, the body is compressed and covered in red-yellow or red-white

Fig. 953 *Pterois radiata*

transverse stripes. When handling Firefishes, one must be very careful. Being pricked by the D-spines can be very painful and dangerous, as poison will be introduced into the wound. Immediately treat the damaged part of the body by bathing it in very hot water; in severe cases, a doctor must be called in and cobra antiserum will have to be applied. Firefishes are often described as being 'fearless'—they hardly ever resort to flight, and simply protect themselves by erecting their poison spines if threatened. *Pterois* is active at dusk. It has been noticed in *P. volitans* that the large Ps are used during a battle. In the course of the hunt, the prey-fish are driven into a corner, from where they often swim up to a transparent part of the P lying near the body, and *Pterois* then finds it easy to snap them into its much enlarged mouth. Firefishes survive well in the aquarium. They need enormous quantities of food; to start with, they will only accept live fish, but usually they will become adapted to dead food. The metabolism is intense, so filtration must be very efficient and the water must be changed regularly. It is possible to keep Firefishes in a community tank with larger species of fish. Various species are often imported, especially:

– *P. volitans* (LINNAEUS, 1758); Butterfly Cod, Caribbean Scorpion, Cobra Fish, Common Dragonfish, Peacock Lionfish, Red Firefish, Scorpionfish, Turkeyfish. Indo-Pacific, Red Sea. Length to 35 cm. In younger specimens, the P is more than the length of the body, and like the tentacles above the eyes, it becomes relatively shorter with age.

Pterolebias GARMAN, 1895. G of the Cyprinodontidae*. Contains about 5 species distributed in S America, in fact from the lowland areas east of the Andes, from Venezuela to Bolivia. Most closely related to the Ga *Austrofundulus** and *Rachovia**. Elongated fishes, up to 10 cm long, that are distinguished from the above-mentioned Ga mainly by the small D (which is in contrast to the strongly developed An). The unpaired fins in the male are larger than in the female; the D and An are pointed at the rear and the C is long and looks as if it is double-pointed or fan-like. In the female all the fins are rounded and of normal size.

As annual species, *Pterolebias* is confined to waters that only contain water some of the time (compare Cyprinodontidae). Their life-cycle is adapted to this seasonal way of life, and in the aquarium this has to be borne in mind (ie the brevity of life, the extremely quick attainment of the ability to reproduce and the dry phase which is necessary for the eggs to develop normally.) In line with their tropical origins, *Pterolebias* species need more warmth than, say, the *Cynolebias** species, which are found further south; on the other hand, too high a temperature will shorten *Pterolebias'* lifespan. Therefore, choose a temperature of around 23–24 °C. A good substrate for *Pterolebias* is well boiled and well rinsed leaf-mould. It has a loose enough consistency to allow the fish to dig into it when spawning, at the same time it makes it easy to deal with the eggs as necessary (compare G *Cynolebias*). Before any attempt at breeding, it is recommended that the sexes are separated and fed vigorously. When brought together again, the fishes often spawn directly afterwards and spontaneously. The spawning period lasts several weeks, and in between times the partners should be separated again from time to time. As with all annual S American fish, *Pterolebias* also requires a dry phase for its eggs, lasting several months (in contrast to the African annual species, where a few weeks is sufficient); indeed, the eggs can remain dry for years, without any adverse effect upon their ability to hatch out. To breed successfully, simply sieve the mould and the eggs and thus let the water run away. Then dry off the contents of the sieve on old newspapers, until there is just a touch (which is absolutely necessary!) of damp left; now place in plastic bottles or containers. Every few weeks, it is advisable to let some air in. After several months of being kept like this, give the mould a good soak (clean rainwater is best), and then some of the embryos will hatch out.

Treat the remainder as before—ie dry them again and a few months later, some more will be ready to hatch out. The fry will be able to eat nauplii (Nauplius* and rotifers (Rotatoria*), as well as other similar food, immediately. After 6–8 weeks, they are already sexually mature.

– *P. longipinnis* GARMAN, 1895; Longfin, Longfin Killie. Brazil. Length to 10 cm. Males brown with a marked metallic shimmer. Flanks with loose rows of iridescent dots; in the shoulder region there is a deep-black blotch, into which are embedded some shining red scales. Unpaired fins (especially C and An) fan-like. Females of similar background colouring, likewise with bright iridescent dots, but smaller and with normal fins (fig. 954).

– *P. peruensis* MYERS, 1954; Peruvian Longfin. Peru. Length to 9 cm. Pale transverse stripes on a brown background. The stripes on the D are at an angle. C in the males double-pointed. The An has 14–17 fin rays.

Fig. 954 *Pterolebias longipinnis*

– *P. zonatus* MYERS, 1935. Venezuela. Length to 9 cm. Very similar to the previous species, but with 22 fin rays in the An.

Pterophyllum HECKEL, 1840; Long-finned Angelfishes. G of the Cichlidae* but with few species. Native to tropical S America (the river basins of the Orinoco and the Amazon). A synonym for this G is *Plataxoides* CASTELNAU, 1855. The body is compressed like a disc and very tall. The C and An are also very elongated, so that the fish looks triangular. V also very long, bent like a sabre. The D has 11–13 hard rays and 19–29 soft ones; the An has 5–7 hard rays and 19–32 soft ones. (These figures are species-specific within a particular distribution area.) In colour, these Angelfishes are a gleaming silver with dark transverse bands. This G is among the best-known and most popular of all ornamental fish, namely the species *P. scalare* (including *P. eimekei*). In the past few decades, it has been used in breeding experiments by aquarists and today, the bred forms are more and more ousting the original form. Types that now exist include long-finned mutants, the so-called Veil-tail Scalare, as well as colour varieties such as 'Black Scalares' (melanistic mutants), Smoke Scalares (the result of cross-breedings between the original form and the Black Scalare), Marbled varieties, Albinos etc. But it is universally true that the bred forms are not only expensive, but are also significantly more delicate.

Pterophyllum species should be kept in large, tall tanks with plenty of vegetation that should reach up to the water surface. Large-leaved water plants, such as *Echinodorus** and Giant Vallisnerias *(Vallisneria* gigantea)* are most suitable. The substrate can be fine-grained, as the members of the G *Pterophyllum* do not stir it up (unlike many other Cichlids). In fact, *Pterophyllum* is a very peaceable genus and the fish move calmly round the tank. As far as possible, the temperature and water conditions should correspond to those of the natural habitat – ie 23–25 °C and soft water. A frequent partial exchange of water or water preparation* is advisable. If kept under the wrong conditions, the cornea often clouds over and the skin tends to develop fungus. Sickly, apathetic fish usually result from too low a temperature and too high an electrolyte* content. If this happens, do not bother with any medicaments, as most *Pterophyllum* species are extremely sensitive to them. Instead, raise the temperature to 32 °C, ventilate vigorously and provide a varied selection of live food; at the same time, try to get the living conditions up to their best possible. Also take care not to feed too much to the fish, or to feed them too much of one thing – the fish may start refusing food as a result. Provided the specimens are healthy, breeding is not a problem. It is best to let the pair select themselves from amongst a shoal of youngish specimens, as partners forced on one another do not always get on. As typical open-water spawners, the sexes are not easy to tell apart – even the form of the genital papilla is no sure sign with *Pterophyllum*. The spawning ground likely to be chosen is formed by large aquatic plant leaves, and occasionally wood and rocks. After cleaning the spot painstakingly, the female attaches the spawn granules one by one, then the male smears it with his genital papilla to fertilise them. Up to 1,000 eggs may be laid. Both parents take part in spawn and brood care, alternately fanning fresh water over the spawn, removing any fungi-ridden eggs and also helping to chew the fry out of the egg-case. The fry are spat out onto a leaf where they remain for some time, attached there by a

Fig. 955 *Pterophyllum scalare* (marbled form)

thread secreted from a gland on the head. Following that, they are finally taken to shallow hollows in the substrate. For this reason, the tank should contain little mould. After 4–5 days, the young fish start trying to swim, and lastly the whole shoal is led through the tank by the parents. Naturally, even with well-practised pairs, it sometimes happens that the spawn or brood are eaten, particularly if the pair is disturbed. That is why breeders prefer to rear artificially. A leaf from the Parlour Palm *(Aspidistra elatior)* offers the best spawning substrate, as Long-finned Angelfishes like to spawn on it. Afterwards, the leaf, together with the attached spawn, is put into a separate, hygienic rearing tank* (no substrate and no plants). By means of gentle ventilation, a flow of water is continually sent across the eggs and the newly hatched brood. In the beginning, keep the lighting in the tank fairly subdued, and regularly remove any eggs that become attacked by fungus. The fry are easy to rear as they will eat micro-organisms directly after they can swim freely.

– *P. altum* PELLEGRIN, 1903; Deep Angelfish, Long-finned Angelfish. Orinoco. Length to 12 cm. Body very tall. A gleaming silver colour with black transverse bands and dark flecks on the flanks.

– *P. leopoldi* (GOSSE, 1963). Amazon Basin. Length to 12 cm. It may have been imported already and confused with the following species. In contrast to the following, it has only 19–22 soft rays in the An.

– *P. scalare* (LICHTENSTEIN, 1823); Angelfish, Scalare. Amazon Basin. Length to 15 cm. A gleaming silver colour with black transverse bands, the band between the D and An being the widest. Iris red. The An has 24–28 soft rays. A well-known and popular species, with many bred forms. *P. eimekei* AHL, is a synonym (figs. 955 and 956).

Fig. 956 *Pterophyllum scalare*

Pteropoda. Sea Butterflies. O of the Opisthobranchiata, Sub-Cl Euthyneura*. Plankton animals from the sea with see-through bodies. They swim by rapidly beating their wing-like parapodia. A mass collection of Pteropoda forms an important food source for fish and

whales. Not important in aquarium-keeping, however.

Pteropsida, Cl, *see* Pteridophyta

Pufferfish types, O, *see* Tetraodontiformes

Pufferfishes, F, *see* Tetraodontidae

Pulmonata, Sup-O, *see* Euthyneura

Pulmonate Snails, O, *see* Basommatophora

Pulmonate Snails, Sup-O, *see* Euthyneura

Pumpkinseed Sunfish *(Lepomis gibbosus) see Lepomis*

Pumps are used in aquarium-keeping to supply air and water. Various electrically-driven aerators are available. Diaphragm pumps, which work by alternating

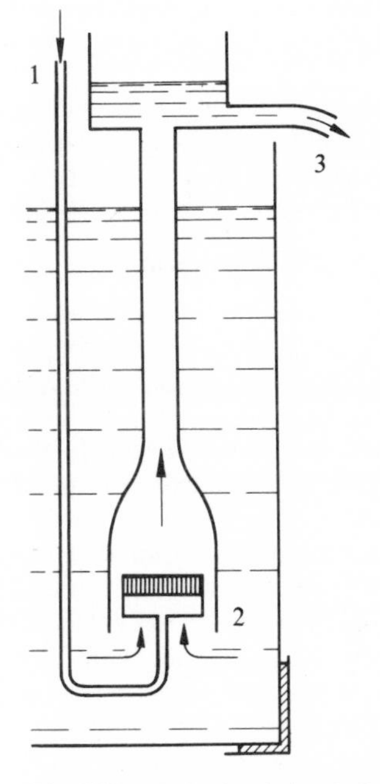

Fig. 957 Methods of circulating water.
a) Air-lifting. 1 Air inlet 2 Diffuser 3 Water outlet

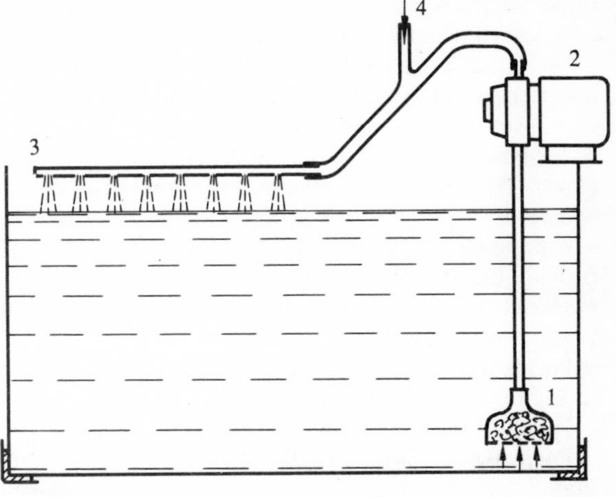

b) By circulatory pump. 1 Water inlet 2 Water impeller circulation pump 3 Spray bar tube 4 Aeration inlet nozzle

current on the principle of the wheelwright's hammer, are being used increasingly in small and medium-sized tanks. Even small aerators are very efficient, make little noise, and can be left to run continuously without the need for any attention. Larger establishments require piston pumps or compressors. If pumps are used that are not specifically designed for the aquarium, care must be taken that no oil leaks out (Air Filtration*). By fitting pressure reservoirs with reducing valves in between, an equal pressure is guaranteed as air is alternately taken in. To ventilate (Ventilation*) an aquarium, the compressed air is blown in via a diffuser directly into the aquarium, or into filter systems (Filtration*), and if necessary, the air can be enriched with ozone*. Compressed air can also be used to supply water. For this, a finely distributed air-flow is introduced from below in an ascending pipe standing vertically in the water. There develops a slight air-water mixture, that is raised above the normal water level (supply height) and that sucks water in after it from below. The relatively low supply height is inconvenient, but it is sufficient nevertheless to conduct water over the edge of the aquarium into filter systems. The efficiency is determined by the ratio of the diameter of the ascending pipe and the distribution and amount of air blown in. The air should be introduced in fine amounts via a diffuser (fig. 957a).

In addition, other water pumps are available that work according to the principle of centrifugal pumps. It is important, particularly in marine tanks, that the water in the pump does not come into contact with pieces of metal or unsuitable packing material, nor must it be contaminated with oil. Such pumps are extremely efficient and can be used in conjunction with filter systems or to circulate the water. An appropriate arrangement of suction pipes and diffuser openings produce a strong water flow or water circulation. Water discharged via a radiating tube or air-suction nozzle fixed on the outlet tube greatly enriches the air in the water. To protect the pumps, the water must always be sucked in via a gauze filter to keep the pump system free of coarse particles (fig. 957b).

Pungitius COSTA (1846). G of the Gasterosteidae, Sub-O Gasterosteoidei*. These fish live in fresh and sea water and come from the northern hemisphere. Similar to *Gasterosteus**, but without armour, and with 9–11 spines in front of the D. They build nests in aquatic plants. The following species survives well in a habitat aquarium:

– *P. pungitius* (LINNAEUS, 1758); Nine-spined Stickleback. N Europe, N America, where it lives in fresh and brackish water. Length to 6 cm. Olive-brown. Males black at spawning time.

Purple Sea-urchin *(Psammechinus miliaris) see Psammechinus*

Purple Star *(Echinaster sepositus) see Echinaster*

Purple-headed Barb *(Barbus nigrofasciatus) see Barbus*

Purple-striped Gudgeon *(Mogurnda mogurnda) see Mogurnda*

Purse Minnow *(Cyprinodon variegatus) see Cyprinodon*

Pygidiidae, F, name not in current usage, *see* Trichomycteridae

Pygidium itatiayae *see* Trichomycteridae

Pygmy Angel *(Centropyge argi) see Centropyge*

Pygmy Angelfishes, G, *see Centropyge*

Pygmy Barb *(Barbus phutunio) see Barbus*

Pygmy Corydoras *(Corydoras hastatus) see Corydoras*

Pygmy Courami *(Trichopsis pumilus) see Trichopsis*

Pygmy Rasbora *(Rasbora maculata) see Rasbora*

Pygmy Seahorse *(Hippocampus zosterae) see Hippocampus*

Pygmy Sunfish *(Elassoma evergladei) see Elassoma*

Pygmy Swordtail *(Xiphophorus pygmaeus) see Xiphophorus*

Pygocentrus, G, name not in current usage, *see Serrasalmus*

Pygoplites FRASER-BRUNNER, 1933. G of the Chaetodontidae*, Sub-F Pomacanthinae. Contains a single, unmistakable species, *P. diacanthus* (BODDAERT, 1772), that lives on the coral reefs of the tropical Indo-Pacific, including the Red Sea. It is a solitary fish. Its body is orange and is crossed by blue-white transverse bands enclosed by a black colour; the markings on the head are blue, the D is purply, and the An has violet-pink stripes. *Pygoplites* is sensitive and very difficult to feed in the aquarium; some will accept Bivalve flesh.

Pyjama Cardinalfish *(Apogon nematopterus) see Apogon*

Pyrrhulina CUVIER and VALENCIENNES, 1846. G of the Lebiasinidae*, Sub-F Pyrrhuliniae. Small South American Characins, up to 10 cm long. Shoaling fish that like to stand near plants and search there for small food animals. The males usually have deeper coloured unpaired fins and sometimes they are elongated, too. *Pyrrhulina* is closely related to the Ga *Copeina** and *Copella**, although it is distinguished from the former by having only one row of teeth (in *Copeina* there are 2). When the G *Copella* was established, many species were removed from the G *Pyrrhulina* and placed there; this was because differences in the maxillary bones in the males had been detected (see under *Copella*). Besides, *Copella* species are slimmer and more elongate than those species left behind in the G *Pyrrhulina*. The tip of the D and the upper lobe of the C in the males are also particularly elongate. The body is usually a little compressed, but only moderately so from side to side. The D is short, not elongated into a point, and there is no adipose fin. C moderately forked; An short an rounded, P's and Vs well developed. In the majority, a black band extends from the mouth, through the eye and over the operculum to the body. As a rule, *Pyrrhulina* is not difficult to rear. The fish spawn on leaves lying horizontally in the water; beforehand, they have been cleaned by the males. After the act of spawning is over, the eggs are watched over until the fry hatch out, which takes place after 25–30 hours. The fry are very small, and they are reared most successfully if fed on Rotatoria* for the first few days of life. All the species are ideal colourful aquarium fish. However, particularly with imported fish, if specimens are kept in water that is too hard and lacking in peat, they seem to be more frail.

– *P. brevis* STEINDACHNER, 1875; Short Pyrrhulina, Dotted-scale Pyrrhulina. From the Amazon Basin, Rio Negro. Length to 9 cm. Flanks brownish with a bluish sheen, dorsal surface brown, throat and ventral surface silvery, often with reddish tones. A black band runs from the mouth, through the eye to the level of the D insertion. There are 4 longitudinal rows of scales on the flanks, with red dots. In the front part of the D there is a black blotch. In the male, the fins are fire-red, the D, An and Vs with black edges. The black D-blotch is outlined in white. The upper caudal lobe is a bit longer than in the females, whose fins are yellowish and not edged in black. A very beautiful species, unfortunately rarely imported, but bred successfully by FRANKE (fig. 958).

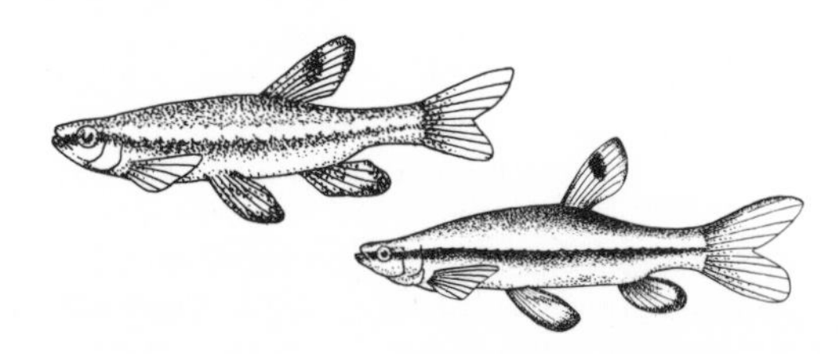

Fig. 958 *Pyrrhulina rachowiana* and *Pyrrhulina brevis*

– *P. laeta* (COPE, 1871); Half-banded Pyrrhulina. From the Peruvian part of the Amazon Basin, the Rio Ucayali near Jenaro Herrera. Length to 8 cm. The shape of the body is strong, a lot like that of *Copeina* guttata*. Upper surface dark-grey, sides a mousey-grey with a violet-coloured shimmer, ventral surface whitish. A blackish band stretches from the tip of the snout, through the eye at an angle downwards and ending at the level of the An. Fins a delicate yellow-reddish; the D has a large, black fleck in the middle. This is a very attractive, robust species, peaceable and hardy in the aquarium, but unfortunately only one or two examples have ever reached us, as part of a supplementary catch (Unintentional Catches*). Breeding unknown *P. semifasciata* STEINDACHNER and *P. maxima* EIGENMANN and EIGENMANN are probably both alternative names for this species that are no longer in use (fig. 959).

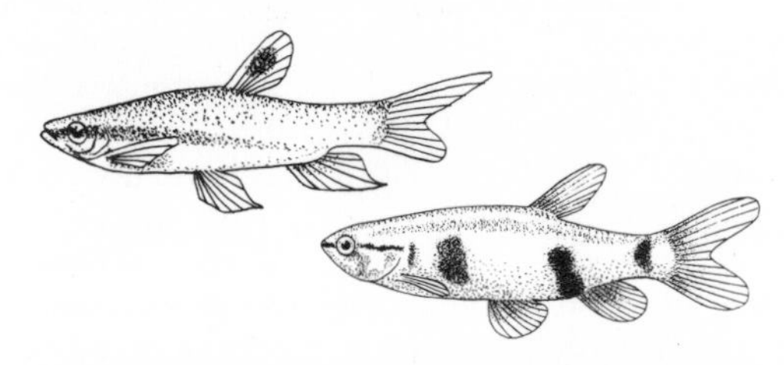

Fig. 959 *Pyrrhulina laeta* and *Pyrrhulina vittata*

– *P. rachowiana* MYERS, 1926; Rachow's Pyrrhulina, Fanning Characin. From the lower Rio Parana and La Plata. Length to 5 cm. Flanks a mousey-grey; in the anterior third of the body there are 2 longitudinal rows of red dots on the scales. Dorsal surface a gleaming dark-brown, ventral surface yellowish-white. A golden gleaming longitudinal band extends from the head to the insertion of the C. On the ventral side this band is bounded by a brown band that gradually merges into the yellow-white ventral colouring. There is a pale green shining blotch on the operculum. Fins yellowish-green with a pale blue edge, D with an oval black blotch. The females are paler in colour. The lobes of the C are of equal length; the An and the Vs in the males have brick-red edges. Spawning takes place on leaves or in shallow hollows in the substrate. This species used to be common in the aquarium, but nowadays it is no longer found there (fig. 958).

– *P. vittata* REGAN, 1912; Striped Pyrrhulina, Headband Pyrrhulina. From the Amazon Basin near Santarem, Rio Tapajoz. Length to 4 cm. Flanks a matt silver with a blue or green sheen, dorsal surface grey-green, ventral surface whitish, often with reddish overtones. A black streak extends from the lower jaw, through the eye to shortly behind the gill-covering. On the flanks there are 3 blackish flecks arranged at an angle to each other and looking a bit like transverse band markings. Only the middle one reaches the edge of the belly. The first one lies between the Ps and Vs, the second one goes from the D to the An, and the third is located above the caudal peduncle just before the C insertion. The fins are colourless or a weak red, the Vs and An have a bluish edge and the D has a black blotch outlined in a paler colour. Only in the males are the lobes of the C slightly elongated. This is a delightful species that is not difficult to breed (fig. 959).

Pyrrhulina nigrofasciata *see Copella metae*
Pyrrhulininae, Sub-F, *see* Characoidei

Quarantine. The holding in isolation of sick organisms or those that are suspected of being ill. Newly acquired fish, especially imports and those caught from the wild, should be kept in quarantine for 4 weeks before being placed in a tank that already has specimens in it. The same goes for invertebrate animals. Quarantine should go hand in hand with a thorough check on state of health, so that latent diseases are detected and if necessary cured. Specimens that have been weakened by the processes of being caught, transported, underfed and generally kept in less than ideal conditions, should be brought back up to a good state of health during quarantine. For specimens caught in the wild, the quarantine period is also useful for getting them accustomed to conditions in captivity. Quarantine tanks should offer the best conditions possible. To be able to have a good view of the specimens, spacious tanks are used with no substrate and no plants. Hiding-places are provided by stones that can be cleaned easily, flowerpots or plastic tubes. The water must have a known composition and be biologically weighed out. Technically, quarantine tanks must be kitted out well and safely. It is essential that quarantine specimens are given extra special care, that the food offered is varied and that they are observed constantly. As a precautionary measure (Prophylaxis*) against diseases, medicaments with a broad effective spectrum can be applied (Trypaflavin*, Copper Sulphate*, Antibiotics*). When the animals are transferred to the already occupied aquaria, great care must be taken and the behaviour of the specimens must be regarded strictly. For importer, breeders and dealers alike, quarantine regulations with a strict and regular health check are unavoidable.

Queen Angel *(Holacanthus [Angelichthys] ciliaris) see Holacanthus*

Queen Triggerfish *(Balistes vetula) see Balistes*

Quillback *(Carpiodes cyprinus) see* Catostomidae

Quillwort (Isoetaceae, F), *see* Pteridophyta

Quillwort *(Isoetes lacustris) see Isoetes*

Quinine is an alkaloid obtained from tropical Bedstraw plants (Rubiaceae) which reacts with acids to form water-soluble salts such as Quinine Sulphate, Quinine Hydrochloride and the like. It is poisonous to one-celled organisms even in very weak concentrations and is used in human medicine for treating malaria, influenza and the like. In the aquarium quinine is used to combat *Ichthyopthirius**, *Cryptocaryon (Cryptocaryon-disease*), Oodinium** and other skin parasites of fish. The fish should be kept in a solution of 1–1.5 g to 100 litres if water for 3 days and this should be repeated if necessary. Before repeating the dose, however, the remains of the quinine should be completely removed from the aquarium water by filtering it over activated charcoal for 24 hours.

Even in small concentrations quinine is deadly for invertebrates such as Actiniaria* and some algae such as *Caulerpa**.

Quintana HUBBS, 1934. G of the Poeciliidae*, Sub-F Poeciliinae. Cuba. Lives in still and slow-flowing waters. So far, the only known species is:

– *Q. atrizona* (HUBBS, 1934); Black-barred Livebearer. Cuba. Male up to 2.5 cm, female up to 4 cm. Body compressed, and very flat from side to side. Yellowish-green with up to 9 distinct dark transverse bands. In direct light, the body takes on a bluish shimmer. There is a dark blotch at the base of the D; the C is yellowish.

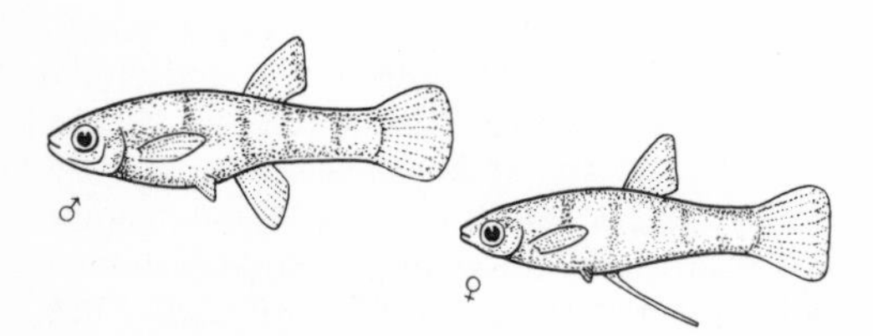

Fig. 960 *Quintana atrizona*

In the female, the V and An are encased in blue. The aquarium should be thickly vegetated in places, and the temperature should be 23–28 °C. Omnivore, with live food and algae being preferred. Sensitive to disturbances (fig. 960).

Rabaut's Corydoras *(Corydoras rabauti) see Corydoras*

Rabbitfish *(Chimaera monstrosa) see* Chondrichthyes

Rabbitfishes (Chimaeriformes, O) *see* Chondrichthyes

Rabbitfishes (Siganidae, F) *see* Acanthuroidei

Rachovia MYERS, 1927. G of the Cyprinodontidae*. Only a few species, all of which live in the savannah regions of Colombia. Closely related to the Ga *Pterolebias** and *Austrofundulus**. To the layman, *Rachovia* is indistinguishable from *Austrofundulus. Rachovia* species are about 7 cm long, moderately elongated and a bit like the well-known African G *Nothobranchius** in shape; however, the colouring is less striking. The D is not so well developed as the An, although both fins are similarly shaped and located opposite one another (a contrast to *Pterolebias*, where the An is always far more strikingly developed than the D). The D and An in the male are pointed at the rear, and the C has distinct twin peaks; in the female all the fins are rounded. *Rachovia* lives in bogs or marshes that occasionally dry up and is extraordinarily adapted in life-cycle to such biotopes. They are annual fish (compare Cyprinodontidae) as are the related Ga already mentioned; the dry periods are survived as an egg-phase. When water flows once again, the fry hatch out and within a few weeks, they are capable of reproduction. *Rachovia* species are so markedly adapted to this seasonal way of life that in the aquarium they are also extremely short-lived and the spawn must experience a dry phase if it is to develop normally. Very small tanks are suitable for care and breeding, with scalded, well watered leaf-mould for a substrate. The sexes should be separated before breeding, then brought together again; often the pair will spawn immediately when this happens. The eggs are laid in the substrate. To encourage the spawn to hatch out, follow the instructions as given for the G *Cynolebias*.

– *R. brevis* (REGAN, 1912). From W Colombia. Length to 6.5 cm. Body and fins blue-grey to blue-green. Edges of the scales and fin markings blue-violet. The sides of the head have lovely green iridescent blotches (fig. 961).

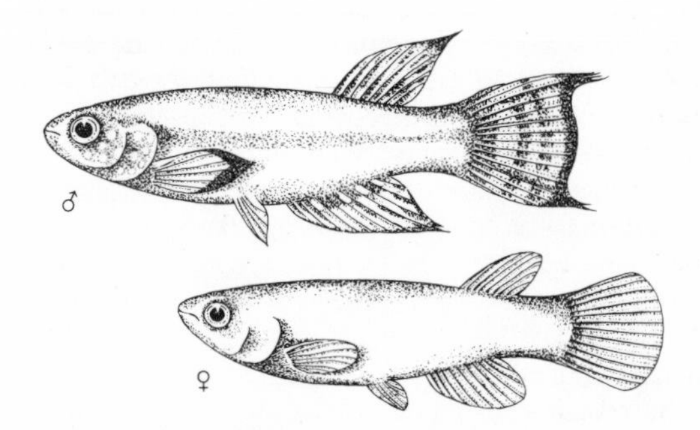

Fig. 961 *Rachovia brevis*

Rachow's Gambusia *(Gambusia rachowi) see Gambusia*

Rachow's Pyrrhulina *(Pyrrhulina rachowiana) see Pyrrhulina*

Racial Group *see* Concept of Species

Radianthus KWIETNIEWSKI, 1896. G of tropical Sea-anemones (Actiniaria*). From the Indo-Pacific reef areas, often living in communities with Anemonefishes*. In contrast to *Stoichactis** and *Discosoma*, the large *Radianthus* species have longer, usually pointed tentacles (some in fact are slightly spindle-shaped). The mouth is off-white to olive, and the column is yellow, green, blue or violet. Some species are easier to keep in the aquarium than *Stoichactis* and *Discosoma* species. Besides live food, even food tablets are often accepted.

Radiolaria; Radiolarians. Cl of the Rhizopoda*. The Radiolaria are microscopically small single-celled organisms Protozoa*) of the marine plankton; they are characterised in particular by their complicated skeleton made of silicic acid or strontium sulphate. They have thread-like pseudopodia stretched out like rays for catching food. Reproduction is often by means of an alternation of generations*. The Radiolaria are among the oldest and best-known fossils of all.

Radiolarians, Cl, *see* Radiolaria

Radix, G, *see Lymnaea*

Ragged Filefish *(Monacanthus spinosissimus) see Monacanthus*

Rainbow Cichlids, G, *see Pelvicachromis*

Rainbow Copella *(Copella vilmae) see Copella*

Rainbow Fish *(Thalassoma lunare) see Thalassoma*

Rainbow Trout *(Salmo gaidneri) see* Salmonoidei

Rainbow Wrasse *(Coris julis) see Coris*

Rainbowfishes (Melanotaenidae, F) *see* Atherinoidei

Rainwater *see* Water, for Breeding Purposes

Raja (various species) *see* Rajidae

Rajidae; True Rays (fig. 962). F of the Chondrichthyes*, O Rajiformes. This F contains a large number of species with typical features of a very flattened, broad, rhombic body and a fairly thin caudal peduncle. The large Ps form the side parts of the body itself and also reach forwards along the head right to the very front of the snout, which in turn may be stumpy or very pointed. These fins are the main source of propulsion for the fish, which they perform with wave-like fin movements. The front of the Vs are covered by the Ps.

On the caudal peduncle there are 2 Ds, and the C is small. The dentition consists of numerous rows of small, mostly rounded teeth. As with all rays, the gill slits lie on the underside of the body. The skin is rough, ie it contains placoid scales. Between the normal placoid scales, many species also have very large placoid scales, many species also have very large placoid scales. The males are usually smaller, and their Vs are sometimes transformed into copulatory organs (myxopterygia), a feature that is perfectly normal for cartilaginous fishes. In the Rajidae, the eggs are fertilised in the oviduct and then laid in horny capsules, the form of which looks like a cushion with very long corners. It is not uncommon to find some among seaweeds that get washed up on the beach. Depending on species and on temperature, the embryos take between 4 and 14 months to develop. Many species, perhaps all of them, have weak electric organs in the caudal peduncle. Rays are typical bottom-fish, and they occur mainly in the shallow water zones along the coast in temperate and cold seas. They feed on molluscs, crustaceans, bottom-dwelling fish and other animals. They are not dangerous to man. The Roker or Thornback Ray *(Raja clavata)* (fig. 963) is common along the Atlantic coasts of Europe, including the Mediterranean. It is a brownish colour and has a stumpy snout. The females can be up to 120 cm long, but the males remain much smaller. Large placoid scales are scattered over the upperside of the body. They look like drawing-pins that have been left sticking in the body, which is where the common name 'Thornback' comes from. The Skate *(Raja batis)* (fig. 964) has similar distribution area but does not reach quite so far south. It has thorns on the caudal peduncle only. At least sometimes, the Skate is caught at depths of 350 m. On rare occasions, the females can reach lengths of over 2 m. Like the Roker, Skates are important in the fishing industries of the North Sea and around Norway, because they occur in sufficient numbers and their meat is tasty. Over the counter, smoked Skate meat is sold as 'Sea Trout'. Crab salad or Lobster salad often consists mainly of Ray-meat. Other species of note include

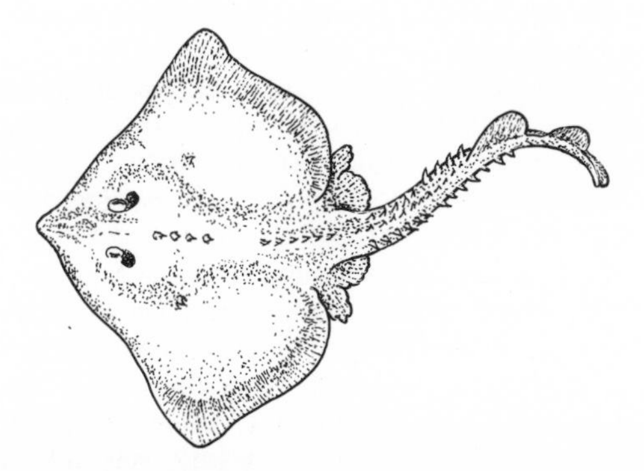

Fig. 962 Rajidae-type

Fig. 963 *Raja clavata*

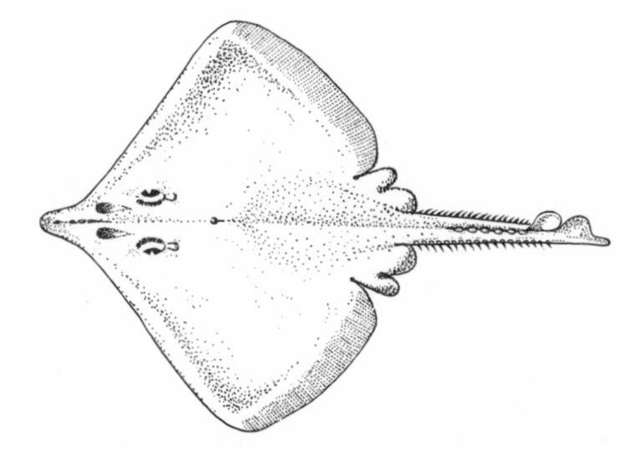

Fig. 964 *Raja batis*

the smaller *Raja radiata* from the eastern N Atlantic and the Hedgehog or Little Skate *(Raja erinacea)* from the western N Atlantic. This last species only grows up to 50 cm long.

Rajiformes, O, *see* Chondrichthyes

Ramirez's Dwarf Cichlid *(Apistogramma ramirezi) see Apistogramma*

Ramming. Term used in ethology. A type of aggressive fighting behaviour* in which the attacker tries to use the full force of his attack to hit the flanks of his opponent. As he does so, the operculum is held open away from the body and the mouth is opened wide. If both fighters meet each other front-on in this attitude, then a mouth-to-mouth fight* may ensue as a consequence of the ramming action. Alternatively ramming circles may be delineated as they take it in turns to swim very quickly round and round a spot trying to ram the other's caudal fin. Ramming is particularly common among Cichlids and Egg-laying Tooth-carps as well as Characins, but it may also be observed among members of all the other fish families.

Ramming Circles *see* Ramming

Ramshorn Snails, Ga, *see Helisoma, Planorbis, Planorbarius*

Ranatra Fabricius, 1795 (fig. 965). G of the Bugs (Heteroptera*). Closely related to the well-known Water Scorpion (*Nepa**) and belonging to the F Nepidae. In shape, *Ranatra* resembles a Walking-stick Insect and reaches a length of about 80 mm, of which about half makes up the body itself and the other half a breathing tube sticking out from the end of the abdomen. Its shape and unremarkable brown colouring makes *Ranatra* marvellously camouflaged; it positions itself near plants in still waters and lies in wait there for suitable prey to come along. When a victim approaches close to the front legs, it is snapped up very quickly and held firmly with the feet pressing against the side of the

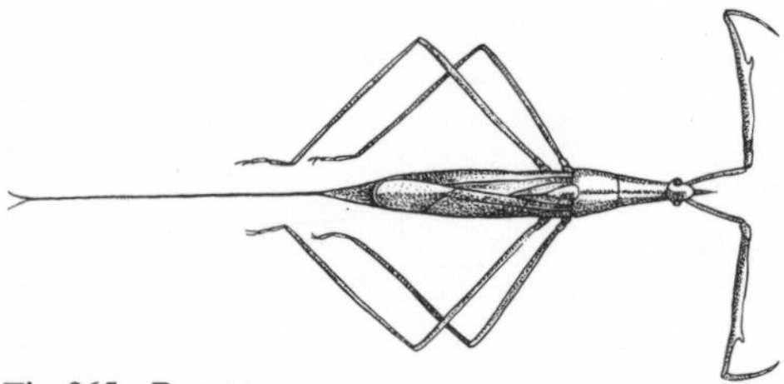

Fig. 965 *Ranatra*

limb (like a jack-knife). A prick with the proboscis paralyses the prey and then it is sucked dry. *Ranatra* makes an excellent object of observation in pond-aquaria, provided there are enough water plants around that still allow the Bugs to breathe air through their breathing-tubes as they sit quietly in the water.

Ranunculaceae; Crowfoot Family. F of the Magnoliatae* with 50 Ga and 2,000 species. Distributed worldwide but mainly in the northern hemisphere. Most are perennials and annuals. The leaves are alternate, pinnate or simple. Flowers hermaphroditic, poly- or monosymmetrical; the perianth comes in a multitude of shapes. The Ranunculaceae colonise a great variety of habitats from low-lying land to the high mountains, but only a few species are marsh and water plants. In spring, on nutrient-rich marshlands, in springs, streams and ditches, you will find *Caltha palustris* (Marsh Marigold), which is distributed throughout the northern hemisphere. Some *Ranunculus* species also grow in reeds, whilst others, usually with white flowers, grow submerged. Well-known ones are *Ranunculus aquatilis* (Water Crowfoot) with soft, pinnate aquatic leaves shaped like a hand, and *Ranunculus circinalis* (Stiff-leaved Water Crowfoot) with stiff, pinnate aquatic leaves. There are hardly any reliable experiences with cultivation.

Rapana Schumacher, 1817. G of Indo-Pacific Prosobranches (Prosobranchia*, Gasteropoda*). The species live a predatory life in the littoral. In World War II, one species was introduced into the Black Sea, since when it has reproduced in enormous numbers because there are no natural competitors or enemies there: *R. thomasiana* Crosse. This snail is native to the Japanese sea and is normally about 10 cm long, sometimes bigger. Its shell is compressed and its last coil is very large. *Rapana* feeds on Mussels *(Mytilus*)*; it can open their shells. In the aquarium, these snails make undemanding, hardy specimens if fed a varied diet of animal food (Bivalve meat and fish meat) (see also *Murex*).

Raphiodontinae, Sub-F, *see* Characoidei

Rasbora Bleeker, 1860. G of the Cyprinidae*, Sub-F Rasborinae. Tropical SE Asia, Sri Lanka (Ceylon), Sumatra, Thailand, the Greater Sunda Islands, the Malaysian Archipelago, with a few species also in E Africa. *Rasbora* species are very small to small fishes, sometimes found in large shoals, that live in flowing and still waters with varied water conditions and ecological factors. Within the G are species of very slim body-shape and species with a stocky shape. They are all very flat laterally. The mouth is terminal or slightly superior. LL complete, incomplete or absent altogether. Rasboras are distinguished from other Ga by having a knob (symphyseal knob) on the lower jaw which dovetails into a recess of the upper lip. They also have no barbels and a short An supported by only 5 rays. Caring for Rasboras is not difficult. They are peaceful fish that require large tanks with enough swimming room, clumps of vegetation round the edges and a dark substrate, and the tank should stand in not too bright a spot. Always keep them as a shoal and they can also be kept together with similar-sized, peaceful fish. Temperature 24–26 °C. Preferably, the water should be soft and slightly acidic. Depending on the size of the fish, feed them on small crustaceans, insects and their larvae, and dried food. There have been successes with breeding many kinds of Rasboras in the aquarium. However, the individual species have different requirements of water quality for breeding purposes. Some species will reproduce in soft, neutral or slightly acidic fresh water; for others, the water must be extremely soft (1.5–2° total hardness) and slightly acidic (pH 5.5–5.7). Filtration through peat is helpful. Often the choice of partner is a crucial issue; it is best to choose from a fairly large number of specimens). The breeding tank selected must not be too small. The sun shining in often stimulates the spawning act. After forceful driving, the male lures the female between some plants where they nestle tightly against one another. The eggs are not very sticky and sink to the bottom. The parents do not bother to care for the brood, and after spawning they should be removed. 24–30 hours later, the fry hatch out and after 3–5 days they will be swimming freely. Feed them with the finest of micro-organimsms to begin with; they grow very quickly. It is a good idea to add some fresh water regularly when rearing.

– *R. argyrotaenia* (BLEEKER, 1850); Silver Rasbora. Japan, China, Thailand, Malaysian Archipelago. Length to 15 cm. An elongated species. Dorsal surface grey-olive, flanks a gleaming silver with a blue-green longitudinal band bounded on top by yellow. Fins yellowish, the C with a black border. Not yet bred in captivity.

– *R. borapetensis* SMITH, 1934; Red-tailed Rasbora. Thailand. Length to 5 cm. Elongated. Dorsal surface olive, flanks green-yellow with a silver iridescence and a black longitudinal band bounded on top by a gold-green line. The dorsal surface and base of the An each have a black stripe. The D and C are reddish. Breed in a low level of water and place the tank in a dark spot. During mating, the male entwines himself round the female (fig. 966).

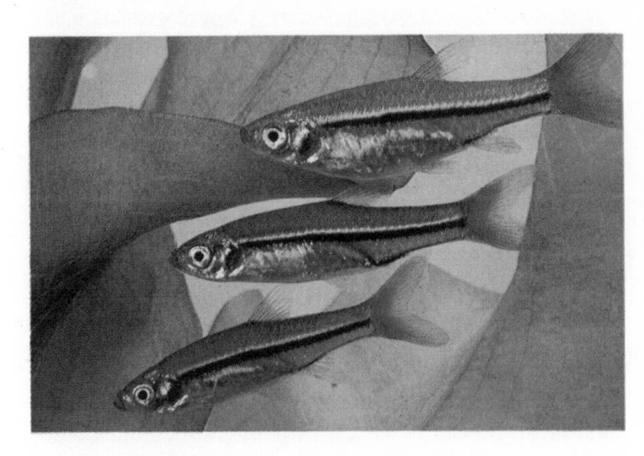

Fig. 966 *Rasbora borapetensis*

– *R. caudimaculata* VOLZ, 1903; Giant Scissortail, Greater Scissor-tail. Malaysian Archipelago, Sumatra, Thailand. Length to 12 cm. Slim. Dorsal surface olive-green, flanks with a silver sheen, ventral surface white. The scales are dark-edged. Along the body runs a greenish-golden longitudinal band that gleams. There is a black band with a red-brown colour below on each of the caudal lobes. There are only a few details about breeding.

– *R. cephalotaenia* (BLEEKER, 1852); Porthole Rasbora. Kalimantan (Borneo), Malaysia, Sumatra. Length to 12 cm. Elongated. Dorsal surface grey-green, flanks copper-brown with a black longitudinal band bounded above by a shining gold colour and running through the middle of the body. It stretches from the Ps to the An. This species prefers the top water levels. There is a sub-species, described as *R. cephalotaenia tornieri* AHL, 1922.

– *R. chrysotaenia* AHL, 1937; Gold-striped Rasbora. Malaysian Archipelago, Sumatra. Length to 10 cm. Elongated. Dorsal surface olive-brown, flanks with a pale silvery sheen, ventral surface with a longitudinal band in a gleaming white, green or red-golden colour and bounded below by a blue-black stripe. Have plenty of vegetation in the tank.

– *R. daniconius* (HAMILTON-BUCHANAN, 1822); Slender Rasbora. From West India, Sri Lanka (Ceylon), Burma, Thailand, the Greater Sunda Islands. Length to 20 cm. Very slim. Dorsal surface bronze-coloured, flanks with a mother-of-pearl silver sheen, ventral surface white. There is a deep blue longitudinal band, occasionally interrupted, and flanked by narrow gold stripes. The D, An and C are yellow-orange. A large tank is needed. Medium-hard water for breeding (8 to 10° total hardness) (fig. 967).

Fig. 967 *Rasbora daniconius*

– *R. dorsiocellata* DUNCKER, 1904, 1904; Eye-spot Rasbora, High-spot Rasbora. Malaysian Archipelago, Sumatra. Length to 6 cm. A bit stocky. Dorsal surface grey-yellowish or blue-violet shimmer. Ventral surface white. There are 2 black longitudinal bands, the lower one of which breaks up into flecks in part. Fins yellowish; the D has a black blotch, outlined in white. Provide cushions of finely pinnate plants in the breeding tank. There is a sub-species, *R. dorsiocellata macrophthalma* MEINKEN, 1951 (Emerald-eye Rasbora). Its length is up to 3.5 cm. Its stripe markings are less distinct and beneath the eye there is an intense blue-green iridescent zone.

– *R. einthoveni* (BLEEKER, 1851); Brilliant Rasbora, Blue-line Rasbora. East India, Malaysian Archipelago, Thailand, Indonesia. Length to 9 cm. Elongated. Dorsal surface yellow-brown to green-brown, flanks a silvery iridescent grey-blue, ventral surface white. There is a black longitudinal band with a greenish iridescence, and bounded above by a golden gleaming stripe. Breeding tank should be fairly large and the water must be free of Infusoria; total hardness about 15–18°.

– *R. elegans* (VOLZ, 1903); Elegant Rasbora, Two-spot Rasbora. Malaysian Archipelago, Kalimantan (Borneo), Sumatra. Length to 13 cm. Stocky build. Dorsal surface brown-olive, flanks brown with a rusty-red to greenish sheen, ventral surface pale brown to yellow with a silver sheen. The scales on the dorsal surface are edged in black. There is a black fleck at the height of the D and on the caudal peduncle. The D, C and An are yellowish.

– *R. gerlachi* (AHL, 1928); Gerlach's Rasbora. Africa: Cameroun. Dorsal surface brown-yellow to olive, flanks yellow with an intense blue longitudinal band, bending downwards and bounded on top by a gleaming yellow stripe. Not yet bred in captivity.

– *R. hengeli* MEINKEN, 1956; Hengel's Rasbora. Sumatra. Length to 3.5 cm. Stocky build. Dorsal surface grey-brown, flanks a translucent grey with a reddish shimmer. A dark blue longitudinal band begins as a triangle at the height of the D; it quickly becomes narrower. Below this band is an orange-yellow zone. Above the P is a red blotch. The fins are yellowish with dark streaking. During spawning, the fish stick their eggs to the underside of broad-leaved aquatic plants; as they do so, both partners turn onto their backs.

– *R. heteromorpha* DUNCKER, 1904; Harlequin Fish, Red Rasbora. Malaysian Archipelago, Thailand, Indonesia, widely distributed. Length to 4.5 cm. Stocky build. Dorsal surface brownish, body a delicate reddish or violet with a silver-grey sheen, ventral surface pale. A blue-black, wedge-shaped blotch begins beneath the D and ends on the caudal peduncle. D and C red. The mating pair lie on their backs to lay their eggs on the underside of aquatic plant leaves. As this happens, the male entwines himself round the female (fig. 968).

Fig. 968 *Rasbora heteromorpha*

– *R. jacobsoni* WEBER and DE BEAUFORD, 1916; Jacobson's Rasbora. Sumatra. Length to 7 cm. Dorsal surface and flanks reddish-brown, ventral surface yellowish-white. A dark longitudinal band is bounded above by a gold-coloured stripe. A second, less distinct band runs from the P to the An. Fins brownish.

– *R. kalochroma* (BLEEKER, 1850); Clown Rasbora, Big-spot Rasbora, Iridescent Rasbora. Sumatra. Kalimantan (Borneo), Malaysian Archipelago. Length to 5 cm. Dorsal surface brown-red, flanks reddish with a pale green gleaming stripe. On this stripe, at both the levels of the P and D, is a dark fleck joined together by speckles. The fins are red; and the An has a black tip. Keep this species in medium-hard water. Probably not yet bred in captivity (fig. 969).

Fig. 969 *Rasbora kalochroma*

– *R. lateristriata* (VAN HASSELT, 1823); Side-striped Rasbora. Sumatra, Malaysian Archipelago. Length to 12 cm. Dorsal surface olive-brown with a gold-green sheen. Scales edged in a dark colour. A dark longitudinal band is bounded above by a red-golden stripe. Parallel with the An is a narrow, black streak. The and C are rust-red, the An and V reddish.

– *R. leptosoma* (BLEEKER, 1855); Copper-striped Rasbora. Thailand, Malaya, Sumatra. Length to 10 cm. Dorsal surface brownish, flanks yellow-brown, ventral surface white. A red-golden longitudinal band is bounded below by a dark band which broadens out towards the rear. Between the V and An is a blackish zone. The D and C are reddish.

– *R. maculata* DUNCKER, 1904; Spotted Rasbora, Pygmy Rasbora. East India, Malaysian Archipelago, Sumatra. Length to 2.5 cm. A stocky build. Dorsal surface greenish-brown, flanks a brick-red, ventral surface yellowish-red to yellow. There are blue-black flecks of varying size above the P, on the base of the An and on the caudal peduncle. Fins red-yellow, the D and An with black blotches. Soft, slightly acidic water. During mating, the female is entwined by the male. Open-water spawner (fig. 970).

Fig. 970 *Rasbora maculata*

– *R. meinkeni* DE BEAUFORT, 1931; Meinken's Rasbora. Sumatra. Length to 7 cm. Dorsal surface olive-brown, flanks and ventral surface brassy-yellow. A golden-yellow stripe runs above a black longitudinal band and

parallel to it. There is a black stripe above the base of the An. Fins brownish.
– *R. nigromarginata* MEINKEN, 1957; Red Rasbora. Sri Lanka (Ceylon). Length to 5 cm. A stocky species. Dorsal surface brown-red, flanks coppery, ventral surface pale. Fins ochre-yellow. The first fin rays of the D, V and Ps are black, and the upper fin lobe of the C is blue-green. A sensitive species. Not yet bred in captivity.
– *R. pauciperforata* WEBER and DE BEAUFORT, 1916; Red-striped Rasbora, Glowlight Rasbora. Malaysian Archipelago, Sumatra. Length to 7 cm. Of slim build. Dorsal surface yellow-olive and brownish, flanks a gleaming silver with a turquoise shimmer, ventral surface silver-white. Some of the scales are dark-edged. Beneath a red longitudinal band runs a parallel, narrow, blue-black band. The lower edge of the caudal peduncle is dark. Fins a translucent yellow colour. So far, has only been bred successfully on a few occasions (fig. 971).

Fig. 971 *Rasbora pauciperforata*

– *R. paucisquamis* AHL, 1935; Large-scaled Rasbora. Malaysian Archipelago. Length to 6 cm. A slim species with strikingly large scales. Dorsal surface and flanks olive-brown, ventral surface pale with a bluish shimmer. A brassy-yellow longitudinal band is divided at the level of the D by a dark longitudinal band that broadens out towards the rear. There is another, less distinct band above the base of the An. Only isolated successes with breeding.
– *R. philippinica* GÜNTHER, 1880; Philippine Rasbora. Length to 7 cm. Dorsal surface greenish-grey, flanks a pale olive with a silvery sheen, ventral surface silver-white. An indistinct, gleaming gold longitudinal band is bounded below by a black line. There is a narrow dark stripe at the base of the An. Fins yellowish. Also tolerates a temperature of around 18 °C.
– *R. rasbora* (HAMILTON-BUCHANAN, 1822); Rasbora. Malaysian Archipelago, Burma, Thailand. Length to 10 cm. Dorsal surface a dark brown-yellow, flanks a paler brown-yellow with a golden iridescence; ventral surface yellow. A black, iridescent blue longitudinal band extends across the whole body to the C. Fins yellowish.
– *R. somphongsi* MEINKEN, 1958; Somphongs' Rasbora. S Thailand. Length to 3 cm. Of stocky build. Dorsal surface and flanks yellow-brown. There is a broad, gleaming golden longitudinal band that is joined by a black band at the level of the D and which continues to the C. At the end of the caudal peduncle there is a dark blotch above and below. Fins yellow-brown. The eggs are laid on the underside of leaves, to do which the fish have to lie on their backs.
– *R. taeniata* (AHL, 1922); Black-striped Rasbora. Malaysian Archipelago, Sumatra. A slim species. Length to 8 cm. Dorsal surface greenish, flanks a pale olive, ventral surface silvery with a reddish shimmer. There is a dark longitudinal band edged on top by a gleaming red-golden stripe. The base of the An has a dark line. C yellow-red.
– *R. trilineata* STEINDACHNER, 1870; Scissortail Rasbora, Three-line Rasbora. Malaysian Archipelago, Sumatra, Kalimantan (Borneo). Length to 15 cm. An elongated form. Body translucent. Dorsal surface olive with a dark longitudinal streak, flanks a silver iridescent colour, ventral surface silver. A narrow dark band begins at the level of the V and continues into the C. On each of the C lobes there is a black transverse band. Breeding is not totally simple. The water must be soft and contain no Infusoria (fig. 972).

Fig. 972 *Rasbora trilineata*

– *R. urophthalma* AHL, 1922; Exclamation Mark Rasbora, Miniature Rasbora. Sumatra. Length to 2.5 cm. Stocky build. Dorsal surface and flanks red-brown, yellowish-white towards the ventral surface, ventral surface silver. A dark-blue longitudinal band that becomes very narrow behind the D ends in a golden-yellow-edged blotch on the C-root. Above, this band borders onto a narrow, gold-red stripe that follows the same route. Fins brownish. The front edges of the D and An and the upper edge of the C black. The mating pair typically entwine and lay their spawn in open water.
– *R. vaterifloris* DERANIYAGALA, 1930; Pearly Rasbora, Ceylonese Fire Rasbora. Sri Lanka (Ceylon). Length to 4 cm. Stocky species. Dorsal surface olive-green, body grey-green with a silvery violet sheen, ventral surface pale orange. The D, An and the base of the C are orange to red. Breed in tanks that are stood in a dark spot. The mating pair nestle closely against each other and lay the spawn among plants.

– *R. wijnbergi* MEINKEN, 1963; Wijnberg's Rasbora. Kalimantan (Borneo). Length to 4 cm. Stocky species. Dorsal surface olive-green, flanks a gleaming silver with dark-edged scales. There is the suggestion of a narrow longitudinal band on the caudal peduncle. The D and An are ochre-yellow with black tips. The base of the C is red, and the C lobes are ochre-yellow. Not yet bred in captivity.

Rasborichthys BLEEKER, 1860. G of the Cyprinidae*, Sub-F Abraminae. Small to medium-sized shoaling fish that are spread throughout SE Asia. They have a stocky appearance, similar to Barbs, and as they get older they become fairly tall. Much flattened laterally. Between the Vs and An, the belly is keeled, and the An is longer than the D. No barbs around the terminal mouth. LL complete. They are peaceable fish that should be kept as a shoal in spacious tanks with a moderate amount of vegetation. Water quality is not a decisive factor. Temperature 22–24 °C. Live and dried food. Very few details available regarding reproductive behaviour and breeding in the aquarium.

– *R. altior* REGAN, 1913. From Singapore. Length to 9 cm. Dorsal surface grey-green, flanks a gleaming silver with a greenish-blue shimmer, ventral surface silver-white. At the base of every scale there is a dark blotch, which produces the effect of rows of dark dots. Fins colourless.

Rasborinae, Sub-F, *see* Cyprinidae

Ratfish *(Chimaera monstrosa) see* Chondrichthyes

Ratfish *(Gonorynchus gonorynchus) see* Gonorynchiformes

Ratfishes (Chimaeridae, F) *see* Chondrichthyes

Ratfishes (Chimaeriformes, O) *see* Chondrichthyes

Rat-tails (Macrouridae, F) *see* Gadiformes

Ray-fins (Actinoptherygia, Sub-Cl) *see* Osteichthyes

Rays (collective term) *see* Chondrichthyes

Rays, Electric, F, *see* Torpedinidae

Rays, True, F, *see* Rajidae

Ray-types (Rajiformes, O) *see* Chondrichthyes

Razorfishes (Centriscidae, F) *see* Aulostomioidei

Rearing Tank. Shortly after the young fish are swimming around freely – or in the case of a numerically very large brood, immediately after the eggs hatch – they must be transferred from the breeding tank to a larger fearing tank. This will minimise losses. Tanks used for rearing must be roomy and transparent enough to be able to keep a constant check on the state of health of the growing youngsters. The substrate can be dispensed with, there should be adequate filtration and aeration of the water and part of the water should be changed at frequent intervals to ensure successful rearing. At the beginning, the mix of the water in the rearing tank should as far as possible correspond to that of the breeding tank, ie soft and slightly acidic. With each change of water, harder, neutral to slightly alkali, fresh water can gradually be added to get the young fish used to it and toughen them up.

Rearing Young Fish *see* Fry, Rearing of

Recessive *see* Heredity

Red Algae, D, *see* Rhodophyta

Red Algae, G, *see* Cystoseira

Red Comb Starfish *(Astropecten aurantiacus) see Astropecten*

Red Emperor *(Lutjanus sebae) see Lutjanus*

Red Fire Fish *(Pterois volitans) see Pterois*

Red Giant Vallisneria *(Vallisneria neotropicalis) see Vallisneria*

Red Gurnard *(Trigla hirundo) see* Scorpaenoidei

Red Hermit Crab *(Dardanus megistos) see Dardanus*

Red Killi *(Aphyosemion bivittatum) see Aphyosemion*

Red Pencilfish *(Nannostomus beckfordi aripirangensis) see Nannostomus*

Red Phantom Tetra *(Megalamphodus sweglesi) see Megalamphodus*

Red Piranha *(Serrasalmus nattereri) see Serrasalmus*

Red Rasbora *(Rasbora heteromorpha) see Rasbora*

Red Rasbora *(Rasbora nigromarginata) see Rasbora*

Red Ruby Tetra *(Axelrodia riesei) see Axelrodia*

Red Sea. A neighbouring sea of the Indian Ocean*, characterised by a very high salt content (about 40 ‰) and a high water temperature. Even at depths of 2,000 m, the water still has a temperature of 21.5 °C, whilst at similar depths in the Indian Ocean measurements have indicated a temperature of only 3.5 °C. Thus, tropical organisms are able to penetrate considerable depths in the Red Sea. At an earlier time in history, the Red Sea was very probably joined to the Mediterranean* and had a very similar fauna. During the Ice Age, this link was largely destroyed. Today's abundant tropical fauna stems from the post-glacial species that migrated there from the Indian Ocean; among them are many well-known coral-fish. On the other hand, the Red Sea also has a number of endemic species* as a consequence of temporary isolation.

Red Sea-squirt *(Halocynthia papillosa) see Halocynthia*

Red Snapper *(Lutjanus sebae) see Lutjanus*

Red Soldierfish *(Myripristis murdjan) see Myripristis*

Red Spiny Starfish *(Protoreaster lincki) see Protoreaster*

Red Squirrelfish *(Holocentrus [Adioryx] rubra) see Holocentrus*

Red Soldierfish *(Holocentrus [Adioryx] rubra) see Holocentrus*

Red Tetra from Rio *(Hyphessobrycon flammeus) see Hyphessobrycon*

Red Top Zebra *(Labeotropheus trewavasae) see Labeotropheus*

Red Water-lily *(Nymphaea rubra) see Nymphaea*

Red Wrasse *(Coris gaimardi) see Coris*

Red-bellied Piranha *(Serrasalmus nattereri) see Serrasalmus*

Redbelly Dace *(Chrosomus erythrogaster) see Chrosomus*

Redbreast Sunfish *(Lepomis auritus) see Lepomis*

Red-dotted 'Pyrrhulina' *(Copella metae) see Copella*

Red-eyed Tetra *(Moenkhausia sanctaefilomenae) see Moenkhausia*

Redfang Trigger *(Odonus niger) see Odonus*

Red-finned Shark *(Labeo erythrurus) see Labeo*

Redfish *(Sebastes marinus) see* Scorpaenoidei

Red-lined Triggerfish *(Balistapus undulatus) see Balistapus*

Redox Potential. The term 'redox potential' when broken down means reduction, oxidation and potential. By reduction we mean the surrender of oxygen, the acceptance of hydrogen and electrons; oxidation is the opposite, ie the acceptance of oxygen, and the surrender of hydrogen and electrons. In nature, nearly all substances can occur in both a reduced and oxidised form – the properties of the substance depend greatly on which form it is in. For example, sulphur in its reduced state as hydrogen sulphide is poisonous even in the tiniest amounts, whilst in its oxidised state as sulphate (eg in the water hardness*) it can be tolerated in quite considerable amounts. Potential is best translated as capacity – with a high redox potential it is the particular capacity for oxidation; with a low redox potential it is the particular capacity for reduction. A high redox potential leads in the end to a complete mineralisation of all dead organic matter that can be broken down; so it is typical of very pure stretches of water, a mountain stream perhaps, or a coral reef. Conversely, a low redox potential will lead to the accumulation of decaying matter and the build-up of ammonia and hydrogen sulphide. It is also of importance to note that there is a significantly higher redox potential in the open-water zones than in the substrate. Although it is always the case in clean, natural waters that oxidising tendencies are the rule, in the aquarium the rules are reversed. However, with appropriate measures of care, the tempo at which the redox potential falls can at least be slowed down. It is generally of benefit, for example, to include the substrate in the water circulation, and indeed the substrate should not act as a filter and therefore as a collector of dirt, but more as a diffuser. Unfortunately, compromises have to be made regarding redox potential, since the different plants and animals necessarily prefer varying degrees of high redox potential. Green algae, for example, flourish particularly well with a high redox potential, *Cryptocoryne* on the other hand with a low redox potential. Because of this, we also have here a barometer of the redox potential in the aquarium. In the marine aquarium, a good growth of green algae is a sign of its high redox potential and therefore the stability of the tank. A smeary algal growth, on the other hand, means the redox potential is quite low and some care is needed. An exact measurement of the redox potential needs a lot of expensive equipment and is only rarely carried out in aquarium-keeping. The level of redox potential is given by the measure rH value*.

Red-spotted Aphyosemion *(Aphyosemion cognatum) see Aphyosemion*

Red-spotted Copeina *(Copeina guttata) see Copeina*

Red-spotted Hawkfish *(Amblycirrhites pinos) see Amblycirrhites*

Red-spotted Panchax *(Epiplatys longiventralis) see Epiplatys*

Red-striped Rasbora *(Rasbora pauciperforata) see Rasbora*

Red-striped Soldierfish *(Myripristis murdjan) see Myripristis*

Red-tailed Black Shark *(Labeo bicolor) see Labeo*

Red-tailed Rasbora *(Rasbora borapetensis) see Rasbora*

Red-tailed Tetra *(Hemigrammus caudovittatus) see Hemigrammus*

Red-throated Rainbow Wrasse *(Coris angulata) see Coris*

Red-toothed Triggerfish *(Odonus niger) see Odonus*

Reduction Division *see* Cell Division

Reed or Common Reed *(Phragmites communis) see* Poaceae

Reed Mace *(Typha, G) see* Typhaceae

Reed Maces, F, *see* Typhaceae

Reedfish *(Calamoichthys calabaricus) see Calamoichthys*

Reef *see* Coral Reef

Reef Angelfishes, Ga, *see* Pomacentridae

Reef Plate *see* Coral Reef

Refuge *see* Refugium

Refugium, also refuge. The area to which plants and animals retreated and survived during the Ice Age. As a consequence, therefore, it is also the area from which they spread again in the post-glacial period. Such places of refuge are to be found, for example, in S Europe, in the Black Sea area, in the river basins of the Amur, and in Mexico; even today, they are conspicuous by their abundance of species and endemic organisms.

In a wider sense, a refugium is any place of retreat and survival – ie not necessarily in relation to the Ice Age.

Regal Angelfish *(Pygoplites diacanthus) see Pygoplites*

Regalescidae, F, *see* Lampridiformes

Regan, Charles Tate (1878–1943). English ichthyologist. Studied at Cambridge University. In 1901 he became an assistant at the British Museum and a student of G. A. Boulenger. He developed into an exact systematician and set about organising the extensive fish collections of the Museum along modern systematic lines. He also had published a large number of publications. They include the *Revision of the Fishes of the American Cichlid genus Cichlasoma* (1905), *The Freshwater Fishes of the British Isles* (1911), *Morphology of the Cyprinodont Fishes of the Sub-family Phallostethinae* (1916), and others. A result of his extensive researches was his own comprehensive system of fish.

Regeneration. The ability of organisms to replace damaged, dead, or lost parts of the body. Plants usually have a very good regenerative capacity (vegetative propagation through cuttings). Animals, in the main, have only slight regenerative powers. Physiological regeneration enables cell and tissue elements to be replaced, particularly those like blood and skin cells that naturally wear out a lot. Through reparative regeneration, damaged parts of the body, or parts that have been ripped off, can be replaced. The Coelenterata* and Asteroidea*, for example, are capable of regenerating a completely new animal from parts of their own organism. Fish can restore parts of fins that have been lost (in fact, the new parts are often elongated). Wounds to the body normally heal quickly in fish, also through regeneration of the damaged tissue. Usually, this causes no complications.

Regularia, Sub-Cl, *see* Echinoidea

Reitzig's Dwarf Cichlid *(Apistogramma reitzigi) see Apistogramma*

Releaser. Term used in ethology. Releasers work as carriers of stimuli that act upon a release mechanism*. The key stimuli* emitted by the releaser set in motion specific modes of behaviour in the fish, depending on the species. Colour, shape or movement can all be releasers, as well as signs of alarm or sound waves. From the multitude of possible examples, I have selected the following: the yellow-black spawning coloration and the spread of the fins in a female *Apistogramma agassizi* is the key stimulus for the male to release his courtship behaviour. Alarm substances released into the water by a minnow caught by a predator are the releasers for the immediate flight of all other minnows from the area. The fin beats of an approaching predator produce sound waves which are detected by the lateral line organ of the prey and induce its flight reaction.

Relict Species. An ancient animal or plant species, or higher systematic category, that has long since died out in other places and only survives in a particular place of refuge (Refugium*). Among the fish, these include the Stump-fins *(Latimeria)*, the Gar Pikes (Lepisosteidae) and the Bony Tongues (Osteoglossidae).

Other species are also called relicts, however: they are those species that have remained in narrowly delineated habitats, even though the environmental conditions have altered vastly in the meantime which has meant that other types of flora and fauna prevail. In such a case, these do not have to be ancient forms. For example, in highly industrialised places today, many freshwater animals show the distribution of relict species, caused by universal, extensive water pollution.

Remilegia australis *see* Echeneidae

Remoras, F, *see* Echeneidae

Reproduction. The process of producing new individuals from existing ones, thus ensuring the continued existence of the organism concerned. Two types of reproduction are distinguished: asexual and sexual.

Asexual reproduction is the splitting off of body cells which are capable of developing into a new organism. The simplest type is where single-celled organisms split in two. This has developed into multiple division, particularly in parasitic sporozoa. It is in this way that a single malarial pathogen *(Plasmodium)* can produce up to 10,000 new individuals from just one cell. The formation of spores in plants is also asexual reproduction through single cells. Certain kinds of plants can break up completely or partially into multi-celled sections (eg blue algae or the marine alga *Caulerpa**); or they can lose parts of their shoots (eg water-weed). Higher plants often produce runners (eg *Cryptocoryne**, *Echinodorus**) or brood buds on the roots or leaves (eg *Ceratopteris**). Asexual reproduction is not unusual among invertebrate animals either. It is particularly common as fission or gemmation in coelenterates (eg *Hydra**) and various worms.

Sexual reproduction involves sexually different germinative cells (gametes*). During sexual reproduction a male germ cell and a female one become united; the hereditary material they each contain fuses together (*see* Fertilisation). This method of reproduction is common throughout the plant and animal kingdoms. There is one great advantage that sexual reproduction has over asexual reproduction: because the hereditary material can combine in continually new ways within a group of organisms, it allows those organisms (as a group) to adapt or develop further over a period of time (*see* Evolution*). It is not uncommon for a species to be able to reproduce in several different ways – sometimes the methods are alternated in particular cycles (*see* Reproductive Cycles). Parthenogenesis* is an exceptional case – an unfertilised egg-cell develops into a new individual.

The product in sexual reproduction does not always leave the body at the same stage of development. Most animals lay eggs (oviparity). When fertilisation takes place internally the foetus may already have begun to develop by the time the egg is laid; or development may already be well advanced (ovoviparity, eg sharks, birds and reptiles). Viviparity means that the young are born live. For information about reproduction in fish *see* Embryology.

Reproductive Cycles; Alternation of Generations. The existence of different methods of reproduction in successive generations, whereby a certain sequence of reproductive methods is determined either by heredity or by environmental factors. Clearly differentiated sexual and asexual phases occur in single-celled organisms and in the lower plants, whilst the sexual generation in the Spermatophyta no longer appears, at least externally. A sexual and asexual alternation of generations *(metagenesis)* also occurs in a large number of invertebrate animals (Coelenterata, Bryozoa, Tunicata etc). For example, medusae usually break off from sessile polyps while the polyps are the product of the eggs of the medusae. With *heterogenesis*, bisexual reproduction alternates with unisexual reproduction (Parthenogenesis*). In the case of water-fleas, for example, the summer generation consists entirely of females that reproduce parthenogenetically. In the autumn, males occur again, and they fertilise the winter eggs. This particular alternation is caused seasonally by temperature and water conditions.

Reproductive Period *see* Breeding Season

Reptantia; Lobsters, Hermit Crabs; True Crabs. Sub-O of the ten-footed crustaceans (Decapoda*). They deviate in a number of different ways from the more primitive shrimps (Natantia*): their abdominal feet, for example, are not used for swimming, and the forehead process (rostrum) is usually small or absent altogether. The body is also flattened and the walking legs are powerful. Nevertheless, any division between Reptantia and Natantia is arbitrary and open to dispute. Members of the Reptantia include the long-tailed, pincerless Palinura (eg *Palinurus**, *Scyllarus**); the long-tailed, but pincered Astacura (eg *Astacus**, *Cambarus**, *Homa-*

*rus**, *Nephrops**); the Anomura, characterised by a soft-skinned, reduced abdomen that is usually asymmetrical; the Paguridea*; the Galatheidea with the Fs Galatheidae* and Porcellanidae*. There are also the Brachyura*, in which the abdomen is much shortened and tongue-like; it is folded forwards beneath the cephalothorax, so that it cannot be seen from above.

Reptiles, Cl, *see* Reptilia

Reptilia. The reptiles are a Cl of vertebrate animals (Vertebrata*) that comprises about 6,000 species. Those alive today are divided into 4 Os, the tortoises and turtles (Testudines), the tuataras (Rhynchocephalia), the crocodiles and alligators (Crocodilia) and the lizards and snakes (Squamata). The Tuatara *(Sphenodon punctatus)* is the only living species of the Rhynchocephalia; biologically and phylogenetically it is a very interesting animal, and is found only in New Zealand. By far the largest order is the Squamata.

The reptiles reached their zenith in the Mesozoic period, about 200 million years ago. The dinosaurs that were living at that time were the largest land animals ever. With the exception of the polar and sub-polar regions and the upper reaches of high mountain ranges, the reptiles are represented worldwide in almost every type of biotope. The largest of them measure about 9 m long and include the Nile Crocodile and giant snakes like the Anaconda and the Reticulated Python.

Modern reptiles are poikilothermal creatures and usually warmth-loving. Those that live in temperate zones usually endure a winter resting period, whereas those from warm regions often endure a summer resting phase. Although very varied in appearance, reptiles do have certain anatomical features in common, such as skin containing few glands, that is protected by horny scales or plates. In the turtles and crocodiles especially, these dermal plates form a complete armour. Apart from this, various peculiarities of feature have developed that are only of importance for some species or systematic groups: autotomy is common among geckos and lizards (ie the ability to discard their tails); in some snakes (Poisonous Adders, Sea-snakes, Vipers, Rattlesnaker) and in two North American lizards, the salivary glands have developed into poison glands; other reptiles (eg chameleons) are able to change colour. There are also differences in reproductive techniques among the reptiles: turtles lay eggs with calcareous shells, many snakes and lizards are livebearing (ovoviviparous or truly viviparous like many chameleons). Some reptiles can emit sounds, like the geckos and the crocodiles (which in many respects are more highly developed than other reptiles). The majority of reptiles are predatory, but tortoises and others are planteaters.

Large numbers of reptiles live on or in water, and so are important in aquarium-keeping, although many species can only be kept in zoological gardens. Some of the most rewarding specimens in this group are the many kinds of marsh-turtles (Emydae). Other reptiles are specialists of one form or another – for example, the South American Mata-mata *(Chelus fimbriatus)* that lives on the substrate of inland waters, the Asiatic Gavials (O Crocodylia) that hunt fish exclusively, or the Marine Iguanas of the Galapagos *(Amblyrhynchus cristatus)* that eat marine plants. The Sea-snakes are marine-dwelling creatures (Hydropheidae); one livebearing species, Yellow-bellied, or Pelagic, Sea-snake, *Pelamis platurus*, has conquered the high seas of the Indian and Pacific Oceans. Such snakes are adapted to *thirius**, *Oodinium**, the instigator of Discus Disease* their aquatic way of life by having an oar-like tail flattened laterally, and nasal openings that can be sealed off. The Sea Turtles (Cheloniidae) and the Leather-back Turtles (Dermochelyidae) also only visit the shore to lay their eggs at certain nesting sites, eg along the Gulf of Mexico. They are adapted to their way of life by having a flat body shape, a reduced bony armour and limbs transformed into kinds of fin.

For a very long time, many kinds of reptile have provided man with food or raw materials, so much so that many species are now endangered. Examples of reptiles that are used in these ways include the meat of turtles, snakes, skinks, turtle eggs, the leathery skin of crocodiles etc. The poison of many poisonous snakes is important in medicine.

Resistance. This term is taken to mean all those factors which enable an organism to defend itself against damaging influences (Parasitism*, Temperature* etc). Such factors are rooted in the genetic make-up of the organism, and are of a mechanical or physiological nature. Certain structural features, such as the formation of cell walls, can in themselves often prevent a parasite from entering the organism (passive resistance). But there are also specific defence mechanisms that are only brought into play in response to a parasite or other damaging factor (active resistance). Fish have both a natural and acquired resistance (Immunity*) that is best helped by providing optimum living conditions in the aquarium so that diseases are prevented (Prophylaxis*). Because of their high rate of propagation, many parasites are predestined to form strains that are resistant to particular medicaments. The occurrence of such resistance, when preparations (Antibiotics*) that have been effective for years suddenly become ineffective, has been observed in *Ichthyophthirius**, *Oodinium**, the instigator of Discus disease*, and in many other strains of bacteria.

Respiration. Gaseous exchange of living beings, during the course of which oxygen is taken in and carbon dioxide is given out. In animals, there are different types of respiration depending on where the external gaseous exchange takes place – lungs, gills, tracheae, through the intestine or skin etc. During the process oxygen dissolved in the air or water passes over the broad surface of respiratory organs* or through the external surface, and as a result of falling concentrations the oxygen is taken into the bloodstream. As the blood circulates through the body it takes the oxygen with it. Particular pigments in the blood (eg haemoglobin) increase the receptive capacity for oxygen and carbon dioxide through temporary chemical combining. The body cells receive the oxygen from the blood and through biological oxidation* they obtain the necessary energy for all life's processes. During this gradual process which is regulated by enzymes*, the oxygen combines with the hydrogen in organic substances to make water. Derivatives of carbohydrates* and, in part, fats* and proteins* (which originally stemmed from the food) yield the hydrogen. The carbon dioxide that results from the breaking down of these combinations is passed through the blood and respiratory organs and out of the body. Plants either take their respiratory oxygen requirements from their surroundings or obtain it metabolically.

Respiratory Organs. Those specialised parts of an animal organism involved in gaseous exchange. From an evolutionary point of view they are formations on the upper surface of the body, such as the skin and the first and last segments of the digestive tract. In many animals such as sponges, coelenterates, many worms and small crustaceans (eg copepods) there are no well defined respiratory organs. Breathing is to a large extent carried out along the entire upper surface of the body. With the development of respiratory organs, gaseous exchange increases because the respiratory surface gets bigger. Moreover, efficiency increases often as a result of movement mechanisms which renew the external medium, and as a result of more intense contact with the body's circulating fluids. In essence there are 3 different types of respiratory organs: tracheae, lungs and gills.

– *Tracheae* are the typical respiratory system of the Tracheata (insects, centipedes and millipedes), as well as other arthropods such as scorpions and spiders. The tracheal system consists mostly of very fine tubes which begin at the upper surface of the body with an opening (stigma) that can be shut off. They form a delicate mesh round the organs of the body, which allows gaseous exchange with the tissues to take place through diffusion. Specialised forms are the tracheal sacs of the Chironomidae* or the tracheal gills (fig. 973) of numerous aquatic insect larvae (eg Mayflies or Libellae).

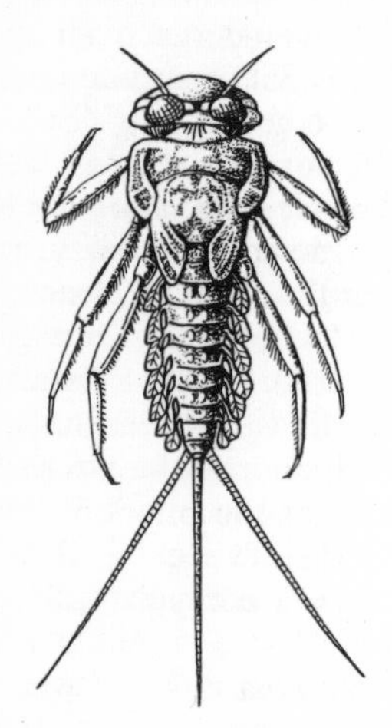

Fig. 973 Tracheal gills on the abdomen of a Mayfly larva

– *Lungs* are the respiratory organs of air-breathing vertebrates and man. They are arranged as a pair of projections from the anterior part of the intestine, down towards the stomach. Therefore the additional respiratory organs of the Lungfishes (Dipnoi*) can be regarded as true lungs. The lungs of some invertebrates, eg Euthyneura*, are merely analogous organs.

– *Gills* (fig. 974) are the respiratory organs of the fish, amphibian larvae and numerous invertebrates. Their basic function is gaseous exchange with water. Fish gills are similarly formations of the front part of the intestine. This can be seen clearly in primitive Chordata*. The tunicates (Tunicata*) and the *Acrania* have branchial baskets. Water swirls into the throat and back out again via numerous slits in the side of the upper part of the intestine. Food particles are retained and gaseous exchange takes place. Bony fish only have 5 gill slits

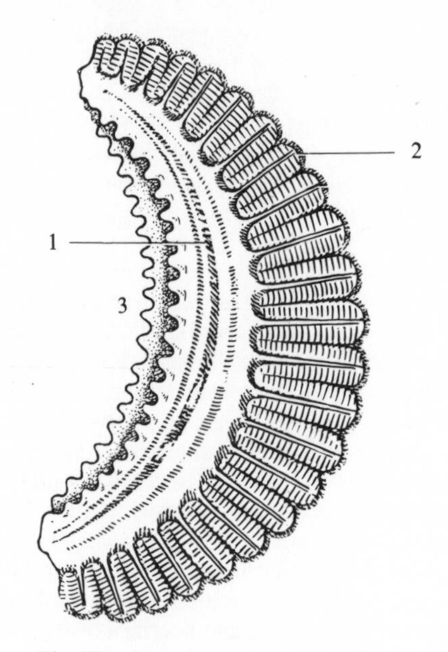

Fig. 974 Structure of a fish gill. 1 Gill arch skeleton 2 Gill membrane 3 Raker teeth

running down towards the belly, and they do not filter their food by this means. Above all, fish gills are used for breathing. The connecting tissue between the gill slits, the 4 gill arches, possess differently shaped comb-like rakers, which point inwards. These prevent food particles from entering. There are blood vessels and skeletal parts in the gill arches, on the outside of which lie the gills. The upper surface of the gills is enlarged by numerous gill plates which are arranged in 2 rows (hemibranchi) in bony fish. The plates in turn are divided into smaller lamellae. After water has been breathed in and has passed over the gills, it returns outside again via the openings in the upper body surface. In sharks and rays these are slits bounded by gill walls; in bony fish, the gill cavity is sealed off by a movable gill covering (operculum), which is also of great importance in respiratory movements. Respiration in bony fish is a suck in and press out action (fig. 975). When breathing in, the mouth and gill cavity are enlarged. Then the mouth is opened. A suction is created and the water streams in through the mouth, as during this phase the gill opening is sealed off by the operculum and a fold of skin known as the operculum membrane. Once the mouth is shut again, 2 folds of skin are laid across just inside the mouth. The mouth cavity is constricted by a special muscular system, producing a pressure which sends the water between the gills. Meanwhile the operculum is lifted away and the water passes out of the body. After the gills, the skin is most important in a fish's respiration. This is particularly apparent in fish that can live for lengthy periods out of the water in damp places, eg in the Mudskipper *Periophthalmus,* in the Eel *Anguilla* or in the Climbing Perch *Anabas.*

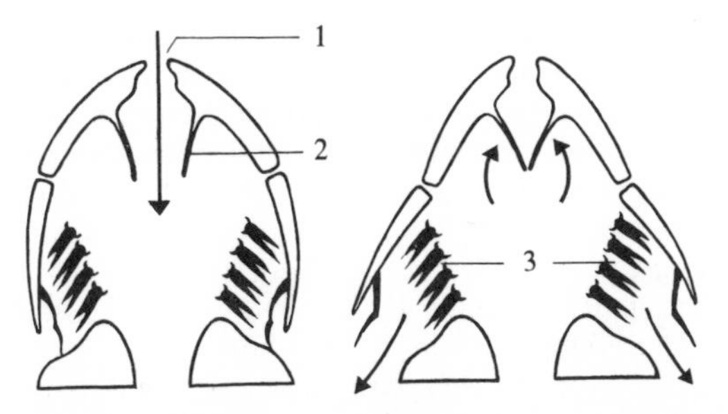

Fig. 975 Respiration of a bony fish. Left, breathing in; right, breathing out. 1 Opening to mouth 2 Palatal folds 3 Gill slits

Many fish possess accessory respiratory organs which chiefly help them to breathe in air. Such organs enable fish to live in waters with a low oxygen content – too low for breathing through gills. (Oxygen levels may fall because of water stagnation, the inability of the sun's rays to penetrate the water or because of severe pollution.) It is with these organs that the Lungfish *Lepidosiren* and *Protopterus,* the Reedfish *Calamoichthys* and the Climbing Perch manage to survive dry periods buried in the mud. The mouth cavity itself can be the simplest additional respiratory organ of all; it may be enlarged by bag-like extensions (eg *Electrophorus, Channa, Periophthalmus*). Often, the intestine, particularly the hind gut, takes on the additional function of breathing. Air from the atmosphere is swallowed and its oxygen content removed before being given off into the water again through the anus (eg in the Weatherfish *Misgurnus* and in the Catfishes *Callichthys* and *Hoplosternum*). Carbon dioxide is largely released through the gills. The labyrinth organ (fig. 976) found in the Anabantoidei is an extension of the gill cavity into which projects a branched outgrowth of the fourth gill arch. The labyrinth may stretch beneath the vertebral

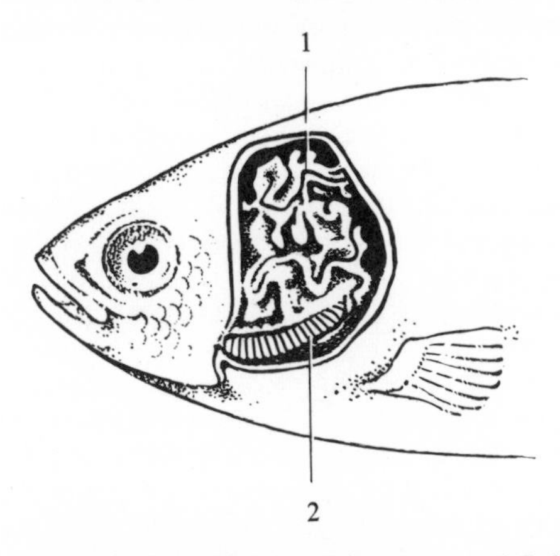

Fig. 976 Accessory breathing organs. 1 Labyrinth organ of the Anabantidae 2 Gill arches

column and far into the body cavity eg in *Betta.* At intervals of a few minutes the labyrinth is filled with fresh air breathed in at the water surface. Oxygen in the air is released into the labyrinth's mucus membrane which contains many blood vessels. Similar organs are possessed by the Catfishes (Clariidae), the Pikeheads *(Luciocephalus)* and the African Characin genus *Citharinus.* The swimbladder* may also serve a function in the breathing of air, when combined with the intestine, eg, in *Amia, Erythrinus, Gymnarchus* and *Umbra.* In these cases the wall of the swimbladder is particularly well supplied with blood vessels and its inner surface is often enlarged by folds or a sponge-like structure. Additional breathing through lungs is fairly rare in fish. Apart from the Lungfishes (Dipnoi), whose lungs are chambered or are structured like a sponge, *Polypterus* and *Calamoichthys* still possess relatively simple lung-like swimbladders. The juveniles of many air-breathing species have larval respiratory organs. In addition to the usual larval system of breathing through the blood vessels of the yolk-sac and the fins, external feathery gills provide extra oxygen *(Lepidosiren, Protopterus, Polypterus, Calamoichthys).*

Respiratory Trees *see* Holothuroidea

Reticulated Loach *(Botia lohachata) see Botia*

Reticulated Rabbitfish *(Siganus vermiculatus) see* Acanthuroidei

Retina *see* Sight, Organs of

rH value. The type of measurement given for redox potential*, ie the capacity for reduction or oxidation in a solution. By analogy with the definition of the pH value*, the rH value is likewise a decadic, negative logarithm – in fact it is the hydrogen loading with which a platinum electrode must be charged, in order to bring about an analogous reducing effect. The rH-scale goes from nil (the strongest reducing effect) to 42 (the strongest oxidising effect). The important area of the scale as far as aquarium-keeping is concerned is between rH 27 and rH 32 – ie with a slight to moderate oxidising capacity. Just as the redox potential is important in aquarium-keeping (eg for plants to grow at their best), it is just as difficult to determine exactly what the rH value is. A precision millivoltmeter with a very high internal resistance to the shutting down of electrolysing processes is needed, together with special electrodes. Usually, measurements are done with a platinum electrode in relation to a calomel electrode. As the rH value is very much dependent on the pH value, at the same time the rH is being calculated, it must also be possible to make an exact determination of the pH. It is not usually worthwhile for an individual to acquire all this equipment, so such calculations will normally only be possible in laboratories.

Rhabdaethiops, Sub-G, *see Neolebias*

Rhamphichthyidae, F, *see* Gymnotoidei

Rhamphocottidae, F, name not in current usage, *see* Cottoidei

Rhenanida, Sub-Cl, *see* Aphetohyoidea

Rheotaxis. An active change of site caused by the flow of water. For example, young eels (*see* Anguilliformes)

always swim against the current. Rheotaxis is a widespread phenomenon among animals that live in flowing waters, as it is by such means that they can oppose the tendency to drift with the current.

Rhinecanthus SWAINSON, 1835. G of the Balistidae*, Sub-F Balistinae. From the tropical Indo-Pacific and the Red Sea. Life-style and form are typical of the family. In this G there is no furrow in front of the eyes, the cheeks are fully covered in scale, the 3rd D spine is small, the caudal peduncle short and crossed by 2–4 rows of small spines. The colouring of these fish is striking. In the aquarium, feed them on animal food of all kinds. They should survive very well, and can even be put in the same tank as smaller or peaceful fish or invertebrates. The following species is imported most often.

– *R. aculeatus* (LINNAEUS, 1758); Humuhumu-nuku-nuku-a-puaa, Picasso Trigger, White-barred Trigger-fish. Tropical Indo-Pacific. Length to 30 cm. A whitish colour with stripes on the head edged in yellow, or sometimes blue; on the rear part of the body there are blackish markings (fig. 977).

Fig. 977 *Rhinecanthus aculeatus*

Rhinichthys AGASSIZ, 1849. G of the Cyprinidae*, Sub-F Cyprininae. There are about 12 well-known species and numerous sub-species distributed in N America. They are small to medium-sized fish that live in clear streams, rivers and lakes. The body is elongated, slim and only slightly flattened laterally. The scales are very small. The ventral profile is more or less straight. The mouth is slightly inferior and is characterised by a slightly elongated snout and small barbels. The D is inserted behind the V. LL complete. Provide a largish tank for *Rhinichthys* with plenty of room for swimming, a clean sandy substrate, and good ventilation. At regular intervals, the tank must be topped up with fresh water. Temperature 15–20 °C, 4–8 ° in winter. They are omnivores that accept every kind of live and dried food as well as plant matter. Also suitable fish for keeping in small garden ponds. After a cool over-wintering, it is possible to breed them in the aquarium. For this, build up the substrate to one side of the tank so that you create a shallow water zone. The mating pair push forwards into the shallow water during the spawning act, the male entwining the female with his caudal peduncle. The breeding season lasts many weeks. Breeding is possible in ponds that have shallow edges. The fry are easy to rear. The individual species are difficult to tell apart.

– *R. atratulus atratulus* (HERMANN, 1804); American Black-nosed Dace. From eastern N America, S Canada, the region around the Great Lakes to Alabama. Length to 12 cm. Dorsal surface olive to blackish, flanks a gleaming silver with a blue shimmer, ventral surface white. A broad, black, longitudinal band stretches the entire length of the body; it is bounded either side by a gold-coloured stripe. Fins yellowish, V and An reddish. At spawning time, the ventral side and fins are orange-coloured or red (fig. 978).

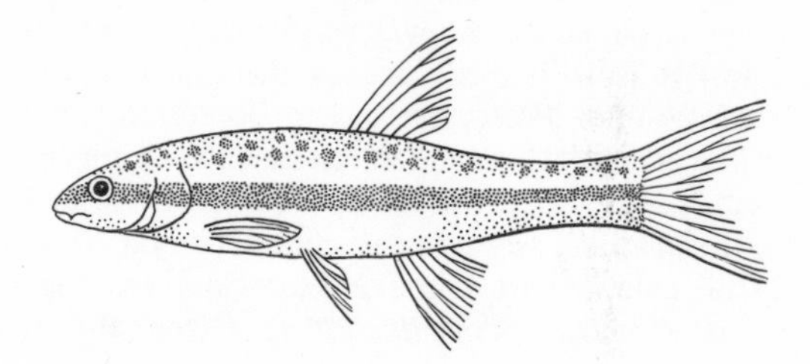

Fig. 978 *Rhinichthys atratulus*

Rhiniodontidae; Whale Sharks. F of the Chondrichthyes*, O Orectolobiformes. There is only one species, but all the same it is a very interesting one *(Rhiniodon typus)* (fig. 979). It has a fairly bulky, cylindrical body, up to 18 m long. The head is truncated, conical, and the mouth is at the front (unlike any other kinds of shark). The eyes are small, the Ps very large and like a sickle. *Rhiniodon* is dark grey or reddish and is covered with a lot of white or yellow dots. Although it is a giant fish, it is in fact a sluggish and completely harmless plankton-

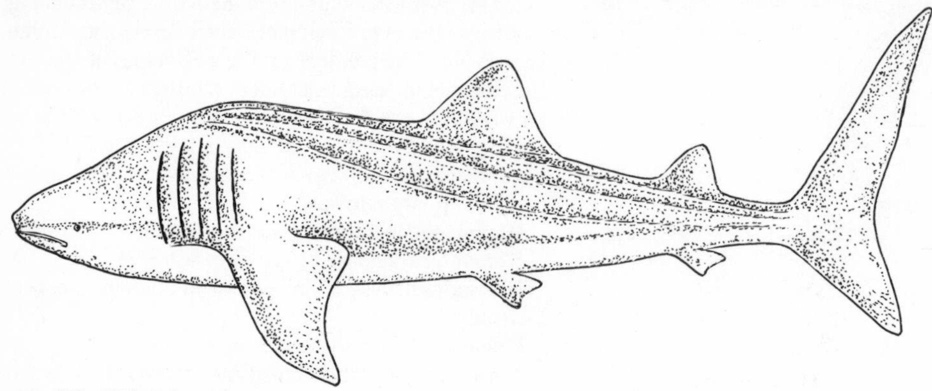

Fig. 979 *Rhiniodon typus*

eater, that filters crustaceans, small fish, squids or other animals from the water with its gill rakers. When doing this, *Rhiniodon* sometimes stands vertical in the water, and occasionally it suns itself at the surface. It has only been confirmed recently that *Rhiniodon* is oviparous and that its eggs are laid in egg-capsules about 30 cm long and 12 cm wide. The young sharks that hatch out are about 35 cm long. The Rhiniodontidae are found in all tropical seas and tales about these monsters are as proliferous as they are highly imaginative.

Rhinobatidae; Guitarfishes (fig. 980). F of the Chondrichthyes*, O Rajiformes. These Rays look most like the Angelsharks (Squatinidae) with their elongated body, flattened on the ventral side. The Guitarfishes also have their Ps joined along a good stretch of the body, but they are still much smaller than those of Rays belonging to other Fs. The effect of this is not so much a disc-shape as a triangle (produced in conjunction with the long, pointed snout), and it is clearly distinct from the Vs. The caudal peduncle is very powerful and has 2 well developed Ds on it. As with all Rays, the gill slits are on the underside. The dentition is made up of numerous rows (40–70) of very small, pointed teeth. It is possible that all the species give birth to live young. Guitarfishes occur in all tropical and subtropical seas

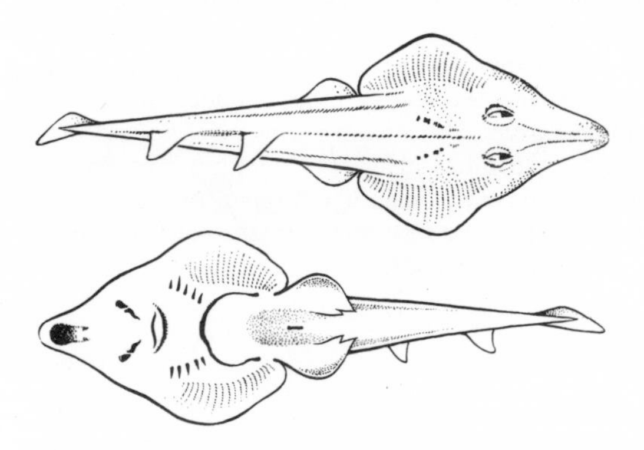

Fig. 980 Rhinobatidae-type. Upper and lower sides

where they are found primarily in the shallow water regions at the foot of the continental blocks. They swim over the substrate, often in small groups, constantly looking for food. Their prey consists mainly of crustaceans, as well as bottom-dwelling fish and molluscs. Most species remain quite small. Nevertheless, the largest one, the Shovel-nose or Sand Shark *(Rhynchobatus djiddensis)*, can reach a length of 3 m. It comes from the Indo-Pacific. However, it does not belong to the Rhinobatidae, but to one of the 2 other Guitarfish Fs. The Common Guitarfish or Shovel-nosed Ray *(Rhinobatos rhinobatos)* lives in the Mediterranean as well as the E Atlantic. Guitarfishes have little commercial value. A good leather is produced in some places from the skin, and the meat of the Shovel-nose is eaten in parts of India. Guitarfishes can be fished quite easily with a rod (the fish do not offer much resistance). No species is dangerous to man.

Rhinobatos rhinobatos *see* Rhinobatidae

Rhinomuraena Garman, 1888; Ribbon Eel. G of the Muraenidae, O Anguilliformes*. There are 2 species from the coral reefs of the tropical Indo-Pacific. They are predatory fish, over 1 m long, and active at night. Their body is very slender, very flat laterally, and with a pointed snout. A characteristic feature are the leaf-like outgrowths on the front nasal openings. Rhinomuraenas are not easy to care for, as they are sensitive to competition for food and often difficult to get used to food at all. To start with, offer them live shrimps or gobies, and later on they should accept bivalve meat as well. In recent years, both the Blue Ribbon Eel (*R. amboinensis* Barbour, 1908) as well as the Black Ribbon Eel (*R. quaesita* Garman, 1889) (fig. 981) have been imported.

Fig. 981 *Rhinomuraena quaesita*

Rhipidogorgia, G, *see* Gorgonaria

Rhithral, also rhithron. The region of a flowing stretch of water that has the character of a mountain stream – ie the water flows fast, is cool and is rich in oxygen. This kind of flowing water type is found all over the world, but is particularly common in the mountains and in regions near the poles. The ecological factors in this kind of habitat are very pronounced, and as a result the shape and life-style of the animals that live there are astonishingly similar, although they come from various parts of the world. The rhithral used to be called the Salmon-type region, because of the main type of fish that are found there – Salmons and Graylings. But because these species sometimes have a very restricted distribution, the term was by and large given up. So today the upper Salmon-region corresponds to the epirhithral, the lower Salmon-region to the metarhithral and the Grayling-region to the hyporhithral. Compare crenal and potamal.

Rhithron *see* Rhithral

Rhizocephala. O of the Barnacles (Cirripedia*), internal parasites mainly belonging to the marine ten-footed crustaceans (Decapoda*). The Rhizocephala are so altered to suit their parasitical way of life that their crustacean nature was only discovered first of all from the larva. The larva is a typical Cirripedia-nauplius. The cypris larva develops from it which attaches itself firmly to a host. The parasite bores through and forms a fine web round the internal organs of the host crustacean, through which it can draw nutrients. After some time, the parasite breaks through the host's epidermis where it forms a brood sac that contains the gonads. The host is well able to survive having the parasite – for example, *Carcinides** and its parasite *Sacculina carcini* THOMPSON. In this case, the brood sac appears beneath the Crab's abdomen. Hermit Crabs are often attacked by *Peltogaster paguri* RATHKE, 1843.

Rhizome. A rhizome or root-stock is a fairly long, thickened piece of shoot with compressed internodes in which reserve substances are stored. Rhizomes occur commonly among strong perennial plants (such as the larger kinds of *Echinodorus* species) that do not necessarily have to experience a resting period. The rhizome extends into a point where it is expanded by new sections, and it gradually dies off from behind. The individual annual sections remain functioning for several years.

Rhizopoda; Rhizopods (fig. 982). Ph of single-celled animals (Protozoa*) descended from the Flagellates (Flagellata*) and characterised by having plasma feet (pseudopodia). Most types are microscopically small, the largest ones being just visible to the naked eye. Well-known Rhizopods include the Amoebae, with or without shells, that can be found in still inland waters among Water-lily leaves, reed stems and in floating Sphagnum-moss mats. Other kinds, such as the Heliozoa, are free-swimming and live in the clear water of forest ponds; yet others, like the Radiolaria* and Foraminifera*, are typical inhabitants of warm seas and sometimes occur in very large numbers of individuals as well as species and forms.

The An is inserted at the same level as the D. Within the distribution areas of the individual species, you will always find Bivalves (Bivalvia*) of the G *Unio** or *Anodonta**, as the fish are reliant on them for their special method of reproduction. During the spawning season, the female develops a very elongate laying-tube, with the aid of which she lays the eggs in the gill cavity of a Bivalve chosen by the male. He then deposits his sperm freely into the water, directly on top of the Bivalve; as the Bivalve draws in some respiratory water, some sperm goes in with it and the current directs it to the egg-cells. 4–5 weeks later, the Bivalve ejects the young fish (which are already capable of swimming) along with the outflow of respiratory water. *Rhodeus* species are peaceful and can be kept as a shoal in the same tank as other cold water fish. They thrive best if the tanks are well planted and the temperature is up to 22 °C. They are omnivores, that like small crustaceans and Midge larvae most of all, as well as plant food. If allowed a cool overwintering, the interesting method of reproduction can also be observed in the aquarium. For this to happen some large *Unio* or *Anodonta* species must be placed in the tank too. After leaving the Bivalves, the young fish should be reared on the finest of micro-organisms. There are about 7 species and sub-species in Asia, and 1 species in Europe.

– *R. sericeus amarus* BLOCH, 1782; Bitterling. Central and E Europe, Asia Minor. Length to 9 cm. A very attractive species, highly recommended for the cold water aquarium. Dorsal surface grey, flanks and ventral surface a gleaming silver colour with a blue shimmer. A gleaming green longitudinal band, beginning at the level of the D, extends as far as the caudal peduncle. Fins yellowish to reddish. At spawning time the male is

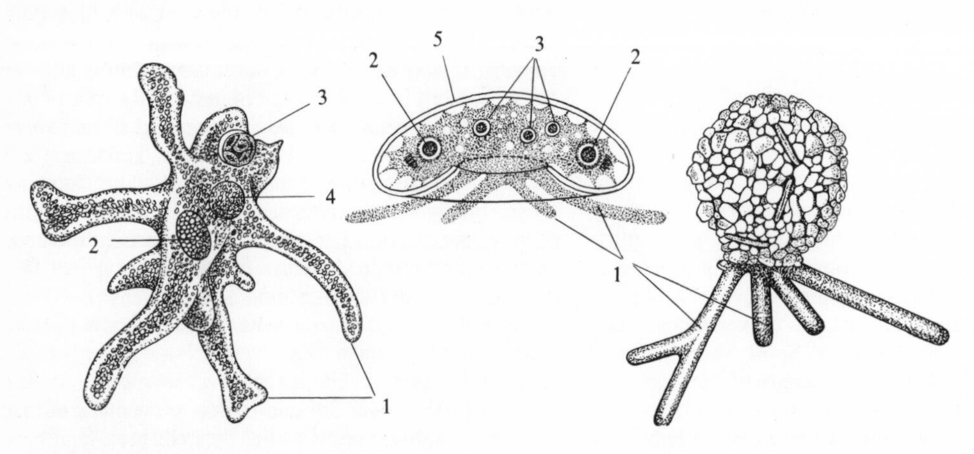

Fig. 982 Rhizopoda. Left, G *Amoeba*; Middle, G *Arcella*; Right, G *Difflugia*. 1 Pseudopodia 2 Cell nucleus 3 Food vacuole 4 Pulsating vacuole 5 Shell

Rhizopods, Ph, *see* Rhizopoda

Rhodeinae, Sub-F, *see* Cyprinidae

Rhodeus AGASSIZ, 1835. G of the Cyprinidae*, Sub-F Rhodeinae. From Central and E Europe, E Asia. These are small fish that live in loosely-knit shoals in fairly calm waters with vegetation and a sandy or muddy substrate. In shape, they are moderately elongated with equally arched ventral and dorsal profiles, and flattened laterally. Caudal peduncle short. The mouth is terminal or inferior; it is small and has no barbels. LL complete.

coloured thus: dorsal surface olive-green, flanks a shimmering blue-green to violet, throat and ventral surface red. The D and An are a strong red colour with a black edge. There are spawning tubercles between the upper lip and the eyes. The female is pale yellow with a laying-tube 45–50 mm long.

Rhodophyta; Red Algae. D of the Algae with 4,000 species. These Algae are in fact red, red-brown or violet in colour with chlorophyll a and sometimes d, which are covered with red phycoerythrin; the chromatophores are lenticular or lobed. There are no cells that move along through flagella. The vast majority of the Rhodophyta are species of the marine benthos*. They prefer fairly warm seas and still appear at depths of 200 m. A few live in fresh water or on firm land. Single-celled Red Algae are decidedly rare; branching threads (forked or feathery) are more common. *Compsopogon*, a species that is occasionally kept in aquaria, belongs in this category, for example. Larger varieties of form have ribbon-shaped or flat-lobed thalli. The most complicated forms stick to the substrate with basal discs and their thalli are divided into shoot- and leaf-like sections, although they are not made up of differentiated tissues. In some species, certain parts of the thallus drop off in the autumn, just like the foliage of summer-green trees. An example of this is *Delesseria**. In many Rhodophyta, the cell walls become calcareous. Fossil forms of this type were important as builders of rock formations. Various polysaccharides are obtained from the cell walls of some species, and these are used for pharmaceutical and technical purposes. One species is cultivated on plantations in E Asia as a food. Ga that are important in marine aquarium-keeping include *Ceramium**, *Furcellaria** and *Peyssonellia**.

Rhodsiinae, Sub-F, *see* Characoidei

Rhyacophila PICTET, 1834 (fig. 983). G of the Caddis Flies (Trichoptera*). There are more than 70 species in Europe alone. The body measures about 2.5 cm in length. *Rhyacophila* species are characteristic inhabitants of fast-flowing waters, in which the larvae wander around freely. These kinds of Caddis Flies do not form caddis cases or trap nets. In contrast to *Hydropsyche**, the larvae have only one horny plate on the anterior part of the thorax. They make suitable food animals and observation specimens in the aquarium.

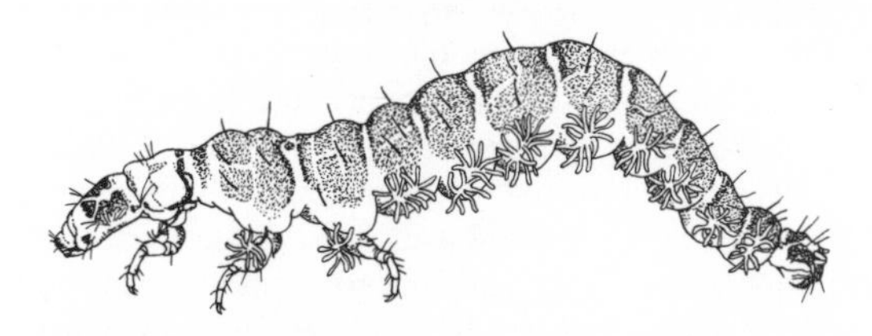

Fig. 983 *Rhyacophila*. Larva

Rhynchobdellidae, F, name not in current usage, *see* Mastacembeloidei

Rhynchocephala, Ph, *see* Nemertinea

Ribbon Worms, Ph, *see* Nemertinea

Ribbonbearers (Eutaeniophoridae, F) *see* Cetomimiformes

Ribbonfish *(Equetus lanceolatus) see Equetus*

Ribbonfishes (Regalescidae, F) *see* Lampridiformes

Ribbonfishes (Trachipteridae, F) *see* Lampridiformes

Riccia LINNÉ, 1753; Thallose Liverwort. G of the Bryophyta*, comprising 20 species, distributed almost throughout the world. *Riccia* has an undifferentiated, branching vegetative body. Reproduction through spores. The various species of *Riccia* are mainly land-dwelling Liverworts that live in a big variety of habitats. Some are able to tolerate flooding for some or all of the time. Thus, the species described below is popular and

undemanding in the aquarium; it floats just below the surface. It provides a good spawning substrate for fish that lay their eggs at the surface, especially Egg-laying Tooth-carps.

– *R. fluitans* LINNÉ; Floating Liverwort. Distributed almost throughout the world.

Ricciocarpus, G, *see* Bryophyta

Rice, G, *see Oryza*

Rice Eel *(Monopterus albus) see* Synbranchiformes

Ricefish *(Chologaster cornutus) see* Percopsiformes

Ricefish *(Oryzias latipes) see Oryzias*

Ricefishes, G, *see Oryzias*

Riddle's Pristella *(Pristella riddlei) see Pristella*

Ring-tailed Pike Cichlid *(Crenicichla saxatilis) see Crenicichla*

Rio de la Plata (River Plate). This river is 300 km long and flows through subtropical S America. It comes into being through the confluence of the Rio Uruguay and the Rio Paraná, and discharges into the S Atlantic. The Rio Paraná comes from tropical Brazil, together with its tributaries the Rio Paranaiba and the Rio Uruguya; these rivers are the homes of many well-known aquarium-fish, such as the Hatchetfish, *Gasteropelecus* sternicla* and the Red-eyed Tetra, *Moenkhausia* sanctaefilomenae*.

Rio Madeira *see* Amazon

Rio Negro *see* Amazon

Rio Xingu *see* Amazon

Ritualised Behaviour. Term used in ethology. Types of behaviour that have lost their original significance in the course of phylogenetic evolution and that usually belonged to another sphere of function* originally. Nowadays, compared to the original form, the ritualised mode of behaviour is very exaggerated, and often occurs as a part of courtship, mating and brood care behaviour. In addition, such ritualised behaviour is often associated with alterations to the stimulus threshold of the innate release mechanism* (IRM) to which it belongs. Very often, well-established actions become ritualised, such as locomotion, grooming, the acquisition of food, the regulation of warmth etc. When this happens, the instinctive movement concerned is released from its controlling central-nervous mechanism and is often coordinated afresh. This is particularly common if the movement acquires a new function-value as a signal (Expressive Behaviour*), and so, for this new purpose, it must differ from its original kinetic rhythm. Intention movements* and displacement activities* are also often ritualised. An example might be the way in which two fish with neighbouring territories swim alternately up to the boundary point. These movements of intent originally suggested attack and flight, but through ritualisation the meaning has been changed, and they are now interpreted as threatening gestures by the opponents.

Rivanol®. A yellow, acridine dyestuff that is used in an aqueous solution to disinfect wounds. In aquarium-keeping, Rivanol® can be used as a prolonged bath (Bathing Treatments*) in a concentration of 0.1 g to

40–50 litres of water, in the treatment of skin parasites (Skin Diseases*, Cloudiness of the Skin*). A more concentrated aqueous solution (100 mg dissolved in 100 ml of hot water and allowed to cool) is used as a preventative measure to disinfect skin wounds and to combat stubborn parasites, especially Fungi (Mycoses*). For this treatment, the affected areas are daubed with the solution, out of water, and then the fish are placed back in the tank immediately. Rivanol® can be removed from the water with activated charcoal*. It is a mitotic poison (Cell Division*) and if too much is applied it can produce sterility* in fish.

River Snails, G, *see Viviparus*

Riverine. Term used to describe organisms that live in flowing waters (ie in the various zones of flowing waters – crenal*, rhithral*, potamal* – or in the tidal zone of lakes*). Riverine forms have need of a high oxygen content. They are also more sensitive to temperature differences in contrast to lacustrine* organisms. Their body structure has also been adapted to offset the effect of being swept away with the current (eg sucking discs as in the Blepharoceridae*).

Rivulinae, Sub-F, *see* Cyprinodontidae

Rivulus POEY, 1858. G of the Cyprinodontidae*. There are precisely 50 species widely distributed in S and Central America, reaching north to around Florida and also found in Cuba. Most of the members of the G live in these places in plant-rich flowing waters, namely the streams of the lowlands and the mountains. Very elongate fish, between 3.5 and 10 cm long. The body is hardly flattened at all laterally. The D is usually smaller than the An and located well towards the tail region; both these fins are round as a rule, in rare cases a bit pointed. C fan-shaped. Mouth superior. The markings and patterns are very simple, usually in the form of longitudinal rows of dots. Other species have a dark longitudinal band or have an irregular marbled effect (eg *R. ocellatus*); narrow transverse stripes also occur now and again *(R. dorni)*. A distinction that *Rivulus* has is the so-called 'Rivulus fleck', a dark fleck, outlined in a paler colour, and located near the upper base of the C; this appears in the juveniles, and is retained for life in the females of many species, but disappears in most males (exception, eg *R. ocellatus*). *Rivulus* species also have the very unusual habit of remaining out of the water for a while; they like to sun themselves on floating leaves and can even cross short stretches of land by jumping. So, if you decide to keep *Rivulus* in the aquarium, it is essential that an excellent lid is provided, as they will use the smallest of cracks to escape. Specimens should be kept in shallow, well-planted tanks which can receive a lot of sunshine or which are well illuminated. The chemistry of the water is insignificant, provided it is changed regularly. As a rule, a temperature of 19–22 °C will suffice. With specimens from tropical countries 23–26 °C. Feeding is no problem, but plenty of variety must be provided. *Rivulus* species are typical attached spawners (compare Cyprinodontidae); there are no annual species in this G. The spawn is laid in clumps of plants, the eggs develop continuously and take about 8–14 days to reach hatching-out point, depending on water temperature. The spawning period itself can last for several weeks. To rationalise the breeding technique, the spawning substrate, together with the attached eggs, can be transferred every few days to a separate breeding tank.

– *R. agilae* HOEDEMAN, 1954. Surinam. Length to 4.5 cm. Yellow-brown to orange-coloured with a metallic blue shimmer and longitudinal rows of indistinct dark flecks. Fins sometimes with orange patterns.

– *R. cylindraceus* POEY, 1861; Green Rivulus, Brown Rivulus, Cuban Rivulus. Found in the mountain streams of Cuba. Length to 5.5 cm. Metallic green, underside reddish. There is a broad, dark band between the snout and the C, with irregular red flecks below.

– *R. harti* (BOULENGER, 1890); Hart's Rivulus, Giant Rivulus. E Colombia, Venezuela and offshore islands. Length to 10 cm. Ground colour brownish-green with longitudinal rows of red dots on top. Unpaired fins green, some with orange borders (fig. 984).

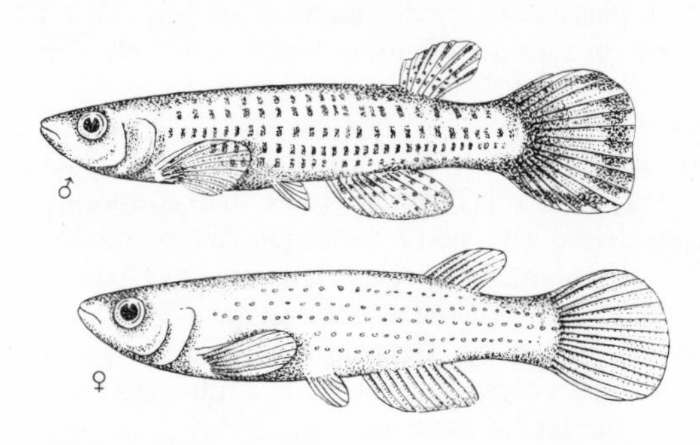

Fig. 984 *Rivulus harti*

– *R. milesi* FOWLER, 1941; Fire-tailed Rivulus. Colombia. Length to 6 cm. Brownish-green with irregular red dots. Operculum a gleaming blue colour. Unpaired fins orange-yellow on the outside, especially the C.

– *R. urophthalmus* GÜNTHER, 1866; Golden Rivulus, Green Rivulus. Peru and W Brazil, and also Guyana. Length to 6 cm. Upperside a metallic blue, underside with violet tones. Body and fins covered with fine red dots. The An has a yellow edge, and the C is brownish.

Roach *(Rutilus rutilus) see Rutilus*

Robber-Crab *(Birgus latro) see* Paguridea

Rock Bass *(Ambloplites rupestris) see Ambloplites*

Rock Beauty *(Holacanthus tricolor) see Holacanthus*

Rock Cods, G, *see Cephalopholis*

Rock Prawn *(Palaemon squilla) see Palaemon*

Rocket Killie *(Epiplatys annulatus) see Epiplatys*

Rocks, Stones *see* Decorative Material

Rock-sucker *(Chorisochismus dentex) see* Gobiesociformes

Roeboides GÜNTHER, 1864. G of the Characidae*, Sub-F Characinae. Peaceful, American Characins from the lower levels of water. Length up to 10 cm. The species of this G have a typical way of holding their bodies at an angle, pointing downwards. They are closely related to the G *Charax**. The body is elongated and tall, very compressed laterally and glassily transparent. The D1 is short and is located roughly in the middle of the body; in the males it is pointed, and in the

females rounded. D2 present. The C is forked, the lower lobe is longer, and the An is very long. Head fairly short, eyes large, LL complete. Not very sensitive to fluctuations in temperature. Will eat all kinds of live food. *Roeboides* likes to have the sun shining into the tank occasionally. Very productive species; the spawning act takes place during a stormy driving session, amongst finely pinnate plants. The fry hatch out after about 24 hours and grow quickly; they are easy to rear on micro-organisms. *Roeboides* species were at one time very popular, but today they are rarely kept in the aquarium.
– *R. guatemalensis* (GÜNTHER, 1864); Guatemalan Glass Characin (fig. 985). Central America. Length to 10 cm. Flanks a transparent yellowish colour, becoming brownish-yellow towards the dorsal surface, and paler towards the ventral surface. There is a dark longitudinal band beneath the LL, which becomes paler the nearer it gets to the ventral surface; it is also bounded back and front by a dark blotch. Above it, there is a narrow band, faintly silver, often turquoise-blue when illuminated. On the flanks there are tiny gleaming dots. The fins are yellowish or colourless, and their tips are sometimes a weak red colour. The D1 in the males is drawn out into a point; in the females it is rounded. *R. microlepis* (RHEINHARDT, 1849), the Small-scaled Glass Characin, is very similarly built and coloured. It comes from the western Amazon Basin.

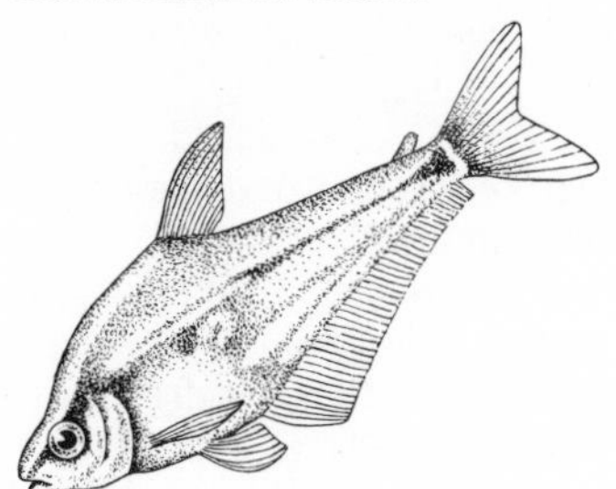

Fig. 985 *Roeboides guatemalensis*

Roker *(Raja clavata) see* Rajidae
Roloffia CLAUSEN, 1966. G of the Cyprinodontidae*. At present comprising exactly 20 species. Their centre of distribution is the tropical rain forests of W Africa (Guinea, Sierra Leone, Liberia and the Ivory Coast). They live in these places in small, shady riverlets mainly, in the cool part near the river's source; only a few forms penetrate savannah waters. *Roloffia* species are particularly closely related to the G *Aphyosemion**, which represents *Roloffia* in the more easterly parts of the rain forests and which lives under very similar conditions. The two Ga are scarcely distinguishable externally; they can be told apart from the arrangement of the sense depressions near the nose that are used to detect current, and from the number of chromosomes present. In *Roloffia*, the sense depressions nearest the front lie very close to one another, forming a triangle with the ones that follow; in *Aphyosemion*, these depressions form a quadrilateral shape. These differences of feature can be seen quite clearly under a magnifying glass. The chromosome number in *Roloffia* is 20–23 n (haploid amount), whereas in *Aphyosemion* it is 9–20 n. *Roloffia* species are elongated fish, only slightly flattened laterally. Length 4–9 cm. The mouth is superior, whereas the upper jaw (as with many other Cyprinodontids except for *Epiplatys** and *Aplocheilus**) is strikingly short. The D and An are built similarly – they are pointed and lie opposite each other; however, the D has less fin rays. Male *Roloffia* species have a C that is characterised by short tips at most (in *Aphyosemion*, it often has long tips). In the females, the C never has an elongated middle lobe. The Vs are noticeably small. The small *Roloffia* species live in very small watercourses; at times of drought, they sometimes seek out larger rivers as a protection against the water drying up. These forms are attached spawners and non-annual (compare Cyprinodontidae). In contrast, the large species *(R. occidentalis, R. toddi, R. monroviae)* are never found in rivers, always in stretches of water that do dry up for certain periods. Their life-cycle is highly adapted to these conditions – the old fish die off, and the eggs survive in the mud till the next rainy season, safe in their hard, watertight shells. These forms are bottom spawners, and annual, although not in such an extreme way as some South American Ga *(Cynolebias*, Rachovia*, Pterolebias*)*, whose eggs have to undergo an obligatory dry phase before they are able to hatch out. When keeping *Roloffia* in the aquarium, the natural conditions must be replicated as far as possible. The small, non-annual species require cool, oxygen-rich water, at a temperature of 20–23 °C. Under natural conditions, it is true, the robust annual species live in waters that are at times very warm up to 35 °C), but such a high temperature reduces the already short lifespan still further, and so should be avoided. An upper limit of 24 °C is perfectly adequate, even for breeding purposes and for rearing the young. Above all, a tank suitable for *Roloffia* must not be stood in too bright a spot; a dark substrate, thick plant vegetation in clumps, root-stumps and floating plants help subdue the light. These conditions must be achieved if *Roloffia* is to show its natural desire for swimming and only then will it go off in search of food. In nature too, *Roloffia* restricts its activity to around dusk. Many natural habitats where *Roloffia* lives contain very soft water, but they are also able to tolerate up to about 10 °dH, and they can even breed under such conditions. More important than the hardness and pH value* is a regular water preparation or change of water, in order to keep the levels of waste and excretory products low. Other than that, *Roloffia* and related Ga *(Aphyosemion, Nothobranchius*)* are very susceptible to fish tuberculosis* and *Oodinium**. If breeding, follow the advice given for the Ga *Aphyosemion* and *Nothobranchius*. The eggs of the small, attached-spawners should be transferred, together with the spawning substrate to shallow rearing dishes, into which they should be shaken. Of course, you can also use a pipette to remove them from the breeding tank. If the young fish do not hatch out their own accord after 2–3 weeks, try to get them to hatch by changing their surroundings – for example, by adding some fresh water, or by adding some dried food, or by placing some aquatic plants in the dishes and then illuminating them. The eggs of the

annual species of *Roloffia* should be kept for at least 4–6 weeks in water dishes or even in damp peat; only then should you try to get the embryos to hatch out. Sometimes it can take months for all the eggs of the annual species to reach the hatching stage. The young fish are capable of accepting food from the first day; to begin with, they should be fed on micro-organisms (nauplii*, Rotatoria*), and then the breeder should work towards as varied a diet as possible. In nature, the small species live mainly on food that drops onto the surface (tiny insects), the large species also tackle Cladocera*, insect larvae and other live food of appropriate size. The colour descriptions given for the individual species relate to the males. The females are much more plainly coloured, very similar to one another and usually much smaller. Doubt has recently been cast once again upon the genus.

– *R. bertholdi* (ROLOFF, 1965). Sierra Leone. Length to 5 cm. Irregular red dots on a metallic blue background. The unpaired fins are flecked in red at the bottom only; the outside edges are crossed by red bands and blue borders, and there is never any yellow marking. Usually, a non-annual.

– *R. geryi* (LAMBERT, 1958). Guinea and Sierra Leone. Length to 4.5 cm. A gleaming blue to green; the red fin dots are concentrated into a distinct longitudinal band. The C is gold-yellow above and below, bounded on the inside by red bands. Usually, a non-annual.

– *R. guineensis* (DAGET, 1954). Guinea, Sierra Leone, and Liberia. Length to 7 cm. Body a strong blue or green, almost black at spawning time. Red markings usually absent. The C has a very pale (white or yellowish) under-edge. A non-annual.

– *R. liberiensis* (BOULENGER, 1908); Calabar Lyretail. W Liberia, where it lives in still waters. Length to 6 cm. A strong blue-green colour; the red dots are united distinctly into sloping lines or into a zigzag band. The C has a yellow edge above and below. Sometimes annual (fig. 986).

Fig. 986 *Roloffia liberiensis*

– *R. occidentalis* (CLAUSEN, 1965); Golden Pheasant. Sierra Leone. Length to 9 cm. Body and fins orange-red or golden, with irregular dark-red dots and stripes. The D, An and C are developed in lovely colours, and are fringe-like. The C has red bands above and below. An annual species that used to be known as *Aphyosemion sjoestedti* (ARNOLD, 1911) in aquarium-keeping (fig. 987).

Fig. 987 *Roloffia occidentalis*

– *R. petersi* (SAUVAGE, 1882). The Ivory Coast, Ghana. Length to 5 cm. Bronze-coloured or brown with very fine gleaming golden or gleaming green dots. The C has a lower yellow edge. A non-annual.

– *R. roloffi* (AHL, 1938). Sierra Leone. Length to 5 cm. Similar in colouring to *R. bertholdi*, but greener. Sometimes an annual species.

– *R. toddi* (CLAUSEN, 1965). As opposed to *R. occidentalis*, this species has red reticulated markings and a horseshoe-shaped band in the C.

– *R. viridis* LADIGES and ROLOFF, 1973. Liberia. Length to 6 cm. Brownish with a pale green or dark green shimmer and very fine red dots. Fins usually transparent, more rarely with bands (as is typical of the genus).

Roloff's Barb *(Barbus roloffi) see Barbus*

Root (fig. 988). The root, like the stem axis* and leaf*, is a component part of the vegetative body (corm*) of Ferns (Pteridophyta*) and seed-plants (Spermatophyta*). It has the character of an axis, but it lacks the leaves found on the stem axis. As a rule, the root develops in the ground where it anchors the vegetative body and also serves to take in water and minerals; many roots also store reserve substances. Roots can, however, develop similarly in water or in air.

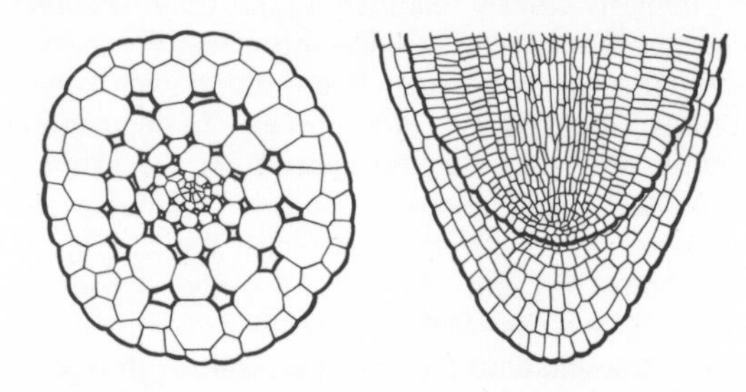

Fig. 988 Root anatomy. Left, transverse section; right, longitudinal section

The *radicle* develops from the root portion of the embryo inside the seed*. In some dicotyledonous plants (Magnoliatae*), the whole root system of the plant concerned develops from this radicle by branching. But many dicotyledonous plants form additional so-called *stem-born roots* from the stem axes. In these cases, the radicle may sooner or later die. With the monocotyledonous plants (Liliatae*) the radicle is always short-lived and is replaced early on by stem-born roots. Ferns only have stem-born roots. The ability to form stem-born roots is very important in aquarium-keeping.

If plants are propagated by cuttings, these root by means of stem-born roots. All new roots that form after the planting of rhizome or tuberous plants are also stem-born. In some plants, eg *Microsorium**, shoots may develop on roots, and then these will grow into adventitious plants*.

The root grows by means of a vegetative cone* which is protected by a special sheath called the root-cap. Just behind the vegetative cone (in earth roots), the root hairs form. These do not live long at all; they join up intimately with the surrounding substrate and further the intake of water and minerals. Many marsh and water plants do not have root hairs. The newest sections of a root are sealed on the outside by an epidermis known as the *rhizodermis*. This has very thin cell walls, no cuticle and no stomata. The rhizodermis is short-lived and is soon replaced by a secondary surface tissue that develops through cork being laid in the cell walls of the outside cortex layers. As in the stem axis, the central cylinder joins up with the cortex on the inside. The cortex is made up of ground tissue, and in marsh and aquatic plants it can also be traversed by air canals. Because of the corkiness of the cell walls, the innermost layer of cortex cells stands out clearly as a dividing line between it and the central cylinder. Through branching, lateral roots form from the outside layers of cells of the central cylinder, ie in contrast to the branching of the stem, the lateral roots form within the mother root and so have to break through the cortex. The conductive tissue of the root is contained in the central cylinder in a central vascular bundle*. Most of the roots in dicotyledonous plants have the capacity for a secondary growth in thickness*.

Some aquatic plants, such as *Ceratophyllum**, *Utricularia** and *Wolffia** have no roots.

Root Hair *see* Root

Root-cap *see* Vegetative Cone

Rootless Duckweed *(Wolffia arrhiza) see Wolffia*

Root-stock *see* Rhizome

Rorippa SCOPARIUS, 1760; Marsh Cress. G of the Brassicaceae*. 15 species. From temperate regions of the northern hemisphere. Small perennials with rosettes and elongated stem axes, on which the leaves are arranged alternately. The leaf blades are very variable, undivided, divided into pinnules or pinnate. The flowers are hermaphroditic and small. Double perianth. Calyx and corolla free-standing, 4-part. Stamens 6. Carpels 2, fused, superior. The fruit is a siliqua. The various species of the G colonise the bank of stretches of water and substrata that have very fluctuating levels of water. They can tolerate lengthy periods of flooding (at least some species). The following is an undemanding, very decorative aquarium plant.

– *R. aquatica* (EATON) PALMER (*Armoracia aquatica*, name not in current usage); Water Cress. Eastern N America. Propagation by means of leaves that have become separated, on the stalks of which adventitious plants will form (fig. 989).

Fig. 989 *Rorippa aquatica*

Rose-coloured Curimatopsis *(Curimatopsis saladensis) see Curimatopsis*

Rosettes *see* Leaf Arrangement

Rossmässler, Emil Adolf (1806–67). Naturalist and writer. Born in Leipzig. By 1825 he was studying theology in his home town, and later he was take up natural sciences there, too. In 1830 he became Professor at the Forest Academy in Tharandt, where he wrote his 3-volume work *Ikonographie der Land- und Süßwassermollusken Europas* ('Iconography of the Land and Freshwater Molluscs of Europe'). From 1850 onwards he was a private tutor in Leipzig, and it was here that he wrote various popular works that contributed to the spread of natural scientific knowledge. His article 'Der See im Glase' ('The Lake in the Glass') appeared in 1856 in the periodical *Die Gartenlaube*, and in 1857 his book *Das Süßwasseraquarium* ('The Freshwater Aquarium') appeared. It was the latter that was to make Rossmässler the father of aquarium-keeping in Germany. He undertook various lecture tours which helped to spread and popularise natural scientific knowledge. Together with A. E. BREHM he published *Tiere des Waldes* ('Animals of the Forest').

Rosy Barb *(Barbus conchonius) see Barbus*

Rosy Feather-star *(Antedon mediterranea) see Antedon*

Rotala LINNÉ, 1771; Rotala. G of the Lythraceae*, comprising 40 species. Found in the tropics and subtropics of the whole world. Small perennial and annual plants with elongated stem axes. Leaf arrangement decussate or in whorls. Leaves sedentary, simply built, rounded, pointed or double-pointed. Flowers hermaphroditic, small, polysymmetrical. Perianth consists of

circles in sixes or threes, doubled or corolla missing in part or completely. Stamens 6–1. Carpels 4–2, fused, superior. Capsule-fruit. The species of the G are small marsh plants that to a large extent can tolerate flooding well. In recent years, some species have been kept in the aquarium with varying degrees of success. They have proven sensitive to small durations of illumination, Propagation is possible through cuttings.
– *R. indica* (WILLDENOW) KOEHNE; Indian Rotala. S Asia, also as a weed in European rice-fields. Leaf arrangements decussate. Aquatic leaves lanceolate. Rarely kept in aquaria. *R. rotundifolia* is often offered commercially by this name.
– *R. macranda* KOEHNE: Broad-leaved Rotala. E Asia. Leaf arrangement decussate. Aquatic leaves ovate.
– *R. rotundifolia* (ROXBURGH) KOEHNE; Round-leaved Rotala. S Asia. Leaf arrangement decussate. Aquatic leaves very variable. The most commonly cultivated species (fig. 990).

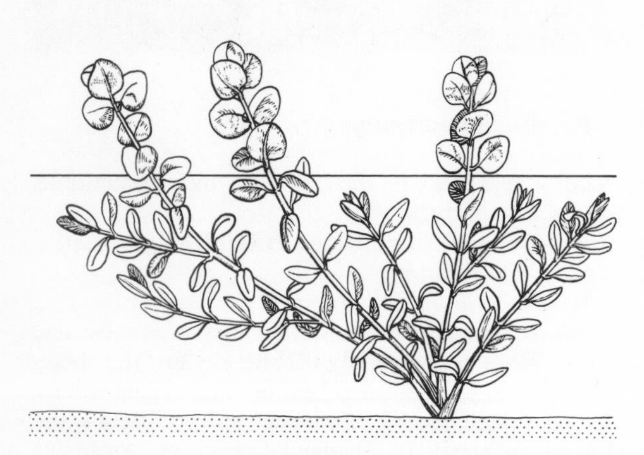

Fig. 990 *Rotala rotundifolia*

– *R.* sp.; Mayacca-like Rotala. Further India. Leaf arrangement whorled. Aquatic leaves narrowly linear.

Rotatoria; Rotifers (fig. 991). Cl of the Nemathelminthes*. There is a large number of species and a multitude of forms, mainly represented in freshwater systems, from the largest lake to the smallest pool of water. Rotifers are microscopically small and certainly not always worm-shaped, as one might deduce from their systematic position. Typical of most Rotatoria is a crown of cilia around the mouth area. From the physiological view-point, they are almost unbelievably resistant to unfavourable living conditions, such as drought, cold etc. Those forms that belong to the plankton* often have long, floating processes and are inclined to accumulate in vast numbers. Other rotifers creep along the bottom or among plants. Many plankton-dwelling Rotatoria are very important as a high-value powdered food (micro-organisms) for rearing aquarium-fishes. Many young fish in the wild also feed exclusively on Rotatoria.

Rotiferans, Cl, *see* Rotatoria

Round Scales *see* Scales

Round-leaved Rotala *(Rotala rotundifolia) see Rotala*

Round-tailed Guppy (*Poecilia reticulata,* standards) *see Poecilia*

Round-tailed Paradisefish *(Macropodus chinensis) see Macropodus*

Roundworms, Ph, *see* Nemathelminthes

Royal Blue Discus *(Symphysodon discus) see Symphysodon*

Royal Blue Triggerfish *(Odonus niger) see Odonus*

Royal Empress Angelfish *(Pygoplites diacanthus) see Pygoplites*

Royal Gramma *(Gramma loreto) see Gramma*

Ruby Shark *(Labeo erythrurus) see Labeo*

Rudd *(Scardinius erythrophthalmus) see Scardinius*

Rudderfish *(Naucrates ductor) see* Carangidae

Rudderfishes, F, *see* Kyphosidae

Rudimentary Organs. Organs that have regressed during the course of phylogenesis, eg the eyes of (blind) Cave-dwelling Fishes. They often undergo a normal embryonal development, nevertheless.

Ruffe *(Acerina cernua) see Acerina*

Ruffe-Flounder Region *see* Potamal

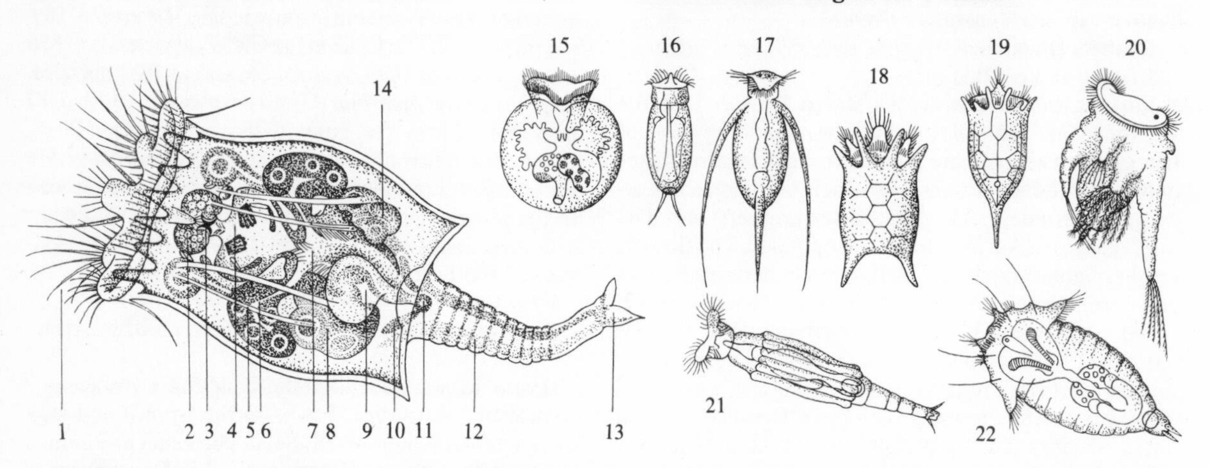

Fig. 991 Rotatoria. Anatomy of a rotifer (1–14) and various Ga of Rotatoria. 1 Ciliated corona 2 Brain 3 Eye 4 Masticatory stomach 5 Retractor muscle 6 Stomach glands 7 Mid-gut 8 Yolk-sac 9 Bladder 10 Foot gland 11 Anus 12 Foot 13 Toes 14 Ovary 15 *Pterodina* 16 *Cephalodella* 17 *Filina* 18 and 19 *Keratella* 20 *Pedalion* 21 *Philodina* 22 *Synchaeta*

Rummy-nosed Tetra *(Hemigrammus rhodostomus) see Hemigrammus*

Runner. A shoot that grows with its long internodes pressed against the substratum and which has normal foliage leaves on its nodes and roots that have normally developed from the shoot. Runners occur in many *Ludwigia* species, for example, during the emerse phase. Cataphyllary leaves* occur at the node. The runner is a means of asexual reproduction and distribution, eg in *Vallisneria, Echinodorus tenellus* or *Potamogeton gayi.*

Runula JORDAN and BOLLMAN, 1890; Sabre-toothed Blenny. G of the Blennidae, Sub-O Blennioidei*. Contains a few species, about 15 cm long, that live in tropical parts of the Indo-Pacific. They live on the bottom in tubes and they swim in a sinuous manner. *Runula* species are highly predatory and feed on pieces of skin and fins that they tear from other fish, as well as on polychaete tentacles. The fish are slender with long, bent sabre-teeth in the lower jaw. They are very fussy eaters and must on no account be put in the same tank as any other fish – either like species or unlike species. Because of a certain similarity with the Cleanerfishes *(Labroides*)*, the following species is imported now and again.

– *R. rhinorhynchus* (BLEEKER, 1852); Blue-lined Blenny, Sabre-toothed Blenny. From the Indo-Pacific. Length to 15 cm. Blue-black with a yellow longitudinal band.

Russian Catfish *(Leiocassis brashnikowi) see Leiocassis*

Rust Disease *see Oodinium*

Rutilus RAFINESQUE, 1820. G of the Cyprinidae*, Sub-F Leuciscinae. Europe, Central Asia. These are large shoaling fish, and together with some sub-species and local races they make up the commonest fish in still and slowly flowing waters in these areas. In the youthful phase, *Rutilus* keeps to the thickly vegetated bank regions; older fish keep to the deeper levels of the water.

Like anadromous fishes*, also found in brackish water. Body oblong or compact, very flat laterally; becomes taller with age. The mouth is small and either terminal or inferior; no barbels. LL complete, curving downwards. Scales large. The C is deeply inward-curving. In contrast to the G *Scardinius**, the D is inserted over the V, and the belly is rounded between the V and An. *Rutilus* often spawn in large shoals in shallow areas near the bank. The males have spawning tubercles. Their diet consists mainly of small crustaceans, worms, insect larvae, but also plant food. Only young specimens are suitable for the aquarium. Provide a large tank with a moderate amount of plants and sufficient room for swimming. Temperature 18–22 °C. An occasional top up with fresh water is advisable.

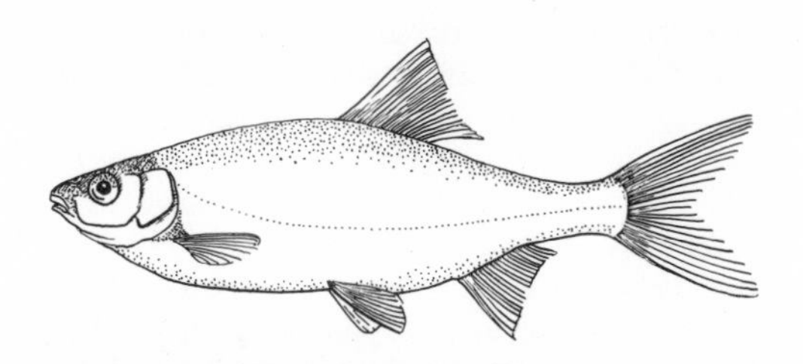

Fig. 992 *Rutilus rutilus*

– *R. rutilus* (LINNAEUS, 1758); Roach. Found in almost the whole of Europe, Central Asia. Length to 45 cm. Dorsal surface olive to blue-green, flanks a gleaming silver colour, ventral surface silver-white. The Ps and Vs are yellowish to red, the D and C are reddish-grey, and the An is red (fig. 992).

Ruvettus pretiosus *see* Scombroidei

Sabella LINNAEUS, 1767. G of the sessile Bristleworms (Polychaeta*), with a soft leathery tube and a two-part crown of tentacles that is not wound in a spiral (unlike *Spirographis**). *Sabella* feeds on suspended matter, and lives in the upper coastal waters of cool and warm seas. The tube is stuck in a vertical position between rocks or in the substrate; The method of catching food has been the subject of a detailed study in the species, *S. pavonina*: the tentacles surrounding the mouth have lateral pinnules on them (pinnulae). These in turn are covered with cilia which create a whirlpool into which the suspended particles are pulled. The cilia form a kind of deep groove along which the food particles are driven as if on a conveyor belt to the mouth. As the groove becomes gradually narrower towards the bottom, this means that the particles are graded according to size – coarse pieces are flushed out of the centre of the funnel, medium-sized ones are stored in special sacs for building the tube, and only the finest particles are transported right to the bottom of the groove and into the mouth. Various species are suitable for keeping in aquaria, eg:

– *S. pavonina* SAVIGNY; Peacock Worm. From the Mediterranean. Length to 25 cm. The crown of tentacles is ringed in brown.

Sabellastarte KRÖYER, 1856. G of the sessile Bristleworms (Polychaeta*) from the tropical Indo-Pacific with an almost circular crown of tentacles. They build grey, pliable tubes, 20–30 cm long, on firm substrata. The crown of tentacles can reach over 10 cm in diameter; it is brown-white, yellowish or red-violet in colour. *Sabellastarte* survives well in marine aquaria with a good algal growth.

Sabre-toothed Blennies, Ga, *see Aspidontus, Ranula*

Saccopharyngoidei, Sub-O, *see* Anguilliformes

Sacculina carcini *see* Rhizocephala

Sacramento-Smelt *(Spirinchus theleichthys) see* Salmonoidei

Saddle Anemonefish *(Amphiprion ephippium) see Amphiprion*

Saddle-spotted Anemonefish *(Amphiprion polymnus) see Amphiprion*

Sagartia GOSSE, 1855. G of small Sea-anemones (Actiniaria*) with stinging threads (acontia) which, when stimulated, can be extended through the mouth and the pores of the body column. There are sucking discs all the way up the column. In contrast to *Cereus**, the number of tentacles is below 200. The Anemones are only a few cm large and are extremely variable in

colour, even within the one species. They reproduce oviparously, sometimes viviparously, or through laceration. They present no problems when kept in the aquarium. A species frequently kept is the following.
– *S. troglodytes* (PRICE, 1847). North Sea to the Mediterranean, where it lives in cracks and crevices, also on rocks in the mud flats. Around 3 cm in size. White, reddish, brown or blackish. Livebearing.

Sagitta, G, *see* Chaetognatha

Sagittaria LINNÉ, 1753 (Lophotocarpus, name not in current usage); Arrowhead. G of the Alismataceae*, comprising 25 species. Mainly in America; originally only 3 species on the other continents, but numerous others have been introduced, however. Perennials of very varied sizes with rhizomes, tubers and runners. Leaves in basilar rosettes, the submerged ones linear and of varying widths; the emersed leaves are stalked, with very variable, lanceolate or sagittate blades. Flowers unisexual, usually monoecious, but also dioecious, polysymmetrical. Perianth doubled, free-standing: calyx and corolla 3-part, petals white. Stamens 9, up to many. Carpels numerous, free-standing, superior. The fruit is a collective nut.

The species of the G are marsh and aquatic plants. Among the large species, linear aquatic leaves occur only for a short time during the youthful phase. Among many medium-sized species, which often experience resting periods, such leaves occur at the start of each vegetation. Among the smallest species, the linear leaves are the dominant leaf form. In the monoecious plants, the female flowers in the inflorescences are located below, and above them are the male flowers (which usually occur in larger numbers). Starch-rich *Sagittaria* tubers are used as a food in some areas. Depending on origin, the various species have very different requirements for cultivation. For practical purposes, the species are divided into 3 groups:

1) Large marsh plants that can only be cultivated in botanical gardens.
– *S. lancifolia* LINNÉ. From Central and S America. A rhizome perennial with broadly lanceolate leaf blades.
– *S. montevidensis* CHAMISSO et SCHLECHTENDAL; Argentinian Arrowhead. America. A rhizome perennial with sagittate leaf blades.

2) Medium-sized and small species that are suitable for keeping in an aquarium.
– *S. graminea* MICHAUX; 'Chinese Arrowhead'. N America. Forms runners. Similar to *S. plathyphylla*, except that the broadly lanceolate aquatic leaves are usually darker green. Aerial leaves appear more rarely, and their leaf blades are narrowly lanceolate (fig. 993).
– *S. plathyphylla* (ENGELMANN) SMITH; Broad-leaved Arrowhead. N America; has also been introduced in places in Central America and SE Asia. Forms runners. Aquatic leaves broadly linear, pale green. Under long-day conditions will often form aerial leaves, whose blades are broadly lanceolate.
– *S. subulata* (LINNÉ) BUCHENAU. America. Forms runners. Aquatic leaves narrowly linear. More rarely floating leaves with lanceolate blades. Aerial leaves are like the aquatic leaves. The inflorescences stream at the surface. 3 varieties are usually distinguished: var. *subulata*, aquatic leaves up to 30 cm long and 6 mm broad—this is the commonest aquarium species; var. *gracillima*, aquatic leaves over 30 cm long, very narrow; var. *kurtziana*, aquatic leaves 7–14 mm wide.

Fig. 993 *Sagittaria graminea*

3) Medium-sized species with winter resting periods that are suitable for open-air ponds.
– *S. latifolia* WILLDENOW. N America. Has runners and overwintering tubers. Aerial leaves have very variable blades, lanceolate or sagittate.
– *S. sagittifolia* LINNÉ. Europe, W Asia. Has runners and over-wintering tubers. Aerial leaves sagittate.

Sailfin Characin *(Crenuchus spilurus) see Crenuchus*

Sailfin Leaf-fish *(Taenionotus triacanthus) see Taenionotus*

Sailfin Molly *(Poecilia latipinna) see Poecilia*

Sailfin Shiner *(Notropis hypselopterus) see Notropis*

Saithe *(Pollachius virens) see* Gadiformes

Salamander (Salamandridae, F) *see* Amphibia

Salamanders (Caudata, O) *see* Amphibia

Salamanders (collective term) *see* Amphibia

Salientia, O, *see* Amphibia

Salinity Damage can occur among water animals that are suddenly transferred to water with a very different salt concentration. In aquarium-keeping, whereas slight alterations in the total salt content of fresh water do not play a significant role in the non-physiological, osmotic burden of an organism, variations in the salt content (salinity) of sea water *do* have a decisive effect upon animals and plants kept there. Invertebrate marine animals consist in the main of so-called stenohaline organisms—ie organisms that can only compensate to a very limited extent for any changes in salinity levels. Only a few invertebrates are euryhaline and able to exist under very different salinity levels by means of various physiological mechanisms; they can in fact leave the sea and penetrate brackish waters (*Arenicola**, *Asterias**, *Mytilus**, *Carcinus** and others). Fish have an osmoregulation at their disposal. Here too, there is a distinction between stenohaline forms with a limited capacity for regulation and euryhaline groups, that can live under a great diversity of salinity levels. A large number of so-called freshwater fish also enter brackish water, eg *Perca**, *Gasterosteus**. Other species alternate between fresh and sea water (Migration*) at different periods of their lives. Under certain circumstances, such fish can tolerate being transferred directly from fresh water into sea water without suffering any damage, eg some Salmonidae*. As a general rule, however, marine fish kept in aquaria are limited in the salinity levels they can tolerate and will only become accustomed to changes in salt content very slowly. Coral-fishes are fairly limited in their tolerance levels. Sudden changes always produce a damaging effect. If the fish are placed directly in water with a much higher salt concentration, they hang around at the surface making rocking swimming movements and their breathing rate increases; conversely, if the salinity level drops quickly, the fish will lie at the bottom finding it difficult to breathe. Extreme changes in salt content can cause shock reactions, in which the fish sink to the bottom, fins splayed open and the operculum held away from the body. Specimens affected in this way must be transferred immediately into water that is normal for them; you can also try moving the operculum manually to overcome the benumbed breathing action.

Salmo, G, *see* Salmonoidei

Salmon *(Salmo salar) see* Salmonoidei

Salmon *(Salmonidae, F) see* Salmonoidei

Salmon, related species, Sub-O, *see* Salmonoidei

Salmon Discus *(Poptella orbicularis) see Poptella*

Salmon Herring *(Chanos chanos) see* Gonorynchiformes

Salmonidae, F, *see* Salmonoidei

Salmoniformes; Salmon-types. O of the Osteichthyes*, Coh Teleostei. GREENWOOD and colleagues (1966) have united a whole series of fish groups in the O Salmoniformes, that used to be partly attributed to the Isospondyli and partly to the Clupeiformes. They also included in the O some fish groups that used to be regarded as Os in their own right – Galaxiiformes, Haplomi, Xenomi, Iniomi, Scopeliformes and Myctophiformes – and gave some of them the rank of a Sub-O. The reasons for this new approach are of an extremely specialist nature and to understand them a grasp of detailed morphological knowledge is needed. It is not possible to consider even the most important pronouncements here; where some details are discussed, you will find them under the the Sub-O entries. Such an attitude to the subject is further compounded by the fact that some of the individual Sub-Os, at least externally, comprise very different fish groups, and secondly, the placing of the separate Sub-Os under the Salmoniformes is not necessarily totally adequate. Thus, it is only possible to characterise the Salmoniformes very generally as follows: They are usually slim fish with primitive features. Many oceanic specimens have luminescent organs. Adipose fins often present, and the swimbladder, if formed, is openly connected to the intestine. The head and gill skeleton have primitive characters, and there are usually more than 24 vertebrae. The following Sub-Os belong to the O: Esocoidei*, Galaxioidei*, Myctophoidei*, Salmonoidei*, Stomiatoidei*.

Salmonoidei; Salmon, related species. Sub-O of the Salmoniformes*, Sup-O Protacanthopterygii. Fairly primitive fish with an elongated body shape, laterally compressed. Fins normally not very big, and only supported by soft rays. Almost always there is an adipose fin between the D and C; it is only a lobe of skin with no fin rays. The body is covered with cycloid scales. Other features, some of which are quite usual for primitive Teleostei*, are as follows: the upper edge of the mouth is defined by a toothed premaxillary and maxillary bone; there are 2 supramaxillary bones above the posterior end of the maxilla; and the swimbladder is connected to the intestine (physostomes). Special features include the lack of an oviduct (at least in most cases), the lack of fish bones in the musculature on the ventral side and unusual vertebral features. Fossil Salmonoidei date back to the lower Eocene period. The modern forms are found mainly in fresh water.

– *Osmeridae*; Smelts. In the Smelts, the last vertebrae in the root of the tail do not form an arch open at the top (which contrasts with the Salmonidae). They are fairly small, elongated coastal fish from the N Atlantic and N Pacific; often, they come up-river at spawning time and form pure freshwater forms regionally. The adipose fin is located opposite the An. The European Smelt *(Osmerus eperlanus)* (fig. 994) is up to 30 cm long; it is elongated in a spindle-shape and hardly flattened at all laterally. Its area of distribution stretches from the coast of N Spain to the Gulf of Bothnia in the Baltic Sea.

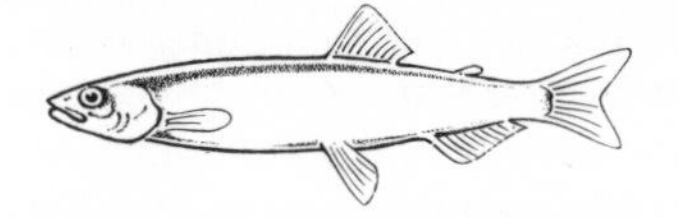

Fig. 994 *Osmerus eperlanus*

The larger marine Smelt migrates to the rivers to spawn, whereas the smaller inland Smelt is a pure freshwater form and is found mainly east of the Oder. At first, the marine Smelts are plankton-eaters, later on they become predators. They spawn in March and April and then large numbers of them die. The small Sacramento-Smelt *(Spirinchus theleichthys)* is famous because it occurs in such large numbers. When dried, the large, oil-rich Candlefish or Eluachon *(Thaleichthys pacificus)* is used by the Indians along the N American W coast as a torch. Even by the beginning of our century *Osmerus mordax* had been introduced into large N American lakes where it has developed into an important useful fish. Because Smelts occur in such large numbers, they are important as food for various predatory fish. They are fished commercially, partly for manufacturing into fertiliser and pig food, and partly for preparing as a food (salted or fried).

– *Salmonidae*; Salmon. Fish of the well-known trout type, that with the exception of the G *Retropinna* (New Zealand) occur only in the northern hemisphere. Most are freshwater species, although some are anadromous. In all the Salmonidae, the last vertebrae in the root of the tail form an arch open at the top. Included in the F are the Thymallidae (Graylings and the Coregonidae (Whitefishes), both of which were often regarded as independent Fs in older systems. Most members of the

F are very important in the fishery industry. The F takes its name from the Salmon *(Salmo salar)* (fig. 995), a species that used to play a dominant role in the inland fishing industry because of the excellence of its meat.

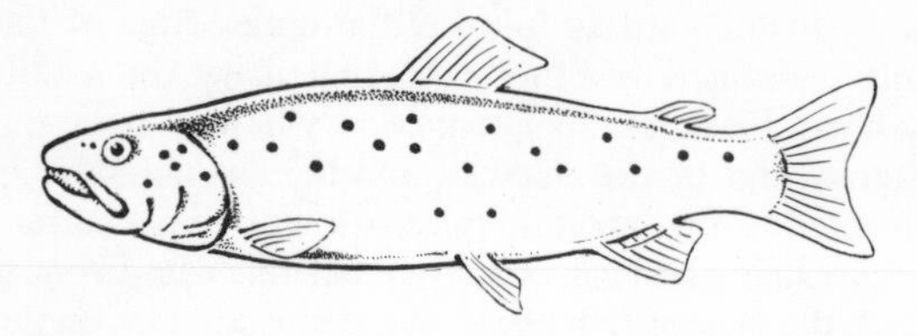

Fig. 995 *Salmo salar*

In the last quarter of the nineteenth century, on average 50,000 salmon were caught in the Lower Rhine every year. After 1900 the river became polluted so rapidly that catches were made only by chance, and now the Rhine Salmon has died out. The situation is similar in many other parts of its distribution area which extends in Europe from the Pechora river in the north to the Duoro river in Portugal. In N America, at least in the northern parts of its distribution area, the situation is a little better. Salmon are actually sea fish, but their life begins in fresh water. The young salmon that hatch out of the 6 mm long eggs remain for 2–5 years in fresh water (they are known as parr at this stage). They have a typical youthful colouring and patterning – particularly striking are the 7–10 dark transverse bands and the red dots. After migrating to the sea, another 1–3 years go by while the fish feed intensively; then they become sexually mature. During this period salmon are bluegreen on top, off-white on the ventral surface, and silvery everywhere else (they are now known as smolt). With the onset of sexual maturity, the salmon cease feeding and in late summer they begin their great spawning migration (anadromous migration). It is now known that, guided by certain scents, the salmon come up-river to the exact spot where their own lives began as fertilised eggs. Each day, they travel about 30–60 km. The reproductive colours of migrating males are unusually striking. The upperside is a shimmering blue with rows of red flecks on the head, whereas the underside can be lemon-yellow to deep black. The females too have much more intense fawn-brown or red-brown tones. In the males, however, other features are altered too. The upper and lower jaws bend into a hook shape at the end, the teeth fall out and are replaced by a strong, but functionless denture; the skin, particularly on the fins, becomes thicker. The ability of the salmon to jump over obstacles as they migrate is well known. At the spawning ground, which is usually a clear-water stream with a gravelly substrate, the females make large troughs by beating their tails vigorously; after spawning, they fan the eggs with sand and gravel. In contrast to the Pacific Salmon, *Salmo salar* can migrate back to the sea again after spawning – ie they are able to spawn for several years one after the other. However, usually the energy reserves stored in the form of red fat (hence the red meat) are so exhausted after the spawning act that the fish die during the catadromous migration – only about 5% ever spawn a second time. Some local races in Sweden and England have become completely adapted to life in fresh water. The so-called Pacific Salmon (G *Oncorhynchus*, 6 species) are closely related to *Salmo salar*. Their natural distribution area stretches from Taiwan to Kamtschatka and the Polar Sea north of Siberia and from Alaska to California. They were often transposed to other areas and are now found in the southern hemisphere, eg around New Zealand. The Pacific Salmon are now the main type of salmon to be fished commercially. In many countries, particularly the USSR, they are bred artificially. The most important species economically are the Chum Salmon *(Oncorhynchus keta)* (fig. 996) which grow up to 1 m long, the Sockeye Salmon *(Oncorhynchus nerka)* (fig. 997) and the Chinook Salmon *(Oncorhynchus tschawytscha)*.

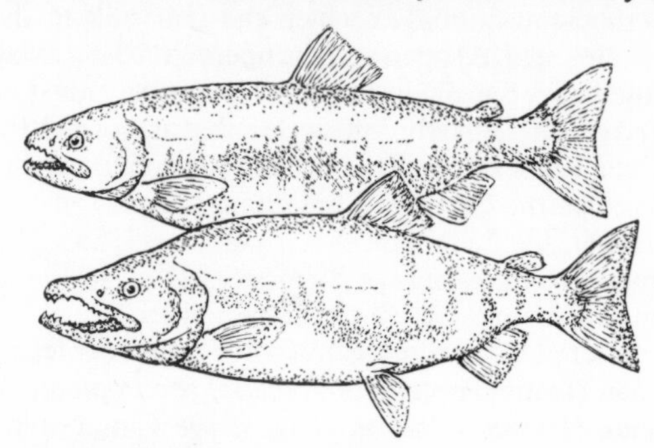

Fig. 996 *Oncorhynchus keta*

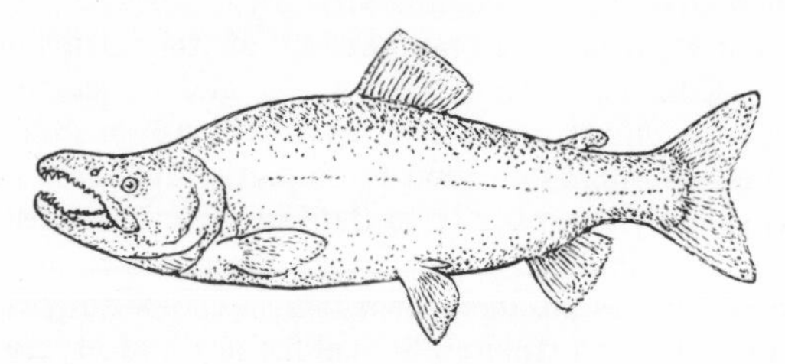

Fig. 997 *Oncorhynchus nerka* (male)

The Huchen *(Hucho hucho)*, which can grow up to 1.5 m long but which usually remains much smaller, is related to the salmon; it is a non-migrating form of salmon from the Danube. In Europe today, trout are much better known than salmon. However, this is less true of the native Brown Trout *(Salmo trutta fario)* (fig. 998), which

Fig. 998 *Salmo trutta fario*

is very rare in many places, and is applied rather to the Rainbow Trout *(Salmo gaidneri)* (fig. 999) which is reared and fattened in large breeding stations. It originally came from America and was introduced to Europe in 1882; it is better suited to intensive fish culture. The

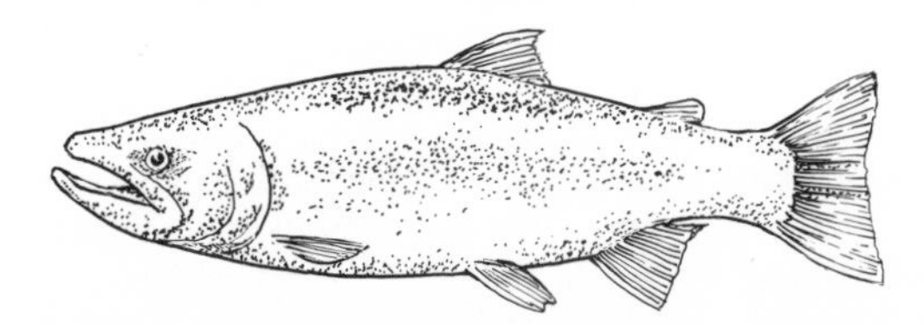

Fig. 999 *Salmo gaidneri*

Brown Trout is distributed right across Europe to N Africa and Asia Minor. In addition, it has been introduced into places with the right living conditions – ie fairly cool, fast-flowing waters. It remains true to its habitat, and is normally about 25–30 cm long, although much bigger specimens are known. It feeds mainly on insect larvae and smaller kinds of fish, including its own species; it will also snap greedily at anything that drops onto the surface, and sport-fishermen have learned to use this behaviour to their advantage. Young trout have dark blotches on the flanks, arranged like transverse bands; adult ones are easy to recognise from their black and red dots, the latter of which usually have blue edges. A distinction is made between the smaller Forest Trout and the larger Rock Trout of the lower river courses. Very dark specimens are also known as Black Trout. The Brown Trout spawns in winter, sexual maturity being reached in the 2nd or 3rd year. One female can produce 500–1,500 eggs a year. A special trout from the Alps and Alpine lakes is known as the Lake Trout *(Salmo trutta lacustris)*; it grows up to 90 cm long. The Salmon or Sea Trout *(Salmo trutta trutta)* is about 1.3 m long and is found near rivermouths along the European coasts; like the salmon, it comes up-river to spawn, in late autumn or early winter. The young fish remain in fresh water for some years. The systematic demarcations of all the named sub-species are unclear. Many people hold the view that they are simply different forms for different habitats. In the high mountain ranges of western N America is found the Golden Trout *(Salmo aquabonita)*; *Salmo nelsoni* from Mexico prefers waters with a temperature of 27 °C, which would bring death to any other species of the G.

Much rarer in Europe than the trout are the chars, that are among the most delightful of any freshwater fish from the temperate zone. The Char *(Salvelinus salvelinus)* is found in the clear Alpine lakes, but also occurs in N Europe, N Asia and Alaska. The upper surface is moss-green and provides a lovely contrast to the yellowy under surface that is bright red at spawning time. The body is also covered with yellow dots, and the front edges of the paired fins are white. The American 'Brook Trout' *(Salvelinus fontinalis)* (fig. 1000) comes from eastern N America; it too has lovely colouring and is normally about 25 cm long. Lake Trout *(Salvelinus namaycush)* was, until the 1940s, the main Salmonid catch in North America's Great Lakes. It was virtually wiped out by a sub-species of the parasitical Sea Lamprey *(Petromyzon marinus)*. Throughout Europe,

Fig. 1000 *Salvelinus fontinalis*

despite its many hard scales, the Grayling *(Thymallus thymallus)* (fig. 1001) was at one time considered a great delicacy. It usually grows to around 30 cm. It is a habitat fish found below the trout region in clean flowing waters

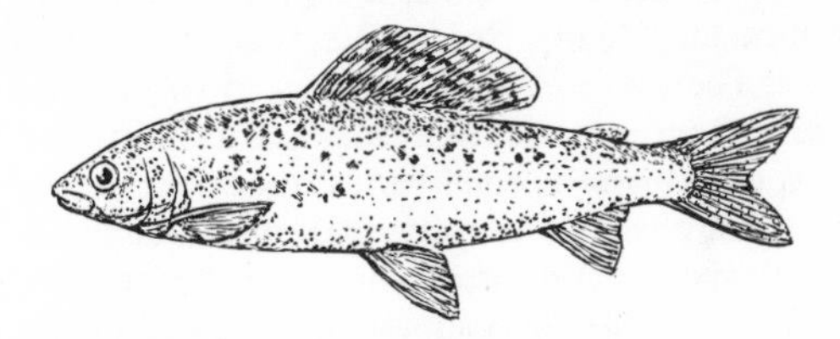

Fig. 1001 *Thymallus thymallus*

(the so-called grayling region). Apart from the easily recognisable matt golden-green scales, the grayling also has a large, fan-like D that has green stripes on a red background. Grayling numbers have been considerably reduced in recent years by pollution. Other species of the G are distributed in N Asia and N America.

The members of the G *Coregonus* that we are also going to consider here are often regarded as an F in their own right. In contrast to the Salmonids so far discussed, *Coregonus* species eat plankton primarily. The denture is therefore only weakly developed, whereas the gill filters are much better developed. All the species have a narrow mouth, large scales and are a fairly uniform silver colour with a dark dorsal side. Their systematics are not obvious. SVÄRDSON (1949–58) and DOTRENS (1959) divide them into small and large Whitefishes. The former are represented by the *Coregonus albula* (fig. 1002) group which are found throughout N Europe

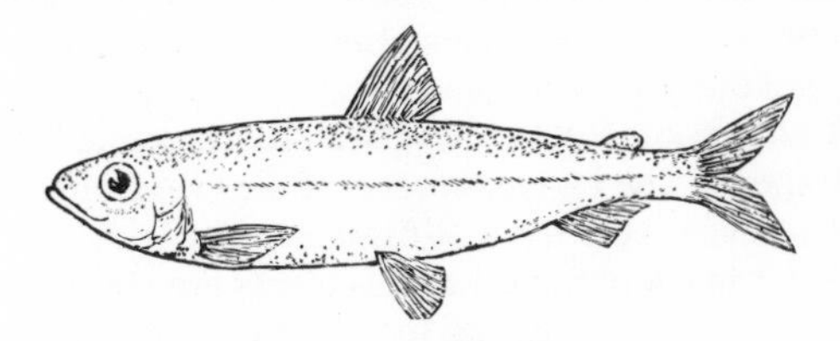

Fig. 1002 *Coregonus albula*

and in some Mecklenburg lakes. The *Coregonus lavaratus* group and others make up the large Whitefishes; they usually have 30–34 gill raker processes, and are found throughout the palaearctic realm. They include such well-known species as the Blaufelchen from Lake

Constance and other lakes of the Lower Alps; also the *Coregonus oxyrhynchus* (fig. 1003) group with 36–44 gill raker processes with a distribution area that stretches from the British Isles, across Europe and Siberia to Alaska. The common names for this group

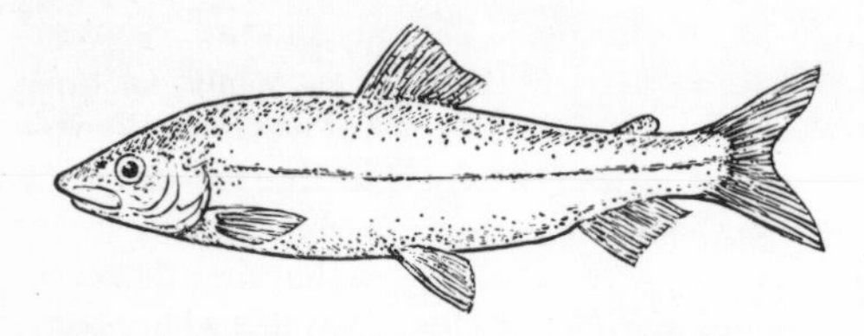

Fig. 1003 *Coregonus oxyrhynchus*

are numerous, but one, the Houting, has a rather protuberant snout like a nose; it is a migrating fish from the south-eastern North Sea coasts and the western Baltic Sea. There are in total 7 groups of forms, and almost all of them comprise both stationary and migrating forms, dwarf and normal forms, that have sometimes been described in scientific literature as separate species, sub-species and forms. All *Coregonus* species, particularly those that form shoals, are important commercially as their meat tastes good and they have no fish bones. They are also widely bred artificially and introduced into particular areas.

Unfortunately, the Salmonidae are not very suitable for aquarium-keeping. Their requirements can normally only be met in special large aquaria. Most importantly, they require cool, oxygen-rich and flowing water. Young trout are easiest to accustom to life in captivity.

Salmon-type Region *see* Flowing Waters

Salmon-types, O, *see* Salmoniformes

Salmopercae, O, name not in current usage, *see* Percopsiformes

Salvelinus, G, *see* Salmonoidei

Salvinaceae, F, *see* Pteridophyta

Salvinia SEGUIER, 1754; Floating Moss. G of the Pteridophyta*, comprising 10 species. Tropical-subtropical, radiating out into temperate regions. Small annual or perennial plants that reproduce in great numbers asexually, with no roots. The stem axes are elongated and lie horizontally. The leaves come in 3-part whorls, two of which unfold horizontally as assimilation leaves, and the third of which acts as a root and is very pointed (Anisophylly*). Complicated reproduction through spores. The various species of the G live on the surface of calm, still waters – often in rice fields, for example, or in damp mud. *Salvinia* plants are not very suitable for the aquarium, as they shut out light from above. On the other hand, they are very good for providing natural shade in greenhouses. Many species are endangered if they are illuminated for less than 12 hours. Well-known species are:

– *S. auriculata* AUBLET. Central and S America. A perennial. Very variable in size. Isolated hairs on the leaf upperside (fig. 1004).

Fig. 1004 *Salvinia auriculata*

– *S. natans* (LINNÉ) ALLINONI. Europe to S Asia. An annual in Europe, overwintering with spores. Bushy hairs on the leaf upperside.

Samolus LINNÉ, 1753; Water Pimpernel. G of the Primulaceae* comprising 15 species. Distributed throughout the world, mainly in the southern hemisphere. *Samolus* species are small perennial plants with rosettes or perennials with horizontal stem axes and an alternate leaf arrangement; a few are annual plants. Flowers hermaphroditic, polysymmetrical, small. Double perianth, calyx and corolla 5-part, fused at the base. Stamens 5. Carpels 5, fused, semi-inferior. Capsule-fruit. The species of the G grow in wet places and in shallow water; some occur in brackish water or on saline substrata. The following species is kept in aquaria.

– *S. valerandi* LINNÉ; Common Water Pimpernel. Africa, W Asia. In rosettes or with gently elongated stem axes. Suitable for cold and heated aquaria.

Sand *see* Substrate, Filtration

Sand Divers (Trichonotidae, F) *see* Trachinoidei

Sand Dollars, G, *see Clypeaster*

Sand Eel *(Ammodytes tobianus) see Ammodytes*

Sand Eel *(Gonorynchus gonorynchus) see* Gonorynchiformes

Sand Eel types, O, *see* Gonorynchiformes

Sand Eels (Ammodytidae, F) *see* Ammodytoidei

Sand Eels, G, *see Ammodytes*

Sand Eels (Gonorynchidae, F) *see* Gonorynchiformes

Sand Eels, related species, Sub-O, *see* Ammodytoidei

Sand Hermit Crab *(Diogenes pugilator) see Diogenes*

Sand Lance *(Ammodytes tobianus) see Ammodytes*

Sand Perch *(Diplectrum formosum) see Diplectrum*

Sand Shark *(Odontaspis taurus) see* Odontaspididae

Sand Sharks, F, *see* Odontaspididae

Sand Shrimp *(Crangon crangon) see Crangon*

Sand Smelts (Atherinidae, F) *see* Atherinoidei

Sand Stargazers (Dactyloscopidae, F) *see* Trachinoidei

Sandelia CASTELNAU, 1861. G of the Anabantidae, Sub-O Anabantoidei*. S Africa. Medium-sized fish. Body elongated, flattened laterally. Head and mouth

large. 2 nasal orifices on both sides. Operculum covered in scale, smooth edge. The D and An are elongated, with a fairly large hard-rayed section. C rounded. V not elongated like a thread. LL interrupted and broken up into 2 rows of scales, located deeper in the body, towards the posterior end of the body. As a whole, *Sandelia* species are very similar to the elongated members of the G *Ctenopoma*, although they are distinguished from them by having a smooth edge to the operculum and by having a more simply constructed labyrinth. Basic colour olive-brown to yellowish with dark markings. There is a distinct fleck on the operculum. Detailed information about care and breeding in the aquarium is largely inadequate. No bubble-nest is built. At present, the following species are known:
– *S. bainsii* CASTELNAU, 1861 (fig. 1005).
– *S. capensis* (CUVIER and VALENCIENNES, 1831).

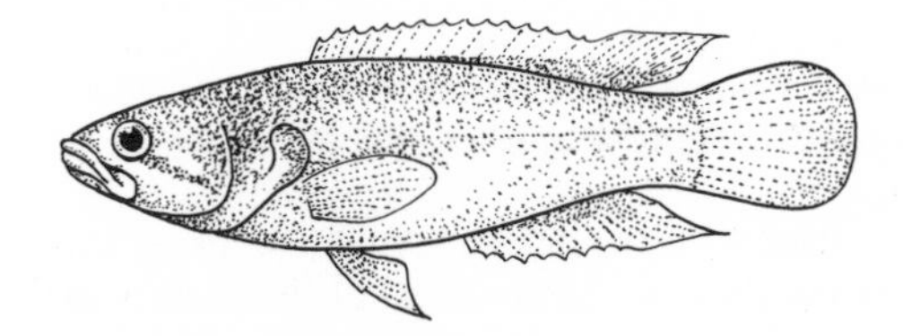

Fig. 1005 *Sandelia bainsii*

Sandfishes (Trichodontidae, F) *see* Trachinoidei

Sanguinicola *see* Trematodes

Saprobionts; saprobes. Organisms that only occur in foul water, and so they are a barometer of the state of the water. *Tubifex** is a well-known saprobe, for example. Depending on the degree of decaying matter in a stretch of water, characteristic saprobe communities will develop, and these can be used to classify the water (biological water analysis). The majority of saprobes are microscopically small, and by recognising them, the aquarist can also tell a great deal about the state of his aquarium water. However, saprobes are difficult to analyse and a lot of experience in identifying them is needed.

Saprolegnia, G, *see* Saprolegniacea

Saprolegniacea (fig. 1006). Fungi that grow on dead, organic matter. In aquarium-keeping, important Ga are *Achlya* and *Saprolegnia* both of which have several species. Saprolegniacea develop on fish and their spawn (ie on living matter) only if they are already damaged. Infections will often be observed after fish have been kept at too cool a temperature after bacterial infections*, *Ichthyosporidium** attacks, fish tuberculosis* or on open wounds. The fins are attacked most of all, but the fungi will appear on almost every part of the body, including the eyes and mouth; in severe cases, the fungi can even penetrate into the musculature. Healthy fish are not attacked. The fungi forms white, cotton-wool growths which disintegrate out of water and which are made up of microscopically thin threads (hypha). The fungi spreads through actively mobile spores which are concentrated at the end of the hypha; they disperse from here to alight on a suitable substrate, where they develop once more into hypha. A cure for Saprolegniacea comes in many forms: try raising the temperature, transferring the infected fish to clean old water, bathing treatments* or localised treatment with potassium permanganate*, Rivanol*, copper sulphate*, Griseofulvin* or daubing the areas with tincture of iodide*. In every case, however, the first priority is to identify and clear up the primary cause of damage.

Fig. 1006 Saprolegniacea

Sarcoma *see* Tumour Diseases

Sarcopterygia, Sub-Cl, *see* Osteichthyes

Sarcopterygians (Sarcopterygia, Sub-Cl) *see* Osteichthyes

Sardina pilchardus *see* Clupeiformes

Sardine *(Sardina pilchardus) see* Clupeiformes

Sargasso Fish *(Histrio histrio) see Histrio* and Lophiiformes

Sargus annularis, name not in current usage, *see Diplodus*

Sarotherodon RÜPPEL, 1853. A taxon* of a rank that is much disputed (sub-genus? genus?) belonging to the Cichlidae*. Very closely related to *Tilapia** in a narrow sense. *Sarotherodon* comprises those *Tilapia* species in which the skull bones, mesethmoid and vomer, do not touch; they are also mouth-brooders. Eg *S. galilaeus (= Tilapia galilaea), S. niloticus (= Tilapia nilotica), S. spilurus, S. hunteri.*

Sauries (Scomberesocidae, F) *see* Exocoetoidei

Saurita endosquamis *see* Myctophoidei

Saururaceae; Lizard's Tail Family. F of the Magnoliatae* with 4 Ga and 5 species. N and Central America, E Asia. Small to medium-sized perennials with runners and upright elongated shoots. Leaves alternate, stalked. Flowers very small and inconspicuous, united into inflorescences, some of which have striking hysophyllary leaves at the base.

The Saururaceae are marsh plants that can tolerate occasional flooding. Species that are sometimes kept in aquaria include the North American *Saururus cernuus* LINNÉ (American Lizard's Tail) and the East Asian *Houttuynia cordata* THUNBERG (Asiatic Lizard's Tail) which is often sold (wrongly) under the same name. Neither makes a very good aquarium plant. Some are propagated in an emersed state and then planted in the aquarium as cuttings; they will then survive for a while.

Saw Sharks, F, *see* Pristiophoridae

Sawfish *(Pristis pristis) see* Pristidae
Sawfishes, F, *see* Pristidae
Scad *(Trachurus trachurus) see Trachurus*
Scads, F, *see* Carangidae
Scalare *(Pterophyllum scalare) see Pterophyllum*
Scale Formula (fig. 1007). In fish systematics, apart from the description of the shape, arrangement and size of the scales, a scale count from certain parts of the body is also important. The scale formula of a species usually consists of the number of scales contained in the LL or the middle longitudinal row between the gill opening and the C insertion (eg 66–70), plus the number of scales in a transverse row from the front insertion point of the D to the longitudinal row just mentioned (eg 8–9), together with the scale count of a transverse row from the insertion of the Vs to this same longitudinal row (eg 12–13). These figures are then written down as follows: $66\text{–}70 \frac{8\text{–}9}{12\text{–}13}$.

If there is no LL, then usually a count is made right through from the dorsal ridge to the middle of the belly. If the LL is incomplete, then the scale count is taken from the longitudinal row of scales that is broken up by the LL organ. In addition, the number of scales from the circumference of the caudal peduncle may be counted, or the scaling on the head, and so on.

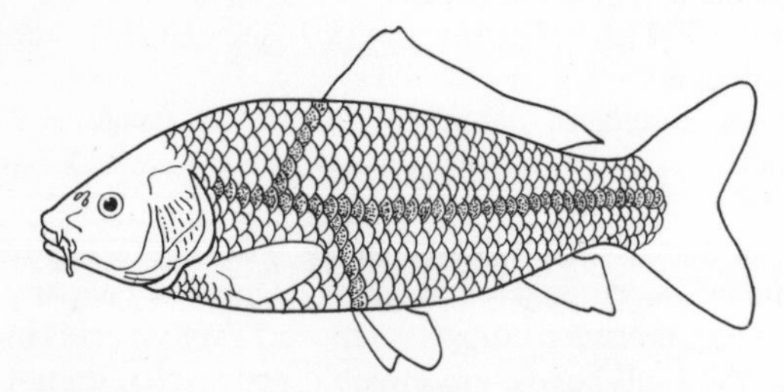

Fig. 1007 Scale formula. The rows of scales to be counted are shaded

Scale Protrusion. Occurs in fish as an accompaniment to stomach dropsy*, *Ichthyosporidium** attack and fish tuberculosis*. Very common are swellings of the digestive tract (Intestinal Diseases*, Liver, Diseases of*) as a result of incorrect feeding* or poisoning* – the outward sign of this is a protrusion of the scales. Sometimes, along with the scale protrusion, a few scales may look bloodshot; this is often the result of a bacterial infection* of the skin and it can then be treated with copper sulphate*, Chloromycetin*, Streptomycin* or Tetracyclin*.

Scaled Blennies (F Clinidae) *see* Blennioidei
Scaleless Blennies (Blenniidae, F) *see* Blennioidei
Scales (fig. 1008). Elements of the dermoskeleton in fish. Scales lend the surface of a fish's body a significant stability and at the same time sufficient mobility. There are 3 basic types of scale.

1) The *placoid scales* of sharks and rays. These are tooth-like formations that are fixed on a chalky base plate lying in the corium (dermis). They consist of a cone of dentine formed from the corium and a tough coating of enamel, which stems from the epidermis. Placoid scales can only go up to a certain size; as the size of the body increases, the scales increase in number.

2) The *elasmoid scales* of bony fish that are formed only from the corium. They are usually gently arched and consist of fine bony lamellae layered one on top of the other, and a thin top-coat of hyalodentine or similar substances. Bony scales dig deep into the scale pockets of the corium causing the rear half of one scale to overlap the front half of the following scale, and so on like a series of roof-tiles. There are 2 types of elasmoid scales. With comb-edged or *ctenoid scales* the rear edge is toothed or the rear portion has spines on it (eg as in Perches). The rear end of round or *cycloid scales* is smooth (eg as in Characins and Barbs). Sometimes, both types of scale occur on the one fish (eg as in many Flatfishes). Throughout a fish's life, elasmoid scales grow in size as the fish's body grows, so growth marks may be detectable (providing a method of telling the fish's age). Elasmoid scales may be microscopically small or be absent altogether (eg as in many Catfishes).

3) *Ganoid scales* are phylogenetically speaking a very old form and they are forerunners of the bony scales. They have a fairly thick coat of enamel (Ganoin*) and a rhombic shape. At the present time, they are only found in relict species, such as the Garpikes and the Polypterids. The bony spines, hooks or plates of various kinds of fish are dermal bones* corresponding to scales.

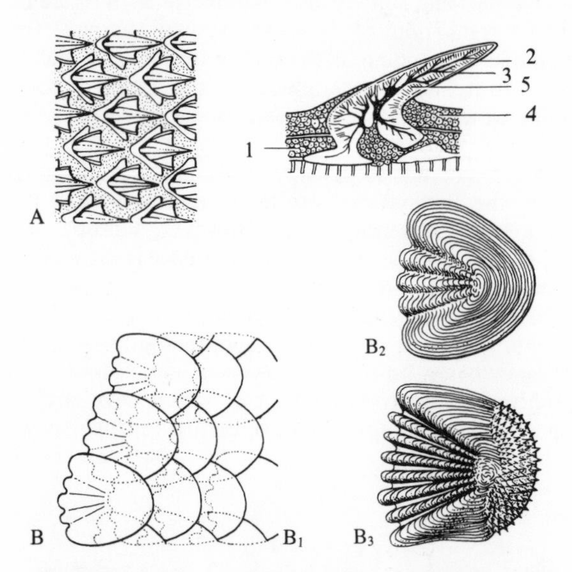

Fig. 1008 Scales. A Position of placoid scales and a longitudinal cut through a placoid scale: 1 Corium 2 Ganoid covering 3 Dentine 4 Epidermis 5 Pulp cavity. B 1 Position of the scales on a bony fish, B 2 Cycloid scale, B 3 Ctenoid scale

Scallops, G, *see Pecten*
Scapanorhynchidae; Goblin or Elfin Sharks. F of the Chondrichthyes*, O Lamniformes. The only species, *Scapanorhynchus owstoni* (Goblin Shark) (fig. 1009) is a deep-sea Shark from the western Pacific and Indian Ocean. It has a long, shovel-shaped 'nose' and a protrusible mouth beset with sharp teeth. There are 2 Ds

roughly the same size, and the lower tail fin lobe is only slightly discernible. Similar species lived in the Cretaceous period.

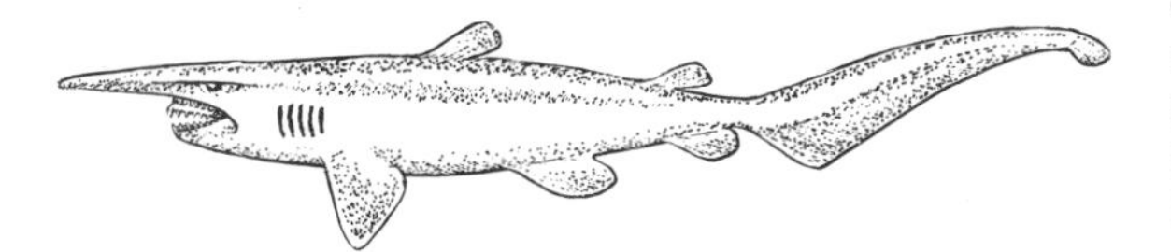

Fig. 1009 *Scapanorhynchus owstoni*

Scaphopoda; Tusk-shells. Cl of the Mollusca* with a tube-shaped shell open at both ends. They are marine creatures that lie on sand. Special snaring threads (captacula) enable them to catch small animals. The best-known Tusk-shells are the 'Elephant-tusk' of the G *Dentalium* LINNAEUS, 1758. There are no aquarium observations.

Scardinius BONAPARTE, 1837. G of the Cyprinidae*, Sub-F Leuciscinae. Europe, Asia. These are large fish that live as small shoals in ponds, lakes and stretches of water that flow slowly. The habitats must also have a soft substrate and an abundant plant growth. The body is stocky, becoming increasingly tall with age, and very flat laterally. The mouth is small and terminal, the mouth-slit being at an angle and reaching upwards. No barbels. LL complete, bending gently downwards. *Scardinius* can be distinguished from the very similar G *Rutilus** in the following features: the D is inserted clearly behind the V and the belly is keeled between the V and An. When young, *Scardinius* is well suited to large domestic aquaria with plenty of swimming room and robust vegetation. Provide good-quality old water that should be replaced occasionally be the addition of fresh water. Temperature 18–22 °C, higher for short spells. Feed with live and dried food, additional plant food is necessary. At spawning time, the males have spawning tubercles on their head and dorsal surface. Eggs are laid in shallow water on aquatic plants.

– ***S. erythrophthalmus*** (LINNAEUS, 1758); Rudd (fig. 1010). Europe. Up to 40 cm. Dorsal surface olive-brown to olive-grey, flanks greenish-yellow, ventral surface silver. Fins red, Vs, D, An and C with a dark base. The species has the tendency to form hybrids with the Ga *Rutilus**, *Blicca** and *Alburnus**.

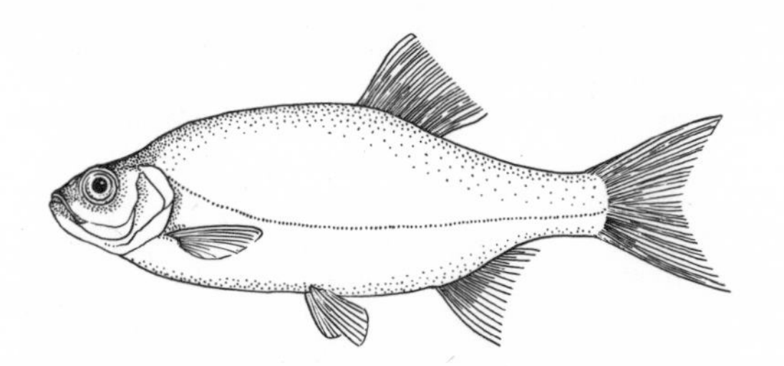

Fig. 1010 *Scardinius erythrophthalmus*

Scaridae, F, *see* Labroidei

Scarlet Lobelia *(Lobelia cardinalis) see Lobelia*

Scarus GRONOVIUS, 1763 (fig. 1011). G of the Scaridae, Sub-O Labroidei*. Numerous species, circum-tropical. Length 20–60 cm. *Scarus* species live a social existence on reefs, rocks and in seaweed mats; they live in schools

or in smallish groups that often consist of one male and several females. Their diet consists of animal and plant growth, and also of pieces of madrepore coral which they cut off with their beak-like denture and grind into fine coral sand. The individual species may be food specialists of different kinds. The powerful body is a long oval shape in side-view and, as with all Scarids, it is covered with large scales. On the cheek beneath the eye, in this G, there is a row of scales. When the mouth is shut, the upper row of theeth is in front of the lower. In contrast to *Sparisoma**, many *Scarus* species wrap themselves in a mucus cocoon at night. Various species have been kept successfully in the aquarium. All of them need additional plant food and some also take dog biscuits (DE GRAAF).

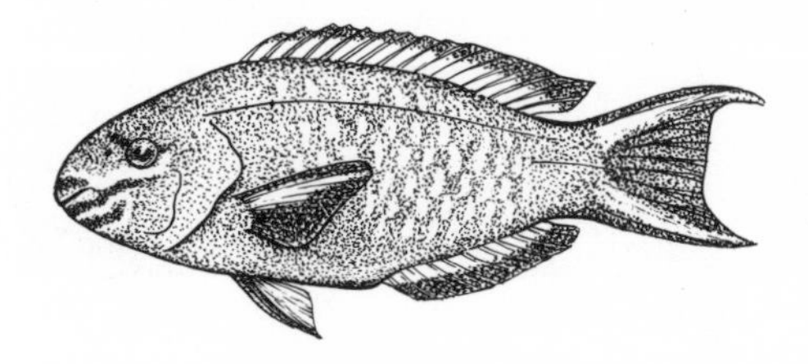

Fig. 1011 *Scarus vetula*

Scatophagidae; Scats (fig. 1012). F of the Perciformes*, Sub-O Percoidei*. Tall fish, laterally flattened, that are found in the coastal zones and on the reefs of tropical parts of the Indo-Pacific. Mouth small, with fine teeth arranged in bands. Small ctenoid scales

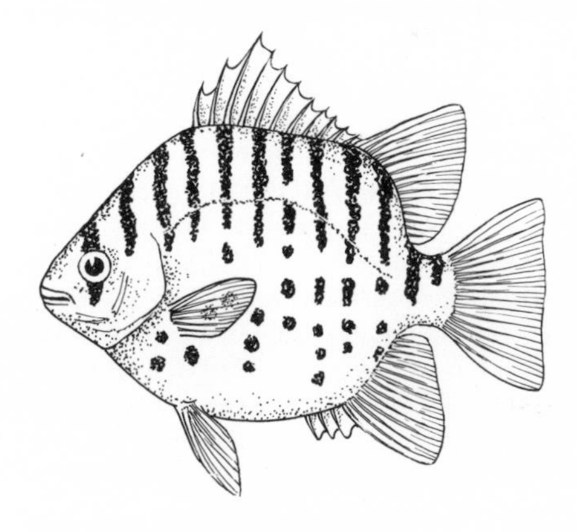

Fig. 1012 Scatophagidae-type

that not only cover the body and head but also extend onto the D and An. An indentation in the D almost divides it into a hard-rayed and a soft-rayed section; the latter section is shorter and often broadens out into a kind of lobe. The An has 4 spines. Young Scatophagidae have bony plates on the head and a large spine on both sides of the nape (tholichthys-stage). Fairly young Scats are found mainly in the brackish water of estuaries. The older ones stay around port areas and have often become specialised as omnivores; they lie in wait for food in shoals near the drainage exit points (Scatophagidae = coprophage or dung-eater). The Argus Fish *(Scatophagus argus)* (fig. 1013) is often kept in marine aquaria; young specimens are attractive and will also flourish in fresh water. Apart from animal food, also give lettuce, spinach or rolled oats. Large Scats are very greedy and pollute the water greatly. Scats have hardly any economic value.

Fig. 1013 *Scatophagus argus*

Scatophagus argus *see* Scatophagidae

Schilbeidae; Glass Catfishes. F of the Siluriformes*, Sup-O Ostariophysi. Naked Catfishes from Africa and S and SE Asia, very variable in size. The earliest fossils date back to the Neogene period. Their body shape is reminiscent of the Siluridae*, although there are characteristic differences between the two Fs. The D is very short and stands well forwards with a powerful spine at the front; however, the fin is always fairly tall (there is no D in *Physailia*, for example). In almost all the species, and in contrast to the Siluridae, there is a small, pointed adipose fin. The C is indented, and always separated from the very long An. The first fin rays of the Ps are formed as spines. The number of barbels varies, as does their length, depending on Ga. Usually, there are 3 pairs of barbels, more rarely 4; the upper jaw barbels are almost always the longest and most powerful pair. Apart from large, commercially valuable fish, this F also comprises some small species that are free-swimming and active by day; they also occur in shoals or groups, eg *Eutropiellus**. The term 'Glass Catfish' is only appropriate for the small species or the young fish. Their body is often translucent, although never as see-through as the body of Glass Catfishes belonging to the Siluridae *(Kryptopterus*)*. When disturbed, many Schilbeidae have an unusual reaction. They pretend to be dead and lie on their sides on top of the substrate, eg *Physailia*. In their homelands, the large species are prized edible fish and recently have been caught in large numbers.

Various species are occasionally kept as unusual aquarium specimens. See under *Eutropiellus, Physailia*.

Schistocephalus, G, *see* Cestodes

Schizobionta. All non-nucleated organisms with a celled structure (*see also* Prokaryota). In contrast to the Eukaryota*, these extremely small primitive cells do not have a true cell nucleus separated off by a membrane. The genetic material is embedded in the central plasma. Other characteristics of the Schizobionta concern the structure of the flagella (the cells can actively move around) and the fine structure of the cell wall. The assimilation pigments (if there are any) are not concentrated in special chromatophores, but are dispersed in the outer plasma layers. Propagation is through simple division. Sexual reproduction is unknown. The Schizobionta are single-celled organisms, they can form loosely knit colonies or a thread-like structure. They are divided into the 2 divisions of the Blue Algae (Cyanophyta*) and the Bacteria (Bacteriophyta*).

Schneider *(Alburnoides bipunctatus) see Alburnus*

Scholze's Tetra *(Hyphessobrycon scholzei) see Hyphessobrycon*

Schomburgk's Leaf Fish *(Polycentrus schomburgki) see Polycentrus*

Schraetzer *(Acerina schraetzer) see Acerina*

Schwartz's Neon *(Hyphessobrycon simulans) see Hyphessobrycon*

Sciaena LINNAEUS, 1758. G of the Sciaenidae*. There are species in warmer parts of the Indo-Pacific, the E Atlantic and the Mediterranean. *Sciaena* species are sociable fish, feeding on small animals, and living near the coast and in rivermouths. They are like a Perch in build, and are 20–70 cm long. Their most striking feature is a very short lower jaw with a short, thick barbel attached to it. No aquarium observations are recorded. Some *Sciaena* species are tasty edible fish. The following species lives along European coasts:

– *S. cirrhosa* LINNAEUS, 1758; Fringed Croaker or Drum. E Atlantic, Mediterranean, Black Sea, Red Sea. Up to 70 cm. Has sloping, wavy, gold stripes on a silver background (fig. 1014).

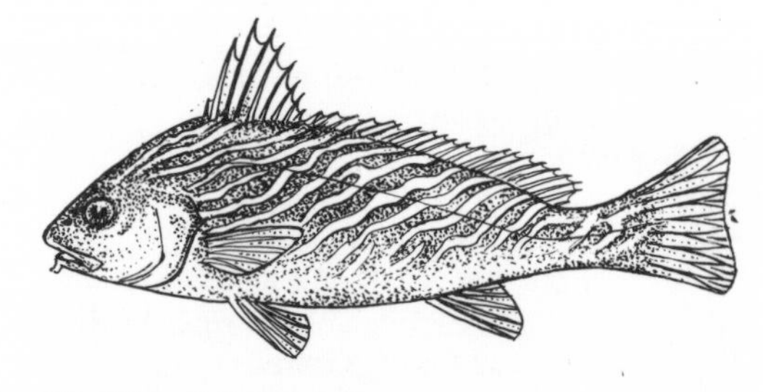

Fig. 1014 *Sciaena cirrhosa*

Sciaenidae; Croakers or Drums (fig. 1015). F of the Perciformes*, Sub-O Percoidei*. Found in tropical and subtropical seas, more rarely in cooler regions; a few species live in brackish water and a very few in fresh water. Croakers are predatory fish with a moderately or strongly elongate body, flattened laterally. The body is covered with scale and it has broad mucous channels. Mouth terminal, snout and chin often with pores or slits. The jaw teeth can vary; there are no teeth on the palate or on the tongue. 2 flat tips on the operculum. D1 short, spiny-rayed, inserted far forwards. D2 directly behind it and very long. The C is never forked, the An is short and has 1–2 spiny rays at the front edge. The Vs are positioned beneath or almost beneath the Ps. LL complete and extending into the tail fin. Ctenoid scales with slight serrations, that also extend onto those parts of the unpaired fins that are near the body. Croakers are particularly well-known for their ability to produce loud species-specific rhythmical noises – tapping, croaking, grunting, snoring or drumming kinds of sounds. This is where their common names, Croakers or Drums, come from. Of the 160 species, only a few are not able to drum. It is often only the males that can produce these

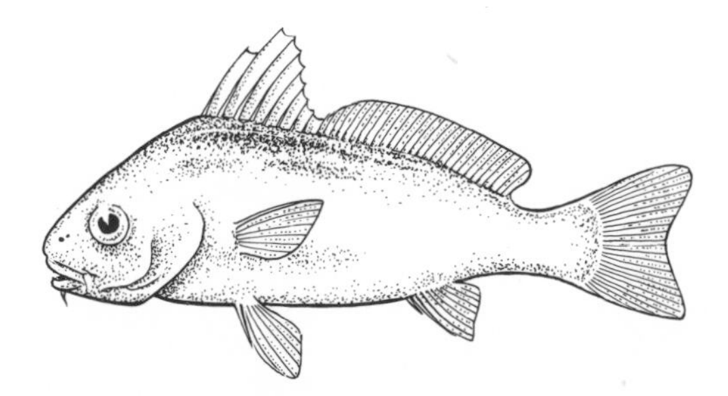

Fig. 1015 Sciaenidae-type

noises, which means that they are probably used in courtship or as part of threatening behaviour. The large swimbladder and the ventral musculature are involved in the production of the sounds. In almost all the species, the colours are inconspicuous and the reproductive technique holds no unusual features. Croakers often appear in large shoals and prefer shallow water regions with a sandy substrate. Many species are very important commercially on account of their highly valued, tasty flesh. In Florida, the hard scales are used to make flower jewellery ('fish-scale jewellery'). For the rod-fisherman, the Sciaenidae are renowned as superb fighters. The large species reach lengths of 1–2 m, as does *Johnius hololepidotus* (fig. 1016) from central and southern parts of the E Atlantic, including the Mediterranean, and around S Africa. The

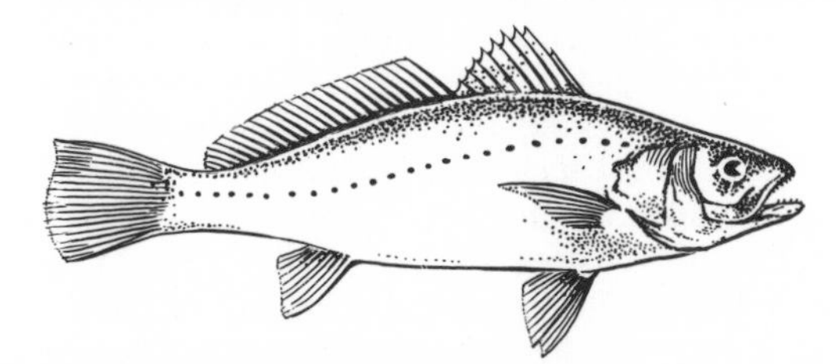

Fig. 1016 *Johnius hololepidotus*

Teraglin or Cape Salmon *(Atractoscion aequidens)* from S Africa, the Black Drum *(Pogonias cromis)* from the W Atlantic and the Indian Croaker *(Sciaena diacantha)* from the Indian Ocean are all further examples of large species. The Freshwater Drum *(Aplodinotus grunnieus)* is found in the fresh water of the North American Great Lakes and in various rivers as far as Texas. Important Ga in aquarium-keeping include *Corvina**, *Equetus** and *Sciaena**.

Scientific Name. Latin or latinised name, the use of which in the ranks of species, genus or family is subject to internationally agreed rules, making that name unequivocal as well as universally understood (Zoological Nomenclature*, Botanical Nomenclature*). The use of scientific names is of fundamental necessity when checking out a biological observation and it is likewise indispensable for any scientific use. In aquarium-keeping too, the common name* of a fish should always be followed by the exact scientific name, particularly in any publications and in the fish trade.

Scissortail Rasbora *(Rasbora trilineata) see Rasbora*

Scleropages, G, *see* Osteoglossiformes

Scleroparei, O, name not in current usage, *see* Scorpaeniformes

Scomberesocidae, F, *see* Exocoetoidei

Scomberesox saurus *see* Exocoetoidei

Scombridae, F, *see* Scombroidei

Scombroidei; Mackerels, related species. Sub-O of the Perciformes, Sup-O Acanthopterygii. The Scombroidei are distributed worldwide; they are pelagial fish of the oceans and deep seas. Their body shape can vary a great deal. A typical characteristic of all the species is the fusion of the premaxilla and the maxilla to form a bony kind of beak. Greenwood and colleagues (1966) include in this Sub-O the Trichiuroidei (Fs Gempylidae, Trichiuridae) which are often regarded as a Sub-O in their own right. The various species of the Scombroidei belong to the fairly primitive Perciformes*, and the oldest fossils known date back to the upper Cretaceous period.

– *Gempylidae*; Snake Mackerels. Elongated, eel-like fish from tropical and temperate seas. The head is long with a powerful, pointed snout; the lower jaw often protrudes. Long, conical fang-teeth are located on the jaws – together they form a formidable denture. D1 very long, with powerful spines. There are usually some small finlets behind the short D2 and the similarly formed An opposite. C deeply indented. The Vs are small, inserted beneath the Ps. Scales very small. 0–2 LLs. There are no keels on the sides of the caudal peduncle. Approximately 20 species of Gempylidae that are found mainly in deep water; a few live pelagically, however. An example is the 1 m long Barracouta or Snoek *(Thyrsites atun)* (fig. 1017) which is distributed around S Africa and Australia. It is among the most

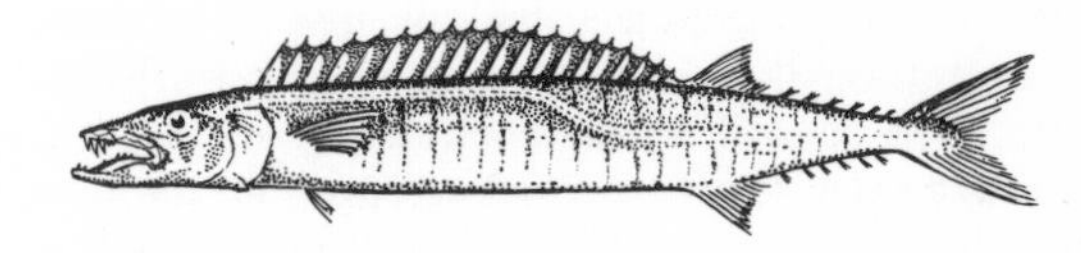

Fig. 1017 *Thyrsites atun*

useful of all the species belonging to the F. The Oilfish or Escolar *(Ruvettus pretiosus)* (fig. 1018), at 1.5 m long, is like an elongated Tuna-fish; it is found throughout the world at depths of 400–600 m.

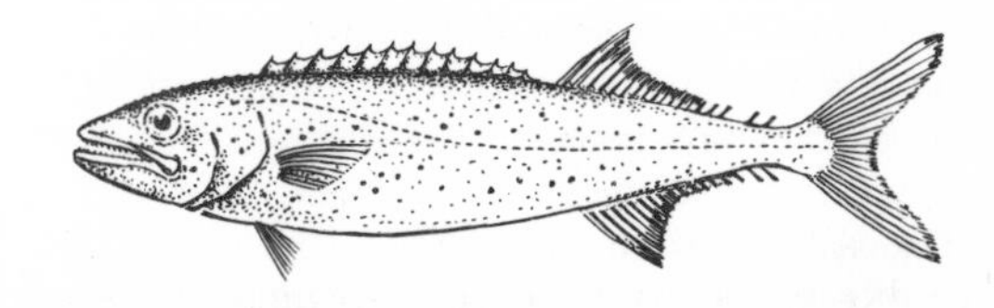

Fig. 1018 *Ruvettus pretiosus*

– *Trichiuridae*; Hairtails, Scabbardfishes or Cutlassfishes (fig. 1019). Very elongate, ribbon-like fish from fairly warm seas (roughly 20 species). The head is a pointed cone-shape with a deeply slit mouth and a powerful, often very powerful, denture made up of fang-teeth. D very long, stretching from the head to the small, deeply forked C. In the G *Trichiurus*, there is no C, the tail narrowing instead into a whip-like appendage. An long, Vs very small or absent altogether. No scales. Nearly all the species are a silvery colour. Hairtails are found both in shallow coastal waters, where they usually spawn, as well as in depths of up to 600 m. Various species become very large and their powerful dentition can make them look very frightening. An example is the 1.5 m long Atlantic Cutlassfish *(Trichiurus lepturus)* from tropical parts of the Atlantic and western parts of the Indo-Pacific. It has a delicious taste, although it is caught only rarely with nets or through ground-fishing.

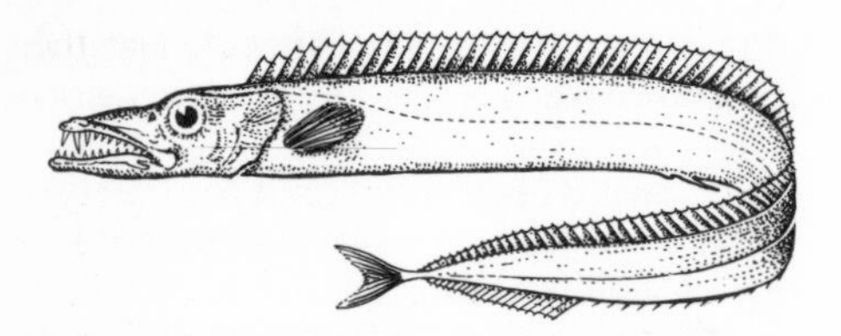

Fig. 1019 Trichiuridae-type

– *Scombridae*; Mackerels. Mainly pelagial fish from fairly warm and temperate seas. They are a spindle-shape, ideally suited to fast swimming, the pointed, conical head and thin caudal peduncle fitting harmoniously in with the body. At high swimming speeds, the Mackerels are able to reduce the currents they create (and therefore the friction) because they have developed special adaptations for this. For a start, they can lower the Ps, Vs and D1 into shallow grooves to become more streamlined. The D1 is short or reaches to the D2, which in its turn is also short and located exactly or nearly opposite a similarly shaped An. Usually, several small finlets follow on from both the D2 and An. The large C is broad and deeply forked; it is moved like a ship's propeller, and so represents the main motive power of the fish. Many species of Scombridae have stabilising bumps on the caudal peduncle. The skin is smooth, elastic with small round scales or no scales at all. There is usually no swimbladder. For brief spells, many species can reach speeds of 80–90 km/h. As young fish, at least, the Scombridae occur in large shoals; later, they may appear in packs. They undergo lengthy migrations relating to reproduction and food requirements. Most species have an excellent-tasting meat. As a result, Mackerels are fished intensively with drag-nets, lines and other pieces of equipment. The greatest catches are made in the Pacific. The best-known species, the Common Mackerel *(Scomber scombrus)* (fig. 1020) grows to about 50 cm long and weighs about 1.5 kg. Its upperside is a greenish colour covered with dark, wavy saddle-lines that do not quite reach the middle of the flank. There is no tail keel and it is found in the whole of the N Atlantic and neighbouring seas. Its diet consists mainly of crustaceans, and after spawning the Mackerel will also eat other fish. The spawning

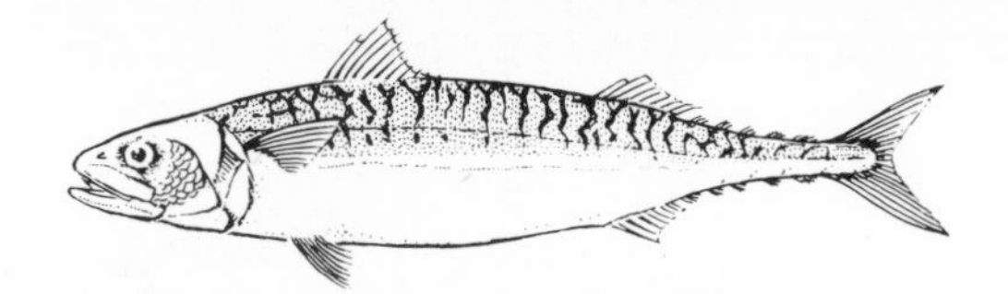

Fig. 1020 *Scomber scombrus*

grounds are mainly near the coast; each female can produce up to about 450,000 pelagic eggs. In contrast to the G *Scomber*, the Mackerels of the mainly Pacific G *Pneumatophorus* have a swimbladder. The giants among the Scombridae are the Tunas. The Blue-fin or Atlantic Tuna or Tunny *(Thunnus thynnus)* (fig. 1021) is found in the Atlantic, and may grow up to 5 m long and weigh 500 kg. However, the average length is 1–2 m.

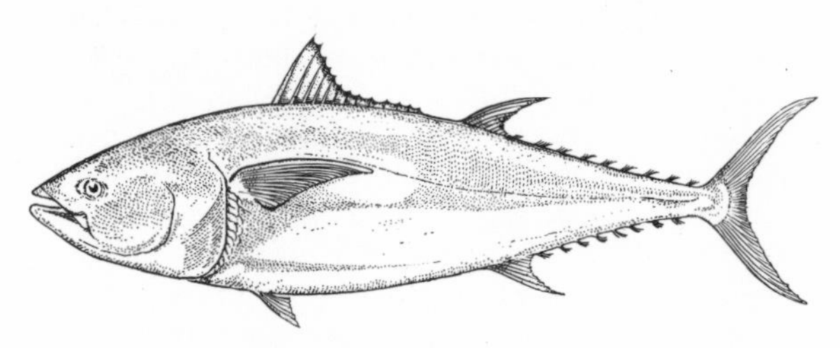

Fig. 1021 *Thunnus thynnus*

The blue-black upperside of the fish suddenly gives way to a grey-silver tone on the sides, ventral surface white. Its blood temperature is a few degrees above the water temperature. Diet consists mainly of fish, especially Mackerels. The centre of distribution is the Mediterranean. In Europe alone, about 20,000 tonnes of tuna is caught a year. The Yellowfin Tuna *(Thunnus albacora)* is found worldwide in fairly warm seas; it measures 2.5 m in length. Other Atlantic species, particularly important in the fishing industry, include the 1 m long Skipjack Tuna *(Euthynnus pelamis)* and the 75 cm long Atlantic Bonito *(Sarda sarda)* (fig. 1022). All the larger kinds of species of this F make excellent sport for fishermen, as they fight hard on the rod.

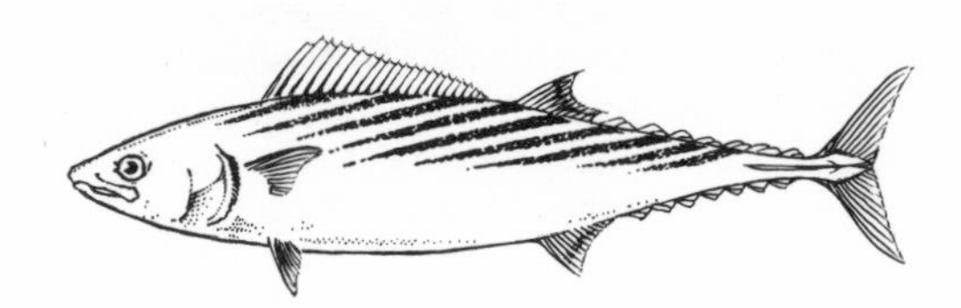

Fig. 1022 *Sarda sarda*

– *Istiophoridae*; Billfishes, Marlins (fig. 1023). The Istiophoridae are distinguished from the Swordfishes that follow because they have a stronger elongated body; the 'sword' is round in cross-section; the D1 is very long and often broad like a sail; and there are 2 lateral keels in front of the C. Sometimes there are long

Vs, located at the breast (eg G *Istiophorus*) or there may be none (eg G *Istiompax*). Like the Xiphiidae, the Istiophoridae live in the oceans and feed mainly on fish; like them also, they are excellent swimmers. Well-known

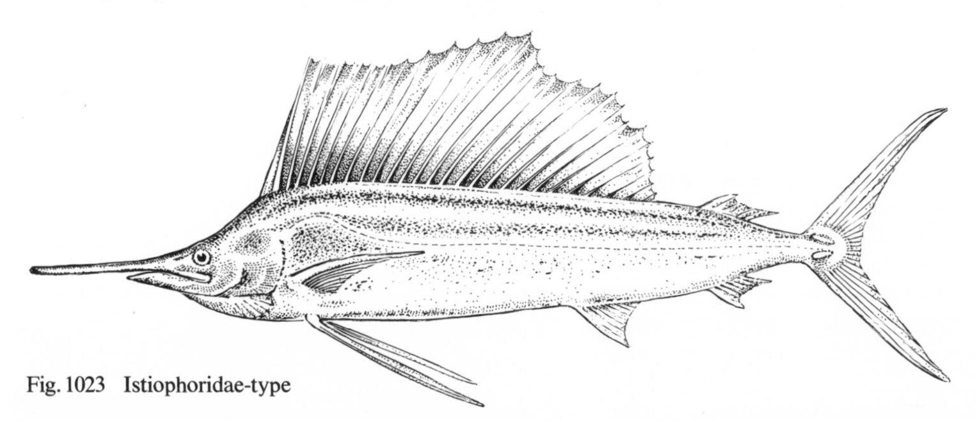

Fig. 1023 Istiophoridae-type

species include the 4 m long Black Marlin *(Istiompax marlina)* from tropical parts of the Indo-Pacific, the not much smaller Blue Marlin *(Makaira audax)* (fig. 1024) from the Atlantic and the Pacific, and the 3.5 m long Sailfish *(Istiophorus orientalis)* which has a very tall D1. Like the Swordfish, all the species are excellent food-fish and are by far and away the best sportfish. Catching one is the dream of many an ocean angler.

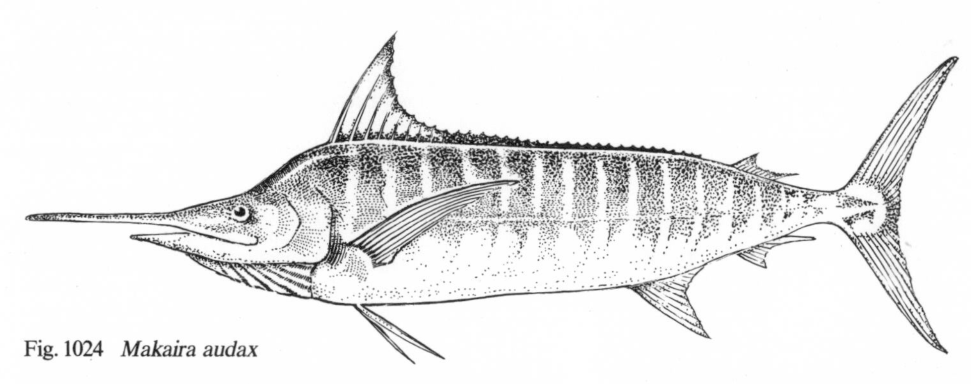

Fig. 1024 *Makaira audax*

– *Xiphiidae*; Swordfishes. The only species, the 4 m long Swordfish *(Xiphias gladius)* (fig. 1025), weighing 250–300 kg, is practically a tuna-fish with a very long, sword-like (ie flattened) rostrum. It is found throughout the world. The rostrum is formed from the nasal bones

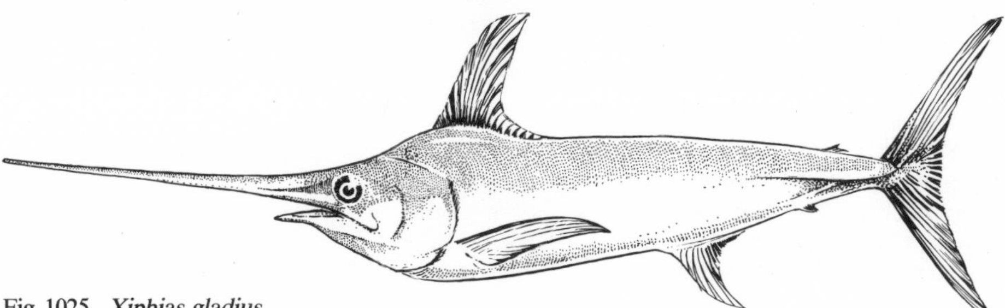

Fig. 1025 *Xiphias gladius*

and the premaxillary bones. The D1 is inserted immediately behind the head – it is large and like a pointed sail. D2 very small and located just in front of the large, sickle-shaped C. The An is divided in two. No Vs. There is a lateral keel in front of the C. No scales on the skin, but there are small dermal scales. In colour, the Swordfish is dark red-brown or bronze on the dorsal surface, paler on the flanks and on the ventral side. It hunts in the open sea alone or in small groups, both near the surface and at greater depths. It chases after fish and Cephalopods mainly. The sword is probably used to stun his prey. If a Swordfish feels under threat, it will attack immediately and ram its sword deep into its adversary, perhaps a whale, but also boats as well. *Xiphias* makes one of the best sportfish. Its meat is very good and its liver contains an unusually high amount of vitamin A. Isolated species occur in the North Sea and in the Baltic, but in the Mediterranean it is by no means rare.

Scophthalmus (various species) *see* Pleuronectiformes

Scorpaena; Scorpionfishes. G of the Scorpaenidae, Sub-O Scorpaenoidei*. Crepuscular fish from fairly warm seas that lie in wait for prey. They are stocky, large-headed fish with a D that has strong spines, at the base of which are located poison glands (so too on the operculum spines). A prick with one of these can be painful. A good camouflage is provided by numerous outgrowths on the head and a pattern of flecks that seem to break up the body. Interestingly, *Scorpaena* moults regularly. Length up to 50 cm. Keeping *Scorpaena* in the aquarium is easy. However, only food that is moving will be eaten; if no live food is available, the aquarist must either drop the food just in front of the fish or move it himself. A partial water change is needed frequently. Several species live along the southern European coasts. One example is the Large-scaled Scorpionfish (*S. scrofa* LINNAEUS, 1758) that grows up to 50 cm long and is red-black in colour. The following species is better suited for aquarium-keeping:

– *S. porcus* LINNAEUS, 1758. Small-scaled Scorpionfish. Mediterranean. Length to 30 cm. Patterned in brown or reddish-brown (fig. 1026).

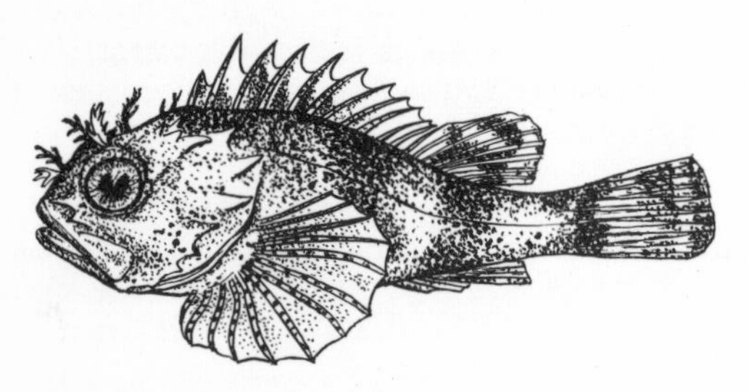

Fig. 1026 *Scorpaena porcus*

Scorpaenidae, F, *see* Scorpaenoidei

Scorpaeniformes; Scorpionfish-types. O of the Osteichthyes*, Coh Teleostei. The fish groups placed in this O largely correspond to the Sub-O Scleroparei (= armoured-cheeks) in the REGAN system, and the Sub-O Cottoidei in the BERG system. Occasionally, the names Cataphracti or Loricati are used. The term 'armoured-cheeks' relates to a feature virtually all the Scorpionfish-types (6 Sub-Os, 21 Fs) have in common, namely a bony connection between the row of bones beneath the eyes (infraorbitalia) and the preoperculum, thus creating a bony covering of the cheek area. However, other parts of the body too are covered in bony plates that often have spines. The mouth is deeply slit, toothed, and the upper mouth-edge is bounded by the premaxilla only. The fins have strong spines. There may or may not be scales on the body. The gill slits are wide, and the gill membrane is not joined to the isthmus. The oldest fossils known date back to the early Tertiary period. The Scorpaeniformes are often monster fish, but the majority are fairly small bottom-dwelling species that are found mainly in warm seas, although some also extend into more temperate regions. Some species are from the deep sea, and a lot of the species are important useful fish. Of the 6 Sub-Os, only the Scorpaenoidei* and Cottoidei* with their important Fs are given separate entries.

Scorpaenoidei; Scorpionfishes, related species. Sub-O of the Scorpaeniformes*, Sup-O Acanthopterygii. Perch-like fish with a very large head and well developed fins. The Scorpaenoidei can be told apart from the other Sub-Os of the O Scorpaeniformes* mainly because they have extremely strong dermal bones around the head area and strong fin spines. Other features of the Sub-O concern details of the skull skeleton – 1–3 infraorbitalia are normally fused, for example. The pelvic bones are directly in contact with the shoulder girdle (cleithrum). They are bottom-dwelling fish of the coastal regions found worldwide in warm and temperate seas. The oldest known fossils are from the Eocene period

– *Caracanthidae*; Orbicular Velvetfishes (fig. 1027). Found on coral reefs in tropical parts of the Pacific. In contrast to all other members of the Sub-O, the Caracanthidae are tall and almost like a coin in side-view. They are mentioned here only because of their unusual skin that has a velvet quality caused by lots of fine processes. This kind of velvet feel is virtually unknown among any other fish.

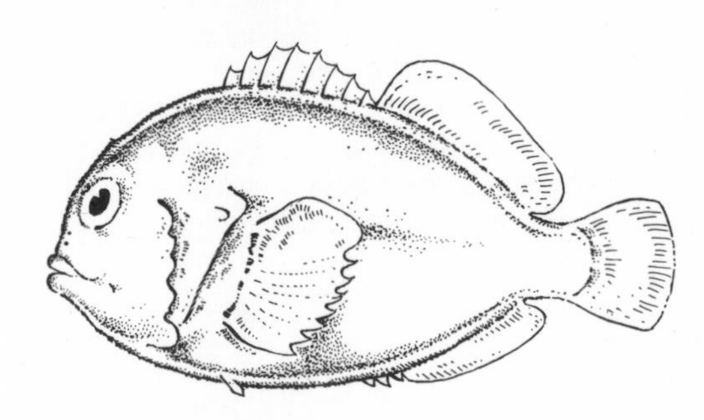

Fig. 1027 Caracanthidae-type

– *Scorpaenidae*; Scorpionfishes. Body moderately elongated, only clearly flattened laterally in the region of the caudal peduncle. Head large, like a bulldog's, with a large number of bony plates and spines. The mouth is deeply slit; it is toothed, including the palatal bone. The fins are large and carry strong spines. The Vs are close to one another and always have 1 spine and 5 soft rays. Broad gill slits; gill membranes not fused with the isthmus. There are scales on the body. The most important species economically is the Redfish *(Sebastes marinus)* (fig. 1028) that comes from the northern parts of the N and W Atlantic. It can grow up to 50 cm long and weighs 2 kg. When alive, the fish are a lovely red colour and they occur as 2 ecological races. The more valuable race, the Deep-sea Redfish, is caught at depths

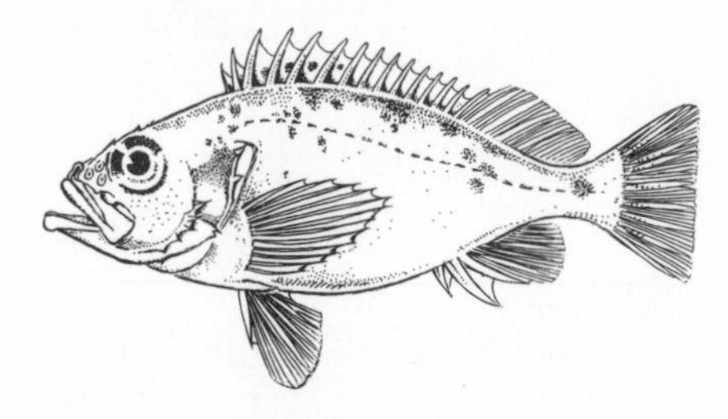

Fig. 1028 *Sebastes marinus*

of 300–600 m, whereas the true, strong red variety lives at depths of 200–300 m. Both races occur as far-flung shoals. Larger specimens eat mainly herring, cod and

crustaceans. The Redfish is livebearing; to begin with, the young fish live in the surface waters. Other species of the F have some regional economic importance. Many species are well-known aquariumfish. See under the Ga *Dendrochirus, Pterois, Scorpaena, Taeniotus.*
– *Synancejidae*; Stonefishes (figs. 1029 and 1030). These are excellently camouflaged fish from the Indo-Pacific – the camouflage is provided by wart-like or

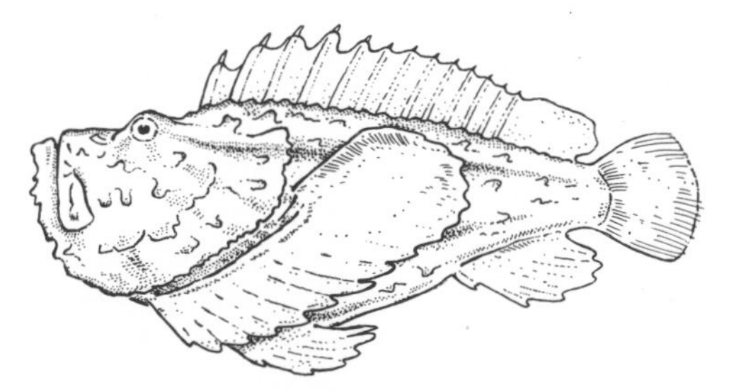

Fig. 1029 Synancejidae-type

spiny knobs on the skin and by the fact that the body is robust and stocky in shape. Often Stonefishes burrow so far into the sandy substrate that only the upper part of the head containing the eyes protrudes. Large head,

Fig. 1030 *Synanceja* sp.

Ps enlarged like wings. The D has 7–15 spines, the An 2–3 spines. Despite their fairly meagre size, the Stonefishes are among the most poisonous of all fish. A nerve poison flows down the spine grooves and disturbs the musculature of the heart and respiration. There have

Fig. 1031 *Inimicus* sp.

been incidences of fatalities because respiration has been totally paralysed. The best-known species is the 30 cm long Stonefish *(Synanceja verrucosa)* which is occasionally exhibited in public aquaria. It can be distinguished from other members of the F by its flattened head. Other species often imported belong to the G *Inimicus* (fig. 1031).
– *Triglidae*; Gurnards or Sea Robins (fig. 1032). Elongated fish, slightly flattened on the ventral side, that live primarily on the bottom of warm and temperate seas. They can immediately be recognised from their Ps, which are divided into a large, wing-like upper section, used for swimming, and a lower section. This latter part consists of stilt-like fin rays, pointing in a downwards

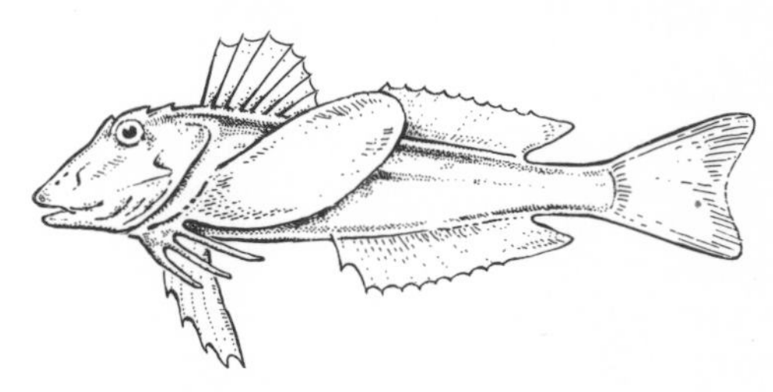

Fig. 1032 Triglidae-type

direction, with no skin membrane in between. They can be moved separately. They are used not only when moving along the substrate, but also as sense organs for feeling and smelling. Sea Robins can produce purring or cracking noises with the aid of their abdominal muscles and large swimbladder. Such sounds are always heard when the fish is picked out of the water. The head is more or less encased in bony plates, and the body is covered with scales or bony plates. D1 spiny-rayed, tall and relatively short. D2 long, located opposite the long An. C indented. The Triglidae eat mainly invertebrate bottom-dwelling animals and smallish bottom-dwelling fish. Many species have a lovely colouring. The 70 cm long Red Gurnard *(Trigla hirundo)* (fig. 1033) is found

Fig. 1033 *Trigla hirundo*

throughout the E Atlantic and adjoining seas. During the summer, a few individuals are even caught in western parts of the Baltic. The spawn is laid from April to August; it floats in the water, and the young fish live pelagically to start with. The smaller Grey Gurnard *(Trigla gurnardus)* is found from Murmansk to the Mediterranean. All Gurnards are good-quality edible fish.

Scorpion *(Dendrochirus brachypterus) see Dendrochirus*

Scorpion Bugs, G, *see Nepa*

Scorpionfish *(Pterois volitans) see Pterois*

Scorpionfish, related species, Sub-O, *see* Scorpaenoidei

Scorpionfishes (Scorpaenidae, F) *see* Scorpaenoidei

Scorpionfishes, G, *see Pterois*

Scorpionfishes, G, *see* Scorpaena

Scorpionfish-types, O, *see* Scorpaeniformes

Scrawled Cowfish *(Acanthostracion quadricornis) see Acanthostracion*

Scribbled Spinefoot *(Siganus vermiculatus) see* Acanthuroidei

Scrophulariaceae; Figwort Family. F of the Magnoliatae*, with 200 Ga and 3,000 species. Found worldwide. In rare cases shrubs or lianas, but usually perennials; also some annuals. Leaf arrangement alternate or whorled. Flowers of varying sizes, monosymmetrical, with a corolla that is usually brightly coloured. The corolla can come in a variety of shapes and it is often two-lipped. Some Scrophulariaceae are feeding-specialists, eg semi-parasites.

Fig. 1034 *Micranthemum micranthemoides*

Well-known ornamental plants belong to the Scrophulariaceae, such as the Slipperwort and the Snapdragon. They can colonise virtually every type of habitat from the low-lying lands to the mountains, and to a certain extent also marsh and aquatic areals. For example, various *Gratiola* species (Hedge Hyssop) grow in and on the waters of N America, as does a small *Lindernia* species and the brilliant yellow flowering *Mimulus guttatus* which has also been introduced into Europe and other areas. Some species of the G *Veronica* (Speedwell) are also adapted to a submerged life. *Veronica beccabunga*, for example, which is common in the northern hemisphere, used to be used quite frequently in cold-water aquaria when aquarium-keeping was in its infancy. In heated aquaria, the Ga *Bacopa**, *Limnophila**, *Hydrotriche**, *Limosella** and *Micranthemum* (fig. 1034) are of importance.

Scyliorhinus (various species) *see* Scyliorhynidae

Scyliorhynidae; Cat Sharks or Dogfishes (fig. 1035). F of the Chondrichthyes*, O Carcharhiniformes. Small sharks, often with a lovely colouring and patterning, found in shallow, coastal waters. The slender shape is emphasised by the tail elongated backwards, not arched

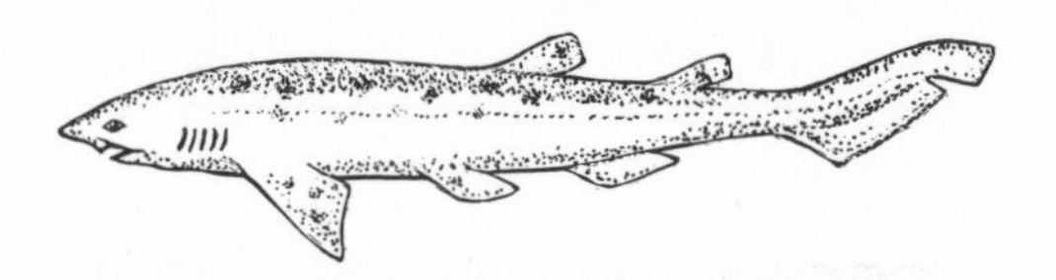

Fig. 1035 Scyliorhynidae-type

upwards. Snout usually truncated. In the light, the eye has a vertical, slit-like pupil, and there is no nictitating membrane. Spray-holes present, and usually 2 Ds. Dogfish feed mainly on crustaceans and molluscs. After internal fertilisation, all the species lay their eggs in horny capsules. These oblong capsules end in 2 very long threads on each of the long sides of the capsule; they are used for anchoring them to seaweed or to rocks. The embryos take many months to develop. The 1 m long Lesser-spotted Dogfish *(Scyliorhinus canicula)* (fig. 1036) occurs along the coasts of Europe, especially around England. It has a yellow-grey to reddish-grey

Fig. 1036 Head of *Scyliorhinus canicula*

upperside with numerous dark blotches on top. Ventral surface pale. In the Mediterranean, you will also find the 1.5 m long Nurse Hound *(Scyliorhinus stellaris)*. The species of the G *Cephaloscyllium* (fig. 1037) swallow a large amount of water or air if molested (eg when caught in a net), and this blows them up like a Pufferfish. The

Skaamoogs (G *Haploblepharus*) along the coasts of S Africa will lie on the bottom if danger threatens, and cover their eyes with their tail fins. Many Cat Sharks can be kept in fairly large marine tanks if fed on fish meat, Crabs and Bivalves.

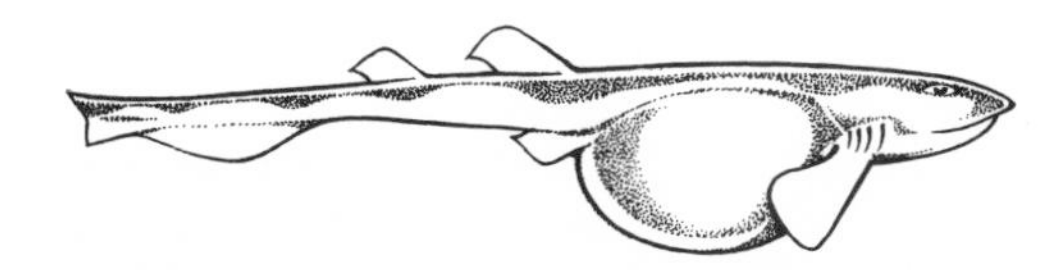

Fig. 1037 *Cephaloscyllium sufflans*

Scyllarus FABRICIUS, 1775; Slipper Lobster. G of the Decapoda*, Sub-O Reptantia*, F Scyllaridae. Various species are found in the littoral* of warm seas, and sometimes also around coral reefs. Related to the Spiny Lobsters (eg *Palinurus**), *Scyllarus* species have no pincers and are relatively bulky, broad and flattened. The antennae have been transformed into broad, serrated plates that are used both for defence and to hold onto food. Length 20 cm. *Scyllarus* is a colourful animal that will accept various kinds of animal food. In the aquarium, they often have moulting difficulties. The rust-red coloured species, *S. arctus* (LINNAEUS, 1758) is found on both sides of the Atlantic and in the Mediterranean. It grows up to 10 cm long and makes a lovely aquarium specimen. The brown-coloured *Scyllarides latus* (LATREILLE) from the same areas, plus the Red Sea, can grow up to 45 cm long.

Scyphomedusae, Cl, *see* Scyphozoa

Scyphopolyps (collective term) *see* Scyphozoa

Scyphozoa; True Jellyfish. Cl of the Coelenterates (Coelenterata*), Sub-Ph of Stinging Animals (Cnidaria*). The large oceanic jellyfish that live on plankton and larger animals belong to the Scyphozoa. Jellyfish have an alternation of generations; the tiny scyphopolyps produce medusae asexually, which in the larger species can measure up to 2 m. The medusae produce egg and sperm cells; a ciliated planula larva will develop from a fertilised egg, and then a polyp once again. The type of medusa-formation on the polyp is interesting; various discs are formed one on top of the other like a stack of coins, then the top-most one at any given time detaches itself to become an immature medusa (ephyra); this is known as strobilation (fig. 1038). Some jellyfish have a sharp sting, such as those of the G *Cyanea* which are common in northern seas. The tropical Box Jellyfishes (Cubomedusae*) release a fatal poison, whilst the well-known *Aurelia** are harmless. Jellyfish are usually difficult to keep in the aquarium. There is virtually no point at all in introducing larger kinds of medusae from the sea into the aquarium. More joy will be had from rearing scyphoplyps that are often accidentally placed in the tank. *Aurelia*, for example, has been reared to sexual maturity by feeding it on small crustaceans, Whiteworms, crustacean and bivalve meat. Larger animals can be fed to it one by one. A tank containing jellyfish must have no diffuser, otherwise air will get under the bell. Tropical *Cassiopeia* (fig. 1039) from mangrove areas can also be kept and bred in the aquarium; they are free-swimming only in the youthful phase; later on they become sessile, their indented bell stuck upside-down on the substrate, while they prey on small animals. Finally, there are the small, sessile Stauromedusae, eg *Lucernaria* (fig. 1040), which are also suitable for the aquarium. These species are stalked.

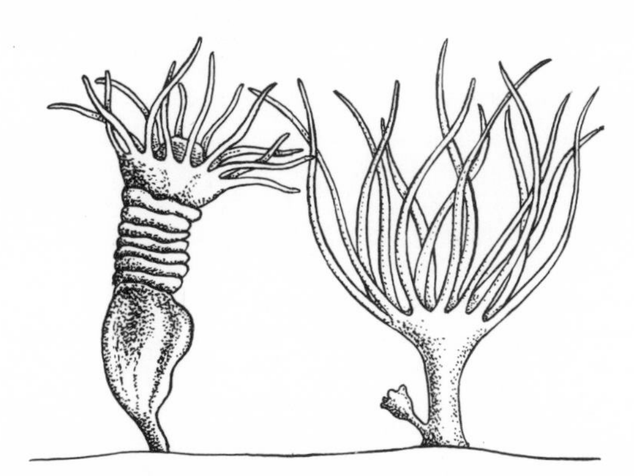

Fig. 1038 Scyphopolyp. Left, in strobilation

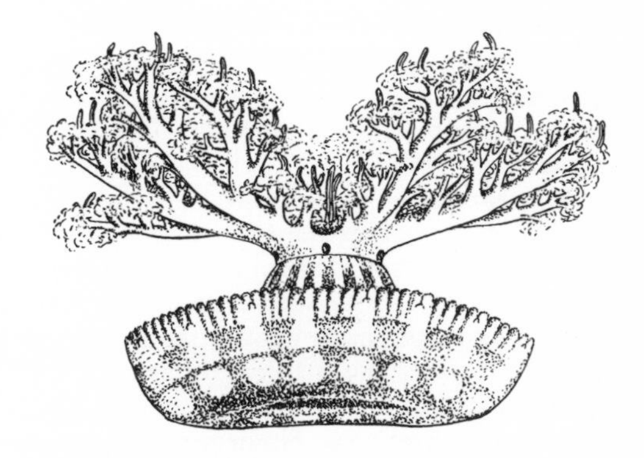

Fig. 1039 *Cassiopeia* sp.

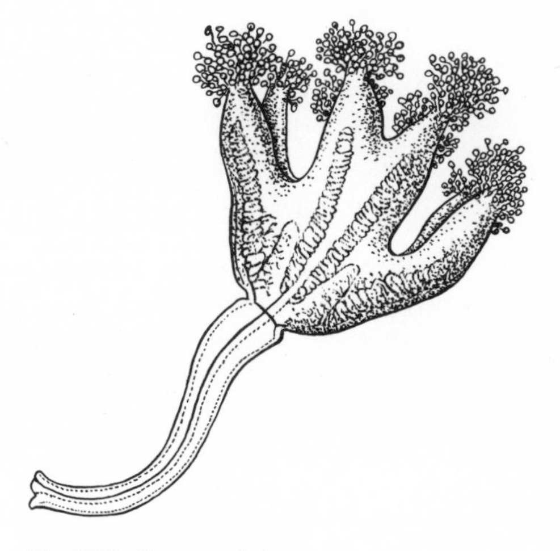

Fig. 1040 *Lucernaria* sp.

Sea Basses, F, *see* Serranidae

Sea Bream *(Sparus auratus) see Sparus,* also Sparidae

Sea Breams, F, *see* Sparidae
Sea Butterflies, O, *see* Pteropoda
Sea Catfishes, F, *see* Ariidae
Sea Crabs, G, *see* Brachyura
Sea Cucumbers, Cl, *see* Holothuroidea
Sea Gooseberries, Cl, *see* Ctenophora
Sea Gooseberries, Ph, *see* Tentaculata
Sea Hair, G, *see Sertularia*
Sea Hares, G, *see Aplysia*
Sea Hen *(Cyclopterus lumpus) see Cyclopterus*
Sea Lamprey *(Petromyzon marinus) see* Cyclostomata
Sea Lettuce, G, *see Ulva*
Sea Moth types, O, *see* Pegasiformes
Sea Moths (Pegasidae, F) *see* Pegasiformes
Sea Mouse *(Aphrodita aculeata) see Aphrodita*
Sea Pens, O, *see* Pennatularia
Sea Robins (Triglidae, F) *see* Scorpaenoidei
Sea Salt Baths *see* Sodium Chloride Baths
Sea Snail (Liparidae, F, name not in current usage) *see* Cottoidei
Sea Snail *(Liparis liparis) see* Cottoidei
Sea Stickleback *(Spinachia vulgaris) see Spinachia*

Sea Water. In respect of dissolved electrolytes*, in comparison with fresh water and brackish water, sea water is much more uniformly put together and has a total salt content above 30‰. In the oceans, the salt content is between 32 and 38‰, but in the Red Sea* and the Persian Gulf it is over 41‰. The density at 36‰ and a temperature of 23 °C is 1.026 g/ml, and the pH value* 8.3–8.4

Sea water can only be used in the aquarium in a very few instances; usually it is made up artificially. The different recipes for this vary in detail, but essentially the relation of the individual ions* remains the same in all of them.

CZENSNY puts forward the following list of ingredients to be mixed with 100 litres of water (distilled):

2,816 g sodium chloride
65 g potassium chloride
550 g crystalised magnesium chloride
338 g magnesium sulphate
129 g calcium chloride
50 g sodium hydrocarbonate

plus traces of iron, manganese, aluminium, phosphorus, silicon, iodine, silver, lead and copper.

Another recipe, put forward by WIDEMANN and KRAMER and modified by HÜCKSTEDT, suggests the following for 100 litres water:

2,765 g sodium chloride
692 g crystallised magnesium sulphate
551 g crystallised magnesium chloride
165 g crystallised calcium sulphate
65 g potassium chloride
25 g sodium hydrocarbonate
10 g sodium bromide
1.5 g strontium chloride
0.5 g potassium iodide

plus a mixture of trace salts, small quantities of hydrogen phosphate and iron citrate.

The calcium sulphate should be dissolved first. If the pH does not reach a level of 8.3 despite ventilation, add some grams of soda. If spring water is available with a high carbonate hardness, the sodium hydrocarbonate and soda can be dispensed with.

To calculate the pH value of sea water, HÜCKSTEDT recommends a special indicator, because CZENSNY's indicator does not show a clear colour change in this respect. Dissolve 1 g alpha-naphthol-phthaline in 400 ml 96% alcohol then add a solution of 2% alcoholic phenolphthaline. The colour changes in relation to the pH value are as follows:

below 7.5	almost colourless
7.5–7.8	green
7.9–8.1	turquoise
8.2–8.3	sky blue
8.4–8.5	dark blue
8.6–8.7	blue-violet
above 8.7	dark violet

If, despite obtaining a good blue colour at a pH value of 8.3–8.4, you want to work with a standard solution, dissolve 8.4 g of sodium hydrocarbonate in 1 litre of distilled water, ventilate it, and a pH value of 8.4 will be arrived at.

A problem that occurs with sea water especially is the proportion of nitrogen compounds it contains and the tendency to revert into nitrogen compounds. In order to set the nitrification process in motion in newly set-up sea water, ie the conversion of ammonia into nitrate, 'inoculate' it with 'old' sea water. If none is available, stir up some sea water with garden soil, let it settle, then pour the remainder into the tank. The nitrate content in sea water can only be allowed to rise to a certain level too. To prevent it building up, only keep a small amount of animals in the tank and feed them sparingly. From time to time, change some of the water. The specific removal of nitrates, using exchange resins, is possible in theory, but at the present time it is much more expensive than making up a fresh lot of sea water.

Sea-anemones, O, *see* Actiniaria
Sea-cows (Sirenia, O) *see* Mammalia
Sea-ears, G, *see Haliotis*
Sea-horse, G, *see Hippocampus*
Sea-horse Sickness *see Glugea*
Sea-horses (Syngathidae, F) *see* Syngnathoidei
Sea-lilies, Cl, *see* Crinoidea
Seals (Pinnipedia, O) *see* Mammalia
Sea-purse *(Codium bursa) see Codium*
Search Behaviour *see* Appetence Behaviour
Search Swimming *see* Appetence Behaviour
Sea-tangs, G, *see Fucus*
Sea-squirts (Ascidiacea, Cl) *see* Tunicata
Sea-squirts, Sub-Ph, *see* Tunicata
Sea-urchins, Cl, *see* Echinoidea
Seavark *(Abalistes stellaris) see Abalistes*
Sea-water Ichthyophthirius *see* Cryptocaryon Disease
Seaweeds, G, *see Fucus*
Sebastes marinus *see* Scorpaenoidei

Secondary Growth in Thickness. The considerable thickening up which stems and roots of perennial plants belonging to the Dicotyledons (Magnoliatae*) may undergo, is based on secondary growth in thickness. This starts from a meristematic tissue that lies between the xylem and phloem in the vascular bundle; after some time, it forms a closed ring when viewed along the axis. The cells deposited on the inside are called wood, and those on the outside bast. Among the Monocotyledons (Liliatae*), secondary growth in thickness is extremely rare and it takes place in a different way.

Secret Brooders *see* Cichlidae

Sedentaria, Sub-Cl, name not in current usage, *see* Polychaeta

Sedges, F, *see* Cyperaceae

Seed. The organ of dispersal of the seed-plants (Spermatophyta*). It is usually formed after a successful fertilisation from an ovule. The ripe seed consists of an embryo that is temporarily resting, nutritive tissue and a seed-envelope. The embryo is made up of a short stem axis with a vegetative cone, radicle and seed-leaves (cotyledons). In a lot of cases, primary leaves are already present too. There may be no nutritive tissue, in which case the nutrients are stored in the seed-leaves.

Seed-plants, D, *see* Spermatophyta

Seed-sowing *see* Propagation of Aquarium Plants

Selachiformes, O, name not in current usage, *see* Chondrichthyes

Selachii, O, name not in current usage, *see* Chondrichthyes

Selection. *Natural selection* is a choice mechanism, effective at all times, through which the offspring of certain individuals of a group of organisms are continued into the next generations. This leads to the outward form of the species changing and eventually to the development of new species (Darwin's theory of selection). For the process of selection to be able to take place, the hereditary material must be repeatedly altered (Mutation*), so that new characters emerge that bring some advantage to the species, eg better chances of survival under extreme environmental conditions, better protection against enemies or diseases, or advantages relating to food competition or sexual partners.

Artificial selection is a basic principle of breeding*; it means that only particularly suitable specimens are selected for reproduction.

Selection of Breeding Fish. Those species that remain as a couple for a short or long period during the breeding season, eg Cichlids, can be left to find their own partners among a shoal of growing youngsters; such partners usually make for successful breeding. In species that do not remain faithful, such as Characins and Barbs, it is best to isolate a pair that has already mated successfully and bearing in mind their breeding season*, place them together again with a view to breeding. It is also recommended that unwilling partners should be separated and another combination tried.

Selenichthyes, O, name not in current usage, *see* Lampridiformes

Self-pollination *see* Pollination

Semi-circular Swimming Display *see* Display Behaviour

Senegal Cichlid *(Tilapia heudeloti) see Tilapia*

Sense Organs. Parts of a multi-celled animal's body that are receptive to stimuli. The essential components of a sense organ are the sense cells which conduct specific information via a nerve connection to the central nervous system*. Various kinds of sense organs are distinguished – light-sensitive organs (Sight, Organs of*); olfactory organs*; organs of taste (Taste, Organs of*); organs with an acoustic sense (Hearing*); balancing organs*; organs that sense temperature; organs of touch (Touch, Organs of*); and, particularly among fish, organs that can sense current (Lateral Line System*). Those sense organs that have a similar function in vertebrates and invertebrates are made up of the same basic elements. The external forms may be very different, or very similar. For example, the eyes of Cephalopods and mammals have the same structure (Convergence*) although they develop from completely different embryonal hereditary factors.

Sepal *see* Flower

Sepia LINNAEUS, 1758. G of the 10-armed Cephalopods (Cephalopoda*, O Decabrachia). Contains a large number of species widely distributed in the continental shelf areas of the oceans on gravelly and on sandy substrates. They often rest on the bottom, perhaps buried in the sand, until prey comes along (crustaceans, fish), then they stretch out their 2 trapping tentacles and ensnare the victim. *Sepia* has a bulky ovate body with a fringe of fin along the sides. It measures between 10 and 40 cm long. The body is normally camouflage colours – when excited, *Sepia* can change colour very quickly. A male *S. officinalis* is striped like a zebra during courtship. The spermatozoa are stored by the female in sperm sacs, so that she only fertilises the 5–7 mm long eggs as she lays them. The eggs are coloured black because of an inky secretion; they are shaped like a lemon and are stuck onto plants, worm-tubes etc, in clusters. Many *Sepia* species are pleasant to eat. The cuttlebone is a good source of calcium for cage-birds, and the contents of the ink sac in this genus provides us with the dyestuff known as sepia. There are a great many problems with keeping *Sepia* species in captivity. Essentials are as follows: large aquaria, the very best water quality, a very good amount of filtration and a continual supply of suitable food animals. Large specimens are the most difficult to adapt to life in the aquarium – it is far better to keep young specimens reared from eggs. To start with, they will eat small crustaceans (also *Daphnia*); later on, they will accept shrimps, crabs and fish. Be careful if you touch *Sepia* – they can give a nasty bite! The commonest species along European coasts is:

– *S. officinalis* LINNAEUS, 1758; Common Cuttlefish. E Atlantic, North Sea and Mediterranean. 30 cm. Dorsal surface yellow-brownish, ventral surface bluish-white. Eggs laid from spring to autumn.

Sergeant Major *(Abudefduf saxatilis) see Abudefduf*

Serpae Tetra *(Hyphessobrycon serpae) see Hyphessobrycon*

Serpula LINNAEUS, 1758. G of the marine Bristle-worms (Polychaeta*). They are sessile creatures that feed on matter in suspension. They often form colonies and have a chalky tube, often round in cross-section, out of which they extend their colourful crown of tentacles (up to 1 cm wide). A typical feature is the stalked lid with which they can shut their shell. *Serpula* species are hardy aquarium specimens that do not require any particular feeding. *S. vermicularis* LINNAEUS, 1758 is found all over the world. Its tube is up to 5 mm thick and its tentacles are a reddish colour.

Serranellus JORDAN and EIGENMANN, 1890. G of the Serranidae*. From tropical to temperature seas. Territorial fish, 10–40 cm long, that feed on crustaceans, molluscs and small fish. They are similar to *Serranus**, but usually have 2 spines on the main operculum and a scaleless lower jaw. *Serranellus*, like many Serranids, is an hermaphrodite. *S. subligarius* (COPE) from the Caribbean Sea can even fertilise its eggs itself, even when it is the only specimen in the tank. Under natural conditions, several fish spawn together. These are prettily patterned species that are excellent for the aquarium. Eg:

– *S. scriba* (LINNAEUS, 1758); Belted Sandfish, Banded Sea Perch. Mediterranean, E Atlantic. 25 cm. There are 5–6 reddish-brown transverse bands on the body, a pale blue blotch on the belly, and scroll markings on the head, like hieroglyphics.

Serranidae; Sea Basses (fig. 1041). F of the Perciformes*, Sub-O Percoidei*. Found mainly in tropical and subtropical seas, more rarely in brackish water and only very occasionally in fresh water. The Serranidae are predators and have an elongate, laterally flattened body that is robust. The head is large with a protrusible, often deeply slit mouth. Usually, the head is completely covered with scale. The jaws contain rows or bands of pointed teeth, and the palatal bone is normally toothed. 2 nasal orifices on each side. The operculum has 1–3 flat spines. The edge of the preoperculum is serrated. D consists of a long, hard-rayed part that continues immediately into a shorter, soft-rayed section, or via a slight indented area first. C straight or rounded, rarely indented. An short, with at least 3 hard rays on the front edge. The Vs are normally well to the front, often directly below the Ps. LL complete; very small to medium-sized, firmly rooted ctenoid scales. The Serranidae are closely related to the True Perches (Percidae*) – the oldest fossils come from the Cretaceous period. They live mainly in shallow water zones above the foot of the continental blocks; many are found around coral reefs. Many species have a lovely colouring, and in many places they are considered extremely tasty. Some species grow quite huge, such as the Stone Bass *(Polyprion americanus)* from warm parts of the Atlantic and the Mediterranean, which can grow to over 2 m. Even bigger is the Giant Sea Bass *(Stereolepis gigas)* which can weigh as much as 350 kg. There have been observations in the aquarium of many Serranid species, see under the Ga *Anthias**, *Cephalopholis**, *Chorististium**, *Chromileptis** (fig. 1042), *Epinephelus**, *Grammistes**, *Seranellus**, *Serranus**.

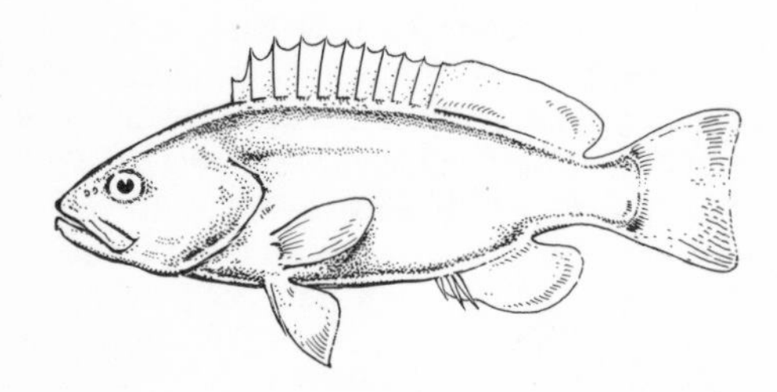

Fig. 1041 Serranidae-type

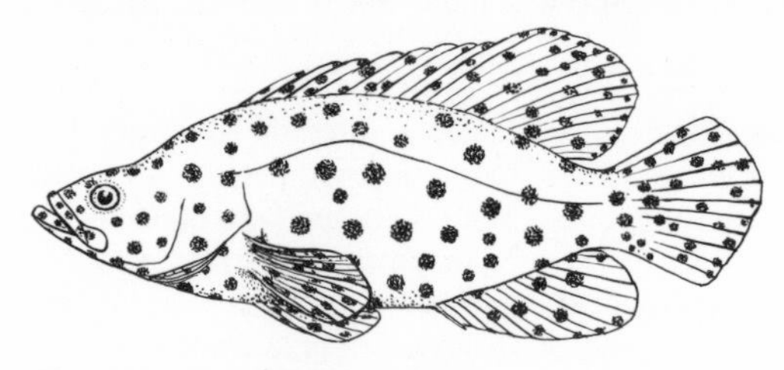

Fig. 1042 *Chromileptis altivelis*

Serranus CUVIER, 1844. G of the Serranidae*. Numerous species found in tropical to temperate seas where they live on rocky and coral coasts. They are territorial predatory fish, living on fish, crustaceans and molluscs. *Serranus* has a typical perch shape, only 1 D and the main operculum has 3 spines; however, there is no longitudinal ridge, and the lower jaw is scaled *(see Serranellus)*. There is much dispute as to whether many of the species in this G *do* belong here. They are hermaphrodites, and measure 10–140 cm long. The larger species are important as food and as sportfish, but uncontrolled harpooning in some places has made them scarce sometimes. The smaller species are suitable for the aquarium. Eg:

– *S. tigrinus* (BLOCH); Harlequin Bass. Caribbean Sea. 15 cm. A whitish-yellow background colour with a brown-black patterning.

Serrasalmidae; Piranhas (fig. 1043). F of the Cypriniformes*, Sub-O Characoidei. Tall to disc-shaped Characoidei, very flat from side to side, and found in central and northern S America. Characteristic features of the F are as follows: the premaxillae have 1 row of very sharp teeth (eg G *Serrasalmus*) or 2 rows of incisor or grinding teeth, or 1 row of incisor teeth and 2 rows of grinding teeth (fig. 1044); the maxilla is reduced and does not carry teeth; the lower jaw has usually 1 row of teeth, and the pterygoid can also be toothed (eg *Serrasalmus*). The keel of the belly in front of the C reaches well forwards, and has serrated teeth on it. All fins are present; the D has 16–17 fin rays normally, more in only a few cases (eg G *Myleus*). On rare occasions, there is a single, spine-like soft ray in front of the

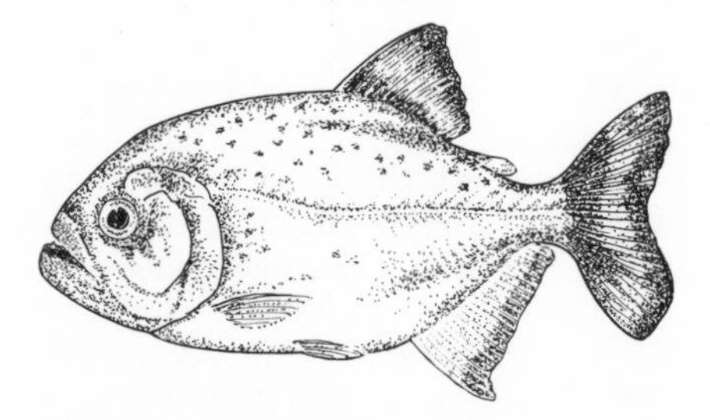

Fig. 1043 Serrasalmidae-type

D; in all species the adipose fin is distinct; C is often only slightly indented; An much longer than the D, occasionally bowing forwards or elongated into a tip; V fairly small. Cycloid scales, very small, and only fixed in loosely. The best-known members of the F are the Piranhas of S America – their legendary ferociousness is not matched by any other freshwater fish. They often provide an episode in many an adventure yarn, although, in reality, the story is somewhat different. Firstly, some species are dangerous, others are not.

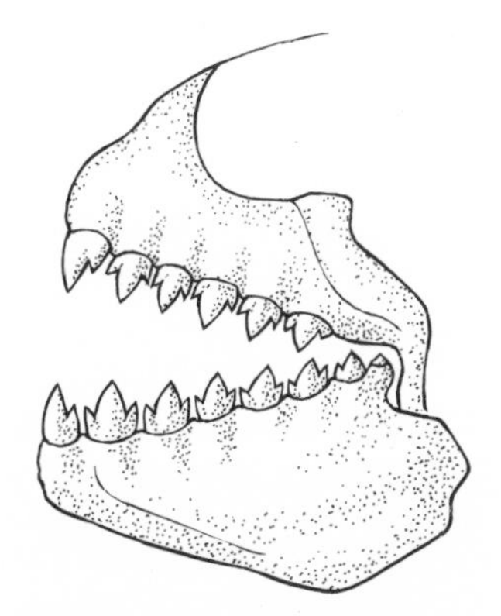

Fig. 1044 Piranha dentition

There are around 16 species, and only 4 are really dangerous. As adults, these kind of Piranhas are easiest to recognise from the swollen, protruding chin. With their unusually sharp and powerful dentition, they are able to tear out a large chunk of flesh with one bite. As Piranhas go around in large packs, and as they are extremely fast and agile fish that rush upon their prey in unison, it is easy to see that very soon there is not going to be much left of the victim. However, their main food consists of sick fish, and so Piranhas are, if you like, a way of policing health in a stretch of water. In reality, the danger is very slight for any mammals or humans that do not have open wounds. Normally, Indians bathe quite happily in stretches of water where dangerous Piranhas are known to live, and in many places they are caught on wire ropes or speared, as their flesh is said to taste excellent. HARALD SCHULTZ, the anthropologist and ichthyologist, travelled through the Amazon region for many years, and he once parodied the exaggerated danger of Piranhas thus: 'When my father was 15 years old, he had to make his escape from attacking Indians in a small, unsteady log canoe. It overturned. So he tried to escape by swimming, but it was as a skeleton only that he was able to climb out of the water! Such a thing never happened to him again.' (quoted from LÜLING, 1973). Normally, only young Piranhas are kept in aquaria; they have been bred successfully, *see under* the G *Serrasalmus*. Various other Ga are interesting shoaling fish and are caught as food in their natural habitats. They also make interesting subjects for an aquarium. *See Catoprion, Metynnis, Mylossoma.*

Serrasalminae, Sub-F, *see* Characoidei

Serrasalmus LACÉPÈDE, 1803. G of the Serrasalmidae*; Piranha. South-American Characins. Length to 40 cm. This G comprises 16 species, some of which are dangerous and greedy predators, (the Piranhas) that live in large shoals in the rivers and streams of the Amazon Basin. With their knife-sharp dentition borne on short, muscular jaws, Piranhas are capable of overpowering large prey, such as deer and tapirs, or even humans, and reducing them in a very short space of time to skeletons. As they do this, they can effortlessly bite through tendons and so separate the bones from one another. However, as a rule, only animals that are already wounded as they enter the water will be attacked – the Piranhas are attracted towards them by the blood from the wound spreading out over the water. There are also fairly harmless species of Piranha, as well as very dangerous ones. Their main source of food is said to be fish. In captivity, most Piranhas are very timid. Only rarely can several specimens be kept together in the same tank for any length of time, as they often start attacking each other and eating each other. Probably the main reason why this happens is the confined nature of the tank. A Piranha's body is a tall elliptical shape, much compressed laterally. The D is inserted at the highest point on the dorsal ridge; it is right-angled and has only the suggestion of a tip. Adipose fin well developed. The broad C is only slightly forked, the An is fairly long and unlobed, usually with a clear tip in both sexes. Vs small, Ps short but powerful. Eyes very large. The jaws are powerful and contain a row of multi-pointed teeth very close to one another, in adult specimens. In captivity, *Serrasalmus* can be fed on lean mammal flesh (heart), fish meat, worms and large water insect larvae. It is extremely difficult to distinguish the species, especially when young. So far, 2 species have been bred: SCHMIDT has submitted a report that one species, probably *S. rhombeus* (LINNAEUS, 1766), was bred successfully in a large public aquarium in Duisburg. The eggs, about 4 mm large, were laid on water moss *(Fontinalis*)* and they hatched out after 2 days. 8–9 days later, the fry were swimming freely and, fed on *Artemia* nauplii, they grew very quickly. SWEGLES bred another species, probably *S. nattereri* (KNER, 1859), in a pond at the Rainbow Aquarium in Chicago. The fish produced about 5,000 large, gold-coloured eggs that were very sticky. They were laid above plants, and hatched out after 9 days. The parent fish did not eat the spawn and even watched over the spawning ground. After 14 days, they spawned again and were fed in the meantime with mice. The fry required large amounts of food (type of food not described in any clearer detail) as they grew – this was only so that they did not try to tear out each other's eyes to eat. Only young Piranhas can even be considered for the aquarium – adult fishes are extremely attractive exhibits for large, public aquaria.

– *S. nattereri* (KNER, 1859); Red Piranha, Natterer's Piranha, Red-bellied Piranha. Widely distributed in the Amazon, Orinoco and Paraná. Length to 30 cm. Flanks pale brown or olive with a strong silvery iridescence and a large number of metallic green glittering spots. Dorsal surface blue-grey; throat, ventral surface, Ps, Vs and An a shining cinnabar-red or blood-red. D and C dark, the latter with a see-through area. The An has a broad black edge. 24–31 saw-teeth on the keel of the belly. The

colouring varies enormously during the various stages of growth. For information on breeding, see under generic description. A very aggressive species.
– *S. piraya* CUVIER, 1819; Piraya, Piranha. Rio San Francisco. Length to 35 cm. Very similar in colour to the species *S. natteri*, except the flanks are more blue-green. 22–24 saw-teeth on the keel of the belly. A characteristic feature of older specimens (those measuring more than about 12 cm) is the frayed adipose fin, that remains round like a lobe in all the other species. A very dangerous species. *Pygocentrus piraya* (CUVIER, 1819) MÜLLER and TROSCHEL, 1844, is another name for this species, but it is no longer used (fig. 1045).

Fig. 1045 *Serrasalmus piraya*

– *S. rhombeus* (LINNAEUS, 1766); White Piranha, Spotted Piranha. From the Amazon and north-east S America. Length up to 38 cm. Body shape more elongated than the other species. Head elongated, forehead profile slightly concave. Dorsal surface and upper half of the body dark-grey to olive; lower part of the body a dirty white with a silver sheen. Large numbers of small, dark dots are scattered over the flanks, and behind the operculum there is a dark shoulder fleck. Fins grey, C blackish with a paler to reddish middle part. Old fish often become a monotone black. The sexes can be distinguished – in contrast to other species of the G – by the elongated An in the males (either an arch shape or a peak shape). In the females, this fin is straight-edged. 37–38 saw-teeth on the keel of the belly. For breeding information see under generic description. This species is said to be less aggressive (fig. 1046).

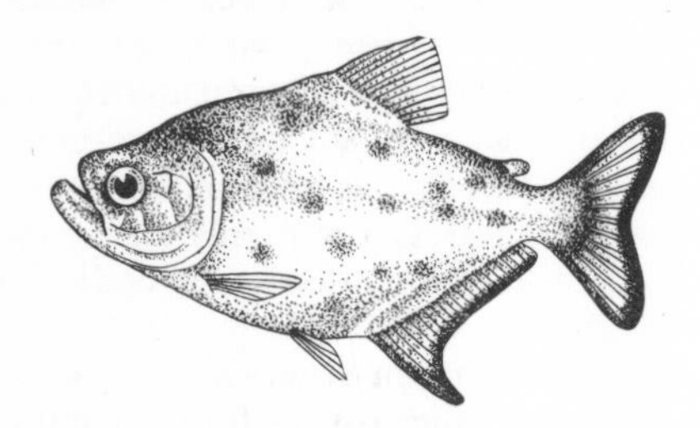

Fig. 1046 *Serrasalmus rhombeus*

Serrivomeridae, F, *see* Anguilliformes

Sertularia LINNAEUS, 1758; Sea Hair. G of the Hydrozoa*. Forms richly branching colonies that are supported by the horny external skin (periderm). Sea Hairs are found especially in northern seas on a firm substrate where they often form very regular dense mats. Usually, only the dead periderm is used as a decoration in the marine aquarium, although it is not very sturdy.

Sessile; sedentary, attached. Eg Acorn Barnacles *(Balanus*)* on rocks and breakwaters, and Sea-anemones *(Ceriantharia*)* in sand etc.

Sessile Jellyfish (Stauromedusae, O) *see* Scyphozoa

Sex Cells *see* Gametes

Sex Determination. The mechanics of determining the sex of offspring. In general, sex determination is brought about by fertilisation (*genotypic* sex determination); it is rare for external factors to be responsible for it (*phenotypic* sex determination).

The sex-determining hereditary factors are located in the cell nucleus, often in two special sex chromosomes (*see* Cell) known as the X and Y chromosome. In humans, the female has two X chromosomes (XX), the male an X and a Y chromosome (XY). If two X chromosomes unite during fertilisation, a female will develop from the egg-cell, whereas an X and a Y combination will produce a male. Similar circumstances exist in many fish. For example, in the Tooth-Carps there are the following groupings of the sex chromosomes.

	Male	*Female*
Poecilia nigrofasciata	XY	XX
Poecilia reticulata	XY	XX
Xiphophorus xiphidium	XY	XX
Xiphophorus variatus	XY	XX
Xiphophorus maculatus (Mexican phylum)	XY	XX
Xiphophorus maculatus (phylum from Honduras)	XX (= ZZ)	XY (= WZ)
Xiphophorus helleri	No sex chromosomes	

Many species of fish (eg *Xiphophorus helleri*) have no sex chromosomes – their sex-determining genes are located in the normal chromosomes. In such cases, both male and female differential sex genes are present, usually one being more dominant than the other, thus producing the final sexual characteristics.

Sex Differences *see* Sexual Dimorphism

Sex Division in Higher Plants. Among the higher plants (angiosperms), the typical flower is hermaphroditic, ie it has both stamens and carpels. However, particularly among plants that are pollinated by the wind or by water, single-sex flowers occur that contain only stamens or carpels (this happens because one sex becomes involuted). Thus, the flowers are either male or female. If both are found on the one plant, it is described as monoecious *(Sagittaria)*. A dioecious plant is one where single-sex flowers grow on separate plants *(Vallisneria, Elodea)*. In the latter case, therefore, there exist male and female plants.

Sex Organs (fig. 1047). In animals it is important to distinguish between *internal* sex organs (genital glands, gonads) which produce and release sex cells (Gametes*) and *external* sex organs which perform the function of procreation. In bony fish, the gonads lie in the intestinal region of the body cavity. In the female sex, the egg-cells form in a pair of *ovaries* which are sometimes fused. After the multiplication phase, the ovification (oogenesis) is concluded in the oviducts. These ducts

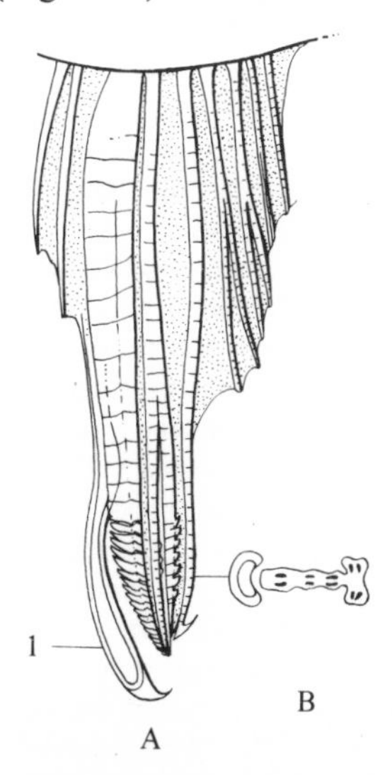

Fig. 1047 Sex organs. A Gonopodium of the Guppy *(Poecilia reticulata)*, 1 Hook B Cross-section through the gonopodium at the point indicated

open to the outside via paired or unpaired sexual openings. Some species, such as the Salmonidae, do not have oviducts that go right through; instead, the eggs drift about freely in the abdominal cavity. In the male sex, the seeds are formed (spermiogenesis) in paired elongated *testes*. The ripe sperm are collected mainly in the deferent ducts (more rarely in the seminal receptacles), which likewise open to the outside of the body in the anal region. The gonads in adult fish too vary in size depending on the reproductive cycle. Livebearing (viviparous) fish have special *brood sacs* on the ovaries or as parts of the oviducts. It is in these sacs that the embryonic development* takes place after the eggs are fertilised. These fishes and the egg-laying species exhibiting internal fertilisation (ovoviviparous species) have *copulatory organs* in the male sex. In bony fish (Poeciliidae, Goodeidae, Hemirhamphidae etc) they appear on the anal fin after complicated transformations. On the gonopodium of the Guppy, for example, the 3rd–5th fin rays form lateral grooves along which flow pockets of sperm (spermatophores) when they are introduced into the female sexual opening. The groove-like, thickened parts (pterygopodia) on the longer inner edge of the ventral fins have the same function in the Sharks and Rays. The *genital papilla* found in the Cichlids for example, is the easily extendable mouth of the oviducts or of the deferent ducts. Its purpose is to help lay the eggs and get them fertilised.

Sex Reversal. The formation of opposite sexual features in sexually different individuals. Sex reversal is possible in one sex because of the presence of male and female hereditary material (bisexual potential). In vertebrates, the ratio of male and female sex hormones (expecially androgen and oestrogen) is a decisive factor in sex determination. Experiments have been carried out to induce a sex reversal by treating the individual with male or female sex hormones. Among the fish, this has been carried out repeatedly on Guppies and on Swordtails.

Natural sex reversal occurs, for example, in the Serranid fishes. In this particular case, the age of the fish determines whether it is male or female. Spontaneous sex reversal has been observed among aquarium fish, especially among the Livebearing Tooth-carps (eg *Xiphophorus**, *Poecilia**, *Heterandria**). It is mostly fully-grown females that develop completely masculine sexual features. The genetic reasons for this are usually very complicated to explain.

Sexual Dimorphism (fig. 1048). A very common occurrence in the animal kingdom, whereby sexually separate species differ not only in their male and female sex organs *(primary sexual dimorphism)* but also in other anatomical features *(secondary sexual dimorphism)*. Sexual dimorphism is determined through heredity (Sex Determination*) in the majority of cases. In fish, secondary sexual dimorphism concerns size, body shape, fin arrangement and colouring in the main, but there may also be differences in the skeleton and in the teeth. Females are often bigger when fully grown – this can clearly be seen, for example, in the Livebearing Toothcarps. An extreme case is the dwarf male Deep-sea Anglerfish which grows firmly onto the female. In the Dwarf Cichlids, however, the males are larger. Colouring is usually more intense in the male; sometimes he may even have different colour patterns. The unpaired fins of the male are usually bigger and drawn out into a point in many species, eg in Labyrinth fishes. Sexual dimorphism is often apparent only during the breeding season, eg in the form of courtship colouring or spawning tubercles.

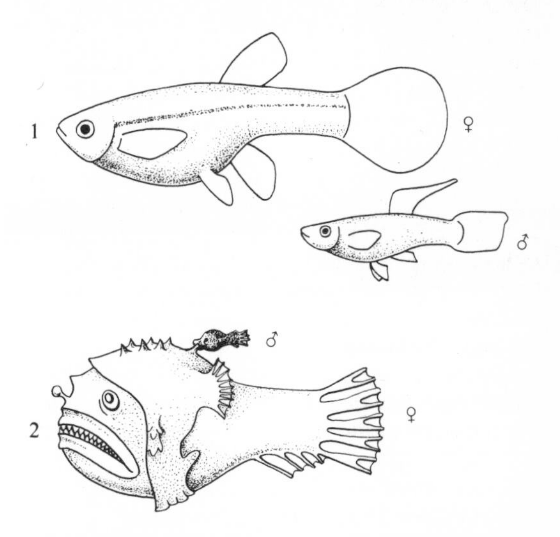

Fig. 1048 Sexual dimorphism. 1 In Livebearing Tooth-carps 2 Dwarf male in Deep-sea Anglerfishes

Sexual Isolation. The complete separation of species brought about because it is impossible for the populations* concerned to exchange hereditary material. Such sexual isolation can be caused by many things – perhaps by anatomical specialisation of the copulatory organs, different ecological needs, different behaviour, different reproductive periods, cytogenetic factors or geographical separation. It is this last-mentioned factor that contributes greatly to the formation of species.

Sexual Reproduction *see* Reproduction

Sharks (Selachiformes, O) *see* Chondrichthyes

Sharksucker *(Echeneis naucrates) see* Echeneidae

Sharksuckers, F, *see* Echeneidae

Sharp-nosed Climbing Perch *(Ctenopoma oxyrhynchus) see Ctenopoma*

Sharp-toothed Tetra *(Acestrorhamphus hepsetus) see Acestrorhamphus*

Sheepshead Acara *(Aequidens curviceps) see Aequidens*

Sheepshead Minnow *(Cyprinodon variegatus) see Cyprinodon*

Shelford's Coolie Loach *(Acanthophthalmus shelfordi) see Acanthophthalmus*

Shipworm *(Teredo navalis) see* Boring Molluscs

Shoal Spawners *see* Selection of Breeding Fish

Shoreline is the point at which an open water merges into dry land. Usually, the shoreline is formed from the layering of substrate material and dead plant parts from the bottom upwards. In relation to the depth of the water a certain kind of plant zonation is formed, therefore, known as shoreline vegetation. If a stretch of water has a low pH value, a covering of vegetation may also spread onto the water surface and the material that is dying is then pressed downwards. Then the shoreline will form from above downwards.

Short Pyrrhulina *(Pyrrhulina brevis) see Pyrrhulina*

Short-bodied Catfish *(Brochis splendens) see Brochis*

Short-day Plant *see* Photoperiodism

Short-term Baths *see* Bathing Treatments

Shovel-nosed Catfish *(Sorubim lima) see Sorubim*

Shovel-nosed Ray *(Rhinobatos rhinobatos) see* Rhinobatidae

Shrimps and Prawns, Sub-O, *see* Natantia

Shubunkin *(Carassius auratus auratus) see Carassius*

Siamese Fighting Fish *(Betta splendens) see Betta*

Siamese Twins *see* Twins

Sickle Characin (*Hyphessobrycon* sp.) *see Hyphessobrycon*

Side-striped Rasbora *(Rasbora lateristriata) see Rasbora*

Siebold, Karl Theodor Ernst von (1804–85). Doctor, zoologist and ichthyologist. Born in Würzburg. Studied medicine in Berlin and Göttingen. By 1840, he was Professor of Physiology and Comparative Anatomy at Erlangen, later at Breslau and Munich, where he was appointed Professor of Zoology in 1854. He made extensive and exact studies of central-European fish fauna, the results of which appeared in the book entitled *Die Süßwasserfische von Mitteleuropa* ('The Freshwater Fishes of Central Europe') (1863). Even today, his descriptions are among the best ever made.

Siganidae, F, *see* Acanthuroidei

Siganus vermicularis *see* Acanthuroidei

Sight, Organs of. Organs found in countless invertebrates and vertebrates that enable them to orientate themselves in the special sense, to seek food, or to recognise like-species and enemies. When light-sensitive cells are dispersed over the body or on parts of the body, there exists the simplest orientation method – the distinction between light and dark (eg as found in many annelids and echinoderms). Many invertebrates are able to see in particular directions by individual sense cells or groups of sense cells being directed in a particular way towards a light source, at the same time being screened off on one side by pigment cells (eg the pigment-cup eyes of the turbellarians or the optic vesicles of many snails). Organs of sight may also be equipped with lenses (eg as in many Medusae, molluscs, Bristle-worms and arthropods). The compound eyes of arthropods and the eyes of vertebrate animals (in conjunction with a nervous system) enable them to see in pictures. Crustaceans, insects and vertebrates can also see in colour.

The vertebrate eye (fig. 1049) consists of a cone-shaped optic cup, into the opening of which light falls (it has to cross through the cornea and the lens that lies behind it). The inner cavity of the eye is largely made up of a watery vitreous body. The inner wall of the optic cup is made up of the light-sensitive retina whose sense cells (rods and cones) receive the picture projected through the lens. From there outwards there follow a layer of pigment cells, the choroid (arachnoid membrane) and the sclerotic membrane next to which the eye muscles are located. The amount of light let into the eye is regulated by a structure in front of the lens known as the iris. The range of view (focus) is adjusted by the lens. A fish's eye, unlike that of a mammal, is adjusted to close vision when at rest as befits conditions for seeing in water. For long-distance vision the spherical lens is pulled towards the retina. In mammals the shape of the lens is altered.

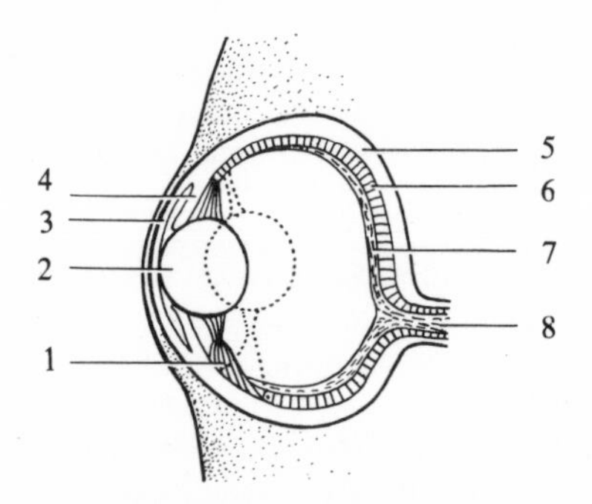

Fig. 1049 Eye of a bony fish. 1 Retractor muscle 2 Lens 3 Cornea 4 Iris 5 Sclerotic membrane 6 Arachnoid membrane 7 Retina 8 Optic nerve

The information received by the eye is conducted to the vision centre of the brain via the optic nerve.

Sign Stimulus *see* Key Stimulus

Signal Movement *see* Expressive Behaviour

Silicon Rubber *see* Aquarium Sealants

Siluridae; True Catfishes (fig. 1050). F of the Siluriformes*, Sup-O Ostariophysi. Naked Catfishes first appeared in the early Miocene period, and today they are found in Europe, Africa and Asia. The body is broad at the front, like an oblique oval in cross-section, but increasingly flattened laterally towards the rear end. D very short, small and with no spine; adipose fin absent, C small, in some Ga fused with the unusually long An which is like a fin fringe. V small, in rare cases absent, P with a powerful spine. Eyes small and usually covered by a transparent skin. Locomotion is achieved mainly by the long An. Most Catfishes are crepuscular or night-time fish – included here is the Wels or European Catfish *(Siluris glanis)* which reputedly grows up to 3 m long, although most of them are around 1 m long. It is one of the tastiest of all fish from Central and E Europe.

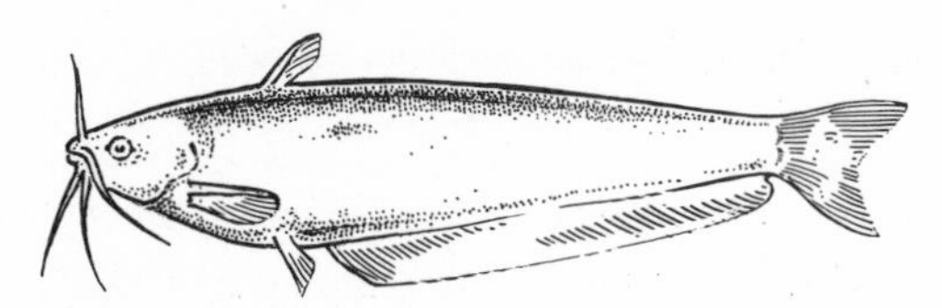

Fig. 1050 Siluridae-type

But also in this F are species that are decidedly active by day; some of them even live in large shoals, eg *Kryptopterus**. All the larger kinds of Catfishes are edible fish, usually eaten fresh. Some Ga, because of their general importance, or because they are important in aquarium-keeping, are treated under separate headings: *Ageniosus, Kryptopterus, Silurus.*

Siluriformes; Catfish-types. O of the Osteichthyes*, Coh Teleostei. With roughly 2,000 species extant today, the Siluriformes are one of the most abundant O of the Osteichthyes. The oldest fossils date from the Palaeogene period. Together with the Cypriniformes they form the Sup-O Ostariophysi, and in common with them the Siluriformes also have the characteristic feature of the Weberian apparatus (*see under* Cypriniformes). They are differentiated by the following features: specialisation is further advanced in almost all the 30 or so Fs than it is in the majority of Cypriniformes; in the skull there is no parietal bone, no symplecticum and no subopercular bone, some parts of the pharyngeal skeleton are also missing; the vomer, palatinum and pterygoid are toothed; usually, there are no fish-bones; the 2nd–4th vertebrae in the Weberian apparatus are fused into one; the swimbladder is divided into a front and a rear section (not simply a narrow constriction!); skin naked or covered with bony plates, never true scales – in some species, there are small placoid scales.

Catfish-types are found in fresh waters worldwide. Only some Ariidae* (Sea Catfishes) and the Plotosidae* (Coral Catfishes) live in the sea, in tropical and subtropical coastal waters. For the freshwater forms, the main area of distribution lies in S America and Africa. The most primitive Catfishes are the Diplomystidae – they still have a toothed maxilla and a primitive Weberian apparatus. But primitive features are also found among the Loricariidae* and the Callichthyiidae*. Most Catfishes are bottom-dwelling fish, active at dusk and live a solitary existence; only a few species swim around freely in the main, are active by day, and form schools or shoals. Adaptations to a night-time life-style include the partial regression of the eyes and the formation of an efficient system of barbels – this enables the Catfish to orientate itself in the dark and to recognise food. Many species are important in the fishing industries of their natural areas of distribution, and they are the tastiest of all fish. Many small and very small species are popular aquarium-fish that are always fascinating objects to observe. For information on the Fs comprising the Siluriformes, *see under* Osteichthyes. Each F is treated separately under its own entry.

Silurus LINNAEUS, 1758. G of the Siluridae*. Naked Catfishes distributed in the Mediterranean and E Europe, as well as W Asia. To 3 m. Predatory fish, active at night, that rest by day in hiding-places along the bank, or on the substrate of rivers and lakes. They feed on fish and even waterfowl. Body elongate, head massive and flattened with a mighty mouth slit, small eyes and 3 pairs of barbels, of which the longest pair extends from the upper jaw (the other 2 pairs project from the lower jaw). D1 very small and inserted far forwards, no D2, C fairly small and not lobed. When swimming, the very long An is moved in wavy lines. Vs and Ps well developed. The spawning period is in May and June, and up to 100,000 eggs are laid in shallow hollows, previously carved out with the fins. These hollows are usually among plants. The spawning ground is then watched over by the parents. The young Catfishes grow quickly and to start with they are a uniform black colour. Only small young fish are suitable for the aquarium – they are best kept alone. Larger specimens are popular as exhibits in zoo aquaria.

– *S. glanis* LINNAEUS, 1758; Wels, European or Danube Catfish. Central and E Europe, W Asia. To 3 m, although most are about 1 m long. Body colouring murky, upperside olive-brown, paler on the flanks with a reddish-brown sheen and a dark marbly effect. Underside a dirty-white. There are also monotone blue-black or completely pale specimens too. Fins violet-brown. *S. glanis* is a well-loved edible fish (fig. 1051).

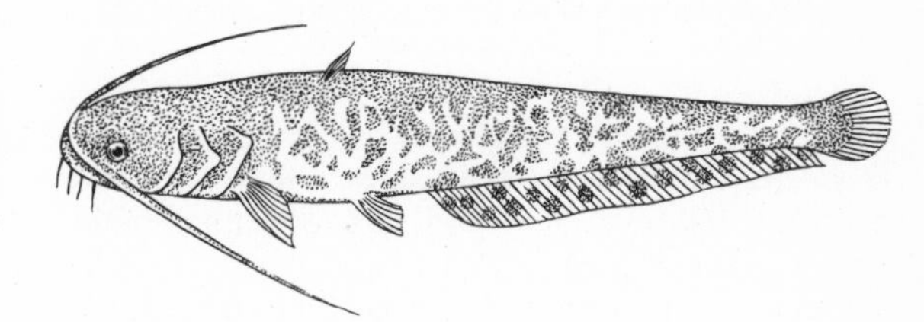

Fig. 1051 *Silurus glanis*

Silver Barb *(Barbus viviparus) see Barbus*

Silver Beetles, G, *see Hydrous*

Silver Bream *(Blicca bjoerkna) see Blicca*

Silver Chromide *(Etroplus suratensis) see Etroplus*

Silver Gambusia *(Gambusia affinis) see Gambusia*

Silver Hatchetfish *(Thoracocharax securis) see Thoracocharax*

Silver Hatchetfish *(Thoracocharax stellatus) see Thoracocharax*

Silver Rasbora *(Rasbora argyrotaenia) see Rasbora*

Silver Shark *(Balantiocheilus melanopterus) see Balantiocheilus*

Silver Tetra *(Ctenobrycon spilurus) see Ctenobrycon*

Silver Tetra *(Gephyrocharax atracaudatus) see Gephyrocharax*

Silver Tip *(Hemigrammus nanus) see Hemigrammus*

Silver-bellied Climbing Perch *(Ctenopoma argentoventer) see Ctenopoma*

Silver-green Fanwort *(Cabomba caroliniana* var. *tortifolia) see Cabomba*

Silver-leaf Fin *(Monodactylus argenteus) see* Monodactylidae

Silversides (Atherinidae, F) *see* Atherinoidei

Silver-tipped Tetra *(Hemigrammus nanus) see Hemigrammus*

Simochromis BOULENGER, 1898 (fig. 1052). G of the Cichlidae*. 4 species only found in Lake Tanganyika*. These fish have an unusual head and body shape, and are up to 20 cm long. Mouth terminal, small. Up to the level of the eyes, the naso-forehead profile climbs vertically and then arches in a gentle curve to the insertion of the D. Chin area likewise arched. The shape of the head thus becomes very truncated and short; the body shape is like a droplet taken in total. D inserted along a long stretch of the body, but low. The Ps are best developed; C clearly indented.

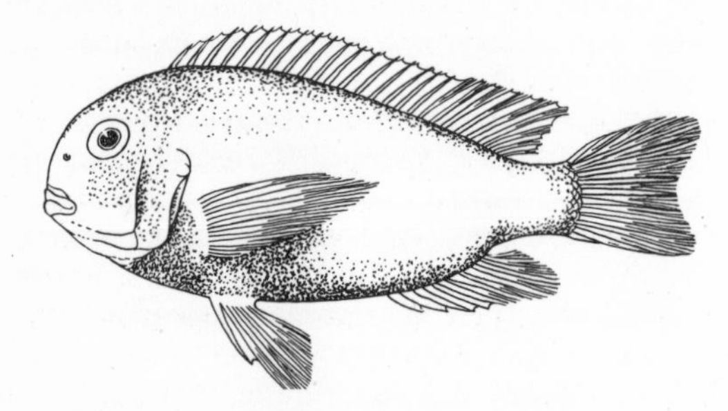

Fig. 1052 Representative of the G *Simochromis*

Simochromis is kept in the aquarium as per instructions given for *Lamprologus**, *Julidochromis** and *Tropheus**. Specimens need large tanks with rocky structures and caves, plus hard, slightly alkali water and a temperature of around 27 °C.

Simulium LATREILLE, 1802; Black Flies (fig. 1053). G of the Midges (Nematocera*). Represented in Europe by about 40 species. In the larval and pupal stages, *Simulium* prefers flowing waters and at times they can be found there in large numbers on firm substrata, such as rocks, plants and wood. The larvae anchor themselves firmly with the rear end of the abdomen, leaving their body dangling in the current. A typical characteristic are the 2 fan-like crowns of hair on the head which form a complicated filter apparatus for trapping food. The pupal stage is spent in a cone-shaped shell which they spin for themselves. Black Flies themselves are small, hunch-backed insects that swarm in bright sunshine and attack large kinds of mammal. Their stings are painful and may also transmit pathogens. Large

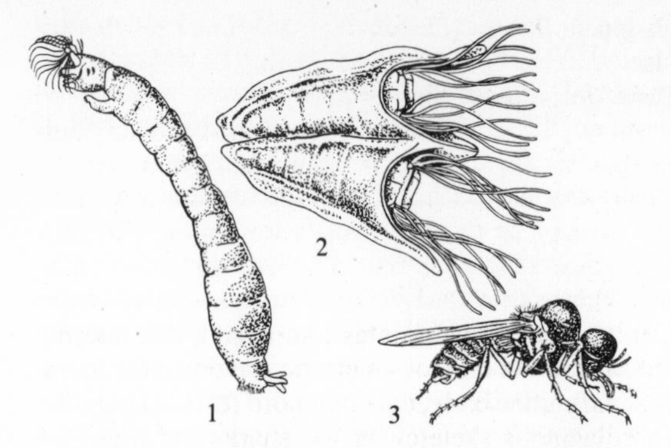

Fig. 1053 *Simulium*. 1 Larva 2 Pupae 3 Imago

numbers of stings can damage the circulation, and in some cases paralysis of the heart has even been known.

Singapore Angel *(Monodactylus argenteus) see* Monodactylidae

Single-celled Animals, Ph, *see* Protozoa

Single-rooted Duckweed *(Lemna trisulca) see Lemna*

Siphamia WEBER, 1909 (fig. 1054). G of the Apogonidae*. Comprises several species from the tropical Indo-Pacific. These fish live in groups and feed on small animals. They are small, compact, large-scaled fish that have silvery luminescent glands extending from the tongue, down both sides of the belly to the An. Some species have become known as 'Sea-urchin Fishes'. During the day, they remain between the long spines of the Diadem Sea-Urchin, thus affording themselves protection. *S. versicolor* SMITH and RADCLIFF cleans the upper surface of the Sea-urchin. No reports of any aquarium observations.

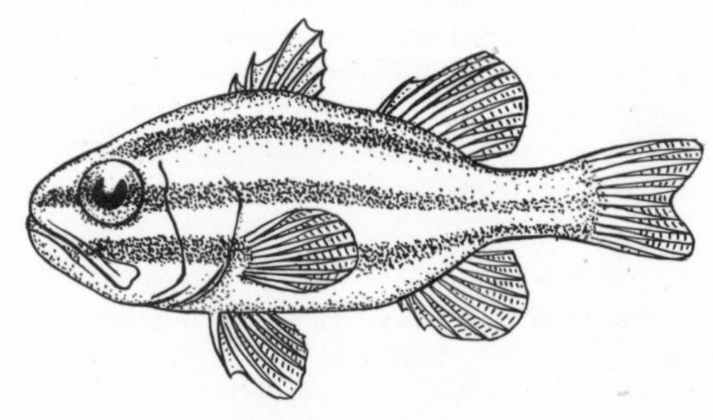

Fig. 1054 Representative of the G *Siphamia*

Siphonophora, O, *see* Hydrozoa

Sipunculida; Sipunculoids. A marine animal Ph. Unsegmented, cylindrical animals that live in soft substrata. They have tentacles arranged in a star-shape on an extensible proboscis, and they use them to catch small animals as well as detritus. Occasionally, they are inadvertently introduced into the marine aquarium.

Sirenia, O, *see* Mammalia

Sium, G, *see* Apiaceae

Six-banded Barb *(Barbus pentazona hexazona) see Barbus*

Six-banded Sergeant Major *(Abudefduf sexfasciatus) see Abudefduf*

Six-barred Panchax *(Epiplatys sexfasciatus) see Epiplatys*

Six-gilled Shark *(Hexanchus griseus) see* Hexanchidae

Skate *(Raja batis) see* Rajidae

Skeleton (fig. 1055). A formation found in many animals that mainly serves as an anchor point for the musculature* and at the same time offers stability to the body. In contrast to the invertebrates, many of which have formed external skeletons (eg a chitin or chalky armour), the vertebrates always have an internal skeleton. In the Cephalochordates (Acrania*), this internal skeleton consists only of an elastic rod down the longitudinal axis of the body, the notochord *(Chorda)*. Unlike the cartilaginous skeleton of the sharks and rays (*see* Chondrichthyes), in the bony fish, the skeleton is by and large made of bone. The following details are restricted in the main to the skeleton of the bony fish (*see also* Chondrichthyes, Gnathostomata, Osteichthyes). Along the body axis lies the *vertebral column*, which consists of a large number of vertebrae arranged one after the other in series. The cylinder-like vertebral bodies are sunken-in like funnels at the front and rear side. The gaps thus produced and the central vertebral canal are filled out by the Chorda throughout life. Towards the rear end, each vertebral body has a pair of neural arches which envelop the spinal cord and which unite to form a long, pointed, spiny process. The large blood vessels lying under the vertebral column are only protected in the tail region by similar ventral formations, the haemal arches. In the main body vertebrae, these processes are short and linked by joints to the *ribs*. The ribs themselves have free ends on the ventral side, as fish do not have a breastbone. In many species, parts of the vertebrae at the front may be in contact with the swimbladder (*see* Weberian apparatus). A jointed cervical region is not apparent, which means that the vertebral column is fixed fairly rigidly to the *skull*. Its upper part, the neurocranium, encloses the brain and protects the olfactory organs, eyes and balancing organs. It is joined together like a capsule made up of a large number of individual bones. The lower part of the skull (viscerocranium) is moveable and comprises the mandibular arches with the upper and lower jaw, the hyoid arch and 4–5 gill arches as well as the operculum bones. All the parts are divided up further.

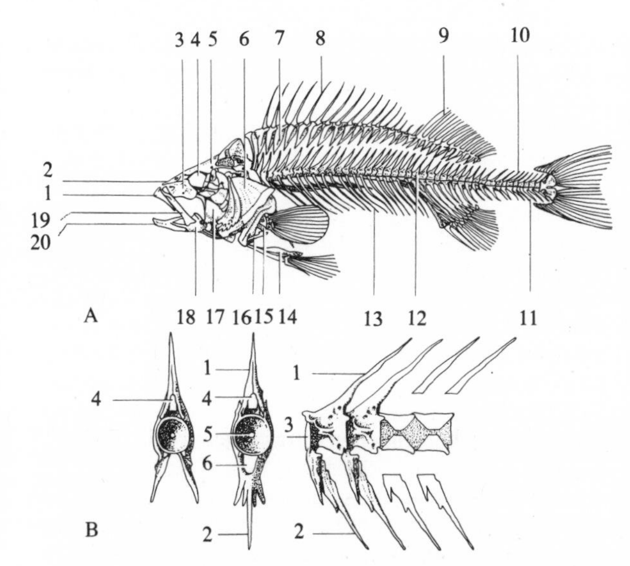

Fig. 1055 A skeleton of a bony fish. 1 Premaxilla 2 Nasal bone 3 Lacrimal bone 4 Orbit 5 Parietal bone 6 Operculum 7 Fin ray carrier 8 Hard rays 9 Soft rays 10 Upper thorn processes 11 Lower thorn processes 12 Vertebra 13 Ribs 14 Pelvic girdle 15 Shoulder girdle 16 Coracoid 17 Quadratum 18 Articular bone 19 Upper jaw 20 Lower jaw

B Left, torso vertebra from the front; right, tail vertebra from the front, from the side and in longitudinal section. 1 Upper thorn processes 2 Lower thorn processes 3 Vertebral bodies 4 Spinal canal 5 Vertebral canal 6 Haemal canal

The supportive elements of the fins* are the *fin rays*. In the D and An, these rays rest on *fin ray carriers* that are often slotted in between the spiny processes and the vertebrae. The rays of the C can spring directly from the vertebral column. The Ps are connected to a simple *shoulder girdle* made up of some dermal bones, shoulder-blades and coracoid bones. The shoulder-girdle is often fixed to the skull. The *pelvic or hip girdle* consists of only 2 clasps of bone that provide a surface of attachment for the musculature of the Vs.

The scales* in bony fish are skeletal elements of the skin.

Skeleton Shrimps, G, *see Caprella*

Skin; integument, cutis (fig. 1056). The external covering of the body in multi-celled animals. It has many functions, eg as a protection from mechanical and chemical effects and infections; and it plays a part in respiration*, excretion*, water and salt economy and temperature regulation. The skin also contains a number of sense receptors, such as those of touch and temperature – a fish's skin also contains taste receptors as well as receptors that detect currents. The characteristic colouring* of an animal is usually produced by component parts of the skin.

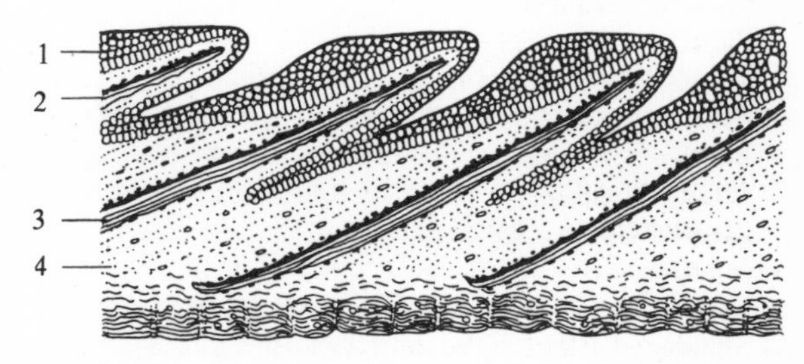

Fig. 1056 Skin of a bony fish. 1 Epidermis 2 and 3 Scales 4 Corium (dermis)

In vertebrate animals, the skin consists of 2 layers: the outer layer *(epidermis)* and the leathery skin beneath it called the *corium*. The outer layers of skin are constantly sloughing off and being replaced by cells from beneath. Unlike that of other vertebrates, the epidermis in fish is horny only in exceptional cases (eg Spawning Tubercles*), but it contains a number of gland cells which secrete slime in particular, as well as alarm substances and poisons. The corium in fish consists mainly of connective tissue with which are mingled blood vessels, nerves and chromatophores*. Bony scales, ganoid scales (*see* Scales) and fish teeth are all derived from skin. So too are the horny scales of reptiles, bird feathers and mammalian hairs.

Skin Cloudiness *see* Skin Diseases

Skin Diseases. Fish skin is an important respiratory organ and defence system. Extreme environmental factors, especially unfavourable water conditions, alter the skin and reduce its efficiency. Unsuitable pH values (Alkali Sickness*, Acid Sickness*), jumps in temperature (Temperature Damage*), lack of oxygen* and nutrition damage (Feeding*) are all examples of factors which can cause visible changes in the skin. At the very least they weaken the fish, thus creating the right circumstances for skin parasites to attack (*see* Parasitism). Signs of such an attack include cloudiness of the epidermis, skin chafing, unsteady swimming and increased frequency of breathing.

Skin damage or parasites can often be detected with the naked eye, but a categorical diagnosis is only possible by examining skin smears of living fish (*see* Examination Technique) under a microscope. In this way, single-celled parasites like *Ichthyophthirius**, *Trichodina**, *Ichthyobodo**, *Chilodonella**, *Cryptocaryon** and *Oodinium** can be detected. Fungi (Mycoses*) often follow a parasitical attack. It is difficult to prove the existence of bacterial infections* of the skin, which can produce inflammation*, scale protrusion*, spot diseases* etc. Apart from all these things, the skin can also be subject to attack from 'Worm' parasites (Helminthiasis*) and parasitical crustaceans (Copepod Attack*, Argulidae* etc). Serious fish diseases often start at the skin or produce characteristic skin symptoms (Stomach Dropsy*, Fish Tuberculosis*).

Skin Smear *see* Examination Technique

Skipper *(Scomberesox saurus) see* Exocoetoidei

Skippers (Scomberesocidae, F) *see* Exocoetoidei

Skull *see* Skeleton

Skunk Catfish *(Corydoras arcuatus) see Corydoras*

Skunk Loach *(Botia horae) see Botia*

Skunk Loach *(Botia sidthimunki) see Botia*

Slate-pencil Urchins, G, *see Heterocentrotus*

Sleep Movement (in plants) *see* Nastic Movements

Sleeper Gobies (Eleotrinae, Sub-F, name not in current usage) *see* Gobioidei

Sleeper Trout *(Mogurnda mogurnda) see Mogurnda*

Slender Cichlid *(Pseudotropheus elongatus) see Pseudotropheus*

Slender Pimelodella *(Pimelodella gracilis) see Pimelodella*

Slender Rasbora *(Rasbora daniconius) see Rasbora*

Slim Catfishes, F, *see* Pangasiidae

Slim Characins, F, *see* Lebiasinidae

Slim Fighting Fish *(Betta bellica) see Betta*

Slime Gland *see* Skin

Slimies, F, *see* Leiognathidae

Slimy Myersi *(Acanthophthalmus myersi) see Acanthophthalmus*

Slipper Lobster *(Scyllarides latus) see Scyllarus*

Slipper Lobster *(Scyllarus arctus) see Scyllarus*

Smallmouth Bass *(Micropterus dolomieu) see Micropterus*

Small-scaled Glass Characin *(Roeboides microlepis) see Roeboides*

Small-scaled Scorpionfish *(Scorpaena porcus) see Scorpaena*

Smell, Sense of, *see* Olfactory Organs

Smelt *(Osmerus eperlanus) see* Salmonoidei

Smelts (Osmeridae, F) *see* Salmonoidei

Smolt-Salmon *(Salmo salar) see* Salmonoidei

Smooth Hammerhead Shark *(Sphyrna zygaena) see* Sphyrnidae

Smooth Hound *(Mustelus mustelus) see* Triakidae

Smooth-head Unicornfish *(Naso lituratus) see Naso*

Snails, Cl, *see* Gastropoda

Snake Blenny *(Lumpenus lampretiformis) see* Blennioidei

Snake Eels (Ophichthidae, F) *see* Anguilliformes

Snake Mackerels (Gempylidae, F) *see* Scombroidei

Snake-head Goby *(Ophiocara porocephala) see Ophiocara*

Snake-head types, O, *see* Channiformes

Snake-heads (Channidae, F) *see* Channiformes

Snakelocks Anemone *(Anemonia sulcata) see Anemonia*

Snakeskin Gourami *(Trichogaster pectoralis) see Trichogaster*

Snappers, F, *see* Lutjanidae

Snapping Shrimps, G, *see Alpheus*

Snipe Eels (Nemichthyidae, F) *see* Anguilliformes

Snipefishes (Macrorhamphosidae, F) *see* Aulostomioidei

Snook *(Centropomus undecimalis) see* Centropomidae

Snowflake Moray *(Echidna nebulosa) see Echidna*

Soapies, F, *see* Leiognathidae

Social Behaviour. Term used in ethology. The tendency of animals that live socially to allow certain or every mode of behaviour to run their course within the group or in relation to it. The following types of social behaviour are distinguished (as far as fish are concerned): like-species to like-species, male to female, male to male, female to female, juvenile to male, juvenile to female, juvenile to juvenile, individual to shoal, shoal to shoal and social behaviour between different species. Special social behavioural tendencies include the instinctive way in which many like individuals stick together or the way in which individuals of different species do this (the herd or shoal instinct), maintaining the shoal quite independently of any mode of behaviour that may be activated in the individual sense. Other examples include the instinct to return to the shoal, the influence of the single individual by the shoal (shoal effect, transmittance of mood), co-operative achievements (among predators, the chasing of prey), the formation of social hierarchies (Order of Rank*) etc. The maintenance of the social behaviour serves to provide various means of communication.

Social Hierarchy *see* Order of Rank

Sockeye Salmon *(Oncorhynchus nerka) see* Salmonoidei

Sodium Chloride Baths. Sodium chloride (cooking salt) and sea salt (sea water*), in small concentrations

(15–20 g to 10 litres of water), have a favourable stimulating effect on the epidermis and on the gills of freshwater fish. This causes a general heightening of the metabolism and an increase in the body's readiness to defend itself against diseases. Many parasites tolerate the higher salt concentrations less well than fish. Short-term baths (bathing treatments*) in solutions of 10–15 g cooking salt (NaCl) in 1 litre of water are therefore suitable in the treatment of skin parasites (skin diseases*). The same concentrations of sea salt are, in general, tolerated better by fish. Treatment is applied in a separate glass container for up to 20 minutes. NaCl baths can be used against *Chilodonella**, *Trichodina**, *Gyrodactylus* and *Dactylogyrus* (Trematodes*), as well as against Copepod attacks* and Saprolegniacea*.

Soft Corals, O, *see* Alcyonaria

Soft Water is lacking in calcium and magnesium salts. The most obvious examples are rainwater and spring water from primitive rock, but soft water is also found in many tropical streams and rivers. Water with a hardness value (Water Hardness*) of up to 5 °dH is described as very soft, up to 10 °dH as soft. Many organisms occur only in soft water, sometimes because their stages of development are very sensitive towards calcium and magnesium salts. One of the best known examples of this is the Neon Tetra *(Paracheirodon* kinnesi)* which needs water with a hardness value of less than 2 °dH for breeding.

Softening (of water) *see* Desalination

Solaster FORBES, 1839; Sunstars. G of predatory Starfishes (Asteroidea*) with species in the cold seas of the northern hemisphere found at depths of 0–1,200 m. 8–14 fairly short arms spring from a broad mouth area – the diameter of the arms may reach as much as 40 cm. Sunstars like the cold and they are sensitive, particularly to temperatures above 15 °C. *S. papposus* LINNAEUS, 1758, loves eating starfish *(Asterias*)* but also bivalves. It is only possible to keep *Solaster* species in the aquarium if the water is kept constantly cool.

Soldier Flies, G, *see Stratiomys*

Soldierfishes (Holocentridae, F) *see* Beryciformes

Sole *(Solea solea) see* Pleuronectiformes

Solea solea *see* Pleuronectiformes

Soleidae, F, *see* Pleuronectiformes

Solenichthyes, O, name not in current usage, *see* Gasterosteiformes

Solenostomidae, F, *see* Syngnathoidei

Soles (Soleidae, F) *see* Pleuronectiformes

Somniosus microcephalus *see* Squalidae

Somphongs' Rasbora *(Rasbora somphongsi) see Rasbora*

Sorubim SPIX, 1829. G of the Pimelodidae*. Naked Antennae-Catfishes from the river areas of the Amazon, Rio de la Plata and Rio Magdalena as well as their tributaries. They have a bizarre shape. Length up to 60 cm. Only one species. *Sorubim* is active at dusk and at night, although it will also leave its day-time hiding-places (among pieces of root, broad-leaved plants etc) at times, particularly if food is available. However, they return to the hiding-place immediately afterwards. Body elongate, cylindrical at the front, flattened laterally at the back. Head flattened with a very long snout that looks like a spatula from above and which overlaps the inferior mouth. D1 narrow but extending into a long point. D2 well developed. The upper lobe of the deeply forked C ends in a point and is longer than the rounded, lower lobe. The An, Vs and Ps are fairly small. 3 pairs of barbels that are held splayed outwards, at an angle upwards or downwards. *Sorubim* is well suited to a large Catfish aquarium, and can be kept in the same tank as robust species such as *Pimelodus, Synodontis, Plecostomus* and others. It is a very greedy fish *(Tubifex,* earthworms, grated meat, insect larvae). Sex differences and breeding unknown.

– *S. lima* (BLOCH and SCHNEIDER, 1806); Shovel-nosed Catfish, Cucharon. Flanks silver-grey to greenish-grey, often with a brassy gleam. Dorsal surface covered with dark marbling, ventral surface pure white. A dark longitudinal band beginning at the tip of the snout and continuing into the dark middle area of the C, is narrow to start with, getting broader towards the posterior end. Other fins colourless. An extremely interesting Catfish, but unfortunately it has only been imported on rare occasions so far (fig. 1057).

Fig. 1057 *Sorubim lima*

Sounds, Production and Use of. A form of expression found in some invertebrates (insects, crustaceans) and in many vertebrates. It can be associated with particular modes of behaviour (eg defence, threatening behaviour, seeking a mate) or it can serve the purpose of general understanding between the individuals of a species.

Fish do not have a larynx to produce voices, but the utterance of sounds is widespread. Noises and tones of various kinds can be produced, for example, by releasing swallowed air through the anus (Cobitidae), by rubbing pharyngeal teeth against one another *(Balistes, Mola, Trachurus)* or by rubbing skeletal elements of the fins together *(Acanthurus, Capros, Monacanthus, Gasterosteus).* The swimbladder* is often used to amplify the sounds. This is achieved either through vibrations of external muscles that lie alongside the swimbladder or that are connected to it via tendons (*Micropogon*, Sciaenidae); or by means of bones that lie alongside it *(Rhinecanthus, Pseudauchenipterus)*; or by means of the muscles in the swimbladder itself (Triglidae).

Source (of river) *see* Crenal

South America. The fourth largest land mass in the world, covering almost exactly 18 million km^2. From N to S it measures over 7,500 km, from W to E over 5,000 km. The geographical formations at the surface contrast sharply between W and E. In the West, the young Andes Mountains, which were formed in the Tertiary period, rise up to 7,000 m; the middle and eastern parts of South America, on the other hand, are made up to old mountain strata that have largely levelled off with huge river systems (Orinoco, Amazon*, Rio de la Plata*) gouged out of them, up to 2,000 m deep. The widest areas of South America lie within the tropical belt, and it is here that the greatest expanse of tropical rain forest in the world is found. To the north and south this area is bounded by bush-steppes and grasslands. The rain forest areas have an extremely varied flora and fauna, and among the freshwater fish species those related to the Characins (Characoidei*) and the Catfish-types (Siluriformes*) have a special role to play. Together with the Knifefish-types (Gymnotoidei*), from these 3 groups alone South America is home to 17 endemic Fs. Other South American fish families with an abundant selection of species include the Cichlids (Cichlidae*) and the Livebearing Tooth-carps (Poeciliidae*).

Finally, another fish family that should be mentioned here, the Cyprinodontidae* (Egg-laying Tooth-carps), has successfully conquered extreme habitats in South America. With their life-cycle, some of the Ga in this family were able to adapt completely to living in savannah waters that only have water in them for some of the time (so-called annual species). The Sub-F Orestiinae forms part of the fauna of the high mountain ranges of the Andes. Zoogeographically, South America belongs to the neotropic realm*.

South American Frogbit *(Limnobium laevigatum) see Limnobium*

South American Lungfish *(Lepidosiren paradoxa) see* Ceratodiformes

South American Lungfishes (Lepidosirenidae, F) *see* Ceratodiformes

South American Molly *(Poecilia caucana) see Poecilia*

Southern Blue Eye *(Pseudomugil signifer) see Pseudomugil*

Sp. Abbreviation of 'species'.

Spadefishes, F, *see* Ephippidae

Spade-tail *(Poecilia reticulata*, standard species) *see Poecilia*

Spanish Killi *(Aphanius iberus) see Aphanius*

Spanish Minnow *(Aphanius iberus) see Aphanius*

Spanish Toothcarp *(Valencia hispanica) see Valencia*

Spanner Barb *(Barbus lateristriga) see Barbus*

Sparganiacea; Sparganium or Bur-reed. F of the Liliatae*. 1 G, 20 species. Found in northern, temperate and cold regions, rare in the southern hemisphere. Forms runners, most species are sturdy perennials. Leaves distichous, linear. Flowers unisexual and monoecious, inconspicuous, with a greenish perianth. The species of the one G *Sparganium* are plants that grow in marshes or along the banks of stretches of water; some also grow submerged and have floating leaves. They are also suitable for planting in open-air ponds.

Sparganium, G, *see* Sparganiaceae

Sparidae; Sea Breams (fig. 1058). F of the Perciformes*, Sub-O Percoidei*. This F contains a large number of species, all of which are found mainly in tropical and subtropical seas. Sea Breams are fairly tall fish, very compressed laterally, with a perch-like shape.

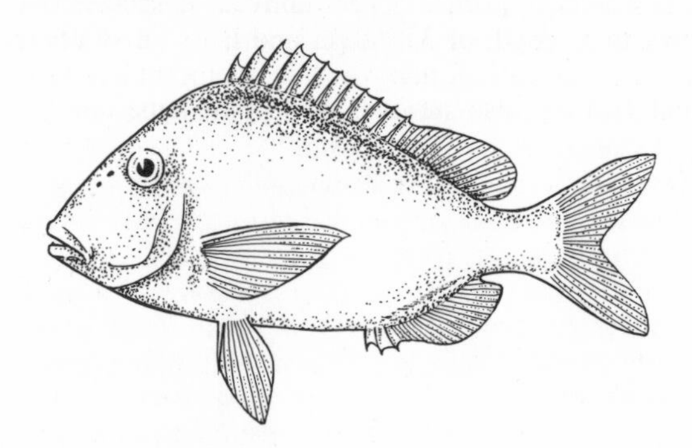

Fig. 1058 Sparidae-type

Head relatively small, mouth not very large. The edge of the operculum is never serrated. The front of the jaws usually contain sharply conical teeth with grinding teeth at the sides. The premaxilla has a typical groove into which the upper jaw slots. D in one piece, usually with 10–13 spines and 10–15 soft rays. C indented; An about as long as the soft-rayed part of the D and located opposite it, with 3 spines. Ctenoid scales, snout area never covered in scale. A particularly interesting biological feature is the hermaphroditism of some Sea Breams. Those concerned are either protoandrous or protogynous, ie they are male to start with, and become female, or vice versa. The corresponding sex glands develop after one another. Most of the species live in shoals in shallow coastal waters; many also enter estuaries. They feed on fish, small animals and algae. Many species are some of the most valuable of all fish in the fishing industry, and some are highly prized as sport-fishes. One of the largest species is the Dentex *(Dentex dentex)* which can grow up to 1 m long and weigh 10 kg. It is a predator from the Mediterranean and neighbouring Atlantic regions, and is coloured a delicate reddish-silver with black dots on top. *Sparus auratus* from the Mediterranean used to be kept and fattened in ancient Greece, because its meat tasted so good. As juveniles, many Sparidae are very good in a marine aquarium, see under the Ga *Boops**, *Diplodus**, *Sparus**.

Sparisoma (SWAINSON, 1839). G of the Scaridae, Sub-O Labroidei*. Numerous species in the tropical and subtropical parts of the Atlantic. They live alone or in loose-knit small groups on reefs and among carpets of seaweed. They feed on pieces of coral as well as other animal and plant growths. *Sparisoma* species are brightly coloured fish with a row of scales below the eyes. Length 20–50 cm. When the mouth is closed, the lower dental plate lies in front of the upper one. Unlike other Scarids, the *Sparisoma* species do not sleep in a protective mucus cocoon. Very few aquarium observa-

tions so far recorded. One species also lives on southern European coasts:
– *S. cretense* (LINNAEUS, 1758 ; Parrotfish. Atlantic coast, W Africa, Canaries, Mediterranean. Length to 40 cm. Purple to red-brown, flanks violet, P and V orange. C violet with a white edge.

Sparrman's Cichlid *(Tilapia sparrmani) see Tilapia*

Sparus Linnaeus, 1758. G of the Sparidae*. These fish feed on molluscs and growths. They are found in warmish seas. In this G, the front jaw teeth are in the form of small, conical fangs; there are 3–5 rows of grinding teeth. The head profile is very convex. A species that grows to a length of 35–60 cm and lives on southern European coasts is *S. auratus*. It is important as a tasty edible fish and also survives well in the aquarium:
– *S. auratus* (LINNAEUS, 1758). Silver with a gold band between the eyes and a dark blotch on the upper corner of the gill slit. This species causes damage to oyster cultures.

Spathe. A hypsophyllary leaf that more or less encases or surrounds an inflorescence. Sometimes, its only function is to offer protection, and then it is inconspicuous, as in the Hydrocharitaceae*. However, sometimes it also serves to attract pollinators, in which case the spathe will be brightly coloured, as in the Araceae*, for example. In aquarium-keeping, the spathe of flowering Cryptocorynes is well-known; it is known as the 'inflorescence'.

Spathiphyllum, G, *see* Araceae

Spathodus BOULENGER, 1900. G of the Cichlidae*. Comprises 2 species confined to Lake Tanganyika*. Elongated fish, up to 10 cm long, with a large head. Particularly in the males, the forehead juts forwards, swollen up. Mouth very thick-lipped and with large, spatula-shaped teeth. D has 21–23 hard rays, An 3. C fan-shaped. Life-style and care as for *Eretmodus**. Possibly a mouth-brooder.
– *S. erythrodon* BOULENGER, 1900. To 10 cm. Male, at the front, yellowish-brown, darker at the back. Head covered in dark-brown bands above the snout and between the eyes, as well as in the nape region. The An has egg-traps. Female pale brown on top, silver below; no head bands and no egg-trap (fig. 1059).

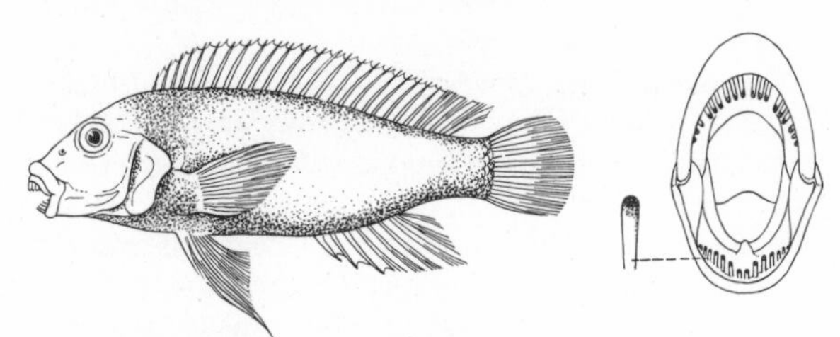

Fig. 1059 *Spathodus erythrodon* (right, jaw with spatula-teeth)

– *S. marlieri* POLL, 1950. To 9 cm. Brownish, head and fins darker to blackish.

Spawn *see* Gametes

Spawn Care Behaviour *see* Parental Care Behaviour

Spawn, Eating of. In particular the females of many fish species that do not practise brood care are guilty of eating their own spawn. Sometimes, they will try to eat it during the spawning act itself, but more usually directly after it. In the wild, this is less worrying because the egg-cells are spread across a wide area due to several matings taking place, thus only some of the eggs will be located and eaten. However, in a breeding aquarium*, the fish are forced to lay all the spawn in a very small area. This, of course, encourages the spawn to be eaten. One effective way round this problem with species that produce non-sticky or only slightly sticky egg-cells is to place a spawning grill in the aquarium. This is made of a plastic gauze and is placed about 1–2 cm above the bottom. The spawn will fall through it and so it will be protected from the female's clutches. If the egg-cells are very sticky, it has been found that very little spawn is eaten if there is thick vegetative cover and partial darkness in the tank during the spawning act. A large number of Characin species are noted as eaters of spawn, in particular the Ga *Nannostomus, Hemigrammus, Hyphessobrycon, Aphyocharax*, also every *Brachydanio* and *Danio* species, as well as most species of Barb. In captivity, even some species that practise brood care (especially Cichlids) will often eat their spawn after starting off with every sign of good care. The reasons for such atypical behaviour normally lie in the breeding aquarium having been set up wrongly or in the absence of a hostility factor* etc.

Spawning Act *see* Spawning Behaviour

Spawning Behaviour, Mating Behaviour. Terms used in ethology to describe the persistent courtship behaviour* between fish until the female has released her ripe eggs. The type of behaviour varies enormously, so only a few important examples can be described here. The display behaviour of the male becomes less obvious with the onset of the spawning behaviour. Substrate spawners* (fig. 1060) use swimming movements that lead to the spawning site with increasing determination. After mating, the eggs are dispersed among the substrate medium by fin movements. In *Rasbora** *hetero-*

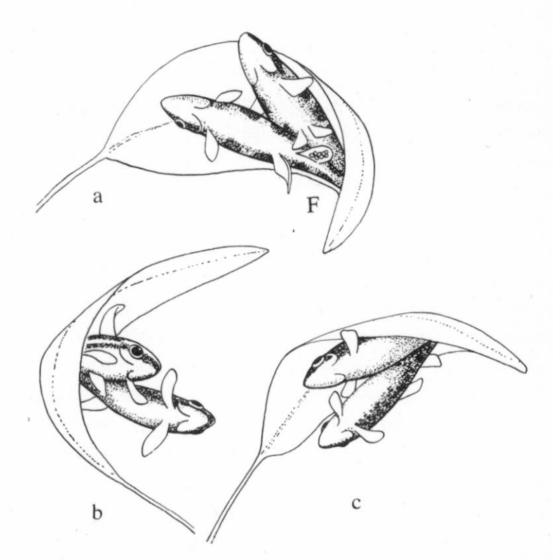

Fig. 1060 Spawning Behaviour. Mating positions of a) *Nannostomus espei*, b) *N. marginatus*, c) *N. eques* (F anal fin of the male folded into a fin sac)

morpha, they are stuck to the leaves of plants. Random spawners*, including many marine fish, such as Herring, mate in open water or near plants. The fertilised eggs sink to the bottom or in among the plants where they develop. In shoaling spawners (*see* Selection of Culture Fish), several males may mate one after the other with ripe females which may result in aggressive behaviour* between the males. In several types of fish the eggs are laid and fertilised without bodily contact between the male and female, for example in tropical Cichlids. The eggs are laid first on the spawning site then fertilised by the male. In Glandulocaudinae, the female is fertilised internally by the male. Then she lays her eggs on the leaves of plants unaccompanied by the male. Spraying Characins (*see Copella* sp. 1) leap several cm out of the water during spawning in order to lay their eggs on aerial leaves of plants. Egg-laying Tooth-carps of the genera *Nothobranchius**, *Pterolebias** etc lay their eggs in the substrate where they must survive very dry periods lasting many months. Female Pipefishes (*see* Syngnathidae) place their eggs in a brood pouch on the ventral side of the male.

Spawning Colours. Many fish species adopt spawning colours at spawning time that can be very different from the normal colouring because the colour intensity is so much greater. These colours, together with developmental processes in the sex glands and specific modes of behaviour, indicate to a mate that the fish is ready to spawn. They are released through the action of certain hormonal glands. In our climatic conditions, these glands react to abiotic factors such as changes in water temperature, length of periods of light or its intensity, atmospheric conditions, whereas in the tropics they react to the onset of the rainy season. In particular, it is the males in species with well developed eyesight that have the most striking spawning colours. A good example is the Three-spined Stickleback *(Gasterosteus aculeatus)* with its blood-red ventral surface and vivid blue-green eyes. The colours have very specific signalling functions that are interpreted correspondingly by a mate or rival (they are also emphasised by particular modes of behaviour). A nesting male Stickleback, for example, will furiously attack any fish with a red underside, since, to him, it signifies a rival. The female Stickleback, on the other hand, will be unable to find the entrance to the nest if the male does not show her the way with his blue-green, shining eyes next to her. Among the Cichlids belonging to the Ga *Apistogramma* and *Nannacara*, it is the females who have the most striking spawning colours – this is because the females look after the brood alone. Usually, their colours consist of an intense yellow body with deep black bands running across it, and this will indicate several different things: it will tell the male that she is ready to mate; it will also act as warning colours to others of its species, telling them not to stray too close to her brood; and it also helps to announce to the young fish where she is, which in turn helps them to orientate themselves (it is not the colours alone that do this, but the colours together with species-specific instinctive movements [*see* Fixed Action Pattern]). In cases where both the parents practise brood care, either at the same time or alternately (eg *Cichlasoma, Symphysodon* etc), there is no difference in spawning colours between the sexes.

This is because the young fish will have to orientate themselves according to the signals first from the male, then from the female, and so on.

Spawning Period *see* Breeding Season

Spawning Tank *see* Breeding Aquarium

Spawning Tubercles. Horny patches, like rinds of skin or like nodules, that appear in the epidermal (*see* Skin) layer of the head and flanks in many species of fish during the spawning period (eg *Rutilus**). Usually, only the males have spawning tubercles.

Spear-leaved Water Plantain *(Alisma lanceolatum) see Alisma*

Species. The correct scientific term for a type or kind of organism (Concept of Species*), one of the systematic categories*. Used in conjunction with adjectives to denote status, eg

species nova (spec. nov., sp. n)	= new species
species bona	= good species
species dubia	= doubtful species

Specimen Plants are plants with a rosette leaf arrangement that grow very large. They are usually from the Ga *Aponogeton** and *Echinodorus**, and in the aquarium they need, inevitably, a large area to grow in. The term is only used in aquarium-keeping.

Sperm *see* Gametes

Spermatophores. Capsules for transferring sperm-cells (Gametes*) to the female sex partner. Fertilisation via spermatophores is known among various kinds of Eddy-worms, Molluscs, ametabolous insects, Scorpions, Millipedes and Salamanders. In many cases, the complicatedly structured spermatophores are received by the female without copulation.

Among the bony fish, the Livebearing Tooth-carps have simple spermatophores. The packets of sperm, in which several thousand sperm threads may be packed tightly, are introduced into the female Tooth-carp's sexual orifice by the copulatory organ (Gonopodium*).

Spermatophyta; Seed-plants. D of the Plants. 230,000 species. The Spermatophyta are the highest developed D within the plant kingdom and they represent the largest part of the Earth's flora. They are characterised by the formation of seeds*. Descended from the Ferns (Pteridophyta*), seed-plants have in common with them a vegetative body made up of stem axis, leaf and root, ie it is a corm*. Like the Pteridophytes too, they have an alternation of generations*, but it is no longer recognisable as such because the generations have been much simplified and united in the one body. With this, the organisms of the Spermatophyta represent the asexual generation of the Pteridophytes. The Spermatophyta were originally, and still are, primarily land plants. A small number have secondarily gone over to an aquatic existence in the course of development. Most of our aquarium plants today belong to them. The Spermatophyta are divided into 2 Sub-Ds:

– *Gymnospermae* (naked seeds). 1,000 species. The seeds develop openly on the carpel. Woody plants only with cone flowers. Our conifers belong here. No aquatic plants belong to this Sub-D.

– *Angiospermae;* (enclosed seeds). 220,000 species. The seeds develop in an envelope formed by the carpels. These are woody plants and herbaceous plants with true flowers. Marsh and aquatic plants are found in many different Fs. The two Cls of the Angiospermae are the Dicotyledons (Magnoliatae*) and Monocotyledons (Liliatae*).

Sphaerechinus DESOR, 1856, G of the Sea-urchins (Echinoidea*). *S. granularis* LAMARCK, the Violet or White-tipped Sea-urchin, lives on soft substrata in the Mediterranean and the Atlantic coast. *Sphaerechinus* is high-arched, with a diameter of up to 13 cm, and it has short violet spines with white tips. It likes to camouflage itself. *Sphaerechinus* eats various types of plant and animal food. It has also been observed in the aquarium catching fairly large crustaceans. This species is well suited to life in an aquarium.

Sphaerichthys CANESTRINI, 1860; Chocolate Gourami. G of the Belontiidae, Sub-O Anabantoidei*. Sumatra, Malaysian Peninsula. Found in weedy streams, rivers, ditches and ponds whose bottoms may be covered with dead parts of plants. Length to 6 cm. Body relatively tall, laterally flattened. Head pointed, mouth small. An elongated, covered in scale at the bottom. V has a strong hard ray and its first soft ray is elongated into a thread. Scales large, very regular order. Ground colour brown, with several, pale yellow to white transverse bands on the body, crossing at irregular intervals. In juveniles, there is a distinct longitudinal band. D and An brown, and especially the An has pale markings and a white border area. C translucent. *Sphaerichthys* likes warmth, and soft, slightly acidic water is recommended. If necessary, filter through peat. Temperature 26–30 °C. Live food is essential, particularly smallish midge larvae, *Daphnia* and flying insects *(Drosophila)*. Artificial food leads to fatty degeneration all too easily. The only well-known species is:

Fig. 1061 *Sphaerichthys osphromenoides*

– *S. osphromenoides* CANESTRINI, 1860; Chocolate Gourami. The sexes are very similar to one another. In the male, the D and An may be slightly drawn into a point. Mouth-brooder. According to available reports, the eggs are laid on the bottom and usually gathered together by the male, or in conjunction with the female, and allowed to brood in the mouth for about 2 weeks. About 20–50 fry in a brood (fig. 1061).

Sphagnum LINNÉ, 1753; Sphagnum Moss, Bog Moss. G of the Bryophyta*. 300 species. Distributed throughout the world. Spongy Mosses with sparse main branches and subsidiary branches arranged like bushes. Leaves very small. Reproduction through spores. The species of the G prefer acid, chalk-free and nutrient-rich habitats. Found growing mainly in marshes and moors. They form thick pads and mats, continually growing on the surface, whilst the deeper layers die off and change into peat*.

Sphenisciformes, O, *see* Aves

Sphenopsida, Cl, *see* Pteridophyta

Sphere of Function. Term used in ethology. Spheres of function relate to animal behaviour patterns of great variety, eg aggression, courtship, parental care; they also include types of behaviour that are caused metabolically, such as the search for food, sleep etc. To be able to describe a sphere of function it is necessary to understand not only all the behavioural elements in the ethogram*, but also all the abiotic and biotic factors, such as lighting, temperature, air pressure, time of day and of year, fresh water supply, changes in food supply, hostility factors etc, all of which can influence how the sphere of function is activated. All spheres of function are subject to a hierarchical division* of their behavioural elements, but they do not need to be prefaced by appetitive behaviour* to trigger them off.

Sphinx Mudfish *(Blennius sphinx) see Blennius*

Sphyraena, G, *see* Sphyraenoidei

Sphyraenidae, F, *see* Sphyraenoidei

Sphyraenoidei; Barracudas, related species. Sub-O of the Perciformes*, Sup-O Acanthopterygii. The Sub-O is only represented by the F Sphyraenidae (fig. 1062) containing the G *Sphyraena*, to which at most 20 species belong. The Sphyraenoidei have been known since the Neogene period, and are now found primarily in tropical and subtropical waters. They have an elongate, extremely slender build. That is why they are sometimes called Arrow-pikes. Head ends in a long point; large eyes and a very deeply rent mouth, rigid with teeth. The teeth themselves are chisel-shaped, extremely sharp and sit inside alveoli (tooth-sockets). D1 has only 5 spines and is inserted well in front of a short D2, which in its turn lies opposite a similar-looking An. C broad, gently indented. Scales small, LL complete. The females are very fertile and spawn in rationed amounts. Eggs pelagic. The Sphyraenoidei belong to the largest predators among the fish, and may even be dangerous

Fig. 1062 Sphyraenidae-type

for humans, particularly divers and fishermen. Barracuda-types swim extremely fast, are very agile and, as young fish, they hunt other fish in packs. The large sexually mature fish are mostly solitary creatures that lie in wait for prey. Many species, such as the Great Barracuda *(Sphyraena barracuda)* from the Atlantic and Pacific, normally grow to a length of 1.5 m, more rarely up to 3 m. The European Barracuda *(Sphyraena sphyraena)* from the Atlantic, Mediterranean and Black Sea can reach a length of 1 m. It is usually found as part of a shoal. Almost all the species make excellent sport fish and are highly prized for their meat. In S California, a Barracuda is a much sought-after market fish. Occasionally when people have enjoyed a meal of Barracuda meat, fish poisoning has been known.

Sphyrion, G, *see* Copepod Infestation

Sphyrna zygaena *see* Sphyrnidae

Sphyrnidae; Hammerhead Sharks (fig. 1063). F of the Chondrichthyes*, O Carcharhiniformes. These sharks have a very spread out head, which looks like a two-headed hammer from above. Eyes are located at the outside ends of the head lobes, and further inside the nasal openings. The head lobes are probably some kind

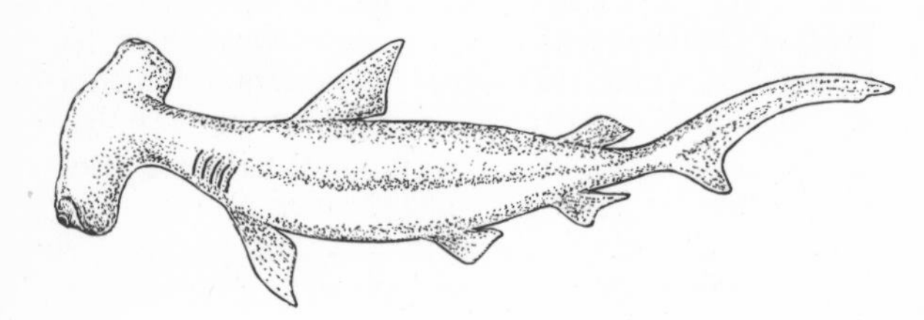

Fig. 1063 Sphyrnidae-type

of stabilising organ. There is no nictitating membrane and no spray holes. 2 Ds. Upper tail fin lobe much bigger than the lower one. Hammerheads are found mainly in tropical seas, with only a few isolated specimens entering temperate areas. They feed on fish and crustaceans mainly. The large species give birth to up to 40 young, and these have a normally shaped head to start with. The larger species, particularly the 4.5 m long Smooth Hammerhead Shark *(Sphyrna zygaena)*, can be dangerous to man. They are reported to attack unprovoked. Hammerhead meat is eaten a lot in Japan, and the skin makes a good leather.

Spider Crabs, F, *see* Brachyura

Spider Crabs, G, *see* Macropodia

Spiders, Cl, *see* Arachnida

Spike Rushes, G, *see Eleocharis*

Spiked Water Milfoil *(Myriophyllum spicatum) see Myriophyllum*

Spikelet. Part of an inflorescence*. On a short shaft are the sessile awns amongst which are the inconspicuous flowers. Spikelets occur, for example, in grasses and in the G *Cyperus**.

Spike-tail Molly *(Poecilia petenensis) see Poecilia*

Spike-tailed Paradise-fish *(Macropodus cupanus cupanus) see Macropodus*

Spike-tailed Paradise-fish *(Macropodus cupanus dayi) see Macropodus*

Spinachia CUVIER, 1817 (fig. 1064). G of the Gasterosteidae, Sub-O Gasterosteoidei*. Contains the species, *S. spinachia* (LINNAEUS, 1758) – the Sea Stickleback – which is found along European sea coasts from the Bay of Biscay to the North Cape, particularly near carpets of seaweed. It is a very elongate, up to 17 cm long, olive-brown fish and has 14–17 spines in front of the D. The male builds a plant-nest. Can be kept in cool, calm aquaria; feed on small crustaceans and small fish. A fairly sensitive species, short-lived.

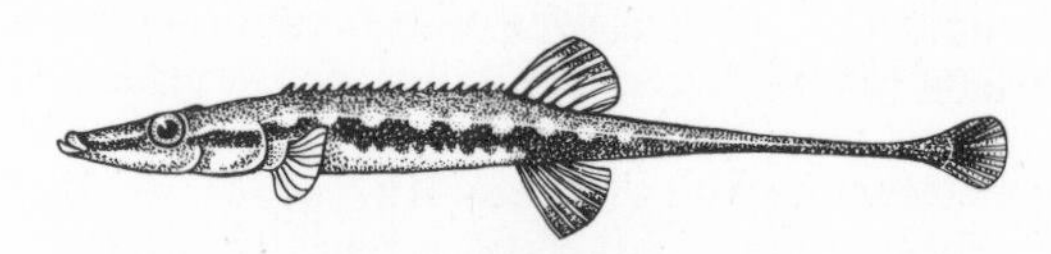

Fig. 1064 *Spinachia spinachia*

Spinal Cord *see* Nervous System

Spine-cheeked Anemonefish *(Premnas biaculeatus) see Premnas*

Spined Loach *(Cobitis taenia) see Cobitis*

Spiny Boxfish *(Chilomycterus schoepfi) see Chilomycterus*

Spiny Burrfish *(Chilomycterus schoepfi) see Chilomycterus*

Spiny Catfish *(Acanthodoras spinosissimus) see Acanthodoras*

Spiny Catfishes, F, *see* Bagridae

Spiny Cockle *(Cardium aculeatum) see Cardium*

Spiny Crab *(Eriphia spinifrons) see Eriphia*

Spiny Dogfish *(Squalus acanthias) see* Squalidae

Spiny Eels, G, *see Mastacembelus*

Spiny Eels (Mastacembelidae, F) *see* Mastacembeloidei

Spiny Fins (Acanthopterygii, Sup-O) *see* Osteichthyes

Spiny Lobsters, G, *see Palinurus*

Spiny Starfish *(Marthasterias glacialis) see Marthasterias*

Spiracle *see* Spray Hole

Spiral Intestine *see* Digestive Organs

Spiranthes cernua *see* Orchideaceae

Spirinchus theleichthys *see* Salmonoidei

Spirodela SCHLEIDEN, 1839; Great Duckweed. G of the Lemnaceae* comprising 3 species distributed almost throughout the world. Very small, perennial plants with a much simplified vegetative body. Daughter members of the plant develop from 2 lateral sacs on each individual component plant, and every member plant has several roots. Flowers unisexual and monoecious, very small. There are 2 male flowers and 1 female flower inside a spathe. No perianth. Male flower: 1 stamen. Female flower: 1 carpel, superior. Fruit like a berry. The various species of Great Duckweed grow in still, nutrient-rich waters, floating on the surface. They do not grow as luxuriantly as the species of the G *Lemna** when kept in aquatic plant gardens. The commonest species is:

– *S. polyrrhiza* (LINNÉ) SCHLEIDEN: Many-rooted Duckweed. Found almost everywhere on Earth. The member plants are roundish.

Spirographis VIVIANI, 1805. G of the marine Bristle-worms (Polychaeta*). These are sessile creatures that feed on matter in suspension. Their tube is leathery and the crown of tentacles is wound in a spiral.
– *S. spallanzanii* VIVIANI, 1805. Very common on firm substrates in shallow coastal waters of the Mediterranean and European Atlantic. Length 20–30 cm. The tentacles are usually patterned in an orange-brown-white colour. Quite frequently, the crown of tentacles is cast off, but with strong specimens it quickly grows back again. In the aquarium, the tube should be fixed between some rocks, such that it rises up unencumbered into the water, but take care not to squash the animal. Unfortunately, *Spirographis* is not very hardy in the aquarium (fig. 1065).

Fig. 1065 *Spirographis spallanzanii*

Spironucleus, G, *see Hexamita*

Spirorbis DAUDIN, 1800; Calcareous Tube-worm. A widely distributed G of sessile marine Bristle-worms (Polychaeta*). They eat matter in suspension and have a chalky tube wound in a spiral. Only a few mm long. Often, they manage to arrive in the marine aquarium attached to firm substrate or algae; large numbers may then start to appear.

Spleen or Milt. An organ, located in the stomach-intestine region, that makes blood cells. Among the lower vertebrates, that contain little bone marrow, it is not only white blood cells that are produced here, but a large amount of the red blood corpuscles as well. The spleen can also expand to be able to store blood cells and red corpuscles are also broken down there.

Splendid Rainbowfish *(Halichoeres marginatus) see Halichoeres*

Sponge Crab (*Dromia*, G), *see* Brachyura

Sponges, Ph, *see* Porifera

Spongilla LAMARCK, 1815 (fig. 1066). A freshwater G of Sponges (Porifera*). There are several species in Europe, and some have a widespread distribution (Africa, N America). *Spongilla* lives like the related G *Ephydatia** in the shallow areas of ponds, lakes and slowly flowing water, but in contrast to it *Spongilla* tends to grow in the shape of antlers or bushes.

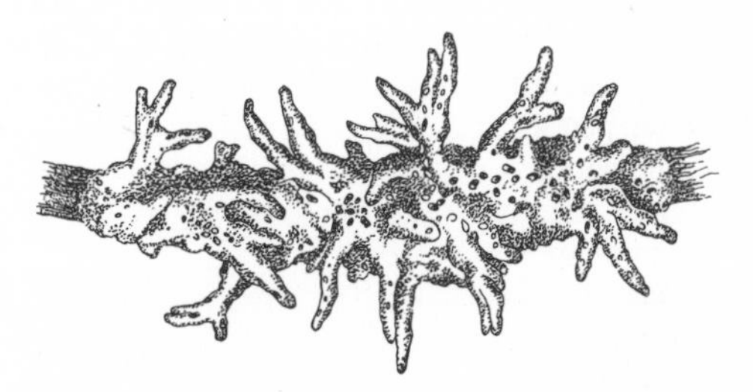

Fig. 1066 *Spongilla lacustris*

Spontaneous Overflow Action. Term used in ethology. Actions that reveal themselves suddenly without any special stimulus after a great amount of unreleased action-specific energy for a particular sphere of function* has been stored up. In the moment such actions take place they appear incongruous, since they are a part of another sphere of function, not the present one. However, they do serve to break up the accumulation of action-specific energy that has become superfluous with the change in situation. For example, *Nannostomus marginatus* standing calmly on the spot after taking in some food will bend its body at regular intervals to aid digestion. This same movement of the sphere of function relating to metabolically induced behaviour is performed again by the males of this species during spawning, only this time it is a ritualised form of behaviour*. It occurs when the female suddenly disappears from his field of vision and as a result a high pent-up energy for reproductive behaviour develops. But as this has become inappropriate it is broken up by overflow action.

Spoonbills (Polyodontidae, F) *see* Acipenseriformes

Sporangium (spore container) *see* Angium

Spore. An asexual reproductive cell found in plants – opposite to gamete*. In Algae* and Fungi*, spores can develop from normal cell division or from maturation division; they may be immobile or mobile by means of cilia. In Mosses and Liverworts (Bryophyta*) and Ferns (Pteridophyta*), spores always develop from maturation division; they are also immobile, the same size, and when they germinate they form the hermaphroditic, sexual generation in the alternation of generations*. Among the Water Ferns (under Pteridophyta*), small microspores are formed, from which the male sexual generation develops. Larger megaspores are also formed that develop into the female sexual generation. *See also* Winter-spore.

Spore Animals, Cl, *see* Sporozoa

Spore Container (Sporangium) *see* Angium

Sporocyst (spore container) *see* Cyst

Sporophyte (asexual generation) *see* Alternation of Generations

Sporozoa; Spore Animals. Cl of parasitical single-celled organisms with a complicated reproduction, usually including an alternation of generations*. The formation of an encapsulated stage (spores) is characteristic – these spores are hardy and infectious (fig. 1067). There are many fish-pathogenic Sporozoans, that live as parasites in the skin and gills, as well as in the internal organs. They produce small, white to yellowish nodules containing large amounts of spores. In the fish industry, *Eimeria cyprini* PLEHN, which causes enteritiscoccidiosis, is one of the most feared Sporozoans. Others include *Myxosoma cerebralis* HOFER and PLENN which causes giddiness (whirling or tumbling disease) among the Salmonidae*, and *Myxobolus pfeifferi* THELOHAN, 1870, the cause of swellings (boil disease) among the Cyprinidae*. In aquarium-keeping, Sporozoan diseases are rare, although they are occasionally introduced into the aquarium along with catches made in the wild. One exception is *Pleistophora* hyphessobryconis* which is extremely common, and causes the so-called Neon disease. Sea-horses *(Hippocampus*)* and Sticklebacks *(Gasterosteus*)* are also known to be attacked by *Glugea**. Sporozoan infections are not possible to cure. Where any fish is infected, the whole batch must be destroyed. The aquarium, pond and pieces of equipment must be thoroughly disinfected (Disinfection*) afterwards.

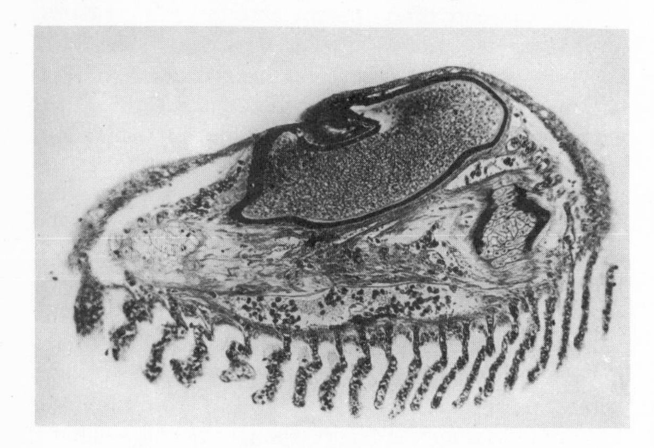

Fig. 1067 Spore cyst in the gill epithelium

Spot Diseases. This term embraces a whole series of diseases that occur in fish. The causes are many and various, but the symptoms are similar. Red, inflamed spots appear on the skin and the fins, and eventually these expand into abscesses that penetrate the muscular system. Bacteria belonging to the Ga *Pseudomonas, Aeromonas, Vibrio* etc may be the cause. Fin rot*, boils and swellings may often accompany the diseases.

Spot diseases are particularly prevalent among freshwater fish. *Vibrio anguillarum* can cause an epidemic among many types of marine fish, and in the aquarium it can be extremely virulent. One way of combating the disease in salt water is to add copper* sulphate and zinc* sulphate. However, more success seems to be met with if the food is mixed with aureomycin* and streptomycin*, and the infected specimens are bathed.

Spotted Climbing Perch *(Ctenopoma acutirostre) see Ctenopoma*

Spotted Danio *(Brachydanio nigrofasciatus) see Brachydanio*

Spotted Gambusia *(Gambusia affinis) see Gambusia*

Spotted Hatchetfish *(Gasteropelecus maculatus) see Gasteropelecus*

Spotted Hawkfish *(Cirrhitichthys aprinus) see Cirrhitichthys*

Spotted Hawkfish *(Cirrhitichthys aureus) see Cirrhitichthys*

Spotted Headstander *(Chilodus punctatus) see Chilodus*

Spotted Leporinus *(Laporinus maculatus) see Leporinus*

Spotted Livebearer *(Phalloceros caudimaculatus reticulatus) see Phalloceros*

Spotted Metynnis *(Metynnis maculatus) see Metynnis*

Spotted Panchax *(Epiplatys macrostigma) see Epiplatys*

Spotted Phago *(Phago maculatus) see Phago*

Spotted Piranha *(Serrasalmus rhombeus) see Serrasalmus*

Spotted Rasbora *(Rasbora maculata) see Rasbora*

Spotted Sea-horse *(Hippocampus kuda) see Hippocampus*

Spotted Shark *(Mustelus punctulatus) see* Triakidae

Spotted Sucker *(Minytrema melanops) see* Catostomidae

Sprat *(Sprattus sprattus) see* Clupeiformes

Sprattus sprattus *see* Clupeiformes

Spray Hole; spiracle. An opening found in most sharks, rays and sturgeons which admits respiratory water. There is a spray hole on each side behind the eyes, and this corresponds to the 1st gill slit (no longer formed among the bony fish). Among air-breathing vertebrates, the embryonal matter that forms a spiracle develops instead into the middle-ear and the Eustachian tube which connects the middle ear to the pharynx.

Spraying Characin *(Copella 'arnoldi') see Copella*

Springtails, O, *see* Collembola

Spurdog *(Squalus acanthias) see* Squalidae

Squalidae; Spiny Dogfishes. F of the Chondrichthyes*, O Squaliformes. Various Fs from older systems have been placed in this F, such as the Dalatiidae. The Squalidae have a typical Shark shape, and various Sub-Fs have a spine in front of both Ds. In other Fs there are no spines here, or only one D has a spine. None of the species has an An. Spray-holes present. There are 5 gill slits in front of the Ps, and the C is asymmetrical. Nearly all the Squalidae are livebearing, some are prettily patterned and coloured. The best-known Sub-F is the Squalinae, which contains species that have a very distinct spine in front of each D. An example is the Spiny Dogfish or Spurdog *(Squalus acanthias)* (fig. 1068) which is grey with white dots. It grows up to between 70–90 cm normally, although a few reach 120 cm; the males remain smaller. The female undergoes an unusually long pregnancy (18–22 months), then gives birth to 8–12 live young. Spiny Dogfishes are found in temperate seas, and from there they penetrate

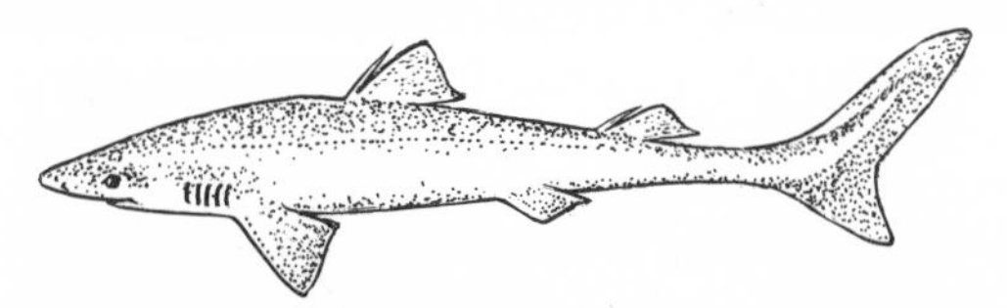

Fig. 1068 *Squalus acanthias*

into Arctic and equatorial regions; occasionally some are spotted in the Baltic. The Spurdog hunts fish in packs; it can cause great damage among fish stocks and to fishing nets. This particular species also has a poison gland which comes out at the spine. As a result, a prick from the spine is very painful. Other species do not have this gland. In the fishing industry, the Spiny Dogfish is made use of mainly in Europe. Strips of skin cut from the ventral side are sold smoked, and Spurdog meat in aspic is also sold on the market. The Sub-F Etmopterinae includes the G *Etmopterus* which contains various deep-sea forms, most of which have luminescent organs. One example is *Etmopterus hillianus*, which with a maximum length of 30 cm, is the smallest living Shark. The Sharks that make up the Sub-F Semniosinae (Ice Sharks) do not have a spine in front of the Ds. The best-known species, the livebearing Greenland or Ice Shark *(Somniosus microcephalus)* (fig. 1069) lives in the Arctic far from land and usually at depths of 150-600 m. It grows to a length of 6 m, sometimes bigger.

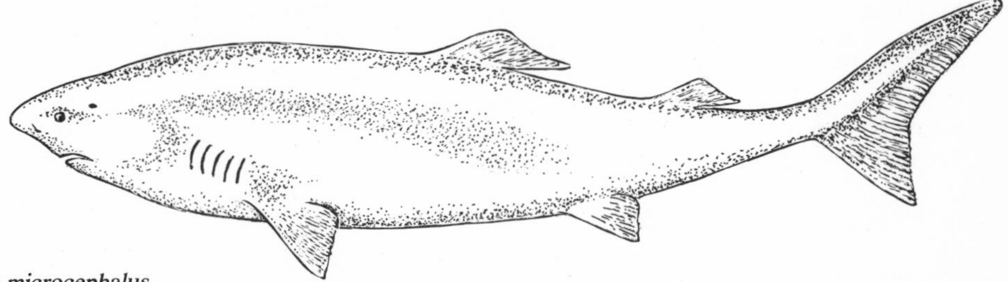

Fig. 1069 *Somniosus microcephalus*

It feeds on carrion, fish and marine mammals, such as seals. It also attacks whales, particularly Narwals, tearing pieces of flesh from the body. Ice Sharks often gather round whaling vessels in large packs, waiting to catch anything that falls their way. No attacks upon humans have been recorded. It is remarkable to note that Ice Sharks caught with fishing tackle give themselves up quite lethargically. They have an enormous liver which yields a good cod-liver oil. Their meat contains a poison that is destroyed when the meat is dried. The skin makes a good leather (so-called Coralshark leather). Another Ice Shark lives in the Antarctic and yet another in the Mediterranean. The rare Luminous Shark *(Isistius brasiliensis)* belongs to the Sub-F Dalatiinae. It is a small deep-sea form from tropical seas, that has occasionally been found at the surface.

Squalus acanthias *see* Squalidae

Squat Lobsters, F, *see* Galatheidae

Squatina squatina *see* Squatinidae

Squatinidae; Angelsharks. F of the Chondrichthyes, O Squatiniformes. These sharks have an atypical body form reminiscent of rays, particularly the Guitarfishes (Rhinobatidae*). Body flattened ventrally and broad. Head rounded, mouth almost terminal and beset with pointed teeth. Each nasal opening has a short, strong barbel. Eyes on the upperside of the head with the large, half-moon shaped spiracles right behind them. Gill slits in front of the wing-like Ps; however, the Ps are not fused with the head. The two Ds are located well to the

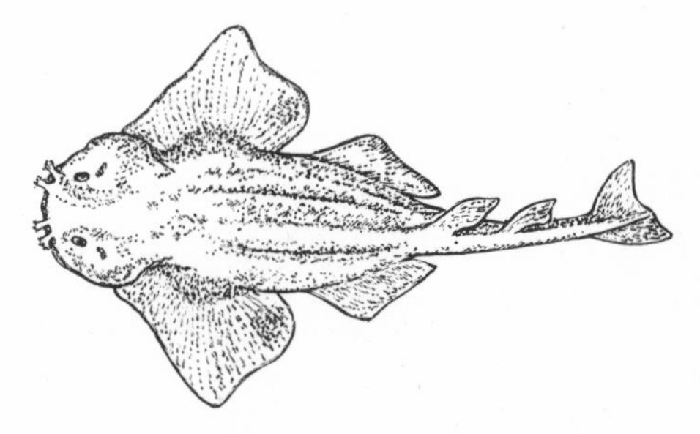

Fig. 1070 *Squatina squatina*

rear on the caudal peduncle, C small. The Angelsharks live in shallow coastal waters in temperate zones, where they feed on bottom-dwelling fish mainly and on molluscs. They are livebearing. The largest and best-known species, the Monkfish *(Squatina squatina)* (fig. 1070), grows to a maximum 2.5 m and lives in the NE Atlantic and the Mediternanean. During the cold seasons of the year, it keeps itself far from land at fairly deep water levels, but in summer it will be found near the coast. The females give birth to 25–35 young. Its skin used to be used for various industrial purposes.

Squids (*Loligo*, G) *see* Cephalopoda

Squilla mantis *see* Stomatopoda

Squirrelfishes (Holocentridae, F) *see* Beryciformes

Stages in Development *see* Ontogenesis

Stamen *see* Flower

Staminodium; staminode. Sterile stamen that does not produce any pollen. It may be inconspicuous or like petals.

Standard Water. If standard water is used it makes water preparation* easier. You start with water that has had all the salts removed (*see* Desalination), to which are then added certain salts that satisfy the needs of the majority of tropical freshwater aquarium fish. At the same time standard water will also guarantee minimal pH value fluctuations and good plant growth.

Characteristics of standard water are a low salt content and a very low carbonate hardness (less than 1° of total carbonate hardness). In theory such water has a very low buffer* effect, but in practice it is sufficiently increased through ventilation and the constant introduction of carbonic acid that this creates. HÜCKSTEDT and EUL recommend the following recipe for standard water:

Dilute the following in a litre of water:
32.73 mg crystallised calcium sulphate
11.73 mg crystallised magnesium sulphate
8.73 mg crystallised calcium chloride
13.2 mg sodium hydrocarbonate
3.33 mg potassium hydrophosphate
0.05 mg Cheloplex-III or Komplexon-III
1.0 mg iron citrate

A mixture of trace elements should also be added. If more humic acids* are required to make the standard water more acidic (eg for fish that occur naturally in dystrophic waters), small amounts of peat may be added. This works because peat contains minimal amounts of hydro-carbons, making its alkalinity* low as well as its ability to form acidic compounds. For animals and plants that require a higher salt content, more calcium and magnesium sulphates may be added in the ratios given above until a sulphate hardness of 10–15° is achieved. For every litre of water also add 100–200 mg cooking salt.

Standing Waters *see* Still Waters

Starfishes, Cl, *see* Asteroidea

Stargazer *(Uranoscopus scaber) see* Trachinoidei

Stargazers (Uranoscopidae, F) *see* Trachinoidei

Star-head Top Minnow *(Fundulus notatus) see Fundulus*

Starry Flounder *(Platichthys stellatus) see* Pleuronectiformes

Starry Moray *(Echidna nebulosa) see Echidna*

Starry Triggerfish *(Abalistes stellaris) see Abalistes*

Stars and Stripes Toad *(Arothron hispidus) see Arothron*

Stauromedusae, O, *see* Scyphozoa

Steatocranus BOULENGER, 1899. G of the Cichlidae*. Native to the Congo area and particularly closely related to the G *Leptotilapia.* It is not yet definite whether there are 1 or two species. The fishes are elongate, up to 10 cm long, and only flattened very little from side to side. A particular characteristic is the lump of fat on the forehead which can reach considerable heights in old males. Other Cichlid Ga (eg *Lamprologus**) also have this fatty lump, but it always stays much smaller. Mouth thick-lipped. The D has a long base and 19–20 hard rays. The An has only 3 hard rays. Soft ray parts are elongated into tips, especially in old males. Colouring inconspicuous. *Steatocranus* has an interesting biology. Unlike most other Cichlids, *Steatocranus* is very peaceful – in fact, it hardly does any damage to plants. When kept in captivity, it is important to provide caves into which the fish can withdraw. A mating pair, when making spawning preparations, defend their brood-cave and also make changes to it to a certain extent (sand is piled up quite a way in front of the entrance, for example). So when the fish spawn it is often not possible to observe them, and the brood also remains in the cave for a further 14 days at least, completely hidden. It is probably unnecessary to feed the fry with powdered food to rear them, as the parents chew the food up beforehand.

To reflect the natural habitat, *Steatocranus* should be kept in oxygen-rich water that is not too warm, at a temperature of 23–24 °C. There are no particular requirements of the water quality, ie the water can be medium-hard, and the pH value should be around 7.

– *S. casuarius* POLL, 1939; Lumphead, African Blockhead. From central and lower Congo where it lives in fast-flowing water. Male to 9 cm, female smaller. Colour a monotone grey-brown to olive-green, occasionally with dark transverse stripes. Iris a brilliant emerald green (fig. 1071).

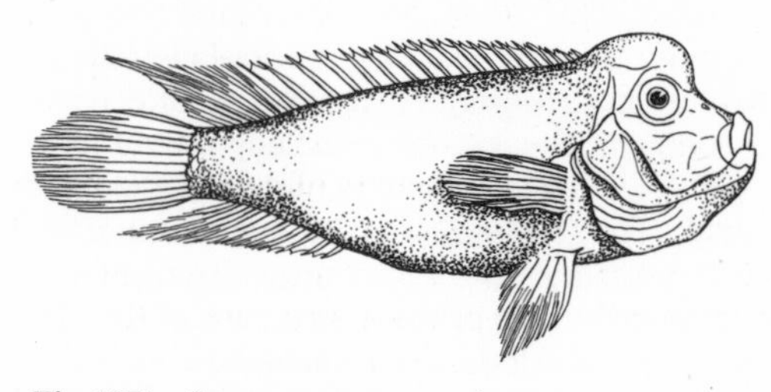

Fig. 1071 *Steatocranus casuarius*

– *S. gibbiceps* BOULENGER, 1888. Only distinguished externally from the above by the smaller lump on the head. Should this species prove to be identical with *S. casuarius,* BOULENGER's name for it becomes the valid one. (This is because of the rules of nomenclature, where the older name holds true.)

Steel Blue Limia *(Poecilia caudofasciata) see Poecilia*

Stem Axis (fig. 1072). Like the leaf* and root*, the stem axis is a part of the vegetative body (corm*) of Ferns (Pteridophyta*) and seed-plants (Spermatophyta*). In contrast to the root, the stem axis has leaves

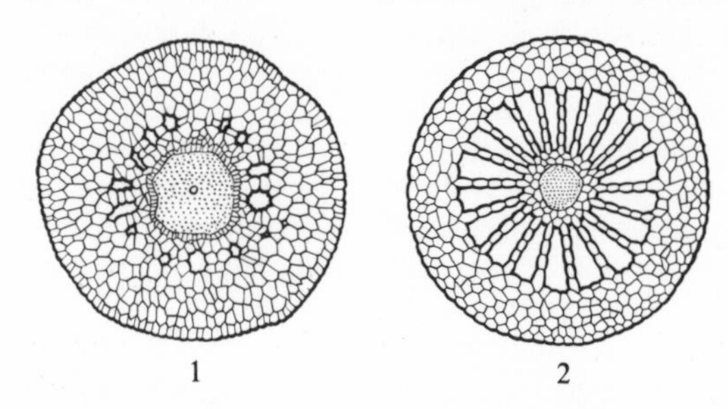

Fig. 1072 Stem anatomy. 1 *Elodea canadensis* 2 *Myriophyllum spicatum*

along its sides. Leaves and stem axis develop from a common vegetative cone*. The stem develops primarily in air, but also in the ground and in water. It grows upwards but may follow a vertical or horizontal course; it is divided into nodes, where the leaves are inserted, and the areas between the nodes known as internodes. These may be cylindrical or angular in shape. The length and thickness of the internodes vary and this results in the formation of very different kinds of stem axes; this, together with the direction of growth of the stem has a profound influence upon the plant's life-style. For example, many aquarium plants have elongate and thin internodes, eg *Hygrophila**, *Ludwigia** or *Limnophila**. If the internodes are very short, however, rosette-growth develops, as in *Vallisneria**. When the internodes are very thick, you have rhizomes* or tubers*.

The young stem axis has a fairly uniform, primary anatomical structure. The outer part is made up of epidermis. Next comes the primary cortex made up of basic tissue – a layer found in land plants that need to have great bending strength. The outer layers of cortex cells contain chlorophyll. The inner part is taken up with pith. Embedded in this is the vascular bundle* which is normally arranged in a ring shape among the Magnoliatae* but scattered along the cross-section in the Liliatae*. In the centre of the pith there may be a cavity of varying size. Pith, pith cavity and vascular bundle are also known by the umbrella term 'central cylinder'. In elongated, submerged stem axes that require tensile strength, the vascular bundle is usually united into a central line. Here, an examination of cortex and pith is not normally possible, and the whole internal area is threaded through by ventilation canals arranged regularly or irregularly. The primary structure of the stem axis can be greatly altered by secondary growth in thickness*.

Stem System. A stem and all its side-stems that have developed from it by branching*. The stem system is known as a *monopodium* when the main stem continues its growth for a long time and all side-stems are subordinate to it. A *sympodium* is where the main stem stops growing after a while and becomes overgrown with side-stems. There are several types of sympodial stem systems.

Stem-developed Roots *see* Root

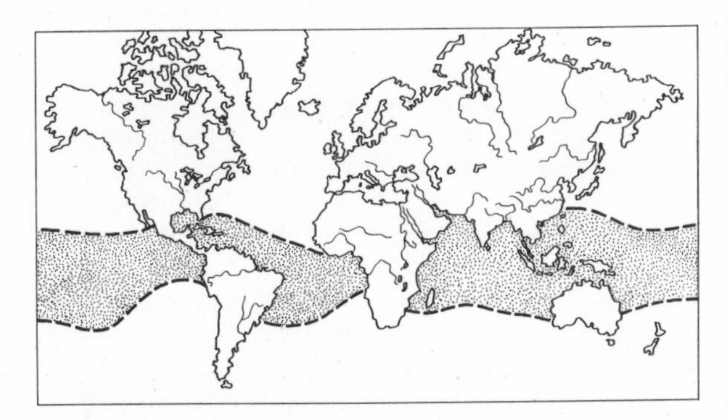

Fig. 1073 Coral reefs are stenothermic organisms and only occur in a narrow temperature range. Note the way the areal shrinks along the west coast of South America because of the influence of the Humboldt current.

Sten- (fig. 1073). In word compounds, means 'narrow'. Eg: stenohaline = only occurring within a narrow range of salt concentration; stenothermic = only occurring in a narrow temperature range.

Stenopus LATREILLE, 1819. G of the Shrimps (Natantia*). There are species in warm parts of the W Atlantic, in the Mediterranean and in the Indo-Pacific. They live on coral reefs and on rocks. *Stenopus* species are long-legged, graceful creatures that live quite a predatory existence. The third pair of walking legs are enlarged into pincers, and it is with these that *Stenopus* attacks small fish and severs Sea-anemone tentacles; even members of their own G are sometimes attacked during moulting, which takes place every 6–8 weeks. The following is the most commonly imported tropical shrimp:

– *S. hispidus;* Banded Coral Shrimp, Cleaner Shrimp. From tropical parts of the W Atlantic. Body length about 8 cm. Eats various types of animal food. Also known to clean fish. Its body is covered with red-white bands and it has very long, shining white antennae. In nature, it is thought to live in pairs.

Stenorhynchus DESMAREST, 1823. G of the Spider Crabs (Brachyura*, F Majidae). They eat small animals and growths, and live in tropical seas. The body is slender and the forehead long; very elongate, slim legs. *Stenorhynchus* species are harmless, peaceful aquarium specimens that will accept animal food, but also algae and detritus. *S. seticornis* from the Caribbean is regularly imported – this species, unlike many other Spider Crabs, does not camouflage itself (fig. 1074).

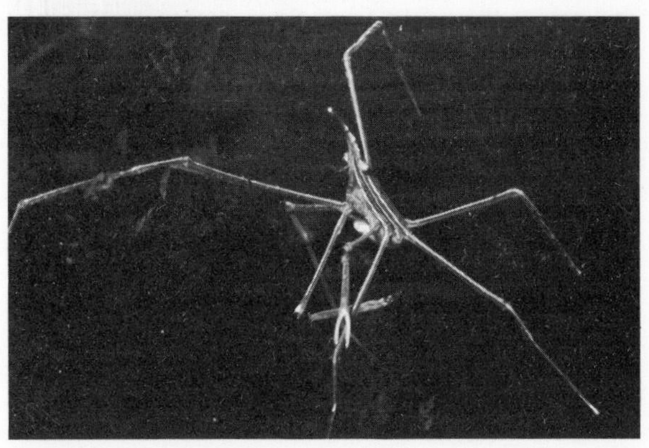

Fig. 1074 *Stenorhynchus seticornis*

Stephanolepis, G, *see Monacanthus*

Stereolepis gigas *see* Serranidae

Sterilisation *see* Disinfection

Sterility; infecundity. Plants and animals that are sterile can no longer produce fertilisable sex cells (Gametes*). Sterility can be caused organically – ie the sex organs may be damaged by infections or other external influences. Infertile individuals can also occur because of inherited damage to the genetic material *(genetic sterility)* or because of the combination of dissimilar hereditary matter, eg when cross-fertilising species with a dissimilar chromosome count *(chromosomal sterility). See* Hybrid.

Sternarchidae, F, *see* Gymnotoidei

Sternoptychidae, F, *see* Stomiatoidei

Stethaprioninae, Sub-F, *see* Characoidei

Stevardia GILL, 1858. G of the Characidae*, Sub-F Glandulocaudinae. Small Characins, 6–8 cm long, found in northern S America, especially Venezuela, and in Trinidad. They are shoaling fish from the middle and upper levels of water and require plenty of room for free swimming and a sunny spot at a temperature of 24–27 °C. *Stevardia* species are elegant fish, elongate and very flat from side to side. The unpaired fins of the male are elongated into a pennant-shape, thus making the lower lobe of the C more drawn out than the upper lobe. In the female, the fins are a normal shape. No D2, LL complete. In aquarium literature, the species *S. riisei* GILL, 1858, has for a long time been included under the names *Stevardia albipinnis* or *Corynopoma riisei.* The operculum of the males extends into a thread-like process that widens out like a spoon at the end and reaches to the An; during courtship (the fluttering dance*), this is held splayed out on the side of the male that is turned towards the female. The spoon-like part gleams patent black during this period. The female snaps at this, thinking it to be morsels of food. Then, at that moment, the male swims very fast to the side of his partner and introduces a sperm packet sealed inside a capsule into her oviduct. The sperm are stored and will fertilise the eggs for several spawning periods. In most cases, a single copulation is enough for a lifetime. The eggs are attached beneath plant leaves by the female, without the male being there. The fry hatch out after 24–26 hours and are easy to rear. Breeding is also possible in medium-hard water.

– *S. riisei* GILL, 1858; Swordtail Characin. Trinidad and northern Venezuela. Length to 7 cm. Body almost transparent; it shimmers in a soft bronze or blue colour. Dorsal surface greenish, ventral surface silver. Fins colourless to pale reddish (for shape, *see above*). In recent years, this original form has almost been ousted completely by a xanthic mutant (fig. 1075).

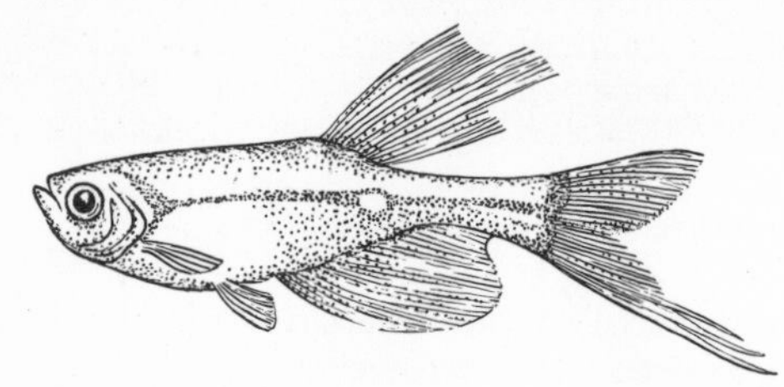

Fig. 1075 *Stevardia riisei*

Stichaeidae, F, *see* Blennioidei

Sticklebacks (Gasterosteidae, F) *see* Gasterosteoidei

Sticklebacks, related species, Sub-O, *see* Gasterosteoidei

Stickleback-types, O, *see* Gasterosteiformes

Stiff-leaved Water Crowfoot *(Ranunculus circinalis) see* Ranunculaceae

Stigma (part of the carpel) *see* Flower

Stigmatogobius BLEEKER, 1874. G of the Gobiidae, Sub-O Gobioidei*. S Asia, Malaysian Archipelago, Philippines, New Guinea. Lives in sea and brackish water of the inter-tidal zone, as well as in fresh water. Small fish, 6–8.5 cm long. Body elongate, cylindrical at the front, flattened laterally at the back. The first ray of both Ds is not bony. The scale behind the eye is enlarged. Front part of the operculum (preoperculum) not covered in scale. There are several rows of lower jaw teeth. Ground colour varying shades of grey or yellow, body and fins often covered with dark flecks. Female's colouring less striking. Can be kept in captivity in slightly brackish water (1–2 tablespoons of sea salt to 10 litres of water). 22–26 °C. Feed with small live food of every sort. Interesting species are:

– *S. hoeveni* (BLEEKER, 1851); Celebes Goby. To 6 cm. Upper side of the body grey-yellow or green-yellow, often a light red-brown on the sides with irregular transverse bands. Lower surface pale. Scales have dark edges. There are 2 dark flecks arranged one above the other at the insertion of the C; these are more distinct in the male. The middle area of the D1 is blue to black, bounded by white. Fin border blue. The front rays of the D1 are often elongated as threads.

– *S. sadanundio* (HAMILTON-BUCHANAN, 1822); Spotted Goby, Knight Goby. To 8.5 cm. Body relatively stocky, with large scales. Ground colour blue-grey to yellowish. Flanks crossed by one or several rows of roundish, black flecks. There are small black dots on the middle area of the C. Also, black flecks and rows of white flecks particularly predominant on the soft parts of the D2 and An. In the D1, the 3rd and 4th rays are clearly elongate. In the female, the fins are smaller.

Still Waters. Inland stretches of water that in contrast to flowing waters do not have any current in any one direction, caused by inclines. The best-known types of still waters are pools*, ponds*, garden ponds* and lakes*; the differences between them are largely defined by the way in which they developed, how long they exist for and their depth. The size of the stretch of water is of less importance.

Stimulus. Term used in physiology. A whole host of external factors which have an effect upon an organism such that its state of being is altered. General protoplasmic stimuli can be chemical, osmotic, thermal, mechanical or electrical. These are in contrast to acoustic stimuli and light stimuli, for the detection of which the animal organism has usually developed special sense organs* in the form of ears and eyes. They transform these special stimuli into general plasma stimuli. The term 'adequate stimulus' is used to describe a stimulus that is typical for a sense organ or any other biological structure that acts as a 'stimulus receiver'. This is as opposed to an inadequate stimulus which also produces a state of excitement in the organ concerned, but at the same time leaves behind some slight structural damage.

Stimulus Threshold. The minimum size of stimulus needed to release a reaction. Usually, stimuli have an effect upon specific receptors, eg light, smell or touch sense cells. The stimulus threshold in these cases is very low. On the other hand, stimuli with a high intensity can cause often unspecific reactions, eg the conduction of excitement in a nerve cell. Below-threshold stimuli can accumulate into threshold stimuli and so the over-all stimulus threshold will be broken through. Different types of such accumulations can occur: the simultaneous stimulation of neighbouring sense cells that are connected spatially with one another, or the accumulation may take place over a period of time, the stimulus being repeated. The stimulus threshold and the functional state of a biological object stand in a reciprocal relationship to each other; this then means that the stimulus threshold can be applied as a characterisation of the object.

Sting *see* Stinging Capsules

Stinging Capsules (Cnida, Nematocysts). Complicatedly constructed, microscopically small cell products that occur in the Cnidaria*. They are located in cells of the ectoderm, especially those of the tentacles and special stinging organs (Acontia*). They contain a rolled-up thread of attachment or stinging thread and, sometimes, special stiletto-like structures which penetrate an enemy or prey when the capsule is fired. As this happens, a stinging poison is set free. The capsules are made to explode when the cnidocils (little sense rods that tower above the cell surface) are touched. About 20 different kinds of stinging capsules have been identified, in *Hydra** there are 3 types. Although most Cnidaria are not a danger to Man, the stinging poison of the tropical Cubomedusae* is fatal. The Portuguese Man o' War (G *Physalia* LAMARCK) which belongs to the Hydrozoa* (Siphonophora), is also very dangerous. The stinging capsules of the common aquarium Sea-anemones (Actiniaria*) normally cause fairly harmless skin irritations only. For illustration see Coelenterata.

Stinging Jellyfish, G, *see* Scyphozoa

Stingray *(Dasyatis pastinaca) see* Dasyatidae

Stingrays, F, *see* Dasyatidae

Stizostedion RAFINESQUE, 1820. G of the Percidae*, Sub-F Luciopercinae. Europe, eastern N America. These are large or very large predatory fishes that live in the open water of large oxygen-rich rivers and lakes that have a firm substrate. The body is elongated like a pike's and only slightly flattened laterally. The pointed head ends in a large mouth containing a lot of teeth; the mouth can be extended far forwards. D divided, supported in the front part by hard rays only. LL complete. *Stizostedion* spawns in late spring along deep parts of the bank where there are overhanging branches and roots. The eggs are large and very sticky, and are guarded by the male. Only young specimens can be kept in captivity; they need a well ventilated domestic aquarium, as this G requires plenty of oxygen. The water must be partly renewed on a regular basis with non-chlorinated, clean fresh water. Temperature not greater than 21 °C. Even specimens 10 cm long need a regular supply of live fry as food – meat is not accepted. There are 3 species in Europe and Asia, and 2 species in eastern N America:

– *S. lucioperca* (LINNAEUS, 1758); Zander, Pike-perch. From central and E Europe. Length to 130 cm, but most are 40–70 cm. Dorsal surface dark olive, flanks grey-green. Young specimens have 8–11 dark, irregular transverse bands. Fins yellowish-grey. D covered with oblong dark flecks, C with dot-flecks. An important food fish (fig. 1076).

– *S. vitreum* (MITCHILL, 1818); Walleye. N America, the area around the Great Lakes, east to Pennsylvania. Length to 90 cm. Dorsal surface dark olive, flanks paler with brassy-yellow flecks, and dark, arched lines in the head area. D yellowish-green, D1 with a brown-black fleck at the rear edge, D2 and C yellowish with dark dots. Strikingly large, translucent eyes.

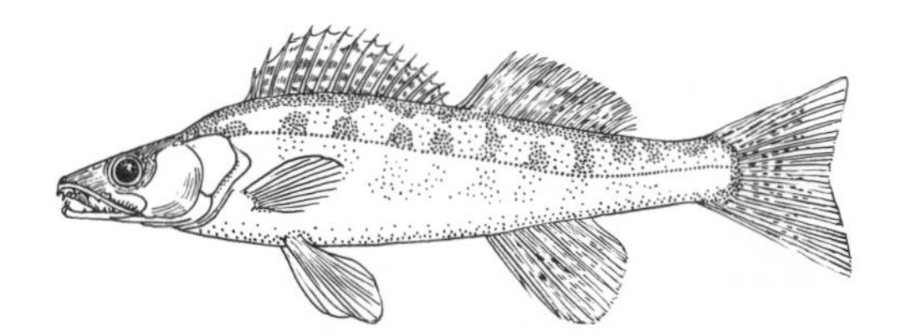

Fig. 1076 *Stizostedion lucioperca*

Stoichactis HADDON, 1898; Giant Sea-anemone. G of the tropical Sea-anemones (Actiniaria*). Lives in the littoral* from about 2 m deep downwards where the sun still penetrates. Normally, Anemonefishes* live inside them. The mouth area can reach a diameteter of 1.5 m, but the tentacles are fairly short with truncated or club-shaped ends; they only have a weak sting. Some *Stoichactis* species are very colourful. Inside are symbiotic Algae (Zooxanthellae) and so that is why the Anemone needs bright light. But in the aquarium *Stoichactis* may be damaged by too high a red light content, and if kept in the dark for any length of time the Zooxanthellae will die. The Giant Anemones are not easy to feed either; large pieces of food are not accepted and various kinds of pulpy food as a replacement for plankton have not proved themselves particularly good. Thus, *Stoichactis* will only survive in the aquarium for a limited period, which is very sad, as they are lovely to look at and are imported often. The same is also true of the very similar Anemones belonging to the G *Discosoma* LEUCKART, 1828.

Stoma (pl stomata). A microscopically small gap in the epidermis (Skin*) of the above-ground green parts of higher plants. Stomata are employed in the process of gaseous exchange and in the giving off of water vapour. The actual gap is bounded by guard cells that are usually bean-shaped; because of their structure, and depending on water content, these cells can alter their shape so that the gap becomes bigger or is shut completely. Unlike the surrounding epidermal cells, the guard cells contain grains of chlorophyll. Occasionally, other nearby cells of the epidermis, auxiliary cells, are also engaged in the function of the stoma. Stomata are found mainly on the underside of leaves, although they can appear on the upperside too (eg as in floating leaves); usually, there are no stomata at all on submerged plant parts.

Stomach *see* Digestive Organs

Stomach Dropsy. Infectious stomach dropsy is one of the most common diseases found among Carps in fish economy. In tropical aquarium fishes it occurs rarely, and mostly as a result of unsuitable conditions in the aquarium or a previous illness. A bacterium *Pseudomonas punctata* or a so far unidentified virus is held responsible for causing the disease. Acute stomach dropsy is recognised by the accumulation of great quantities of liquid in the body cavity (*see* Ascites), the fraying of the fins and a high proportion of deaths. The chronic stage is characterised by large dark swellings in the skin with a white ring in the centre or by soft tumours. Often, not visible from the outside, the liver turns yellow-green and there are extreme deformities of the internal organs. Stomach dropsy is a debilitating illness. Once fish fall ill with the disease, however, highly virulent (infectious) bacterial strains build up, which also attack strong, healthy fish. In the early stage, a cure can be attempted by administering injections* of Streptomycin* or Chloramphenicol*. It is critical to increase the resistance of the fish to prevent the disease. Usually after stomach dropsy has occurred in the aquarium, it is necessary to kill off all the fish and plants as well as a thorough disinfection* of the tank. Fisheries take the precaution of immunising against the disease.

Stomatopoda; Mantis Shrimps. O of the Crustaceans (Crustacea*, Malacostraca), which at the same time forms its own Sup-O (Hoplocarida) with around 200 species. These are predators that usually live hidden in holes and cracks in warm seas. Their highly mobile eyes, their shape and, in particular, their strong predatory feet (metamorphosed 2nd maxillary feet) make them look very like the well-known insects, the Praying Mantis (Mantidae). The Stomatopoda move forwards in a very pliable way, and when trying to escape they shoot backwards, beating their tail. Many Stomatopoda are very colourful crustaceans, 1–33 cm long. They recognise their prey (crustaceans, fish, molluscs, worms) by eyesight; many stalk up to their prey, then kill it quickly with their predatory legs. Then the victim is held fast with the trapping legs and the following three pairs of legs that carry pincers whilst it is devoured. *Gonodactylus* (LATREILLE, 1828) smashes hard-shelled molluscs for himself with his predatory legs; in the aquarium, he goes around hammering at empty bivalve shells (it can be heard quite distinctly). *Squilla mantis* LATREILLE from the Mediterranean appears to hunt crustaceans and annelids mainly; from aquarium ob-

Fig. 1077 *Odontodactylus scyllaris*

servations at least, it appears to clean fish. The females keep the eggs in a clump for weeks between the jaw-feet; they do not take any food until the planktonic larvae have hatched. The Stomatopoda are very worthwhile objects to study. But because they are such great predators, they are best kept on their own. They accept various kinds of animal food but otherwise they have no special requirements. Being struck by the trapping legs can be very painful for the aquarist sometimes. An Indo-Pacific tropical species that is often imported belongs to the G *Odontodactylus* BIGELOW, 1895; it is a lovely green-gold-blue colour (fig. 1077).

Stomiatoidei; Wide-mouths, related species. Sub-O of the Salmoniformes*, Sup-O Protacanthopterygii. Small deep-sea fish, often monstrous-looking, with a series of luminescent organs on the body. There is usually an adipose fin behind the D, and some species have another adipose fin in front of the An. The upper mouth edge is bounded by the premaxilla and maxilla. Jaw apparatus often much enlarged. Various skull bones and the Ps may be absent. Species belonging to the Stomiatoidei are common and are regularly caught in deep-sea fishing nets. The stomachs of large, bathypelagic fish sometimes contain only members of this Sub-O. The many varied shapes of the Wide-mouths, like those of other deep-sea fish groups, can only be understood if one bears in mind that their phylogenetic development was largely determined by a lack of light and low density of food at such great depths. All mutative changes that brought about advantages in these two areas enabled the fish at the same time to penetrate deeper and deeper. Conversely, any natural selective factors geared towards more surface-water zones came to play a less and less significant role in relation to the drive into the depths. There is no doubt that among these natural selective factors is the one that forces almost all surface-dwelling fish into certain types of body shape. So, the often hideous deep-sea forms have arisen because the selectively effective factors were reduced to the bare necessities, while at the same time control through selection was also reduced. Another way of putting it is to say that the deep-sea forms are the results of specialisations and deviations during phylogenesis. There are 9 Fs in the Stomiatoidei, 2 of which are looked at here.

– *Gonostomatidae;* Bristlemouths. Small deep-sea fish, mostly less than 7 cm long. Externally, they look like herrings, except there are rows, usually a double row, of luminescent organs. These are thought to send out red and green light. Bristlemouths are distinguished from the other Fs by peculiarities of the organ of balance and the construction of the pharynx. This can be expanded quite dramatically into a kind of pointed ice-cream cone, whereby the jaws form the entrance and the pharynx the tip. On each gill arch there are a series of long, bristle-like rake-teeth which go to make up a kind of weir-basket. Thus, the whole apparatus can function like a coarse cone-filter and it is by this means that food is obtained. Bristlemouths eat Arrow Worms mainly. There are approximately 32 species and recently some have been able to be studied more closely. AHLSTRÖM and COUNTS for example, have been able to show that larvae develop from the tiny, floating eggs of the east-Pacific species, *Vinciguerria lucetia*, that at first are very similar to the Sardine larvae. The luminescent organs only develop after the metamorphosis into the young fish. Several members of the G *Gonostoma* are found in the Atlantic, each of which lives at different depths. *Gonostoma denudatum* lives at depths of 100–500 m; it is a pale-coloured species with very large eyes. *Gonostoma elongatum* lives at depths of 500–1,500 m; it is black and its eyes are smaller. This is the same for *Gonostoma bathyphilum* (fig. 1078), a species that is found at still greater depths. Comparison

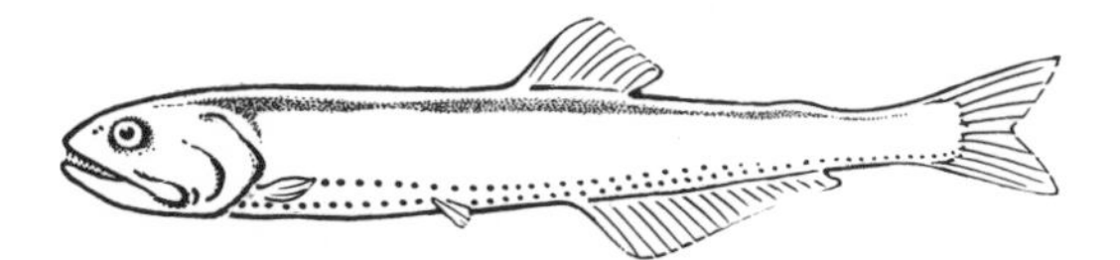

Fig. 1078 *Gonostoma bathyphilum*

studies have shown that with increasing depths the width of the pupils increases, although the ability to see does not improve. In fact, various details seem to suggest that seeing in picture-form is reduced at great depths to seeing only dark and light. The commonest species belong to the G *Cyclothone*; they occur in vast numbers, and so are the main source of food for many other deep-sea fish.

– *Sternoptychidae;* Marine Hatchetfishes (fig. 1079). Small, compact, hatchet-shaped fish from the deep sea. They have luminescent organs on the sides of the belly edge. They live in the deep waters of warm and temperate seas. So far, we know of about 15 species. The G

Fig. 1079 Sternoptychidae-type

Argyropelecus has telescopic eyes. Other interesting members of the Sub-O include the Malacosteidae with the anatomically interesing species, *Malacosteus niger* (fig. 1080), which has an enormous trapping mouth. Also

Fig. 1080 *Malacosteus niger*

the eel-shaped Idiacanthidae with their unusual larvae (the eyes are on very long stalks). BEEBE (1934) was the first to prove that the Stalk-eye *'Stylophthalmus paradoxus'* was not a species is its own right (as originally thought), but the larva of *Idiacanthus fasciola* (fig. 1081).

Stomphia, G, *see* Actiniaria

Stone Bass *(Polyprion americanus) see* Serranidae

Stone Corals, O, *see* Madreporaria

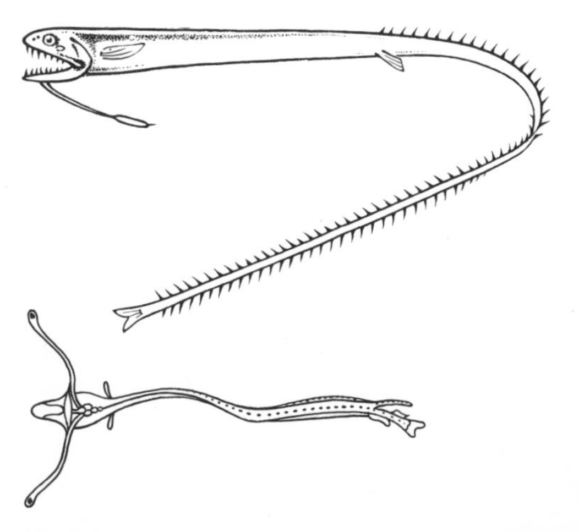

Fig. 1081 Idiacanthus-type and larva of *Idiacanthus fasciola*

Stone Flies, O, *see* Plecoptera

Stone Loach *(Noemacheilus barbatulus) see Noemacheilus*

Stonefish *(Synanceja verrucosa) see* Scorpaenoidei

Stonefishes (Synancejidae, F) *see* Scorpaenoidei

Stony Corals (*Fungia,* G) *see* Madreporaria

Stratiomys FABRICIUS, 1775; Soldier Flies (fig. 1082). G of the Flies (Brachycera*). There are about a dozen species in Europe. The larvae of *Stratiomys* live in weedy pool biotopes, hanging vertically beneath the

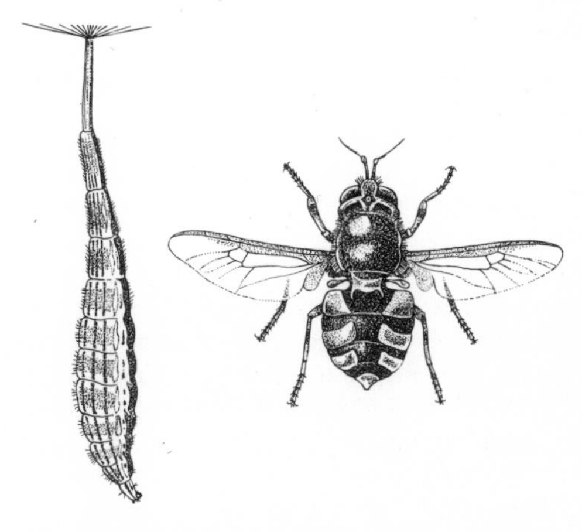

Fig. 1082 *Stratiomys.* Left, larva; right, imago

surface. They are striking because of their size (up to about 50 mm) and because of the fine circlet of hair at the end of the abdomen. The common name comes from the 2 spines on the rear part of the thorax in the adult insect (imago) that look a bit like soldiers' weapons. Another feature of the adult fly is the flattened, broad abdomen coloured black-yellow.

Stratiotes LINNÉ, 1753; Water Soldier; Crab's Claw; Water Aloe. G of the Hydrocharitaceae*. Only one species. Found in Europe, W Asia. Water Soldiers are rosette perennials with runners. Leaves linear, pointed,

with strong serrations, dark green. Flowers unisexual and dioecious, developing within a spathe. Double perianth, with a three-part green calyx and a 3-part white corolla. Male flower: up to 12 stamens. Female flower: 3 carpels, fused, inferior. Fruit like a berry. *Stratiotes* grows at the water surface, its leaves either rising into the air or remaining submerged at varying depths. During the winter, all the plants dive below the surface. The roots have their ends dangling freely in the water or they push into the substrate, thus giving the plant a firm anchorage. Water Soldiers propagate by means of runners and usually grow in large clumps. Often, only one sex is found in a particular area. The species can be cultivated in open-air ponds, but look out for the sharp leaf edges as they can cut.
– *S. aloides* LINNÉ (fig. 1083).

Fig. 1083 *Stratiotes aloides*

Strawberry Beadlet Anemone *(Actinia equina) see Actinia*

Streak-scale Barb *(Barbus roloffi) see Barbus*

Streptomycin. An antibiotic* that is obtained as a metabolic product of the fungus, *Streptomyces griseus*. It is effective against aerobic* bacteria, especially tubercular bacteria. Streptomycin is available in several compounds that dissolve well in water, and often it is used in conjunction with other antibiotics or sulphonamides*. With fish, Streptomycin is used against bacterial infections*. It is given in the form of prolonged baths (Bathing Treatments*) lasting 1–2 days in separate tanks (20–50 mg Streptomycin to 1 litre of aquarium water). If necessary, after a change of water, apply again at intervals of 2 days. Larger kinds of fish can be injected in the body cavity with an aqueous solution of Streptomycin (1–2.5 mg solution to 100 g body weight). *See* Fish Disease Therapy Injection.

Striped Anostomus *(Anostomus anostomus) Anostomus*

Striped Barb *(Barbus fasciatus) see Barbus*

Striped Barb *(Barbus lineatus) see Barbus*

Striped Damsel *(Dascyllus aruanus) see Dascyllus*

Striped Fighting Fish *(Betta fasciata) see Betta*

Striped Flying Barb *(Esomus lineatus) see Esomus*

Striped Headstander *(Abramites hypselonotus) see Abramites*

Striped Leporinus *(Leporinus fasciatus fasciatus) see Leporinus*

Striped Leporinus *(Leporinus striatus) see Leporinus*

Striped Mullet *(Mugil cephalus) see* Mugiloidei

Striped Panchax *(Aplocheilus lineatus) see Aplocheilus*

Striped Panchax *(Epiplatys fasciolatus) see Epiplatys*

Striped Pyrrhulina *(Pyrrhulina vittata) see Pyrrhulina*

Striped Slimefish *(Petroscirtes temmincki) see Petroscirtes*

Striped Surgeon *(Acanthurus lineatus) see Acanthurus*

Striped Wrasse *(Labrus bimaculatus) see Labrus*

Striped-faced Unicornfish *(Naso lituratus) see Naso*

Strobilation *see* Scyphozoa

Stroemer *(Leuciscus souffia) see Leuciscus*

Stromateoidei; Butterfishes, Driftfishes, Rudderfishes, related species (fig. 1084). Sub-O of the Perciformes*, Sub-O Acanthopterygii. Small to medium-sized fish from warmer seas. They are elongate to fairly tall in shape, flattened laterally to a greater or lesser extent. The jaws contain a simple row of conical teeth.

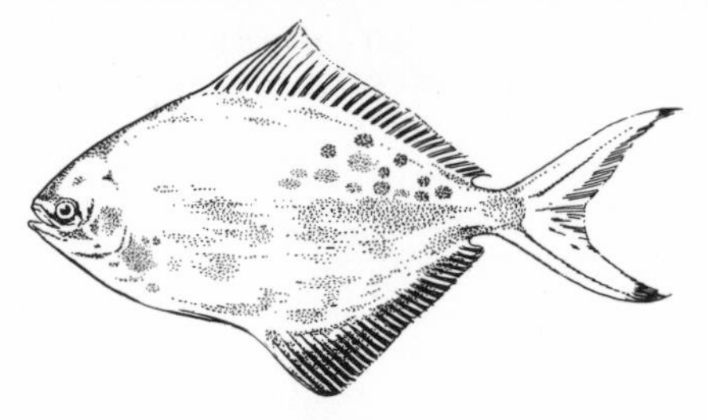

Fig. 1084 Stromateoidae-type

D very long, the front, spiny part in some Ga being separate, C indented. An similarly long. The Vs can fold down into a groove; between the inner edge of the Vs and the belly there is always a membrane. With age, the Vs sometimes become reduced and disappear altogether (F Stromatidae). Small scales, usually fixed loosely. There is often no swimbladder. Particularly typical of the Stromateoidei are the sac-like enlargements of the oesophagus which is lined with length-wise folds and papillae. In the Nomeidae, these folds and papillae are supported by bone and actually carry teeth. What function this structure performs is unknown. The Stromateoidei are found both in shallow coastal waters as well as in the open sea and at deeper water levels. Various species during their youthful phase, or indeed throughout their lives, live in the protection of Jellyfish tentacles. The young of the Pacific

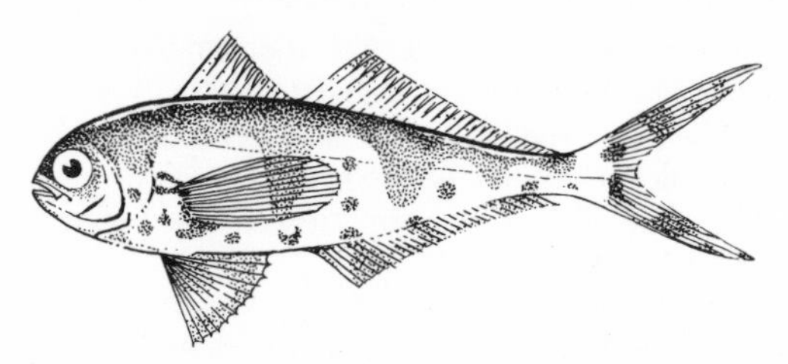

Fig. 1085 *Nomeus albula*

Pompano *(Peprilus simillimus)*, for example, which is up to 28 cm long and an iridescent green colour, is occasionally found in the trail of tentacles in the Portuguese Man o'War *(Physalia)*. A more typical Jellyfish-fish is *Nomeus albula* (fig. 1085) which is blue-violet in colour, about 15 cm long, and has elongated Vs and drawn-out C lobes. This species is found in all warm parts of the open seas. Some Harvestfishes (F Stromatidae) are regionally economically important – their flesh is said to taste very good.

Stump-fins (Crossopterygii, Sup-O) *see* Coelacanthiformes

Sturgeons (Acipenseridae, F) *see* Acipenseriformes

Sturgeon-types, O, *see* Acipenseriformes

Stygicola dendatus *see* Gadiformes

Style (part of the carpel) *see* Flower

Stylocidaris, G, *see* Echinoidea

Stylonychia, G, *see* Ciliata

Suberites, G, *see Paguristes*

Submersed. Those plants that grow completely under water and all those part of marsh plants that grow in water may be described as submersed. The opposite is emersed*.

Submissive Behaviour *see* Inferiority Behaviour

Submissive Colouring *see* Inferiority Colouring

Sub-species *see* Concept of Species, Systematic Categories

Substrate. The decorative function of the substrate in an aquarium is of less importance than its biological function. It is the place where numerous organisms live that have a decisive effect upon the total metabolism of an aquarium because they break down organic products. Such organisms also play a crucial role in the formation of a biological balance capable of life. A substrate made up of too fine sand or one that is too muddy hinders water circulation and ventilation, thus creating a lack of oxygen. When this happens, the breakdown processes proceed largely without the involvement of oxygen (*see* Anaerobes) and highly toxic, oxygen-consuming metabolic products build up, eg hydrogen sulphide, ferrous sulphide. Under such circumstances the substrate behaves like a reducing system, and as a result mud deterioration, decaying plant roots, stagnating plant growth and a general lowering of the redox potential* can be observed in the aquarium water. If the substrate is included in the water circulation* (*see* Filtration) and enough oxygen is provided, anaerobic breakdown processes are prevented. The relationships of plants and animals to the substrate can vary greatly. Aquatic plants which naturally grow as marsh plants obtain a large part of their mineral needs via the roots (*Cryptocoryne**, *Echinodorus**, *Nymphaea**, *Aponogeton** etc) and so require a nutrient-rich substrate. Many aquatic plants only anchor themselves in the substrate and obtain their minerals via the leaves direct from the water. Sessile animals need to attach themselves to the substrate or burrow into it. Fish place the most varied demands upon the condition of the substrate. A dark, non-reflecting surface layer suits all fish, even strictly surface fish or open water fish which usually have no direct relationship with the substrate. Bottom-dwelling fish and animals that like to burrow are suited to a soft, sandy substrate. For some species a layer of humus is necessary, even essential, for spawning. Other kinds of fish need a coarse, gravelly bottom with lots of holes and crevices offering places of refuge. In aquarium-keeping there usually has to be a compromise between these various demands on the substrate. However, it may be said that putting plants and animals with different requirements indiscriminately together in the same tank usually meets with lack of success.

If keeping fish that require soft, slightly acidic water, make sure the substrate contains no hardness agents. The most suitable substances are inert quartz gravel or ground down parent rock of varied grain size. Parent rocks often have dark hues and slowly release trace elements* into the water. To decalcify sand with a small amount of hardness-forming agents treat with 5–10 % hydrochloric acid. This sterilises the sand at the same time. Wash thoroughly under running water. Calcareous sand, or mussel and coral sand consisting of pure chalk, can be used in the marine aquarium. However, it will not be possible to achieve a buffer action against too high a content of carbonic acid or organic acids in this way (pH Value*, Carbon Dioxide*).

Before adding a substrate to the aquarium, arrange all other decorations*, such as large stones, corals and roots. This prevents stagnation in any sand layers trapped under the decorations, and also prevents them from being upturned by over-active fish. The sand should be washed thoroughly, piled up when still damp and pressed lightly. By tipping it up slightly, any rotting material will later gather at the lowest point which makes it easier to remove. Excess water can be soaked up with a sponge. If keeping aquatic plants that have need of a lot of nutrient from the substrate, add some pure, well broken up loam to the lowest layer of sand. Soil or mud from a pond is not suitable. Place a layer of clean, rinsed sand on top of the loam. With fish that burrow a lot or make depressions in the substrate, the whole substrate should consist of carefully rinsed sand. If plants are also added, they must be placed in shallow plant pots. Fish that require a layer of humus can be supplied with a layer of peat*. The sand must not be too coarse for animals that burrow. In order still to be able to provide adequate water circulation* and ventilation, the substrate layer should be kept shallow. A sub-layer of coarse gravel is unsuitable, as the fine sand will slip through the gaps in the gravel after a while, and will no longer meet the requirements of the animals.

Substrate Filter *see* Filtration

Substrate Spawners. Species of fish that lay their spawn on parts of plants, on rocks, roots or in the substratum. There is still a distinction, however, between those species that entrust their eggs to the safety of the substrate, and those that practise brood care. Usually in the latter case the number of eggs is much smaller than with open-water spawners*, as the egg cells are better protected against loss, whether they are protected simply by being put in a safe place or whether both spawn and fry are guarded by parent fish.

Sub-threshold Stimuli *see* Stimulus Threshold

Subularia LINNÉ, 1753. G of the Brassicaceae* comprising one species. Found in temperate regions of the northern hemisphere, E Africa (where there could well be a second species). Very small rosette perennials. Leaves awl-shaped. Flowers hermaphroditic, very small. Perianths doubled. Calyx and corolla free-standing, 4-part. Stamens 6. Carpels 2, fused, superior. The fruit is a pod. *Subularia* lives in shallow, clear, cool waters and along the bank. It is suitable as a foreground plant in a cold water aquarium. Can only be propagated from seed. This normally happens of its own accord, since submerged plants can form seeds even without the flowers opening.
– *S. aquatica* LINNÉ.

Suckerfish *(Echeneis naucratis) see* Echeneidae

Suckers, F, *see* Catostomidae

Sucking Barbs (Garrinae, Sub-F) *see* Cyprinidae

Sucking Discs *see* Organs of Attachment

Suction Holders. Sucking discs made of plastic or rubber which are used to support or anchor thermometers*, heaters, filter tubes, air ducts and other pieces of technical equipment in the aquarium; depending on use, the holders may have clamps on them. By using them, unintentional rearrangements of the technical equipment are largely avoided and the threat of accidents is greatly reduced (particularly with electrically driven apparatus). Care has to be taken that the materials* used are neutral and absolutely poison-free.

Sufflamen JORDAN, 1916. G of the Balistidae*, Sub-F Balistinae. From tropical parts of the Indo-Pacific. Has a life-style and build typical of the family. Generic characteristics similar to *Balistoides**, but with a straight head profile and a straight-edged or slightly elongated C.

Sulphate Hardness *see* Permanent Hardness

Sulphonamides. Antibacterial substances, some of which have the effect of a fungicide. The main component is p-aminobenzol-sulphonamide. There are about 20 different medical compounds that are relatively unpoisonous for higher organisms. Among fish, sulphonamides are used against bacterial infections* and occasionally against fungal diseases (Mycoses*). Sulphonamides are difficult to dissolve in water and have to be dissolved in water at 40–50 °C. In aquarium-keeping, Sulphanilamide and Sulphathiazol are usable – they are applied in the form of a prolonged bath (Bathing Treatments*) lasting 3–4 days, the concentration being 100–250 mg per 1 litre of water. If treatment is repeated, the water must be changed completely for a new dose – further doses are not recommended.

Sumatra Barb *(Barbus tetrazona tetrazona) see Barbus*

Sumatra Fern *(Ceratopteris thalictroides) see Ceratopteris*

Sumatra Loach *(Acanthophthalmus kuhli sumatranus) see Acanthophthalmus*

Sundews, F, *see* Droseraceae

Sunfish *(Mola mola) see* Molidae

Sunfishes, F, *see* Centrarchidae

Sunfishes, F, *see* Molidae

Sunset Platy *(Xiphophorus variatus) see Xiphophorus*

Sunstars, Ga, *see Heliaster, Solaster*

Surfperches, F, *see* Embiotocidae

Surgeonfish related species, Sub-O, *see* Acanthuroidei

Surgeonfishes (Acanthuridae, F) *see* Acanthuroidei

Surgeonfishes, G, *see Acanthurus*

Swallowers (Saccopharyngoidei, Sub-O) *see* Anguilliformes

Swamp Barb *(Barbus chola) see Barbus*

Swamp Eels, O, *see* Synbranchiformes

Swamp Eels (Synbranchidae, F) *see* Synbranchiformes

Swampfish *(Chologaster cornutus) see* Percopsiformes

Swayfish *(Taenionotus triacanthus) see Taenionotus*

Sweet Flag, G, *see Acorus*

Sweetlips, F, *see* Pomadasyidae

Swegles' Tetra *(Megalamphodus sweglesi) see Megalamphodus*

Swelling of the Body can occur in fish as a result of stomach dropsy* or fish tuberculosis*. More rarely it can be caused by the parasite *Ichthyosporidium**. In addition tumours* in the thyroid gland, kidney and liver can increase the girth of the body.

Swellings *see* Tumour Diseases

Swimbladder (fig. 1086). A protruberance of the intestinal canal in fishes, located on the dorsal side. It is filled with gases and is usually unpaired. In some groups there is no swimbladder, eg the Cyclostomes, Sharks and Rays, and among the bony fish, the Trachinidae, Cottidae, Cyclopteridae and many Gobiidae. The Pleuronectidae and Uranoscopidae have a swimbladder during the larval stage only. In deep-sea fish, the swimbladder may contain oil.

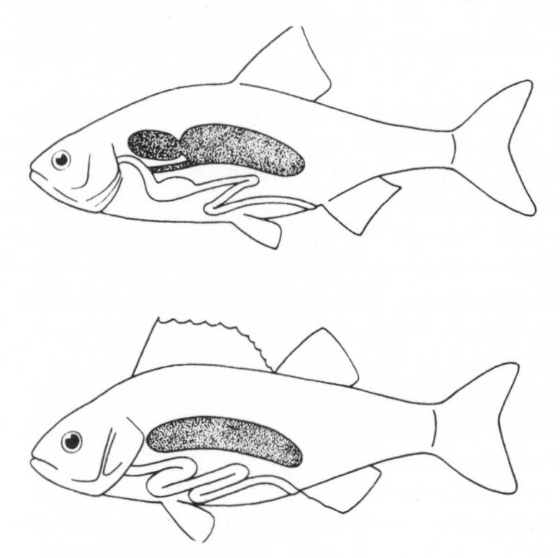

Fig. 1086 Types of swimbladder. Top, Physostomes; below, Physoclists

The swimbladder is mainly a hydrostatic organ. By altering the volume of gas inside it, the specific gravity of the body can be balanced out with the surrounding water. This enables the fish to float, saving its strength, and also preventing it from sinking steadily. Therefore, species with no swimbladder are usually fast swimmers or they are bottom-dwelling fish. More rarely, the swimbladder serves to amplify sounds the fish produce (Sounds, Production and Use of*) and to receive sound waves (Hearing*), or it may become an accessory respiratory organ*.

The swimbladder lies between the body cavity and the vertebral column. In many cases, it consists of a simple, elongated compartment (Acipenseridae, Salmonidae); usually, however, it has 2 or 3 sections separated by constrictions. The embryonal connective duct between intestine and swimbladder may be retained throughout life as an *air duct* (Ductus pneumaticus) or it may regress completely. Based on this criterion, fish with swimbladders can be divided into 2 large groups: *physostomes* (eg Carp-types) and *physoclists* (phylogenetically, mostly more highly developed fish such as Perch-types). With the physostomes, the gas pressure of the swimbladder is regulated solely or partly by the passage of air. The physoclists, on the other hand, raise the pressure by using *gas glands* (red bodies) that are rich in blood vessels and located in the wall of the swimbladder. These glands transfer carbon dioxide from the blood to the swimbladder. Other regions (eg the *oval* in the rear part of the swimbladder) work in opposition and produce a lowering of the pressure. The composition of the gas varies with different fish species and fluctuates within certain limits. For the Perch *(Perca fluviatilis)*, for example, levels of 2.5% carbon dioxide, 19.4% oxygen and 78.1% hydrogen have been recorded.

The swimbladders of many fish are used commercially, eg to obtain fish-glue (mainly those of Sturgeons and Catfishes) and isinglass, an organic substance for keeping beer and wine clear. In many parts of China, swimbladders are considered a delicacy.

Swimbladder Inflammation. Occurs in fish as an accompaniment to stomach dropsy*, but bacterial* and viral* infections, as well as metabolic disturbances (Deficiency Diseases*, Feeding*) can be the cause of diseased changes to the swimbladder. Fish suffering in this way will have an abnormal swimming behaviour or they will lie on the bottom, incapable of swimming; but as these can both be indications of diseases of the nervous system, the fish will have to be examined. With swimbladder inflammation, the bladder will be partially enlarged and filled with a mucus-blood fluid. A treatment that can be tried at least is to raise the temperature (28–30 °C) and apply antibiotics*; if not successful, the fish affected should be killed.

Swordfish *(Xiphias gladius) see* Scombroidei

Swordfishes (Xiphiidae, F) *see* Scombroidei

Swordtail *(Xiphophorus helleri) see Xiphophorus*

Swordtail Characin *(Stevardia riisei) see Stevardia*

Swordtail Platy *(Xiphophorus xiphidium) see Xiphophorus*

Symbiosis. Various kinds of organisms living together for mutual benefit. There are symbiotic relationships among micro-organisms, plants and animals. One of the best-known examples of symbiosis is that between the parasitic Anemone *Calliactis parasitica** and various Hermit Crabs. The Anemone profits from the food remains of the Crab, on top of whose shell it sits, and also from the fact that the Crab enables the Anemone

to move around. The Crab for its part enjoys protection because of the stinging cells on the Anemone's tentacles. Often, an apparent form of symbiosis takes place, *commensalism*, whereby the food of one animal is also used by another. It is not uncommon for this to lead to *parasitism*, in which only one of the partners benefits at the expense of the other.

Symbolic Fighting Behaviour (fig. 1087). Term used in ethology. In contrast to aggressive fighting behaviour*, symbolic fighting behaviour is purely a display contest

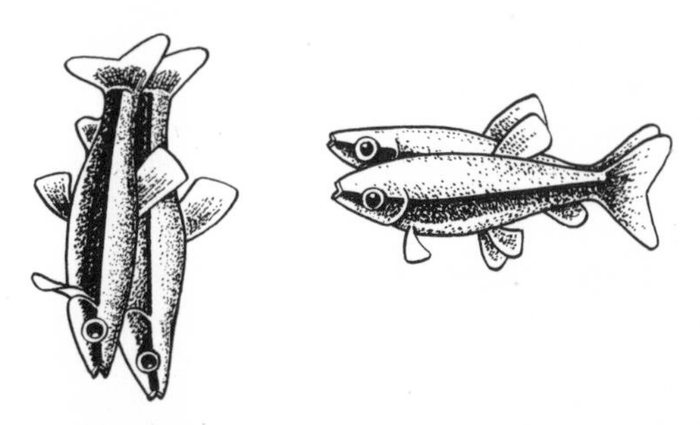

Fig. 1087 Symbolic fighting behaviour of *Nannostomus beckfordi aripirangensis.* Left, vertical fighting; right, horizontal fighting

and never ends up with the adversary being hurt or even killed. OHM (1956) has observed symbolic fighting behaviour among Cichlids and FRANKE (1972) among Nannostomids. The following forms of symbolic fighting behaviour are typical: in many species of fish, the tail fin is beaten in the direction of the head and body of the opponent, but never actually touching him; instead, a wave of water is simply sent towards the adversary. There are never any injuries either during horizontal fighting behaviour, in which the adversaries stand opposite each other, bodies virtually touching (in a similar way to parallel display [Display Behaviour*]), and make beating movements at each other with the head and C. Particularly in *Nannostomus beckfordi aripirangensis*, this jousting behaviour increases in intensity to become vertical fighting behaviour. This position is achieved by the opponents slowly, and at the same time turning round from the horizontal position into the vertical until they are standing, head-down, next to one another, tail and head beating.

Symbranchii, O, name not in current usage, *see* Synbranchiformes

Sympatry. The occurrence of distinguishable populations* of a species in the same area, ie the areals* concerned overlap. Unrelated to the size of the overlapping area, sympatry (when it involves very similar forms) is always a sure sign that isolation mechanisms are at work, and that in the course of time these will lead to the formation of species. The same conclusion cannot be drawn in the case of allopatric distribution (Allopatry*).

Symphysodon HECKEL, 1840; Discus. G of the Cichlidae*. Native to the Amazon and tributaries; along with *Pterophyllum**, one of the few highly specialised (Apomorphic*) Cichlid Ga of S America. Body disc-shaped, much flattened laterally; size up to 20 cm. Many species have a lovely colouring. Fin formulae differ from those of other Cichlid Ga: soft-rayed parts of the D and An are particularly wide (D has 8–10 hard rays and 29–34 soft rays, An 7–9 hard rays and 26–32 soft ones); both fins are very similar in shape. C fan-shaped. V very long, bent like a sabre. Head and mouth, in comparison to the body, very small. In their natural habitats, many *Symphysodon* species are caught by the Indians for their good-tasting flesh.

At the moment, the opinion is being voiced increasingly loudly that all the known members of *Symphysodon* belong to a single species, although that species is, to be sure, very variable. Following the rules of zoological nomenclature*, the specific name *discus* HECKEL, 1840, takes precedence over *aequifasciata* PELLEGRIN, 1903.

Since it was first introduced in 1921, *Symphysodon* has been a real problem fish for the freshwater aquarium. Today, where the ecological data of the natural waters is known and imitated successfully, taking care of these beautiful fish is no longer as tricky. Very much adapted to their biotope, Discus-fishes can tolerate only small changes to their environment. Their natural water habitats have extremely low amounts of electrolytes, the total hardness being less than 1 °dH. The high amount of humic acids* and tannins (Tannic Substances*) mean that there is a low amount of bacteria and fungi. The pH value* is about 6, so the water reacts slightly acid (Dystrophic Waters*). To look after *Symphysodon* successfully, water must be prepared in the necessary way (Water Preparation*) and also checked at intervals of 2–3 weeks, otherwise the electrolyte content in the aquarium will climb too high. Apart from the lowering of water hardness, it is especially necessary to achieve the properties of dystrophic waters by adding humic acids and tannins, either by periodic filtration through peat or by adding peat extracts. By these methods, the necessary slightly acidic pH value will be arrived at too. Increasing acidity with phosphoric acid is not advisable because of the sensitivity of *Symphysodon*; moreover, the buffer* properties of the extremely soft water are slight. A tank containing Discus-fishes should always be very large so that the specimens have enough room for swimming, Temperature 26–29 °C, as befits their equatorial origin. Large water plants (eg *Echinodorus**) and leached root stumps create the necessary cover and lodging places. Feed specimens on a varied diet, but do not overfeed – eg use black and white Mosquito larvae *(Aedes*, Culex*, Corethra*)*, small Mayfly larvae *(Baetis*, Cloeon*)* and crustaceans of appropriate size *(Daphnia**, large *Artemia*)* Although *Symphysodon* species like to gather food from the substrate and are also very adept at washing worms from the bottom, *Tubifex** and Red Mosquito larvae *(Chironomus, see* Chironomidae) should not be given. These kinds of food animals live in the mud of very polluted waters and therefore contain many pathogens and toxins.

If kept under inadequate conditions, Discus-fishes react very quickly; they refuse food, the digestive system becomes disturbed (viscous white faeces), and they are attacked by the dreaded 'Discus disease'*. Although Discus-fishes are very peaceful towards other fish, they must be kept exclusively in a species-aquarium. Only there are they at their best and undisturbed by other fish.

As open-water spawners (*see* Cichlidae), the G is very primitive in its method of reproduction (Plesiomorphic*), but their form of brood care is highly specialised, to a degree that is comparable with only a few fish. After swimming free, the brood feeds exclusively at first on a special secretion from the parents' skin. The lack of success with rearing Discus-fishes in the beginning could be traced to the fact that aquarists did not know of this form of feeding the fry; probably for safety reasons, the parents were always removed from the tank and the fry were reared 'artificially' (but successfully).

In order to get Discus-fishes to spawn in the aquarium, it is best to let a pair choose themselves from a shoal of young fish. Discus-fishes are intelligent enough in fact personally to recognise individuals in a small group, and as a result they can also pick out their chosen partner from amongst other like species. Once a pair has selected itself, they will leave the other inmates of the tank in peace, although they will restrict the others' freedom. A suitable substrate (perhaps an *Echinodorus* leaf) is chosen for spawning and is painstakingly cleaned by both partners. The eggs are deposited and fertilised without any visible sign of excitement on behalf of the pair. The fry hatch after 2 days and are partly chewed out of their egg-cases by the parents. After that, the parents deposit their youngsters on aquatic plant leaves where they hang from short secreted threads (compare *Pterophyllum*) and eat up the yolk supply. After a further two-and-a-half days, the brood can swim freely; they attach themselves to their parents and start feeding on the skin secretion. The partners take it in turns to tend the fry – certain types of behaviour patterns help them in this. Very soon after they can swim freely, the young fish will also be eating powdered food and so they gradually lose their dependence on the parental care. To begin with, the fry are still elongate; the typical disc shape develops after 3 months, and the final colouring after 7–9 months.

– *S. discus* HECKEL, 1840 (True Discus); Many varieties:

– *Blue Discus.* From the Amazon near Letitia and Benjamin Constant. Ground colour brown, head a shimmering purple. Body as in the other forms crossed by oblique bands. Face, dorsal, ventral and tail areas, plus D and An, are all crossed by lovely gleaming blue wavy lines. Iris red.

– *Brown Discus* or Common Discus. From the Amazon near Belem, Rio Urubu. Yellow-brown to rusty-coloured; the 9 bands are narrow and only have an intense colour some of the time. D and An dark-brown on the outside edges with fine rust-red markings. V a green iridescent colour at the front, the other parts rust-red. Head and dark fin areas crossed by pale blue or bright green iridescent lines.

– *True Discus.* From the Middle Amazon and tributaries (Rio Negro and Rio Xingu). Colour similar to the

Royal Blue Discus; it is a delightful species. The 5th oblique band is especially wide and striking. Unpaired fins sky-blue. Other authors describe the True Discus as inconspicuous (fig. 1088).

– *Green Discus.* From the Amazon near Santarem and Tefé. Has 9 dark-brown, equally intense oblique bands. D and An blackish at the base, outside edge pale olive-green with pale markings. V dark-green. Cheeks crossed by pale blue stripes. Operculum crossed by 3 vertical pale blue stripes.

Fig. 1088 *Symphysodon discus*

– *Royal Blue Discus.* From the middle Amazon Basin area (Rio Purus). Ground colour olive-brown, oblique bands broad and very intense. The wavy horizontal lines are a lovely mid-blue, uninterrupted (fig. 1089).

Fig. 1089 *Symphysodon* 'Royal Blue'

– *Red Discus.* Body and base part of the D and An deep-red, otherwise similarly coloured to the Brown Discus (fig. 1090).

Sympodium *see* Stem System
Synanceja verrucosa *see* Scorpaenoidei
Synancejidae, F, *see* Scorpaenoidei
Synaphobranchidae, F, *see* Anguilliformes
Synapta, G, *see* Holothuroidea
Synascidea, Cl, *see* Tunicata
Synbranchidae, F, *see* Synbranchiformes

Fig. 1090 *Symphysodon* 'Red'

Synbranchiformes; Swamp Eels. O of the Osteichthyes*, Coh Teleostei. Eel-shaped fish from tropical parts of S America and Africa, although they are not related to the True Eels (Anguilliformes*). The D, C and An form a very low, often almost completely reduced fin fringe. Paired fins missing. Only the members of the F Alabetidae (Australia and Tasmania) make an exception here – they have both a distinct fringe of fin as well as very small Vs standing at the throat. Mouth large, dentition complete. Small eyes covered by the skin. No swimbladder. Gills often reduced, and skin usually naked. No fossil forms are known. The Synbranchiformes have developed different forms of additional respiration, and this has enabled them to live in marshy, oxygen-poor biotopes that sometimes dry up. They come regularly to the surface to breathe or they stand vertically, resting on their tails, and in this position they can breathe air. Membranes in the pharynx that are rich in blood vessels, lung-type sacs in the gill cavity or the hind-gut are all used as air-breathing organs. Many species survive the dry seasons buried in the mud and by reducing respiration (aestivation or summer dormancy). *Synbranchus marmoratus* (F Synbranchidae) (fig. 1091), which grows up to 1.5 m long, has an un-

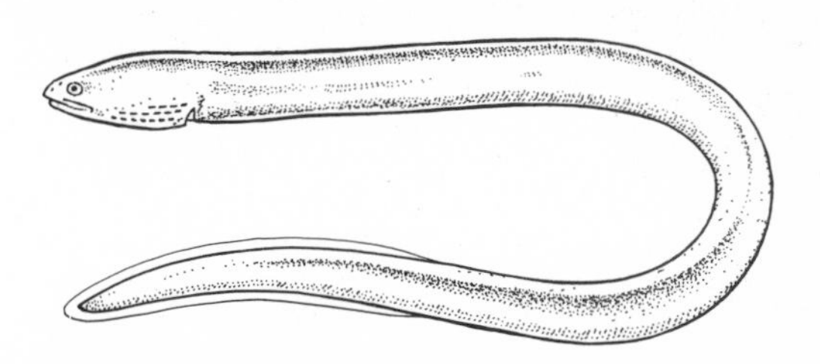

Fig. 1091 *Synbranchus marmoratus*

divided gill cavity enlarged like a balloon for its accessory breathing organ, and 4 gills that function well. The gill cavity opens out on the throat side by means of a slit sited at an angle. In the 90 cm long *Monopterus albus* (F Synbranchidae) the gill cavity is divided – there are 3 pairs of rudimentary gills and 2 gill slits at the throat. This particular species is distributed from N China to Burma and on the islands of the Malaysian Archipelago. *Amphipnous cuchia* (F Amphipnoidae) grows up to 70 cm long and has 2 large air sacs like lungs that are connected to the gill cavity. These sacs reach as far as beneath the skin of the nape. The gills are simplified, and the species also has small scales. All the members of the O Synbranchiformes are very greedy predators, and many are considered excellent-tasting fish. *Monopterus* puts its eggs in a bubble-nest that is guarded by the male. Smaller kinds of specimens have also been kept in the aquarium (many have a very pretty colouring), but because of their predatory nature they must be kept on their own. However, they are active at night, and so the aquarist will see them rarely.

Synbranchii, O, name not in current usage, *see* Synbranchiformes

Synbranchoid Eels, O, *see* Synbranchiformes

Synbranchus marmoratus *see* Synbranchiformes

Synchiropus GILL, 1860 (fig. 1092). G of the Callionymidae, Sub-O Callionymoidei*. A few species only found in the tropical Indo-Pacific. They live on the bottom and feed on small animals. Length 5–20 cm. Body cylindrical and fairly stout. D1 with 4 spines; last ray of the V joined to the P by a membrane. Mouth small. The D in the male is bigger than in the female. Eggs pelagic. *Synchiropus* species are extremely colourful and very intolerant of like-species. They require peaceful tanks with no lively food competitors and a rich variety of small animal food. The following fantastically coloured species is imported quite often:

– *S. splendidus* (HERRE, 1927); Mandarin Fish. W Pacific. Patterned in a shining green-blue and coppery red. The picture shows the rare Mandarin Fish (*S. picturatus* PETERS, 1876).

Fig. 1092 *Synchiropus picturatus*

Synecology. A branch of ecology* dealing with the inter-relationships between the members of a living community (Biocenesis*) and the inter-relationships between living community and environment*. Important aspects of synecology are the circulation of materials in nature, the flow of energy within an ecosystem, the productivity of ecosystems (production biology) and the mutual relationships of the species. Within synecology different branches have been established relating to the broad concept of different habitats – ecology of the sea (oceanology), inland waters (limnology) and land (epeirology).

Synentognathis, O, name not in current usage, *see* Atheriniformes

Syngnathidae, F, *see* Syngnathoidei

Syngnathiformes, O, name not in current usage, *see* Gasterosteiformes

Syngnathoidei; Pipefishes, related species. Sub-O of the Gasterosteiformes*, Sup-O Acanthopterygii. Fish of the type of the well-known Pipefishes and Sea-horses belong here. Main features of the Sub-O are: tufty gills, a tube snout that functions as a sucking tube for catching things, a sturdy armour made of dermal bones, the fusion of the first 3 vertebrae and peculiarities of the reproductive organs and reproduction itself. No teeth. In the REGAN system, this fish group is represented by its own O (Lophobranchii), BERG placed them as a Sub-O, together with the Sub-O Aulostomoidei, in the O Syngnathiformes. Most of the species live along the coasts of warm and temperate seas, and some are also found in fresh water. The oldest fossils date from the lower Eocene.

– *Solenostomidae*; Ghost Pipefishes (fig. 1093). Small, bizarre fish from the coastal region of tropical parts of the Indo-Pacific. Head very large, shaped as in the Pipefishes; body short with lateral extensions at the level of the tall D1 and D2. C and Vs large, elongated like pennants. The star-shaped bony covering on the skin is ripply. In contrast to the species of the F that is described below, it is the female Ghost Pipefish that has a breast-pouch (formed from the ventral fins). Little is known about Ghost Pipefish biology.

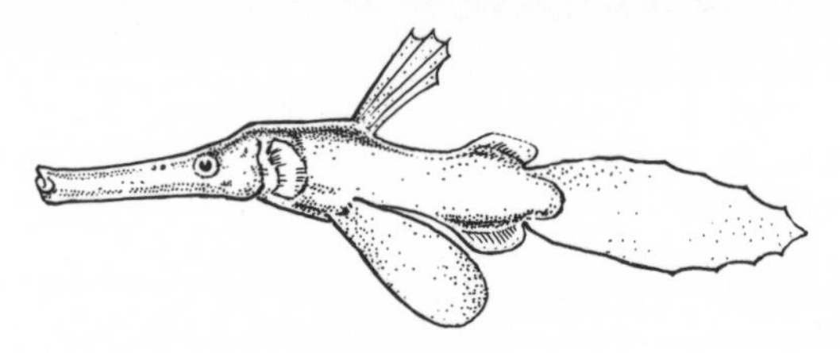

Fig. 1093 Solenostomidae-type

– *Syngnathidae*; Pipefishes, Sea-horses (fig. 1094). Despite their lack of similarity in appearance, both these types of fish have so many things in common that it is justified to group them into one F. Such features include the tube-snout, the complete body-armour, the prehensile tail, the simple D functioning as a organ of propulsion, the lack of Vs and the tiny An. There are other commonly held oddities of feature in the skeleton, kidneys and gonads. The males carry the eggs freely on the ventral side that will have changed to a spongy texture, or they form folds of skin around the belly area or on the underside of the tail, which overlap at the sides to make a breast-pouch. Most Syngnathidae live in shallow parts of the sea, although a very few enter fresh water. A great number of species are popular for the aquarium – see under the Ga *Corytoichthys**, *Dunckerocampus*, *Hippocambus**, *Nerophis**, *Syngnathus**. The F has no economic importance, although in many places, dried Sea-horses are sold as souvenirs.

Fig. 1094 Syngnathidae-representatives

Syngnathus LINNAEUS, 1758. G of the Syngnathidae, Sub-O Syngnathoidei*. Contains numerous species in all seas, some also in brackish and fresh water. They live among groups of plants and feed mainly on small crustaceans. *Syngnathus* is 20–50 cm long, extremely slim, and square or hexagonal in cross-section. They have a small C. The females display in front of the males by shaking and turning along their axis. The sealed breast-pouch in the males is located behind the anus on the underside of the caudal peduncle. The head profile continues straight into the snout. *Syngnathus* is not difficult to keep in the aquarium and it can be very rewarding. They need calm aquaria with plenty of vegetation, and no over-active food competitors or Sea-anemones or larger kinds of crustaceans. Feed with live small crustaceans. Well-known species are:

– *S. acus* LINNAEUS, 1758; Great Pipefish. Indian Ocean to the E Atlantic, including the North Sea. 40 cm. Brown with 12–13 oblique bands.

– *S. pulchellus* BOULENGER, 1917; African Freshwater Pipefish. Congo, Ogowe, where it lives in fresh and brackish water. To 15 cm. Grey to dark-brown, belly reddish.

– *S. typhle* LINNAEUS, 1758; Deep-snouted, Broad-nosed Pipefish. From the Baltic to the Black Sea. To 30 cm. Camouflage colours, brown to green. Snout very large and laterally compressed. The most common species along European coasts.

Synnema BENTHAM, 1845. G of the Acanthaceae*, comprising 12 species. (Placed under the G *Hygrophila** by many botanists.) From tropical Africa and Asia. They are small perennials with horizontal and vertical stems. Leaf arrangement decussate. Aerial and aquatic leaves undivided to pinnate. Flowers located in dense inflorescences at the leaf-axils, hermaphroditic, monosymmetrical. Double perianth. Calyx 5-part, virtually free-standing. Corolla 2-lipped, lower lip flat; corolla tube not sealed. Stamens 5. Carpels 2, fused, superior. Capsule-fruit. The various species are marsh plants almost all of which can tolerate lengthy periods of flooding. The species described below is one of the prettiest and most popular of all aquarium plants (occasionally, a mutant with very narrow leaves is offered for sale). It needs a lot of light to develop its decorative, feathery aquatic leaves. Propagation normally through cuttings. Adventitious plants can also be obtained, however, if you allow leaves that have become separated to float on the surface.

– *S. triflorum* (NEES) O. KUNTZE (*Hygrophila difformis*, Name not in current usage). Unhappily, also known as 'Water Wisteria'. From SE Asia (fig. 1095).

Fig. 1095 Heterophylly. *Synnema triflorum*

Synodontidae, F, *see* Myctophoidei

Synodontis CUVIER, 1817. G of the Mochokidae*. Very uniformly built Catfishes, native to Africa. Length to 45 cm. Active only at night – during the day, they stay in gaps in the rock, often with the head pointing downwards or belly upwards. Other day-time refuges include root-works or among large plant leaves. *Synodontis* species are peaceful fish, but at night they swim around in gangs searching for food, and this can disturb other specimens in the aquarium. Whilst doing this, many species often adopt a ventral position either for some or all of the time. *S. nigriventris* is best-known for this – its ventral side is coloured dark-brown to give it better protection against discovery by water birds. This species is also said to rest during the day directly beneath the surface, floating freely. The body is scaleless, compact and only very slightly flattened laterally.

The dorsal surface (with the exception of females full of spawn) is more strongly arched than the ventral profile. Head only moderately flattened, with 3 pairs of barbels held outstretched, a pair on each of the upper jaw, lower jaw and chin. The upper jaw pair is longest and only rare, the remaining 2 pairs are usually feathery. D1 large, triangular with a strong spine. D2 has a large surface and is rounded; C twin-lobed, deeply forked with long drawn-out tips. Vs and Ps well developed, the latter with a strong spine. Eyes large. A long shoulder process extends behind the operculum in a triangle into the lower half of the body. *Synodontis* requires subdued light and roomy, well planted tanks. It is an omnivore that prefers worm food. So far, only *S. nigriventris* has been bred, and even then the spawning behaviour of the fish was not able to be observed. PINTER found the spawn stuck to the darkest corner of the aquarium. The eggs were 3 mm long and the fry hatched out after 7–8 days; after a further 4 days, they were eating *Artemia* and nauplii and only adopted the 'back position' after 10 weeks. In SCHMITT's case, this species spawned in a branching piece of root-work in a large well planted community tank. The fry were only noticed after they were a few weeks old. They grew under the protection of the roots among a multitude of larger Catfishes. When the parents were ready to spawn, this state was advertised by a brightening of the body colouring. Sometimes, the females become very full of spawn, and then their bellies look very round indeed. In recent years, several species have been imported, although many have virtually been ignored because they are not particularly striking in colour.

– *S. angelicus* SCHILTHUIS, 1891; Polka-dot African Catfish. From tropical W Africa, Congo. Length to 20 cm. Young specimens very attractively coloured:

Fig. 1096 *Synodontis angelicus* (top) and *S. flavitaeniatus*

flanks and fins red-violet with large, round, pastel-white dots, arranged in oblique bands on the fins. With older specimens, the body turns dark-violet to grey-violet, and the white dots reddish-yellow to dark brown-red. The most attractive species of the G, but only rarely imported unfortunately (fig. 1096).

– *S. flavitaeniatus* BOULENGER, 1906. Tropical W Africa. Probably can grow up to 16 cm. Ground colour yellowish-pink and 2 broad, irregular, chocolate-brown bands run along the flanks – the upper one encircles the eyes. The lower band is often divided into several sections. Both bands continue into the tips of the upper and lower C lobe. The dorsal ridge too is traversed by a broad, chocolate-brown band that includes the base of the D1 and the whole of the D2. The D1, An and Vs are a basic yellow colour crossed by several brown, wavy lines. An extremely attractive species, also fairly peaceful, that only reaches a length of 10 cm in the aquarium. Unfortunately, only imported very rarely. Hiding-places beneath roots or in rock crevices essential if this species is to feel at home (fig. 1096).

– *S. melanostictus* BOULENGER, 1906. Zambezi, Lake Tanganyika. To 32 cm. Flanks and fins pale to dark-brown with a large number of fine black dots, distributed irregularly. Underside bright (fig. 1097).

Fig. 1097 *Synodontis melanostictus*

– *S. nigrita* (CUVIER and VALENCIENNES, 1840). White Nile, Senegal, Niger, River Gambia. To 17 cm. Young fish coffee-brown to brown-violet with a large number of largish, black dots in the upper half of the body, often forming oblique rows. Characteristic of this species is an irregular, dark, oblique band marking on the C, as well as a pale band, edged in a darker colour, extending from the eye to the snout. Older specimens are a monotone dark-brown with a greenish or violet shimmer and dark fleck markings.

– *S. nigriventris* DAVID, 1936; Upside-down Catfish, Congo Backswimmer. Congo Basin. To 8 cm. Body colouring beige to pale brown with dark-brown, often cloudy flecks that may be grouped into oblique bands. The transparent fins are covered in dark dots. The best-known species, which, because of its meagre size, is well suited for the aquarium. It is striking because of its unusual way of swimming.

– *S. notatus* VAILLANT, 1893; Spotted Synodontis. Congo Basin. To 22 cm. Body colouring grey to silver-grey, dorsal area dark-grey. Near the middle of the body is a large, round blotch, deep black in colour; it may be followed by some small blotches. Belly off-white, fins colourless to pale grey (fig. 1098).

– *S. nummifer* BOULENGER, 1899. Congo Basin. To 18 cm. Very similar in colouring to *S. notatus*, but distinct from it in various ways – the upper jaw barbels are longer (reaching to behind the head), the D2 is longer and the eyes are slightly oval. Also, this species usually has 3 black dots – the largest in the middle of the body behind the D1, the second beneath the D2, and the third on the base of the C (fig. 1098).

Fig. 1098 *Synodontis notatus* (top) and *Synodontis nummifer*

– *S. schall* (BLOCH and SCHNEIDER, 1801). Niger, Senegal, Lake Chad. To 40 cm. Juveniles have a pale brown or olive brown background colour with a fine dark-brown marbling effect. A good contrast is provided by the wavy band markings on the snout. Older specimens become a monotone dark-grey or brown, with no flecking. Underside bright, fins blackish. Only suitable when young for the aquarium (fig. 1099).

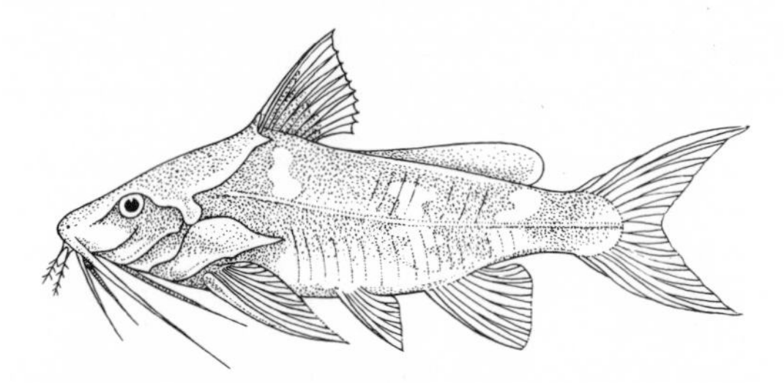

Fig. 1099 *Synodontis schall*

Synodus synodus *see* Myctophoidei

Synonym. 2 or more names given to the same taxon*. It is quite common for species and other systematic groups to be given (unknowingly) several different names. For example: The names *Pterophyllum scalare* (LICHTENSTEIN, 1823) and *Pterophyllum eimekei* AHL, 1928 are one and the same species, which means that they are synonyms. According to the rules of zoological nomenclature only one of the synonyms can be valid, usually the elder one (in the case mentioned above, *Pterophyllum scalare*).

System. A survey of all the world's organisms, arranged hierarchically according to systematic categories*, and worked out by systematics* or taxonomy*. 3 types of systems are distinguished: artificial, natural and phylogenetic. *Artificial systems* are based on just one or a few features, and so they can provide simply an overall view of the diversity of living matter. Such systems first came into being out of a need for a certain ordering, and even today artificial systems can be justified inasmuch as they can help to identify plants and animals quickly and without complication. Thus, most identification tables are basically nothing other than artificial systems compiled according to polyphyletic* groups.

Natural systems are based on the use of as many features as possible, and so they reflect graduated similarities between groups of organisms. With this method, typological groups are encompassed, eg the paraphyletic* group of the fish. *Phylogenetic systems* are not content with typological groupings. Instead, they aim to reflect as accurately as possible the actual succession of generations and, with that, the tangible phylogenesis. To do this, forked, branching diagrams are used (either expressed numerically or graphically), in which those extant species that stem from one and the same original form (Monophyletic* Group) are brought together in groups.

Systematic Categories. The graduated divisions within systematics* which add up in total to a system* constructed on a hierarchical basis. The basic systematical categories, in declining order of rank, are:

Kingdom (Regnum)
Phylum (Phylum)
Class (Classis)
Order (Ordo)
Family (Familia)
Genus (Genus)
Species (Species)

A high degree of different ranks is attained by the use of prefixes like super-and sub-, as well as other systematic categories such as division and tribe. The use of systematic categories does not mean they are equivalent, however; for example, an order of plants, insects and fish cannot be compared in rank without further investigation. There have been attempt to correlate the systematic categories with a particular geological age, but such a procedure assumes the same rate of development for all the groups of organisms.

Systematics. An ordered representation of something – in biology, it is the ordering of living organisms into a system. As the term 'systematics' is very general, increasingly in biological circles the term 'taxonomy' is being used.

Tachypleus, G, *see* Xiphosura

Tadpole Mad Tom *(Noturus gyrinus) see* Ictaluridae

Tadpoles *see* Amphibia

Taenioididae, F, name not in current usage, *see* Gobioidei

Taenionotus LACÉPÈDE, 1802. G of the Scorpaenidae, Sub-O Scorpaenoidei*. There is only one species from tropical parts of the Indo-Pacific: *T. triacanthus* LACÉPÈDE, 1802, the Sailfin Leaf-fish or Swayfish (fig. 1100). This is a predatory fish that lies in wait for its prey among seaweeds; it has quite a fantastic appearance. Length to 15 cm. Usually has camouflage colours

Fig. 1100 *Taenionotus triacanthus*

(yellowish), very flat from side to side, D and P large. It lurks between plants, making swaying movements that get stronger as the fish gets excited–it also stalks its prey with this swaying action. Now and again, the top-most layer of skin is covered entirely with stripes. *Taenionotus* is quite delicate in the aquarium; it will only accept live food and must not be put in the same tank as over-active food competitors.

Tail Fin *see* Fins

Tail Rot *see* Fin-rot

Tail-spot Climbing Perch *(Ctenopoma kingsleyae) see Ctenopoma*

Tail-spot Corydoras *(Corydoras caudimaculatus) see Corydoras*

Talking Catfish *(Acanthodoras spinosissimus) see Acanthodoras*

Tangs (Acanthuridae, F) *see* Acanthuroidei

Tanichthys LIN SHU YEN, 1932. G of the Cyprinidae*, Sub-F Rasborinae. SE Asia, from the White Cloud Mountains near Canton, and the area around Hong Kong. *Tanichthys* species are small shoaling fish that live in mountain streams and small lakes where the winter temperature can drop to 14 °C and the summer temperature can climb to 26 °C. Their body is elongate and only slightly flattened laterally. Mouth slightly superior, no barbels. LL absent. *Tanichthys* is distinguished from other Ga of the Sub-F mainly because it has horny tubercles on either side along the edge of the snout and because the front and rear nasal openings are joined to form a long furrow. In the aquarium, specimens are highly mobile and undemanding; keep them as a shoal in long tanks with plants located along the edges. Regularly adding fresh water helps the fish a great deal. Temperature not too high, in summer 20–22 °C, in winter 16–18 °C. They are omnivores, feeding on live and dried food. Easy to breed. The pairs spawn between plants, and the fry that hatch out after 48 hours should be fed on fine powdered food. Only one species:

– *T. albonubes* LIN SHU YEN, 1932; White Cloud Mountain Minnow. To 4 cm. Colouring varies depending on origin. Dorsal surface dark-brown with a greenish shimmer; flanks paler, ventral surface white. A blue-black stripe ends in a blotch on the C root and is bounded above by a silvery, golden, green or blue-green iridescent band, and below by a broad, brown zone. Base of the D red, with a yellow band towards the outside, and silver-blue at the edge itself. In aquarium literature, the Venus Fish (*Aphyocypris pooni*, LIN SHU YEN, 1937) has often been mistakenly confused with the form described by LIN SHU YEN in 1937, but it is not identical with it; because a new description proved necessary, it is now known as *Hemigrammocypris* lini* WEITZMAN and CHAN, 1966. In fact the form described by LIN SHU YEN in 1937 is a colour deviant (local variety) of *T. albonubes* and is found in the area around Hong Kong. Both local races can cross-breed fruitfully, and today, it is scarcely possible to get pure phylogenetic forms any more (fig. 1101).

Fig. 1101 *Tanichthys albonubes*

Tannic Substances. Astringent vegetable substances. In aquarium-keeping, this term is used in its narrowest sense to mean tannic acids, its main representative being tannin. They are chemically non-uniform substances of vegetable origin which are particularly prevalent in oak bark, horse-chestnut and peat*. They have the property of being soluble in water as colloids (Colloid*), reacting as a weak acid, behaving as an antiseptic (impeding bacteria and fungi) and as an astringent. Some waters have large concentrations of tannic substances. This causes the water to turn brown (Dystrophic Waters*, Black Water Rivers). In aquarium-keeping, the antiseptic, astringent and acidic properties of tannic substances are utilised, since fish from black water rivers, such as *Symphysodon**, *Aphyosemion** and *Roloffia**, are adapted to such conditions. In contrast, other fish (Poeciliidae*) can react badly to tannic substances. If these substances are to be added to the aquarium water, it should be filtered for some time through peat. Pure tannic substances or extracts should be used with great care.

Tapetails (Eutaeniophoridae, F) *see* Cetomimiformes

Tapeworm Infection *see* Cestodes

Tarpon *(Megalops atlanticus) see* Elopiformes

Tarpons (Elopidae and Megalopidae, Fs) *see* Elopiformes

Tarpon-types (Elopomorpha, Sup-O) *see* Osteichthyes

Tarpon-types, O, *see* Elopiformes

Taste, Organs of. In fish, the organs of taste are concentrated as taste buds in the mouth, on the lips and on the barbels (eg in the palatal organ of the carp), or, in addition, they are scattered over the body. They consist of rod-like sense cells arranged in clusters that are so deeply embedded in the epidermis that they sit directly on top of the corium. The four taste distinctions of sweet, sour, salty and bitter can be detected. The sense stimuli reach the central nervous system in the form of information received via the nerves.

Tatia, G, *see Centromochlus*

Taxis. The free movement of plant cells or cell colonies in the direction of a stimulus. Examples are single-celled Algae actively swimming towards a light source (phototaxis) or the movement of male gametes from a Fern to the egg-cell, which is brought about by chemical stimulants (chemotaxis). *See also* Nastic Movements and Tropism.

Taxis Components; norm of reaction. Term used in ethology. In an animal, the turning movement towards the source of a stimulus after it has orientated itself. If, for example, a frog has noticed a resting insect at the edge of his field of vision, he will turn his body towards that insect, and it is this movement that is called the taxis component, or norm of reaction. The rigidly fixed snapping movement that follows (for catching the prey) is the consummatory act, and it is a fixed action pattern* or norm of movement.

Taxocenosis. All the species of a group of organisms (Taxon*) that live together in a particular habitat (Biotope*) – perhaps the fish in a coral reef, the Dragonflies of a moor etc. To that extent, then, taxocenosis is that part of the living community (Biocenosis*) that is restricted to the taxon. The reason for confining oneself to taxocenoses is because it is only possible in a very few cases for the individual author to comprehend the whole spectrum of species in a living community.

Taxon, plural taxa. Each systematic category* that has a valid, or even an invalid, name, eg the order Perciformes, the family Cichlidae, the genus *Pterophyllum*, the species *Pterophyllum scalare.* In the family category and below (genus, species, sub-species), each taxon is baded on a type*.

Taxonomic Units *see* Systematic Categories

Taxonomy. Branch of biology concerned with the ordering of living organisms into a hierarchically divided system*, and for that reason often called systematics*. It is through comparison of features that organisms can be organized in this way. As much evidence as possible is taken into account, eg factors to do with morphology, anatomy, physiology and biochemistry, ethology and palaeontology. There are various methods of approaching taxonomy. *Numerical taxonomy* tackles mathematically all the features that can be treated numerically, and in this way similarities can be concluded. What is not taken into account by this method is the possibility of a feature developing independently (Analogy, Convergence*) and the different value of features. For these reasons, numerical taxonomy can only make limited assertions. The study of homologies* plays a central role in taxonomy by providing clues to a common descent and this leads to a natural system of organisms. With *phylogenetic taxonomy*, different values are placed on features since a distinction is made between the original formation of the feature and one that has been derived from it (Plesiomorphic*, Apomorphic*). With this method, the aim is to achieve a system as laid down by homological studies, that faithfully reflects the phylogenetic relationships of the individual groups of organisms.

Tealia Gosse, 1858. G of large Sea-anemones (Actiniaria*) from cold seas. They eat coarse-grained animals which they catch in their strong but short tentacles. There are sucking warts on the body column. *T. felina* (Linnaeus, 1758), the Dahlia Anemone, is found along northern European coasts; it can reach a diameter of 30 cm, and is coloured a lovely blue, green or red. This Anemone will not tolerate temperatures over 20 °C and so is difficult to keep in the aquarium.

Teddy *(Neoheterandria elegans) see Neoheterandria*

Teeth (figs. 1102 and 1103). Most fish have teeth. The way they are formed is very much dependent on what they eat. With pointed teeth the food is often only held and swallowed without being broken up. Hard-shelled Bivalves, Snails and Crustaceans can be broken up by massive grinding or pavement teeth. Many fish species have incisor teeth or the teeth are fused into tooth ridges (Pufferfishes, Porcupinefishes). Teeth are homologous with the placoid scales of the sharks (Homologous Organs*). Like them, they have a coat of enamel on the outside. The main mass of the teeth in bony fish is made of dentine or similar substances which surround an area (pulp cavity) filled with blood vessels, nerves and connective tissue.

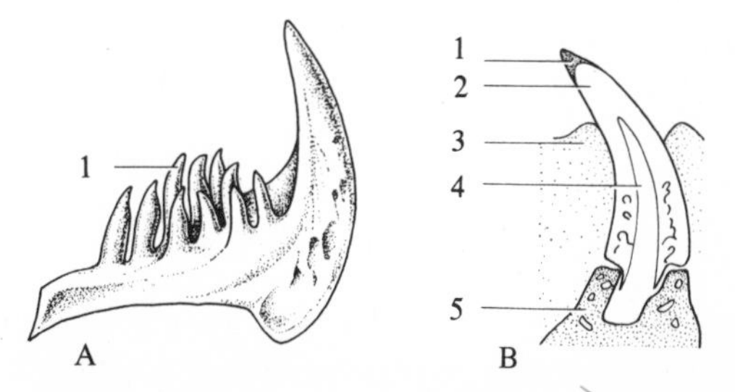

Fig. 1102 Teeth. A Pharyngeal bones with pharyngeal teeth (1) of a bony fish. B Longitudinal section through the tooth of a bony fish. 1 Ganoid coating 2 Dentine 3 Gum 4 Pulp cavity 5 Jaw bone

Whereas sharks and rays only have teeth on the jaws, bony fish may have teeth on all the dermal bones of the viscero-cranium. There may be teeth on the jaws, on the roof of the mouth, on the tongue bones, as well as *pharyngeal teeth* on the pharyngeal bones (upper and lower gill arch bones).

Fish are constantly replacing their teeth. In sharks and rays there are rows of developing teeth behind the old ones, and in bony fish they develop next to or below the old ones.

Dentition is an important feature for the systematics of fish. It can be expressed in terms of a *tooth formula.*

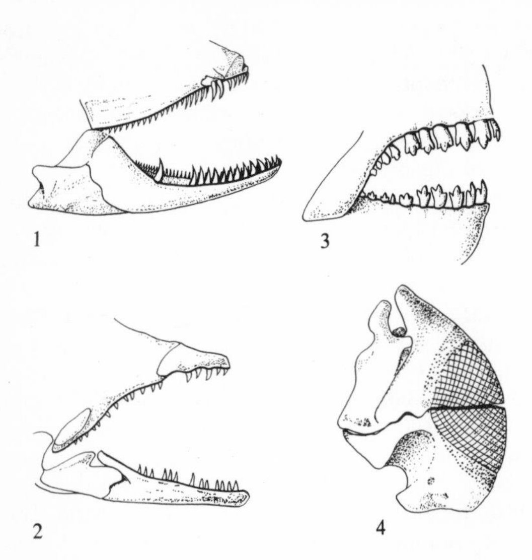

Fig. 1103 Examples of various types of dentition (jaw teeth). 1 *Amia calva* 2 *Salmo salar* 3 *Psalidodon* 4 *Pseudoscarus*

Teleogramma BOULENGER, 1899. G of the Cichlidae* with few species. Native to tropical Africa where the fish live in fast-flowing stretches of water. Their body is very much adapted to this biotope, in shape it is very elongate and cylindrical. Swimbladder no longer functional. Other identification features of this G are the undivided LL continuing as far as the tail, and the large number of hard rays in the D (XXIV). The An has 5 hard rays and 10 soft ones. C rhomboid. Colouring, particularly in the male, lacking in contrast – pale sandy colours, olive or blackish. The females have a broad white edge to the D, likewise the C at the top; also, during the spawning period, a broad red ring-band behind the Ps. Black-white ring markings appear when the fish is frightened.

In the aquarium, *Teleogramma* must encounter conditions that are like its natural habitat. Therefore, long shallow tanks with clear water and strong turbulence (strong ventilation or circulation pump); water temperature around 22 °C. As *Teleogramma* is highly territorial and lives in holes in the rock, the tank should only contain a few specimens, each of which must have its own shelter. As with most Cichlids, reproduction takes place according to a certain pattern of behaviour. With *Teleogramma*, the female displays her red ring-band to the male who reacts by making his head and the fins by the side of the body 'shiver'. The spawn is laid in a hole in the rock, and the female takes care of the brood. Usually, the spawn consists of a few (10–30) large, oval eggs, yellow in colour. They take about 3 days to mature. For the first 2–3 weeks, the brood remains in the hole, and when they do emerge the fry are already colourful and quite big. By this time too they show the same territorial behaviour as their parents.

– *T. brichardi* POLL, 1959; Brichard's African Dwarf Cichlid. The lower reaches of the Congo, in fast-flowing currents. To 12 cm. Colouring as described above.

Teleostei, Coh, *see* Osteichthyes

Telescope Fishes (Giganturidae, F) *see* Cetomimiformes

Telescope-eyed Veil-tail *(Carassius auratus auratus) see Carassius*

Telmatherina BOULENGER, 1917. G of the Atherinidae*. SE Asia, Sulawesi (Celebes). These are small fish that live as a shoal in oxygen-rich and fast-flowing mountain streams, although they are also found in stretches of water near the coast. Body elongate with equally arched dorsal and ventral profiles, flattened laterally. D divided and like the An inserted well to the rear. In contrast to the G *Pseudomugil**, the An begins just behind the D1, and the start of the D1 lies in front of the V. In the males, the fin rays of the D2 and An are very long. Small, superior mouth. No LL. *Telmatherina* species are good swimmers, peaceful shoaling fish and they need fairly large tanks therefore, with a moderate amount of plants and plenty of room for swimming. They particularly like oxygen-rich, clear water that is moving. Total hardness 12 ° dH or above. A partial water exchange should be carried out frequently. Temperature 20–25 °C. *Telmatherina* is an omnivore, taking food preferably from the surface, as well as dried food. For breeding purposes, medium-hard fresh water with a neutral pH value is required. The eggs are laid on fine-leaved plants. The spawning period itself may last several weeks (Breeding Season*). At a temperature of 24 °C, the fry hatch out after 12–14 days and must be fed immediately with powdered food. Growth fairly slow at the beginning.

– *T. ladigesi* AHL, 1936; Celebes Rainbow Fish. Sulawesi, near Makassar. To 7 cm. Dorsal surface, upper side of the caudal peduncle and belly yellow. Flanks yellowish with a blue-green iridescence. A blue-green longitudinal band runs from the middle of the body to the root of the tail. Fins glassy. In the male, the front parts of the D2 and An have black fin rays, yellow at the back. C yellowish with white flecks at the ends of the fin lobes (fig. 1104).

Fig. 1104 *Telmatherina ladigesi*

Telmatochromis BOULENGER, 1898. G of the Cichlidae*. 5 species confined to Lake Tanganyika*. These are elongated fish, up to 12 cm long, with a large, bullish head, as the forehead protrudes in a lump in a similar way to *Steatocranus** and *Cyphotilapia**. Mouth beset with strong teeth, thick lips. Along with *Lamprologus**

and *Julidochromis**, *Telmatochromis* belongs to those African Cichlid Ga that have more than 3 hard rays in the An. D with 18–22 hard rays. Soft-rayed parts of the D and An pointed, C ovate or gently inward-curving. As with other Lake Tanganyika Cichlids, *Telmatochromis* should be kept in hard, neutral to slightly alkali water at a temperature of around 27 °C. The natural habitat is best imitated by providing algae-covered structures (rocks) with plenty of holes, and a sandy substrate. In contrast to the G *Julidochromis* already mentioned, *Telmatochromis* is fairly peaceful, despite its predatory appearance. The species can be kept in quite small tanks and can be bred there too. If you have a fairly large aquarium, it is best to offer several females to one male, as these fishes are polygamous. It is also possible, of course, to make up a community tank with other Cichlids from Lakes Tanganyika and Malawi.

Provided brood-holes are provided, *Telmatochrmis* is not difficult to breed. The females prefer small, very narrow holes, into which the male cannot even follow. So, the spawn can only be fertilised if the male releases his sperm just in front of the hole. There are usually several hundred eggs, and the female practises an intensive form of brood-care. About 2 weeks go by before the fry make their first appearance at the entrance to the hole and start snapping after food. During the next few weeks of their lives, they remain exclusively on the bottom of the aquarium, largely unnoticed because of their unremarkable, fleck markings. Only as they become sexually mature do the fry become more colourful, eg shining flecks on the head and unpaired fins, as well as colourful border areas to the D and An. Two species that have been introduced and bred successfully are *T. caninus* POLL (fig. 1105) and *T. temporalis* BOULENGER, 1904.

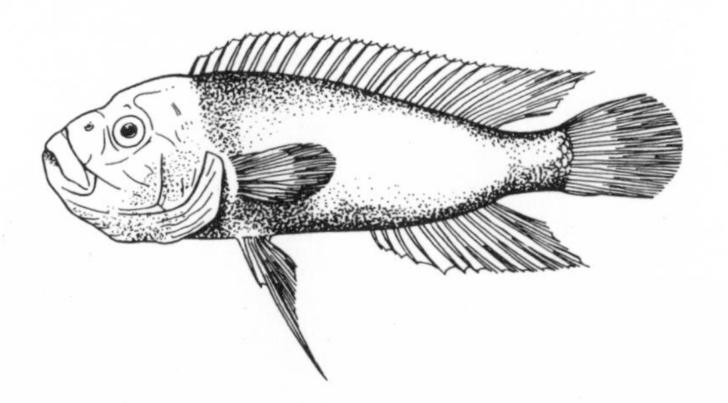

Fig. 1105 *Telmatochromis caninus*

Temperature Control in the Aquarium. All organisms are dependent for their life-style on a particular optimal temperature that is determined genetically. Temperature is a factor, together with light (Lighting*), gas content of the water (Oxygen*, Carbon Dioxide*) and other factors, that has an important effect upon all the metabolic processes going on in the aquarium and it can have a decisive influence upon the water quality. Like other aquarium specimens, fish are poikilothermal* animals, which means that as the outside temperature rises so does their body temperature, and their metabolism intensifies in comparison to lower temperatures. Naturally, animals and plants are subjected to more or less rhythmical fluctuations in temperature, and these will correspond to the characteristic optimal temperature range. If these external temperatures fluctuate too much, damage will occur (Temperature Damage*). The temperature in tropical areas fluctuates little, and lies between 20 and 28 °C. The geographical location alone does not determine the temperature requirements of aquarium plants and animals, however; the ecology of the waters where they live is also a deciding factor. For example, in tropical seas, especially coral seas, and in tropical rain forests, there are only very slight temperature differences; but in shallow tropical savannah pools where the sun's rays can penetrate extremely high fluctuations have been measured between day and night temperatures. Tropical fish from high-up mountain streams and cave waters are usually kept at too high a temperature in the aquarium *(Tanichthys*, Anoptichthys*)* and as a result they do not survive long. Fish from temperate areas can tolerate far greater temperature fluctuations, from between 4 and 20 °C. Apart from the daily night-day rhythm, in temperate latitudes fish also experience a marked seasonal cycle between summer and winter. During the summer months, deep waters only heat up at the surface, so that even in relatively shallow depths of water marked temperature zonation can occur. Animals from temperate areas must have a winter resting period and a reduced metabolism if their lives are to unfold naturally. It is only if such demands are taken into account that the specimens will spawn in the aquarium. But, for tropical fish too, giving regard to temperature rhthyms and other factors, such as feeding, lighting and changes to the water composition, is very worthwhile. In most cases it is necessary to heat the aquarium. In aquarium-keeping today, electric heaters made of glass, enamelled metal or flexible cable are used almost exclusively. Cable heaters work economically as the heating element can be laid as a flexible cable on or in the entire substrate and assists the water circulation. It is vital that the heaters hung in the water are unbreakable and that the contacts are absolutely water-tight. For sea-water tanks, glass heaters only are suitable, or pieces of equipment made of materials that are impervious to sea water. Normal rubber sealants are quickly decomposed by sea water. To prevent unfortunate accidents, the tank should be earthed. The power of the heater must be adapted to the size of the tank and the average room temperature (*see* table). Graded heaters are good because they can be set at different levels of performance to adjust to changes in the outside temperature. Thermostatically controlled heaters need no attention at all – they can maintain a selected temperature electronically or by means of a bi-metallic strip, a mercury relay and a contact thermometer. By adjusting the settings morning and night, a day-night rhythm can be approximated, and this is very beneficial for plants and animals. With one thermostat, several tanks can be operated provided that calculations show that each tank needs the same amount of water and the same wattage. If different wattages are worked out for the same amount of water, several tanks can still be operated from one thermostat if they have different temperature requirements. If the thermostat is to work without causing any disturbance, it must be protected securely from moisture in the air. With an insulated housing, the warmth of an aquarium lost to the outside can be greatly reduced and the current consomption kept low. Places that house a lot of tanks are best heated by space or room heating. For sea and freshwater fish from temperate zones, there is always the danger that the water will heat up too much during the summer months. Apart from a cooling coil which is kept cool constantly by tap water (which can mean it is expensive to operate), fully automatic pieces of equipment are available that will cool or heat the aquarium water as necessary. Because the heating of cooling element only affects the temperature of the water immediately surrounding it, the tank must have good circulation (Water Circulation*) to ensure that the temperature is distributed evenly through the tank. Even with fully automated temperature control, a well calibrated thermometer is needed so that any fault in the technical apparatus will be noticed during regular checks (Aquarium, Care of*). For the same reasons, do not choose too high a power for the heater even when a thermostat is involved, just in case the automatic control goes wrong and the aquarium 'boils'.

fixed temperature boundaries. Among fish, the dependency on a certain water temperature may also change with species-specific life cycles (Reproduction*). Most tropical aquarium fish have an optimum temperature range of 20–28 °C; temperatures up to 38 °C can usually be tolerated for short periods, but below 18 °C severe damage usually occurs. However, many endemic Cyprinidae* *(Cyprinus* carpio, Carassius* auratus auratus)* can slowly become used to ranges of between 6 and 40 °C. Marine fish from Arctic waters that live permanently at temperatures of around −1.8 °C and breed there have adapted to such conditions in an extreme manner, and even at a few degrees higher than this they can no longer adapt. Corals, whose metabolism is genetically bound to constant high temperatures, will die from cold if the temperature drops below +18 °C. It is known that some *Tilapia** species can reproduce at a water temperature of up to 42 °C, and *Cyprinodon** *macularis* lives in pools that may reach temperatures of nearly 49 °C. Here too, the ability to tolerate lower temperatures is limited. In the aquarium, sudden temperature changes are always damaging; with small fish, especially, this can lead to temperature shock and system and the fish have been plunged into extreme even to death. If there has been a fault in the heating temperature ranges, they must only be brought back to their normal temperature again very slowly. High temperatures for fish mean increased metabolic rate, and as a result they will require more oxygen. But as

Tank contents in litres	Heated to °C (temperature differential)														
	1	2	3	4	5	6	7	8	9	10	11	12	13	14	15
10	2	5	7	9	11	13	16	18	20	22	24	27	29	31	33
20	4	8	12	16	20	24	28	32	35	39	44	47	51	55	59
30	6	11	16	22	28	33	38	44	49	55	60	66	71	77	82
40	7	14	20	27	34	40	47	54	60	67	74	80	87	93	100
50	8	16	23	31	39	47	54	62	69	77	85	93	100	108	115
60	9	18	26	34	42	51	59	68	76	85	93	102	110	119	128
70	9	18	28	37	46	55	64	73	82	91	101	110	119	128	137
80	10	19	29	38	48	57	67	77	86	96	105	115	124	134	144
90	10	20	30	40	50	59	69	79	89	98	108	118	128	138	148
100	10	20	30	40	50	60	70	80	90	100	110	120	130	140	150

Required wattage of electrical aquarium heater

Temperature Damage. Like invertebrate organisms, fish are poikilothermal* and so depend for their existence on a certain surrounding temperature. The optimal temperature range in each case is determined through heredity and any deviations from this range can only be tolerated within certain limits if no damage is to result. Spawn, fish embryos and fry, as well as most fish species and invertebrates from tropical and Arctic waters, are usually bound to a very narrowly defined optimal range of temperature (stenothermic) and their ability to adapt (Adaptation*) to changed circumstances is slight. In contrast, forms from temperate zones, for example, can usually exist in far greater temperature zones (eurythermic). Here, depending on ecological origin, individuals with different preferences of temperature can develop from one species which can slowly become displaced by changes to the genetically

temperature increases, the oxygen content of the water drops rapidly, so that acute oxygen deficiency* may also be a limiting factor here.

Temperature Increase *see* Fish Disease Therapy

Temporary Hardness. That part of water hardness* that can be removed by boiling, ie the calcium and magnesium bound to carbonate and hydrocarbonate. That is why temporary hardness is sometimes called carbonate hardness (bicarbonate hardness, hydrocarbonate hardness). The temperature dependency of the temporary hardness rests on the fact that carbon dioxide does not dissolve well in warm water, and the chalk-carbonic acid balance present in the water shifts towards chalk (to the left in the equation):

$$CaCO_3 + CO_2 + H_2O \rightleftharpoons Ca(HCO_3)_2$$

Chalk + carbon dioxide (difficult to dissolve) + water ⇌ Calcium hydrocarbonate (easy to dissolve)

The same occurs with the magnesium carbonate-carbonic acid balance. As a result of temperature increase, therefore, calcium and magnesium carbonate precipitate. A well-known example of this process in action is the scale that is deposited in hot-water heaters; under certain circumstances it also occurs with aquarium heating apparatus. Although it is quite easy to be rid of temporary hardness by boiling the water, increasingly ion exchangers* are being used today. By choosing this method, you will be using a more elegant, less energy-wasting and more effective means, especially as the permanent hardness* and the rest of the salt content is also removed by it (Water Preparation*).

Reference should finally be made to the connection between alkalinity* and temporary hardness which is used to determine the temporary hardness. Calcium and magnesium carbonate or hydrocarbonate exchange with hydrochloric acid and (when converted) 2.8 mg CaO reacts with 1 ml of a 0.1 hydrochloric acid. The temporary hardness can therefore be arrived at by multiplying the alkalinity number by the factor 2.8, given in degrees of German (carbonate) hardness (°KH), ie alkalinity (ability to bind acids) times 2.8 = carbonate hardness.

Tench *(Tinca tinca) see Tinca*

Tenebrio. G of the Darkling Beetles belonging to the F Tenebrionidae. There are almost 20,000 species in the F, most of which are found in tropical dry steppe areas and desserts; some cause damage to food supplies. Apart from *Tenebrio*, the main damager of supplies is *Tribolinus*. *Tenebrio* beetles are almost all a monotone black colour with fairly long legs. In aquarium-keeping and bird-keeping, the larvae of *Tenebrio molitor* (known by the name mealworm) are very important as food animals. They can also be bought from shops (fig. 1106). The larvae are a horn colour, a long cylindrical shape, and they eat decaying matter. When free, they live in the pulp underneath the bark of trees or in tree sponges

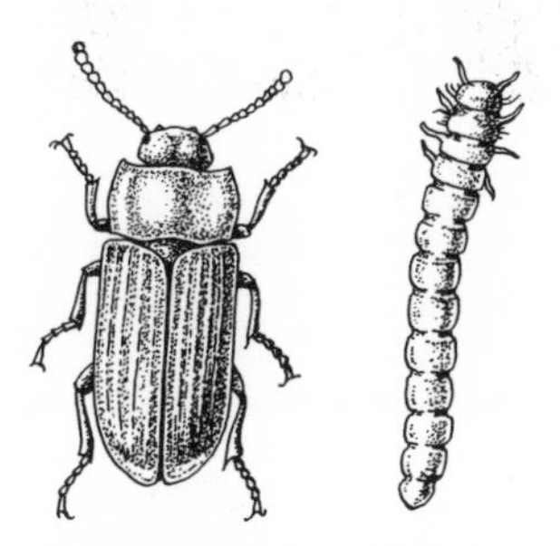

Fig. 1106 *Tenebrio molitor.* Right, larva

(Fungi). As pests, they are found in warehouses, mills or bakeries where they feed on flour, bran, cereals or flour products. The damage is usually slight, as they take about $1^1/_2$ years to develop at a temperature of 18–20 °C. At higher temperatures, a generation can develop in 6 months.

Tenebrio can be cultivated in fairly large glass vessels or clay pots that are shut with a ventilated lid. Flour, bran or rolled oats (all dry) should be offered as a nutrient substrate. Loosened up with saw-dust, shredded bits of paper, wool fibre etc. FREY recommends dried leaves (eg birch) as a food, as the floury contents of the gut prevent many fish from feeding. The best larvae to feed to your fish are those that are soft and newly moulted. It is advisable to press down the head and the front part of the body containing the 3 pairs of legs with some pincers. For feeding to smaller kinds of fish, the larvae should be chopped up. Mealworms are rich in fats and carbohydrates and so should only be fed sparingly to the fish.

Ten-footed Crustaceans, O, *see* Decapoda

Ten-spotted Livebearer *(Cnesterodon decemmaculatus) see Cnesterodon*

Tentaculata. An exclusive aquatic animal Ph, to which the Cl Bryozoa* belongs, including the mussel-like Brachyopoda and the Horseshoe Worms (Phoronida). The body is characteristically divided into three parts, the head, the thorax and the abdomen. The thorax has a tentacle. Most of the Tentaculata live in colonies, and they are very similar to microplankton.

Teredo navalis *see* Boring Molluscs

Terramycin® (Hydroxytetracyclin). An antibiotic* formed from *Streptomyces rimosus* that is effective against many bacteria (Schizomycetes*) as well as viruses (Viral Infections*) and Protozoa*, without causing much toxic side-effects. Terramycin® is applied as a sodium salt which dissolves at 0.025% in water, but which dissolves well in alcohol. Among fish, Terramycin is used in the treatment of bacterial infections*, especially stomach dropsy*. For a prolonged bathing treatment* 10–20 mg are dissolved for every 1 litre of water. Larger fish can be injected in the body cavity with 3 mg for every 150–400 g body weight (intraperitoneal injection). *See under* Fish Disease Therapy, Injection.

Terranatus, G, *see Austrofundulus*

Terrarium-keeping. The observation, care and rearing of amphibious or terrestrial reptiles and amphibians based on keeping the specimens in terraria that provide as near natural conditions as possible. Along with aquarium-keeping, the setting up of a terrarium is a very important hobby, and it has added greatly to our scientific knowledge, especially in the fields of systematics, distribution, ethology and reproductive biology of reptiles and amphibians.

Territorial Behaviour *see* Inter-territorial Fighting

Territorial Boundary. Term used in ethology. The dividing line between the territories of neighbouring individuals who may or may not live there on a permanent basis (some may live there only during the breeding season). The boundary itself is normally defined by particular landmarks, such as groups of plants, rock formations, watercourses. Mammals define their territories additionally by spraying secretions or urine along the boundary, and these marks are respected by their neighbours. For the observer, fish territorial boundaries are often not easy to determine, because such noticeable ways of marking out their areas do not exist. If a rival crosses over the boundary line into another's territory, it is almost inevitable that aggressive fighting behaviour* will break out; usually, the occupier of the territory will emerge as the victor.

Testis *see* Sex Organs

Tetra Ulreyi *(Hemigrammus ulreyi) see Hemigrammus*

Tetrabranchiata, Sub-Cl, *see* Cephalopoda

Tetracyclin. A highly effective antibiotic* produced from various fungi belonging to the G *Streptomyces*. Like Aureomycin*, it can be used against bacterial, viral and protozoan infections. Tetracyclin is contained in many preparations on the market – some are water-soluble. In aquarium-keeping, concentrations of 10–15 mg/l litre fresh water or 15–20 mg/l litre sea water are applied as a prolonged bath (Bathing Treatments*).

Tetragonopterinae, Sub-F, *see* Characoidei

Tetraodon LINNAEUS, 1758. G of the Tetraodontidae*. Contains species that live in the fresh waters and coastal areas of tropical Africa, SE Asia, Australia and the Philippines. Some live in pure fresh water (*T. mbu, T. schoutedeni, T. miurus,* etc) whereas others also live in brackish water (eg *T. cutcutia, T. fluviatilis, T. palembangensis*). *Tetraodon* species are territorial fish, prone to biting. They eat small animals, using their strong dental plates to crush molluscs, crustaceans and insect larvae; *T. miurus* especially likes to prey on smaller fish. The species of *Tetraodon* have a typical family build and they are between 8 and 75 cm long. Many have small soft spines in their leathery skin which they can erect by puffing themselves up (*T. fluviatilis, T. palembangensis, T. leiurus* etc). An important recognition feature is the number and shape of the tentacles around the nose. Within the G, there are both substrate spawners *(T. cutcutia, T. fluviatilis, T. leiurus brevirostris)* and open-water spawners like *T. schoutedeni*, in which 1–2 males bite firmly onto the ventral side of the female during the spawning act. Pufferfishes are extremely charming and long-lived members of an aquarium – they even get to know their keeper 'personally' sometimes. Some individual *T. fluviatilis* specimens have been known to live for 9 years in the aquarium. Apart from the pure freshwater species from Africa that require water that is not too hard and a bit peaty, most of the others are happiest in hard, slightly alkali water or even brackish water. A particular problem with this G is that many of the species are intolerant of one another, and some also of other species. They can give quite a painful nip, and despite their plumpish build they are quite agile. They should be fed with all kinds of animal food, especially hard-shelled food like snails, plus mealworms and earthworms; usually, feeding should present no problems. A lack of desire to eat, discoloration (the black colour noticeable on the ventral side), continual curving round of the tail and emaciation are typical disease symptoms. In a lot of cases, the specimens are cured by putting them into brackish water. In many specimens the teeth start to grow abnormally; they can be cut off carefully using a sharpened pair of fine nail clippers, or similar. Some of the species have been bred successfully *(T. cutcutia, T. leiurus brevirostris, T. fluviatilis, T. schoutedeni)*. The substrate spawners lay their eggs on rocks. There are about 200–300 eggs with *T. cutcutia*, 300–500 with *T. leiurus brevirostris*. As far as we know, only the males carry out brood care. With *T. leiurus* care ends with the hatching of the fry, but with the other species mentioned, the males still guard the youngsters and keep them safely in shallow grooves. Infusorians, nauplii *(Cyclops, Artemia)*, rotifers and microworms *(Turbatrix*)* etc, are suitable foods for rearing them on. The young fish are still tolerant of one another, unlike the older specimens that form territories. The young even form groups sometimes, eg *T. leiurus brevirostris*. The following species are imported fairly often:

– *T. cutcutia* (HAMILTON-BUCHANAN, 1822); Common Pufferfish, Globe Fish. From SE Asia, Malaysian Archipelago, where it lives in fresh and brackish water. Length to 8 cm. No spines on the skin. Nasal tentacles undivided. Grey-yellow with fairly dark flecks and a fine honeycomb patterning.

– *T. fluviatilis* (HAMILTON-BUCHANAN, 1812); Green Puffer, Figure Eight Puffer. From SE Asia, Malaysian Archipelago, the Philippines where it lives in fresh and brackish water. Length to 17 cm. Short forked nasal tentacles. Yellow-green with very variable dark dots and bands. Mostly fairly peaceful. Leaves small fishes unmolested (fig. 1107).

Fig. 1107 *Tetraodon fluviatilis*

– *T. leiurus brevirostris* (BENL, 1957). From Thailand (?). Length to 12 cm. Snout short and truncated. Each nasal slit bears a tentacle that is divided at the end into 2 lips. Spiny skin. Grey-yellow with a dense covering of dark flecks. During the spawning act, the flecks are pale on a dark background. Prone to biting. A subspecies of *T. leiurus leiurus* BLEEKER, 1850.

– *T. mbu* BOULENGER, 1899; Gold-ringed Puffer. From the Congo, only found in fresh water. Length to 75 cm. Fairly elongate. 2 nasal tentacles in each slit, divided like a fork. Head and body covered with fine spines. Dorsal surface and flanks covered in dark, worm-like lines on a yellow to orange background colour. When young, has large dots.

– *T. miurus* BOULENGER, 1902; Valise Puffer. From the Congo, only found in fresh water. Length to 15 cm. Eyes small, directed upwards. Has a striking capacity for changing colour, usually patterned in brownish/sandy

colours. Buries itself into the sand up to its eyes. Very prone to biting. Eats small fish.
– *T. palembangensis* BLEEKER, 1852; Figure Eight Puffer. From Thailand, Sumatra, Kalimantan (Borneo), where it lives in fresh and brackish water. Length to 20 cm. There is a nasal tentacle on both sides. Upper side of body dark with yellow or green patterns of stripes that form a closed ring below the D. Survives better in brackish and sea water than it does in pure fresh water.
– *T. schoutedeni* PELLEGRIN, 1926; Leopard Puffer. From the lower Congo, only in fresh water. Female 10 cm, male smaller. Similar to *T. fluviatilis*. 2 long, forked nasal tentacles on both sides. Dark dots on an ochre-coloured background. One of the most peaceful of Pufferfishes!

Tetraodon-poisoning *see* Poisonous Fish

Tetraodontidae; Pufferfishes (fig. 1108). F of the Tetraodontiformes*, Sub-O Tetraodontoidei. Found in warm seas, in brackish water, and a very few in fresh water. Pufferfishes have an oblong shape and the caudal peduncle is elongated to a greater or lesser extent. Mouth small with ridges of teeth on the jaws which add up to form a kind of beak that always protrudes slightly;

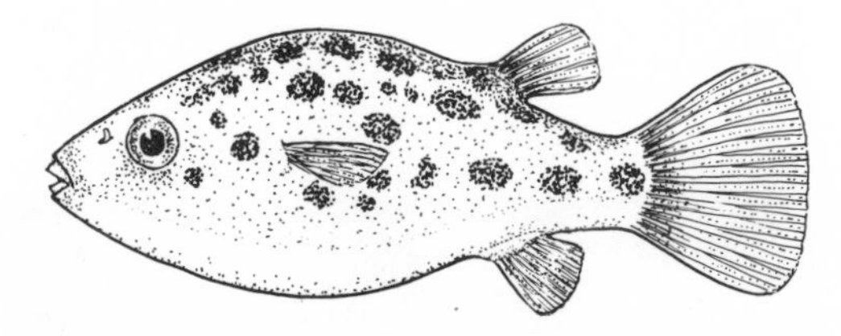

Fig. 1108 Tetraodontidae-type

the beak consists of teeth that are fused. There is a suture between the right and left tooth ridge of the upper and lower jaw, which is where the name Tetraodontidae comes from (= four-toothed). Pufferfishes move by using their Ps like a propeller, sometimes also the short D and An. The rounded C is only used for steering, in conjunction with the caudal peduncle. No Vs. Skin naked or covered with triaxial spines which are laid down, pointing backwards, when at rest. The Tetraodontidae glide peacefully through the water and are very agile; they can also turn on the spot or swim backwards. All Pufferfishes have a special extension of the stomach which spreads to below the skin of the breast and belly areas; when danger threatens, it can be filled with water or air, and at the same time, the spines become erect. This makes the fish look bigger in the eyes of a potential attacker, and it can also blow the water it has taken in towards its enemy. Pufferfishes often blow themselves up like this when they are trapped in a net. The flesh of many species contains the well-known tetrodotoxin – a strong nerve poison. Pufferfishes are very important in aquarium-keeping. *See under* the Ga *Arothron**, *Canthigaster**, *Carinotetraodon**, *Colomesus**, *Tetraodon**. You will also find further details there about their biology.

Tetraodontiformes; Pufferfish-types. O of the Osteichthyes*, Coh Teleostei. Also known as the O Plectognathi, this O embraces a whole series of Fs, some of which at first sight seem to have no similarities.

Nevertheless, they do form a relatively uniform group. This is difficult to prove from any external features and demands comparison studies of the skeleton, particularly of the skull. Here, a fusion of the premaxilla and maxilla is only hinted at, there are no parietal bones and the number of vertebrae is small. From the outside, features to note are the very small, non-protrusible mouth and small, hole-shaped gill slits. The individual groups also have certain parallels in behaviour patterns. The jaws may carry individual teeth or beak-like teeth-ridges. Often, the Vs are missing and with them also the pelvic bones. The skin may be scaled or naked, or it may be covered with large spines or bony plates. Nearly all Tetraodontiformes live in tropical and subtropical seas, with only a few species from fresh water areas. The Tetraodontiformes are related to the Acanthuroidei*, and the oldest fossils date from the early Tertiary period. GREENWOOD and colleagues (1966) distinguish between 2 Sub-Os, the Balistoidei containing the Fs Balistidae*, Ostraciontidae*, Triacanthidae* and the Tetraodontoidei containing the Fs Diodontidae*, Molidae*, Tetraodontidae* and Triodontidae*. *See also* the system under the heading Osteichthyes.

Tetrodotoxin *see* Poisonous Fish

Tetrosomus SWAINSON, 1834. G of the Ostraciontidae*. From tropical parts of the Indo-Pacific. Way of life and form typical of the family. This G has a downward-pointing spine above the eyes, and there is another spine, this time strong and flattened, on the edge of the dorsal surface towards the rear; along the belly edge there are also 5 short spines. The body is triangular in cross-section. For instructions on how to care for *Tetrosomus* in the aquarium, see under *Ostracion** and *Lactoria**. The only known species is:
– *T. gibbosus* (LINNAEUS, 1758); Black-blotched Turretfish, Camel Trunkfish, Hovercraft, Pyramid Trunkfish. Suez Canal to the Philippines. Length to 30 cm. Yellow to olive-brown with dark flecks.

Texas Cichlid *(Cichlasoma cyanoguttatum) see Cichlasoma*

Thalassoma (SWAINSON, 1839). G of the Labridae, Sub-O Labroidei*. Contains numerous species in tropical and subtropical parts of the Indo-Pacific, Atlantic and in the Mediterranean, where they live alone near the coasts or in loose bands, searching for small animal food on the bottom. At night they lie buried in the substrate. Most reach sizes of around 20–30 cm only. They are slim, agile fish that are usually a lovely colour. The D is long and contains 7–8 spines, and in older specimens, the C is often extended into a sickle-shape. Males, females and juveniles can often by very different in colour and are often regarded as separate species. During the youthful phase, some *Thalassoma* species act as cleaners, others are Anemone-fishes* for a while. They do not take care of their brood and the eggs are pelagic. In the aquarium *Thalassoma* survives very well and will accept any kind of animal food. However, they are often intolerant towards like-species. As they swim around constantly throughout the day, *Thalassoma* species require plenty of free swimming room and a

sandy surface for disappearing into at night. Various species that are frequently imported include:
– *T. bifasciatum* (BLOCH, 1791); Bluehead. Tropical W Atlantic. To 15 cm. Male has a blue-violet head and a green body, with a broad, white band, bounded by black, in between. Female and juveniles yellowish with a dark longitudinal band.
– *T. lunare* (LINNAEUS, 1758); Green Parrot Wrasse, Lyretail Wrasse, Moon Wrasse, Rainbow Fish. Indo-Pacific. To 30 cm. C long drawn-out, sickle-shaped. A shining green colour with red and blue stripes on the head. Each scale has a vertical red streak.
– *T. pavo* (LINNAEUS, 1758); Peacock Wrasse. E Atlantic, Mediterranean. To 20 cm. A bronze-green body, some of it crossed by paler transverse bands; blue stripe markings on the head. In mature males, the C is elongated. A very pretty aquarium fish (fig. 1109).

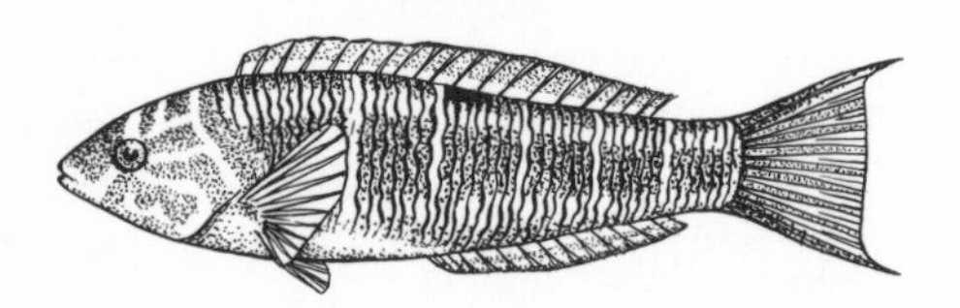

Fig. 1109 *Thalassoma pavo*

Thaleichthys pacificus *see* Salmonoidei

Thaliacea, Cl, *see* Tunicata

Thallus. A multi-celled or multi-nucleated vegetative body that is not divided into stem axes*, leaves* and roots*. The thallus can occur as a freely mobile cell colony or it can consist of unbranched or branched threads, or of simple or multi-layered surfaces. Another kind of thallus may be even more differentiated and can begin to look like higher plants with their stems and leaves. A thallus occurs in many Blue-algae (Cyanophyta*), many Algae* and Fungi*, in Mosses and Liverworts (Bryophyta*) and in the sexual generation of Ferns (Pteridophyta*). Opposites are protophyte* and corm*.

Thayeria EIGENMANN, 1908. G of the Characidae*, Sub-F Tetragonopterinae. These are peaceable shoaling fish of the surface zone that live in the Amazon Basin and its tributaries. Length to 7 cm. Body elongate, much flattened laterally and held in a slight sloping position, head up. This position is emphasised optically by a broad black band, which in the well-known species extends into the elongated lower lobe of the caudal fin. D1 short, triangular, D2 present. C deeply forked, An elongate and extending into a tip in the front third. The base of the C is covered with scale. Eyes relatively large. LL incomplete. In habits, *Thayeria* species are very like members of the Ga *Hemiodopsis** and *Hemiodus**, with whom they also share a skilful way of swimming and a delight in jumping. Omnivores. The female is easy to distinguish from her fuller ventral area. The only species that has so far been bred successfully is *T. boehlkei*, which was wrongly thought to be *T. obliqua* for a long time. After violent driving behaviour, and turning the body very fast, the mating pair laid their spawn just as dusk was falling beneath the surface, in open water. At 25 °C, the fry hatched out after about 20 hours from the olive-green eggs. After a further 4–5 days, they can swim freely and grow very quickly if fed with live food suitable for rearing fry (Fry, Feeding of*). One brood can contain more than a thousand fry. After spawning, it is recommended that the breeding water be largely renewed. Established breeding pairs should not be separated. *Thayeria* species are undemanding and long-lived Characins. *T. sanctaemariae* LADIGES, 1949, and *T. ifati* GÉRY, 1959, have so far not been imported alive (at least, it is not thought so).
– *T. boehlkei* WEITZMAN, 1957; Penguinfish. Amazon Basin. Length to 6 cm. Flanks a gleaming silver. Dorsal surface brownish-olive. A broad, deep-black longitudinal band runs from the operculum through the body to the lower C-lobe, accompanied below by a greenish-golden band. The front rays of the lower caudal fin lobe and the An are yellowish-white; all the other fins are colourless (fig. 1110).

Fig. 1110 *Thayeria obliqua* and *Thayeria boehlkei* (below)

– *T. obliqua* EIGENMANN, 1908; Penguin Fish. Amazon Basin. Length to 7 cm. The body and fin colouring largely corresponds to *T. boehlkei.* The black band, however, begins first on the caudal peduncle and runs from here downwards into the lower caudal fin lobe. In the years following 1949, this species was imported in quite large numbers (only under the incorrect name of *T. sanctaemariae*) but no offspring resulted (fig. 1110).

The Trehira *(Hoplias malabaricus) see Hoplias*

Therapon jarbua *see* Theraponidae

Theraponidae; Tigerfishes. F of the Perciformes*, Sub-O Percoidei*. Native to the tropical Indo-Pacific. Tigerfishes are fairly small and they are related to the Sea Perches (Serranidae*). Body like a perch's, D elongated, almost divided in two and containing 12–14 spiny rays. C usually indented. There are fine bands of teeth on the jaws and often on the palate, too. Preoperculum serrated at the rear end, occasionally with spines. Several dark longitudinal bands show up on most species. Young specimens of the Three-striped Tigerfish *(Therapon jarbua)* (fig. 1111) are often kept in the marine aquarium; this species grows up to 35 cm long. It is found from E Africa to N Australia and the Philippines, and often enters brackish water. When young, this species is territorial, defending small hollows in the tidal zone. Breeding has already been successful in the

Fig. 1111 *Therapon jarbua*

aquarium – the spawn is laid on rocks and the male watches over it. Almost all Theraponidae are timid. Various species have regional importance as good edible fish.

Theria, Sub-Cl, *see* Mammalia

Thermometers in the aquarium are indispensable for keeping a constant check on water temperature (Temperature Control*). You can buy either a floating thermometer which is fixed to the side of the aquarium by a suction holder, or one that you stick into the substrate to anchor it firmly. Site the thermometer as far away as possible from the heating or cooling system in the tank. Thermometers filled with alcohol have the advantage over mercury thermometers that should they break no poisonous mercury will leak into the tank. Always make sure the thermometer is calibrated accurately, as inaccuracies of up to 5 °C are not uncommon. Centigrade (°C), Fahrenheit (°F) and Réaumur (°R) scales are all used.

Thick Topshell *(Monodonta turbinata) see Monodonta*

Thick-lipped Gourami *(Colisa labiosa) see Colisa*

Thick-lipped Wrasse *(Hemigymnus melapterus) see Hemigymnus*

Tholichthys-stage *see* Scatophagidae

Thollon's Tilapia *(Tilapia tholloni) see Tilapia*

Thoracica, O, *see* Cirripedia

Thoracocharax FOWLER, 1907. G of the Gasteropelecidae*. Small surface shoaling fish from S America. Length to 9 cm. Can be distinguished from the Ga *Carnegiella** and *Gasteropelecus** by the number and structure of the scales in the middle longitudinal row. Here too, the breast-belly line bends out particularly strongly. The dorsal profile is straight as far as the insertion of the broad D; the small D2 sits well back on the caudal peduncle. C deeply forked. An very long, Vs small. Ps, especially in *T. securis* very long and reaching back as far as the beginning of the D. Like all the other Gasteropelecidae, *Thoracocharax* species are good jumpers (tank must have a lid, therefore); they require free swimming room at the water surface and subdued lighting. They are omnivores that like to eat food that happens to drop onto the surface, as well as dried food. Breeding habits unknown. Rarely imported.

– *T. securis* (FILIPPI, 1853); Silver Hatchetfish. From Central S America, east of the Andes, the Amazon Basin, Rio Paraná. Length to 9 cm. Flanks a strong gleaming silver with a shimmering bluish or greenish band reaching from the operculum to the insertion of the C. Fins colourless. In adults, the body is often as tall as it is long. The bottom of the An is covered by 5–6 rows of scales (fig. 1112).

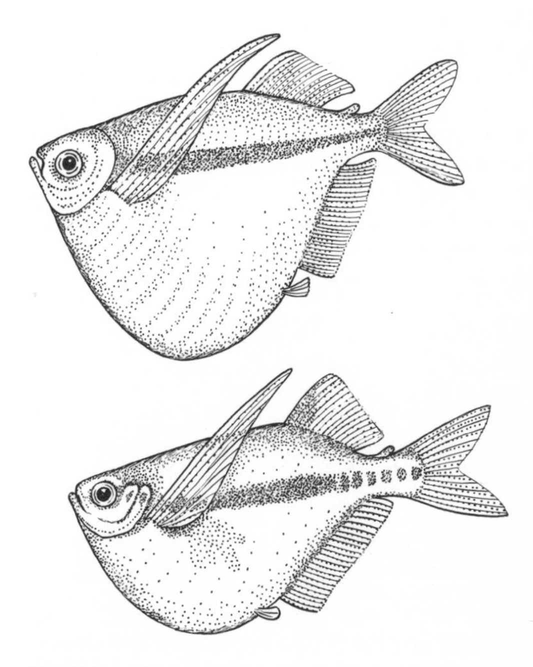

Fig. 1112 *Thoracocharax securis* (top) and *Thoracocharax stellatus*

– *T. stellatus* (KNER, 1859); Silver Hatchetfish. From central Brazil to Argentina. Length to 7 cm. Similar in colour to *T. securis*, but distinguished from it by having a less bulging breast-belly line and only 2–3 rows of scales at the base of the An. In addition, the bottom parts of the first rays of the D are a dark colour (fig. 1112).

Thoracostei, O, name not in current usage, *see* Gasterosteiformes

Thornback Ray *(Raja clavata) see* Rajidae

Thorny Catfishes, F, *see* Doradidae

Threadfin Breams (Nemipteridae, F) *see* Lutjanidae

Thread-fin Butterfly-fish *(Chaetodon auriga) see Chaetodon*

Threadfins, related species, Sub-O, *see* Polynemoidei

Threatening Behaviour *see* Display Behaviour

Three-banded Pencilfish *(Nannostomus trifasciatus) see Nannostomus*

Three-fin Blenny *(Tripterygion tripteronotus) see Tripterygion*

Three-line Rasbora *(Rasbora trilineata) see Rasbora*

Three-spined Stickleback *(Gasterosteus aculeatus) see Gasterosteus*

Three-spot Anostomus *(Anostomus trimaculatus) see Anostomus*

Three-spot Damsel *(Dascyllus trimaculatus) see Dascyllus*

Three-spot Damselfish *(Pomacentrus tripunctatus) see Pomacentrus*

Three-spot Gourami *(Trichogaster trichopterus trichopterus) see Trichogaster*

Three-striped Damselfish *(Dascyllus aruanus) see Dascyllus*

Three-striped Tigerfish *(Therapon jarbua) see* Theraponidae

Thresher Shark *(Alopias vulpinus) see* Alopiidae

Thresher Sharks, F, *see* Alopiidae

Threshold Stimulus *see* Stimulus Threshold

Thunnus (several species) *see* Scombroidei

Thymallus thymallus *see* Salmonoidei

Thyroid Gland Tumours *see* Tumour Diseases

Thyrsites atun *see* Scombroidei

Tiger Barb *(Barbus tetrazona tetrazona) see Barbus*

Tiger Botia *(Botia macracantha) see Botia*

Tiger Cowrie *(Cypraea tigrina) see Cypraea*

Tiger Fish *(Hoplias malabaricus) see Hoplias*

Tiger Loach or Botia *(Botia hymenophysa) see Botia*

Tiger Shark *(Galeocerdo cuvieri) see* Carcharhinidae

Tigerfishes, F, *see* Theraponidae

Tilapia SMITH, 1840 (fig. 1113). G of the Cichlidae* with a large number of species. Widely distributed in Africa, with the exception of the southern part, and in the north penetrating into Asia Minor. Some species have been introduced successfully into other tropical parts of the world as useful fish. Tilapias live in all sorts of inland waters and are also found in brackish water, sometimes. Recently, the G has been revised on several occasions by THYS VAN DEN AUDENAERDE and TREWAVAS. But as the results differ greatly, even within different revisions made by the one author, we have decided to consider the G *Tilapia* in its widest sense. It is also taken to include members of the taxon* *Sarotherodon** RÜPPELL, 1853 – ie mouth-brooding forms like *T. galilaea* and *T. nilotica*.

So, in the broadest context, *Tilapia* species are fairly large, tall fish of a robust appearance. Some species grow up to 40–50 cm long, eg *T. galilaea* and *T. nilotica*.

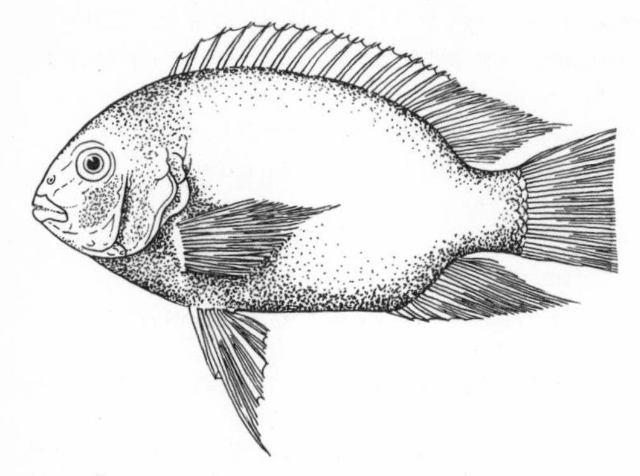

Fig. 1113 Representative of the G *Tilapia*

The mouth is large and thick-lipped. Forehead sunken in in front of the eyes, especially in older specimens: arching nape. Scales large. The hard-ray and soft-ray count of the D is very variable, but normally around XV–XVII/11–13; the An always has 3 hard rays, soft-ray count variable. C fan-shaped, truncated or slightly outward-curving. In old fish, the tips of the D and An are also only slightly elongate. Outside the spawning period, the colouring is usually simple – silver-grey and yellow-grey tones predominantly. Dark fleck markings or transverse band markings often show up, particularly in excited fish. Included in the fleck markings are the operculum (gill-covering process) and the iris. During the spawning period, the males of some species are very highly coloured, eg shining dark-blue to deep black *(T. mossambica, T. macrochir)*.

There are several forms of brood care within the G. The most primitive (Plesiomorphic*) are the open-water spawners (*see* Cichlidae) that form definite partners (eg *T. mossambica* and *T. tholloni*). Other species, such as *T. galilaea* and *T. heudeloti*, have gone over to mouth-brooding; they also form partners only before spawning, and it is the male who takes care of the brood. Particularly apomorphic* forms do not form pairs at all any more – the female in fact spawns with any males that happen to come along. In these cases, the females are the mouth-brooders; as each portion of spawn is laid, she snaps it into her mouth immediately, and only then is it fertilised in the mouth. With this type of reproduction, the male exhibits 'dummy' eggs which the female tries to snap into her mouth, when in fact she is taking in clouds of sperm. The dummy eggs may consist of egg-spots on the An *(T. sparrmani)*, particularly intensely coloured genital papillae *(T. nilotica)*, or the genital papillae themselves may look like eggs on account of their knotty or grape-like shape *(T. karomo)*.

In the aquarium, *Tilapia* species are very undemanding, at least as regards water chemistry, choice of food and temperature. Species from N Africa and Asia Minor are particularly indifferent to temperature and can also tolerate temperatures of 15–17 °C for lengthy periods without damage. In line with their size, Tilapias need very spacious tanks – as a result, certain limitations are placed upon keeping them in normal domestic aquaria, unlike public display aquaria where they are common specimens. *Tilapia* will also rearrange the tank to suit its own taste – a trait not particularly liked by many aquarists. They dig out a lot of spawning hollows, for example, often causing stone structures to collapse, and they also uproot plants (if they have not in fact eaten them beforehand). So, a tank full of *Tilapia* specimens is hardly going to be a centrepiece of the room unless you use very hardy marsh plants in flowerpots, allowing the leaves to rise up above the surface, as well as decorative root-work and shored up rock structures. Feeding is no problem, but make sure there is plenty of variety. Apart from animal food of every sort, also supply enough plant food, such as lettuce, pondweed and cooked rolled oats. Artificial food is also accepted readily. Because they are so greedy, a good filter system must be installed if the water is to remain clear. Partly exchanging the water on a regular basis is also very important. Breeding causes few problems, provided you have a sufficiently large tank available for this purpose.

For the mouth-brooding species that make spawning hollows, coarse-grained, well-washed sand makes the best substrate; substrate spawners need suitable spawning areas, such as large stones that are not stood in too bright a spot. The number of eggs in a brood depends very much on the type of reproduction – open-water spawners may lay up to 1,000 eggs, whereas the mouth-brooders will produce a maximum 250 eggs (usually much less). When any *Tilapia* species reproduced in the aquarium, particularly the mouth-brooding forms, it makes for very interesting glimpses of their behaviour patterns. The depositing of the spawn itself is always preceded by a certain genetically fixed ritual, which serves to make the potential partners aware of one another and to synchronise both of them. A behaviour pattern will unfold step by step until it culminates in mutual spawning. The partner that takes care of the brood (be it male or female) should be kept well away from any source of disturbance – either make sure there is a safe hiding-hole for the fish or remove all the other specimens from the tank. The eggs take just about 2 weeks to mature, although the fry will remain in the parent's mouth for a further few days after hatching. After the brood can swim freely, the parent fish guides the fry shoal with certain movements, and takes them all back into the mouth again should danger threaten.

– *T. galilaea* (ARTEDI, 1758); Galilee Cichlid. Asia Minor and the northern half of Africa. Up to 40 cm long, sexually mature from 12 cm. Silver with a blue shimmer; blotch on operculum blue to black. Male mouth-brooder, female carries out the courtship.

– *T. guinasana* TREWAVAS, 1936; Guianas Cichlid. Northern SW Africa (Lake Guiana). To 14 cm. Colour very variable, either a pale cobalt blue or velvet black on the underside, upperside a gleaming bronze. Open-water brooder.

– *T. guineensis* (BLEEKER, 1863); Guineas Cichlid. Tropical Africa. To 25 cm. Grey-silver; at spawning time, the male has a darker, sometimes patent black underside. As befits their natural habitat, needs warmth. Open-water brooder.

– *T. heudeloti* (DUMERIL, 1858); Senegal Cichlid (including *T. macrocephala* [BLEEKER, 1863]). Tropical W Africa; also found in brackish water. To 30 cm. Grey-silver, with a dark shoulder fleck. At spawning time, the male has a velvet black throat, breast and base areas of P and V. Female mouth-brooder.

– *T. lepidura* BOULENGER, 1899; Pearl Tilapia. Tropical W and Central Africa; also found in brackish water. To 20 cm. Dotted in red and with 6–8 strong, dark oblique bands. Needs warmth. A peaceful species.

– *T. macrochir* BOULENGER, 1912. Southern E Africa (Zambezi and Great Lake areas). To 35 cm. Pale yellow. Male black at spawning time, D and C with a milky white border area and iris red. Female mouth-brooder.

– *T. mariae* BOULENGER, 1899; Tiger Cichlid, Zebra Cichlid. W Africa. To 15 cm. Pale yellow with a longitudinal row of dark flecks. Lives mainly as a vegetarian. Substrate-brooder.

– *T. mossambica* (PETERS, 1852); Mozambique Cichlid. E Africa; also found in brackish water. To 40 cm. Green-silver with a dark blotch on the operculum. Male blue at spawning time. Female mouth-brooder (fig. 1114).

Fig. 1114 *Tilapia mossambica*

– *T. nilotica* (LINNAEUS, 1766 ; Nile Mouthbrooder. Widely distributed in Africa, in the north as far as Asia Minor; also in brackish water. To 50 cm. Grey-silver; at spawning time the male has a dark-red throat, P and V. Sensitive to temperature. Female mouth-brooder.

– *T. sparrmani* SMITH, 1840; Sparrman's Cichlid. Southern part of Africa. To 20 cm. Green, scales dotted in orange; fins dotted in red and with red borders. A peaceful species. Open-water brooder.

– *T. tholloni* (SAUVAGE, 1884); Thollon's Tilapia. Tropical W Africa. To 30 cm. Greenish, with dark markings made up of 2 longitudinal bands and several transverse bands. Blotch on operculum a shiny black colour. An intolerant species. Lives mainly as a vegetarian. Open-water brooder.

– *T. zilli* (GERVAIS, 1848); Zill's Tilapia. Northern Africa and Asia Minor. To 30 cm. Grey-silver with dark band markings, at least when excited. Becomes a lovely colour during spawning period: dorsal surface a gleaming olive-green, underside red, belly black, transverse bands show up clearly. Not troubled by low temperatures.

Tillaea recurva, name not in current usage, *see Crassula*

Tinca CUVIER, 1817. G of the Cyprinidae*, Sub-F Leuciscinae. Europe, Asia, Siberia. These are large fish that live in waters that flow sluggishly, or in warm lakes and ponds with dense plant growth and a muddy substrate; but some also enter brackish water and the rhithral* region of streams. The body is powerful with a fairly deep caudal peduncle, flattened laterally. The mouth is terminal and has one pair of short barbels. Scales small and located deep below the thick, slimy skin. LL complete. D inserted clearly behind the Vs. C very gently indented. Tenches are very peaceful fish, active at night. They stay around the bottom mainly and feed on small animals, Snails and insect larvae, plus plant remains. In the spring, large shoals form for reproduction. The eggs are deposited on plants. Younger specimens are ideal for keeping in the domestic aquarium. The substrate should be covered with a layer of washed sand, as *Tinca* likes to dig around in the bottom. Specimens will feel particularly at home if robust aquatic plants and some hiding-places are provided. Temperature 18–22 °C, cooler in winter. Only one species:

– *T. tinca* (LINNAEUS, 1758); Tench. To 70 cm. Dorsal surface dark brown-green, flanks greenish-brown with a golden sheen. Belly golden-yellow, fins non-transparent, greenish-brown to grey. Judged a very good edible fish. There is a xanthic variety of this species (Colouring*), which is yellowish-red on top, and golden-yellow below. Dark dots are scattered all over the body and the fins (fig. 1115).

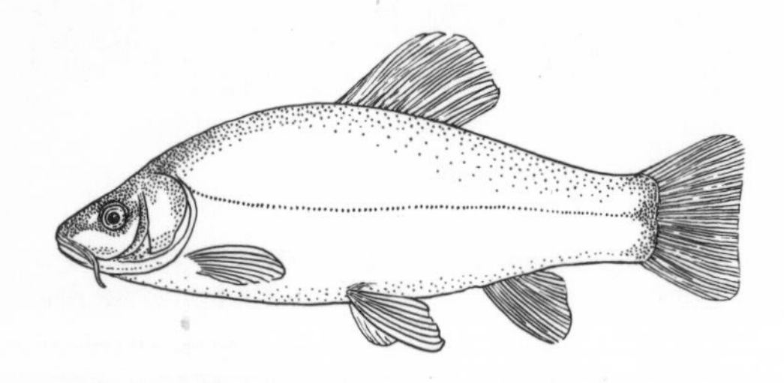

Fig. 1115 *Tinca tinca*

Tincture of Iodide. Iodine (I) has strong disinfectant properties and so is suitable in solution as a preventive treatment for wounds to the skin in fish. The following solutions are used: 2 g iodine (I) + 2.4 g sodium iodide (NaI) dissolved in 100 ml 70% alcohol, or, 10 g iodine (I) dissolved in 100 ml 100% ethyl alcohol, or, chloroform. The wound is dabbed with the solution (fish out of the water), before returning the fish immediately to the aquarium.

Tinfoil Barb *(Barbus schwanenfeldi) see Barbus*

Tipulidae; Craneflies or Daddy-long-legs (fig. 1116). F of the Flies (Nematocera*). Craneflies are widespread. There are several thousand species, around 180 of which are found in Central Europe. They are medium-sized to very large, two-winged insects. In Europe, the largest native species is *Tipula maxima* with a body length of 40 mm and a wingspan of over 50 mm. Craneflies have long, very thin legs on which they can get around as if on stilts. Their mouth parts are brought forward like a snout, but are only used for sucking up water, nectar and plant juices, not for stinging. At rest, the wings are held at an angle pointing backwards, and they are often flecked in brown or grey. Craneflies need damp ground for laying their eggs, and so they are often found near stretches of water. As larvae, some species live purely aquatically, eg the above-mentioned species, *Tipula maxima*, lives in streams. Cranefly larvae are cylindrical in shape; they have no feet-like appendages, and at the anus they have protrusible anal tubes, the function of which is unknown. Craneflies are suitable as food for larger kinds of fish. They are also interesting to observe in a pond-aquarium, perhaps the emergence of the imagines, copulation and laying of the eggs.

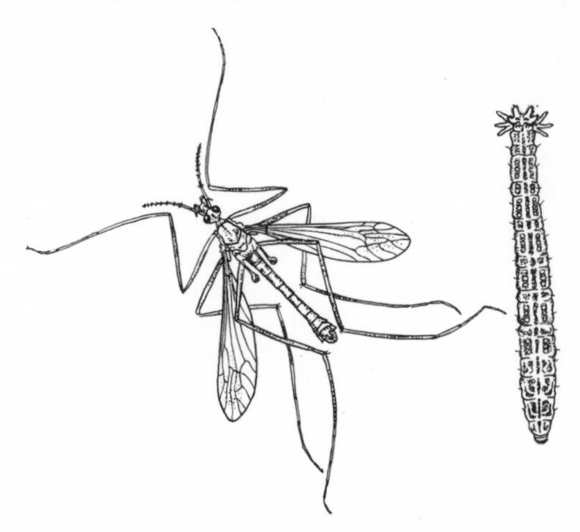

Fig. 1116 Tipulidae. Left, imago; right, larva

Tissue. Groups of cells found in multi-celled organisms that are specialised in some way for particular functions. For example, in animals you will find connective tissue, muscular tissue and nerve tissue; in plants, meristematic tissue, surface tissue, conductive tissue and supportive tissue.

Titriplex *see* Chelaplex

Toadfishes (Antennariidae, F) *see* Lophiiformes

Toadfishes (Batrachoididae, F) *see* Batrachoidiformes

Toadfishes, related species, O, *see* Batrachoidiformes

Toadfishes or Frogfishes, G, *see Antennarius*

Toads (and Frogs) (Salientia, O) *see* Amphibia

Toads (collective term) *see* Amphibia

Toads (Discoglossidae, F) *see* Amphibia

Tomato Clown *(Amphiprion ephippium) see Amphiprion*

Tomeuridae, F, name not in current usage, *see* Poeciliidae

Tongue *see* Digestive Organs

Tongue-soles (Cynoglossidae, F) *see* Pleuronectiformes

Tooth-carps *see* Cyprinodontidae or Poeciliidae

Tooth-carps, related species, Sub-O, *see* Cyprinodontoidei

Tope *(Galeorhinus galeus) see* Triakidae

Torpedinidae; Electric Rays (fig. 1117). F of the Chondrichthyes*, O Torpediniformes. In Electric Rays, the head and front of the body, together with the pectoral fins, are joined to form a flat, elliptical or round disc that continues into the rear part of the body and caudal peduncle (which is usually keeled). The nasal openings, the extensible and slightly protruding mouth with its plaster-like pointed teeth, and the hole-shaped gill openings all lie on the underside of the disc, whilst the eyes and the spray holes are on the upperside. In some species the eyes are reduced or absent altogether. There are usually 2 Ds. No placoid scales on the skin.

But the most interesting feature of the Electric Ray is its very large electric organ which may take up anything from $^1/_6$ to $^1/_4$ of the mass of the body. It is a paired organ lying in the body disc where it forms a kidney-shaped complex on both the right and left sides. If the organ is viewed from above after the removal of the skin, it would look like a somewhat irregular honeycomb. Each area of honeycomb is the upper layer of a column, but the total number of columns varies with different species. On one side it can vary between 150 and 1,000. Each column is made up of a large number of small plates arranged one on top of the other, and it is these that correspond to electric elements. The organ has a positive charge on the top side and negative on the ventral side. The voltage is normally around 60–80,

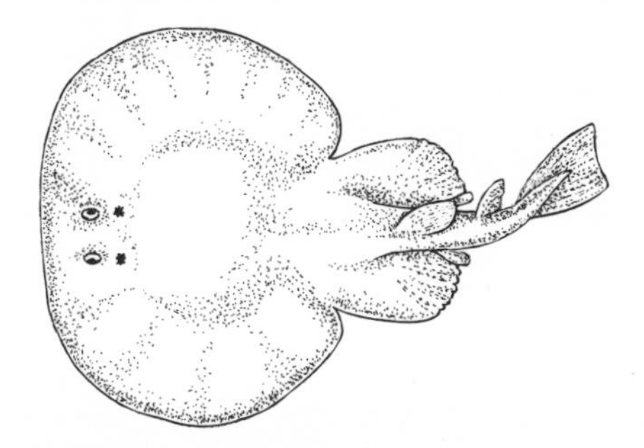

Fig. 1117 Torpedinidae-type

but in some species it can be as high as 150–200. The shocks can come very quickly in succession, although with each surge of current the effect is weaker. After wards, the Electric Rays need several days to regenerate. Ancient peoples knew of this bioelectricity, and indeed they used it for medicinal purposes. The Rays themselves use their electricity to kill their prey (especially crustaceans and fish), to defend themselves against their enemies, and with weak surges of current

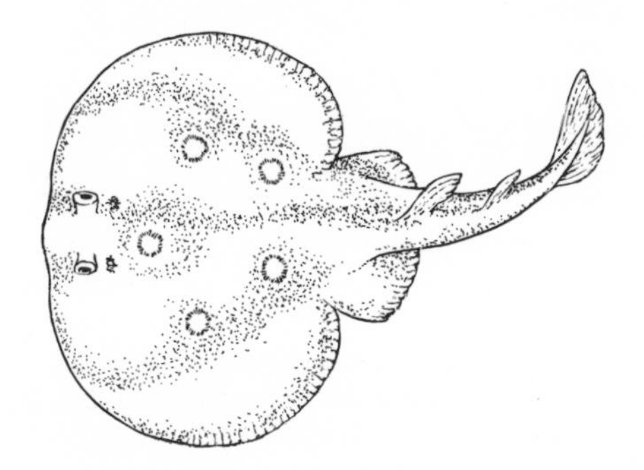

Fig. 1118 *Torpedo torpedo*

they can perhaps orientate themselves (a form of radar) or keep in contact with like-species. Electric Rays are found mainly in shallow coastal waters, more rarely in deep water of tropical and subtropical seas. They are livebearing. The young Rays have thread-like external gills that hang out of the gill openings. The largest species, the Electric Ray or Torpedo *(Torpedo nobiliana)*, grows up to 1.5 m long, and is found throughout the E Atlantic, including the Mediterranean, from Scotland to W Africa, and in the W Atlantic in the north to Florida. Also found in the Mediterranean and E Atlantic are *Torpedo torpedo* (fig. 1118) and the marbled *Torpedo marmorata* (fig. 1119). Both only enter the North Sea very rarely, whereas they are common in the Mediterranean.

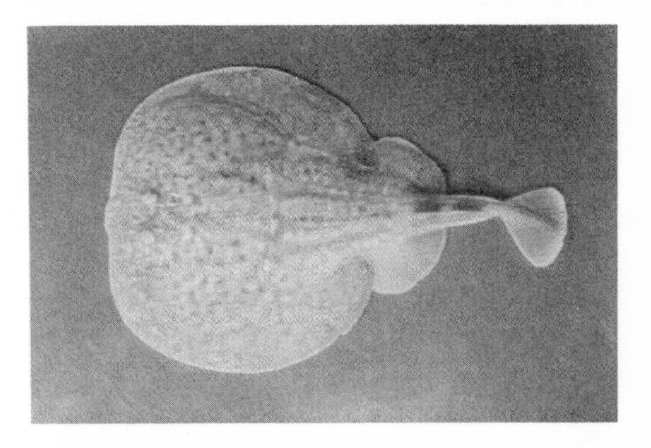

Fig. 1119 *Torpedo marmorata*

Torpedo *(Torpedo nobiliana) see Torpedo*
Torpedo (several species) *see* Torpedinidae
Torpedo Rays, F, *see* Torpedinidae
Total Hardness *see* Water Hardness

Touch, Organs of. In fish, the organs of touch are neural arches sensitive to pressure that are found in some parts of the body that are free of scales. In particular, these areas include the lips and barbels (especially in species active at night, cave-dwelling fish and deep-sea fish), as well as elongated fin rays (eg in the long-finned Gouramis *Colisa* and *Trichogaster*).

Toxins. Substances that are produced by organisms (bacteria, animals and plants) and which can cause specific or unspecific poisoning in other organisms. All toxins have antigene properties, ie in higher animals and in man they release a protective reaction – the formation of antibodies. Toxins can serve to protect their own species, to acquire food or perform other biofunctions. Many bacterial toxins are metabolic end products which are released from the bacterial body. Other bacterial toxins are only set free when the bacteria die. Chemically, toxins are mainly made of proteins, protein and carbohydrate compounds or other kinds of substances. The effects caused by toxins vary greatly – in higher vertebrates, toxins often break up the blood or block the activity of the nervous system.

For fish, bacterial toxins are particularly important. Parasitic forms are often pathogenic simply because of the toxins they release. In aquarium water, disturbed metabolic processes can also lead to a very high multiplication rate of certain putrefactive bacteria that break down proteins. Their toxic metabolic products can in a short time reach concentrations that are fatal for fish. Occasionally, reports of fish deaths caused by algal toxins are known—normally it is attributed to Blue Algae (Cyanophyta*).

Toxopneustes, G, *see* Echinoidea

Toxotes jaculatrix *see* Toxotidae

Toxotidae; Archerfishes. F of the Perciformes*, Sub-O Percoidei*. Typical surface-dwelling fish from coastal areas, especially the mangrove estuaries of the whole of S and SE Asia, including the Malaysian Archipelago. In side-view, the body looks like a boat, eyes large, mouth deeply slit. D and An almost the same size and virtually opposite one another. Ps very powerful. There is one G with 5 species. What the Archerfishes are well-known for is their ability to spit a jet of water very accurately at insects scrabbling about on vegetation above the water surface. To do this, the mouth cavity is suddenly compressed, and the water shot out through a channel formed by the tongue and roof of the mouth (the mouth is held slightly above the water surface at the time). Large Archerfishes are able to hit targets up to 1.5 m away (fig. 1120). Young fish learn to spit when they are 2–3 cm long. At this stage of development, they are also distinguished by having pale

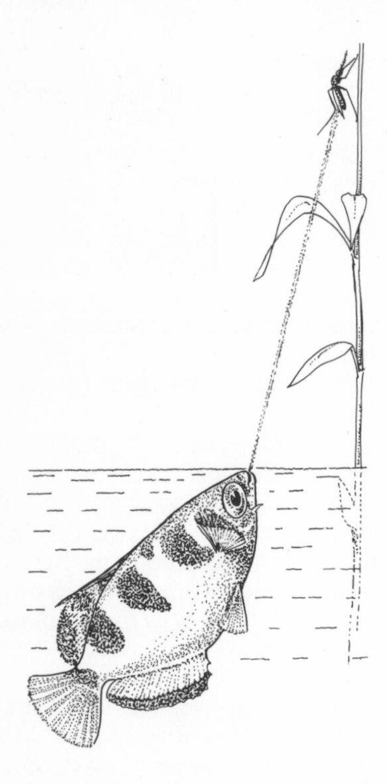

Fig. 1120 *Toxotes* firing jet of water

green iridescent saddle-markings—a feature which makes it easier for the young fish to keep in contact in the usually murky mangrove waters. The sexually mature fish penetrate further into the sea. The 25 cm long Archerfish *(Toxotes jaculatrix)* is often imported. It has black saddle bands on a pale brown to yellow-green background colour and broad black edges to the fins. It is best kept in slightly brackish water. Feed with live food, especially insects (fig. 1121).

Fig. 1121 *Toxotes jaculatrix*

Trace Elements (fig. 1122). In particular iron, manganese, aluminium, phosphorus and silicon, but also including iodine, silver, lead, copper and other elements that are dissolved in small or very small amounts in natural waters. They have a physiological importance for the animals and plants that live there. For example, the small concentration of inorganic phosphate in most stretches of water is the deciding factor for algal growth. If the phosphate content increases, perhaps by vegeable manure being washed in, the algae will often grow luxuriantly, and a state known as water-bloom* will result.

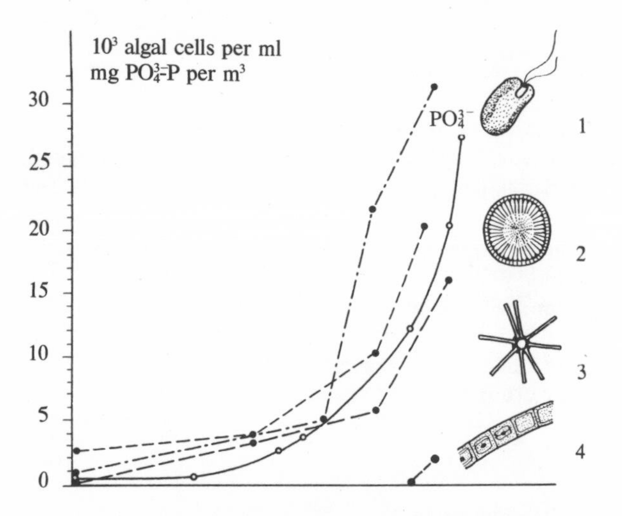

Fig. 1122 Trace elements. How the density of algae is related to the phosphate content in a lake. 1 *Rhodomonas* (Green Algae) 2 *Cyclotella* 3 *Asterionella* 4 *Mougeotia*

When preparing water (Water Preparation*) or setting up sea water*, trace elements must be added in the correct proportions. As few people are able to have the use of sensitive analytical balances and fine chemicals, it is best to buy ready-made mixtures that should be used as per the instructions. With the addition of chelatores*, especially Chelaplex-III, the effect of these salt mixtures will be increased, as the metal ions of the trace elements are kept in solution in this way.

Trachinidae, F, *see* Trachinoidei

Trachinoidei; Weeverfishes, related species. Sub-O of the Perciformes*, Sup-O Acanthopterygii. Fairly small to very small bottom-dwelling fish found mainly on tropical and subtropical sandy and muddy substrata. The relationships between all the Fs in the Sub-O are not yet fully understood. In the system described here by GREENWOOD and colleagues (1966), the Sub-O Trachinoidei is a collective group uniting several Fs (as in BERG's system). In other systems, these same Fs are sometimes given different positions within the system. Almost all Trachinoidei are fish that like to burrow into the substrate, some even build tubes where they lie in wait for prey. Exceptions are some deep-sea forms, eg the members of the F Chiasmodontidae. They are small, free-swimming Trachinoidei, usually black in colour, and with a giant, highly extensible mouth rigid with teeth; they can overcome prey fish that are bigger than themselves. Various members of the Sub-O are considered good edible fish and are caught either in a ground net or on the rod. From the aquarist's point of view, some species are very interesting, although they are kept only rarely. For details see under the following family descriptions:

– *Dactyloscopidae*; Sand Stargazers (fig. 1123). Fairly small fish, of the well-known Uranoscopidae type, that live along the tropical and subtropical coasts of America. The body is a bit more elongate, the Vs at the throat have only 1 spine and 3 soft rays, the D is undivided and there are no electric organs.

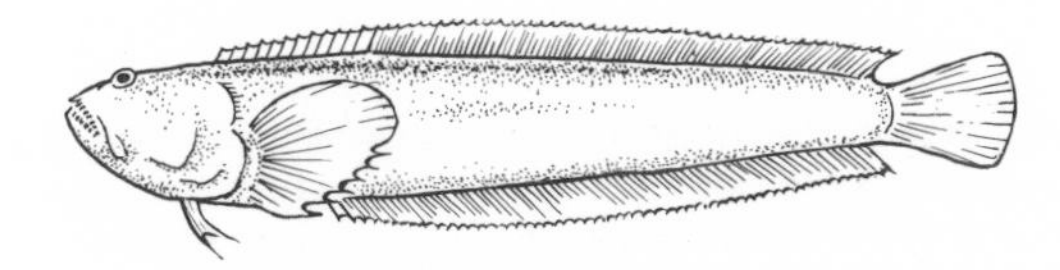

Fig. 1123 Dactyloscopidae-type

– *Opisthognathidae*; Jawfishes, Smilers (figs. 1124 and 11125). Small, elongate fish with a bullish head and found in tropical seas. Mouth unusually large, slit to behind the eyes. Jaws contain fine teeth. The D reaches from the nape to the C. C rounded, An long. Scales small; LL incomplete. Jawfishes stay true to a particular area, building tubes with their mouths in the mud, which they often make stronger with shells and small stones.

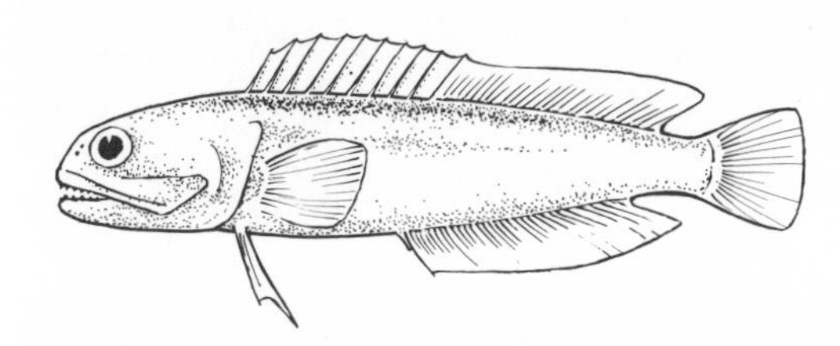

Fig. 1124 Opisthognathidae-type

In open water, they usually stand up vertically on their angular C or at an angle (Tail-standers). Some species, perhaps all of them, are mouth-brooders. With their large mouths, capable of being greatly expanded, the Jawfishes can overpower quite large prey-animals, usually fishes. Some have an interesting colour pattern.

– *Trachinidae*; Weeverfishes (fig. 1126). 4–5 species from the E Atlantic and Mediterranean. Body elongate,

Fig. 1125 *Opisthognathus aurifrons*, below with eggs in its mouth

laterally compressed, head bullish, mouth slit large, either at an angle or pointing steeply upwards. Fine teeth on the jaws. There is a spine pointing backwards on the main operculum, which is also the outlet of poison glands. D1 short, with 5–6 spines at the front

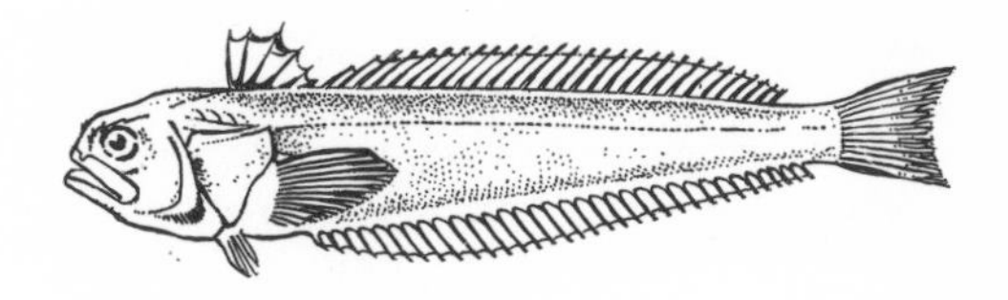

Fig. 1126 Trachinidae-type

poison glands; D2, like the An, unusually long. C straight. Vs located at the throat. Scales round; LL complete, arched at the front. The 45 cm long Greater Weever *(Trachinus draco)* (fig. 1127) lives in the Mediterranean and neighbouring Atlantic, north as far as

Norway. It is found on sandy substrate in which it normally lies buried up to its eyes during the day. It is a yellow-brown colour with oblique stripes or flecks running across it. Like all the other species, the Greater Weever is very poisonous. Injuries caused by the spines

Fig. 1127 *Trachinus draco*

on the D1 or operculum can cause long-lasting, severe pain about 30–40 minutes after being pricked. The poison has an effect on the blood and the activity of the nerves. No-one is sure whether or not injuries can cause death. Weeverfishes eat Shrimps and small bottom-dwelling fishes mainly. Their eggs, like their larvae, are pelagic. Since their flesh is tasty, Weaverfishes are valuable marketable commodities. The Greater Weever is normally not very attractive for an aquarium.

– *Trichodontidae*; Sandfishes (fig. 1128). 2 species from the northern Pacific. Body elongate, almost round at the front. Head large with a large mouth-slit, positioned vertically, and with clear fringes on both lips.

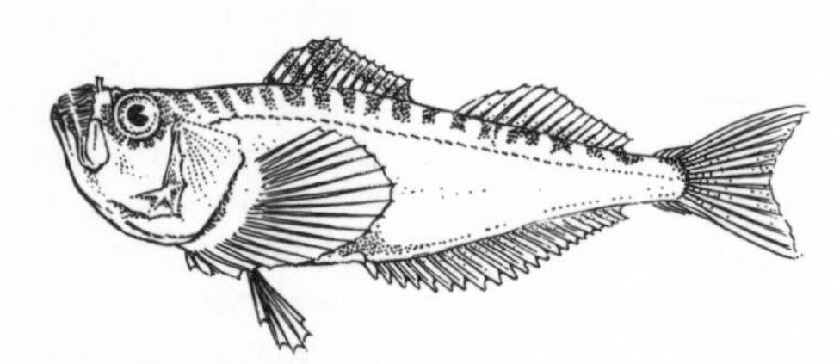

Fig. 1128 Trichodontidae-type

2 Ds, C indented. An long, Ps large. Sandfishes live at depths of 100–200 m; they lie in wait buried in the sand hoping for prey to pass by. They lay their spawn in shallow water. The 30 cm long species *Arctoscopus japonicus* is a tasty edible fish in Japan.

– *Trichonotidae*; Sand Divers (fig. 1129). Small, very elongate, cylindrical fish live on sandy substrate in tropical seas (particularly coral sand). They are able to burrow head-first into the sand very quickly. Head pointed, mouth deeply slit, lower jaw protruding. Eyes highly mobile – in some of the 50 or so species, they can be elongated like a telescope. D and An very long, almost taking in the entire body. The first rays of the

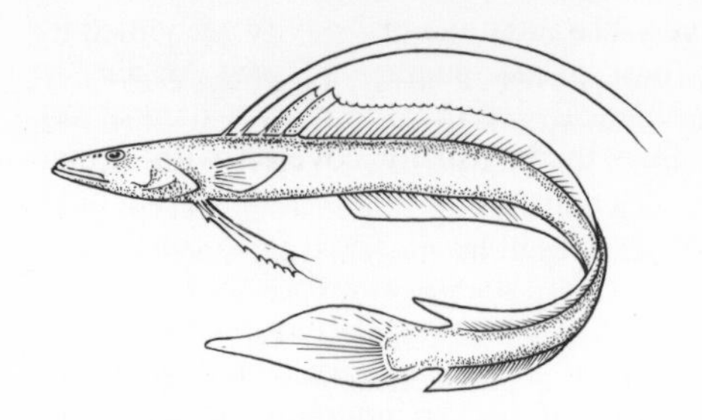

Fig. 1129 Trichonotidae-type

D are elongated, often free-standing. C pointed, Vs elongated. Colouring usually unremarkable. Sandfishes feed on small animals. Their main area of distribution is the tropical W Pacific. Closely related Fs are the Creediidae (fig. 1130) and Limnichthyidae.

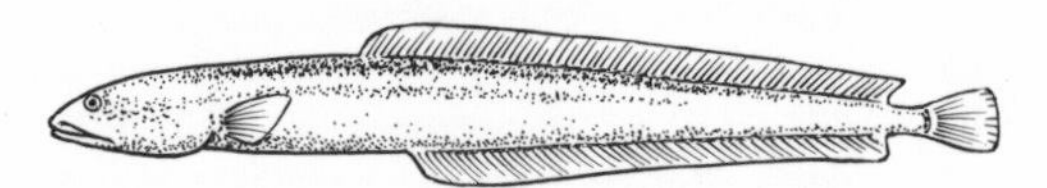

Fig. 1130 Creediidae-type

– *Uranoscopidae*; Stargazers. Bottom-dwelling fish that live both in the depths and in shallow-water zones of tropical and temperate seas. There are about 25 species. The Stargazers are similar to the Trachinidae described above, except their head is bigger, broader and has an armour of bony plates. The large mouth-slit is vertical, and the eyes lie on the upper side of the head. In the shoulder region, there is a strong spine, pointing backwards, which is sometimes described as poisonous, sometimes not. Many species have a thread-like or tree-like process at the chin which acts as a lure for prey-animals. A weak electric organ may be formed

Fig. 1131 *Uranoscopus scaber*

behind the eyes. The nose and oral cavity are joined to one another (naso-palatal passage). D and An almost equally long. Vs located at the throat, with 1 spine and 5 soft rays. Skin thick, usually covered with strong scales. Stargazers burrow so far into the sand that only their eyes and mouth-slit show. Their food consists of crustaceans and small fish which are slurped in with a quick movement. The Stargazer *(Uranoscopus scaber)* (fig. 1131) comes from the E Atlantic, Mediterranean and Black Sea where it lives on muddy substrata. It can grow up to 30 cm long and is covered in pale flecks on a grey-brown background. In the W Atlantic, it is represented by the similar-sized species *Ariscopus iburius* and *Astroscopus guttatus*. Stargazer meat is very tasty.

Trachinus draco *see* Trachinoidei

Trachipteridae, F, *see* Lampridiformes

Trachipterus trachipterus *see* Lampridiformes

Trachurus GÜNTHER, 1854. G of the Carangidae* comprising various species found in warm to temperate parts of the Atlantic and neighbouring seas. *Trachurus* species are shoaling fish from open waters, as well as coastal areas. Length 40–60 cm. Silvery fish with a long-oval shape. In contrast to *Caranx**, the LL is accompanied by shield scales. Like other related Ga and their species, young *Trachurus trachurus* live beneath the bell and in the genital cavities of large Scyphozoa* (Jellyfishes) (*Cyanea, Rhizostoma*, and others). No observations have been made in the aquarium. An important edible fish is:

– *T. trachurus* (LINNAEUS, 1758); Scad, Horse Mackerel, Maasbanker. E Atlantic, Mediterranean, Black Sea, also rarely in the Baltic. 40 cm.

Translucent Bloodfin *(Prionobrama filigera) see Prionobrama*

Transpiration, in contrast to guttation*, is a physically induced release of water vapour by plant organs. The water vapour is given off via the stomata*, as well as through the external walls of the epidermal cells. It can be actively regulated to a certain extent by altering the width of the stomata. Transpiration takes place mainly on the leaves. There is no transpiration in parts of plants that are submerged.

Transportation of Aquatic Animals. The transporting of live aquatic animals is inevitably linked with a worsening of their living conditions, and so any time in transit must be as little as possible. Dispatching specimens bred in captivity that are adapted to living conditions in the aquarium is normally simpler than transporting animals caught in the wild. In transport containers with a fairly small amount of water and a lot of animals, the gaseous exchange is disturbed and it is all too easy for oxygen to become scarce and carbon dioxide to build up. Poisonous metabolic products quickly accumulate and the temperature can fluctuate enormously. All these undesirable factors can be compounded by a shortage of food and a lack of ecosystem. It is very rare for transport containers to have ventilation*, filtration* and temperature control*—usually, methods have to be improvised. Firstly, only healthy, strong and well fed specimens should be selected for transportation. Sick or weak specimens die after a short while and then they will make the living conditions deteriorate for the healthy animals too, and so the whole batch will be endangered. Before dispatch, specimens should be starved for 2–3 days so that the contents of the gut are emptied out, which will mean that excreta in the water in the container will be kept to a minimum. How many animals should be transported in a certain amount of water will depend on their needs. In the final analysis, it is always better to have too few animals than too many. Large tanks are only rarely available as a means of transport. The transport tins (made of metal) that used to be popular are more and more being replaced by light and inert plastic containers. Increasingly, aquatic animals are being sent in plastic pouches with rounded corners. To guard against too high temperature differences, insulation materials made of foam, especially Polystyrol (Styropor), are a good idea. Thermos-flask containers with good insulation are usually heavy and break too easily. The containers should be filled only a third full with suitable quality fresh water. The volume of air above it will ensure sufficient gaseous exchange. Plastic pouches are often pumped up with pure oxygen, which means that there is then an increased internal pressure. If the specimens are removed from the transport container, the balancing out of the gas must be allowed to take place slowly to protect the animals from gas bubble disease*. With the addition of a small amount of anaesthetics, the metabolism of the specimens is greatly reduced and so more animals can be placed in the same volume of water. Animals that are to be transported together must be selected with care. Intolerant species or predatory fish on no account belong in the same container as smaller species. Especially with animals caught from the wild, it is important to keep them separate for a while and to allow the animals several days' adaptation time. Plants must not be transported together with animals; sea-anemones, crustaceans, molluscs, fish and other groups must also be placed in separate containers. Molluscs, crustaceans and some sea-anemones, like most aquatic plants, normally travel better without water, simply wrapped in some damp material. If animals are being transported by post, by train, ship or aircraft, the containers must be shut very firmly and stickers should be added saying 'Handle with care–live animals!'. Adding some information about temperature is often advisable too. When the transported animals reach their destination, they should gradually be accustomed to conditions in the aquarium.

Trapa LINNÉ, 1753; Water Chestnut. Only G of the Trapacea*. 3 species, perhaps more. From Europe, N Africa, Asia. Annual plants with floating leaves in rosettes. The leaves have pumped up stalks, leaf blades rhomboid. Flowers hermaphroditic, polysymmetrical. Perianth consists of calyx and corolla, 4-part. Stamens 4. Carpels 2, fused, standing in the middle. A complicated fruit with horns that develop from the sepals.

The species of *Trapa* colonise still, nutrient-rich stretches of water with little calcium-content that warm up nicely in the summer. They make rewarding and attractive specimens in open-air ponds. If the water is

drained away, the ripe fruits must be gathered in the autumn and kept in cold water throughout the winter. They contain a lot of starch and were formerly used as a food. In some parts of Asia, they still are today. The best-known species is:
– *T. natans* LINNÉ; Water Chestnut. Europe, Asia (fig. 1132).

Fig. 1132 *Trapa natans*

Trapaceae (Hydrocaryaceae, name not in current usage); Water Chestnut Family, F of the Magnoliatae*. The only G is *Trapa**.

Trematocara BOULENGER, 1899 (fig. 1133). G of the Cichlidae*. There are about 8 species only found in Lake Tanganyika*. They are small to medium-sized

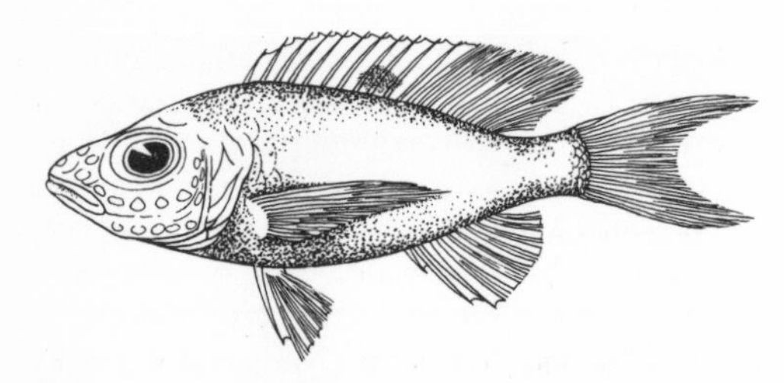

Fig. 1133 Representative of the G *Trematocara*

Cichlids, measuring between 4.5 and 15 cm. Shape normally very elongate, only a few species are more deep. Looks very predatory. Mouth deeply slit, slightly superior. Eyes very large. Head pointed, forehead and chin very flat. D has 8–12 hard rays and 10–13 soft ones. The An has 3 hard rays and 9–12 soft ones. The rear sections of both these fins are pointed to some extent. P and C long, and so too are the Vs, sometimes. C has pointed tips, curving in a semicircular shape. The members of the G should be cared for as per instructions given under *Lamprologus** and *Julidochromis**.

Trematodes. Cl of the Plathelminthes*. In the adult stage Trematodes live parasitically – some also during the larval stage. The Monogena group are hermaphrodites (Hermaphroditism*) which develop directly without alternation of generations. Numerous species live as ectoparasites (Parasitism*) on the skin and gills of freshwater and sea-water fish. A mass attack will only be observed among weakened animals in the aquarium. The Ga *Dactylogyrus* DIESING, 1850, and *Monocoelium* WEGENER, 1909, are gill parasites that anchor themselves in the gill epithelium by means of an attachment disc and 2 *(Dactylogyrus)* or 4 *(Monocoelium)* clasping hooks. The front end of the 1 mm long *Dactylogyrus* species (fig. 1134) has 4 tips, whilst *Monocoelium* has no tips. Both Ga are characterised by 4 or more pigment spots. All the species lay eggs, and the larvae that hatch out attach themselves once again to the fish's gills. Species belonging to the G *Gyrodactylus* NORDMANN, 1832 (fig. 1134) are skin parasites that only occasionally attach themselves to gills. The front end

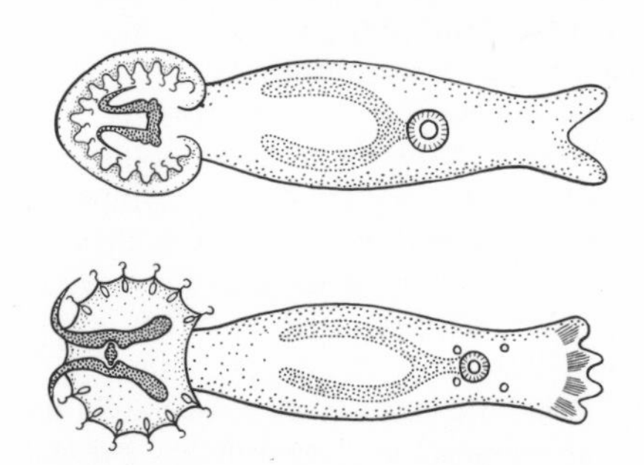

Fig. 1134 Trematodes. 1 *Gyrodactylus* 2 *Dactylogyrus*

is double-pointed and there are no pigment spots; the rear disc of attachment has 2 clasping hooks. These parasites are 0.2–0.8 mm long, and they are livebearing. Fish that are suffering from a Monogena attack often show no symptoms. Only if there is a high number of parasites will there be swellings on the gills and shortage of breath, or white-grey coatings on the skin. Short-term sodium chloride baths* or formalin* baths are recommended as treatment. Formalin is also suitable for treating sea fish, along with copper sulphate* and zinc sulphate*. Prolonged baths using Trichlorophon* are very effective. Livebearing *Gyrodactylus* species will die after 4–6 days if the tank is kept free of fish; if the egg-laying varieties of parasite appear, the tank must be disinfected (Disinfection*). *Diplozoon* NORDMANN, 1832, is interesting from the biological point of view; it is a gill parasite that only occurs occasionally in the aquarium; even in the larval stage, 2 specimens always fuse together (fig. 1135).

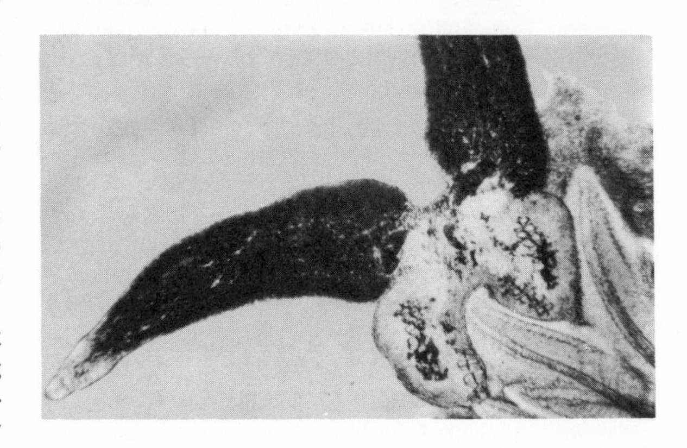

Fig. 1135 *Diplozoon tetragonopterini* sitting on the gills

The Digena are characteristic endoparasites, the development of which takes place over several larval stages via 1–2 intermediate hosts. Adult parasites live in vertebrate animals, often as intestinal and blood parasites in fish. The Blood Fluke (*Sanguinicola* PLEHN, 1905) (fig. 1136) is found in free-living Cyprinidae*, more rarely in aquarium fish. It lives as a parasite in the adult form in the blood vessel system of the gills. Its cap-shaped eggs 40–70 μm) get into all the organs via the blood stream, block up the capillaries, and so cause considerable damage to the areas affected. The larvae that hatch out bore their way to the outside in the gill region, but they can only infect other fishes via Snails as the intermediate host. Fish are also intermediate hosts of Trematodes, giving refuge to their larval stages in these cases. The adult parasites often live in the intestine of fish-eating birds. The eggs are released along with the bird's droppings and develop in the water into small larvae (miracidia), which actively bore into snails. Here they undergo metamorphic and reproductive processes, then leave the snail as cercaria larvae that penetrate the fish skin. Here, they are encapsulated in tissue as metacercaria and are visible as dark nodules in the skin and in the internal organs. A bad attack can cause the organ to shrink, or it can cause paralysis and metabolic disturbances. The metacercaria

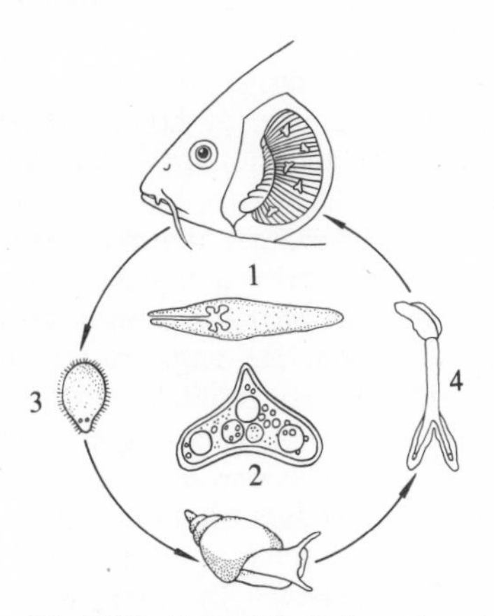

Fig. 1136 *Sanguinicola-inermis*-cycle. 1 Adult animal 2 Egg 3 Miracidium 4 Cercaria

of *Proalaria spathaceum* LA RUE attach themselves to a fish's eye, causing the so-called worm cataract. Only when sick fish are eaten by birds does the adult parasite develop once again in the bird's intestine. Cercaria and metacercaria diseases usually occur in the aquarium only with imported fish. Any slight attack normally goes unnoticed and hardly damages the fish at all. New infections are only possible via infected Mollusca*. Attempts at curing attacked fish are useless.

Triacanthidae; Triple-spines (fig. 1137). F of the Tetraodontiformes*, Sub-O Balistoidei. The most primitive F of the O, premaxilla and maxilla not yet fused, mouth therefore slightly protruding (compare Tetraodontiformes*). Fairly tall, laterally flattened fishes with a pointed or very elongated, tube-like snout. D1 has a few, very strong and long spines. D2 short, standing opposite a similar An. C rounded. The Vs consist of a large spine that can be locked in position and a small soft ray. The 9 Ga are found mainly in fairly deep parts of the Atlantic and Indo-Pacific. A well-known species is the 12 cm long *Halimochirurgus triacanthus*. Triple-spines are not important economically or in the aquarium.

Fig. 1137 Triacanthidae-type

Triaenodon obesus *see* Triakidae

Triakidae; Smooth Dogfishes (fig. 1138). F of the Chondrichthyes*, O Charcharhiniformes. There are about 30 species, none of which is dangerous. Leopard Sharks are found mainly around the coast, and often occur in large numbers. Normally, they are about 1 m long. They have a typical Shark build, 2 spiracles and they always have an An. Placoid scales small; 2 Ds. Denture consists of plaster teeth with which they can break open-shelled food animals such as bivalves, snails and crustaceans. In addition, smaller kinds of fish and worms are eaten. Leopard Sharks are livebearing. The Smooth Hound, for example *(Mustelus mustelus)*, gives birth to about 20 offspring after a pregnancy lasting many months. This particular shark comes from the W and E coasts of the Atlantic as well as the seas bordering it. Before birth, the young are fed by the mother via a pseudo-placenta, and they are attached by an umbilical cord. The lovely Leopard Shark *(Triakis semifasciata)* with its saddle markings is found along the Pacific coast of N America. It is often an exhibit in public aquaria.

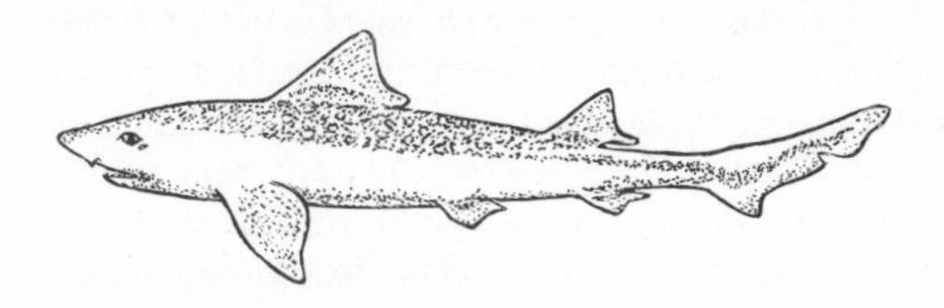

Fig. 1138 Triakidae-type

Triaenodon obesus has white tips on its dorsal fins. It is distributed from the Red Sea east to Polynesia and west to Panama. One of the largest species, the 1.8 m long Spotted Shark *(Mustelus punctulatus)* is found in the Mediterranean and along the coasts of S Africa. Just as big is the Tope *(Galeorhinus galeus)* (fig. 1139) which is mainly found in the areas around the bottom of the continental land masses in the E Atlantic, from Scandinavia to S Africa. It is grey-green on the upper surface and white on the lower. It will often circle round swimmers and divers, only out of curiosity, and never attacks. The Triakidae have a rather tough, bland meat and so they are scarcely ever eaten.

Fig. 1139 Juvenile *Galeorhinus galeus*

Triakis semifasciata *see* Triakidae

Triangle Cichlid *(Uaru amphiacanthoides) see Uaru*

Tribe *see* Systematic Categories

Trichiuridae, F, *see* Scombroidei

Trichiurus lepturus *see* Scombroidei

Trichlorophon (2.2.2-trichloro-1-hydroxyethylene phosphoric acid-0.0-dimethylester). A highly effective insecticide that can also be applied against worm parasites*. It is a white, crystalline powder that dissolves in water (1:10); for fish, it is applied as a prolonged bath (Bathing Treatments*) for 3–4 days in concentrations of 2–3 mg to every 10 litres of water. Trichlorophon can be removed from the water again by means of activated charcoal*. Trichlorophon is the active ingredient of many common insecticides, but it can only be used in aquarium-keeping when it is not mixed with anything else.

Trichodina EHRENBERG, 1832 (fig. 1140). G of the Ciliata*. Several species of *Trichodina* are encountered as fish parasites in fresh and sea water. They have 2 crowns of cilia which enable them to be highly mobile. *Trichodina* is bell-shaped with a ring of attachment beset with hooks (diameter 45–50 μm), with which it anchors itself to skin and gills, or more rarely the urinary bladder. *Trichodina* is a typical debilitating parasite

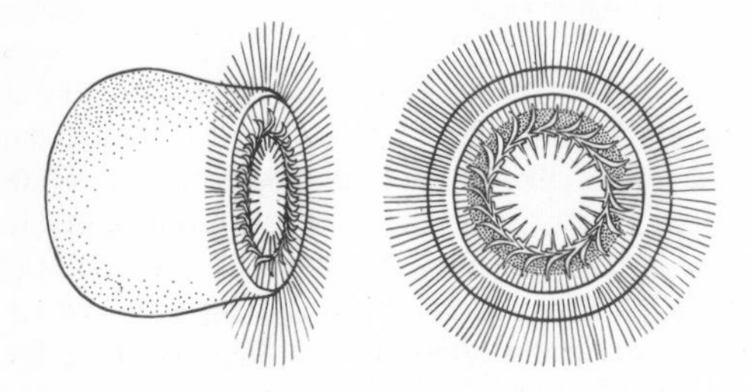

Fig. 1140 *Trichodina*

(Parasitism*), and, as such, can only reproduce in any great numbers if the host fish is already damaged in some way. Single parasites are harmless, but a concentrated attack (trichodiniasis) may destroy the epithelium of the skin, and this will then appear a murky bluish-white colour. When this happens, the parasite also penetrates into the skin tissue where it digs deep passageways. *Trichodina* can survive a long time without a host. Its presence can only be detected under a microscope from fresh skin smears. Various kinds of prolonged bathing treatments* are recommended as a cure – sodium chloride*, formalin*, potassium permanganate*, malachite green*, methylene blue*, and for freshwater fish Trypaflavine*, Chinin*, copper sulphate* and zinc sulphate* can also be used. Any treatment will only have lasting success, however, if the primary cause of damage is removed.

Trichodontidae, F, *see* Trachinoidei

Trichogaster STERKI, 1878. G of the Belontiidae, Sub-O Anabantoidei*. SE Asia: Near East and Further India, Malaysian Archipelago, Greater Sunda Islands. *Trichogaster* is found mostly in streams, river areas, ponds and lakes – wherever there is rich vegetation. They are medium-sized fishes, 11–25 cm long. Body usually strong, elongate to a greater or lesser extent and at the same time fairly tall. Mouth small. LL irregular. In contrast to the G *Colisa*, the D is short, ending before the C, and in the male it is pointed. The An begins beneath the base of the P; it is usually rounded at the rear end. V short and inserted in front of the P; the first ray is elongated into a long thread. Colouring variable, sometimes very bright.

Trichogaster species should be kept in well planted aquaria at a temperature of 24–30 °C. Floating plants are a good idea. The fish require fine live food, but they also accept dried and aritificial food. The eggs are laid in bubble-nests – *Trichogaster* is very productive. Often over 1,000 fry will hatch out, and in order to rear them they should be graded according to size. In their homelands, the various species are sought-after edible fish. Interesting and popular species in the aquarium are:

– *T. leeri* (BLEEKER, 1852); Pearl Gourami, Lace Gourami, Mosaic Gourami. From Thailand, Malaysian Peninsula, Sumatra, Kalimantan (Borneo). To 11 cm. Ground colour brownish; net markings on the flanks that are made up of a large number of gleaming silver dots that often shimmer a bluish-violet colour. A broken up, dark brown longitudinal band extends from the mouth to the root of the tail where it ends in a blotch. There is a soft mosaic pattern on the D, An and C. The front section of the An is red. In the male the breast and throat are red; in fact, the male is more intensely coloured altogether. This species reaches sexual maturity fairly late (fig. 1141).

– *T. microlepis* (GÜNTHER, 1861); Moonlight Gourami. From Thailand. To 15 cm. Body slim. Upper side of the body slightly concave at the front. Scales very small. Colour bluish-silver. Younger specimens often have a dark longitudinal band. In the male, the V is orange, yellowish in the female. The D in the male is noticeably elongate. This species generally is a little timid. It spawns mainly in the evening and at night. Builds a bubble-nest with lots of plant parts. Up to 5,000 eggs may be laid.

Fig. 1141 *Trichogaster leeri*

– *T. pectoralis* (REGAN, 1910); Snakeskin Gourami. From S Vietnam, Thailand, Malaysian Peninsula. To 25 cm. Ground colouring grey-green to olive-green. Irregular, yellowish transverse bands cross the body at an angle, and there is also one incomplete, dark longitudinal band. In the male especially, the tips of the hard-rayed part of the An are a stronger yellow colour.
– *T. trichopterus trichopterus* PALLAS, 1777); Three-spot Gourami. From the Near East, Thailand, Kampuchea, S Vietnam, Malaysian Archipelago, Greater Sunda Islands. To 15 cm. Colour variable. Dorsal surface olive, otherwise the body is silvery with a weak violet tinge. There is a dark fleck on the middle of the flanks and on the root of the tail. Sometimes, brownish oblique bands show up, and occasionally there is also an incomplete longitudinal band. The D, An and C are greenish to grey with white to orange-coloured blotches. This species has one sub-species and several ornamental forms:
– *T. trichopterus sumatranus* LADIGES, 1933; Opaline Gourami. From Sumatra. To 13 cm. This is a blue-coloured sub-species of *T. trichopterus* with darker oblique bands that show up particularly when excited and at the breeding period. Soft, slightly acidic water is recommended. Specimens are somewhat intolerant. Ornamental forms are 'Cosby' (fig. 1142), which is blue with dark to black marble markings, the Gold Gourami which is a strong gold-yellow ground colour with irregular, dark oblique bands, and the Silver Gourami with its green-silver marbly effect.

Fig. 1142 *Trichogaster trichopterus sumatranus* 'Cosby'

Trichomycteridae; Parasitic Catfishes (fig. 1143). F of the Siluriformes*, Sup-O Ostariophysi. Small to very small, worm-shaped or Loach-like naked Catfishes from S America. The D and An are short and have no spines, and none of the species has an adipose fin. Barbels very short or almost completely regressed. The swimbladder too is very much reduced – it is encapsulated in a bony housing. The operculum has numerous spines on it, pointing backwards. Many Trichomycteridae are free-living, and as typical bottom-dwelling fish they dig their food from the soft substrate. A few have become parasites, among which a very few may also be dangerous to Man and mammals. *Stegophilus insidiosus*, for example, lives in the gill cavity of

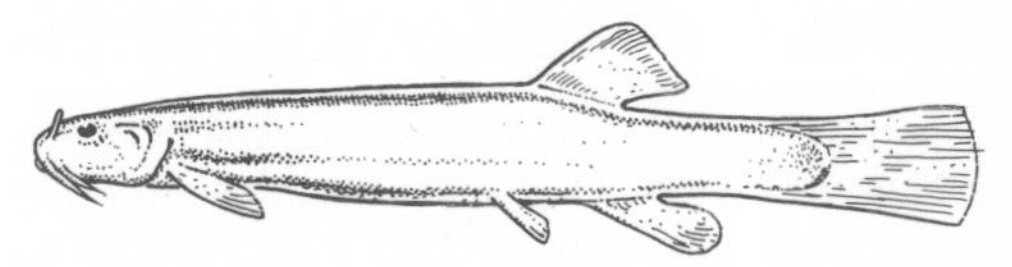

Fig. 1143 Trichomycteridae-type

large Mailed Catfishes (Callichthyidae*) where it nourishes itself on gill hooks or sucks blood. *Vandelia* species (only 2.5–6 cm long and 3–5 mm thick) enter the urethra of men and women who are bathing. There, they use the spines on the operculum to hook themselves in firmly and this can cause extreme pain and inflammation. Usually, they can only be removed if the victim is operated on. It is probable that the penetration of the urethra is not intentional parasitism, but a reaction to the current caused by people passing water. The fish swim against this current and so enter the urinary ducts. In the areas that this 'Candiru' is known to exist, the Indians wear protective gear when bathing. Non-parasitical Trichomycteridae have been imported very rarely – in particular *Pygidium itatiayae*, a 15 cm long Catfish from E Brazil. These crepuscular fish were also very active during the day in the aquarium; they also accepted any kind of food and proved to be very hardy.

Trichonotidae, F, *see* Trachinoidei

Trichopsis CANESTRINI, 1860. G of the Belontiidae, Sub-O Anabantoidei*. SE Asia: Thailand, S Vietnam, Malaysian Peninsula, Greater Sunda Islands. Found in ponds, weedy streams and rivers. Length 4–6.5 cm. Body elongate, flattened laterally. Head pointed, lower jaw protruding. D short, with 2–4 hard rays and 6–8 soft ones. An long, with up to 28 soft rays. D, An and C, particularly in the male, clearly elongated at the rear or drawn out into a point. The first soft ray of the V is elongated in a thread, although it is shorter than in the related G *Trichogaster* and thinner than in the similarly built *Betta* species. Ground colour of the body brownish with rows of dark flecks or longitudinal bands. Colourful dots sprinkled over the unpaired fins. Both sexes, but especially rival males at spawning times, can produce purring noises, probably by means of the labyrinth. Tanks should have plenty of vegetation and a temperature of 22–25 °C. Temperature fluctuations should be avoided. Food should preferably be live and small: *Cyclops*, small *Daphnia* etc. For breeding purposes, the level of water in the tank should only be about 10 cm, and the temperature around 26–30°C. *Trichopsis* builds a small bubble-nest usually, laid between plants; *T. pumilus* and *T. schalleri* often build it away from the surface. Open-water spawning is also possible. How productive the species are can vary, in line with their size – *T. vittatus* may produce 60 to over 200 fry. The species known so far and recommended for the aquarium are:

– *T. pumilus* (ARNOLD, 1936); Dwarf Gourami. Vietnam, Thailand, Greater Sunda Islands. To 4 cm. C elongated in the middle, but rounded. Ground colouring olive, dorsal area darker. Belly and caudal peduncle greenish-white. A row of dark flecks or an irregular band runs from the tip of the snout to the root of the tail. There are also red flecks on the flanks and scattered blue or blue-green dots. D, An and C bluish with rows of dark dots and red edges. Female less conspicuous; An rounded at the rear and (fig. 1144).

Fig. 1144 *Trichopsis pumilus*

– *T. schalleri* (LADIGES, 1962). From Thailand. To 6 cm. Colouring similar to *T. pumilus*, but with fairly strong longitudinal bands. The rear sections of the D, An and C clearly pointed. Fins paler in the female.

– *T. vittatus* (CUVIER and VALENCIENNES, 1831). From Thailand, S Vietnam, Malaysian Peninsula, Greater Sunda Islands. To 6.5 cm. D and An pointed; middle rays of the C, particularly in the male, elongated. Ground colour of the body yellowish to brownish with a greenish gleam. Along the flanks run 2–4 brown longitudinal bands, and here too is a greenish-coloured shoulder blotch. D, An and C reddish with blue-green and red flecks or dots. An red with a blue-white edge area. Female's colouring weaker, fins smaller, breast and belly paler.

Trichoptera; Caddis Flies (fig. 1145). O of the Insecta*. Closely related to the Butterflies, but unlike them the Caddis Flies are usually small, dull in colour and have hairy rather than scaly wings. The imagines (adult insects) are active at dusk and during the day they normally rest among the vegetation on the banks of the stretch of water where they live. That is why hardly anyone pays them much attention. Their larvae and pupae live in a wide variety of inland waters (the water channels of springs, streams, rivers, lakes and moorland pools). A very few species have largely gone over to a terrestrial way of life. The common name Caddis Flies refers to the ability of many larvae to cement together with a salivary secretion pieces of stone, sand, plant material and other matter to form ingenious cases (eg *Limnephilus** and *Phryganea**). The larva is protected in its case and it always carries it around wherever it goes. Other Caddis Fly larvae build complicated webs in which they live and also trap food drifting along in the moving water (eg *Hydropsyche**). The larvae of some species build neither cases nor webs (eg *Rhyacophila**). Caddis Flies are highly recommended for the pond-aquarium where all their activities can be observed: construction of caddis cases, feeding, locomotion, pupation and hatching. In addition, all the stages of Caddis Fly can also be used as fish food.

Tridacna BRUGUIERE, 1797; Giant Clams. G of tropical marine Bivalves (Bivalvia*, O Eulamellibranchiata) that bore mechanically with their vertebral column foremost into the coral. Their shell can be up to 175 cm across, and clams can hold their shells shut so tightly that any diver who happens to get trapped in one cannot free himself. Symbiotic algae (Zooxanthellae), which provide additional food, live in the colourful edges of the mantle and in other parts of the bivalve's body that receives light. Apart from this, Giant Clams feed on

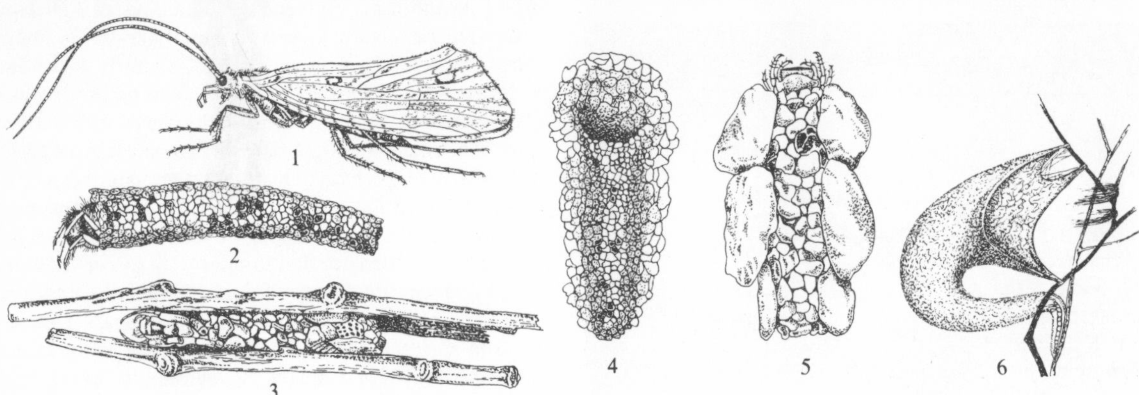

Fig. 1145 Trichoptera. 1 Imago 2–5 Various larva with caddis cases. 2 *Seriacostoma* 3 *Anabolia* 4 *Molanna* 5 *Silo* 6 Net-building Caddis Fly larva of *Neureclipsis*

plankton and detritus only. Young specimens can be kept in the aquarium, and because of the Zooxanthellae they need bright light.

Triggerfishes, F, *see* Balistidae

Trigla (several species) *see* Scorpaenoidei

Triglidae, F, *see* Scorpaenoidei

Triodon bursaria *see* Triodontidae

Triodontidae; Three-toothed Puffer. F of the Tetraodontiformes*, Sub-O Tetraodontoidei. The only species of this F, which lives in the Indo-Pacific, is *Triodon bursaria* (fig. 1146), a 40 cm long Pufferfish with a very large belly pouch. On the side of this pouch is a large, round fleck bounded by a yellow colour. In the upper jaw there is a right and a left edge to the teeth, in the lower jaw, the two edges are fused. A rare fish.

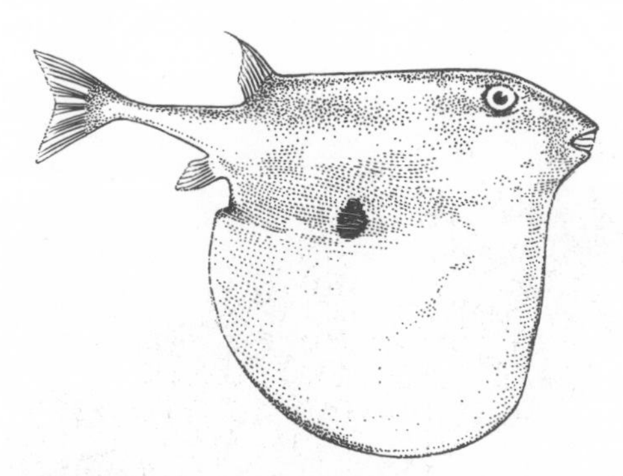

Fig. 1146 *Triodon bursaria*

Triops, G, *see* Triopsidae

Triopsidae; Fairy Shrimps. F of the Phyllopoda*. 2 Ga, *Lepidurus* LEACH, 1819, and *Triops* SCHRANK, 1803, are very widely distributed in the northern hemisphere. The Triopsidae are a very ancient group of crustaceans which have survived into the present time with only a few species. They live a seasonal existence, living in temporary bodies of fresh water. *Lepidurus*, for example, appears only directly after the winter snows have melted, whilst *Triops* is a definite summer form (fig. 1147). The hard-shelled eggs of the Triopsidae can lie dry for years, and in fact they may have to experience a dry period before they are capable of developing

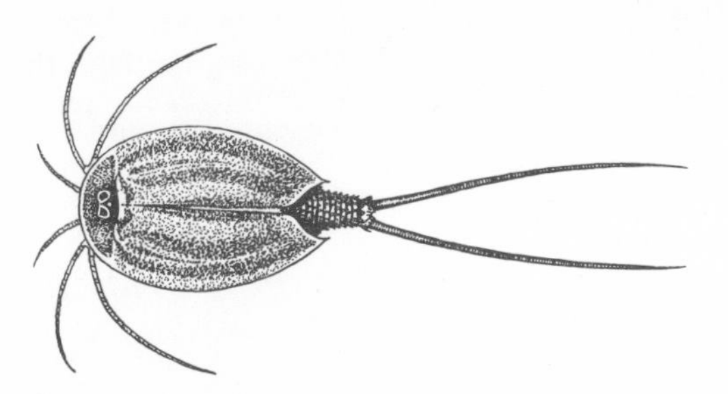

Fig. 1147 *Triops cancriformis*

properly under suitable environmental conditions. The crustaceans grow very quickly and are already several cm long within a few weeks (*Lepidurus* up to 5 cm, *Triops* up to 10 cm). A characteristic of the Triopsidae is a large dorsal shell which covers the body like a shield, with the exception of the last abdominal segments. The last segment of all carries a long forking tail (furca).

The Triopsidae are found in very localised areas, but in such places they are usually common. To find specimens is always a great thrill for any nature-lover. They can be kept in the pond-aquarium for some time. They should not be used as a fish food because they are rare.

Triple-spines, F, *see* Triacanthidae

Triple-tails, F, *see* Lobotidae

Tripneustes AGASSIZ, 1841 (fig. 1148). G of tropical Sea-urchins (Echinoidea*). *Tripneustes* is a grazer and opportune predator, like an apple to look at, with short spines and powerful poisonous pincers. It is similar to the related G *Sphaerechinus**. The diameter of the body is about 15 cm. Various species are frequently imported and have proved to be hardy in the aquarium.

Fig. 1148 *Tripneustes gratilla*

Tripterygiidae, F, *see* Blennioidei

Tripterygion RISSO, 1826. G of the Clinidae, Sub-O Blennioidei*. Often regarded as its own F Tripterygiidae. There are various species that live in the rocky littoral of warm seas. They eat small creatures. *Tripterygion* is scarcely 10 cm long, but it has 3 Ds. It is an elongated fish, with a pointed mouth; when at rest, it props itself on its large Ps and the Vs that are located at the throat. During the breeding season the male is territorial, and it is he who practises brood care. *Tripterygion* species make charming aquarium fish, but they are more delicate than most *Blennius** species, and can rarely be bought anyway. The most common European species is:

– *T. tripteronotus* RISSO, 1826; Three-fin Blenny. Mediterranean. 7 cm. Females and young have a yellowish-brown pattern. Females ready to spawn are a shining red with a black head. Can change colour quickly.

Trissocles setifer *see* Clupeiformes

Tritonium, G, name not in current usage, *see Charonia*

Trochophora *see* Polychaeta

Tropheus BOULENGER, 1898. G of the Cichlidae*. 2 species endemic to Lake Tanganyika*. These are oval-shaped fish, very compressed laterally, up to 12 cm long. Their basic colour is velvet-black. D has a very long base and contains about 20 hard rays; the soft-rayed part, like that of the An, is not elongate. C slightly inward-curving. V very well developed, shaped like a pennant. *Tropheus* has a very interesting life-style and behaviour pattern. The 2 species live around rocks and they live according to a marked social structure. Both species form groups whose individual members know each other 'personally'. Strangers to the group are chased away relentlessly and in the narrow confines of an aquarium they may be bitten to death. In the wild, however, *Tropheus* bands live much more peaceably together. They have an inferior mouth with which they rasp algal growths containing large numbers of tiny creatures from the underwater rocks. Although *Tropheus* species are highly specialised mouth-brooders, the sexes are the same size and colour – probably because this is an adaptation to the social structure just mentioned. The highest ranking male performs a typical shaking ritual (similar to *Haplochromis**) when courting a female from his troop that is ready to spawn. The female lays only a few eggs, although they are very large (diameter about 7 mm); often, she takes them into her mouth even as they are sinking to the bottom. After this, she seeks out the genital region of the male with her mouth, then the sperm that she sucks into her mouth along with her respiratory water will fertilise the eggs. About 4 weeks go by before the eggs are fully developed, but the fry that hatch out are already 15 mm long and completely formed, so that they are already able to swim. If *Tropheus* is to be cared for and bred successfully, it is important to take into account the natural living conditions of the species. The tanks must be very large with plenty of rock structures. The normal kind of vegetation can be dispensed with; light can be shielded off with floating plants, however (this appears to be good for the fish); otherwise, subdue the overhead lighting! *T. duboisi* can be put in the same tank as foreign specimens if you like, but with *T. moorei*, it is best to keep one pair to begin with, as this particular species is very intolerant of its own species in the aquarium. However, they do not much molest their own offspring, so in this way it is possible to acquire a *T. moorei* shoal. In line with their natural habitat, *Tropheus* will only feel happy in extremely hard water. If none is available, the waters of Lake Tanganyika can be imitated by adding about 35 g magnesium sulphate and 2 g sodium chloride to every 100 litres of water. With sodium carbonate, the pH value can be pushed up to 8. Temperature 27 °C.

– *T. duboisi* MARLIER, 1959. From Lake Tanganyika where it lives above rocky ground up to 12 m deep. Length to 10 cm. Velvet black. As a juvenile, it has irregular white flecks which gradually disappear with the onset of sexual maturity (especially in the male). Depending on mood, there is a yellow-brown or reddish transverse stripe behind the Ps which varies in intensity.

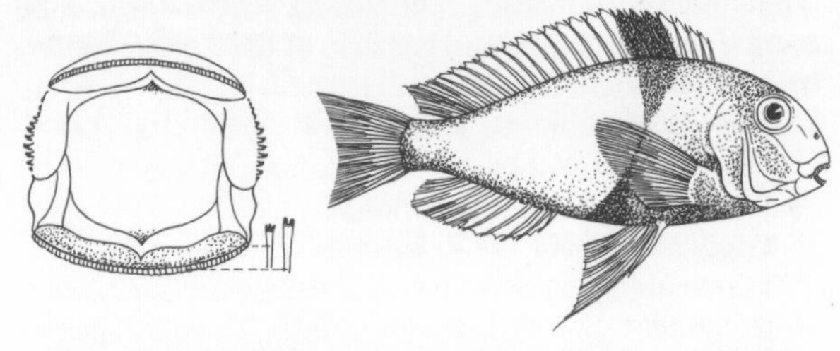

Fig. 1149 *Tropheus moorei* (left, jaw with spatula-teeth)

– *T. moorei* BOULENGER, 1898. From Lake Tanganyika. Found above rocky ground up to 5 m deep. Length to 12 cm. Velvet black. There is a yellow-red saddle marking stretching from the D to the belly which becomes more or less intense as mood changes. Inferiority colouring* a monotone grey (fig. 1149 and 1150).

Fig. 1150 *Tropheus moorei*

Tropism is a movement of parts of sessile plants either towards the source of a stimulus or away from it. The bending of shoots towards a light source, for example, is known as positive phototropism. *See also* Nastic Movements and Taxis.

Trout *(Salmo trutta trutta) see* Salmonoidei

Trout (collective term) *see* Salmonoidei

Trout Region *see* Rhithral

Trout-perches (Percopsidae, F) *see* Percopsiformes

True Carps (Cyprininae, Sub-F) *see* Cyprinidae

True Catfishes, F, *see* Siluridae

True Characins, F, *see* Characidae

True Eels (Anguillidae, F) *see* Anguilliformes

True Flies, O, *see* Diptera

True Perches, F, *see* Percidae

True Rays, F, *see* Rajidae

Trumpetfishes (Aulostomidae, F) *see* Aulostomoidei

Trumpetfishes, G, *see Aulostoma*

Trumpetfishes, related species, Sub-O, *see* Aulostomoidei

Trumpet-snails, G, *see Charonia*

Trunkfishes, F, *see* Ostraciontidae

Trypaflavin. A yellow acridine dyestuff that dissolves easily in water and has good disinfectant properties. In aquarium-keeping, Trypaflavin is used to good effect against skin* and gill diseases* caused by Protozoa*, against ectoparasitical Trematodes* and against external bacterial infections*. Parasitical Fungi (Mycoses*) are only slightly impeded by Trypaflavin. The normal way of applying Trypaflavin is as a prolonged bath (Bathing Treatments*) in concentrations of 0.1–1 g per 100 litres of water. Plants are damaged by Trypaflavin. The dye can be removed from the water again by filtering it through activated charcoal*. Trypaflavin is a mitotic poison (Cell Division*), and if given in overdoses it can lead to sterility* because the sex cells (which actively divide) are damaged.

Trypanoplasma-disease *see* Blood Flagellates

Trypanosoma *see* Blood Flagellates

Tube-mouthed Pencilfish *(Nannostomus eques) see Nannostomus*

Tuber. A roundish, compressed section of the stem axis in which reserve substances are stored. Found mainly among plants that have a dormant period, eg in *Nymphaea* species that grow in waters that dry up at certain times. In every period of growth, the old tuber is used up and a new one formed.

Tuberculosis *see* Fish Tuberculosis

Tubifex LAMARCK, 1816 (fig. 1151). G of the Oligochaeta*. In Europe there are 7 species, some of which have a cosmopolitan* distribution. They are most commonly found in the mud of very polluted waters where they accumulate in large numbers. *Tubifex* worms are up to 8 cm long, very thin and coloured red (caused by the red colouring matter of the blood). The rear end of the worms sticks up freely into the water and undulates gently for respiratory exchange. The front end, on the other hand, sits vertically in mud tubes which have a kind of chimney-stack on the outside. If disturbed, *Tubifex* can disappear completely into its tube. *Tubifex* is very important in aquarium-keeping as a fish food.

Fig. 1151 *Tubifex*. Various phases of burrowing

Samples are sieved from the mud, transported damp, then kept under running water for a few days. Only then should they be fed to your fish. They can be kept alive for several weeks in the fridge if kept in shallow dishes, just covered with water (change the water frequently).

Tubificids, G, *see Tubifex*

Tubipora LINNAEUS, 1758; Organ-pipe Coral. G of the Leathery Corals (Alcyonaria*) which are found on the coral reefs of the Indo-Pacific. The firm skeleton is a shining red colour and consists of tubes standing next to one another. These tubes are joined together by transverse walls stood in tiers. The top-most tier is occupied by the brownish-green polyps. *Tubipora* skeletons are popular as decorative material in the marine aquarium.

Tubularia LINNAEUS, 1758. G of the Hydrozoa*. The polyps are colourful and up to 1 cm long, and with two circlets of tentacles. They sit individually on high,

unbranched stalks which spring from a common creeping root. As the very pretty polyps eat only micro-plankton, they are very difficult to care for in the aquarium. The colonies are also easily overwhelmed by Thread-algae. It is best to keep *Tubularia* in its own small aquarium.

Tubularians, G, *see Tubularia*

Tule Perch *(Hysterocarpus traski) see* Embiotocidae

Tumbling Disease, Whirling Disease, *see* Sporozoa

Tumour Diseases (fig. 1152) are the result of an abnormally large growth of certain tissue cells. What causes tumours to form in fish is largely unclear. Two types of tumour can form: benign, slow-growing tumours of limited extent, and malignant tumours. The latter sort infiltrate neighbouring tissues and destroy them. Tumour cells can be transported by the blood, thus allowing new tumours to form in other organs (metastases). Different names are given to benign tumours, depending on what kind of tissue is affected. Epithelioma are tumours of the skin, myoma are muscle tumours and adenoma affect glands. Malignant tumours are referred to as sarcoma. Fish often have tumours of the thyroid gland, which, if they spread, can obstruct the mouth and gills. Lack of iodine can cause these changes to the thyroid gland. The tumours in such cases are benign and can be treated by bathing the fish in potassium iodide*. If the tumour does not respond to the treatment, then it is malignant and untreatable. Melanosis is also untreatable. This is a tumour disease of the black pigment cells which often begins in the skin of species that have a large amount of melanin *(Poecilia*)* Many tumour diseases can be traced back to a hereditary condition (Hereditary Diseases*, Heredity).

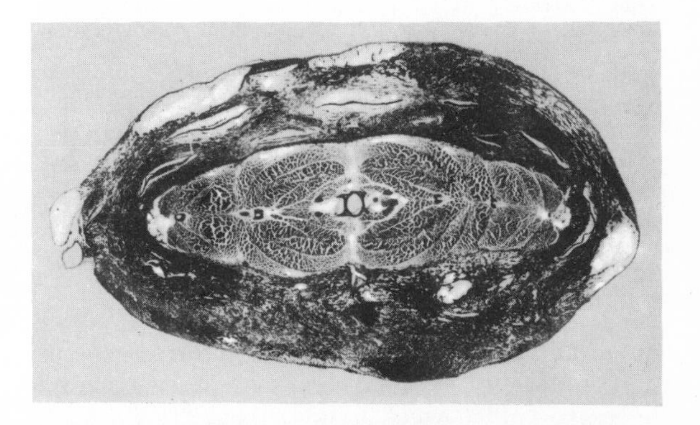

Fig. 1152 Melanotic sarcoma on the caudal peduncle of *Xiphophorus*

Tunicata; Tunicates, Ascidians. Sub-Ph of the Chordata*. All Tunicates live a purely marine life-style. The best-known Tunicates are probably the Sea-squirts (Cl Ascidiacea) which lead a sessile existence on various kinds of substrata. Others include the pelagic, planktonic members of the Cl Thaliacea (with no tails) and another planktonic Cl known as the Copelata or Appendicularia (with tails). The essential chordate features of the Tunicata are the Chorda dorsalis and a nerve tube lying above it on the dorsal side, (which only appear in the larval stage in the Sea-squirts and Thaliacea *(Salpa)* or which have disappeared altogether), plus the foregut (pharynx) which is broken up by gill slits. As the Chorda only appears in the tail region, the Tunicata are also known as Urochordata. All Tunicates are filter-feeders. The Thaliacea *(Salpa)* are biologically very interesting, but unfortunately they cannot be considered in any detail here. It should be mentioned, however, that the poet and naturalist CHAMISSO (1821) was the first to describe the alternation of generations in these animals. Another group of Tunicates belonging to the Thaliacea form hose-shaped colonies that live in warm seas – they shine the most brilliantly of all marine organisms. The Appendicularia have an unusual feature in the very complicated housing that surrounds them. It contains a filter apparatus by means of which very fine plankton (nannoplankton) is held firmly and steered towards the mouth. These particular Tunicates can evacuate their casing through a built-in opening and then they build a new casing within a short while. Only Sea-squirts can be considered for the aquarium. Many have lovely colouring. Sea-squirts are tube-shaped organisms with an inlet opening (ingestion opening) at the front and to the side of this an outlet opening (egestion opening). They anchor themselves firmly to substrata by means of root-like processes and are often difficult to shift. The mantle may be delicate and transparent *(Clavelina, Ciona*)*, tough like leather *(Halocynthia*)* or gristly *(Phallusia)*. It is not unusual for a Sea-squirt to be covered over and over with other organisms (eg *Microcosmus*). Large Sea-squirts are eaten in many places.

Various kinds of Sea-squirt are distinguished – solitary ones (Monascidea, O, eg *Ascidia**, *Ciona**, *Halocynthia**), social ones that form loose-knit groups joined together by runners (stolons) (eg *Clavelina*), as well as the O Synascidea (eg *Botryllus**) in which groups of individual animals are united in a common mantle into compact colonies. These forms do not represent any naturally related groups, however: modern systematics is concerned in the first instance with anatomical features (structure and position of the gonads, of the pharynx etc). Ascidians are hermaphrodites that normally release their eggs and spermatozoa into the water. Many species can also reproduce asexually by budding. Water flows in through the ingestion opening into the pharynx which takes up the largest part of the Ascidian body. Above the fairly large gill slits lies an extraordinarily fine filter band made of mucus. It is formed continually in a ventral furrow (hypobranchial furrow), shifted sideways and upwards, then in the so-called epibranchial furrow on the dorsal side it is rolled up with the food particles contained there into a sausage and finally transported backwards into the stomach. The filtered water goes into a peribranchial chamber surrounding the pharynx and from there via a cloaca (which is also the exit point of the intestine and sex organs) through the egestion opening to the outside. The water current is produced by the cilia of the pharynx. In a domestic aquarium, as with many filter-feeders, feeding is often left to chance, and although Sea-squirts are hardy in themselves, they gradually starve. The best preparations to make are to provide large tanks with plenty of algae. In addition, very fine food in suspension should be added, as well as chopped-up, soaked artificial food. You may also like to try producing your own suitable plankton by developing it in small glasses stood in a bright spot. In large aquaria, many Ascidians reproduce regularly.

Tunicates (Ascidiacea, Cl) *see* Tunicata

Tunicates (Thaliacea, Cl) *see* Tunicata

Tunny *(Thunnus thynnus) see* Scombroidei

Turbatrix. G of the Nematodes. The Vinegar Worm *(T. aceti)* is a nourishing food for fry (Fry, Feeding of*). These worms live in the earth and are up to 2.5 mm long. They feed on fermenting bacteria. *T. aceti* is livebearing and reproduces very quickly under favourable conditions. At room temperature (18–20 °C), they are best cultivated in sealable glass jars, the bottom of which should bow inwards. The culture medium (rolled oats with milk, meal pap, beer that has been left standing) should be placed in the jar such that the raised middle of the bottom stays free. A small shallow dish or tin lid can also be used if placed in the middle. When some Vinegar Worms are placed into the culture medium, they will reproduce in large numbers within a few days. Most will collect on the free surface of the bottom and at the edge above the culture medium. They can then be lifted off easily with a paintbrush or a damp cloth. After some time the culture ceases to be any good because it becomes too acidic, so a new one will have to be set up and inoculated. It is a good idea to use several jars and continually to renew them.

Before feeding to your fish, the worms must be washed if necessary. To do this, the worms are put in a glass of water which is then poured out after the worms have fallen to the bottom. *Turbatrix* also sinks to the bottom of the aquarium quite quickly. Although they can survive up to 8 hours in the water, they are often not eaten any more if they are lying around. To prevent rotting matter from building up, small amounts of worms should be put in the aquarium at fairly short intervals. This also suits the constant food requirements of young fish. But it is ill-advised to feed fish *only* with *Turbatrix*, as the worms are quite fatty.

Turbellaria; Eddy Worms (fig. 1153). Cl of the Platyhelminthes*. There are several thousand species, some microscopically small and others several cm long; most are flattened. The majority of Eddy Worms are a part of the bottom-dwelling fauna of the seas and fresh waters (Benthos*), and they lead a hidden life-style. Some tropical species have conquered very moist habitats on land. A characteristic feature of the Turbellaria is their covering of cilia which help them to move around and to guide food towards them. Larger kinds of Eddy Worms also creep around like naked snails. They lead a partially predatory existence. Various Eddy Worms digest their food at least partly outside the body – ie they shower the prey with digestive juices and suck in the half-digested food-pap through their strong proboscis. Many marine-dwelling Turbellaria are attractively coloured and for this reason they are occasionally kept in the aquarium.

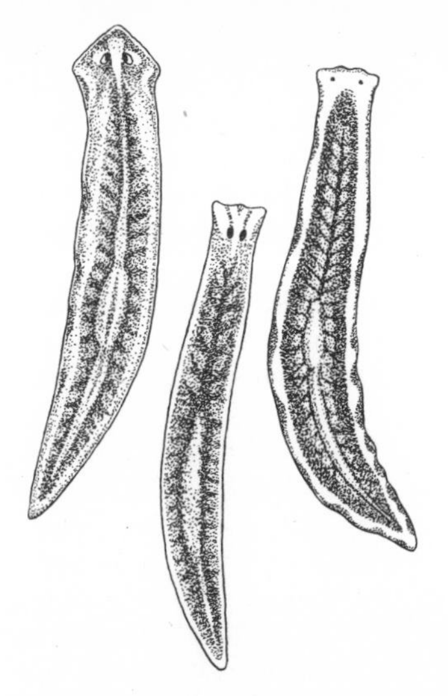

Fig. 1153 Various representatives of the Turbellaria. Left to right: *Planaria gonocephala, Planaria alpina, Dendrocoelum lacteum*

Turbot *(Scophthalmus maximus) see* Pleuronectiformes

Turgor is the state of tension in a plant cell; ie the pressure of the cell contents on the cell wall. Lack of water, disturbances to the water intake or vigorous tranpiration* can reduce the turgor and eventually, parts of the plant, or the whole plant will wilt*.

Turkeyfish *(Pterois volitans) see Pterois*

Tusk-shells, Cl, *see* Scaphopoda

Tusk-shells (*Dentalium*, G) *see* Scaphopoda

Tuxedo Swordtail *(Xiphophorus helleri helleri) see Xiphophorus*

Tuxedo-Platy *(Xiphophorus maculatus) see Xiphophorus*

Twaite Shad *(Alosa fallax) see* Clupeiformes

Twins. In plants, animals and in man, 2 individuals can develop from one egg-cell as a result of premature division of the germ cell. The hereditary material and phenotype of these 2 individuals are completely identical *(monozygotic twins)*. Monozygotic twins have also been produced experimentally by splitting up amphibian eggs in this way. *Siamese twins* are partly joined together at birth because the germ cell did not completely separate in two. *Dizygotic twins* are the simplest form of multiple birth that is normal in most animals. 2 or more egg-cells are fertilised at the same time.

Twinspot Wrasse *(Coris angulata) see Coris*

Two-banded Anemonefish *(Amphiprion bicinctus) see Amphiprion*

Two-banded Pencilfish *(Nannostomus bifasciatus) see Nannostomus*

Two-coloured Angelfish *(Centropyge bicolor) see Centropyge*

Two-coloured Slimefish *(Escenius bicolor) see Escenius*

Two-spot Livebearer *(Poecilia parae) see Poecilia*

Two-Spotted Cichlid *(Cichlasoma bimaculatum) see Cichlasoma*

Two-spotted Climbing Perch *(Ctenopoma nigropannosum) see Ctenopoma*

Two-striped Aphyosemion *(Aphyosemion bivittatum) see Aphyosemion*

Two-striped Cichlid *(Chaetobranchiopsis bitaeniatus) see Chaetobranchiopsis*

Type. The objective, defined, unalterable name-bearer and datum point of a taxon*: a type-specimen of a species, a type-species of a genus, a type-genus of a family. In contrast, the circumspection of a taxon is subjective and, as such, it can be altered. Every zoological taxon is in the last analysis based on a type, either real or potential, and *not* (as is often wrongly assumed) on its description. Whereas a type-species and type-genus need only be named in a new description of the taxon concerned, the type-specimen of a species should be brought to a museum by the original describer so that comparison examinations can be made at any time. If the describer of a species has defined only one type-specimen this is called a *holotype*. If there is a series of type-specimens, each one is known as *syntype*, and this is so even if the specimens belong to one or several species. In the latter case, to make things clearer, when the next author comes along, he must declare one specimen as a type from amongst the syntypes. Types determined at a later date like this are known as *lectotypes*. A *paratype* is each specimen from a type-series that is not the holotype; if a lectotype is chosen, then the remaining specimens are *paralectotypes*. The term *allotype* or *allolectotype* is used to describe a paratype which is opposite to the holotype or lectotype – perhaps the female version of a male specimen or the development stage (larva) of an animal in comparison to the sexually mature animal (imago). If the type-specimen of a species should become lost for some reason (during wartime for example), a *neotype* can be defined for important reasons provided attention is paid to certain rules of nomenclature.

Typha, G, *see* Typhaceae

Typhaceae; Reed Mace Family. F of the Liliatae*, comprising 1 G and 15 species. Widely distributed in the tropics and in northern temperate regions. Most Reed Maces are strong perennials that form runners. Leaves distichous, linear. Flowers unisexual and monoecious, inconspicuous, much simplified and with no perianth. The species of the G *Typha* (Reed Mace) with their characteristic brownish, club-shaped infructescences are plants from amongst the marsh vegetation and around the water's edge. They are suitable for planting round the edges of open-air ponds and in shallow spots. Young shoots and the starch-rich runnners are eaten in many places.

Uaru HECKEL, 1840. G of the Cichlidae*. There is one species found in the black water rivers of tropical S America. It lives in the same areas as *Pterophyllum** and *Symphysodon**. *Uaru* species are large, deep-bodied fish, very flat from side to side – they are reminiscent of tall *Cichlasoma** species and they are particularly closely related with this G. The forehead in older specimens has a fatty lump. Mouth relatively small. In the youthful phase, the soft-rayed parts of the D and An are lobed and clear as glass (as in the F Nandidae*), but as they get older these fins become pointed as is typical of Cichlids and they then reach to the end of the C. The body colouring is also dependent on age. *Uaru* species are peaceful, calm Cichlids and are only liable to be quarrelsome during the spawning season. Even water plants are hardly touched. Nevertheless, there are considerable difficulties with care and breeding (more details of which appear under *Symphysodon*). The water needs to be very soft, slightly acidic and enriched with peaty substances (Humic Acids*, Tannic Substances*). Temperature between 27 and 30 °C. Apart from being a suitable size, the tank must also be furnished with places of refuge, such as holes in the rock or root-work, and the lighting should be subdued. A good variety of food is also necessary if care is to be successful. Courtship and spawning behaviour is similar to that found in the G *Pterophyllum*, although *Uaru* always lays its eggs in the darkest corner of the tank (cave).

– *U. amphiacanthoides* HECKEL, 1840; Triangle Cichlid. From the Amazon Basin, Guyana. Length to 28 cm. Up to a length of 4 cm, this species is dark-brown to deep-black in colour, later on a dirty yellow or pale brown, with irregular bright flecks. Colouring in adult specimens yellow-brown with a black blotch behind each of the eyes, another behind the P insertion (this one is particularly large and is in the shape of a droplet lying on its side), and another on the caudal peduncle. Iris bright red (fig. 1154).

Fig. 1154 *Uaru amphiacanthoides*

Uca LEACH, 1830; Fiddler Crab (fig. 1155). Amphibious Crabs (Brachyura*, F Ocypodidae) that build hiding-holes and live in the tidal zones of warm seas. Large numbers of them colonise mud banks and coastal swamps. *Uca* Crabs mainly eat things from the muddy slime. The angular carapace is only 0.8–3.5 cm in width and its edges are smooth. The males have a giant signalling pincer which can make up as much as a half of the body weight, and a small pincer for eating with. Tropical species of *Uca* can be very colourful. The eyes are perched on top of long stalks, and these enable the animal to spot anything that moves at a distance of 3–8 cm; if they do, they slip very quickly back into their home-tube. They also retreat here during a flood, sealing the tube from the inside with slime. Food is taken in at low tide: small clumps of mud are wedged between the mouth appendages with the eating pincer and the particles of food contained in the mud are washed out. *Uca* also eats dead fish, algae, fruits that have been washed up etc. Male *Uca* Crabs make waving signals to attract females and also as threatening gestures to other males. Copulation normally takes place in the tubes. Breeding has also been successful in the aquarium. The larvae that hatch from the eggs are placed

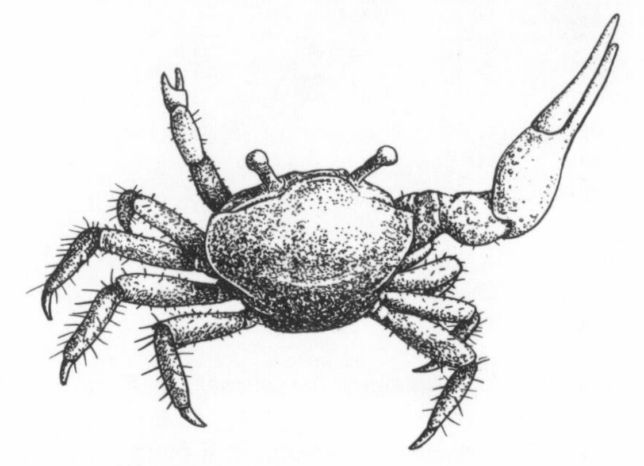

Fig. 1155 *Uca* sp.

in the water by the females. They can be reared in algae-covered dishes stood in a bright spot, or in aquaria which include a land area. *Uca* Crabs make interesting specimens and are fairly easy to care for in well-lit marine aqual-terraria. They can be fed with mud from a pond, artificial food, fish meat and pieces of fruit. An ideal place to keep these crabs is where it is subject to high and low tide. The well-known Ghost Crabs (G *Ocypode* WEBER, 1795) are very similar, although they do not have a signalling pincer. They are found along the coasts of warm seas above the water line.

Udotea. G of marine Green Algae (Chlorophyta*). *Udotea* has a creeping axis and stalked, dark-green thalli with an irregular leaf-shape, which grow in shaded areas or in fairly deep water. *U. petiolata* (TURRA) BÖRGES is common in the Mediterranean, and it also grows well sometimes in the aquarium.

Ulrey's Tetra *(Hemigrammus ulreyi) see Hemigrammus*

Ultra-violet Lamps. In aquarium-keeping, lights with a high proportion of ultraviolet (UV) radiation are used to sterilise and clarify the water. The UV light is surrounded by a watertight glass casing through which water can be passed by means of a pump*. Depending on the power of the lamp, the speed of the water flowing through and the quality of the water, a strong influence can be exerted on the state of the water. Single-celled organisms, floating algae, bacteria, fungi, proteins and other colloids are killed off or denatured by UV radiation. They then stick together to form larger units that flake off and remain behind in the filter. In this way the water can be sterilised or cloudiness caused by the proliferation of single-celled organisms will be quickly removed. If, in sea water, there is a foam filter* operating at the same time, it will be ineffective, as the protein bodies necessary to form the foam will be flaked out by the UV radiation. Depending on its reaction partner,

UV radiation can have an oxidising or reducing effect and thus decisively alter the redox potential* in the water. By denaturing and flaking off the protein bodies that accrue during metabolism, much less free ammonia will develop. The nitrogen compounds are present chiefly as nitrite, as nitrate (particularly in the alkali range) is once again reduced to nitrite. If copper sulphate* is added for healing purposes, the bivalent copper ions are reduced very quickly to univalent ions, whereby catalysis will cause oxygen deficiency*. Iodides which are contained in sea water are oxidised by UV radiation into non-combined, very poisonous iodine. Both positive and negative reactions can be expected whenever UV radiators are installed, so it is important to have a specific kind of light for a specific purpose and only controlled doses should be applied. Continual UV radiation or too high dosages are harmful. But a UV light working for 2–3 hours a day with a power of 6 watts in a medium-sized tank will produce positive effects that are useful in aquarium-keeping. Before it reaches the UV radiator, the water must flow through a pre-filter, otherwise the pieces of equipment will quickly become polluted by coarse particles of dirt and then become ineffective. Fish and other aquarium specimens must never be subjected directly to a strong UV radiation. If ever medicaments are applied, the UV radiation must be switched off, because catalytic reactions will take place that are impossible to control. Very little is known about the efficacity of weak UV radiation (which is also present under natural conditions) on aquatic organisms (Illumination*).

Ulva; Sea Lettuce. G of marine Green Algae (Chlorophyta*) with a lobed, wavy thallus which consists of a double cell layer (there is only one cell layer in the very similar G *Monostroma*). *Ulva* is bright green and grows near the surface. It also flourishes in the aquarium if illuminated strongly.

Umbelliferae, F, name not in current usage, *see* Apiaceae

Umbra krameri *see* Esocoidei

Undulate Triggerfish *(Balistapus undulatus) see Balistapus*

Undulating Display Swimming *see* Courtship Swimming

Unicornfishes, G, *see Naso*

Unintentional Catches, Make-weight Catches. This term is used to describe single specimens or small numbers of rarely imported species of fish, which makes them little known to the aquarist. They crop up from time to time among a large shipment of different kinds of particularly popular ornamental fish, usually unbeknown to the dealer (eg *Cheirodon axelrodi, Petitella georgiae, Carnegiella strigata*). Usually, one is talking about rather inconspicuously coloured species, which are almost without exception discarded upon capture, as their value is slight to the trade. Occasionally, upon the special request of large import concerns, whole collections of such fish do reach Europe and N America. For many aquarists, who are interested in the care and breeding of little known species of fish, such imports are of as incalculable value as they are for ichthyologists working on systematics or ethology. It is not unusual to find among them scientifically unknown species or species that had been imported once, decades ago, and then only as a few specimens which did not breed and were only known from literature.

Unio RETZIUS, 1783; River Mussels (fig. 1156). G of the Bivalves (Bivalvia*). There are about 5 species in Europe, some distributed in the palaearctic realm. They grow up to 8 cm long and live in the substrate of rivers and clear lakes. Unlike the Pond Mussels *(Anodonta*)*, *Unio* species have a notched seal to their shell. *Unio* are well suited to keeping as individuals in the pond-aquarium, and they sometimes serve as a host for the eggs or embryos of the Bitterling *(Rhodeus)*.

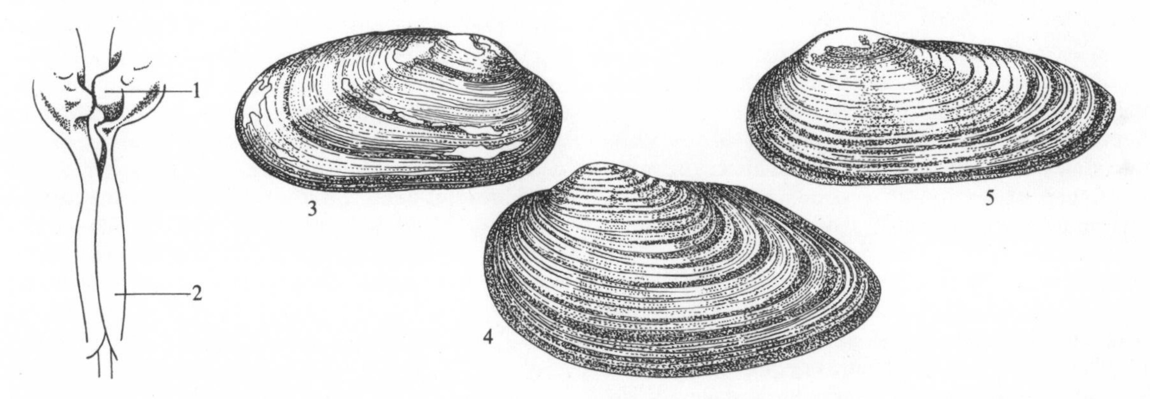

Fig. 1156 *Unio.* 1 Main teeth 2 Ancillary teeth, together forming the seal 3 *Unio crassus* 4 *U. tumidus* 5 *U. pictorum*

Umbrella Palm *(Cyperus alternifolius) see* Cyperaceae

Umbrella Plant *(Eleocharis vivipara) see Eleocharis*

Umbridae, F, *see* Esocoidei

Underwater Banana Plant *(Nymphoides aquatica) see Nymphoides*

Unisexual *see* Sexual Division in Higher Plants

Unstriped Red Mullet *(Mullus barbatus) see Mullus*

Upeneus CUVIER and VALENCIENNES, 1829. G of the Red Mullets (Mullidae*). There are species in tropical and subtropical parts of the Indo-Pacific and W Pacific, where they live on soft substrata. They feed mainly on small creatures and have a build that is typical of the

F. Length 15–30 cm. *Upeneus* are characterized by having a scaly soft D and An; the upper and lower jaws contain fine teeth. They are lively fish in the aquarium but usually they are not easy to look after. In a community tank they often go short on food. Nevertheless, *U. tragula* RICHARDS, 1845, is one species that is imported now and again.

Upside-down Catfish *(Synodontis nigriventris) see Synodontis*

Uranoscopidae, F, *see* Trachinoidei

Uranoscopus scaber *see* Trachinoidei

Urethane *see* Anaesthetics

Urinary Organs (fig. 1157). A system of organs which serve to expel water and salts (Osmoregulation*) as well as to render harmless toxic matter and metabolic end products (Excretion*). In vertebrates, the urinary

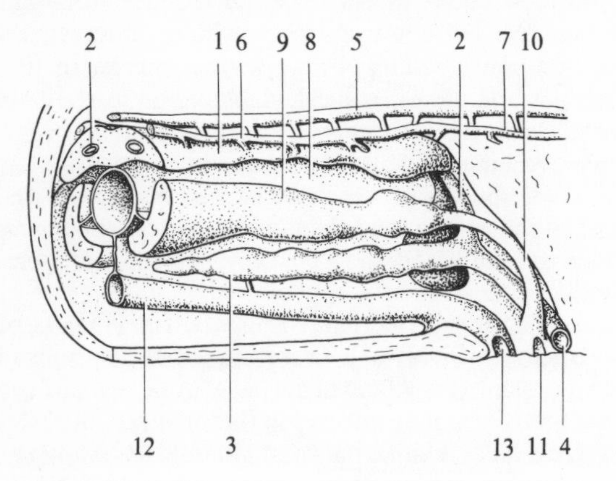

Fig. 1157 Location of the urinary and sex organs in a male Pike. 1 Kidney 2 Ureter 3 Urinary bladder 4 Urinary opening 5 Aorta dorsalis 6 Rear cardinal vein 7 Tail vein 8 Swimbladder 9 Testis 10 Spermatic duct 11 Genital aperture 12 Intestine 13 Anus

system is made up of the *kidneys* (the urine-forming organs), the *ureter* (the ducts which lead away from the kidneys), and the *urinary bladder* (where the urine is temporarily stored). Since the urinary organs have undergone changes in the phylogenetic history of vertebrates, there exist some differences in the structure and embryonic development of these organs in today's systematic groupings. For example, the Cyclostomes and some bony fish still have a segmented, fore-kidney (pronephros) in the anterior part of their bodies – this is an old feature in terms of evolutionary history. Usually, it becomes a lymphatic organ and only fulfils the function of a kidney in the early stages of development. The cartilaginous fish and most bony fish have two kinds of kidneys – the mesonephros and the metanephros, which together are called the opisthonephros. They lie in pairs just beneath the vertebral column, and are dark-red in colour and usually elongated in shape. They are connected with the sex organs only in cartilaginous fish. The ureters in bony fish open out behind the sexual openings and may be paired or unpaired. In some cases there is no urinary bladder.

These essential structural elements of all types of kidney function more or less in the same way. Blood plasma passes through the walls of fine arterial glomeruli ultrafiltered and protein-free, and flows as primary urine into the coils of kidney tubules. With kidneys of the pronephros, and more rarely of the mesonephros types, the fluid of the body cavity is also fed to the kidney tubules. Large amounts of water are removed from the primary urine as it passes through the various sections of the tubules. At the same time those substances needed by the body (eg glucose) are retrieved through reabsorption. It is mainly the end products of protein metabolism that are left – particularly urea in fish. The secondary urine passes form the kidney tubules into the ureter and the urinary bladder. A peculiarity of marine bony fish are kidneys without glomeruli or with very simple glomeruli. This is connected with the small amount of water that these fish expel (Osmoregulation*).

Utricularia LINNÉ: Bladderwort. G of the Lentibulariaceae*. 150 species. Found in tropical to temperate regions. Small perennials with elongate or truncated stem axes. Leaf arrangement alternate, opposite, whorled or in rosettes. No roots. Leaves come in a variety of shapes, eg peltate or their blades are kidney-shaped, linear, awl-shaped, pinnate, with snapping bladders. Flowers hermaphroditic, monosymmetrical. Perianth doubled. Calyx 2 part, corolla 5-part with 2 lips fused. Stamens 2. Carpels 2, fused, superior. Capsule-fruit. The species of the G grow mainly in moors, marshes or in water, but there are also some in other kinds of habitats (for example, there are epiphytes). The aquatic forms, which usually have only the inflorescences reaching up into the air, float in still waters or grow attached to the bottom (often fixed to rocks) in flowing waters. All Bladderworts are food-specialists: they use the ovate, roundish or lens-shaped snapping bladders found on the leaves to trap and digest small water animals. Because they do this, they also acquire for themselves an extra source of nitrogen, although they survive perfectly well without animal food. The bladders are sealed with a lid, and inside there is a weak vacuum. When prey animals touch the bristles at the opening to the bladder, the lid snaps shut inwards, and the victim is sucked in. Digestion takes place by means of glands that secrete proteolytic enzymes and at the same time aid resorption. Each bladder can take effect several times. So, for these reasons, the aquatic species of *Utricularia* make interesting biological objects for observation. But so far, only a few tropical and temperate species have been kept in the aquarium,

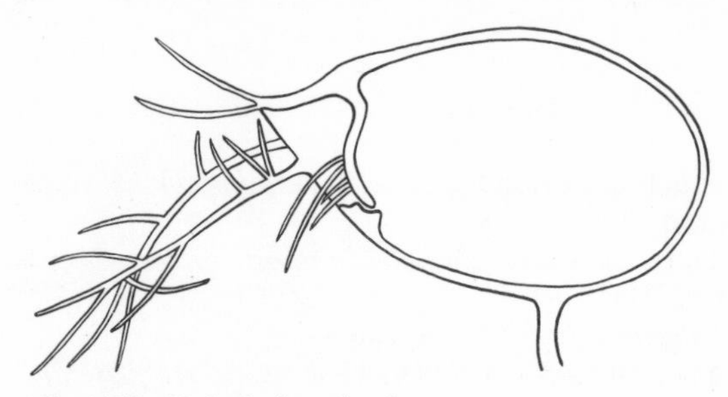

Fig. 1158 *Utricularia vulgaris*

with varying degrees of success. As a result, no reliable information can be given from aquarium observations. Well-known species are:

– *U. gibba* LINNÉ; Dwarf Bladderwort. Found in all tropical areas, and in eastern N America. Very small. Aquatic leaves simple with one snapping bladder. Suitable for heated aquaria.

– *U. vulgaris* LINNÉ; Greater Bladderwort. From the northern hemisphere. Large. Aquatic leaves very pinnate and with many snapping bladders. A cold water plant with winter buds (fig. 1158).

Vagile; having the ability to move about (as opposed to sessile*).

Valencia MYERS, 1928. G of the Cyprinodontidae*. One species found in S and E Spain (provinces of Catalonia and Valencia), on the island of Corfu (Greece) and probably also in Albania. *Valencia* lives in still waters near the coast. Particularly closely related to the G *Aphanius**. These fish have a pike's elongated shape and can measure up to 8 cm long; they are moderately flattened laterally but all in all they look very compact. D and An similarly shaped, rounded, positioned at the tail end. C fan-shaped and rounded. Caudal peduncle very short and tall. Mouth slightly superior.

Valencia is a good specimen for the less experienced aquarist because it makes no particular demands regarding temperature, water quality or food. The fish survive well in fairly large, well-planted tanks, stood in a sunny position, and are even simple to breed. In rooms with a normal temperature, it is possible to do without heating for the aquarium, and a cool overwintering (10–15 °C) increases the fish's readiness to reproduce. During the summer months, *Valencia* is also suitable for open-air ponds and such like. It is a very intolerant fish however, so enough hiding-places and refuges (plants, root-work etc) must be on hand. During the spawning period, the males drive hard, the spawning act itself taking place in clumps of water plants in the way typical of attached spawners (compare Cyprinodontidae). As the parent fish are likely to eat their own spawn, the eggs are best reared in separate tanks. Simply transfer the plants with the spawn attached to them. Roughly 2 weeks are needed for the eggs to mature, and if fed abundantly the brood grows very quickly.

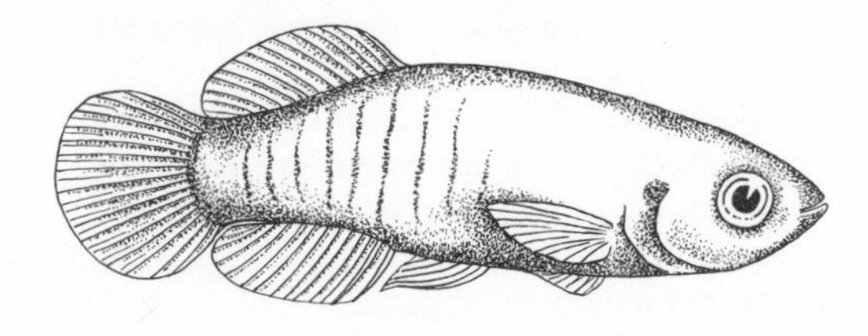

Fig. 1159 *Valencia hispanica*

– *V. hispanica* (CUVIER and VALENCIENNES, 1846); Spanish Tooth-carp. To 8 cm. Brownish with a metallic green shimmer and indistinct dark transverse bands. There is a clear dark shoulder fleck above the P. Edges of the scales darker than the bottoms of the scales. Unpaired fins yellow, with dark borders (fig. 1159).

Valencia Characin *(Gephyrocharax valencia) see Gephyrocharax*

Valenciennes, Achille (1794–1864). French ichthyologist. Pupil and colleague of CUVIER. In 1828, in Paris, the first two volumes of a work entitled *Histoire Naturelle des Poissons* ('Natural History of Fishes') was published as the joint effort of Cuvier and Valenciennes. After Cuvier's death in 1832, Valenciennes worked alone on the 9th to the 20th volume and the work in total still counts as one of the most genial and comprehensive books ever written on ichthyology.

Valentini Mimic *(Paraluteres prionurus) see Paraluteres*

Vallisneria LINNÉ, 1753; Vallisneria, Wild Celery, Eel Grass. G of the Hydrocharitaceae* with 8 species. Tropical to temperate, but only in certain areas. They are perennial plants with truncated stem axes and a great number of runners. Leaf arrangement in rosettes. Leaves long-linear, flat or twisting like a screw, green to reddish in colour. Flowers unisexual and dioecious, inconspicuous. Double perianth, free-standing leaves, 3-part. Calyx greenish or brownish. Corolla scaly. Male flower: stamens 1–2. Female flower: carpels 3, fused, inferior. Fruit like a berry. *Vallisneria* species are water plants with an interesting flower biology. The female flowers develop on their own and are enveloped in a translucent spathe; they also have long stalks that grow up to the water surface. The male flowers form in large numbers in inflorescences at the base of the plants, and are similarly encased by a spathe. They are gradually released and climb up to the surface. Once there, they throw off their sepals and then they can float along to the female flowers. They are less than 1 mm in size. In pollinated female flowers, the long flower stalk often rolls together in a spiral and pulls the ripening fruit beneath the water. All *Vallisneria* species like light, but in all other respects they are undemanding. Propagation can take place by separating off the plants that form from the runners. *V. spiralis* has been cultivated in the aquarium for about the longest period of all plants. Widespread species are:

– *V. gigantea* GRAEBNER, Giant Vallisneria. From SE Asia. Leaves broad and green.

– *V. neotropicalis* MARIE-VICTORIN; Red Giant Vallisneria. From south-eastern N America, Cuba. Leaves broad and reddish.

– *V. spiralis* LINNÉ; Tape Grass, Eel Grass. From S Europe. Leaves narrow, flat or spiralling (fig. 1160).

Vandellia-species, G, *see* Trichomycteridae

Variety. A group of organisms within a species that have hereditary features in common that are different from those of the basic form. But this cannot be attributed to any geographical distribution, and so a variety does not represent a geographical race or sub-species*. The variety, which is often developed to extremes in the course of artificial breeding (whence it is known as a variety, strain or cultivar), has no status within nomenclature and so the name it is given is not subject to the rules of nomenclature. In aquarium-keeping, particular varieties are striven for by means of breeding, particu-

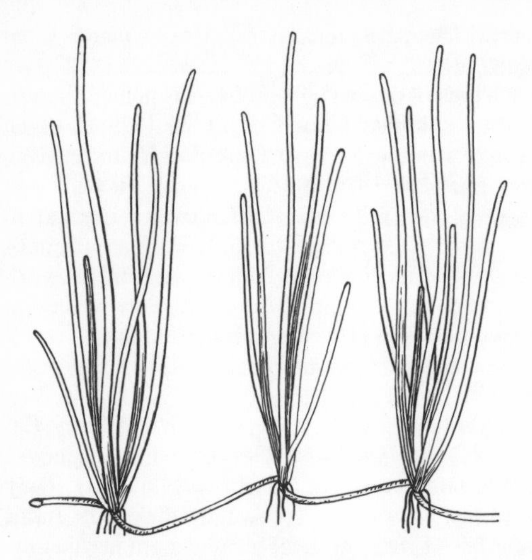

Fig. 1160 *Vallisneria spiralis*

larly among the Livebearing Tooth-carps (Poeciliidae*). Examples are the 'double sword' of the Swordtail and the numerous varieties of Platy *(Xiphophorus*).*

Varkenvis *(Abalistes stellaris) see Abalistes*

Vascular Bundle. Tissue systems found in pteridophytes (Pteridophyta*) and seed-plants (Spermatophyta*) that run through the stems, leaves and roots for the purpose of transporting matter. Their essential components are therefore connective tissue *(see* Tissue), but also included in their structure is ground tissue and supportive tissue, as well as meristematic tissue in places. In cross-section, vascular bundles are circular or elliptical. Water and minerals are transported by a portion of the bundle known as the xylem, whereas organic matter is transported through a sieve-like portion known as the phloem. Anything growing underwater will often have much simplified vascular bundles, the xylem sections being more reduced than the phloem.

Vaucheria. A G of the Yellow Algae (Xanthophyta*) which occur primarily in fresh waters. *Vaucheria* has fine, unbranched threads that are not divided into individual cells. In healthy marine aquaria, these Algae form thick mats, usually. But care must be taken that sessile animals of the shaded zone, such as Horn Corals (Gorgonaria*), and Red Algae are not smothered.

Vegetation. The covering of plants in a geographical area that is made up of the individual plant communities*. Essentially, vegetation is determined by climate* and results from the geographical latitude – on large land masses, distance from the sea and height of the land also play a part. Other influential factors include conditions in the ground and the history of the region. So it is easy to see how a complicated mosaic of vegetation builds up. In most parts of the world today, instead of the original vegetation you will normally find a secondary vegetation which has been profoundly influenced by man. Broadly speaking, the following climatic and vegetation zones can be picked out starting from the equator and going towards the poles:

– *Tropical Zone*. The equatorial regions stretching to the 10° north and 10° south latitudes are characterised by a warm, permanently damp climate. Annual fluctuations in temperature are slight, and seasons are scarcely distinguishable; every day there are stormy downpours. The highest amounts of rainfall occur when the sun is at its highest point. Typical of the tropical zone is the evergreen tropical rain forest. To the north and south lie regions of changeable tropical climate. The rain linked with the sun reaching its highest altitude is broken up by a dry period that can vary in length, but never lasts more than 8 months. The forest is a semi-evergreen seasonal forest, in which some species lose their foliage in the dry period; or, if the dry period is longer, the forest will be a rain-green, dry forest, in which all wood lands lie leafless during the dry period. Depending on the structure of the ground, the woodlands may be suppressed and grasses will take their place (various kinds of savannah lands).

– *Subtropical Zone*. In the damp parts of the subtropics in which the bulk of the rain falls in the summer months (influenced by the monsoons), the large-leaved laurel forest predominates. Temperatures are fairly even. In the winter-rain regions affected by cyclones, small-leaved and hard-leaved forests are typical, that are likewise evergreen. In these regions there are occasional frosts, but there is no real cold season. Summers are dry and hot. If the dry period lasts 8–10 months, spongy and low, rain-green, dry bushes and succulents are characteristic. Of the rainfall recedes further, scrubby semi-deserts and succulent semi-deserts appear and these gradually give way to true deserts where sometimes no rain at all falls for years. In the subtropical dry regions the annual and daily fluctuations in temperature (caused by the very slight air moisture and heat reflection) are very high. Night frosts are known.

– *Temperate Zone*. Characterised by cyclonal rain during all seasons, whereby the annual precipitation declines with distance from the sea. Towards the interior of the continents the summers also become hotter and the winters colder. Provided precipitation and air moisture reach a certain level and the winters are not too long or too cold, then summer-green deciduous forest is characteristic of the temperate zone. In the more continental areas, however, treeless steppes predominate. In the southern hemisphere there is a temperate zone only in S America, as the other continents do not reach far enough south.

– *Boreal Zone*. This zone too is marked by cyclonal rain in all seasons. The summers are cool and damp, and the winters over 6 months long – fairly mild in areas near the oceans, cold on the continents. Evergreen coniferous forests and moors are typical.

– *Arctic Zone*. Slight precipitation spread throughout the year, falling mostly as snow. The summers are short and cool and are characterised by unbroken daylight. The deep-frozen ground only thaws at the surface, and because it cannot flow away large marshy areas develop down below. The real vegetative period only lasts 3–4 months. The winters are long, fairly mild close to the oceans, extremely cold into the continent. Treeless tundra vegetation is typical, consisting of dwarf shrubs and low, herb-like plants.

Vegetative Cone (fig. 1161). The tip of the shoot or root in Ferns (Pteridophyta*) and Seed-plants (Spermatophyta*). It consists of meristematic tissue in which intensive cell division takes place during growth. In most ferns, cell divisions start from a single apical cell; in seed-plants from a group of equivalent initial cells. The shoot vegetative cone divides at its rearward part into the rudimentary leaves and lateral shoot buds. The young leaves grow quickly and envelope the vegetative cone protectively. The root vegetative cone is protected by a special root-cap.

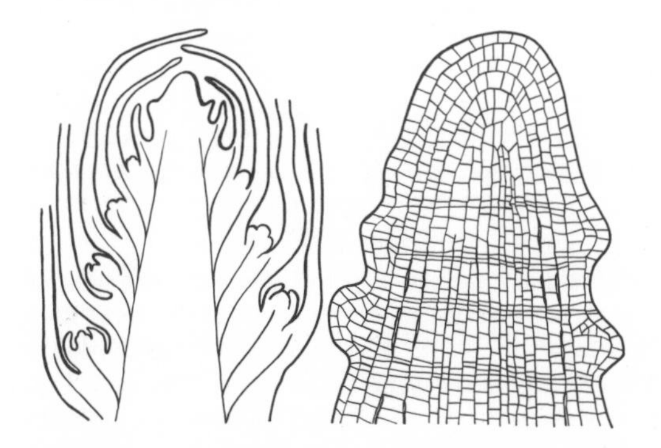

Fig. 1161 Vegetative cone

Vegetative Reproduction. An asexual form of reproduction*. Opposite: generative (sexual) reproduction.

Veil-tail *(Carassius auratus auratus) see Carassius*

Veil-tail (*Poecilia reticulata*, standard species) *see Poecilia*

Vein *see* Blood Vascular System

Velia LATREILLE, 1804; Water Cricket (fig. 1162). F of the Bugs (Heteroptera*); water-loving land Bugs. *Velia* has a very contrasty colour and is also able to move around with great skill on the water surface. These two factors make it a striking inhabitant of flowing waters. Even on mild winter days Water Crickets are sometimes active.

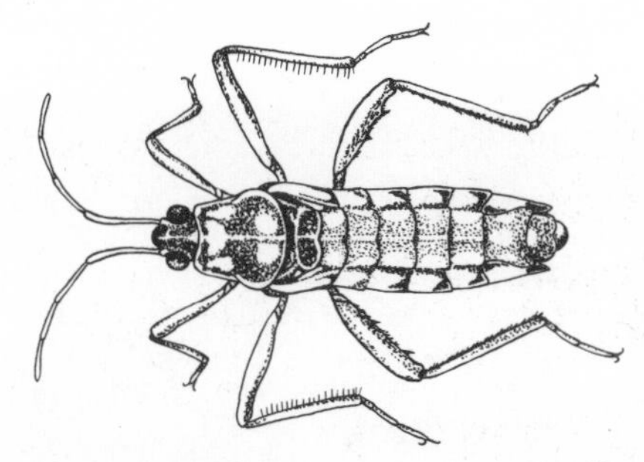

Fig. 1162 *Velia* sp.

Velvet Disease *see Oodinium*

Ventilation. Gases in the atmosphere are dissolved in water, depending on its temperature, pressure and salinity. Of great importance is oxygen, which is needed by most living things and which is produced under the influence of light by water plants during photosynthesis processes. Also important is carbon dioxide, which is released during respiration* in animals and plants. In an aquarium there exist completely different conditions from those found naturally. As a rule water in an aquarium tends to have a low redox potential which leads to a lack of oxygen and an excess of carbon dioxide. With a moderate amount of plants and animals in the tank, it is only possible in a few cases to maintain an equilibrium for the fish and plants and all their metabolic processes, without the need for ventilation. Firstly, introducing air into aquarium water produces a circulation* of water, by means of which excess carbon dioxide is driven out and oxygen from the air is taken in. Ventilation helps carry water from the bottom up to the surface where a rapid and unimpeded exchange can take place between the air and the aquarium water, which is rich in CO_2 and lacking in oxygen. The direct assimilation of oxygen by the introduced bubbles of air is slight as they do not remain long enough in the water for this purpose. The efficiency of a ventilation system is quite emphatically dependent on water quality. For example, organically overconcentrated water or water with a lot of metal ions has much reduced characteristics (Redox Potential*) and cannot be enriched with oxygen even with a lot of ventilation. To get the ventilation working, air is compressed by electric pumps*. For smaller aquaria diaphragm pumps are sufficient, for larger tanks piston pumps or compressors are necessary. It is important with these types of pumps to clean the air of traces of oil (*see* Filtration of Oil). Airflow in piston pumps is generally in regular pulses – to smooth out these pulses an air reservoir tank is used which results in smooth continuous air pressure. In the event of a stoppage or failure by the pump, to stop the aquarium water siphoning back through the air tubing, the pumps must stand above the surface of the water. Adequate non-return valves must be built into the airline. To distribute the air into individual tanks, ducts made out of metal, plastic, glass or rubber are used. The flow is regulated by hose-clips or stop-cocks. The air comes out under the water via a diffuser block; porous materials made of clay, wood etc, the size of the pores determining how big or small the air bubbles will be. All materials placed in water should be inert, capable of withstanding sea water if necessary (*see* Raw Materials). The diffusers (airstones) are always placed directly above the substrate, in order to achieve a complete circulation of water and give the maximum length for the air bubbles to travel through. If air in filtration systems is used as a means of moving the water, efficiency is increased manyfold by the use of an airstone. Filters (Filtration*), foam filters* or ventilation systems which work with ozone-enriched air must be made of ozone resistant material.

Ventral; towards the belly. From the Latin *venter*, meaning stomach. Term used in anatomy* in order to describe the position of parts of the body. Opposite: dorsal*.

Ventral Display Swimming *see* Courtship Swimming

Ventral Fin *see* Fins

Venus Fan (*Rhipidogorgia*, G) *see* Gorgonaria

Venus's Girdle (*Cestus*, G) *see* Ctenophora

Vermetus BRUGUIÈRE, 1792. G of the Gastropoda*, Prosobranchia*. Found in fairly warm seas. *Vermetus* is a sessile plankton-eater. They are related to the Turret Snails (eg *Cerithium**), but the shell, which grows on firm substrata, is not coiled (apart from the beginning part which is not normally visible), and it is thus reminiscent of worm tubes, eg that of *Protula**. *Vermetus* catches its prey with a net of mucus: it produces fine mucus threads with its foot gland, and all sorts of small animals become entangled in them. After a while, the net and all its contents are eaten. *Vermetus* species, eg *V. arenarius* (LINNAEUS, 1758) from the Mediterranean, survive well in the aquarium for years. If small crustaceans are introduced or if the sides of the aquarium are cleaned, *Vermetus* reacts by building a mucus net.

Vermiculated Rabbitfish or Spinefoot *(Siganus vermiculatus) see* Acanthuroidei

Vermiculated Triggerfish *(Balistapus undulatus) see Balistapus*

Veronica, G, *see* Scrophulariaceae

Vertebral Column *see* Skeleton

Vertebrata; Vertebrates. Term used in comparative anatomy and physiology. All Chordata* are defined as vertebrates if they have rudimentary vertebrae or true vertebral bodies. So all the members of the Sub-Pha Agnatha* and Gnathostomata* are regarded as vertebrates. In some extant Agnatha (Petromyzoniformes) only tiny remains of vertebral bodies still exist well to the front behind the skull–the specimens from the Cambrian and Silurian periods sometimes had very well-developed vertebrae. In some primitive Gnathostomata too, the vertebrae are developed only in the form of their processes; in some deep-sea fish they have virtually disappeared as a secondary development phase. The individual vertebrae together form the vertebral column (*see* Skeleton). Often, all other animal Pha, including the invertebrate Chordata (Tunicata*, Acrania*), are described as 'invertebrates' as opposed to the vertebrates.

Vertical Fighting *see* Symbolic Fighting Behaviour

Verticillate *see* Leaf Arrangement

Vesicularia MÜLLER, 1896; Java Moss. G of the Bryophyta*. 100 species. Tropical-subtropical. These mosses are very branching plants with densely packed leaflets. Reproduction through spores. *Vesicularia* species colonise a wide variety of habitats. They grow on the ground, on rocks, on trees, and some flourish in a partial or permanent state of submersion in flowing and still waters. The following species is a popular aquarium plant, as it is very decorative under subdued lighting growing on wood or on rocks:

– *V. dubyana* (MÜLLER) BROTHERUS; Common Java Moss. From the islands of southern Asia.

Vibrio anguillarum *see* Spot Diseases

Vicariance. The mutual occurrence of closely related animal or plant forms in neighbouring geographical realms. The Crayfish Ga *Astacus** (Europe) and *Cambarus** (N America) are vicariads, for example. All geographical sub-species (Concept of Species*) live vicariously with one another, as they inhabit neighbouring geographical realms and do not overlap in area of distribution (Areal*) (Allopatry*).

Victoria LINDLEY, 1837; Queen Victoria Water-lily. G of the Nymphaeaceae*. 2 species. From S America. These are rhizome perennials with very large floating leaves arranged in rosettes. These floating leaves have a characteristic upturned edge which is missing in the leaves of young plants. Flowers hermaphroditic, large, polysymmetrical, with 4 sepals and many petals. Many stamens and many carpels which are fused and inferior. Fruit like a berry and green. The species of *Victoria* are found in calm inlets and in old waters of large rivers. They are cultivated in every large botanical garden and always attract the visitor. They are cared for in specially built Victoria-houses with large tanks of water; in warmer places, they may also be kept in open-air ponds which are partly heated. In their natural habitats, *Victoria* species are hardy perennials, but in Europe and N America they are usually cultivated as annuals, because lack of light in the winter season is bad for them and besides the Victoria-houses usually have to be used for other purposes during the winter. The seeds are sown in February. The first leaves are small and submerged; they have arrow-shaped leaves. But, soon afterwards, simple floating leaves appear. As the plants gain in strength, they are planted in position, at the very latest at the beginning of May, in very nutrient-rich earth. Growth is now very rapid, and by July the plants are capable of flowering. The leaves of fairly old specimens, which may reach up to a diameter of 2 m, have strong, forward-springing nerves on the underside. If loaded evenly they are able to carry a weight of up to 50 kg as a result. Water that happens to get onto the upperside can flow away through tiny pores in the leaf-blades. *Victoria* flowers at night. The individual flowers which follow very quickly after one another in the summer open twice. The first night they are a shining white, the second night pink. Then they dive beneath the water and the seeds mature here.

– *V. amazonica* (PÖPPIG) SOWERBY (*V. regia*, name not in current usage). From the river system of the Amazon. Leaf edges up to 10 cm. Sepals spiny.

Fig. 1163 *Victoria cruziana*

– *V. cruziana* D'ORBIGNY. From the river systems of the Paraná and the Paraguay. Leaf edges to 15 cm. Sepals bare (fig. 1163).

Vinciguerria lucetia *see* Stomiatoidei

Vinegar-fly, G, *see Drosophila*

Violet Sea Urchin *(Sphaerechinus granularis) see Sphaerechinus*

Viral Infections (fig. 1164). Viruses are non-cellular living beings that can only reproduce in the cells of bacteria, plants or animals. Very little knowledge about viruses in fish is certain. Epidemics feared in fish economy, including septicaemia, pancreas-necrosis, infectious swellings of the kidneys, are probably all caused by viruses. Numerous tumour formations may also have their cause in some kind of virus. In aquarium fish the only definite virus known is lymphocystis*.

Viruses *see* Viral Infections

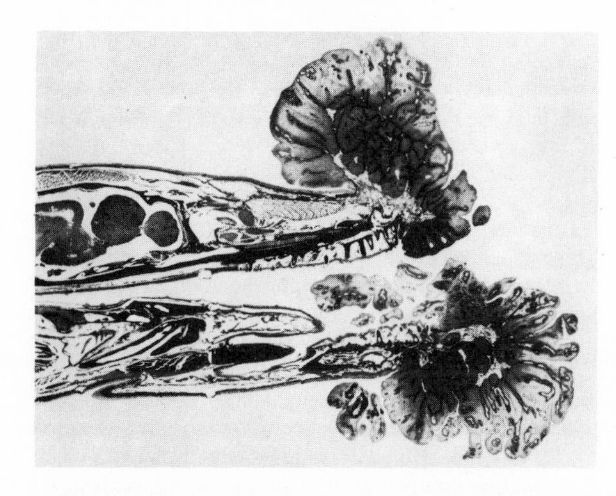

Fig. 1164 Viral infections. Cauliflower abscesses on the edges of the lips on an Eel

Viscera, Fatty Degeneration of, *see* Degeneration

Visceral Examination *see* Examination Technique

Vitamin Deficiency *see* Deficiency Diseases

Vitamins. Organic substances that are important for life. Animals and humans are either incapable of producing any vitamins for themselves or they produce them in insufficient quantities. So, vitamins or their direct precursors (provitamins) must be introduced along with food. The original source of vitamins are the plants, so a vitamin food chain (Nutrition*) is also in operation. Plant-eaters store vitamins in the body and thus also provide predatory species with their necessary amounts of vitamins. The daily requirement only amounts to micro and milligrams. Lack of vitamins can have serious consequences (avitaminosis). In youthful organisms this can be seen in retarded growth: in adults, deficiency diseases* set in.

Most vitamins are built into enzymes* that regulate metabolic processes. Vitamin A aids growth and is in particular a component part of the visual purple of the retina; the vitamins of the B group have, among other things, functions in energy and proteometabolism; Vitamin D helps the reception of calcium from the food and thus helps to firm up the bones; Vitamin E is an antisterility factor; Vitamin H is mainly a participant in fat metabolism and Vitamin K is important in the production of clotting proteins of the blood. Vitamin C, which works in conjunction with various processes (eg the breakdown of certain proteins or the formation of adrenal hormones) is produced by most animals themselves. Humans, however, have to get their Vitamin C from their food. The chemical and physical properties of vitamins are not uniform. Some are water-soluble (eg B_1, B_2, B_6, B_{12}, C), others are fat-soluble (eg A_1, A_2, D, E, K). Vitamins A and D therefore are found especially in fat tissue and in organs that contain fat, eg a fish's liver (which is why cod-liver oil is important in medicine). There is hardly any information at all on the special vitamin needs of aquarium fish. But care should be taken to ensure that the complete spectrum of vitamins is supplied by providing a good variety of foods. DE GRAAF identified the following fish foods as the most important sources of vitamins:

– *Vitamin A and provitamin carotin,* found in crustacean-types, arthropods, egg-yolk, algae, lettuce, spinach, water plants, cow's liver and fish liver.

– *Vitamin* B_1 found in algae, especially Bacillariophyta*, lettuce, spinach, yeast, egg-yolk, cow's heart and fish meat, bivalves.

– *Vitamins* B_2 *and* B_6, found in crustacean-types, beef, cow's liver, fish meat, bivalves, chicken eggs, spinach, lettuce, yeast.

– *Vitamin* B_{12}, *pantothene acid and nicotine acid,* found in green algae, red seaweeds, brown seaweeds, lettuce, bivalves, yeast, beef, cow's liver, egg-yolk.

– *Vitamin D and provitamin,* found in Earthworms, Mealworms, *Tubifex,* egg-yolk, snails, fish liver, Water-fleas, Shrimps.

– *Vitamin E,* found in green algae, lettuce, spinach, egg-yolk.

– *Vitamin H* (biotin), found in yeast, cow's liver, egg-yolk.

– *Vitamin K,* found in cow's liver, lettuce, spinach, Water-fleas.

Viviparity *see* Reproduction

Viviparous Blenny (Zoarcidae, F), *see* Gadiformes

Viviparus MONTFORT, 1810; River Snail (fig. 1165). G of the Prosobranchs (Prosobranchia*). There are about 6 species in Europe, some of which are found in the Palaearctic* realm. The shell is a roundish cone-shape up to 4 cm tall and 3 cm across. It is covered with brown bands and there is a closeable lid. River Snails are interesting objects of study in the pond-aquarium.

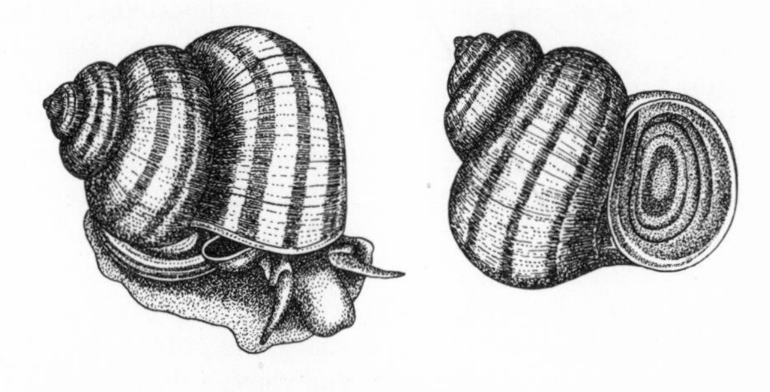

Fig. 1165 *Viviparus.* Right, closed shell

Wagtail Platy *(Xiphophorus maculatus) see Xiphophorus*

Wagtail Sword(tail) *(Xiphophorus helleri helleri) see Xiphophorus*

Wash-bottle *see* Air Filtration

Water. Chemical combination of hydrogen (symbol H) and oxygen (symbol O) in the atomic ratio 2:1 and with a formula H_2O. In relation to the temperature, several water molecules come together into complex structures known as aggregates; it is this that causes water to be not easily volatised. Aggregates form because of a certain displacement of the electrical charge within the water molecule. This also explains why water is able to dissolve many bases*, acids* and salts so well. Nothing can live unless it is in the presence of water, be it on a molecular, cellular or organismal plane. At the same time water is also a habitat for an unbelievably large number of organisms, be it fresh, brackish or sea water. Every natural water has a certain electrolyte* content (salt content, Alkalinity*, Acidity*) which, among other things, is reflected in the conductivity. Chemically pure water has a damaging effect on almost all organisms, a fact that must be borne in mind when preparing water (Water Preparation*). It is interesting to note that in almost all natural waters the composition of the dissolved electrolytes, especially of the salts, is fairly homogeneous, ie in most cases, the relations of the types of ions* to one another are similar (standard ion-combination, *see* Standard Water). In the aquarium, however, certain imbalanced accumulations of individual types of ion occur and this can lead to injuries to the organisms. This is true of the nitrogen compounds, for example (Ammonia*, Nitrite Content*, Nitrate Content*).

Water Analysis. Water analysis should give information about the quality of the water, especially its content of particular electrolytes (Electrolyte Content*). For aquarium-keeping purposes approximate values will do, and these can be arrived at using chemicals and pieces of equipment that cost relatively little. If water analysis is to be successful, it is important to have some basic chemical knowledge; never carry out any procedures according to 'recipes' the meaning of which you do not understand. In practical aquarium-keeping, just a few chemical data are enough to give you an estimate of the water* in front of you and likewise to alter it with a particular aim in mind (Water Preparation*). Apart from constantly checking temperature, colour, smell and cloudiness, the ascertainment of pH value* and water hardness* is very important. By determining the alkalinity*, the carbonate content is arrived at and this gives you the ratio of temporary to permanent hardness*. Conductivity is a measure of the total content of electrolytes*, and, finally, individual ion concentrations can be ascertained quantitively (eg ammonium, nitrate, chloride, calcium and magnesium). Some special procedures do involve a lengthy methodical approach. From the consumption of permanganate* the aquarist can obtain a general view of the content of oxidisable substances, and these point to how much organic decomposable material there is in the water (which gets into the water mainly through feeding and from animal excreta). Increasingly important in aquarium-keeping is the determination of the rH value* which provides information about the ratio of oxidisable substances to reduceable ones, and thus is a measure of the redox potential*.

Water Beetle *(Dytiscus marginalis) see Dytiscus*

Water Beetles, F, *see* Dytiscidae

Water Bloom is a cloudiness of water, usually in varying shades of green. It is caused by the proliferation of single-celled, free-moving algae*.

Water Boatmen, popular name for *Notonecta**

Water Boatmen, G, *see Corixa*

Water, for Breeding Purposes. Usually very soft (1–4 °dH) and slightly acidic (pH value somewhere between 6.0 and 6.8) *natural* or *artificial water,* often enriched with peat extracts. The composition of the water should correspond as nearly as possible to the native waters of the breeding fish concerned in order to ensure that a high percentage of the spawn will hatch out. It is not possible everywhere to create such water in the form of spring, stream or tap water, or to have it available. So the aquarist will have to dilute very hard water by adding distilled water – 1 litre of tap water with a hardness level of 10 °dH mixed with 1 litre of distilled water will give 2 litres of water at 5 °dH. In recent years, removing the hardness of water chemically has found increasing use in aquarium-keeping; for fish-breeding purposes, partly or fully desalinated water is suitable. However, care must always to taken to ensure that in both cases some of the original water is put back in (depending on its hardness in proportion to the breeding water desired). But, as it is expensive to make or buy desalinating equipment, it is best to obtain such water from industrial manufacturers. In the correct mixtures, these manufactured waters have proved just as valuable as natural waters. Good results with breeding problem fish (eg *Paracheirodon* innesi*) have been achieved with clean *rainwater.* But, because of the high content of sulphurous acids, H_2SO_3, it is important not to catch rain from town areas, or if you do, it must be left standing for several weeks. As the high oxygen potential of *fresh water* stimulates the reproductive drive of most fish, it is beneficial to use the breeding water just a few hours after it has been set up. Only a few fish species, eg all Atherinidae (Atherinoidei*) require hard or very hard water for successful breeding and care.

Water Bugs, G, *see Naucoris*

Water, Changing of. A partial change of water in the aquarium is one of the simplest and surest methods of ensuring that poisonous metabolic products do not build up and that substances important for life (especially trace elements) remain present in sufficient quantities. Depending on the size of the aquarium, how many fish it contains and the capacity of the filter, every 15–20 days between $^1/_5$ and $^1/_4$ of the water in the tank should be sucked out and replaced by fresh water at the correct temperature. This partial water exchange is carried out most easily as a part of the aquarium care* routine (eg when sucking out mould from the substrate). In a breeding tank*, it is often necessary to change part of the water a short while after spawning has taken place. In rearing tanks, if fry are to develop healthily

(Fry, Rearing of*), a regular partial change of water, every 1–2 days, is necessary.

Water Chestnuts, F, *see* Trapaceae

Water Chestnuts, G, *see Trapa*

Water Circulation (fig. 1166). A continual circulation of the aquarium water, ie the rotating of water at the surface and in the depths, including the water system between the particles of the substrate (Filtration*), is essential if a biological equilibrium is to be maintained in the aquarium. Water circulation helps promote an even temperature (Temperature Control*), as a heating or cooling system usually operates from small pieces of equipment with a localised effect. Secondly, it helps maintain an even distribution of oxygen and carbon

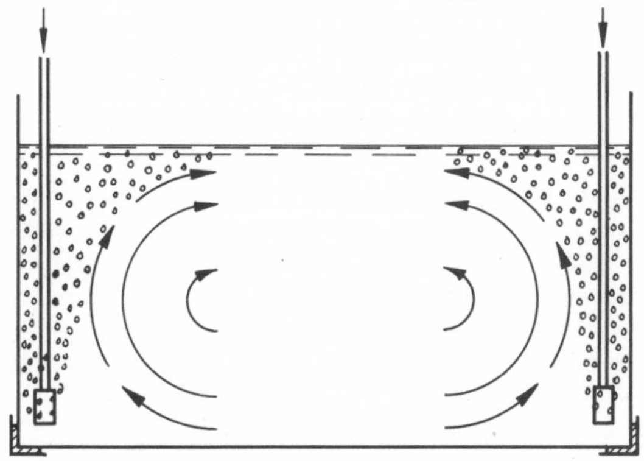

Fig. 1166 Water circulation

dioxide* as well as promoting their unimpeded exchange at the water surface (Ventilation*). In the third place, water circulation is necessary to bring nutrients continually towards plants and animals that obtain them directly from their surrounding water, as well as to take away from them breakdown products that have been released. Algae, Coelenterata* and other more or less sessile organisms thrive in areas of strong water circulation or they actively seek out places with marked water movement near diffusers or near the entry and exit points of filters. It is important to include the water system of the substrate with its countless micro-organisms in the water circulation, as this will largely prevent the formation of still water zones with accumulating poisonous metabolic products. A good water circulation is particularly important for the sea water aquarium, as the animals kept here mostly come from the coastal zone where there is strong water movement. To produce an adequate water circulation in the simplest way, set up a ventilation system with the diffuser on the substrate if possible and on the sides of the aquarium (this will ensure full circulation). In the same way, the intake and exit tubes of a filter system can be arranged similarly, and this too will provide an efficient water circulation (Filtration*). Circulatory pumps* work best of all; these actively suck out water from the aquarium and pass it through appropriate filter systems, before it is returned under pressure to the tank. To achieve good gaseous exchange, the returning water can be sprayed over the surface from a radiating tube or extra air can be sucked in through nozzles. If the substrate is included in the water circulation, the returning water is led back into the tank via a water chamber beneath the substrate.

Water Circulation Pump *see* Pumps

Water Cress *(Rorippa aquatica) see Rorippa*

Water Crickets, G, *see Velia*

Water Crowfoot *(Ranunculus aquatilis) see* Ranunculaceae

Water Fern, G, *see Bolbitis*

Water Ferns (Parkeriaceae, F) *see* Pteridophyta

Water Ferns, Sub-Cl, *see* Pteridophyta

Water Flea, G, *see Daphnia*

Water Fleas, tribe, *see* Cladocera

Water Goldfinch *(Pristella riddlei) see Pristella*

Water Hardness. The term 'water hardness' has been coined because it has been noticed that some natural waters feel soft and slippery to the touch when using soap whereas others feel hard. This difference is caused by the different amounts of calcium and magnesium salts in the water, the so-called hardness-forming substances. Hard water contains many of these salts, soft water only a few. It is interesting to note that most tropical region waters contain small amounts of electrolytes* and as a result they also have very little hardness-forming substances. This is particularly true of dystrophic waters* (black water rivers, etc). Famous exceptions, such as the East African drainage lakes (Lakes Tanganyika* and Malawi*) have been referred to. So, it can be concluded that many inhabitants of tropical waters are very definitely bound to soft water – this is especially true of the so-called problem fish and many water plants. If care and propagation are to be successful in the aquarium, therefore, the water has to be prepared (Water Preparation*) in a lot of cases. The hardness of the water is made up of various factors. By temporary hardness* we mean that part of the water that can be removed by boiling – ie, the magnesium and calcium bound to hydrogen carbonate or carbonate. The remainder (usually by far the greater part) is called permanent hardness*. This, together with the temporary hardness, produces the total hardness. In order not to have to take into account the different kinds of anions (Ions*), reference is always made to the calcium or magnesium oxide (CaO or MgO); the Ca^{++} or Mg^{++} values that are ascertained are also converted into CaO or MgO. In this country we measure in degrees of German (total-) hardness (°dH) whereby 1 dH is defined as the content of 10 mg CaO in 1 litre of water. There is an equivalent value of 7.19 mg per litre for the lighter magnesium. If you wish to convert the MgO into CaO, the MgO content must therefore by multiplied by a factor of 1.4.

As a general rule, the following hardness values are classified thus:

0–5° dH very soft
5–10 °dH soft
10–20 ° medium-hard
20–30 °dH hard
over 30 °dH very hard

The hardness of water (Water Hardness, Determination of*) is determined chemically and can even be carried out with an adequate degree of accuracy by the layman. A bit more knowledge is needed to get rid of hardness-forming substances. The most elegant method is using ion exchangers* (Desalination*, Partial Desalination*, Neutral Exchange*), and many aquarists have already mastered its theory and application, especially as this method also regulates the salt proportions (which does not have any effect on the water hardness itself).

Water Hardness, Determination of. 1 Determination of *total hardness* (Water Hardness*) with soap solutions. The method relies on the fact that hardness-forming substances (calcium and magnesium ions) form insoluble precipitations with the palmitate anion of the soap. Only after a sufficient amount has been precipitated out will lasting soap bubbles be formed. So, the amount of soap solution needed to form the soap bubbles is directly proportional to the total hardness. It is best to buy everything you need (stoppered bottle, suitable soap solution, special burette) from a shop that deals in such apparatus. To measure the water hardness, fill the bottle with the water to be tested up to a particular calibration, then add the soap solution drop by drop from the burette. As soon as lasting bubbles form when you shake the bottle vigorously (ie no crackling as the bubbles burst!), read off from the burette how much soap solution has been used up. This equals the water hardness. It should be noted that this method only gives sufficiently accurate readings up to a total hardness value of 12. When testing harder water, the sample should be diluted with a known volume of distilled water *before* the test is conducted, and the resulting hardness figure multiplied by the dilution factor to give the true hardness figure (eg a 1:1 dilution produces a multiplication factor of 2; 1:3 dilution produces a multiplication factor of 4, etc).

2) Determination of *temporary hardness**. This can be calculated by multiplying the alkalinity* value by 2.8.

3) Determination of *permanent hardness**. This can be calculated from the difference between the total hardness and the temporary hardness.

More sophisticated methods, such as titration methods to determine calcium and magnesium content with Chelaplex* are not used in aquarium-keeping because of the prohibitive cost of the chemicals. In most cases, such methods are totally unnecessary anyway.

Water Hemlock *(Cicuta virosa) see* Apiaceae

Water Houseleek *(Crassula aquatica) see Crassula*

Water Hyacinths, F, *see* Pontederiaceae

Water Hyacinths, G, *see Eichhornia*

Water Hyssop (several species) *see Bacopa*

Water Lettuce *(Pistia stratiotes) see Pistia*

Water Measurer, G, *see Hydrometra*

Water Milfoils, F, *see* Haloragaceae

Water Mites, various Fs, *see* Acari

Water Orchid *(Spiranthes cernua) see* Orchidaceae

Water Pennyworts, G, *see Hydrocotyle*

Water Pimpernel *(Samolus valerandi) see Samolus*

Water Plant or Aquatic Plant. A water plant grows permanently or primarily in a submerged state or it has floating leaves. Many water plants can also exist for shorter or longer periods on land, but they are never at their best here. *See also* Marsh Plant.

Water (Aquatic) Plant Horticulture. A specialised industry the aim of which is different depending on geographical location. In tropical and subtropical areas, this form of horticulture is concerned with the collection of aquarium plants in the wild or acquiring them from endemic collectors. The plants are then prepared for export, although sometimes certain species are propagated in open-air cultures. Aquatic plant horticulture in temperate regions is concerned with importing aquarium plants and getting them ready for the wholesale and retail trades. But a fairly large amount is also propagated for the inland market and for export. For this, culture tanks are set up in heated greenhouses, and during the winter months artificial lighting may also have to be installed. Some of the activities of water plant horticulture are linked with the collection, propagation and trade in other plant groups (eg Orchids), or with those of aquarium and terrarium animals, so this leads to broader-based industries developing. Modern horticultural plants are often connected up to thermal springs or to power stations.

Water Plantain *(Alisma plantago-aquatica) see Alisma*

Water Plantain Family *see* Alismataceae

Water Plants, Transportation of. Water Plants are usually transported without water, regardless of the distance to be covered or the length of time it will take. The plants are either placed directly in plastic bottles or wrapped in damp paper first. In exceptional cases, water plants are transported in water-filled containers.

Water Pollination *see* Pollination

Water Preparation. Raw water, particularly tap water, has to be prepared for aquarium-keeping purposes in many areas. The aim in mind with water preparation is the removal of any chlorine* mixed in with it, as well as the altering of the salt content (Electrolyte Content*). Lengthy, vigorous ventilation drives out chlorine and carbonic acid. Unwanted salts (Water Hardness*) are best removed by ion exchange (Ion Exchangers*, Partial Desalination*, Desalination*, Neutral Exchange*). Finally, under certain circumstances, the pH value* must be corrected, and if necessary the water should also be enriched with tannins (Tannic Substances*) and humic acids*, as well as chelatores*. All these forms of water preparation must be repeated from time to time because the composition of water in an aquarium can alter quickly, eg through evaporation, food animals dying and excretion products.

Before any water is prepared, a good water analysis* must be carried out and the aquarist must also have in mind a very clear idea of the water quality he wants to achieve. It must also be said, however, that even before there was a great deal of knowledge about water chemistry, many people were perfectly successful with caring for and breeding aquarium animals. But it is true to say that most successes were with freshwater aquaria, rather less so with sea water aquaria. If no suitable water was available, in many places people simply helped themselves to spring water and rainwater. Today's approach of preparing water chemically, however, is beneficial for 2 reasons: firstly, it makes it much easier to prepare fairly large amounts of water, and secondly, the nature of the water can be altered very specifically. If water is prepared using ion exchangers, it is important to remember that a particular ratio of ions* must be retained (standard ion-combination, *see* Standard Water) and that a certain minimum salt content is not gone below. So the lowest level of hardness should be about 2 °dH. These conditions can either be achieved by adding in raw water or standard water* is produced using chemical procedures.

Water Purslane, G, *see Peplis*

Water Renewal see Aquarium, Installation of, Aquarium, Care of

Water Scorpion *(Nepa cinerca) see Nepa*

Water Shield *(Brasenia schreberi) see Brasenia*

Water Soldier *(Stratiotes aloides) see Stratiotes*

Water Spiders, G, *see Argyroneta*

Water Sprite *(Ceratopteris thalictroides) see Ceratopteris*

Water Sprites, G, *see Ceratopteris*

Water Starwort *(Callitriche vernalis) see Callitriche*

Water Starworts, F, *see* Callitrichaceae

Water Starworts, G, *see Callitriche*

Water Stick Insect, G, *see Ranatra*

Water Trumpet, G, *see Cryptocoryne*

Water Vascular System *see* Echinodermata

'Water Wisteria' *(Synnema triflorum) see Synnema*

Water-lilies, F, *see* Nymphaeaceae

Water-lily, G, *see* Nymphaea

Watermeal *(Wolffia columbiana) see Wolffia*

Water-pepper, G, *see Elatine*

Water-peppers, F, *see* Elatinaceae

Wax Moths. Small Butterflies with a worldwide distribution, the larvae of which often live as parasites in beehives. Occasionally they are used as a fish food. The best-known species are: the Large Wax Moth *(Galleria mellonella)* and the Small Wax Moth *(Achroia grisella).*

Weatherfish *(Misgurnus fossilis) see Misgurnus*

Weatherfishes (collective term) *see Misgurnus*

Weberian Apparatus (fig. 1167). A characteristic apparatus for the Sup-O Ostariophysi of the Osteichthyes*, which transmits changes in pressure detected by the swimbladder* or sound waves to the outside fluid space of the inner ear. The Weberian apparatus consists of 3–4 pairs of bones (stapes, claustrum, incus, malleus) lying one behind the other (they stem from vertebrae). The last one of these bones, the malleus, is in contact with the swimbladder. (*See also* Cypriniformes and Siluriformes.)

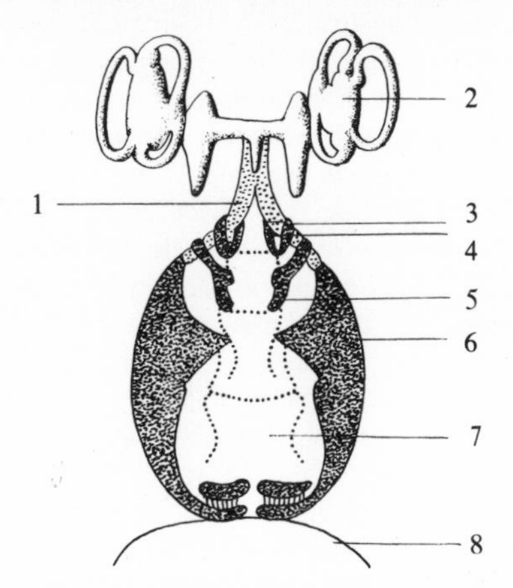

Fig. 1167 Weberian apparatus. 1 Sinus impar 2 Labyrinth 3 Claustrum 4 Stapes 5 Incus 6 Malleus 7 Vertebral body 8 Swimbladder

Wedgetailed Blue Tang *(Paracanthurus hepatus) see Paracanthurus*

Weeverfish, related species, Sub-O, *see* Trachinoidei

Weeverfishes (Trachinidae, F) *see* Trachinoidei

Wels *(Silurus glanis) see Silurus*

Wendt, Albert (1887–1958). German aquarist. He was acknowledged as one of the greatest experts on endemic and tropical aquatic plants and contributed greatly to the spread of aquarium-keeping with the publication of numerous works. His book *Aquarienpflanzen in Wort und Bild* ('Aquarium Plants in Words and Pictures') was the leading work of its day on this subject.

Western Mosquito *(Gambusia affinis) see Gambusia*

Whale Sharks, F, *see* Rhiniodontidae

Whalefish-types, O, *see* Cetomimiformes

Whalefishes (Cetomimidae, F) *see* Cetomimiformes

Whales, O, *see* Cetacea

Whiptail Catfish *(Loricaria filamentosa) see Loricaria*

Whirligig Beetles, F, *see* Gyrinidae

Whirling Disease, Tumbling Disease, *see* Sporozoa

White Bream *(Blicca bjoerkna) see Blicca*

White Cloud Mountain Minnow *(Tanichthys albonubes) see Tanichthys*

White Piranha *(Serrasalmus rhombeus) see Serrasalmus*

White Seaperch *(Phanerodon furcatus) see* Embiotocidae

White Shark *(Carcharodon carcharias) see* Lamnidae

White Water-lily *(Nymphaea alba) see Nymphaea*

White-barred Triggerfish *(Rhinecanthus aculeatus) see Rhinecanthus*

White-breasted Surgeonfish *(Acanthurus leucosternon) see Acanthurus*

White-cheeked Surgeonfish *(Acanthurus glaucopareius) see Acanthurus*

White-fishes (*Coregonus*, G) *see* Salmonoidei

White-fishes (Leuciscinae, Sub-F) *see* Cyprinidae

White's Cynolebias *(Cynolebias whitei) see Cynolebias*

White's Pearlfish *(Cynolebias whitei) see Cynolebias*

White-spot Disease *see Ichthyophthirius*

White-tailed Damsel *(Dascyllus aruanus) see Dascyllus*

White-tailed Footballer *(Dascyllus aruanus) see Dascyllus*

White-tipped Sea-urchin *(Sphaerechinus granularis) see Sphaerechinus*

Whiteworm *(Enchytraeus buchholtzi) see Enchytraeus*

Whiting *(Merlangus merlangus) see* Gadiformes

Whorled *see* Leaf Arrangement

Whorled Pennywort *(Hydrocotyle verticillata) see Hydrocotyle*

Whorled Water Milfoil *(Myriophyllum verticillatum) see Myriophyllum*

Whorled Waterwort *(Elatine alsinastrum) see Elatine*

Wide-mouths, related species, Sub-O, *see* Stomiatoidei

Wiesbaden Swordtail *(Xiphophorus helleri helleri) see Xiphophorus*

Wijnberg's Rasbora *(Rasbora wijnbergi) see Rasbora*

Wilting is a drooping of plant parts or of the whole plant, which can result in the plant dying. Plants wilt because there is a reduction in the turgor*, and so it occurs when water is released at a greater rate than it is received.

Wimple Piranha *(Catoprion mento) see Catoprion*

Wimplefish *(Heniochus acuminatus) see Heniochus*

Wind Pollination *see* Pollination

Winged Piranha *(Catoprion mento) see Catoprion*

Winter Bud. A form of overwintering found among many plants that come from temperate and cold regions. It consists of a stunted stem axis which is packed tight with cataphyllary leaves or, at least in comparison with the summer leaves, with differently shaped foliage leaves. These are enriched with reserve substances. Winter buds develop from the tips of stems, from lateral buds or from the tips of runners. They are found among the Ga *Elodea**, *Hydrocharis**, *Myriophyllum**, *Utricularia**.

Winter-spore. A strong-walled, usually single-celled dormant stage that occurs in Schizobionta*, algae Phycophyta*) and fungi*. In this state the spore can survive adverse environmental conditions.

Wobbegong *(Orectolobus barbatus) see* Orectolobidae

Wolf Herring *(Chirocentrus dorab) see* Clupeiformes

Wolf Herrings (Chirocentridae, F) *see* Clupeiformes

Wolffia HORKEL and SCHLEIDEN, 1844; Watermeal, Rootless Duckweed. G of the Lemnaceae*, with 10 species. Found in the tropics and subtropics but also radiating out into temperate regions. Very small, perennial plants with much simplified vegetative bodies. Daughter members of the plant develop from a backward-pointing pocket of the individual members. No roots. Flowers unisexual and monoecious, very small. There is one male flower and one female flower in each inflorescence and no spathe. No perianth. Male flower: 1 stamen. Female flower: 1 carpel, superior. Fruit is like a berry. The species of the G grow in still waters where they float on the water surface or live submerged. In cultures of the temperate regions they grow most vigorously (without any additional light) in the light-scarce winter months. They are suitable as a food for many plant-eating fish and can also be prepared as dried food. The most widespread species are:

– *W. arrhiza* (LINNÉ) HORKEL and WIMMER; Watermeal, Rootless Duckweed. From tropical and subtropical parts of Asia and Africa. Occasionally reaches temperate regions when brought there by migrating water fowl. Floats on the surface of the water.

– *W. columbiana* KARSTEN. From Central America and eastern N America. Grows submerged.

Wolf-fish *(Anarrhichas lupus) see* Blennioidei

Wolf-fishes (Anarrhichadidae, F) *see* Blenniodidei

Wolterstorff's Pearlfish *(Cynolebias wolterstorffi) see Cynolebias*

Wood *see* Decorative Material

Woodlice, O, *see* Isopoda

Worm Cataract *see* Trematodes

Worm Eels (Moringuidae, F) *see* Anguilliformes

Worm Leech *(Piscicola geometra) see Piscicola*

Worm Parasites *see* Helminthiasis

Worm Pipefishes, G, *see Nerophis*

Wrasses (Labridae, F) *see* Labroidei

Wrasses, related species, Sub-O, *see* Labroidei

Wray's Gambusia *(Gambusia wrayi) see Gambusia*

Wreckfish *(Anthias squamipinnis) see Anthias*

Wreckfish *(Polyprion americanus) see* Serranidae

Wrestling Half-beak *(Dermogenys pusillus) see* Exocoetoidei

Xantho LEACH, 1830. G of the Crabs (Brachyura*, F Xanthidae). These crabs have a broad carapace, a smooth forehead, short walking legs and strong pincers. They are found in warm seas. The carapace, between 2 and 4 cm wide, is more or less camouflaged in colour, and the fingers of the pincers are usually black. *Xantho* species are sluggish creatures that remain hidden much of the time, but in the aquarium they are very hardy if fed on animal food. However, they easily manage to upset rock structures in the aquarium, because they are prone to digging around in the substrate. The Stone Crab (*X. hydrophilus* HERBST) is very common beneath rocks in shallow waters along the southern European coasts.

Xanthophyta; Yellow Algae. D of the Algae* with 200 species. Green Algae with chlorophyll a, c and e; the chromatophores are usually lens-shaped. The mobile cells have two flagella of different lengths, located slightly to the side. The Xanthophyta live in fresh water as well as on damp earth. The simplest forms are mobile or sessile single-celled organisms, which join together into colonies or they build unbranched and branched threads. In many Xanthophyta the thallus is not divided into cells, eg as in the G *Vaucheria**.

Xenarchi, O, name not in current usage, *see* Percopsiformes

Xenentodon cancila *see* Exocoetoidei

Xenoberyces, O, name not in current usage, *see* Beryciformes

Xenomystus nigri *see* Mormyriformes

Xiphias gladius *see* Scombroidei

Xiphiidae, F, *see* Scombroidei

Xiphophorus HECKEL, 1848; Swordtail. G of the Poeciliidae*, Sub-F Poeciliinae. From Central America, especially the Atlantic coasts of Mexico, Guatemala and

Honduras. They live mainly in flowing waters, in the strong current of mountain rivers to the rivermouths, but also in lakes, marshes and lagoons. Not found in brackish water. *Xiphophorus* species are small to medium-sized fish, males 3–10 cm long, females 4–12 cm. 3 species groups can be distinguished according to body structure and colouring:

1) Helleri-group (*X. clemenciae, X. helleri* with sub-species). C drawn out at the bottom into a long kind of sword-shape. Body fairly slim with a red or brown stripe along the middle of the flanks.

2) Maculatus-group *X. couchianus, X. maculatus, X. variatus, X. xiphidium* and sub-species). C has no sword or there is only the suggestion of any elongation. Body deep, more compact. Platy-type. No longitudinal band.

3) Montezumae-group (*X. milleri, X. montezumae, X. pygmaeus* and sub-species). Sometimes, the C has a short sword. Body relatively elongate, with a dark, distinct, zig-zag band on the flanks.

The fish should be kept in sufficiently large tanks. Filtration and ventilation systems are both recommended. Requirements as far as heat is concerned depends on the fish's origin, but it is usually around 22–26 °C. Omnivores. Feed with a good variety of live food, supplemented with plant food (algae, lettuce leaves) as well as dried and artificial foods. For a productive breeding, pregnant females should be kept isolated.

Some of the most popular of all aquarium fish belong to this G. *X. helleri* and *X. maculatus* have various-sub-species with a large number of natural and ornamental races. Important members of the G are:

– *X. clemenciae* (ALVAREZ, 1959); Yellow Swordtail. From Mexico. Male to 4.5 cm not including the sword. Female to 5 cm. Body yellowish-brown. Sword edged in black. D has distinct rows of red to black dots.

– *X. couchianus couchianus* (GIRARD, 1859); Northern Platy. From Mexico. Male to 3.5 cm, female to 6 cm. Body brownish on the upper side, paler below, no colourful longitudinal stripe. There are some small dark flecks on the caudal peduncle.

– *X. helleri helleri* HECKEL, 1848; Swordtail. From S Mexico, Guatemala. Male up to 10 cm long, not including the sword; female to 12 cm. Body olive, upper side green. Sword outlined in black, inside green or yellow to red. D covered in brownish rows of dots. The sub-species *X. h. alvarezi, X. h. guentheri* and *X. h. strigatus* differ somewhat in body shape and in the structure of the gonopodium. There are over 30 ornamental forms* of which the following are important to mention:

'Berlin Cross-breed' (species cross, green helleri-males with red maculatus-females). No stripes on the body, red with black flecks. Male more yellowish-red, short sword and outlined in a weak black colour.

'Hamburg Cross breed'. Black helleri (species cross, green helleri-females with red maculatus-males). Scales a gleaming green colour with black edges or completely black. Fins pale. Sword black.

'Wiesbaden Cross-breed' (helleri ornamental form). Lower parts of the body black, colour otherwise red or green. The offspring of the black-red form are always pure black and red specimens.

'Wagtail-Helleri'. Fin rays and sword black. Body red, green or yellow.

'Tuxedo-Helleri'. Lower parts of the body and the fin rays black. Rest of body red or green (fig. 1168).

Fig. 1168 *Xiphophorus helleri* Lyre-tail Tuxedo (male)

'Double-sword Helleri'. A second sword is formed from the upper rays of the C.

'Delta-High-Fin'. D tall, like a sail.

'Lyre-tail'. D tall, C extending into a point at the top. Gonopodium very long. There are also albino and xanthic (yellow) forms. For information about the genetic basis of cross-breeding and mutations, turn to Heredity. Sex reversal* is of particular interest in this species (it occurs frequently) (fig. 1169).

Fig. 1169 *Xiphophorus helleri* Red Lyre-tail (pair)

– *X. maculatus* (GÜNTHER, 1866); Platy. From S Mexico, Guatemala, Honduras. Male to 4 cm, female to 6 cm. Colour very variable. Basic form: body olive brown to grey on the upper side. Flanks have a bluish or greenish shimmer in direct light. Throat and belly pale. The body is covered with irregular black speckles, and on the caudal peduncle there are 1–2 fairly large black flecks. Other natural colour varieties are more red; others have stronger black flecks or are partially black, predominantly blue or the form of the caudal peduncle flecks differ. There are over 30 ornamental forms

Fig. 1170 *Xiphophorus maculatus* (Moon Platy)

(figs. 1170, 1172 and 1174), of which the following are important to mention:

'Gold Platy'. Body yellow on the upper side, belly pale. D red.

'Red Platy'. Body and fins red. There exist races of this and the previous form with a black half-moon blotch or a round moon blotch on the caudal peduncle (fig. 1171).

Fig. 1171 *Xiphophorus maculatus* red (Platy)

'Velvet-black Platy'. Dorsal surface brown, otherwise the body is predominantly black.

'Tuxedo-Platy'. Lower half black, rest of body either red or greenish. Fins more or less black.

'Wagtail-Platy'. Fin rays black. Body either red, yellow, bluish or black.

There are also albino forms and ornamental races with altered fin shapes.

– *X. montezumae* JORDAN and SNYDER, 1900; Montezuma Swordtail. From Mexico. Male to 4.5 cm, female to 6 cm. Body olive-brown, belly pale. D and upper parts of the body covered in black flecks. D and C yellowish. Sword outlined in a dark colour. There are some faint bands running parallel to the dark brown zigzag band, but these are less clear in the sub-species *X. m. cortezi.*

– *X. pygmaeus* HUBBS and GORDON, 1943; Pygmy Swordtail. From Mexico. Male and female up to 4 cm long. Body less robust. Only the suggestion of a sword, a bit longer in the sub-species *nigrensis*. As is typical of the Montezumae group, the body has a zigzag band and distinct net-markings. A shimmering blue colour on the flanks. In the male the D has dark flecks and dark edges. In the sub-species *X. p. nigrensis* only, the sword has a dark edge below. Should be kept at a temperature of 25–27 °C.

– *X. variatus* (MEEK, 1904); Sunset Platy. From Mexico. Male to 5.5 cm, female to 7 cm. Colouring very variable. Male yellowish at the front, greenish to blue at the rear. Sides covered in irregular black or brown flecks, sometimes appearing as longitudinal lines. Sometimes 3–6 transverse bands make as appearance. C yellowish to reddish. D yellow with dark markings and a black edge. There are often 2 black blotches on the caudal peduncle. Female paler.

Fig. 1172 *Xiphophorus* var. (Bleeding Heart Platy) (male)

The sub-species *X. v. evelynae* is more simply coloured, although it has 8–12 black transverse bands on the flanks.

Various ornamental forms exist, in which blue, yellow to red, or black colours predominate. As with *X. helleri* and *X. maculatus* there is a Tuxedo-form and a Delta-hi-fin (fig. 1173).

Fig. 1173 *Xiphophorus* var. (Hi-fin form) (male)

– *X. xiphidium* (HUBBS and GORDON, 1932 ; Swordtail Platy. From Mexico. Male to 4 cm, female to 5 cm. In older males, the lower part of the C is pointed at the rear. Body yellowish, paler below. The male has clear black

Fig. 1174 *Xiphophorus* var. (blue bred form)

oblique streaks, which are particularly noticeable in the shoulder area. Both male and female have 2 black flecks, one over the other, on the caudal peduncle.

Xiphosura; King, or Horseshoe, Crabs. These are ancient marine creatures, like Crustaceans, which together with the Spiders represent the Sub-Ph Chelicerata. The Xiphosura live on soft substrata in the littoral*, rummaging about in the ground for worms and molluscs which they swallow whole.

The front and middle section (prosoma and mesosoma) of the body form a firm shield, and on the last part of the body (metasoma) is a long telson spine. Two large compound eyes are located on the prosoma. There are 5 species still alive today. The West Atlantic species, *Limulus polyphemus* LINNAEUS, 1758, can reach a length of 60 cm (fig. 1175). The Moluccan crabs of the G *Tachypleus* LEACH, 1819, are native to warm parts of the Indo-Pacific. Horseshoe Crabs swim on their back, turning right side up after 'landing' with the aid of their tail spine. When cared for in the aquarium, PROBST has reported that the specimens should be taken out every two days and bivalve meat (or similar) stuffed between their legs. They usually eat food presented to them in this way immediately.

Fig. 1175 *Limulus polyphemus*

X-ray Fish *(Pristella riddlei) see Pristella*

Xyrichthys CUVIER, 1815. G of the Labridae, Sub-O Labroidei*. There are species in the Atlantic Ocean where they live above sandy substrata in which they can bury themselves deep. Elongated fish, up to 15 cm long, with a strikingly steep forehead area and very compressed laterally. In contrast to the G *Hemipteronotus**, the cheeks are hardly covered in scale at all. The following species lives along European coasts:

– *X. novacula* (LINNAEUS, 1758). From the E Atlantic, Mediterranean. To 20 cm. Young fish pink, later grey to green with blue markings on the head.

Yellow Algae, D, *see* Xanthophyta

Yellow and Black Angelfish *(Centropyge bicolor) see Centropyge*

Yellow Belly *(Girardinus falcatus) see Girardinus*

Yellow Congo Characin *(Alestopetersius caudalis) see Alestopetersius*

Yellow Dwarf Cichlid *(Apistogramma reitzigi) see Apistogramma*

Yellow Flag *(Iris pseudacorus) see* Iridaceae

Yellow Limia *(Poecilia nicholsi) see Poecilia*

Yellow Long-nosed Butterfly-fish *(Forcipiger longirostris) see Forcipiger*

Yellow Perch *(Perca flavescens) see Perca*

Yellow Sea-horse *(Hippocampus kuda) see Hippocampus*

Yellow Swordtail *(Xiphophorus clemenciae) see Xiphophorus*

Yellow Tetra *(Hyphessobrycon bifasciatus) see Hyphessobrycon*

Yellow Water Lily *(Nuphar lutea) see Nuphar*

Yellow-banded Tetra *(Moenkhausia sanctaefilomenae) see Moenkhausia*

Yellow-faced Angelfish *(Euxiphipops xanthometopon) see Euxiphipops*

Yellow-fin Tuna *(Thunnus albacora) see* Scombroidei

Yellow-green Algae, D, *see Chrysophyta*

Yellow-spotted Triggerfish *(Pseudobalistes fuscus) see Pseudobalistes*

Yellowtail Damselfish *(Microspathodon chrysurus) see Microspathodon*

Yellow-tail Wrasse *(Coris formosa) see Coris*

Yellowtailed Tang *(Zebrasoma xanthurus) see Zebrasoma*

Yolk-sac (fig. 1176). An organ found in embryos or larvae that contains the yolk supply of the growing youngster. It comprises a number of blood vessels which unite to form the vitelline vein. These vessels convey the nutrients of the liquefied yolk substances to the developing embryo and because they are located near the upper surface they are also important in respi-

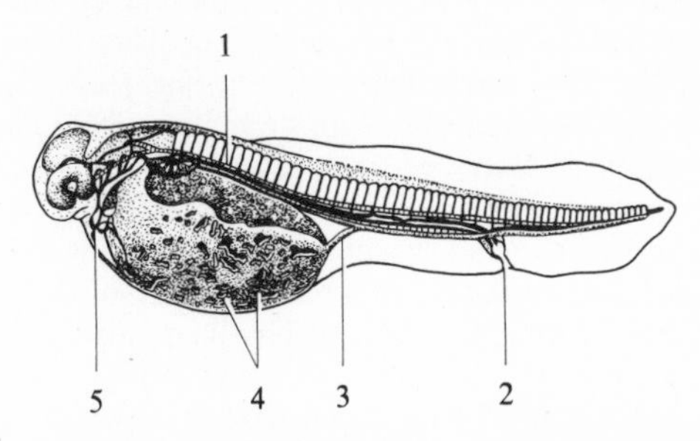

Fig. 1176 Bony fish embryo with yolk-sac. 1 Aorta 2 Anus 3 Vitelline vein 4 Vascular network of the yolk-sac 5 Heart

ration. Part of the yolk passes from the yolk-sac through to the intestine via the yolk duct to be digested. As the yolk gets used up the yolk-sac gradually shrivels away. The size of the yolk-sac depends on the yolk content of the eggs (*see* Gametes). Cephalopods, amphibians, reptiles and birds all possess a relatively large yolk-sac. Livebearing mammals either have no yolk-sac or a very small one. In fish, even after the spawn has hatched out, the yolk-sac usually remains for several days, weeks or months until the young fish can aquire food for itself.

Yolk-sac Dropsy; Blue Sac Disease (fig. 1177). This disease occurs in newly hatched fish embryos or in embryos that are a few days old. A translucent liquid builds up in the yolk-sac making it swell to a great size;

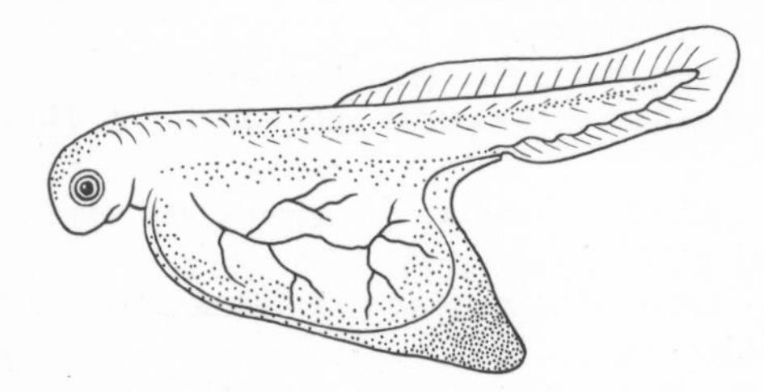

Fig. 1177 Fry with yolk-sac dropsy

usually the infected fish dies. As yet the causes of the disease can only be guessed at. It is probably inherited from the parents, and a combination of other factors such as nutrient errors, the effect of chemicals, precocious females giving birth at too young an age, hybrid formation etc may be the prime causes. Occasionally, bacteria (*Diplobacillus liquefaciens piscium* VAN BETEGH) infect the yolk-sac; this may be regarded as a secondary infection of the diseased embryos. Any females that produce a high percentage of embryos with yolk-sac dropsy should not be allowed to breed.

Yucatan Gambusia *(Gambusia yucatana) see Gambusia*

Zambezi Shark *(Carcharhinus zambezensis) see* Carcharhinidae

Zanclidae, F, name not in current usage, *see* Acanthuroidei

Zanclus CUVIER and VALENCIENNES, 1831; Moorish Idols, Tobies. G of the Acanthuridae, Sub-O Acanthuroidei*. One species, *Z. cornutus* (LINNAEUS, 1758) known as the Moorish Idol, lives in tropical parts of the Indo-Pacific. They form shoals among the reefs and rocky coasts. Characteristic features are the protruding snout, the greatly flattened (laterally), very deep body and the very elongate 3rd dorsal spine. The young fish pass through a so-called acronurus-stage which is very similar to that of young Surgeonfishes (Acanthuridae*). The second *Zanclus* species that has been described, *Z. canescens* LINNAEUS, 1758, may be the youthful form of *Z. cornutus* (fig. 1178); as it gets older it is characterised by having 2 little horns in front of the eyes. *Zanclus* grows to a length of 25 cm, and has 2 broad oblique bands on a yellowish-white background. The D, C and An are yellow at the rear. They are very sensitive food specialists that require large, bright aquaria and abundant food supplies.

Zander *(Stizostedion lucioperca) see Stizostedion*

Zannichelliaceae; Horned Pondweeds. F of the Liliatae* with 5 Ga and 25 species. Liberally scattered throughout the world, chiefly in Australia. Horned Pondweeds are small perennials with elongated stem axes and narrowly linear or ribbon-shaped leaves. Flowers unisexual, either monoecious or dioecious, very small and simplified.

The Zannichelliaceae live in fresh, brackish and sea water. They are all water-pollinated. From the G *Zannichellia* (Horned Pondweed) which contains 2 species, *Zannichellia palustris* is found in some places in brackish and fresh water throughout the northern hemisphere. No experiences with cultivation.

Zantedeschia aethiopica *see* Araceae

Zebra Barb *(Barbus fasciatus) see Barbus*

Zebra Barb *(Barbus lineatus) see Barbus*

Zebra Danio *(Brachydanio rerio) see Brachydanio*

Zebra Firefish *(Dendrochirus zebra) see Dendrochirus*

Zebra Killi *(Fundulus heteroclitus) see Fundulus*

Zebra Lionfish *(Dendrochirus zebra) see Dendrochirus*

Zebra Mussels, G, *see Dreissena*

Zebra Scorpion *(Dendrochirus zebra) see Dendrochirus*

Zebrasoma SWAINSON, 1839. G of the Acanthuridae, Sub-O Acanthuroidei*. There are several species in the tropical Indo-Pacific where they live in shallow waters. They eat Algae and small creatures. Length 20–40 cm. *Zebrasoma* species are tall fish, very flat laterally and

Fig. 1178 *Zanclus cornutus*

have a remarkably large D and An. Snout pointed. On the caudal peduncle there are moveable stiletto spines. Younger specimens especially can become accustomed to life in captivity if the tank is roomy and if plant food is provided. The following is a typical example:
– *Z. xanthurum* BLYTH, 1852; Yellowtailed Tang. From the Red Sea, Indo-Pacific. 20 cm. Blue to brownish-yellow with a yellow C (fig. 1179).

Fig. 1179 *Zebrasoma xanthurum*

Zeidae, F, *see* Zeiformes

Zeiformes; John Dory types. O of the Osteichthyes*, Coh Teleostei. Known in the REGAN system as the Zeomorphi, this O is a small but very typical fish group. The oldest fossils date from the early Tertiary period. Various features point to their being related to the Perciformes*. The body is usually tall and very flat from side to side. Head large with a deeply slit mouth that can be extended and protruded considerably. The spiny-rayed and soft-rayed parts of the D and An are almost completely separated by an indentation. The soft-rayed sections of both fins are long, stand opposite each other and reach almost to the C. The Vs are inserted beneath the Ps. Ctenoid scales. All Zeiformes live in temperate and warm seas, and several Fs are found only in the greater depths of the oceans (bathypelagic).
– *Zeidae;* John Dories. Deep-bodied, laterally very flat fish. D has 6–10 very strong spines, the An 1–4. A well-known representative of the F is the John Dory *(Zeus faber)* (fig. 1180). It lives in the high seas and is found along the coasts of Norway and Great Britain to E Africa, although the species is also caught quite often in the Mediterranean. Length up to 70 cm, although most remain smaller. Its main characteristics are: thread-like elongations of the fin membranes between the 9th and 10th dorsal fin spines; eyes positioned high up; a large, black blotch on the flanks outlined in a pale colour; elongated Vs. On both sides of the base of the D and An, and on the edge of the belly, there is a row of strong spiny plates. John Dory sails calmly through the water, propelled along by the soft-rayed parts of the D and An, but it is also able to put in short bursts of

Fig. 1180 *Zeus faber*

speed. Spawning takes place at a depth of 100–200 m, and the eggs float in the water. Small troops of John Dory follow along after shoals of smaller kinds of herrings, and that is why they are often caught along with them. In certain regions John Dory meat is highly prized. *Zeus japonicus* lives in the Indo-Pacific.
– *Caproidae;* Boarfishes (fig. 1181). Similarly very tall fish, laterally very flat, but in side-view they tend to have a more rhomboid appearance. Forehead profile concave. The mouth can be protruded even further. There are no bony plates at the base of the An nor on the belly. The 15 cm long Boarfish *(Capros aper)* comes from the Mediterranean, although it also wanders into the NE Atlantic. The Antigone-Boarfishes of the Atlantic and Pacific are shining red. They are often regarded as a F (Antigonidae) in their own right.

Fig. 1181 Caproidae-type

Zeomorphi, O, name not in current usage, *see* Zeiformes

Zeus faber *see* Zeiformes

Zigzag-banded Copella (*Copella* sp.) *see Copella*

Zill's Tilapia *(Tilapia zilli) see Tilapia*

Zinc Poisoning *see* Poisoning

Zinc Sulphate ($ZnSO_4 \cdot 7H_2O$). Forms colourless crystals with good water-solubility. Like copper sulphate*, it works like a poison on lower organisms and is used in marine aquarium-keeping in particular to combat skin parasites found in fish (Fin-rot*, Oodinium*, Cryptocaryon-disease*, Trichodina*, Helminthiasis* and

Mycoses* of the skin). Zinc sulphate has the advantage over copper sulphate that it precipitates in sea water only very slowly, and so it is retained as an effective concentration over a longer period of time. A prolonged bathing treatment* in separate tanks uses 1 ml of an original solution (4 g $ZnSO_4 \cdot 7H_2O$ to 1 litre of water) to every 2–4 litres of sea water. After 4 days the bath has to be renewed. Further doses are not advisable.

Zirfaea, G, *see* Boring Molluscs

Zoantharia. O of the Anthozoa*. There are species in both warm and cold seas. Usually, these Anemones are colonial, the polyps being only a few cm long and growing from a communal crusty base or joined together by runners. The majority of Zoantharia are particle-eaters. They do not form their own skeleton but often take sponge spicules or grains of sand into the mesogloea which gives them some stability. Many grow on the shells of Hermit Crabs, on Horny Corals or on Sponges (as is the case with the *Parazoanthus axinellae* O. SCHMIDT, 1862, fig. 1182) which is quite common in the Mediterranean). Like many Zoantharians, this species has proved very durable in the aquarium. It is important to feed them daily with small crustaceans *(Cyclops, Artemia). Parazoanthus axinellae* grows in shaded spots and is easily overwhelmed by algae; tropical Zoantharians on the other hand, such as the frequently imported G *Palythoa* LAMOUROUX, 1816, thrive best if illuminated very brightly.

Fig. 1182 *Parazoanthus axinellae*

Zoarces viviparus *see* Gadiformes

Zoarcidae, F, *see* Gadiformes

Zobel *(Abramis sapa) see Abramis*

Zoochlorellae *see* Zooxanthellae

Zoogamy *see* Pollination

Zoogeographical Regions (fig. 1183). The areas into which the Earth's surface is divided according to criteria of animal distribution. On the one hand, by comparing areals* certain areas of overlap are recognised, and on the other, it becomes clear that certain animal groups are excluded from appearing. And so a zoogeographical region is characterised both by the fact that certain systematic taxa (Taxon*) occur together and by the fact that certain others are absent. Since animal distribution is not only dependent on geographical conditions, but also on ecological and historical factors, the boundaries of the zoogeographical regions do not always (and sometimes never) coincide with the geographical land boundaries. The dividing line between the nearctic (North American) and neotropic (South American) fauna, for example, runs roughly from California through Mexico to Florida. The relationships of the two sets of fauna are less strong than those of N America and temperate Eurasia. A similar situation exists for the north of Africa which is grouped not with the Ethiopian* realm but with the palaearctic*.

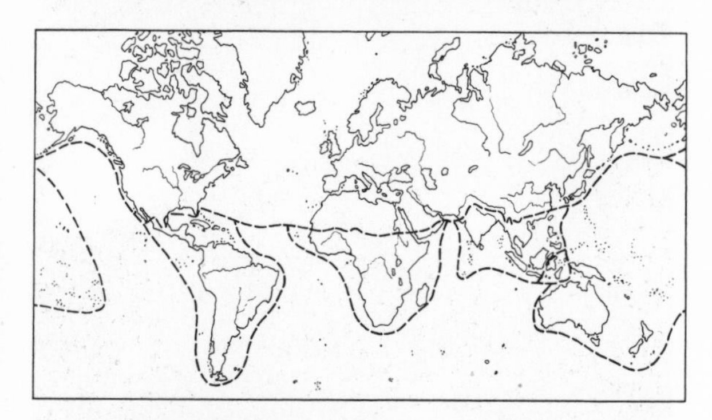

Fig. 1183 Zoogeographical realms. Holarctic (N America, N Africa, Eurasia), Neotropic (Central, S America), Ethiopian (Africa), Oriental (S Asia), Australis (Australia, Pacific island groups)

The broad divisions of the two major habitats, land and inland waters, into zoogeographical regions is represented by the picture. It also extends in part into including the fauna of the littoral*, ie the animal world of the coastal regions. In contrast, the zoogeographical division of the oceanic pelagial* fauna is fundamentally different – it is essentially divided into zones corresponding to climates. So far we known much too little about the abyssal (Abyss*) fauna to be able to divide it up zoogeographically.

Zoogeography. A branch of biogeography* that is concerned with the distribution of animals on Earth and the governing factors that lie behind it. Apart from the chorological aspect, ie understanding and describing the areal* picture of a particular animal group, zoogeography also tries to uncover the relationships between areal and ecology*; in addition, it also tries to explain the areal picture by historical-geological processes. Comparative studies of the areals have led to the Earth being divided up into zoogeographical regions* which do not always coincide with the geographical divisions.

Zoological Nomenclature. Every taxon* has a Latin or latinised name which may not be reused in the animal kingdom, so that it will only ever have one meaning. The formation and use of the name is subject to internationally agreed rules which are enforced and presided over by an international panel, the Nomenclature Commission.

The scientific name* above the level of species consists of a single word (uninominal), that of a species is made up of two words (binominal) and that of a subspecies of three words (trinominal), eg

Genus	*Symphysodon*
Species	*Symphysodon aequifasciata*
Sub-species	*Symphysodon aequifasciata axelrodi*

If a sub-genus is introduced, the name of a species is nevertheless binominal, that of a sub-species trinominal, eg *Neolebias (Micraethiops) ansorgei* is binominal. Family names end in -idae and must be taken from a genus included in the family, eg Cichlidae, taken from the genus *Cichla*. For a name to be valid, it must fulfil certain conditions. These are fixed at present in 87 articles in the International Rules for Zoological Nomenclature. For example, a name published not before 1758 and after 1960 may not be published anonymously etc. With an exact description, after the name there follows the author, then separated off by a comma, the year of publication, eg

Symphysodon HECKEL, 1840

which means that the generic name *Symphysodon* was first published by HECKEL in 1840. Specific names are applied in the same way. As a result of new studies, it may happen that the species must be transferred to another genus. This is evident from brackets being placed round the author's name and the year, eg

Pterophyllum scalare (LICHTENSTEIN, 1823)

which means that the species described by LICHTENSTEIN in 1823 was originally placed in another genus. If a name is inadvertently given several times (Homonym*), for reasons of clarity of meaning, the more recent homonym must be given a new name. Sometimes, several different names are given to one and the same taxon* (Synonym*). In order to preserve the right of priority, it is always the oldest name which is valid, unless the rules or the Commission state otherwise.

The introduction of binominal nomenclature by the Swedish researcher CARL LINNÉ* (1708–78) was an important advance for biology, since as many taxa as required can be indicated briefly but without the possibility of dual meaning. Any renamings that may be necessary in order to preserve priority, or because of synonyms, homonyms and revisions (unpopular as they may be) must be accepted if the unequivocal meaning of a name is to be preserved.

Zooxanthellae. Symbiotic single-celled algae with a yellow-brown pigment which live in the cells of many Protozoans, Coelenterates, Turbellarians and Molluscs. It is certain that the oxygen and products of assimilation produced by the Zooxanthellae when illuminated are of advantage to those animals that house them. They are found in many Sea-anemones (Actiniaria*), among other things, as well as among tropical coral reefs. The latter are sometimes able to live with the aid of Zooxanthellae completely without organic food. All animals containing Zooxanthellae need bright light in the aquarium. The same goes for the green-coloured Zoochlorellae.

Zope *(Abramis ballerus) see Abramis*

Zostera; Grass Wrack. G of the marine Wracks (Potamogetonaceae*). They form expansive mats at shallow depths on soft substrata. Leaves pale green, narrowly linear. All attempts at introducing these plants into the marine aquarium have so far proved unsuccessful.

Zosterella SMALL, 1913. G of the Pontederiaceae* with 2 species. Subtropical to northern temperate America. Small perennials with elongated stem axes and an alternate leaf arrangement. Leaves linear. Flowers hermaphroditic, polysymmetrical. Perianth consists of a 6-part yellow perigonium. 3 stamens. 3 carpels, superior, fused. Capsule-fruit. The species live mostly submerged, although they are sometimes found as creeping plants in marshes. The following species is kept in the aquarium:

– *Z. dubia* (JACQUIN) SMALL *Heteranthera dubia*, name not in current usage). From America.

Zygogaster, Sub-G, *see Astyanax*

Zygote. The product obtained by the fusion of a female and a male sex cell (Gametes*). The uniting of the hereditary material of the gametes (or fertilisation*) takes place in the zygote. The zygote is the first stage of embryonal development in multi-celled organisms.

Bibliography

Axelrod, H. et al, *Breeding Aquarium Fishes*, Vols 1–5 (T.F.H. 1967, 1971, 1973, 1976, 1978)

Axelrod, H. et al, *Exotic Tropical Fishes* (T.F.H. 1981)

Dal Vesco et al, *Life in the Aquarium* (Octopus 1975)

Duijn, C. van, *Diseases of Fishes*, 3rd edition (Butterworth Group 1973)

Favre, H., *Dictionary of the Freshwater Aquarium* (Ward Lock 1977)

Federation of British Aquatic Societies:
National Show Fish Sizes (1981)
National Show Guide to Cultivated Fishes (1981)
Dictionary of Common/Scientific Names of Freshwater Fishes (1982)
Scientific Names of Fishes and their Meanings (1980)

Frank, Dr. S., *Illustrated Encyclopaedia of Aquarium Fish* (Octopus 1980)

George, D. and J., *Marine Life* (Harrap 1979)

Gilbert, J. (editor), *Complete Aquarist's Guide to Freshwater Tropical Fish* (Ward Lock 1970, Peter Lowe 1982)

Goldstein, R., *Cichlids of the World* (T.F.H. 1980)

Géry, Dr., *Characoid Fishes* (T.F.H. 1973)

de Graaf, F., *Marine Aquarium* (Pet Library 1973)

Hargreaves, V.C., *The Tropical Marine Aquarium* (David & Charles 1978)

Hervey and Hems, *A Guide to Freshwater Aquarium Fishes* (Hamlyn 1973)

Hoedeman, J. J., *Naturalist's Guide to Freshwater Aquarium Fishes* (Sterling Pub. Inc. 1974)

Hunman, Milne and Stebbing, *The Living Aquarium* (Ward Lock 1981)

Jacobsen, N., *Aquarium Plants* (Blandford Press 1979)

Lythgoe, J. and G., *Fishes of the Sea* (Blandford Press 1971)

Madsen, J. M., *Aquarium Fishes in Colour* (Blandford Press 1975)

Mills, D., *Know the Game: Aquaria* (EP Publishing 1978)

Mills, D., *Aquarium Fishes* (Kingfisher Books 1980, John Wiley, Canada, 1980)

Mills, D., *Illustrated Guide to Aquarium Fishes* (Kingfisher Books/ Ward Lock 1981)

Mühlberg, H., The Complete Guide to Water Plants (EP Publishing 1982)

Perry, Fances, *The Water Garden* (Ward Lock 1981)

Ramshorst, Dr J. D. van, *Complete Aquarium Encyclopedia of Tropical Freshwater Fish* (Elsevier Phaidon 1978)

Rataj and Horeman, *Aquarium Plants* (T.F.H. 1977)

Schiotz and Dahlstrom, *Collins Guide to Aquarium Fishes and Plants* (Collins 1972)

Singleton, V., and Mills, D., *How About Keeping Fish* (EP Publishing 1979)

Spotte, S., *Fish and Invertebrate Culture* (Wiley Interscience 1970, 1979)

Sterba, G., *Freshwater Fishes of the World* (Studio Vista 1966)

Sterba, G., *Aquarium Care* (Studio Vista 1967)

Sterba, G., *Dr Sterba's Aquarium Handbook* (Pet Library 1973)

Thabrew, Dr. V. de, *Popular Tropical Aquarium Plants* (Thornhill Press 1981)

Vevers, G., *Pocket Guide to Aquarium Fishes* (Mitchell Beazley 1980)

Whitehead, P., *How Fishes Live* (Elsevier Phaidon 1975)

Picture Credits

Barth, Dessau: Fig. 121
Birnbaum, Halle: Fig. 1132
Brückner, Leipzig: Figs. 33, 35, 232, 631
Fach, Dessau: Fig. 225
Foersch, Munich: Figs. 59, 576, 590
Franke, Gera: Figs. 89, 292, 300, 355, 356, 367, 372, 405, 548, 583, 640, 642, 645, 647, 648, 653, 711, 713, 759, 761, 811, 817, 891, 1045
Hansen, W. Berlin: Figs. 24, 389, 479, 587, 1125, 1154
Hertel, Bad Langensalza: Figs. 122, 124, 463, 466, 824, 825
Jacob, Leipzig: Figs. 198, 564, 635, 875
Kahl, Stuttgart: Figs. 3, 7, 9, 14, 15, 32, 37, 88, 94, 102, 104, 110, 115, 116, 119, 130, 131, 159, 161, 166, 168, 169, 170, 171, 172, 175, 183, 208, 209, 210, 211, 213, 214, 215, 240, 244, 254, 284, 296, 317, 318, 319, 320, 321, 322, 349, 350, 373, 388, 400, 417, 422, 425, 426, 429, 436, 448, 449, 474, 487, 495, 498, 500, 509, 510, 511, 526, 558, 565, 567, 595, 596, 598, 601, 607, 615, 620, 624, 630, 669, 670, 671, 672, 676, 678, 682, 695, 708, 762, 774, 803, 807, 808, 847, 848, 859, 861, 881, 884, 885, 893, 899, 912, 919, 942, 948, 949, 951, 952, 955, 956, 966, 967, 968, 970, 972, 986, 987, 998, 1033, 1089, 1090, 1096, 1097, 1100, 1104, 1107, 1111, 1121, 1141, 1150, 1170, 1171, 1174, 1178
Knaack, Vienna: Figs. 103, 105, 106, 107, 108, 109, 112, 114, 185, 414, 415, 496, 818, 829, 830, 873, 954, 1088, 1168, 1169, 1173
Kormann, Berlin: Figs. 28, 42, 70, 92, 136, 160, 186, 196, 234, 265, 266, 303, 326, 352, 364, 384, 443, 627, 683, 747, 753, 821, 833, 862, 863, 1065, 1182
Marcuse, Munich: Figs. 13, 22, 45, 55, 65, 138, 189, 207, 258, 382, 412, 424, 431, 457, 460, 491, 533, 542, 586, 660, 677, 706, 716, 719, 751, 851, 905, 946, 950, 963, 1013, 1031, 1036, 1074, 1114, 1127, 1131, 1139
Müller, Leipzig Fig. 52
Naumann, Leipzig: Figs. 386, 445, 522, 657, 659, 839, 900, 1006, 1067, 1135, 1152, 1164
Nieuwenhuizen, Hemstede: Figs. 53, 60, 237, 256, 257, 276, 277, 278, 280, 298, 363, 402, 428, 430, 433, 461, 497, 572, 580, 581, 621, 680, 685, 723, 728, 741, 748, 771, 798, 813, 843, 844, 853, 864, 910, 925, 926, 927, 934, 953, 981, 1030, 1077, 1092, 1148, 1179
Paysan, Stuttgart: Fig. 441
Richter, Leipzig: Figs. 178, 399, 744, 745, 746, 971, 1061, 1142, 1144
Rodemann, Halle: Figs. 95, 120, 153, 154, 206, 223, 264, 387, 393, 394, 416, 464, 468, 471, 472, 634, 721, 722, 739, 797, 834, 858, 874, 894, 989, 993, 1004, 1034, 1163
Schmitt, Gera: Fig. 1057
Sterba, Leipzig: Figs. 71, 184, 261, 503, 504, 516, 585, 705, 737, 780, 781, 851, 852, 867, 963, 1000, 1101
Zukal, Brno: Figs. 113, 177, 494, 508, 544, 552, 831, 872, 884, 885, 913